To the Student: How to Use This Book

Find important information in this book by focusing on the Gold Thread.

These black swan chicks are being introduced to the world by their mother, who trails close behind them. Likewise, this chapter introduces the key ideas that launched biological science as a discipline.

Biology and the Tree of Life 1

In essence, biological science is a search for ideas and observations that unify our understanding of the diversity of life, from bacteria living in rocks a mile underground to hedgehogs and humans. Chapter 1 is an introduction to this search.

The goals of this chapter are to introduce the nature of life and explore how biologists go about studying it. The chapter also introduces themes that will resonate throughout this book: (1) analyzing how organisms work at the molecular level, (2) understanding organisms in terms of their evolutionary history, and (3) helping you learn to think like a biologist.

Let's begin with what may be the most fundamental question of all: What is life?

1.1 What Does It Mean to Say That Something Is Alive?

An **organism** is a life-form—a living entity made up of one or more cells. Although there is no simple definition of life that is endorsed by all biologists, most agree that organisms share a suite of five fundamental characteristics.

- *Energy* To stay alive and reproduce, organisms have to acquire and use energy. To give just two examples: plants absorb sunlight; animals ingest food.
- *Cells* Organisms are made up of membrane-bound units called cells. A cell's membrane regulates the passage of materials between exterior and interior spaces.
- *Information* Organisms process hereditary or genetic information, encoded in units called genes, along with information they acquire from the environment. Right now cells throughout your body are using genetic information to make the molecules that keep you alive; your eyes and brain are decoding information on this page that will help you learn some biology.

KEY CONCEPTS

- Organisms obtain and use energy, are made up of cells, process information, replicate, and as populations evolve.
- The cell theory proposes that all organisms are made of cells and that all cells come from preexisting cells.
- The theory of evolution by natural selection maintains that species change through time because individuals with certain heritable traits produce more offspring than other individuals do.
- A phylogenetic tree is a graphical representation of the evolutionary relationships between species. These relationships can be estimated by analyzing similarities and differences in traits. Species that share distinctive traits are closely related and are placed close to each other on the tree of life.
- Biologists ask questions, generate hypotheses to answer them, and design experiments that test the predictions made by competing hypotheses.

✓ When you see this checkmark, stop and test yourself. Answers are available in Appendix B. 1

Key Concepts

Start with Key Concepts on the first page of every chapter. Read these gold key points first to familiarize yourself with the chapter's big ideas.

MORE! Bulleted Lists

Take note of bulleted lists that "chunk" information and ideas. This will help you manage the information that you are learning in the course.

Gold Highlighting

Watch for important information highlighted in gold. Gold highlighting is always a signal to slow down and pay special attention.

Gold Key

Material related to Key Concepts will be signaled with a gold key.

As you read the text and view the figures, practice with the Blue Thread.

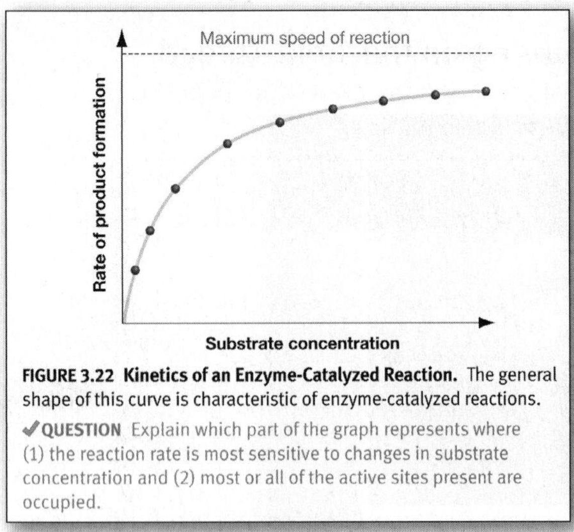

FIGURE 3.22 Kinetics of an Enzyme-Catalyzed Reaction. The general shape of this curve is characteristic of enzyme-catalyzed reactions.

✔**QUESTION** Explain which part of the graph represents where (1) the reaction rate is most sensitive to changes in substrate concentration and (2) most or all of the active sites present are occupied.

Blue Thread Questions

Many figures include Blue Thread Questions or Exercises to help you check your understanding of the material they present.

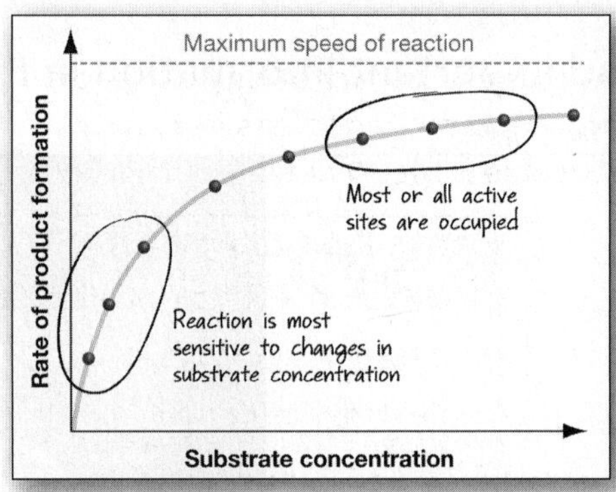

Drawing Exercises

Some Blue Thread Questions contain artwork from the textbook that you will be asked to draw on or modify.

NEW! Suggested Answers

Suggested answers for the Blue Thread Questions and Exercises are provided in Appendix B.

"You Should be Able To" Exercises

Text passages flagged with blue type and the words "you should be able to" offer exercises on concepts that professors and students have identified as most difficult. These are the topics most students struggle with on exams.

To read this graph, put your finger on the *x*-axis at time 0. Then read up the *y*-axis, and note that kernels averaged about 11 percent protein at the start of the experiment. Now read the graph to the right. Each dot is a data point, representing the average kernel protein concentration in a particular generation. (A generation in maize is one year.) The lines on this graph simply connect the dots, to make the pattern in the data easier to see. At the end of the graph, after 100 generations of selection, average kernel protein content is about 29 percent. (For more help with reading graphs, see **BioSkills 2** in Appendix A.)

This sort of change in the characteristics of a population, over time, is evolution. Humans have been practicing artificial selection for thousands of years, and biologists have now documented evolution by *natural* selection—where humans don't do the selecting—occurring in thousands of different populations, including humans.

To practice applying the principles of artificial selection, go to the online study area at *www.masteringbiology.com*.

(MB) **Web Activity** Artificial Selection

Evolution occurs when heritable variation leads to differential success in reproduction. ✔If you understand this concept, you should be able to describe how protein content in maize kernels changed over time, using the same *x*-axis and *y*-axis as in Figure 1.3, when researchers selected individuals with *lowest* kernel protein content to be the parents of the next generation. (This experiment was actually done, starting with the same population at the same time as selection for high protein content.)

FITNESS AND ADAPTATION Darwin also introduced some new terminology to identify what is happening during natural selection.

CHECK YOUR UNDERSTANDING

If you understand that . . .

- Natural selection occurs when heritable variation in certain traits leads to improved success in reproduction. Because individuals with these traits produce many offspring with the same traits, the traits increase in frequency and evolution occurs.
- Evolution is a change in the characteristics of a population over time.

✔ **You should be able to . . .**

On the graph you just analyzed, describe the average kernel protein content over time in a maize population where *no* selection occurred.

Answers are available in Appendix B.

1.4 The Tree of Life

Section 1.3 focused on how individual populations change through time in response to natural selection. But over the past several decades, biologists have also documented dozens of cases in which natural selection has caused populations of one species to diverge and form new species. This divergence process is called **speciation**.

Research on speciation has two important implications: All species come from preexisting species, and all species, past and present, trace their ancestry back to a single common ancestor.

The theory of evolution by natural selection predicts that biologists should be able to reconstruct a **tree of life**—a family tree

CHAPTER 1 Biology and the Tree of Life 5

Make Learning Part of the Grade®

It's possible that your professor will include these Blue Thread Questions in a graded assignment at www.masteringbiology.com.

Check Your Understanding

The blue half of the Check Your Understanding boxes asks you to do something with the information in the top half. If you can't complete these exercises, go back and re-read that section of the chapter.

...is to practice. Here's how.

NEW! Bulleted Summary of Key Concepts

The succinct Summary of Key Concepts reviews important concepts in short, manageable bullet points.

Blue Thread Exercises

End-of-chapter Blue Thread Exercises help you review the major themes of the chapter and synthesize information.

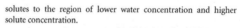

CHAPTER 6 REVIEW *For media, go to the study area at www.masteringbiology.com* 

Summary of Key Concepts

Phospholipids are amphipathic molecules—they have a hydrophilic region and a hydrophobic region. In solution, phospholipids spontaneously form bilayers that are selectively permeable—meaning that only certain substances cross them readily.

- The plasma membrane forms a physical barrier between the internal and external environment—often between life and nonlife.
- The basic structure of plasma membranes is created by a phospholipid bilayer.

solutes to the region of lower water concentration and higher solute concentration.

- Osmosis is a passive process driven by an increase in entropy.

✔You should be able to imagine a beaker with solutions separated by a plasma membrane, and then predict what will happen after addition of a solute to one side if the solute (1) crosses the membrane readily or (2) is incapable of crossing the membrane.

(MB) **Web Activity** Diffusion and Osmosis

Make Learning Part of the Grade®

Visit www.masteringbiology.com for practice quizzes, 3-D animations, the eText, and more.

BioFlix™

BioFlix™ 3-D animations are included in MasteringBiology's Study Area and are available as automatically graded assignments.

Analyze: Can I recognize underlying patterns and structure?

Evaluate: Can I make judgements on the relative value of ideas and information?

Synthesize: Can I put ideas and information together to create something new?

Apply: Can I use these ideas in a new situation?

Explain: Can I explain this concept in my own words?

Remember: Can I recall the key terms and ideas?

Bloom's Taxonomy

Bloom's Taxonomy categorizes six levels of learning competency. The Blue Thread Questions and Exercises in the textbook test on the higher levels of the scale—Explain, Apply, Analyze, Evaluate, and Synthesize—to help you develop critical thinking skills and prepare you for exams.

Steps to Understanding

End of Chapter questions are scaled along Bloom's Taxonomy.

✔TEST YOUR KNOWLEDGE

Begin by testing your knowledge of new facts.

✔TEST YOUR UNDERSTANDING

Once you're confident in your knowledge of the material, demonstrate your understanding by answering the Test Your Understanding questions.

✔APPLYING CONCEPTS TO NEW SITUATIONS

Challenge yourself even further by applying your understanding of the concepts to new situations.

Learn to think like a scientist. Here's how.

A unique emphasis on the process of scientific discovery and experimental design teaches you how to think like a scientist as you learn fundamental biology concepts.

MasteringBIOLOGY®

Make Learning Part of the Grade®

www.masteringbiology.com

Experiment Boxes

Study Experiment Boxes to help you understand how experiments are designed and give you practice interpreting data.

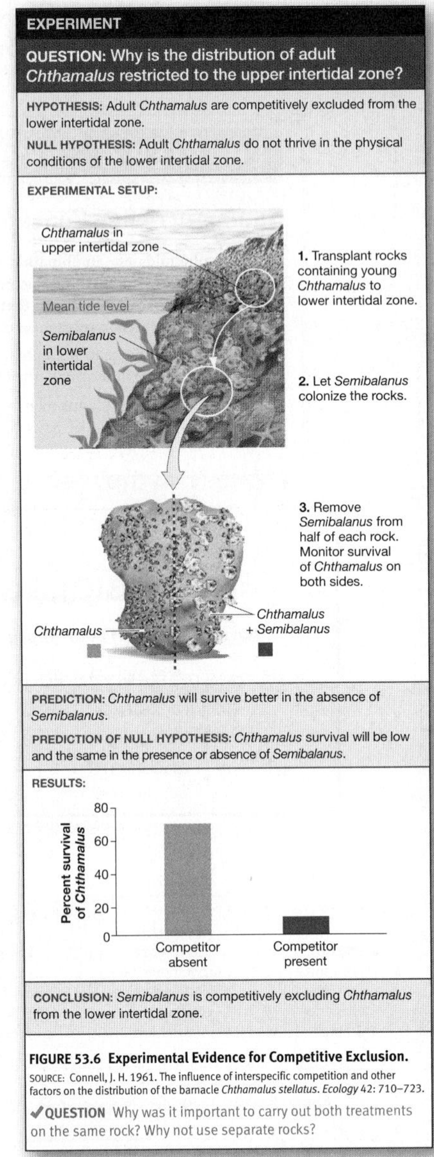
EXPERIMENT
QUESTION: Why is the distribution of adult Chthamalus restricted to the upper intertidal zone?
HYPOTHESIS: Adult Chthamalus are competitively excluded from the lower intertidal zone.
NULL HYPOTHESIS: Adult Chthamalus do not thrive in the physical conditions of the lower intertidal zone.
EXPERIMENTAL SETUP:
Chthamalus in upper intertidal zone
Mean tide level
Semibalanus in lower intertidal zone
1. Transplant rocks containing young Chthamalus to lower intertidal zone.
2. Let Semibalanus colonize the rocks.
3. Remove Semibalanus from half of each rock. Monitor survival of Chthamalus on both sides.
Chthamalus
Chthamalus + Semibalanus
PREDICTION: Chthamalus will survive better in the absence of Semibalanus.
PREDICTION OF NULL HYPOTHESIS: Chthamalus survival will be low and the same in the presence or absence of Semibalanus.
RESULTS:
Percent survival of Chthamalus
80
60
40
20
0
Competitor absent
Competitor present
CONCLUSION: Semibalanus is competitively excluding Chthamalus from the lower intertidal zone.

FIGURE 53.6 Experimental Evidence for Competitive Exclusion.

SOURCE: Connell, J. H. 1961. The influence of interspecific competition and other factors on the distribution of the barnacle *Chthamalus stellatus*. *Ecology* 42: 710–723.

✔**QUESTION** Why was it important to carry out both treatments on the same rock? Why not use separate rocks?

NEW! Experimental Inquiry Tutorials

Experimental Inquiry Tutorials based on some of biology's most seminal experiments can be found on **www.masteringbiology.com**. Your instructor may assign these. They will give you practice analyzing the experimental design and data, and help you understand reasoning that led scientists from the data they collected to their conclusions.

Some of the topics include:

- The Process of Science
- Engelmann's Photosynthesis and Wavelengths of Light
- Morgan's Cross with White-Eyed Males
- Meselson-Stahl's Semiconservative Replication
- Steinhardt et al and Hafner et al's Polyspermy
- Grant's Changes in Finch Beak Size
- Went's Phototropism and Auxin Distribution
- Coleman's Obesity Gene
- Connell's Competition in Barnacles
- Bormann, Likens et al's Nutrient Cycling in Hubbard Brook Forest

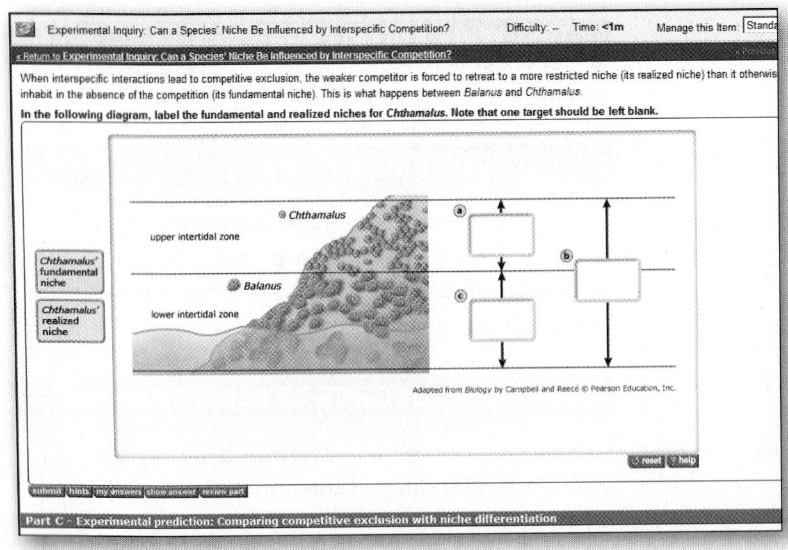
Experimental Inquiry: Can a Species' Niche Be Influenced by Interspecific Competition?
Difficulty: –   Time: <1m   Manage this Item:
« Return to Experimental Inquiry: Can a Species' Niche Be Influenced by Interspecific Competition?
When interspecific interactions lead to competitive exclusion, the weaker competitor is forced to retreat to a more restricted niche (its realized niche) than it otherwise would inhabit in the absence of the competition (its fundamental niche). This is what happens between Balanus and Chthamalus.
In the following diagram, label the fundamental and realized niches for Chthamalus. Note that one target should be left blank.
Chthamalus' fundamental niche
Chthamalus' realized niche
Chthamalus
upper intertidal zone
Balanus
lower intertidal zone
a
b
c
Adapted from Biology by Campbell and Reece © Pearson Education, Inc.
reset
help
submit
hints
my answers
show answer
review part
Part C - Experimental prediction: Comparing competitive exclusion with niche differentiation

NEW! Source Citations

Each Experiment Box now cites the original research paper, encouraging you to extend your learning by exploring the primary literature.

NEW! Experiment Box Questions

Each Experiment Box now includes a question that asks students to analyze the design of the experiment.

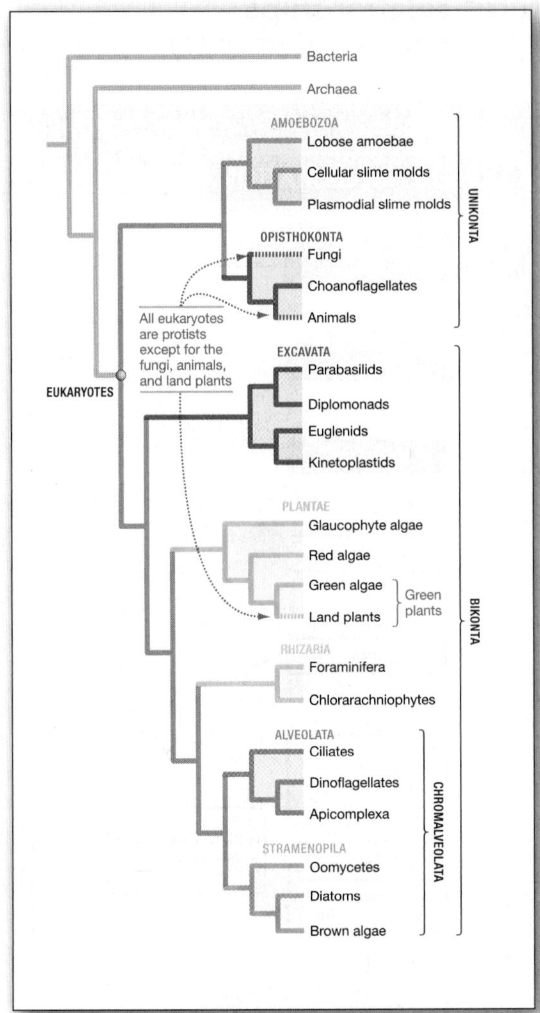

NEW! Redesigned Phylogenetic Trees

Practice "tree thinking" using these newly redesigned phylogenetic trees. Their U-shaped, top-to-bottom format is consistent with the way such trees are most commonly depicted in the scientific literature.

Expanded BioSkills Appendix

BIOSKILLS

Build skills that will be important to your success in future courses. At relevant points in the text, you'll find references to the expanded BioSkills Appendix that will help you learn and practice the following foundational skills:

- NEW! The Metric System
- Reading Graphs
- Reading a Phylogenetic Tree
- NEW! Some Common Latin and Greek Roots Used in Biology
- Using Statistical Tests and Interpreting Standard Error Bars
- Reading Chemical Structures
- Using Logarithms
- Making Concept Maps
- Separating and Visualizing Molecules
- Biological Imaging: Microscopy and X-Ray Crystallography
- NEW! Separating Cell Components by Centrifugation
- NEW! Cell Culture Methods
- Combining Probabilities
- NEW! Model Organisms

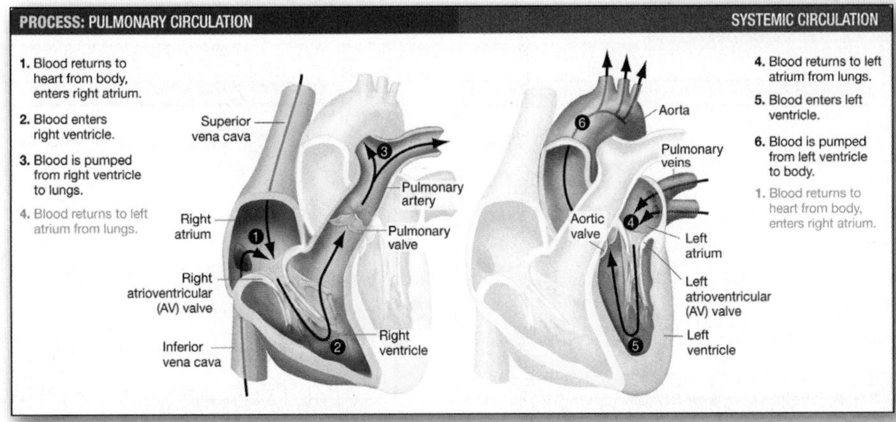

Informative Figures

Think through complex biological processes with figures that clearly define concepts.

Concept maps help you to keep sight of "big picture" relationships among biological concepts.

NEW! Big Picture Concept Maps

Four remarkable Big Picture concept maps help you synthesize information across the chapters on energy, genetics, evolution, and ecology.

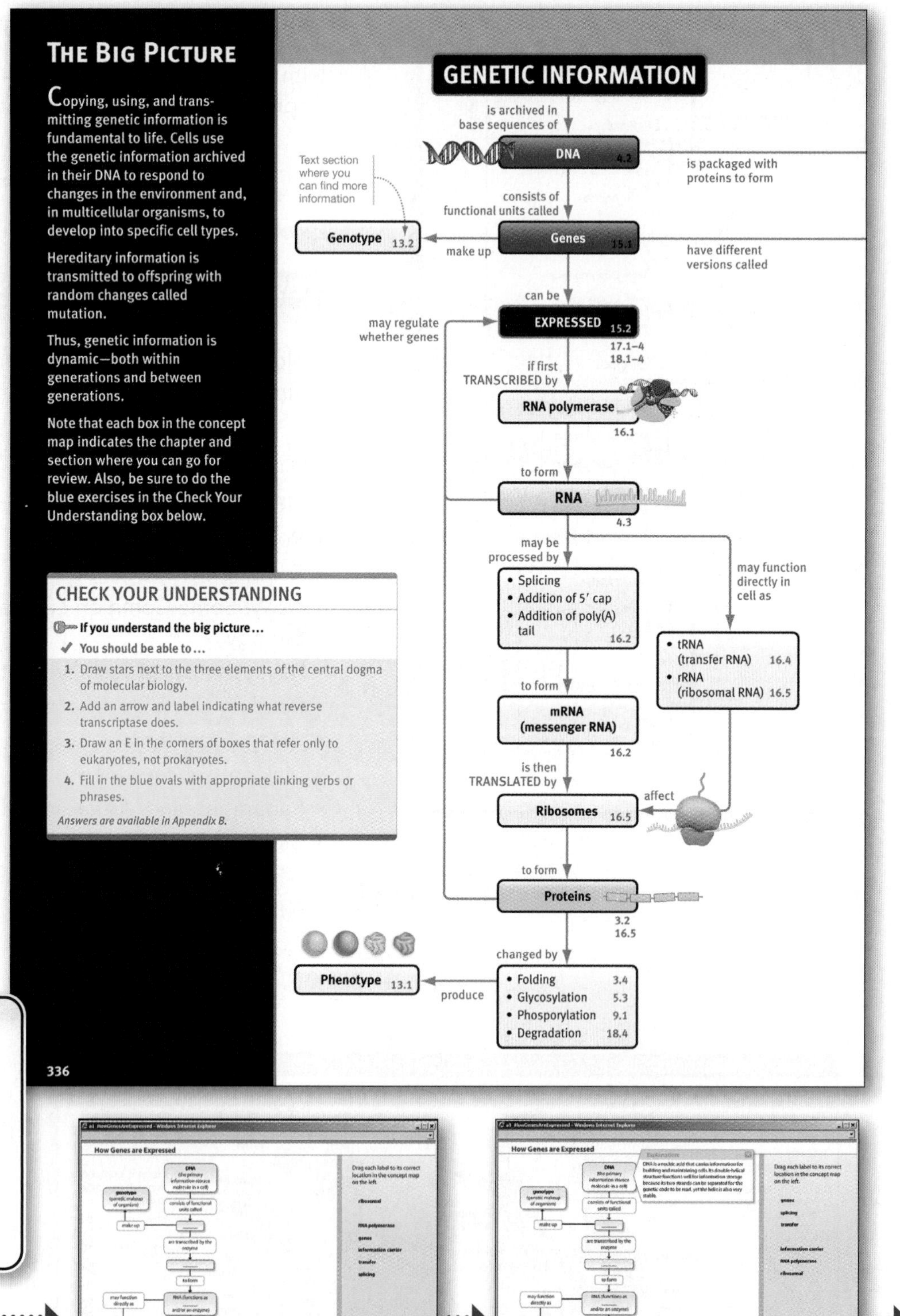

THE BIG PICTURE

Copying, using, and transmitting genetic information is fundamental to life. Cells use the genetic information archived in their DNA to respond to changes in the environment and, in multicellular organisms, to develop into specific cell types.

Hereditary information is transmitted to offspring with random changes called mutation.

Thus, genetic information is dynamic—both within generations and between generations.

Note that each box in the concept map indicates the chapter and section where you can go for review. Also, be sure to do the blue exercises in the Check Your Understanding box below.

CHECK YOUR UNDERSTANDING

If you understand the big picture . . .

You should be able to . . .

1. Draw stars next to the three elements of the central dogma of molecular biology.
2. Add an arrow and label indicating what reverse transcriptase does.
3. Draw an E in the corners of boxes that refer only to eukaryotes, not prokaryotes.
4. Fill in the blue ovals with appropriate linking verbs or phrases.

Answers are available in Appendix B.

Check Your Understanding

Check your understanding of these big picture relationships by answering the Blue Thread Questions.

MasteringBIOLOGY®

Make Learning Part of the Grade®

Your professor may assign interactive Big Picture concept map exercises at www.masteringbiology.com.

More Big Picture activities are available in the study area at www.masteringbiology.com (MB)

Chromatin 18.2 → ( ) → Chromosomes 11.1, 18.2

Chromosomes — may change due to →
- Breakage
- Duplication or deletion due to errors in meiosis
- Damage by radiation or other agents

12.4, 14.5, 15.4

— causing → Mutation 15.4 — can be → TRANSMITTED 11.1

Alleles 13.2 — are → COPIED 14.3 — and can be → TRANSMITTED 11.1

COPIED 14.3 — by → DNA polymerase 14.3 — occasionally makes errors, causing → MUTATION 15.4 — can be → TRANSMITTED 11.1

Chromosomes — can be → TRANSMITTED 11.1 (12.1, 13.1–4)

TRANSMITTED — to somatic cells by → MITOSIS 11.1

TRANSMITTED — to germ cells by → MEIOSIS 12.1

MEIOSIS — includes →
- Independent assortment
- Recombination

12.2, 13.3–4

MITOSIS — starts with → Parent cell (2n) — ends with → 2n, 2n — Two daughter cells with the same genetic information as the parent cell (unless mutation has occurred). — occurs during → GROWTH and ASEXUAL REPRODUCTION 11.0 — result in → Low genetic diversity

MEIOSIS — starts with → Parent cell (2n) — ends with → n, n, n, n — Four daughter cells with half the genetic information as the parent cell. — occurs during → SEXUAL REPRODUCTION 12.3 — results in → High genetic diversity

Independent assortment / Recombination → ( ) → High genetic diversity

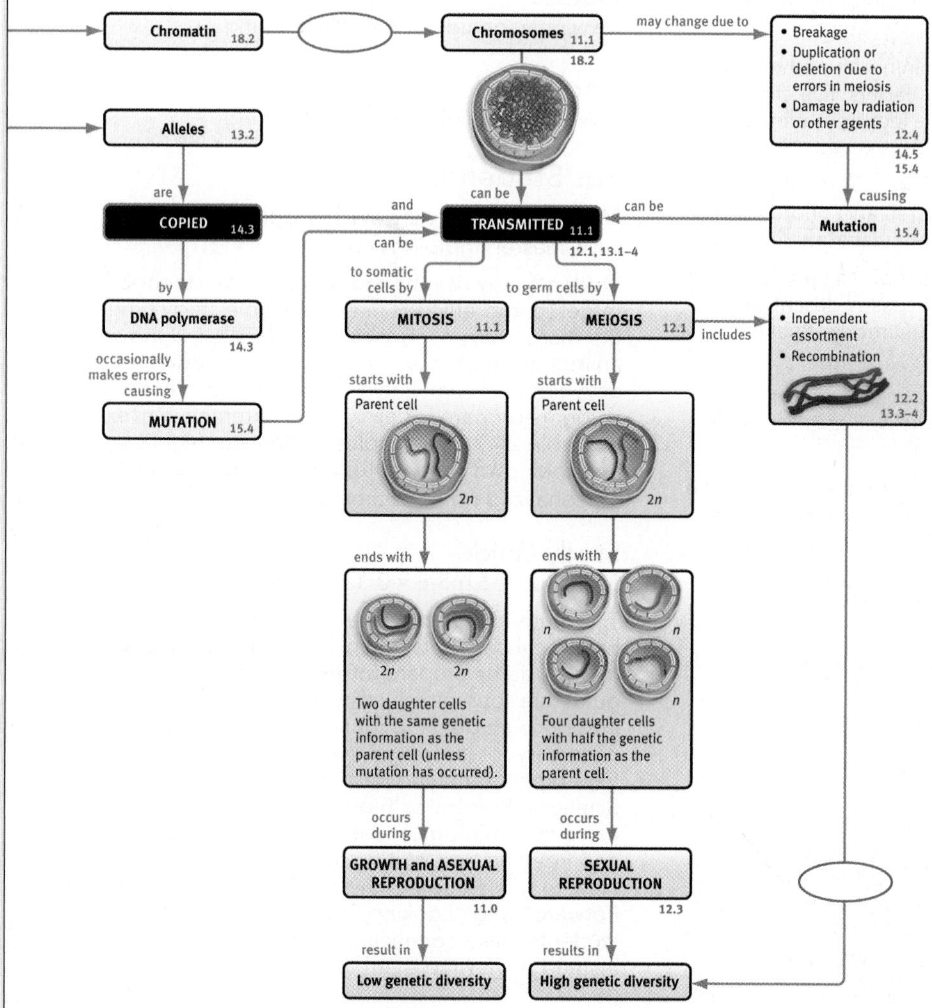

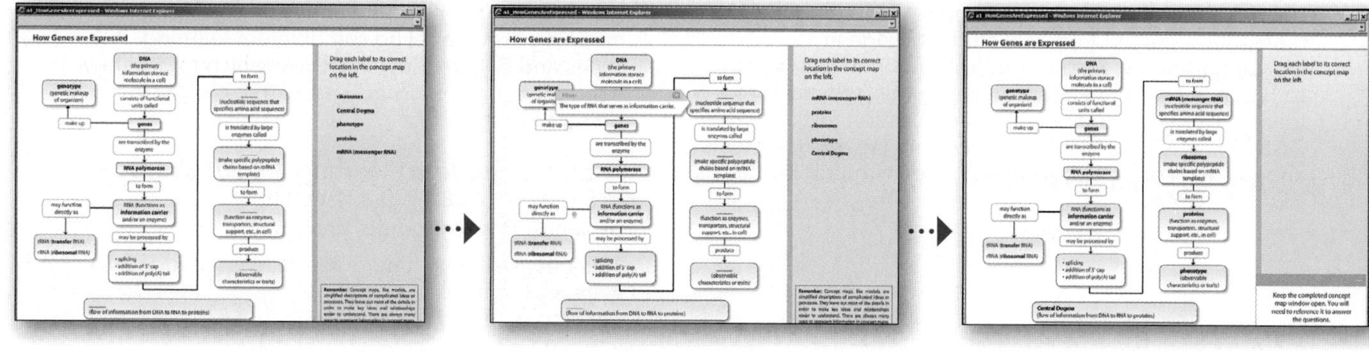

For Instructors

Instructor Resource CD/DVD-ROM

978-0-321-61351-6 • 0-321-61351-1

Everything instructors need for lectures is in one place, including video segments that demonstrate how to incorporate active-learning techniques into your own classroom. The Instructor Resource CD/DVD-ROM includes:

- PowerPoint® Lecture Tools containing all of the figures and photos, which have editable labels; five clicker questions per chapter; pre-made lecture outlines containing select images from the text with embedded animations
- JPEG images of all textbook figures and photos including printer-ready transparency acetate masters
- Over 300 animations and videos that accurately depict complex topics and dynamic processes described in the book; five new BioFlix™ 3-D movie-quality animations
- Instructor's Guides for *Biological Science* and *Practicing Biology* are available as well as a list of all primary literature citations
- The full test bank for *Biological Science* and the *Active Learning Workshop DVD* also come included with the IR-DVD

MasteringBiology® with Pearson eText

www.masteringbiology.com

Assign dynamic homework into your course with automatic grading and adaptive tutoring. Choose from a wide variety of stimulating activities, including visually stunning and scientifically accurate tutorials, ranking questions, 3-D animations, and test bank questions. The powerful gradebook compiles all your favorite teaching diagnostics—the hardest concept, class grade distribution, which students are spending the most or the least time on homework—with the click of a button. Instructors are empowered to customize the eText for themselves and students, including highlighting key text, annotating with comments, adding weblinks, and hiding chapters.

TestGen®

978-0-321-60533-7 • 0-321-60533-0

All of the exam questions in the test bank have been rigorously peer reviewed and revised using the metadata collected from real student usage in MasteringBiology. Test questions have been correlated to Bloom's Taxonomy of cognitive learning domains to identify which level of learning the question tests on. The Test Bank is also available in course management systems and in Microsoft® Word format on the Instructor Resources CD/DVD-ROM.

Course Management Options

CourseCompass™

www.pearsonhighered.com/elearning

This course management system contains preloaded content such as testing and assessment question pools.

WebCT

www.pearsonhighered.com/elearning

Blackboard

www.pearsonhighered.com/elearning

For Students

MasteringBiology with Pearson eText

www.masteringbiology.com

Students may use the Study Area in MasteringBiology for targeted and efficient use of valuable study time. Some of the many study tools include BioFlix™ 3-D movie-quality animations that focus on the toughest topics, engaging activities and cumulative chapter quizzes that help students prepare for exams. The interactive eText is available 24/7 and enables students to highlight text, add their own study notes, and review their Instructor's personalized notes at their convenience.

Study Guide

978-0-321-56168-8 • 0-321-56168-6

The Study Guide presents a breakdown of key biological concepts, difficult topics, and quizzes to help students prepare for exams. Unique to this study guide are four introductory, stand-alone chapters that introduce students to foundational ideas and skills necessary for classroom success: Introduction to Experimentation and Research in the Biological Sciences, Presenting Biological Data, Understanding Patterns in Biology and Improving Study Techniques, and Reading and Writing to Understand Biology. New to this edition of the Study Guide are "Looking Forward" and "Looking Back" sections that help students make connections across the chapters instead of viewing them as discrete entities.

Practicing Biology: A Student Workbook

978-0-321-61264-9 • 0-321-61264-7

This workbook focuses on key ideas, principles, and concepts that are fundamental to understanding biology. A variety of hands-on activities such as mapping and modeling suit different learning styles and help students discover which topics they need more help on. Students learn biology by doing biology.

A Short Guide to Writing About Biology

978-0-321-66838-7 • 0-321-66838-3

by Jan A. Pechenik, Tufts University

This best-selling writing guide teaches students to write and think as biologists.

BIOLOGICAL SCIENCE

VOLUME 1 The Cell, Genetics, & Development

Black Swan, *Cygnus atratus*
Male and female black swans have identical coloration—an all-black body with white flight feathers at the tips of the wings. Although the significance of their orange-red beak coloration is unknown, experiments with other species have shown that individuals with particularly bright beaks or feathers are in exceptionally good health and are attractive to potential mates. To explore how you might test this hypothesis in black swans, see Chapter 25.

BIOLOGICAL SCIENCE

FOURTH EDITION

SCOTT FREEMAN

University of Washington

Benjamin Cummings

Boston Columbus Indianapolis New York San Francisco Upper Saddle River
Amsterdam Cape Town Dubai London Madrid Milan Munich Paris Montréal Toronto
Delhi Mexico City São Paulo Sydney Hong Kong Seoul Singapore Taipei Tokyo

VP, Editor-in-Chief, Biology: Beth Wilbur
Acquisitions Editor: Becky Ruden
Executive Director of Development, Biology: Deborah Gale
Editorial Project Manager: Sonia DiVittorio
Development Editors: Alice Fugate, Moira Lerner-Nelson, William O'Neal, Susan Teahan
Art Editor: Kelly Murphy
Assistant Editor: Brady Golden
Editorial Assistant: Leslie Allen
Senior Media Producer: Laura Tommasi
Director of Editorial Content, Mastering Biology: Tania Mlawer
Developmental Editor, Mastering Biology: Sarah Jensen
Director of Marketing: Christy Lawrence
Executive Marketing Manager: Lauren Harp
Director of Production, Science: Erin Gregg
Managing Editor, Biology: Michael Early
Production Supervisor: Lori Newman
Media Production Supervisor: James Bruce
Supplements Production Supervisor: Jane Brundage
Production Management and Composition: S4Carlisle Publishing Services
Design Manager and Interior Designer: Marilyn Perry
Cover Designer: Riezebos Holzbaur Design Group
Illustrators: Kim Quillin, Imagineering Media Services
Photo Researcher: Maureen Spuhler
Manufacturing Buyer: Michael Penne
Cover Printer: Phoenix Color Corp.
Printer and Binder: Courier, Kendallville

Cover Photo Credits: Black Swan—*Cygnus atratus,* © Eric Isselée/Fotolia (front cover); Black swan among white swans, Hokkaido, Japan, North-East Asia © Keren Su/Getty Images, Inc. (back cover)

Credits and acknowledgments borrowed from other sources and reproduced with permission in this textbook appear on the appropriate page within the text or beginning on page C:1 of the backmatter.

Library of Congress Cataloging-in-Publication Data

Freeman, Scott
Biological science / Scott Freeman.—4th ed.
p. cm.
Includes index.
ISBN 978-0-32-159820-2 (student ed.)—ISBN 978-0-32-159819-6 (professional copy)—
ISBN 978-0-32-161347-9 (v. 1 : the cell, genetics, and development—ISBN 978-0-32-160530-6
(v. 2 : evolution, diversity, and ecology)—ISBN 978-0-32-157676-7 (v. 3 : how plants and animals work)
1. Biology—Textbooks. I. Title.

QH308. 2. F73 2011
570—dc22 2009047825

ISBN 10: 0-32-159820-2; ISBN 13: 978-0-32-159820-2 (Student edition)
ISBN 10: 0-32-159819-9; ISBN 13: 978-0-32-159819-6 (Professional copy)
ISBN 10: 0-32-161347-3; ISBN 13: 978-0-32-161347-9 (Volume 1)
ISBN 10: 0-32-160530-6; ISBN 13: 978-0-32-160530-6 (Volume 2)
ISBN 10: 0-32-157676-4; ISBN 13: 978-0-32-157676-7 (Volume 3)

3 4 5 6 7 8 9 10—CRK—14 13 12 11

Benjamin Cummings
is an imprint of

www.pearsonhighered.com

Brief Contents: Volume 1

Detailed Contents

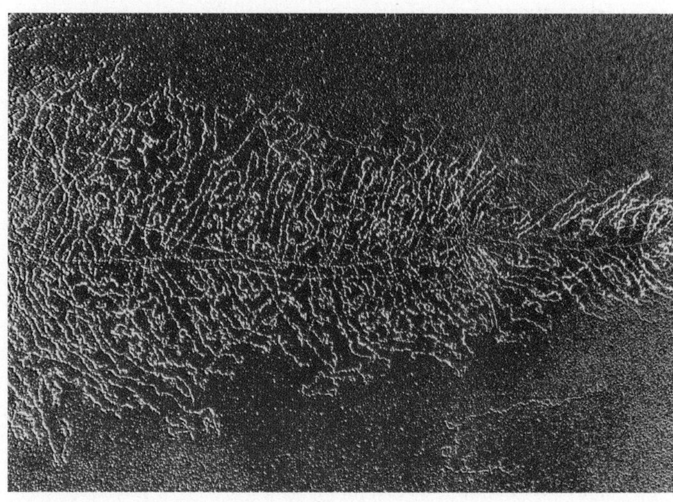

About the Author

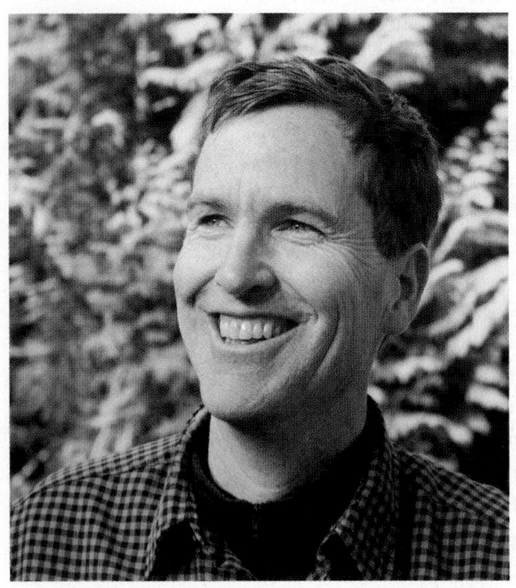

SCOTT FREEMAN received his Ph.D. in Zoology from the University of Washington and was subsequently awarded an Alfred P. Sloan Postdoctoral Fellowship in Molecular Evolution at Princeton University. His current research focuses on the scholarship of teaching and learning—specifically (1) how active learning and peer teaching techniques increase student learning and improve performance in introductory biology, and (2) how the levels of exam questions vary among introductory biology courses, standardized postgraduate entrance exams, and professional school courses. He has also done research in evolutionary biology on topics ranging from nest parasitism to the molecular systematics of the blackbird family. Scott teaches introductory biology for majors at the University of Washington and is coauthor, with Jon Herron, of the standard-setting undergraduate text *Evolutionary Analysis.*

Unit Advisors

Twelve cherished colleagues guided the revision process by synthesizing reviews, providing citations for recent high-impact publications, and drawing on their extensive teaching experience and subject-matter expertise to advise Scott on hundreds of questions, ranging from what to include to which analogies might communicate best to students. It is hard to overstate just how critical these people were to this revision. Through their own teaching and research and their work on this book, they are having a profound effect on how biology is taught.

Jason Flores, University of North Carolina, Charlotte (Unit 1)

Lisa Elfring, University of Arizona, Tucson (Unit 1)

Suzanne Simon-Westendorf, Ohio University (Unit 2)

Gregory Podgorski, Utah State University (Unit 3)

Kathleen Marrs, Indiana University–Purdue University, Indianapolis (Units 4, 6, and 7)

Jon Monroe, James Madison University (Units 4, 6, and 7)

Warren Burggren, University of North Texas (Units 4 and 8)

Joan Sharp, Simon Fraser University (Units 5 and 6)

Michael Black, California Polytechnic State University, San Luis Obispo (Units 6 and 8)

Kathleen Hunt, University of Portland (Units 6 and 9)

Emily Taylor, California Polytechnic State University, San Luis Obispo (Unit 8)

Fred Wasserman, Boston University (Unit 9)

Illustrator

KIM QUILLIN combines expertise in biology and information design to create lucid visual representations of biological principles. She received her B.A. in Biology at Oberlin College and her Ph.D. in Integrative Biology from the University of California, Berkeley (as a National Science Foundation Graduate Fellow), and taught undergraduate biology at both schools. Students and instructors alike have praised Kim's illustration programs for *Biological Science*, as well as *Biology: A Guide to the Natural World* by David Krogh and *Biology: Science for Life* by Colleen Belk and Virginia Borden, for their success in the visual communication of biology. Kim is a lecturer in the Department of Biological Sciences at Salisbury University.

Preface to Instructors

This book is a response to calls—from the National Academy of Sciences, the Howard Hughes Medical Institute and American Association of Medical Colleges, and the National Science Foundation—for changes in the way introductory biology is taught. Reports like *Biology 2010*, *Scientific Foundations for Future Physicians*, and *Vision and Change* are asking that introductory students not only learn the language of biology and understand fundamental concepts, but begin to apply those concepts in new situations, analyze experimental design, synthesize results, and evaluate hypotheses and data.

I wrote this book for instructors who embrace this challenge—who want to help their students learn how to think like a biologist. The essence of higher education is to promote higher-order thinking. Our job is to help students understand biological science at all six levels of Bloom's taxonomy of learning.

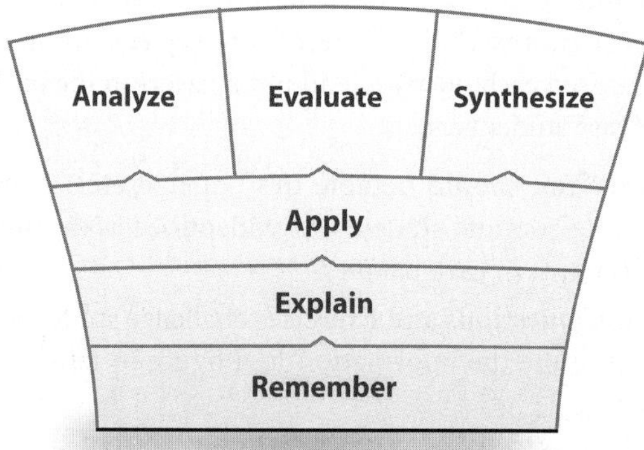

Bloom's Taxonomy. An annotated version of this graphic can be found in "To the Student: How to Use This Book" at the front of this book.

The Evolution of a Textbook

Evolution can be extremely fast in populations with short generation times and high mutation rates. Biology textbooks are no exception. Generation times have to be short because the pace of research in biology and student learning is so fast. This book, in particular, evolves quickly because it incorporates so many new ideas with each edition. Some of these "alleles" are novel mutations, but most arrive via lateral transfer—from advisors, reviewers, friends, students, and the literature.

In comparing the first edition of *Biological Science* with the fourth, the thing that jumps out is what educational researchers call "scaffolding"—tools that help professors and students teach and learn at the level demanded by the NAS, HHMI, AMC, and NSF. It's essential to have high expectations of our students, but we also need to provide the help and practice they need to meet those expectations.

What's New in This Edition

This revision was about making the book a better teaching and learning tool. To help students manage the mass of information and ideas that is contemporary biology, I broke long paragraphs into shorter paragraphs, made liberal use of numbered lists and bulleted lists to "chunk" information and ideas, and broke out dozens of new sections and subsections. I could almost hear my mother's voice: "Take small bites, and chew them well." I also made the book over 100 pages shorter by following Thoreau's dictum to "Simplify, simplify, simplify"—focusing on the most critical concepts that an introductory student needs to master.

In addition, the book team and I came up with a long list of new or expanded features.

- **The Big Picture** These new, two-page spreads are meant to help students see the forest for the trees. They are concept maps that focus on particularly critical areas—Energy, Genetic Information, Evolution, and Ecology. Each synthesizes content and concepts from an array of chapters and includes exercises for students to complete. You'll recognize these pages readily—their edges are colored black (for example, see The Big Picture: Energy on pages 192–193). In addition, the book's MasteringBiology® website has ten new concept map activities based on Big Picture content that will allow you to explore the concepts and their connections with your students during lecture.
- **BioSkills** *Biology 2010*, *Scientific Foundations*, and *Vision and Change* all place a premium on skills—the ability to read a graph, interpret an equation, understand the bands on a gel. The third edition of *Biological Science* introduced a series of appendixes focused on key skills for introductory biology students. Instructors and students found them extraordinarily helpful. New in this edition are BioSkills on using the metric system, common Latin and Greek roots, techniques for isolating and visualizing cell components, cell and tissue culture methods, and model organisms. BioSkills are located in Appendix A.
- **Answer Key** New to the Fourth Edition are suggested answers to all questions and exercises in the textbook. Students asked us to make this important change between editions to make the book a more complete study tool. The answer key will allow them to self-check their understanding while reading, and when reviewing for exams. Answers are in Appendix B.
- **Experiment Boxes** This text's hallmark has always been its emphasis on experimental evidence—on teaching how we know what we know. In the second edition, key experiments were converted to a boxed format so students could easily

navigate through the logic of the question, hypothesis, and test. In this edition, I added a new question to every experiment box to encourage students to analyze some aspect of the experiment's design.

- **Art Program** Recent research shows that students are more likely to interpret phylogenetic trees correctly if the trees are designed with U-shaped branches instead of Y-shaped branches. We responded by redesigning every phylogenetic tree in the text. To make other subject areas more accessible to visual learners, we enlarged figures, replaced hundreds of photos with clearer images, and strove to streamline labels and graphics across the board. (More on improvements to the art program below.)
- **MasteringBiology Quizzes** MasteringBiology gives students round-the-clock access to quizzes. We developed 550 new assignable questions based on the book's "Blue Thread" questions (more on the "Blue Thread" and its evolution below). We also developed 1100 new quiz questions, along with a cumulative practice test to simulate what a real exam might be like. To help students keep up with their reading, we created 55 new reading quizzes—one for each chapter—that you can assign through Mastering Biology.
- **MasteringBiology Experimental Inquiry Tutorials** The call to teach students about the process of science has never been louder. In response, a team lead by Tom Owens of Cornell University developed 10 new interactive tutorials on classic scientific experiments—ranging from Meselson–Stahl on DNA replication to the Grants' work on Galápagos finches and Connell's work on competition. Students who use these interactive tutorials should be better prepared to think critically about experimental design and evaluate the wider implications of the data—preparing them to do the work of real scientists in the future.
- **MasteringBiology BioFlix Animations and Tutorials** BioFlix™ are movie-quality, 3-D animations available on MasteringBiology. They focus on the most difficult core topics and are accompanied by in-depth, online tutorials that provide hints and feedback to help guide student learning. Thirteen BioFlix were available with the third edition of *Biological Science.* Five new BioFlix 3-D animations and tutorials have been developed for this edition—on mechanisms of evolution, homeostasis, gas exchange, population ecology, and the carbon cycle.

Changes to Gold Thread Scaffolding

The third edition introduced a dramatically expanded set of tools designed to help with a chronic problem for novice learners: picking out important information. Novices highlight every line in the text and try to memorize everything mentioned in lecture; experts instinctively home in on the key unifying ideas.

For students to make the novice-to-expert transition, we have to help them with features like:

1. **Key concepts** that are declared at the start of each chapter, highlighted with a key icon within the chapter, and reviewed at the end of the chapter.
2. **In-text highlighting**, in gold, that directs their attention to particularly important ideas.
3. **Check Your Understanding boxes**, at the end of key sections, with a bulleted list of key points.
4. **Summary tables** that pull information together in a compact format that is easy to review and synthesize.

The book team and I scrubbed every aspect of the gold thread—reviewing, revising, and re-revising. The third edition was a coming-out party for the gold thread; in the Fourth Edition we wanted it to dance.

Changes to Blue Thread Scaffolding

Each edition of this text has added tools to help students with metacognition—understanding what they do and don't understand. Novices like to receive information passively, and easily persuade themselves that they know what's going on. Experts are skeptical—they want to solve some problems before they're convinced that they know and understand an idea.

In the third edition, we formalized the metacognitive tools in *Biological Science* as a "Blue Thread" set of questions; in this edition, we revised each question and put answers in the back of the book for easy student access.

1. **In-text "You should be able to's"** offer exercises on topics that professors and students have identified as the most difficult concepts in each chapter.
2. **Caption Questions and Exercises** challenge students to critically examine the information in a figure or table—not just absorb it.
3. **Check Your Understanding boxes** present two to three tasks that students should be able to complete in order to demonstrate a mastery of summarized key ideas.
4. **Chapter Summaries** include "You should be able to" problems or exercises related to each of the key concepts declared in the gold thread.
5. **End-of-Chapter Questions** are organized around Bloom's taxonomy of learning, so students can test their understanding at the knowledge, comprehension, and application levels.

The fundamental idea is that if students really understand a piece of information or a concept, they should be able to do something with it. How do you get to Carnegie Hall? Practice.

As students mature as biologists-in-training and start taking upper-division courses, most or all of this scaffolding can disappear. By the time our students are juniors and seniors, they should have enough expertise to construct a high-level understanding on their own. But if a well-designed scaffold isn't there to get them started in their first and second years, when they are novices, most will flounder. We have to help them learn how to become good students.

Supporting Visual Learners

Figures can help students, especially visual learners, at all levels of Bloom's taxonomy—not only to understand and remember the material, but also to exercise higher levels of critical thinking. The overall goal of the Fourth Edition art revision was to hone the figures for accessibility to help novice learners recognize and engage with important visual information. In addition to re-designing the previously mentioned phylogenetic trees, Kim Quillin led the effort to enhance virtually every other aspect of the visual-teaching program.

- **Art and Photos** Kim enlarged art and photographs in figures throughout the book to increase clarity by making details physically easier to see. She also reduced the amount of detail in labels and graphics to simplify, simplify, simplify.
- **Color Use** Kim continues to use color strategically to draw attention to important parts of the figures. In this revision, she boosted color contrast in many figures to make the art more vibrant and the details easier to see.
- **Molecular Icons** Kim redesigned many molecular icons to simplify their shapes. The overall contours are based on molecular coordinates, when available, to accurately represent size and geometry, but she smoothed the textures for a simpler appearance—one that is more memorable and pleasing.
- **Molecular Models** New molecular models have been introduced to help students visualize structure-function relationships. In Chapter 5, for example, redesigned 2-D line drawings of sugars are now paired with 3-D ball-and-stick models.
- **"Pointers"** The Fourth Edition figures still use pointer annotations as a "whisper in the ear" to guide students in interpreting figures, but Kim has replaced the hand with an arrow to be more precise.

Serving a Community of Teachers

I love students, but I love teachers even more. There is nothing I like better than to sit in a workshop with a bunch of biology instructors, steal some good ideas, and come home to try them out—in some cases in a framework where I can collect data and test the hypothesis that the new approaches are improving student learning.

Research on biology education is gathering momentum, trying to catch up on the trail blazed by physics education researchers, bringing the same level of rigor to our classrooms that we bring to our lab benches and field sites. I try to bring the spirit and practice of evidence-based teaching into this textbook, and welcome your comments, suggestions, and questions.

Thank you for considering this text, and for your work on behalf of your students. We have the best jobs in the world.

SCOTT FREEMAN
University of Washington

Content Highlights of the Fourth Edition

As discussed in the preface, a major focus of this revision is to enhance the pedagogical utility of *Biological Science.* Another major goal is to ensure that the content reflects the current state of science and is accurate. In addition, every chapter has been rigorously evaluated for discussions that, in the previous edition, may have been too complex or overly detailed. As a result of this scrutiny, certain sections in every chapter have been simplified, content has been pruned judiciously, and the approach to certain topics has been re-envisioned to enhance student comprehension. In this section, some of the key content improvements to the textbook are highlighted.

Unit 1 The Molecules of Life

Chapter 1 A new experiment on ant navigation and discussions of tree-based naming systems and artificial selection in maize has been added. Coverage is expanded on the definition of life and the nature of science and religion.

Chapter 2 The descriptions of bond angles and the geometry of simple molecules are simplified. Added is a discussion on the hot-start hypothesis as well as a new Key Concept on the nature of chemical energy.

Chapter 3 This chapter has been streamlined by eliminating discussion of optical isomers/chirality and reducing coverage of enzyme kinetics and reaction rates.

Chapter 4 The discussion of RNA is expanded to include recently discovered roles for RNAs in cells. Also added is a new summary table (**Table 4.1**) comparing DNA and RNA structure.

Chapter 5 A stronger emphasis on the link between electronegativity of atoms and potential energy in C–C, C–H, and C–O bonds is developed. New ball-and-stick models are added to clarify the differences in location and orientation of functional groups.

Chapter 6 Coverage of secondary active transport has been expanded. Also included in this chapter is current research on the "first cell" and a discussion of non-random distribution of membrane proteins and phospholipids.

Unit 2 Cell Structure and Function

Chapter 7 New research on bacterial cell structure has been included, and a more explicit connection between lysosomes and the endomembrane system is emphasized. Centrifugation is moved to **BioSkills 11** in Appendix A.

Chapter 8 New sections on quorum sensing in bacteria and cross-talk among signal-transduction pathways have been added.

Chapter 9 The discussions of mitochondrial structure, ATP yield from glucose oxidation, and the role of GDP in the citric acid cycle have been updated. The introductory section on cellular respiration has been simplified.

Chapter 10 A new section on regulation (inhibition) has been added. The sections on C_4 and CAM photosynthesis now emphasize the role of these pathways in increasing CO_2 concentrations versus water conservation.

Chapter 11 Micrographs have been added to the phases of mitosis figure (**Figure 11.5**). The discussion on the role of activated MPF has been updated to include the triggering M phase of the cell cycle. Animal-cell culture methods are moved to **BioSkills 12** in Appendix A.

Unit 3 Gene Structure and Expression

Chapter 12 The discussions of recombination rates and aneuploidy rates in humans are updated. New micrographs have been added to the phases of meiosis figure (**Figure 12.7**).

Chapter 13 The linkage discussion and notation in fly crosses have been simplified. Sex-linkage is moved to the Mendelian section (Section 13.4 The Chromosome Theory of Inheritance), and mapping is now covered in Box 13.1 Quantitative Methods: Linkage. A new summary table (**Table 13.3**) presenting basic vocabulary used in Mendelian genetics has been added.

Chapter 14 A new space-filling model of DNA has been added to **Figure 14.4**.

Chapter 15 Discussions on mutation in the melanocortin receptor (link to mouse-coat-color camouflage) and karyotypes of cancerous cells have been added.

Chapter 16 The sections on transcription in bacteria and eukaryotes are now combined. The structure of the translation initiation complex in bacteria has been updated to reflect current science; snRNAs have been added to the discussion of RNA splicing.

Chapter 17 The chapter was streamlined with the removal of discussions of DNA fingerprinting and the structure of the operator and DNA-binding proteins. Treatment of catabolite repression/positive control has been trimmed.

Chapter 18 Included in this chapter is a new summary table (**Table 18.1**) comparing control of gene expression in bacteria and eukaryotes. Also added are discussions on ubiquitination and protein degradation, the importance of epigenetic inheritance (chromosome structure), and the histone code hypothesis.

Chapter 19 Southern/Northern/Western blots have moved to **BioSkills 9** in Appendix A. The discussions on golden rice, the impact of GM crops, and SNP association studies for human diseases have been updated with the most recent research. Notes on "next generation" sequencing technologies have been included.

Chapter 20 Human health applications now emphasize the use of genomics and microarrays to study cancer. Several datasets are updated, including sequencing database totals. New notes on miRNA genes, metagenomics, and the definition of the gene have been added.

Unit 4 Developmental Biology

Chapter 21 The discussions of *bicoid* and regulatory gene cascades are simplified. New material on auxin as a master regulator in early development and the importance of apoptosis has been added.

Chapter 22 The discussion about sea urchin fertilization and variation has been streamlined.

Chapter 23 A new section introducing basic concepts in angiosperm gametogenesis is added.

Unit 5 Evolutionary Processes and Patterns

Chapter 24 A section on the internal consistency of diverse data as evidence for evolution, including a new phylogeny and timeline of whale evolution, has been added. **Figure 24.6**, depicting the evolution of the Galápagos mockingbird, and long-term data on ground finches (**Figure 24.17**) are updated to reflect the most current science. There is a new graph on the evolution of drug resistance in pathogenic bacteria (**Figure 24.14**).

Chapter 25 The genetic drift example has changed from breeding in a small population on Pitcairn Island to coin flips simulating mating in a single couple (using data from the author's classroom). The prairie lupine gene flow example is replaced by recent work on an island population of *Parus major.* Notes on balancing selection and interactions among evolutionary forces have been included.

Chapter 26 The speciation-by-vicariance example has been changed from ratites to snapping shrimp, and the sympatric speciation example featuring soapberry bugs has been changed to apple/hawthorn flies.

Chapter 27 The sections on adaptive radiation and mass extinction have been completely reorganized. A new hypothesis for the cause of the Cambrian explosion is included, and detail on the "new genes, new bodies" hypothesis has been removed. Presentation of "Life's Timeline" has been significantly overhauled (see **Figures 27.8, 27.9,** and **27.10**).

Unit 6 The Diversification of Life

The model organisms have been moved to **BioSkills 14** in Appendix A. Phylogenetic trees have been redrawn to reflect a horizontal orientation with U-shaped branches for easier comprehension.

Chapter 28 New information on mechanisms of pathogenicity is added. Extensive updates include new notes on archaeon-eukaryote polymerases, the discovery of extensive biomass in the marine subfloor, an archaeon associated with a human disease, discovery of N-fixation and nitrification in archaea, and bacteriorhodopsin's role in phototrophy.

Chapter 29 A stronger emphasis on endosymbiosis as a theme in protist diversification has been threaded throughout this chapter.

Chapter 30 New content on green algae as a grade and on convergence in vascular tissue in mosses-vascular plants and gnetophytes-angiosperms has been added.

Chapter 31 The dynamic nature of mycelia, the importance of glomalin in soil, the role of mating types, and the discovery of "multigenomic" asexual glomales all have new supporting material.

Chapter 32 The treatment of embryonic tissues, developmental patterns, the coelom, and body symmetry have been updated to reflect the latest scientific thinking. A shift in emphasis to the origin of the neuron and cephalization has been implemented.

Chapter 33 New commentary on the independent transitions to land as well as a clarified discussion on the nature of the ecdysozoan-lophotrochozoan split are included. The discussion of annelids is updated to reflect recent results.

Chapter 34 The coverage of the echinoderm endoskeleton has been expanded and a phylogeny of early tetrapods has been added to the fin-to-limb transition figure (**Figure 34.16**). New data have been incorporated in the evolution-of-fishes timeline (**Figure 34.11**). The treatment of hagfish-lamprey, evolution-of-the-jaw (**Figure 34.14**) and *H. sapiens* migration (**Figure 34.40**) also include the most recent data available. The emphasis on the adaptive significance of the amniotic egg has changed from watertightness to increased size and support. Emphasis in the discussion of viviparity has changed to the adaptive advantage of embryo portability and temperature control. The recent analysis of *Ardipithecus ramidus* as the first hominin, with data on estimated body mass and braincase volume, has been included.

Chapter 35 The material on HIV phylogeny has been moved to the section on emerging viruses.

Unit 7 How Plants Work

Chapter 36 Surface area-to-volume ratios have been added as a theme in root and shoot systems. New information on contractile roots in *Ficus* and bulbs is incorporated into this chapter.

Chapter 37 New content on aquaporins and the transmembrane route to root xylem has been added, and coverage of why air has such low water pressure potential has been expanded.

Chapter 38 The description of nitrogen fixation has been clarified.

Chapter 39 **Figure 39.8** on the acid-growth hypothesis has been redesigned, and the discussion of polar auxin transport is simplified. New commentary on the role of brassinosteroids in growth regulation and on "talking trees" is included. The coverage of the receptors for GA, auxin, ABA, and brassinosteroids, and MeSA's role in the SAR has been updated with the most current research. Plant-tissue culture methods have been moved to **BioSkills 12** in Appendix A.

Chapter 40 Comments on day-length sensing and on pollination syndromes are new to this chapter.

Unit 8 How Animals Work

Chapter 41 New details on tissue types (especially connective tissue) have been incorporated. The discussion of thermoregulation has been completely reorganized for a more logical flow.

Chapter 42 The sections on the shark rectal gland and the mammalian loop of Henle have been revised to improve focus.

Chapter 43 A description of incomplete digestive systems is now included, and coverage of comparative aspects of digestive tract structure and function has been expanded.

Chapter 44 Information on the types of circulatory systems and types of blood vessels has been consolidated. Details on surface tension and lung elasticity have been removed while new content on countercurrent exchange in fish gills has been added.

Chapter 45 The chapter and section introductions have been rewritten to introduce a comparative context and to make the neuron-to-systems chapter organization more transparent. New content on interspecific variation in nervous systems has been added.

Chapter 46 The chapter has been shortened and its focus sharpened by the removal of nonessential information.

Chapter 47 New material on EPO abuse in athletes has been included.

Chapter 48 The section on sperm competition includes new data from experiments on seed beetles.

Chapter 49 The discussion of the V regions of BCRs and antibodies and recombination in BCR/TCR genes has been simplified. New content on autoimmune disorders and diseases associated with immunosuppression, allergies, and immunodeficiency diseases has been added. The discussion of vaccination has been expanded.

Unit 9 Ecology

Chapter 50 New information on the importance of nutrient availability in aquatic ecosystems, with details on lake turnover and ocean upwelling, is included. A new section on the Wallace line has also been added.

Chapter 51 The content in this chapter has been completely reorganized to increase cohesiveness. It is presented as a series of questions in behavioral ecology, with each question addressed at the proximate and ultimate levels with separate case studies. Material on modes of learning, innate behavior, bat-moth interactions, sex change in wrasses, and acoustic and visual signaling in red-winged blackbirds has been trimmed or dropped. New content on child abuse in humans is added.

Chapter 52 Discussion of the hare–lynx-cycle field experiment has been reorganized for clarity, with new supporting "Results" data added to accompanying **Figure 52.12**.

Chapter 53 New content has been added on species richness and resistance of communities to invasion, the use of predators or parasites as biocontrol agents, and character displacement in finches. The discussion of succession in Glacier Bay is reorganized and simplified. The discussion of alternative hypotheses to explain the latitudinal gradient in species richness has been expanded and clarified.

Chapter 54 The chapter was rewritten and reorganized to sharpen its focus on human impacts. Sections on trophic cascades and biomagnification have been added, as have recent data on human appropriation of NPP, sources of nutrient gain and loss, and the impact of ocean acidification on coral growth.

Chapter 55 New content on the impact of global climate change and a new section on ways to preserve biodiversity are now included. Two new boxes on quantitative methods have been added: one on estimating species numbers and species losses and the other on population viability analysis.

Acknowledgments

Reviewers

The peer review system is the key to quality and clarity in science publishing. In addition to providing a filter, the investment that respected individuals make in vetting the material—catching errors or inconsistencies and making suggestions to improve the presentation—gives authors, editors, and readers confidence that what they are publishing and reading meets rigorous professional standards.

Peer review plays the same role in textbook publishing. The time and care that this book's reviewers have invested is a tribute to their professional integrity, their scholarship, and their concern for the quality of teaching. Virtually every paragraph in this edition has been revised and improved based on insights from the following individuals.

Ann Aguanno, *Marymount Manhattan College*
Adrienne Alaie, *City University of New York, Hunter College*
John Alcock, *Arizona State University*
Sylvester Allred, *Northern Arizona University*
Suzanne Alonzo, *Yale University*
Dan Ardia, *Franklin and Marshall College*
Peter Armbruster, *Georgetown University*
David Asch, *University of Pennsylvania*
Andrea Aspbury, *Texas State University, San Marcos*
Nicanor Austriaco, *Providence College*
Mitchell Balish, *Miami University*
Elizabeth Balko, *State University of New York, Oswego*
Ralston Bartholomew, *Warren County Community College*
Christine Barton, *Centre College*
Robert Bauman, *Amarillo College*
Wayne Becker, *University of Wisconsin, Madison*
Peter Berget, *Carnegie Mellon University*
Ethan Bier, *University of California, San Diego*
Michael Black, *California Polytechnic State University, San Luis Obispo*
Anthony Bledsoe, *University of Pittsburgh*
James Bottesch, *Brevard Community College*
Scott Bowling, *Auburn University*
Robert Boyd, *Auburn University*
Ronald Breaker, *Yale University*
Diane Bridge, *Elizabethtown College*
Andrew Brower, *Middle Tennessee State University*
Rebecca Brown, *College of Marin*
Mark Browning, *Purdue University*
Arthur Buikema, *Virginia Polytechnic Institute and State University*
Warren Burggren, *University of North Texas*
Jennifer Carbrey, *Duke University*
Dale Casamatta, *University of North Florida*
Gregory Chandler, *University of North Carolina, Wilmington*
Deborah Chapman, *University of Pittsburgh*
Curt Coffman, *Vincennes University*
Patricia Colberg, *University of Wyoming*
Kathleen Cornely, *Providence College*
Elizabeth Cowles, *Eastern Connecticut State University*
Jason Curtis, *Purdue University, North Central*
Karen Curto, *University of Pittsburgh*
Farahad Dastoor, *University of Maine*
Robin Lee Davies, *Sweet Briar College*
Charles Delwiche, *University of Maryland, College Park*
Jean DeSaix, *University of North Carolina, Chapel Hill*
Hudson DeYoe, *University of Texas, Pan American*
Sunethra Dharmasari, *Texas State University, San Marcos*
Lisa Elfring, *University of Arizona, Tucson*
Eric Engstrom, *College of William and Mary*
Jean Everett, *College of Charleston*
Brent Ewers, *University of Wyoming*
Andrew Fabich, *Tennessee Temple University*
Gary Firestone, *University of California, Berkeley*
Ryan Fisher, *Salem State College*
David Fitch, *New York University*
Jason Flores, *University of North Carolina, Charlotte*
Andrew Forbes, *University of Notre Dame*
Edward Freeman, *St. John Fisher College*
Caitlin Gabor, *Texas State University, San Marcos*
George Gilchrist, *College of William and Mary*
Lynda Goff, *University of California, Santa Cruz*
Elliot Goldstein, *Arizona State University*
Andrew Goliszek, *North Carolina Agricultural & Technical State University*
Margaret Goodman, *Wittenberg University*
Joyce Gordon, *University of British Columbia*
Steve Gorsich, *Central Michigan University*
Susan Michele Green, *Texas State University, San Marcos*
Paul Greenwood, *Colby College*
Stanley Guffey, *University of Tennessee, Knoxville*
Wendy Hanna-Rose, *Pennsylvania State University*
Kiki Harbitz, *Gustavus Adolphus College*
Jeffrey Hardin, *University of Wisconsin, Madison*
Jana Henson, *Georgetown College*
Helen Hess, *College of the Atlantic*
Tracey Hickox, *University of Illinois, Urbana-Champaign*
Sara Hoot, *University of Wisconsin, Milwaukee*
Laurie Host, *Harford Community College*
Kelly Howe, *University of New Mexico, Valencia*
Kathleen Hunt, *University of Portland*
Christine Janis, *Brown University*
Eric Jellen, *Brigham Young University*
Warren Johnson, *University of Wisconsin, Green Bay*
Cindy Johnson-Groh, *Gustavus Adolphus College*
Greg M. Kelly, *University of Western Ontario*
Paul King, *Massasoit Community College*
Joel Kingsolver, *University of North Carolina, Chapel Hill*
Roger Koeppe, *Oklahoma State University*
David Kooyman, *Brigham Young University*

Jocelyn Krebs, *University of Alaska, Anchorage*
John Krenetsky, *Metro State College of Denver*
Patrick Krug, *California State University, Los Angeles*
Holly Kupfer, *Central Piedmont Community College*
Mary Rose Lamb, *University of Puget Sound*
John Lammert, *Gustavus Adolphus College*
Hans Landel, *Edmonds Community College*
Dominic Lannutti, *El Paso Community College*
Janet Lanza, *University of Arkansas, Little Rock*
Georgia Lind, *Kingsborough Community College*
Debra Linton, *Central Michigan University*
Curtis Loer, *University of California, San Diego*
Frank Logiudice, *University of Central Florida*
Barbara Lom, *Davidson College*
James Maller, *University of Colorado, Denver*
James Manser, *Harvey Mudd College (retired)*
Kathleen Marrs, *Indiana University–Purdue University, Indianapolis*
Jennifer Martin, *University of Colorado, Boulder*
Kathy Martin-Troy, *Central Connecticut State University*
John H. McDonald, *University of Delaware*
Robert McLean, *Texas State University, San Marcos*
Michael Meighan, *University of California, Berkeley*
Mariana Melo, *North Essex Community College*
Philip Meneely, *Haverford College*
Dennis Minchella, *Purdue University*
Alan Molumby, *University of Illinois, Chicago*
Vertigo Moody, *Santa Fe Community College*
Elizabeth Morgan, *Lonestar College Kingswood*
Nancy Morvillo, *Florida Southern College*
Mike Muller, *University of Illinois, Chicago*
Dennis Nyberg, *University of Illinois, Chicago*
Amanda N. Orenstein, *Centenary College of New Jersey*
Stephanie Scher Pandolfi, *Michigan State University*
Lisa Parks, *North Carolina State University*
Robert Paul, *Kennesaw State University*
Andrew Pease, *Stevenson University*
Andrew Pekosz, *Johns Hopkins University*
Nancy Pelaez, *Purdue University*
Roger Persell, *City University of New York, Hunter College*
John Peters, *College of Charleston*
Debra Pires, *University of California, Los Angeles*
Gregory Podgorski, *Utah State University*
Robert Podolsky, *College of Charleston*
Therese Poole, *Georgia State University*
Harvey Pough, *Rochester Institute of Technology*
Vanessa Quinn, *Purdue University North Central*
Stephanie Randell, *McLennan Community College*
Clifford Ross, *University of North Florida*
Michael Rutledge, *Middle Tennessee State University*
James Ryan, *Hobart and William Smith Colleges*
Margaret Saha, *College of William and Mary*
Mark Sandheinrich, *University of Wisconsin, La Crosse*
Terry Saropoulos, *Vanier College*
Thomas Sasek, *University of Louisiana, Monroe*
Jon Scales, *Midwestern State University*
Gregory Schmaltz, *Great Basin College*
Oswald Schmitz, *Yale University*
Joan Sharp, *Simon Fraser University*
Michele Shuster, *New Mexico State University*
Suzanne Simon-Westendorf, *Ohio State University*
David Skelly, *Yale University*
Meredith Somerville-Norris, *University of North Carolina, Charlotte*
Sally Sommers-Smith, *Boston University*
Eric Stavney, *Pierce College*
Scott Steinmaus, *California Polytechnic State University, San Luis Obispo*
Janet Steven, *Sweet Briar College*
Kirk Stowe, *University of South Carolina*
Christine Strand, *California Polytechnic State University, San Luis Obispo*
Cynthia Surmacz, *Bloomsburg University*
Jackie Swanik, *North Carolina Central University*
Jerilyn Swann, *Maryville College*
Brad Swanson, *Central Michigan University*
Emily Taylor, *California Polytechnic State University, San Luis Obispo*
Eric Thobaben, *Carroll University*
Ken Thomas, *North Essex Community College*
Briana Timmerman, *University of South Carolina*
Alexandru M. F. Tomescu, *Humboldt State University*
James Traniello, *Boston University*
William Velhagen, *New York University*
Sara Via, *University of Maryland, College Park*
Susan Waaland, *University of Washington*
Frederick Wasserman, *Boston University*
Elizabeth Weiss-Kuziel, *University of Texas, Austin*
Jason Wiles, *Syracuse University*
Kelly P. Williams, *Virginia Polytechnic Institute and State University*
Elizabeth Willott, *University of Arizona, Tucson*
Clifford Wilson, *Kennedy King College*
Charles Wimpee, *University of Wisconsin, Milwaukee*
Leslie Wooten-Blanks, *College of Charleston*
Todd Yetter, *University of the Cumberlands*

Correspondents

One of the most enjoyable interactions I have as a textbook author is correspondence or conversations with researchers and teachers who take the time and trouble to contact me to discuss an issue with the book, or who respond to my queries about a particular data set or study. I'm always amazed and heartened by the generosity of these individuals. They care, deeply.

Julie Aires, *Florida Community College, Jacksonville*
Göran Arnqvist, *Uppsala University*
Terry Bidleman, *Environment Canada*
Brian Buchwitz, *University of Washington*
Helaine Burstein, *Ohio University*
Curt Coffman, *Vincennes University*
Mark Cooper, *University of Washington*
Scott Creel, *Montana State University*
Laura DiCaprio, *Ohio University*
Mary Durant, *Lone Star College*
Victoria Finnerty, *Emory University*
Kathleen Foltz, *University of California, Santa Barbara*
Kathy Gillen, *Kenyon College*
Peter Grant, *Princeton University*
Rosemary Grant, *Princeton University*
Takato Imaizumi, *University of Washington*
Kathleen Janech, *College of Charleston*
Paul King, *Massosoit Community College*

John Lammert, *Gustavus Adolphus College*
Curtis Loer, *University of San Diego*
Scott Meissner, *Cornell University*
Philip Meneely, *Haverford College*
Diarmaid Ó Foighil, *University of Michigan*
John Roth, *University of California, Davis*
Brian Spohn, *Florida Community College, Jacksonville*
Fayla Schwartz, *Everett Community College*
Elizabeth Willott, *University of Arizona, Tucson*
Dan Wulff, *University at Albany, State University of New York*
Glenn Yasuda, *Seattle University*

Supplements Contributors

Instructors depend on an impressive array of support materials—in print and online—to design and deliver their courses. The student experience would be much weaker without the study guide, test bank, activities, animations, quizzes, and tutorials written by the following individuals.

Brian Bagatto, *University of Akron*
Michael Black, *California Polytechnic State University, San Luis Obispo*
Jay L. Brewster, *Pepperdine University*
Warren Burggren, *University of North Texas*
Patricia Colberg, *University of Wyoming*
Clarissa Dirks, *Evergreen State College*
Lisa Elfring, *University of Arizona, Tucson*
Brent Ewers, *University of Wyoming*
Miriam Ferzli, *North Carolina State University*
Cheryl Frederick, *University of Washington*
Cindee Giffen, *University of Wisconsin, Madison*
Kathy M. Gillen, *Kenyon College*
Mary Catherine Hager
Christopher Harendza, *Montgomery County Community College*
Laurel Hester, *University of South Carolina*
Jean Heitz, *University of Wisconsin, Madison*
Tracey Hickox, *University of Illinois, Urbana-Champaign*
Kathleen Hunt, *University of Portland*
Barbara Lom, *Davidson College*
Jennifer Nauen, *University of Delaware*
Chris Pagliarulo, *University of Arizona, Tucson*
Stephanie Scher Pandolfi, *Michigan State University*
Debra Pires, *University of California, Los Angeles*
Gregory Podgorski, *Utah State University*
Carol Pollock, *University of British Columbia*
Jessica Poulin, *University at Buffalo, the State University of New York*
Vanessa Quinn, *Purdue University North Central*
Eric Ribbens, *Western Illinois University*
Joan Sharp, *Simon Fraser University*
Suzanne Simon-Westendorf, *Ohio University*
Emily Taylor, *California Polytechnic State University, San Luis Obispo*
Fred Wasserman, *Boston University*
Cindy White, *University of Northern Colorado*

Book Team

People who watch film credits won't be surprised to learn how many talented people are involved in making a textbook happen. Ruth Steyn provided incisive comments on the earliest drafts of the revised manuscript, which were then implemented by Development Editors Alice Fugate, Moira Lerner-Nelson, Bill O'Neal, and Susan Teahan; the final version of the text was copyedited by Michael Rossa and proofread by Pete Shanks. This editorial team was directed by Executive Director of Development Deborah Gale.

As in the first three editions, the book's figures were designed by Dr. Kim Quillin. Kim's artistic sensibilities, scientific training, and teaching talent are what make the figures in this book sing. If imitation really is the sincerest form of flattery, then Kim should be blushing—we've started seeing her style copied in textbook figures and scientific illustrations by other illustrator/designers. Kim's designs were rendered by Imagineering Media Services; the final art manuscript was vetted by Art Editor Kelly Murphy and rendered art expertly proofread by Frank Purcell. Maureen Spuhler did a thorough review of the photography program and researched the hundreds of images new to the Fourth Edition.

The book's clean, elegant design is the brainchild of Marilyn Perry; the text and art were set in Marilyn's design by S4Carlisle Publishing Services. The book's production was supervised by Lori Newman and Mike Early.

The extensive supplements program was managed by Assistant Editor Brady Golden. All of the individuals I've mentioned—and more—were supported by Editorial Assistant Leslie Allen.

Creating MasteringBiology® tutorials and activities requires a talented team of people playing many different roles. Media content development was overseen by Tania Mlawer and Sarah Jensen who benefited from the program expertise of Caroline Power and Karen Sheh. Julia Henderson and Laura Tommasi worked together as Media Producers with the leadership of Senior Media Producer Deb Greco. Lauren Fogel (VP, Director, Media Development), Stacy Treco (VP, Director, Media Product Strategy), and Laura ensured that the complete media program that accompanies the Fourth Edition, including MasteringBiology, will meet the needs of the students and professors who use our offerings.

Pearson's talented Sales Reps, who listen to professors, advise the editorial staff, and get the book in students' hands, are supported by tireless Executive Marketing Manager Lauren Harp and Director of Marketing Christy Lawrence. The marketing materials that support the outreach effort were produced by Lillian Carr and her colleagues in Pearson's MarCom group.

The vision and resources required to run this entire enterprise are the responsibility of Vice President and Editor-in-Chief Beth Wilbur, Senior Vice President and Editorial Director Frank Ruggirello, President of Pearson Science Paul Corey, and President of Pearson Science & Math Linda Davis.

Finally, the driving forces behind this edition were Project Manager Sonia DiVittorio and Acquisitions Editor Becky Ruden. Sonia's razor-sharp mind, depth of heart, and dedication to excellence shine in every page. Becky's energy and belief in this book surmounted obstacles that would have felled any other editor. I thank the two of them from the bottom of my heart.

Media Guide

Students who purchase a new copy of the text receive free access to MasteringBiology® *(www.masteringbiology.com),* which contains valuable videos, animations, and practice quizzes to help students learn and prepare for exams.

THE BIG PICTURE New to the Fourth Edition, The Big Pictures are interactive concept maps based on four overarching topics in biology that help students synthesize information across broad concepts and not get lost in the details.

Energy (Chapters 9 and 10)

- How Photosynthesis Yields Sugar
- How Cellular Respiration Yields ATP
- How Photosynthesis Relates to Cellular Respiration

Genetic Information (Chapters 12–18)

- How Genes Are Expressed
- How Genetic Information Is Copied and Transmitted
- How Genetic Information Changes

Evolution (Chapters 24–27)

- How Species Evolve
- How Species Form the Tree of Life

Ecology (Chapters 50–55)

- How Organisms Interact in Their Environment
- How Energy and Nutrients Flow through Ecosystems

BIOFLIX™ BioFlix are 3-D movie-quality animations with carefully constructed student tutorials, labeled slide shows, study sheets, and quizzes which bring biology to life.

WEB ACTIVITIES Web Activities help students learn biological concepts via simple, cartoon-style animations and contain pre-quizzes and post-quizzes to test student's understanding of biology's dynamic processes and concepts.

DISCOVERY VIDEOS Brief videos from the Discovery Channel on 29 different biology topics are available for student viewing along with a corresponding video quiz.

VIDEOS Additional molecular and microscopy videos provide vivid images of processes of the cell.

BIOSKILLS BioSkills (in Appendix A) provide background on key skills and techniques for introductory biology students. New to the Fourth Edition are online questions that give students practice building their skill set.

GRAPHIT! Graphing tutorials show students how to plot, interpret, and critically evaluate real data.

Chapter 1

- An Introduction to Graphing

Chapter 50

- Animal Food Production Efficiency and Food Policy
- Atmospheric CO_2 and Temperature Changes

Chapter 52

- Age Pyramids and Population Growth

Chapter 53

- Species Area Effect and Island Biogeography

Chapter 55

- Forestation Change
- Global Fisheries and Overfishing
- Municipal Solid Waste Trends in the U.S.
- Global Freshwater Resources
- Prospects for Renewable Energy
- Global Soil Degradation

WORD STUDY TOOLS New to the Fourth Edition are Latin and Greek root word flash cards to help students practice the language of biology. In addition, an audio glossary provides correct pronunciation to help students learn key terms introduced in the book.

CUMULATIVE TEST Every chapter offers 20 Practice Test questions that students can pool from different chapters into a Cumulative Test to simulate a practice exam.

RSS FEEDS Real Simple Syndication directly links breaking news from four important sources: National Public Radio, *Scientific American, Science Daily News,* and *BioScience.* Current articles reinforce the dynamic nature of science in our daily lives.

eTEXT The eText of *Biological Science,* Fourth Edition, is available online 24/7 for students' convenience. New annotation, highlighting, and bookmarking tools allow students to personalize the material for efficient review.

MEDIA AT A GLANCE

	BIOFLIX	WEB ACTIVITIES	DISCOVERY VIDEOS	VIDEOS	BIOSKILLS
1 Biology and the Tree of Life		Artificial Selection; Introduction to Experimental Design	Cells; Charles Darwin; Early Life; Bacteria		The Metric System; Reading Graphs; Reading a Phylogenetic Tree; Some Common Latin and Greek Roots Used in Biology
Unit 1 The Molecules of Life					
2 Water and Carbon: The Chemical Basis of Life		The Properties of Water			Reading Chemical Structures; Using Logarithms; Making Concept Maps; Reading Graphs
3 Protein Structure and Function		Condensation and Hydrolysis Reactions; Activation Energy and Enzymes		An Idealized Alpha Helix (A); An Idealized Alpha Helix (B); An Idealized Beta-Pleated Sheet (A); An Idealized Beta-Pleated Sheet (B)	
4 Nucleic Acids and the RNA World		Structure of RNA and DNA		Stick Model of DNA; Surface Model of DNA	Separating and Visualizing Molecules; Biological Imaging: Microscopy and X-Ray Crystallography
5 An Introduction to Carbohydrates		Carbohydrate Structure and Function			
6 Lipids, Membranes, and the First Cells	Membrane Transport	Diffusion and Osmosis; Membrane Transport Proteins		Space-filling Model of Cholesterol; Stick Model of Cholesterol; Space-filling Model of Phosphatidylcholine; Stick Model of a Phosphatidylcholine	Biological Imaging: Microscopy and X-Ray Crystallography; Separating and Visualizing Molecules
Unit 2 Cell Structure and Function					
7 Inside the Cell	Tour of an Animal Cell; Tour of a Plant Cell	Transport into the Nucleus; A Pulse-Chase Experiment	Bacteria; Cells	Confocal vs. Standard Fluorescence Microscopy; Cytoplasmic Streaming; Crawling Amoeba	Separating Cell Components by Centrifugation; Biological Imaging: Microscopy and X-Ray Crystallography; Separating and Visualizing Molecules
8 Cell-Cell Interactions			Cells	Connexon Structure	Separating and Visualizing Molecules
9 Cellular Respiration and Fermentation	Cellular Respiration	Redox Reactions; Glucose Metabolism		Space-filling Model of ATP (adenosine triphosphate); Stick Model of ATP (adenosine triphosphate)	
10 Photosynthesis	Photosynthesis	Chemiosmosis; Photosynthesis; Strategies for Carbon Fixation	Space Plants	Space-filling Model of Chlorophyll	
11 The Cell Cycle	Mitosis	The Phases of Mitosis; Four Phases of the Cell Cycle	Fighting Cancer	Mitosis	Separating and Visualizing Molecules; Cell and Tissue Culture Methods

MEDIA AT A GLANCE *(continued)*

	BIOFLIX	WEB ACTIVITIES	DISCOVERY VIDEOS	VIDEOS	BIOSKILLS
Unit 3 Gene Structure and Expression					
12 Meiosis	Meiosis	Meiosis; Mistakes in Meiosis			Combining Probabilities; Using Statistical Tests and Interpreting Standard Error Bars
13 Mendel and the Gene		Mendel's Experiments; The Principle of Independent Assortment			Model Organisms; Combining Probabilities; Reading Graphs
14 DNA and the Gene: Synthesis and Repair	DNA Replication	DNA Synthesis			Separating Cell Components by Centrifugation; Cell and Tissue Culture Methods; Using Logarithms; Reading Graphs
15 How Genes Work		The One-Gene One-Enzyme Hypothesis; The Triplet Nature of the Genetic Code			
16 Transcription, RNA Processing, and Translation	Protein Synthesis	RNA Synthesis; Synthesizing Proteins		A Stick-and-Ribbon Rendering of a tRNA	
17 Control of Gene Expression in Bacteria		The *lac* Operon		Cartoon Model of the *lac* Repressor from *E. coli*	
18 Control of Gene Expression in Eukaryotes		Transcription Initiation in Eukaryotes		Cartoon Model of the DNA-Binding Portion of TATA-Binding Protein Interacting with DNA; Cartoon Model of the GAL4 Transcription Factor from the yeast *S. cerevisiae*	Biological Imaging: Microscopy and X-Ray Crystallography; Separating and Visualizing Molecules
19 Analyzing and Engineering Genes		Producing Human Growth Hormone; The Polymerase Chain Reaction	Transgenics	Cartoon Model of the BamH1a Endonuclease	Separating and Visualizing Molecules
20 Genomics		Human Genome Sequencing Strategies	DNA Forensics		Model Organisms; Using Logarithms
Unit 4 Developmental Biology					
21 Principles of Development		Early Pattern Formation in *Drosophila*	Cloning	A Cartoon and Stick Model of the Homeodomain of the *Engrailed* Protein from *Drosophila* Interacting with DNA	Model Organisms; Cell and Tissue Culture Methods
22 An Introduction to Animal Development		Early Stages of Animal Development			
23 An Introduction to Plant Development					Model Organisms
Unit 5 Evolutionary Processes and Patterns					
24 Evolution by Natural Selection		Natural Selection for Antibiotic Resistance	Charles Darwin; Antibiotics		Reading a Phylogenetic Tree; Model Organisms; Reading Graphs

MEDIA AT A GLANCE *(continued)*

	BIOFLIX	WEB ACTIVITIES	DISCOVERY VIDEOS	VIDEOS	BIOSKILLS
25 Evolutionary Processes	Mechanisms of Evolution	The Hardy-Weinberg Principle; Three Modes of Natural Selection			Combining Probabilities; Using Statistical Tests and Interpreting Standard Error Bars; Reading Graphs
26 Speciation		Allopatric Speciation; Speciation by Changes in Ploidy			Reading a Phylogenetic Tree
27 Phylogenies and the History of Life		Adaptive Radiation	Mass Extinctions		Reading a Phylogenetic Tree
Unit 6 The Diversification of Life					
28 Bacteria and Archaea		The Tree of Life	Bacteria; Antibiotics; Molds; Tasty Bacteria; Early Life		Reading a Phylogenetic Tree; Model Organisms
29 Protists		Alternation of Generations in a Protist		A Crawling Amoeba	Biological Imaging: Microscopy and X-Ray Crystallography; Model Organisms
30 Green Algae and Land Plants		Plant Evolution and the PhylogeneticTree	Colored Cotton; Space Plants; Plant Pollination		
31 Fungi		Life Cycle of a Mushroom	Leafcutter Ants; Molds		
32 An Introduction to Animals		The Architecture of Animals	Leafcutter Ants; Invertebrates		
33 Protostome Animals		Protostome Diversity			Model Organisms
34 Deuterostome Animals		Deuterostome Diversity	Invertebrates		
35 Viruses		The HIV Replicative Cycle	Vaccines; Emerging Diseases		Biological Imaging: Microscopy and X-Ray Crystallography; Separating and Visualizing Molecules
Unit 7 How Plants Work					
36 Plant Form and Function		Plant Growth			
37 Water and Sugar Transport in Plants	Water Transport in Plants	Solute Transport in Plants		Plasmolysis of Plant Cells	
38 Plant Nutrition		Soil Formation and Nutrient Uptake			
39 Plant Sensory Systems, Signals, and Responses		Sensing Light; Plant Hormones; Plant Defenses			Cell and Tissue Culture Methods
40 Plant Reproduction		Reproduction in Flowering Plants; Fruit Structure and Development	Colored Cotton; Plant Pollination		
Unit 8 How Animals Work					
41 Animal Form and Function		Surface Area/Volume Relationships; Homeostasis	Blood; Human Body		Using Logarithms
42 Water and Electrolyte Balance in Animals		The Mammalian Kidney			

MEDIA AT A GLANCE *(continued)*

	BIOFLIX	WEB ACTIVITIES	DISCOVERY VIDEOS	VIDEOS	BIOSKILLS
43 Animal Nutrition	Homeostasis: Regulating Blood Sugar	The Digestion and Absorption of Food; Understanding Diabetes Mellitus	Nutrition		Biological Imaging: Microscopy and X-Ray Crystallography; Separating and Visualizing Molecules
44 Gas Exchange and Circulation	Gas Exchange	Gas Exchange in the Lungs and Tissues; The Human Heart	Blood		
45 Electrical Signals in Animals	How Neurons Work; How Synapses Work	Membrane Potentials; Action Potentials	Novelty Gene; Teen Brains	The Acetylcholine Receptor	Using Logarithms
46 Animal Sensory Systems and Movement	Muscle Contraction	The Vertebrate Eye; Structure and Contraction of Muscle Fibers	Muscles & Bones	The Acetylcholine Receptor	
47 Chemical Signals in Animals		Endocrine System Anatomy; Hormone Actions on Target Cells	Endocrine System	Cartoon Model of the DNA Binding Motif of a Zinc Finger Transcription Factor Binding to DNA	Separating Cell Components by Centrifugation
48 Animal Reproduction		Human Gametogenesis; Human Reproduction			Using Logarithms; Reading a Phylogenetic Tree
49 The Immune System in Animals		The Inflammatory Response; The Adaptive Immune Response	Emerging Diseases; Vaccines	Chemotaxis of a Neutrophil	
Unit 9 Ecology					
50 An Introduction to Ecology		Tropical Atmospheric Circulation	Rain Forests; Trees; Introduced Species		
51 Behavioral Ecology		Homing Behavior in Digger Wasps	Novelty Gene		
52 Population Ecology	Population Ecology	Modeling Population Growth; Human Population Growth and Regulation			
53 Community Ecology		Life Cycle of a Malaria Parasite; Succession	Leafcutter Ants		
54 Ecosystems	The Carbon Cycle	The Global Carbon Cycle			
55 Biodiversity and Conservation Biology		Habitat Fragmentation	Rain Forests		Using Logarithms

These black swan chicks are being introduced to the world by their mother, who trails close behind them. Likewise, this chapter introduces the key ideas that launched biological science as a discipline.

Biology and the Tree of Life 1

In essence, biological science is a search for ideas and observations that unify our understanding of the diversity of life, from bacteria living in rocks a mile underground to hedgehogs and humans. Chapter 1 is an introduction to this search.

The goals of this chapter are to introduce the nature of life and explore how biologists go about studying it. The chapter also introduces themes that will resonate throughout this book: (**1**) analyzing how organisms work at the molecular level, (**2**) understanding organisms in terms of their evolutionary history, and (**3**) helping you learn to think like a biologist.

Let's begin with what may be the most fundamental question of all: What is life?

KEY CONCEPTS

- Organisms obtain and use energy, are made up of cells, process information, replicate, and as populations evolve.
- The cell theory proposes that all organisms are made of cells and that all cells come from preexisting cells.
- The theory of evolution by natural selection maintains that species change through time because individuals with certain heritable traits produce more offspring than other individuals do.
- A phylogenetic tree is a graphical representation of the evolutionary relationships between species. These relationships can be estimated by analyzing similarities and differences in traits. Species that share distinctive traits are closely related and are placed close to each other on the tree of life.
- Biologists ask questions, generate hypotheses to answer them, and design experiments that test the predictions made by competing hypotheses.

1.1 What Does It Mean to Say That Something Is Alive?

An **organism** is a life-form—a living entity made up of one or more cells. Although there is no simple definition of life that is endorsed by all biologists, most agree that organisms share a suite of five fundamental characteristics.

- *Energy* To stay alive and reproduce, organisms have to acquire and use energy. To give just two examples: plants absorb sunlight; animals ingest food.
- *Cells* Organisms are made up of membrane-bound units called cells. A cell's membrane regulates the passage of materials between exterior and interior spaces.
- *Information* Organisms process hereditary or genetic information, encoded in units called genes, along with information they acquire from the environment. Right now cells throughout your body are using genetic information to make the molecules that keep you alive; your eyes and brain are decoding information on this page that will help you learn some biology.

✓ When you see this checkmark, stop and test yourself. Answers are available in Appendix B.

- *Replication* One of the great biologists of the twentieth century, François Jacob, said that the "dream of a bacterium is to become two bacteria." Almost everything an organism does contributes to one goal: replicating itself.
- *Evolution* Organisms are the product of evolution, and their populations continue to evolve.

You can think of this text as one long exploration of these five traits. Here's to life!

1.2 The Cell Theory

Two of the greatest unifying ideas in all of science laid the groundwork for modern biology: the cell theory and the theory of evolution by natural selection. Formally, scientists define a **theory** as an explanation for a very general class of phenomena or observations. The cell theory and theory of evolution address fundamental questions: What are organisms made of? Where do they come from?

When these concepts emerged in the mid-1800s, they revolutionized the way biologists think about the world. They established two of the five attributes of life: Organisms are cellular, and their populations change over time.

Neither insight came easily, however. The cell theory, for example, emerged after some 200 years of work. In 1665 Robert Hooke used a crude microscope to examine the structure of cork (a bark tissue) from an oak tree. The instrument magnified objects to just 30× (30 times) their normal size, but it allowed Hooke to see something extraordinary. In the cork he observed small, pore-like compartments that were invisible to the naked eye. These structures came to be called cells.

Soon after Hooke published his results, Anton van Leeuwenhoek succeeded in developing much more powerful microscopes, some capable of magnifications up to 300×. With these instruments, Leeuwenhoek inspected samples of pond water and made the first observations of human blood cells, and of sperm cells, shown in **Figure 1.1**.

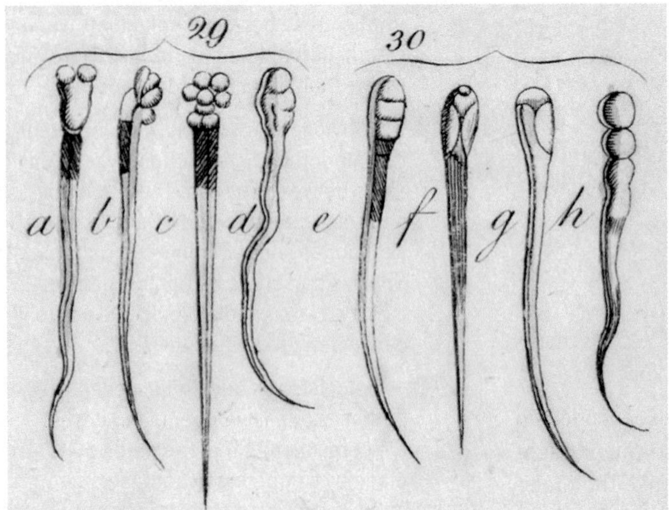

FIGURE 1.1 First Images of Cells. This drawing, by Anton van Leeuwenhoek, shows sperm cells.

In the 1670s a researcher who was studying the leaves and stems of plants with a microscope concluded that these large, complex structures are composed of many individual cells. By the early 1800s enough data had accumulated for a biologist to claim that *all* organisms consist of cells.

Are *All* Organisms Made of Cells?

The smallest organisms known today are bacteria that are barely 80 nanometers wide, or 80 *billionths* of a meter. (See **BioSkills 1** in Appendix A to review the metric system and its prefixes.[1]) It would take 12,500 of these organisms lined up end to end to span a millimeter. This is the distance between the smallest hash marks on a metric ruler.

In contrast, sequoia trees can be over 100 meters tall. This is the equivalent of a 20-story building. Bacteria and sequoias are composed of the same fundamental building block, however—the cell. Bacteria consist of a single cell; sequoias are made up of many cells.

Today, a **cell** is defined as a highly organized compartment that is bounded by a thin, flexible structure called a plasma membrane and that contains concentrated chemicals in an aqueous (watery) solution. The chemical reactions that sustain life take place inside cells. Most cells are also capable of reproducing by dividing—in effect, by making a copy of themselves.

The realization that all organisms are made of cells was fundamentally important, but it formed only the first part of the cell theory. In addition to understanding what organisms are made of, scientists wanted to understand how cells come to be.

Where Do Cells Come From?

Most scientific theories have two components: The first describes a pattern in the natural world; the second identifies a mechanism or process that is responsible for creating that pattern. Hooke and his fellow scientists articulated the pattern component of the cell theory. In 1858 Rudolph Virchow added the process component by stating that all cells arise from preexisting cells.

The complete **cell theory** can be stated as follows: All organisms are made of cells, and all cells come from preexisting cells.

TWO HYPOTHESES The cell theory was a direct challenge to the prevailing explanation of where cells come from, called spontaneous generation. At the time, most biologists believed that organisms arise spontaneously under certain conditions. For example, the bacteria and fungi that spoil foods such as milk and wine were thought to appear in these nutrient-rich media of their own accord—springing to life from nonliving materials. Spontaneous generation was a **hypothesis**: a proposed explanation.

The all-cells-from-cells hypothesis, in contrast, maintained that cells do not spring to life spontaneously but are produced only when preexisting cells grow and divide.

Biologists usually use the word theory to refer to proposed explanations for broad patterns in nature and prefer hypothesis to refer to explanations for more tightly focused questions.

[1]BioSkills are located in the first appendix at the back of the book. They focus on general skills that you'll use throughout this course. More than a few students have found them to be a life-saver. Please use them!

AN EXPERIMENT TO SETTLE THE QUESTION Soon after the all-cells-from-cells hypothesis appeared in print, Louis Pasteur set out to test its predictions experimentally. A **prediction** is something that can be measured and that must be correct if a hypothesis is valid.

Pasteur wanted to determine whether microorganisms could arise spontaneously in a nutrient broth or whether they appear only when a broth is exposed to a source of preexisting cells. To address the question, he created two treatment groups: a broth that was not exposed to a source of preexisting cells and a broth that was.

The spontaneous generation hypothesis predicted that cells would appear in both treatment groups. The all-cells-from-cells hypothesis predicted that cells would appear only in the treatment exposed to a source of preexisting cells.

Figure 1.2 shows Pasteur's experimental setup. Note that the two treatments are identical in every respect but one. Both used glass flasks filled with the same amount of the same nutrient broth. Both were boiled for the same amount of time to kill any existing organisms such as bacteria or fungi. But because the flask pictured in Figure 1.2a had a straight neck, it was exposed to preexisting cells after sterilization by the heat treatment. These preexisting cells are the bacteria and fungi that cling to dust particles in the air. They could drop into the nutrient broth because the neck of the flask was straight.

In contrast, the flask drawn in Figure 1.2b had a long swan neck. Pasteur knew that water would condense in the crook of the swan neck after the boiling treatment and that this pool of water

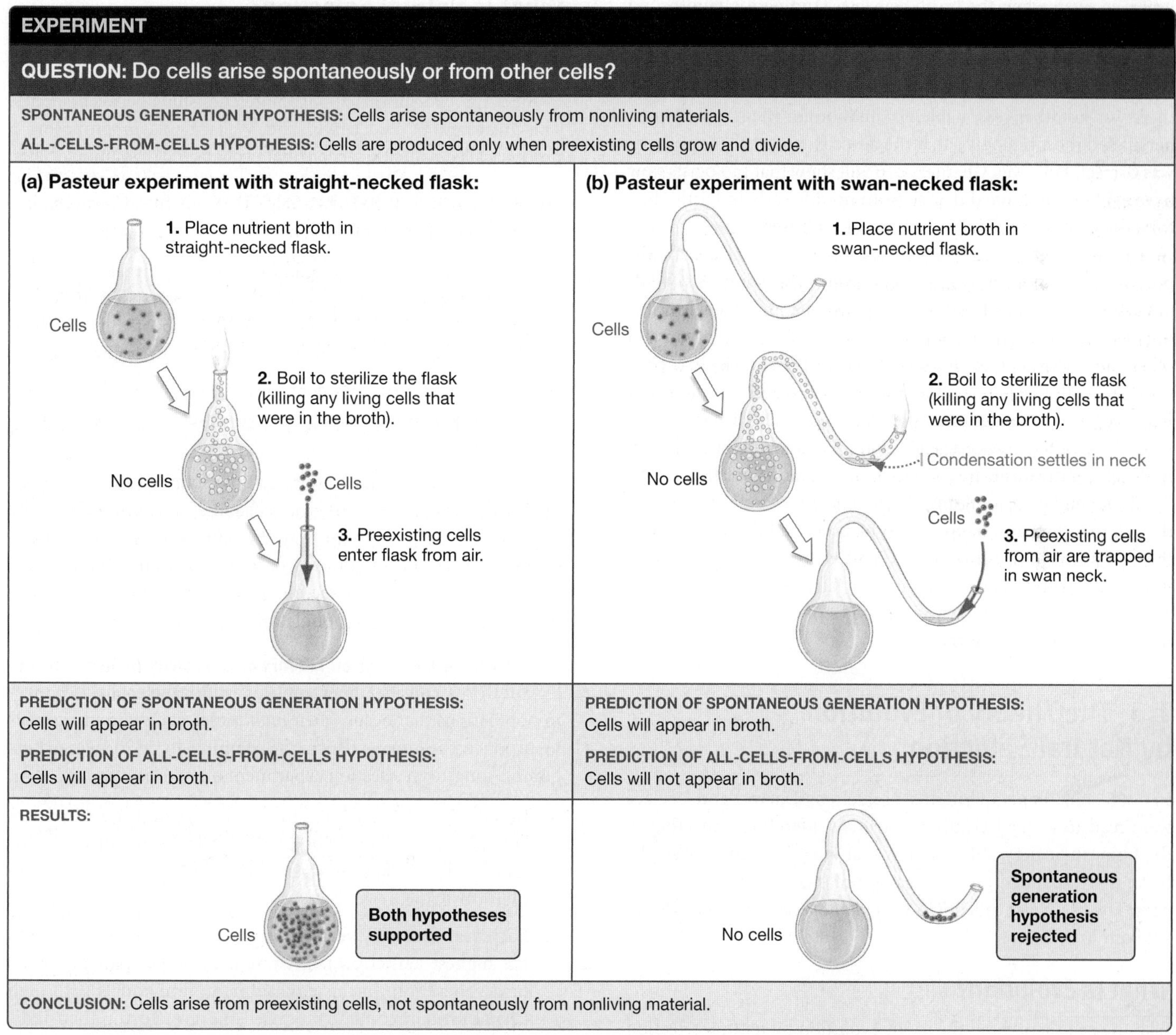

EXPERIMENT
QUESTION: Do cells arise spontaneously or from other cells?
SPONTANEOUS GENERATION HYPOTHESIS: Cells arise spontaneously from nonliving materials.
ALL-CELLS-FROM-CELLS HYPOTHESIS: Cells are produced only when preexisting cells grow and divide.
(a) Pasteur experiment with straight-necked flask:
1. Place nutrient broth in straight-necked flask.
Cells
2. Boil to sterilize the flask (killing any living cells that were in the broth).
No cells
Cells
3. Preexisting cells enter flask from air.
(b) Pasteur experiment with swan-necked flask:
1. Place nutrient broth in swan-necked flask.
Cells
2. Boil to sterilize the flask (killing any living cells that were in the broth).
Condensation settles in neck
No cells
Cells
3. Preexisting cells from air are trapped in swan neck.
PREDICTION OF SPONTANEOUS GENERATION HYPOTHESIS: Cells will appear in broth.
PREDICTION OF ALL-CELLS-FROM-CELLS HYPOTHESIS: Cells will appear in broth.
PREDICTION OF SPONTANEOUS GENERATION HYPOTHESIS: Cells will appear in broth.
PREDICTION OF ALL-CELLS-FROM-CELLS HYPOTHESIS: Cells will not appear in broth.
RESULTS:
Cells
Both hypotheses supported
No cells
Spontaneous generation hypothesis rejected
CONCLUSION: Cells arise from preexisting cells, not spontaneously from nonliving material.

FIGURE 1.2 The Spontaneous Generation and All-Cells-from-Cells Hypotheses Were Tested Experimentally.

✔**QUESTION** What problem would arise if Pasteur had (1) put different types of broth in the two treatments, (2) heated them for different lengths of time, or (3) used a ceramic flask for one treatment and a glass flask for the other?

would trap any bacteria or fungi that entered on dust particles. Thus, the contents of the swan-necked flask was isolated from any source of preexisting cells even though still open to the air.

Pasteur's experimental setup was effective because there was only one difference between the two treatments and because that difference was the factor being tested—in this case, a broth's exposure to preexisting cells.

ONE HYPOTHESIS SUPPORTED And Pasteur's results? As Figure 1.2 shows, the treatment exposed to preexisting cells quickly filled with bacteria and fungi. This observation was important because it showed that the heat sterilization step had not altered the nutrient broth's capacity to support growth.

The treatment in the swan-necked flask remained sterile, however. Even when the broth was left standing for months, no organisms appeared in it. This result was inconsistent with the hypothesis of spontaneous generation.

Because Pasteur's data were so conclusive—meaning that there was no other reasonable explanation for them—the results persuaded most biologists that the all-cells-from-cells hypothesis was correct. However, Chapters 2–6 will show that biologists now have evidence that life did arise from nonlife early in Earth's history, through a process called chemical evolution.

The success of the cell theory's process component had an important implication: If all cells come from preexisting cells, it follows that all individuals in an isolated population of single-celled organisms are related by common ancestry. Similarly, in you and other multicellular individuals, all the cells present are descended from preexisting cells, tracing back to a fertilized egg. A fertilized egg is a cell created by the fusion of sperm and egg—cells that formed in individuals of the previous generation. In this way, all the cells in a multicellular organism are connected by common ancestry.

The second great founding idea in biology is similar, in spirit, to the cell theory. It also happened to be published the same year as the all-cells-from-cells hypothesis. This was the realization, made independently by Charles Darwin and Alfred Russel Wallace, that all species—all distinct, identifiable types of organisms—are connected by common ancestry.

1.3 The Theory of Evolution by Natural Selection

In 1858 short papers written separately by Darwin and Wallace were read to a small group of scientists attending a meeting of the Linnean Society of London. A year later, Darwin published a book that expanded on the idea summarized in those brief papers. The book was called *The Origin of Species.* The first edition sold out in a day.

What Is Evolution?

Like the cell theory, the theory of evolution by natural selection has a pattern and a process component. Darwin and Wallace's theory made two important claims concerning patterns that exist in the natural world.

- Species are related by common ancestry. This contrasted with the prevailing view in science at the time, which was that species represent independent entities created separately by a divine being.
- In contrast to the accepted view that species remain unchanged through time, Darwin and Wallace proposed that the characteristics of species can be modified from generation to generation. Darwin called this process "descent with modification."

Evolution is a change in the characteristics of a population over time. It means that species are not independent and unchanging entities, but are related to one another and can change through time.

What Is Natural Selection?

This pattern component of the theory of evolution was actually not original to Darwin and Wallace. Several scientists had already come to the same conclusions about the relationships between species. The great insight by Darwin and Wallace was in proposing a process, called **natural selection**, that explains *how* evolution occurs.

TWO CONDITIONS OF NATURAL SELECTION Natural selection occurs whenever two conditions are met.

1. Individuals within a population vary in characteristics that are **heritable**—meaning, traits that can be passed on to offspring. A **population** is defined as a group of individuals of the same species living in the same area at the same time.
2. In a particular environment, certain versions of these heritable traits help individuals survive better or reproduce more than do other versions.

If certain heritable traits lead to increased success in producing offspring, then those traits become more common in the population over time. In this way, the population's characteristics change as a result of natural selection acting on individuals. This is a key insight: Natural selection acts on individuals, but evolutionary change occurs in populations.

SELECTION ON MAIZE AS AN EXAMPLE To clarify how natural selection works, consider an example of **artificial selection**—changes in populations that occur when *humans* select certain individuals to produce the most offspring. Beginning in 1896, researchers began a long-term selection experiment on maize (corn).

1. In the original population, the percentage of protein in maize kernels was variable among individuals. Kernel protein content is a heritable trait—parents tend to pass the trait on to their offspring.
2. Each year for many years, researchers chose individuals with the highest kernel protein content to be the parents of the next generation. In this environment, individuals with high kernel protein content produced more offspring than individuals with low kernel protein content.

Figure 1.3 shows the results. Note that this graph plots generation number on the *x*-axis, starting from the first generation

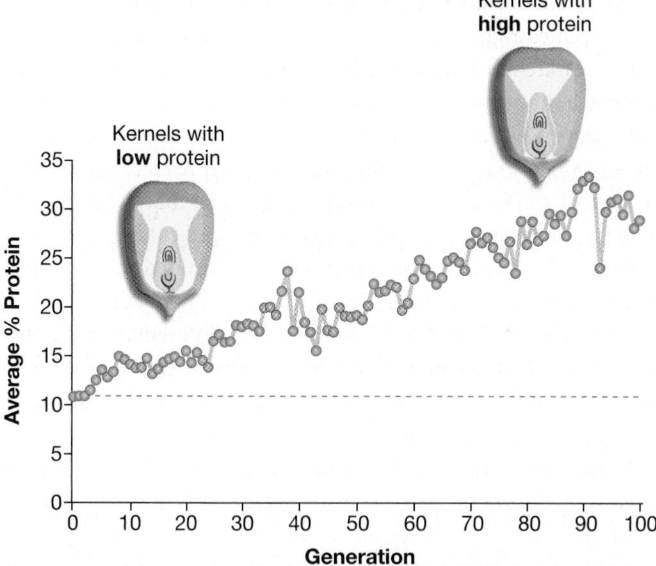

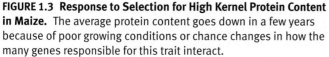

FIGURE 1.3 Response to Selection for High Kernel Protein Content in Maize. The average protein content goes down in a few years because of poor growing conditions or chance changes in how the many genes responsible for this trait interact.

("0") and continuing for 100 generations. The average percentage of protein in a kernel among individuals in this population is plotted on the *y*-axis.

To read this graph, put your finger on the *x*-axis at time 0. Then read up the *y*-axis, and note that kernels averaged about 11 percent protein at the start of the experiment. Now read the graph to the right. Each dot is a data point, representing the average kernel protein concentration in a particular generation. (A generation in maize is one year.) The lines on this graph simply connect the dots, to make the pattern in the data easier to see. At the end of the graph, after 100 generations of selection, average kernel protein content is about 29 percent. (For more help with reading graphs, see **BioSkills 2** in Appendix A.)

This sort of change in the characteristics of a population, over time, is evolution. Humans have been practicing artificial selection for thousands of years, and biologists have now documented evolution by *natural* selection—where humans don't do the selecting—occurring in thousands of different populations, including humans.

To practice applying the principles of artificial selection, go to the online study area at *www.masteringbiology.com.*

MB **Web Activity** Artificial Selection

Evolution occurs when heritable variation leads to differential success in reproduction. ✔If you understand this concept, you should be able to describe how protein content in maize kernels changed over time, using the same *x*-axis and *y*-axis as in Figure 1.3, when researchers selected individuals with *lowest* kernel protein content to be the parents of the next generation. (This experiment was actually done, starting with the same population at the same time as selection for high protein content.)

FITNESS AND ADAPTATION Darwin also introduced some new terminology to identify what is happening during natural selection.

- In everyday English, fitness means health and well-being. But in biology, **fitness** means the ability of an individual to produce offspring. Individuals with high fitness produce many surviving offspring.
- In everyday English, adaptation means that an individual is adjusting and changing to function in new circumstances. But in biology, an **adaptation** is a trait that increases the fitness of an individual in a particular environment.

Once again, consider kernel protein content in maize: In the environment of the experiment graphed in Figure 1.3, individuals with high kernel protein content produced more offspring and had higher fitness than individuals with lower kernel protein content. In this population and this environment, high kernel protein content was an adaptation that allowed certain individuals to thrive.

Note that during this process, the amount of protein in the kernels of any individual maize plant did not change within its lifetime—the change occurred in the characteristics of the population over time.

Together, the cell theory and the theory of evolution provided the young science of biology with two central, unifying ideas:

1. The cell is the fundamental structural unit in all organisms.
2. All species are related by common ancestry and have changed over time in response to natural selection.

CHECK YOUR UNDERSTANDING

If you understand that . . .

- Natural selection occurs when heritable variation in certain traits leads to improved success in reproduction. Because individuals with these traits produce many offspring with the same traits, the traits increase in frequency and evolution occurs.
- Evolution is a change in the characteristics of a population over time.

✔ **You should be able to . . .**

On the graph you just analyzed, describe the average kernel protein content over time in a maize population where *no* selection occurred.

Answers are available in Appendix B.

1.4 The Tree of Life

Section 1.3 focused on how individual populations change through time in response to natural selection. But over the past several decades, biologists have also documented dozens of cases in which natural selection has caused populations of one species to diverge and form new species. This divergence process is called **speciation**.

Research on speciation has two important implications: All species come from preexisting species, and all species, past and present, trace their ancestry back to a single common ancestor.

The theory of evolution by natural selection predicts that biologists should be able to reconstruct a **tree of life**—a family tree of organisms. If life on Earth arose just once, then such a diagram would describe the genealogical relationships between species with a single, ancestral species at its base.

Has this task been accomplished? If the tree of life exists, what does it look like?

Using Molecules to Understand the Tree of Life

One of the great breakthroughs in research on the tree of life occurred when Carl Woese (pronounced *woes*) and colleagues began analyzing the chemical components of organisms, as a way to understand their evolutionary relationships. Their goal was to understand the **phylogeny** of all organisms—their actual genealogical relationships. Translated literally, phylogeny means "tribe-source."

To understand which organisms are closely versus distantly related, Woese and co-workers needed to study a molecule that is found in all organisms. The molecule they selected is called small subunit ribosomal RNA (rRNA). It is an essential part of the machinery that all cells use to grow and reproduce.

Although rRNA is a large and complex molecule, its underlying structure is simple. The rRNA molecule is made up of sequences of four smaller chemical components called ribonucleotides. These ribonucleotides are symbolized by the letters A, U, C, and G. In rRNA, ribonucleotides are connected to one another linearly, like boxcars of a freight train (**Figure 1.4**).

ANALYZING rRNA Why might rRNA be useful for understanding the relationships between organisms? The answer is that the ribonucleotide sequence in rRNA is a trait that can change during the course of evolution. Although rRNA performs the same function in all organisms, the sequence of ribonucleotide building blocks in this molecule is not identical among species.

In land plants, for example, the molecule might start with the sequence A-U-A-U-C-G-A-G. In green algae, which are closely related to land plants, the same section of the molecule might contain A-U-A-U-G-G-A-G. But in brown algae, which are not closely related to green algae or to land plants, the same part of the molecule might consist of A-A-A-U-G-G-A-C.

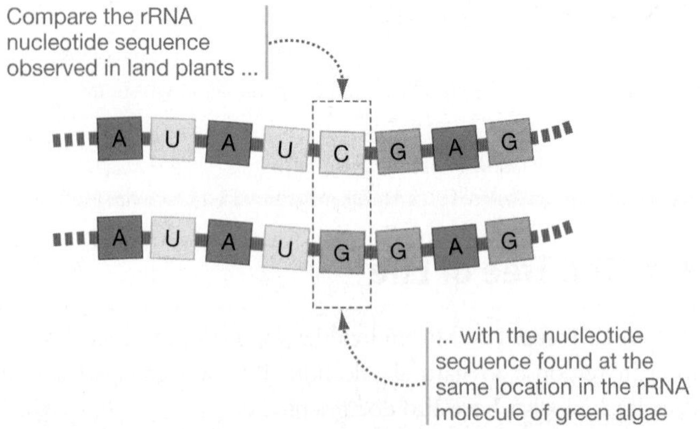

The research program that Woese and co-workers pursued was based on a simple premise: If the theory of evolution is correct, then rRNA sequences should be very similar in closely related organisms but less similar in organisms that are less closely related. Species that are part of the same evolutionary lineage, like the plants, should share certain changes in rRNA that no other species have.

To test this premise, the researchers determined the sequence of ribonucleotides in the rRNA of a wide array of species. Then they considered what the similarities and differences in the sequences implied about relationships between the species. The goal was to produce a diagram that described the phylogeny of the organisms in the study.

A diagram that depicts evolutionary history in this way is called a phylogenetic tree. Just as a family tree shows relationships between individuals, a phylogenetic tree shows relationships between species. On a phylogenetic tree, branches that share a recent common ancestor represent species that are closely related; branches that don't share recent common ancestors represent species that are more distantly related.

Compare the rRNA nucleotide sequence observed in land plants ...

A U A U C G A G

A U A U G G A G

... with the nucleotide sequence found at the same location in the rRNA molecule of green algae

FIGURE 1.4 RNA Molecules Are Made Up of Smaller Molecules. The complete small subunit rRNA molecule contains about 2000 ribonucleotides; just 8 are shown in this comparison.

✔**QUESTION** Suppose that in the same portion of rRNA, molds and other fungi have the sequence A-U-A-U-G-G-A-C. According to these data, are fungi more closely related to green algae or to land plants? Explain your logic.

THE TREE OF LIFE ESTIMATED FROM AN ARRAY OF GENES To construct a phylogenetic tree, researchers use a computer to find the arrangement of branches that is most consistent with the similarities and differences observed in the data.

Although the initial work was based only on the sequences of ribonucleotides observed in rRNA, biologists now use data sets that include sequences from a wide array of genes. **Figure 1.5** shows a recent tree produced by comparing these sequences. Because this tree includes such a diverse array of species, it is often called the universal tree, or the tree of life. (For help in learning how to read a phylogenetic tree, see **BioSkills 3** in Appendix A.)

The tree of life implied by rRNA and other genetic data established that there are three fundamental groups or lineages of organisms: (**1**) the Bacteria, (**2**) the Archaea, and (**3**) the Eukarya. In all **eukaryotes**, cells have a prominent component called the nucleus (**Figure 1.6a**). Translated literally, the word eukaryotes means "true kernel." Because the vast majority of bacterial and archaeal cells lack a nucleus, they are referred to as **prokaryotes** (literally, "before kernel"; see **Figure 1.6b**). The vast majority of bacteria and archaea are unicellular ("one-celled"); many eukaryotes are multicellular ("many-celled").

When these results were first published, biologists were astonished. For example:

- Prior to Woese's work and follow-up studies, biologists thought that the most fundamental division among organisms was between prokaryotes and eukaryotes. The Archaea were virtually unknown—much less recognized as a major and highly distinctive branch on the tree of life.
- Fungi were thought to be closely related to plants. Instead, they are actually much more closely related to animals.
- Traditional approaches for classifying organisms—including the system of five kingdoms divided into various classes, or-

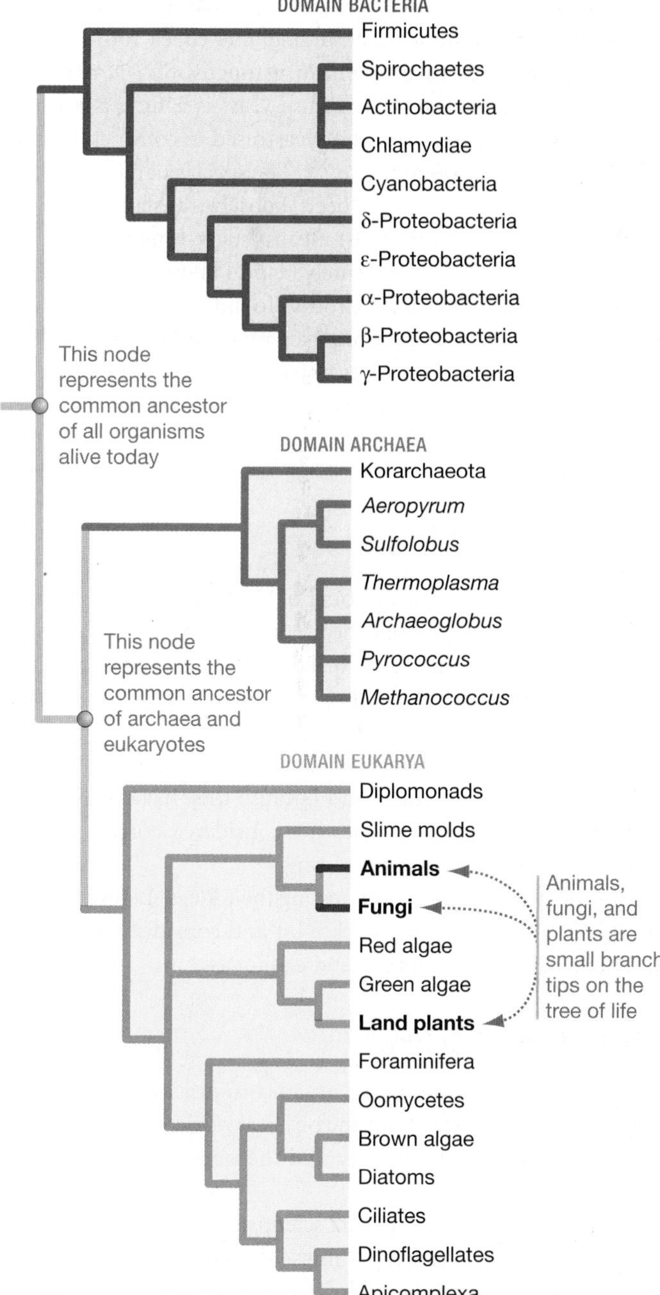

FIGURE 1.5 The Tree of Life. "Universal tree" estimated from a large amount of gene sequence data. The three domains of life revealed by the analysis are labeled. Common names are given for most lineages in the domains Bacteria and Eukarya. Genus names are given for members of the domain Archaea, because most of these organisms have no common names.

ders, and families that you may have learned in high school—are inaccurate in many cases, because they do not reflect the actual evolutionary history of the organisms involved.

THE TREE OF LIFE IS A WORK IN PROGRESS Just as researching your family tree can help you understand who you are and where you came from, so the tree of life helps biologists understand the relationships between organisms and the history of species. The discovery of the Archaea and the accurate placement of lineages

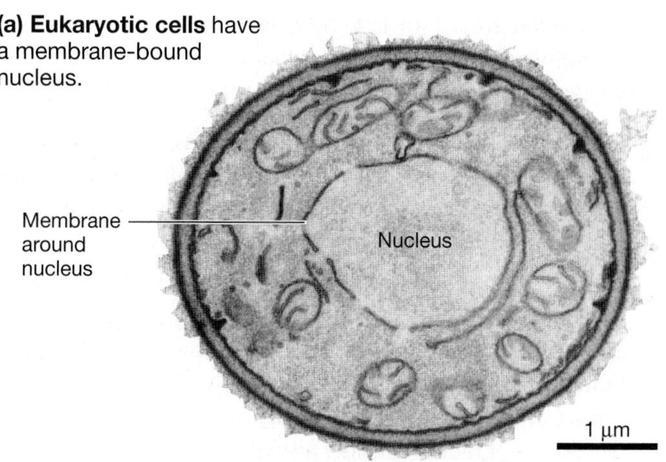

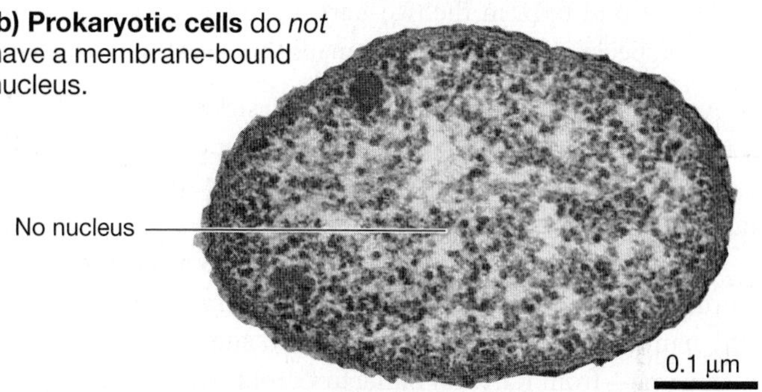

FIGURE 1.6 Eukaryotes and Prokaryotes.

✔**QUESTION** How many times larger is the eukaryotic cell in this figure than the prokaryotic cell? (Hint: study the scale bars.)

such as the fungi qualify as exciting breakthroughs in our understanding of evolutionary history and life's diversity.

Work on the tree of life continues at a furious pace, however, and the location of certain branches on the tree is hotly debated. As databases expand and as techniques for analyzing data improve, the shape of the tree of life presented in Figure 1.5 will undoubtedly change. Our understanding of the tree of life, like our understanding of every other topic in biological science, is dynamic.

How Should We Name Branches on the Tree of Life?

In science, the effort to name and classify organisms is called **taxonomy**. Any named group is called a **taxon** (plural: **taxa**). Currently, biologists are working to create a taxonomy, or naming system, that accurately reflects the phylogeny of organisms.

For example, Woese proposed a new taxonomic category called the **domain**. The three domains of life are the Bacteria, Archaea, and Eukarya.

Biologists often use the term **phylum** (plural: **phyla**) to refer to major lineages within each domain. Although the designation is somewhat arbitrary, each phylum is considered a major branch on the tree of life. For example, within the lineage called animals, biologists name about 35 phyla—each of which is distinguished by distinctive aspects of its body structure as well as by distinctive gene sequences. The mollusks (clams, squid, octopuses) constitute a phylum, as do chordates (the vertebrates and their close relatives).

Because the tree of life is so new, though, naming systems are still being worked out. One thing that hasn't changed for centuries, however, is the naming system for individual species.

SCIENTIFIC (SPECIES) NAMES In 1735, a Swedish botanist named Carolus Linnaeus established a system for naming species that is still in use today. Linnaeus created a two-part name unique to each type of organism.

- The first part indicates the organism's **genus** (plural: **genera**). A genus is made up of a closely related group of species. For example, Linnaeus put humans in the genus *Homo*. Although humans are the only living species in this genus, at least five extinct organisms, all of which walked upright and made extensive use of tools, were later also assigned to *Homo*.
- The second term in the two-part name identifies the organism's species. Linnaeus gave humans the species name *sapiens*.

An organism's genus and species designation is called its **scientific name** or Latin name. Scientific names are always italicized. Genus names are always capitalized, but species names are not—for instance, *Homo sapiens*.

Scientific names are based on Latin or Greek word roots or on words "Latinized" from other languages. Linnaeus gave a scientific name to every species then known, and also Latinized his own name—from Karl von Linné to Carolus Linnaeus.

Linnaeus maintained that different types of organisms should not be given the same genus and species names. Other species may be assigned to the genus *Homo*, and members of other genera may be named *sapiens*, but only humans are named *Homo sapiens*. Each scientific name is unique.

SCIENTIFIC NAMES ARE OFTEN DESCRIPTIVE Scientific names and terms are often based on Latin or Greek word roots that are descriptive. For example, *Homo sapiens* is derived from the Latin *homo* for "man" and *sapiens* for "wise" or "knowing." The yeast that bakers use to produce bread and that brewers use to brew beer is called *Saccharomyces cerevisiae*. The Greek root *saccharo* means "sugar," and *myces* refers to a fungus. *Saccharomyces* is aptly named "sugar fungus" because yeast is a fungus and because the domesticated strains of yeast used in commercial baking and brewing are often fed sugar. The species name of this organism, *cerevisiae*, is Latin for "beer." Loosely translated, then, the scientific name of brewer's yeast means "sugar-fungus for beer."

Most biologists find it extremely helpful to memorize some of the common Latin and Greek roots. To aid you in this process, new terms in this text are often accompanied by a reference to their Latin or Greek word roots in parentheses, and a glossary of common root words with translations and examples is provided in **BioSkills 4** in Appendix A.

CHECK YOUR UNDERSTANDING

If you understand that . . .

- A phylogenetic tree shows the evolutionary relationships between species.
- To infer where species belong on a phylogenetic tree, biologists examine the characteristics of the species involved. Closely related species should have similar characteristics, while less closely related species should be less similar.

✔ You should be able to . . .

Examine the following sequences and draw a phylogenetic tree showing the relationships between species A, B, and C that these data imply:

Species A: A A C T A G C G C G A T
Species B: A A C T A G C G C C A T
Species C: T T C T A G C G G T A T

Answers are available in Appendix B.

1.5 Doing Biology

This chapter has introduced some of the great ideas in biology. The development of the cell theory and the theory of evolution by natural selection provided cornerstones when the science was young; the tree of life is a relatively recent insight that has revolutionized our understanding of life's diversity.

These theories are considered great because they explain fundamental aspects of nature, and because they have consistently been shown to be correct. They are considered correct because they have withstood extensive testing.

How do biologists go about testing their ideas? Before answering this question, let's step back a bit and consider the types of questions that researchers can and cannot ask.

The Nature of Science

Biologists ask questions about organisms, just as physicists and chemists ask questions about the physical world or geologists ask questions about Earth's history and the ongoing processes that shape landforms.

No matter what their field, all scientists ask questions that can be answered by measuring things—by collecting data. Conversely, scientists cannot address questions that can't be answered by measuring things.

This distinction is important. It is at the root of continuing controversies about teaching evolution in publicly funded schools. In the United States and in Turkey, in particular, some Christian and Islamic leaders have been particularly successful in pushing their claim that evolution and religious faith are in conflict. Even though the theory of evolution is considered one of the most successful and best-substantiated ideas in the history of science, they object to teaching it.

The vast majority of biologists and religious leaders reject this claim; they see no conflict between evolution and religious faith. Their view is that science and religion are compatible because they address different types of questions.

- Science is about formulating hypotheses and finding evidence that supports or conflicts with those hypotheses.

- Religious faith addresses questions that cannot be answered by data. The questions addressed by the world's great religions focus on why we exist and how we should live.

Both types of questions are seen as legitimate and important.

So how do biologists go about answering questions? Let's consider two issues currently being addressed by researchers.

Why Do Giraffes Have Long Necks? An Introduction to Hypothesis Testing

If you were asked why giraffes have long necks, you might say that long necks enable giraffes to reach food that is unavailable to other mammals. This hypothesis is expressed in African folktales and has traditionally been accepted by many biologists. The food competition hypothesis is so plausible, in fact, that for decades no one thought to test it.

In the mid-1990s, however, Robert Simmons and Lue Scheepers assembled data suggesting that the food competition hypothesis is only part of the story. Their analysis supports an alternative hypothesis—that long necks allow giraffes to use their heads as effective weapons for battering their opponents.

How did biologists test the food competition hypothesis? What data support their alternative explanation? Before attempting to answer these questions, it's important to recognize that hypothesis testing is a two-step process:

1. State the hypothesis as precisely as possible and list the predictions it makes.
2. Design an observational or experimental study that is capable of testing those predictions.

If the predictions are accurate, the hypothesis is supported. If the predictions are not met, then researchers do further tests, modify the original hypothesis, or search for alternative explanations.

THE FOOD COMPETITION HYPOTHESIS: PREDICTIONS AND TESTS

The food competition hypothesis claims that giraffes compete for food with other species of mammals. When food is scarce, as it is during the dry season, giraffes with longer necks can reach food that is unavailable to other species and to giraffes with shorter necks. As a result, the longest-necked individuals in a giraffe population survive better and produce more young than do shorter-necked individuals, and average neck length of the population increases with each generation.

To use the terms introduced earlier, long necks are adaptations that increase the fitness of individual giraffes during competition for food. This type of natural selection has gone on so long that the population has become extremely long necked.

The food competition hypothesis makes several explicit predictions. For example, the food competition hypothesis predicts that

1. neck length is variable among giraffes;
2. neck length in giraffes is heritable; and
3. giraffes feed high in trees, especially during the dry season, when food is scarce and the threat of starvation is high.

The first prediction is correct. Studies in zoos and natural populations confirm that neck length is variable among individuals.

The researchers were unable to test the second prediction, however, because they studied giraffes in a natural population and were unable to do breeding experiments. As a result, they simply had to accept this prediction as an assumption. In general, though, biologists prefer to test every assumption behind a hypothesis.

What about the prediction regarding feeding high in trees? According to Simmons and Scheepers, this is where the food competition hypothesis breaks down.

Consider, for example, data collected by a different research team on the amount of time that giraffes spend feeding in vegetation of different heights. **Figure 1.7a** plots the height of vegetation versus the percentage of bites taken by a giraffe, for males and for females from the same population in Kenya. The dashed line on each graph indicates the average height of a male or female in this population.

Note that the average height of a giraffe in this population is much greater than the height where most feeding takes place. In

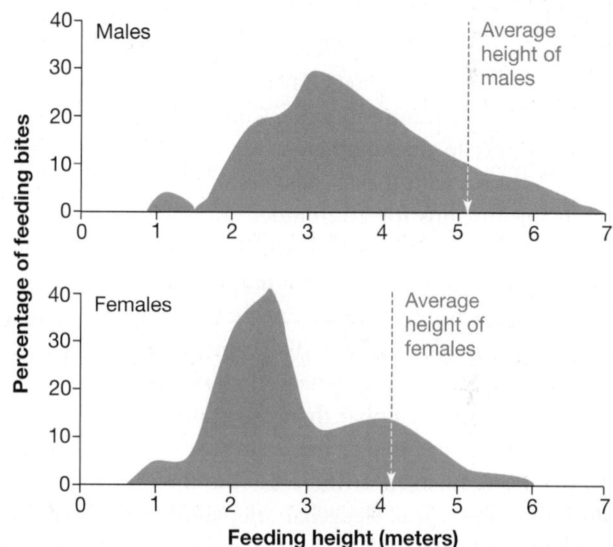

FIGURE 1.7 Giraffes Do Not Usually Extend Their Necks to Feed.

this population, both male and female giraffes spend most of their feeding time eating vegetation that averages just 60 percent of their full height. Studies on other populations of giraffes, during both the wet and dry seasons, are consistent with these data. Giraffes usually feed with their necks bent (**Figure 1.7b**).

These data cast doubt on the food competition hypothesis, because one of its predictions does not appear to hold. Biologists have not abandoned this hypothesis completely, though, because feeding high in trees may be particularly valuable during extreme droughts, when a giraffe's ability to reach leaves far above the ground could mean the difference between life and death. Still, Simmons and Scheepers have offered an alternative explanation for why giraffes have long necks. The new hypothesis is based on the mating system of giraffes.

THE SEXUAL COMPETITION HYPOTHESIS: PREDICTIONS AND TESTS Giraffes have an unusual mating system. Breeding occurs year round rather than seasonally. To determine when females are coming into estrus or "heat" and are thus receptive to mating, the males nuzzle the rumps of females. In response, the females urinate into the males' mouths. The males then tip their heads back and pull their lips to and fro, as if tasting the liquid. Biologists who have witnessed this behavior have proposed that the males taste the females' urine to detect whether estrus has begun.

Once a female giraffe enters estrus, males fight among themselves for the opportunity to mate. Combat is spectacular. The bulls stand next to one another, swing their necks, and strike thunderous blows with their heads. Researchers have seen males knocked unconscious for 20 minutes after being hit and have cataloged numerous instances in which the loser died. Giraffes are the only animals known to fight in this way.

These observations inspired a new explanation for why giraffes have long necks. The sexual competition hypothesis is based on the idea that longer-necked giraffes are able to strike harder blows during combat than can shorter-necked giraffes. In engineering terms, longer necks provide a longer "moment arm." A long moment arm increases the force of the impact. (Think about the type of sledgehammer you'd use to bash down a concrete wall—one with a short handle or one with a long handle?)

The idea here is that longer-necked males should win more fights and, as a result, father more offspring than shorter-necked males do. If neck length in giraffes is inherited, then the average neck length in the population should increase over time. Under the sexual competition hypothesis, long necks are adaptations that increase the fitness of males during competition for females.

Although several studies have shown that long-necked males are more successful in fighting and that the winners of fights gain access to estrous females, the question of why giraffes have long necks is not closed. With the data collected to date, most biologists would probably concede that the food competition hypothesis needs further testing and refinement and that the sexual selection hypothesis appears promising. It could also be true that both hypotheses are correct. For our purposes, the important take-home message is that all hypotheses must be tested rigorously.

In many cases in biological science, testing hypotheses rigorously involves experimentation. Experimenting on giraffes is difficult. But in the case study considered next, biologists were able to test an interesting hypothesis experimentally.

How Do Ants Navigate? An Introduction to Experimental Design

Experiments are a powerful scientific tool because they allow researchers to test the effect of a single, well-defined factor on a particular phenomenon. Because experiments testing the effect of neck length on food and sexual competition in giraffes haven't been done yet, let's consider a different question: When ants leave their nest to search for food, how do they find their way back?

The Saharan desert ant lives in colonies and makes a living by scavenging the dead carcasses of insects. Individuals leave the burrow and wander about searching for food at midday, when temperatures at the surface can reach 60°C (140°F) and predators are hiding from the heat.

Foraging trips can take the ants hundreds of meters—an impressive distance when you consider that these animals are only about a centimeter long. But when an ant returns, it doesn't follow the same long, wandering route it took on its way away from the nest. Instead, individuals return in a straight line. How do they do this?

THE PEDOMETER HYPOTHESIS Early work on navigation in desert ants showed that they use the Sun's position as a compass—meaning that they always know the approximate direction of the nest relative to the Sun. But how do they know how far to go?

After experiments had shown that the ants do not use landmarks to navigate, Matthias Wittlinger and co-workers set out to test a novel idea. The biologists proposed that Saharan desert ants know how far they are from the nest by integrating information from leg movements.

According to this pedometer hypothesis, the ants always know how far they are from the nest because they track the number of steps they have taken and their stride length. The idea is that they can make a beeline back to the burrow because they integrate information on the angles they have traveled *and* the distance they have gone—based on step number and stride length.

TESTING THE HYPOTHESIS To test their idea, Wittlinger's group allowed ants to walk from a nest to a feeder through a channel—a distance of 10 m. Then they caught ants at the feeder and created three test groups, each with 25 individuals:

- *Stumps* By cutting the lower legs of some individuals off, they created ants with shorter-than-normal legs.
- *Normal* Some individuals were left alone, meaning that they had normal leg length.
- *Stilts* By gluing pig bristles onto each leg, the biologists created ants with longer-than-normal legs.

Next they put the ants in a different channel and recorded how far they traveled before starting the characteristic set of 180° turns that ants make when they are looking for the nest hole. To see the data they collected, look at the graph on the left side of the "Results" section in **Figure 1.8**.

- *Stumps* The ants with stumps stopped short, by about 5 m, before starting to look for the nest opening.
- *Normal* The normal ants walked the correct distance—about 10 m.
- *Stilts* The ants with stilts walked about 5 m too far before starting to look for the nest opening.

To check the validity of this result, they put the test ants back in the nest and recaptured them one to several days later, when they had walked to the feeder on their stumps, normal legs, or stilts. Now when the ants were put into the other channel to "walk back," they all traveled the correct distance—10 m—before starting to look for the nest (see the graph on the right side of the "Results" section in Figure 1.8).

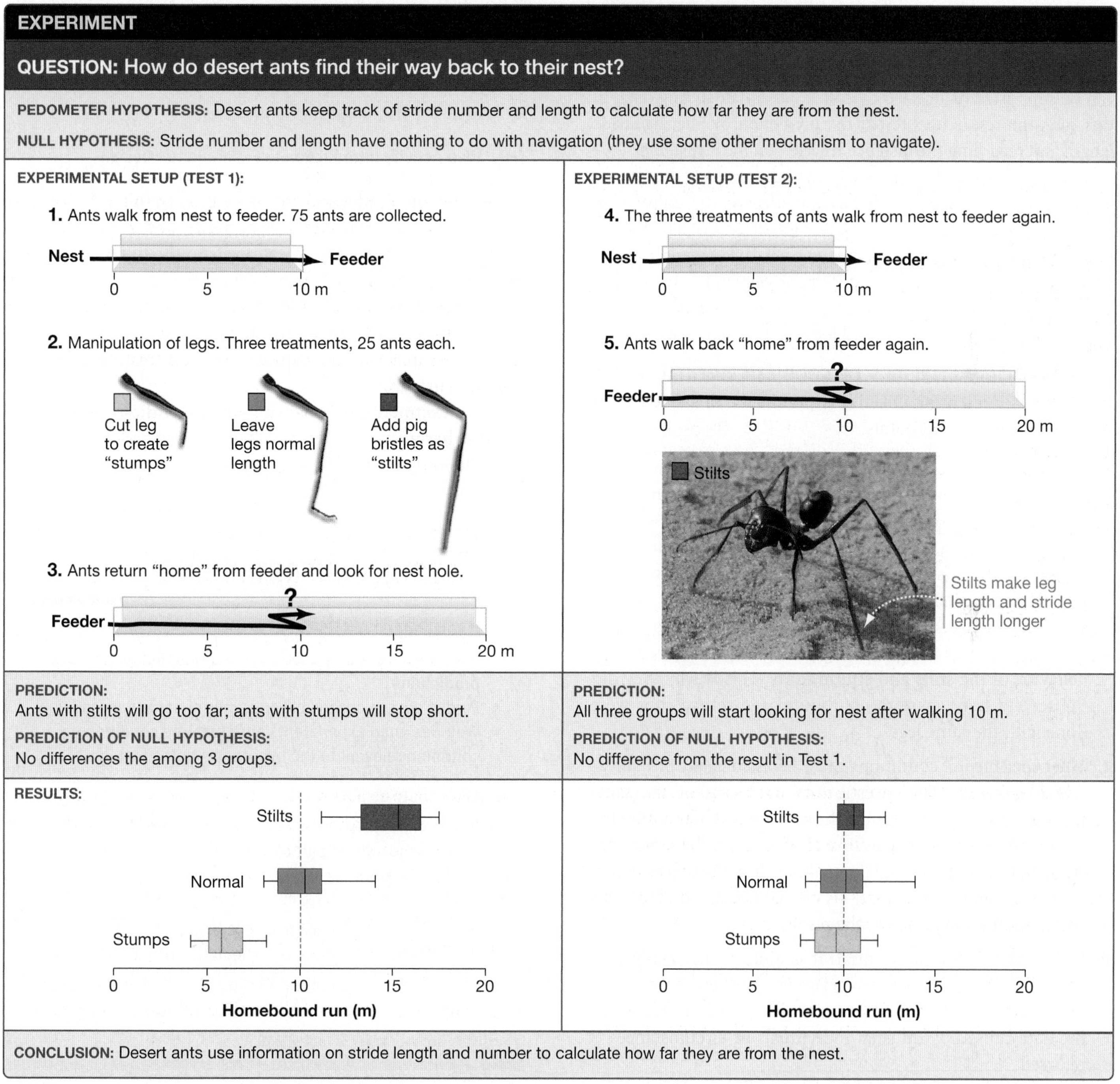

EXPERIMENT
QUESTION: How do desert ants find their way back to their nest?
PEDOMETER HYPOTHESIS: Desert ants keep track of stride number and length to calculate how far they are from the nest.
NULL HYPOTHESIS: Stride number and length have nothing to do with navigation (they use some other mechanism to navigate).
EXPERIMENTAL SETUP (TEST 1):
1. Ants walk from nest to feeder. 75 ants are collected.
Nest
Feeder
0
5
10 m
2. Manipulation of legs. Three treatments, 25 ants each.
Cut leg to create "stumps"
Leave legs normal length
Add pig bristles as "stilts"
3. Ants return "home" from feeder and look for nest hole.
?
Feeder
0
5
10
15
20 m
EXPERIMENTAL SETUP (TEST 2):
4. The three treatments of ants walk from nest to feeder again.
Nest
Feeder
0
5
10 m
5. Ants walk back "home" from feeder again.
?
Feeder
0
5
10
15
20 m
Stilts
Stilts make leg length and stride length longer
PREDICTION:
Ants with stilts will go too far; ants with stumps will stop short.
PREDICTION OF NULL HYPOTHESIS:
No differences the among 3 groups.
PREDICTION:
All three groups will start looking for nest after walking 10 m.
PREDICTION OF NULL HYPOTHESIS:
No difference from the result in Test 1.
RESULTS:
Stilts
Normal
Stumps
0
5
10
15
20
Homebound run (m)
Stilts
Normal
Stumps
0
5
10
15
20
Homebound run (m)
CONCLUSION: Desert ants use information on stride length and number to calculate how far they are from the nest.

FIGURE 1.8 An Experimental Test: Do Desert Ants Use a "Pedometer"?

SOURCE: Wittlinger, M., R. Wehner, and H. Wolf. 2006. The ant odometer: Stepping on stilts and stumps. *Science* 312: 1965–1967.

✔**QUESTION** How would you interpret the experiment if the researchers had used just one ant in each group instead of 25?

The graphs in the "Results" display "box-and-whisker" plots. Each box indicates the range of distances where 50 percent of the ants stopped to search for the nest; the whiskers indicate the range where 95 percent of the ants stopped to search. The vertical line inside each box indicates the median—meaning that half the ants stopped above this distance and half below. For more details on how biologists report averages and indicate the variability and uncertainty in data, see **BioSkills 5** in Appendix A.

INTERPRETING THE RESULTS The pedometer hypothesis predicts that an ant's ability to walk home depends on the number and length of steps taken on its outbound trip. Recall that a prediction specifies what we should observe if a hypothesis is correct. Good scientific hypotheses make testable predictions—predictions that can be supported or rejected by collecting and analyzing data. In this case, the researchers tested the prediction by altering stride length and recording the distance traveled on the return trip.

If the pedometer hypothesis is wrong, however, then stride length and step number should have no effect on the ability of an ant to get back to its nest. This latter possibility is called a **null hypothesis**. A null hypothesis specifies what we should observe when the hypothesis being tested isn't correct. Under the null hypothesis in this experiment, all the ants should have walked 10 m in the first test before they started looking for their nest.

IMPORTANT CHARACTERISTICS OF GOOD EXPERIMENTAL DESIGN In relation to designing effective experiments, this study illustrates several important points:

- It is critical to include **control** groups. A control checks for factors, other than the one being tested, that might influence the experiment's outcome. In this case, there were two controls. Including a normal, unmanipulated individual controlled for the possibility that switching the individuals to a new channel altered their behavior. In addition, the researchers had to control for the possibility that the manipulation itself—and not the change in leg length—affected the behavior of the stilts and stumps ants. This is why they did the second test, where the outbound and return runs were done with the same legs.
- The experimental conditions must be as constant or equivalent as possible. The investigators used ants of the same species, from the same nest, at the same time of day, under the same humidity and temperature conditions, at the same feeders, in the same channels. Controlling all the variables except one—leg length in this case—is crucial because it eliminates alternative explanations for the results.
- Repeating the test is essential. It is almost universally true that larger sample sizes in experiments are better. By testing many individuals, the amount of distortion or "noise" in the data caused by unusual individuals or circumstances is reduced.

✔You should be able to explain: (1) What issue would arise if in the first test, the normal individual had not walked 10 m on the return trip before looking for the nest; and (2) What you would conclude if the stilts and stumps ants had not navigated normally during the second test.

From the outcomes of these experiments, the researchers concluded that desert ants use stride length and number to measure how far they are from the nest. They interpreted their results as strong support for the pedometer hypothesis.

 Biologists practice evidence-based decision-making. They ask questions about how organisms work, pose hypotheses to answer those questions, and use experimental or observational evidence to decide which hypotheses are correct.

To review the principles of experimental design, go to the study area at *www.masteringbiology.com*.

(MB) **Web Activity** Introduction to Experimental Design

The data on giraffes and ants are a taste of things to come. In this text you will encounter hypotheses and experiments on questions ranging from how water gets to the top of 100-meter-tall sequoia trees to why the bacterium that causes tuberculosis has become resistant to antibiotics. As you work through this book, you'll get lots of practice thinking about hypotheses and predictions, analyzing the nature of control treatments, and interpreting graphs.

A commitment to tough-minded hypothesis testing and sound experimental design is a hallmark of biological science. Understanding their value is an important first step in becoming a biologist.

CHECK YOUR UNDERSTANDING

If you understand that . . .

- Hypotheses are proposed explanations that make testable predictions.
- Predictions are observable outcomes of particular conditions.
- Well-designed experiments alter just one condition—a condition relevant to the hypothesis being tested.

✔ **You should be able to . . .**

Design an experiment to test the hypothesis that desert ants feed during the hottest part of the day because it allows them to avoid being eaten by lizards. Then answer the following questions about your experimental design:

1. How does the presence of a control group in your experiment allow you to test the null hypothesis?
2. How are experimental conditions controlled or standardized in a way that precludes alternative explanations of the data?

Answers are available in Appendix B.

CHAPTER 1 REVIEW

For media, go to the study area at www.masteringbiology.com 

Summary of Key Concepts

Organisms obtain and use energy, are made up of cells, process information, replicate, and as populations evolve.

- There is no single, well-accepted definition of life. Instead, biologists point to five characteristics that organisms share.

✔You should be able to explain why the cells in a dead organism are different from the cells in a live organism.

The cell theory proposes that all organisms are made of cells and that all cells come from preexisting cells.

- The cell theory identified the fundamental structural unit common to all life.

✔You should be able to describe the evidence that supported the pattern and the process components of the cell theory.

The theory of evolution by natural selection maintains that species change through time because individuals with certain heritable traits produce more offspring than other individuals do.

- The theory of evolution states that all organisms are related by common ancestry.
- Natural selection is a well-tested explanation for why species change through time and why they are so well adapted to their habitats.

✔You should be able to explain why the average protein content of seeds in a natural population of a grass species would increase over time, if seeds with higher protein content survive better and grow into individuals that produce many seeds with high protein content when they mature.

MB **Web Activity** Artificial Selection

A phylogenetic tree is a graphical representation of the evolutionary relationships between species. These relationships can be estimated by analyzing similarities and differences in traits. Species that share distinctive traits are closely related and are placed close to each other on the tree of life.

- The cell theory and the theory of evolution predict that all organisms are part of a genealogy of species, and that all species trace their ancestry back to a single common ancestor.
- To reconstruct this phylogeny, biologists have analyzed the sequence of components in rRNA and other molecules found in all cells.
- A tree of life, based on similarities and differences in these molecules has three major lineages: the Bacteria, Archaea, and Eukarya.

✔You should be able to explain how biologists can determine whether newly discovered species are members of the Bacteria, Archaea, or Eukarya by analyzing their rRNA or other molecules.

Biologists ask questions, generate hypotheses to answer them, and design experiments that test the predictions made by competing hypotheses.

- Biology is a hypothesis-driven, experimental science.

✔You should be able to explain (1) the relationship between a hypothesis and a prediction and (2) why experiments are convincing ways to test predictions.

MB **Web Activity** Introduction to Experimental Design

Questions

✔TEST YOUR KNOWLEDGE

Answers are available in Appendix B

1. Anton van Leeuwenhoek made an important contribution to the development of the cell theory. How?
   a. He articulated the pattern component of the theory—that all organisms are made of cells.
   b. He articulated the process component of the theory—that all cells come from preexisting cells.
   c. He invented the first microscope and saw the first cell.
   d. He invented more powerful microscopes and was the first to describe the diversity of cells.
2. What does it mean to say that experimental conditions are controlled?
   a. The test groups consist of the same individuals.
   b. The null hypothesis is correct.
   c. There is no difference in outcome between the control and experimental treatment.
   d. Physical conditions are identical for all groups tested.
3. What does the term *evolution* mean?
   a. The strongest individuals produce the most offspring.
   b. The characteristics of an individual change through the course of its life, in response to natural selection.
   c. The characteristics of populations change through time.
   d. The characteristics of species become more complex over time.
4. What does it mean to say that a characteristic of an organism is heritable?
   a. The characteristic evolves.
   b. The characteristic can be passed on to offspring.
   c. The characteristic is advantageous to the organism.
   d. The characteristic does not vary in the population.
5. In biology, to what does the term *fitness* refer?
   a. The degree of training and muscle mass an individual has, relative to others in the same population
   b. An individual's slimness, relative to others in the same population
   c. The longevity of a particular individual
   d. An individual's ability to survive and reproduce
6. Could *both* the food competition hypothesis and the sexual selection hypothesis explain why giraffes have long necks? Why or why not?
   a. No. In science, only one hypothesis can be correct.
   b. No. Observations have shown that the food competition hypothesis cannot be correct.
   c. Yes. Long necks could be advantageous for more than one reason.
   d. Yes. All giraffes have been shown to feed at the highest possible height and fight for mates.

✓TEST YOUR UNDERSTANDING

Answers are available in Appendix B

1. What would researchers have to demonstrate to convince you that they had discovered life on another planet?
2. It was once thought that the deepest split between life-forms was between two groups: prokaryotes and eukaryotes. Draw and label a phylogenetic tree that represents this hypothesis. Then draw and label a phylogenetic tree that shows the actual relationships between the three domains of organisms.
3. Why was it important for Linnaeus to establish the rule that only one type of organism can have a particular genus and species name?
4. What does it mean to say that a species is adapted to a particular habitat?
5. Explain how selection occurs during natural selection. What is selected, and why?
6. The following two statements explain the logic behind the use of molecular sequence data to estimate evolutionary relationships:

   "If the theory of evolution is true, then rRNA sequences should be very similar in closely related organisms but less similar in organisms that are less closely related."

   "On a phylogenetic tree, branches that share a recent common ancestor represent species that are closely related; branches that don't share recent common ancestors represent species that are more distantly related."

   Is the logic of these statements sound? Why or why not?

✓APPLYING CONCEPTS TO NEW SITUATIONS

Answers are available in Appendix B

1. A scientific theory is a set of propositions that defines and explains some aspect of the world. This definition contrasts sharply with the everyday usage of the word theory, which often carries meanings such as "speculation" or "guess." Explain the difference between the two definitions, using the cell theory and the theory of evolution by natural selection as examples.
2. Turn back to the tree of life shown in Figure 1.5. Note that Bacteria and Archaea are prokaryotes, while Eukarya are eukaryotes. On the simplified tree below, draw an arrow that points to the branch where the structure called the nucleus originated. Explain your reasoning.

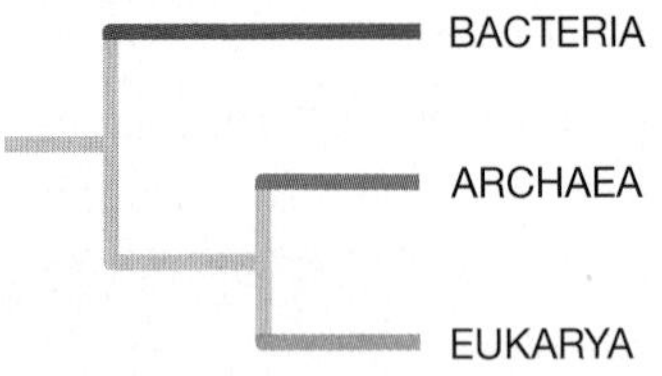

3. The proponents of the cell theory could not "prove" that it was correct in the sense of providing incontrovertible evidence that all organisms are made up of cells. They could state only that all organisms examined to date were made of cells. Why was it reasonable for them to conclude that the theory was valid?
4. Some humans have heritable traits that make them resistant to infection by HIV. In areas of the world where HIV infection rates are high, are human populations evolving? Explain your logic.

The mounds in the foreground, called stromatolites, are teeming with bacterial and archaeal cells. The fossil record includes 3.42-billion-year-old stromatolites that are virtually identical with the ones pictured here.

Water and Carbon: The Chemical Basis of Life 2

Chapter 1 introduced a classic experiment on spontaneous generation, which tested the idea that life arises from nonliving materials. This work helped build a consensus that spontaneous generation does not occur. But for life to exist, spontaneous generation must have occurred at least once, early in Earth's history.

How did life begin? This simple query has been called "the mother of all questions."

This chapter examines a theory, called **chemical evolution**, that is the leading scientific explanation for the origin of life. Like the theories introduced in Chapter 1, the theory of chemical evolution has a pattern component that makes a claim about the natural world and a process component that explains that pattern.

- *The pattern component* Early in Earth's history, simple chemical compounds combined to form more complex carbon-containing substances that accumulated in the oceans.
- *The process component* Complex carbon-containing compounds formed because energy in sunlight and extremely hot water was converted to chemical energy in the form of new chemical bonds.

The theory maintains that continued inputs of energy led to the formation of increasingly complex carbon-containing substances, culminating in a compound that could make a copy of itself, or self-replicate.

At this point, there was a switch from chemical evolution to biological evolution. As the original molecule multiplied, the process of evolution by natural selection took over. Eventually a descendant of the original self-replicating molecule became surrounded by a membrane. When this occurred, the five attributes of life discussed in Chapter 1 were fulfilled. Life had begun.

At first glance, the theory of chemical evolution may seem implausible. But is it? What evidence do biologists have that chemical evolution occurred? Let's start with the fundamentals: the atoms and molecules that would have combined to get chemical evolution started.

KEY CONCEPTS

- Molecules form when atoms bond to each other. Chemical bonds are based on electron sharing. The degree of electron sharing varies from nonpolar covalent bonds, to polar covalent bonds, to ionic bonds.
- Water is essential for life. Water is highly polar and readily forms hydrogen bonds. Hydrogen bonding makes water an extremely efficient solvent.
- Energy is the capacity to do work or supply heat, and can be (1) a stored potential or (2) an active motion. Chemical energy is a form of potential energy, stored in chemical bonds.
- Chemical reactions tend to be spontaneous if they lower potential energy and increase entropy (disorder). An input of energy is required for nonspontaneous reactions to occur.
- Most of the important compounds in organisms contain carbon. Key carbon-containing molecules formed early in Earth's history.

✓ When you see this checkmark, stop and test yourself. Answers are available in Appendix B.

2.1 Atoms, Ions, and Molecules: The Building Blocks of Chemical Evolution

Just four types of atoms—hydrogen, carbon, nitrogen, and oxygen—make up 96 percent of all matter found in organisms today. Many of the molecules found in your cells contain thousands, or even millions, of these atoms bonded together. But early in Earth's history, these elements existed only in simple substances such as water and carbon dioxide, which contain just three atoms apiece.

Two questions are fundamental to understanding how elements could have evolved into the more complex substances found in living cells:

1. What is the physical structure of the hydrogen, carbon, nitrogen, and oxygen atoms found in living cells?
2. What is the structure of the simple molecules—water, carbon dioxide, and others—that acted as the building blocks of chemical evolution?

The focus on structure follows from one of the most central themes in biology: Structure affects function. To understand how a molecule affects your body or the role it played in chemical evolution, you have to understand how it is put together.

Basic Atomic Structure

Figure 2.1a shows a simple way of depicting the structure of an atom, using hydrogen and carbon as examples. Extremely small particles called electrons orbit an atomic nucleus made up of larger particles called protons and neutrons. Protons have a positive electric charge, neutrons are electrically neutral, and electrons have a negative electric charge. Opposite charges attract; like charges repel. When the number of protons and the number of electrons in an atom (or molecule) are the same, the charges balance and the atom is electrically neutral. **Figure 2.1b** provides a sense of scale at the atomic level.

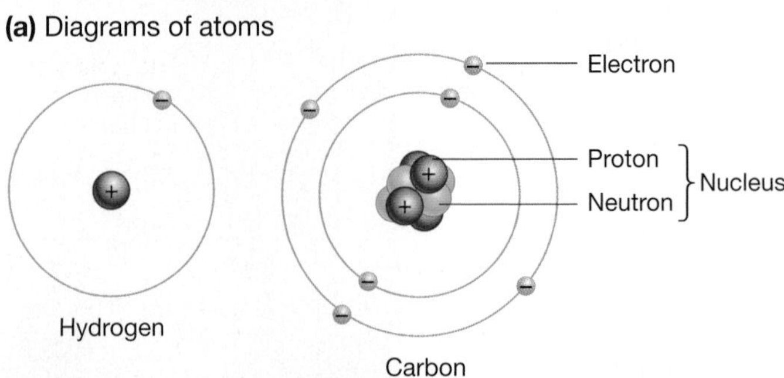

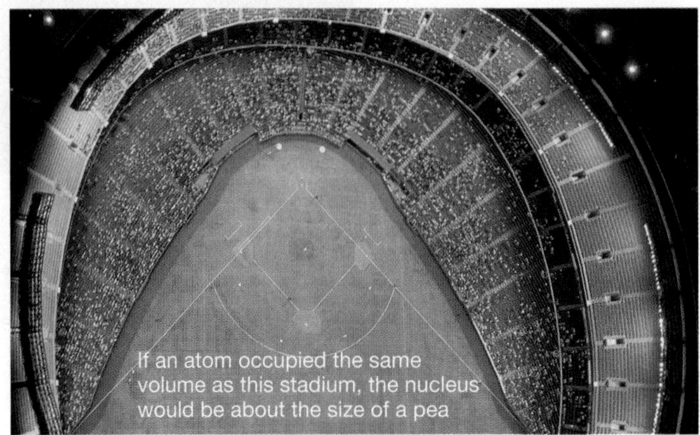

(b) Most of an atom's volume is empty space.
If an atom occupied the same volume as this stadium, the nucleus would be about the size of a pea

FIGURE 2.1 Parts of an Atom. The atomic nucleus, made up of protons and neutrons, is surrounded by orbiting electrons. In reality, electrons do not orbit the nucleus in circles; their actual orbits are complex.

Figure 2.2 shows a segment of the periodic table of the elements. Notice that each atom of an **element** contains a characteristic number of protons, called its **atomic number**. The atomic number is given as a subscript of each element symbol in the table. The sum of the protons and neutrons in an atom is called its **mass number**, given as a superscript of each symbol in Figure 2.2.

Mass number (number of protons + neutrons)

Atomic number (number of protons)

$^{1}_{1}\mathrm{H}$							$^{4}_{2}\mathrm{He}$
$^{7}_{3}\mathrm{Li}$	$^{9}_{4}\mathrm{Be}$	$^{11}_{5}\mathrm{B}$	$^{12}_{6}\mathrm{C}$	$^{14}_{7}\mathrm{N}$	$^{16}_{8}\mathrm{O}$	$^{19}_{9}\mathrm{F}$	$^{20}_{10}\mathrm{Ne}$
$^{23}_{11}\mathrm{Na}$	$^{24}_{12}\mathrm{Mg}$	$^{27}_{13}\mathrm{Al}$	$^{28}_{14}\mathrm{Si}$	$^{31}_{15}\mathrm{P}$	$^{32}_{16}\mathrm{S}$	$^{35}_{17}\mathrm{Cl}$	$^{40}_{18}\mathrm{Ar}$

FIGURE 2.2 A Portion of the Periodic Table. Each element has a unique atomic number and is represented by a unique one- or two-letter symbol. The mass numbers given here are the most common for each element.

The number of protons in an element does not vary—if the atomic number of an atom changes, then one element is transformed to another element. The number of neutrons present in an element can vary, however. Forms of an element with different numbers of neutrons are known as **isotopes** (literally, "equal-places"). For example, all atoms of the element carbon have 6 protons. But naturally occurring isotopes of carbon can have either 6 or 7 neutrons, giving them a total of 12 or 13 protons and neutrons, respectively. Isotopes have different masses.

Although the masses of protons, neutrons, and electrons can be measured in grams, the numbers involved are so small that biologists prefer to use a special unit called the **dalton**. The masses of protons and neutrons are virtually identical and are routinely rounded to 1 dalton. A carbon atom that contains 6 protons and 6 neutrons has a mass of 12 daltons and a mass number of 12, while a carbon atom with 6 protons and 7 neutrons would have a mass number of 13. These isotopes would be written as ^{12}C and ^{13}C, respectively. The mass of an electron is so small that it is normally ignored.

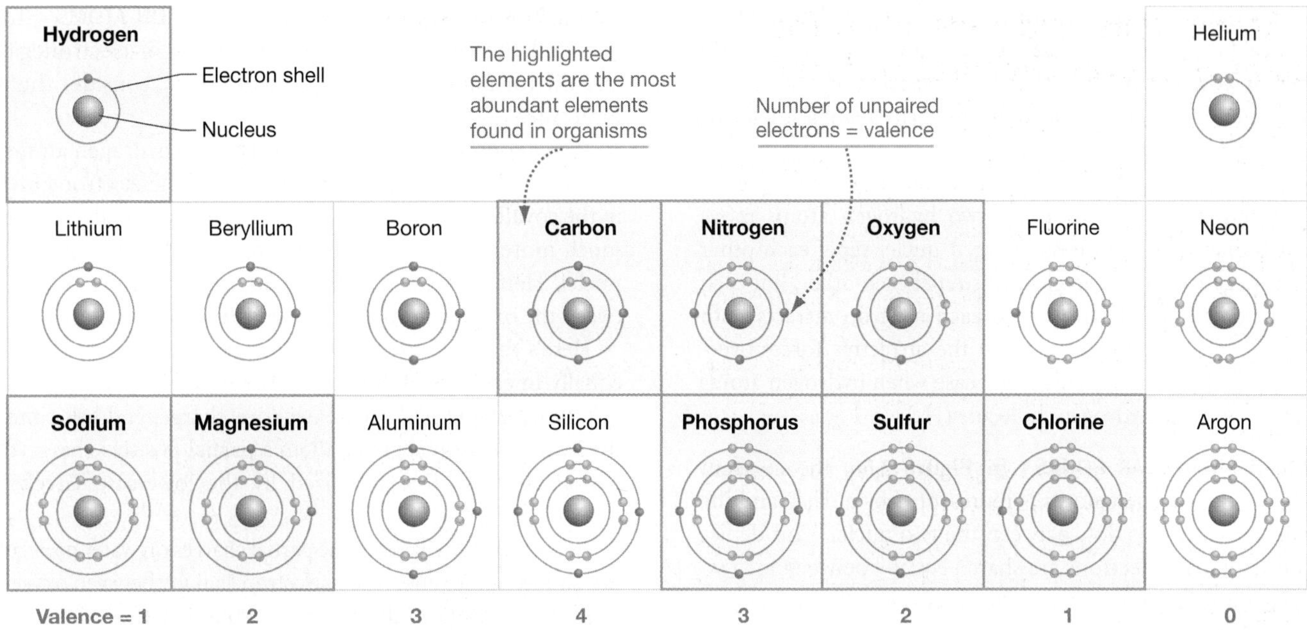

FIGURE 2.3 The Structure of Atoms Found in Organisms.

To understand how the atoms involved in chemical evolution behave, focus on how electrons are arranged around the nucleus:

- Electrons move around atomic nuclei in specific regions called **orbitals.**
- Each orbital can hold up to two electrons.
- Orbitals are grouped into levels called **electron shells**.
- Electron shells are numbered 1, 2, 3, and so on, to indicate their relative distance from the nucleus, with smaller numbers closer to the nucleus.
- Each electron shell contains a specific number of orbitals. An electron shell comprising a single orbital can hold up to two electrons; a shell with four orbitals can contain up to eight electrons.
- The electrons of an atom fill the innermost shells first, before filling outer shells.

To understand how the structures of atoms differ, take a moment to study **Figure 2.3**. This chart highlights the atoms that are most abundant in living cells. The gray ball in the center of each box represents a nucleus, and the orange circle or circles represent the electron shells around that nucleus. The small orange balls on the circles indicate how electrons are distributed in the shells of each element.

Now focus on the outermost shell of each atom. This is the element's valence shell. The electrons found in this shell are referred to as **valence electrons**. Two observations are important:

1. In each of the highlighted elements, the outermost electron shell is not full—not all the orbitals in the valence shell have two electrons. The highlighted elements have at least one unpaired valence electron.
2. The number of unpaired electrons in the valence shell varies among elements. Carbon has four unpaired electrons in its outermost shell; hydrogen has one. The number of unpaired electrons found in an atom is called its **valence**. Thus, carbon's valence is four, and hydrogen's valence is one.

These observations are significant because an atom is most stable when its valence shell is filled. One way that shells can be filled is through the formation of **chemical bonds**—strong attractions that bind atoms together.

How Does Covalent Bonding Hold Molecules Together?

To understand how atoms can become more stable by making chemical bonds, consider hydrogen. The hydrogen atom has just one electron, which resides in a shell that can hold two electrons.

Because it has an unpaired valence electron, the hydrogen atom is not very stable. But when two atoms of hydrogen come into contact, the two electrons become shared by the two nuclei (**Figure 2.4**). Both atoms now have a completely filled shell. Together, the hydrogen atoms are more stable than the two individual hydrogen atoms.

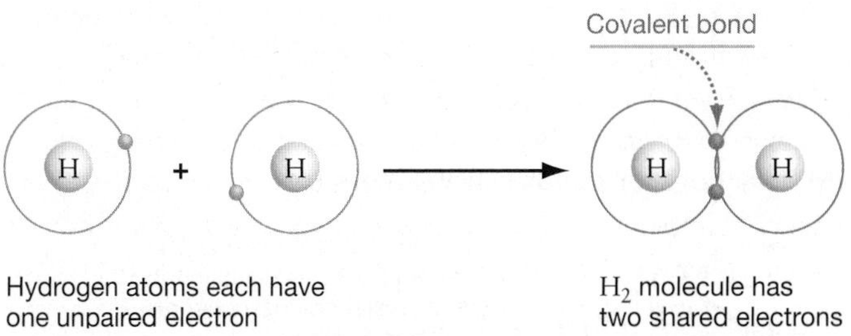

FIGURE 2.4 Covalent Bonds Result from Electron Sharing. When two hydrogen atoms come into contact, their electrons are attracted to the positive charge in each nucleus. As a result, their orbitals overlap, the electrons are shared by each nucleus, and a covalent bond forms.

Shared electrons "glue" two hydrogen atoms together in a type of chemical bond called a **covalent bond**. Substances that are held together by covalent bonds are called **molecules**. In the case of two hydrogen atoms, the bonded atoms form a single molecule of hydrogen, written as H—H or H_2.

It can also be helpful to think about covalent bonding as electrical attraction and repulsion. As two hydrogen atoms move closer together, their positively charged nuclei repel each other and their negatively charged electrons repel each other. But each proton attracts both electrons, and each electron attracts both protons. Covalent bonds form when the attractive forces overcome the repulsive forces. This is the case when hydrogen atoms interact to form the hydrogen molecule (H_2).

NONPOLAR AND POLAR BONDS In **Figure 2.5a**, the covalent bond between hydrogen atoms is represented by a dash, and the electrons are drawn as dots between the two nuclei. This depiction shows that the electrons are shared equally between the two hydrogen atoms, resulting in a nonpolar covalent bond—a covalent bond that is symmetrical.

It's important to note, though, that the electrons participating in a covalent bond are not always shared equally between the atoms involved. This happens because some atoms hold the electrons in covalent bonds much more tightly than do other atoms. Chemists call this property an atom's **electronegativity**.

Oxygen is among the most electronegative of all elements: It attracts covalently bonded electrons more strongly than does any other atom commonly found in organisms. Nitrogen's electronegativity is somewhat lower than oxygen's. Carbon and hydrogen, in turn, have relatively low and approximately equal electronegativities. Thus, the electronegativities of the four most abundant elements in organisms are related as follows: $O > N > C \cong H$.

Because carbon and hydrogen have approximately equal electronegativity, the electrons in a C—H bond are shared equally. The result is a **nonpolar covalent bond**. In contrast, asymmetric sharing of electrons results in a **polar covalent bond**. The electrons in a polar covalent bond spend most of their time close to the nucleus of the more electronegative atom. Why is this important?

(a) Nonpolar covalent bond in hydrogen molecule

H—H

Electrons are shown to be superimposed on the bond to indicate that they are halfway between the two atoms, shared equally

(b) Polar covalent bonds in water molecule

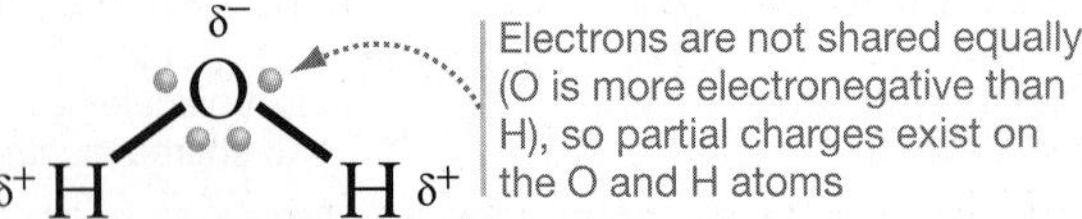
δ−
O
δ+ H
H δ+
Electrons are not shared equally (O is more electronegative than H), so partial charges exist on the O and H atoms

FIGURE 2.5 Electron Sharing and Bond Polarity. Delta (δ) symbols in polar covalent bonds refer to partial positive and negative charges that arise owing to unequal electron sharing.

POLAR BONDS PRODUCE PARTIAL CHARGES ON ATOMS To understand the consequences of differences in electronegativity and the formation of polar covalent bonds, consider the water molecule.

Water consists of oxygen bonded to two hydrogen atoms, and is written H_2O. As **Figure 2.5b** illustrates, the electrons involved in the covalent bonds in water are not shared equally but are held much more tightly by the oxygen nucleus than by the hydrogen nuclei. Hence, water has two polar covalent bonds—those between the oxygen atom and each of the hydrogen atoms.

Here's the key observation: Because electrons are shared unequally in each O—H bond, they spend more time near the oxygen atom, giving it a partial negative charge, and less time near the hydrogen atoms, giving them a partial positive charge. These partial charges are symbolized by the lowercase Greek letter delta, δ.

As Section 2.2 will show, the partial charges on water molecules—due simply to the difference in electronegativity between oxygen and hydrogen—are one of the primary reasons that life exists.

Ionic Bonding, Ions, and the Electron-Sharing Continuum

Ionic bonds are similar in principle to covalent bonds, but instead of sharing electrons between two atoms, the electrons in ionic bonds are completely transferred from one atom to the other. The electron transfer occurs because it gives the resulting atoms a full outermost shell.

Sodium atoms (Na), for example, tend to lose an electron, leaving them with a full second shell. This is a much more stable arrangement, energetically, than having a lone electron in their third shell (**Figure 2.6a**). The atom that results has a net electric charge of +1, because it has one more proton than it has electrons.

An atom or molecule that carries a charge is called an **ion**. The sodium ion is written Na^+ and, like other positively charged ions, is called a **cation**.

Chlorine atoms (Cl), in contrast, tend to gain an electron, filling their outermost shell (**Figure 2.6b**). The ion has a net charge of −1, because it has one more electron than protons. This negatively charged ion, or **anion**, is written Cl^- and is called chloride.

When sodium and chlorine combine to form table salt (sodium chloride, NaCl), the atoms pack into a crystal structure consisting of sodium cations and chloride anions (**Figure 2.6c**). The electrical attraction between the ions is so strong that salt crystals are difficult to break apart.

This discussion of covalent and ionic bonding supports an important general observation: The degree to which electrons are shared in chemical bonds forms a continuum, from equal sharing in nonpolar covalent bonds, to unequal sharing in polar covalent bonds, to the transfer of electrons in ionic bonds.

As the left-hand side of **Figure 2.7** shows, covalent bonds between atoms with exactly the same electronegativity—for exam-

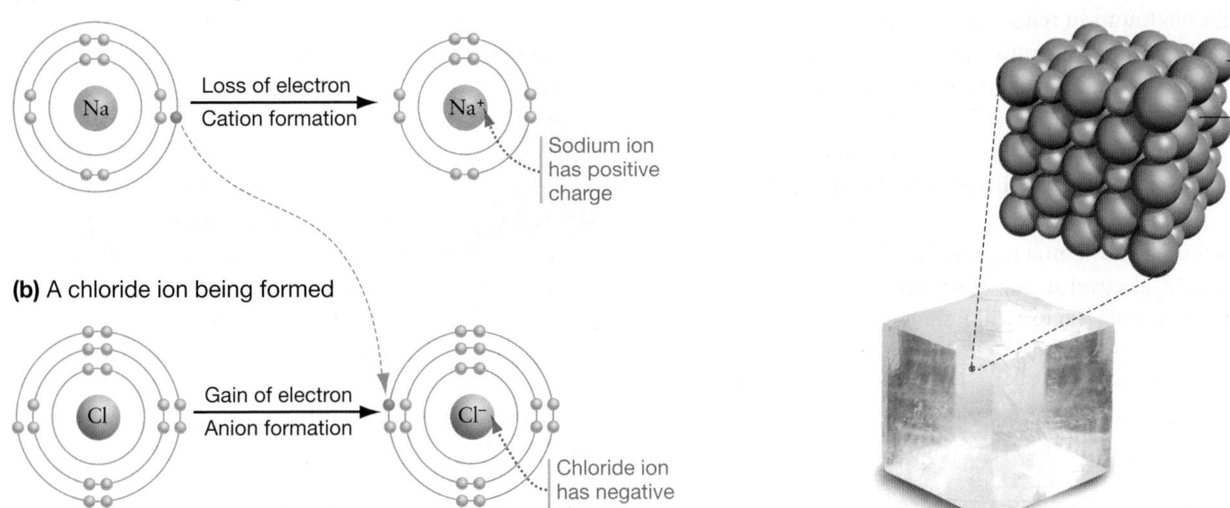

FIGURE 2.6 Ion Formation and Ionic Bonding. The sodium ion (Na^+) and the chloride ion (Cl^-) are stable because they have full valence shells. In table salt (NaCl), sodium and chloride ions pack into a crystal structure held together by electrical attraction between their positive and negative charges.

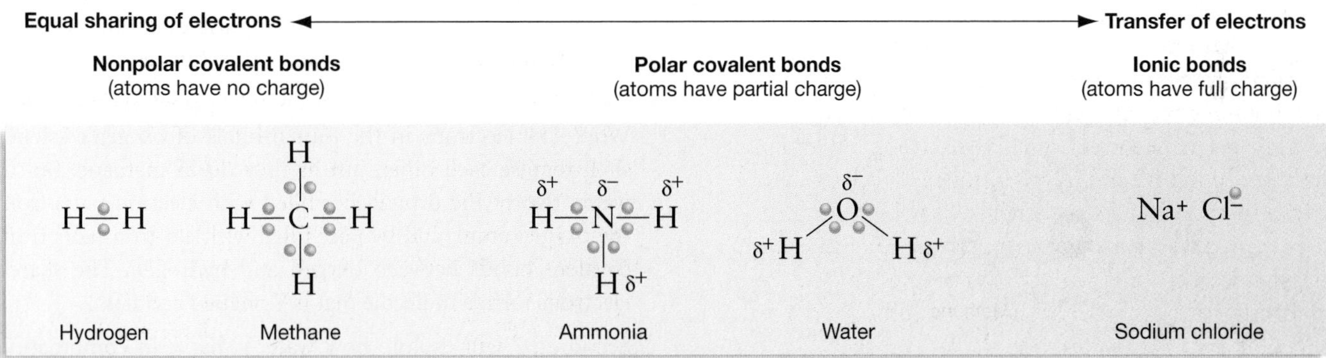

FIGURE 2.7 The Electron-Sharing Continuum. The degree of electron sharing in chemical bonds can be thought of as a continuum, from equal sharing in nonpolar covalent bonds to no sharing in ionic bonds.

✔**QUESTION** Why do most polar covalent bonds involve nitrogen or oxygen?

ple, between the atoms of hydrogen in H_2—represent one end of the continuum. The electrons in these nonpolar bonds are shared equally.

In the middle of the continuum are bonds where one atom is much more electronegative than the other. In these asymmetric bonds, substantial partial charges exist on each of the atoms. These types of polar covalent bonds occur when a highly electronegative atom such as oxygen or nitrogen is bound to an atom with a lower affinity for electrons, such as carbon or hydrogen. Ammonia (NH_3) and water (H_2O) are examples of molecules with polar covalent bonds.

At the right-hand side of the continuum are molecules made up of atoms with extreme differences in their electronegativities. In this case, electrons are transferred rather than shared, the atoms have full charges, and the bonding is ionic. Common table salt, NaCl, is a familiar example.

Most chemical bonds that occur in biological molecules are on the left-hand side and the middle of the continuum; in the molecules found in organisms, ionic bonding is rare.

Some Simple Molecules Formed from C, H, N, and O

Look back at Figure 2.3 and count the number of unpaired electrons in the valence shells of carbon, nitrogen, oxygen, and hydrogen atoms. Each unpaired electron in a valence shell can make up half of a covalent bond. It should make sense to you that a carbon atom can form a total of four covalent bonds; nitrogen can form three; oxygen can form two; and hydrogen, one.

When each of the four unpaired electrons of a carbon atom covalently bonds with a hydrogen atom, the molecule that results is written CH_4 and is called methane (**Figure 2.8a**). Methane is the most common molecule found in natural gas. When a nitrogen atom's three unpaired electrons bond with three hydrogen atoms, the result is NH_3, or ammonia. Similarly, an atom of oxygen can form covalent bonds with two atoms of hydrogen, resulting in a water molecule (H_2O). As Figure 2.4 showed, a hydrogen atom can bond with another hydrogen atom to form hydrogen gas (H_2).

In addition to forming more than one single bond, atoms with more than one unpaired electron in the valence shell can form double bonds or triple bonds. **Figure 2.8b** shows how carbon forms double bonds with oxygen atoms to produce carbon dioxide (CO_2). Triple bonds result when three pairs of electrons are shared. **Figure 2.8c** shows the structure of molecular nitrogen (N_2), which forms when two nitrogen atoms establish a triple bond.

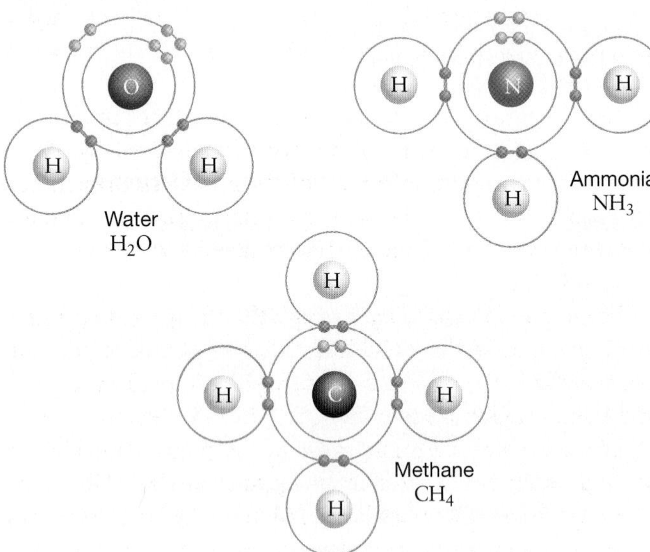

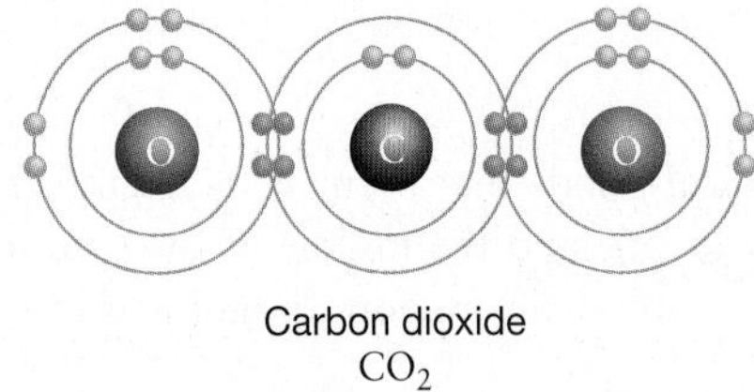

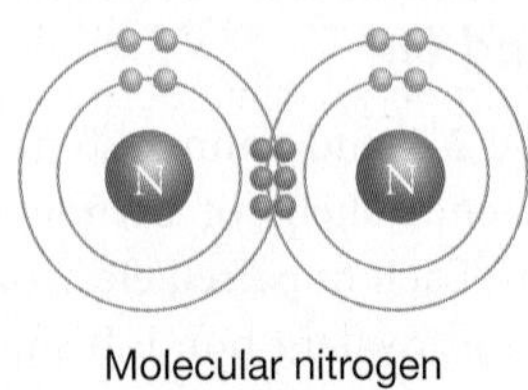

FIGURE 2.8 Unpaired Electrons in the Valence Shell Participate in Covalent Bonds. Covalent bonding is based on sharing of electrons in the outermost shell. Covalent bonds can be **(a)** single, **(b)** double, or **(c)** triple.

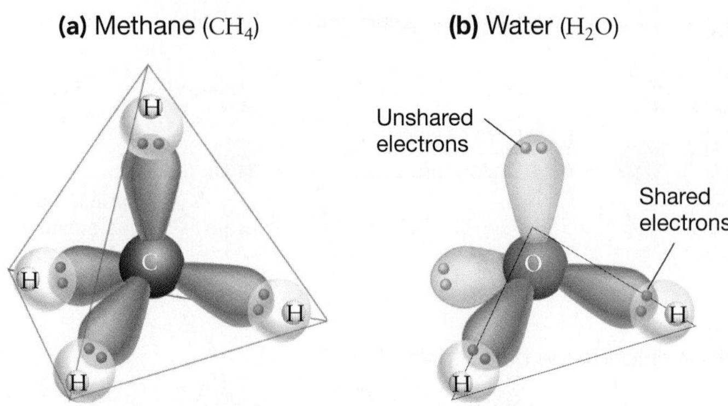

FIGURE 2.9 The Geometry of Methane and Water.

The Geometry of Simple Molecules

In many cases, the overall shape of a molecule dictates how it behaves. In chemistry and in biology, function is based on structure.

The shapes of the simple molecules you've just learned about are governed by the geometry of their bonds.

- Methane (CH_4) is tetrahedral—a structure with four triangular surfaces (**Figure 2.9a**). The tetrahedron forms because the electrons in the four C—H bonds repulse each other equally. The electron pairs are as far apart as they can get.
- Water is bent and two-dimensional, or planar (**Figure 2.9b**). Why? The electrons in the four orbitals of oxygen's valence shell repulse each other, just as they do in methane. But in water, two of the orbitals are filled with electron pairs from the oxygen atom, and two are filled with electron pairs from covalent bonds between oxygen and hydrogen. The shared electrons form a molecule that is V-shaped and flat.

Section 2.2 will explore how water's shape, in combination with the partial charges on the oxygen and hydrogen atoms, make it the most important molecule on Earth.

Representing Molecules

Molecules can be represented in a variety of increasingly complex ways—only some of which reflect their actual shape. Each method has advantages and disadvantages.

- **Molecular formulas** are compact but don't contain a great deal of information—they indicate only the numbers and types of atoms in a molecule (**Figure 2.10a**).
- **Structural formulas** indicate which atoms in a molecule are bonded together. Single, double, and triple bonds are represented by single, double, and triple dashes, respectively. Structural formulas also indicate geometry in two dimensions (**Figure 2.10b**). This is useful for planar molecules such as water and O_2.
- **Ball-and-stick models** take up more space than structural formulas, but provide information on the three-dimensional shape of molecules and indicate the relative sizes of the atoms involved (**Figure 2.10c**).

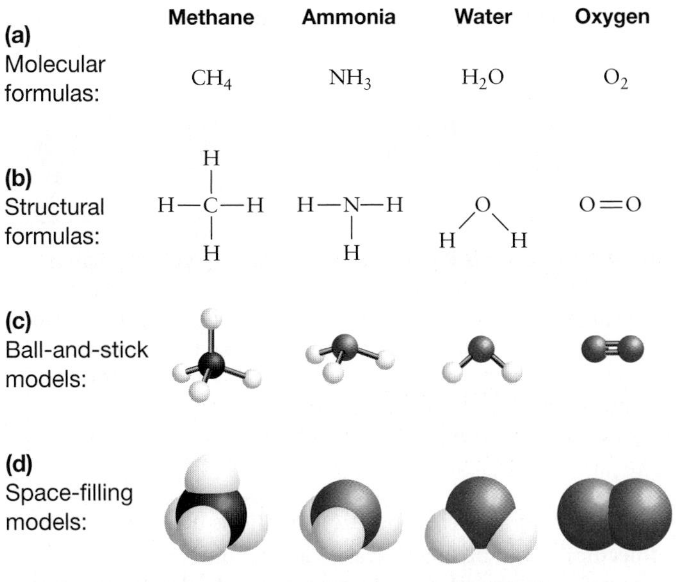

	Methane	Ammonia	Water	Oxygen
(a) Molecular formulas:	CH_4	NH_3	H_2O	O_2
(b) Structural formulas:	H—C—H (with H above and below C)	H—N—H (with H below N)	H—O—H (bent)	O=O
(c) Ball-and-stick models:				
(d) Space-filling models:				

FIGURE 2.10 Molecules Can Be Represented Several Ways. Each method of representing a molecule has particular advantages.

- **Space-filling models** are more difficult to read than ball-and-stick models but more accurately depict the spatial relationships between atoms (**Figure 2.10d**).

In both ball-and-stick and space-filling models, biologists use certain colors to represent certain atoms. A black ball, for example, always symbolizes carbon. For more information on interpreting chemical structures, see **BioSkills 6** in Appendix A.

Basic Concepts in Chemical Reactions

The molecules you've just learned about—CH_4, NH_3, H_2O, CO_2, and N_2—are found in volcanic gases on Earth and in the atmospheres of nearby planets. Based on these observations, researchers claim that they were important components of Earth's ancient atmosphere and ocean. If so, then they provided the building blocks for chemical evolution.

The question now is: How did these simple building blocks combine to form more complex products, early in Earth's history?

When a **chemical reaction** occurs, one substance is combined with others or broken down into another substance. Atoms are rearranged in molecules; in most cases, chemical bonds are broken and new bonds form.

Chemicals react in simple whole-number combinations. For example, one molecule of acetic acid (a prominent component of vinegar) reacts with one molecule of ethanol (the active ingredient in alcoholic beverages) to form one molecule of ethyl acetate.

Now suppose that you wanted to set up this reaction in an experiment. How would you know how much acetic acid and ethanol to add? The problem is that there is no simple way of counting the numbers of molecules present in a sample. Researchers solve this problem using the mole concept.

A **mole** refers to the number 6.022×10^{23}—just as the unit called the dozen refers to the number 12 or the unit million refers to the number 1×10^6. The mole is a useful unit because the mass of one mole of any molecule is the same as its molecular weight expressed in grams. **Molecular weight** is the sum of the mass numbers of all the atoms in a molecule.

For example, look back at Figure 2.2 and note that hydrogen has a mass number of 1 and oxygen has a mass number of 16. To get the molecular weight of H_2O, you sum the mass numbers of two atoms of hydrogen and one atom of oxygen, giving $1 + 1 + 16$, or a total of 18. Because mass number can also be measured in grams, it follows that if you weighed a sample of 18 grams of water, it would contain 6.022×10^{23} water molecules, or 1 mole of water molecules.

Now let's get back to your experiment. If you wanted to react one mole of acetic acid and one mole of ethanol, you'd use a balance to weigh out a number of grams of acetic acid and ethanol equal to their respective molecular weights and make up a **solution**—a homogenous (uniform) mixture of one or more substances dissolved in a liquid.

When substances are dissolved in liquid, their concentration is expressed in terms of molarity (symbolized by "M"). **Molarity** is the number of moles of the substance present per liter of solution. A 1-molar solution of acetic acid in water, for example, means that 1 mole of acetic acid is contained in 1 liter of solution.

The mole and molarity are concepts that allow you to connect what is going on in the invisible world at the atomic level to the visible world of the lab bench—where you can perform experiments with compounds such as acetic acid and ethanol.

Researchers postulate that most of the critical reactions in chemical evolution occurred in an aqueous, or water-based, solution. To understand what happened and why, let's delve into the properties of water, then start analyzing the reactions that triggered chemical evolution.

CHECK YOUR UNDERSTANDING

If you understand that . . .

- Covalent bonds are based on electron sharing. Electron sharing allows atoms to fill all the orbitals in their valence shell, making them more stable.
- Covalent bonds can be polar or nonpolar, depending on whether the electronegativities of the two atoms involved are the same or different.

✓ You should be able to . . .

Draw the structural formulas of methane (CH_4) and ammonia (NH_3) and add dots to indicate the relative locations of the covalently bonded electrons, based on the relative electronegativities of C, H, and N.

Answers are available in Appendix B.

2.2 The Early Oceans and the Properties of Water

Life is based on water. In a typical living cell, water comprises over 75 percent of the volume (**Figure 2.11**). You can survive for weeks without eating, but you aren't likely to live more than 3–4 days without drinking.

Water is vital for a simple reason: It is an excellent **solvent**—an agent for getting substances into solution. The chemical reactions occurring inside your body right now, and the reactions that caused chemical evolution some 3.5 billion years ago, depend on direct, physical interaction between the reactants. Substances are most likely to come into contact and react as **solutes**—meaning, when they are dissolved. The formation of Earth's first ocean, about 3.8 billion years ago, was a turning point in chemical evolution because it gave the process a place to happen.

Why Is Water Such an Efficient Solvent?

To understand why water is such an effective solvent, recall that:

1. Both of the O–H bonds in the molecule are polar, owing to oxygen's high electronegativity. As a result, the oxygen atom has a partial negative charge and each hydrogen atom has a partial positive charge.
2. The molecule is bent. Consequently, the partial negative charge on the oxygen atom sticks out, away from the partial positive charges on the hydrogen atoms (**Figure 2.12a**).

Figure 2.12b illustrates how water's structure affects its interactions with other water molecules. When two liquid water molecules approach each other, the partial positive charge on hydrogen attracts the partial negative charge on oxygen. This weak electrical attraction forms a **hydrogen bond** between the molecules.

✔If you understand how water's polarity makes hydrogen bonding possible, you should be able to (1) draw a fictional version of Figure 2.12b that shows water as a linear (not bent) molecule, with partial charges on the oxygen and hydrogen atoms; and (2) explain why electrostatic attractions between such water molecules would be much weaker, as a result.

FIGURE 2.11 Water Is the Most Abundant Molecule in Organisms. Fruits shrink when they are dried because they consist primarily of water.

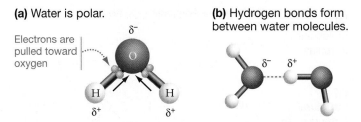

(a) Water is polar.

Electrons are pulled toward oxygen

δ⁻ O H H δ⁺ δ⁺

(b) Hydrogen bonds form between water molecules.

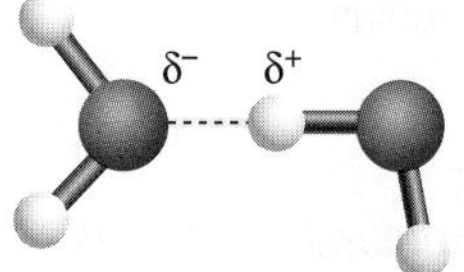

FIGURE 2.12 Water Is Polar and Participates in Hydrogen Bonds. **(a)** Because of oxygen's high electronegativity, the electrons that are shared between hydrogen and oxygen spend more time close to the oxygen nucleus, giving the oxygen atom a partial negative charge and the hydrogen atom a partial positive charge. **(b)** The electrical attraction that occurs between the partial positive and negative charges on water molecules forms a hydrogen bond.

In an aqueous solution, hydrogen bonds also form between water molecules and other polar molecules. Similar interactions occur between water and ions. Ions and polar molecules stay in solution because of their interactions with water's partial charges (**Figure 2.13a**). Substances that interact with water in this way are said to be **hydrophilic** ("water-loving").

In contrast, compounds that are uncharged and nonpolar do not interact with water through hydrogen bonding and do not dissolve in water. Because their interactions with water are minimal or nonexistent, they are forced to interact with each other (**Figure 2.13b**). Substances that do not interact with water are said to be **hydrophobic** ("water-fearing"). Hydrogen bonding makes it possible for almost any charged or polar molecule to dissolve in water.

Although individual hydrogen bonds are not nearly as strong as covalent or ionic bonds, many of them occur in a solution. Hydrogen bonding is extremely important in biology owing to the sheer number of hydrogen bonds that form between water and other molecules that are polar or carry a charge.

How Does Water's Structure Correlate with Its Properties?

Water's small size, bent shape, highly polar covalent bonds, and overall polarity are unique among molecules. Because the structure of molecules routinely correlates with their function, it's not surprising that water has some remarkable properties, in addition to its extraordinary capacity to act as a solvent.

COHESION, ADHESION, AND SURFACE TENSION Binding between like molecules is called **cohesion**. Water is cohesive—meaning that it stays together—because of the hydrogen bonds that form between individual molecules.

Binding between unlike molecules, in contrast, is called **adhesion**. Adhesion is usually analyzed in regard to interactions between a liquid and a solid surface. Water adheres to surfaces that have any polar or charged components.

Cohesion and adhesion are important in explaining how water moves from the roots of plants to their leaves (see Chapter 37).

(a) Polar molecules and ions dissolve readily in water.

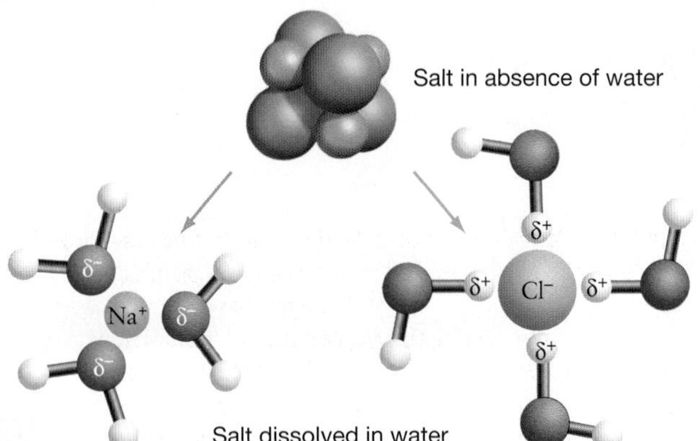

(b) Nonpolar molecules do not dissolve in water.

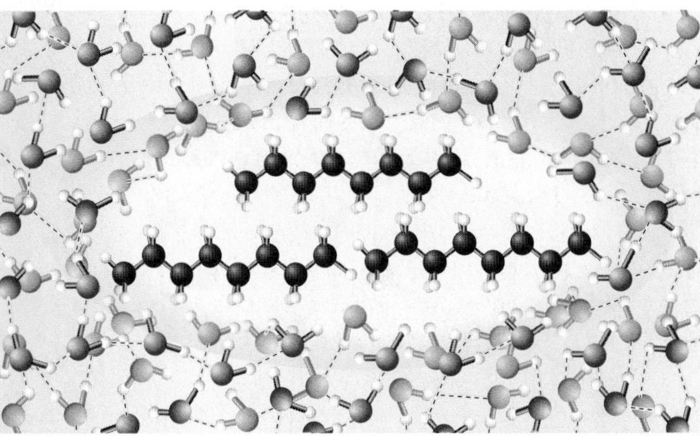

FIGURE 2.13 Water Interacts with Other Polar Molecules. (a) Water's polarity makes it a superb solvent for polar molecules and ions. **(b)** In aqueous solution, nonpolar molecules and compounds are forced to interact with each other. This occurs because water is much more stable when it interacts with itself rather than with the nonpolar molecules.

✔ **QUESTION** Explain the physical basis of the expression, "Oil and water don't mix."

But you can also see them in action in the concave surface, or meniscus, that forms in a glass of water (**Figure 2.14a**). A meniscus forms as a result of two forces:

1. Water molecules at the surface hydrogen-bond with water molecules below them, so they experience a net downward pull.
2. Water molecules at the surface also adhere to the glass, allowing them to resist the downward pull.

Cohesion is also instrumental in the phenomenon known as surface tension. Because hydrogen bonding exerts a pulling force, or tension, at the surface of any body of water, water molecules are not stable there—they are constantly being pulled away from the surface. A body of water is most stable when this source of instability is minimized—meaning that its total surface area is minimized.

This fact has an important consequence: Water resists any force that increases its surface area. More specifically, any force that depresses a water surface meets with resistance. This resistance makes a water surface act as if it had an elastic membrane (**Figure 2.14b**)—a property called **surface tension**.

All liquids have a surface tension. Water's surface tension is extraordinarily high because of the extensive hydrogen bonding that occurs between molecules. In water, the "elastic membrane" at the surface is stronger than it is in other liquids.

WATER IS DENSER AS A LIQUID THAN AS A SOLID When factory workers pour molten metal or plastic into a mold and allow it to cool to the solid state, the material shrinks. When molten lava pours out of a volcano and cools to solid rock, it shrinks. But when you fill an ice tray with water and put it in the freezer to make ice, the water expands.

Unlike most substances, water is denser as a liquid than it is as a solid. In other words, there are more molecules of water in a given volume of liquid water than there are in the same volume

(a) A meniscus forms where water meets a solid surface, as a result of two forces.

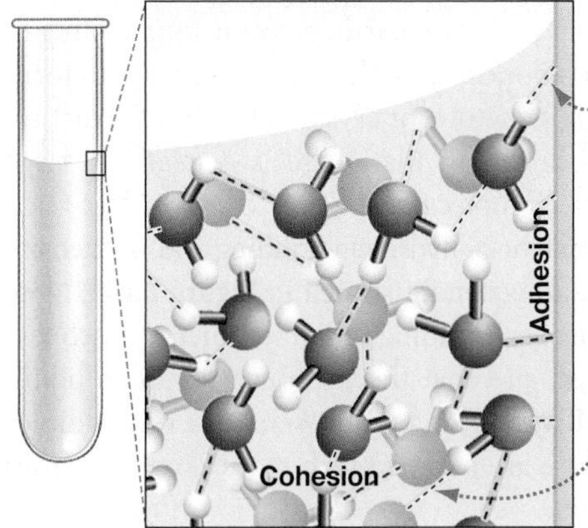

(b) Water has high surface tension.

Because of surface tension, light objects do not fall through the water's surface

FIGURE 2.14 Cohesion, Adhesion, and Surface Tension. (a) Meniscus formation is based on hydrogen bonding. **(b)** Water resists forces—like the weight of an insect—that increase its surface area. The resistance is great enough that light objects do not break the surface.

(a) In ice, water molecules form a crystal lattice.

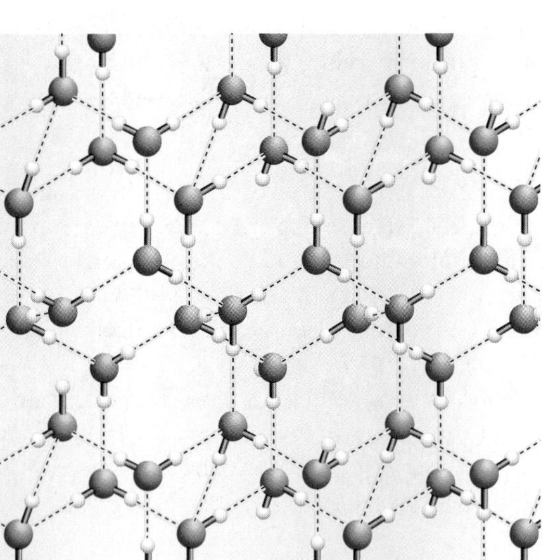

(b) In liquid water, no crystal lattice forms.

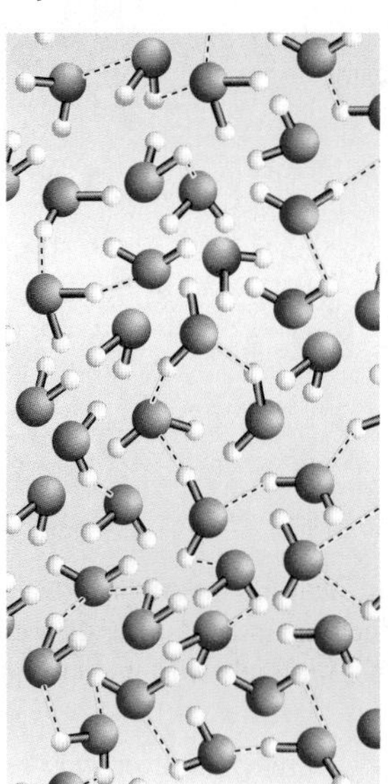

(c) Liquid water is denser than ice. As a result, ice floats.

FIGURE 2.15 Hydrogen Bonding Forms the Crystal Structure of Ice. In ice, each molecule can form four hydrogen bonds at one time. Each oxygen atom can form two; each hydrogen atom can form one.

of solid water, or ice. **Figure 2.15a** illustrates why this is so. Note that in ice, each water molecule participates in four hydrogen bonds. These hydrogen bonds cause the water molecules to form a regular and repeating structure, or crystal. The crystal structure of ice is fairly open, meaning that there is a relatively large amount of space between molecules.

Now compare the extent of hydrogen bonding and the density of ice with that of liquid water, illustrated in **Figure 2.15b**. Note that the extent of hydrogen bonding in liquid water is much less than that found in ice, and that the hydrogen bonds in liquid water are constantly being made and broken. As a result, molecules in the liquid phase are packed much more closely together than in the solid phase.

Normally, heating a substance causes it to expand because molecules begin moving faster and colliding more often and with greater force. But heating ice causes hydrogen bonds to break and the open crystal structure to collapse. In this way, hydrogen bonding explains why water is denser as a liquid than as a solid.

This property of water has an important result: Ice floats (**Figure 2.15c**). If it did not, ice would sink to the bottom of lakes, ponds, and oceans soon after it formed. The ice would stay frozen in the cold depths. If water weren't so unusual, it is almost certain that Earth's oceans would have frozen almost solid before life had a chance to start.

WATER HAS A HIGH CAPACITY FOR ABSORBING ENERGY Hydrogen bonding is also responsible for another of water's remarkable physical properties: Water has a high capacity for absorbing energy.

Specific heat, for example, is the amount of energy required to raise the temperature of 1 gram of a substance by 1°C. Water has an extremely high specific heat because when a source of energy hits it, hydrogen bonds must be broken before heat can be transferred and the water molecules begin moving faster. As **Table 2.1** indicates, it takes an extraordinarily large amount of energy to change the temperature of water and other molecules where extensive hydrogen bonding occurs.

Similarly, it takes a large amount of energy to break the hydrogen bonds in liquid water and change the molecules from the liquid phase to the gas phase. Water's **heat of vaporization**—the energy required to change 1 gram of it from a liquid to gas—is higher than that of most molecules that are liquid at room temperature. As a result, water has to absorb a great deal of energy to evaporate. Water's high heat of vaporization is the reason that sweating or dousing yourself with water is an effective way to cool off on a hot day. Water molecules absorb a great deal of energy from your body before they evaporate, so you lose heat.

TABLE 2.1 **Specific Heats of Some Liquids**

The specific heats reported in this table were measured at 25°C and are given in units of joules per gram of substance per degree Celsius. (The joule is a unit of energy.)

Liquids with extensive hydrogen bonding	Specific Heat
Ammonia (NH_3)	4.70
Water (H_2O)	4.18
Liquids with some hydrogen bonding	
Ethanol (CH_3CH_2OH)	2.44
Ethylene glycol ($HOCH_2CH_2OH$; used in antifreeze)	2.22
Liquids with little or no hydrogen bonding	
Benzene (C_6H_6)	1.80
Xylene (C_8H_{10})	1.72
Sulfuric acid (H_2SO_4)	1.40

SOURCE: J. Murray and R. C. Fay 2004. *Chemistry,* 4th ed., Prentice-Hall, Table 8.1

Water's ability to absorb energy is critical to the theory of chemical evolution. Because compounds that were important in chemical evolution dissolve readily in water, they would have formed in the ocean or rained out of the atmosphere into the ocean. Once these compounds were in aqueous solution, they were well-protected from sources of energy that could break them apart, such as intense sunlight. As a result, they would have persisted and slowly increased in concentration over time, making them more likely to react and continue the process.

Acid–Base Reactions Involve a Transfer of Protons

One other aspect of water's chemistry is important to understanding chemical evolution and how organisms work today. Water is not a completely stable molecule. In reality, water molecules continually undergo a chemical reaction with themselves. This "dissociation" reaction can be written as follows:

$$H_2O \rightleftharpoons H^+ + OH^-$$

The double arrow indicates that the reaction proceeds in both directions.

The substances on the right-hand side of the expression are the **hydrogen ion** (H^+) and the **hydroxide ion** (OH^-). A hydrogen ion is simply a proton. In reality, however, protons never exist by themselves. In water, for example, protons associate with water molecules to form the hydronium ion (H_3O^+). Thus, the dissociation of water is more accurately written as:

$$H_2O + H_2O \rightleftharpoons H_3O^+ + OH^-$$

One of the water molecules on the left-hand side of the expression has given up a proton, while the other water molecule has accepted a proton.

Substances that give up protons during chemical reactions and raise the hydrogen ion concentration of water are called **acids**; molecules or ions that acquire protons during chemical reactions and lower the hydrogen ion concentration of water are called **bases**. Most acids act only as acids, and most bases act only as bases; but water can act as both an acid and a base.

A chemical reaction that involves a transfer of protons is called an acid–base reaction. Every acid–base reaction requires a proton donor and a proton acceptor—an acid and a base, respectively.

Water is an extremely weak acid—very few water molecules dissociate to form hydronium ions and hydroxide ions. In contrast, strong acids like the hydrochloric acid in your stomach readily give up a proton when they react with water.

$$HCl + H_2O \rightleftharpoons H_3O^+ + Cl^-$$

Strong bases readily acquire a proton when they react with water. For example, sodium hydroxide (NaOH, commonly called lye) dissociates completely in water to form Na^+ and OH^-. The hydroxide ion produced by that reaction then accepts a proton from a hydronium ion in the water, forming two water molecules.

$$NaOH(aq) \longrightarrow Na^+ + OH^-$$
$$OH^- + H_3O^+ \rightleftharpoons 2\,H_2O$$

The "*aq*" in the expression indicates that NaOH is in aqueous solution.

To summarize, adding an acid to a solution increases the concentration of protons; adding a base to a solution lowers the concentration of protons. Water is both a weak acid and a weak base.

pH INDICATES THE CONCENTRATION OF PROTONS In a solution, the tendency for acid–base reactions to occur is largely a function of the number of protons present. Chemists can measure the concentration of protons in a solution directly. In a sample of pure water at 25°C, the concentration of H^+ is 1.0×10^{-7} M, or 1 ten-millionth molar.

Because the concentration of protons in water is such a small number, exponential notation is cumbersome. So chemists and biologists prefer to express the concentration of protons in a solution, and thus its acidity or alkalinity, with a logarithmic notation called **pH.**[1]

By definition, the pH of a solution is the negative of the base-10 logarithm, or log, of the hydrogen ion concentration:

$$pH = -\log[H^+]$$

(The square brackets are a standard notation for indicating "concentration of" a substance in solution.) Taking antilogs gives:

$$[H^+] = \text{antilog}(-pH) = 10^{-pH}$$

pH, then, is a convenient way to indicate the concentration of protons in a solution. If the concentration of H^+ in a sample of water is 1.0×10^{-7} M, then its pH is 7. If the solution changes to pH 9, then it has become 100 times more basic. To review logarithms, see **BioSkills 7** in Appendix A.

BUFFERS PROTECT AGAINST DAMAGING CHANGES IN pH Cells are extremely sensitive to changes in pH. Changes in hydrogen ion concentration affect the structure and function of polar or charged substances, as well as the tendency of acid–base reactions to occur.

Compounds that minimize changes in pH are called **buffers** because they buffer a solution against the damaging effects of pH change. Buffers are important in maintaining relatively constant conditions, or **homeostasis**, in cells and tissues. In cells, a wide array of naturally occurring molecules act as buffers.

Most buffers are weak acids, meaning that they are somewhat likely to give up a proton in solution. To see how they work, consider the disassociation of acetic acid in water to form acetate ions and protons.

$$CH_3COOH \rightleftharpoons CH_3COO^- + H^+$$

When acetic acid and acetate are present in about equal concentrations in a solution, they function as a buffering system. If the concentration of protons increases slightly, the protons react with acetate ions to form acetic acid and pH does not change. If the concentration of protons decreases slightly, acetic acid gives up protons and pH does not change.

[1]The term *pH* is derived from the French *puissance d'hydrogène*, or "power of hydrogen."

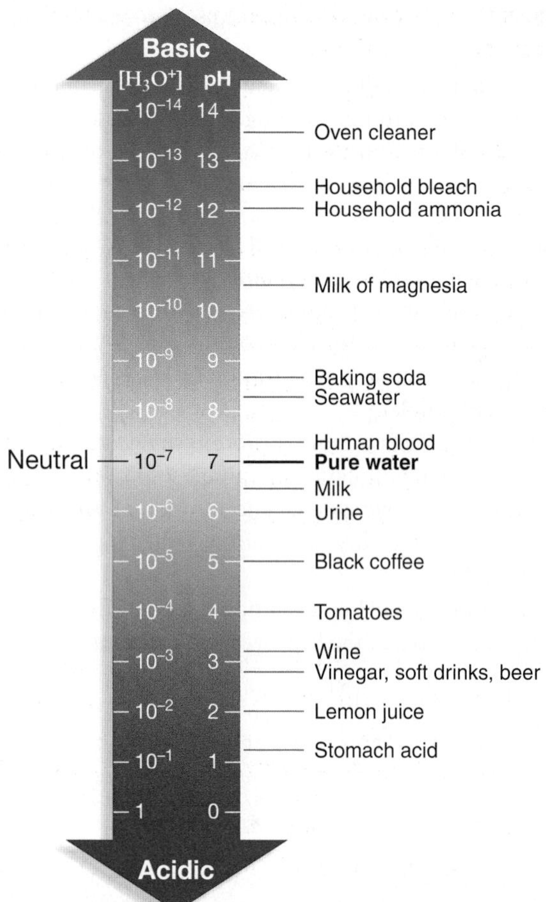

FIGURE 2.16 The pH Scale. Because the pH scale is logarithmic, a change in one unit of pH represents a change in the concentration of hydrogen ions equal to a factor of 10. Coffee has a hundred times more H^+ than pure water has.

✔**QUESTION** How is the pH of black coffee affected if you add milk?

THE pH SCALE MEASURES THE ACIDITY OR ALKALINITY OF A SOLUTION **Figure 2.16** shows the pH scale and reports the pH of some common solutions. Note that the pH of pure water at 25°C is 7. Pure water is used as a standard, or point of reference, on the pH scale.

Solutions that contain acids have a proton concentration larger than 1×10^{-7} M and thus a pH < 7. This is because acidic molecules tend to release protons into solution. In contrast, solutions that contain bases have a proton concentration less than 1×10^{-7} M and thus a pH > 7. This is because basic molecules tend to accept protons from solution. ✔If you understand this concept, you should be able to explain the difference in hydrogen ion concentration between a slightly basic solution with pH 8 and a slightly acidic solution with pH 6.

Solutions with a pH of 7 are considered neutral—neither acidic nor basic. The solution inside living cells is about 7. Many of today's lakes also have a nearly neutral pH, which probably mimics the conditions in the early oceans—it would have taken hundreds of millions of years for rain and wind to erode rocks and send dissolved ions to the ocean, increasing its salinity and raising its pH to the current level of about 8.

As chemical evolution began, then, water provided the physical environment for key reactions to take place. In some cases, water also acted as an important reactant. Although acid–base reactions were not critical to the initial stages of chemical evolution, they became extremely important once the process was under way.

Table 2.2 summarizes some of the key properties of water, for you to use as a reference. You can also go to the study area at *www.masteringbiology.com* and review how hydrogen bonding gives water such unusual properties.

(MB) **Web Activity** The Properties of Water

Water provided a medium for chemical evolution to occur. Now let's consider what happened in solution, some 3.5 billion years ago.

SUMMARY TABLE 2.2 **Properties of Water**

Property	Cause	Biological Consequences
Solvent for charged or polar compounds	______________	Most chemical reactions important for life take place in aqueous solution.
Denser as a liquid than a solid	As water freezes, each molecule forms a total of four hydrogen bonds, leading to the formation of the low-density crystal structure called ice.	______________
High specific heat	Water molecules must absorb lots of heat energy to break hydrogen bonds and experience increased movement (and thus temperature).	Oceans absorb and release heat slowly, moderating coastal climates.
High heat of vaporization	______________	Evaporation of water from an organism cools the body.

✔**EXERCISE** You should be able to fill in the missing cells in this table. You should also be able to make a concept map relating water's structure to the properties listed here. (For an introduction to concept mapping, see **BioSkills 8** in Appendix A.) Your concept map should include the following terms or phrases: polar covalent bonds, polarity (of the water molecule), hydrogen bonding, high heat of vaporization, high specific heat, less dense as a solid, effective solvent, unequal sharing of electrons, high energy input required to break bonds, high electronegativity of oxygen.

2.3 Chemical Reactions, Chemical Evolution, and Chemical Energy

Proponents of the theory of chemical evolution contend that simple molecules present in the atmosphere and ocean of ancient Earth participated in chemical reactions that eventually produced larger, more complex molecules—such as the proteins, nucleic acids, sugars, and lipids introduced in Chapters 3–6. Currently, researchers are investigating two environments where these reactions could have occurred.

1. The atmosphere, which was probably dominated by gases ejected from volcanoes. Carbon dioxide, water vapor, and nitrogen are the dominant gases ejected from volcanoes today; a small amount of molecular hydrogen (H_2), methane (CH_4), and carbon monoxide (CO) may also be present.
2. Deep-sea vents, where extremely hot rocks from the Earth's interior contact the seafloor. In addition to the gases listed above, today's deep-sea vents are rich in metals such as iron and nickel.

When volcanic gases are put together and allowed to interact, however, very little happens. The simple molecules do not suddenly link together to create large, complex substances like those found in living cells. Instead, their bonds remain intact. How, then, did chemical evolution occur?

How Do Chemical Reactions Happen?

Chemical reactions are written in a format similar to mathematical equations, with the initial, or **reactant**, atoms or molecules shown on the left and the resulting reaction **product(s)** shown on the right. For example, the most common reaction in the mix of gases and water that emerge from volcanoes is:

$$CO_2(g) + H_2O(l) \rightleftharpoons H_2CO_3(aq)$$

This expression indicates that carbon dioxide (CO_2) reacts with water (H_2O), forming carbonic acid (H_2CO_3). The state of each reactant and product is indicated as gas (g), liquid (l), in aqueous solution (aq), or solid (s).

Note that the expression is balanced; that is, 1 carbon, 3 oxygen, and 2 hydrogen atoms are present on each side of the expression. Note also that the expression contains a double arrow, meaning that the reaction is reversible. When the forward and reverse reactions proceed at the same rate, the quantities of reactants and products remain constant, although not necessarily equal. A dynamic but stable state such as this is termed a **chemical equilibrium**.

A chemical equilibrium can be disturbed by changing the concentration of reactants or products. For example, adding CO_2 to the mixture would drive the reaction to the right, creating more H_2CO_3 until the equilibrium proportions of reactants and products are reestablished. Adding H_2CO_3 instead would drive the reaction to the left. Removing CO_2 would also drive the reaction to the left; removing H_2CO_3 would drive it to the right.

A chemical equilibrium can also be altered by changes in temperature. For example, the water molecules in this set of interacting elements, or **system**, would be present as a combination of liquid water and water vapor:

$$H_2O \rightleftharpoons H_2O(g)$$

If liquid water molecules absorb enough heat, they transform to the gaseous state. This is called an **endothermic** ("within heating") process because heat is absorbed during the process. In contrast, the transformation of water vapor to liquid water releases heat and is called **exothermic** ("outside heating"). Raising the temperature of this system drives the equilibrium to the right; cooling the system drives it to the left.

In relation to chemical evolution, though, these reactions and changes of state are not particularly interesting. Carbonic acid is not an important intermediate in the formation of more complex molecules. Interesting things do begin to happen, though, when large amounts of energy are added to mixtures of volcanic gases.

What Is Energy?

Energy can be defined as the capacity to do work or to supply heat. This capacity exists in one of two ways—as a stored potential or as an active motion.

Stored energy is called **potential energy**. An object gains or loses its ability to store energy because of its position. An electron that resides in an outer electron shell will, if the opportunity arises, fall into a lower electron shell closer to the positive charges on the protons in the nucleus. Because of its position farther from the positive charges in the nucleus, an electron in an outer electron shell has more potential energy than does an electron in an inner shell (**Figure 2.17**).

Kinetic energy is the energy of motion. Molecules have kinetic energy because they are constantly in motion.

- The kinetic energy of molecular motion is called **thermal energy**.
- The **temperature** of an object is a measure of how much thermal energy its molecules possess. If an object has a low temperature, its molecules are moving slowly. (We perceive this as "cold.") If an object has a high temperature, its molecules are moving rapidly. (We perceive this as "hot.")

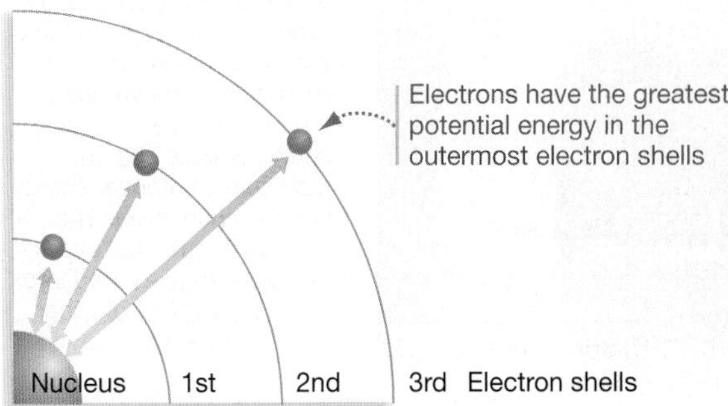

FIGURE 2.17 Potential Energy as a Function of Electron Shells. Electrons in outer shells have more potential energy than do electrons in inner shells, because the negative charges of the electrons in outer shells are farther from the positive charges of the protons in the nucleus. Each shell represents a distinct level of potential energy.

- When two objects with different temperatures come into contact, thermal energy is transferred between them. This transferred energy is called **heat**.

There are many forms of potential energy and kinetic energy, and energy can change from one form into another. To drive this point home, consider a water molecule sitting at the top of a waterfall, as in **Figure 2.18a**.

Step 1 The molecule has potential energy (E_p) because of its position.

Step 2 As the molecule passes over the waterfall, its potential energy is converted to the kinetic energy (E_k) of motion.

Step 3 When the molecule reaches the rocks below, it experiences a change in potential energy because it has changed position. The change in potential energy is transformed into an equal amount of energy in other forms: mechanical energy, which tends to break up the rocks; heat (thermal energy), which raises the temperature of the rocks and the water itself; and sound.

An electron in an outer electron shell is analogous to the water molecule at the top of a waterfall (**Figure 2.18b**). If the electron falls to a lower shell, its potential energy is converted to the kinetic energy of motion. After the electron occupies the lower electron shell, it experiences a change in potential energy. As panel 3 in Figure 2.18b shows, the change in potential energy

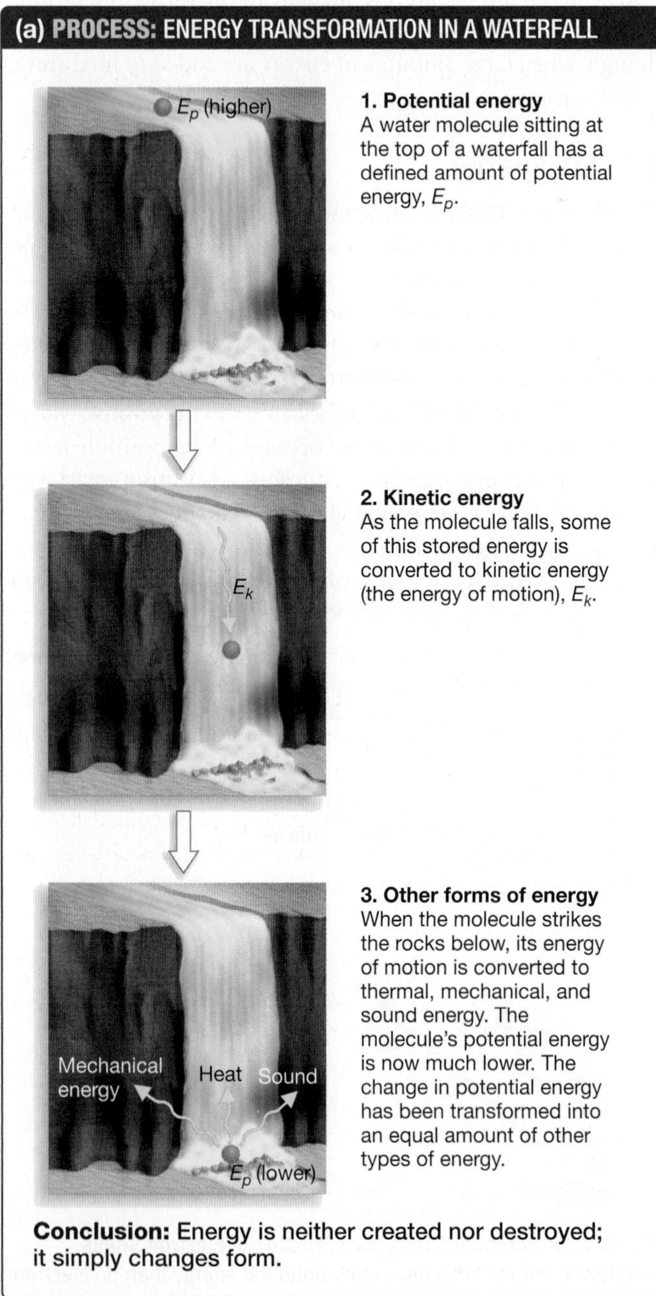

(a) PROCESS: ENERGY TRANSFORMATION IN A WATERFALL
E_p (higher)
1. Potential energy
A water molecule sitting at the top of a waterfall has a defined amount of potential energy, E_p.
E_k
2. Kinetic energy
As the molecule falls, some of this stored energy is converted to kinetic energy (the energy of motion), E_k.
Mechanical energy
Heat
Sound
E_p (lower)
3. Other forms of energy
When the molecule strikes the rocks below, its energy of motion is converted to thermal, mechanical, and sound energy. The molecule's potential energy is now much lower. The change in potential energy has been transformed into an equal amount of other types of energy.
Conclusion: Energy is neither created nor destroyed; it simply changes form.

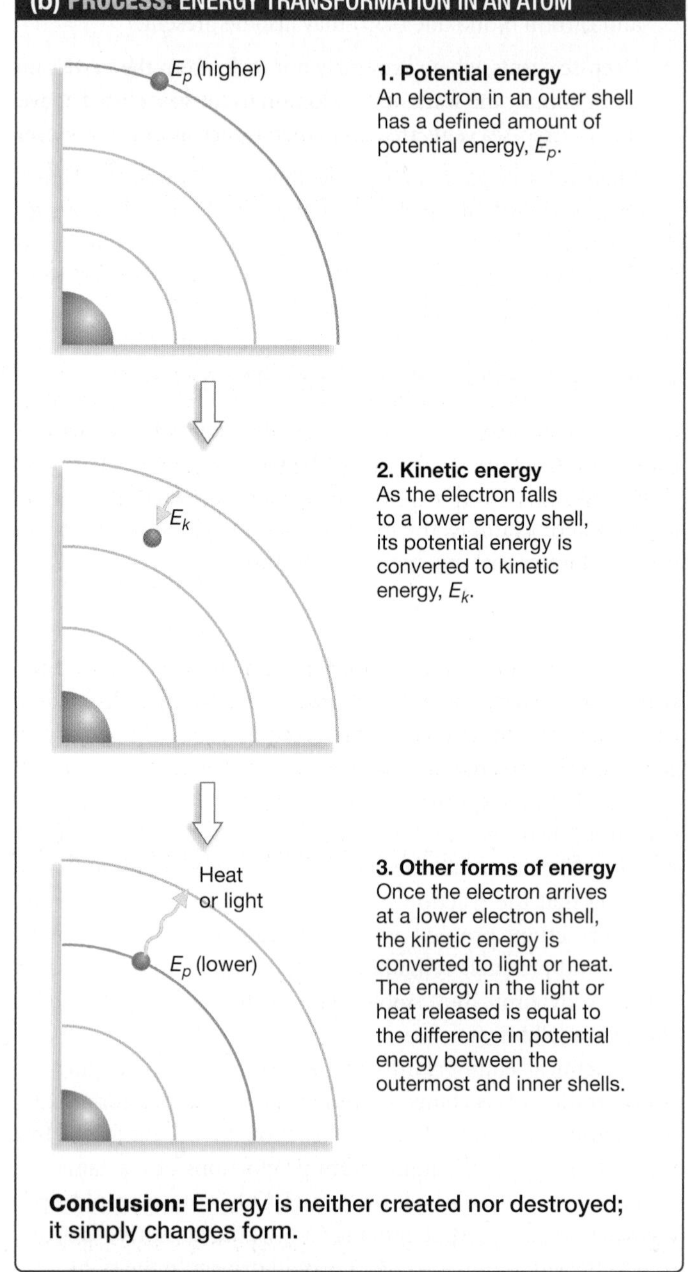

(b) PROCESS: ENERGY TRANSFORMATION IN AN ATOM
E_p (higher)
1. Potential energy
An electron in an outer shell has a defined amount of potential energy, E_p.
E_k
2. Kinetic energy
As the electron falls to a lower energy shell, its potential energy is converted to kinetic energy, E_k.
Heat or light
E_p (lower)
3. Other forms of energy
Once the electron arrives at a lower electron shell, the kinetic energy is converted to light or heat. The energy in the light or heat released is equal to the difference in potential energy between the outermost and inner shells.
Conclusion: Energy is neither created nor destroyed; it simply changes form.

FIGURE 2.18 Energy Transformations. During an energy transformation, the total amount of energy in the system remains constant.

is transformed into an equal amount of energy in other forms—usually thermal energy, but sometimes light.

These examples illustrate the **first law of thermodynamics**, which states that energy is conserved. Energy cannot be created or destroyed, but only transferred and transformed.

Energy transformation is the heart of chemical evolution. According to the best data available, molecules that were part of the young Earth were exposed to massive inputs of energy. Part of this energy was in the form of heat, from the gradually cooling molten mass that initially formed the planet. Part of the energy was in the form of high-energy radiation from the Sun, which today is screened out by molecules in the atmosphere. How would the application of large amounts of heat and radiation affect the course of chemical evolution?

Chemical Evolution: A Model System

To determine how energy inputs affected the simple molecules present in the early oceans and atmosphere, researchers have constructed computer models to simulate the reactions that can occur between carbon dioxide, water, ammonia, and hydrogen molecules.

The goal of one such study was quite specific: The researchers wanted to determine whether a molecule called formaldehyde (H_2CO) could be produced. Along with hydrogen cyanide (HCN), formaldehyde is a key intermediate in the creation of the larger, more complex molecules found in cells. Forming formaldehyde and hydrogen cyanide is the critical first step in chemical evolution—a trigger that could set the process in motion.

The research group began by proposing that the following reaction could take place:

$$CO_2(g) + 2\,H_2(g) \longrightarrow H_2CO(g) + H_2O(g)$$

This reaction, however, doesn't occur spontaneously. Let's consider why it doesn't occur without an input of energy. We can then look at where the necessary energy input might have originated, and how conditions found on early Earth, such as temperature and the concentration of reactants, influenced the reaction.

WHAT MAKES A CHEMICAL REACTION SPONTANEOUS? When chemists say that a reaction is spontaneous, they have a precise meaning in mind: Chemical reactions are spontaneous if they proceed on their own, without any continuous external influence such as added energy. Two factors determine whether a reaction is spontaneous or nonspontaneous:

1. Reactions tend to be spontaneous if the products have lower potential energy than the reactants. If the electrons in the reaction products are held more tightly than the electrons in the reactants, then they have lower potential energy. Recall that highly electronegative atoms such as oxygen and nitrogen hold electrons much more tightly than do atoms with a lower electronegativity, such as carbon and hydrogen. For example, when natural gas burns, methane reacts with oxygen gas to produce carbon dioxide and water:

$$CH_4(g) + 2\,O_2(g) \longrightarrow CO_2(g) + 2\,H_2O(g)$$

The electrons involved in the C=O and O–H bonds of carbon dioxide and water are held much more tightly than they were in the C–H and O=O bonds of methane and oxygen (**Figure 2.19**). As a result, the products have much lower potential energy than the reactants. The difference in potential energy between reactants and products is given off as heat, so the reaction is exothermic. In chemical reactions, the difference in potential energy between the products and the reactants is symbolized by ΔH. (The uppercase Greek letter Δ delta, is often used in chemical and mathematical notation to represent change.) When a reaction is exothermic, ΔH is negative.

2. Reactions tend to be spontaneous when the product molecules are less ordered than the reactant molecules. TNT, or dynamite, has a highly ordered chemical structure. But when TNT explodes, gases like carbon dioxide, carbon monoxide, various nitrogen oxides, and small particulates are given off (**Figure 2.20**). These molecules are much less ordered than the reactant molecules in TNT. The amount of disorder in a group of molecules is called its **entropy**, which is symbolized by S. When the products of a chemical reaction are less ordered than the reactant molecules are, entropy increases and ΔS is positive. Reactions tend to be spontaneous if they increase entropy. The **second law of thermodynamics**, in fact, states that entropy always increases in an isolated system.

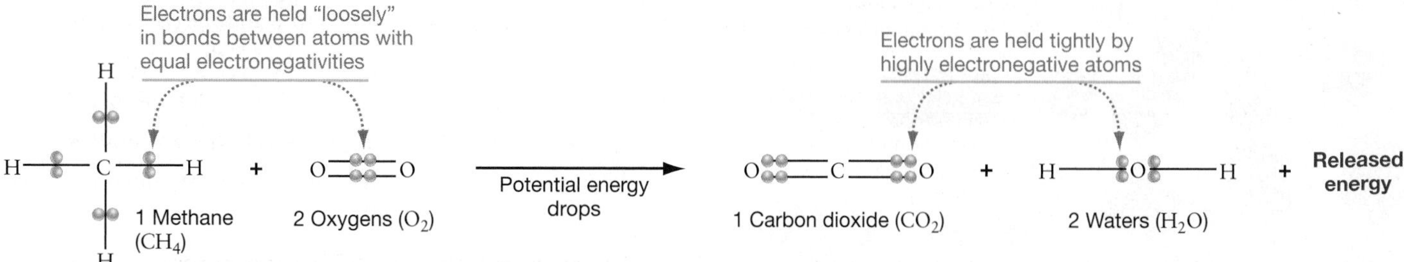

FIGURE 2.19 Potential Energy and/or Entropy May Change during Chemical Reactions. When methane burns, the products have much lower potential energy than the reactants.

✔ **EXERCISE** Label which electrons have relatively low potential energy and which electrons have relatively high potential energy.

FIGURE 2.20 Entropy May Change during Chemical Reactions. When TNT explodes, it produces carbon dioxide, water vapor, smoke, and other compounds that are much less ordered than the original system. The reaction results in an increase in entropy.

In general, physical and chemical processes proceed in the direction that results in lower potential energy and increased disorder (**Figure 2.21**). If potential energy drops, then ΔH is negative; if entropy increases, then ΔS is positive.

In the case of exploding TNT, the reaction is exothermic *and* results in higher entropy—less-ordered products. However, an increase in entropy does not always accompany a drop in potential energy. When methane burns, for example, ΔH is negative but ΔS is essentially 0.

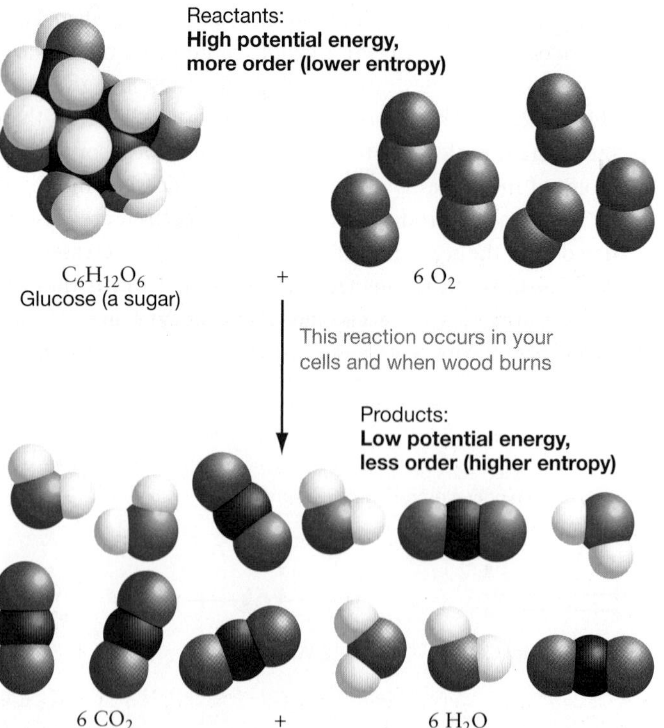

FIGURE 2.21 Spontaneous Processes Result in Lower Potential Energy, Increased Disorder, or Both.

To determine whether a chemical reaction is spontaneous, it's necessary to assess the combined contributions of changes in heat and disorder. Chemists do this with a quantity called the **Gibbs free-energy change**, symbolized by ΔG:

$$\Delta G = \Delta H - T\Delta S$$

Here, T stands for temperature measured on the Kelvin scale (see **BioSkills 1**, in Appendix A). Water freezes at 273.15 K and boils at 373.15 K.

In words, the free-energy change in a reaction is equal to the change in potential energy minus the change in entropy multiplied by the temperature. The $T\Delta S$ term simply means that entropy becomes more important in determining free-energy change as the temperature of the molecules increases. The faster molecules are moving, the more important entropy becomes in determining the overall free-energy change.

Chemical reactions are spontaneous when ΔG is less than zero. Such reactions are said to be **exergonic**. Reactions are nonspontaneous when ΔG is greater than zero. Such reactions are termed **endergonic**. When ΔG is zero, reactions are at equilibrium. ✔**If you understand these concepts, you should be able to explain (1) why the same reaction can be nonspontaneous at low temperature but spontaneous at high temperature, and (2) why some exothermic reactions are nonspontaneous.**

Free energy changes when the potential energy and/or entropy of substances changes. Spontaneous chemical reactions run in the direction that lowers the free energy of the system. Exergonic reactions are spontaneous and release energy; endergonic reactions are nonspontaneous and require an input of energy to proceed.

THE ROLES OF TEMPERATURE AND CONCENTRATION IN CHEMICAL REACTIONS Even if a chemical reaction occurs spontaneously, it may not happen quickly. The reactions that convert iron to rust or sugar molecules to carbon dioxide and water are spontaneous, but at room temperature they occur very slowly, if at all. For most reactions to proceed, one chemical bond has to break and another one has to form. For this to happen, the substances involved must collide in a specific orientation that brings the electrons involved near each other.

The number of collisions occurring between the substances in a mixture depends on the temperature and the concentrations of the reactants:

- When the concentration of reactants is high, more collisions should occur and reactions should proceed more quickly.
- When their temperature is high, reactants should move faster and collide more frequently.

Higher concentrations and higher temperatures should tend to speed up chemical reactions.

Figure 2.22 provides data from a set of experiments performed by students from Parkland College in Champaign, Illi-

EXPERIMENT

QUESTION: Do chemical reaction rates increase with increased temperature and concentration?

RATE INCREASE HYPOTHESIS: Chemical reaction rates increase with increased temperature. They also increase with increased concentration of reactants.

NULL HYPOTHESIS: Chemical reaction rates are not affected by increases in temperature or concentration of reactants.

EXPERIMENTAL SETUP:

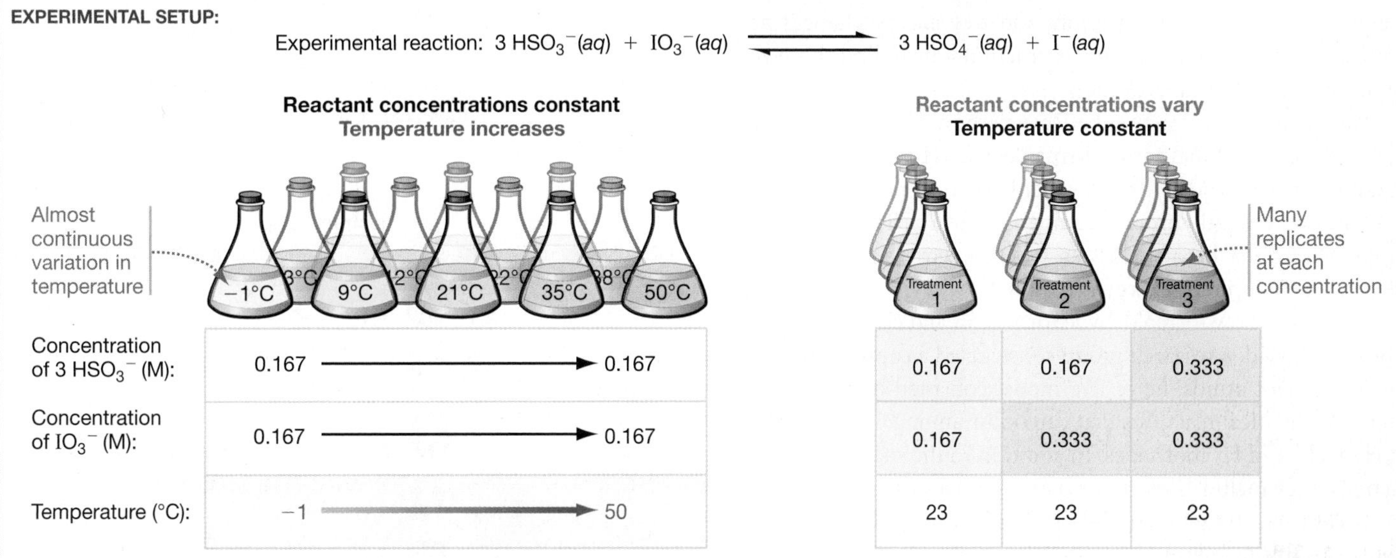

PREDICTION: Reaction rate, measured as 1/(time for reaction to go to completion), will increase with increased concentrations of reactants and increased temperature of reaction mix.

PREDICTION OF NULL HYPOTHESIS: No difference in reaction rates among treatments in each setup.

RESULTS:

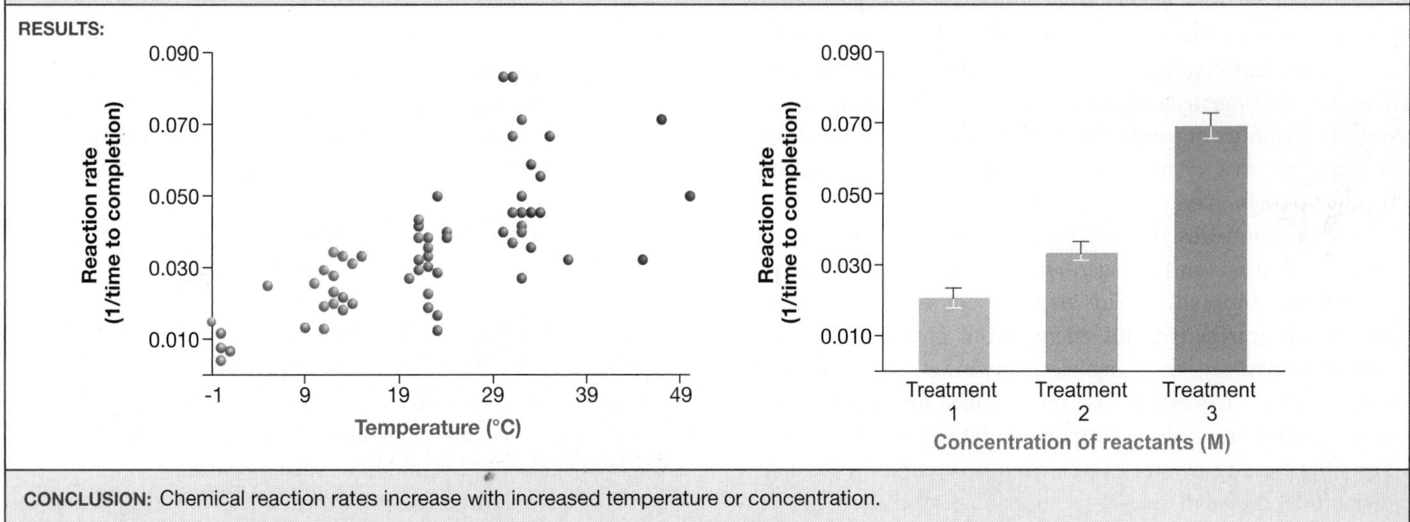

CONCLUSION: Chemical reaction rates increase with increased temperature or concentration.

FIGURE 2.22 Testing the Hypothesis That Reaction Rates Are Sensitive to Changes in Temperature and Concentration.

✔ **QUESTION** In the graph on the right side of the "Results" section, what do the heights of the orange bars

nois, to test these predictions. Pay special attention to the two graphs in the "Results" section:

- *Temperature versus reaction rate* The graph on the left is based on experiments where the concentration of the reactants was the same, but the temperature varied. Each data point represents one experiment. Note that the points rise from left to right—meaning, in this case, that the reaction rate speeded up when the temperature of the reaction mixture was higher.
- *Concentration versus reaction rate* The graph on the right is based on experiments where the temperature was constant, but the concentration of reactants varied. Each bar represents the average reaction rate over many replicates of each treatment, or

set of concentrations. The thin lines at the top of each bar indicate the standard error of the average—a measure of variability (see **BioSkills 2** in Appendix A). The take-home message of this graph is that reaction rates are higher when substrate concentrations are higher.

ENERGY INPUTS AND THE START OF CHEMICAL EVOLUTION Temperature, substrate concentration, and free energy changes are critical to analyzing any chemical reaction—including those responsible for chemical evolution.

For example, the reaction between carbon dioxide (CO_2) and hydrogen gas (H_2) that forms formaldehyde (H_2CO) and water is endergonic. Formaldehyde and water have more potential energy and are more highly ordered than CO_2 and H_2. The reaction is nonspontaneous because ΔG is positive. For the reaction to occur, a large input of energy is required.

To explore how formaldehyde formation and chemical evolution got started, a research group constructed a computer model of the ancient atmosphere. The model consisted of a list of all possible chemical reactions that can occur among CO_2, H_2O, N_2, NH_3, CH_4, and H_2 molecules. In addition to the spontaneous reactions, they included reactions that occur when these molecules are struck by sunlight. This was crucial because sunlight represents a source of energy.

The sunlight that strikes Earth is made up of packets of light energy called **photons**. The amount of light energy contained in a photon can vary widely. Today, most of the higher-energy photons in sunlight never reach Earth's lower atmosphere. Instead, they are absorbed by a molecule called ozone (O_3) in the upper atmosphere. But if Earth's early atmosphere was filled with volcanic gases released as the molten planet cooled, it is extremely unlikely that appreciable quantities of ozone existed. On the basis of this logic, researchers infer that when chemical evolution was occurring, large quantities of high-energy photons bombarded the planet.

To understand why this energy source was so important, recall that the atoms in hydrogen and carbon dioxide molecules have full outermost shells. This arrangement makes these molecules largely unreactive. But energy from photons can break molecules apart by knocking electrons away from the outer shells of atoms. The atoms that result, called **free radicals**, have unpaired electrons and are extremely reactive (**Figure 2.23**). To mimic the conditions on early Earth more accurately, the computer model included several reactions that produce highly reactive free radicals.

To understand which of the long list of possible reactions would actually occur, and to estimate how much formaldehyde could have been produced in the ancient atmosphere, the researchers needed to consider the effects of two additional factors: temperature and concentration.

THE FIRST REACTIONS IN CHEMICAL EVOLUTION To model the behavior of simple molecules in the ancient atmosphere, the researchers working on the formaldehyde synthesis reaction needed to specify temperature and the concentration of each

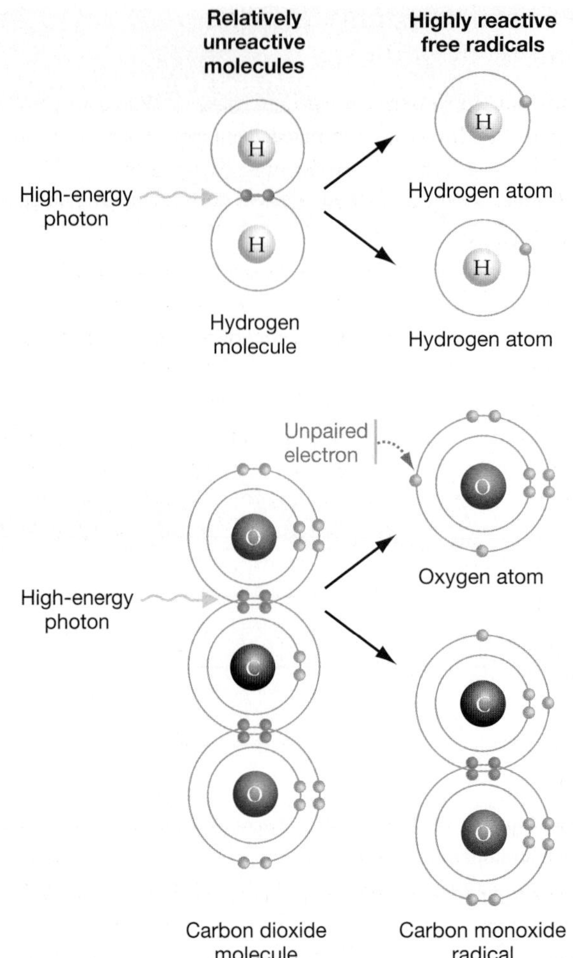

FIGURE 2.23 Free Radicals Are Extremely Reactive. When a high-energy photon strikes a hydrogen or carbon dioxide molecule, free radicals can be created. Formation of free radicals is thought to be responsible for some key reactions in chemical evolution.

molecule. In experiments like those in Figure 2.22, they measured the actual reaction rates observed at controlled temperatures and concentrations. They were then able to assign a rate to each of the reactions listed in their model.

Their result? They calculated that, under temperature and concentration conditions accepted as reasonable approximations of early Earth conditions by most scientists, appreciable quantities of formaldehyde would have been produced.

Using a similar model, other researchers have shown that significant amounts of hydrogen cyanide (HCN) could also have been produced in the ancient atmosphere. According to this research, large quantities of the critical intermediates in chemical evolution would have formed in the ancient atmosphere.

Once formaldehyde and hydrogen cyanide formed in the atmosphere, they would have rained out into the ocean. Similar reactions occur at deep-sea vents, where extremely hot water provides energy to drive endergonic reactions. As a result, organic compounds with relatively high potential energy were bubbling up from below. Once these types of compounds began to accumulate, the groundwork was in place for chemical evolution to take off.

How Did Chemical Energy Change during Chemical Evolution?

The initial products of chemical evolution are important, for a simple reason: They have more potential energy than the reactant molecules have. When formaldehyde is produced, an increase in potential energy occurs because the electrons that bond CO_2 and H_2 together are held more tightly than they are in H_2CO or H_2O. This form of potential energy—the potential energy stored in chemical bonds—is called **chemical energy**.

This observation gets right to the heart of chemical evolution: The energy in sunlight was converted to chemical energy—potential energy in chemical bonds. This energy transformation explains how chemical evolution was possible. When small, simple molecules absorb energy, chemical reactions can occur that transform the external energy into potential energy stored in chemical bonds. More specifically, the energy in sunlight was converted to chemical energy in the form of formaldehyde and hydrogen cyanide.

The complete reaction that results in the formation of formaldehyde is written as:

$$CO_2(g) + 2\,H_2(g) + \text{sunlight} \longrightarrow H_2CO(g) + H_2O(g)$$

Notice that the reaction is balanced in terms of the atoms *and* the energy involved. The sunlight on the left side of the expression balances the higher chemical energy in formaldehyde and water.

This result makes sense intuitively. Energy is the capacity to do work, and it seems logical that building larger, more complex molecules requires work to be done. Here's another way to say the same thing: The reactions involved in chemical evolution are endergonic, so inputs of energy were required. Now the question is, what happened to these first building blocks of chemical evolution?

CHECK YOUR UNDERSTANDING

If you understand that . . .

- Chemical reactions tend to be spontaneous if they lead to lower potential energy and higher entropy (more disorder).
- The combined effects of potential energy and entropy changes are summarized in the equation for the Gibbs free-energy change.

✓ You should be able to . . .

1. Write out the Gibbs equation and define each of the components.
2. When is the overall free energy change in a reaction most likely to be negative (meaning that the reaction is exergonic)?

Answers are available in Appendix B.

2.4 The Importance of Carbon

Life has been called a carbon-based phenomenon, and with good reason. Except for water, almost all of the molecules found in organisms contain this atom. Molecules that contain carbon are called **organic** molecules. (Other types of molecules are referred to as *inorganic* compounds.)

Carbon has great importance in biology because it is the most versatile atom on Earth. Because of its four valence electrons, it can form many covalent bonds. With different combinations of single and double bonds, an almost limitless array of molecular shapes are possible.

You have already examined the tetrahedral structure of methane and the linear shape of carbon dioxide. When molecules contain more than one carbon atom, these atoms can be bonded to one another in long chains, as in the component of the gasoline called octane (C_8H_{18}; **Figure 2.24a**), or in a ring, as in the sugar glucose ($C_6H_{12}O_6$; **Figure 2.24b**). Carbon atoms

(a) Carbons linked in a linear molecule

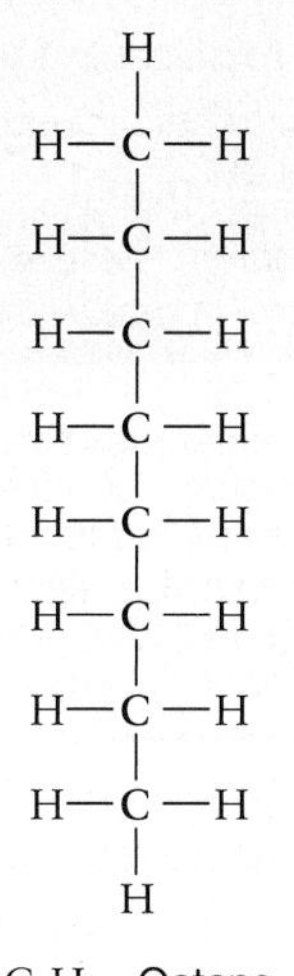

C_8H_{18} Octane

(b) Carbons linked in a ring

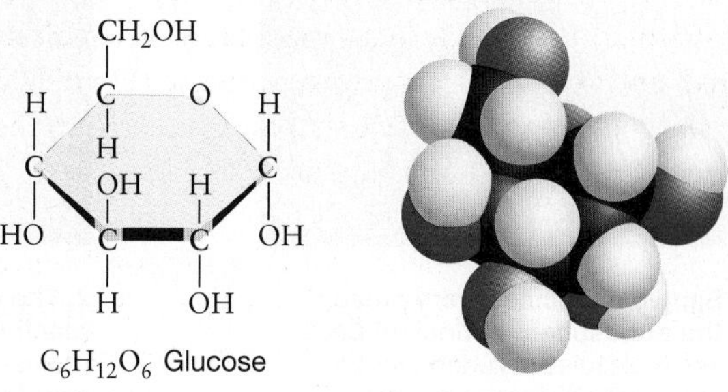

FIGURE 2.24 The Shapes of Carbon-Containing Molecules. **(a)** Octane is a linear molecule, and one of the primary ingredients in gasoline. **(b)** Glucose is a sugar that can form the ring-like structure.

provide the structural framework for virtually all the important compounds associated with life.

Linking Carbon Atoms Together

The formation of carbon-carbon bonds was an important event in chemical evolution: It represented a crucial step toward the production of the types of molecules found in living organisms.

Once organic compounds such as formaldehyde and hydrogen cyanide had formed, continued chemical evolution could occur by the addition of heat alone. For example, when molecules of formaldehyde are heated, they react with one another to form a molecule called acetaldehyde. Acetaldehyde contains a carbon-carbon bond. With continued heating, reactions between formaldehyde and acetaldehyde molecules can produce the larger organic compounds called sugars.

Figure 2.25 summarizes these steps: the formation of building block compounds like formaldehyde and HCN, followed by reactions that resulted in small molecules with carbon-carbon bonds—including molecules that are found in organisms living today.

Functional Groups

In general, the carbon atoms in an organic molecule furnish a skeleton that gives the molecule its overall shape. But the chemical behavior of the compound—meaning the types of reactions that it participates in—is dictated by groups of H, N, or O atoms that are bonded to one of the carbon atoms in a specific way.

The critically important H-, N-, and O-containing groups found in organic compounds are called **functional groups**. The composition and properties of six prominent functional groups recognized by organic chemists are summarized in **Table 2.3**. To understand the role that organic compounds play in organisms, it is important to analyze how these functional groups behave.

- In solution, the *amino and carboxyl functional groups* tend to attract or drop a proton, respectively. Amino groups function as bases, while carboxyl groups act as acids. During chemical evolution and in organisms today, the most important types of amino- and carboxyl-containing molecules are the amino acids, which Chapter 3 analyzes in detail. Amino acids contain both an amino group and a carboxyl group. (It's common for organic compounds to contain more than one functional group.) Amino acids can be linked together by covalent bonds that form between amino- and carboxyl-containing groups. In addition, both of these functional groups participate in hydrogen-bonding.
- The *carbonyl group* is found on aldehyde and ketone molecules such as formaldehyde, acetaldehyde, and acetone. This functional group is the site of reactions that link these molecules into larger, more complex compounds, like the sugar ribose illustrated in Figure 2.25.

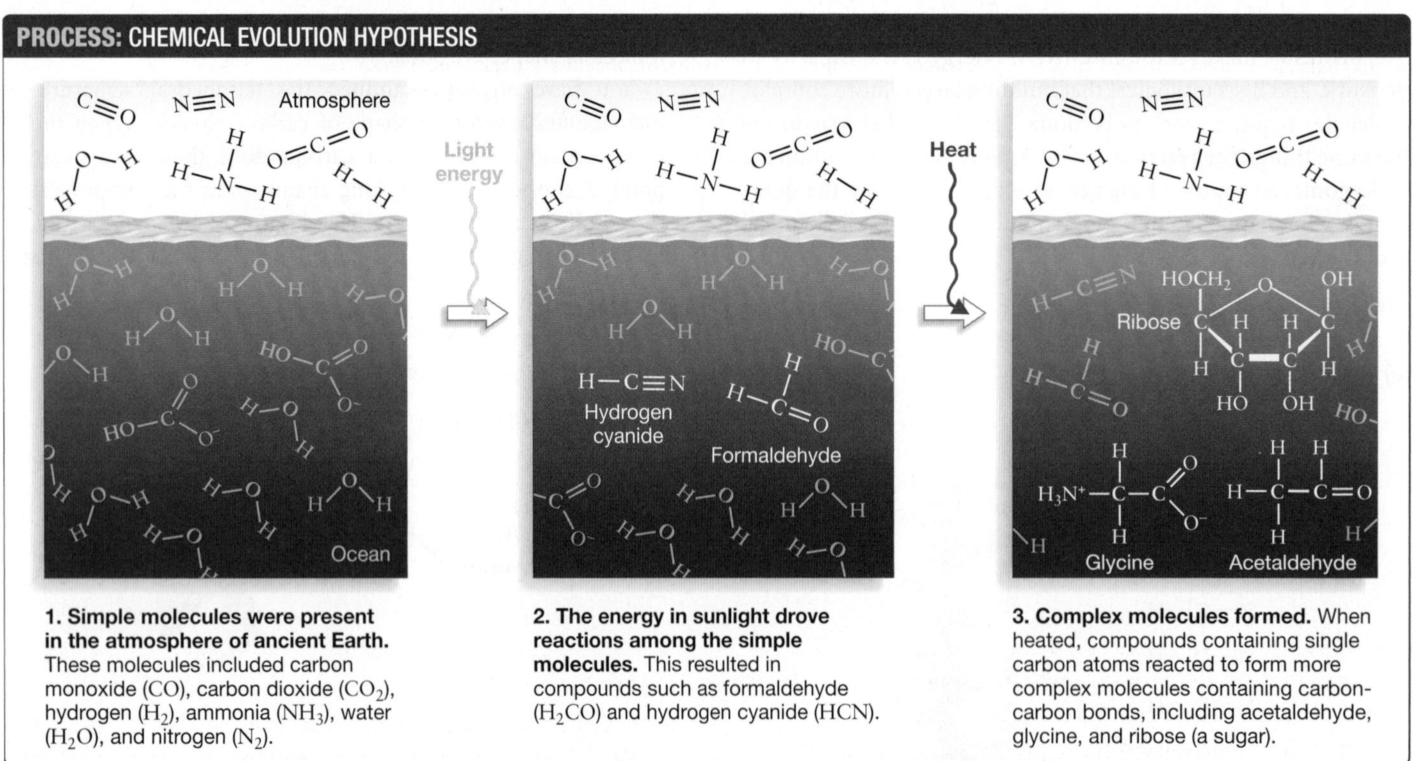

PROCESS: CHEMICAL EVOLUTION HYPOTHESIS
Atmosphere
Ocean
Light energy
Heat
H—C≡N
Hydrogen cyanide
Formaldehyde
Ribose
Glycine
Acetaldehyde
1. Simple molecules were present in the atmosphere of ancient Earth. These molecules included carbon monoxide (CO), carbon dioxide (CO_2), hydrogen (H_2), ammonia (NH_3), water (H_2O), and nitrogen (N_2).
2. The energy in sunlight drove reactions among the simple molecules. This resulted in compounds such as formaldehyde (H_2CO) and hydrogen cyanide (HCN).
3. Complex molecules formed. When heated, compounds containing single carbon atoms reacted to form more complex molecules containing carbon-carbon bonds, including acetaldehyde, glycine, and ribose (a sugar).

FIGURE 2.25 The Start of Chemical Evolution—An Overview. During chemical evolution, simple molecules containing C, H, O, and N reacted to form organic compounds with higher potential energy in the form of carbon-carbon bonds. The process was triggered by an energy source such as sunlight, or by the heat released from a volcanic eruption.

SUMMARY TABLE 2.3 **Six Functional Groups Commonly Attached to Carbon Atoms**

Functional Group	*Formula	Family of Molecules	Properties of Functional Group	Example
Amino	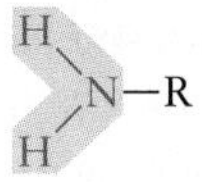 $H_2N{-}R$	Amines	Acts as a base—tends to attract a proton to form: 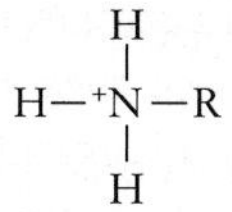 $H{-}^{+}NH_2{-}R$	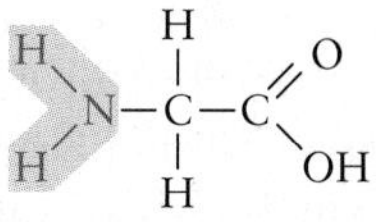 $H_2N{-}CH_2{-}COOH$ Glycine (an amino acid)
Carbonyl	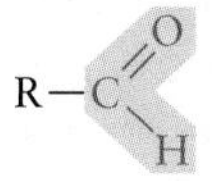 $R{-}C(=O){-}H$	Aldehydes	Aldehydes, especially, react with certain compounds to produce larger molecules with form: 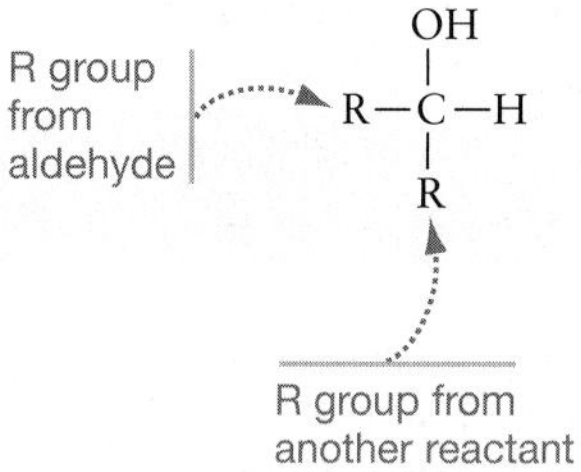 $R{-}C(OH)(H){-}R$ (R group from aldehyde; R group from another reactant)	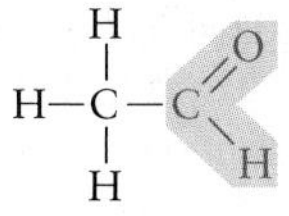 $H_3C{-}C(=O){-}H$ Acetaldehyde
	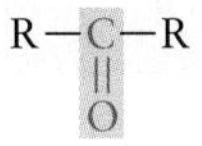 $R{-}C(=O){-}R$	Ketones		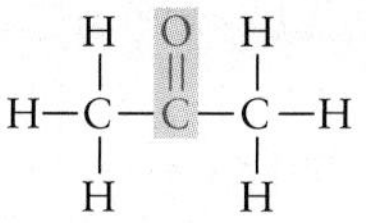 $H_3C{-}C(=O){-}CH_3$ Acetone
Carboxyl	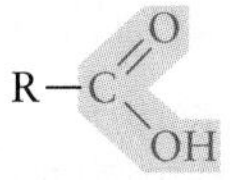 $R{-}C(=O){-}OH$	Carboxylic acids	Acts as an acid—tends to lose a proton in solution to form: 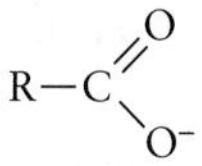 $R{-}C(=O){-}O^-$	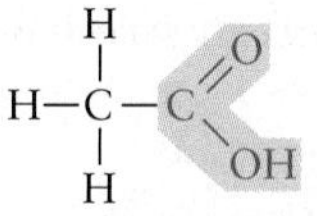 $H_3C{-}C(=O){-}OH$ Acetic acid
Hydroxyl	$R{-}OH$	Alcohols	Highly polar, so makes compounds more soluble through hydrogen bonding with water; may also act as a weak acid and drop a proton	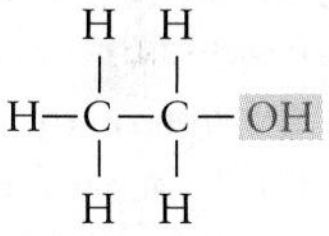 $H_3C{-}CH_2{-}OH$ Ethanol
Phosphate	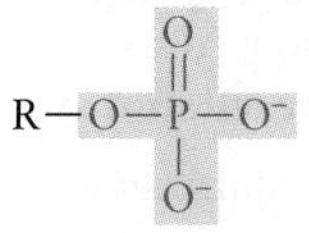 $R{-}O{-}P(=O)(O^-){-}O^-$	Organic phosphates	When several phosphate groups are linked together, breaking O–P bonds between them releases large amounts of energy	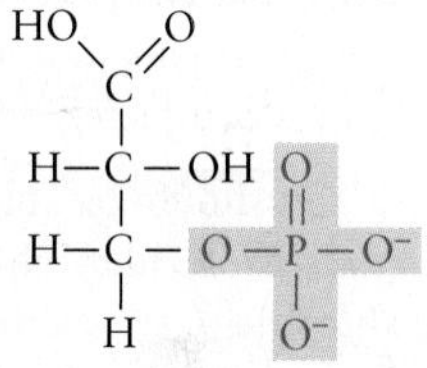 $HO{-}C(=O){-}CH(OH){-}CH_2{-}O{-}P(=O)(O^-){-}O^-$ 3–Phosphoglyceric acid
Sulfhydryl	$R{-}SH$	Thiols	When present in proteins, can form disulfide (S–S) bonds that contribute to protein structure	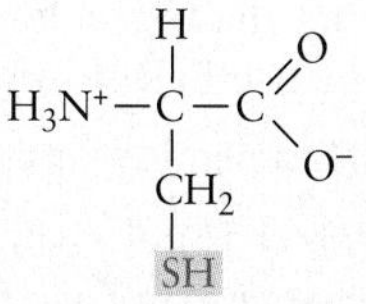 $H_3N^+{-}CH(CH_2{-}SH){-}C(=O){-}O^-$ Cysteine

*In these structural formulas, "R" stands for the rest of the molecule.

✔ **EXERCISE** On the basis of the electron negativities of the atoms involved, predict whether each functional group is polar or nonpolar.

- *Hydroxyl groups* are important because they act as weak acids. In many cases, the protons involved in acid–base reactions that occur in cells come from hydroxyl groups on organic compounds. Because hydroxyl groups are polar, molecules containing a number of hydroxyl groups will form hydrogen bonds and be highly soluble in water.
- *Phosphate groups* carry two negative charges. When phosphate groups are transferred from one organic compound to another, the change in charge often dramatically affects the recipient molecule.
- *Sulfhydryl groups* consist of a sulfur atom bonded to a hydrogen atom. They are important because sulfhydryl groups on different molecules can link together via disulfide (S–S) bonds.

To summarize, functional groups make things happen. The number and types of functional groups attached to a framework of carbon atoms imply a great deal about how that molecule is going to behave.

When you encounter an organic compound that is new to you, it's important to do two things: Examine its overall size and shape, and locate any functional groups. Understanding these two features will help you understand the molecule's role in chemical evolution and today's cells.

Once carbon-containing molecules with functional groups had appeared early in Earth's history, what happened next? For chemical evolution to continue, two things had to happen. First, reactions between relatively small and simple organic compounds had to produce the building blocks of the large molecules found in living cells. Second, these building blocks had to link together to form the large, complex compounds found in organisms—proteins, nucleic acids, and carbohydrates.

The next three chapters focus on how these events occurred and how proteins, nucleic acids, and carbohydrates function in organisms today. The atoms that make up your body were present in the ocean before life began, some 3.5 billion years ago.

CHAPTER 2 REVIEW

For media, go to the study area at www.masteringbiology.com 

Summary of Key Concepts

Molecules form when atoms bond to each other. Chemical bonds are based on electron sharing. The degree of electron sharing varies from nonpolar covalent bonds, to polar covalent bonds, to ionic bonds.

- When atoms participate in chemical bonds to form molecules, the shared electrons give the atoms full valence shells and thus contribute to the atoms' stability.
- The electrons in chemical bonds may be shared equally or unequally, depending on the relative electronegativities of the two atoms involved.
- Nonpolar covalent bonds result from equal sharing; polar covalent bonds are due to unequal sharing. Ionic bonds form when an electron is completely transferred from one atom to another.

✔You should be able to draw the electron-sharing continuum and place molecular oxygen (O_2), carbon dioxide (CO_2), and calcium chloride ($CaCl_2$) on it.

Water is essential for life. Water is highly polar and readily forms hydrogen bonds. Hydrogen bonding makes water an extremely efficient solvent.

- The chemical reactions required for life take place in water.
- Water is polar—meaning that it has partial positive and negative charges—because it is bent and has two polar covalent bonds.
- Polar molecules and charged substances, including ions, interact with water and stay in solution via hydrogen bonding and electrostatic attraction.
- Water's ability to participate in hydrogen bonding also gives it an extraordinarily high capacity to absorb heat and cohere to other water molecules.

✔You should be able to explain how water interacts with amino, carboxyl, and hydroxyl functional groups in solution.

MB **Web Activity** The Properties of Water

Energy is the capacity to do work or supply heat, and can be (1) a stored potential or (2) an active motion. Chemical energy is a form of potential energy, stored in chemical bonds.

- Energy comes in different forms. Although energy cannot be created or destroyed, one form of energy can be transformed into another.
- Experiments suggest that early in Earth's history, the energy in sunlight and hot water drove chemical reactions between simple molecules, resulting in the formation of more complex compounds with higher potential energy. In this way, energy in the form of sunlight or heat was transformed into chemical energy.

✔You should be able to explain how the heat released when methane burns represents an energy transformation.

Chemical reactions tend to be spontaneous if they lower potential energy and increase entropy (disorder). An input of energy is required for nonspontaneous reactions to occur.

- Spontaneous reactions do not require an input of energy to occur.
- The Gibbs free energy change, ΔG, summarizes the combined effects of changes in energy and entropy during a chemical reaction.
- Spontaneous reactions have a negative ΔG and are said to be exergonic; nonspontaneous reactions have a positive ΔG and are said to be endergonic.

✔You should be able to explain why cells cannot stay alive without inputs of energy.

Most of the important compounds in organisms contain carbon. Key carbon-containing molecules formed early in Earth's history.

- Organic molecules are critical to life because they have complex shapes provided by a framework of carbon atoms, along with complex chemical behavior due to the presence of functional groups.
- The first step in chemical evolution was the formation of the small organic compounds called formaldehyde and hydrogen cyanide from molecules such as ammonia (NH_3), methane (CH_4), molecular hydrogen (H_2), and carbon dioxide (CO_2). These reactions occur readily when a source of intense energy, such as the radiation in sunlight, is present.

✔You should be able to explain why molecules with carbon-carbon bonds have more potential energy and lower entropy than carbon dioxide.

Questions

✓TEST YOUR KNOWLEDGE

Answers are available in Appendix B

1. Which of the following occurs when a covalent bond forms?
   a. The potential energy of electrons drops.
   b. Electrons in valence shells are shared between nuclei.
   c. Ions of opposite charge interact.
   d. Polar molecules interact.

2. If a reaction is exothermic, then which of the following statements is true?
   a. The products have lower potential energy than the reactants.
   b. Energy must be added for the reaction to proceed.
   c. The products have lower entropy (are more ordered) than the reactants.
   d. It occurs extremely quickly.

3. If a reaction is exergonic, then which of the following statements is true?
   a. The products have lower free energy than the reactants.
   b. Energy must be added for the reaction to proceed.
   c. The products have lower entropy (are more ordered) than the reactants.
   d. It occurs extremely quickly.

4. What is thermal energy?
   a. a form of potential energy
   b. the temperature increase that occurs when any form of energy is added to a system
   c. mechanical energy
   d. the kinetic energy of molecular motion, measured as heat

5. What determines whether a chemical reaction is spontaneous?
   a. if it increases the disorder, or entropy, of the substances involved
   b. if it decreases the potential energy of the substances involved
   c. the temperature only—reactions are spontaneous at high temperatures and nonspontaneous at low temperatures
   d. the combined effect of changes in potential energy and entropy

6. Which of the following is *not* an example of an energy transformation?
   a. A shoe drops, converting potential energy to kinetic energy.
   b. A chemical reaction converts the energy in sunlight into the chemical energy in formaldehyde.
   c. The electrical energy flowing through a lightbulb's filament is converted into light and heat.
   d. Sunlight strikes a prism and separates into distinct wavelengths.

✓TEST YOUR UNDERSTANDING

Answers are available in Appendix B

1. Consider the reaction between carbon dioxide and water, which forms carbonic acid:

   $$CO_2(g) + H_2O(l) \rightleftharpoons H_2CO_3(aq)$$

   In aqueous solution, carbonic acid immediately dissociates to form a proton and the bicarbonate ion, as follows:

   $$H_2CO_3(aq) \rightleftharpoons H^+(aq) + HCO_3^-(aq)$$

   Does this second reaction raise or lower the pH of the solution? If an underwater volcano bubbled additional CO_2 into the ocean, would this sequence of reactions be driven to the left or the right? How would this affect the pH of the ocean?

2. When chemistry texts introduce the concept of electron shells, they emphasize that shells represent distinct potential energy levels. In introducing electron shells, this chapter also emphasized that they represent distinct distances from the positive charges in the nucleus. Are these two points of view in conflict? Why or why not?

3. Draw a ball-and-stick model of the water molecule, and explain why this molecule is bent. Indicate the location of the partial electric charges on it. Why do these partial charges exist?

4. Hydrogen bonds form because the opposite, partial electric charges on polar molecules attract. Covalent bonds form as a result of the electrical attraction between electrons and protons. Covalent bonds are much stronger than hydrogen bonds. Explain why, in terms of the electrical attractions involved.

5. Explain why extensive hydrogen bonding gives water an extraordinarily high specific heat.

6. Explain the relationship between the carbon framework in an organic molecule (the "R" in Table 2.3) and the functional groups on the same molecule.

✓APPLYING CONCEPTS TO NEW SITUATIONS

Answers are available in Appendix B

1. Why isn't CO_2 bent and polar, like H_2O? Why is H_2O much more likely to participate in chemical reactions than CO_2?

2. Oxygen is extremely electronegative, meaning that its nucleus pulls in electrons shared in covalent bonds. Explain the changes in electron position that are illustrated in Figure 2.19 based on oxygen's electronegativity.

3. When nuclear fusion reactions take place, some of the mass in the atoms involved is converted to energy. The energy in sunlight is created during nuclear fusion reactions on the Sun. Explain what astronomers mean when they say that the Sun is burning down and that it will eventually burn out.

4. Why do coastal regions tend to have climates with moderate temperatures and lower annual variation in temperature than do inland areas at the same latitude?

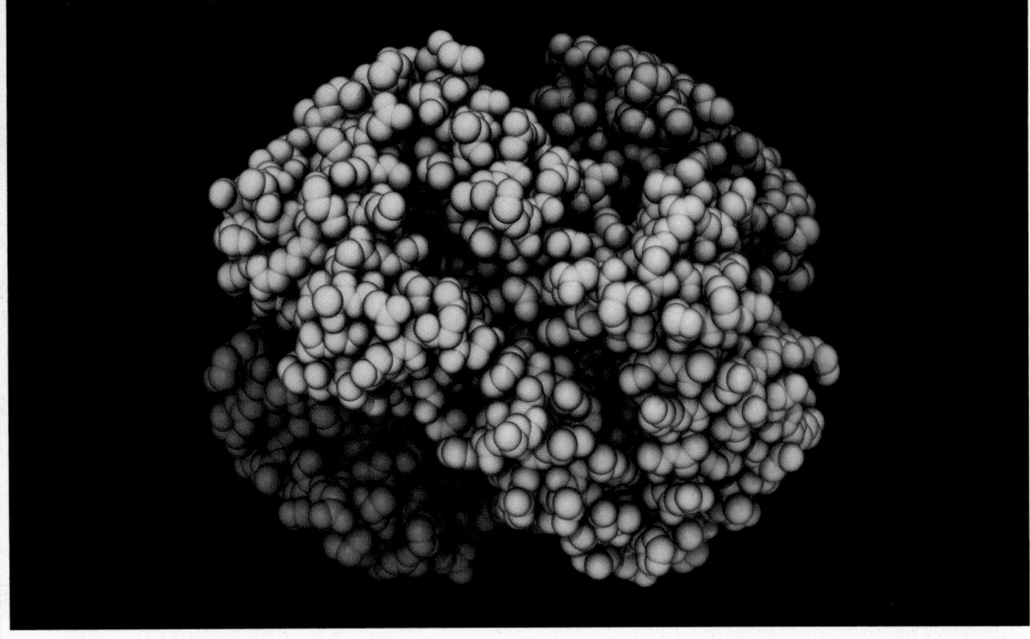

A space-filling model of hemoglobin—a protein that is carrying oxygen in your blood right now.

3 Protein Structure and Function

KEY CONCEPTS

- Most cell functions depend on proteins.
- Amino acids are the building blocks of proteins. Amino acids vary in structure and function because their side chains vary in composition.
- Proteins vary widely in structure. The structure of a protein can be analyzed at four levels that form a hierarchy—the amino acid sequence, substructures called α-helices and β-pleated sheets, interactions between amino acids that dictate a protein's overall shape, and combinations of individual proteins that make up larger, multiunit molecules.
- In cells, most proteins are enzymes that function as catalysts. Chemical reactions occur much faster when they are catalyzed by enzymes. During enzyme catalysis, the reactants bind to an enzyme's active site in a way that allows the reaction to proceed efficiently.

Chapter 2 introduced the hypothesis that chemical reactions in the atmosphere and ocean of ancient Earth led to the formation of the first complex carbon-containing compounds. This idea, called chemical evolution, was first proposed in 1923 by Alexander I. Oparin. The hypothesis was published again—independently and six years later—by J. B. S. Haldane.

Today, the Oparin-Haldane proposal can best be understood as a formal scientific theory. As Chapter 1 noted, scientific theories typically have two components: a statement about a pattern that exists in the natural world and a proposed mechanism or process that explains the pattern.

In the case of chemical evolution, the pattern is that increasingly complex carbon-containing molecules formed in the atmosphere and ocean of ancient Earth. The process responsible for this pattern was the conversion of energy, from sunlight and other sources, into chemical energy in the bonds of large, complex molecules.

Scientific theories are continuously refined as new information comes to light, and many of Oparin and Haldane's original ideas have been revised. In its current form, the theory can be broken into four steps, each requiring an input of energy:

1. Chemical evolution began with the production of small organic compounds such as formaldehyde (H_2CO) and hydrogen cyanide (HCN), from reactants such as H_2, CO_2, CH_4, and NH_3. Chapter 2 focused on this step.
2. Formaldehyde, hydrogen cyanide, and other simple organic compounds reacted to form the mid-sized molecules called amino acids, nitrogenous bases, and sugars. Amino acids are introduced in this chapter, nitrogenous bases are analyzed in Chapter 4, and sugars are discussed in Chapter 5. According to the Oparin-Haldane theory these molecules accumulated in the shallow waters of the ancient ocean, forming a complex solution called the **prebiotic soup.**

✔ When you see this checkmark, stop and test yourself. Answers are available in Appendix B.

3. Mid-sized, building-block molecules linked to form the types of large molecules found in cells today, including proteins, nucleic acids, and complex carbohydrates. Each of these large molecules is made up of distinctive chemical subunits that join together: Proteins are composed of amino acids, nucleic acids are composed of nucleotides, and complex carbohydrates are composed of sugars. Proteins, nucleic acids, and carbohydrates are featured in Chapters 3–5, respectively.
4. Life became possible when one of these large, complex molecules made a copy of itself. This self-replicating molecule began to multiply by means of chemical reactions that it controlled. At that point, life had begun. Chemical evolution gave way to biological evolution.

Which type of molecule was responsible for the origin of life? Answering this question is a recurring theme in this and the subsequent three chapters.

Researchers in several labs are pushing proteins as the candidate for the molecule that was the initial spark of life. Why?

3.1 Early Origin-of-Life Experiments

In 1953 a graduate student named Stanley Miller performed a breakthrough experiment in the study of chemical evolution. Miller wanted to answer a simple question: Can complex organic compounds be synthesized from the simple molecules present in Earth's early atmosphere and ocean? In other words, is it possible to re-create the first steps in chemical evolution by simulating ancient Earth conditions in the laboratory?

Miller's experimental setup (**Figure 3.1**) was designed to produce a microcosm of ancient Earth. The large glass flask represented the atmosphere and contained the gases methane (CH_4), ammonia (NH_3), and hydrogen (H_2), all of which have high free energy. This large flask was connected to a smaller flask by glass tubing. The small flask held a tiny ocean—200 milliliters (mL) of liquid water.

To connect the mini-atmosphere with the mini-ocean, Miller boiled the water constantly. This added water vapor to the mix of gases in the large flask. As the vapor cooled and condensed, it flowed back into the smaller flask, where it boiled again. In this way, water vapor circulated continuously through the system. This was important: If the molecules in the simulated atmosphere reacted with one another, the "rain" would carry them into the simulated ocean, forming a simulated version of the prebiotic soup.

Had Miller stopped at merely boiling the molecules, little or nothing would have happened. Even at the boiling point of water (100°C), the starting molecules used in the experiment are stable. They do not undergo spontaneous chemical reactions, even at high temperatures.

Something did start to happen in the apparatus, however, when Miller sent electrical discharges across the electrodes he'd inserted into the atmosphere. These miniature lightning bolts added a crucial element to the reaction mix—pulses of intense electrical energy. After a day of continuous boiling and sparking,

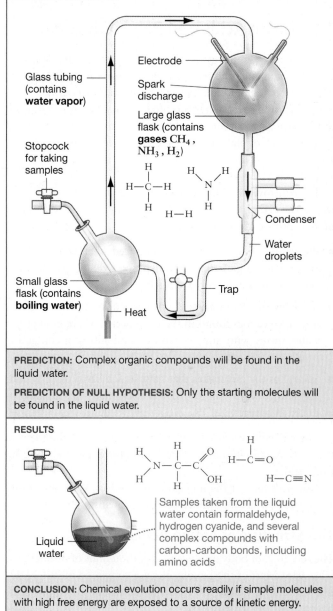

EXPERIMENT

QUESTION: Can simple molecules and kinetic energy lead to chemical evolution?

HYPOTHESIS: If kinetic energy is added to a mix of simple molecules with high free energy, reactions will occur that produce more complex molecules, perhaps including some with C-C bonds.

NULL HYPOTHESIS: Chemical evolution will not occur, even with an input of energy.

EXPERIMENTAL SETUP:

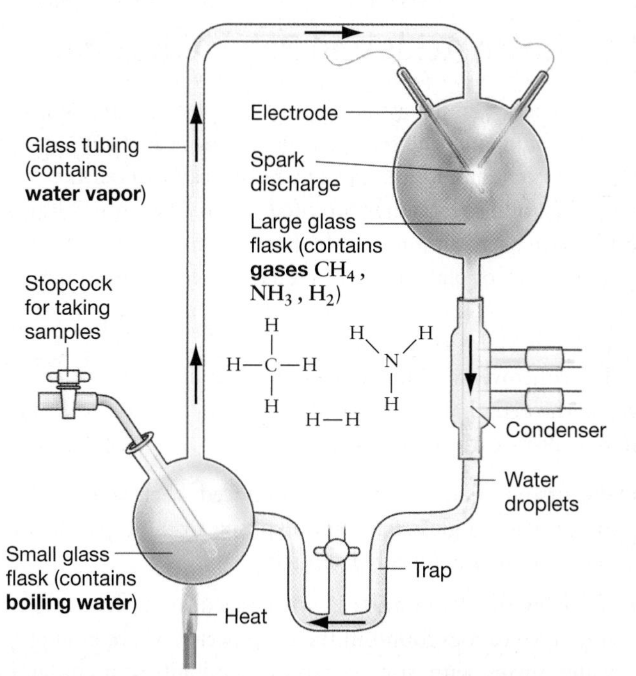

PREDICTION: Complex organic compounds will be found in the liquid water.

PREDICTION OF NULL HYPOTHESIS: Only the starting molecules will be found in the liquid water.

RESULTS

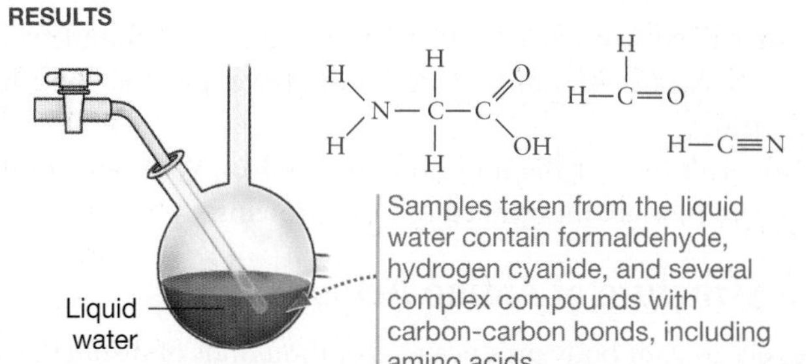

CONCLUSION: Chemical evolution occurs readily if simple molecules with high free energy are exposed to a source of kinetic energy.

FIGURE 3.1 Miller's Spark-Discharge Experiment. The arrows in the "Experimental setup" diagram indicate the flow of water vapor or liquid. The condenser is a jacket with cold water flowing through it.

SOURCE: Miller, S. L. 1953. A production of amino acids under possible primitive Earth conditions. *Science* 117: 528–529.

✔**QUESTION** Which parts of the apparatus mimic the ocean, atmosphere, rain, and lightning?

the solution in the boiling flask began to turn pink. After a week, it was deep red and cloudy.

When Miller analyzed samples from the mini-ocean, he found large quantities of hydrogen cyanide (HCN) and formaldehyde (H_2CO). Chapter 2 claimed that these were key compounds in chemical evolution, because they are required for reactions that lead to the synthesis of more complex organic molecules. Indeed, some of these more complex compounds were actually present in the miniature ocean. The sparks and heating had led to the synthesis of compounds that are the building blocks of proteins: amino acids.

3.2 Amino Acids and Polymerization

The presence of amino acids supported Miller's claim that his experiment simulated the second stage in chemical evolution—the formation of a prebiotic soup. The results came under fire, however, when other researchers pointed out that because volcanism would have been common as the Earth's crust began to cool and form the crust of plates observed today, the early atmosphere was dominated by volcanic gases like CO, CO_2, and H_2, not the CH_4 and NH_3 in Miller's experiment.

The controversy stimulated a series of follow-up experiments, which showed that amino acids can also be produced under more realistic early-Earth conditions:

- Precursors to amino acids are produced when volcanic gases are put into a glass flask and exposed to the types of high-energy radiation found in sunlight.
- An array of amino acids is synthesized in experiments that mimic volcanoes found in the deep ocean, where boiling-hot water mixes with simple carbon- and nitrogen-containing molecules and metal ions.
- Several of the meteorites that have struck Earth recently contain amino acids—indicating that these molecules can be produced in outer space.

The current consensus is that amino acids were abundant in the early oceans—raining down from above and bubbling up from below.

Now let's look at the molecules themselves. What are amino acids, and how are they linked to form proteins?

The Structure of Amino Acids

The cells in your body produce tens of thousands of distinct proteins. Most of these molecules are composed of just 20 different building blocks, called **amino acids.** All 20 amino acids have a common structure.

To understand how amino acids are put together, recall that carbon atoms have a valence of four—they form four covalent bonds. Amino acids have a carbon atom that forms bonds to the four groups or atoms diagrammed in **Figure 3.2a**:

1. NH_2—the amino functional group
2. COOH—the carboxyl functional group

(a) Non-ionized form of amino acid

$$\mathrm{H_2N{-}C(H)(R){-}C({=}O)OH}$$

(b) Ionized form of amino acid

$$\mathrm{H_3N^{+}{-}C(H)(R){-}C({=}O)O^{-}}$$

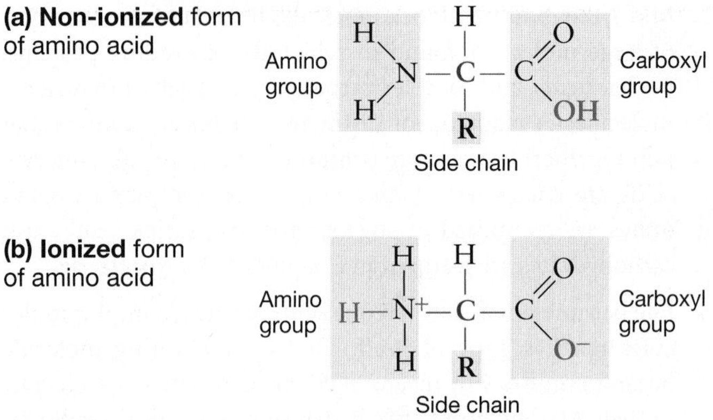

FIGURE 3.2 All Amino Acids Have the Same General Structure. The central carbon is shown in red.

3. H—a hydrogen atom
4. an "R-group"—an atom or group of atoms called a side chain

The combination of amino and carboxyl groups not only inspired the name amino acid, but is key to how these molecules behave. In water at pH 7, the concentration of protons causes the amino group to act as a base. It attracts a proton to form NH_3^+ (**Figure 3.2b**). The carboxyl group, in contrast, is acidic because its two oxygen atoms are highly electronegative. They pull electrons away from the hydrogen atom, which means that it is relatively easy for this group to lose a proton to form COO^-.

The charges on these functional groups are important for two reasons: (1) They help amino acids stay in solution, where they can interact with one another and with other solutes, and (2) they alter amino acid chemical reactivity.

The Nature of Side Chains

What about the R-group? The R-groups on amino acids vary from a single hydrogen atom to large structures containing carbon atoms linked into rings. All amino acids have a carbon atom bonded to an amino group, a hydrogen atom, and a carboxyl group, but each R-group is unique. The properties of amino acids vary because their R-groups vary.

Figure 3.3 highlights the R-groups on the 20 most common amino acids found in cells.[1] As you examine these side chains, ask yourself two questions: Is this R-group likely to participate in chemical reactions? Will it help this amino acid stay in solution?

FUNCTIONAL GROUPS AFFECT REACTIVITY Several of the side chains found in amino acids contain the carboxyl, sulfhydryl, hydroxyl, or amino functional groups introduced in Chapter 2, Table 2.3. Under the right conditions, these functional groups can participate in chemical reactions. For example, amino acids

[1]There are actually 22 amino acids found in proteins that occur in organisms, but two are rare.

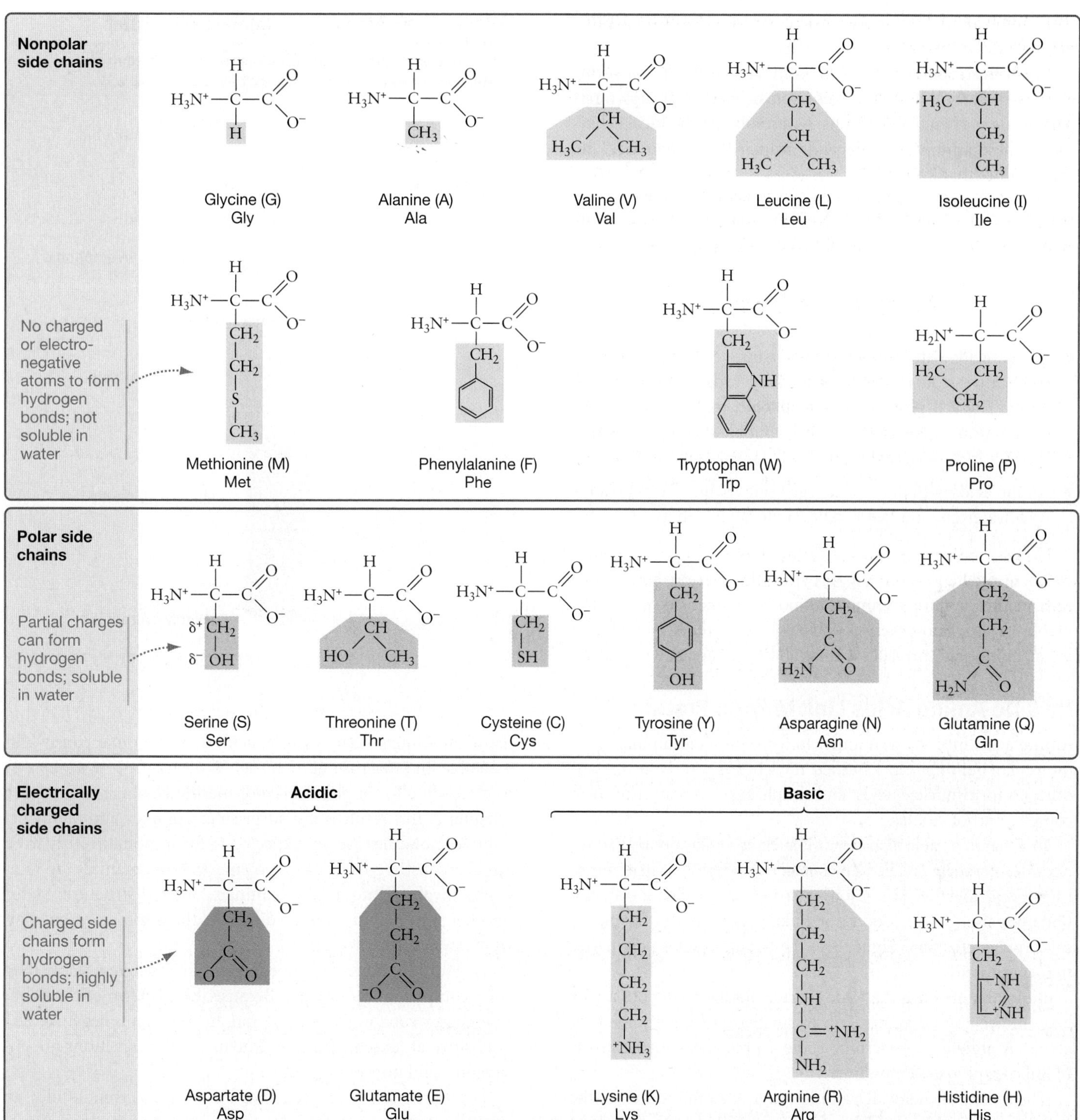

FIGURE 3.3 The 20 Major Amino Acids Found in Organisms. At the pH (about 7.0) found in cells, the 20 major amino acids found in organisms have the structural formulas shown here. The side chains are highlighted, and standard single-letter and three-letter abbreviations for each amino acid are given. For clarity, the carbon atoms in the ring structures of phenylalanine, tyrosine, tryptophan, and histidine are not shown; each bend in a ring is the site of a carbon atom. The hydrogen atoms in these structures are also not shown. A double line inside a ring indicates a double bond.

✔**EXERCISE** Explain why the green R-groups are nonpolar and why the lavender R-groups are polar, based on the relative electronegativities of O, N, C, and H. Note that sulfur (S) has an electronegativity almost equal that of carbon, and slightly higher than that of hydrogen.

with sulfur atoms (S) in their side chains can form bonds that help link different parts of large proteins.

In contrast, some amino acids contain side chains consisting entirely of carbon and hydrogen atoms. These R-groups rarely participate in chemical reactions. As a result, the chemical behavior of these amino acids depends primarily on their size and shape rather than reactivity.

THE POLARITY OF SIDE CHAINS AFFECTS SOLUBILITY The nature of the R-group affects how soluble the amino acid is in water.

Figure 3.3 emphasizes this point by sorting the amino acids according to whether their side chain is nonpolar, polar, or electrically charged.

- Amino acids with nonpolar side chains lack charged or electronegative atoms capable of forming hydrogen bonds with water. These R-groups are **hydrophobic**, meaning that water does not interact with them. Hydrophobic side chains tend to coalesce in aqueous solution.
- Amino acids with polar or charged side chains interact readily with water and are **hydrophilic**. Hydrophilic amino acids dissolve in water easily.

These predictions are supported by the data on how readily each of the 20 most common amino acids interacts with water (**Table 3.1**). In almost every case, the polarity of the R-group found on an amino acid correlates with its ability to interact with water.

TABLE 3.1 **How Amino Acids Interact with Water**

20 amino acids are ranked according to how likely they are to interact with water. Color codes are based on Figure 3.3.

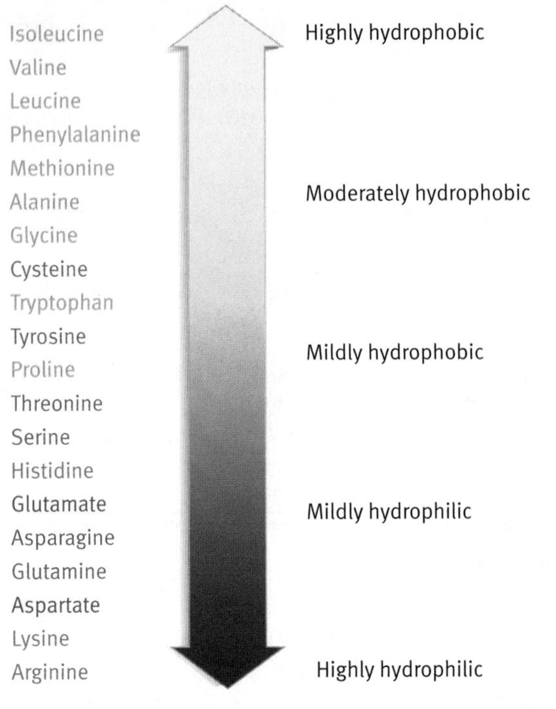

Amino acid	Interaction with water
Isoleucine	Highly hydrophobic
Valine	
Leucine	
Phenylalanine	
Methionine	
Alanine	Moderately hydrophobic
Glycine	
Cysteine	
Tryptophan	
Tyrosine	Mildly hydrophobic
Proline	
Threonine	
Serine	
Histidine	
Glutamate	Mildly hydrophilic
Asparagine	
Glutamine	
Aspartate	
Lysine	
Arginine	Highly hydrophilic

How Do Amino Acids Link to Form Proteins?

Amino acids link to one another to form proteins. Similarly, the molecular building blocks called nucleotides attach to one another to form nucleic acids, and simple sugars connect to form complex carbohydrates.

In general, a molecular subunit such as an amino acid, a nucleotide, or a sugar is called a **monomer** ("one-part"). When monomers bond together, the resulting structure is called a polymer ("many-parts"). The process of linking monomers together is called **polymerization** (**Figure 3.4**). Thus, amino acids polymerize to form proteins.

Biologists also use the word **macromolecule** to denote a very large molecule that is made up of smaller molecules joined together. A **protein** is a macromolecule—a polymer—that consists of linked amino acid monomers.

The theory of chemical evolution states that monomers in the prebiotic soup polymerized to form proteins and other types of macromolecules found in organisms. This is a difficult step, because monomers such as amino acids do not spontaneously self-assemble into macromolecules such as proteins.

According to the second law of thermodynamics reviewed in Chapter 2, this result is not surprising. Complex and highly organized molecules are not expected to form spontaneously from simpler constituents, because polymerization organizes the molecules involved into a more complex, ordered structure. Stated another way, polymerization decreases the disorder, or entropy, of the molecules involved.

In addition, polymers are energetically much less stable than their component monomers. Because the ΔH term of the Gibbs free-energy equation is positive and the ΔG term is negative, $T\Delta S$ is positive at all temperatures. Polymerization reactions are endergonic and nonspontaneous.

For monomers to link together and form macromolecules, an input of energy is required. How could this have happened during chemical evolution?

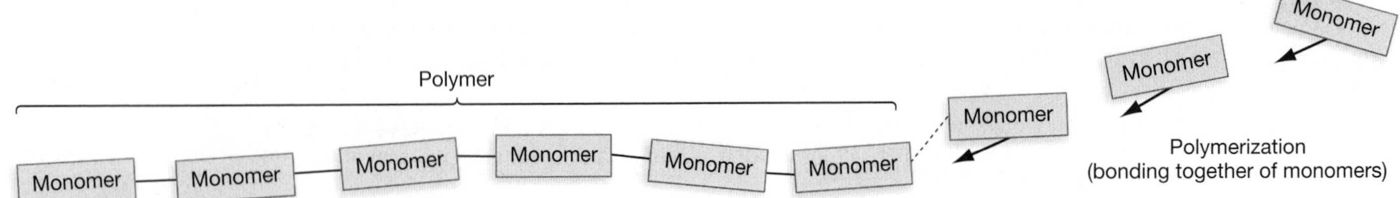

FIGURE 3.4 Monomers Are the Building Blocks of Polymers.

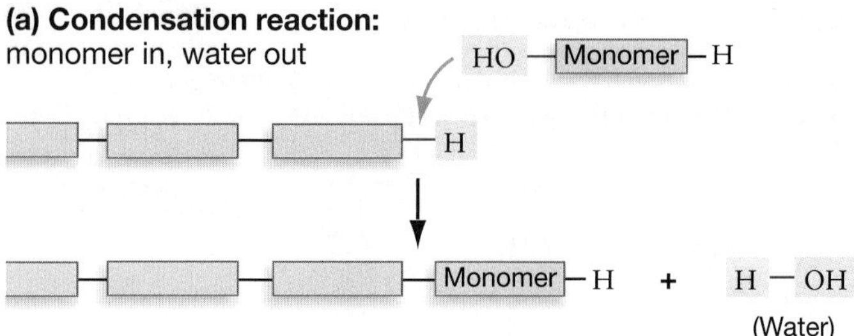

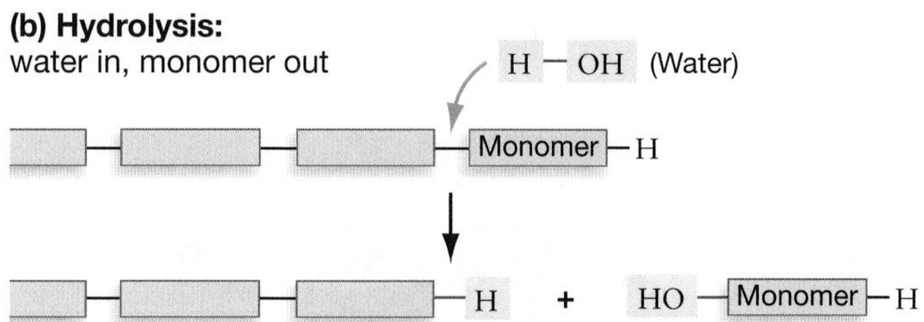

FIGURE 3.5 Polymers Can Be Extended or Broken Apart.

COULD POLYMERIZATION OCCUR IN THE ENERGY-RICH ENVIRONMENT OF EARLY EARTH? Monomers polymerize through **condensation reactions**, also known as **dehydration reactions**. These reactions are aptly named because the newly formed bond results in the loss of a water molecule (**Figure 3.5a**). The reverse reaction, called **hydrolysis**, breaks polymers apart by adding a water molecule (**Figure 3.5b**). The water molecule reacts with the bond linking the monomers, separating one monomer from the polymer chain.

In a solution such as the prebiotic soup, condensation and hydrolysis represent the forward and reverse reactions of a chemical equilibrium. Hydrolysis dominates because it is exergonic—it increases entropy and is favorable energetically.

According to recent experiments, though, there are several ways that amino acids could have polymerized early in chemical evolution.

- Researchers have been able to create stable polymers by mixing monomers with a source of chemical energy and tiny mineral particles—the size found in clay or mud. Apparently, growing macromolecules are protected from hydrolysis if they cling, or adsorb, to a mineral surface. One experiment with clay particles produced polymers that were 55 amino acids long.
- In conditions that simulate the hot, metal-rich environments of undersea volcanoes, researchers have observed not only amino acid formation but also their polymerization.
- Amino acids have also joined into polymers in experiments in cooler water if a carbon- and sulfur-containing gas—one that is commonly ejected from undersea volcanoes—is present.

The current consensus is that several mechanisms led to polymerization reactions between amino acids, early in chemical evolution. What kind of bond is responsible for linking these monomers?

THE PEPTIDE BOND As **Figure 3.6** shows, amino acids polymerize when a bond forms between the carboxyl group of one amino acid and the amino group of another. The C—N bond that results from this condensation reaction is called a **peptide bond**. Because water is lost in the condensation reaction, the carboxyl group of the amino acid is converted to a carbonyl functional group in the polymer.

To study peptide bond formation in more detail and review the general principles of polymerization reactions, go to the study area at *www.masteringbiology.com.*

MB **Web Activity** Condensation and Hydrolysis Reactions

Peptide bonds are unusually stable because the electrons involved are partially shared between the peptide bond and the neighboring carbonyl functional group. The degree of electron sharing is great enough that peptide bonds actually have some of the characteristics of a double bond. For example, the peptide bond is planar.

When amino acids are linked by peptide bonds into a chain, the amino acids are referred to as residues and the resulting molecule is called a **polypeptide**.

Figure 3.7a shows how the chain of peptide bonds in a polypeptide gives the molecule a structural framework, or a "backbone." There are three key points to note about the peptide-bonded backbone of a polypeptide:

1. ***R-group orientation*** The side chains present in each residue extend out from the backbone, making it possible for them to interact with each other and with water.
2. ***Directionality*** There is an amino group ($-NH_3^+$) on one end of every polypeptide chain and a carboxyl group ($-COO^-$) on the other. By convention, biologists always write amino acid sequences in the same direction. The end of the sequence that has the free amino group is placed on the left and is called the N-terminus, or amino-terminus, and the

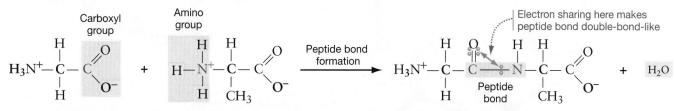

FIGURE 3.6 Peptide Bonds Form When the Carboxyl Group of One Amino Acid Reacts with the Amino Group of a Second Amino Acid.

✔**QUESTION** Is a peptide bond a hydrogen bond, a nonpolar covalent bond, a polar covalent bond, or an ionic bond?

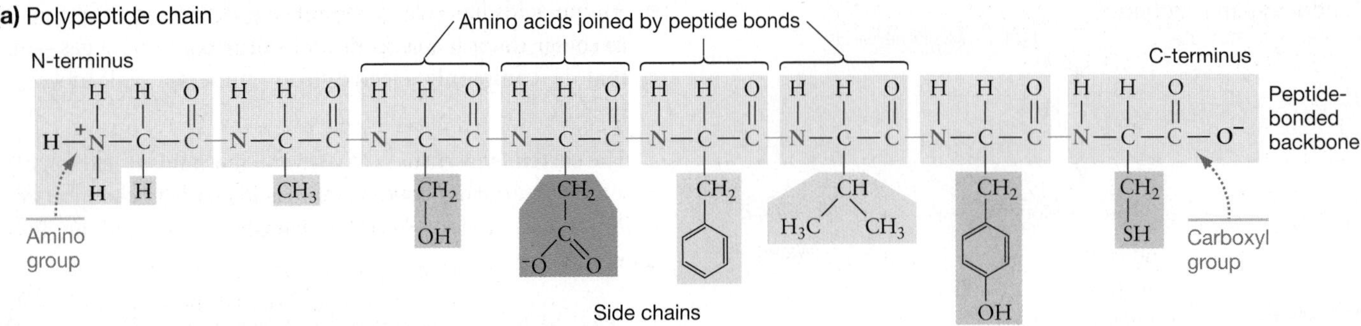

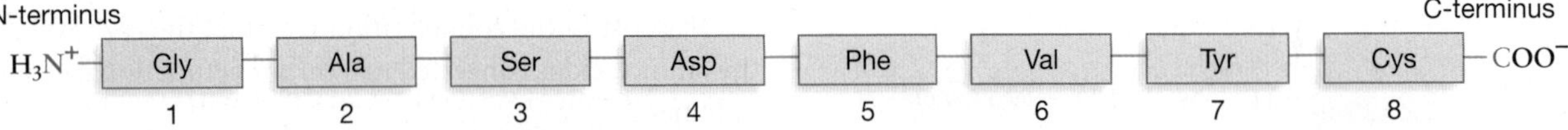

FIGURE 3.7 Amino Acids Polymerize to Form Polypeptides.

end with the free carboxyl group appears on the right-hand side of the sequence and is called the C-terminus, or carboxy-terminus. The amino acids in the chain are always numbered starting from the N-terminus (**Figure 3.7b**), because the N-terminus is the start of the chain when proteins are synthesized in cells.

3. ***Flexibility*** Although the peptide bond itself cannot rotate because of its double-bond nature, the single bonds on either side of the peptide bond can rotate. As a result, the structure as a whole is flexible (**Figure 3.8**).

When fewer than 50 amino acids are linked together in this way, the resulting polypeptide is called an **oligopeptide** ("few peptides") or simply a **peptide**. Polypeptides that contain 50 or more amino acids are formally called **proteins**. Proteins may consist of single polypeptides or multiple polypeptides that are bonded to one another.

Proteins are the stuff of life. Let's take a look at how they are put together, and then at what they do.

CHECK YOUR UNDERSTANDING

If you understand that . . .

- Amino acids are small molecules with a carbon atom bonded to a carboxyl group, an amino group, a hydrogen atom, and a side chain called an R-group.
- Each amino acid has distinctive chemical properties because each has a unique R-group.
- Polypeptides are polymers made up of amino acids.
- When the carboxyl group of one amino acid reacts with the amino group of another amino acid, a strong covalent bond called a peptide bond forms. Small polypeptides are called oligopeptides, and large polypeptides are called proteins.

✔ You should be able to . . .

Draw the structural formulas of two glycine residues (glycine's R-group is an H) linked by a peptide bond, and label the amino- and carboxy-terminus.

Answers are available in Appendix B.

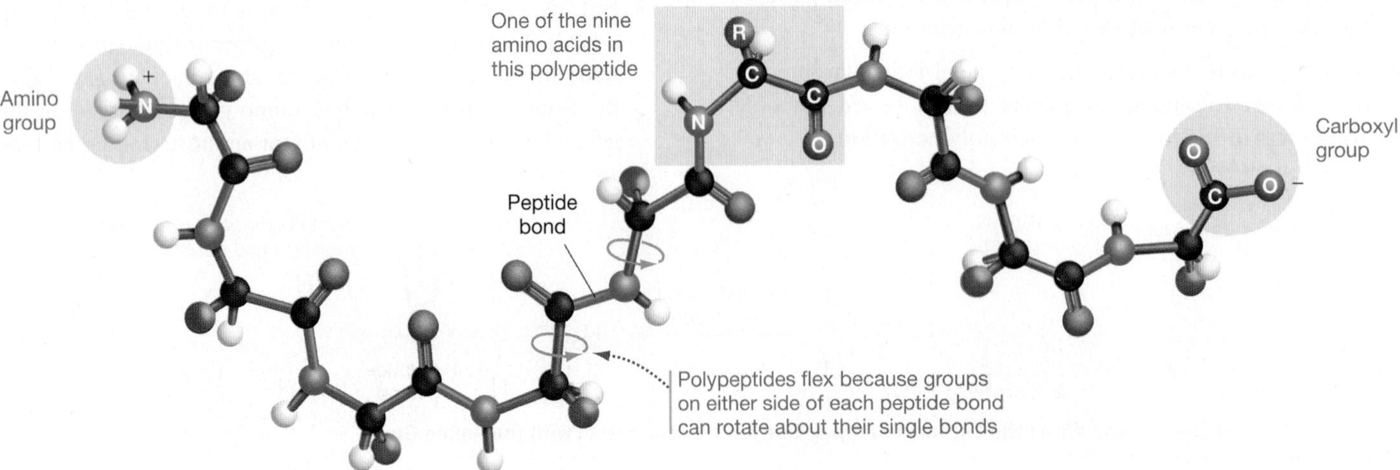

FIGURE 3.8 Polypeptides Are Flexible.

3.3 Proteins Are the Most Versatile Large Molecules in Cells

As a group, proteins perform more types of cell functions than any other type of molecule does. It makes sense to hypothesize that life began with proteins, simply because proteins are so vital to the life of today's cells.

Consider the red blood cells that are moving through your arteries right now. Each of these cells contains about 300 million copies of the protein called hemoglobin. Hemoglobin carries oxygen from your lungs to cells throughout the body. But every red blood cell also has thousands of copies of a protein called carbonic anhydrase, which is important for moving carbon dioxide from cells back to the lungs, where it can be breathed out. Other proteins form the cell's internal skeleton or act as identification badges on its membrane.

Proteins are crucial to most tasks required for cells to exist. These tasks include:

- ***Catalysis*** Many proteins are specialized to **catalyze**, or speed up, chemical reactions. A protein that functions as a catalyst is called an **enzyme**. The carbonic anhydrase molecules in red blood cells are catalysts. So is the protein called salivary amylase, found in your mouth. Salivary amylase helps begin the digestion of starch and other complex carbohydrates into simple sugars. Most chemical reactions that make life possible depend on enzymes.
- ***Defense*** Proteins called antibodies and complement proteins attack and destroy viruses and bacteria that cause disease.
- ***Movement*** Motor proteins and contractile proteins are responsible for moving the cell itself, or for moving large molecules and other types of cargo inside the cell. As you turn this page, for example, specialized proteins called actin and myosin will slide past one another as they work to flex or extend muscle cells in your fingers and arm.
- ***Signaling*** Proteins are involved in carrying and receiving signals from cell to cell inside the body. If sugar levels in your blood are low, a small protein called glucagon will bind to receptor proteins on your liver cells, triggering enzymes inside to release sugar into your bloodstream.
- ***Structure*** Structural proteins make up body components such as fingernails and hair, and define the shape of individual cells. Structural proteins keep red blood cells flexible and in their normal disc-like shape.
- ***Transport*** Proteins allow particular molecules to enter or exit cells, and carry specific compounds throughout the body. Hemoglobin is a particularly well-studied transport protein, but virtually every cell is studded with membrane proteins that control the passage of specific molecules and ions.

Each of these aspects of protein function will be explored in detail later in the text. The immediate task, though, is different. Could functioning proteins be produced by chemical evolution?

3.4 What Do Proteins Look Like?

Figure 3.9 illustrates some of the diverse sizes and shapes observed in proteins. In the case of the TATA-box binding protein in Figure 3.9a and the protein called porin in Figure 3.9b, the shape of the molecule has obvious clear correlation with its function. The TATA-box binding protein has a groove where DNA molecules fit; porin has a hole that forms a pore. The groove in the TATA-box binding protein interacts with specific regions of DNA inside cells, while porin fits in cell membranes and allows certain hydrophilic molecules to pass through. But most of the proteins found in cells function as enzymes and are globular like

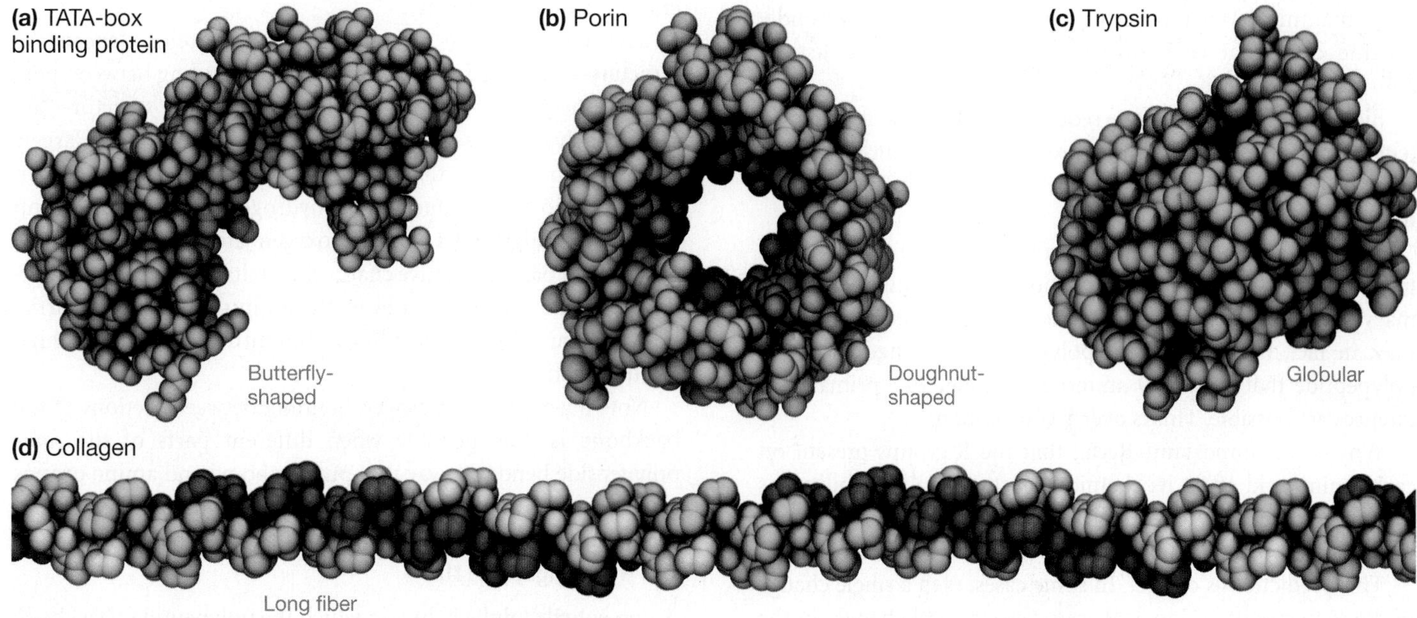

FIGURE 3.9 In Overall Shape, Proteins Are the Most Diverse Class of Molecules Known.

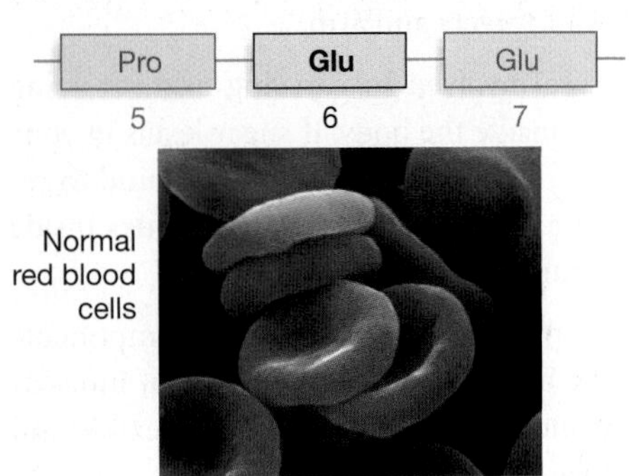

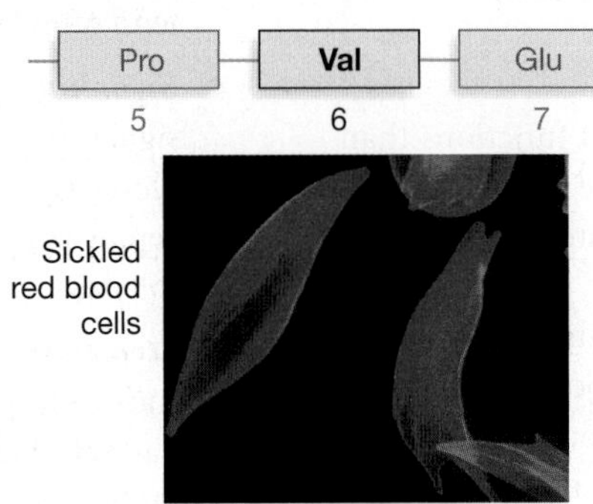

FIGURE 3.10 Changes in Primary Structure Affect Protein Function. Compare the primary structure of normal hemoglobin **(a)** with that of hemoglobin molecules of people with sickle-cell disease **(b)**. The single amino acid change causes red blood cells to change from their normal disc shape in (a) to a sickled shape in (b) when oxygen concentrations are low. Each red blood cell contains about 300 million hemoglobin molecules.

the example in Figure 3.9c. Proteins that provide structural support for cells or tissues, such as the collagen protein in Figure 3.9d, often form long fibers.

The unparalleled diversity of proteins—in size, shape, and other aspects of structure—is important because function follows from structure. Proteins can serve diverse functions in cells because they are diverse in size and shape as well as in the chemical properties of their amino acids.

How can biologists make sense of this diversity of protein size and shape? Initially, the amount of variation seems overwhelming. Fortunately, it is not. No matter how large or complex a protein may be, its underlying structure can be broken down into just four basic levels of organization.

Primary Structure

Every protein has a unique sequence of amino acids. That simple conclusion was the culmination of 12 years of study by Frederick Sanger and co-workers during the 1940s and 1950s. Sanger's group worked out the first techniques for determining the amino acid sequence of a protein and published the completed sequence of the hormone insulin, a protein that helps regulate sugar concentrations in the blood of humans and other mammals. When other proteins were analyzed, it rapidly became clear that each protein has a definite and distinct amino acid sequence.

Biochemists call the unique sequence of amino acids in a protein the **primary structure** of that protein. The sequence of amino acids in Figure 3.7, for example, defines that polypeptide's primary structure.

With 20 types of amino acids available and size ranging from two amino acid residues to tens of thousands, the number of primary structures that are possible is practically limitless. There may, in fact, be 20^n different polypeptides of length n. For a polypeptide that is just 10 amino acids long, 20^{10} primary sequences are possible. This is over 10,000 billion.

Why is this important? Recall that the R-groups present on each amino acid affect its chemical reactivity and solubility. It's therefore reasonable to predict that the R-groups present in a polypeptide will affect that molecule's properties and function.

This prediction is correct. In some cases, even a single change in the sequence of amino acids can cause radical changes in the way the macromolecule as a whole behaves.

As an example, consider the hemoglobin protein of humans. In some individuals, hemoglobin has a valine instead of a glutamate at the amino acid numbered 6 in a strand of 146 amino acids (**Figure 3.10a**). Valine's side chain is radically different from the R-group in glutamate. The change in R-group composition produces a protein that tends to crystallize instead of staying in solution when oxygen concentrations in the blood are low. When hemoglobin crystallizes, the red blood cells that carry the protein adopt a sickled shape (**Figure 3.10b**). Sickled red blood cells get stuck in the small blood vessels called capillaries. In people whose hemoglobin contains this single amino acid change, cells downstream of blocked capillaries become starved for oxygen. A debilitating illness called sickle-cell disease results.

A protein's primary structure is fundamental to its function. Primary structure is also fundamental to the higher levels of protein structure: secondary, tertiary, and quaternary.

Secondary Structure

Even though variation in the amino acid sequence of a protein is virtually limitless, it is only the tip of the iceberg in terms of generating structural diversity.

The next level of organization in proteins—**secondary structure**—is created in part by hydrogen bonding between portions of the peptide-bonded backbone. Secondary structures are distinctively shaped sections of proteins that are stabilized largely by hydrogen bonding that occurs between the carbonyl oxygen of one amino acid residue and the hydrogen on the amino group of another (**Figure 3.11a**). The oxygen atom in the carbonyl group has a partial negative charge due to its high electronegativity, while the hydrogen atom in the amino group has a partial positive charge because it is bonded to nitrogen, which has high electronegativity.

Note a key point: Hydrogen bonding between sections of the backbone is only possible when different parts of the *same* polypeptide bend in a way that puts carbonyl and amino groups close together. In most proteins, the bending that aligns parts of the backbone and allows these bonds to form occurs in one of two ways (**Figure 3.11b**):

1. an **α-helix** (alpha helix), in which the polypeptide's backbone is coiled; or

(a) Hydrogen bonds form between peptide chains.

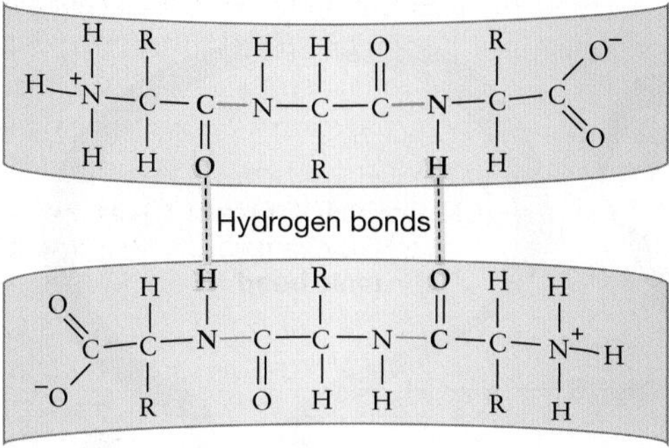

(b) Secondary structures of proteins result.

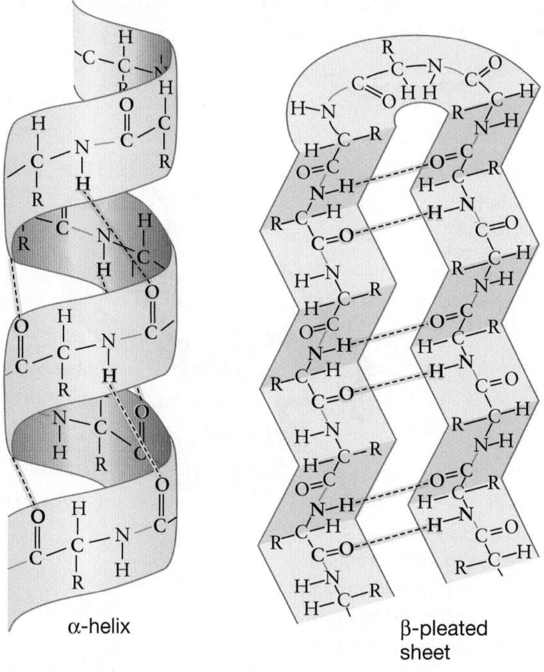

(c) Ribbon diagrams of secondary structure

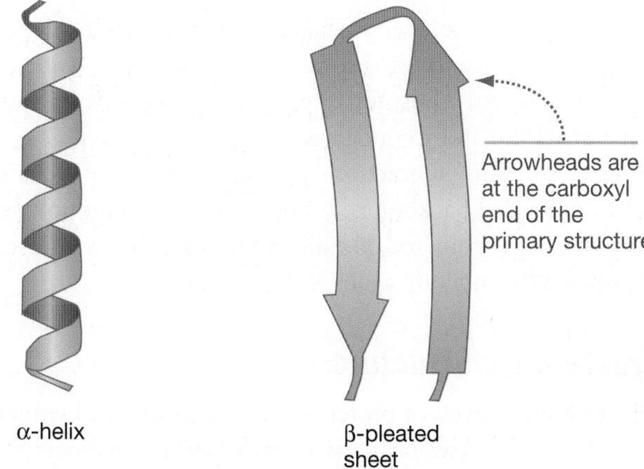

FIGURE 3.11 Secondary Structures of Proteins. A polypeptide chain can coil or fold in on itself when hydrogen bonds form between amino groups and carbonyl groups on its peptide-bonded backbone.

2. a **β-pleated sheet** (beta-pleated sheet), in which segments of a peptide chain bend 180° and then fold in the same plane.

When biologists use illustrations called ribbon diagrams to represent the shape of a protein, α-helices are shown as coils while β-pleated sheets are shown by groups of arrows in a plane (**Figure 3.11c**).

In most cases, secondary structure consists of α-helices and β-pleated sheets. Which one forms, if either, depends on the molecule's primary structure—specifically, the geometry and properties of the amino acids in the sequence. Methionine and glutamic acid, for example, are much more likely to be involved in α-helices than in β-pleated sheets. The opposite is true for valine and isoleucine. Proline, in contrast, is unlikely to be involved in either type of secondary structure.

This is a key point: Which secondary structures form depends on the protein's primary structure. A protein may have different types of secondary structure at different points along its sequence.

Although each of the hydrogen bonds in an α-helix or a β-pleated sheet is weak relative to a covalent bond, the large number of hydrogen bonds in these structures makes them highly stable. As a result, they increase the stability of the molecule as a whole and help define its shape. In terms of overall shape and stability, though, a protein's tertiary structure is even more important.

Tertiary Structure

Interactions between components of a protein's peptide-bonded backbone are responsible for α-helices and β-pleated sheets. In contrast, most of the overall shape, or **tertiary structure**, of a polypeptide results from interactions between R-groups or between R-groups and the peptide backbone. As **Figure 3.12a** shows, side chains can be involved in a wide variety of bonds and interactions. Because each contact between R-groups causes the peptide-bonded backbone to bend and fold, each contributes to the distinctive three-dimensional shape of a polypeptide.

Five types of interactions involving side chains are particularly important:

1. ***Hydrogen bonding*** Hydrogen bonds form between hydrogen atoms and the carbonyl group in the peptide-bonded backbone, and between hydrogen and atoms with partial negative charges in side chains.
2. ***Hydrophobic interactions*** In an aqueous solution, water molecules interact with hydrophilic side chains and force hydrophobic side chains to coalesce. When hydrophobic portions of proteins come together, the surrounding water molecules form more hydrogen bonds, increasing their stability. As a result, the hydrophobic side chains of proteins tend to form globular masses.
3. ***van der Waals interactions*** Once hydrophobic side chains are close to one another, they are stabilized by electrical attractions known as **van der Waals interactions**. These weak

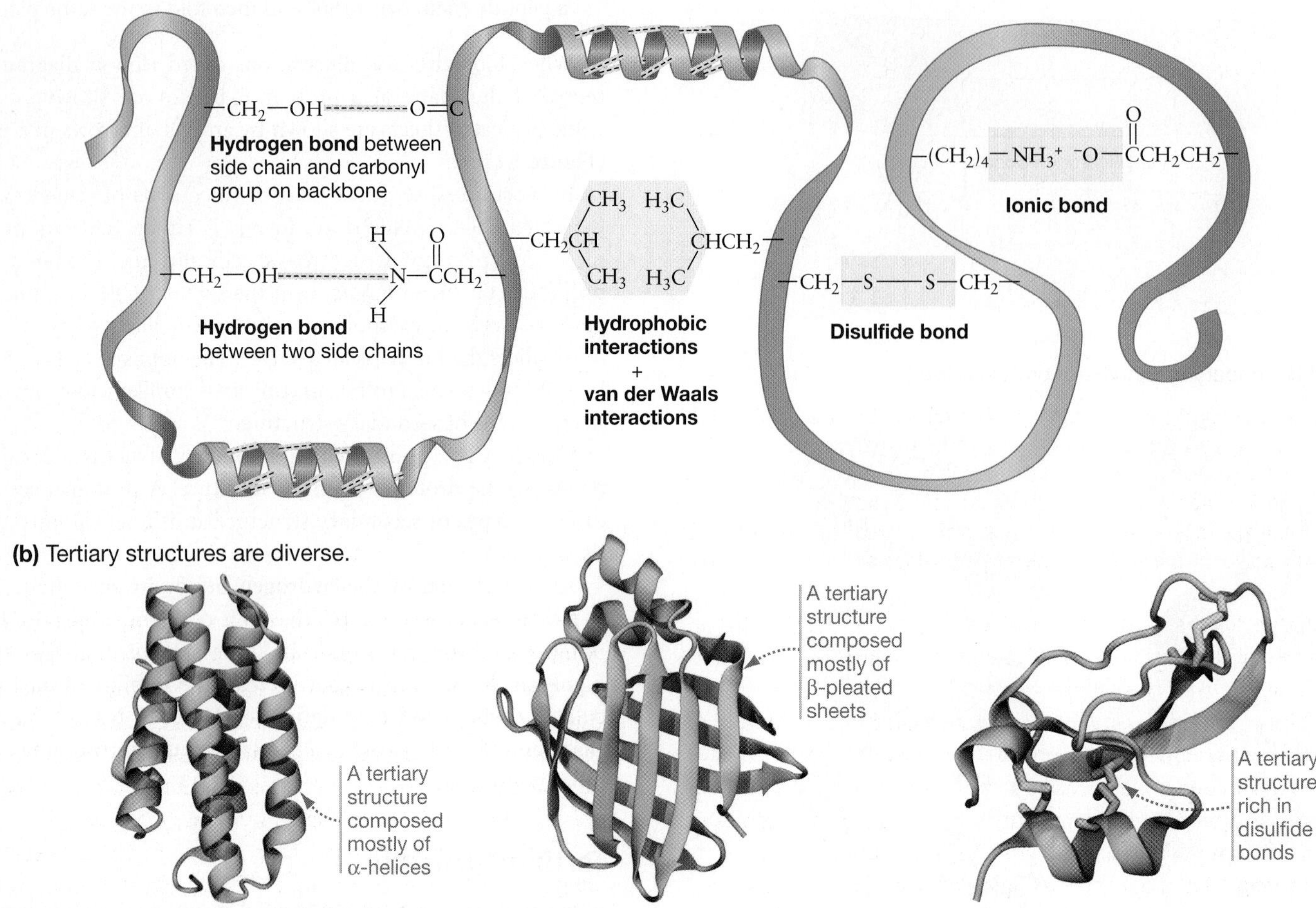

FIGURE 3.12 Tertiary Structure of Proteins Results from Interactions Involving R-Groups. (a) Each protein has a unique overall shape called its tertiary structure. Tertiary structure is created by bonds and other interactions that cause proteins to fold in a precise way. **(b)** The tertiary structure of these proteins includes interactions between α-helices and β-pleated sheets. The polypeptide chains are color-coded so that you can follow the chain from the amino-end (dark blue) to the carboxy-end (red).

attractions occur because the constant motion of electrons gives molecules a tiny asymmetry in charge that changes with time. If molecules get extremely close to each other, the minute partial charge on one molecule induces an opposite partial charge in the nearby molecule and causes an attraction. Although the attraction is weak relative to covalent bonds or even hydrogen bonds, a large number of van der Waals interactions can occur in a polypeptide when many hydrophobic residues congregate. The result is a significant increase in stability.

4. ***Covalent bonding*** Covalent bonds can form between sulfur atoms when a reaction occurs between the sulfur-containing R-groups of two cysteines. These **disulfide** ("two-sulfur") **bonds** are frequently referred to as bridges because they create strong links between distinct regions of the same polypeptide.
5. ***Ionic bonding*** Ionic bonds form between groups that have full and opposing charges, such as the ionized amino and carboxyl functional groups highlighted on the right in Figure 3.12a.

In addition, the overall shape of many proteins depends in part on the presence of secondary structures like α-helices and β-pleated sheets. Thus, tertiary structure depends on both primary structure and secondary structure.

With so many interactions possible between side chains and peptide-bonded backbones, it's not surprising that polypeptides vary in shape from rod-like filaments to ball-like masses. (See **Figure 3.12b,** and look again at Figure 3.9.)

Quaternary Structure

The first three levels of protein structure involve individual polypeptides. But many proteins contain several distinct polypeptides that interact to form a single structure. The combination of polypeptide subunits gives proteins a **quaternary structure**. The individual polypeptides may be held together by bonds or other interactions among R-groups or sections of their peptide backbones.

In the simplest case, a protein with quaternary structure can consist of just two subunits that are identical. The Cro protein

(a) Cro protein, a dimer

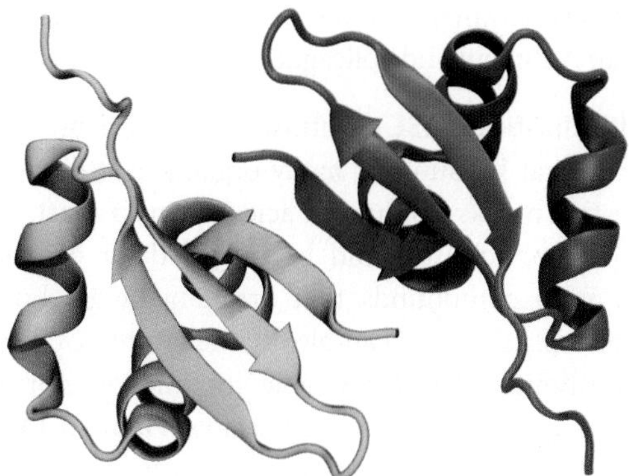

(b) Hemoglobin, a tetramer

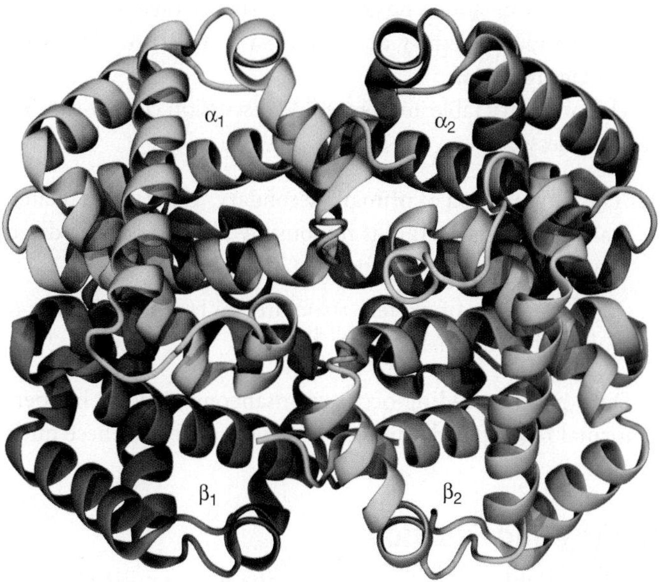

FIGURE 3.13 Quaternary Structures of Proteins Are Created by Multiple Polypeptides. These diagrams represent the primary sequence as a ribbon. **(a)** The Cro protein is a dimer—it consists of two polypeptides. The polypeptides are identical in this case but are colored yellow and green here. **(b)** Hemoglobin is a tetramer—it consists of four polypeptides. The yellow and purple polypeptides are identical; so are the blue and green polypeptides.

found in a virus called bacteriophage λ (pronounced *LAMB-da*) is an example (**Figure 3.13a**). Proteins with two polypeptide subunits are called dimers ("two-parts").

More than two polypeptides can be linked into a single protein, however, and the polypeptides involved may be distinct in primary, secondary, and tertiary structure. Hemoglobin, for example, is a tetramer ("four-parts"). It consists of two copies of two different polypeptides (**Figure 3.13b**).

Proteins that consist of a single polypeptide lack quaternary structure. Quaternary structure is common, however. Many proteins comprise several polypeptide subunits.

In addition, most cells contain at least one **multienzyme complex**: a group of enzymes, each of which catalyzes one reaction, that are physically joined to each other. The enzymes in a multi-enzyme complex are all involved in the same task. The nitrogenase complex found in certain bacteria, for example, contains the enzymes required to convert molecular nitrogen (N_2) to amine groups ($-NH_2$) used in amino acid synthesis.

Table 3.2 summarizes the four levels of protein structure, using hemoglobin as an example. The key thing to note is that protein structure is hierarchical. Quaternary structure is based on tertiary structure, which is based in part on secondary structure.

SUMMARY TABLE 3.2 **Protein Structure**

Level	Description	Stabilized by	Example: Hemoglobin
Primary	The sequence of amino acids in a polypeptide	Peptide bonds	Gly – Ser – Asp – Cys
Secondary	Formation of α-helices and β-pleated sheets in a polypeptide	Hydrogen bonding between groups along the peptide-bonded backbone; thus, depends on primary structure.	One α-helix
Tertiary	Overall three-dimensional shape of a polypeptide (includes contribution from secondary structures)	Bonds and other interactions between R-groups, or between R-groups and the peptide-bonded backbone; thus, depends on primary structure.	One of hemoglobin's subunits
Quaternary	Shape produced by combinations of polypeptides (thus, combinations of tertiary structures)	Bonds and other interactions between R-groups, and between peptide backbones of different polypeptides; thus, depends on primary structure.	Hemoglobin, which consists of four polypeptide subunits

All three of the higher-level structures are based on primary structure. ✔**You should be able to describe elements of the primary, secondary, tertiary, and quaternary structure of the protein in Figure 3.13a.**

The summary table and preceding discussion convey two important messages:

1. The combination of primary, secondary, tertiary, and quaternary levels of structure is responsible for the fantastic diversity of sizes and shapes observed in proteins.
2. Most elements of protein structure are based on folding of polypeptide chains.

Does protein folding occur spontaneously? What happens if normal folding is disrupted? Let's take a look at these questions next.

Folding and Function

If you were able to synthesize one of the polypeptides in hemoglobin from individual amino acids, and if you placed the resulting chain in water, it would spontaneously fold into the shape of the tertiary structure shown in Table 3.2.

This result seems counterintuitive. Because an unfolded protein has many more ways to move about, it has much higher entropy than the folded version. Folding *is* spontaneous in some cases, however, because the bonds, hydrophobic interactions, and van der Waals interactions that occur make the folded molecule more stable energetically than the unfolded molecule. Thus, folding may release enough free energy to be exergonic and occur spontaneously.

Folding is also crucial to the function of a completed protein. This point was hammered home in a set of classic experiments by Christian Anfinson and colleagues during the 1950s.

FOLDING IN RIBONUCLEASE Anfinson studied a protein called ribonuclease that is found in many organisms. Ribonuclease is an enzyme that breaks ribonucleic acid polymers apart. Anfinson found that ribonuclease could be unfolded, or **denatured**, by treating it with compounds that break hydrogen bonds and disulfide bonds. (Proteins can also be denatured by heat.) The denatured ribonuclease was unable to function normally—it could no longer break apart nucleic acids (**Figure 3.14**). This is not surprising, given that the function of a protein depends on its structure.

When Anfinson removed the denaturing agents, however, the molecule refolded and began to function normally again. These experiments confirmed that ribonuclease folds spontaneously and that folding is essential for normal function.

More recent work has shown that in cells, folding is often facilitated by specific proteins called **molecular chaperones.** Many molecular chaperones belong to a family of molecules called the heat-shock proteins. These compounds are produced in large quantities after cells experience high temperatures or other treatments that make proteins lose their tertiary structure. Heat-shock proteins speed the refolding of other proteins into their normal shape after denaturation has occurred.

FOLDING IN PRIONS In 1982, Stanley Prusiner published what may be the most surprising result to emerge from research on protein folding: Some improperly folded proteins act as infec-

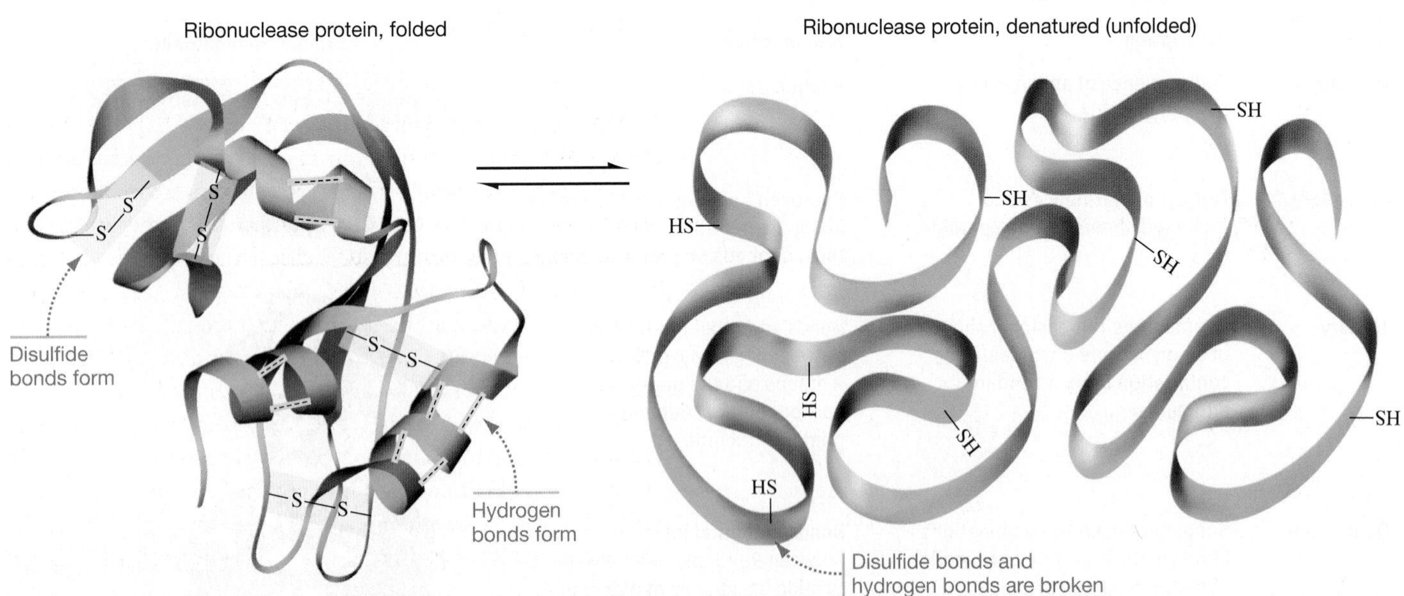

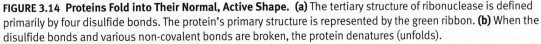

FIGURE 3.14 Proteins Fold into Their Normal, Active Shape. (a) The tertiary structure of ribonuclease is defined primarily by four disulfide bonds. The protein's primary structure is represented by the green ribbon. **(b)** When the disulfide bonds and various non-covalent bonds are broken, the protein denatures (unfolds).

tious, disease-causing agents. The proteins involved are called **prions** (pronounced *PREE-ons*), or proteinaceous infectious particles.

Follow-up work showed that prions are improperly folded forms of normal proteins that are present in healthy individuals. The infectious and normal forms do not necessarily differ in amino acid sequence, however. Instead, their *shapes* are radically different. Further, the infectious form of a protein can induce other normal protein molecules to change their shape to the altered form.

Figure 3.15 illustrates the shape differences observed in the normal and infectious forms of the first prion ever described. The molecule in Figure 3.15a is called the prion protein (PrP) and is a normal component of mammalian cells. Mutant versions of this protein, like the one in Figure 3.15b, are found in a wide variety of species and cause a family of diseases known as the spongiform encephalopathies—literally, "sponge-brain-illnesses." Hamsters, cows, goats, and humans afflicted with these diseases undergo massive degeneration of the brain. Cattle suffer from "mad cow disease"; sheep and goats acquire scrapie (so called because the animals itch so badly that they scratch off their wool or hair); humans develop kuru or Creutzfeldt-Jakob disease.

Although some spongiform encephalopathies can be inherited, in many cases the disease is transmitted when individuals eat tissues containing the infectious form of PrP. All the prion illnesses are fatal.

Prions are particularly dramatic examples of how a protein's function depends on its shape, and how the final shape of a protein depends on folding.

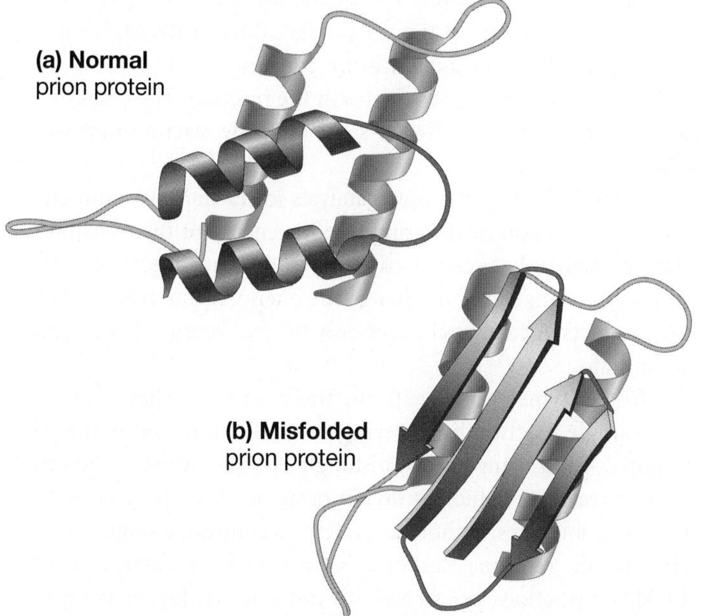

FIGURE 3.15 Prions Are Improperly Folded Proteins. Ribbon model of **(a)** a normal prion protein and **(b)** the misfolded form that causes mad cow disease in cattle. Secondary structure is represented by coils (α-helices) and arrows (β-pleated sheets).

CHECK YOUR UNDERSTANDING

If you understand that . . .

- Proteins have up to four levels of structure.
- Primary structure consists of the sequence of amino acids.
- Secondary structure results from bonds between atoms in the peptide-bonded backbone of the same polypeptide. These bonds produce structures such as α-helices and β-pleated sheets.
- Tertiary structure is the overall shape of a polypeptide. Most tertiary structure is a consequence of bonds or other interactions between R-groups or between R-groups and the peptide-bonded backbone.
- Quaternary structure occurs when multiple polypeptides interact to form a single protein.

✔ **You should be able to . . .**

Explain why secondary, tertiary, and quarternary structure depend on a polypeptide's primary structure.

Answers are available in Appendix B.

3.5 Enzymes: An Introduction to Catalysis

Of all the functions that proteins perform in cells, catalysis may be the most important. The reason is speed. Life, at its most basic level, consists of chemical reactions. But most of the chemical reactions in cells don't occur fast enough to support life unless a catalyst is present.

Even when life began, the first molecule that could make a copy of itself would have had to catalyze the polymerization reactions involved. Catalysis is at the heart of life's origin, and all life today. Let's dig in.

Enzymes Help Reactions Clear Two Hurdles

To appreciate how enzymes work, recall from Chapter 2 that reactions take place when reactants (1) collide in a precise orientation and (2) have enough kinetic energy to overcome repulsion between electrons that come into contact as a bond forms. Part of the reason enzymes are such effective catalysts is that they bring reactant molecules—called **substrates**—together in a precise orientation, so that the electrons involved in the reaction can interact. But enzymes can also affect the amount of kinetic energy reactants must have for a reaction to proceed.

To understand why catalysts affect the energy required for a reaction, it's critical to realize that a collision between reactants creates a combination of old and new bonds. This intermediate condition is called a **transition state**. The amount of free energy required to reach the transition state is called the **activation energy** of the reaction.

Reactions happen when reactants have enough kinetic energy to reach the transition state. The kinetic energy of molecules, in turn, is a function of their temperature. This is why

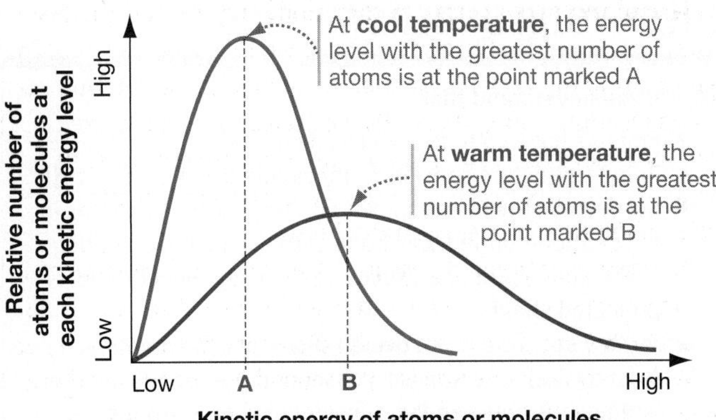

FIGURE 3.16 Atoms and Molecules Have More Kinetic Energy at High Temperatures than at Low Temperatures.

✔**EXERCISE** Add a vertical line two-thirds of the way along the horizontal axis. Label this line "Activation energy for reaction." At low and high temperatures, which portion of the respective curves represents molecules with enough kinetic energy for this reaction to proceed?

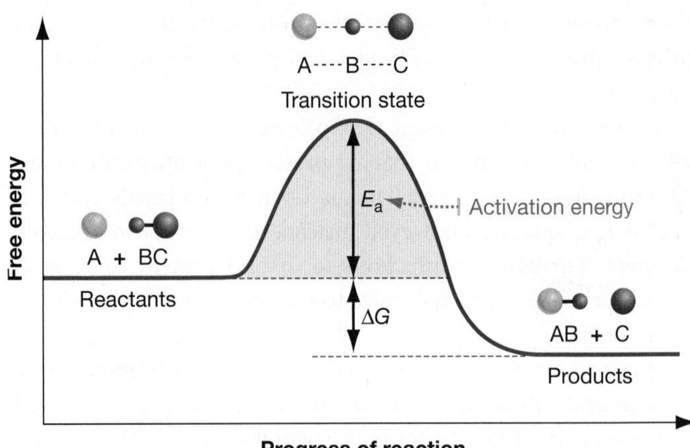

FIGURE 3.17 Changes in Free Energy during a Chemical Reaction. The changes in free energy that occur over the course of a hypothetical reaction between an atom A and a molecule containing atoms B and C. The overall reaction would be written as A + BC → AB + C. E_a is the activation energy of the reaction, and ΔG is the overall change in free energy.

chemical reactions tend to proceed faster at higher temperatures, as noted in Chapter 2.

To make sure you understand these points, study **Figure 3.16.** The *x*-axis plots the kinetic energy of a molecule in a cell or a solution; the *y*-axis plots the proportion of all the molecules present that have a particular amount of kinetic energy. The blue curve shows the distribution of kinetic energy in molecules at cool temperature; the red curve shows the distribution of energy in molecules at high temperature. Even if the activation energy for a reaction involving these molecules is low, more of them will have enough kinetic energy to react if they are at high temperature versus low temperature.

Catalysts don't change the temperature of a solution, though. How do they fit in?

GRAPHING THE ENERGY OF THE TRANSITION STATE **Figure 3.17** graphs the changes in free energy that take place during the course of a chemical reaction. As you read along the *x*-axis from left to right, note that a dramatic rise in free energy occurs when the reactants combine to form the transition state—followed by a dramatic drop in free energy when products form. The free energy of the transition state is high because the bonds that existed in the substrate are destabilized—it is the transition point between breaking old bonds and forming new ones.

The ΔG label on the graph indicates the overall change in free energy in the reaction—that is, the energy of the products minus the energy of the reactants. In this particular case, the products have lower potential energy than the reactants, meaning that the reaction is exothermic. But because the activation energy for this reaction, symbolized by E_a, is high, the reaction would proceed slowly—even at high temperature.

This is an important point: The more unstable the transition state, the higher the activation energy and the less likely a reaction is to proceed quickly.

Reaction rates, then, depend on both the kinetic energy of the reactants and the activation energy of the particular reaction—meaning the free energy of the transition state. If the kinetic energy of the participating molecules is high, then molecular collisions are likely to result in completed reactions. But if the activation energy of a particular reaction is also high, then collisions are less likely to result in completed reactions.

ENZYMES LOWER THE ACTIVATION ENERGY In many cases, the electrons in the transition-state molecule can be stabilized when they interact with another ion, atom, or molecule. When this occurs, the activation energy required for the reaction drops and the reaction rate increases.

A substance that lowers the activation energy of a reaction and increases the rate of the reaction is called a **catalyst.** It's important to note that a catalyst is not consumed in a chemical reaction, even though it participates in the reaction. The composition of a catalyst is exactly the same after the reaction as it was before.

Figure 3.18 diagrams how catalysts lower the activation energy for a reaction by lowering the free energy of the transition state. Note that the presence of a catalyst does not affect the overall energy change, ΔG, or change the energy of the reactants or the products. A catalyst changes only the free energy of the transition state.

Most enzymes are specific in their activity—they catalyze just a single reaction by lowering the activation energy that is required—and many are astonishingly efficient. Most of the important reactions in biology do not occur at all, or else proceed at imperceptible rates, without a catalyst. In contrast, a single molecule of the enzyme carbonic anhydrase can catalyze over 1,000,000 reactions *per second.* It's not unusual for enzymes to speed up reactions by a factor of a million; some enzymes make reactions go many *trillions* of times faster than they would without a catalyst.

To understand why catalysts make reactions happen so much faster than uncatalyzed reactions, go back to Figure 3.16. After

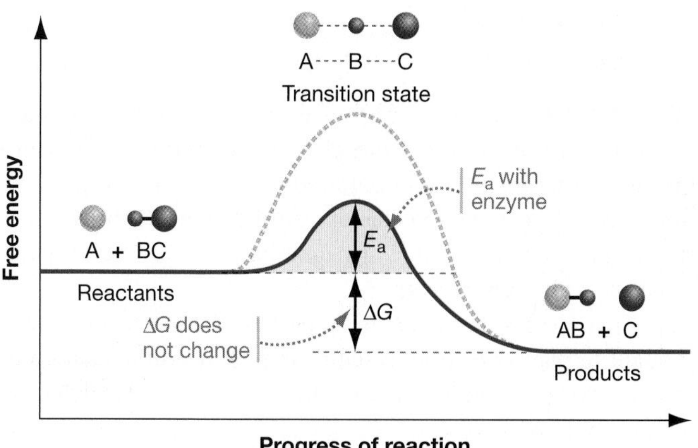

FIGURE 3.18 A Catalyst Changes the Activation Energy of a Reaction. The energy profile for the same reaction diagrammed in Figure 3.17, but with a catalyst present. Even though the energy barrier to the reaction, E_a, is much lower, ΔG does not change.

✔**QUESTION** Can a catalyst make a nonspontaneous reaction occur spontaneously? Explain why or why not.

you've done the original exercise in the caption, add another vertical line one-third of the way along the *x*-axis. Label this line "Activation energy for reaction with catalyst." Label the portion of each curve that has enough energy to undergo a reaction with a catalyst but not without a catalyst. ✔**If you understand what catalysts do, you should be able to determine whether more reactions occur with a catalyst or without, and explain why.**

To summarize, enzymes (1) bring reactants together in precise orientations and (2) stabilize transition states. Protein catalysts are important because they speed up the chemical reactions that are required for life. How do they do what they do?

How Do Enzymes Work?

The initial hypothesis for how enzymes work was proposed by Emil Fischer in 1894. According to Fischer's lock-and-key model, enzymes are rigid structures analogous to a lock. The keys are substrates that fit into the lock and then react.

Several important ideas in this model have stood the test of time. For example, Fischer was correct in proposing that enzymes bring substrates together in a precise orientation that makes reactions more likely. His model also accurately explained why most enzymes can catalyze only one specific reaction. Enzyme specificity is a product of the geometry and chemical properties of the sites where substrates bind.

As researchers began to test and extend Fischer's model, the location where substrates bind and react became known as the enzyme's **active site**. The active site is where catalysis actually occurs.

When techniques for solving the three-dimensional structure of enzymes became available, it turned out that enzymes tend to be very large relative to substrates and roughly globular, and that the active site is in a cleft or cavity within the globular shape. The enzyme hexokinase, which is at work in most cells of your body now, is a good example. (Many enzymes have names that end with *-ase.*) As the left-hand side of **Figure 3.19** shows, the active site in hexokinase is a small notch in an otherwise large, crescent-shaped enzyme.

Fischer's model had to be modified, however, as research on enzyme action progressed. Perhaps the most important realization was that enzymes are not rigid and static, but flexible and dynamic. In fact, many enzymes undergo a significant change in shape, or conformation, when reactant molecules bind to the active site. You can see this conformational change, called an **induced fit**, in the hexokinase molecule on the right of Figure 3.19. As hexokinase binds its substrate—the sugar glucose—the enzyme rocks forward over the active site.

In addition, recent research has clarified the nature of Fischer's key. When one or more substrate molecules enter the active site, they are held in place through hydrogen bonding or other electrical interactions with amino acids in the active site. Once the substrate is bound, one or more R-groups in the active site come into play. The degree of interaction between the substrate and enzyme increases and reaches a maximum when the transition state is formed. Thus, Fischer's key is actually the transition state.

Interactions with R-groups at the active site stabilize the transition state and thus lower the activation energy required for the reaction to proceed. At the atomic level, R-groups that line the

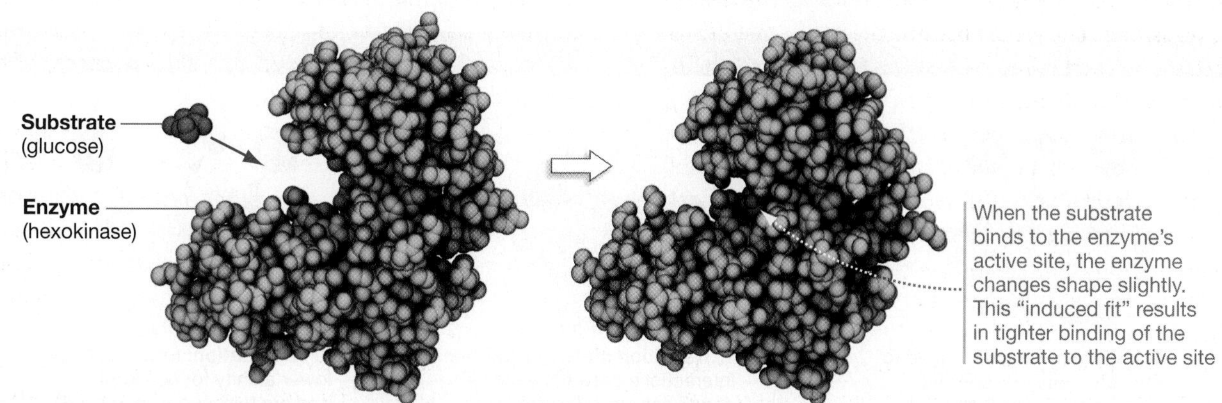

FIGURE 3.19 Reactant Molecules Bind to Specific Locations in an Enzyme. The reactant molecule, shown in red, fits into a precise location, called the active site, in the green enzyme. In this enzyme and in many others, the binding event causes the protein to change shape.

active site may form short-lived covalent bonds that assist with the transfer of atoms or groups of atoms from one reactant to another. More commonly, the presence of acidic or basic R-groups allows the reactants to lose or gain a proton more readily.

The reaction products have a much lower affinity for the active site than do either the reactants or the transition state, however. As a result, they are released from the enzyme once they form.

Figure 3.20 summarizes these principles. Enzyme catalysis can be analyzed as a three-step process:

1. **Initiation**: Instead of reactants occasionally colliding in a random fashion, enzymes orient reactants precisely as they bind at specific locations within the active site.
2. **Transition state facilitation**: The act of binding induces the formation of the transition state. In some cases the transition state is stabilized by a change in the enzyme's shape. The interaction between the substrate and R-groups in the enzyme's active site lowers the activation energy required for the reaction. Inside a catalyst's active site, then, more reactant molecules have sufficient kinetic energy to reach this lowered activation energy. Thus, the catalyzed reaction proceeds much more rapidly than the uncatalyzed reaction.
3. **Termination**: The reaction products have less affinity for the active site than the transition state does. Binding ends, the enzyme returns to its original conformation, and the products are released.

To make sure you understand how catalysts affect transition states and activation energies, go to the study area at *www.masteringbiology.com.*

(MB) **Web Activity** Activation Energy and Enzymes

✔If you understand the basic principles of enzyme catalysis, you should be able to complete the following sentences: (1) Enzymes speed reaction rates by ________ and lowering activation energy. (2) Activation energies drop because enzymes may destabilize bonds in the reactant, stabilize the ________, make acid–base reactions more favorable, and change the reaction mechanism through a covalent bonding interaction. (3) Enzyme specificity is a function of the active site's shape and the chemical properties of the ________ at the active site. (4) In enzymes, as in many molecules, function follows from ________.

DO ENZYMES ACT ALONE? The answer to this question, in many cases, is no. Atoms or molecules that are not part of an enzyme's primary structure are often required for an enzyme to function normally. These enzyme **cofactors** can be either **(1)** metal ions such as Zn^{2+} (zinc) or Mg^{2+} (magnesium) or **(2)** small organic molecules called **coenzymes**.

In many cases, the cofactor is part of the active site and is thought to play a key role in stabilizing the transition state during the reaction. The presence of the cofactor is therefore essential for catalysis.

To appreciate why this is important, consider that many of the vitamins in your diet are required for the production of enzyme cofactors. Vitamin deficiencies cause enzyme-cofactor deficiencies. Lack of cofactors, in turn, disrupts normal enzyme function and causes disease. For example, thiamine (vitamin B_1) is required for the production of an enzyme cofactor called thiamine pyrophosphate, which is required by three different enzymes. Lack of thiamine in the diet dramatically reduces the activity of these enzymes and causes an array of nervous system and heart disorders collectively known as beriberi.

MOST ENZYMES ARE REGULATED In addition to requiring a cofactor for activity, most enzymes associate with other molecules that regulate them. These regulatory molecules change the protein's structure in some way, and their presence either activates or inactivates the enzyme.

Most enzyme regulation occurs in one of two ways:

- Catalysis is inhibited when a molecule that is similar in size and shape to a substrate binds to the active site. This event is called **competitive inhibition**, because the molecule involved competes with the substrate for access to the enzyme's active site (**Figure 3.21a**).
- Regulatory molecules may bind at a location other than the active site. This type of regulation is called **allosteric** ("different-structure") **regulation**, because the molecule involved does not affect the active site directly. Instead, the binding event

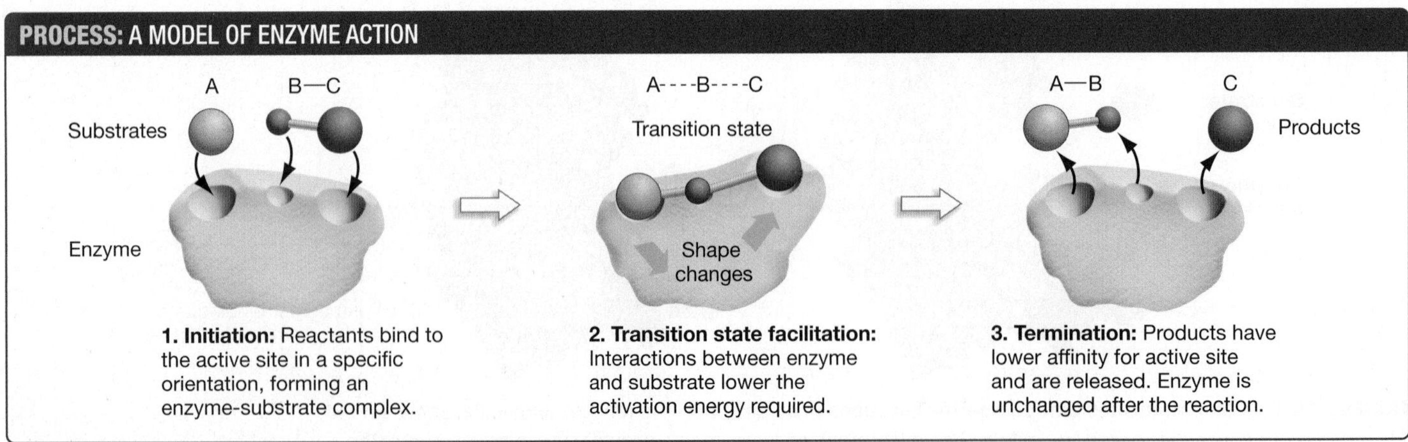
PROCESS: A MODEL OF ENZYME ACTION
A
B—C
Substrates
Enzyme
A- - -B- - -C
Transition state
Shape changes
A—B
C
Products
1. Initiation: Reactants bind to the active site in a specific orientation, forming an enzyme-substrate complex.
2. Transition state facilitation: Interactions between enzyme and substrate lower the activation energy required.
3. Termination: Products have lower affinity for active site and are released. Enzyme is unchanged after the reaction.

FIGURE 3.20 Enzyme Action Can Be Analyzed as a Three-Step Process.

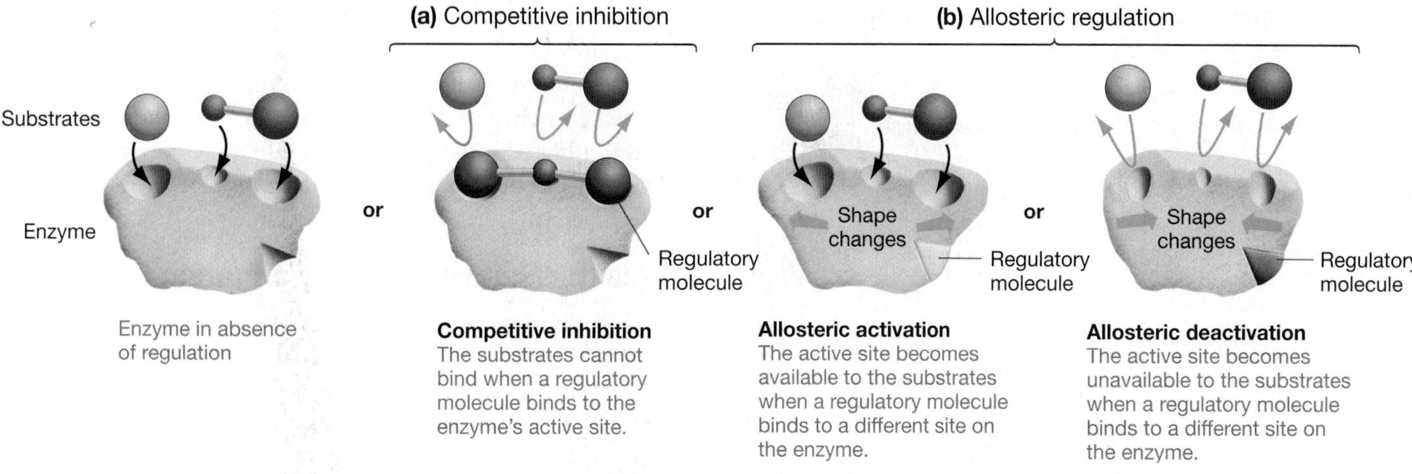
(a) Competitive inhibition
(b) Allosteric regulation
Substrates
Enzyme
or
or
or
Regulatory molecule
Shape changes
Regulatory molecule
Shape changes
Regulatory molecule
Enzyme in absence of regulation
Competitive inhibition
The substrates cannot bind when a regulatory molecule binds to the enzyme's active site.
Allosteric activation
The active site becomes available to the substrates when a regulatory molecule binds to a different site on the enzyme.
Allosteric deactivation
The active site becomes unavailable to the substrates when a regulatory molecule binds to a different site on the enzyme.

FIGURE 3.21 An Enzyme's Activity Is Precisely Regulated. Enzymes are turned on or off when specific molecules bind to them.

changes the shape of the enzyme in a way that makes the active site accessible or inaccessible (**Figure 3.21b**).

Allosteric regulation is much more common than competitive inhibition; later chapters will provide detailed examples of both processes.

WHAT LIMITS THE RATE OF CATALYSIS? For several decades after Fischer's model was published, most research on enzymes focused on rates of enzyme action, or what biologists call enzyme kinetics. Researchers observed that, when the amount of product produced per second—indicating the speed of the reaction—is plotted as a function of substrate concentration, a graph like that shown in **Figure 3.22** results.

In this graph, each data point represents an experiment where reaction rate was measured when substrates were at various concentrations. As you read the curve from left to right, note that it has three basic sections:

1. When substrate concentrations are low, the speed of an enzyme-catalyzed reaction increases in a linear fashion.

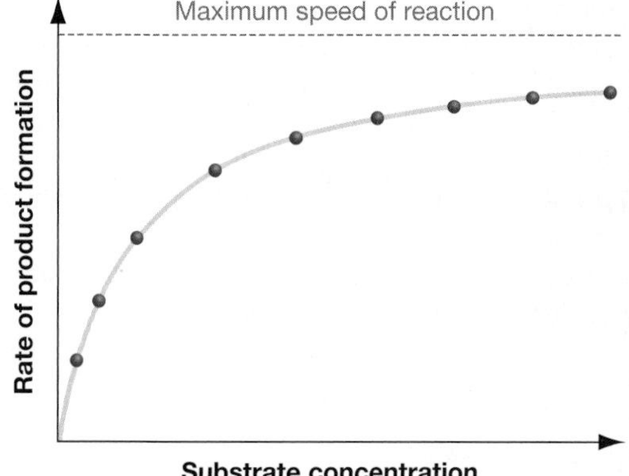
Maximum speed of reaction
Rate of product formation
Substrate concentration

FIGURE 3.22 Kinetics of an Enzyme-Catalyzed Reaction. The general shape of this curve is characteristic of enzyme-catalyzed reactions.

✔**EXERCISE** Label the parts of the graph represents where (1) the reaction rate is most sensitive to changes in substrate concentration and (2) most or all of the active sites present are occupied.

2. At intermediate substrate concentrations, the increase in speed begins to slow.
3. At high substrate concentration, the reaction rate plateaus at a maximum speed.

This pattern is in striking contrast to the situation for uncatalyzed reactions, in which reaction speed tends to show a continuing linear increase with substrate concentration. The "saturation kinetics" of enzyme-catalyzed reactions were taken as strong evidence that the enzyme-substrate complex proposed by Fischer actually exists. The idea was that, at some point, active sites cannot accept substrates any faster, no matter how large the concentration of substrates gets. Stated another way, reaction rates level off because all available enzyme molecules are being used.

HOW DO PHYSICAL CONDITIONS AFFECT ENZYME FUNCTION? Given that an enzyme's structure is critical to its function, it's not surprising that an enzyme's activity is sensitive to conditions that alter its structure. In particular, the activity of an enzyme often changes drastically as a function of temperature and pH.

Temperature affects the movement of the enzyme as well as the kinetic energy of the substrates. pH affects the makeup and charge of amino acid side chains with carboxyl or amino groups, and the active site's ability to participate in proton-transfer or electron-transfer reactions.

Do data support these assertions? **Figure 3.23a** shows how the activity of an enzyme, plotted on the *y*-axis, changes as a function of temperature, plotted on the *x*-axis. Data for the enzyme glucose-6-phosphatase, which is helping to produce usable energy in your cells right now, are shown for two species of bacteria. In this graph, each data point represents the enzyme's relative activity—meaning, the rate of the enzyme-catalyzed reaction, scaled relative to the highest rate observed—in experiments conducted under conditions that were identical except for differences in temperature.

Note that, in both bacterial species, the enzyme has a distinct optimum or peak—a temperature at which it functions best. One of the bacterial species lives inside your gut, where the temperature is about 40°C, while the other lives in hot springs, where

(a) Enzymes from different organisms may function best at different temperatures.

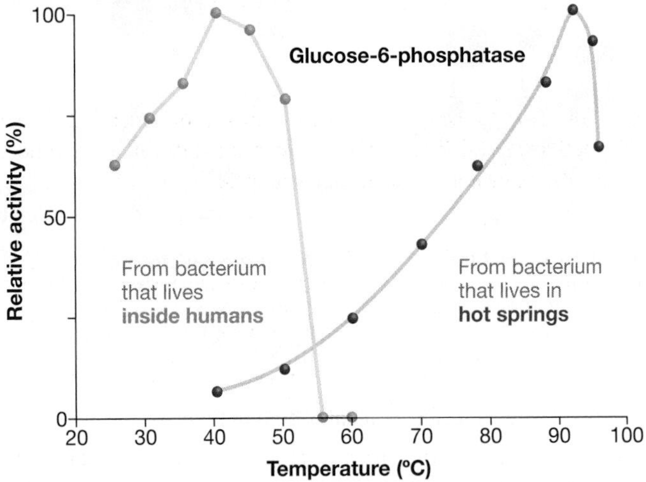

(b) Enzymes from different organisms may function best at different pHs.

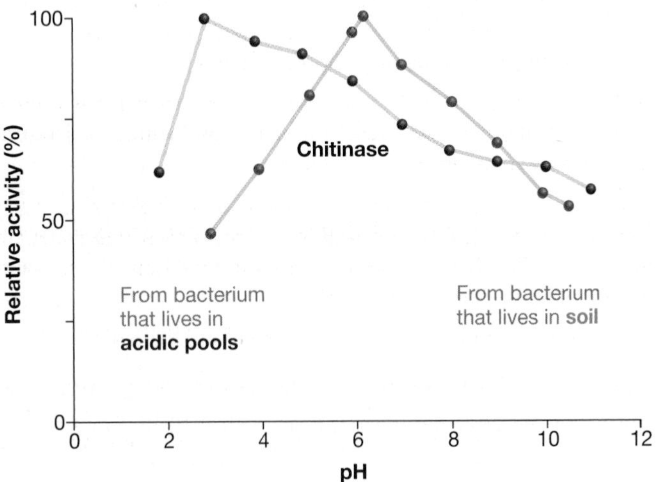

FIGURE 3.23 Enzymes Have an Optimal Temperature and pH. Enzymes are sensitive to changes in temperature **(a)** and pH **(b)**.

temperatures can be close to 100°C. The temperature optimum for the enzyme reflects these environments.

The two types of bacteria have different versions of the enzyme that differ in primary structure. Natural selection—the process introduced in Chapter 1—has favored different structures that have different functions. The enzymes are adaptations that allow each species to thrive at different temperatures.

Figure 3.23b makes the same point for pH. The enzyme in this graph, called chitinase, protects bacterial cells by digesting a molecule found in the cell walls of fungi that use bacteria as food. The data come from a species of bacterium that lives in acidic pools and a bacterial species that lives in the soil under palm trees.

Note that the organism that thrives in an acidic environment has a version of the enzyme that performs best at low pH; the organism that lives near palms has a version of the enzyme that functions best near neutral pH. Each enzyme is sensitive to changes in pH, but each species' version of the enzyme has a structure that allows it to function best at the pH of its environment.

To summarize, the rate of an enzyme-catalyzed reaction depends not only on substrate concentration and the enzyme's intrinsic affinity for the substrate but also on temperature and pH (and other factors). Temperature affects the movement of the substrates and enzyme; pH affects the enzyme's shape and reactivity.

Was the First Living Entity a Protein Catalyst?

Several observations argue that the answer to this question is yes. Experimental studies have shown that amino acids were likely to be abundant in the prebiotic soup, and that they could have polymerized to form small proteins. In addition, proteins are the most efficient catalysts known, and a self-replicating molecule had to act as a catalyst during the assembly and polymerization of its copy.

These observations support the hypothesis that the self-replicating molecule was a polypeptide. Indeed, several laboratories that are currently working to create life have focused on synthesizing a self-replicating protein.

To date, however, attempts to simulate the origin of life with proteins have not been successful. Although it is much too early to make definitive conclusions, most origin-of-life researchers are increasingly skeptical that life began with a protein. Their reasoning is that to make a copy of something, a mold or template is required. Proteins cannot furnish this information. Nucleic acids, in contrast, can. How they do so is the subject of Chapter 4.

CHECK YOUR UNDERSTANDING

If you understand that . . .

- Most proteins are enzymes, which make specific chemical reactions occur rapidly.
- Enzymes catalyze reactions by means of a three-step mechanism:
  Step 1 Binding of reactants in a precise orientation inside the active site.
  Step 2 Facilitation of the transition state, thus lowering the activation energy required for the reaction. This step often involves a shape change in the enzyme, resulting in an "induced fit" between active site and substrate. Enzyme cofactors are often required at this step.
  Step 3 Release of products, which do not bind tightly to the active site.
- The activity of enzymes is controlled through allosteric regulation or competitive inhibition.
- Different enzymes work at different rates, and enzymes are sensitive to changes in temperature and pH.

✓ You should be able to . . .

1. Explain why the precise orientation of reactants in the active site is important.
2. Explain why the shape change that occurs during allosteric regulation can change an enzyme's activity.

Answers are available in Appendix B.

CHAPTER 3 REVIEW

For media, go to the study area at www.masteringbiology.com 

Summary of Key Concepts

Most cell functions depend on proteins.

- In organisms, proteins function in catalysis, defense, movement, signaling, structural support, and transport of materials.
- Proteins can have diverse functions in cells because they have such diverse structures and chemical properties.

✔**You should be able to give an example of how a specific protein's structure is correlated with its function.**

Amino acids are the building blocks of proteins. Amino acids vary in structure and function because their side chains vary in composition.

- Amino acids have a central carbon bonded to an amino group, a hydrogen atom, a carboxyl group, and an R-group.
- The structure of the R-group affects the chemical reactivity and solubility of the amino acid.
- In proteins, amino acids are joined by a peptide bond between the carboxyl group of one amino acid and the amino group of a second amino acid.

✔**You should be able to explain why some R-groups make amino acids soluble versus insoluble in water.**

MB **Web Activity** Condensation and Hydrolysis Reactions

Proteins vary widely in structure. The structure of a protein can be analyzed at four levels that form a hierarchy—the amino acid sequence, substructures called α-helices and β-pleated sheets, interactions between amino acids that dictate a protein's overall shape, and combinations of individual proteins that make up larger, multiunit molecules.

- A protein's primary structure, or sequence of amino acids, is responsible for most of its chemical properties.
- Interactions that take place between carbonyl and amino groups in the same peptide-bonded backbone create secondary structures, which are stabilized primarily by hydrogen bonding.
- Interactions between R-groups found in the same polypeptide and between R-groups and the peptide-bonded backbone allow the protein to fold into a characteristic overall shape—its tertiary structure.
- In many cases, a complete protein consists of several different polypeptides, bonded together. The combination of polypeptides represents the protein's quaternary structure.

✔**You should be able to make a concept map (see BioSkills 8 in Appendix A) that relates the four levels of protein structure to one another and to how proteins function as catalysts. Your map should include the following boxed terms: Primary structure, secondary structure, tertiary structure, quaternary structure, transition state, activation energy, amino acid sequence, active site, R-groups, helices and sheets, 3D shape.**

In cells, most proteins are enzymes that function as catalysts. Chemical reactions occur much faster when catalyzed by enzymes. During enzyme catalysis, the reactants bind to an enzyme's active site in a way that allows the reaction to proceed efficiently.

- Enzymes are protein catalysts that lower activation energy by stabilizing the transition state of the reaction. This speeds reaction rates. The enzyme itself is unchanged by the reaction.
- Catalysis takes place at the enzyme's active site, which has unique chemical properties and a distinctive size and shape. As a result, most enzymes catalyze a specific reaction.
- As a group, enzymes are able to catalyze many types of reactions because the chemical and physical structures of their active sites are so diverse. This diversity is due to the variety of amino acids and the four levels of protein structure.
- Many enzymes function only with the help of cofactors.
- Virtually all enzyme activity in cells is regulated. In most cases, regulation occurs when molecules bind at the active site or at locations on the protein that induce a change in the size or shape of the active site.
- The rate at which enzymes work depends on substrate concentration, their affinity for the substrate, temperature, and pH.

✔**You should be able to explain why a catalyst changes only the activation energy for a reaction—not the overall free energy change.**

MB **Web Activity** Activation Energy and Enzymes

Questions

✔TEST YOUR KNOWLEDGE

Answers are available in Appendix B

1. What two functional groups are present on every amino acid?
   a. a carbonyl group and a carboxyl group
   b. an amino group and a carbonyl group
   c. an amino group and a hydroxyl group
   d. an amino group and a carboxyl group

2. Twenty different amino acids are found in the proteins of cells. What distinguishes these molecules?
   a. the location of their carboxyl group
   b. the location of their amino group
   c. the composition of their side chains, or R-groups
   d. their ability to form peptide bonds

3. What determines the primary structure of a polypeptide?
   a. its sequence of amino acids
   b. hydrogen bonds that form between carboxyl and amino groups on different residues
   c. hydrogen bonds and other interactions between side chains
   d. the number, identity, and arrangement of polypeptides that make up a protein

4. In a polypeptide, what is most responsible for the secondary structure called an α-helix?
   a. the sequence of amino acids
   b. hydrogen bonds that form between carbonyl and amino groups on different residues
   c. hydrogen bonds and other interactions between side chains
   d. the number, identity, and arrangement of polypeptides that make up a protein
5. What is a transition state?
   a. the complex formed as covalent bonds are being broken and re-formed during a reaction
   b. the place where an allosteric regulatory molecule binds to an enzyme
   c. an interaction between reactants with high kinetic energy, due to high temperature
   d. the shape adopted by an enzyme that has an inhibitory molecule bound at its active site
6. By convention, biologists write the sequence of amino acids in a polypeptide in which direction?
   a. carboxy- to amino-terminus
   b. amino- to carboxy-terminus
   c. polar residues to nonpolar residues
   d. charged residues to uncharged residues

✔ TEST YOUR UNDERSTANDING

Answers are available in Appendix B

1. Explain the lock-and-key model of enzyme activity. What was incorrect about this model?
2. Isoleucine, valine, leucine, phenylalanine, and methionine are amino acids with highly hydrophobic side chains. Suppose a section of a protein contains a long series of these hydrophobic residues. How would you expect this portion of the protein to behave when the molecule is in aqueous solution?
3. Compare and contrast competitive inhibition and allosteric regulation.
4. Does it take an input of energy for polymerization reactions to proceed, or do they occur spontaneously? Why or why not?
5. A major theme in this chapter is that the structure of molecules correlates with their function. Use this theme to explain why proteins can perform so many different functions in organisms and why proteins as a group are such effective catalysts.
6. Explain why temperature and pH affect enzyme function.

✔ APPLYING CONCEPTS TO NEW SITUATIONS

Answers are available in Appendix B

1. Researchers are searching for life on other planets. Is it reasonable to expect that protein-based life-forms could exist in extraterrestrial environments that are very hot or highly acidic? Why or why not?
2. Recently, researchers were able to measure movement that occurred in a single amino acid in an enzyme as reactions were taking place in its active site. The amino acid that moved was located in the active site, and the rate of movement correlated closely with the rate at which the reaction was taking place. Discuss the significance of these findings, using the information in Figures 3.19 and 3.20.
3. Researchers can analyze the atomic structure of enzymes during catalysis. In one recent study, investigators found that the transition state included the formation of a free radical, and that a coenzyme bound to the active site donated an electron to help stabilize the free radical. How would the reaction rate and the stability of the transition state change if the coenzyme were not available?
4. Some of the most effective drugs against HIV, the virus that causes AIDS, are competitive inhibitors of an HIV enzyme. But many HIV strains have now evolved resistance to these drugs—their enzymes are not affected by the drugs. Predict which parts of the enzymes changed. Explain your reasoning.

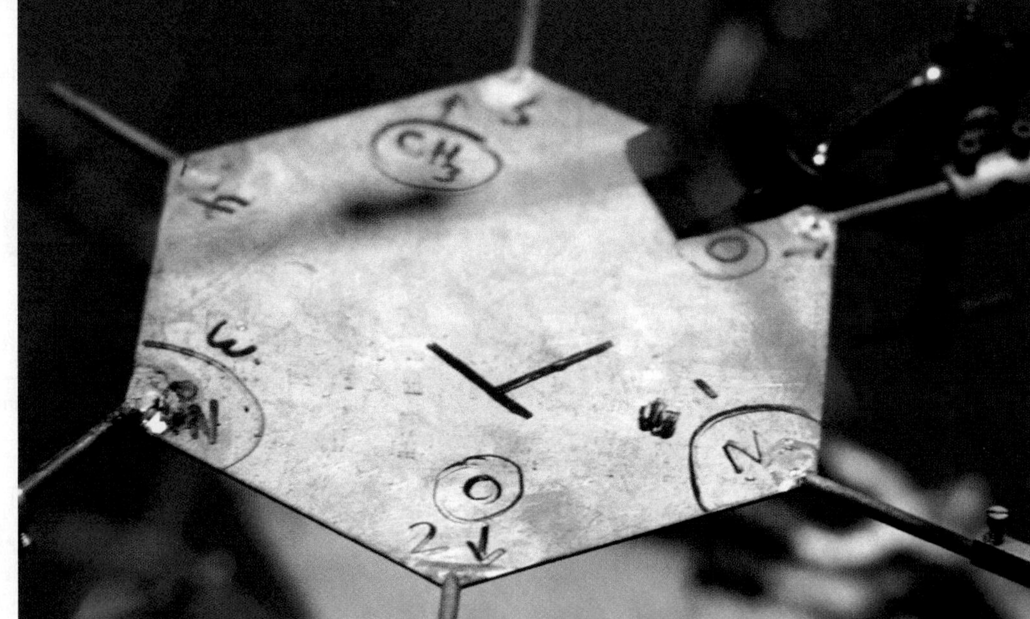

This is part of the sheet-metal-and-wire model that James Watson and Francis Crick used to figure out the secondary structure of DNA. The large "T" stands for the nitrogen-containing base thymine.

Nucleic Acids and the RNA World 4

Chapter 3 began with experimental evidence that chemical evolution produced the monomers called amino acids and their polymers called proteins. But the chapter ended by stating that even though proteins are the workhorse molecules of today's cells, relatively few researchers favor the hypothesis that life began as a protein molecule. Instead, the vast majority of biologists contend that life began as a polymer called a nucleic acid—specifically, a molecule of ribonucleic acid (RNA). This proposal is called the RNA world hypothesis.

The RNA world hypothesis contends that chemical evolution led to the existence of an RNA molecule that could make a copy of itself. Once this self-replicating molecule existed, chance errors in the copying process would create variations that would undergo natural selection—the evolutionary process introduced in Chapter 1. Chemical evolution was over, and biological evolution was off and running.

To test this hypothesis, several labs around the world are working to synthesize a self-replicating RNA molecule in the laboratory. If they succeed, they will have created a life-form in a test tube.

This chapter is focused on the structure and function of nucleic acids. Let's begin with an analysis of nucleic acids as monomers and end with current research efforts to create a self-replicating molecule.

KEY CONCEPTS

- Nucleotides consist of a sugar, a phosphate group, and a nitrogen-containing base. Ribonucleotides polymerize to form RNA. Deoxyribonucleotides polymerize to form DNA.
- DNA's primary structure consists of a sequence of nitrogen-containing bases. Its secondary structure consists of two DNA strands, running in opposite directions, that are held together by complementary base pairing and twisted into a double helix. DNA's structure allows organisms to store and replicate the information needed to grow and reproduce.
- RNA's primary structure consists of a sequence of nitrogen-containing bases. Its secondary structure includes short regions of double helices and structures called hairpins.
- Because RNA molecules can carry information as well as catalyze chemical reactions, it is likely that RNA was the first self-replicating molecule and a forerunner to the first life-form.

4.1 What Is a Nucleic Acid?

Nucleic acids are polymers, just as proteins are polymers. But instead of being made up of monomers called amino acids, **nucleic acids** are made up of monomers called **nucleotides**.

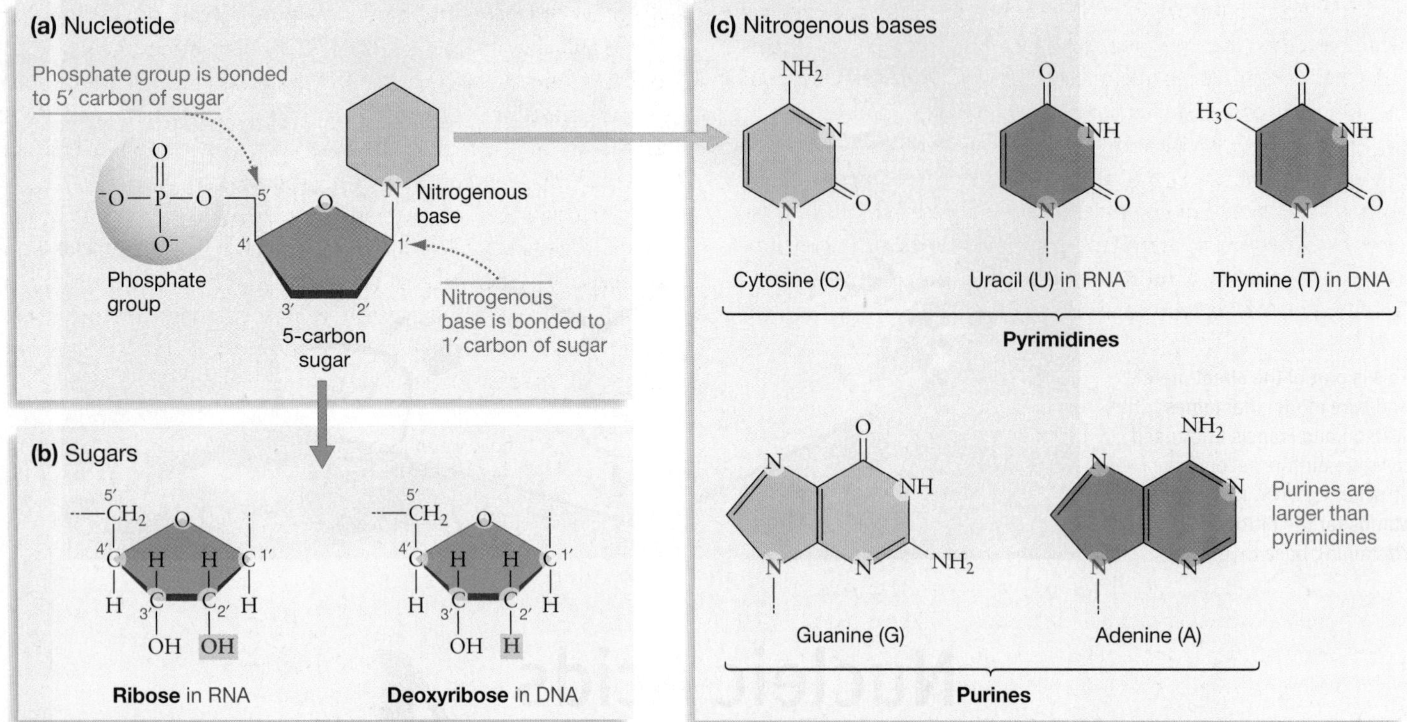

FIGURE 4.1 The General Structure of a Nucleotide. Note that in the bases, the nitrogen that bonds to the sugar is colored blue.

Figure 4.1a diagrams the three components of a nucleotide: **(1)** a phosphate group, **(2)** a sugar, and **(3)** a nitrogenous (nitrogen-containing) base. The phosphate is bonded to the sugar molecule, which in turn is bonded to the nitrogenous base. A **sugar** is an organic compound with a carbonyl group ($\rangle C=O$) and several hydroxyl ($-OH$) groups.

Notice that the prime symbols (′) in Figure 4.1 indicate that the carbon being referred to is part of the sugar—not of the attached nitrogenous base. The phosphate group in a nucleotide is attached to the 5′ carbon.

Although a wide variety of nucleotides are found in living cells, origin-of-life researchers concentrate on two types: **ribonucleotides** and **deoxyribonucleotides**. In ribonucleotides, the sugar is ribose; in deoxyribonucleotides, it is deoxyribose. As **Figure 4.1b** shows, these two sugars differ by a single atom. Ribose has an $-OH$ group bonded to the 2′ carbon. Deoxyribose has an H instead at the same location. Note that deoxy means "lacking oxygen." Deoxyribonucleotides lack oxygen at the 2′ carbon.

Cells today have four different ribonucleotides, each of which contains a different nitrogenous base. These bases, diagrammed in **Figure 4.1c**, belong to structural groups called **purines** and **pyrimidines**. Ribonucleotides include the purines adenine (A) and guanine (G), and the pyrimidines cytosine (C) and uracil (U).

Similarly, four different deoxyribonucleotides are found in cells today and are distinguished by the structure of their nitrogenous base. Like ribonucleotides, deoxyribonucleotides include adenine, guanine, and cytosine. However, instead of uracil, a closely related pyrimidine called thymine (T) occurs in deoxyribonucleotides (Figure 4.1c).

✔You should be able to diagram a ribonucleotide and a deoxyribonucleotide, using a ball for the phosphate group, a pentagon to represent the sugar subunit, and a hexagon to represent the nitrogenous base. Label the **2′**, **3′**, and **5′** carbons on the sugar molecule, and add the atoms that are bonded to the **2′** carbon.

Could Chemical Evolution Result in the Production of Nucleotides?

Based on data presented in Chapter 3, most researchers contend that amino acids were abundant early in Earth's history. As yet, however, no one has observed the formation of a nucleotide via chemical evolution. The problem lies with mechanisms for synthesizing the sugar and nitrogenous base components of these molecules. Let's consider each issue in turn.

Laboratory simulations have shown that many sugars can be synthesized readily under conditions that mimic the prebiotic soup. Specifically, when formaldehyde (H_2CO) molecules are heated in solution, they react with one another to form almost all the sugars that have five or six carbons. (These are called pentoses and hexoses, respectively.)

In modern experiments, the various pentoses and hexoses are produced in approximately equal amounts, but ribose would have had to predominate for RNA or DNA to form in the prebiotic soup. How ribose came to be the dominant sugar during chemical evolution (i.e., what selective process was at work) is still a mystery. Origin-of-life researchers refer to this issue as the "ribose problem."

The origin of the pyrimidines is equally challenging. Simply put, origin-of-life researchers have yet to discover a plausible mechanism for the synthesis of pyrimidines (cytosine, uracil, and thymine) prior to the origin of life. Purines, in contrast, are readily synthesized by reactions among hydrogen cyanide (HCN) molecules. For example, both adenine and guanine have been found in the solutions recovered after spark-discharge experiments.

The ribose problem and the questions about the origin of pyrimidine bases are two of the most serious challenges that remain for the theory of chemical evolution. Research on these issues continues. In the meantime, let's consider the next question: Once nucleic acids formed, how did they polymerize to form RNA and DNA? This question has an answer.

How Do Nucleotides Polymerize to Form Nucleic Acids?

Nucleic acids form when nucleotides polymerize. As **Figure 4.2** shows, the polymerization reaction involves the formation of a bond between the phosphate group of one nucleotide and the hydroxyl group of the sugar component of another nucleotide. The result of this condensation reaction is called a **phosphodiester linkage**, or a phosphodiester bond.

In Figure 4.2, a phosphodiester linkage joins the 5′ carbon on the ribose of one nucleotide to the 3′ carbon on the ribose of the other. When the nucleotides involved contain the sugar ribose, the polymer that is produced is called **ribonucleic acid**, or simply **RNA**. If the nucleotides contain the sugar deoxyribose instead, then the resulting polymer is **deoxyribonucleic acid**, or **DNA**.

DNA AND RNA STRANDS ARE DIRECTIONAL **Figure 4.3** shows how the chain of phosphodiester linkages in a nucleic acid acts as a backbone, analogous to the peptide-bonded backbone found in proteins.

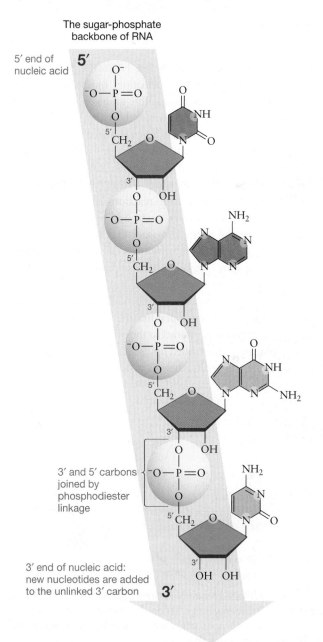

The sugar-phosphate backbone of a nucleic acid is directional, as is the peptide-bonded backbone of a polypeptide. In a strand of RNA or DNA, one end has an unlinked 5′ carbon while the other end has an unlinked 3′ carbon—meaning a carbon that is not linked to another nucleotide. By convention, the sequence of bases found in an RNA or DNA strand is always written in the 5′→3′ direction. (The system is logical because in cells, RNA and DNA are always synthesized in this direction. Bases are added at the 3′ end of the growing molecule.)

The sequence of nitrogenous bases forms the primary structure of the molecule, analogous to the sequence of amino acids in a polypeptide. When biologists write the primary structure of a stretch of DNA or RNA, they simply list the sequence of nucleotides in the 5′ to 3′ direction, using their single-letter abbreviations. For example, a six-base-long DNA sequence might be ATTAGC.

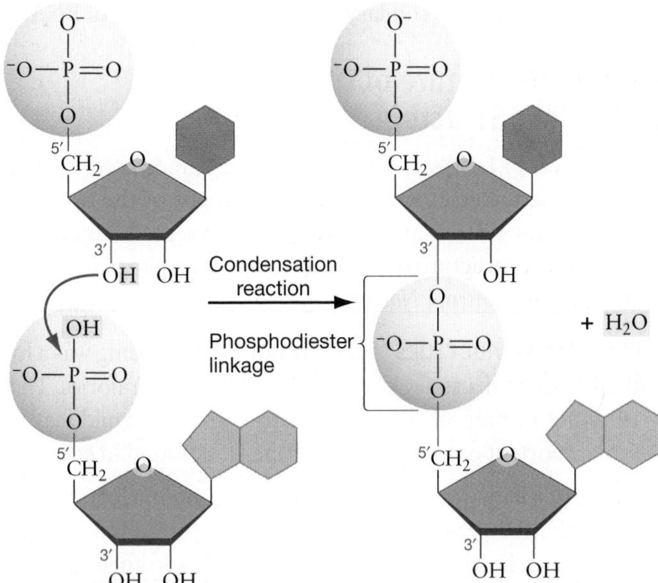

FIGURE 4.2 Nucleotides Polymerize via Phosphodiester Linkages. Ribonucleotides can polymerize via condensation reactions. The resulting phosphodiester linkage connects the 3′ carbon of one ribonucleotide and the 5′ carbon of another ribonucleotide.

FIGURE 4.3 RNA Has a Sugar-Phosphate Backbone.

✔**EXERCISE** Identify the four bases in this RNA strand, using Figure 4.1c as a key. Then write down the base sequence, starting at the 5′ end.

POLYMERIZATION IS AN ENDERGONIC PROCESS In cells, the polymerization reactions that form nucleotides are catalyzed by enzymes. Like other polymerization reactions, the process is endergonic.

Polymerization can take place in cells because the free energy of the nucleotide monomers is first raised by reactions that add two phosphate groups to the ribonucleotides or deoxyribonucleotides, creating nucleoside triphosphates (**Figure 4.4**).[1] Molecules that have phosphate groups attached in this way are said to be **phosphorylated.**

This is a key point, and one that you will encounter again and again in this text: The addition of one or more phosphate groups raises the potential energy of substrate molecules enough to make an otherwise endergonic reaction possible. (Chapter 9 explains how this happens.) In the case of nucleic acid polymerization, researchers refer to the phosphorylated nucleotides as "activated."

COULD NUCLEIC ACIDS FORM IN THE PREBIOTIC SOUP? During chemical evolution, activated nucleotides may have polymerized on the surfaces of clay-sized (very fine grained) mineral particles. In a suite of experiments analogous to those discussed in Chapter 3 for protein synthesis, researchers have produced RNA molecules by incubating activated ribonucleotides with tiny mineral particles.

In one experiment, researchers isolated the clay particles after a day of incubation and then added a fresh batch of activated nucleotides. They repeated this reaction-isolation-reaction sequence for a total of 14 days, or 14 additions of fresh ribonucleotides. At the end of the two-week experiment, they analyzed the mineral particles, using techniques called **gel electrophoresis** and **autoradiography** (see **BioSkills 9** in Appendix A), and found RNA molecules up to 40 nucleotides long.

More recent work has shown that nucleotides can polymerize without being phosphorylated, if heating provides a source of energy and the monomers interact in the presence of nonpolar, non-water-soluble molecules called lipids, introduced in Chapter 6.

Based on these results, there is a strong consensus that if ribonucleotides and deoxyribonucleotides were able to form during chemical evolution, they would be able to polymerize and form DNA and RNA. Now, what would these nucleic acids look like, and what could they do?

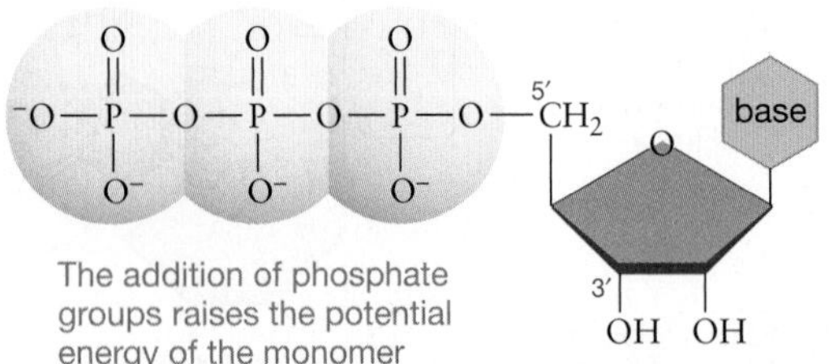
$^{-}O-P(=O)(O^{-})-O-P(=O)(O^{-})-O-P(=O)(O^{-})-O-CH_2$
5′
base
3′
OH OH
The addition of phosphate groups raises the potential energy of the monomer

FIGURE 4.4 Activated Monomers Drive Endergonic Polymerization Reactions. Polymerization reactions involving ribonucleotides are endergonic, but polymerization reactions involving ribonucleoside triphosphates are exergonic.

[1]A molecule consisting of a sugar and one of the bases in Figure 4.1c is called a nucleoside (a nucleotide is a sugar, a base, and one or more phosphate groups). Thus, a sugar attached to a base and three phosphate groups is called a nucleoside triphosphate.

CHECK YOUR UNDERSTANDING

If you understand that . . .

- Nucleotides are monomers that consist of a sugar, a phosphate group, and a nitrogen-containing base.
- Nucleotides polymerize to form nucleic acids through formation of phosphodiester linkages between the 3′ carbon on one nucleotide and the 5′ carbon on another.
- During polymerization, nucleotides are added only to the 3′ end of a nucleic acid strand.

✓ You should be able to . . .

Draw a simplified diagram of the phosphodiester linkage between two nucleotides, indicate the 5′ to 3′ polarity, and mark where the next nucleotide would be added to the growing chain.

Answers are available in Appendix B.

4.2 DNA Structure and Function

The primary structure of nucleic acids is somewhat similar to the primary structure of proteins. Proteins have a peptide-bonded backbone with a series of R-groups that extend from it. DNA and RNA molecules have a sugar-phosphate backbone, created by phosphodiester linkages, and a sequence of any of four nitrogenous bases that extend from it.

DNA and RNA also have secondary structure. But unlike the α-helices and β-pleated sheets of proteins, which are formed by hydrogen bonding between carbonyl and amino groups in the peptide-bonded backbone, the secondary structure of nucleic acids is formed by hydrogen bonding between nitrogenous bases.

Let's analyze the secondary structure and function of DNA first, and then dig into the secondary structure and function of RNA.

What Is the Nature of DNA's Secondary Structure?

The solution to DNA's secondary structure, announced in 1953, ranks among the great scientific breakthroughs of the twentieth century. James Watson and Francis Crick presented a model for the secondary structure of DNA in a one-page paper published in the scientific journal *Nature.*

EARLY DATA PROVIDE CLUES Watson and Crick's finding was a hypothesis based on a series of results from other laboratories. They were trying to propose a secondary structure that could explain several important observations about the DNA found in cells:

- Chemists had worked out the structure of nucleotides and knew that DNA polymerized through the formation of phosphodiester linkages. Thus, Watson and Crick knew that the molecule had a sugar-phosphate backbone.
- By analyzing the nitrogenous bases in DNA samples from different organisms, Erwin Chargaff had established two empirical rules: (1) The number of purines in a given DNA molecule

is equal to the number of pyrimidines, and (2) the number of T's and A's in DNA are equal, and the number of C's and G's in DNA are equal.

- By bombarding DNA with X-rays and analyzing how it scattered the radiation, Rosalind Franklin and Maurice Wilkins had calculated the distances between groups of atoms in the molecule (see **Bioskills 10** in Appendix A for an introduction to this technique, called **X-ray crystallography**). The scattering patterns showed that three distances were repeated many times: 0.34 nanometer (nm), 2.0 nm, and 3.4 nm. Because the measurements repeated, the researchers inferred that DNA molecules had a regular and repeating structure. The pattern of X-ray scattering suggested that the molecule was helical, or spiral, in nature.

FIGURE 4.5 Building a Physical Model of DNA Structure. Watson (left) and Crick (right) represented the arrangement of the four deoxyribonucleotides in a double-helical arrangement, using metal plates and wires with precise lengths and geometries.

Based on this work, understanding DNA's structure boiled down to understanding the nature of the helix involved. What type of helix would have a sugar-phosphate backbone and explain both Chargaff's rules and the Franklin-Wilkins measurements?

DNA STRANDS ARE ANTIPARALLEL Watson and Crick began by analyzing the size and geometry of deoxyribose, phosphate groups, and the nitrogenous bases. The bond angles and measurements suggested that the distance of 2.0 nm probably represented the width of the helix, and that 0.34 nm was likely to be the distance between bases stacked in a spiral.

How could they make sense of Chargaff's rules and the 3.4-nm distance, which appeared to be exactly 10 times the distance between a single pair of bases?

To solve this problem, Watson and Crick constructed a series of physical models such as the one pictured in **Figure 4.5**. The models allowed them to tinker with different types of helical configurations. After many false starts, something clicked:

- They arranged two strands of DNA side by side and running in opposite directions—meaning that one strand ran in the $5' \rightarrow 3'$ direction while the other strand was oriented $3' \rightarrow 5'$. Strands with this orientation are said to be **antiparallel**.
- If the antiparallel strands are twisted together to form a double helix, the coiled sugar-phosphate backbones end up on the outside of the spiral and the nitrogenous bases on the inside.
- For the bases from each backbone to fit in the interior of the 2.0-nm-wide structure, they have to form purine-pyrimidine pairs (see **Figure 4.6a**). This is key: The pairing allows hydrogen bonds to form between certain purines and pyrimidines. Adenine forms hydrogen bonds with thymine, and guanine forms hydrogen bonds with cytosine (**Figure 4.6b**).

(a) Only purine-pyrimidine pairs fit inside the double helix.

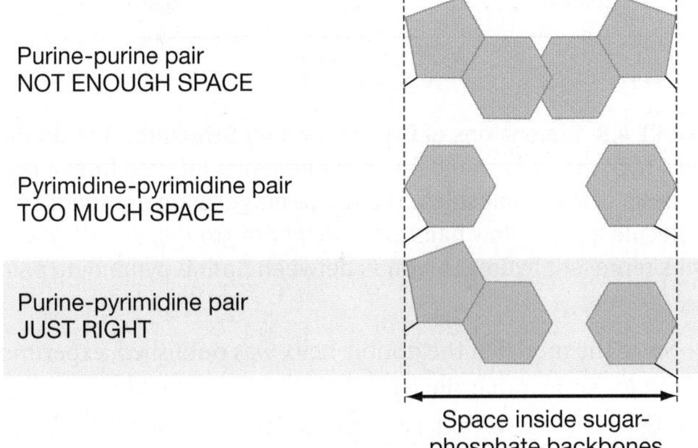

(b) Hydrogen bonds form between G-C pairs and A-T pairs.

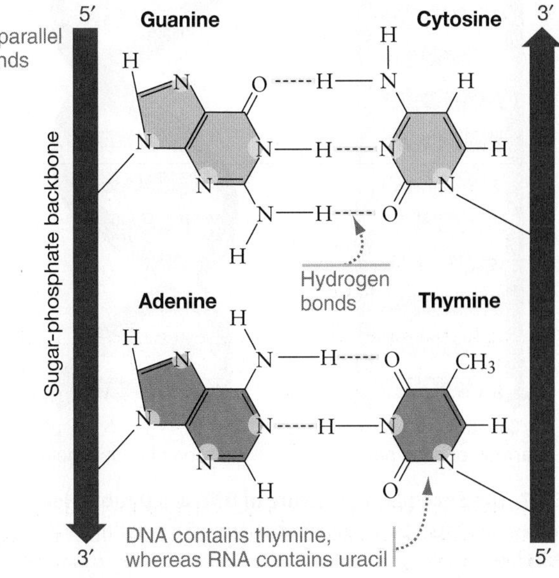

FIGURE 4.6 Complementary Base Pairing Is Based on Hydrogen Bonding.

- The A-T and G-C bases were said to be complementary. Two hydrogen bonds form when A and T pair, but three hydrogen bonds form when G and C pair. As a result, the G-C interaction is slightly stronger than the A-T bond. In contrast, A-C and G-T pairs allow no or only one hydrogen bond.

Watson and Crick had discovered **complementary base pairing**. In fact, the term **Watson-Crick pairing** is now used interchangeably with the phrase complementary base pairing.

THE DOUBLE HELIX **Figure 4.7** shows how antiparallel strands of DNA form when complementary bases line up and form hydrogen bonds. As you study the figure, notice that DNA is put together like a ladder whose ends have been twisted in opposite directions. The sugar-phosphate backbone forms the supports of the ladder; the base pairs represent the rungs of the ladder. The twisting occurs because it allows the nitrogenous bases to line up in a way that makes hydrogen bonding possible between them.

The nitrogenous bases in the middle of the DNA helix also stack tightly on top of each other. This tight packing forms a hydrophobic interior that is difficult to break apart. But the molecule as a whole is hydrophilic and water-soluble because the sugar-phosphate backbones, which face the exterior of the molecule, are negatively charged because of the negative charges on the phosphate groups. As a result, they interact with water.

It's also important to note that the outside of the helical DNA molecule forms two types of grooves. These grooves differ in size. The larger of the two is known as the major groove, and the smaller one is known as the minor groove. **Figure 4.8** highlights these grooves and illustrates how DNA's secondary structure explains the measurements observed by Franklin and Wilkins.

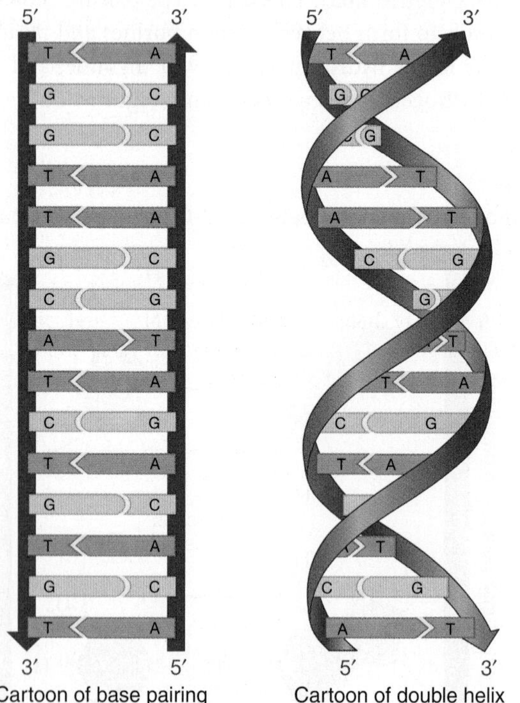

FIGURE 4.7 The Secondary Structure of DNA Is a Double Helix. Complementary base pairing is responsible for twisting DNA into a double helix.

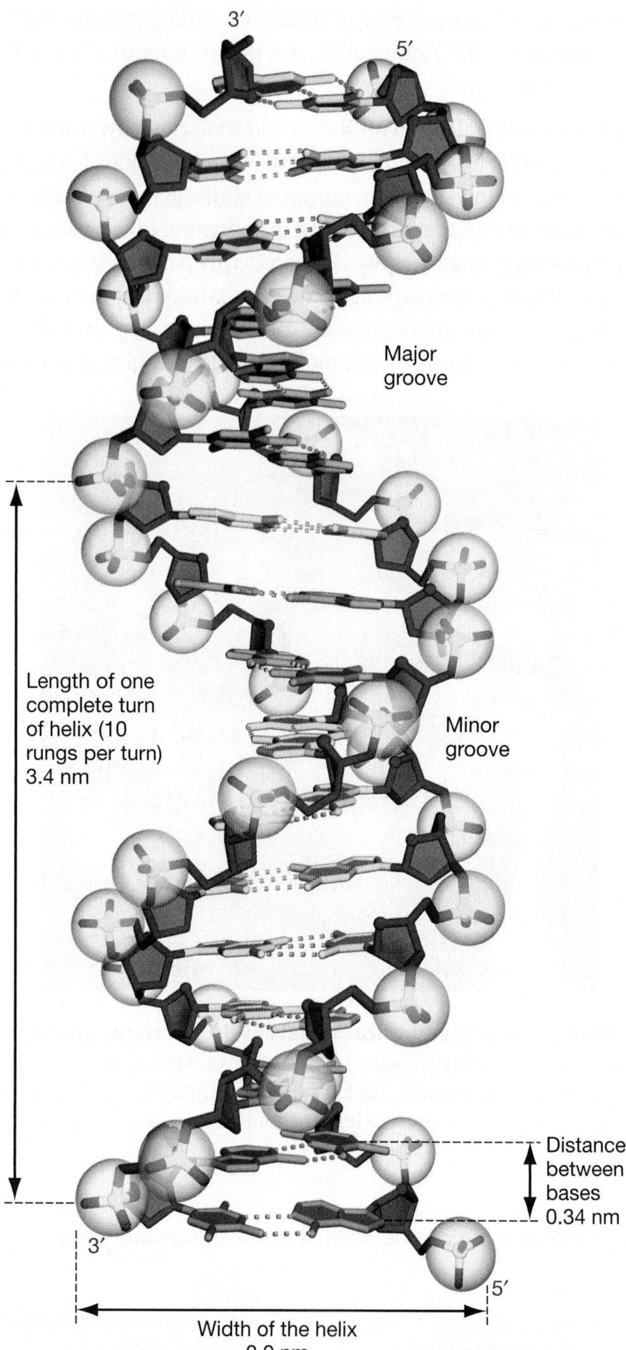

FIGURE 4.8 Dimensions of DNA Secondary Structure. The double-helix hypothesis explains the measurements inferred from X-ray analysis of DNA molecules. The red pentagons are deoxyribose molecules; the yellow balls are phosphate groups, and the dashed lines represent hydrogen bonds between purine-pyrimidine pairs.

Since the model of the double helix was published, experimental tests have shown that the hypothesis is correct in almost every detail. DNA's secondary structure consists of two antiparallel strands twisted into a double helix. The molecule is stabilized by hydrophobic interactions in its interior and by hydrogen bonding between the complementary base pairs A-T and G-C. ✔You should be able to explain why no secondary structure forms when two DNA strands align in a parallel—instead of antiparallel—fashion.

Now the question is, How does this secondary structure affect the molecule's function?

DNA Functions as an Information-Containing Molecule

Watson and Crick's model created a sensation for a simple reason: It revealed how DNA could store and transmit biological information. In literature, information consists of letters on a page. In music, information is composed of the notes on a staff. But inside cells, information consists of a sequence of nucleotides in a nucleic acid. The four nitrogenous bases function like letters of the alphabet. A sequence of bases is like the sequence of letters in a word—it has meaning.

In all cells that have been examined to date, from tiny bacteria to gigantic redwood trees, DNA carries the information required for the organism's growth and reproduction. DNA is the basis for one of the five attributes of life introduced in Chapter 1: storing and processing genetic information. Exploring how hereditary information is encoded and translated into action is the heart of Chapters 15 through 18.

Here, however, our focus is on how life began. The theory of chemical evolution holds that life began as a self-replicating molecule—a molecule that could make a copy of itself. **Figure 4.9** shows how a copy of DNA can be made by complementary base pairing.

Step 1 Heating or enzyme-catalyzed reactions cause the double helix to separate.

Step 2 Free deoxyribonucleotides form hydrogen bonds with complementary bases on the original strand of DNA—also called a **template strand**. As they do, their sugar-phosphate groups form phosphodiester linkages to create a new strand—also called a **complementary strand**. Note that the $5' \rightarrow 3'$ directionality of the complementary strand is opposite that of the template strand.

Step 3 Complementary base pairing allows each strand of a DNA double helix to be copied exactly, producing two daughter molecules.

Watson and Crick ended their paper on the double helix with one of the classic understatements in the scientific literature: "It has not escaped our notice that the specific pairing we have postulated immediately suggests a possible copying mechanism." Here's the key insight: DNA's primary structure serves as a mold or template for the synthesis of a complementary strand. DNA contains the information required for a copy of itself to be made.

To review DNA's primary and secondary structure and the difference between ribonucleotides and deoxyribonucleotides, go to the study area at *www.masteringbiology.com* and view the copying mechanism in action.

Web Activity Structure of RNA and DNA

DNA copying is the basis for a second of the five attributes of life introduced in Chapter 1: replication. But can DNA self-replicate? In today's cells and in laboratory experiments, the answer is no. Instead, the molecule is copied through a complicated series of energy-demanding reactions, catalyzed by a large suite of enzymes. Why can't DNA catalyze these reactions itself?

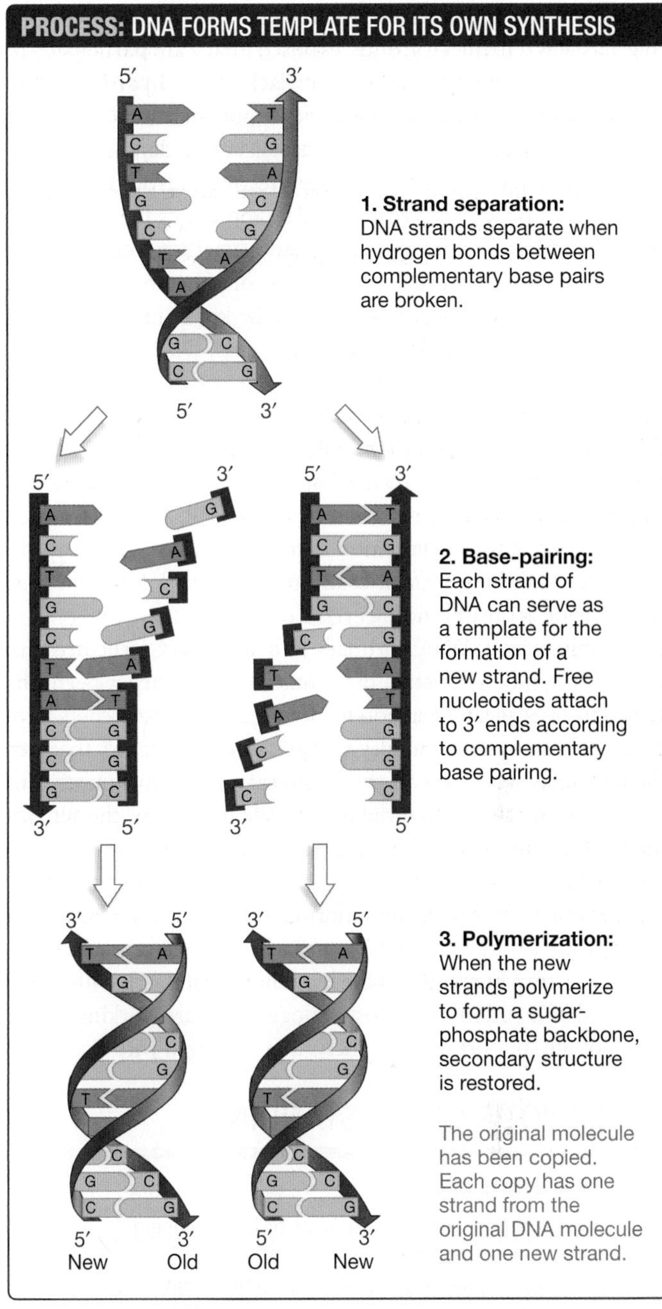

PROCESS: DNA FORMS TEMPLATE FOR ITS OWN SYNTHESIS
5′
3′
1. Strand separation:
DNA strands separate when hydrogen bonds between complementary base pairs are broken.
2. Base-pairing:
Each strand of DNA can serve as a template for the formation of a new strand. Free nucleotides attach to 3′ ends according to complementary base pairing.
3. Polymerization:
When the new strands polymerize to form a sugar-phosphate backbone, secondary structure is restored.
The original molecule has been copied. Each copy has one strand from the original DNA molecule and one new strand.
New
Old
Old
New

FIGURE 4.9 Making a Copy of DNA. If new bases are added to each of the two strands of DNA via complementary base pairing, a copy of the DNA molecule can be produced.

✔**QUESTION** When double-stranded DNA is heated to 95°C, the bonds between complementary base pairs break and single-stranded DNA results. Considering this observation, is the reaction shown in step 1 endergonic or exergonic?

Is DNA a Catalytic Molecule?

The DNA double helix is highly structured. It is regular, symmetric, and held together by hydrogen bonding, hydrophobic interactions, and phosphodiester linkages. In addition, the molecule has few chemical groups exposed that can participate in chemical reactions. For example, the lack of a 2′ hydroxyl group on each deoxyribonucleotide makes the polymer much less reactive than RNA, and thus much more resistant to degradation.

Intact stretches of DNA have been recovered from fossils that are tens of thousands of years old. The molecules have the same sequence of bases as the organisms had when they were alive, despite death and exposure to a wide array of pH, temperature, and chemical conditions. DNA's stability is the key to its effectiveness as a reliable information-bearing molecule. DNA's structure is consistent with its function in cells.

The orderliness and stability that make DNA such a dependable information repository make it extraordinarily inept at catalysis, however. Recall from Chapter 3 that enzyme function is based on a specific binding event between a substrate and a protein catalyst. Thanks to variation in reactivity among R-groups in amino acids and the enormous diversity of shapes found in proteins, a wide array of binding events can occur.

In comparison, DNA's primary and secondary structures are simple. It is not surprising, then, that DNA has never been observed to catalyze any reaction in any organism. Researchers have been able to construct single-stranded DNA molecules that can catalyze some reactions in the laboratory, but the number and diversity of reactions involved is a minute fraction of the activity catalyzed by proteins.

In short, DNA furnishes an extraordinarily stable template for copying itself and for storing information encoded in a sequence of bases. But owing to its inability to act as an effective catalyst, virtually no researchers support the hypothesis that the first life-form consisted of DNA. Instead, most biologists who are working on the origin of life support the hypothesis that life began with RNA.

CHECK YOUR UNDERSTANDING

If you understand that . . .

- DNA's primary structure consists of a sequence of deoxyribonucleotides.
- DNA's secondary structure consists of two DNA molecules that run in opposite orientations to each other. The two strands are twisted into a double helix, and they are held together through hydrogen bonds between A-T and G-C pairs and hydrophobic interactions among atoms inside the helix.
- The sequence of deoxyribonucleotides in DNA contains information, much like the sequence of letters and punctuation on this page. Owing to complementary base pairing, each DNA strand also contains the information required to form its complementary strand.

✔ You should be able to . . .

Make a sketch of a DNA molecule and label the sugar-phosphate backbones, the hydrogen bonds between complementary bases, the location of hydrophobic interactions that stabilize the helix, and the orientation of each strand.

Answers are available in Appendix B.

4.3 RNA Structure and Function

A self-replicating molecule has to perform two key functions: it has to carry information and perform catalysis. At first glance, these two functions appear to conflict. Information storage requires regularity and stability; catalysis requires variation in chemical composition and variation and flexibility in shape.

How is it possible for a molecule to do both? The answer is structure.

Structurally, RNA Differs from DNA

Recall that most proteins have four levels of structure: the primary sequence of amino acids, secondary folds that are stabilized by hydrogen bonding between atoms in the peptide-bonded backbone, tertiary folds that are stabilized by interactions involving R-groups, and an overall or quaternary shape consisting of multiple polypeptides.

DNA has only primary and secondary structure. But RNA, like proteins, can have up to four levels of structure.

PRIMARY STRUCTURE Like DNA, RNA has a primary structure consisting of a sugar-phosphate backbone formed by phosphodiester linkages and, extending from that backbone, a sequence of four types of nitrogenous bases. But it's important to recall two significant differences between these nucleic acids:

1. The sugar in the sugar-phosphate backbone of RNA is ribose, not deoxyribose as in DNA.
2. The pyrimidine base thymine does not exist in RNA. Instead, RNA contains the closely related pyrimidine base uracil.

The first point is critical because the hydroxyl (—OH) group on the 2′-carbon of ribose is much more reactive than the hydrogen atom on the 2′-carbon of deoxyribose. This functional group can participate in reactions that tear the polymer apart. The presence of this —OH group makes RNA much less stable than DNA.

SECONDARY STRUCTURE Like DNA molecules, most RNA molecules have secondary structure that results from complementary base pairing between purine and pyrimidine bases. In RNA, adenine forms hydrogen bonds only with uracil, and guanine again forms hydrogen bonds with cytosine. (Guanine can also bond with uracil, but it does so much less effectively than with cytosine.) Thus, the complementary base pairs in RNA are A-U and G-C. Three hydrogen bonds form between guanine and cytosine, but only two form between adenine and uracil.

How do the secondary structures of RNA and DNA differ? In the vast majority of cases, the purine and pyrimidine bases in RNA undergo hydrogen bonding with complementary bases on the same strand, rather than forming hydrogen bonds with complementary bases on a different strand, as in DNA.

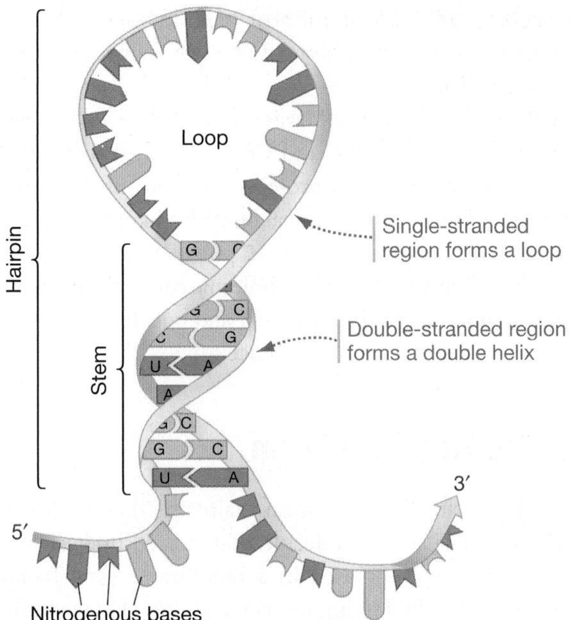

FIGURE 4.10 Complementary Base Pairing and Secondary Structure in RNA: Stem-and-Loop Structures. This RNA molecule has secondary structure. The double-stranded "stem" and single-stranded "loop" form a hairpin. The bonded bases in the stem are antiparallel, meaning that they are oriented in opposite directions.

Figure 4.10 shows how the within-strand pairing works. The key is that when bases on one part of an RNA strand fold over and align with ribonucleotides on another segment of the same strand, the two sugar-phosphate strands are antiparallel. In this orientation, hydrogen bonding between complementary bases results in a stable double helix.

If the section where the fold occurs includes a large number of unbonded bases, then the stem-and-loop configuration shown in Figure 4.10 results. This type of secondary structure is called a **hairpin**. Several other types of RNA secondary structures are possible, each involving a different length and arrangement of base-paired segments.

Like the α-helices and β-pleated sheets observed in many proteins, RNA secondary structures are stabilized by hydrogen bonding and occur spontaneously. Even though hairpins and other types of secondary structure reduce the entropy of RNA molecules, they form spontaneously because the energy released in formation of H-bonds makes the overall process favorable. Hydrogen bond formation is exothermic; overall, the formation of hairpins is exergonic.

TERTIARY AND QUATERNARY STRUCTURES RNA molecules can also have tertiary structure and quaternary structure, owing to interactions that (1) fold secondary structures into complex shapes or (2) hold different RNA strands together. As a result, RNA molecules with different base sequences can have very different overall shapes and chemical properties.

Although they do not begin to rival proteins in overall structural diversity or the presence of various functional groups, RNA molecules are much more diverse in size, shape, and reactivity than DNA molecules are. Structurally and chemically, RNA is intermediate between the complexity of proteins and the simplicity of DNA. **Table 4.1** summarizes the similarities and differences in the structures of RNA and DNA.

RNA's Structure Makes It an Extraordinarily Versatile Molecule

In terms of structure, RNA is intermediate between DNA and proteins. RNA is intermediate in terms of function as well. RNA molecules cannot archive information nearly as efficiently as DNA molecules do, but RNAs do perform key functions in information processing. Likewise, they cannot catalyze as many reactions as proteins do, but as it turns out, the reactions they do catalyze are particularly important.

In cells, RNA molecules function like a jackknife or a pocket tool with an array of attachments: They perform a wide variety of tasks reasonably well. Some of the most surprising results in the last decade of biological science, in fact, involve new insights into the diversity of roles that RNAs play in cells. These molecules process information stored in DNA, help synthesize proteins, and defend against attack by viruses, among other things.

Here let's focus on the roles that RNA could have played in the origin of life—as an information-containing entity and as a catalyst.

RNA Is an Information-Containing Molecule

Because RNA contains a sequence of bases analogous to the letters in a word, it can function as an information-containing molecule. And because hydrogen bonding occurs specifically between A-U pairs and G-C pairs in RNA, it is possible for RNA to furnish the information required to make a copy of itself.

SUMMARY TABLE 4.1 **DNA and RNA Structure**

	DNA	RNA
Primary structure	Sequence of deoxyribonucleotides; bases are A, T, G, C.	Sequence of ribonucleotides; bases are A, U, G, C.
Secondary structure	Two antiparallel strands twist into a double helix, stabilized by hydrogen bonding between complementary bases (A-T and G-C) and hydrophobic interactions.	Most common are hairpins, formed when a single strand folds back on itself to form a double-helix "stem" and a single-stranded "loop."
Tertiary structure	None*	Folds that form distinctive three-dimensional shapes.
Quaternary structure	None	Associations between several RNA molecules.

*In cells, DNA coils around proteins that bind to the double helix. In many cases the DNA-protein complex folds into highly organized, compact structures. But DNA does not form tertiary structure on its own.

The copying process occurs as shown for DNA in Figure 4.9, except that the template RNA exists as a single strand instead of a double strand. When RNA is copied, free ribonucleotides form hydrogen bonds with complementary bases on the original strand of RNA—the template strand. As they do, their sugar-phosphate groups form phosphodiester linkages to produce the complementary strand. Finally, the hydrogen bonds between the strands are broken by heating or by a catalyzed reaction. The new RNA molecule now exists independently of the original strand. If these steps were repeated with the new strand used as a template, the resulting molecule would be a copy of the original. In this way, an RNA's primary sequence serves as a mold.

Although complementary base pairing allows RNA to carry the information required for the molecule to be copied, RNA is not nearly as stable a repository for particular sequences as is DNA. In the prebiotic soup, could an RNA molecule have made a copy of itself before it would have degraded?

RNA Can Function as a Catalytic Molecule

In terms of diversity in chemical reactivity and overall shape, RNA molecules do not begin to match proteins. The primary structure of RNA molecules is much more restricted because there are only four types of nitrogenous bases in RNA versus the 20 types of amino acids found in proteins. Secondary through quaternary structure is more limited as a result, meaning that RNA cannot form the wide array of active sites observed among proteins.

But because RNA has a degree of structural and chemical complexity, it is capable of stabilizing a few transition states and catalyzing at least a limited number of chemical reactions. Sidney Altman and Thomas Cech shared the 1989 Nobel Prize in chemistry for showing that catalytic, enzyme-like RNAs, or **ribozymes**, exist in organisms.

Figure 4.11 shows the RNA molecule that Cech isolated from a single-celled organism called *Tetrahymena*. This ribozyme catalyzes both the hydrolysis and the condensation of phosphodiester linkages. Researchers have since discovered ribozymes that catalyze an array of different reactions in cells. For example, ribozymes catalyze the formation of peptide bonds when amino acids polymerize to form polypeptides. Ribozymes are at work in your cells right now.

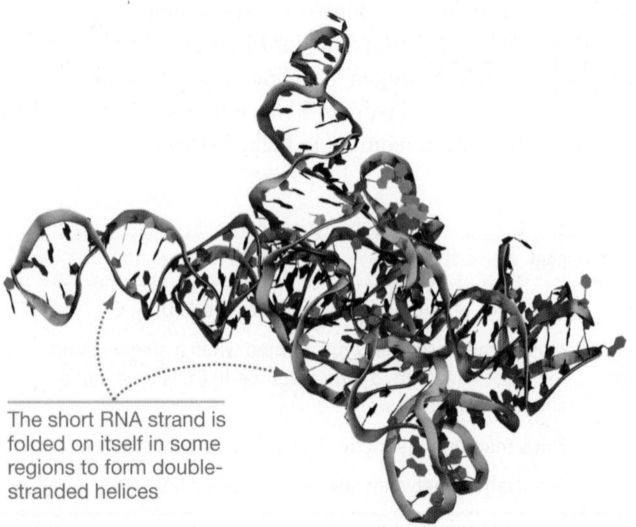

The short RNA strand is folded on itself in some regions to form double-stranded helices

FIGURE 4.11 Tertiary Structure of *Tetrahymena* Ribozyme.

The discovery of ribozymes was a watershed event in origin-of-life research. Before Altman and Cech published their discovery, most biologists thought that proteins were the only type of molecule capable of catalyzing chemical reactions in organisms. But if a ribozyme in *Tetrahymena* catalyzes polymerization reactions, it raises the possibility that an RNA molecule could make a copy of itself. Such a molecule could copy itself and qualify as the first living entity. Does any experimental evidence support this hypothesis?

4.4 The First Life-Form

The theory of chemical evolution maintains that life began as a naked self-replicator—a molecule that existed by itself in solution, without being enclosed in a membrane. To make a copy of itself, that first living molecule had to (1) provide a template that could be copied, and (2) catalyze polymerization reactions that would link monomers into a copy of that template.

Because RNA is capable of both processes, most origin-of-life researchers propose that the first life-form was made of RNA. No self-replicating RNA molecules exist today, however, so researchers test the hypothesis by trying to simulate the RNA world in the laboratory. The eventual goal is to create an RNA molecule that can catalyze its own replication.

To understand how researchers do this work, consider recent experiments by Wendy Johnston and others working in David Bartel's laboratory. This research group's goal was ambitious; they wanted to create, from scratch, a ribozyme that was capable of catalyzing the addition of ribonucleoside triphosphates to an existing strand, via complementary base pairing. Such a ribozyme would be called RNA replicase.

This research program is creating considerable excitement among biologists interested in the origin of life, because adding ribonucleotides to a growing strand is a key attribute of an RNA replicase. The logic of their experiment can be outlined as follows:

Step 1 Synthesize billions of large RNA molecules that contain randomly generated primary sequences.

Step 2 Incubate these newly synthesized, large RNAs with free ribonucleotides—including one that contains a specific molecular tag.

Step 3 Isolate large RNAs that have one of the tagged ribonucleotides attached. These molecules must be ribozymes, because they catalyzed the addition of the tagged ribonucleotide to the end of the growing RNA strand.

Step 4 Copy the ribozymes in a way that introduces a few random changes to their sequence of bases. These copied RNAs have a primary sequence that is similar to the ribozyme, but not identical.

Next Repeat steps 2–4—meaning, re-do the reaction, isolation, and copying sequence—17 times. By the end of the experiment, they had completed a total of 18 rounds of selection for ribozymes that can catalyze the formation of a phosphodiester linkage.

Here's the key to this experiment: When the ribozymes isolated in each round are copied, the molecules that result—and that furnish the starting material for the next round—are not identical in terms of their primary sequence because random changes ("mutations") were introduced during the copying step. Most of these "mutated" RNAs were worse ribozymes than the molecule that was copied, but some were better.

By isolating the best ribozymes from each round and copying them (with more "mutations" that potentially could make them even better) in the next round, the research team continually selected for ribozymes that were more and more efficient. By round 18, the group had a ribozyme that was a much better catalyst than the original molecule.

This experimental protocol was designed to mimic the process of natural selection, introduced in Chapter 1. The population of ribozymes had variable characteristics that could be copied and passed on to the next generation of ribozymes. In addition, the researchers were able to select the most efficient ribozymes to be the "parents" of the next generation.

In effect, evolution was occurring. By round 18, the researchers had a ribozyme that was reasonably proficient at adding ribonucleotides to a growing strand. As this book goes to press, the round-18 ribozyme is the closest biologists have come to creating life.

Thanks to similar efforts at other laboratories around the world, biologists have produced an increasingly impressive set of ribozymes—an array of molecules capable of catalyzing many of the key reactions responsible for replication and metabolism.

Each of these results provides support for the RNA world hypothesis. Each result also brings research teams closer to the creation of an RNA replicase. If this goal is met, human beings will have created a living entity in a test tube.

CHAPTER 4 REVIEW

For media, go to the study area at www.masteringbiology.com 

Summary of Key Concepts

Nucleotides consist of a sugar, a phosphate group, and a nitrogen-containing base. Ribonucleotides polymerize to form RNA. Deoxyribonucleotides polymerize to form DNA.

- Ribonucleotides have a hydroxyl (–OH) group on their 2′ carbon. Deoxyribonucleotides do not.
- Nucleic acids form when nucleotides polymerize. Polymerization occurs via phosphodiester linkages between the 3′ carbon on one nucleotide and the phosphate group on another.
- Nucleic acids have a sugar-phosphate backbone with nitrogenous bases attached.
- Nucleic acids are directional: they have a 5′ end and a 3′ end.
- During polymerization, new nucleotides are added only to the 3′ end.

✓You should be able to compare and contrast proteins and nucleic acids in terms of (1) polymerization, (2) their directionality, and (3) the presence of a "backbone."

DNA's primary structure consists of a sequence of nitrogen-containing bases. Its secondary structure consists of two DNA strands, running in opposite directions, that are held together by complementary base pairing and twisted into a double helix. DNA's structure allows organisms to store and replicate the information needed to grow and reproduce.

- DNA is an extremely stable molecule that serves as a superb archive for information in the form of base sequences.
- DNA is stable because deoxyribonucleotides lack a reactive 2′ hydroxyl group and because antiparallel DNA strands form a secondary structure called a double helix. The DNA double helix is stabilized by hydrogen bonds that form between complementary purine and pyrimidine bases and by hydrophobic interactions between bases stacked on the inside of the spiral.
- DNA's structural stability and regularity make it ineffective at catalysis.
- DNA is readily copied via complementary base pairing. Complementary base pairing occurs between A-T and G-C pairs in DNA.

✓You should be able to explain why DNA molecules with a high percentage of guanine and cytosine are particularly stable.

MB **Web Activity** Structure of RNA and DNA

RNA's primary structure consists of a sequence of nitrogen-containing bases. Its secondary structure includes short regions of double helices and structures called hairpins.

- RNA molecules can have secondary structure because complementary base pairing occurs between A-U and G-C pairs on the same strand. Some tertiary and quaternary structure also occurs, because hydrogen bonding allows RNA molecules to fold in precise ways, and thus interact with each other.
- RNA is versatile. The primary function of proteins is to catalyze chemical reactions, and the primary function of DNA is to carry information. But RNA is an "all-purpose" macromolecule that can do both.

✓You should be able to explain why RNA molecules can have tertiary and quaternary structure, while DNA molecules do not.

Because RNA molecules can carry information as well as catalyze chemical reactions, it is likely that RNA was the first self-replicating molecule and a forerunner to the first life-form.

- To test the RNA world hypothesis, researchers are attempting to synthesize an RNA replicase in the laboratory. They have succeeded in isolating ribozymes that can catalyze formation of phosphodiester linkages.

✓You should be able to explain why, in laboratory experiments designed to produce efficient ribozymes, it is critical to copy the RNA molecules in a way that produces random changes in the primary structure.

Questions

TEST YOUR KNOWLEDGE

Answers are available in Appendix B

1. What are the four nitrogenous bases found in RNA?
   a. uracil, guanine, cytosine, thymine (U, G, C, T)
   b. adenine, guanine, cytosine, thymine (A, G, C, T)
   c. adenine, uracil, guanine, cytosine (A, U, G, C)
   d. alanine, threonine, glycine, cysteine (A, T, G, C)

2. What determines the primary structure of an RNA molecule?
   a. the sugar-phosphate backbone
   b. complementary base pairing and the formation of hairpins
   c. the sequence of deoxyribonucleotides
   d. the sequence of ribonucleotides

3. DNA attains a secondary structure when hydrogen bonds form between the nitrogenous bases called purines and pyrimidines. What are the complementary base pairs that form in DNA?
   a. A-T and G-C
   b. A-U and G-C
   c. A-G and T-C
   d. A-C and T-G

4. By convention, biologists write the sequence of bases in RNA and DNA in which direction?
   a. $3' \rightarrow 5'$
   b. $5' \rightarrow 3'$
   c. N-terminal to C-terminal
   d. C-terminal to N-terminal

5. The secondary structure of DNA is called a double helix. Why?
   a. Two strands wind around one another in a helical, or spiral, arrangement.
   b. A single strand winds around itself in a helical, or spiral, arrangement.
   c. It is shaped like a ladder.
   d. It stabilizes the molecule.

6. In RNA, when does the secondary structure called a hairpin form?
   a. when hydrophobic residues coalesce
   b. when hydrophilic residues interact with water
   c. when complementary base pairing between ribonucleotides on the same strand creates a stem-and-loop structure
   d. when complementary base pairing between two antiparallel strands forms a double helix

TEST YOUR UNDERSTANDING

Answers are available in Appendix B

1. Make a concept map (see **BioSkills 8** in Appendix A) that relates DNA's primary structure to its secondary structure. Your diagram should include deoxyribonucleotides, hydrophobic interactions, purines, pyrimidines, phosphodiester linkages, DNA primary structure, DNA secondary structure, complementary base pairing, and antiparallel strands.
2. Explain why the addition of three phosphate groups to nucleosides allows otherwise-endergonic polymerization reactions to occur.
3. Explain why nucleic acids are directional. In a nucleic acid, what are the 5′ carbons bonded to? What are the 3′ carbons bonded to?
4. Why is DNA such a stable molecule compared with proteins or RNA?
5. Explain how the secondary structures called hairpins form in RNA.
6. A major theme in this chapter is that the structure of molecules correlates with their function. Explain why DNA's secondary structure limits its catalytic abilities compared with that of RNA. Why is it logical that RNA molecules can catalyze a modest but significant array of reactions? Why are proteins the most effective catalysts?

APPLYING CONCEPTS TO NEW SITUATIONS

Answers are available in Appendix B

1. Viruses are particles that infect cells. In some viruses, the genetic material consists of two strands of RNA, bonded together via complementary base pairing. Would these antiparallel strands form a double helix? Explain why or why not.
2. Suppose that experiments like those reviewed in Section 4.4 succeeded in producing a molecule that could make a copy of itself. According to the criteria listed in Chapter 1, would this molecule be considered alive?
3. Before Watson and Crick published their model of the DNA double helix, Linus Pauling offered a model based on a triple helix. If the three sugar-phosphate backbones were on the outside of such a molecule, would hydrogen bonding or hydrophobic interactions be more important in keeping such a secondary structure together?
4. Origin-of-life researcher Robert Crabtree maintains that experiments simulating early Earth conditions are a valid way to test the theory of chemical evolution. Crabtree claims that if scientists working in the field agree that an experiment is a plausible reproduction of early Earth conditions, it is valid to infer that its results are probably correct—that the simulation effectively represents events that occurred some 3.5 billion years ago. Do you agree? Do you find the models and experiments presented in this chapter and previous chapters to be convincing tests of the theory? Explain your answers.

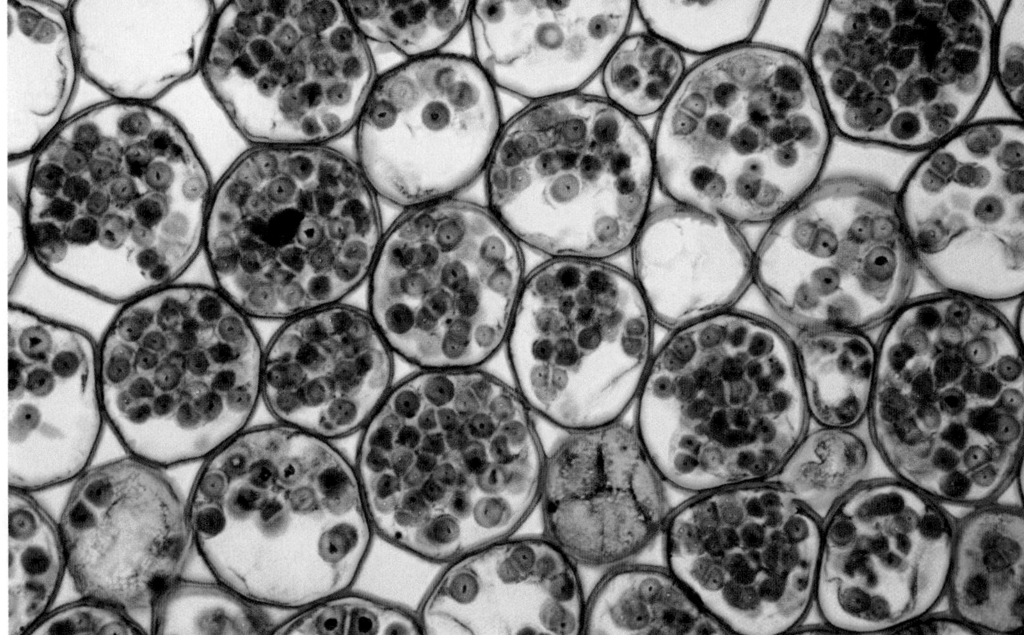

A cross section through a buttercup root. Cellulose-rich cell walls are stained green; starch-filled structures are stained purple. Cellulose is a structural carbohydrate; starch is an energy-storage carbohydrate.

An Introduction to Carbohydrates 5

This unit highlights the four types of macromolecules that are prominent in today's cells: proteins, nucleic acids, carbohydrates, and lipids. Understanding the structure and function of each of these macromolecules is a basic requirement for exploring how life began and how organisms work. Chapters 3 and 4 analyzed the way proteins and nucleic acids are put together and what they do. This chapter focuses on carbohydrates; Chapter 6 will introduce lipids.

The term **carbohydrate, or sugar**, encompasses the monomers called monosaccharides (literally, "one-sugar"), small polymers called oligosaccharides ("few-sugars"), and the large polymers called polysaccharides ("many-sugars"). The name carbohydrate is logical because the chemical formula of many of these molecules is $(CH_2O)_n$, where the n refers to the number of "carbon-hydrate" groups. The name can also be misleading, though, because carbohydrates do not consist of carbon atoms bonded to water molecules. Instead, they are molecules with a carbonyl ($>C{=}O$) and several hydroxyl ($-OH$) functional groups, along with several to many carbon-hydrogen ($C{-}H$) bonds.

Let's begin with monosaccharides, put them together into polysaccharides, then explore how carbohydrates figured in the origin of life and what they do in cells today. As you study this material, be sure to ask yourself the central question of biological chemistry: How does this molecule's stucture relate to its properties and function?

KEY CONCEPTS

- Sugars and other carbohydrates are highly variable in structure.
- Monosaccharides are monomers that polymerize via condensation reactions to form polymers called polysaccharides. The monosaccharides in polysaccharides are joined by different types of glycosidic linkages.
- Carbohydrates have diverse functions in cells. In addition to serving as raw material for synthesizing other molecules, they provide fibrous structural materials, indicate cell identity, and store chemical energy.

5.1 Sugars as Monomers

Sugars are fundamental to life. They provide chemical energy in cells and furnish some of the molecular building blocks required for the synthesis of larger, more complex compounds. Monosaccharides were important during chemical evolution, early in Earth's history. For example, the sugar called ribose is required for the formation of nucleotides. What are sugars, and how do they differ from one another?

✔ When you see this checkmark, stop and test yourself. Answers are available in Appendix B.

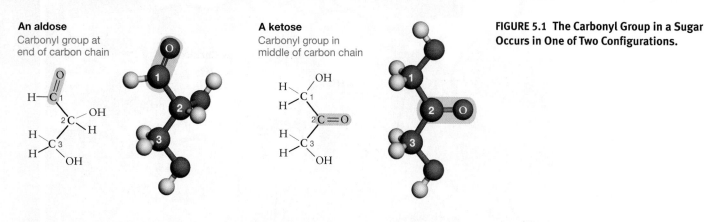

An aldose
Carbonyl group at end of carbon chain

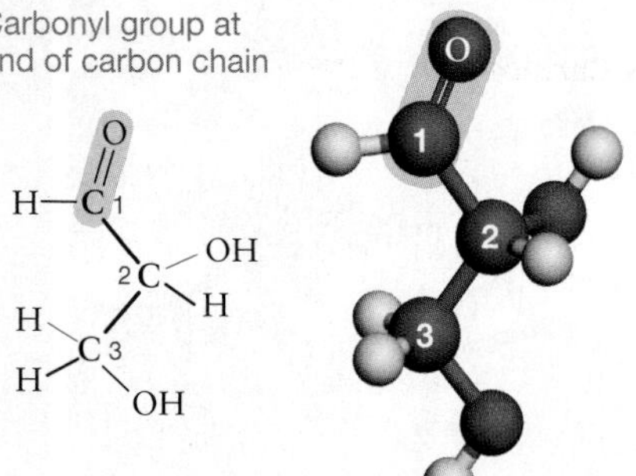

A ketose
Carbonyl group in middle of carbon chain

H
OH
C1
H
2C=O
H
C3
H
OH

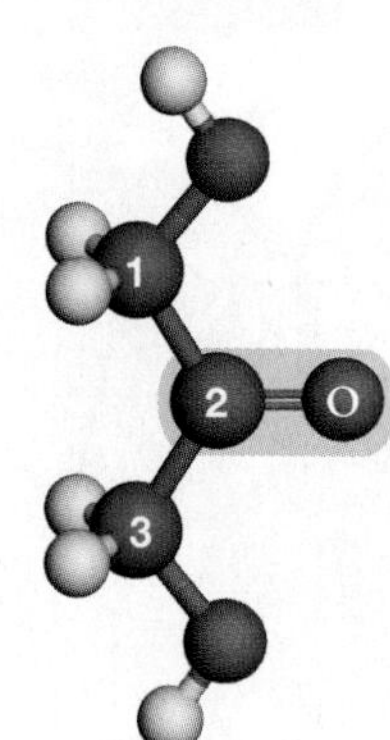

FIGURE 5.1 The Carbonyl Group in a Sugar Occurs in One of Two Configurations.

How Monosaccharides Differ

Figure 5.1 illustrates the structure of the monomer called a **monosaccharide**, or simple sugar. The carbonyl group that serves as one of monosaccharides' distinguishing features can be found either at the end of the molecule, forming an aldehyde sugar (an aldose), or within the carbon chain, forming a ketone sugar (a ketose). The presence of a carbonyl group along with multiple hydroxyl groups provides an array of functional groups in sugars. Based on this observation, it's not surprising that sugars are able to participate in a large number of chemical reactions.

The number of carbon atoms present also varies in monosaccharides. By convention, the carbons in a monosaccharide are numbered consecutively, starting with the end nearest the carbonyl group. Figure 5.1 features three-carbon sugars, or **trioses**. Ribose, which acts as a building block for nucleotides, has five carbons and is called a **pentose**; the glucose that is coursing through your bloodstream and being used by your cells right now is a six-carbon sugar, or a **hexose**.

Besides varying in the location of the carbonyl group and the total number of carbon atoms present, monosaccharides can vary in the spatial arrangement of their atoms. There is, for example, a wide array of pentoses and hexoses. Each is distinguished by the configuration of its hydroxyl functional groups. **Figure 5.2** illustrates glucose and galactose, which are six-carbon sugars. Note that the two molecules have the same chemical formula ($C_6H_{12}O_6$) but not the same structure. Both are aldose sugars with six carbons, but they differ in the spatial arrangement of the hydroxyl group at the carbon highlighted in Figure 5.2.

This is a key point: Because the structures of glucose and galactose differ, their functions differ. In cells, glucose is used as a source of chemical energy that sustains life. But for galactose to be used as a source of energy, it first has to be converted to glucose via an enzyme-catalyzed reaction. This example underscores a general theme: Even seemingly simple changes in structure—like the location of a single hydroxyl group—can have enormous consequences for function. This is because molecules interact in precise ways, based on their shape.

It's rare for sugars to exist in the form of the linear chains illustrated in Figures 5.1 and 5.2, however. In aqueous solution they tend to form ring structures. Glucose serves as the example in **Figure 5.3**. When the cyclic structure forms in glucose, the C-1 carbon (the carbon numbered 1 in the linear chain) forms a bond with the oxygen atom of the C-5 hydroxyl, and its carbonyl group becomes a hydroxyl group. This hydroxyl group can be oriented in two distinct ways: above or below the plane of the ring. The different configurations produce the molecules α-glucose and β-glucose.

To summarize, many distinct monosaccharides exist because so many aspects of their structure are variable: aldose or ketose placement of the carbonyl group, variation in carbon number, different arrangements of hydroxyl groups in space, and alternative ring forms. Each monosaccharide has a unique structure and function.

Glucose

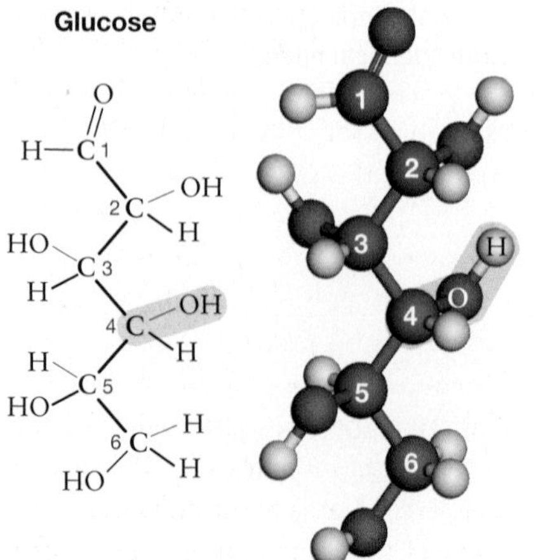

Galactose

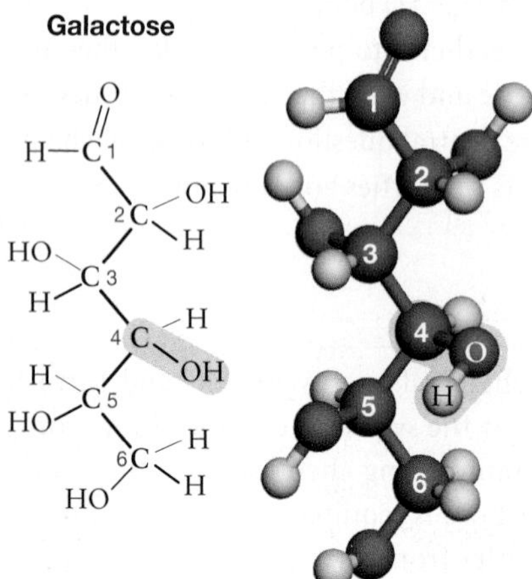

FIGURE 5.2 Sugars May Vary in the Configuration of Their Hydroxyl Groups. The two six-carbon sugars shown here vary only in the spatial orientation of their hydroxyl groups on carbon number 4.

✔ **EXERCISE** Mannose is a six-carbon sugar that is identical with glucose, except that the hydroxyl (—OH) group on carbon number 2 is switched in orientation. Circle carbon number 2 in glucose and galactose, then draw the structural formula of mannose.

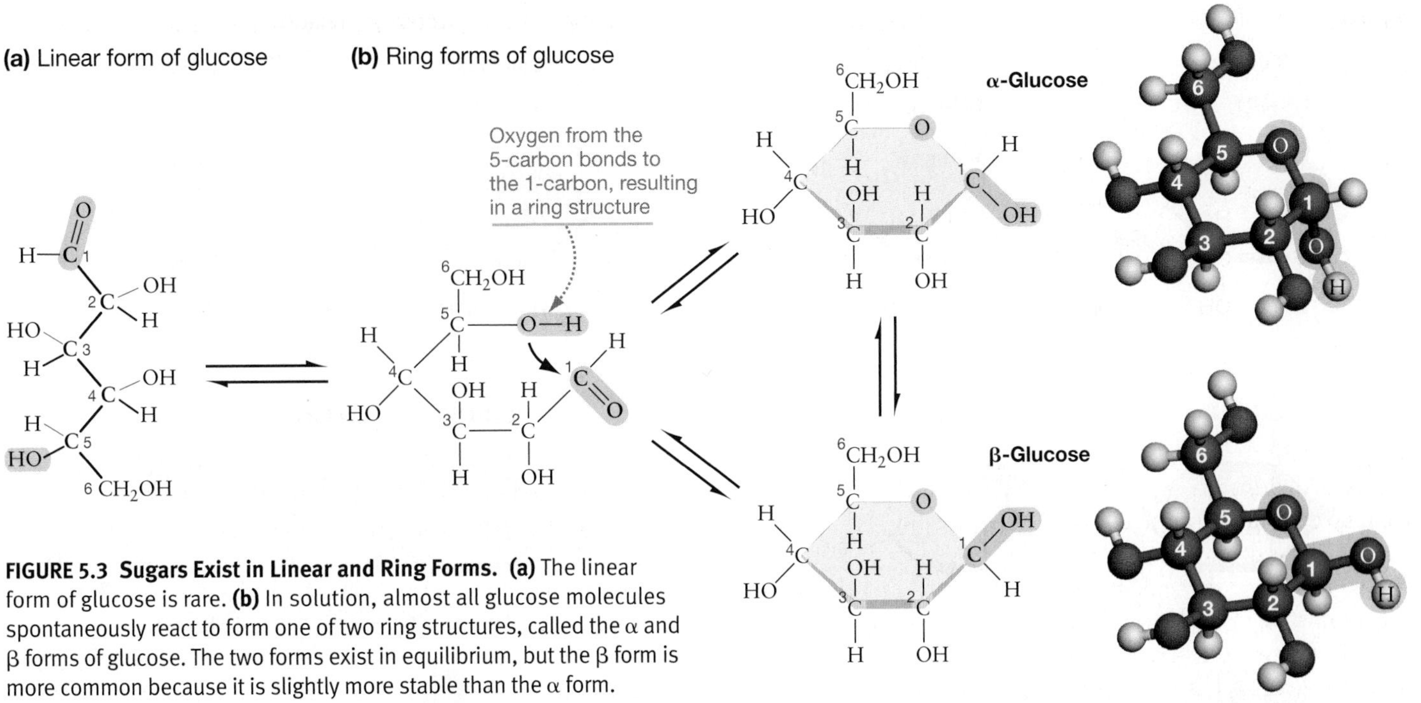

(a) Linear form of glucose

(b) Ring forms of glucose

Oxygen from the 5-carbon bonds to the 1-carbon, resulting in a ring structure

α-Glucose

β-Glucose

FIGURE 5.3 Sugars Exist in Linear and Ring Forms. (a) The linear form of glucose is rare. **(b)** In solution, almost all glucose molecules spontaneously react to form one of two ring structures, called the α and β forms of glucose. The two forms exist in equilibrium, but the β form is more common because it is slightly more stable than the α form.

CHECK YOUR UNDERSTANDING

If you understand that . . .

- Simple sugars differ in three respects:
  1. the location of their carbonyl group,
  2. the number of carbon atoms present, and
  3. the spatial arrangement of their atoms—particularly the relative positions of hydroxyl (–OH) groups.

✔ **You should be able to . . .**

Draw the structural formula of a three-carbon monosaccharide in linear form and then draw other sugars that differ from this one in each of the three aspects listed.

Answers are available in Appendix B.

Monosaccharides and Chemical Evolution

Laboratory simulations have shown that most monosaccharides are readily synthesized under conditions that mimic the prebiotic soup. For example, when formaldehyde (H_2CO) molecules are heated in solution, they react with one another to form almost all the pentoses and hexoses, as well as some seven-carbon sugars. In addition, researchers recently announced the discovery of the three-carbon ketose illustrated in Figure 5.1, along with a wide array of compounds closely related to sugars, on the Murchison meteorite that struck Australia in 1969.

Based on these observations, investigators suspect that sugars are synthesized on dust particles and other debris in interstellar space and could have rained down onto Earth as the planet was forming, as well as being synthesized in the hot water near undersea volcanoes. Most researchers interested in chemical evolution maintain that a wide diversity of monosaccharides existed in the prebiotic soup.

But as Chapter 4 pointed out, it remains a mystery why ribose might have predominated and made the synthesis of nucleotides possible. It also appears highly unlikely that monosaccharides were able to polymerize to form the polysaccharides found in today's cells. Let's explore why.

5.2 The Structure of Polysaccharides

Polysaccharides are polymers that form when monosaccharides are linked together. They are also known as complex carbohydrates. The simplest polysaccharides consist of two sugars and are known as **disaccharides**. The two monomers involved may be identical, as in the two α-glucose molecules that link to form maltose. Or they may be different, as in the combination of a glucose molecule and a galactose molecule that forms lactose—the most important sugar in milk.

Simple sugars polymerize when a condensation reaction occurs between two hydroxyl groups, resulting in a covalent bond called a **glycosidic linkage.** Glycosidic linkages are analogous to the peptide bonds that hold proteins together and to the phosphodiester linkages that connect the nucleotides in nucleic acids. There is an important difference, however. Peptide and phosphodiester bonds always form at the same location in their monomers. But because glycosidic linkages form between hydroxyl groups, and because every monosaccharide contains at least two hydroxyl groups, the location and geometry of glycosidic linkages can vary widely among polysaccharides.

Figure 5.4 shows two of the most common glycosidic linkages, called an α-1,4-glycosidic linkage and a β-1,4-glycosidic linkage. The numbers refer to the carbons on either side of the linkage; the α and β refer to the contrasting orientations of the linkage. As Section 5.3 will show, the differences in orientation are particularly

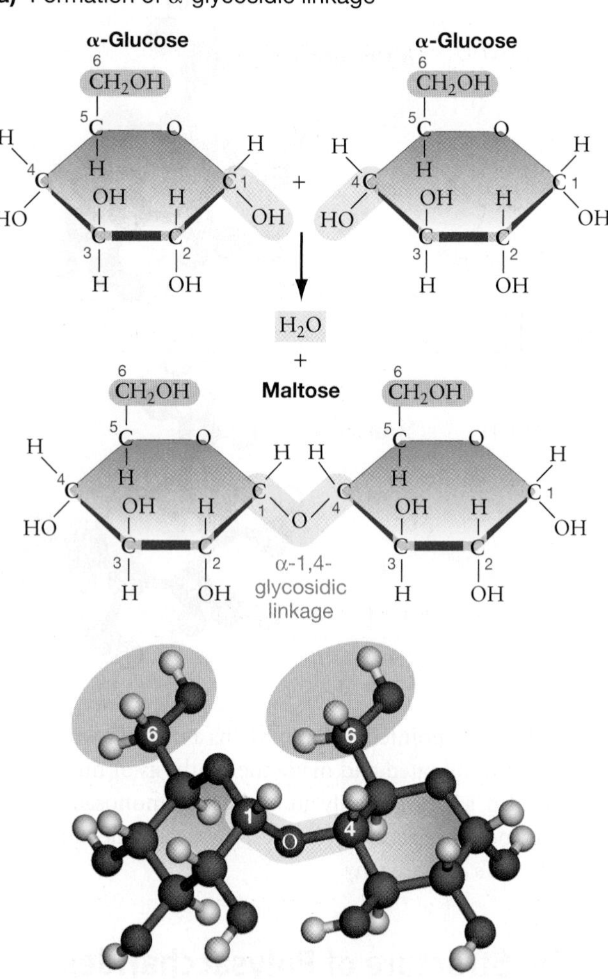

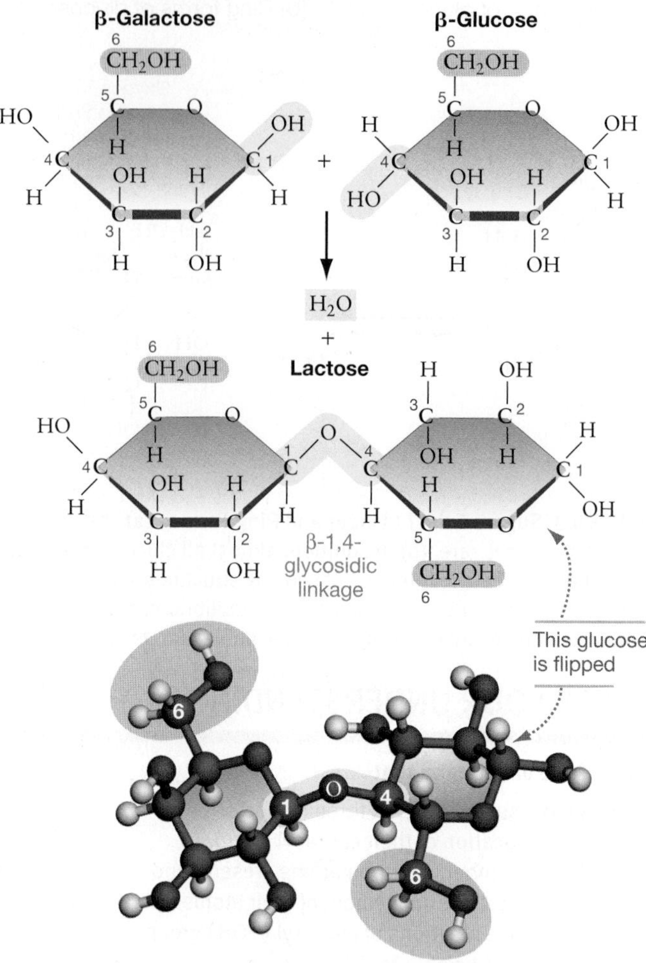

FIGURE 5.4 Monosaccharides Polymerize through Formation of Glycosidic Linkages. A glycosidic linkage occurs when hydroxyl groups on two monosaccharides undergo a condensation reaction to form a bond. Maltose and lactose are disaccharides.

important: α-linkages are easy for enzymes to break; β-linkages are difficult for enzymes to break.

To drive this point home, consider the structures of the most common polysaccharides found in organisms today: starch, glycogen, cellulose, and chitin, along with a modified polysaccharide called peptidoglycan. Each of these macromolecules can consist of a few hundred to many thousands of monomers, joined by glycosidic linkages at different locations.

Starch: A Storage Polysaccharide in Plants

In plant cells, monosaccharides are stored for later use in the form of starch. **Starch** consists entirely of α-glucose monomers that are joined by glycosidic linkages. As the top panel in **Table 5.1** shows, the angle of the linkages between carbons 1 and 4 causes the chain of glucose subunits to coil into a helix. Starch is actually a mixture of two such polysaccharides, however. One is an unbranched molecule called amylose, which contains only α-1,4-glycosidic linkages. The other is a branched molecule called amylopectin. The branching in amylopectin occurs when glycosidic linkages form between carbon 1 of a glucose monomer on one strand and carbon 6 of a glucose monomer on another strand. In amylopectin, branches occur in about one out of every 30 monomers.

Glycogen: A Highly Branched Storage Polysaccharide in Animals

Glycogen performs the same storage role in animals that starch performs in plants. In humans, for example, glycogen is stored in the liver and in muscles. When you start exercising, enzymes begin breaking glycogen into glucose monomers, which are then processed in muscle cells to supply energy. Glycogen is a polymer of α-glucose and is nearly identical with the branched form of starch. However, instead of an α-1,6-glycosidic linkage occurring in about 1 out of every 30 monomers, a branch occurs in about 1 out of every 10 glucose subunits (see Table 5.1).

SUMMARY TABLE 5.1 **Polysaccharides Differ in Structure**

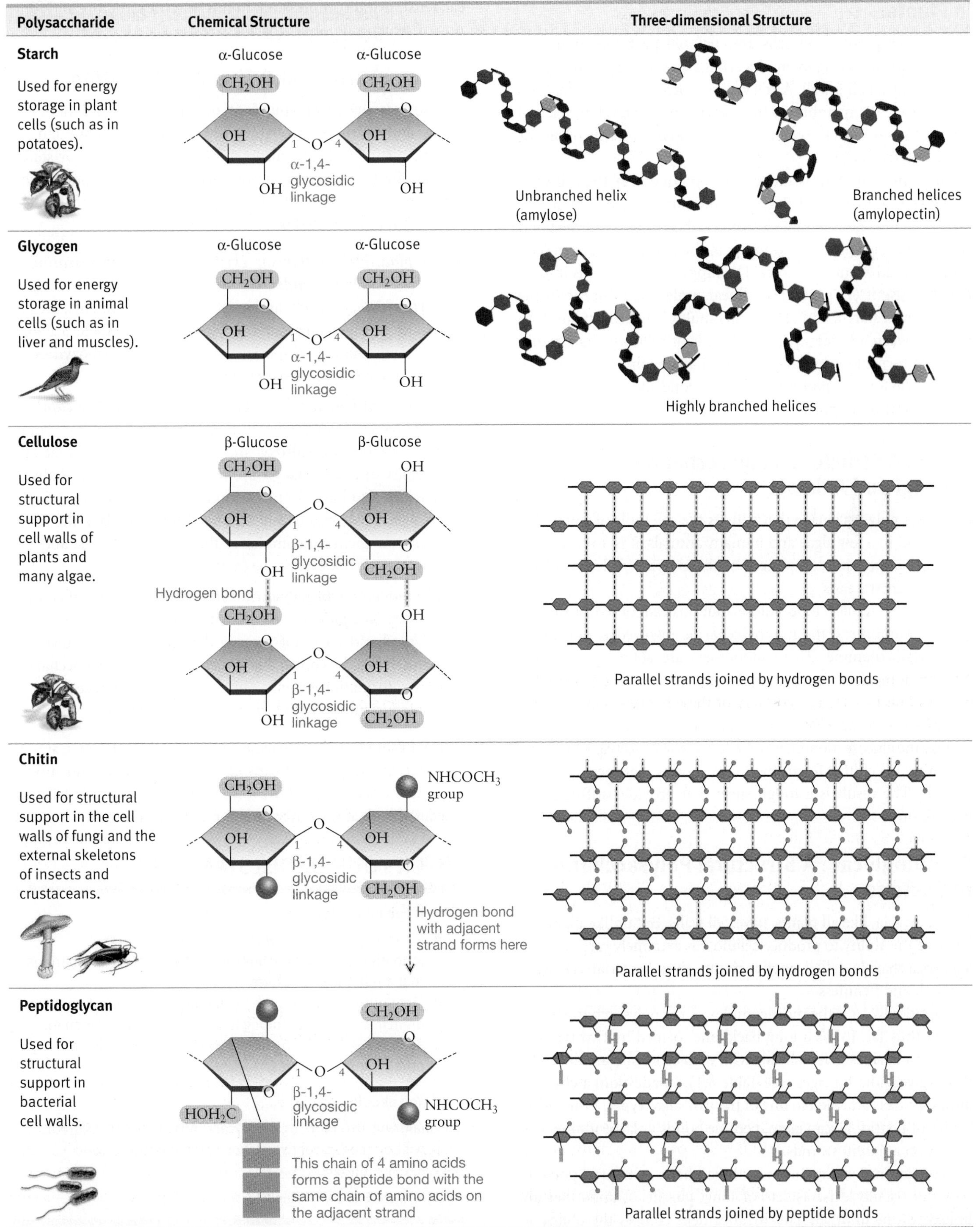

Polysaccharide	Chemical Structure	Three-dimensional Structure
Starch Used for energy storage in plant cells (such as in potatoes).	α-Glucose, α-Glucose; CH_2OH, O, OH, OH; 1, O, 4; α-1,4-glycosidic linkage; CH_2OH, O, OH, OH	Unbranched helix (amylose); Branched helices (amylopectin)
Glycogen Used for energy storage in animal cells (such as in liver and muscles).	α-Glucose, α-Glucose; CH_2OH, O, OH, OH; 1, O, 4; α-1,4-glycosidic linkage; CH_2OH, O, OH, OH	Highly branched helices
Cellulose Used for structural support in cell walls of plants and many algae.	β-Glucose, β-Glucose; CH_2OH, O, OH, OH; O, 1, 4; β-1,4-glycosidic linkage; OH, OH, O, CH_2OH; Hydrogen bond; CH_2OH, O, OH, OH; O, 1, 4; β-1,4-glycosidic linkage; OH, OH, O, CH_2OH	Parallel strands joined by hydrogen bonds
Chitin Used for structural support in the cell walls of fungi and the external skeletons of insects and crustaceans.	CH_2OH, O, OH; O, 1, 4; β-1,4-glycosidic linkage; $NHCOCH_3$ group; OH, O, CH_2OH; Hydrogen bond with adjacent strand forms here	Parallel strands joined by hydrogen bonds
Peptidoglycan Used for structural support in bacterial cell walls.	O, HOH_2C; 1, O, 4; β-1,4-glycosidic linkage; CH_2OH, O, OH; $NHCOCH_3$ group; This chain of 4 amino acids forms a peptide bond with the same chain of amino acids on the adjacent strand	Parallel strands joined by peptide bonds

Cellulose: A Structural Polysaccharide in Plants

As Chapter 1 noted, all cells are enclosed by a membrane. In most organisms living today, the cell is also surrounded by a layer of material called a wall. A **cell wall** is a protective sheet that occurs outside the membrane. In algae, plants, bacteria, fungi, and many other groups, the cell wall is composed primarily of one or more polysaccharides.

In plants, cellulose is the major component of the cell wall. **Cellulose** is a polymer of β-glucose monomers, joined by β-1,4-glycosidic linkages. As Table 5.1 shows, the geometry of the bond is such that each glucose monomer in the chain is flipped in relation to the adjacent monomer. The flipped orientation is important because it (1) generates a linear molecule, rather than the helix seen in starch, and (2) permits multiple hydrogen bonds to form between adjacent, parallel strands of cellulose. As a result, cellulose forms long, parallel strands that are joined by hydrogen bonds. The linked cellulose fibers are strong and provide the cell with structural support.

Chitin: A Structural Polysaccharide in Fungi and Animals

Chitin is a polysaccharide that stiffens the cell walls of fungi. It is also found in a few algae and in many animals; it is, for example, the most important component of the external skeletons of insects and crustaceans.

Chitin is similar to cellulose, but instead of consisting of glucose monomers, the monosaccharide involved is one called *N*-acetylglucosamine. These monomers are abbreviated "NAc." NAc monomers are joined by β-1,4-glycosidic linkages (see Table 5.1). As in cellulose, the geometry of these bonds results in every other residue being flipped in orientation.

Like the glucose monomers in cellulose, the *N*-acetylglucosamine subunits in chitin form hydrogen bonds between adjacent strands. The result is a tough sheet that provides stiffness and protection.

Peptidoglycan: A Structural Polysaccharide in Bacteria

Most bacteria, like all plants, have cell walls. But unlike plants, in bacteria the ability to produce cellulose is extremely rare. Instead, a polysaccharide called **peptidoglycan** gives bacterial cell walls strength and firmness.

Peptidoglycan is the most complex of the polysaccharides discussed thus far. It has a long backbone formed by two types of monosaccharides that alternate with each other and are linked by β-1,4-glycosidic linkages (see Table 5.1). In addition, a chain of amino acids is attached to one of the two sugar types. When molecules of peptidoglycan align, peptide bonds link the amino acid chains on adjacent strands.

Note an important common thread here: Structural polysaccharides usually exist as sets of long, parallel strands that are linked to one another. This arrangement confers the ability to withstand pulling and pushing forces—what an engineer would call tension and compression. In this way, the structure and function of structural polysaccharides are correlated.

Polysaccharides and Chemical Evolution

Cellulose is the most abundant organic compound on Earth today, and chitin is probably the second most abundant by weight. Virtually all organisms manufacture glycogen or starch. But despite their current importance to organisms, polysaccharides probably played little to no role in the origin of life. This conclusion is supported by several observations:

- *No plausible mechanism exists for the polymerization of monosaccharides under conditions that prevailed early in Earth's history.* In cells and in laboratory experiments, the glycosidic linkages illustrated in Figure 5.4 and Table 5.1 form only with the aid of specialized enzymes. No ribozymes are known to catalyze these reactions.
- *To date, no reactions have been discovered that are catalyzed by polysaccharides.* Even though monosaccharides contain large numbers of hydroxyl and carbonyl groups, they lack the diversity of functional groups found in amino acids. Polysaccharides also have simple secondary structures, consisting of linkages between adjacent strands. Thus, they lack the structural and chemical complexity that makes proteins, and to a lesser extent RNA, effective catalysts.
- *The monomers in polysaccharides are not capable of complementary base pairing.* Like proteins, but unlike DNA and RNA, polysaccharides cannot provide the information required for themselves to be copied. As far as is known, no polysaccharides store information in cells. Thus, no one has proposed that the first living entity might have been a polysaccharide.

Even though polysaccharides probably did not play a significant role in the earliest forms of life, they became enormously important once cellular life evolved. In the next section let's take a detailed look at how they function in today's cells.

CHECK YOUR UNDERSTANDING

If you understand that . . .

- Polysaccharides form when enzymes catalyze the formation of glycosidic linkages between monosaccharides that are in the α or β form.
- Most polysaccharides are long, linear molecules, but some branch extensively. Among linear forms, it is common for adjacent strands to be linked by hydrogen bonding or other types of linkages.

✔ **You should be able to . . .**

Explain why two four-sugar polysaccharides can be different, even if both consist of two glucose monomers and two galactose monomers.

Answers are available in Appendix B.

5.3 What Do Carbohydrates Do?

Chapter 4 introduced one of the four basic functions that carbohydrates perform in organisms: serving as a substrate for synthesizing more-complex molecules. Recall that both RNA and DNA contain sugars—the five-carbon sugars ribose and deoxyribose, respectively. In nucleotides, which consist of a sugar, a phosphate group, and a nitrogenous base, the sugar itself acts as a subunit of the larger molecule.

In addition, sugars frequently furnish the raw "carbon skeletons" that are used as building blocks in the synthesis of important molecules. Amino acids are being produced by your cells right now, for example, using sugars as a starting point.

Although the details of how sugars are used in synthesizing amino acids and other complex molecules are beyond the scope of this book, you can delve into the other three major roles of carbohydrates in the rest of this chapter and in the supporting materials in the study area at *www.masteringbiology.com.*

(MB) **Web Activity** Carbohydrate Structure and Function

Carbohydrates have diverse functions in cells: In addition to serving as precursors to larger molecules, they provide fibrous structural materials, indicate cell identity, and store chemical energy.

The Role of Carbohydrates as Structural Molecules

Cellulose and chitin, along with the modified polysaccharide peptidoglycan, are key structural compounds. They form fibers that give cells and organisms strength and elasticity.

To appreciate why cellulose, chitin, and peptidoglycan are effective as structural molecules, recall that they form long strands and that bonds can form between adjacent strands. In the cell walls of plants, for example, a collection of about 80 cellulose molecules are cross-linked by hydrogen bonding to create a tough fiber. These cellulose fibers, in turn, crisscross to form a tough sheet.

Besides being stiff and strong, the structural carbohydrates are durable. Almost all organisms have the enzymes required to break the α-1,4- and α-1,6-glycosidic linkages that hold starch and glycogen molecules together, but only a few organisms have enzymes capable of hydrolyzing the β-1,4-glycosidic linkages in cellulose, chitin, and peptidoglycan. The shape and orientation of β-1,4-glycosidic linkages make them difficult to break, and few enzymes have active sites with the correct geometry and reactive groups to do so. As a result, the structural polysaccharides are resistant to degradation and decay.

Ironically, the durability of cellulose is important for digestion. The cellulose that you ingest when you eat plant cells—what biologists call dietary fiber—passes through your gut without being broken down. The cellulose in dietary fiber, like the cellulose in a paper towel, absorbs water. The presence of dietary fiber adds moisture to the feces. In addition, cellulose adds bulk that helps fecal material move through the intestinal tract more quickly, preventing constipation, and other problems.

The Role of Carbohydrates in Cell Identity

Polysaccharides do not store information in cells, but they do *display* important information. More specifically, polysaccharides act as signage on the outer surface of the plasma membrane that surrounds a cell. (Chapter 6 will describe plasma membranes in detail.)

Figure 5.5 shows how this information display happens. Molecules called glycoproteins project outward from the cell surface into the surrounding environment. A **glycoprotein** is a protein that is covalently bonded to a carbohydrate—usually the relatively short chain of sugars called oligosaccharides.

Glycoproteins are key molecules in what biologists call cell-cell recognition and cell-cell signaling. Each cell in your body has glycoproteins on its surface that identify it as part of your body. Immune system cells use these glycoproteins to distinguish your body's cells from foreign cells, such as bacteria. In addition, each distinct type of cell in a multicellular organism—for example, the nerve cells and muscle cells in your body—displays a different set of glycoproteins on its surface.

In cells, glycoproteins form a "sugar coating" that acts like the magnetic stripe on the back of a credit card or the personal identification number (PIN) that you use to access a bank account—it immediately identifies the individual that bears it. The identification information displayed by glycoproteins helps cells recognize and communicate with each other.

The key point here is to recognize that the enormous number of structurally distinct monosaccharides makes it possible for an enormous number of unique oligosaccharides to exist. As a result, each cell type and each species can display a unique identity.

This is another example where a molecule's structure correlates with its function. Oligosaccharides that function in cell identity vary because their component monomers—and the linkages between them—vary.

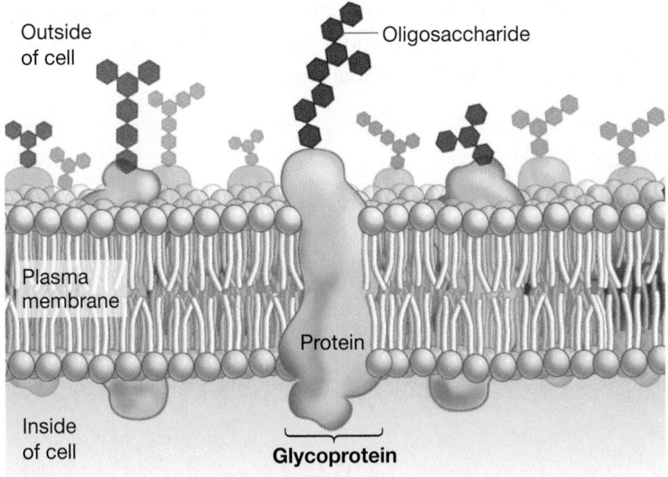

FIGURE 5.5 Carbohydrates Are an Identification Badge for Cells. Glycoproteins contain sugar groups that project outside the cell from the surface of the plasma membrane enclosing the cell. These sugar groups have distinctive structures that identify the type or species of the cell.

The Role of Carbohydrates in Energy Storage

Candy-bar wrappers promise a quick energy boost, and ads for sports drinks claim that their products provide the "carbs" needed for peak activity. If you were to ask friends or family members what carbohydrates do in your body, they would probably say something like "They give you energy." And after pointing out that carbohydrates are also used in cell identity, as a structural material, and as a source of carbon skeletons for the synthesis of other complex molecules, you'd have to agree.

Carbohydrates store and provide chemical energy in cells. What aspect of carbohydrate structure makes this function possible?

CARBOHYDRATES STORE SUNLIGHT AS CHEMICAL ENERGY Recall from earlier chapters that the essence of chemical evolution was the conversion of kinetic energy in sunlight and heat into chemical energy stored in the bonds of molecules such as formaldehyde (H_2CO) and hydrogen cyanide (HCN). Today, the kinetic energy in sunlight is converted to chemical energy stored in the bonds of carbohydrates, via the process known as **photosynthesis**.

Photosynthesis entails a complex set of reactions that can be summarized most simply as follows:

$$CO_2 + H_2O + \text{sunlight} \longrightarrow (CH_2O)_n + O_2$$

where $(CH_2O)_n$ represents a carbohydrate.

Figure 5.6 shows the structural formulas of the molecules involved in the summary reaction of photosynthesis. Note, too, that gold dots represent the relative positions of their covalently bonded electrons. The key to understanding this figure is to compare the positions of the electrons in the reactants and the products.

1. The electrons in the C=O bonds of carbon dioxide and the C–O bonds of carbohydrates are held tightly, because of oxygen's high electronegativity. Thus, they have relatively low potential energy.
2. The covalently bonded electrons in the C–H bonds of carbohydrates are shared equally because the electronegativity of carbon and hydrogen is about the same. Thus, these electrons have relatively high potential energy.
3. Electrons are also shared equally in the carbon-carbon (C–C) bonds of carbohydrates—meaning that they, too, have relatively high potential energy.

C–C and C–H bonds have much higher free energy than C–O bonds have. As a result, carbohydrates have much more free energy than carbon dioxide has.

The essence of photosynthesis, then, is that energy in sunlight is transformed into chemical energy that is stored in the C–H and C–C bonds of carbohydrates. Because carbohydrates are much more highly ordered than carbon dioxide in addition to having higher potential energy, they have much higher free energy (see Chapter 2).

Figure 5.7 summarizes and extends these points. Start by comparing the structure of carbon dioxide in Figure 5.7a with the carbohydrate in Figure 5.7b. The main difference is the presence of hydrogen atoms and C–H bonds in the carbohydrate. Now compare the carbohydrate in Figure 5.7b with the fatty acid—a subunit of a fat molecule—in Figure 5.7c. Compared with carbohydrates, fats contain many more C–C and C–H bonds and many fewer C–O bonds.

This is important. C–C and C–H bonds have high free energy because the electrons are shared equally by atoms with low electronegativities. C–O bonds, in contrast, have low free energy because the highly electronegative oxygen atom holds the electrons so tightly. Both carbohydrates and fats are used as fuel in cells, but fats store twice as much energy per gram compared with carbohydrates. Fats will be discussed in more detail in Chapter 6.

ENZYMES HYDROLYZE CARBOHYDRATES TO RELEASE GLUCOSE Starch and glycogen are efficient energy-storage molecules because they polymerize via α-glycosidic linkages instead of the β-glycosidic linkages observed in the structural polysaccharides. The α-linkages in storage polysaccharides are readily hydrolyzed to release glucose, while the β-linkages in structural polysaccharides resist enzymatic degradation.

The glucose subunits that are hydrolyzed from starch and glycogen are then processed in reactions that result in the pro-

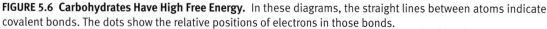

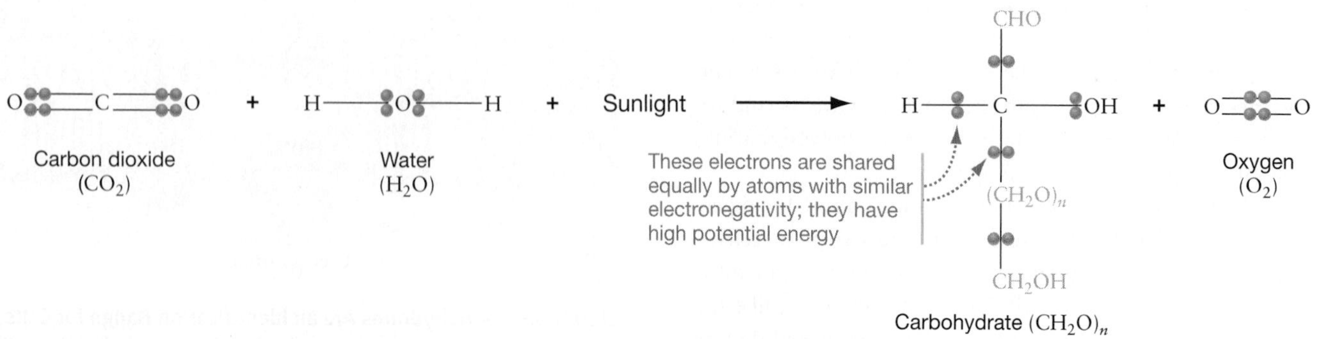

FIGURE 5.6 Carbohydrates Have High Free Energy. In these diagrams, the straight lines between atoms indicate covalent bonds. The dots show the relative positions of electrons in those bonds.

(a) Carbon dioxide

(b) A carbohydrate

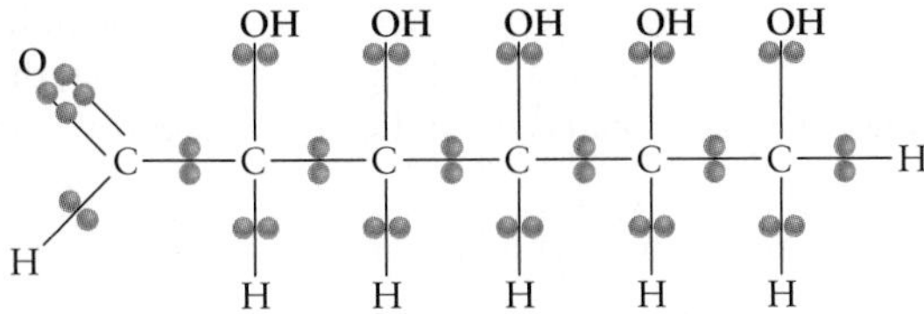

(c) A fatty acid (a component of fat molecules)

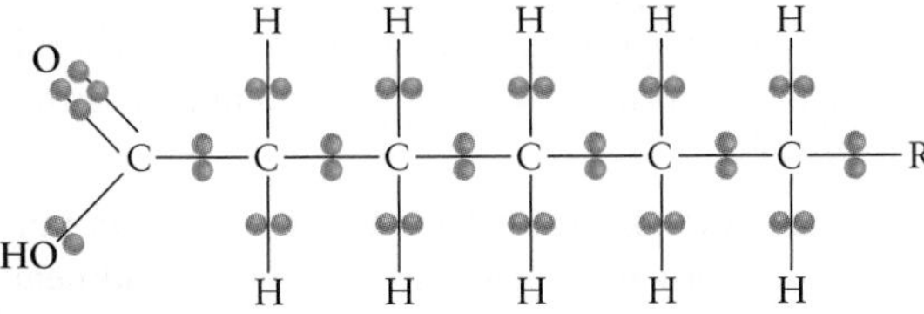

FIGURE 5.7 In Organisms, Free Energy Is Stored in C–H and C–C Bonds. (a) In carbon dioxide, the electrons involved in covalent bonds are held tightly by oxygen atoms. **(b)** In carbohydrates such as the sugar shown here, many of the covalently bonded electrons are held equally between C and H atoms. **(c)** The fatty acids found in fat molecules have more C–H bonds and fewer C–O bonds than carbohydrates do.

✔**EXERCISE** Circle the bonds in this diagram that have high free energy.

duction of chemical energy that can be used in the cell. Starch and glycogen are like a candy bar that has segments, so you can break off chunks whenever you need a boost.

The most important enzyme involved in catalyzing the hydrolysis of α-glycosidic linkages in glycogen is called **phosphorylase.** Most of your cells contain phosphorylase, so they can break down glycogen to provide glucose on demand.

The enzymes involved in breaking the α-glycosidic linkages in starch are called **amylases.** Your salivary glands and pancreas produce amylases that are secreted into your mouth and small intestine, respectively. These amylases are responsible for digesting the starch that you eat.

ENERGY STORED IN GLUCOSE IS TRANSFERRED TO ATP When a cell needs energy, exergonic reactions lead to the breakdown of the glucose and capture of the released energy through synthesis of a molecule called **adenosine triphosphate (ATP).**

More specifically, the energy that is released when sugars are processed is used to synthesize ATP from a precursor called adenosine diphosphate (ADP) plus a free inorganic phosphate (P_i) molecule. The overall reaction can be written as follows:

$$CH_2O + O_2 + ADP + P_i \longrightarrow CO_2 + H_2O + ATP$$

To put this in words, the chemical energy stored in the C–H and C–C bonds of carbohydrate is transferred to chemical energy in the form of the third phosphate group in ATP. The free energy in ATP drives endergonic reactions like polymerization, moves your muscles, and performs other types of work in cells.

Carbohydrates are like the water that piles up behind a dam; ATP is like the electricity, generated at a dam, that lights up your home. Carbohydrates store chemical energy; ATP "spends" it.

Later chapters will analyze in detail how sugars and other carbohydrates are made in organisms, and how these carbohydrates are then broken down to provide cells with usable chemical energy in the form of ATP.

CHECK YOUR UNDERSTANDING

If you understand that . . .

- Carbohydrates provide building blocks for the synthesis of more complex compounds.
- Polysaccharides such as cellulose, chitin, and peptidoglycan form cell walls, which give cells structural strength.
- Glycoproteins project from the surface of cells. They provide a molecular PIN (personal identification number) that identifies the cell's type or species.
- Starch and glycogen store sugars for later use in reactions that produce ATP. Sugars contain large amounts of chemical energy because they contain carbon atoms that are bonded to hydrogen atoms or other carbon atoms, instead of being bonded to oxygen. The C–H and C–C bonds have high free energy because the electrons are shared equally by atoms with low electronegativity.

✔ **You should be able to . . .**

1. Identify two aspects of the structures of cellulose, chitin, and peptidoglycan that correlate with their function as structural molecules.
2. Describe how the carbohydrates you ate during breakfast today are functioning in your body right now.

Answers are available in Appendix B.

CHAPTER 5 REVIEW

For media, go to the study area at www.masteringbiology.com 

Summary of Key Concepts

Sugars and other carbohydrates are highly variable in structure.

- Carbohydrates are organic compounds that have a carbonyl group and several to many hydroxyl groups.
- Carbohydrates can occur as the simple sugars called monosaccharides or as complex polysaccharides.
- Many types of monosaccharide exist, each distinguished by one or more of the following features: (1) the location of their carbonyl group—either at the end of the molecule or within it; (2) the number of carbon atoms they contain—from three to seven, usually; and (3) the orientation of their hydroxyl groups in the linear chain or the ring form.

✔You should be able to explain why a relatively small difference in the location of a carbonyl or hydroxyl group can lead to dramatic changes in the properties and function of a monosaccharide.

Monosaccharides are monomers that polymerize via condensation reactions to form polymers called polysaccharides. The monosaccharides in polysaccharides are joined by different types of glycosidic linkages.

- Monosaccharides can be linked together by covalent bonds, called glycosidic linkages, that join hydroxyl groups on adjacent molecules.
- Polysaccharides are distinguished by the type of monomers involved and the location and orientation of the glycosidic linkages between the monomers.
- The most common polysaccharides in organisms today are starch, glycogen, cellulose, and chitin; peptidoglycan is an abundant polysaccharide that has short chains of amino acids attached.

✔You should be able to explain why different types of glycosidic linkage cause glucose polymers to form a helix (e.g., in amylase) versus a straight chain (e.g., in cellulose).

Carbohydrates have diverse functions in cells. In addition to serving as raw material for synthesizing other molecules, they provide fibrous structural materials, indicate cell identity, and store chemical energy.

- In carbohydrates, as in proteins and nucleic acids, structure correlates with function.
- Cellulose, chitin, and peptidoglycan are polysaccharides that function in support. They are made up of monosaccharides joined by β-1,4-glycosidic linkages, which are difficult for enzymes to degrade. When individual molecules of these polysaccharides align side by side, bonds form between them—resulting in strong, flexible fibers or sheets.
- The oligosaccharides on cell-surface glycoproteins can function as specific signposts or identity tags, because their constituent monosaccharides are so diverse in geometry and composition.
- Both starch and glycogen function as energy-storage molecules. They are made up of glucose molecules that are in the α ring form and that are joined by glycosidic linkages between their first and fourth carbons.
- When cells need energy, enzymes hydrolyze the α-1,4-glycosidic linkages in starch or glycogen, releasing glucose molecules. Glucose and other sugars contain a significant amount of chemical energy because they contain many C–C and C–H bonds. Sugars are processed in reactions that lead to the production ATP; chemical energy in ATP is readily usable by cells.

✔You should be able to describe two key differences in the structure of polysaccharides that function in energy storage versus structural support.

 **Web Activity** Carbohydrate Structure and Function

Questions

✔TEST YOUR KNOWLEDGE

Answers are available in Appendix B

1. What is the difference between a monosaccharide, a disaccharide, and a polysaccharide?
   a. the number of carbon atoms in the molecule
   b. the type of glycosidic linkage between monomers
   c. the spatial arrangement of the various hydroxyl residues in the molecule
   d. the number of monomers in the molecule

2. What type of bond allows sugars to polymerize?
   a. glycosidic linkage
   b. phosphodiester bond
   c. peptide bond
   d. hydrogen bond

3. What holds cellulose molecules together in bundles large enough to form fibers?
   a. the cell wall
   b. peptide bonds
   c. hydrogen bonds
   d. hydrophobic interactions between different residues in the cellulose helix

4. What are the primary functions of carbohydrates in cells?
   a. energy storage, cell identity, structure, and building blocks for synthesis
   b. catalysis, structure, and transport
   c. information storage and catalysis
   d. signal reception, signal transport, and signal response

5. Why is it unlikely that carbohydrates played a large role in the origin of life?
   a. They cannot be produced by chemical evolution.
   b. They are too diverse in terms of structure and function.
   c. More types of glycosidic linkages are possible than are actually observed in organisms.
   d. They do not polymerize without the aid of enzymes.

6. What is a "quick and dirty" way to assess how much free energy an organic molecule has?
   a. Count the number of carbon atoms it contains.
   b. Compare the number of C–H and C–C bonds vs. C–O bonds it contains.
   c. Count the number of functional groups it has.
   d. Determine whether it contains a carbonyl group.

✔TEST YOUR UNDERSTANDING

Answers are available in Appendix B

1. Explain why the structure of carbohydrates supports their function in signaling the identity of a cell.
2. What is the difference between linking glucose molecules with α-1,4-glycosidic linkages versus β-1,4-glycosidic linkages? What are the consequences?
3. Compare and contrast the structures and functions of starch and glycogen. How are these molecules similar? How are they different?
4. Why do the bonds in a carbohydrate store a large amount of chemical energy compared with the chemical energy stored in the bonds of carbon dioxide?
5. What aspects of the structure of cellulose and chitin support their function in protecting and stiffening cells and organisms?
6. Both glycogen and cellulose consist of glucose monomers that are linked end to end. How do the structures of these polysaccharides differ? How do their functions differ?

✔APPLYING CONCEPTS TO NEW SITUATIONS

Answers are available in Appendix B

1. A weight-loss program for humans that emphasized minimal consumption of carbohydrates was popular in some countries in the early 2000s. What was the logic behind this diet? (Note: This diet plan caused controversy and is not endorsed by some physicians and researchers.)
2. Galactosemia is a potentially fatal disease that occurs in humans who lack the enzyme that converts galactose to glucose. To treat this disease, physicians exclude the monosaccharide galactose from the diet. Why does the disaccharide lactose also have to be excluded from the diet?
3. Amylase, an enzyme found in human saliva, catalyzes the hydrolysis of the α-1,4-glycosidic linkages in starch. If you hold a salty cracker in your mouth long enough, it will begin to taste sweet. Why?
4. Lysozyme, an enzyme found in human saliva, tears, and other secretions, catalyzes the hydrolysis of the β-1,4-glycosidic linkages in peptidoglycan. What effect does contact with this enzyme have on bacteria?

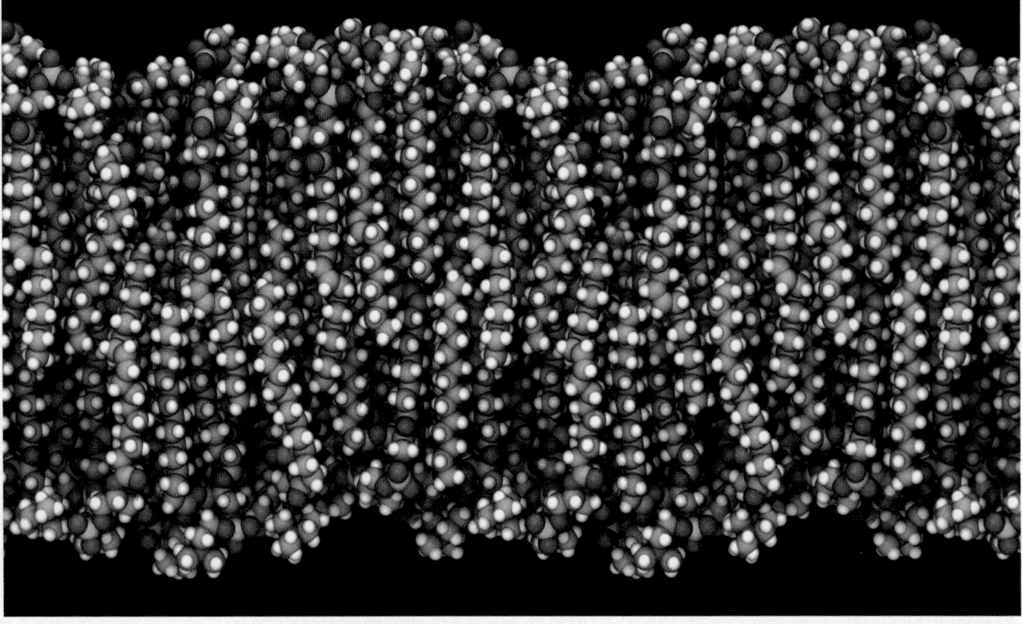

A space-filling model of a phospholipid bilayer from the membrane of a cell. In single-celled organisms, this cluster of molecules forms part of the boundary between life (inside the cell) and nonlife (outside the cell).

6 Lipids, Membranes, and the First Cells

KEY CONCEPTS

- Phospholipids are amphipathic molecules—they have a hydrophilic region and a hydrophobic region. In solution, phospholipids spontaneously form bilayers that are selectively permeable—meaning that only certain substances cross them readily.
- Ions and molecules diffuse spontaneously from regions of high concentration to regions of low concentration. Water moves across lipid bilayers from regions of high water concentration to regions of low water concentration via osmosis—a special case of diffusion.
- In cells, membrane proteins are responsible for the passage of ions, polar molecules, and large molecules that can't cross the membrane on their own because they are not soluble in lipids. Some membrane proteins form channels, some facilitate diffusion by binding to substrates, and some use energy from ATP to actively pump ions or molecules.

Currently, most biologists support the hypothesis that biological evolution began with an RNA molecule that could make a copy of itself. As the offspring of this molecule multiplied in the prebiotic soup, natural selection would have favored versions of the molecule that were particularly stable and efficient at catalysis. A second great milestone in the history of life occurred when a descendant of this replicator became enclosed within a membrane.

Why is the presence of a membrane so important? The **plasma membrane**, or **cell membrane**, separates life from nonlife. It is a layer of molecules that surrounds the cell interior and separates it from the external environment.

- The plasma membrane serves as a selective barrier: It keeps damaging compounds out of the cell and allows entry of compounds needed by the cell.
- Because the plasma membrane sequesters the appropriate chemicals in an enclosed area, reactants collide more frequently—the chemical reactions necessary for life occur much more efficiently.

The first cell functioned as an efficient and dynamic reaction vessel. It was responsible for one of the five key attributes of life introduced in Chapter 1: the presence of a barrier that defines the cell and regulates the passage of materials.

Recent research suggests that the evolution of the first self-replicating molecule and the first membrane could have occurred simultaneously. The experiments focused on nucleotides and on the "oily" or "fatty" compounds called lipids. When researchers heated nucleotides and allowed them to react in the presence of lipids, the nucleotides polymerized to form RNA-like nucleic acids—some of which become enclosed in membrane-like compartments consisting of lipids. Researchers are calling these structures "proto-cells."

Why do lipids form membranes? Which ions and molecules can pass through a membrane formed from lipids and which cannot, and why? These are some of the most fundamental questions in all of biological science. Let's delve into them further.

✓ When you see this checkmark, stop and test yourself. Answers are available in Appendix B.

6.1 Lipids

Lipid is a catch-all term for carbon-containing compounds that are found in organisms and are largely nonpolar and hydrophobic—meaning that they do not dissolve readily in water. (Recall from Chapter 2 that water is a polar solvent.) Lipids do dissolve, however, in liquids consisting of nonpolar organic compounds.

(a) Isoprene

$H_2C{=}C(CH_3){-}CH{=}CH_2$

(b) Fatty acid

HO–C(=O)– (Carboxyl group)

H_2C–CH_2–H_2C–CH_2–H_2C–CH_2–H_2C–CH_2–H_2C–CH_2–H_2C–CH_2–H_2C–CH_2–H_3C (Hydrocarbon chain)

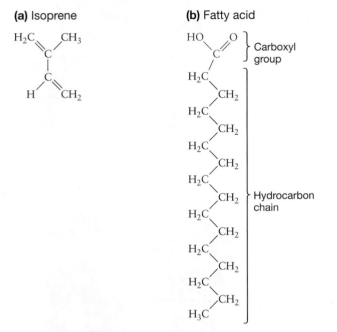

FIGURE 6.1 Hydrocarbon Structure. (a) Isoprene subunits like the one shown here can be linked to each other, end to end, to form long hydrocarbon chains. **(b)** Fatty acids typically contain a total of 14–20 carbon atoms, most found in their long hydrocarbon "tails."

To understand why lipids are insoluble in water, examine the five-carbon compound called isoprene illustrated in **Figure 6.1a**. Note that it consists of carbon atoms bonded to hydrogen atoms. Molecules that contain only carbon and hydrogen are known as **hydrocarbons**. Hydrocarbons are nonpolar because electrons are shared equally in C–H bonds—owing to the approximately equal electronegativity of carbon and hydrogen. This property makes hydrocarbons hydrophobic. Thus, the reason lipids do not dissolve in water is that they have a significant hydrocarbon component.

Figure 6.1b gives the structural formula of a **fatty acid**, which consists of a hydrocarbon chain bonded to a carboxyl (–COOH) functional group. Isoprene and fatty acids are key building blocks of the lipids found in organisms.

A Look at Three Types of Lipids Found in Cells

Unlike amino acids, nucleotides, and monosaccharides, lipids are characterized by a physical property—their solubility—instead of a shared chemical structure. The structure of lipids varies widely. For example, consider the most important types of lipids found in cells: fats, steroids, and phospholipids.

FATS **Fats** are composed of three fatty acids that are linked to a three-carbon molecule called **glycerol**. Because of this structure, fats are also called triacylglycerols or triglycerides.

As **Figure 6.2a** shows, fats form when a dehydration reaction occurs between a hydroxyl group of glycerol and the carboxyl

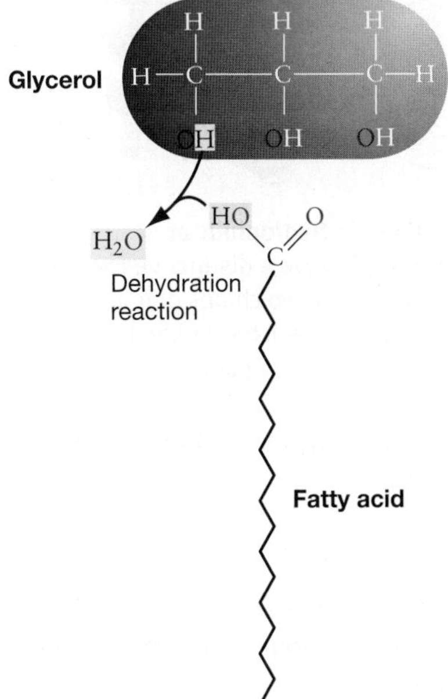

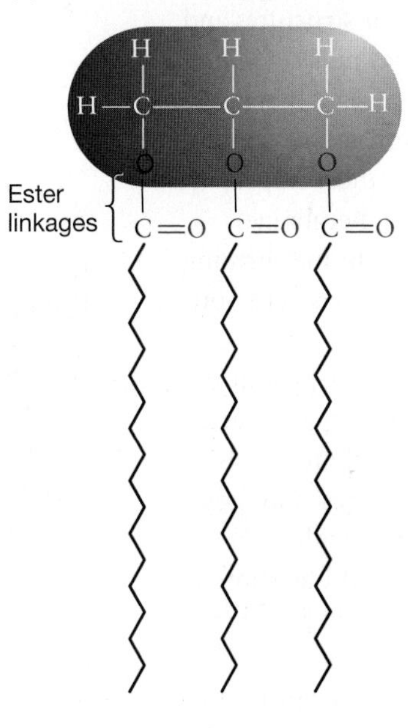

FIGURE 6.2 Fats Are One Type of Lipid Found in Cells. (a) When glycerol and a fatty acid react, a water molecule leaves. The covalent bond that results from this reaction is termed an ester linkage. **(b)** The fat shown here as a structural formula and a space-filling model is tristearin, the most common type of fat in beef.

group of a fatty acid. The glycerol and fatty-acid molecules become joined by an **ester linkage**, which is analogous to the peptide bonds, phosphodiester bonds, and glycosidic linkages in proteins, nucleic acids, and carbohydrates, respectively.

Fats are not polymers, however, and fatty acids are not monomers. As **Figure 6.2b** shows, fatty acids are not linked together to form a macromolecule in the way that amino acids, nucleotides, and monosaccharides are.

✔You should be able to explain why fats store a great deal of chemical energy, and why they are hydrophobic.

STEROIDS **Steroids** are a family of lipids distinguished by the bulky, four-ring structure shown in orange in **Figure 6.3a**. The various steroids differ from one another by the functional groups or side groups attached to those rings—shown as an "R-group" in the schematic diagram in Figure 6.3a. The space-filling model in that figure is cholesterol, which has a hydrophilic hydroxyl group attached to the rings and a hydrocarbon "tail" formed of isoprene subunits. Cholesterol is an important component of plasma membranes in many organisms.

PHOSPHOLIPIDS **Phospholipids** consist of a glycerol that is linked to a phosphate group (PO_4^{3-}) and to either two chains of isoprene or two fatty acids. The phosphate group is also bonded to a small, organic molecule that is charged or polar (**Figure 6.3b**).

Phospholipids with isoprene tails are found in the domain Archaea introduced in Chapter 1; phospholipids composed of fatty acids are found in the domains Bacteria and Eukarya. In all three domains of life, phospholipids are critically important components of the plasma membrane.

The Structures of Membrane Lipids

The lipids found in organisms have a wide array of structures and functions. In addition to storing chemical energy, lipids act as pigments that capture or respond to sunlight, serve as signals between cells, form waterproof coatings on leaves and skin, and act as vitamins used in an array of cellular processes. The most important lipid function, however, is their role in the plasma membrane.

Not all lipids can form membranes. Membrane-forming lipids have a polar, hydrophilic region—in addition to the nonpolar, hydrophobic region found in all lipids.

To better understand this structure, take another look at the phospholipid illustrated in Figure 6.3b. Notice that the molecule has a "head" region containing highly polar covalent bonds as well as positive and negative charges. The charges and polar bonds in the head region interact with water molecules when a phospholipid is placed in solution. In contrast, the long isoprene or fatty-acid tails of a phospholipid are nonpolar and hydrophobic. Water molecules cannot form hydrogen bonds with the hydrocarbon tail, and do not interact extensively with this part of the molecule.

Compounds that contain both hydrophilic and hydrophobic elements are **amphipathic** (literally, "dual-sympathy"). Phospholipids are amphipathic. As Figure 6.3a shows, cholesterol is also amphipathic. Because it has a hydroxyl functional group attached to its rings, it has both hydrophilic and hydrophobic regions.

The amphipathic nature of phospholipids is far and away their most important feature biologically. It is responsible for their presence in plasma membranes.

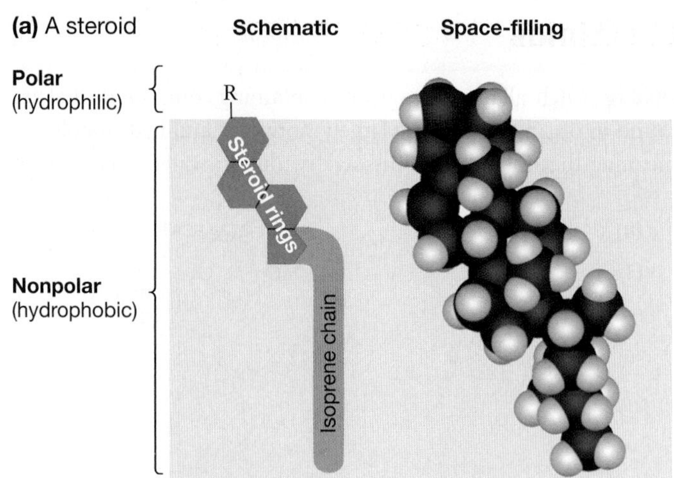

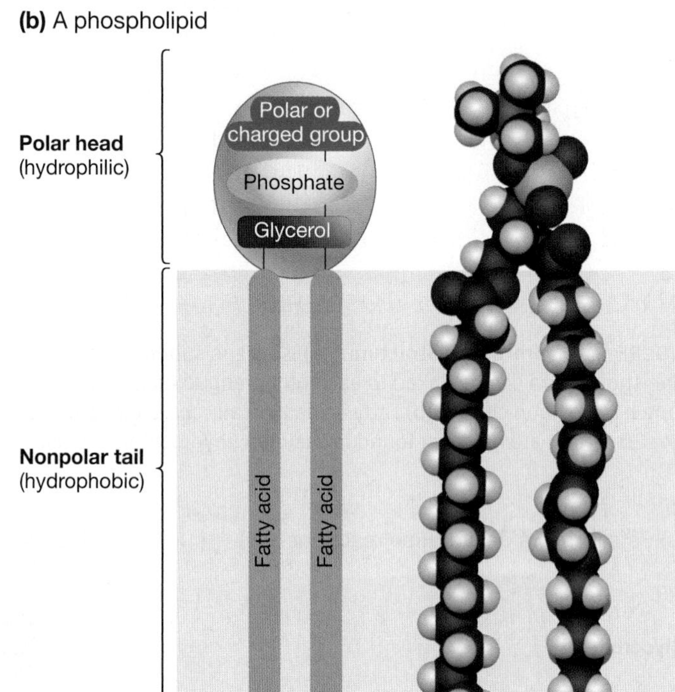

FIGURE 6.3 Amphipathic Lipids Contain Hydrophilic and Hydrophobic Elements. (a) All steroids have a distinctive four-ring structure. **(b)** All phospholipids consist of two chains of isoprene or two fatty acids that are linked to glycerol, which is linked to a phosphate group, which is linked to a small, organic molecule that is polar or charged.

✔**QUESTION** If these molecules were in solution, where would water molecules interact with them?

CHECK YOUR UNDERSTANDING

If you understand that . . .

- Fats, steroids, and phospholipids differ in structure and function.
- Fats are not amphipathic; steroids and phospholipids are.
- Fats store chemical energy; steroids and phospholipids are key components of plasma membranes.

✓ You should be able to . . .

1. Compare and contrast the structure of a fat, a steroid, and a phospholipid.
2. Based on their structure, explain why cholesterol and phospholipids are amphipathic.

Answers are available in Appendix B.

6.2 Phospholipid Bilayers

Phospholipids do not dissolve when they are placed in water. Water molecules interact with the hydrophilic heads of the phospholipids, but not with their hydrophobic tails. The lack of interaction with water drives the hydrophobic tails together.

Instead of dissolving in water, then, phospholipids form one of two types of structures: micelles or lipid bilayers.

- Micelles (**Figure 6.4a**) are tiny droplets created when the hydrophilic heads of phospholipids face the water and the hydrophobic tails are forced together, away from the water.
- Phospholipid bilayers, or simply, **lipid bilayers**, are created when two sheets of phospholipid molecules align. As **Figure 6.4b** shows, the hydrophilic heads in each layer face a surrounding solution while the hydrophobic tails face one another inside the bilayer. In this way, the hydrophilic heads interact with water while the hydrophobic tails interact with one another. These bilayers form spontaneously.

FIGURE 6.4 Phospholipids Form Micelles and Bilayers in Solution. In **(a)** a micelle or **(b)** a lipid bilayer, the hydrophilic heads of phospholipids face out, toward water; the hydrophobic tails face in, away from water. Plasma membranes consist in part of lipid bilayers.

Micelles tend to form from phospholipids with relatively short tails; bilayers tend to form from phospholipids with longer tails.

It's critical to recognize that micelles and phospholipid bilayers form spontaneously—no input of energy is required. This concept can be difficult to grasp, because entropy clearly decreases when these structures form. Micelles and lipid bilayers are much more highly organized than phospholipids floating free in the solution.

The key is to recognize that micelles and lipid bilayers are much more stable energetically than are independent phospholipids in solution. Stated another way, the structures have much lower potential energy than do the independent molecules in solution. Independent phospholipids are unstable in water because their hydrophobic tails disrupt hydrogen bonds that otherwise would form between water molecules. As a result, amphipathic molecules are much more stable in aqueous solution when their hydrophobic tails avoid water, and instead participate in the hydrophobic interactions introduced in Chapter 3.

In this case, the loss of potential energy outweighs the decrease in entropy. Overall, the free energy of the system decreases. Lipid bilayer formation is exergonic and spontaneous.

Artificial Membranes as an Experimental System

When lipid bilayers are agitated by shaking, the layers break and re-form as small, spherical structures. The resulting vesicles have water on the inside as well as the outside, because the hydrophilic heads of the lipids face outward on each side of the bilayer. Artificial membrane-bound vesicles like these are called liposomes.

To explore how plasma membranes work, researchers began creating and experimenting with liposomes (**Figure 6.5**) and

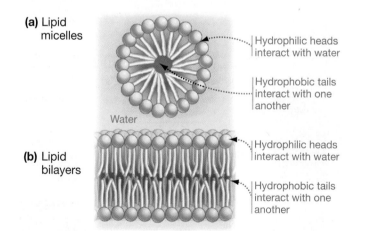

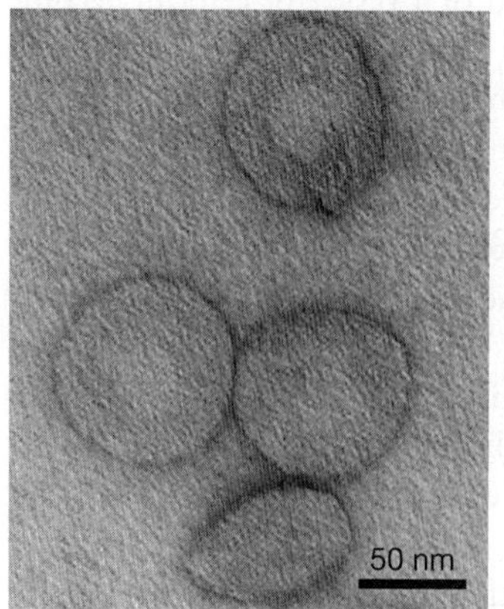

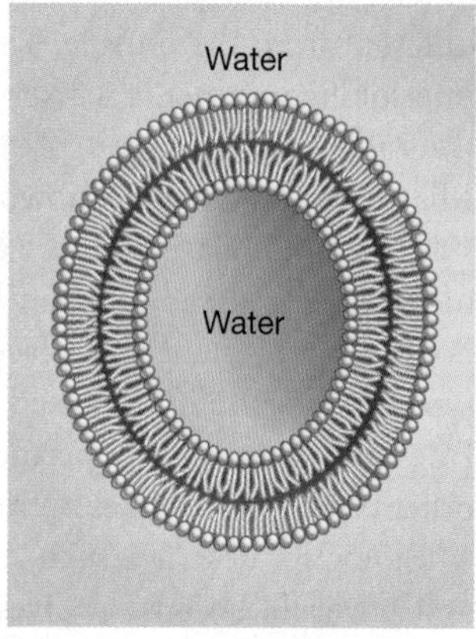

FIGURE 6.5 Liposomes Are Artificial Membrane-Bound Vesicles. Electron micrograph of liposomes in cross section (left) and a cross-sectional diagram of the lipid bilayer in a liposome (right).

(a) Planar bilayers: Artificial membranes

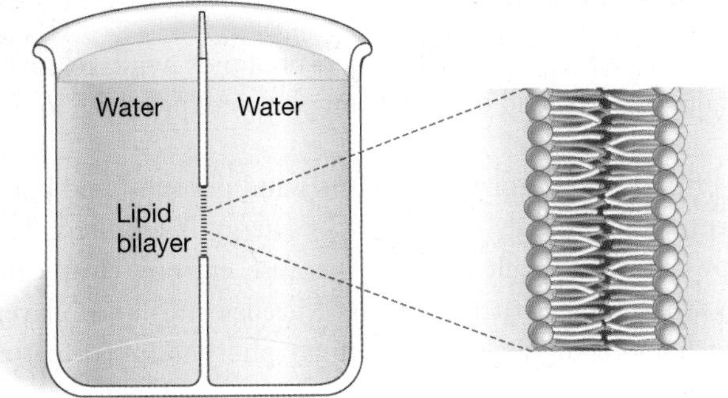

(b) Artificial-membrane experiments

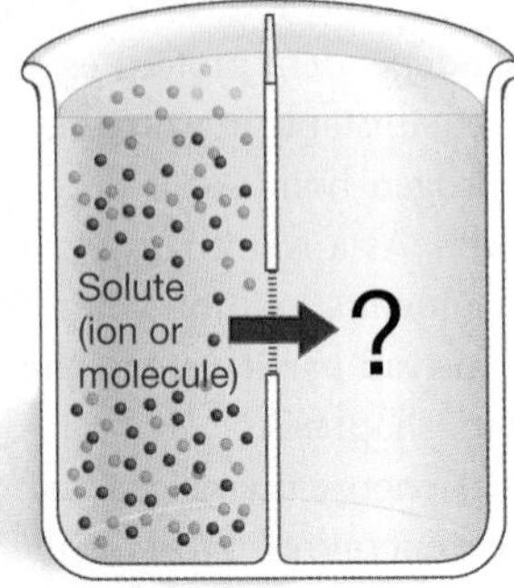

How rapidly can different solutes cross the membrane (if at all) when . . .

1. Different types of phospholipids are used to make the membrane?

2. Proteins or other molecules are added to the membrane?

FIGURE 6.6 Planar Bilayers Are Artificial Membranes. (a) The construction of planar bilayers across a hole in a glass wall separating two water-filled compartments (left), and a close-up sketch of the bilayer (right). **(b)** A wide variety of experiments are possible with planar bilayers; a few are suggested here.

planar bilayers—lipid bilayers constructed across a hole in a glass or plastic wall separating two aqueous solutions (**Figure 6.6a**). Some of the first questions they posed concerned the permeability of lipid bilayers. The **permeability** of a structure is its tendency to allow a given substance to pass across it.

Membrane permeability is a critical issue, because if certain molecules or ions pass through a lipid bilayer more readily than others, the internal environment of a vesicle or cell can become different from the outside. This difference between exterior and interior environments is a key characteristic of cells.

Using liposomes and planar bilayers, researchers can study what happens when a known ion or molecule is added to one side of a lipid bilayer (**Figure 6.6b**). Does the substance cross the membrane and show up on the other side? If so, how rapidly does the movement take place? What happens when a different type of phospholipid is used to make the artificial membrane? Does the membrane's permeability change when proteins or other types of molecules become part of it?

Biologists describe such an experimental system as elegant and powerful because it gives them precise control over which factor changes from one experimental treatment to the next. Control, in turn, is why experiments are such an effective way to explore scientific questions. You might recall from Chapter 1 that good experimental design allows researchers to alter one factor at a time and determine what effect, if any, each has on the process being studied.

Selective Permeability of Lipid Bilayers

When researchers put molecules or ions on one side of a liposome or planar bilayer and measure the rate at which the molecules arrive on the other side, a clear pattern emerges: Lipid bilayers are highly selective.

Selective permeability means that some substances cross a membrane more easily than other substances can. Small, nonpolar molecules move across bilayers quickly. In contrast, large molecules and charged substances cross the membrane slowly, if at all.

According to the data in **Figure 6.7**, small, nonpolar molecules such as oxygen (O_2) move across selectively permeable membranes more than a billion times faster than do chloride ions (Cl^-). In essence, ions cannot cross membranes at all—unless they have "help" in the form of membrane proteins introduced later in the chapter. Very small and uncharged molecules such as water (H_2O) can cross membranes relatively rapidly, even if they are polar. Small, polar molecules such as glycerol have intermediate permeability.

The leading hypothesis to explain this pattern is that charged compounds and large, polar molecules can't pass through the nonpolar, hydrophobic tails of a lipid bilayer. Because of their electrical charge, ions are more stable in solution where they interact with water than they are in the interior of membranes, which is electrically neutral. To test the hypothesis, researchers have manipulated the size and structure of the tails in liposomes or planar bilayers. ✔If you understand this hypothesis, you should be able to predict whether amino acids and nucleotides will cross a membrane readily.

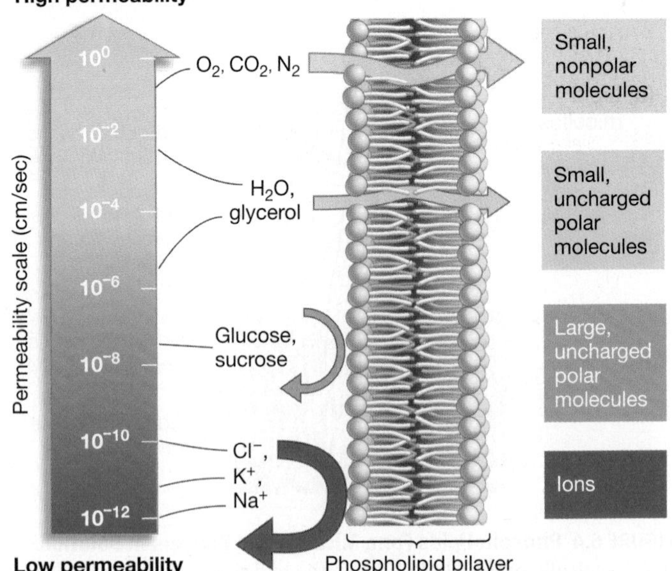

FIGURE 6.7 Lipid Bilayers Show Selective Permeability. Only certain substances cross lipid bilayers readily. Size and charge affect the rate of diffusion across a membrane.

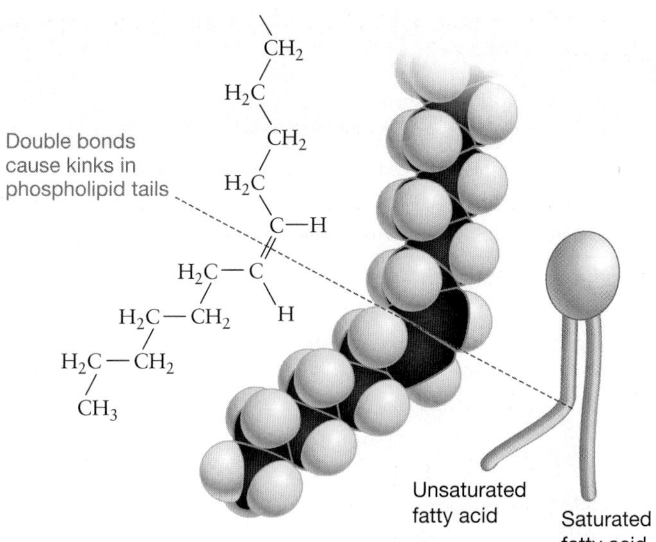

FIGURE 6.8 Unsaturated Hydrocarbons Contain Carbon-Carbon Double Bonds. The icon on the right indicates that one of the hydrocarbon tails in a phospholipid is unsaturated and therefore kinked.

How Does Lipid Structure Affect Membrane Properties?

The correlation between the structure of a molecule and its function in cells may be the most basic theme in biological chemistry. This is certainly true of lipids. The hydrophilic-head-and-hydrophobic-tail structure of phospholipids allows them to form lipid bilayers and membranes. But the number of double bonds in a hydrophobic tail and its overall length, in addition to the number of cholesterol molecules nearby, has a profound influence on how a membrane behaves.

BOND SATURATION IS AN IMPORTANT ASPECT OF LIPID STRUCTURE When two carbon atoms form a double bond, the attached atoms are found in a plane instead of a three-dimensional tetrahedron. The carbon atoms involved are also locked into place. They cannot rotate freely, as they do in carbon-carbon single bonds. As a result, certain double bonds between carbon atoms produce a "kink" in an otherwise straight hydrocarbon chain (**Figure 6.8**).

Hydrocarbon chains without double bonds are said to be **saturated**. But if a hydrocarbon chain contains a double bond, the chain is said to be **unsaturated**. This choice of terms is logical. If a hydrocarbon chain does not contain a double bond, it is saturated with the maximum number of hydrogen atoms that can attach to the carbon skeleton. If it is unsaturated, then fewer than the maximum number of hydrogen atoms are attached.

Because they contain more C–H bonds, which have much more free energy than C=C bonds, saturated fats have more chemical energy than unsaturated fats do. People who are dieting are often encouraged to eat fewer saturated fats, to reduce their energy intake. Foods that contain lipids with many double bonds are said to be polyunsaturated and are advertised as healthier than foods with more-saturated fat.

BOND SATURATION AND HYDROCARBON CHAIN LENGTH CHANGE MEMBRANE FLUIDITY AND PERMEABILITY The degree of saturation in a fat—along with the length of its hydrocarbon tails—affects key aspects of a lipid's behavior in a membrane.

- When hydrophobic tails are packed into a lipid bilayer, the kinks created by double bonds produce spaces among the tightly packed tails. These spaces reduce the strength of hydrophobic interactions between the tails. These interactions are stronger among saturated hydrocarbon tails.
- Hydrophobic interactions also become stronger as saturated hydrocarbon tails increase in length.

These observations have profound impacts on membrane fluidity. Highly saturated fats, such as butter, are solid at room temperature (**Figure 6.9a**). Lipids that have extremely long hydrocarbon tails, as **waxes** do, form particularly stiff solids at room

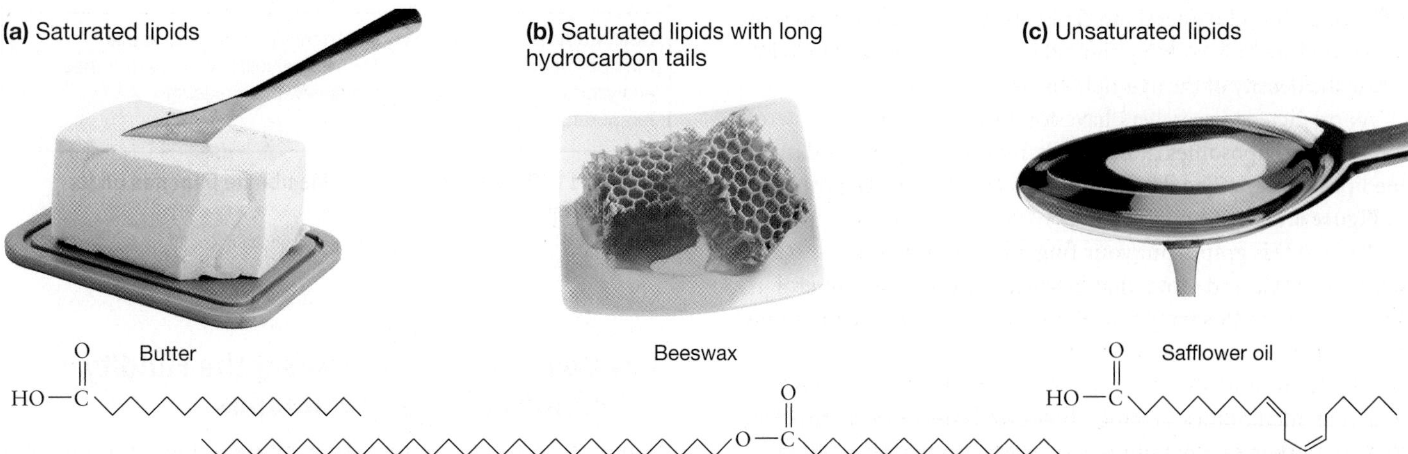

FIGURE 6.9 The Fluidity of Lipids Depends on the Length and Saturation of Their Hydrocarbon Chains. (a) Butter consists primarily of saturated lipids. **(b)** Waxes are lipids with extremely long hydrocarbon chains. **(c)** Oils are dominated by "polyunsaturates"—lipids with hydrocarbon chains that contain multiple double bonds.

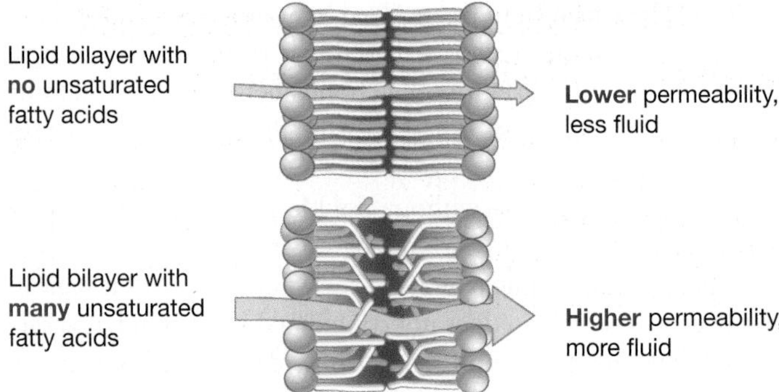

FIGURE 6.10 Fatty-Acid Structure Changes the Permeability of Membranes. Lipid bilayers containing many unsaturated fatty acids have more gaps and should be more permeable than bilayers with few unsaturated fatty acids.

temperature (**Figure 6.9b**). Highly unsaturated fats are liquid at room temperature (**Figure 6.9c**). Liquid triacylglycerols are called **oils**.

Membrane permeability is affected as well—fluidity and permeability are closely related. As **Figure 6.10** shows, lipid bilayers become more permeable as well as more fluid when they consist of short, unsaturated hydrocarbon tails. The membrane allows more materials to pass, because its interior is held together less tightly.

If these hypotheses are correct, then bilayers made of lipids with long, straight, saturated fatty-acid tails should be much less permeable than membranes made of lipids with short, kinked, unsaturated fatty-acid tails. Experiments on liposomes have shown exactly this pattern.

The take-home message is clear: The degree of hydrophobic interactions has a strong impact on the behavior of phospholipid bilayers.

CHOLESTEROL REDUCES MEMBRANE PERMEABILITY In addition to exploring the role of hydrocarbon chain length and degree of saturation on membrane permeability, biologists have investigated the effect of adding cholesterol molecules. Because the steroid rings in cholesterol are bulky, adding cholesterol to a membrane should increase the density of the hydrophobic section.

As predicted, researchers have found that adding cholesterol molecules to liposomes dramatically reduces the permeability of the lipid bilayers. The data behind this conclusion are presented in **Figure 6.11**.

To read this graph, put your finger on the *x*-axis at the point marked 20°C, and note that permeability to the glycerol is much higher at this temperature in membranes that contain no cholesterol versus 20 percent or 50 percent cholesterol. Using this procedure at other temperature points should convince you that membranes lacking cholesterol are more permeable than the other two membranes at every temperature tested in the experiment. Figure 6.11 also shows that for any of the three membranes, regardless of cholesterol content, permeability increases with temperature.

EXPERIMENT

QUESTION: Does adding cholesterol to a membrane affect its permeability?

HYPOTHESIS: Cholesterol reduces permeability because it fills spaces in phospholipid bilayers.

NULL HYPOTHESIS: Cholesterol has no effect on permeability.

EXPERIMENTAL SETUP:

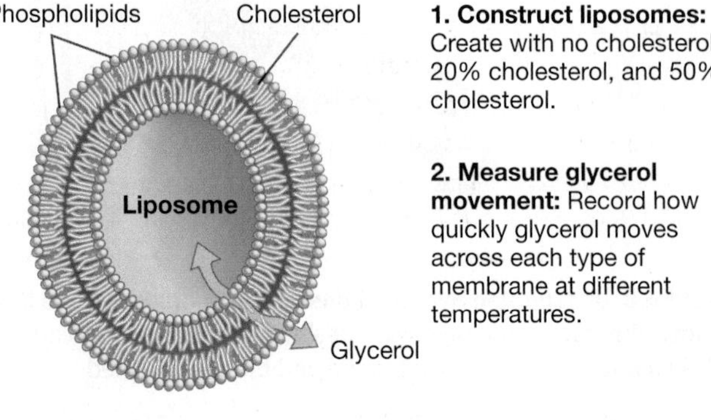

1. Construct liposomes: Create with no cholesterol, 20% cholesterol, and 50% cholesterol.

2. Measure glycerol movement: Record how quickly glycerol moves across each type of membrane at different temperatures.

PREDICTION: Liposomes with higher cholesterol levels will have reduced permeability.

PREDICTION OF NULL HYPOTHESIS: All liposomes will have the same permeability.

RESULTS:

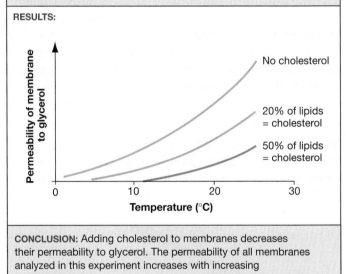

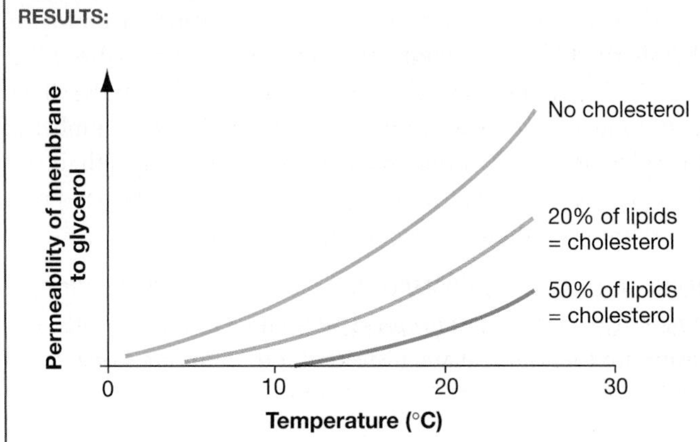

CONCLUSION: Adding cholesterol to membranes decreases their permeability to glycerol. The permeability of all membranes analyzed in this experiment increases with increasing temperature.

FIGURE 6.11 The Permeability of a Membrane Depends on Its Composition.

SOURCE: J. de Gier et al. (1968). Lipid composition and permeability of liposomes. *Biochimica et Biophysica Acta* 150: 666–675.

How Does Temperature Affect the Fluidity and Permeability of Membranes?

At about 25°C—or "room temperature"—the phospholipids found in plasma membranes are liquid, and bilayers have the consistency of olive oil. This fluidity, as well as the membrane's permeability, decreases as temperature decreases. Why?

As temperatures drop, individual molecules in the bilayer move more slowly. As a result, the hydrophobic tails in the interior of membranes pack together more tightly. At very low temperatures, lipid bilayers begin to solidify. As the graph in Figure 6.11 indicates, low temperatures can make membranes impervious to molecules that would normally cross them readily. Put your finger on the *x*-axis just about the freezing point of water (0°C), and note that even membranes that lack cholesterol are almost completely impermeable to glycerol. Indeed, trace any of the three lines in Figure 6.11, and as you move to the right (increasing temperature), you also move up (increasing permeability).

The fluid nature of membranes also allows individual lipid molecules to move laterally within each layer, a little like a person moving about in a dense crowd (**Figure 6.12**). By tagging individual phospholipids and following their movement, researchers have clocked average speeds of 2 micrometers (μm)/second at room temperature. At these speeds, phospholipids could travel the length of a small bacterial cell in a second.

These experiments on lipid and ion movement demonstrate that membranes are dynamic. Phospholipid molecules whiz around each layer, while water and small, nonpolar molecules shoot in and out of the membrane. How quickly molecules move within and across membranes is a function of temperature and the structure of the hydrocarbon tails in the bilayer.

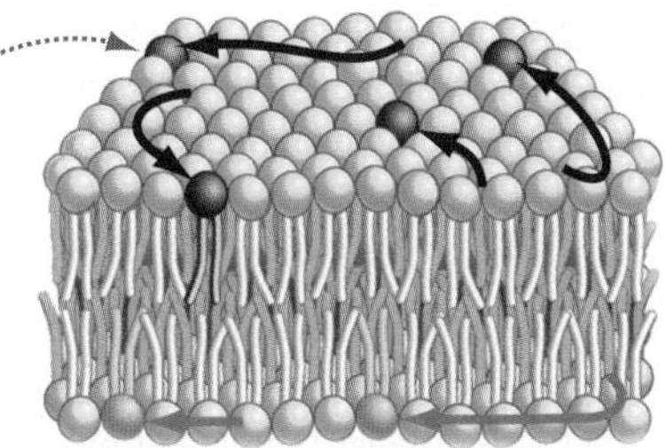

Phospholipids are in constant lateral motion, but rarely flip to the other side of the bilayer

FIGURE 6.12 Phospholipids Move within Membranes. Membranes are dynamic—in part because phospholipid molecules move laterally within each layer in the structure.

CHECK YOUR UNDERSTANDING

If you understand that . . .

- In water, phospholipids form bilayers that are selectively permeable—meaning that some substances cross them much more readily than others do.
- Permeability is a function of the degree of saturation and the length of the hydrocarbon tails in membrane phospholipids, the amount of cholesterol in the membrane, and the temperature.

✓ **You should be able to . . .**

Fill in a chart with rows called Saturation of hydrocarbon tails, Length of hydrocarbon tails, Cholesterol, and Temperature, and columns named Factor, Effect on permeability, and Reason.

Answers are available in Appendix B.

6.3 Why Molecules Move across Lipid Bilayers: Diffusion and Osmosis

Small and uncharged molecules and hydrophobic compounds can cross membranes readily and spontaneously—without an expenditure of energy. The question now is: How is this possible? What process is responsible for the movement of molecules across phospholipid bilayers?

Diffusion

A thought experiment can help explain why molecules and ions can cross membranes spontaneously. Suppose you rack up a set of blue billiard balls on a pool table that contains many white balls, and then begin to vibrate the table.

1. Because of the vibration, the balls will move about randomly. They will also bump into one another.
2. After these collisions, some blue balls will move outward—away from their original position.
3. As movement and collisions continue, the overall or net movement of blue balls will be outward. This occurs because the random motion of the blue balls disrupts their original, nonrandom position. As the blue balls move at random, they are more likely to move away from one another than to stay together.
4. Eventually, the blue billiard balls will be distributed randomly across the table. The entropy of the blue billiard balls has increased.

Recall from Chapter 2 that entropy is a measure of the randomness or disorder in a system. The second law of thermodynamics states that in a closed system, entropy always increases.

This hypothetical example illustrates why molecules or ions located on one side of a lipid bilayer move to the other side spontaneously. The dissolved molecules and ions, or **solutes**, have thermal energy and are in constant, random motion. Movement of molecules and ions that results from their kinetic energy is known as **diffusion**. Because solutes change position randomly owing to diffusion, they tend to move from a region of high concentration to a region of low concentration. A difference in solute concentrations creates a **concentration gradient**.

Molecules and ions move randomly in all directions when a concentration gradient exists, but there is a net movement from regions of high concentration to regions of low concentration. Diffusion along a concentration gradient is a spontaneous process because it results in an increase in entropy.

Once the molecules or ions are randomly distributed throughout a solution, equilibrium is established. For example, consider two aqueous solutions separated by a lipid bilayer. **Figure 6.13** shows how molecules that can pass through the bilayer diffuse to the other side. At equilibrium, molecules continue to move back and forth across the membrane, but at equal rates—simply because each molecule or ion is equally likely to move in any direction. This means that there is no longer a net movement of molecules across the membrane.

PROCESS: DIFFUSION ACROSS A LIPID BILAYER

Lipid bilayer

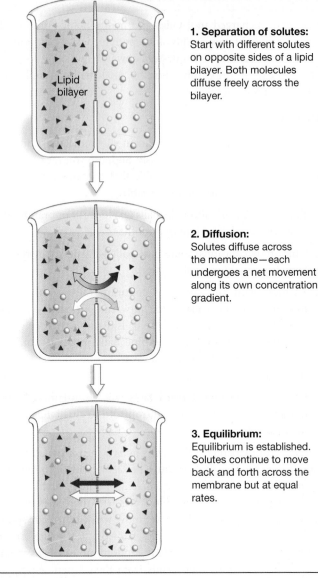

1. Separation of solutes: Start with different solutes on opposite sides of a lipid bilayer. Both molecules diffuse freely across the bilayer.

2. Diffusion: Solutes diffuse across the membrane—each undergoes a net movement along its own concentration gradient.

3. Equilibrium: Equilibrium is established. Solutes continue to move back and forth across the membrane but at equal rates.

FIGURE 6.13 Diffusion across a Selectively Permeable Membrane Establishes an Equilibrium.

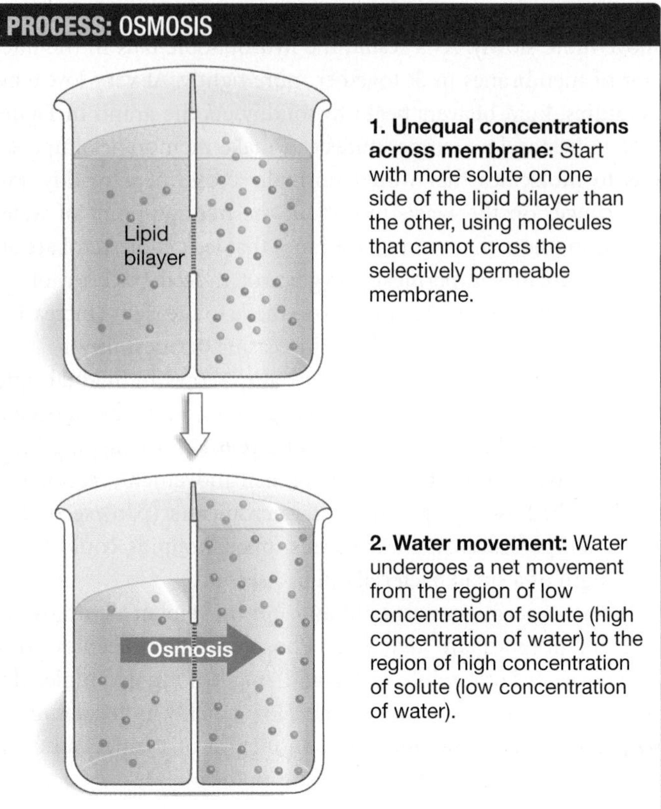

PROCESS: OSMOSIS
Lipid bilayer
1. Unequal concentrations across membrane: Start with more solute on one side of the lipid bilayer than the other, using molecules that cannot cross the selectively permeable membrane.
Osmosis
2. Water movement: Water undergoes a net movement from the region of low concentration of solute (high concentration of water) to the region of high concentration of solute (low concentration of water).

FIGURE 6.14 Osmosis Is the Diffusion of Water.

✔**QUESTION** Suppose you doubled the number of solute molecules on the left side of the membrane (at the start). At equilibrium, would the water level on the left side be higher or lower than what is shown in the second drawing?

Osmosis

What about water? As the data in Figure 6.7 showed, water moves across lipid bilayers relatively quickly. Like other substances that diffuse, water moves along its concentration gradient—from higher to lower concentration. The movement of water is a special case of diffusion that is given its own name: osmosis. **Osmosis** occurs only when solutions are separated by a membrane that is permeable to some molecules but not others—that is, a selectively permeable membrane.

The best way to think about water moving in response to a concentration gradient is to focus on the concentration of solutes in the solution. Let's suppose the concentration of a particular solute is higher on one side of a selectively permeable membrane than it is on the other side (**Figure 6.14**, step 1). Further, suppose that this solute cannot diffuse through the membrane to establish equilibrium. What happens? Water will move from the side with a lower concentration of solute to the side with a higher concentration of solute (step 2). It dilutes the higher concentration and equalizes the concentrations on both sides.

This movement of water is spontaneous. It is driven by the increase in entropy achieved when solute concentrations are equal on both sides of the membrane.

Another way to think about osmosis is to realize that water is at higher concentration on the left side of the beaker in Figure 6.14 than it is on the right side of the beaker. As water diffuses, then, there will be net movement of water molecules from the left side to the right side: from a region of high concentration to a region of low concentration.

The movement of water by osmosis is important because it can swell or shrink a membrane-bound vesicle. Consider the liposomes illustrated in **Figure 6.15**.

- *Left* If the solution outside the membrane has a higher concentration of solutes than the interior has, and the solutes are not able to pass through the lipid bilayer, then water will move out of the vesicle into the solution outside. As a result, the

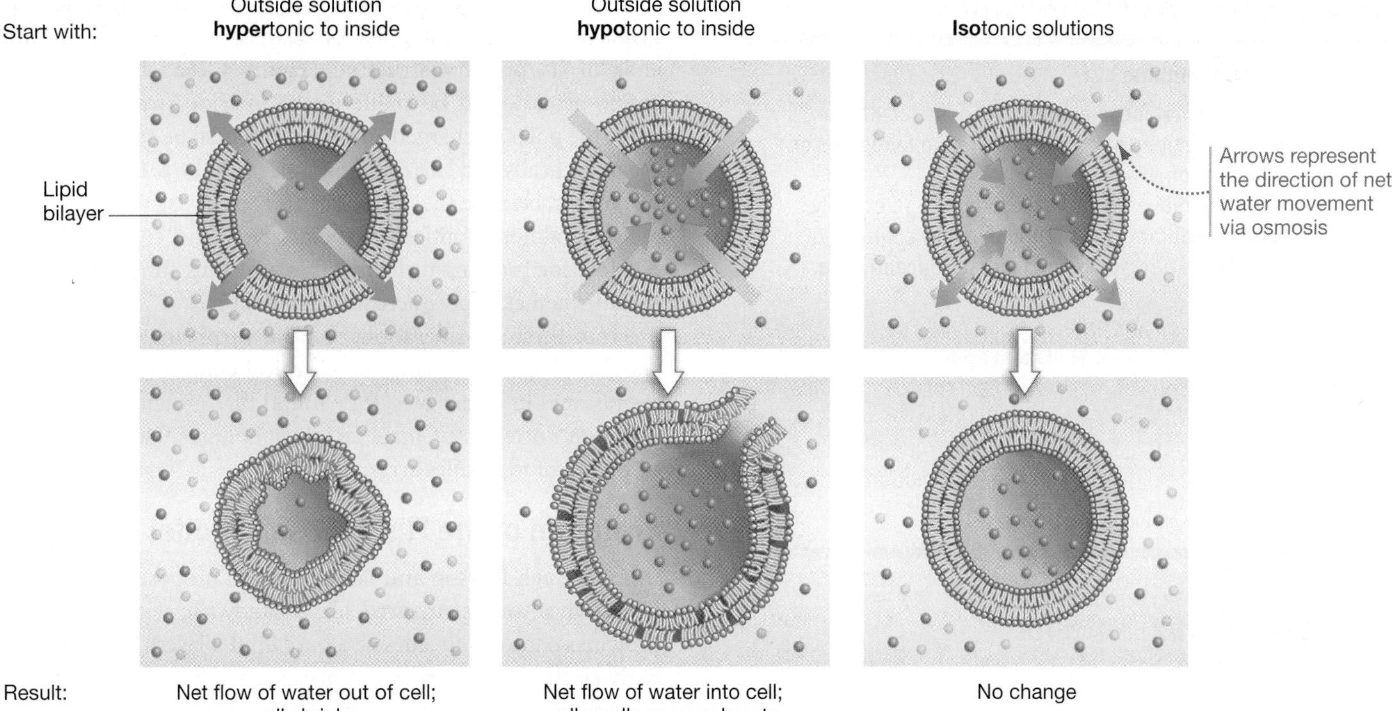

FIGURE 6.15 Osmosis Can Shrink or Burst Membrane-Bound Vesicles.

vesicle will shrink and the membrane shrivel. The solution outside is said to be **hypertonic** ("excess-tone") relative to the inside of the vesicle. The word root hyper refers to the outside solution's containing more solutes than the solution on the other side of the membrane.

- *Middle* If the solution outside the membrane has a lower concentration of solutes than the interior has, water will move into the vesicle via osmosis. The incoming water will cause the vesicle to swell or even burst. The outside solution is termed **hypotonic** ("lower-tone") relative to the inside of the vesicle. Here the word root hypo refers to the outside solution's containing fewer solutes than the inside solution has.
- *Right* If solute concentrations are equal on either side of the membrane, the liposome will maintain its size. When the outside solution does not affect the membrane's shape, that solution is called **isotonic** ("equal-tone").

Note that the terms hypertonic, hypotonic, and isotonic are relative—they can be used only to express the relationship between a given solution and another solution. To make sure that you understand these concepts, review them in the study area at *www.masteringbiology.com.*

(MB) **Web Activity** Diffusion and Osmosis

✔You should be able to (1) modify the drawing on the left side of Figure 6.15, such that the surrounding solution is hypotonic relative to the solution inside the liposome, and (2) modify the drawing in the center of Figure 6.15, such that the surrounding solution is hypertonic relative to the solution inside the liposome.

What does all this have to do with the first membranes floating in the prebiotic soup? Osmosis and diffusion tend to *reduce* differences in chemical composition between the inside and outside of membrane-bound structures. If liposome-like structures were present in the prebiotic soup, it's unlikely that their interiors offered a radically different environment from the surrounding solution. In all likelihood, the primary importance of the first lipid bilayers was simply to provide a container for self-replicating molecules.

It's important to note, though, that ribonucleotides can diffuse across lipid bilayers—meaning that monomers would be available for a ribozyme to make a copy of itself inside a vesicle. Further, experiments have shown that cell-like vesicles grow as additional lipids are added and then divide if sheared by shaking, bubbling, or wave action. On the basis of these observations, it is reasonable to hypothesize that once a self-replicating ribozyme had become surrounded by a lipid bilayer, this simple life-form and its descendants would continue to occupy cell-like structures that grew and divided.

Now let's investigate the next great event in the evolution of life: the formation of a true cell. How could lipid bilayers become a barrier capable of creating and maintaining a specialized internal environment that was conducive to life? How could an effective plasma membrane—one that admits ions and molecules needed by the replicator while excluding ions and molecules that might damage it—evolve in the first cell?

CHECK YOUR UNDERSTANDING

If you understand that . . .

- Diffusion is the net movement of ions or molecules in solution from regions of high concentration to regions of low concentration.
- Osmosis is the movement of water across a selectively permeable membrane, from a region of low solute concentration to a region of high solute concentration.

✓ You should be able to . . .

Make a concept map (see **BioSkills 8** in Appendix A) that includes the boxed terms water molecules, solute molecules, solution, osmosis, diffusion, areas of high-to-low concentration, semipermeable membrane, concentration gradients, hypertonic solutions, hypotonic solutions, and isotonic solutions.

Answers are available in Appendix B.

6.4 Membrane Proteins

What sort of molecule could become incorporated into a lipid bilayer and affect the bilayer's permeability? The title of this section gives the answer away. Proteins that are amphipathic can be inserted into lipid bilayers.

Proteins can be amphipathic because they are made up of amino acids, and because amino acids have side chains, or R-groups, that range from highly nonpolar to highly polar (see Figure 3.3 and Table 3.1). It's conceivable, then, that a protein could have a series of nonpolar amino acids in the middle of its primary structure, but polar or charged amino acids on both ends of its primary structure (**Figure 6.16a**). The nonpolar amino acids would be stable in the interior of a lipid bilayer, while the polar or charged amino acids would be stable alongside the polar heads and surrounding water (**Figure 6.16b**).

Further, because the secondary and tertiary structures of proteins are almost limitless in their variety and complexity, it is possible for proteins to form tubes and thus function as some sort of channel or pore across a lipid bilayer.

From these considerations, it's not surprising that when researchers began analyzing the chemical composition of plasma membranes in eukaryotes, they found that proteins were just as common, in terms of mass, as phospholipids. How were these two types of molecules arranged?

Evolution of the Fluid-Mosaic Model

In 1935 Hugh Davson and James Danielli proposed that plasma membranes were structured like a sandwich, with hydrophilic proteins coating both sides of a pure lipid bilayer (**Figure 6.17a**). Early electron micrographs of plasma membranes seemed to be consistent with the sandwich model, and for decades it was widely accepted.

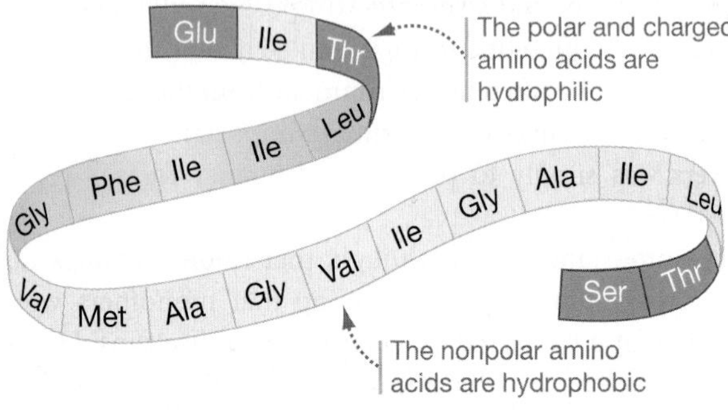

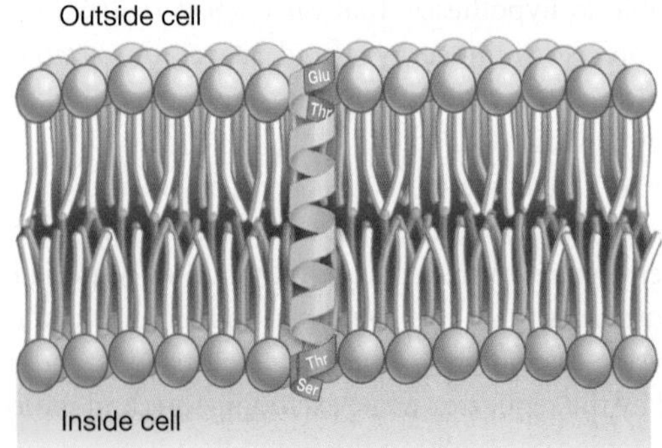

FIGURE 6.16 Amphipathic Proteins Are Stable in Lipid Bilayers.

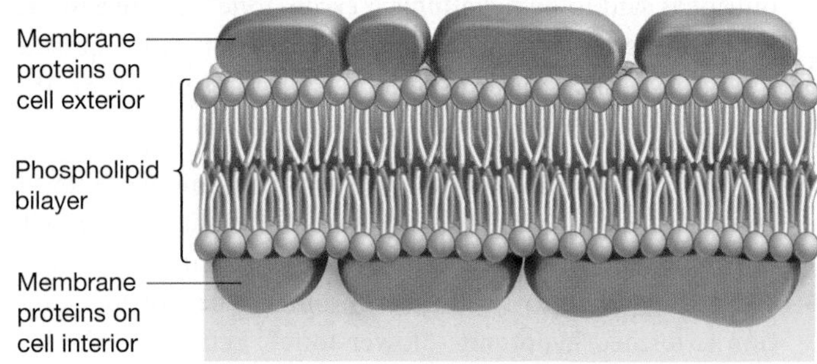

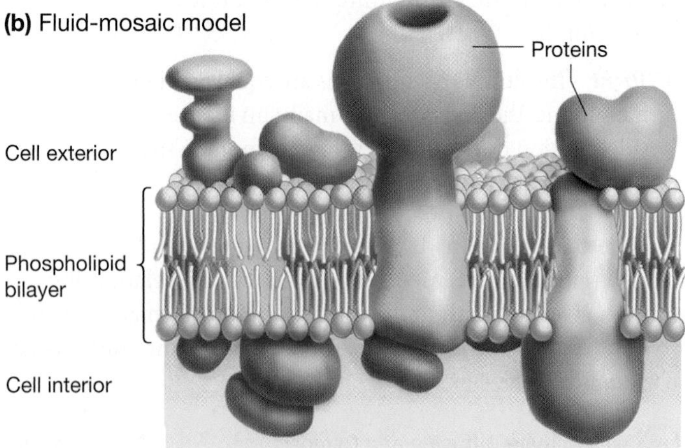

FIGURE 6.17 Past and Current Models of Membrane Structure Differ in Where Membrane Proteins Reside. (a) The protein-lipid-lipid-protein sandwich model was the first hypothesis for the arrangement of lipids and proteins in plasma membranes. **(b)** The fluid-mosaic model was a radical departure from the sandwich hypothesis.

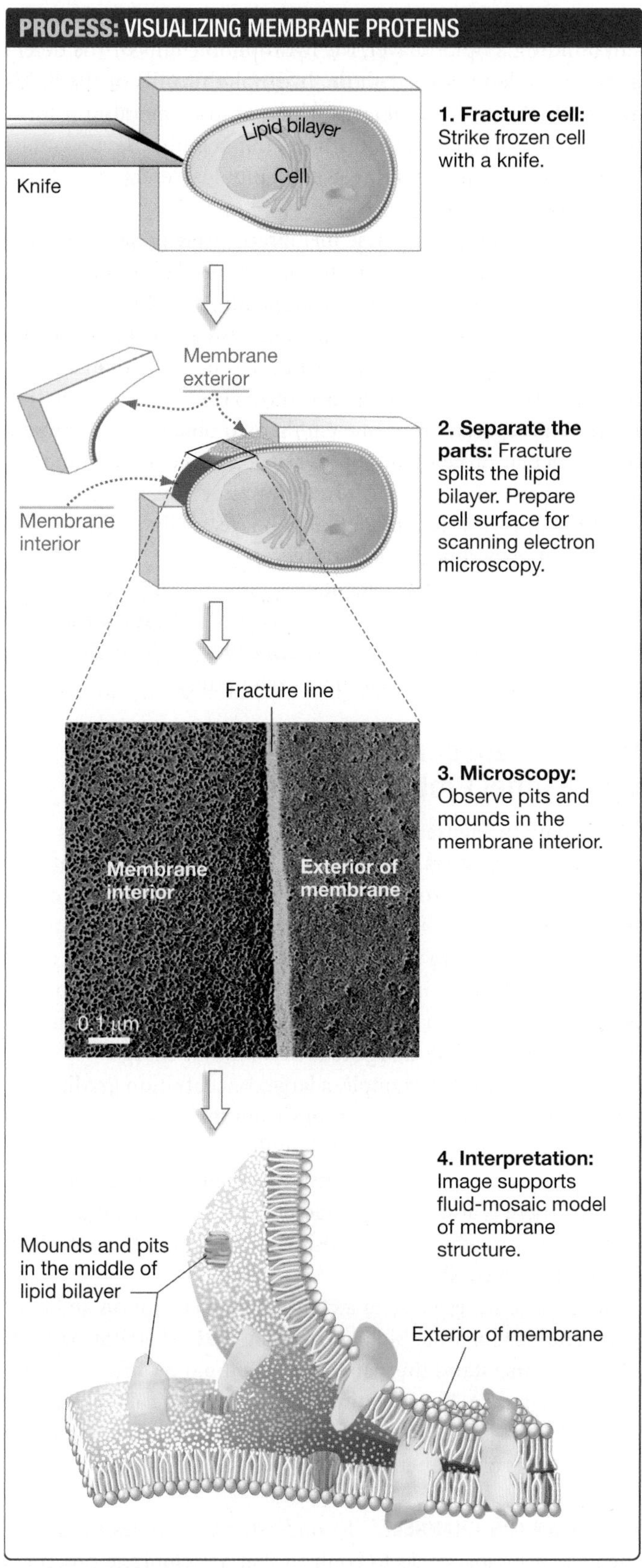
PROCESS: VISUALIZING MEMBRANE PROTEINS
Lipid bilayer
Cell
Knife
1. Fracture cell: Strike frozen cell with a knife.
Membrane exterior
Membrane interior
2. Separate the parts: Fracture splits the lipid bilayer. Prepare cell surface for scanning electron microscopy.
Fracture line
Membrane interior
Exterior of membrane
0.1 μm
3. Microscopy: Observe pits and mounds in the membrane interior.
4. Interpretation: Image supports fluid-mosaic model of membrane structure.
Mounds and pits in the middle of lipid bilayer
Exterior of membrane

FIGURE 6.18 Freeze-Fracture Preparations Allow Biologists to View Membrane Proteins. The scanning electron micrograph in Step 3 shows a freeze-fractured membrane that has been colored to match the drawing in Step 2.

The realization that membrane proteins could be amphipathic led S. Jon Singer and Garth Nicolson to suggest an alternative hypothesis, however. In 1972, they proposed that at least some proteins span the membrane instead of being found only outside the lipid bilayer. Their hypothesis was called the **fluid-mosaic model** (**Figure 6.17b**). Singer and Nicolson suggested that membranes are a mosaic of phospholipids and different types of proteins. The overall structure was proposed to be dynamic and fluid.

The controversy over the nature of the plasma membrane was resolved in the early 1970s with the development of an innovative technique for visualizing the surface of plasma membranes. The method is called freeze-fracture electron microscopy, because the steps involve freezing and fracturing the membrane before examining it with a **scanning electron microscope**, which produces images of an object's surface (see **BioSkills 10** in Appendix A).

As **Figure 6.18** shows, the freeze-fracture technique allows researchers to split plasma membranes and view the middle of the structure. The scanning electron micrographs that result show pits and mounds studding the inner surfaces of the lipid bilayer. Researchers interpret these structures as the locations of membrane proteins. As step 4 in Figure 6.18 shows, the pits and mounds are hypothesized to represent proteins that span the lipid bilayer.

These observations conflicted with the sandwich model but were consistent with the fluid-mosaic model. On the basis of these and subsequent observations, the fluid-mosaic model is now widely accepted.

Figure 6.19 summarizes the current hypothesis for where proteins and lipids are found in a plasma membrane. Note that some proteins span the membrane and have segments facing both the interior and the exterior surfaces. Proteins such as these are called **integral membrane proteins**, or **transmembrane proteins**. Other proteins, called **peripheral membrane proteins**, are found only on one side of the membrane.

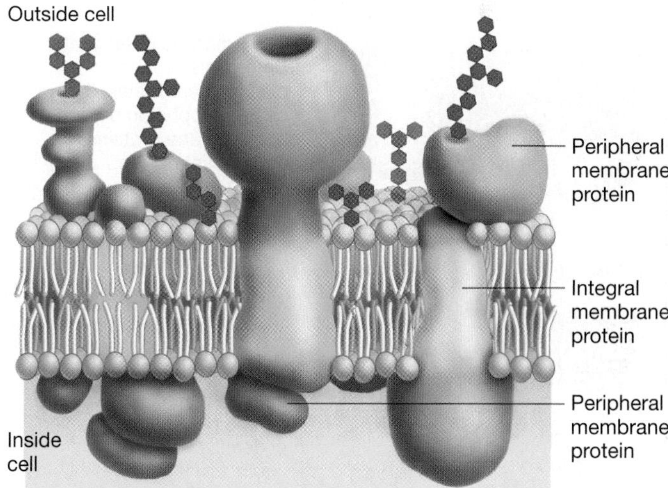
Outside cell
Peripheral membrane protein
Integral membrane protein
Peripheral membrane protein
Inside cell

FIGURE 6.19 Integral and Peripheral Membrane Proteins Individualize Bilayer Surfaces. The interior and exterior surfaces of the plasma membrane are different, because their peripheral proteins differ and the ends of the transmembrane proteins differ.

In most cells, certain peripheral proteins are found only on the interior surface of the plasma membrane and thus inside the cell, while others are found only on the exterior surface and thus face the surrounding environment. Peripheral membrane proteins are often attached to transmembrane membrane proteins.

What do all these proteins do? Later chapters will explore how certain membrane proteins act as enzymes, receive signals from other cells, or make physical connections between cells. Here, let's focus on how transmembrane proteins are involved in the transport of selected ions and molecules across the plasma membrane.

Systems for Studying Membrane Proteins

The discovery of transmembrane proteins was consistent with the hypothesis that proteins affect membrane permeability. The evidence was not considered conclusive, though, because it was also plausible to claim that transmembrane proteins were structural components that influenced membrane strength or flexibility. To test whether proteins actually do affect membrane permeability, researchers needed some way to isolate and purify membrane proteins.

Figure 6.20 outlines one method that researchers developed to separate proteins from membranes. The key to the technique is the use of detergents. A **detergent** is a small, amphipathic molecule. When detergents are added to the solution surrounding a lipid bilayer, the hydrophobic tails of the detergent molecule interact with the hydrophobic tails of the lipids and with the hydrophobic portions of transmembrane proteins. These interactions displace the membrane phospholipids and end up forming water-soluble, detergent-protein complexes.

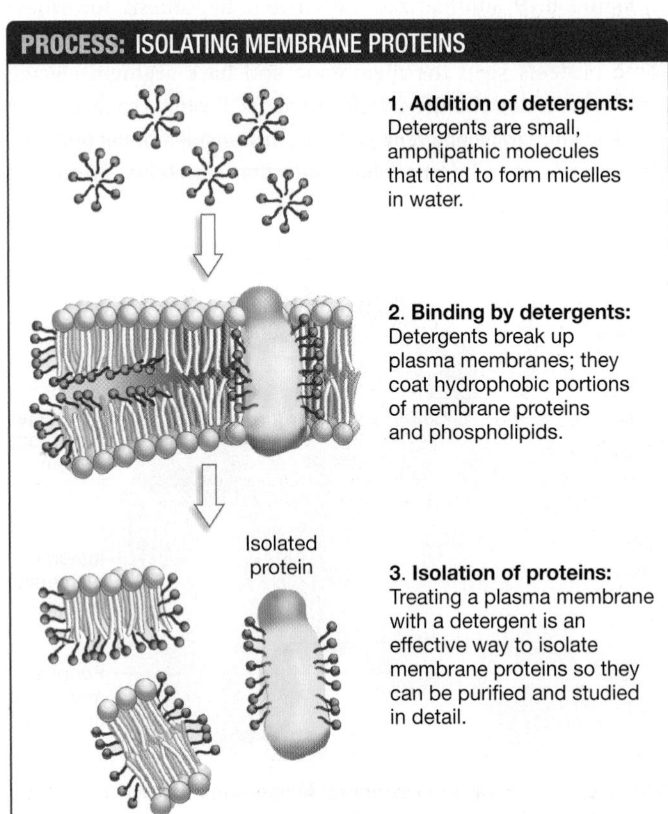

PROCESS: ISOLATING MEMBRANE PROTEINS
1. Addition of detergents: Detergents are small, amphipathic molecules that tend to form micelles in water.
2. Binding by detergents: Detergents break up plasma membranes; they coat hydrophobic portions of membrane proteins and phospholipids.
Isolated protein
3. Isolation of proteins: Treating a plasma membrane with a detergent is an effective way to isolate membrane proteins so they can be purified and studied in detail.

FIGURE 6.20 Detergents Can Be Used to Get Membrane Proteins into Solution.

To isolate and purify these membrane proteins once they are in solution, researchers use the technique called gel electrophoresis, introduced in **BioSkills 9** in Appendix A. When detergent-protein complexes are loaded into a gel and a voltage is applied, the larger protein complexes migrate more slowly than smaller proteins. As a result, the various proteins isolated from a plasma membrane separate from one another. To obtain a pure sample of a particular protein, the appropriate band is cut out of the gel. The gel material is then dissolved to retrieve the protein. Once this protein is inserted into a planar bilayer or liposome, dozens of different experiments are possible.

Since intensive experimentation on membrane proteins began, researchers have identified three broad classes of **transport proteins** that affect membrane permeability: channels, transporters, and pumps. Let's consider each in turn.

Protein Transport I: Facilitated Diffusion via Channel Proteins

As the data in Figure 6.7 showed, ions almost never cross pure phospholipid bilayers on their own. But in cells, ions routinely cross membranes through specialized membrane proteins called **ion channels**.

Ion channels form pores, or holes, in a membrane. Ions move through these pores in a predictable direction: from regions of high concentration to regions of low concentration via diffusion, and from areas of like charge to areas of unlike charge.

In **Figure 6.21**, for example, a large concentration gradient favors the movement of sodium ions from the outside of a membrane to the inside. But in addition, the inside of this cell has a net negative charge while the outside has a net positive charge. As a result, there is also a charge gradient that favors the movement of sodium ions, which have a positive charge, from the outside to the inside of the cell.

Ions move in response to a combined concentration and electrical gradient, or what biologists call an **electrochemical gradient.** ✔If you understand this concept, you should be able to add an arrow to Figure 6.21, indicating the electrochemical gradient for chloride ions—assuming that chloride concentrations are equal on both sides of the membrane.

IS CFTR AN ION CHANNEL? To understand the types of experiments that biologists do to confirm that a membrane protein is an ion channel, consider work on the cystic fibrosis transmembrane conductance regulator (CFTR).

Experiments published in 1983 suggested that cystic fibrosis—the most common genetic disease in humans of Northern

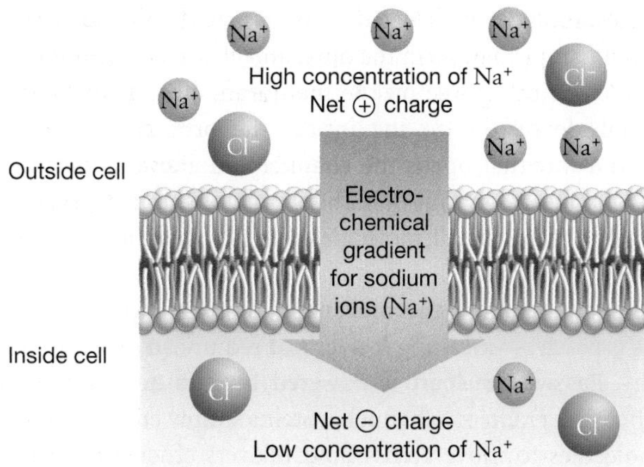

FIGURE 6.21 An Electrochemical Gradient Is a Combined Concentration and Electrical Gradient. Electrochemical gradients are established when ions build up on one side of a membrane.

✔**QUESTION** If you added a large number of sodium ions to the inside of this plasma membrane, what would happen to the electrical gradient, and to the concentration gradient for sodium?

EXPERIMENT

QUESTION: Is CFTR a chloride channel?

HYPOTHESIS: CFTR increases the flow of chloride ions across a membrane.

NULL HYPOTHESIS: CFTR has no effect on membrane permeability.

EXPERIMENTAL SETUP:

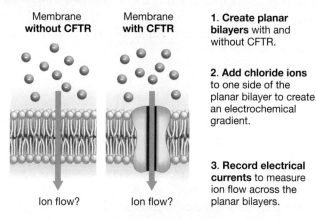

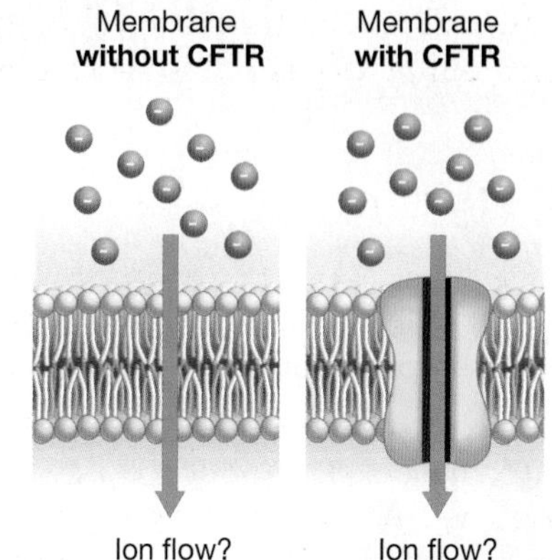

1. **Create planar bilayers** with and without CFTR.
2. **Add chloride ions** to one side of the planar bilayer to create an electrochemical gradient.
3. **Record electrical currents** to measure ion flow across the planar bilayers.

PREDICTION: Ion flow will be higher in membrane with CFTR.

PREDICTION OF NULL HYPOTHESIS: Ion flow will be the same in both membranes.

RESULTS:

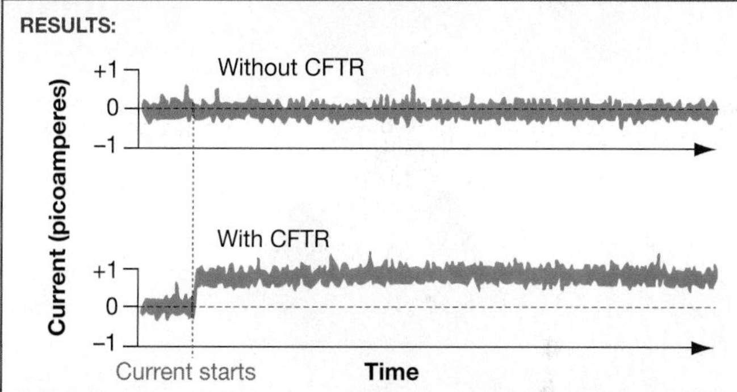

CONCLUSION: CFTR facilitates diffusion of chloride ions along an electrochemical gradient. CFTR is a chloride channel.

FIGURE 6.22 Electric Current Measurements Indicate That Ions Flow through CFTR.

SOURCE: Bear, C. A. et al. (1992). Purification and functional reconstitution of the cystic fibrosis transmembrane conductance regulator (CFTR). *Cell* 68: 809–818.

✔**QUESTION** The researchers repeated the "with CFTR" treatment 45 times, but recorded a current in only 35 of the replicates. Does this observation negate the conclusion? Explain why or why not.

European descent—is caused by defects in a membrane protein that allows chloride ions (Cl^-) to move across plasma membranes. Using techniques introduced in Chapter 19, biologists were also able to (1) find the gene that is defective in people suffering from CF, and (2) use the gene to produce copies of the normal protein, which had come to be called CFTR.

Is CFTR a chloride channel? To answer this question, researchers inserted purified CFTR into planar bilayers and measured the flow of electric current across the membrane. Because ions carry a charge, ion movement across a membrane produces an electric current.

The graphs in **Figure 6.22**, which plot the amount of current flowing across the membrane over time, show their results. Note that when CFTR was absent, no electric current passed through the membrane. But when CFTR was inserted into the membrane, current began to flow. This was strong evidence that CFTR was indeed a chloride channel.

PROTEIN STRUCTURE DETERMINES CHANNEL SELECTIVITY Subsequent research has shown that cells have many different types of pore-like **channel proteins** in their membranes, including the subset called ion channels. Channel proteins are selective. Each channel protein has a structure that allows it to admit a particular type of ion or small molecule.

For example, Peter Agre and co-workers discovered channels called **aquaporins** ("water-pores") that allow water to cross the plasma membrane over 10 times faster than it does in the absence of aquaporins. Aquaporins admit water but not other small molecules or ions.

Figure 6.23a shows a cutaway view from the side of an aquaporin, indicating how it fits in a plasma membrane. Like other channels that have been studied in detail, aquaporins have a pore that is lined with polar regions of amino acids—in this case, functional groups that interact with water. Hydrophobic groups form the outside of channels and interact with the lipid bilayer. A channel's pore is hydrophilic; its exterior is hydrophobic.

(a) Water pores allow only water to pass through.

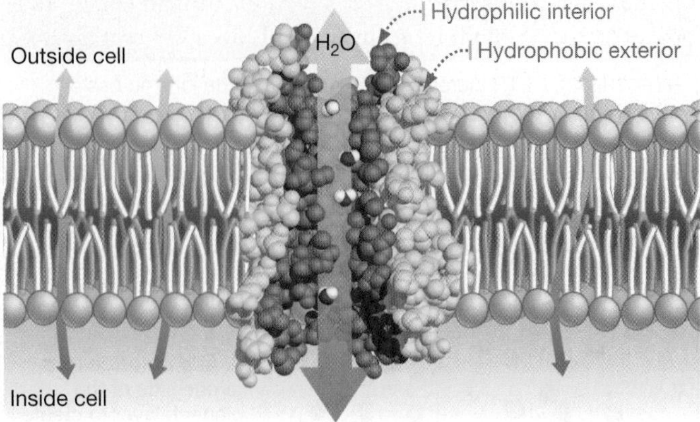

(b) Potassium channels allow only potassium ions to pass through.

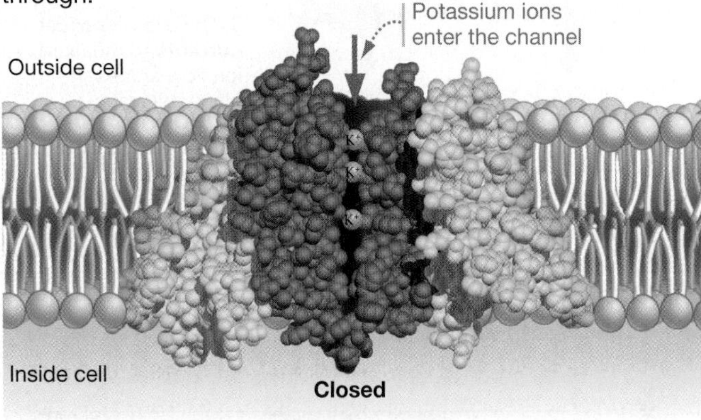

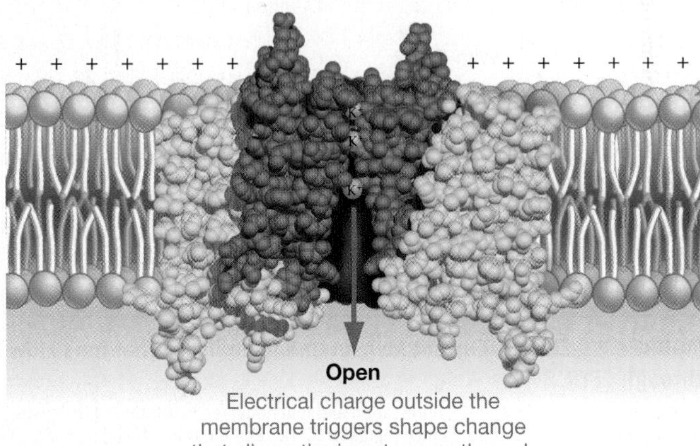

FIGURE 6.23 Membrane Channels Are Highly Selective and Highly Regulated. (a) A cutaway view looking at the inside of an aquaporin—a membrane channel that admits only water. Water moves through its pore via osmosis over 10 times faster than it can move through the lipid bilayer. **(b)** A model of a K^+ channel in the closed and open configurations.

MOVEMENT THROUGH MEMBRANE CHANNELS IS REGULATED Recent research has shown that the aquaporins and ion channels such as CFTR are **gated channels**—meaning that they open or close in response to the binding of a particular molecule or to a change in the electrical charge on the outside of the membrane.

As an example of how gated channels work, **Figure 6.23b** shows a potassium channel in the open and closed configuration. When the electrical charge on the membrane becomes positive on the outside relative to the inside, the protein's structure changes in a way that opens the channel and allows potassium ions to cross. The important point here is that in almost all cases, the flow of ions and small molecules through membrane channels is carefully controlled.

In all cases, however, the movement of substances through channels is passive—meaning it does not require an expenditure of energy. **Passive transport** is powered by diffusion along an electrochemical gradient. Channel proteins simply enable ions or polar molecules to move across lipid bilayers efficiently, in response to an existing gradient.

✔If you understand the nature of membrane channels, you should be able to (1) draw a channel that admits calcium ions (Ca^{2+}) when a signaling molecule binds to it, (2) label hydrophilic and hydrophobic portions of the channel, and (3) add ions to the outside and inside of a membrane containing the channel to explain why an electrochemical gradient favors entry of Ca^{2+}.

To summarize, membrane proteins such as CFTR, aquaporins, and potassium channels circumvent the lipid bilayer's impermeability to small, charged compounds. They are responsible for **facilitated diffusion**: the passive transport of substances that otherwise would not cross a membrane readily.

Protein Transport II: Facilitated Diffusion via Carrier Proteins

Even though facilitated diffusion does not require an expenditure of energy, it is facilitated—aided—by the presence of a specialized membrane protein. Facilitated diffusion can also occur through **carrier proteins**, or **transporters**, that change shape during the process. Perhaps the best-studied transporter is specialized for moving glucose into cells.

THE SEARCH FOR A GLUCOSE TRANSPORTER Next to ribose, the six-carbon sugar glucose is the most important sugar found in organisms. Virtually all cells alive today use glucose as a building block for important macromolecules and as a source of stored chemical energy. But as Figure 6.7 showed, lipid bilayers are only moderately permeable to glucose. It is reasonable to expect, then, that plasma membranes have some mechanism for increasing their permeability to this sugar.

This prediction was supported in experiments on pure preparations of plasma membranes—or "ghosts"—from human red blood cells (see **Figure 6.24**). These plasma membranes turned out to be much more permeable to glucose than are pure lipid bilayers. Why?

After isolating and analyzing many proteins from red blood cell ghosts, researchers found one protein that specifically increases membrane permeability to glucose. When this purified protein was added to liposomes, the artificial membrane transported glucose at the same rate as a membrane from a living cell.

PROCESS: HOW TO MAKE RED BLOOD CELL "GHOSTS"

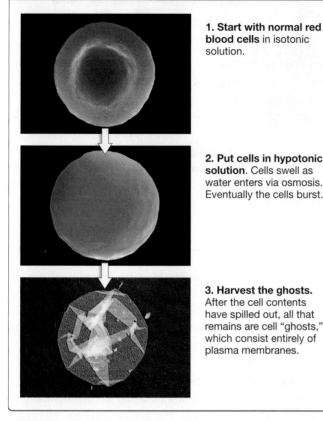

1. Start with normal red blood cells in isotonic solution.

2. Put cells in hypotonic solution. Cells swell as water enters via osmosis. Eventually the cells burst.

3. Harvest the ghosts. After the cell contents have spilled out, all that remains are cell "ghosts," which consist entirely of plasma membranes.

FIGURE 6.24 Rupture of Red Blood Cells Produces "Ghosts." Red blood cell ghosts are simple membranes that can be purified and studied in detail.

This experiment convinced biologists that the membrane protein—now called GLUT-1—was indeed responsible for transporting glucose across plasma membranes.

HOW DOES GLUT-1 WORK? **Figure 6.25** illustrates the current hypothesis for how GLUT-1 works. The idea is that glucose binds to GLUT-1 on the exterior of the membrane, and that this binding induces a conformational change in the protein which transports glucose to the interior of the cell. Recall from Chapter 3 that enzymes frequently change shape when they bind substrates and that such conformational changes are often a critical step in the catalysis of chemical reactions.

What powers the movement of molecules through transporters? The answer is diffusion. When glucose enters a cell via GLUT-1, it does so because it is following its concentration gradient. If the concentration of glucose is the same on both sides of the plasma membrane, then no net movement of glucose occurs even if the membrane contains GLUT-1. A large array of molecules moves across plasma membranes via facilitated diffusion through specific transporter proteins.

Protein Transport III: Active Transport by Pumps

Whether diffusion is facilitated by channel proteins or by transporters, it is a passive process that makes the cell interior and exterior more similar. But it is also possible for today's cells to import molecules or ions *against* their electrochemical gradient. Accomplishing this task requires energy, because the cell must counteract the entropy loss that occurs when molecules or ions are concentrated. It makes sense, then, that transport against an electrochemical gradient is called **active transport**.

In cells, the energy required to move substances against their electrochemical gradient is provided by a phosphate group (HPO_4^{2-}) from adenosine triphosphate (ATP). Recall that ATP contains three phosphate groups (Chapter 5), and that phosphate groups carry two negative charges (Chapter 2). When a phosphate group leaves ATP and binds to a protein, its negative charges are added to the protein. These charges repel other charges on the protein's amino acids. The protein's potential energy increases in response, and its shape usually changes. When a phosphate group leaves ATP, the resulting molecule is adenosine diphosphate (ADP), which has two phosphate groups.

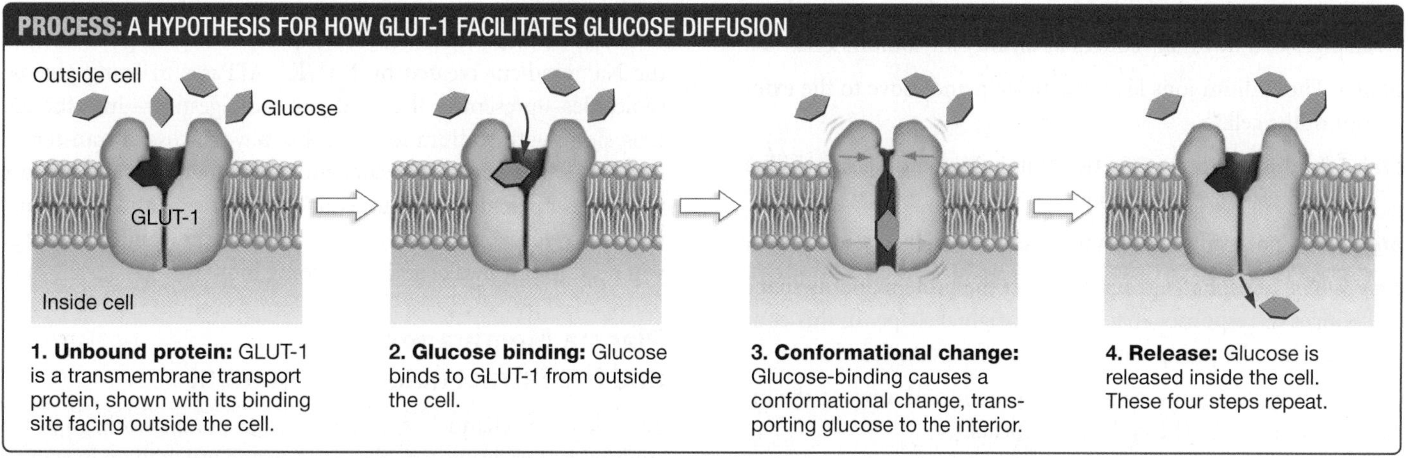

PROCESS: A HYPOTHESIS FOR HOW GLUT-1 FACILITATES GLUCOSE DIFFUSION
Outside cell
Glucose
GLUT-1
Inside cell
1. Unbound protein: GLUT-1 is a transmembrane transport protein, shown with its binding site facing outside the cell.
2. Glucose binding: Glucose binds to GLUT-1 from outside the cell.
3. Conformational change: Glucose-binding causes a conformational change, transporting glucose to the interior.
4. Release: Glucose is released inside the cell. These four steps repeat.

FIGURE 6.25 Membrane Transport Proteins and Enzymes May Work Similarly. This model suggests that the GLUT-1 transporter binds a substrate (in this case, a glucose molecule), undergoes a conformation change, and releases the substrate.

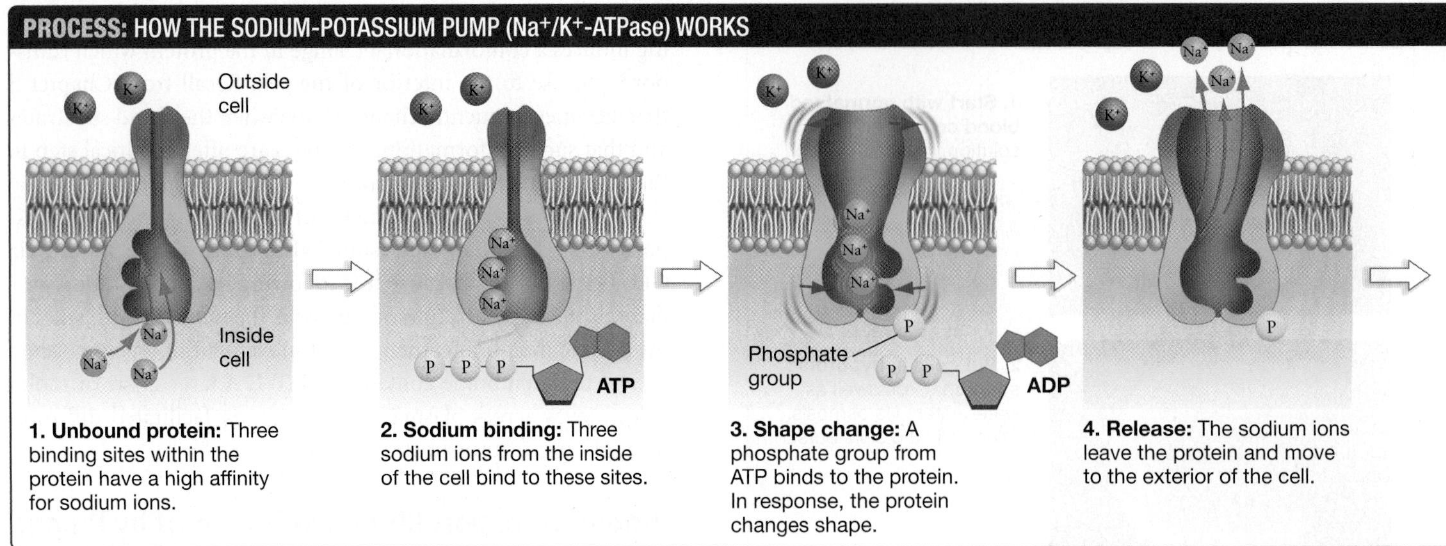

PROCESS: HOW THE SODIUM-POTASSIUM PUMP (Na+/K+-ATPase) WORKS
Outside cell
Inside cell
ATP
Phosphate group
ADP
1. Unbound protein: Three binding sites within the protein have a high affinity for sodium ions.
2. Sodium binding: Three sodium ions from the inside of the cell bind to these sites.
3. Shape change: A phosphate group from ATP binds to the protein. In response, the protein changes shape.
4. Release: The sodium ions leave the protein and move to the exterior of the cell.

FIGURE 6.26 Active Transport Depends on an Input of Chemical Energy Stored in ATP.

THE SODIUM-POTASSIUM PUMP Ions or molecules can move against an electrochemical gradient when membrane proteins called **pumps** change shape.

As an example, **Figure 6.26** highlights the first pump that was discovered and characterized: a protein called the **sodium-potassium pump**, or more formally, Na^+/K^+-ATPase. The Na^+/K^+ part of the name refers to the ions that are transported, ATP indicates that adenosine triphosphate is used, and *–ase* implies that the molecule acts like an enzyme.

As the figure indicates, sodium and potassium ions move in a multistep process:

Step 1 When Na^+/K^+-ATPase is in the conformation shown here, binding sites with a high affinity for sodium ions are available.

Step 2 Three sodium ions from the inside of the cell bind to these sites.

Step 3 A phosphate group from ATP then binds to the pump. When the phosphate group attaches, the pump's shape changes in a way that reduces its affinity for sodium ions.

Step 4 The sodium ions leave the protein and move to the exterior of the cell.

Step 5 In this conformation, the protein has binding sites with a high affinity for potassium ions.

Step 6 Two potassium ions from outside the cell bind to the pump.

Step 7 The phosphate group drops off the protein and its shape changes in response—back to the original shape. In this conformation, the pump has low affinity for potassium ions.

Step 8 The potassium ions leave the protein and move to the interior of the cell. The cycle then repeats.

In an analogous way, other types of pumps move protons (H^+), calcium ions (Ca^{2+}), or other ions or molecules into or out of cells, against electrical or concentration gradients.

Pumps allow cells to concentrate or expel certain substances. As a result, cells can import valuable ions or molecules that are at high concentration inside the cell but at low concentration outside the cell. They can also get rid of damaging molecules or ions, even when a concentration gradient favors diffusion of these substances into the cell.

SECONDARY ACTIVE TRANSPORT In addition to concentrating or expelling certain substances, pumps set up electrochemical gradients. For example, because the Na^+/K^+-ATPase exchanges three sodium ions for every two potassium ions, the outside of the membrane becomes positively charged relative to the inside. In this way, the sodium-potassium pump sets up an electrical gradient and a chemical gradient across the membrane. The electrical gradient favors a flow of anions out of the cell and cations into the cell; the chemical gradient favors a flow of sodium ions into the cell and potassium ions out of the cell.

These gradients are crucial, in part because they make it possible for cells to engage in **secondary active transport**—also known as **cotransport**. When cotransport occurs, a gradient set up by a pump provides the potential energy required to power the movement of a different molecule against its particular gradient.

For example, a transmembrane protein in your gut cells uses the Na^+ gradient created by Na^+/K^+-ATPases to bring glucose molecules—present in the food you are digesting—into the gut cells along with sodium ions. In this way, glucose is transferred into your body against a concentration gradient. The glucose molecules then diffuse into your bloodstream and are transported to your brain, where they provide the chemical energy you need to stay awake and learn some biology.

Plasma Membranes and the Intracellular Environment

To review how channels, transporters, and pumps work, go to the study area at *www.masteringbiology.com* and watch them in action.

(MB) **BioFlix™** Membrane Transport, **Web Activity** Membrane Transport Proteins

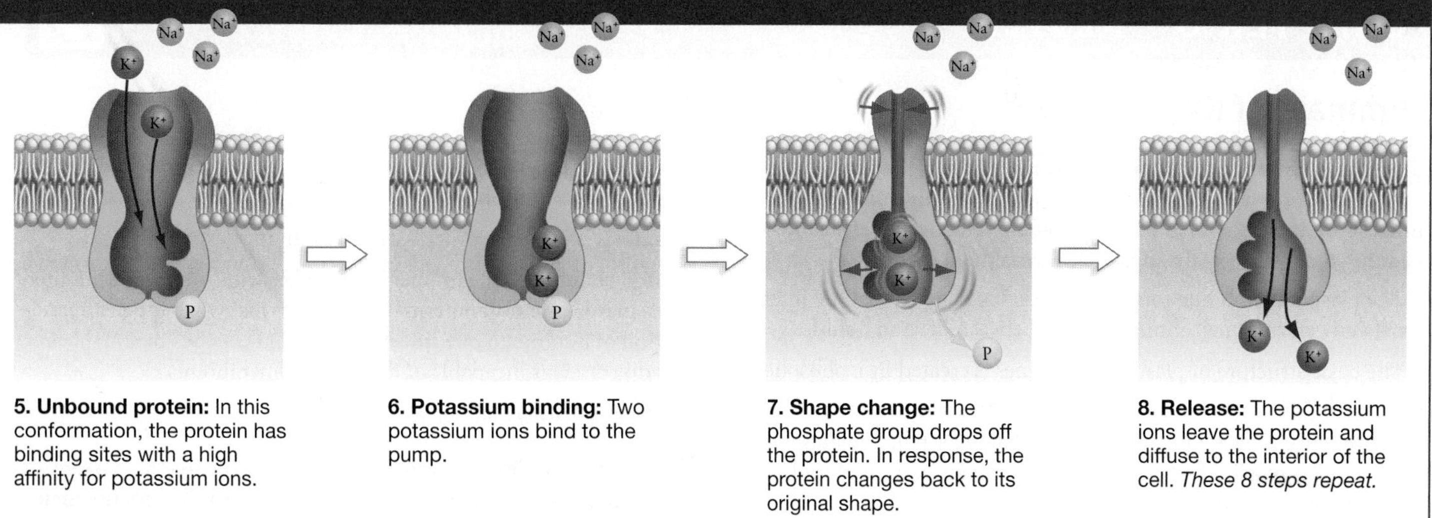

5. Unbound protein: In this conformation, the protein has binding sites with a high affinity for potassium ions.

6. Potassium binding: Two potassium ions bind to the pump.

7. Shape change: The phosphate group drops off the protein. In response, the protein changes back to its original shape.

8. Release: The potassium ions leave the protein and diffuse to the interior of the cell. *These 8 steps repeat.*

Taken together, the selective permeability of the lipid bilayer and the specificity of the proteins involved in passive transport and active transport enable cells to create an internal environment that is much different from the external one (**Figure 6.27**).

With the evolution of membrane proteins, the early cells acquired the ability to create an internal environment that was conducive to life—one that contained the substances required for manufacturing ATP and copying ribozymes. Cells with particularly efficient and selective membrane proteins would be favored by natural selection and would come to dominate the population. Cellular life had begun.

Some 3.5 billion years later, cells continue to evolve. What do today's cells look like, and how do they produce and store the chemical energy that makes life possible? Answering these and related questions is the focus of Unit 2.

CHECK YOUR UNDERSTANDING

If you understand that . . .

- Membrane proteins allow ions and molecules that ordinarily do not readily cross lipid bilayers to enter or exit cells.
- Substances may move across a plasma membrane along an electrochemical gradient, via facilitated diffusion through channel proteins or transport proteins. Or, they may move against an electrochemical gradient in response to work done by pumps.

✓ You should be able to . . .

Explain what is passive about passive transport, what is active about active transport, and what is "co" about cotransport.

Answers are available in Appendix B.

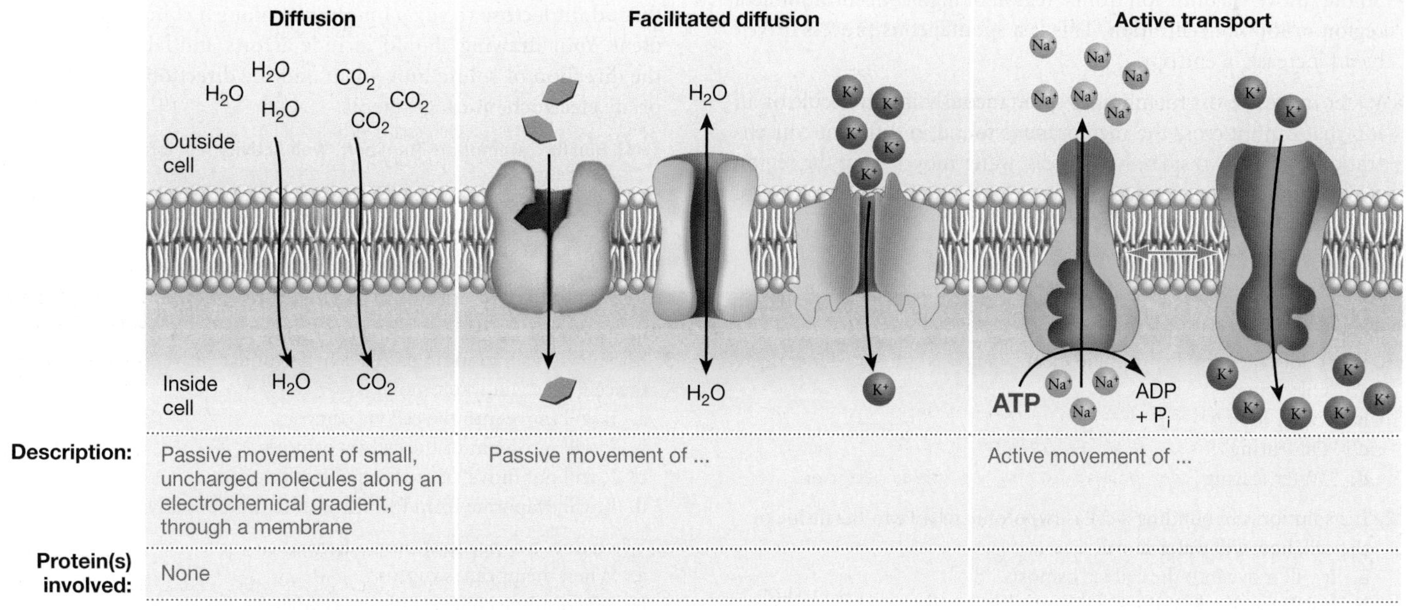

FIGURE 6.27 Summary of the Passive and Active Mechanisms of Membrane Transport.

✓EXERCISE Complete the chart.

CHAPTER 6 REVIEW

For media, go to the study area at www.masteringbiology.com 

Summary of Key Concepts

Phospholipids are amphipathic molecules—they have a hydrophilic region and a hydrophobic region. In solution, phospholipids spontaneously form bilayers that are selectively permeable—meaning that only certain substances cross them readily.

- The plasma membrane forms a physical barrier between the internal and external environment—often between life and nonlife.
- The basic structure of plasma membranes is created by a phospholipid bilayer.
- Phospholipids have a polar head and a nonpolar tail. The nonpolar tail consists of a lipid—usually a fatty acid or an isoprene. Lipids do not dissolve in water.
- Small, nonpolar molecules tend to move across membranes readily; ions and other charged compounds cross rarely, if at all.
- The permeability and fluidity of lipid bilayers depend on temperature and on the types of phospholipids present. Phospholipids that contain long, saturated fatty acids form a dense and highly hydrophobic membrane interior that lowers permeability, relative to phospholipids containing shorter, unsaturated fatty acids.

✔You should be able to describe the structure of phospholipid bilayers that are highly permeable and fluid versus highly impermeable and lacking in fluidity.

Ions and molecules diffuse spontaneously from regions of high concentration to regions of low concentration. Water moves across lipid bilayers from regions of high water concentration to regions of low water concentration via osmosis—a special case of diffusion.

- Diffusion is movement of ions or molecules owing to their kinetic energy.
- Solutes move via diffusion from a region of high concentration to a region of low concentration. This is a spontaneous process driven by an increase in entropy.
- Water moves across membranes spontaneously if a molecule or an ion that cannot cross the membrane is found in different concentrations on the two sides. In osmosis, water moves from the region with a higher concentration of water and lower concentration of solutes to the region of lower water concentration and higher solute concentration.
- Osmosis is a passive process driven by an increase in entropy.

✔You should be able to imagine a beaker with solutions separated by a plasma membrane, and then predict what will happen after addition of a solute to one side if the solute (1) crosses the membrane readily or (2) is incapable of crossing the membrane.

MB **Web Activity** Diffusion and Osmosis

In cells, membrane proteins are responsible for the passage of ions, polar molecules, and large molecules that can't cross the membrane on their own because they are not soluble in lipids. Some membrane proteins form channels, some facilitate diffusion by binding to substrates, and some use energy from ATP to actively pump ions or molecules.

- The permeability of lipid bilayers can be altered significantly by membrane transport proteins.
- Channel proteins provide holes in the membrane and facilitate the diffusion of specific ions into or out of the cell.
- Transport proteins are enzyme-like proteins that facilitate the diffusion of specific molecules into or out of the cell.
- Energy-demanding pumps actively move ions or molecules against their electrochemical gradient.
- In combination, the selective permeability of phospholipid bilayers and the specificity of transport proteins make it possible to create an environment inside a cell that is radically different from the exterior.

✔You should be able to draw and label the membrane of a cell that pumps hydrogen ions to the exterior, has channels that admit calcium ions along an electrochemical gradient, and has carriers that admit lactose (a sugar) molecules along a concentration gradient. Your drawing should include arrows and labels indicating the direction of solute movement and the direction of the appropriate electrochemical gradients.

MB **BioFlix™** Membrane Transport, **Web Activity** Membrane Transport Proteins

Questions

✔TEST YOUR KNOWLEDGE

Answers are available in Appendix B

1. What does the term hydrophilic mean when it is translated literally?
   a. "Oil loving"
   b. "Water loving"
   c. "Oil fearing"
   d. "Water fearing"
2. If a solution surrounding a cell is hypotonic relative to the inside of the cell, how will water move?
   a. It will move into the cell via osmosis.
   b. It will move out of the cell via osmosis.
   c. It will not move, because equilibrium exists.
   d. It will evaporate from the cell surface more rapidly.
3. If a solution surrounding a cell is hypertonic relative to the inside of the cell, how will water move?
   a. It will move into the cell via osmosis.
   b. It will move out of the cell via osmosis.
   c. It will not move, because equilibrium exists.
   d. It will evaporate from the cell surface more rapidly.
4. When does a concentration gradient exist?
   a. When membranes rupture
   b. When solute concentrations are high
   c. When solute concentrations are low
   d. When solute concentrations differ on the two sides of a membrane

5. Which of the following must be true for osmosis to occur?
   a. Water must be at room temperature or above.
   b. Solutions with the same concentration of solutes must be separated by a selectively permeable membrane.
   c. Solutions with different concentrations of solutes must be separated by a selectively permeable membrane.
   d. Water must be under pressure.

6. Why are the lipid bilayers in cells called "selectively permeable"?
   a. They are not all that permeable.
   b. Their permeability changes with their molecular composition.
   c. Their permeability is temperature-dependent.
   d. They are permeable to some substances but not others.

✔TEST YOUR UNDERSTANDING

Answers are available in Appendix B

1. Cooking oil is composed of lipids that consist of long hydrocarbon chains. Would you expect these lipids to form membranes spontaneously? Why or why not? Describe, on a molecular level, how you would expect these lipids to interact with water.
2. Explain why phospholipids form a bilayer in solution, and why the process is spontaneous.
3. Ethanol, the active ingredient in alcoholic beverages, is a small, polar, uncharged molecule. Would you predict that this molecule crosses plasma membranes quickly or slowly? Explain your reasoning.
4. Why can osmosis occur only if solutions are separated by a selectively permeable membrane? What happens in solutions that are *not* separated by a selectively permeable membrane?
5. The text claims that the portion of membrane proteins that spans the hydrophobic tails of phospholipids is itself hydrophobic (see Figure 6.16b). Why is this logical? Look back at Figure 3.3 and Table 3.1, and make a list of amino acids you would expect to find in these regions of transmembrane proteins.
6. Examine the membrane in the accompanying figure. Label the molecules and ions that will pass through the membrane as a result of osmosis, diffusion, and facilitated diffusion. Draw arrows to indicate where each of the molecules and ions will travel.

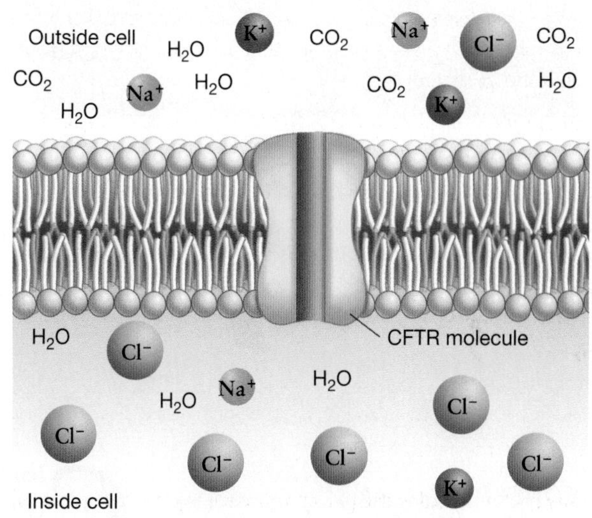

✔APPLYING CONCEPTS TO NEW SITUATIONS

Answers are available in Appendix B

1. When phospholipids are arranged in a bilayer, it is theoretically possible for individual molecules in the bilayer to flip-flop. That is, a phospholipid could turn 180° and become part of the membrane's other surface. From what you know about the behavior of polar heads and nonpolar tails, predict whether flip-flops are frequent or rare. Then design an experiment, using a planar bilayer with one side made up of phospholipids that contain a dye molecule on their hydrophilic head, to test your prediction.
2. Unicellular organisms that live in extremely cold habitats have an unusually high proportion of unsaturated fatty acids in their plasma membranes. Some of these membranes even contain polyunsaturated fatty acids, which have more than one double bond in each hydrocarbon chain. Explain why you agree or disagree with the hypothesis that membranes with unsaturated fatty-acid tails function better at cold temperatures than do membranes with saturated fatty-acid tails. Make a prediction about the structure of fatty acids found in organisms that live in extremely hot environments.
3. When biomedical researchers design drugs that must enter cells to be effective, they sometimes add methyl (CH_3) groups to make the drug molecules more likely to pass through plasma membranes. Conversely, when researchers design drugs that act on the exterior of plasma membranes, they sometimes add a charged group to decrease the likelihood that the drugs will pass through membranes and enter cells. Explain why these strategies are effective.
4. Advertisements frequently claim that laundry and dishwashing detergents "cut grease." The ad writers mean that the detergents surround oil droplets on clothing or dishes, making the droplets water-soluble. When this happens, the oil droplets can be washed away. Explain how this happens on a molecular level.

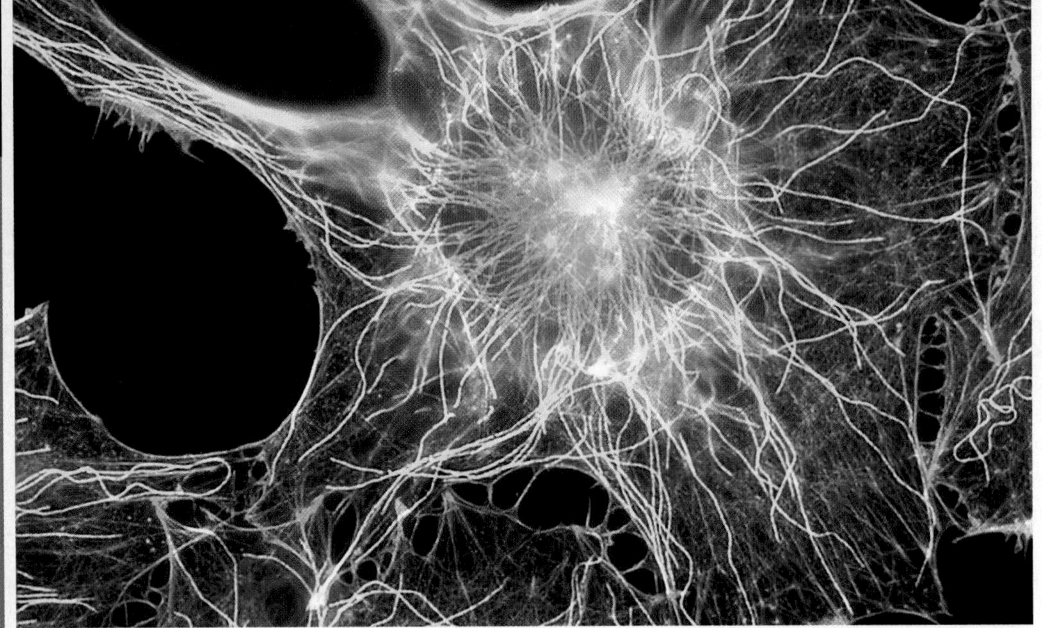

This cell has been treated with fluorescing molecules that bind to its fibrous skeleton. Microtubules (large protein fibers) are yellow; actin filaments (smaller fibers) are blue. The cell's nucleus has been stained green.

7 Inside the Cell

KEY CONCEPTS

- The structure of individual cell components is closely related to their function. The overall shape and composition of a cell is also closely related to its function.
- Inside cells, materials are transported to their destinations with the help of molecular "zip codes."
- The cytoskeleton provides a structural framework inside cells and plays a key role in cell division, movement, and transport.
- Cells are dynamic, highly integrated structures. Thousands of chemical reactions occur each second within cells; molecules constantly enter and exit across the plasma membrane; cell products are shipped along protein fibers; and elements of the cell's internal skeleton grow and shrink.

Chapter 1 introduced the cell theory, which states that all organisms consist of cells and all cells are derived from preexisting cells. Since this theory was initially developed and tested in the 1850s, an enormous body of research has confirmed that the cell is the fundamental structural and functional unit of life. Life on Earth is cellular.

Chapter 6 delved into this fundamental attribute of life by introducing the **plasma membrane**—a structure common to every cell. Thanks to the selective permeability of phospholipid bilayers and the activity of membrane transport proteins, the plasma membrane creates an internal environment that differs from conditions outside the cell.

Our task now is to explore the structures inside the plasma membrane. Let's begin by analyzing how the parts inside a cell function individually, and then explore how they work as a unit. This approach is analogous to studying individual organs in the body, and then analyzing how they work together to form the nervous system or digestive system. As you study this material, keep asking yourself some key questions: How does the structure of this part or group of parts correlate with its function? What problem does it solve?

7.1 Bacterial and Archaeal Cell Structures and Their Functions

In addition to introducing the cell theory, Chapter 1 highlighted a distinction between two fundamental types of cells observed in nature—eukaryotes and prokaryotes. Eukaryotic cells have a membrane-bound compartment called a nucleus, while prokaryotic cells do not.

According to **morphology** ("form-science"), then, species fall into two broad categories: (1) prokaryotes and (2) eukaryotes. But according to **phylogeny** ("tribe-

✔ When you see this checkmark, stop and test yourself. Answers are available in Appendix B.

source"), or evolutionary history, organisms fall into three broad domains called (**1**) Bacteria, (**2**) Archaea, and (**3**) Eukarya. Members of the Bacteria and Archaea are prokaryotic; members of the Eukarya—including algae, fungi, plants, and animals—are eukaryotic.

A Revolutionary New View

For almost 200 years, biologists thought that prokaryotic cells were simple in terms of their morphology, with little structural diversity among species. This was a valid conclusion at the time, given the resolution of the microscopes that were available and the number of species that had been studied.

Things have changed. The structure of archaeal cells is still poorly studied, but recent improvements in microscopy and other research tools have convinced biologists that bacterial cells are highly organized, with an array of distinctive structures found among millions of species. This conclusion represents one of the most exciting discoveries in cell biology over the past 10 years.

To keep things simple at the start, though, **Figure 7.1** offers a low-magnification, stripped-down diagram of a bacterial cell.

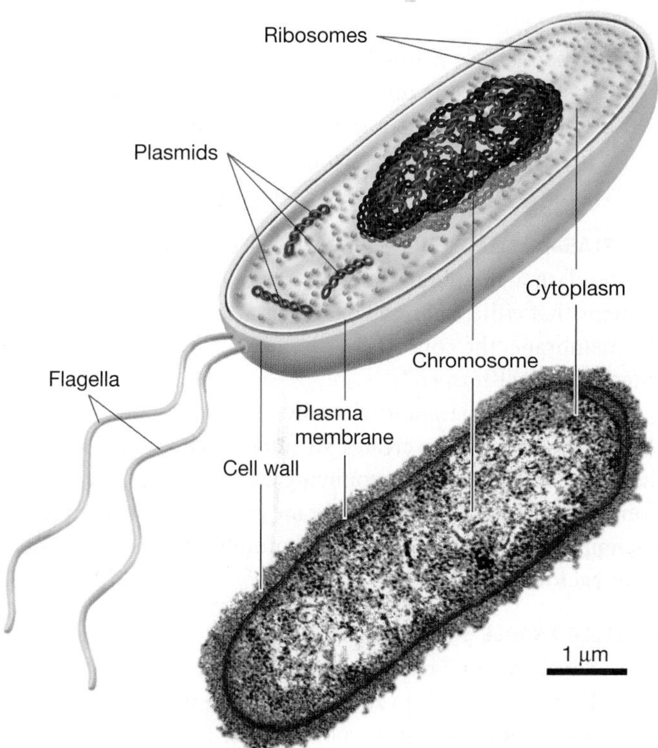

FIGURE 7.1 Overview of a Prokaryotic Cell. Prokaryotic cells are identified by a negative trait—the absence of a membrane-bound nucleus. Although there is wide variation in the size and shape of bacterial and archaeal cells, they all contain a plasma membrane, a chromosome, and protein-synthesizing ribosomes; almost all have a stiff cell wall. Some prokaryotes have flagella used in swimming and/or inner membranes where photosynthesis takes place.

✔**EXERCISE** Label the nucleoid in the drawing and the micrograph.

Prokaryotic Cell Structures: A Parts List

The labels in Figure 7.1 highlight the components common to all or most bacteria studied to date. Let's explore these elements one by one, and also look at more specialized structures found in particular species, starting from the inside and working out.

THE CHROMOSOME IS ORGANIZED IN A NUCLEOID The most prominent structure inside a bacterial cell is the **chromosome**. Most bacterial species have a single, circular chromosome that consists of a large DNA molecule associated with a small number of proteins. The DNA molecule contains information, while the proteins provide structural support for the DNA.

Recall from Chapter 4 that the information in DNA is encoded in its sequence of nitrogenous bases, and a segment of DNA that contains the information for building an RNA molecule or a polypeptide is called a **gene**. Thus, chromosomes contain DNA, which contains genes.

In the well-studied bacterium *Escherichia coli*, the circular chromosome would be over 1 mm long if it were linear—500 times longer than the cell itself. This situation is typical in prokaryotes. To fit into the cell, the DNA double helix coils on itself with the aid of enzymes to form the highly compact, "supercoiled" structure. Supercoiled regions of DNA resemble a string that has been held at either end and then twisted until it coils back upon itself (**Figure 7.2**).

Bacterial chromosomes are found in a localized area of the cell called the **nucleoid** (pronounced *NEW-klee-oyd*). The nucleoid is usually found in the center of the cell and typically represents about 20 percent of the cell's total volume. It's important to note, though, that the genetic material in the nucleoid is not separated from the rest of the cell interior by a membrane. The organization of the nucleoid is currently the subject of intense research.

In addition to one or more main chromosomes, bacterial cells may also contain one to about a hundred small, usually circular, supercoiled DNA molecules called **plasmids**. Plasmids contain genes but are physically independent of the main, cellular

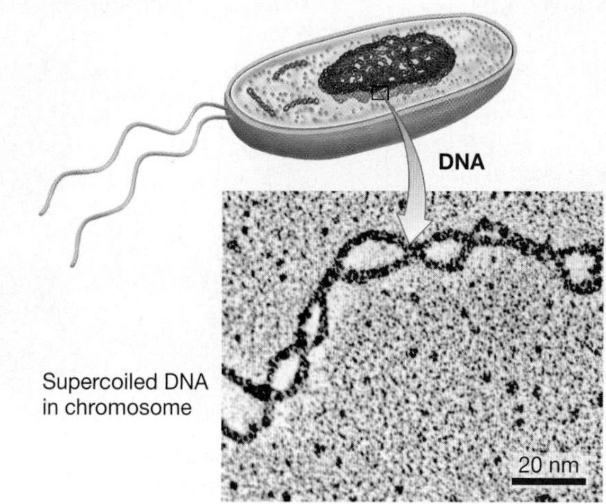

FIGURE 7.2 Bacterial DNA Is Supercoiled. The circular chromosomes of bacteria and archaea must be coiled extensively, into "supercoils," to fit in the cell.

chromosome. In most cases the genes carried by plasmids are not required under normal conditions; instead they help cells adapt to unusual circumstances, such as the sudden presence of a poison in the environment. As a result, plasmids can be considered auxiliary genetic elements.

RIBOSOMES MANUFACTURE PROTEINS **Ribosomes** are observed in all prokaryotic cells and are found throughout the cell interior. It is not unusual for a single cell to contain 10,000 ribosomes, each of which functions as a protein-manufacturing center.

Bacterial ribosomes are complex structures consisting of RNA molecules and proteins. Biologists sometimes refer to ribosomes, along with other multicomponent complexes that perform specialized tasks, as "molecular machines." Chapter 16 analyzes the structure and function of the bacterial ribosome in detail.

INTERNAL MEMBRANE COMPLEXES IN PHOTOSYNTHETIC SPECIES In addition to the nucleoid and ribosomes found in all bacteria and archaea studied to date, it is common to observe extensive internal membranes in prokaryotes that perform photosynthesis. Photosynthesis is the suite of chemical reactions responsible for converting the energy in sunlight into chemical energy stored in sugars.

The photosynthetic membranes observed in prokaryotes contain the enzymes and pigment molecules required for these reactions to occur and develop as infoldings of the plasma membrane. In some cases, vesicles pinch off as the plasma membrane folds in. In other cases, flattened stacks of photosynthetic membrane, like those shown in **Figure 7.3**, form from the infolded sections of the plasma membrane. The extensive surface area provided by these internal membranes makes it possible for more photosynthetic reactions to occur, increasing the cell's ability to make food.

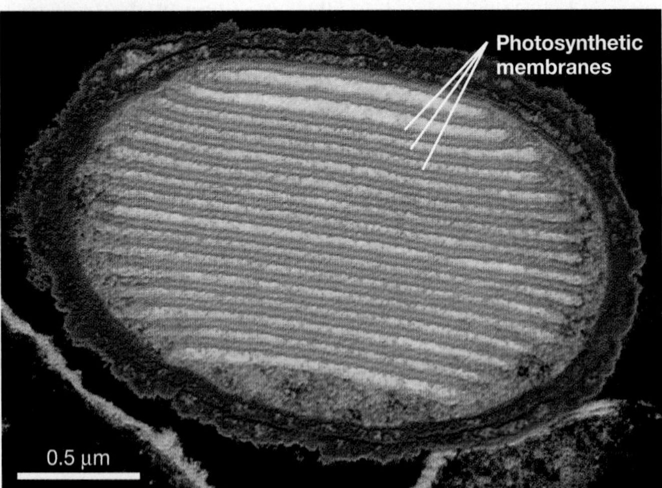

FIGURE 7.3 Photosynthetic Membranes in Bacteria. The green stripes in this photosynthetic bacterium are membranes that contain the pigments and enzymes required for photosynthesis. This photo is a **transmission electron micrograph** that has been colorized (see **BioSkills 10** in Appendix A).

ORGANELLES PERFORM SPECIALIZED FUNCTIONS Recent research indicates that several bacterial species have internal compartments that qualify as **organelles** ("little organs"). An organelle is a membrane-bound compartment inside the cell that contains enzymes or structures specialized for a particular function. Each type of bacterial organelle is found in certain species.

Bacterial organelles perform an array of tasks, including:

- storing calcium ions or other key molecules;
- holding crystals of the mineral magnetite, which function like a compass needle to help cells sense a magnetic field and swim in a directed way;
- organizing enzymes responsible for synthesizing complex carbon compounds from carbon dioxide; and
- sequestering enzymes that generate chemical energy from ammonium ions.

THE CYTOSKELETON STRUCTURES THE CELL INTERIOR Recent research has also shown that bacteria and archaea contain long, thin fibers that serve a structural role inside the cell. All bacterial species, for example, contain protein fibers that are essential for cell division to take place. Some species also have protein filaments that help maintain cell shape. Protein filaments such as these form the basis of the **cytoskeleton** ("cell skeleton").

The discovery of bacterial cytoskeletal elements is so new that much remains to be learned. Currently, researchers are working to understand how the different cytoskeletal elements are localized inside the cell, whether they play a role in transporting materials inside the cell or organizing the cell interior into distinctive regions, and how they move to enable the cell to divide.

THE PLASMA MEMBRANE SEPARATES LIFE FROM NONLIFE The cell, or plasma, membrane consists of a phospholipid bilayer and proteins that either span the bilayer or attach to one side. Inside the membrane, the contents of a cell are collectively termed the **cytoplasm** ("cell-formed").

Because all archaea and virtually all bacteria are unicellular, the plasma membrane creates an internal environment that is distinct from the outside, nonliving environment. In combination with the lipid bilayer, membrane proteins allow the passage of compounds required for life, and prohibit the entry of materials dangerous to life.

FLAGELLA ENABLE SOME SPECIES TO SWIM When **flagella** (singular: **flagellum**) are present in the plasma membrane, their rotation allows aquatic cells to swim through the water. At top speed, flagellar movement can drive a bacterial cell through water at 60 cell lengths per second. In contrast, the fastest land animal—the cheetah—can sprint at a mere 25 body lengths per second.

Bacterial flagella are usually few in number and are located on the surface of the cell. Over 40 different proteins are involved in building and controlling this molecular machine.

THE CELL WALL FORMS A PROTECTIVE "EXOSKELETON" Because the cytoplasm contains a high concentration of solutes, in most

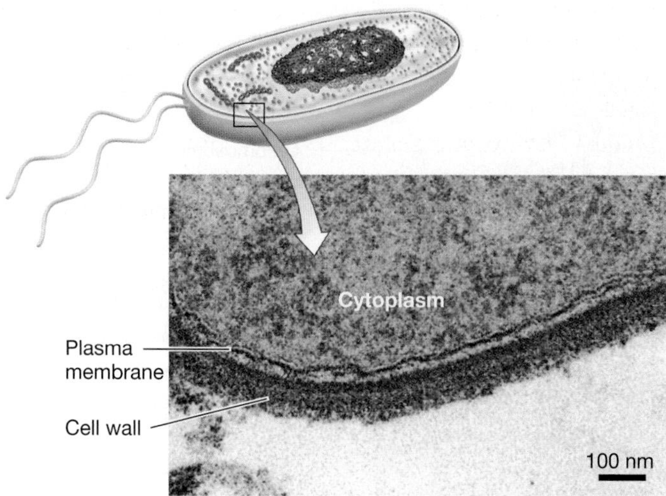

FIGURE 7.4 The Bacterial Cell Wall. In bacteria and archaea, the cell wall consists of peptidoglycan or similar polymers that are cross-linked into tough sheets. The inside of the cell wall contacts the plasma membrane, which pushes up against the wall. The outside of the cell wall makes direct contact with the outside environment, which is almost always filled with competitors and predators.

habitats it is hypertonic relative to the surrounding environment. When this is the case, water enters the cell via osmosis and makes the cell's volume expand. In virtually all bacteria and archaea, this pressure is resisted by a stiff **cell wall** (**Figure 7.4**).

Bacterial and archaeal cell walls are a tough, fibrous layer that surrounds the plasma membrane. In prokaryotes, the pressure of the plasma membrane against the cell wall is about the same as the pressure in an automobile tire.

The cell wall protects the organism and gives it shape and rigidity, much like the exoskeleton (external skeleton) of a crab or insect. In addition, many bacteria have another protective layer outside the cell wall that consists of lipids with polysaccharides attached. Lipids that contain carbohydrate groups are termed **glycolipids**.

The painting in **Figure 7.5** shows a cross section of a bacterial cell, with a close-up view of the cell wall and some of the other structures introduced in this section. One feature that prokaryotic and eukaryotic cells have in common: They are both packed with dynamic, highly integrated structures.

CHECK YOUR UNDERSTANDING

If you understand that . . .

- Each structure in a prokaryotic cell performs a function vital to the cell.

You should be able to . . .

Describe the structure and function of (1) photosynthetic membranes, (2) organelles that contain magnetite, and (3) the cell wall.

Answers are available in Appendix B.

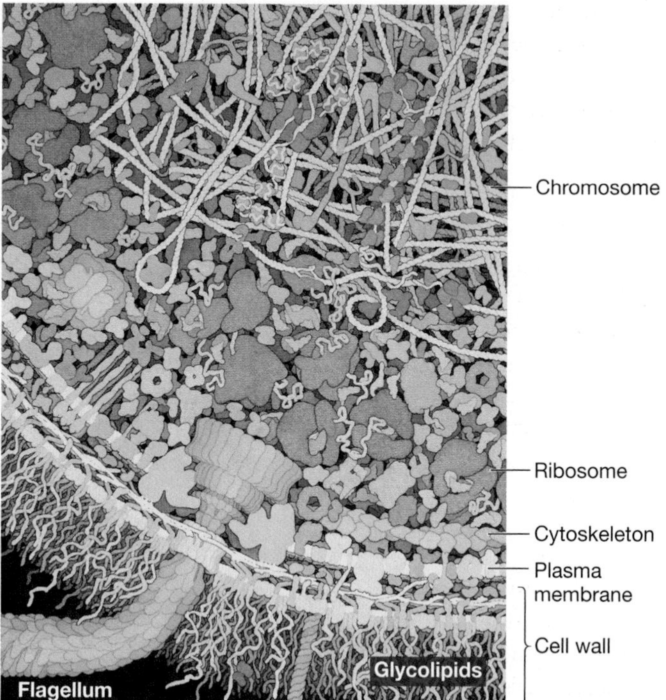

FIGURE 7.5 Close-up View of a Prokaryotic Cell. This painting is an artist's conception of a cross section through part of a bacterial cell. Note that the cell is packed with proteins, DNA, ribosomes, and other molecular machinery.

7.2 Eukaryotic Cell Structures and Their Functions

The lineage called Eukarya includes species that range from microscopic algae to 100-meter-tall redwoods. Brown algae, red algae, fungi, amoebae, and slime molds are all eukaryotic, as are green plants and animals. Although multicellularity has evolved several times among eukaryotes (see Chapter 29), many of these species are unicellular.

The first thing that strikes biologists about eukaryotic cells is how much larger they are on average than bacteria and archaea. Most prokaryotic cells measure 1 to 10 μm in diameter, while most eukaryotic cells range from about 5 to 100 μm in diameter. A micrograph of an average eukaryotic cell, at the same scale as the bacterial cell in Figure 7.3, would fill this page.

In many species of unicellular eukaryotes, large size allows them to make a living by ingesting bacteria and archaea whole. Large size has a downside, however. Ions and small molecules such as adenosine triphosphate (ATP), amino acids, and nucleotides cannot diffuse across a large volume quickly. If the ATP that supplies chemical energy in cells is used up on one side of a large cell, ATP from the other side of the cell would take a long time to diffuse to that location. Prokaryotic cells are small enough that ions and small molecules arrive where they are needed via diffusion.

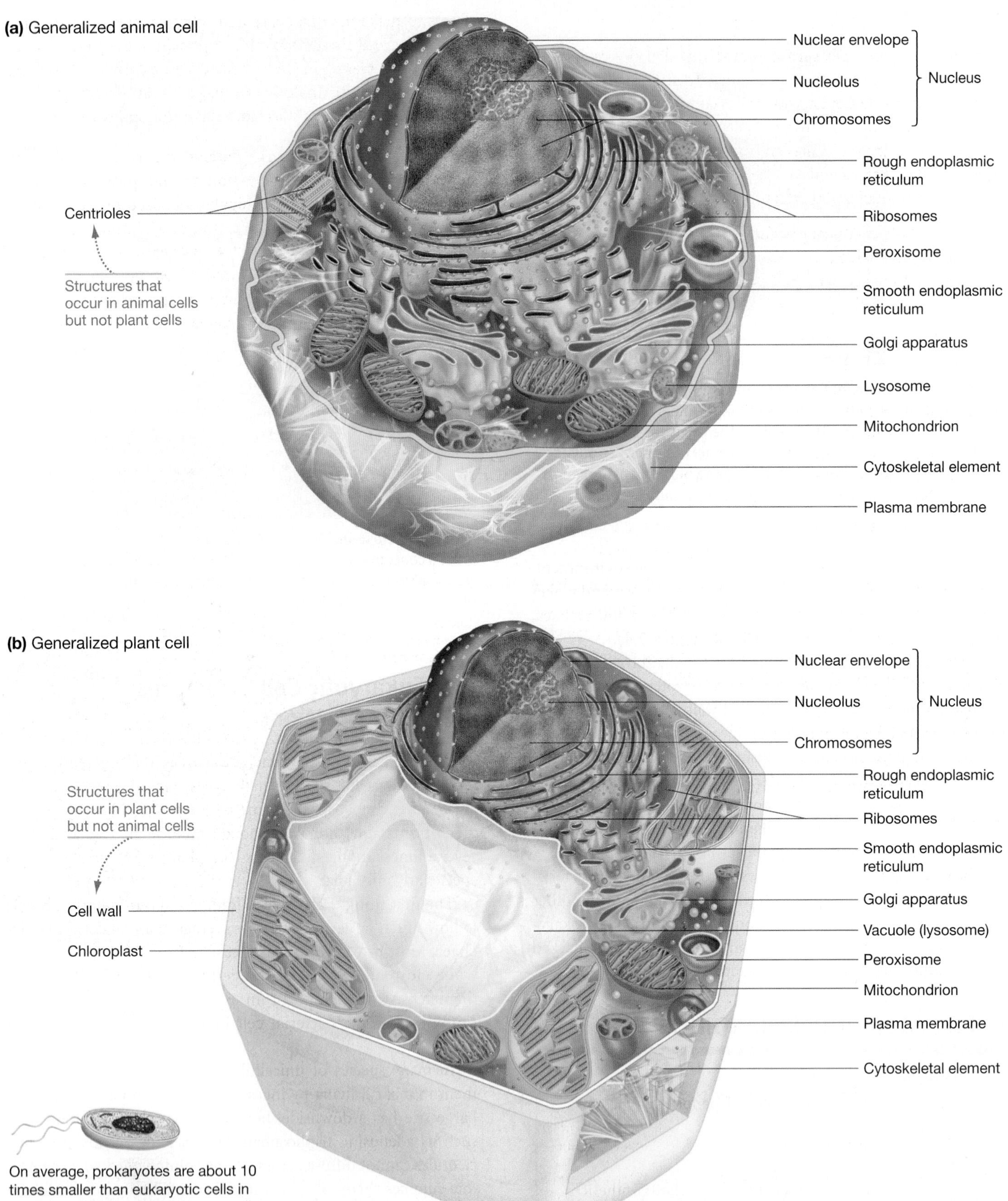

FIGURE 7.6 Overview of Eukaryotic Cells. Generalized, or "typical," **(a)** animal and **(b)** plant cells, color-coded for clarity. Compare with the prokaryotic cell, shown at true relative size at bottom left.

The Benefits of Organelles

How do eukaryotic cells solve the problems that size can engender? The answer lies in their numerous organelles. In effect, the huge volume inside a eukaryotic cell is compartmentalized into a large number of bacterium-sized parts. Because eukaryotic cells are subdivided, the molecules required for specific chemical reactions are often located within a given compartment or organelle.

Compartmentalization offers two key advantages:

- Incompatible chemical reactions can be separated. For example, new fatty acids can be synthesized in one organelle while excess or damaged fatty acids are degraded and recycled in a different organelle.
- Chemical reactions become more efficient. First, the substrates required for particular reactions can be localized and maintained at high concentrations within organelles. Second, if substrates are used up in a particular part of the organelle, they can be replaced by substrates that have only a short distance to diffuse. Third, groups of enzymes that work together can be clustered on internal membranes instead of floating free in the cytoplasm. When the product of one reaction is the substrate for a second reaction catalyzed by another enzyme, clustering the enzymes increases the speed and efficiency of both reaction sequences.

If bacteria and archaea can be compared to small machine shops, then eukaryotic cells resemble sprawling industrial complexes. The organelles and other structures found in eukaryotes are analogous to highly specialized buildings that act as factories, power stations, warehouses, transportation corridors, and administrative centers.

When typical prokaryotic and eukaryotic cells are compared, four key differences identified in **Table 7.1** stand out:

1. Eukaryotic chromosomes are found inside a membrane-bound compartment called the nucleus.
2. Eukaryotic cells are often much larger.
3. Eukaryotic cells contain extensive amounts of internal membrane.
4. Eukaryotic cells feature a particularly diverse and dynamic cytoskeleton.

Eukaryotic Cell Structures: A Parts List

Figure 7.6 provides a simplified view of a typical animal cell and plant cell. The artist has removed most of the cytoskeletal elements to make the organelles and other cellular parts easier to see. As you read about each cell component in the pages that follow, focus on identifying how their structure correlates with their function. Then use **Table 7.2** on page 114 as a study guide. As with bacterial cells, let's start from the inside and work out.

THE NUCLEUS The **nucleus** contains the chromosomes and functions as an information storage and processing center. Among the largest and most highly organized of all organelles (**Figure 7.7**), it is enclosed by a unique structure—a complex double membrane called the **nuclear envelope**. As Section 7.4 will detail, the nuclear envelope is studded with pore-like openings, and its inside surface is linked to fibrous proteins that form a lattice-like sheet called the **nuclear lamina**. The nuclear lamina stiffens the structure and maintains its shape.

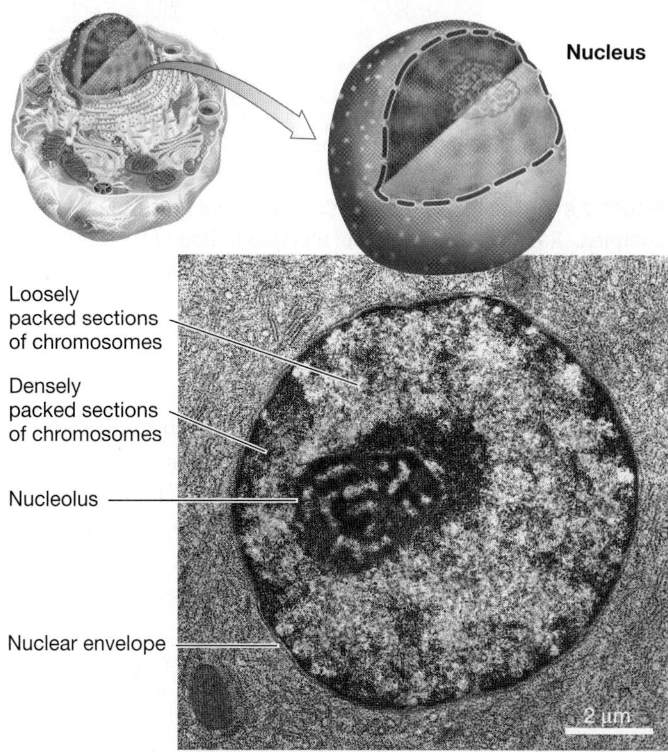

FIGURE 7.7 The Nucleus Is the Eukaryotic Cell's Information Storage and Retrieval Center. The genetic, or hereditary, information is encoded in DNA, which is a component of the chromosomes inside the nucleus.

SUMMARY TABLE 7.1 **How Do the Structures of Prokaryotic and Eukaryotic Cells Differ?**

Cell Type	Location of DNA	Internal Membranes and Organelles	Cytoskeleton	Overall Size
Bacteria and Archaea	In nucleoid (not membrane bound); plasmids also common	Extensive internal membranes only in photosynthetic species; limited types and numbers of organelles	Limited in extent, relative to eukaryotes	Usually small relative to eukaryotes
Eukaryotes	Inside nucleus (membrane bound); plasmids extremely rare	Large numbers of organelles; many types of organelles	Extensive—usually found throughout volume of cell	Most are larger than prokaryotes

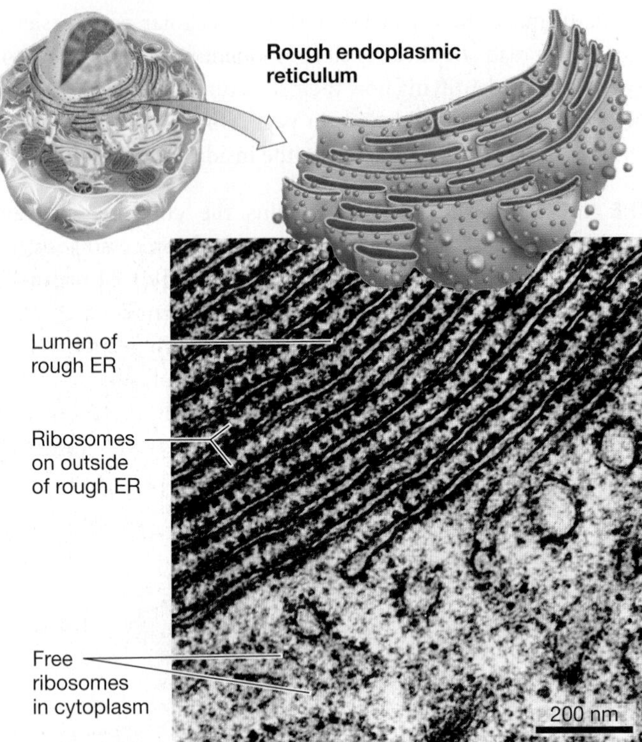

FIGURE 7.8 Rough ER Is a Protein Synthesis and Processing Complex. Rough ER is a system of membrane-bound sacs and tubules with ribosomes attached. It is continuous with the nuclear envelope and with smooth ER.

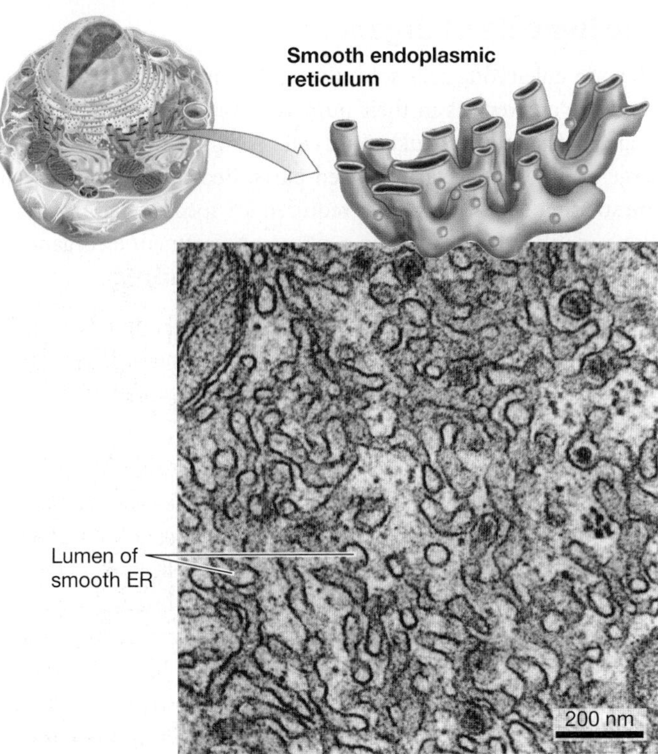

FIGURE 7.9 Smooth ER Is a Lipid-Handling Center and a Storage Facility. Smooth ER is a system of membrane-bound sacs and tubules that lacks ribosomes.

Chromosomes do not float freely inside the nucleus—instead, each chromosome occupies a distinct area and is attached to the nuclear lamina in at least one location. The nucleus also contains specific sites where gene products are processed and includes a distinctive region called the **nucleolus**, where the RNA molecules found in ribosomes are manufactured and the large and small ribosomal subunits are assembled.

The nuclear envelope is continuous with an extensive series of membrane-bound sacs called the **endoplasmic reticulum** (literally, "inside-formed-network"), or ER. As Figure 7.6 shows, the ER extends from the nuclear envelope out into the cytoplasm. Although the ER is a single structure, it has regions that are distinct in structure and function. Let's consider each region in turn.

ROUGH ENDOPLASMIC RETICULUM The **rough endoplasmic reticulum**, or **rough ER (RER)**, is named for its appearance in transmission electron micrographs (**Figure 7.8**). The knobby-looking structures in rough ER are ribosomes that attach to the membrane.

The ribosomes associated with the rough ER synthesize proteins that will be inserted into the plasma membrane, secreted to the cell exterior, or shipped to an organelle. As they are being manufactured by ribosomes, these proteins move to the interior of the sac-like component of the rough ER. The interior of the rough ER, like the interior of any sac-like structure in a cell or body, is called the **lumen**. In the lumen of the rough ER, newly manufactured proteins undergo folding and other types of processing.

The proteins produced in the rough ER have a variety of functions. Some carry messages to other cells; some act as membrane transport proteins or pumps; others are enzymes. The common theme is that rough ER products are packaged into vesicles and transported to various distant destinations—often to the surface of the cell or beyond.

SMOOTH ENDOPLASMIC RETICULUM In electron micrographs, parts of the ER that are free of ribosomes appear smooth and even. Appropriately, these parts of the ER are called **smooth endoplasmic reticulum**, or **smooth ER** (SER; **Figure 7.9**).

The smooth ER membrane contains enzymes that catalyze reactions involving lipids. Depending on the type of cell, these enzymes may synthesize lipids needed by the organism or break down lipids that are poisonous. Smooth ER is also the manufacturing site for phospholipids used in plasma membranes. In addition to processing lipids, smooth ER functions as a reservoir for calcium ions (Ca^{2+}) that act as a signal triggering a wide array of activities inside the cell.

The structure of endoplasmic reticulum correlates closely with its function. Rough ER has ribosomes and functions primarily as a protein-manufacturing center; smooth ER lacks ribosomes and functions primarily as a lipid-processing center.

GOLGI APPARATUS In many cases, the products of the rough ER pass through the Golgi apparatus before they reach their final

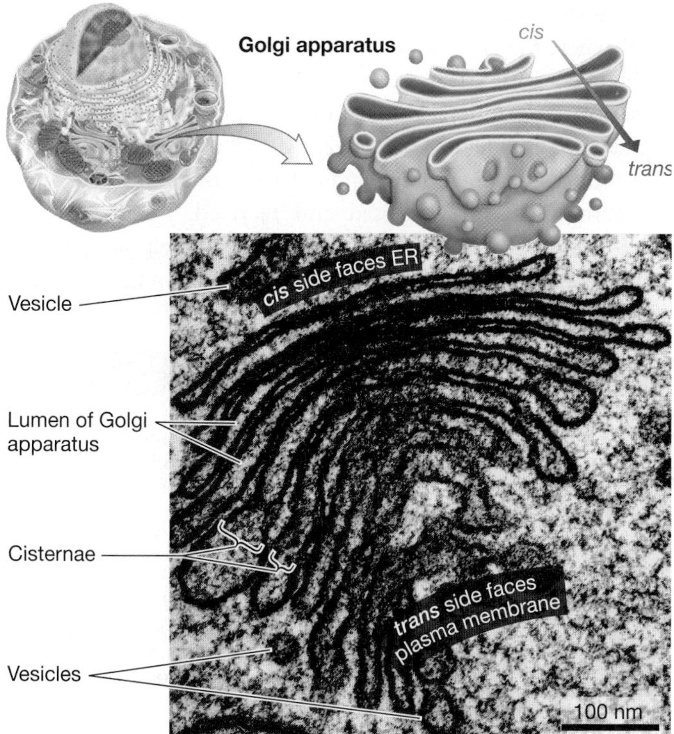

FIGURE 7.10 The Golgi Apparatus Is a Site of Protein Processing, Sorting, and Shipping. The Golgi apparatus is a collection of flattened vesicles called cisternae.

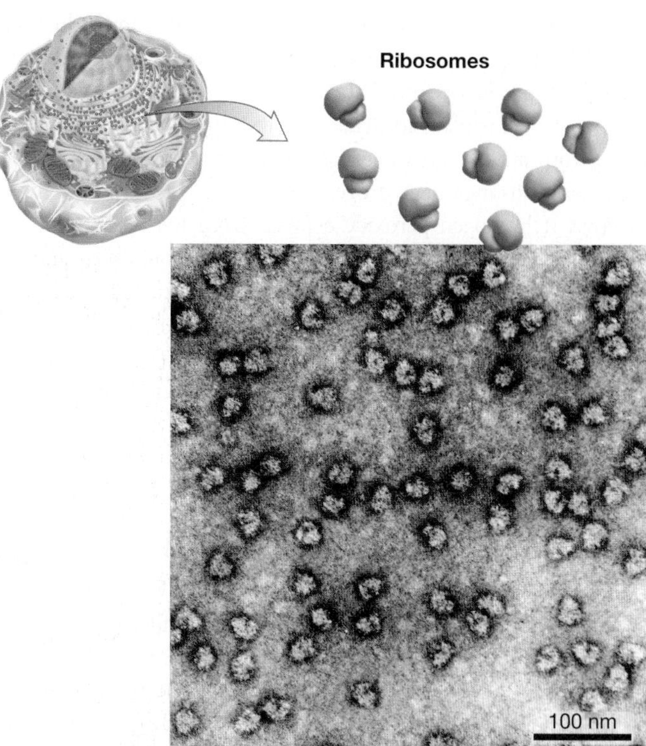

FIGURE 7.11 Ribosomes Are the Site of Protein Synthesis. Eukaryotic ribosomes are similar in structure to bacterial and archaeal ribosomes—though not identical. They are composed of large and small subunits, each of which contains both RNA molecules and proteins.

destination. The **Golgi apparatus** consists of flattened, membranous sacs called **cisternae** (singular: **cisterna**), which are stacked on top of one another (**Figure 7.10**). The organelle also has a distinct polarity, or sidedness. The *cis* ("this side") surface is closest to the rough ER and nucleus, and the *trans* ("across") surface is oriented toward the plasma membrane.

The *cis* side of a Golgi apparatus receives products from the rough ER, and the *trans* side ships them out toward the cell surface. In between, within the cisternae, the rough ER's products are processed and packaged for delivery. Micrographs often show "bubbles" on either side of the Golgi stack. These are membrane-bound vesicles that carry proteins or other products to and from the organelle. Section 7.3 analyzes the intracellular movement of molecules from the rough ER to the Golgi apparatus and beyond in more detail.

RIBOSOMES In eukaryotes, the cytoplasm consists of everything inside the plasma membrane excluding the nucleus; the fluid portion of the cytoplasm is called the **cytosol**. Although many of the cell's millions of ribosomes are attached to the RER, many are scattered throughout the cytosol (**Figure 7.11**).

Like bacterial ribosomes, eukaryotic ribosomes are complex molecular machines that manufacture proteins. They are not classified as organelles because they are not surrounded by a membrane.

PEROXISOMES Virtually all eukaryotic cells contain globular organelles called **peroxisomes** (**Figure 7.12**). These organelles have a single membrane, and originate as buds from the ER.

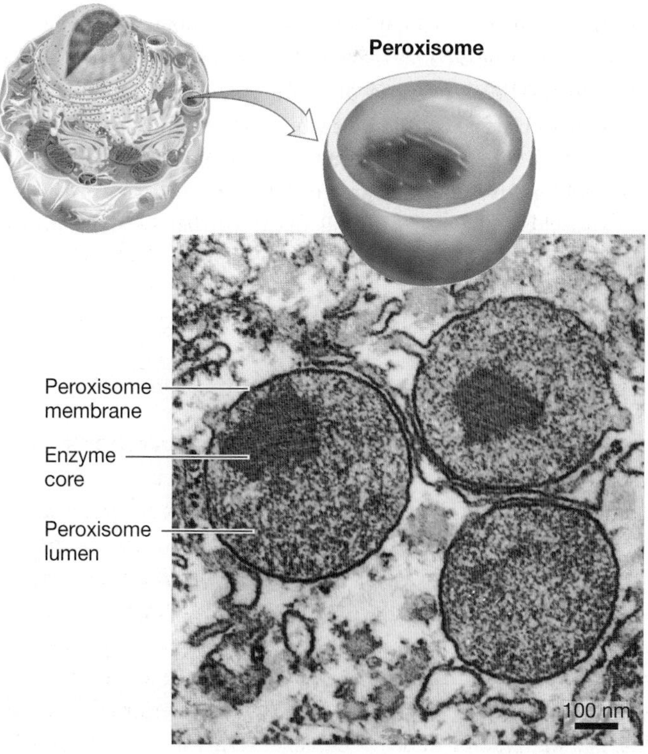

FIGURE 7.12 Peroxisomes Are the Site of Oxidation Reactions. Peroxisomes are globular organelles with a single membrane.

Although different types of cells from the same individual may have distinct types of peroxisomes, these organelles all share a common function: Peroxisomes are centers for oxidation reactions. As Chapter 9 will explain in detail, oxidation reactions remove electrons from atoms and molecules. In many cases the products of these reactions include hydrogen peroxide (H_2O_2), which is highly reactive. If hydrogen peroxide escaped from the peroxisome, the H_2O_2 would quickly react with organelle membranes and the plasma membrane and damage them. This is rare, however. Inside the peroxisome, the enzyme catalase quickly "detoxifies" hydrogen peroxide by converting it to water and oxygen.

Different types of peroxisomes contain different suites of oxidative enzymes. As a result, each is specialized for oxidizing particular compounds. For example, the peroxisomes in your liver cells contain enzymes that oxidize an array of toxins, including the ethanol in alcoholic beverages. The products of these oxidation reactions are usually harmless and are either excreted from the body or used in other reactions.

In plant leaves, specialized peroxisomes called **glyoxysomes** are packed with enzymes that oxidize fats to form a compound that can be used to store energy for the cell. But plant seeds have a different type of peroxisome—one that is packed with enzymes responsible for releasing energy from stored fatty acids. The young plant uses this energy as it begins to grow.

In both animals and plants, there is a clear connection between structure and function: The enzymes found inside the peroxisome make a specialized set of oxidation reactions possible.

LYSOSOMES Animal cells contain organelles called **lysosomes** (**Figure 7.13**) which function as digestive centers. The organelle's interior, or lumen, is acidic because proton pumps in the lysosome membrane import enough hydrogen ions to maintain a pH of 5.0.

Lysosomes also contain about 40 different enzymes, each specialized for breaking up a different type of macromolecule—protein, nucleic acid, lipid, or carbohydrate—into its component monomers. The monomers are then excreted or recycled.

The digestive enzymes are collectively called acid hydrolases because they catalyze hydrolysis reactions that break monomers from macromolecules most efficiently at a pH of 5.0. In the cytosol, where the pH is about 7.2, these enzymes are less active.

Figure 7.14 illustrates two ways that materials are delivered to lysosomes in animal cells: autophagy and phagocytosis. During **autophagy** (literally, "same-eating"), damaged organelles are surrounded by a membrane and delivered to a lysosome. There the components are digested and recycled. In **phagocytosis** ("eat-cell-act"), the plasma membrane of a cell surrounds a smaller cell or food particle and engulfs it, forming a structure

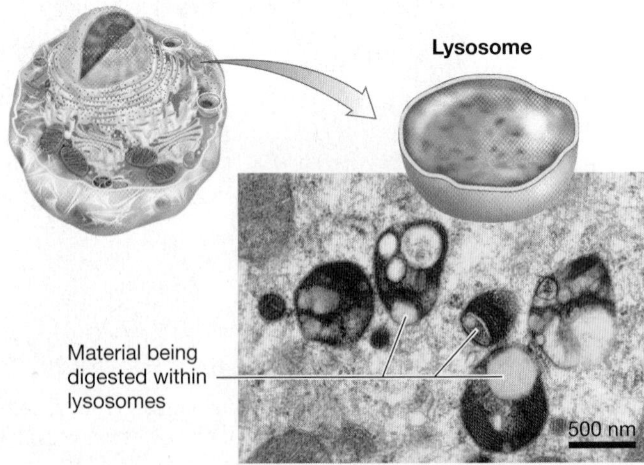

FIGURE 7.13 Lysosomes Are Recycling Centers. Lysosomes are usually oval or globular and have a single membrane.

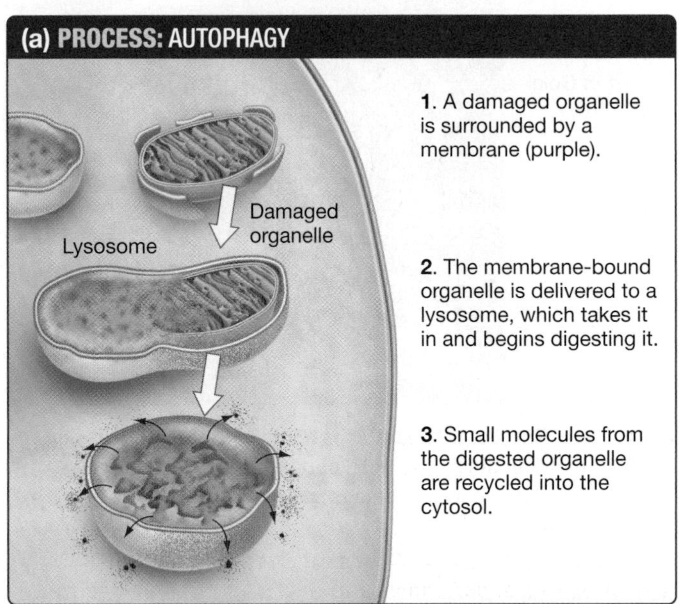

(a) PROCESS: AUTOPHAGY
1. A damaged organelle is surrounded by a membrane (purple).
Damaged organelle
Lysosome
2. The membrane-bound organelle is delivered to a lysosome, which takes it in and begins digesting it.
3. Small molecules from the digested organelle are recycled into the cytosol.

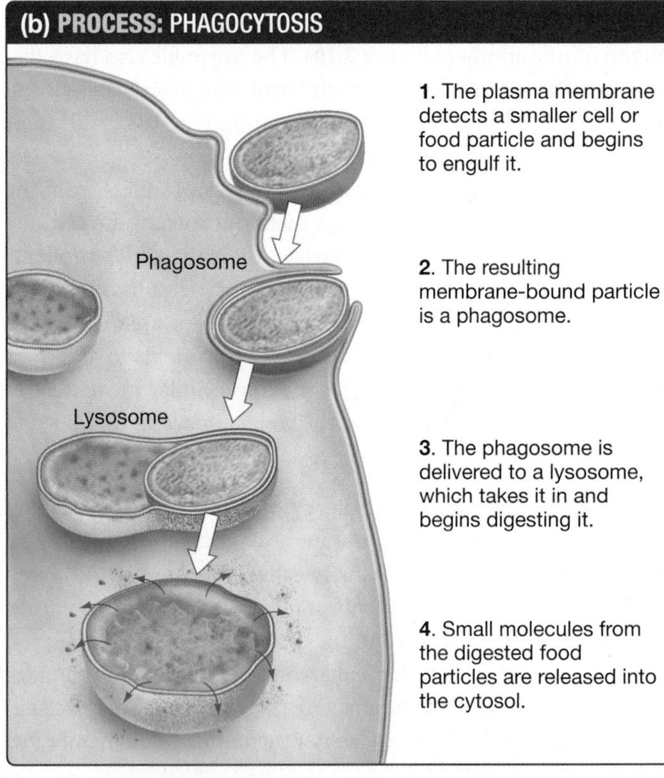

(b) PROCESS: PHAGOCYTOSIS
1. The plasma membrane detects a smaller cell or food particle and begins to engulf it.
Phagosome
2. The resulting membrane-bound particle is a phagosome.
Lysosome
3. The phagosome is delivered to a lysosome, which takes it in and begins digesting it.
4. Small molecules from the digested food particles are released into the cytosol.

FIGURE 7.14 Two Ways to Deliver Materials to Lysosomes. Materials can be transported to lysosomes **(a)** via autophagy or **(b)** after phagocytosis.

PROCESS: RECEPTOR-MEDIATED ENDOCYTOSIS

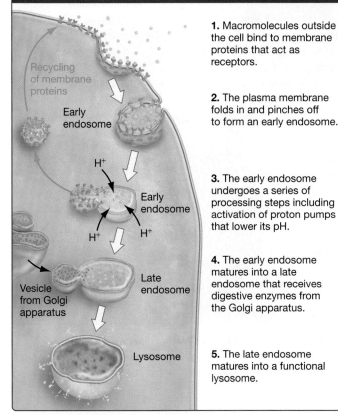

Recycling of membrane proteins

Early endosome

H+

Early endosome

H+ H+

Vesicle from Golgi apparatus

Late endosome

Lysosome

1. Macromolecules outside the cell bind to membrane proteins that act as receptors.

2. The plasma membrane folds in and pinches off to form an early endosome.

3. The early endosome undergoes a series of processing steps including activation of proton pumps that lower its pH.

4. The early endosome matures into a late endosome that receives digestive enzymes from the Golgi apparatus.

5. The late endosome matures into a functional lysosome.

FIGURE 7.15 Receptor-Mediated Endocytosis May Lead to Lysosome Formation. Endosomes created by receptor-mediated endocytosis may mature into lysosomes.

✔**QUESTION** Why is it significant that vesicles from the Golgi apparatus fuse with endosomes?

called a phagosome. This structure is delivered to a lysosome, where it is taken in and digested.

Figure 7.15 illustrates a third way that lysosomes process materials: **receptor-mediated endocytosis.** As its name implies, the sequence of events begins when macromolecules outside the cell bind to membrane proteins that act as receptors. More than 25 distinct receptors have now been characterized, each specialized for responding to a different macromolecule. Receptors are found in specific locations in the cell, where the membrane is underlain by a coating of specialized proteins that include a cage of clathrin molecules.

Once receptor binding occurs, the plasma membrane folds in and pinches off to form a membrane-bound vesicle called an **early endosome** ("inside-body"). The exterior of the endosome is a cage-like structure comprised of clathrins and other proteins.

Early endosomes undergo a series of processing steps that include the activation of proton pumps that gradually lower their pH. When they have matured into a **late endosome**, the structures receive digestive enzymes from the Golgi apparatus and may eventually become fully functioning lysosomes.

Regardless of whether the materials in lysosomes originate via autophagy, phagocytosis, or receptor-mediated endocytosis, the result is similar: Molecules are hydrolyzed. The amino acids, nucleotides, sugars, and other molecules that result from acid hydrolysis leave the lysosome via transport proteins in the organelle's membrane. Once in the cytoplasm, they can be reused.

It is important to note, however, that not all of the materials that are surrounded by membrane and taken into a cell end up in lysosomes. **Endocytosis** ("inside-cell-act") refers to any pinching off of the plasma membrane that results in the uptake of material from outside the cell. In addition to phagocytosis and receptor-mediated endocytosis, endocytosis can occur via **pinocytosis** ("drink-cell-act"). Pinocytosis brings fluid into the cytoplasm via tiny vesicles that form from infoldings of the plasma membrane. The fluid inside these vesicles is not transported to lysosomes, but is used elsewhere in the cell. In addition, most of the macromolecules that collect in early endosomes are selectively removed and used long before the structure becomes a lysosome.

Even though lysosomes are physically separated from the Golgi apparatus and the rough and smooth endoplasmic reticulum, the organelles jointly form a key functional grouping called the **endomembrane system.** The endomembrane ("inner-membrane") system is the primary center for protein and lipid synthesis and processing in eukaryotic cells.

VACUOLES The cells of plants, fungi, and certain other groups lack lysosomes. Instead, they contain a prominent organelle called a vacuole. Compared with the lysosomes of animal cells, the **vacuoles** of plant and fungal cells are large—sometimes taking up as much as 80 percent of a plant cell's volume (**Figure 7.16**).

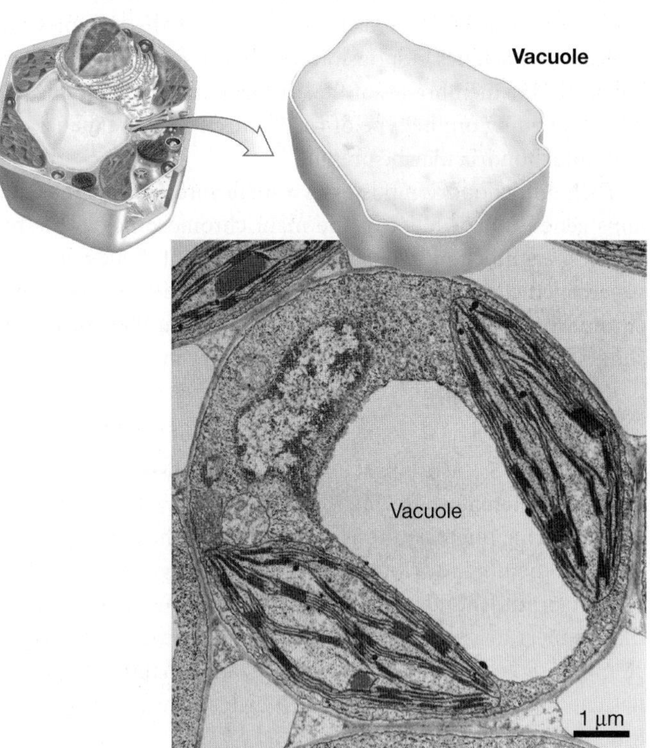

FIGURE 7.16 Vacuoles Are Storage Centers in Plant and Fungal Cells. Vacuoles vary in size and function. Some contain digestive enzymes and serve as recycling centers; most are large storage containers.

✔**QUESTION** Why are toxins like nicotine, cocaine, and caffeine stored in vacuoles instead of the cytosol?

Although some vacuoles contain enzymes that are specialized for digestion, most of the vacuoles observed in plant and fungal cells act as storage depots. In many cases, the stored material is water, which maintains the cell's normal volume, or ions such as potassium (K^+) and chloride (Cl^-). In other cells, vacuoles have more specialized storage functions.

- Inside seeds, cells may contain a large vacuole filled with proteins. When the embryonic plant inside the seed begins to grow, enzymes begin digesting these proteins to provide amino acids for the growing individual.
- In cells that make up flower petals or fruits, vacuoles are filled with colorful pigments.
- Elsewhere, vacuoles may be packed with noxious compounds that protect leaves and stems from being eaten by predators. The type of chemical involved varies by species, ranging from bitter-tasting tannins to toxins such as nicotine, morphine, caffeine, or cocaine.

MITOCHONDRIA The chemical energy required to build all these organelles and do other types of work comes from adenosine triphosphate (ATP), most of which is produced in the cell's **mitochondria** (singular: **mitochondrion**).

As **Figure 7.17** shows, each mitochondrion has two membranes. The outer membrane defines the organelle's surface, while the inner membrane is connected to a series of sac-like **cristae**. The solution inside the inner membrane is called the **mitochondrial matrix**. In eukaryotes, most of the enzymes and molecular machines responsible for synthesizing ATP are embedded in the membranes of the cristae or suspended in the matrix. Depending on the type of cell, from 50 to more than a million mitochondria may be present.

Each mitochondrion possesses a small chromosome that contains genes, independent of the main chromosomes in the nucleus. This mitochondrial DNA is a component of a circular and supercoiled chromosome that is similar in structure to bacterial chromosomes. Mitochondria also manufacture their own ribosomes. Like most organelles, mitochondria can grow and divide independently of nuclear division and cell division.

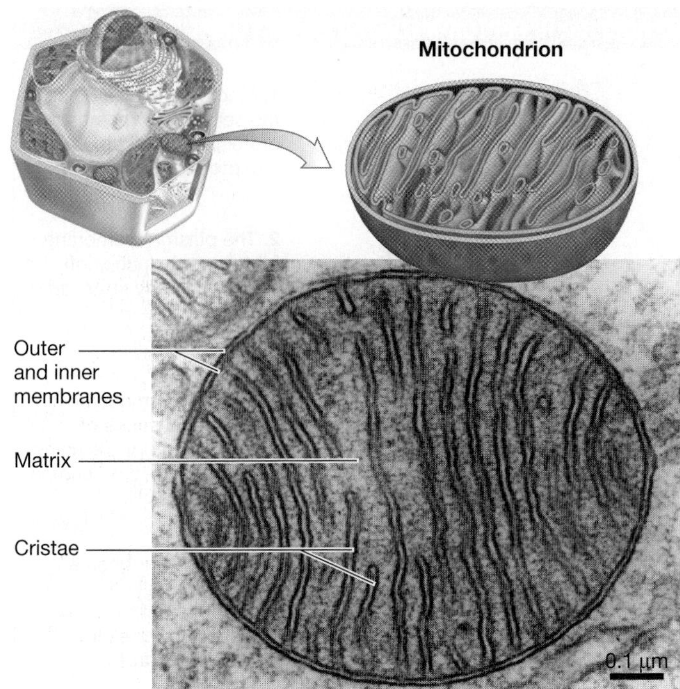

FIGURE 7.17 Mitochondria Are Power-Generating Stations. Mitochondria vary in size and shape, but all have a double membrane with sac-like cristae inside.

CHLOROPLASTS Most algal and plant cells possess an organelle called the **chloroplast**, in which sunlight is converted to chemical energy during photosynthesis (**Figure 7.18**). The chloroplast has a double membrane around its exterior, analogous to the structure of a mitochondrion. Instead of featuring sac-like cristae that connect to the inner membrane, though, the interior of the chloroplast is dominated by hundreds of membrane-bound, flattened vesicles called **thylakoids**, which are independent of the inner membrane.

Thylakoids are stacked like pancakes into piles called **grana** (singular: **granum**). Many of the pigments, enzymes, and molecular machines responsible for converting light energy into carbohydrates are embedded in the thylakoid membranes. Certain

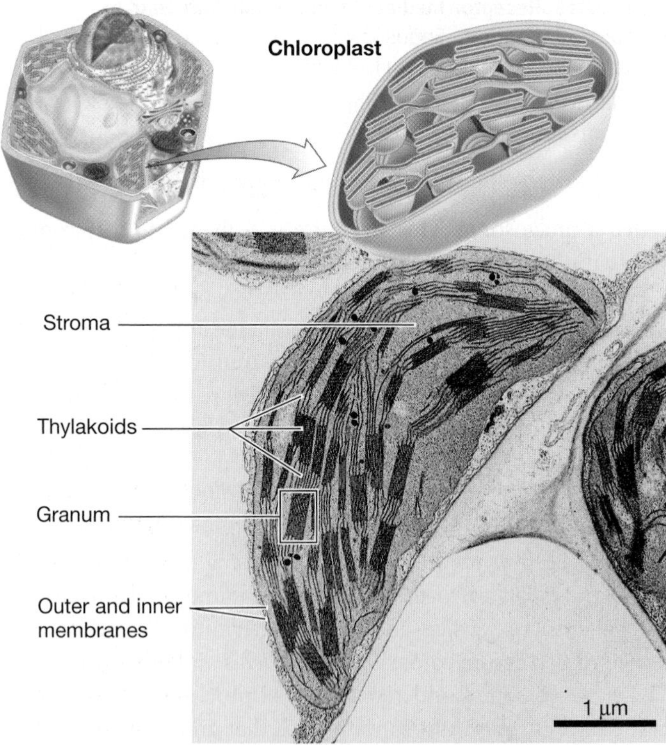

FIGURE 7.18 Chloroplasts Are Sugar-Manufacturing Centers in Plants and Algae. Many of the enzymes and other molecules required for photosynthesis are located in membranes inside the chloroplast. These membranes are folded into thylakoids and stacked into grana.

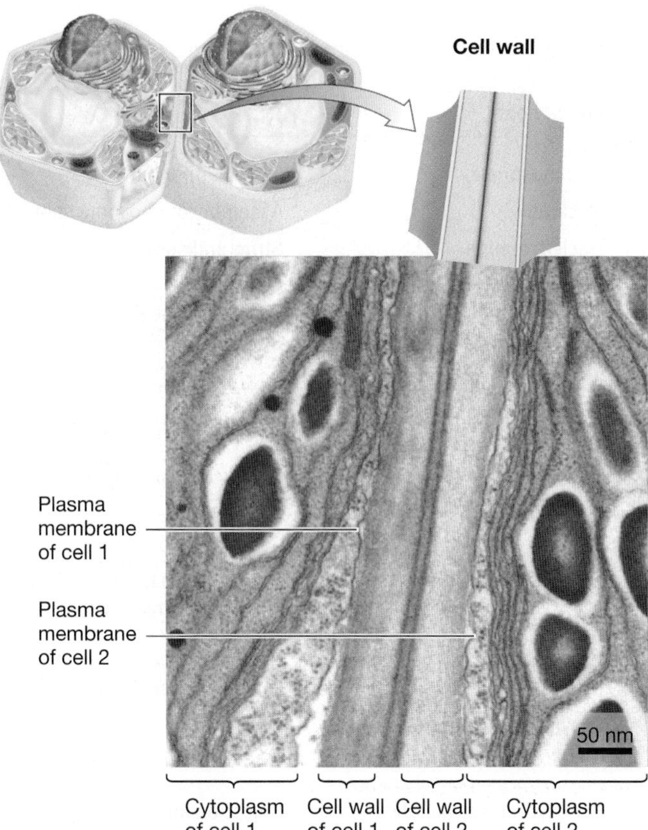

FIGURE 7.19 Cell Walls Protect Plants and Fungi. Plants have cell walls that contain cellulose; in fungi the major structural component of the cell wall is chitin.

✔**QUESTION** Is the cell wall inside or outside the plasma membrane?

critical enzymes and substrates, however, are found outside the thylakoids in the region called the **stroma**.

The number of chloroplasts per cell varies from none to several dozen. Like mitochondria, each chloroplast contains a circular chromosome. Chloroplast DNA is independent of the main genetic material inside the nucleus. Chloroplasts also grow and divide independently of nuclear division and cell division.

THE CELL WALL In fungi, algae, and plants, cells possess an outer cell wall in addition to their plasma membrane (**Figure 7.19**). The cell wall is located outside of the plasma membrane, and furnishes a stiff, outer layer that provides structural support for the cell. The cells of animals, amoebae, and other groups lack a cell wall—their exterior surface consists of just the plasma membrane.

Although the composition of the cell wall varies among species and even between types of cells in the same individual, the general plan is similar: Rods or fibers composed of a carbohydrate run through a stiff matrix made of other polysaccharides and proteins (see Chapter 8 for details).

In addition, some plant cells produce a secondary cell wall that features a particularly tough molecule called lignin. Lignin forms a branching, cagelike network that is almost impossible for enzymes to attack. The combination of cellulose fibers and lignin in secondary cell walls makes up most of the material we call wood.

CYTOSKELETON The final major structural feature that is common to all eukaryotic cells is the cytoskeleton, an extensive system of protein fibers. In addition to giving the cell its shape and structural stability, cytoskeletal proteins are involved in moving the cell itself and moving materials within the cell. In essence, the cytoskeleton organizes all of the organelles and other cellular structures into a cohesive whole. The painting in **Figure 7.20** is a cross section through a small part of a eukaryotic cell. Note the density of cytoskeletal elements—the long tubes colored blue in this painting. Section 7.6 will analyze the structure and functions of the cytoskeleton in detail.

Before moving on to the next section, review animations of animal and plant cells in the study area at *www.masteringbiology.com* and complete the Check Your Understanding box on page 115.

(MB) **BioFlix™** Tour of an Animal Cell, **BioFlix™** Tour of a Plant Cell

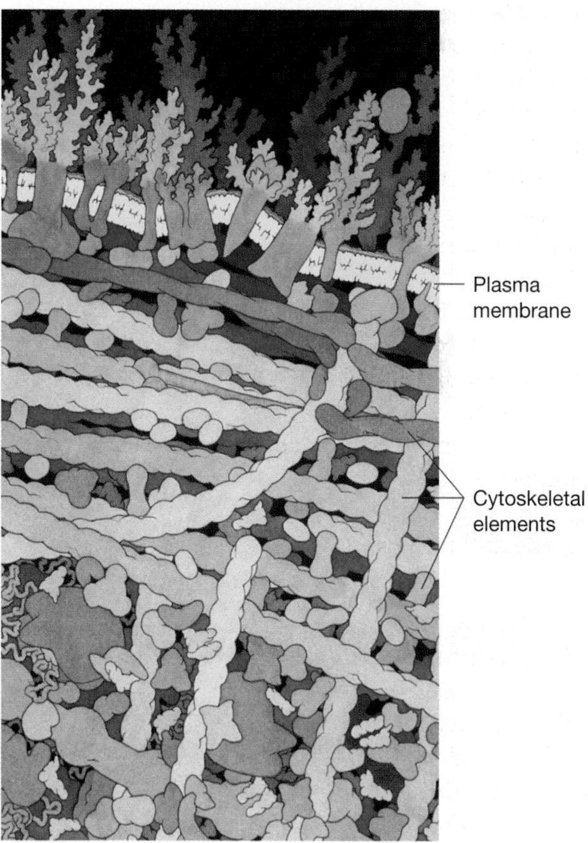

FIGURE 7.20 Close-up View of a Eukaryotic Cell. Cells are highly structured entities packed with organelles, ribosomes, and cytoskeletal elements (colored blue here). Note that the phospholipid bilayer, colored yellow here, is studded with transmembrane proteins—many of which have sugar groups (in green) projecting into extracellular space.

SUMMARY TABLE 7.2 **Eukaryotic Cell Components**

Icons not to scale		Structure		Function
		Membrane	Components	
	Nucleus	Double ("envelope"); openings called nuclear pores	Chromosomes Nucleolus Nuclear lamina	Genetic information Assembly of ribosome subunits Structural support
	Ribosomes	None	Complex of RNA and proteins	Protein synthesis
	Endomembrane system			
	Rough ER	Single; contains receptors for entry of selected proteins	Network of branching sacs Ribosomes associated	Protein synthesis and processing
	Golgi apparatus	Single; contains receptors for products of rough ER	Stack of flattened cisternae	Protein processing (e.g., glycosylation)
	Smooth ER	Single; contains enzymes for synthesizing phospholipids	Network of branching sacs Enzymes for synthesizing lipids	Lipid synthesis
	Lysosomes	Single; contains proton pumps	Acid hydrolases (catalyze hydrolysis reactions)	Digestion and recycling
	Peroxisomes	Single; contains transporters for selected macromolecules	Enzymes that catalyze oxidation reactions Catalase (processes peroxide)	Oxidation of fatty acids, ethanol, or other compounds
	Vacuoles	Single; contains transporters for selected molecules	Varies—pigments, oils, carbohydrates, water, or toxins	Varies—coloration, storage of oils, carbohydrates, water, or toxins
	Mitochondria	Double; inner contains enzymes for ATP production	Enzymes that catalyze oxidation-reduction reactions, ATP synthesis	ATP production
	Chloroplasts	Double; plus membrane-bound sacs in interior	Pigments Enzymes that catalyze oxidation-reduction reactions	Production of ATP and sugars via photosynthesis
	Cytoskeleton	None	Actin filaments Intermediate filaments Microtubules	Structural support; movement of materials; in some species, movement of whole cell
	Plasma membrane	Single; contains transport and receptor proteins	Phospholipid bilayer with transport and receptor proteins	Selective permeability—maintains intracellular environment
	Cell wall	None	Carbohydrate fibers running through carbohydrate or protein matrix	Protection, structural support

CHECK YOUR UNDERSTANDING

If you understand that . . .

- Each structure in a eukaryotic cell performs a function vital to the cell.

You should be able to . . .

1. Explain how the structure of peroxisomes and lysosomes correlates with their function.
2. In Table 7.2, indicate next to each component which component(s) serve as: administrative/information hub, power station, warehouse, large molecule manufacturing and shipping facility (with subtitles for fat factory, protein finishing and shipping line, protein synthesis and folding center, waste processing and recycling center), support beams, perimeter fencing with secured gates, protein factory, food-manufacturing facility, and fatty-acid processing and detox center.

Answers are available in Appendix B.

7.3 Putting the Parts into a Whole

Within a cell, the structure of each organelle and component correlates with its function. In the same way, the overall size, shape, and composition of a cell correlates with its function.

Cells might be analogous to machine shops or factory complexes, but clothing manufacturing centers are very different in layout and composition from airplane production facilities. How does the physical and chemical makeup of a cell correlate with its function?

Structure and Function at the Whole-Cell Level

Inside an individual plant or animal, cells are specialized for certain tasks and have a structure that correlates with those tasks. For example, the muscle cells in your upper leg are extremely long, tube-shaped structures. They are filled with protein fibers that slide past one another as the entire muscle flexes or extends. It is this sliding motion that allows your muscles to contract or extend as you run. Muscle cells are also packed with mitochondria, which produce the ATP required for the sliding motion to occur.

In contrast, nearby fat cells are rounded, globular structures that store fatty acids. They consist of little more than a plasma membrane, a nucleus, and a fat droplet. Neither cell bears a close resemblance to the generalized animal cell pictured in Figure 7.6a.

To drive home the correlation between the overall structure and function of a cell, examine the transmission electron micrographs in **Figure 7.21**.

- The animal cell in Figure 7.21a, located in the pancreas, manufactures and exports digestive enzymes. It is packed with rough ER and Golgi, which make these functions possible.
- The animal cell in Figure 7.21b is from the testis and synthesizes the steroid hormone testosterone. This cell is dominated

(a) Animal pancreatic cell: Exports digestive enzymes.

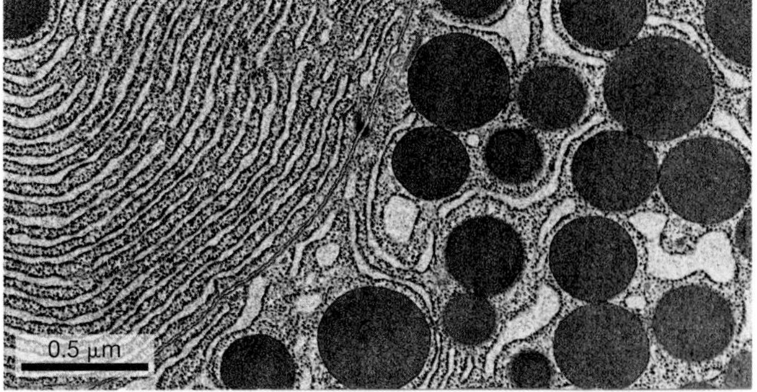

(b) Animal testis cell: Exports lipid-soluble signals.

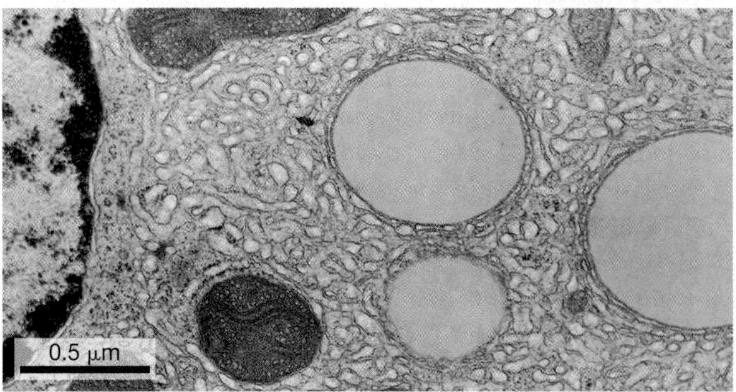

(c) Plant leaf cell: Manufactures ATP and sugar.

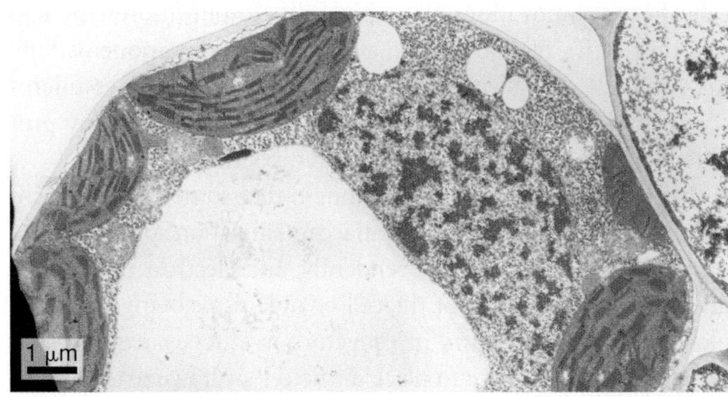

(d) Plant root cell: Stores starch.

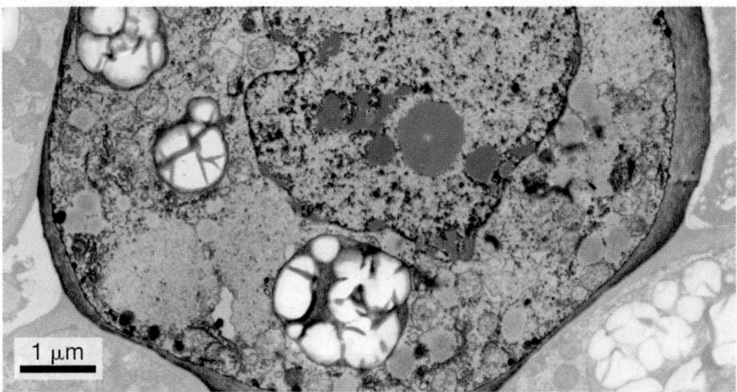

FIGURE 7.21 Cell Structure Correlates with Function.

✔**EXERCISE** In part (a), label the rough ER and secretory vesicles. (They are dark and round.) In (b), label the smooth ER. In (c), label the chloroplasts, vacuole, and nucleus. In (d), label the starch granules.

by smooth ER, where processing of steroids and other lipids takes place.

- The plant cell in Figure 7.21c, from the leaf of a potato, has hundreds of chloroplasts and is specialized for absorbing light and manufacturing sugar.
- The plant cell in Figure 7.21d comes from a potato tuber (part of an underground stem). It has numerous vacuoles that are packed with stored starch—which shows up as white chunks in the micrograph.

In each case, the type of organelles in each cell and their size and number correlate with the cell's specialized function.

The Dynamic Cell

Biologists study the structure and function of organelles and cells with a combination of tools and approaches. For several decades, a technique called **differential centrifugation** was particularly important, because it allowed researchers to isolate particular cell components and analyze their chemical composition. As **BioSkills 11** in Appendix A explains, differential centrifugation is based on breaking cells apart to create a complex mixture and separating components in a centrifuge. The individual parts of the cell can then be purified and studied in detail, in isolation from other parts of the cell.

Historically and currently, however, the most important research in cell biology is based on imaging—simply looking at cells. Recent innovations allow biologists to put fluorescing tags or other types of markers on particular cell components, and then look at them with increasingly sophisticated light microscopes and electron microscopes. Advances in microscopy provide increasingly high magnification and better resolution.

It's important to recognize, though, that some of these techniques have limitations. Differential centrifugation splits cells into parts that are analyzed independently, and electron microscopy gives a fixed "snapshot" of the cell or organisms being observed. Neither technique allows investigators to directly explore how things move from place to place in the cell or how parts interact. The information gleaned from these techniques can make cells seem static. In reality, cells are dynamic.

The amount of chemical activity and the speed of molecular movement inside cells is nothing short of fantastic. Bacterial ribosomes add up to 20 amino acids per second to a growing polypeptide, and eukaryotic ribosomes typically add two per second. Given that there are about 15,000 ribosomes in each bacterium and possibly a million in an average eukaryotic cell, hundreds or even thousands of new protein molecules can be finished each second in every cell.

- In an average second, a typical cell in your body uses an average of 10 million ATP molecules and synthesizes just as many.
- It's not unusual for a cellular enzyme to catalyze 25,000 or more reactions per second; most cells contain hundreds or thousands of enzymes.
- A minute is more than enough time for each membrane phospholipid in your body to travel the breadth of the organelle or cell where it resides.
- The hundreds of trillions of mitochondria inside you are completely replaced about every 10 days, for as long as you live.
- The plasma membrane is fluid, and its composition is constantly changing.

Because humans are such large organisms, it's impossible for us to imagine what life is really like inside a cell. At the scale of a ribosome or an organelle or a cell, gravity is inconsequential. Instead, the dominant forces are the charge- or polarity-based electrostatic attractions between molecules and the kinetic energy of motion. At this level, events take nanoseconds, and speeds are measured in micrometers per second. This is the speed of life.

Contemporary methods for studying cells, including some of the imaging techniques featured in **BioSkills 10** in Appendix A, capture this dynamism by tracking how organelles and molecules move and interact over time. The ability to digitize video images of live cells, or take time-lapse photographs of living cells, is allowing researchers to see and study dynamic processes.

The rest of this chapter focuses on this theme of cellular dynamism and movement. Its goal is to put all the individual pieces of a cell together, and ask how they work as systems to accomplish key tasks.

To begin, let's look at how molecules move into and out of the cell's control center—the nucleus. Then we'll consider how proteins move from ribosomes into the lumen of the rough ER and then to the Golgi apparatus and beyond. The chapter closes by analyzing how cytoskeletal elements help transport cargo inside the cell or move the cell itself.

7.4 Cell Systems I: Nuclear Transport

The nucleus is the information center of eukaryotic cells—a corporate headquarters, design center, and library all rolled into one. Appropriately enough, its interior is highly organized.

The organelle's overall shape and structure are defined by the mesh-like nuclear lamina. The nuclear lamina provides an attachment point for the chromosomes, each of which occupies a well-defined region in the nucleus.

In addition, specific centers exist where the genetic information in DNA is decoded and processed. At these locations, large suites of enzymes interact to produce RNA messages from specific genes at specific times. Meanwhile, the nucleolus functions as the site of ribosome synthesis.

Structure and Function of the Nuclear Envelope

The nuclear envelope separates the nucleus from the rest of the cell. Starting in the 1950s, transmission electron micrographs of cross sections through the nuclear envelope showed that the

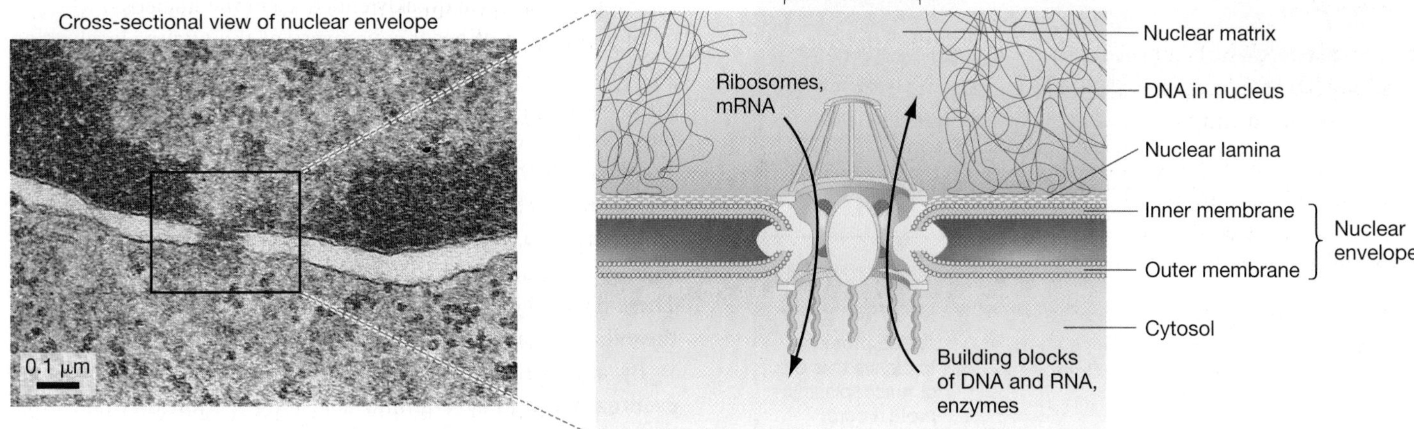

FIGURE 7.22 Structure of the Nuclear Envelope and Nuclear Pore.

structure is supported by the fibrous nuclear lamina and bounded by two membranes, each consisting of a lipid bilayer (**Figure 7.22**).

Micrographs like the one in Figure 7.22 also show that the envelope is broken with openings called **nuclear pores**. Because these gate-like structures extend through both inner and outer nuclear membranes, they connect the inside of the nucleus with the cytosol. Follow-up research showed that each pore consists of over 50 different proteins. As the diagram on the right side of Figure 7.22 shows, these protein molecules form an elaborate structure called the **nuclear pore complex.**

Experiments in the early 1960s showed that molecules travel into and out of the nucleus through the nuclear pore complexes. The initial studies were based on injecting tiny gold particles into cells and then preparing the cells for electron microscopy. In electron micrographs, gold particles show up as black dots. One or two minutes after injection, the micrographs showed that most of the gold particles were in the cytoplasm. A few, however, were closely associated with nuclear pores. Ten minutes after injection, particles were inside the nucleus as well as in the cytoplasm.

These data supported the hypothesis that the pores function as the doors to the nucleus. Follow-up work confirmed that the nuclear pore complex is the only gate between the cytoplasm and the nucleus and only certain molecules go in and out. Passage through a nuclear pore is selective.

What substances traverse nuclear pores? DNA clearly does not—it never leaves the nucleus. But information coded in DNA is used to synthesize RNA inside the nucleus.

Several types of RNA molecules are produced, each distinguished by size and function. For example, most **ribosomal RNAs** are manufactured in the nucleolus, where they bind to proteins to form ribosomes, which are then exported to the cytoplasm. Similarly, molecules called messenger **RNAs (mRNA)** carry the information required to manufacture proteins out to the cytoplasm, where protein synthesis takes place. The outbound traffic, mostly consisting of RNAs, is intense.

Inbound traffic is also impressive. Nucleoside triphosphates that act as building blocks for DNA and RNA enter the nucleus, as do the proteins responsible for copying DNA, synthesizing RNAs, extending the nuclear lamina, assembling ribosomes, or building chromosomes.

To summarize, ribosomal subunits and various types of RNAs exit the nucleus; proteins that are needed inside enter it. In a typical cell, over 500 molecules pass through each of the 3000–4000 nuclear pores every second.

The scale of traffic through the nuclear pores is mind-boggling. How is it regulated and directed?

How Are Molecules Imported into the Nucleus?

The first experiments on how molecules move through the nuclear pore focused on proteins that are produced by viruses. **Viruses** are parasites that use the cell's machinery to make copies of themselves. When a virus infects a cell, certain of its proteins enter the nucleus.

Investigators began studying the transport of viral proteins when they noticed that if a particular amino acid in a viral protein changed, it was no longer able to pass through the nuclear pore. This simple-sounding observation led to a key hypothesis: Proteins that are synthesized by ribosomes in the cytosol but are headed for the nucleus contain a "zip code"—a molecular address tag that marks them for transport through the nuclear pore complex.

The idea was that viral proteins could enter the nucleus only if they carried the same address tag that normal cellular proteins had. Thus, the proteins with the altered amino acid were thwarted. This zip code came to be called the **nuclear localization signal (NLS)**.

A series of experiments on a protein called nucleoplasmin helped researchers understand the nature of the nuclear localization signal. Nucleoplasmin, which plays an important role in the assembly of chromosomes, has a distinctive structure: a globular protein core surrounded by a series of extended protein "tails." When researchers labeled nucleoplasmin with a radioactive atom

EXPERIMENT

QUESTION: Where is the "Send to nucleus" zip code in the nucleoplasmin protein?

HYPOTHESIS: The "Send to nucleus" zip code is in either the tail region or the core region of the nucleoplasmin protein.

NULL HYPOTHESIS: The zip code is not on the nucleoplasmin protein itself, or there is no zip code.

EXPERIMENTAL SETUP:

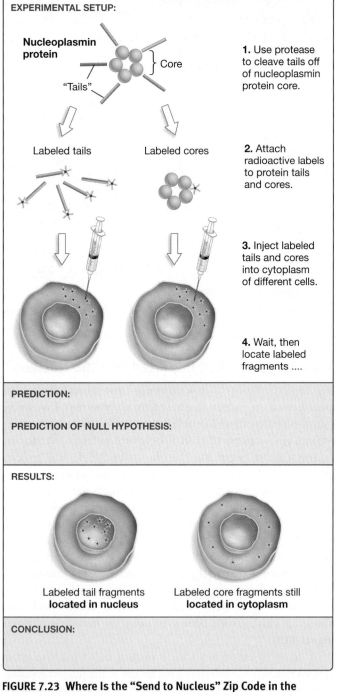

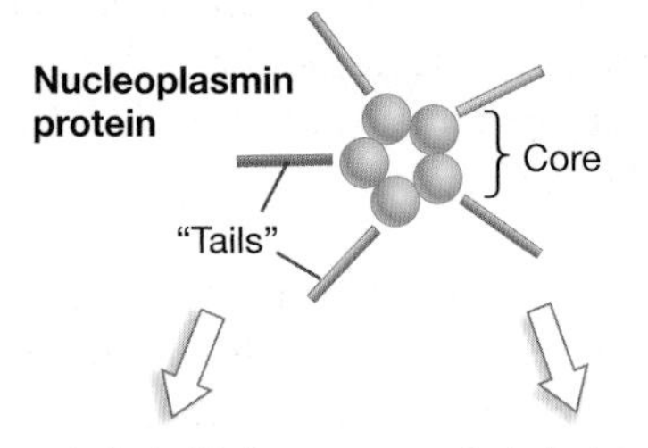

1. Use protease to cleave tails off of nucleoplasmin protein core.

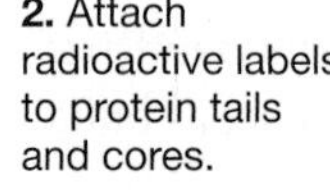

2. Attach radioactive labels to protein tails and cores.

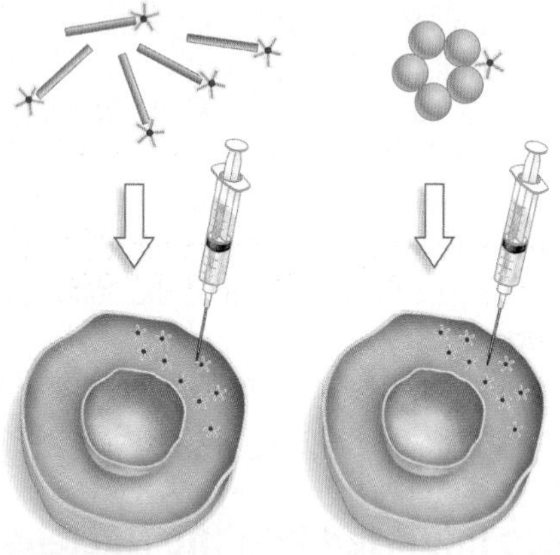

3. Inject labeled tails and cores into cytoplasm of different cells.

4. Wait, then locate labeled fragments ....

PREDICTION:

PREDICTION OF NULL HYPOTHESIS:

RESULTS:

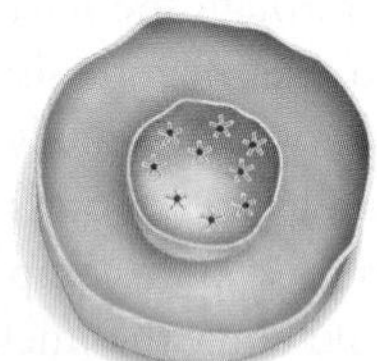

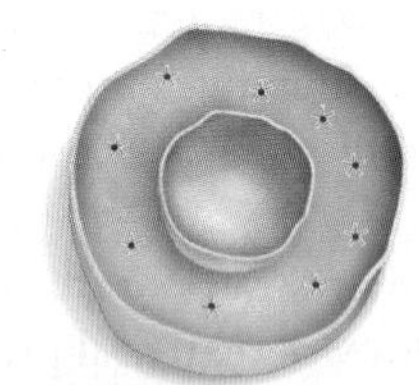

CONCLUSION:

FIGURE 7.23 Where Is the "Send to Nucleus" Zip Code in the Nucleoplasmin Protein?

SOURCES: Mills, A. D. et al. 1980. *Journal of Molecular Biology* 139: 561–568; Dingwall, C. et al. 1982. *Cell* 30: 449–458.

✔**EXERCISE** Without looking at the text, fill in the prediction(s) and conclusion(s) in this experiment.

and injected it into the cytoplasm of living cells, they found that the radioactive signal quickly ended up in the nucleus.

Figure 7.23 outlines the discovery of the nuclear localization signal in nucleoplasmin. First, researchers used enzymes called proteases to separate the core sections of nucleoplasmin from the tails. Then they labeled each component with radioactive atoms and injected them into the cytoplasm of different cells. When they examined the experimental cells with the electron microscope, they found that tail fragments were transported to the nucleus. Core fragments, in contrast, remained in the cytoplasm. These data suggested that the "zip code" must be somewhere in the tail.

By analyzing different stretches of the tail, the biologists eventually found a 17-amino-acid-long section that had to be present to direct proteins to the nucleus. They concluded that the nuclear localization signal consisted of 17 specific amino acids in the tail.

You can review the experimental evidence for the nuclear localization signal in the study area at *www.masteringbiology.com.*

Web Activity Transport into the Nucleus

Follow-up work confirmed that other proteins bound for the nucleus have similar localization signals. More recent research has shown that the movement of proteins and other large molecules into and out of the nucleus is an energy-demanding process that involves transport proteins called importins and exportins. Importins and exportins function like trucks that haul cargo into and out of the nucleus, through the nuclear pore complex.

Currently, biologists are trying to unravel how all this traffic in and out of the nucleus is regulated to avoid backups and head-on collisions. Nuclear transport is a classic research system and a key example of how organelles interact.

7.5 Cell Systems II: The Endomembrane System Manufactures and Ships Proteins

The nuclear membrane is not the only place in cells where cargo moves in a regulated and energy-demanding fashion. For example, Chapter 6 highlighted how specific ions and molecules are pumped into and out of cells or transported across the plasma membrane by specialized membrane proteins.

In addition, proteins that are synthesized by ribosomes in the cytosol for use inside mitochondria or chloroplasts contain special signal sequences, like the nuclear localization signal, that target the proteins for transport to the appropriate organelles. Ions, ATP, amino acids, and other small molecules diffuse randomly throughout the cell, but the movement of proteins and other large molecules is energy demanding and tightly regulated.

If you think about it for a moment, the need to sort proteins and ship them to specific destinations should be clear. Proteins are produced by ribosomes in the cytosol or ribosomes on the ER. Each protein that is synthesized needs to be transported to one of the many compartments inside the eukaryotic cell. Acid hydrolases need to end up in lysosomes, catalase must be shipped to peroxisomes, and ribosomal proteins require transport to the nucleolus. To get to the right location, each protein has to have an address tag and a transport and delivery system.

To get a better understanding of protein sorting and transport in eukaryotic cells, let's consider perhaps the most intricate of all manufacturing and shipping complexes: the endomembrane system. In this system, proteins that are synthesized in the rough ER move to the Golgi apparatus for processing, and from there travel to the cell surface or other destinations.

Studying the Pathway through the Endomembrane System

The idea that materials move through the endomembrane system in an orderly way was inspired by a simple observation. According to electron micrographs, cells that secrete digestive enzymes, hormones, or other products have particularly large amounts of rough ER and Golgi. This correlation led to the idea that these cells have a "secretory pathway" that starts in the rough ER and ends with products leaving the cell (**Figure 7.24**). How does this hypothesized pathway work?

THE LOGIC OF A PULSE-CHASE EXPERIMENT George Palade and colleagues did pioneering research on the secretory pathway using a **pulse-chase experiment**. This strategy is based on two steps:

- *The "Pulse"* Expose experimental cells to a high concentration of a labeled molecule for a short time. For example, if a cell takes in a large amount of radioactively labeled amino acid for a brief time, virtually all of the proteins synthesized during that interval will be labeled.
- *The "Chase"* The pulse of labeled molecule is followed by a chase—large amounts of an unlabeled version of the same molecule, provided for a longer time. If the chase consists of unlabeled amino acid, then the proteins synthesized during the chase period will *not* be labeled.

The idea is to mark a population of molecules at a particular interval and then follow their fate over time. This approach is analogous to adding a small amount of dye to a stream and then following the movement of the dye molecules.

To understand why the chase is necessary in these experiments, imagine what would happen if you added dye to a stream continuously. Soon the entire stream would be dyed—you could no longer tell where a specific population of dye molecules were moving.

In testing the secretory pathway hypothesis, Palade's team focused on pancreatic cells that were growing in **culture**, or in

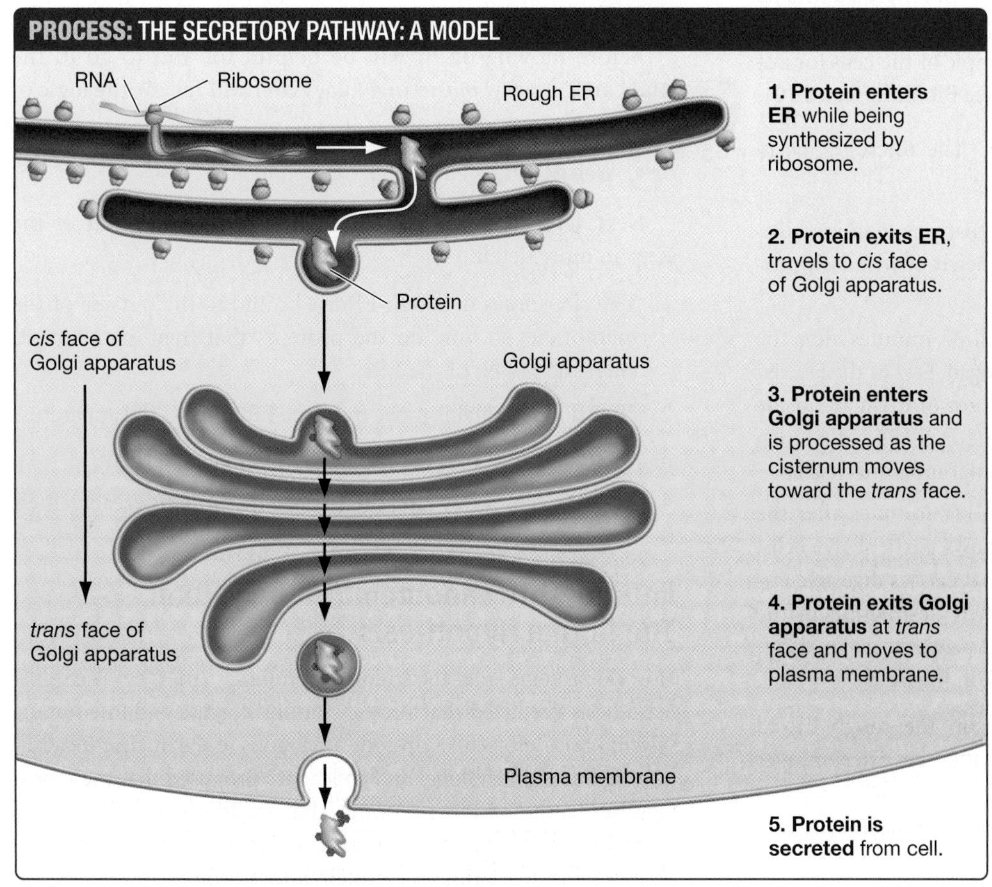
PROCESS: THE SECRETORY PATHWAY: A MODEL
RNA
Ribosome
Rough ER
1. Protein enters ER while being synthesized by ribosome.
2. Protein exits ER, travels to cis face of Golgi apparatus.
Protein
cis face of Golgi apparatus
Golgi apparatus
3. Protein enters Golgi apparatus and is processed as the cisternum moves toward the trans face.
trans face of Golgi apparatus
4. Protein exits Golgi apparatus at trans face and moves to plasma membrane.
Plasma membrane
5. Protein is secreted from cell.

FIGURE 7.24 The Secretory Pathway Hypothesis. The secretory pathway hypothesis proposes that proteins intended for secretion from the cell are synthesized and processed in a highly prescribed set of steps. Note that proteins are packaged into vesicles when they move from the RER to the Golgi and from the Golgi to the cell surface.

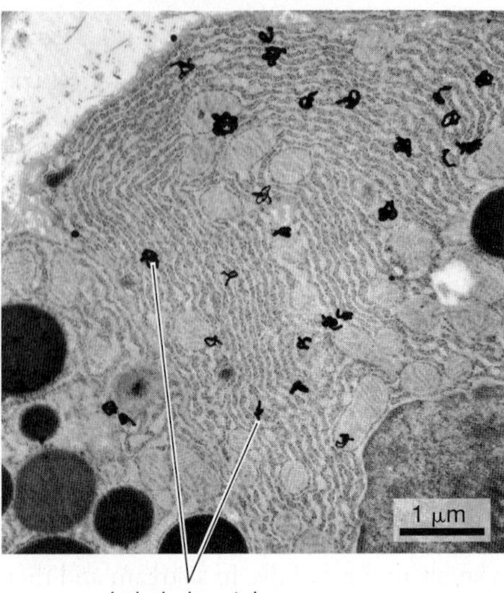

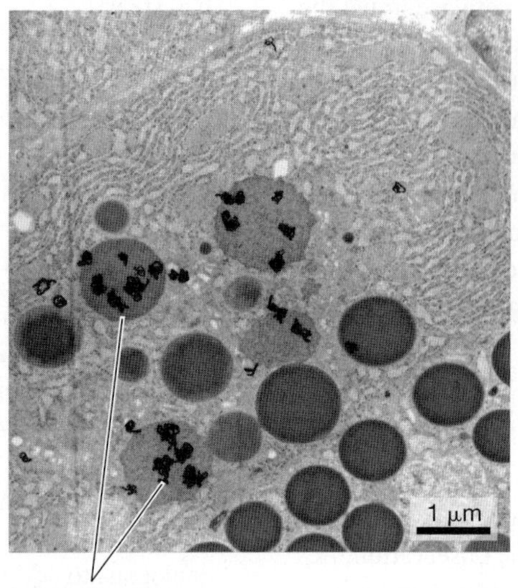

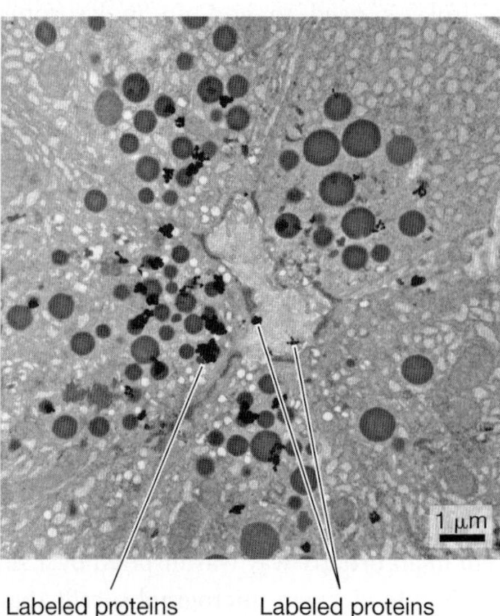

FIGURE 7.25 Results of a Pulse-Chase Experiment. Proteins move **(a)** from the rough ER into **(b)** secretory vesicles before **(c)** being secreted from the cell.

vitro.[1] These cells, specialized for secreting digestive enzymes into the small intestine, are packed with rough ER and Golgi.

The basic experimental approach was to supply the cells with a 3-minute pulse of the amino acid leucine, labeled with a radioactive atom, followed by a long chase with nonradioactive leucine. Because the radioactive leucine was incorporated into all proteins being produced during the pulse, those proteins were labeled. Then the researchers prepared a sample of the cells for autoradiography and electron microscopy (see **BioSkills 9 and 10**).

RESULTS OF A PULSE-CHASE EXPERIMENT The micrographs in **Figure 7.25** show what happened over time.

- Figure 7.25a shows part of a single cell that was examined immediately after the pulse. Most of the newly synthesized proteins are inside this cell's rough ER.
- Figure 7.25b shows part of a single cell 37 minutes after the pulse ended. Now the situation has changed. Few of the labeled proteins are in the rough ER. Instead, most of them are inside structures called secretory vesicles on the *trans* side of a Golgi apparatus (some are found inside the Golgi apparatus itself).
- The micrograph in Figure 7.25c, taken 117 minutes after the pulse, is at lower magnification and shows parts of five cells. The structure in the middle is a duct that carries digestive enzymes from pancreatic cells toward their destination in the small intestine. Note that most labeled proteins are in secretory vesicles or actually outside the cell, in the duct.

Because the labeled proteins move from the rough ER to Golgi apparatus to secretory vesicles to the cell exterior over time, the results support the hypotheses that a secretory pathway exists and that the rough ER and Golgi apparatus function as an integrated endomembrane system.

The data suggest that proteins produced in the rough ER do not drift randomly from organelle to organelle. Instead, traffic through the endomembrane system appears to be highly organized and directed.

Before moving on, it will be helpful for you to go to the study area at *www.masteringbiology.com* and review the logic of a pulse-chase experiment.

(MB) **Web Activity** A Pulse-Chase Experiment

Next, let's break the system down and examine four of the steps in more detail:

1. The ribosomes in rough ER are bound to the outside of the membrane, so how do the proteins that they manufacture enter the lumen of the ER?
2. How do the proteins move from the ER to the Golgi apparatus?
3. Once they're inside the Golgi, what happens to them?
4. And finally, how do the finished proteins reach their destinations?

Entering the Endomembrane System: The Signal Hypothesis

How do proteins enter the endomembrane system? The signal hypothesis predicted that proteins bound for the endomembrane system have a molecular zip code analogous to the nuclear localization signal. Günter Blobel and colleagues proposed that these proteins are synthesized by ribosomes that are attached to the outside of the ER, and the first few amino acids in the growing polypeptide act as a signal that brings the protein into the lumen of the ER.

[1]The term in vitro is Latin for "in glass." Experiments that are performed outside living organisms are done in vitro. The term in vivo, in contrast, is Latin for "in life." Experiments performed with living organisms are done in vivo.

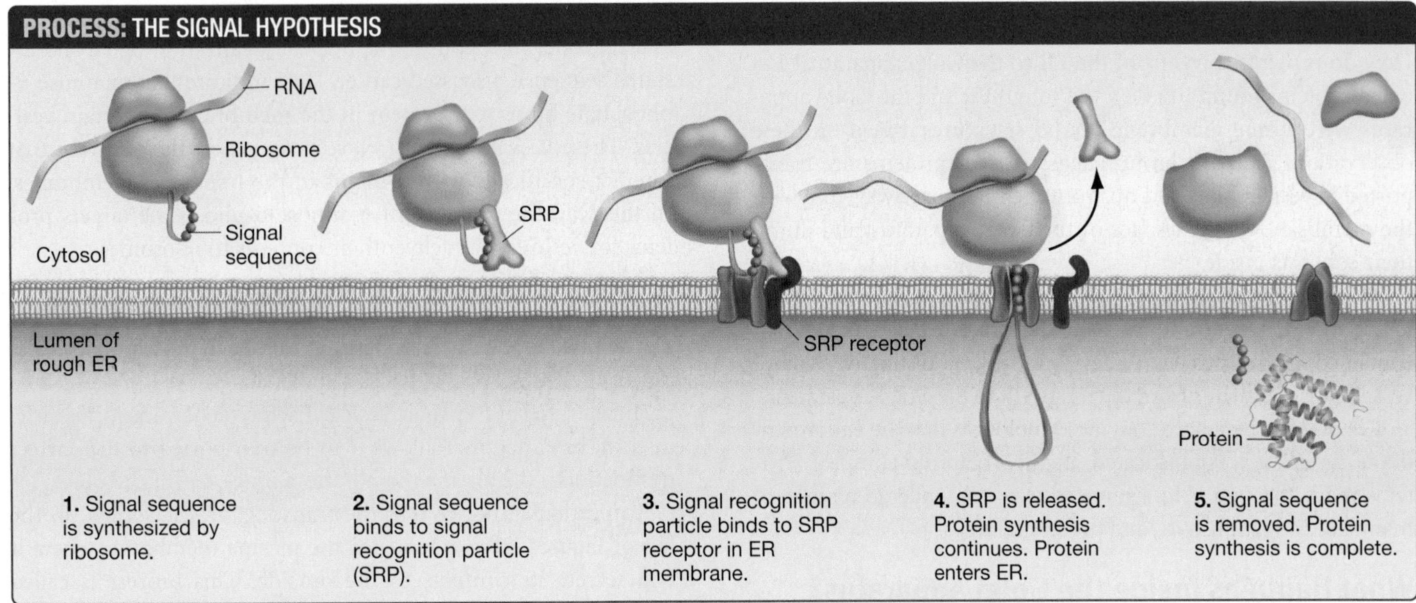
PROCESS: THE SIGNAL HYPOTHESIS
RNA
Ribosome
Signal sequence
Cytosol
SRP
Lumen of rough ER
SRP receptor
Protein
1. Signal sequence is synthesized by ribosome.
2. Signal sequence binds to signal recognition particle (SRP).
3. Signal recognition particle binds to SRP receptor in ER membrane.
4. SRP is released. Protein synthesis continues. Protein enters ER.
5. Signal sequence is removed. Protein synthesis is complete.

FIGURE 7.26 The Signal Hypothesis Explains How Proteins Destined for Secretion Enter the Endomembrane System. According to the signal hypothesis, proteins destined for secretion contain a short stretch of amino acids that interact with a signal recognition particle (SRP) in the cytoplasm. This interaction allows the protein to enter the ER.

This hypothesis received important support when researchers made a puzzling observation: When proteins that are normally synthesized in the rough ER are instead manufactured by isolated ribosomes in vitro—with *no* ER present—they are 20 amino acids longer, on average, than usual.

Blobel seized on these data. He claimed that the extra amino acids are the "send-to-ER" signal, and that the signal is removed inside the organelle. When the same protein is synthesized outside the ER, the signal is not removed.

Blobel's group went on to produce convincing data that supported the hypothesis: They identified the exact series of amino acids in the **ER signal sequence**.

More recent work has documented the mechanisms responsible for receiving the send-to-ER signal and inserting the protein into the rough ER. **Figure 7.26** illustrates the key steps involved.

Step 1 A ribosome synthesizes the ER signal sequence.

Step 2 The signal sequence binds to a **signal recognition particle (SRP)**—a complex of RNA and protein.

Step 3 The ribosome + signal sequence + SRP complex attaches to an SRP receptor in the ER membrane itself. Think of the SRP as a key that is activated by an ER signal sequence. The SRP receptor in the ER membrane is the lock.

Step 4 Once the lock (the receptor) and key (the SRP) connect, the SRP is released.

Step 5 The signal sequence is removed and protein synthesis is completed.

If the finished polypeptide will eventually be shipped to an organelle or secreted from the cell, it enters the lumen of the rough ER. If the finished polypeptide is a membrane protein, it remains in the rough ER membrane while it is being processed.

Once proteins are inside the rough ER or inserted into its membrane, they fold into their three-dimensional shape with the help of chaperone proteins (see Chapter 3). In addition, proteins that enter the lumen interact with enzymes that catalyze the addition of carbohydrate side chains. Because carbohydrates are polymers of sugar monomers, the addition of one or more carbohydrate groups is called **glycosylation** ("sugar-together"). The resulting molecule is a **glycoprotein** ("sugar-protein"; see Chapter 5).

As **Figure 7.27** shows, proteins that enter the ER often gain a specific carbohydrate that consists of 14 sugar subunits. The completed glycoproteins are ready for shipment to the Golgi apparatus.

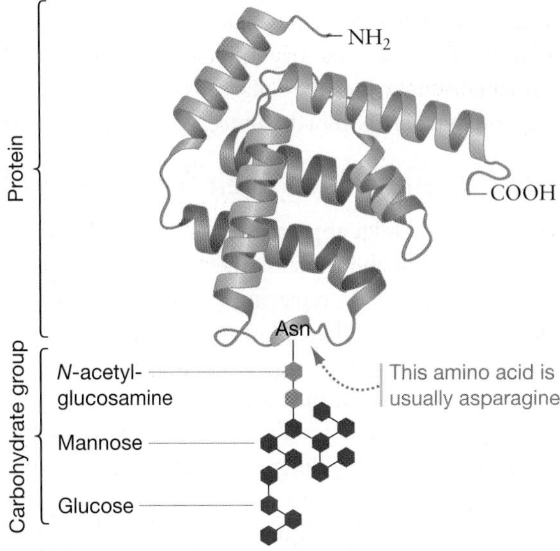
Protein
NH2
COOH
Asn
Carbohydrate group
N-acetyl-glucosamine
Mannose
Glucose
This amino acid is usually asparagine

FIGURE 7.27 Glycosylation Adds Carbohydrate Groups to Proteins. When proteins enter the ER, most acquire the 14 sugar residues shown here. Some of these sugars may be removed or others added as proteins pass through the Golgi apparatus.

Moving from the ER to the Golgi

How do proteins travel from the ER to the Golgi apparatus? Labeled proteins found between the rough ER and the Golgi apparatus were inside membrane-bound structures. Based on these observations, Palade's group suggested that proteins are transported in vesicles that bud off from the ER, move away, fuse with the membrane on the *cis* face of the Golgi apparatus, and dump their contents inside.

This hypothesis was supported when other researchers used differential centrifugation to isolate and characterize the vesicles that contained labeled proteins. They found that a distinctive type of vesicle carries proteins from the rough ER to the Golgi apparatus.

Results like these have convinced biologists that the endomembrane system is a sophisticated, highly organized complex. Proteins move through it in a directed way and undergo a series of manufacturing, transport, and processing steps.

What Happens inside the Golgi Apparatus?

Section 7.2 indicated that the Golgi apparatus consists of a stack of flattened vesicles called cisternae, and that cargo enters one side of the organelle and exits the other. Recent research has shown that the composition of the Golgi apparatus is dynamic. New cisternae constantly form at the *cis* face, while old cisternae break apart at the *trans* face. Once a cisterna forms, it gradually moves toward the *trans* face. As it does, it changes in composition and activity.

By separating individual cisternae and analyzing their contents, researchers have found that cisternae at various stages of maturation contain different suites of enzymes. These enzymes catalyze glycosylation reactions. As a result, proteins undergo further modification as a cisterna matures.

While cisternae are still near the *cis* face, some of the proteins inside have sugar-phosphate groups added. Later, the carbohydrate group that was added in the rough ER is removed. Near the *trans* face, various types of carbohydrate chains are attached that may protect the protein or help it attach to surfaces.

If the rough ER is like a foundry and stamping plant where rough parts are manufactured, then the Golgi can be considered a finishing area where products are polished, painted, and readied for shipping.

How Do Proteins Reach Their Destinations?

The rough ER and Golgi apparatus constitute an impressive assembly line. Some of the proteins they produce stay in the endomembrane system itself, replacing worn-out molecules. But if proteins are processed to the end of the line, they will be sent to one of several destinations, including lysosomes, the plasma membrane, chloroplasts, or the outside of the cell.

How are these finished products put into the right shipping containers, and how are the different containers addressed?

Studies on enzymes that are shipped to lysosomes have provided some answers to both questions. A key finding was that lysosome-bound proteins have a phosphate group attached to a specific sugar subunit on their surface, forming the compound mannose-6-phosphate. If mannose-6-phosphate is removed from these proteins, they are not transported to a lysosome.

This is strong evidence that the phosphorylated sugar serves as a zip code, analogous to the nuclear localization signal and ER signal sequence discussed earlier. Data indicate that mannose-6-phosphate binds to a protein in the membranes of certain vesicles. These vesicles, in turn, have proteins on their surface that interact specifically with proteins in the lysosomal membranes. In this way, the presence of mannose-6-phosphate targets proteins for vesicles that deliver their contents to lysosomes.

Figure 7.28 presents a comprehensive model of how endomembrane system products are loaded into specific vesicles and shipped to their correct destinations. Each protein that comes out of the Golgi apparatus has a molecular tag that places it in a particular type of transport vesicle. Each type of transport vesicle, in turn, has a tag that allows it to be transported to the correct destination—the plasma membrane, a lysosome, or the ER.

In particular, notice that the transport vesicle shown on the left of Figure 7.28 is bound for the plasma membrane, where it will secrete its contents to the outside. This process is called **exocytosis** ("outside-cell-act"). When exocytosis occurs, the vesicle membrane and plasma membrane make contact. As the two membranes fuse, their lipid bilayers rearrange in a way that exposes the interior of the vesicle to the outside of the cell. The vesicle's contents then diffuse into the space outside the cell. This is how cells in your pancreas deliver digestive enzymes to the duct that leads to your small intestine—where food is digested.

Proteins that are synthesized in the cytoplasm also have zip codes directing them to mitochondria, chloroplasts, or other destinations. In general, the proteins produced in a cell have distinctive molecular address labels, which allow proteins to be shipped to the compartments where they function.

If vesicles function like shipping containers for products that move between organelles, do they travel along some sort of road or track? What molecule or molecules function as the delivery truck, and does ATP supply the gas? Let's delve into these questions in Section 7.6.

CHECK YOUR UNDERSTANDING

If you understand that . . .

- In cells, the transport of proteins and other large molecules is energy demanding and tightly regulated.
- Proteins must have the appropriate molecular zip code to enter or leave the nucleus, enter the lumen of the rough ER, or become incorporated into vesicles destined for lysosomes or the plasma membrane.
- In many cases, proteins and other types of cargo are shipped in vesicles that contain molecular zip codes on their surface.

You should be able to . . .

1. Predict what happens to proteins that lack an ER signal sequence.
2. Predict the outcome of an experiment where secreted proteins are placed inside vesicles with a zip code associated with shipment to lysosomes.

Answers are available in Appendix B.

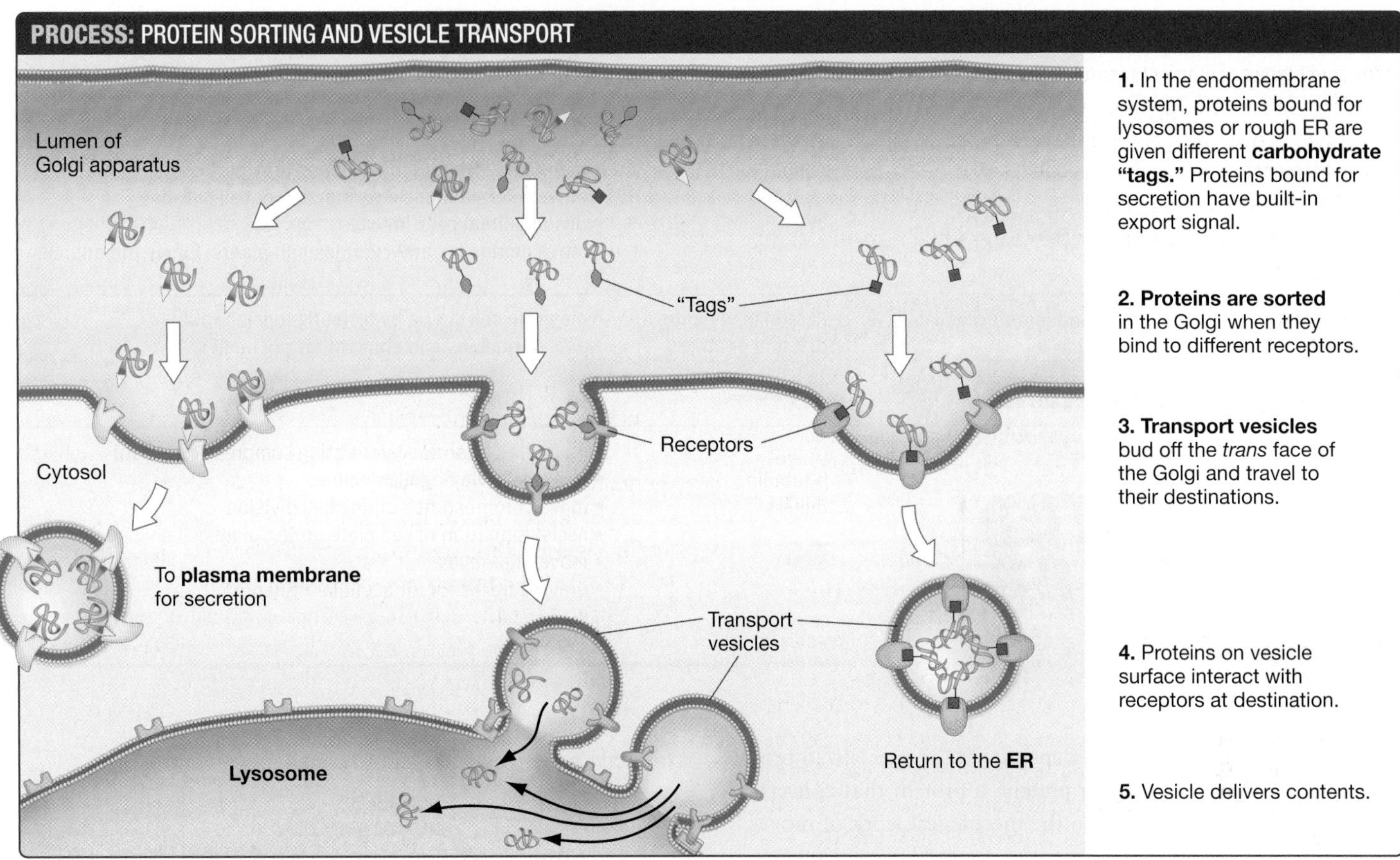

PROCESS: PROTEIN SORTING AND VESICLE TRANSPORT
Lumen of
Golgi apparatus
"Tags"
Receptors
Cytosol
To **plasma membrane**
for secretion
Transport
vesicles
Lysosome
Return to the **ER**
1. In the endomembrane system, proteins bound for lysosomes or rough ER are given different **carbohydrate "tags."** Proteins bound for secretion have built-in export signal.
2. Proteins are sorted in the Golgi when they bind to different receptors.
3. Transport vesicles bud off the *trans* face of the Golgi and travel to their destinations.
4. Proteins on vesicle surface interact with receptors at destination.
5. Vesicle delivers contents.

FIGURE 7.28 In the Golgi Apparatus, Proteins Are Sorted into Vesicles That Are Targeted to a Destination.

7.6 Cell Systems III: The Dynamic Cytoskeleton

The endomembrane system may be the best-studied example of how individual organelles work together in a dynamic, highly integrated way. This integration depends in part on the physical relationship of organelles, which is organized by the cytoskeletal system.

The cytoskeleton is a dense and complex network of fibers that helps maintain cell shape by providing structural support. However, the cytoskeleton is not a static structure like the scaffolding used at construction sites. Its fibrous proteins move and change to alter the cell's shape, shift its contents, and move the entire structure. Like the rest of the cell, the cytoskeleton is dynamic.

As **Table 7.3** shows, there are three distinct cytoskeletal elements in eukaryotic cells: actin filaments (also known as microfilaments), intermediate filaments, and microtubules. Recent research has shown that bacterial cells have cytoskeletal elements that are extremely similar to actin filaments and microtubules.

Each of the three cytoskeletal elements found in eukaryotes has a distinct size, structure, and function. Let's look at each one in turn.

Actin Filaments

Sometimes called **microfilaments** because they are the cytoskeletal element with the smallest diameter, **actin filaments** are fibrous structures made of the globular protein actin (Table 7.3). In animal cells, actin is often the most abundant of all proteins—typically it represents 5–10 percent of the total protein in the cell. Each of your liver cells contains about half a billion of these molecules.

ACTIN FILAMENT STRUCTURE Actin filaments form when individual actin molecules polymerize. Because each actin subunit in the strand is asymmetrical, the structure as a whole has a distinct polarity. The two ends of an actin filament are different and are referred to as plus and minus ends. Each filament grows and shrinks as actin subunits are added to or subtracted from each end of the structure. The addition and deletion of actin subunits is called treadmilling when the rates of addition at one end and deletion at the other are about equal—the fiber stays the same length but individual subunits move as if on a treadmill.

A completed actin filament resembles two long strands that coil around each other. In general, actin filaments tend to grow at the plus end because polymerization occurs fastest there.

In animal cells, actin filaments are particularly abundant just under the plasma membrane, where they are organized into long, parallel bundles or dense, crisscrossing networks, with individual actin filaments linked to one another by other proteins. The reinforced bundles and networks of actin filaments help stiffen the cell and define its shape.

ACTIN FILAMENT FUNCTION In addition to providing structural support, actin filaments are involved in movement. In several

SUMMARY TABLE 7.3 **Cytoskeletal Filaments**

The three types of filaments found in the cytoskeleton are distinguished by their size and structure, and the protein subunit of which they are made.

	Structure	Subunits	Functions
Actin filaments (microfilaments)	Strands in double helix 7 nm – end + end	Actin	• maintain cell shape by resisting tension (pull) • move cells via muscle contraction or cell crawling • divide animal cells in two • move organelles and cytoplasm in plants, fungi, and animals
Intermediate filaments	Fibers wound into thicker cables 10 nm	Keratin or vimentin or lamin or others	• maintain cell shape by resisting tension (pull) • anchor nucleus and some other organelles
Microtubules	Hollow tube 25 nm – end + end	α- and β-tubulin dimers	• maintain cell shape by resisting compression (push) • move cells via flagella or cilia • move chromosomes during cell division • assist formation of cell plate during plant cell division • move organelles • provide tracks for intracellular transport

cases, actin's role in movement depends on the specialized protein myosin. Myosin is a **motor protein**: a protein that converts the chemical energy in ATP into the mechanical work of movement, just as a car's motor converts the chemical energy in gasoline into movement.

Chapter 46 details the interaction between actin and myosin that produces movement. For now, it's enough to recognize that when ATP binds to myosin and is then hydrolyzed to ADP, the "head" region of the myosin molecule binds to actin and moves. The movement of this protein causes the actin filament to slide (**Figure 7.29a**). This type of movement is analogous to a line of people who are passing along a long log or pole. The people are myosin molecules; the log or pole is actin.

As **Figure 7.29b** shows, the ATP-powered interaction between actin and myosin is the basis for an array of cell movements:

- **Cytokinesis** ("cell-moving") is the process of cell division in animals. For these cells to divide in two, actin filaments that are arranged in a ring under the plasma membrane must slide past one another. Because they are connected to the plasma membrane, the movement of the actin fibers pinches the cell in two.
- **Cytoplasmic streaming** is the directed flow of cytosol and organelles around plant cells. The movement occurs along actin filaments and is powered by myosin. It is especially common in large cells, where the circulation of cytoplasm facilitates material transport.

In addition, the movement called **cell crawling** occurs when groups of actin filaments grow, creating bulges in the plasma membrane that extend and move the cell. Cell crawling occurs in a wide range of organisms and cell types, including amoebae, slime molds, and certain human cells.

(a) Actin and myosin interact to cause movement.

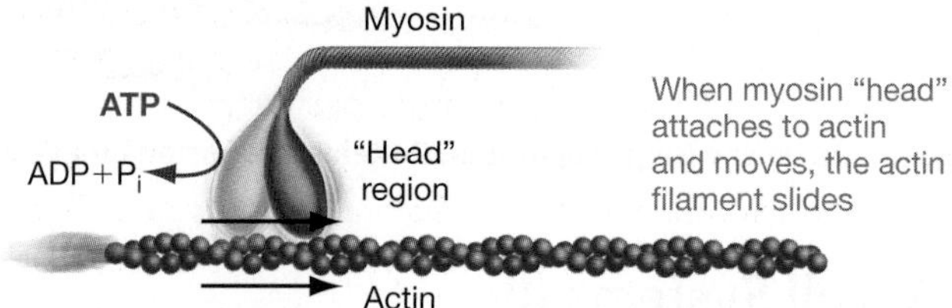
Myosin
ATP
ADP+P_i
"Head" region
Actin
When myosin "head" attaches to actin and moves, the actin filament slides

(b) Examples of movement caused by actin-myosin interactions

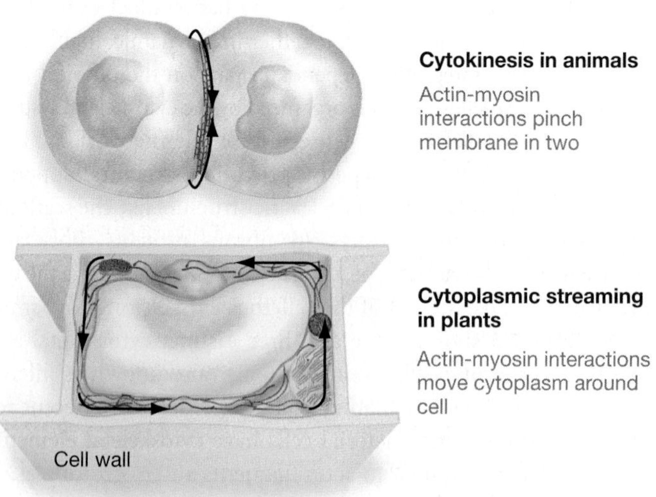
Cytokinesis in animals
Actin-myosin interactions pinch membrane in two
Cytoplasmic streaming in plants
Actin-myosin interactions move cytoplasm around cell
Cell wall

FIGURE 7.29 Many Cellular Movements Are Based on Actin-Myosin Interactions. **(a)** When the "head" region of the myosin protein interacts with ATP, myosin attaches to actin and changes shape. The movement causes the actin filament to slide. **(b)** Actin-myosin interactions can divide cells and move organelles and cytoplasm.

Intermediate Filaments

Although similar in size, many types of **intermediate filaments** exist, each consisting of a different—though structurally similar—type of protein (Table 7.3). Humans, for example, have 70 genes that code for intermediate filament proteins. This is in stark contrast to actin filaments and microtubules, which are made from the same protein subunits in all eukaryotic cells.

In addition, intermediate filaments are not polar; instead, each end of these filaments is identical. As a result, intermediate filaments do not treadmill, and they are not involved in directed movement driven by myosin or related proteins. Intermediate filaments serve a purely structural role in eukaryotic cells.

The intermediate filaments that you are most familiar with belong to a family of molecules called the keratins. The cells that make up your skin and line surfaces inside your body contain about 20 types of keratin. These intermediate filaments provide the mechanical strength required for these cells to resist pressure and abrasion. Skin cells manufacture another 10 distinct forms of keratin. Depending on the location of the skin cell and keratins involved, the secreted filaments form fingernails, toenails, or hair.

Nuclear lamins, which make up the nuclear lamina layer introduced in Section 7.4, also qualify as intermediate filaments. Nuclear lamins form a dense mesh under the nuclear envelope. Recall that in addition to giving the nucleus its shape, they anchor the chromosomes. They are also involved in the breakup and reassembly of the nuclear envelope when cells divide.

Some intermediate filaments project from the nucleus through the cytoplasm to the plasma membrane, where they are linked to intermediate filaments that run parallel to the cell surface. In this way, intermediate filaments form a flexible skeleton that helps shape the cell surface and hold the nucleus in place.

Microtubules

Microtubules are composed of two proteins called α-tubulin and β-tubulin and are the largest cytoskeletal components in terms of diameter (Table 7.3). Molecules of α-tubulin and β-tubulin bind to form **dimers** ("two-parts"), compounds formed by the joining of two monomers.

Tubulin dimers polymerize to form the large, hollow tube called a microtubule. Because each end of a tubulin dimer is different, each end of a microtubule has a distinct polarity. Like actin filaments, microtubules are dynamic and usually grow at their plus end. Microtubules grow and shrink in length as tubulin dimers are added or subtracted.

Microtubules originate from a structure called the **microtubule organizing center** and grow outward, radiating throughout the cell (see the chapter opening photograph on page 102). Although plant cells typically have hundreds of these organizing centers, most animal and fungal cells have just one.

In animals, the microtubule organizing center has a distinctive structure and is called a **centrosome**. As **Figure 7.30** shows, animal centrosomes contain two bundles of microtubules called **centrioles**.

In function, microtubules are similar to actin filaments: They provide stability and are involved in movement. Like steel girders in a skyscraper, the microtubules that radiate from an organizing center stiffen the cell by resisting compression forces. Microtubules may also provide a structural framework for organelles. If microtubules are prevented from forming, the ER no longer assembles in its normal network-like configuration.

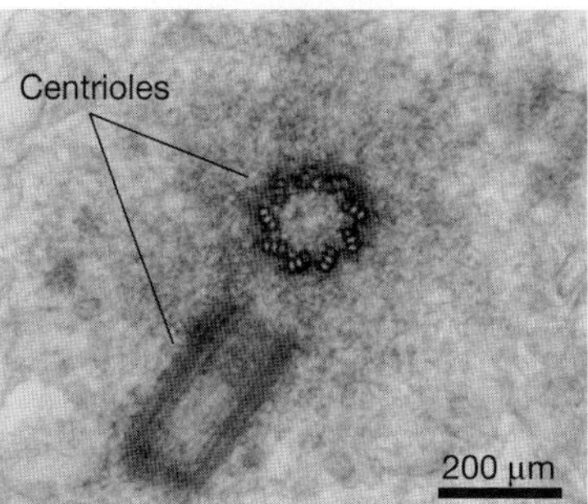

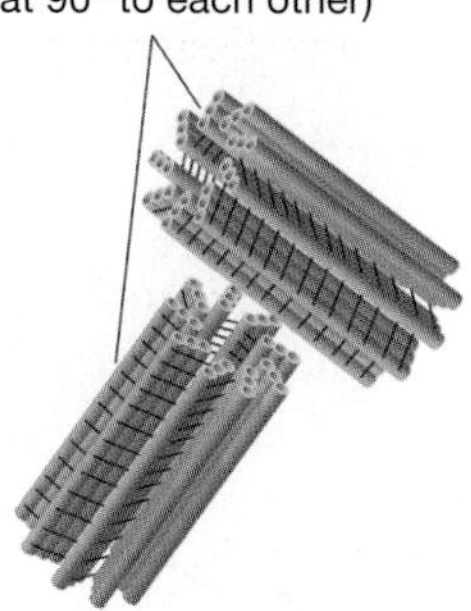

FIGURE 7.30 Centrosomes Are a Type of Microtubule Organizing Center. Microtubules emanate from microtubule organizing centers, which in animals are called centrosomes. The centrioles inside a centrosome are made of microtubules.

During cell division, microtubules from the organizing center move chromosomes from the original cell to each of the two resulting cells (see Chapters 11 and 12). But microtubules are involved in many other types of cellular movement as well. Let's consider their role in moving materials inside cells, then ask how the microtubules in flagella help move an entire cell.

STUDYING VESICLE TRANSPORT Materials are transported to a wide array of destinations inside cells via vesicles. To study how this movement happens, Ronald Vale and colleagues focused on the giant axon, an extremely large nerve cell in squid that runs the length of the animal's body. If the squid is disturbed, the cell signals muscles to contract so it can jet away to safety.

The researchers decided to study this particular cell for three reasons.

1. The giant axon is so large that it is relatively easy to see and manipulate.
2. Large numbers of molecules are synthesized in the cell's ER and then transported in vesicles down the length of the cell, where they are released. As a result, a large amount of cargo moves a long distance.
3. The researchers found that if they gently squeezed the cytoplasm out of the cell, vesicle transport still occurred in the cytoplasmic material. This allowed them to do experiments on vesicle transport without the plasma membrane being in the way.

In short, the squid giant axon provided a system that could be observed and manipulated efficiently in the lab. What did the biologists find out?

MICROTUBULES ACT AS "RAILROAD TRACKS" To watch vesicle transport in action, the researchers mounted a video camera to a

(a) Electron micrograph

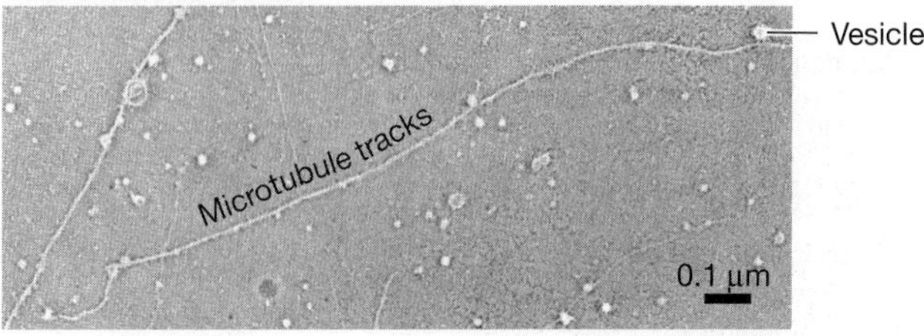

(b) Video image

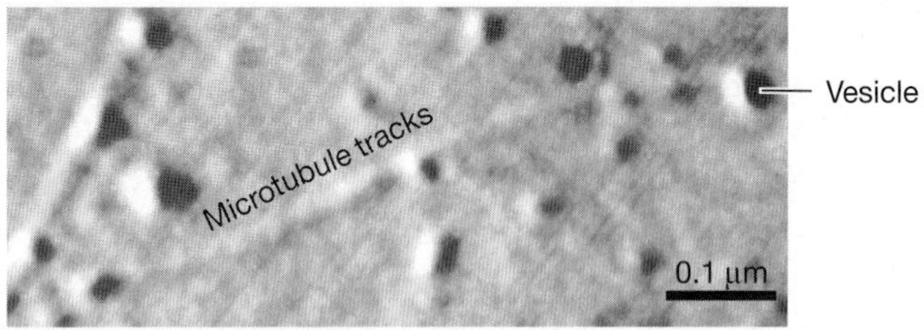

FIGURE 7.31 Transport Vesicles Move along Microtubule Tracks. The images show extruded cytoplasm from a squid giant axon. **(a)** An electron micrograph that allowed researchers to measure the diameter of the filaments and confirm that they are microtubules. In the upper part of this image, you can see a vesicle on a "track." **(b)** A slightly fuzzy but higher-magnification videomicroscope image, in which researchers actually watched vesicles move.

microscope. As **Figure 7.31** shows, this technique allowed them to document that vesicle transport occurred along a filamentous track. A simple experiment convinced the group that this movement is an energy-dependent process: If they depleted the amount of ATP in the cytoplasm, vesicle transport stopped.

To identify the filament involved, the biologists measured the diameter of the tracks and analyzed their chemical composition. Both types of data indicated that the tracks consist of microtubules. Microtubules also appear to be required for movement of materials elsewhere in the cell. For instance, if experimental cells are treated with a drug that disrupts microtubules, the movement of vesicles from the rough ER to the Golgi apparatus is impaired.

The general message of these experiments is that transport vesicles move through the cell along microtubules. How? Do the tracks themselves move, like a conveyer belt, or are vesicles carried along on some sort of molecular truck?

A MOTOR PROTEIN GENERATES MOTILE FORCES To study the way vesicles move along microtubules, Vale's group took the squid axon's transport system apart and put it back together.

To begin, they assembled microtubule fibers from purified α-tubulin and β-tubulin. Then they used differential centrifugation to isolate transport vesicles. But when they mixed purified microtubules and vesicles with ATP, no transport occurred. Something had been left out—but what?

To find the missing element or elements, the researchers purified one subcellular part after another, using differential centrifugation, and added it to the microtubule + vesicle + ATP system. Through trial and error, they found something that triggered movement. After further purification steps, the researchers finally succeeded in isolating a protein that generated vesicle movement. They named the molecule **kinesin**, from the Greek word *kinein* ("to move").

Like myosin, kinesin is a motor protein. Kinesin converts chemical energy in ATP into mechanical energy in the form of movement. More specifically, when ATP is added to kinesin or drops off, the protein moves.

Biologists began to understand how kinesin works when X-ray diffraction studies showed that it has three major regions: a head section with two globular pieces, a tail associated with small polypeptides, and a stalk that connects the head and tail (**Figure 7.32a**). Follow-up studies confirmed that the head region binds to the microtubule, while the tail region binds to the transport vesicle. Recent work has shown that kinesin "walks" along the microtubule when the head region binds to ATP (**Figure 7.32b**).

(a) Structure of kinesin

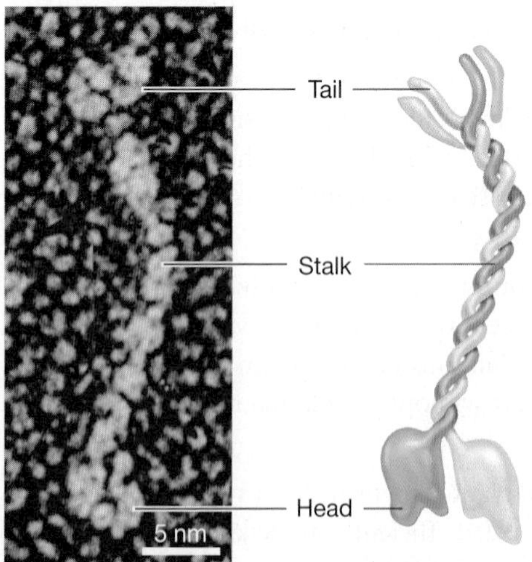

(b) Kinesin "walks" along a microtubule track.

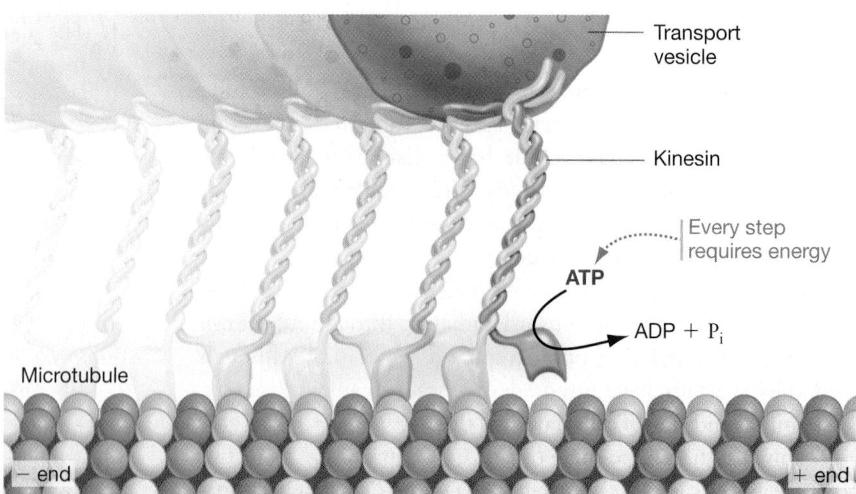

FIGURE 7.32 A Motor Protein Moves Vesicles along Microtubules. (a) Kinesin has three major segments. **(b)** The current model depicting how kinesin "walks" along a microtubule track to transport vesicles. The two head segments act like feet that alternately attach and release in response to the gain or loss of a phosphate group.

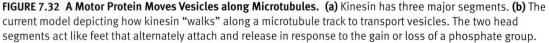

A kinesin molecule is like a delivery truck that carries transport vesicles along microtubule tracks. Cells contain several different kinesin proteins, each specialized for carrying a different type of vesicle. In this way, kinesins move molecular cargo to destinations throughout the cell. How do microtubules move entire cells?

Flagella and Cilia: Moving the Entire Cell

Flagella are long, hairlike projections from the cell surface that function in movement. While many bacteria and eukaryotes have flagella, the structure is completely different in the two groups.

- Bacterial flagella are made of a protein called flagellin; eukaryotic flagella are constructed from microtubules (tubulin).
- Bacterial flagella move the cell by rotating like a ship's propeller; eukaryotic flagella move the cell by undulating—they whip back and forth.
- Eukaryotic flagella are surrounded by plasma membrane and are considered organelles; bacterial flagella are not.

Based on these observations, biologists conclude that the two structures evolved independently, even though their function is similar.

To understand how cells move, let's focus on eukaryotic flagella. Eukaryotic flagella are closely related to structures called **cilia** (singular: **cilium**), which are short, filament-like projections that are also found in some eukaryotic cells (**Figure 7.33**). Flagella are generally longer than cilia, and a cell will typically have just one or two flagella but many cilia. But when researchers examined the two structures with the electron microscope, they found that their underlying organization is identical.

HOW ARE CILIA AND FLAGELLA CONSTRUCTED? In the 1950s, anatomical studies established that most cilia and flagella have a characteristic "9 + 2" arrangement of microtubules. As **Figure 7.34a** shows, nine microtubule pairs, or doublets, surround two central microtubules. The doublets consist of one complete and one incomplete microtubule and are arranged around the periphery of the structure.

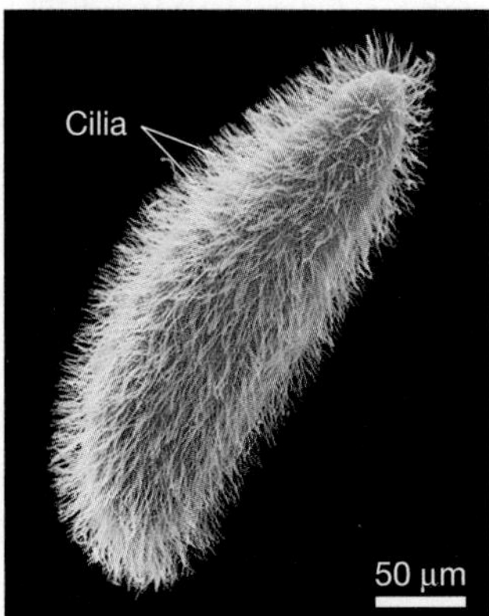

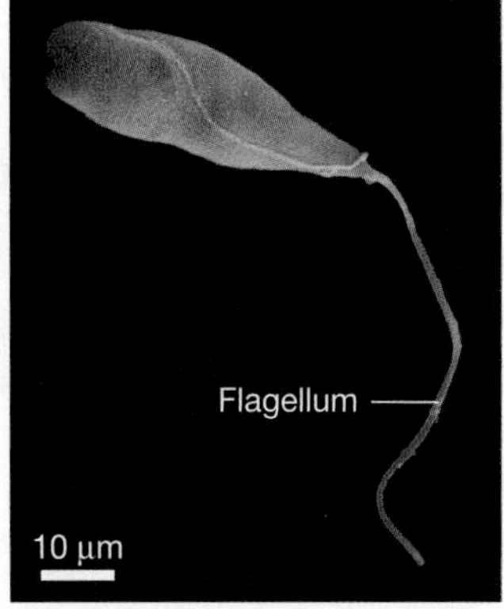

FIGURE 7.33 Cilia and Flagella Differ in Length and Number. The cells in these scanning electron micrographs have been colorized.

(a) Transmission electron micrograph of axoneme

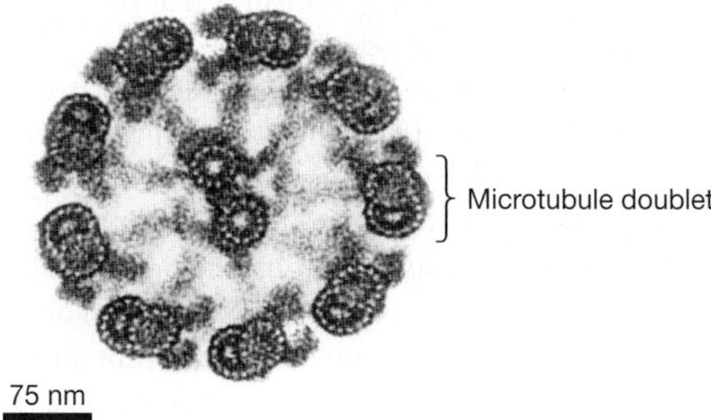

(b) Diagram of axoneme

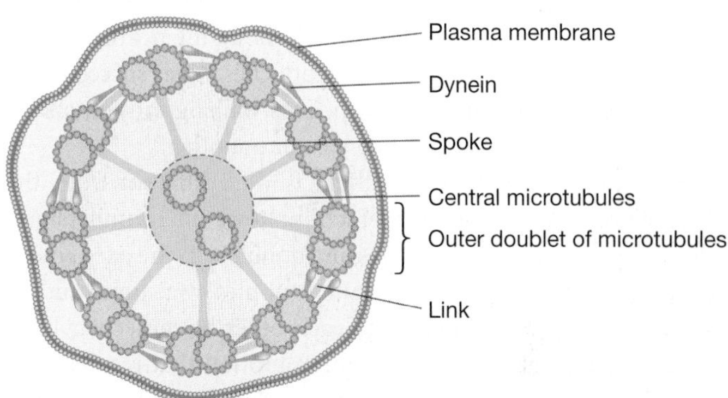

FIGURE 7.34 The Structure of Cilia and Flagella. (a) Transmission electron micrograph of a cross section through an axoneme. **(b)** The major structural elements in cilia and flagella. The microtubules are connected by links and spokes, and the entire structure is surrounded by the plasma membrane.

The entire 9 + 2 structure is called the **axoneme** ("axle-thread"). The axoneme attaches to the cell at a structure called the **basal body**. The basal body is identical in structure to a centriole and plays a central role in the growth of the axoneme.

As electron microscopy improved, biologists gained a more detailed view of the structure. Spoke-like structures connect each doublet to the central pair of microtubules, and molecular links connect the nine doublets to one another (**Figure 7.34b**). Each doublet has a set of arms that project toward an adjacent doublet.

Microtubules are complex. How do their components interact to generate motion?

WHAT PROVIDES THE FORCE REQUIRED FOR MOVEMENT? In the 1960s Ian Gibbons began studying the cilia of a common unicellular eukaryote called *Tetrahymena*, which lives in pond water. Gibbons found that by using a detergent to remove the plasma membrane that surrounds cilia and then subjecting the resulting solution to differential centrifugation, he could isolate axonemes.

These steps gave Gibbons a cell-free system for studying how cilia and flagella work. Cell-free systems are elegant to study because they are isolated—they are relatively easy to manipulate, and no other cell components are present that might confuse experimental results.

Gibbons found that the isolated structures would beat if he supplied them with ATP. This result confirmed that the beating of cilia is an energy-demanding process.

In another early experiment, Gibbons treated the isolated axonemes with a molecule that affects the ability of proteins to bind to one another. The resulting axonemes could not bend or use ATP. When Gibbons examined them in the electron microscope, he found that the arms had fallen off the doublets. This suggested that the arms are required for movement. Follow-up work showed that the arms are made of a large protein that Gibbons named **dynein** (from the Greek word *dyne*, meaning "force").

Like myosin and kinesin, dynein is a motor protein. Dynein changes shape when a phosphate group from ATP attaches to it. This shape change moves the molecule along the nearby microtubule. When the dynein molecule reattaches, it has succeeded in walking up the microtubule a step. This walking motion allows the microtubule doublets to slide past one another.

The outcome of dynein walking is very different from the outcome of a myosin-actin interaction, however. To understand why, remember that each of the nine doublets in the axoneme is connected to the central pair of microtubules by a spoke, and all of the doublets are connected to each other by molecular links (Figure 7.34b). As a result, the sliding motion that results from dynein walking is constrained—if one doublet slides, it transmits force to the rest of the axoneme via the links and spokes (**Figure 7.35**).

If the dynein arms on just one side of the axoneme walk and cause some of the doublets to slide while those on the other side are at rest, this constrained, localized movement results in bending. The bending of cilia or flagella results in a swimming motion. ✔If you understand this concept, you should be able to describe what would happen to the axoneme in Figure 7.35 if (1) the microtubule doublets were not connected by spokes and links, and (2) no ATP molecules were present.

Scaled for size, flagellar-powered swimming can be rapid. In terms of the number of body or cell lengths traveled per second, a sperm cell from a bull moves faster than a human world-record-holder does when swimming freestyle. At the level of the cell, life is fast paced.

Taken together, the data reviewed in this chapter can be summed up in six words: Cells are dynamic, highly integrated structures. Chemical reactions take place at mind-boggling speeds. Actin filaments and microtubules grow and shrink. The endomembrane system synthesizes, sorts, and ships an array of products in a highly regulated manner. How does all this activity inside the cell relate to what is going on outside? This is the issue we'll take up in Chapter 8.

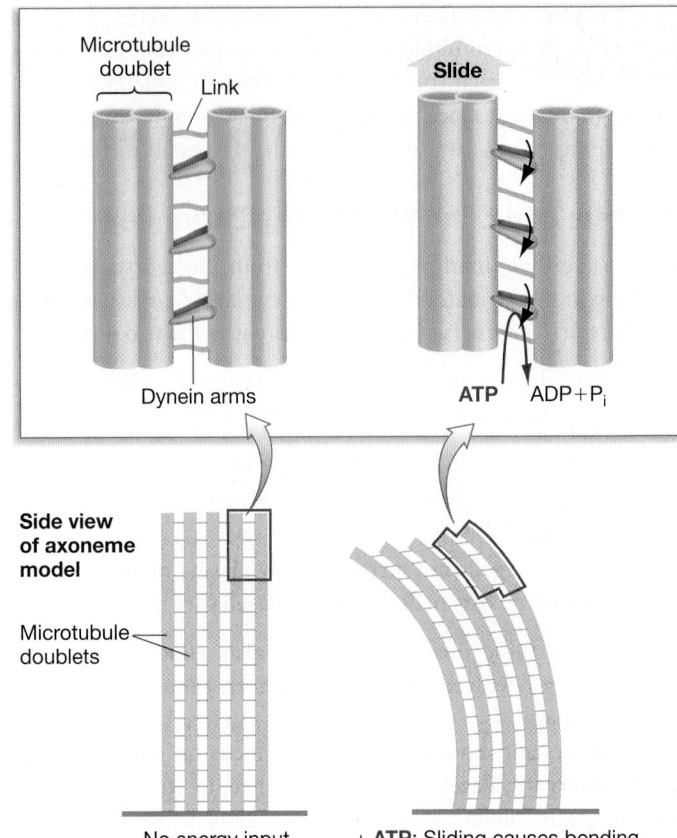

FIGURE 7.35 How Do Flagella Bend? When dynein arms walk along the microtubule doublets on one side of a flagellum, the structure bends.

CHECK YOUR UNDERSTANDING

If you understand that . . .

- Each component of the cytoskeleton has a unique structure and set of functions. In addition to providing structural support, actin filaments and microtubules work in conjunction with motor proteins to move the cell or materials inside the cell. Intermediate filaments provide structural support.
- Most elements of the cytoskeleton are dynamic—they grow and shrink over time.

✔ You should be able to . . .

Predict what will happen to endomembrane system products when experimental cells are treated with drugs that inhibit formation of microtubules.

Answers are available in Appendix B.

CHAPTER 7 REVIEW

For media, go to the Study Area at www.masteringbiology.com 

Summary of Key Concepts

The structure of individual cell components is closely correlated with their function. The overall shape and composition of a cell is also closely related to its function.

- There are two basic cellular designs: prokaryotic and eukaryotic.
- Eukaryotic cells are usually much larger and more structurally complex than prokaryotic cells.
- Most prokaryotic cells consist of a single membrane-bound compartment in which nearly all cellular functions occur.
- Eukaryotic cells contain numerous membrane-bound compartments called organelles. Organelles allow eukaryotic cells to compartmentalize functions and grow to a large size.
- Eukaryotic organelles are specialized for carrying out different functions, and their structure often correlates closely to their function.
- The defining organelle of eukaryotic cells is the nucleus, which contains the cell's chromosomes and serves as its control center.
- Rough ER is named for the ribosomes that attach to it. Ribosomes are protein-making machines, and rough ER is a site for protein synthesis and processing.
- Smooth ER lacks ribosomes because it is a center for lipid synthesis and processing.
- Mitochondria and chloroplasts have extensive internal membrane systems where the enzymes responsible for ATP generation and photosynthesis reside.

✓You should be able to predict what would happen to cells that are exposed to (1) a drug that poisons mitochondria, (2) an enzyme that degrades the cell wall, or (3) a drug that prevents ribosomes from functioning.

MB **BioFlix™** Tour of an Animal Cell, **BioFlix™** Tour of a Plant Cell

Inside cells, materials are transported to their destinations with the help of molecular "zip codes."

- Cells have sophisticated systems for making sure that proteins and other products end up in the right place.
- Traffic across the nuclear envelope occurs through nuclear pores, which contain a multiprotein nuclear pore complex that serves as gatekeeper. Proteins can't pass through the pore and enter the nucleus unless they contain a specific molecular signal.
- Materials move through the endomembrane system inside vesicles. Before products leave the system, they are sorted with molecular "zip codes" that direct them to specific vesicles. The vesicles interact with receptor proteins at the target location so that the contents are delivered correctly.

✓You should be able to explain, using concepts in Chapter 3, why proteins—and not RNA, DNA, carbohydrates, or lipids—are the molecules responsible for "reading" the array of molecular zip codes in cells.

MB **Web Activity** Transport into the Nucleus, **Web Activity** A Pulse-Chase Experiment

The cytoskeleton provides a structural framework inside cells and plays a key role in cell division, movement, and transport.

- The cytoskeleton is an extensive system of fibers that provides:
  (1) structural support and a framework for arranging and organizing organelles and other cell components;
  (2) paths for moving vesicles inside cells; and
  (3) machinery for moving the cell as a whole through the beating of flagella or cilia, or cell crawling.
- Movement often depends on motor proteins, which use chemical energy stored in ATP to change shape or position.
- Myosin causes actin filaments to slide, making cell division and cytoplasmic streaming possible in eukaryotes.
- Kinesin "walks" transport vesicles along microtubule tracks.
- Dynein "walks" along microtubule doublets, bending cilia and flagella so that they beat back and forth, enabling cells to swim or generate water currents.

✓You should be able to compare and contrast the structure and function of actin filaments, intermediate filaments, and microtubules.

Cells are dynamic, highly integrated structures. Thousands of chemical reactions occur each second within cells; molecules constantly enter and exit across the plasma membrane; cell products are shipped along protein fibers; and elements of the cell's internal skeleton grow and shrink.

- Cells are bound by a highly selective membrane and have a tightly organized, dynamic interior.
- Cytoskeletal elements hold the nucleus, components of the endomembrane system, and other organelles in position.
- Materials are transported from one cell compartment to another along well-defined microtubule "tracks."
- Subunits are constantly being added to or removed from actin filaments and microtubules.

✓You should be able to explain how pulse-chase experiments allowed researchers to study cells as dynamic enterprises—looking specifically at how materials move inside cells.

Questions

✔ TEST YOUR KNOWLEDGE

Answers are available in Appendix B

1. Which of the following best describes the nuclear envelope?
   a. It is continuous with the endomembrane system.
   b. It is continuous with the nucleolus.
   c. It is continuous with the plasma membrane.
   d. It contains a single membrane and nuclear pores.

2. Why is "receptor-mediated endocytosis" an appropriate term?
   a. It is the first step in autophagy.
   b. It is the first step in phagocytosis.
   c. It starts when extracellular molecules bind to receptors.
   d. It is the first step in formation of a lysosome.

3. Which of the following is *not* true of secreted proteins?
   a. They are synthesized in ribosomes.
   b. They are transported through the endomembrane system in membrane-bound transport organelles.
   c. They are transported from the Golgi apparatus to the ER.
   d. They contain a signal sequence that directs them into the ER.

4. To find the nuclear localization signal in the protein nucleoplasmin, researchers separated the molecule's core and tail segments, labeled both with a radioactive atom, and injected them into the cytoplasm. Why did the researchers conclude that the signal is in the tail region of the protein?
   a. The protein reassembled and folded into its normal shape spontaneously.
   b. Only the tail segments appeared in the nucleus.
   c. With a confocal microscope, tail segments were clearly visible in the nucleus.
   d. The tail and head segments both appeared in the nucleus.

5. Molecular zip codes direct molecules to particular destinations in the cell. How are these signals read?
   a. They bind to receptor proteins.
   b. They enter transport vesicles.
   c. They bind to motor proteins.
   d. They are glycosylated by enzymes in the Golgi apparatus.

6. What does a motor protein do?
   a. causes actin filaments and/or microtubules to treadmill
   b. catalyzes endergonic reactions
   c. triggers receptor-mediated endocytosis
   d. changes shape in a way that moves another cell structure

✔ TEST YOUR UNDERSTANDING

Answers are available in Appendix B

1. Compare and contrast the structure of a generalized plant cell, animal cell, and prokaryotic cell. Which features are common to all cells? Which are specific to just prokaryotes, or just plants, or just animals?

2. Make a flowchart that traces the movement of a secreted protein from its site of synthesis to the outside of a eukaryotic cell. Identify all of the organelles that the protein passes through. Add notes indicating what happens to the protein at each step.

3. Describe how the motor protein kinesin can move a transport vesicle down a microtubule track. Explain why the movement requires ATP.

4. Actin filaments and microtubules may treadmill and/or undergo dramatic changes in length. How does this observation relate to the claim that the cytoskeleton is dynamic?

5. Explain why cells that are responsible for acquiring ions and other nutrients via specialized membrane proteins would be expected to have a large amount of rough ER.

6. Structurally, what is the difference between microtubules involved in structural support, vesicle movement, and flagella?

✔ APPLYING CONCEPTS TO NEW SITUATIONS

Answers are available in Appendix B

1. In addition to delivering cellular products to specific organelles, eukaryotic cells can take up material from the outside and route it to specific organelles. For example, specialized cells of the human immune system ingest bacteria and viruses and then deliver them to lysosomes for degradation. Suggest a hypothesis for how this material is tagged and directed to lysosomes. How would you test this hypothesis?

2. The enzymes found in peroxisomes are synthesized by ribosomes in the cytosol. Suggest a hypothesis for how the finished proteins find their way to the peroxisomes.

3. Propose a function for cells that contain (a) a large number of lysosomes, (b) a particularly extensive cell wall, and (c) many peroxisomes.

4. Suggest a hypothesis to explain why archaea have cell walls. Suppose that you have mutant archaeal cells that lack a cell wall. How could you use these individuals to test your idea?

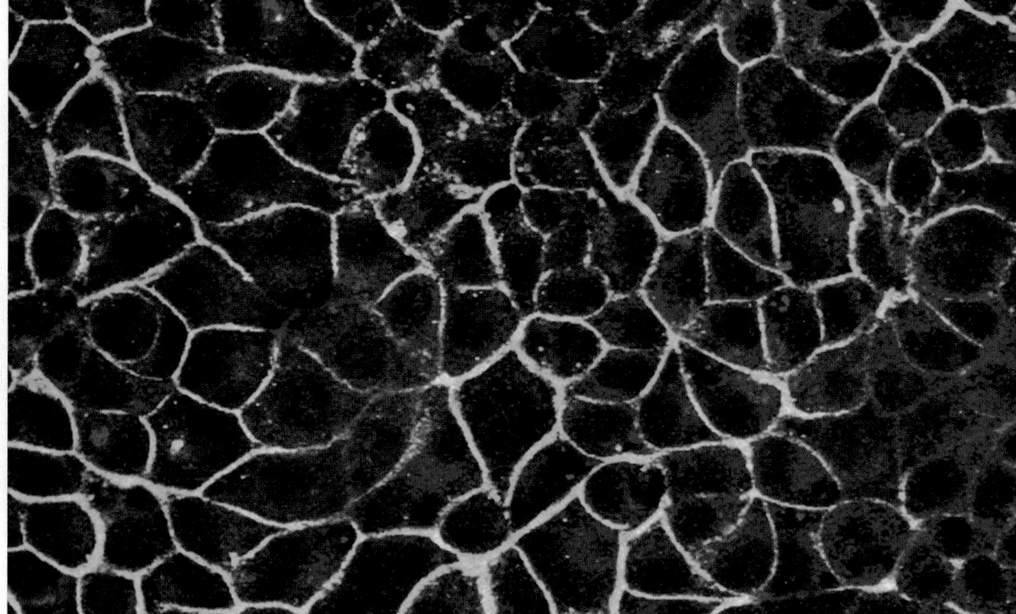

In this micrograph of skin cells, a green dye highlights proteins that help adjacent cells bind to each other. This chapter focuses on how cells in multicellular organisms stick together and communicate.

Cell-Cell Interactions 8

Chapter 6 introduced the structure and function of the plasma membrane—the defining feature of the cell. Chapter 7 surveyed the organelles, molecular machines, and cytoskeletal elements that fill the space inside that membrane and explored how cargo moves from sources to destinations within the cell. Both chapters highlighted the breathtaking speed and diversity of events that take place at the cellular level.

The cell is clearly a bustling enterprise. But it would be a mistake to think that cells are self-contained—that they are worlds in and of themselves. Instead, cells interact with other cells and the surrounding environment.

For most unicellular species, the outside environment is teeming with other organisms. Inside your gut, for example, hundreds of billions of bacterial cells are jostling for space and resources. In addition to interacting with these individuals, every unicellular organism must contend with constant shifts in the physical environment, such as heat, light, ion concentrations, and food supplies. If unicellular organisms are unable to sense these conditions and respond appropriately, they die.

In multicellular species, the environment outside the cell is made up of other cells, both neighboring and distant. The cells that make up a redwood tree, an *Amanita* mushroom, or your body are intensely social. Although biologists often study cells in isolation, an individual tree, fungus, or person is actually an interdependent community of cells. If those cells do not communicate and cooperate, the whole will break into dysfunctional parts and die.

To understand the life of a cell, then, it is critical to analyze how the cell interacts with the world outside its membrane. How do cells obtain information about the world and respond to that information? In particular, how do cells interact with other cells? To answer these questions, let's begin with the cell surface—with the molecules that separate the cell from its environment.

KEY CONCEPTS

- Extracellular material strengthens cells and helps bind them together.
- Cell-cell connections help adjacent cells adhere. Cell-cell gaps allow adjacent cells to communicate.
- Intercellular signals are responsible for creating an integrated whole from many thousands of independent parts.
- In target cells, intercellular signals are received, processed, responded to, and deactivated. If the signal is received at the cell surface, the processing step involves production of an intracellular signal.

✔ When you see this checkmark, stop and test yourself. Answers are available in Appendix B.

8.1 The Cell Surface

Chapter 6 introduced the fluid-mosaic hypothesis, the currently accepted model for the structure of the plasma membrane. Recall that the phospholipid bilayer is studded with membrane proteins that are integral, meaning that they are embedded in the bilayer, or peripheral, meaning that they are attached to one surface.

Chapter 6 also analyzed the primary function of the plasma membrane: to create an environment inside the cell that is different from conditions outside. Ions and molecules move across plasma membranes by direct diffusion through the phospholipid bilayer or via several types of membrane proteins. Transport of materials across the membrane can be energy demanding or passive, but it is always selective.

This picture—of a dynamic, complex plasma membrane that selectively admits or blocks passage of specific substances—is accurate but not complete. The plasma membrane does not exist in isolation. Cytoskeletal elements introduced in Chapter 7 attach to the interior face of the bilayer, and a complex array of extracellular structures exists outside. Let's consider the nature of the material outside the cell and then analyze how the cell interacts with other cells.

The Structure and Function of an Extracellular Layer

It is actually extremely rare for cells to be bounded simply by a plasma membrane. Most cells possess a layer or wall that forms just beyond the membrane. The extracellular material helps define the cell's shape and either attaches it to another cell or acts as a first line of defense against the outside world.

Virtually all types of extracellular structures, in turn—from the cell walls of bacteria, algae, fungi, and plants to the extracellular matrix that surrounds most animal cells—follow the same fundamental design principle. Like reinforced concrete and fiberglass, they are "fiber composites": They consist of a cross-linked network of long filaments embedded in a stiff surrounding material, or ground substance (**Figure 8.1**). The molecules that make up the filaments and the encasing material vary from group to group, but the engineering principle is the same. Why?

- The rods or filaments in a fiber composite are extremely effective at withstanding stretching and straining forces, or tension. The steel rods in reinforced concrete and the cellulose fibers in a plant cell wall are unlikely to break as a result of being pulled or pushed lengthwise.
- The stiff surrounding substance is effective at withstanding the pressing forces called compression. Concrete performs this function in highways, and a gel-forming mixture of polysaccharides achieves the same end in plant cell walls.

Thanks to the combination of tension- and compression-resisting elements, fiber composites are particularly rugged. And in many living cells, the fiber and composite elements are flexible as well as strong.

What molecules make up the rods and ground substance found on the surface of plant and animal cells? How are these extracellular layers synthesized, and what do they do?

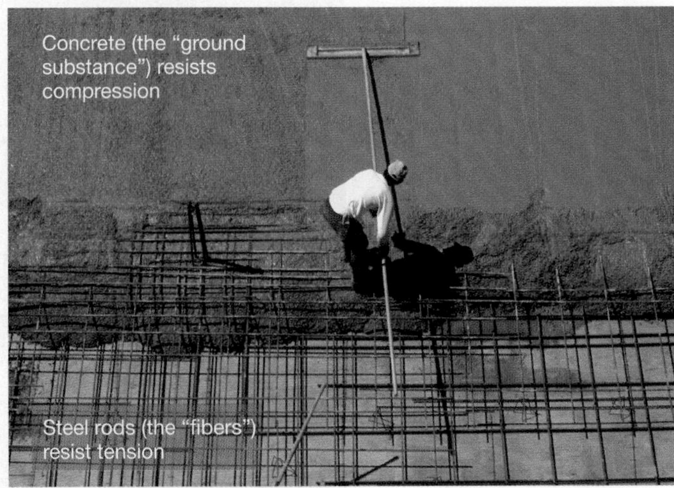
Concrete (the "ground substance") resists compression
Steel rods (the "fibers") resist tension

FIGURE 8.1 Fiber Composites Resist Tension and Compression. Fiber composites such as reinforced concrete consist of a massive ground substance that fills spaces between cross-linked rods.

The Cell Wall in Plants

Virtually all plant cells are surrounded by a cell wall—a fiber composite that is the basis of major industries. The paper in this book, the threads in your cotton clothing, and the wood in your neighborhood's houses are made up primarily of plant cell walls.

Before analyzing the structure of plant cell walls in detail, it's important to note that these structures are dynamic. If they are damaged by attacking insects, they may release signaling molecules that trigger the reinforcement of walls in nearby cells. When a seed germinates (sprouts), the walls of cells that store oils or starch are actively broken down, releasing these nutrients to the growing plant. Cell walls are also degraded in a controlled way as fruits ripen, making the fruits softer and more digestible for the animals that disperse the seeds inside.

PRIMARY CELL WALLS When plant cells first form, they secrete an initial fiber composite designated a **primary cell wall**.

- The fibrous component of the primary cell wall consists of long strands of cellulose, which are cross-linked by other polysaccharide filaments and bundled into stout, cable-like structures termed microfibrils. The microfibrils are synthesized by a complex of enzymes in the plasma membrane, forming a crisscrossed network (**Figure 8.2**).
- The space between microfibrils is filled with gelatinous polysaccharides like **pectins**—the molecules that are used to thicken jams and jellies. Because the polysaccharides in pectin are hydrophilic, they attract and hold large amounts of water to keep the cell wall moist. The gelatinous components of the cell wall are synthesized in the rough endoplasmic reticulum and Golgi apparatus and secreted to the extracellular space.

The primary cell wall defines the shape of a plant cell. Under normal conditions, the nucleus and cytoplasm fill the entire volume of the cell and push the plasma membrane up against the wall. Because the concentration of solutes is higher inside the cell than outside, water tends to enter the cell via osmosis. The in-

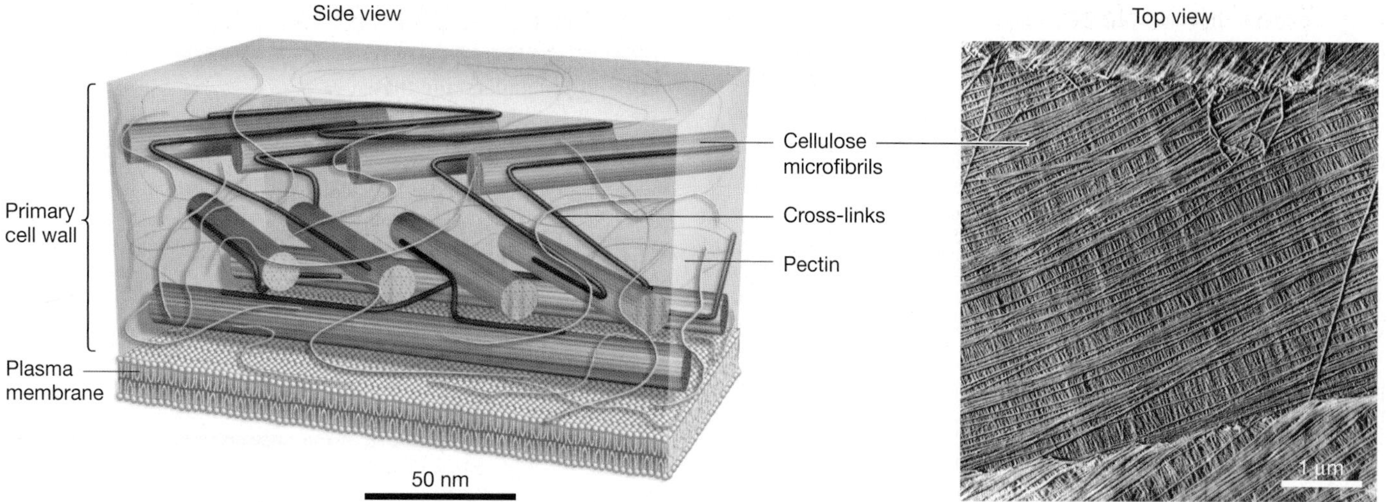

FIGURE 8.2 Primary Cell Walls of Plants Are Fiber Composites. In a plant's primary cell wall, cellulose microfibrils are cross-linked by polysaccharide chains. The spaces between the microfibrils are filled with pectin molecules, which form a gelatinous solid.

coming water inflates the plasma membrane, exerting a force against the wall that is known as **turgor pressure**.

Although plant cells experience turgor pressure throughout their lives, it is particularly important in young cells that are actively growing. Young plant cells secrete enzymes named expansins into their cell-wall matrix. Expansins catalyze reactions that allow the microfibrils in the matrix to slide past one another. Turgor pressure then forces the wall to elongate and expand. The result is cell growth.

SECONDARY CELL WALLS As plant cells mature and stop growing, they may secrete a layer of material—a **secondary cell wall**—inside the primary cell wall. The structure of the secondary cell wall varies from cell to cell in the plant and correlates with that cell's function. Cells on the surface of a leaf have cell walls that are impregnated with waxes that form a waterproof coating; while cells that support the plant's stem have secondary cell walls that contain a great deal of cellulose.

In cells that form wood, the secondary cell wall includes **lignin**, a tough substance that forms an exceptionally rigid network. Cells that have thick secondary cell walls of cellulose and lignin help plants withstand the forces of gravity and wind.

Although animal cells do not make a cell wall, they do form a fiber composite outside their plasma membrane. What is this substance, and what does it do?

The Extracellular Matrix in Animals

Most animal cells secrete a fiber composite called the **extracellular matrix (ECM)**. Like the extracellular materials found in other organisms, one of the ECM's most important functions is structural support.

ECM design follows the same principles observed in the cell walls of bacteria, archaea, algae, fungi, and plants. There is a key difference, however: The animal ECM contains much more protein than a cell wall.

- The fibrous component of animal ECM is dominated by a cable-like protein termed **collagen** (**Figure 8.3a**).

(a) Collagen proteins consist of three polypeptide chains that wind around each other.

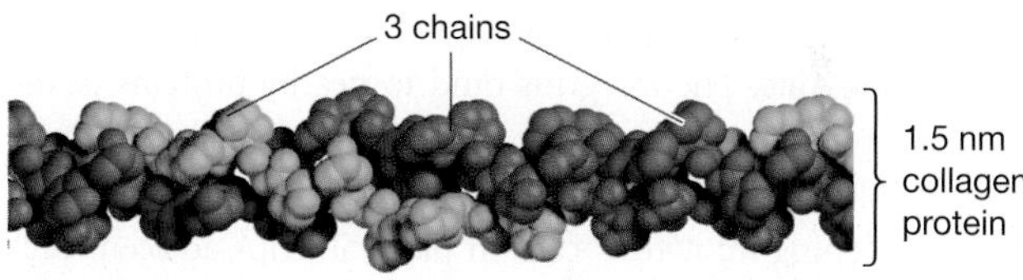

(b) ECM containing many collagen fibrils, each made up of many collagen proteins

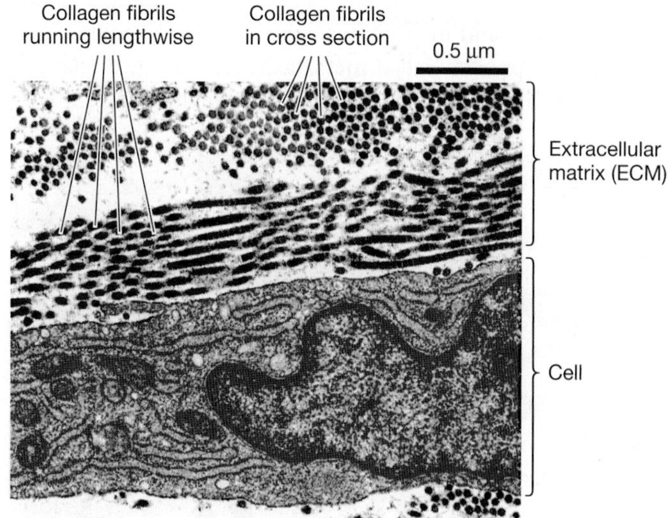

FIGURE 8.3 The Extracellular Matrix Is a Fiber Composite. **(a)** Although several types of fibrous proteins are found in the ECM of animal cells, the most abundant is collagen. Groups of collagen proteins coalesce to form collagen microfibrils, and bundles of microfibrils link to form collagen fibers. **(b)** A cross section of ECM from monkey cartilage tissue, showing a cell surrounded by abundant ECM. The spaces between the collagen fibers are filled with gelatinous polysaccharides.

- The matrix that surrounds collagen and the other fibrous components consists of gel-forming polysaccharides—most of which are attached to a protein core.

Because collagen and the other common ECM proteins are much more elastic and bendable than cellulose or lignin, the structure as a whole is relatively pliable.

Most ECM components are synthesized in the rough ER, processed in the Golgi apparatus, and secreted from the cell via exocytosis. Some of the protein-bound polysaccharides that form the composite material are synthesized by membrane proteins, however.

Even in the same organism, the amount of ECM varies among different types of cells. Bone and cartilage, for example, have relatively few cells but a large amount of ECM (**Figure 8.3b**). Skin cells, in contrast, are packed together with a minimal amount of ECM.

The composition of the ECM also varies among cell types. For example, the ECM of lung cells contains large amounts of a rubber-like protein called elastin, which allows the ECM to expand and contract during breathing. The structure of a cell's ECM correlates with that cell's function.

Although an ECM is not as stiff as a cell wall, it is strengthened by connections to transmembrane proteins. As **Figure 8.4** shows, actin filaments in the cytoskeleton are connected to transmembrane proteins called **integrins**. The integrins bind to nearby proteins in the ECM, including **fibronectins**, which in turn bind to collagen fibers. This direct linkage between the cytoskeleton and ECM is critical. In addition to keeping individual cells in place, it helps adjacent cells adhere to each other via their common connection to the ECM.

If this cytoskeleton-ECM linkage breaks down, cancer can develop. Through mechanisms that are not well understood, cells that are growing in an uncontrolled fashion, forming a tumor, can break away and migrate throughout the body, seeding new tumors. This process, called **metastasis**, can change a relatively benign and localized tumor into a life-threatening cancer.

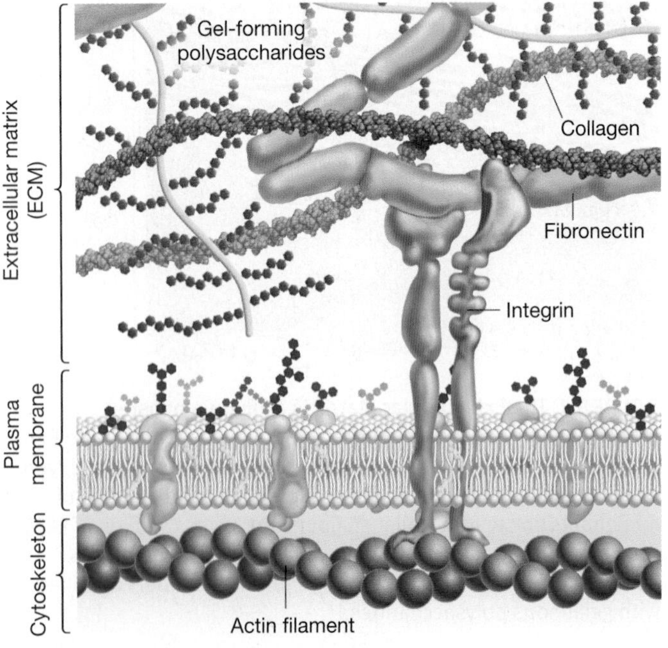

FIGURE 8.4 The Extracellular Matrix Connects to the Cytoskeleton.

CHECK YOUR UNDERSTANDING

If you understand that . . .

- Most cells secrete a layer of structural material that supports the cell and helps define its shape. The extracellular material is usually a fiber composite—a combination of cross-linked filaments surrounded by a ground substance.

✓ You should be able to . . .

Compare and contrast the molecular composition of a plant cell wall and the ECM of animal cells.

Answers are available in Appendix B.

8.2 How Do Adjacent Cells Connect and Communicate?

Although unicellular species may live in close proximity, they usually do not make direct physical connections. Physical connections between cells are the basis of **multicellularity** (**Figure 8.5**). Multicellular organisms are made up of cells that adhere to each other and have distinct structures and functions. The muscle cells in your body, for example, have distinctive shapes and are packed with proteins that are only found in muscle cells.

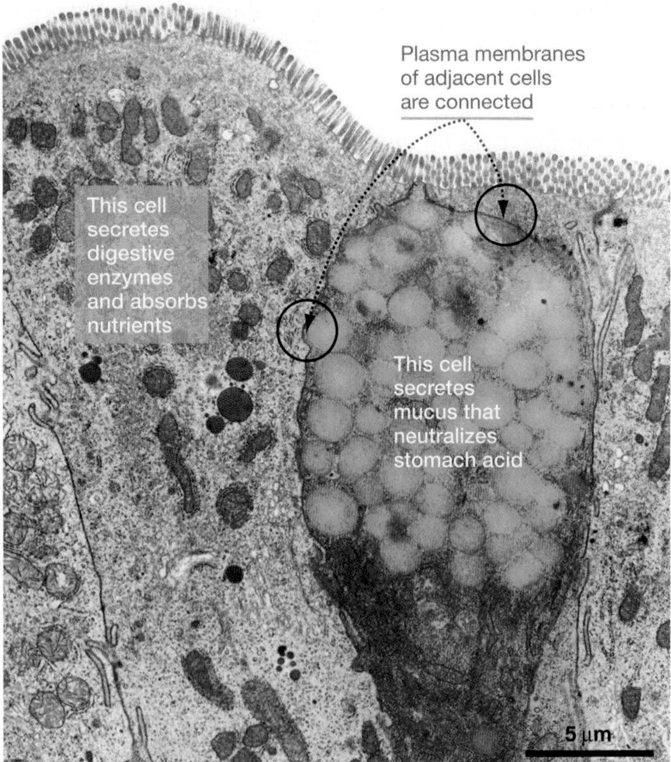

FIGURE 8.5 In Multicellular Organisms, Cells Are Connected. A cross section through cells that line the small intestine of a mammal. This transmission electron micrograph has been colorized to make the contrasting cell types more visible.

Most multicellular organisms also have **tissues**: groups of similar cells that perform a similar function. The individual muscle cells in your body are grouped into muscle tissue that contracts and relaxes to make movement possible.

Let's look first at the structures that attach adjacent cells to each other, then examine the openings that allow nearby cells to exchange materials and information.

Cell-Cell Attachments in Eukaryotes

The structures that hold cells together vary among multicellular organisms. To illustrate this diversity, consider the intercellular connections observed in the best-studied groups of organisms: plants and animals.

The extracellular space between adjacent plant cells comprises three layers (**Figure 8.6**). The primary cell walls of adjacent plant cells sandwich a central layer designated the middle lamella, which consists primarily of gelatinous pectins. Because this gel layer is continuous with the primary cell walls of the adjacent cells, it serves to glue them together. The two cell walls are like slices of bread; the middle lamella is like a layer of peanut butter. If enzymes degrade the middle lamella, as they do when flower petals and leaves detach and fall, the surrounding cells separate.

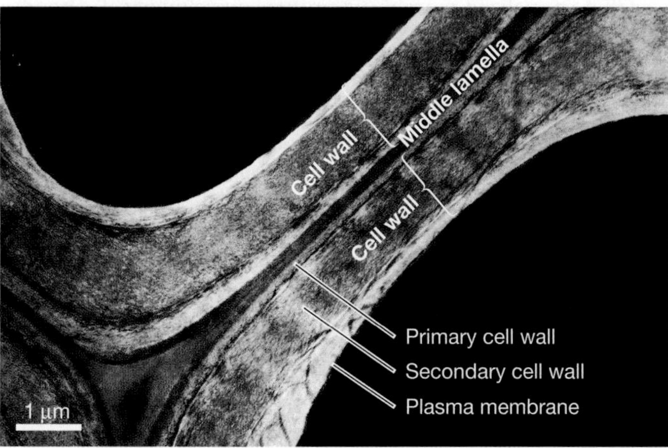

FIGURE 8.6 The Middle Lamella Connects Adjacent Plant Cells. The middle lamella contains gelatinous polysaccharides called pectins.

In many animal tissues, integrins connect the cytoskeleton of each cell to the extracellular matrix (see Section 8.1). A middle-lamella-like layer of gelatinous polysaccharides runs between adjacent animal cells, so cytoskeleton-ECM connections help hold individual animal cells together. In addition, in certain animal tissues the polysaccharide glue is reinforced by cable-like proteins that span the ECM to connect adjacent cells.

Materials and structures that bind cells together are particularly important in **epithelia** (singular: **epithelium**)—tissues that form external and internal surfaces. Epithelial cells form layers that separate organs and other structures. These epithelial layers must be sealed to prevent mixing of solutions from adjacent organs or structures.

A variety of cell-cell attachment structures exist in epithelia and other tissues. Here, however, we'll analyze just two: tight junctions and desmosomes.

TIGHT JUNCTIONS A **tight junction** is a cell-cell attachment composed of specialized proteins in the plasma membranes of adjacent animal cells (**Figure 8.7a**). As the drawing in **Figure 8.7b** indicates, these proteins line up and bind to one another. The resulting structure resembles quilting, with the proteins acting as stitches.

Because tight junctions form a watertight seal, this type of junction is commonly found in cells that form a barrier, such as the epithelial cells lining your stomach and intestines. There, they prevent the ions and molecules in your gut contents from leaking between

(a) Electron micrograph of a tight junction in longitudinal section

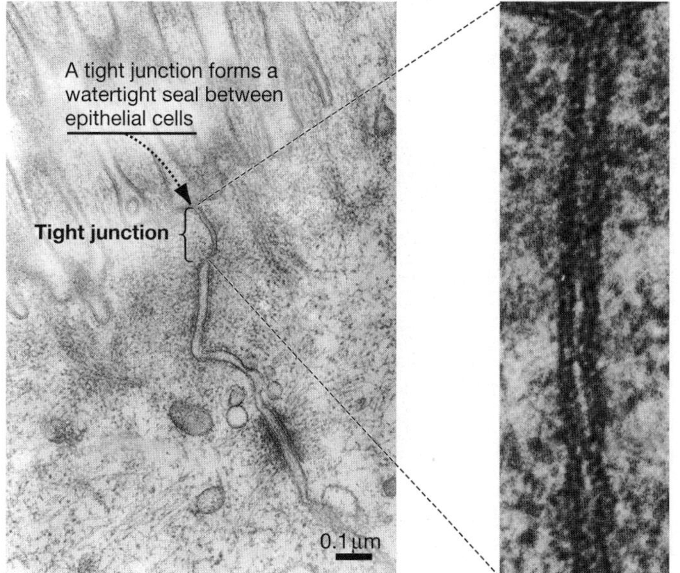

(b) Three-dimensional view of a tight junction

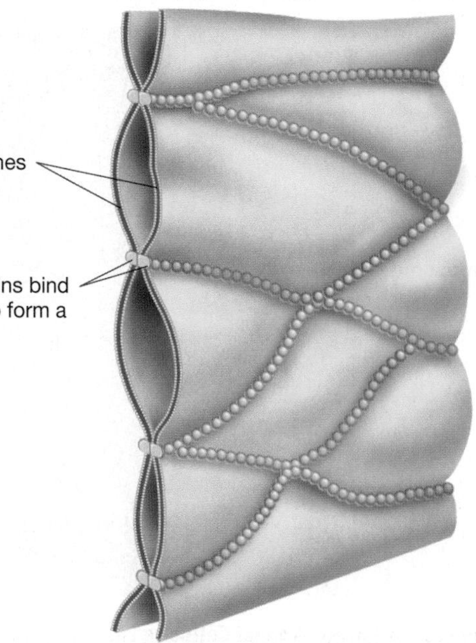

FIGURE 8.7 In Animals, Tight Junctions May Form a Seal between Adjacent Cells.

cells and diffusing into your body. Instead, only selected nutrients enter the cells. These ions and molecules are admitted via specialized transport proteins and channels in the plasma membrane.

Although tight junctions are indeed tight, they are variable. The tight junctions in the cells lining your bladder are much more snug than tight junctions in the cells lining your small intestine, because they consist of different proteins. As a result, small ions can pass through the surface of the small intestine much more easily than through the surface of the bladder—helping you absorb ions in your food and eliminate them in your waste.

Tight junctions are also dynamic. In some cases they may loosen to permit more transport between epithelial cells—for example, in the small intestine after a meal—and then "retighten" later. In this way, tight junctions can open and close in response to changes in environmental conditions.

DESMOSOMES **Figure 8.8a** illustrates **desmosomes**, cell-cell attachments particularly common in animal epithelial cells and certain muscle cells. The structure and function of a desmosome are analogous to the rivets that hold pieces of sheet metal together.

As **Figure 8.8b** indicates, desmosomes are extremely sophisticated cell-cell connections. At their heart are proteins that bind to each other and to larger proteins that anchor intermediate filaments in the cytoskeletons of the two cells. In this way, desmosomes bind together the cytoskeletons of adjacent cells.

What are these cell-attachment proteins? The answer to this question traces back to some of the first experiments conducted on cell-cell interactions.

SELECTIVE ADHESION Long before electron micrographs revealed the presence of desmosomes, biologists realized that some sort of molecule must bind animal cells to each other. This insight grew out of experiments conducted on sponges in the early 1900s.

Sponges are aquatic animals, and the sponge species used in this study consists of just two basic types of cells. When a biologist treated adult sponges with chemicals that made the cells separate from each other, the result was a jumbled mass of individual and unconnected cells. But when normal chemical conditions were restored, the cells gradually began to move and stick to other cells.

As the experiment continued, cells of each type began to combine and aggregate because they adhered to cells of the same tissue type. This phenomenon is now called **selective adhesion**. Eventually the experimental sponge cells re-formed functional adult sponges.

In an even more dramatic experiment, the researcher dissociated the cells of adult sponges from two differently pigmented sponge species and randomly mixed them together in a culture dish. As **Figure 8.9** shows, the cells eventually sorted themselves into two distinct aggregates, each containing cells from only one species and only one cell type. How could this happen?

THE DISCOVERY OF CADHERINS What is the molecular basis of selective adhesion? The initial hypothesis, proposed in the 1970s, was that specialized membrane proteins were involved. The idea was that different types of cells produce different types of adhesion proteins in their membranes, and that the molecules interact in a way that anchors cells of the same type to each other.

This hypothesis was tested through experiments that relied on molecules called antibodies. An **antibody** is a protein that binds specifically to a section of another protein. Because they stick to certain proteins specifically, antibodies can be used to block certain portions of proteins or mark proteins so they can be seen (see **BioSkills 9** in Appendix A).

Figure 8.10 shows how researchers tested the hypothesis that cell-cell adhesion takes place via interactions between membrane proteins:

Step 1 Isolate the membrane proteins from a certain cell type. Produce pure preparations of each protein.

Step 2 Inject one of the membrane proteins into a rabbit. The rabbit's immune system cells respond by creating antibodies to the

(a) Micrograph of desmosome in longitudinal section

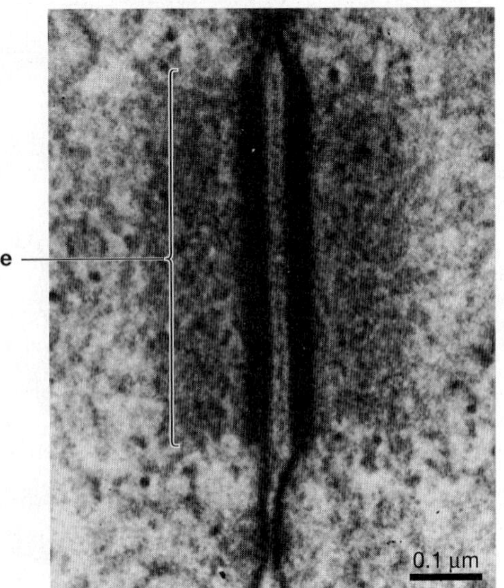

(b) Three-dimensional view of desmosome

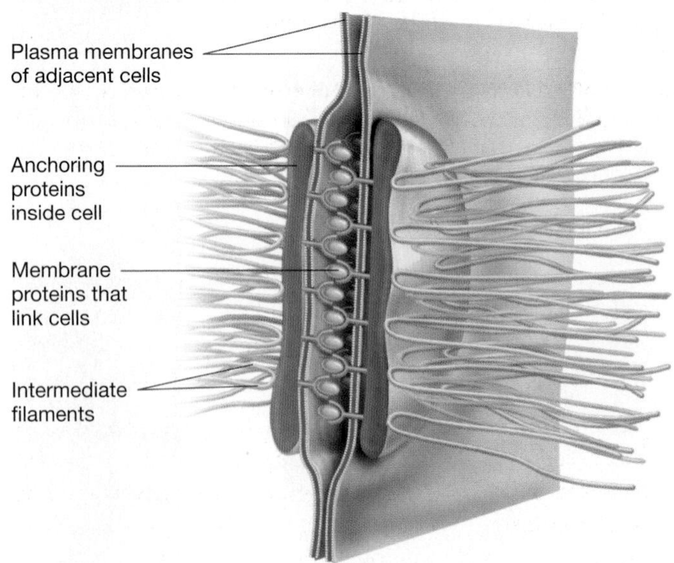

FIGURE 8.8 Adjacent Animal Cells Are Linked by Desmosomes, Which Bind Cytoskeletons Together.

EXPERIMENT

QUESTION: Do animal cells adhere selectively?

HYPOTHESIS: Cells of the same type and from the same species have a mechanism for selectively adhering to each other.

NULL HYPOTHESIS: Cells do not adhere or they adhere to each other randomly, not selectively.

EXPERIMENTAL SETUP:

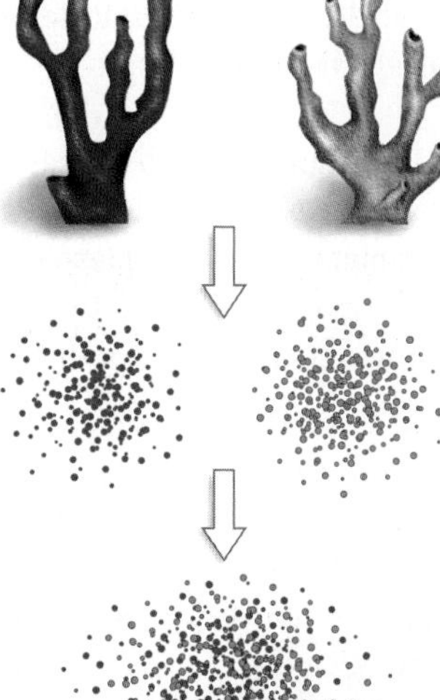

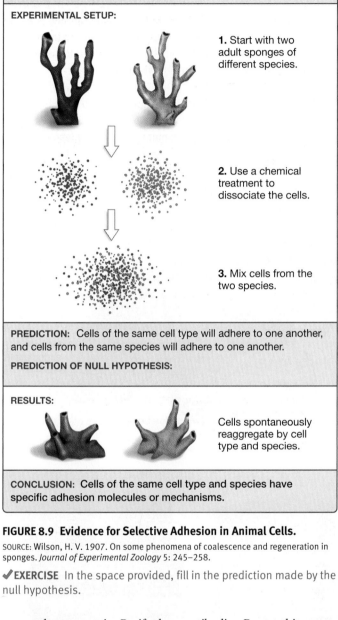

1. Start with two adult sponges of different species.

2. Use a chemical treatment to dissociate the cells.

3. Mix cells from the two species.

PREDICTION: Cells of the same cell type will adhere to one another, and cells from the same species will adhere to one another.

PREDICTION OF NULL HYPOTHESIS:

RESULTS:

Cells spontaneously reaggregate by cell type and species.

CONCLUSION: Cells of the same cell type and species have specific adhesion molecules or mechanisms.

FIGURE 8.9 Evidence for Selective Adhesion in Animal Cells.

SOURCE: Wilson, H. V. 1907. On some phenomena of coalescence and regeneration in sponges. *Journal of Experimental Zoology* 5: 245–258.

✔**EXERCISE** In the space provided, fill in the prediction made by the null hypothesis.

membrane protein. Purify those antibodies. Repeat this procedure for the other membrane proteins that were isolated. In this way, obtain a large collection of antibodies—each of which binds specifically to one (and only one) type of membrane protein.

Step 3 Add one antibody type to the mixture of dissociated cells and observe whether the cells reaggregate normally. Repeat this experiment with each of the other antibody types, one type at a time.

If treatment with a particular antibody prevents the cells from attaching to each other, the antibody is probably bound to an

EXPERIMENT

QUESTION: Do animal cells have adhesion proteins on their surfaces?

HYPOTHESIS: Selective adhesion is due to specific membrane proteins.

NULL HYPOTHESIS: Selective adhesion is not due to specific membrane proteins.

EXPERIMENTAL SETUP:

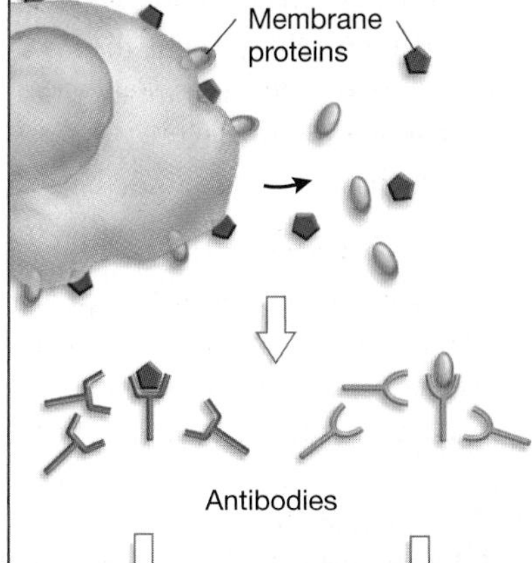

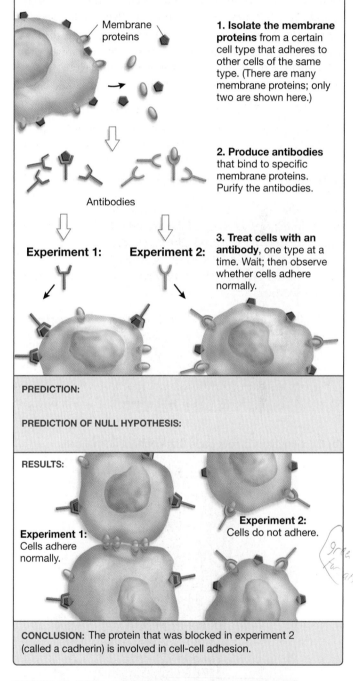

1. Isolate the membrane proteins from a certain cell type that adheres to other cells of the same type. (There are many membrane proteins; only two are shown here.)

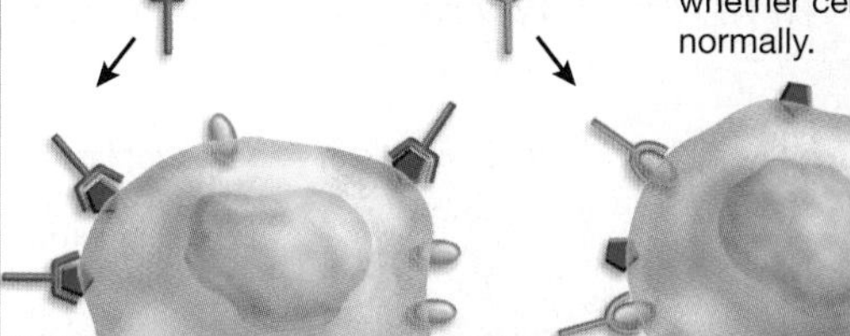

2. Produce antibodies that bind to specific membrane proteins. Purify the antibodies.

3. Treat cells with an antibody, one type at a time. Wait; then observe whether cells adhere normally.

PREDICTION:

PREDICTION OF NULL HYPOTHESIS:

RESULTS:

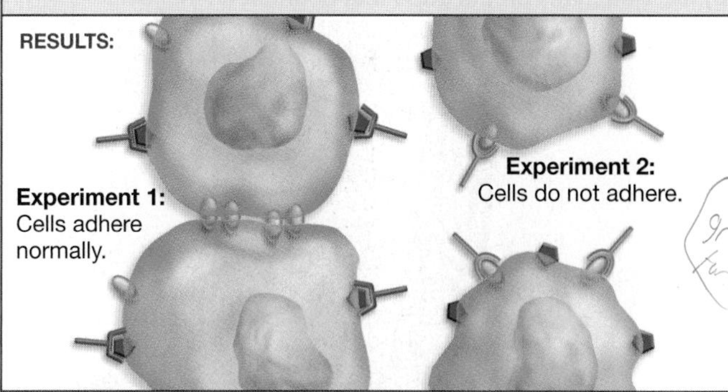

CONCLUSION: The protein that was blocked in experiment 2 (called a cadherin) is involved in cell-cell adhesion.

FIGURE 8.10 Evidence for Adhesion Proteins on Animal Cells.

SOURCES: Hatta, K. and M. Takeichi. 1986. Expression of N-cadherin adhesion molecules associated with early morphogenetic events in chick development. *Nature* 320: 447–449. Also Takeichi, M. 1988. The cadherins: cell-cell adhesion molecules controlling animal morphogenesis. *Development* 102: 639–655.

✔**EXERCISE** Fill in the prediction made by each hypothesis.

adhesion protein. The logic is that if the antibody "shakes hands" with the adhesion protein, the adhesion protein can't "shake hands" with other adhesion proteins and attach the cells to each other. This approach allowed biologists to identify several major classes of cell adhesion proteins, including **cadherins**—the attachment molecules in desmosomes. Different types of cells in the body have different forms of cadherin in their plasma membranes, and each cadherin can bind only to cadherins of the same type. In this way, cells of the same tissue type attach specifically to one another.

To summarize: Animal cells attach to each other in a selective manner because different types of cell adhesion proteins can bind and rivet certain cells together. Cadherins provide the physical basis for selective adhesion in many cells and are a critical component of the desmosomes that join mature cells.

✔If you understand cell-cell attachments, you should be able to predict: (1) whether molecules can pass between adjacent cells through middle lamellae, and (2) what would happen if you treated cells in a developing frog embryo with a molecule that blocked a cadherin observed in muscle tissue.

The presence of a middle lamella, continuous ECM, tight junctions, desmosomes, and cadherins bind adjacent cells to each other. But how do these cells communicate?

Cells Communicate via Cell-Cell Gaps

In both plants and animals, direct connections—holes or gaps—between cells in the same tissue help the cells work in a coordinated fashion. The presence of holes or gaps allows adjacent plant cells and adjacent animal cells to retain their own organelles, proteins, and nucleic acids, but communicate by sharing the ions and small molecules in their cytoplasm.

PLASMODESMATA In plants, gaps in cell walls create direct connections between the cytoplasm of adjacent cells. At these connections, named **plasmodesmata** (singular: **plasmodesma**), the plasma membrane and the cytoplasm of the two cells are continuous. Smooth endoplasmic reticulum (smooth ER) runs through these holes (**Figure 8.11a**).

(a) Plasmodesmata create gaps that connect plant cells.

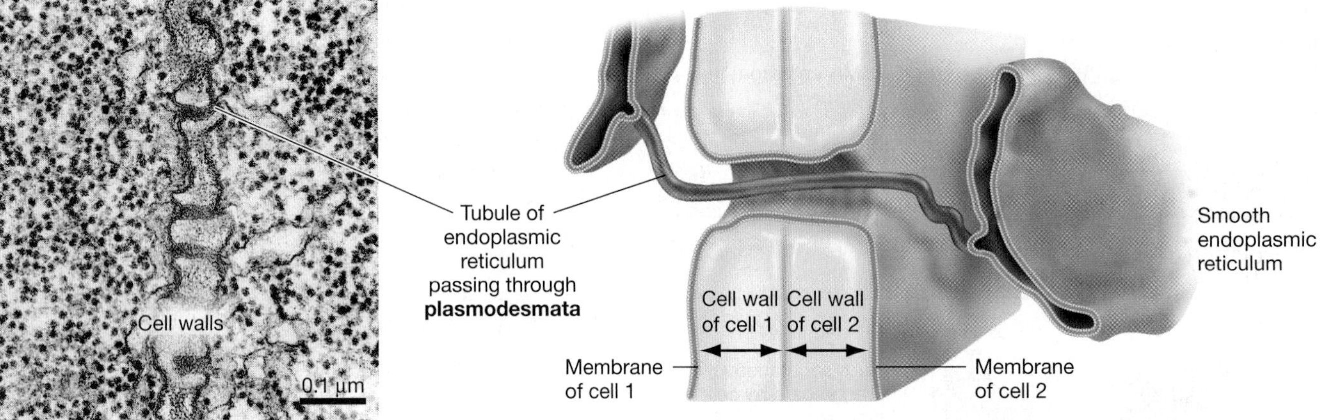

(b) Gap junctions create gaps that connect animal cells.

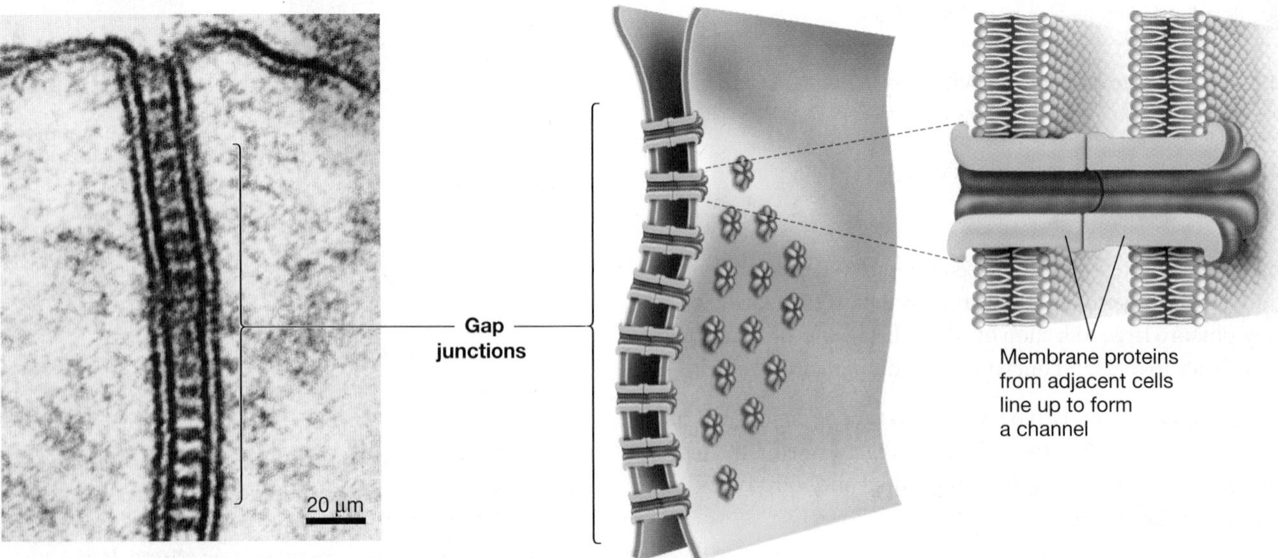

FIGURE 8.11 Adjacent Plant Cells and Adjacent Animal Cells Communicate Directly.

Growing evidence suggests that plasmodesmata also contain proteins that regulate the passage of specific proteins, making the connections similar in function to the nuclear pore complex introduced in Chapter 7. At least some of the proteins that are transported through plasmodesmata are involved in coordinating the activity of adjacent cells. Plasmodesmata are communication portals.

GAP JUNCTIONS In most animal tissues, structures called **gap junctions** connect adjacent cells. The key feature of gap junctions is the specialized proteins that create channels between cells (**Figure 8.11b**). These channels allow water, ions, and small molecules such as amino acids, sugars, and nucleotides to move between adjacent cells.

The flow of small molecules through gap junctions can help adjacent cells coordinate their activities by allowing the rapid passage of regulatory ions or molecules. In the muscle cells of your heart, for example, a flow of ions through gap junctions acts as a signal that coordinates contractions. Without this cell-cell communication, a normal heartbeat would be impossible. A tissue can act as an integrated whole if its component cells are connected by gaps in their extracellular material and in their plasma membranes (**Figure 8.12**).

How do more distant cells in a multicellular organism communicate? For example, suppose that leaf cells in a maple tree are attacked by caterpillars or the muscle cells in your arm are exercising so hard that they run low on sugar. How do these cells signal tissues or organs elsewhere in the body to release materials that are needed to fend off caterpillars or exhaustion? Distant cell communication is the subject of Section 8.3.

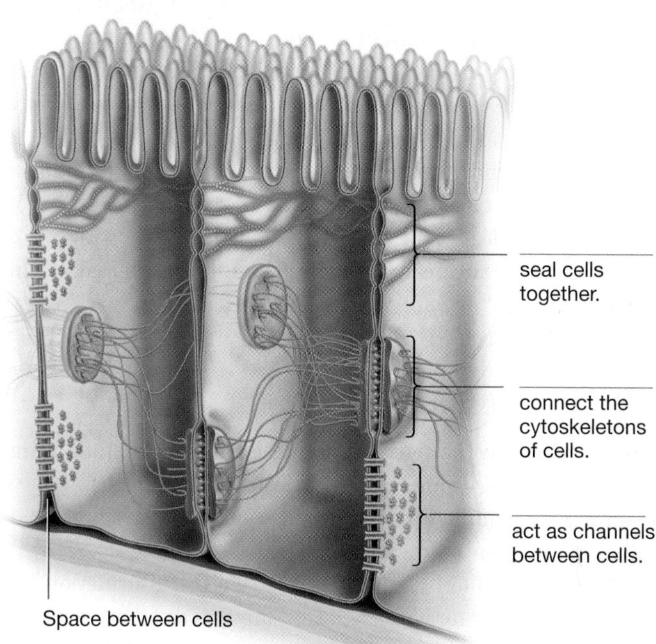

seal cells together.
connect the cytoskeletons of cells.
act as channels between cells.
Space between cells

FIGURE 8.12 An Array of Structures Is Involved in Cell-Cell Adhesion and Communication.

✔**EXERCISE** In the blanks on the right, write the name of each type of cell-cell attachment.

CHECK YOUR UNDERSTANDING

If you understand that . . .

- In plants and animals, adjacent cells are physically connected and communicate with each other through openings in their plasma membranes.

✔ **You should be able to . . .**

1. Compare and contrast the structure and function of the middle lamella of plants and the tight junctions and desmosomes of animals.
2. Describe the structure and function of plasmodesmata and gap junctions.

Answers are available in Appendix B.

8.3 How Do Distant Cells Communicate?

Cells that are not in physical contact communicate with each other. This is true for unicellular organisms, where hundreds or thousands of closely related cells may live in close proximity, as well as for multicellular organisms like humans and maple trees, which typically contain trillions of cells and dozens of tissue types.

Cell-cell communication qualifies as one of the most dynamic and important research areas in biology. Let's begin by analyzing how distant cells in humans and other multicellular eukaryotes exchange information, and close with a brief overview of how bacterial cells from the same species communicate with each other.

Cell-Cell Signaling in Multicellular Organisms

Suppose that cells in your brain sense that you are becoming dehydrated. Brain cells can't do much about the water you lose during urination, but kidney cells can. In response to dehydration, certain brain cells release a signaling molecule that travels to the kidneys. The arrival of the signal activates specialized membrane channels that prevent water from being lost in urine—an important aspect of fighting dehydration.

Thanks to cell-cell signals, the activities of cells in different parts of a multicellular body are coordinated.

Although biologists have classified many different types of signals that keep distant cells in touch, the best-studied may be **hormones**—information-carrying molecules that are secreted from a plant or animal cell, circulate in the body, and act on target cells far from the original cell that sent the signal.

Hormones are usually small molecules and are typically present in minute concentrations. Even so, they have a large impact on the activity of target cells and the condition of the body as a whole. Hormones are like a fleeting scent or whispered phrase from someone you are attracted to—a tiny signal, but one that makes your cheeks flush and your heart pound.

As **Table 8.1** indicates, hormones have a wide array of effects and chemical structures. The important point about a cell-cell signal, though, is not whether it is a gas or peptide or

TABLE 8.1 **Hormones Have Diverse Structures and Functions**

Hormone Name	Chemical Structure	Where is signal received?	Function of Signal
Auxin	Small organic compound	At plasma membrane	Signals changes in long axis of plant body
Brassinosteroids	Steroid	At plasma membrane	Stimulate plant cell elongation
Estrogens	Steroid	Inside cell	Stimulate development of female characteristics in animals
Ethylene	C_2H_4 (a gas)	At plasma membrane	Stimulates fruit ripening, regulates aging
FSH	Glycoprotein	At plasma membrane	Stimulates egg maturation, sperm production in animals
Insulin	Protein, 51 amino acids	At plasma membrane	Stimulates glucose uptake in animal bloodstream
Prostaglandins	Modified fatty acid	At plasma membrane	Perform a variety of functions in animal cells
Systemin	Peptide, 18 amino acids	At plasma membrane	Stimulates plant defenses against herbivores
Thyroxine (T4)	Modified amino acid	Inside cell	Regulates metabolism in animals

glycoprotein, but whether it is lipid soluble. The distinction is crucial because the signal has to be recognized to have an effect on a target cell. Where does this step occur—inside the cell or outside?

- Most lipid-soluble signals diffuse across the plasma membrane and enter the cytoplasm of their target cells.
- Large or hydrophilic cell-cell signals are lipid-insoluble and do not cross the plasma membrane. To affect a target cell, they have to be recognized at the cell surface.

How do cells receive and respond to signals from distant cells? The basic steps are common to all cell-cell signaling systems. Let's consider each in turn.

Signal Reception

Hormones and other types of cell-cell signals deliver their message by binding to receptor molecules. Even though the molecule that carries the message: "We're getting dehydrated—conserve water" is broadcast throughout the body, only certain kidney cells respond because only they have the appropriate receptor. The presence of an appropriate receptor dictates which cells will respond to a particular hormone. Bone and muscle cells don't respond to the "conserve water" message, because they don't have a receptor for it.

Cells in a wide array of tissues may respond to the same signal, though, if they have the appropriate receptor. If you are startled by a loud noise, cells in your adrenal glands secrete a hormone that carries the message: "Get ready to fight or run." In response, your heart rate increases, your breathing rate increases, and cells in your liver release sugars that your muscles can use to power rapid movement. This is the basis of an "adrenalin rush." Heart, lung, and liver cells respond to adrenalin because they have an adrenalin receptor. Identical receptors in diverse cells and tissues allow long-distance signals to coordinate the activities of cells throughout a multicellular organism.

No matter where signal receptors are located, it's critical to note two important points about these proteins:

- Receptors are dynamic. The number of receptors in a particular cell may decline if hormonal stimulation occurs at high levels over a long period of time. The ability of a receptor molecule to bind tightly to a signal may also decline in response to intensive stimulation. As a result, the sensitivity of a cell to a particular hormone may change over time.
- Receptors can be blocked. The drugs called beta-blockers, for example, bind to certain receptors for the hormone adrenalin (also called epinephrine). When adrenalin binds to receptors in heart cells, it stimulates the cells to contract. So if a physician wants to reduce a patient's heart cell contraction as a way to lower pressure, she is likely to prescribe a beta-blocker.

In many cases, the receptors that respond to lipid-soluble signals are located inside the cell, because the signals readily diffuse through the plasma membrane. The majority of signal receptors are located in the plasma membrane, however, where they can bind to signals that cannot or do not cross the membrane.

The most important general characteristic of signal receptors, though, is that their physical conformation—meaning, overall shape—changes when a hormone binds to them. A **signal receptor** is a protein that changes its shape and activity after binding to a signaling molecule.

This is a critical event in cell-cell signaling. The change in receptor structure means that the signal has been received. It's like throwing an "on" switch.

What happens next?

Signal Processing

Once a cell receives a signal, something has to happen to initiate the cell's response. This signal-processing step happens in one of two ways, depending on whether the signal is received inside the cell or at the membrane surface.

When lipid-soluble signals enter a cell, the information they carry is processed directly—without any intermediate steps. For example, steroid hormones such as testosterone and estrogen diffuse through the plasma membrane and enter the cytoplasm, where they bind to a receptor protein. The hormone-receptor complex is then transported to the nucleus, where it triggers changes in the genes being expressed in the cell (**Figure 8.13**).

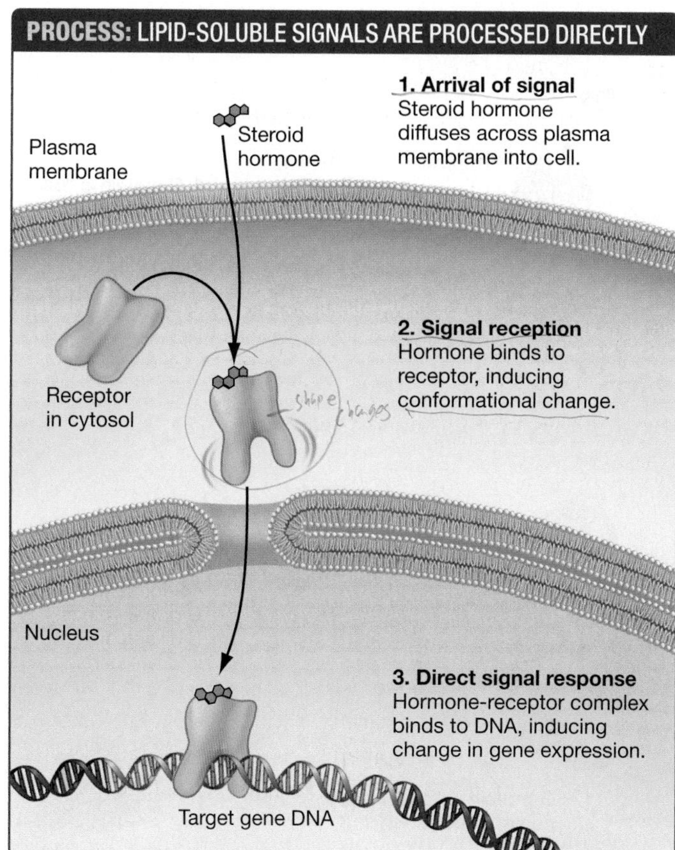

PROCESS: LIPID-SOLUBLE SIGNALS ARE PROCESSED DIRECTLY
1. Arrival of signal
Steroid hormone diffuses across plasma membrane into cell.
Plasma membrane
Steroid hormone
Receptor in cytosol
2. Signal reception
Hormone binds to receptor, inducing conformational change.
Nucleus
3. Direct signal response
Hormone-receptor complex binds to DNA, inducing change in gene expression.
Target gene DNA

FIGURE 8.13 Some Cell-Cell Signals Enter the Cell and Bind to Receptors in the Cytoplasm. Because they are lipids, steroid hormones can diffuse across the plasma membrane and bind to signal receptors inside the cell. The hormone-receptor complex is transported to the nucleus and binds to genes, changing their activity.

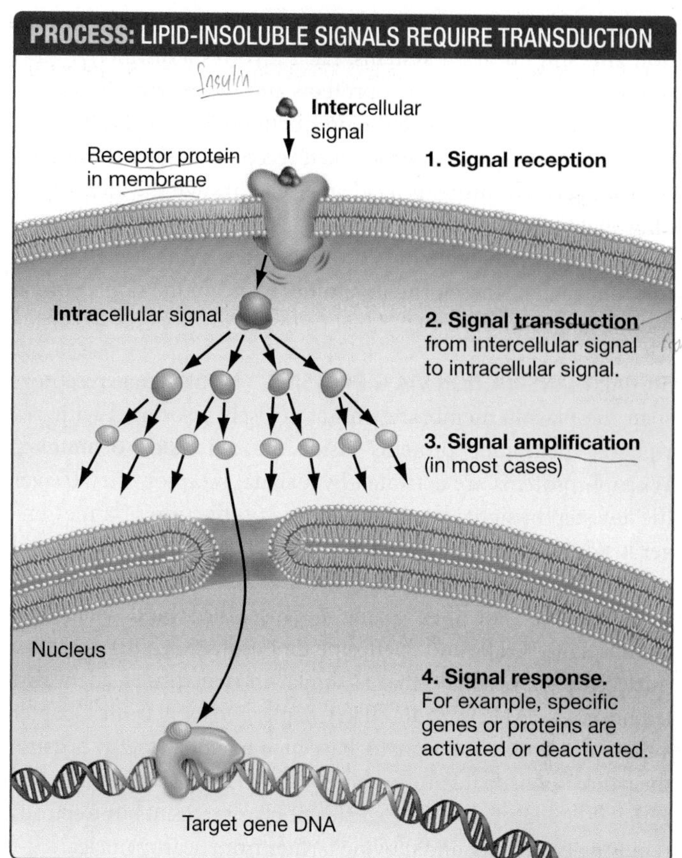

FIGURE 8.14 Signal Transduction Converts an Intercellular Signal to an Intracellular Signal. Signal transduction is a multistep process.

Other types of lipid-soluble hormones bind to receptors that initiate a response by activating certain pumps in the plasma membrane. In each case, the hormone-receptor complex directly initiates the change by binding to the genes or to the pumps.

Hormones that *cannot* diffuse across the plasma membrane and enter the cytoplasm can't change the activity of genes or pumps directly. Instead, the signal that arrives at the surface of the cell has to be changed to an intracellular signal—the signal-processing step is indirect.

When a signal binds at the cell surface it triggers **signal transduction**—the conversion of the signal from one form to another. A long and often complex series of events ensues, collectively called a signal transduction pathway.

Signal transduction is a common occurrence in everyday life. For example, the e-mail messages you receive are transmitted from one computer to another over cables or wireless transmissions. These electronic signals can be transmitted efficiently over long distances, but would be meaningless to you. Software in your computer has to transduce, or convert, the signals into a form that you can understand and respond to, such as words on the screen.

Signal transduction pathways work the same way (**Figure 8.14**). In a cell, signal transduction converts an extracellular signal to an intracellular signal. As in an e-mail transmission, a signal that is easy to transmit is converted to a signal that is easily understood and that triggers a response.

SIGNAL AMPLIFICATION Recall that hormones are present in minuscule concentrations but trigger a large response from cells. Signal amplification is one reason this is possible. When a hormone arrives at the cell surface, the message it transmits may be amplified as it changes form. The amplifier in your portable music player performs an analogous function: Once it is amplified, a tiny sound signal can get a whole roomful of people dancing.

In cells, signal transduction begins at the plasma membrane; amplification occurs inside. Amplification may occur in a variety of ways. In general, the mechanism of amplification correlates with the mechanism of signal transduction. But the general observation is that the arrival of a single signaling molecule may result in a secondary signal that involves many ions or molecules.

For example, one major type of signal transduction system consists of membrane channels that open to allow a flow of ions into the cell, changing the electrical properties of the membrane. Chapter 45 will analyze this type of signal transduction in detail.

Here let's focus on the other major types of signal transduction and amplification systems. Each involves a distinctive class of membrane protein: (1) G proteins and (2) enzyme-linked receptors. G proteins initiate the production of an intracellular or "second" messenger. Enzyme-linked receptors trigger the activation of a series of proteins inside the cell, through the addition of phosphate groups.

There are many types of G proteins and enzyme-linked receptors, but each of these protein families works in the same general way. Let's look at these two signal transduction systems in turn.

SIGNAL TRANSDUCTION VIA G PROTEINS Many signal receptors span the plasma membrane and are closely associated with peripheral membrane proteins inside the cell called **G proteins.** When G proteins are activated by a signal receptor, they trigger the key step in signal transduction: the production of a messenger inside the cell. They link the receipt of an extracellular signal to the production of an intracellular signal.

G proteins got their name because they bind guanosine triphosphate (GTP) and guanosine diphosphate (GDP). GTP is a nucleoside triphosphate that is similar in structure to adenosine triphosphate (ATP; introduced in Chapter 5). Recall from Chapter 4 that nucleoside triphosphates have high potential energy because their three phosphate groups have four negative charges close together (see Figure 4.4). The cluster of charges results in electrons repulsing each other and moving farther from nearby nuclei.

When GTP binds to a protein, the addition of the negative charges alters the protein's shape. Changes in shape produce changes in activity. G proteins are turned on or activated when they bind GTP; they are turned off or inactivated when a phosphate group drops away to form GDP.

To understand how G proteins fit into an overall signal transduction pathway, follow the events in **Figure 8.15**.

Step 1 A hormone arrives and binds to a receptor in the plasma membrane. Notice that the receptor is a transmembrane protein, and that it is coupled to a peripheral G protein on the membrane's inner surface. In response to hormone binding, the receptor changes shape.

Step 2 The shape change is a switch that activates its G protein. Specifically, the G protein releases the GDP molecule that kept it in an inactive state and binds GTP instead. When GTP is attached, the G protein changes shape radically: It splits into two parts.

Step 3 One part of the "split" G protein activates a nearby enzyme that is embedded in the plasma membrane. The enzyme catalyzes the production of a **second messenger**—a nonprotein signaling molecule that elicits a response to the first messenger (the signal that arrived at the cell surface).

Second messengers are effective because they are small and diffuse rapidly to spread the signal throughout the cell. In addition, they can be produced quickly in large quantities. This characteristic is important. Because the arrival of a single hormone molecule can stimulate the production of many second messenger molecules, the signal transduction event amplifies the original signal.

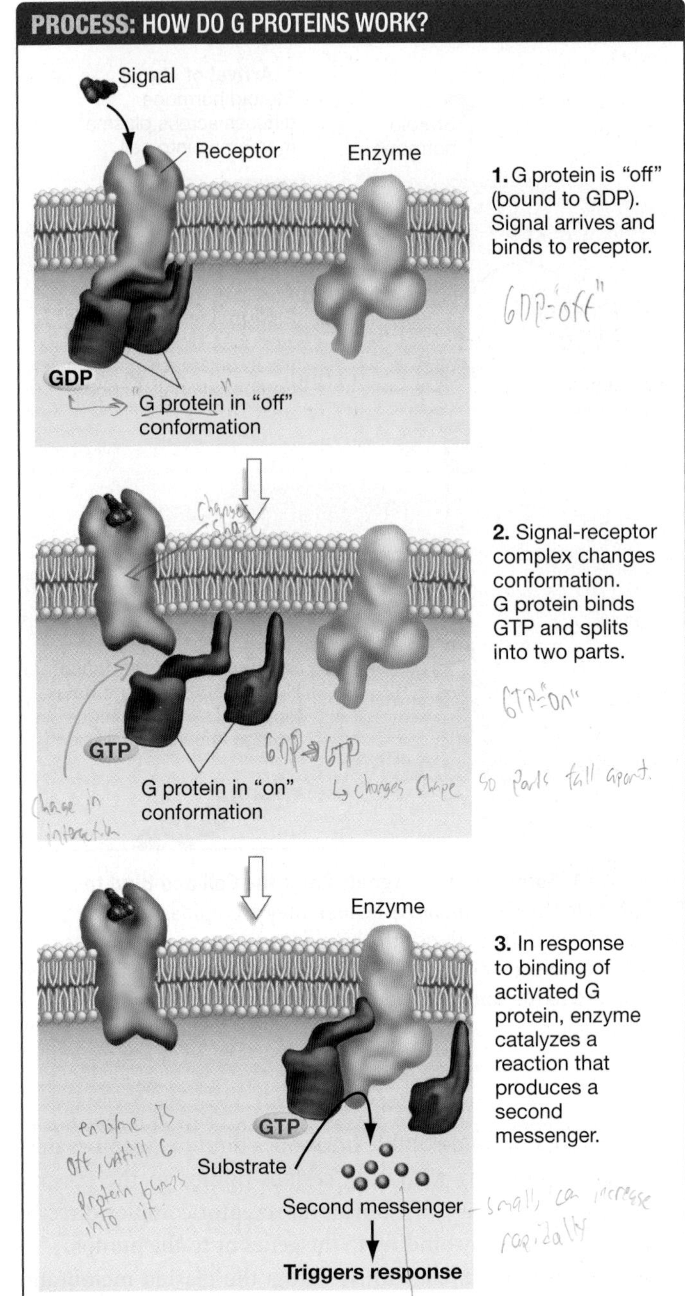
PROCESS: HOW DO G PROTEINS WORK?
Signal
Receptor
Enzyme
1. G protein is "off" (bound to GDP). Signal arrives and binds to receptor.
GDP
G protein in "off" conformation
2. Signal-receptor complex changes conformation. G protein binds GTP and splits into two parts.
GTP
G protein in "on" conformation
Enzyme
3. In response to binding of activated G protein, enzyme catalyzes a reaction that produces a second messenger.
GTP
Substrate
Second messenger
Triggers response

FIGURE 8.15 G Proteins Trigger the Production of a Second Messenger.

Several types of small molecules act as second messengers in cells. **Table 8.2** lists some of the best-studied and provides an example of how cells respond to each second messenger. Note that several second messengers activate **protein kinases**—enzymes that activate or inactivate other proteins by adding a phosphate group to them.

It's also important to note that:

1. Second messengers aren't restricted to a single role or single cell type—the same second messenger can initiate dramatically different events in different cell types; and

TABLE 8.2 **Examples of Second Messengers**

Name	Type of Response
Cyclic guanosine monophosphate (cGMP)	Opens ion channels; activates certain protein kinases
Diacylglycerol (DAG)	Activates certain protein kinases
Inositol triphosphate (IP_3)	Opens calcium channels—mobilizes stored calcium ions
Cyclic adenosine monophosphate (cAMP)	Activates certain protein kinases
Calcium ions (Ca^{2+})	Binds to a receptor called calmodulin; Ca^{2+}/calmodulin complex then activates proteins

2. It is common for more than one second messenger to be involved in triggering a cell's response to the same extracellular signal.

To make sure that you understand how G proteins and second messengers work, imagine the following movie scene: A spy arrives at a castle gate. The guard receives a note from the spy, but he cannot read the coded message. Instead, the guard turns to the queen. She reads the note and summons the commander of the guard, who sends soldiers throughout the castle to warn everyone of approaching danger. ✔You should be able to identify which characters in the scene correspond to the second messenger, G protein, hormone, receptor, and enzyme activated by the G protein.

It's difficult to overstate the importance of signal transduction by G proteins. Biomedical researchers estimate that half of human drugs target signal receptors that are associated with G proteins.

SIGNAL TRANSDUCTION VIA ENZYME-LINKED RECEPTORS Instead of activating a nearby G protein, enzyme-linked receptors transduce the signal from a hormone by directly catalyzing a reaction inside the cell. **Figure 8.16** focuses on the best-studied group of enzyme-linked receptors: the **receptor tyrosine kinases (RTKs)**.

Step 1 A hormone binds to an RTK.

Step 2 The protein forms a dimer. In this conformation, the receptor has a binding site for a phosphate group from ATP inside the cell. Once it is phosphorylated, an RTK becomes an active enzyme.

Step 3 Proteins inside the cell form a bridge between the activated RTK and a peripheral membrane protein called **Ras**, which functions like a G protein. The formation of the RTK bridge activates Ras. Specifically, Ras exchanges its GDP for a GTP.

Step 4 When Ras is activated, it triggers the phosphorylation and activation of another protein.

Step 5 The phosphorylated protein then catalyzes the phosphorylation of other proteins, which phosphorylate yet another population of proteins.

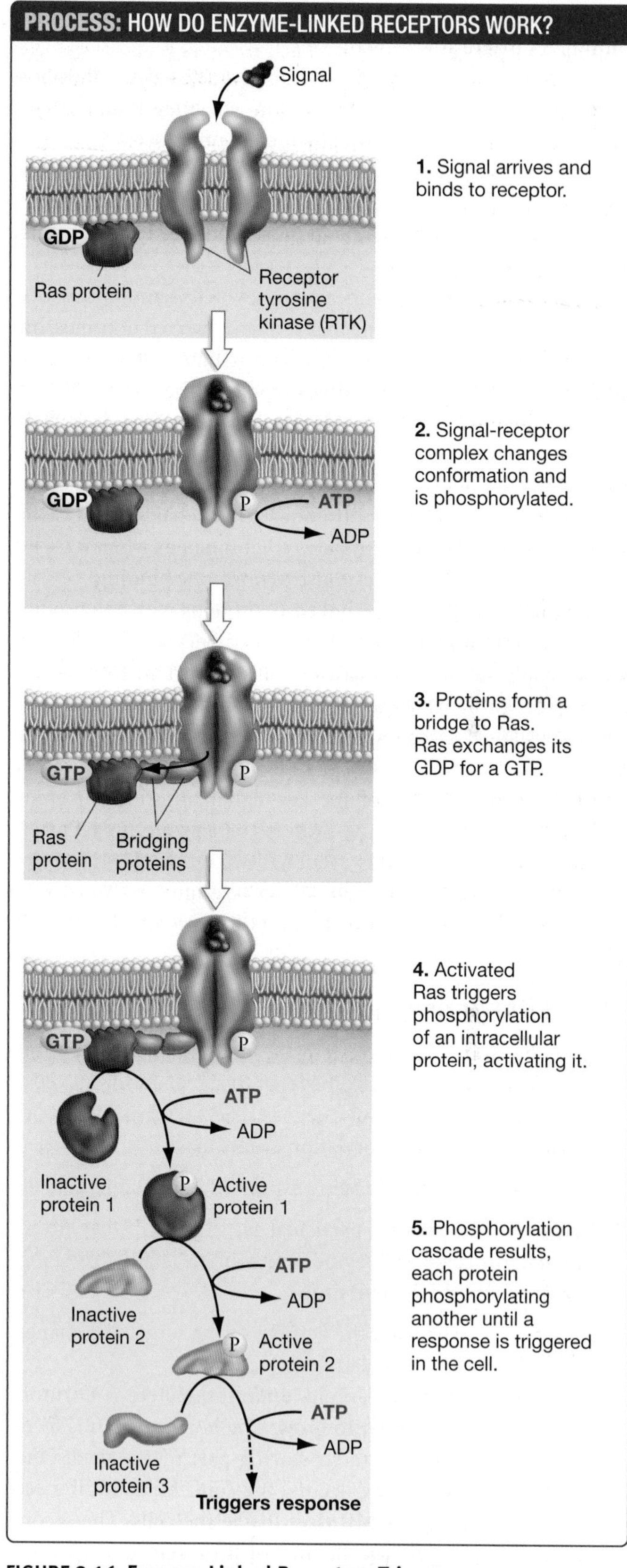
PROCESS: HOW DO ENZYME-LINKED RECEPTORS WORK?
Signal
GDP
Ras protein
Receptor tyrosine kinase (RTK)
1. Signal arrives and binds to receptor.
GDP
P
ATP
ADP
2. Signal-receptor complex changes conformation and is phosphorylated.
GTP
P
Ras protein
Bridging proteins
3. Proteins form a bridge to Ras. Ras exchanges its GDP for a GTP.
GTP
P
4. Activated Ras triggers phosphorylation of an intracellular protein, activating it.
ATP
ADP
Inactive protein 1
P
Active protein 1
ATP
ADP
Inactive protein 2
P
Active protein 2
5. Phosphorylation cascade results, each protein phosphorylating another until a response is triggered in the cell.
ATP
ADP
Inactive protein 3
Triggers response

FIGURE 8.16 Enzyme-Linked Receptors Trigger a Phosphorylation Cascade.

This sequence of events is termed a **phosphorylation cascade**, and culminates in a response by the cell.

In some cases, each enzyme in the cascade catalyzes the phosphorylation of numerous "downstream" enzymes. When this occurs, the original signal is amplified many times over.

In most cases, though, the proteins that take part in a phosphorylation cascade are held in close physical proximity by scaffolding proteins. This structure increases the speed and efficiency of the reaction sequence.

To make sure that you understand how RTKs and phosphorylation cascades work, imagine that you have two red dominos, one black domino, and a large supply of green, blue, yellow, pink, and orange dominos. The red dominos represent the two subunits of an RTK dimer, and the black domino represents Ras. Each of the other colors represents a type of protein in a phosphorylation cascade. ✔You should be able to explain (1) how you would set up the dominos to simulate a phosphorylation cascade, and (2) why tipping the red dominos simulates what happens when RTK becomes phosphorylated in response to hormone binding.

Although this discussion linked G proteins with the production of second messengers and associated enzyme-linked receptors with phosphorylation cascades, it's important to recognize that some G proteins trigger phosphorylation cascades and some phosphorylation cascades result in the production of second messengers.

To summarize: Many of the key signal transduction events observed in cells occur via G proteins or enzyme-linked receptors. The signal transduction event has two results: **(1)** It converts an easily transmitted extracellular message into an intracellular message, and **(2)** in many cases it amplifies the original message many times over.

Signal Response

What is the ultimate response to the messages carried by hormones? The answer varies from signal to signal and from cell to cell, but fall into two general categories. Second messengers or a cascade of protein phosphorylation events may:

1. change which genes are being expressed in the target cell; or
2. activate or deactivate a particular target protein that already exists in the cell—an enzyme, a membrane channel, or a protein that activates certain genes.

Whatever the mechanism, the activity of the target cell changes dramatically after the signal arrives.

In wheat seeds, for example, embryos secrete a hormone called GA_1 when they start to grow. The hormone binds to receptors in cells near the starch-storing part of the seed. Once hormone-receptor binding occurs, the concentration of a second messenger called cGMP rises inside the cells. The second messenger triggers the production of a protein that activates the gene for a starch-digesting enzyme. Large quantities of the starch-digesting enzyme α-amylase enter the cell's endomembrane system.

Vesicles packed with α-amylase can be secreted into the starch storage area because the G protein activated by GA_1 also triggers a rise in intracellular Ca^{2+}. High Ca^{2+} concentrations allow the vesicles to fuse with the plasma membrane and release their contents. The enzymes begin breaking down the starch and releasing sucrose. In this way, the germinating embryo has signaled for the release of the nutrients it needs to grow.

We've analyzed the first three steps of long-distance cell-cell communication: signal reception, signal processing, and response. Now the question is, how is the signal turned off? Consider the flush of testosterone and estrogen that you experienced during puberty, and the morphological changes these hormones induced. Abnormalities would result if these changes continued indefinitely. What limits the response to a cell-cell signal?

Signal Deactivation

Cells have built-in systems for turning off intracellular signals. For example, activated G proteins convert bound GTP to GDP. When this reaction occurs, the G protein's conformation changes. Activation of its associated enzyme stops, and production of the second messenger ceases.

The presence of second messengers in the cytosol is also short lived. For example, pumps in the membrane of the smooth ER return calcium ions to storage, and enzymes called phosphodiesterases convert active cAMP (see Table 8.2) and cGMP to inactive AMP and GMP. When second messengers are cleared from the cytosol, the response stops.

For G proteins to stay activated and continue influencing the behavior of the cell, the extracellular signal has to continue. Otherwise, the signal transduction system quickly shuts down.

Phosphorylation cascades wind down in a similar way. Enzymes called phosphatases are always present in cells, where they catalyze reactions that remove phosphate groups from proteins. If hormone stimulation of a receptor tyrosine kinase ends, phosphatases are able to dephosphorylate enough components of the phosphorylation cascade that the response begins to slow. Eventually it stops.

Although an array of specific mechanisms are involved, here is the general observation: Signal transduction systems trigger a rapid response and can be shut down quickly. As a result, they are exquisitely sensitive to small changes in the concentration of hormones or the number and activity of signal receptors.

It is critical, though, to appreciate what happens when a signal transduction system does not shut down properly. For example, recall that Ras activates a phosphorylation cascade when it binds GTP, but is deactivated once it has broken down GTP to GDP. With surprising frequency, human cells produce defective Ras proteins that no longer convert GTP to GDP once they are activated. As a result, GTP stays bound and the defective Ras proteins stay in the "on" position. They continue stimulating a phosphorylation cascade even when no appropriate signals are present. Cells with defective Ras are likely to keep dividing, which may lead to the development of cancer. An estimated 25–30 percent of all human cancers involve defective Ras proteins. Chapters 11 and 18 explore the family of diseases called cancer in detail.

Cross-Talk: Synthesizing Input from Many Signals

Although the preceding discussion focused on how cells respond to individual signals, it's crucial to realize that every cell has an array of signal receptors on its plasma membrane and in its cytoplasm, and every cell receives an almost constant stream of different signals. You get text messages, e-mails, phone calls, and snail mail about changes in your environment; cells get an array of chemical signals about changes in their environment.

The signal transduction pathways that are triggered by these signals and receptors intersect and connect. In reality, they are not strictly linear like the pathways illustrated in Figures 8.13 through 8.16. Signal transduction pathways form a network. This complexity is important: It allows cells to respond to an array of extracellular signals in an integrated way.

The diverse signals that a cell receives are integrated by what biologists call **cross-talk**—meaning, interactions between signaling pathways (**Figure 8.17**). The key things to note are that:

1. Elements or products from one pathway may inhibit steps in a different pathway—reducing the cell's response, even though the appropriate signal is present.
2. A response from one pathway may stimulate a greater response by a protein in a different pathway, increasing the cell's response to the other signal.
3. The presence of multiple steps in each signaling pathway provides a series of points where cross-talk can regulate the flow of information. These interactions are important, because they allow the cell to respond appropriately to many signals at the same time.

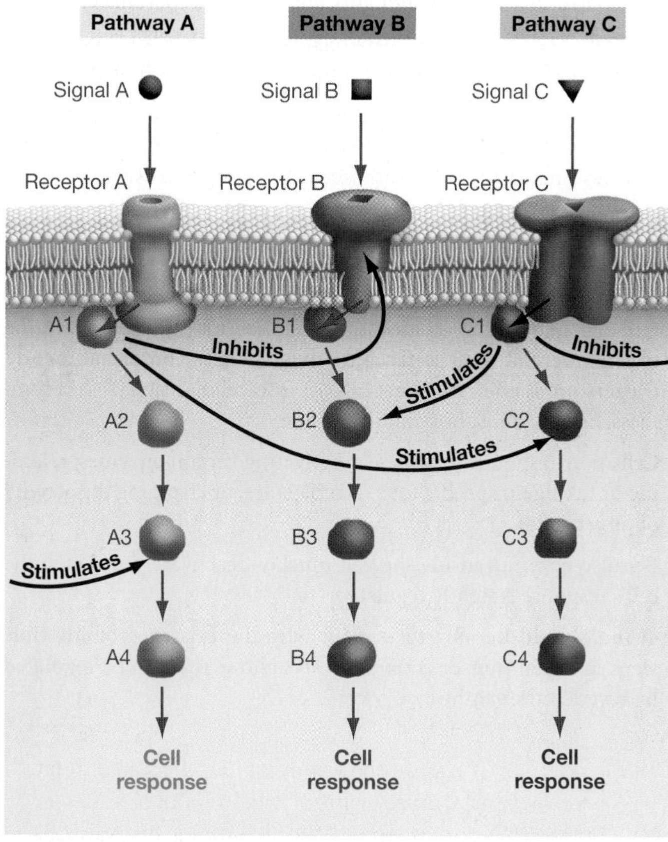

FIGURE 8.17 Signaling Pathways Interact via "Cross-Talk."

Quorum Sensing in Bacteria

In eukaryotes, cell-cell signaling has been one of the hottest research areas in biological science over the past two decades. But in prokaryotes, research on cell-cell signaling is just beginning to gain momentum.

In bacteria, cell-cell communication is called **quorum sensing**. The name was inspired by the observation that bacterial cells of the same species may undergo dramatic changes in activity when their numbers reach a threshold, or quorum.

Quorum sensing is based on species-specific signaling molecules that are secreted by cells and diffuse through the environment. The response to these signals varies dramatically among species and ranges from bioluminescence—or light emission—to secretion of proteins that allow the bacteria to attack multicellular hosts and cause disease, to the production and secretion of **biofilms**—hard, polysaccharide-rich substances that encase the cells and attach them to a surface (**Figure 8.18**).

Biofilm formation has been particularly well-studied because of its medical importance.

- Biofilms that form in the human mouth are called dental plaque and contribute to tooth decay.
- When *Pseudomonas aeruginosa* cells settle on a human lung cell, they secrete a signaling molecule that recruits other *P. aeruginosa* cells to the site and triggers secretion of a biofilm. The biofilm attaches the cells to the lung surface and protects them from attack by immune system cells or antibiotics.

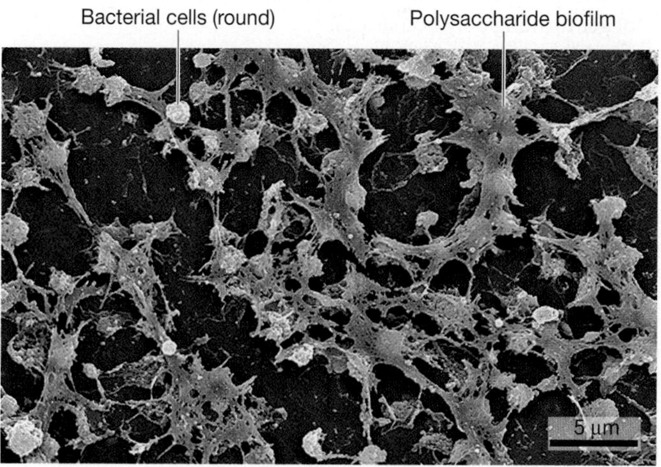

FIGURE 8.18 Groups of Unicellular Organisms May Secrete Biofilms. This micrograph shows *Staphylococcus aureus* cells (the round structures) growing inside a catheter—a tube inserted into a body so that fluids can be withdrawn or injected. The cells have secreted a sticky biofilm that helps them cling to each other and to the walls of the catheter.

In effect, quorum sensing allows closely related bacterial cells to communicate and coordinate activity. When it occurs, these single-celled organisms take on some of the characteristics of multicellular organisms.

This brief introduction to quorum sensing and cross-talk brings us to the frontier of research in cell-cell signaling. It has taken decades of painstaking research to work out each step in individual signaling pathways; now biologists are excited about probing how the major pathways interact, and how similar types of pathways operate in eukaryotes.

Although the number of molecules involved and the complexity of their interactions can seem overwhelming, the punchline is simple: In organisms ranging from bacteria to blue whales, cell-cell signaling helps organisms receive information about their environment and respond to changing conditions in an appropriate way.

CHECK YOUR UNDERSTANDING

If you understand that . . .

- Intercellular signals coordinate the activities of cells throughout the body of a multicellular organism in response to changes in internal or external conditions.
- If intercellular signals do not enter the cell, they bind to a receptor on the plasma membrane. In response, the intercellular signal is transduced to an intracellular signal that the cell responds to.

✓ **You should be able to . . .**

1. Explain why only certain cells respond to particular signals.
2. Explain how some signals are amplified.

Answers are available in Appendix B.

CHAPTER 8 REVIEW

For media, go to the study area at www.masteringbiology.com 

Summary of Key Concepts

Extracellular material strengthens cells and helps bind them together.

- The vast majority of cells secrete an extracellular layer.
- In bacteria, archaea, algae, and plants, the extracellular material is stiff and is called a cell wall. In animals, the secreted layer is flexible and is called the extracellular matrix (ECM).
- Extracellular layers are fiber composites. They consist of cross-linked filaments that provide tensile strength and a ground substance that fills space and resists compression.
- In plants the extracellular filaments are cellulose microfibrils; in animals the most abundant filaments are made of the protein collagen. In both plants and animals, the ground substance is composed of gel-forming polysaccharides.

✓You should be able to predict what happens to animal cells when (1) they are treated with an enzyme that cuts integrin molecules, or (2) collagen fibers degrade.

Cell-cell connections help adjacent cells adhere. Cell-cell gaps allow adjacent cells to communicate.

- In multicellular organisms, molecules in the extracellular layer and plasma membrane mediate interactions between adjacent cells.
- Adjacent cells may be physically bound to one another, by glue-like middle lamella in plants or tight junctions and desmosomes in animals.
- The cytoplasm of adjacent cells is in direct communication through openings called plasmodesmata in plants and gap junctions in animals.

✓You should be able to explain the consequences of loosening the tight junctions between animal epithelial cells.

Intercellular signals are responsible for creating an integrated whole from many thousands of independent parts.

- Distant cells in multicellular organisms communicate through signaling molecules that bind to receptors found in or on specific target cells. As a result, cells and tissues throughout the body can alter their activity in response to changing conditions, and do so in a coordinated way.
- Quorum sensing allows closely related bacterial cells to coordinate changes in their activity when population density is high.

✓You should be able to explain why the hormone adrenalin can stimulate cells in both the heart and liver, yet trigger different responses (increasing heart rate versus releasing glucose).

In target cells, intercellular signals are received, processed, responded to, and deactivated. If the signal is received at the cell surface, the processing step involves production of an intracellular signal.

- Cell-cell signals that are not lipid-soluble bind to receptors in the plasma membrane. The receptor then changes conformation and triggers production of a new type of intracellular signal—a second messenger or phosphorylation cascade.
- Cells may respond to signals by activating certain enzymes, releasing or taking up specific ions or molecules, or changing the activity of target genes.
- Because enzymes inside the cell quickly deactivate the signal, the cell's response is tightly regulated.

✓You should be able to explain why the existence of multiple steps in signal transduction pathways allows them to be regulated by several other pathways.

Questions

✓TEST YOUR KNOWLEDGE

Answers are available in Appendix B

1. Which of the following statements represents a fundamental difference between the fibers found in the extracellular layers of plants and those of animals?
   a. Plant fibers are thicker; they are also stronger because they have more cross-linkages.
   b. Animal fibers consist of proteins; plant fibers consist of polysaccharides instead.
   c. Plant extracellular fibers never move; animal fibers can slide past one another.
   d. Cellulose microfibrils run parallel to each other; collagen filaments crisscross.
2. In animals, where are most components of the extracellular material synthesized?
   a. smooth ER
   b. the rough ER and Golgi apparatus
   c. in the extracellular layer itself
   d. in the plasma membrane
3. Treating dissociated cells with certain antibodies makes the cells unable to reaggregate. Why?
   a. The antibodies bind to cell adhesion proteins called cadherins.
   b. The antibodies bind to the fiber component of the extracellular matrix.
   c. The antibodies bind to receptors on the cell surface.
   d. The antibodies act as enzymes that break down desmosomes.
4. What does it mean to say that a signal is transduced?
   a. The signal enters the cell directly and binds to a receptor inside.
   b. The physical form of the signal changes between the outside of the cell and the inside.
   c. The signal is amplified, such that even a single molecule evokes a large response.
   d. The signal triggers a sequence of phosphorylation events inside the cell.
5. Why are tight junctions found in only certain types of tissues, while desmosomes are found in a wide array of cells?
   a. Tight junctions are required only in cells where communication between adjacent cells is particularly important.
   b. Tight junctions are not as strong as desmosomes.
   c. Tight junctions have different structures but the same functions.
   d. Tight junctions are found only in epithelial cells that must be watertight.
6. What physical event represents the receipt of an intercellular signal?
   a. the passage of ions through a desmosome
   b. the activation of the first protein in a phosphorylation cascade
   c. the binding of a hormone to a signal receptor, which changes conformation in response
   d. the activation of a G protein associated with a signal receptor

✓TEST YOUR UNDERSTANDING

Answers are available in Appendix B

1. Why is it difficult to damage a fiber composite?
2. How is it possible for a phosphorylation cascade to amplify an intercellular signal?
3. Compare and contrast the structure and function of tight junctions and gap junctions. Compare and contrast the structure and function of middle lamellae and plasmodesmata.
4. Animal cells adhere to each other selectively. Summarize experimental evidence that supports this statement. Explain the molecular basis of selective adhesion.
5. Make a flowchart summarizing the reception, processing, response, and deactivation steps for a signal that binds to an intracellular receptor.
6. What is the significance of the observation that many signal transduction pathways intersect or overlap, creating a network?

✓APPLYING CONCEPTS TO NEW SITUATIONS

Answers are available in Appendix B

1. Suppose that an animal and a plant each lacked the ability to secrete an extracellular matrix. What would these organisms look like?
2. Suppose that a cell-cell signal binds to a membrane-receptor, and the cell responds without signal transduction occurring. Compared to signal transduction pathways, how would an event like this affect (a) the types of responses that are possible, (b) amplification, and (c) regulation?
3. In most species of fungi, chitin is a major polysaccharide found in cell walls. Review the structure of chitin as described in Chapter 5, and then make a sketch predicting the structure of the fungal cell wall.
4. Suppose you created an antibody that bound to the receptor illustrated in Figure 8.16. How would the signal transduction pathway be affected? How would the signal transduction pathway be affected by a drug that bound permanently to the receptor?

When table sugar is heated, it undergoes the uncontrolled oxidation reaction known as burning. Burning gives off heat. In cells, the simple sugar called glucose is oxidized through a long series of carefully regulated reactions. Instead of being given off as heat, some of the energy produced by these reactions is used to synthesize ATP.

9 Cellular Respiration and Fermentation

KEY CONCEPTS

- In cells, the endergonic reactions required for life occur in conjunction with an exergonic reaction involving ATP.
- Cellular respiration produces ATP from molecules with high potential energy—often glucose.
- Cellular respiration has four components: (1) glycolysis, (2) pyruvate processing, (3) the citric acid cycle, and (4) electron transport and chemiosmosis.
- Cellular respiration and fermentation are carefully regulated.
- Fermentation pathways allow glycolysis to continue when the lack of an electron acceptor shuts down electron transport chains.

Cells are dynamic. Vesicles move cargo from the Golgi apparatus to the plasma membrane and other destinations, enzymes synthesize a complex array of macromolecules, and millions of proteins pump ions and molecules across the plasma membrane. These cell activities change constantly in response to signals from other cells or the environment.

What fuels all this action? The answer is **adenosine triphosphate (ATP)**. ATP has high potential energy and allows cells to do work. Because staying alive takes work, there is no life without ATP.

This chapter investigates how cells make adenosine triphosphate, starting from sugars and other compounds that have high potential energy. As cells process sugar, the energy that is released is used to transfer a phosphate group to **adenosine diphosphate (ADP)**, generating ATP.

The chapter begins by reviewing fundamental concepts about energy and introducing how it is used in cells. Sections 9.2 through 9.6 delve into the reactions involved in processing **glucose**, the most common fuel used by organisms, and producing ATP. Section 9.7 introduces fermentation, an alternative route for ATP production that occurs in many cells. Section 9.8 examines how cells divert certain carbon-containing compounds away from ATP production and into the synthesis of DNA, RNA, amino acids, and other molecules.

This chapter is about a key attribute of life: the ability to acquire and use energy. It is also your introduction to **metabolism**—all the chemical reactions that occur in living cells.

✔ When you see this checkmark, stop and test yourself. Answers are available in Appendix B.

9.1 The Nature of Chemical Energy and Redox Reactions

Recall from Chapter 2 that chemical energy is a form of potential energy. Potential energy is energy that is associated with position or configuration. In cells, chemical energy is stored in the position of electrons.

The amount of potential energy in an electron is based on its position relative to other electrons and the protons in the nuclei of nearby atoms. If an electron is close to negative charges on other electrons and far from the positive charges in nuclei, it has high potential energy. In general, the potential energy of a molecule is a function of the way its electrons are configured or positioned.

The Structure and Function of ATP

ATP makes things happen in cells because it has a great deal of potential energy. As **Figure 9.1a** shows, four negative charges are confined to a small area in the three phosphate groups in ATP. In part because these negative charges repel each other, the potential energy of the electrons in the phosphate groups is extraordinarily high.

When ATP reacts with water during a hydrolysis reaction, the bond between ATP's outermost phosphate group and its neighbor is broken, resulting in the formation of ADP and inorganic phosphate, P_i, which has the formula $H_2PO_4^-$ (**Figure 9.1b**). This reaction is highly exergonic. Recall that **exergonic** reactions release energy, while **endergonic** reactions require an input of energy. Under standard conditions of temperature and pressure in the laboratory, a total of 7.3 **kilocalories** of energy per mole of ATP (or 7.3 kcal/mol), is released during the reaction. A kcal of energy raises 1 g of water 1°C.

WHY DOES ATP HYDROLYSIS RELEASE ENERGY? You might remember from Chapter 2 that changes in free energy dictate whether a reaction is exergonic or endergonic. Changes in free energy, in turn, depend on the relationship between reactants and products in terms of potential energy and entropy (disorder).

The hydrolysis of ATP is exergonic because the entropy of the product molecules is much higher than that of the reactants, and because there is a large drop in potential energy when ADP and P_i are formed from ATP. The change in potential energy occurs in part because the electrons from ATP's phosphate groups are now spread between two molecules instead of being clustered on one molecule, meaning that there is less electrical repulsion. In addition, the negative charges on ADP and P_i are stabilized much more efficiently by interactions with the partial positive charges on surrounding water molecules than are the charges on ATP.

WHAT HAPPENS WHEN PROTEINS ARE PHOSPHORYLATED BY ATP? If the reaction diagrammed in Figure 9.1b occurred in a test tube, the energy released would be lost as heat. But cells don't lose that 7.3 kcal/mole as heat. Instead, they use it to make things happen. Specifically, things start to happen when ATP is hydrolyzed and the phosphate group that is released is transferred to a protein.

The addition of a phosphate group to a substrate is called **phosphorylation**. Phosphorylation of proteins is exergonic because the electrons in ADP and the phosphate group have much less potential energy than they did in ATP.

When phosphorylation adds a negative charge to a protein, the electrons in the protein change configuration. The molecule's overall shape, or conformation, usually changes as well.

(a) ATP consists of three phosphate groups, ribose, and adenine.

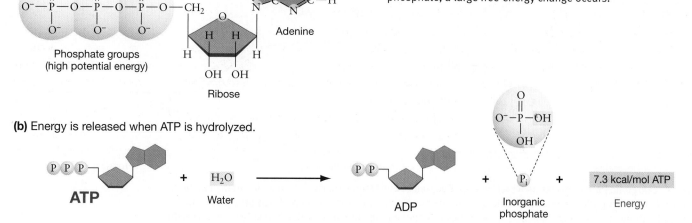

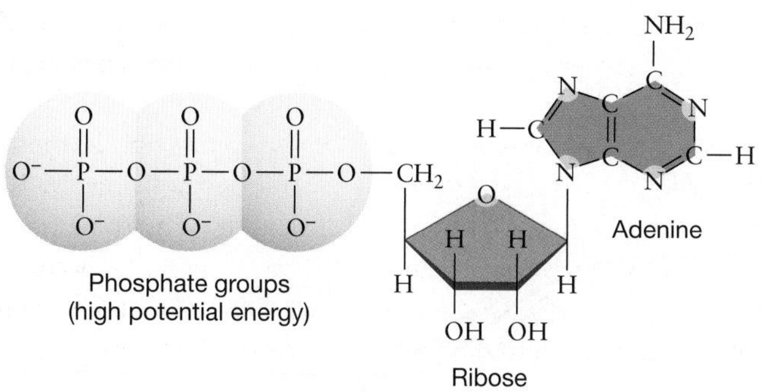

(b) Energy is released when ATP is hydrolyzed.

P P P — ATP + H_2O Water → P P — ADP + P_i Inorganic phosphate + 7.3 kcal/mol ATP Energy

O^-—P—OH (=O, —OH)

FIGURE 9.1 Adenosine Triphosphate (ATP) Has High Potential Energy. (a) ATP's high potential energy results, in part, from the four negative charges clustered in its three phosphate groups. The negative charges repel each other, raising the potential energy of the electrons. **(b)** When ATP is hydrolyzed to ADP and inorganic phosphate, a large free-energy change occurs.

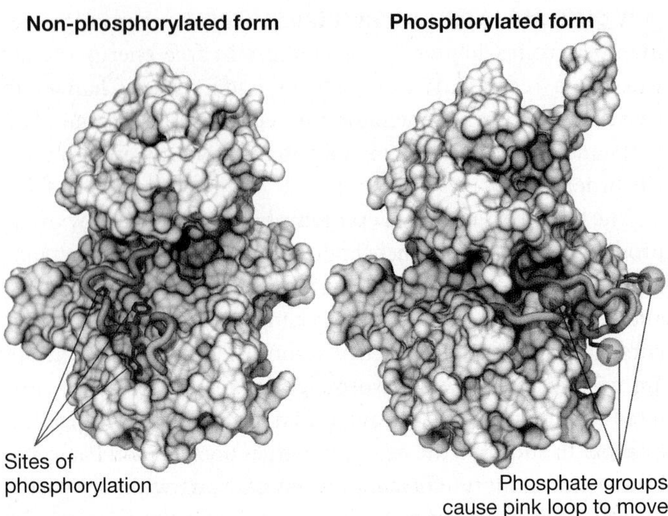

FIGURE 9.2 Phosphorylation Changes the Shape and Activity of Proteins. When proteins are phosphorylated or an ATP molecule binds to them, they often change shape in a way that alters their activity. The figure shows the inactivated and activated forms of an enzyme called insulin receptor tyrosine kinase. Note that three phosphate groups (yellow) activate this molecule.

Part of the protein moves (**Figure 9.2**). Protein movement—either in response to phosphorylation or to binding of an entire ATP molecule—is what transports materials inside cells, powers flagella or cilia, and pumps ions across membranes. It also drives the endergonic reactions required for life.

HOW DOES ATP DRIVE ENDERGONIC REACTIONS? In the time it takes to read this sentence, millions of endergonic reactions have occurred in your cells. This chemical activity is possible because entire ATP molecules or phosphate groups from ATP are being added to reactant molecules or enzymes.

To see how this process works, consider an endergonic reaction between a compound A and compound B that results in a product AB needed by your cells. This reaction can happen only when ATP reacts with the substrate to produce a phosphorylated intermediate molecule and ADP. If the reactant that is phosphorylated is compound B, it is referred to as an activated substrate. Activated substrates contain a phosphate group and have high free energy. This is the critical point: Activated substrates have high enough potential energy that the reaction between compound A and the activated form of compound B is exergonic. The two compounds then go on to react and form the product molecule AB.

In some cases, the enzyme that catalyzes the reaction is phosphorylated instead of a reactant. When either a substrate or an enzyme is phosphorylated, the exergonic phosphorylation reaction is coupled to an endergonic reaction. In many cases, phosphorylation of substrates or enzymes makes the reactions that occur in cells exergonic.

Figure 9.3 graphs how **energetic coupling** between an exergonic and endergonic reaction works. Note, on the far right, that the reaction between A and B to produce the product AB is endergonic—ΔG is positive. But when the exergonic reaction occurs that moves a phosphate group from ATP to B, the free energy of the reactants A and B is now high enough to make the reaction that forms AB exergonic. This is due to "coupling" between phosphorylation reactions and endergonic reactions.

✓If you understand the principles of energetic coupling, consider the following endergonic reaction:

$$\textbf{ribulose} + \textbf{CO}_2 \longrightarrow \textbf{glycerate}$$

Suppose that the enzyme that catalyzes this reaction is next to a ribulose molecule and a CO_2 molecule. You should be able to: (1) Explain why the ribulose does not react with the CO_2, even with the enzyme present. (2) Suppose that 2 ATPs add 2 phosphate groups to the ribulose. Explain why the resulting molecule, called ribulose bisphosphate, is called an activated substrate. (3) The enzyme catalyzes a reaction between the activated substrate and CO_2,

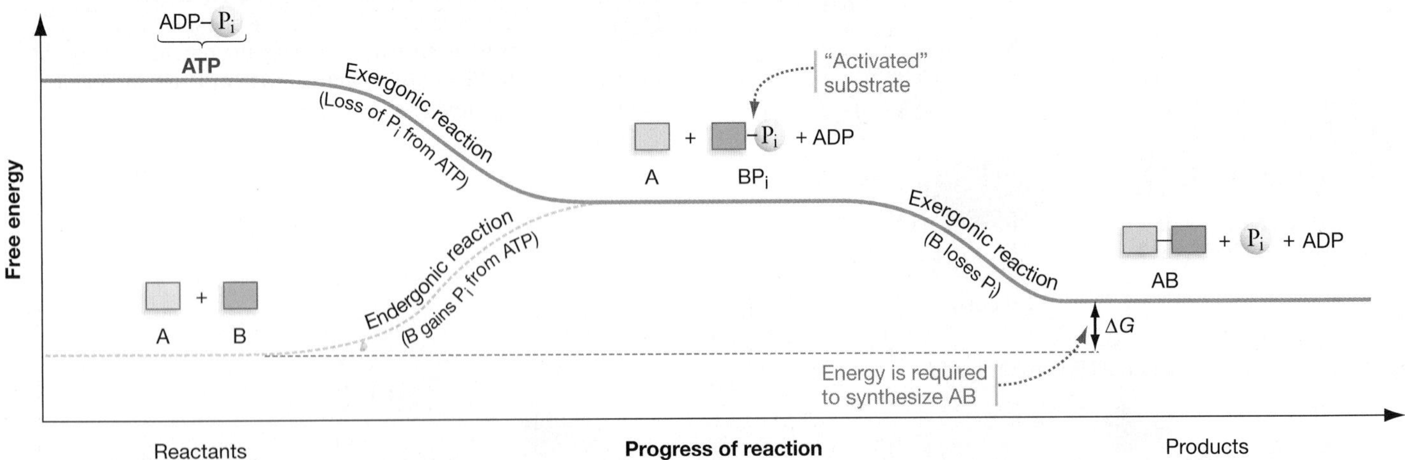

FIGURE 9.3 Exergonic Phosphorylation Reactions Are Coupled to Endergonic Reactions. In cells, many reactions only occur if one reactant (or an enzyme) undergoes phosphorylation—the addition of a phosphate group from ATP. The phosphorylated reactant molecule (or enzyme) has high enough free energy that the subsequent reaction is exergonic.

forming two molecules of glycerate (each of which has a phosphate group attached). Explain why this sequence of events represents the coupling of an exergonic reaction and endergonic reaction.

It is hard to overstate the importance of energetic coupling: Without it, life is impossible. If you ran out of ATP, enzymes and reactants could no longer be phosphorylated and you would die within minutes.

Now the question is, where do cells get ATP in the first place? A great deal of energy is required to synthesize ATP from ADP by adding P_i. Where does this energy come from? The answer is redox reactions.

What Is a Redox Reaction?

Reduction-oxidation reactions, or **redox reactions,** are a class of chemical reactions that involve the loss or gain of one or more electrons. Redox reactions are central in biology because they drive the formation of ATP.

In a redox reaction, the atom that loses one or more electrons is oxidized, and the atom that gains one or more electrons is reduced. To keep these terms straight, chemists use the mnemonic "LEO the lion goes GER"—Loss of Electrons is **Oxidation**; Gain of Electrons is **Reduction**. (An alternative is OIL RIG—Oxidation Is Loss; Reduction Is Gain.)

Oxidation events are always paired with a reduction; if one atom loses an electron, another has to gain it. Stated another way, a reactant that acts as an **electron donor** is always associated with a reactant that acts as an **electron acceptor**.

The gain or loss of an electron can be relative, however. During a redox reaction, an electron can be transferred completely from one atom to another, or an electron can simply shift its position in a covalent bond.

AN EXAMPLE OF REDOX IN ACTION To see how redox reactions work, consider the overall reaction for photosynthesis (**Figure 9.4**). Plants take in carbon dioxide (CO_2) and water (H_2O); and with the aid of sunlight, they synthesize carbohydrate—in this example, the sugar glucose ($C_6H_{12}O_6$)—and release molecular oxygen (O_2) and water. The orange dots in the illustration represent the positions of the electrons involved in covalent bonds.

Now compare the position of the electrons in the first reactant, carbon dioxide, with their position in the first product, glucose. Notice that many of the electrons have moved closer to the carbon nucleus in glucose. This means that carbon has been reduced: it has "gained" electrons. The change occurred because the carbon and oxygen atoms in CO_2 do not share electrons equally, while the carbon and hydrogen atoms in glucose do. In CO_2, the high electronegativity of the oxygen atoms pulled electrons away from the carbon atom.

Now compare the position of the electrons in the reactant water molecules with their position in the O_2 molecules that are produced. In O_2, the electrons have moved farther from the oxygen nuclei than they were in the water molecules, meaning that the oxygen atoms have been oxidized. Oxygen has "lost" electrons. Thus, in photosynthesis, carbon atoms are reduced while oxygen atoms are oxidized.

These shifts in electron position change the amount of chemical energy in the reactants and products. When photosynthesis occurs, electrons are held much more loosely in the product molecules than in the reactant molecules. This means their potential energy has increased. The entropy of the products is also much lower than the reactants. As a result, the reaction is endergonic. It can take place only with an input of energy from sunlight.

ANOTHER APPROACH TO UNDERSTANDING REDOX During the redox reactions that occur in cells, electrons (e^-) are often transferred from an atom in one molecule to an atom in a different molecule. When this occurs, the electron is usually accompanied by H^+.

Molecules that gain a hydrogen (H) atom in this way tend to have high potential energy, because the electrons in C–H bonds are relatively far from the positive charges in a nucleus. This observation should sound familiar, from the introduction to carbohydrates in Chapter 5. Molecules that have a large number of C–H bonds, such as carbohydrates and fats, store a great deal of potential energy.

Conversely, molecules that are oxidized in cells often lose a proton along with an electron. Instead of having many C–H bonds, oxidized molecules in cells tend to have many C–O bonds. Oxidized molecules also have lower potential energy. To understand why, remember that oxygen atoms have extremely

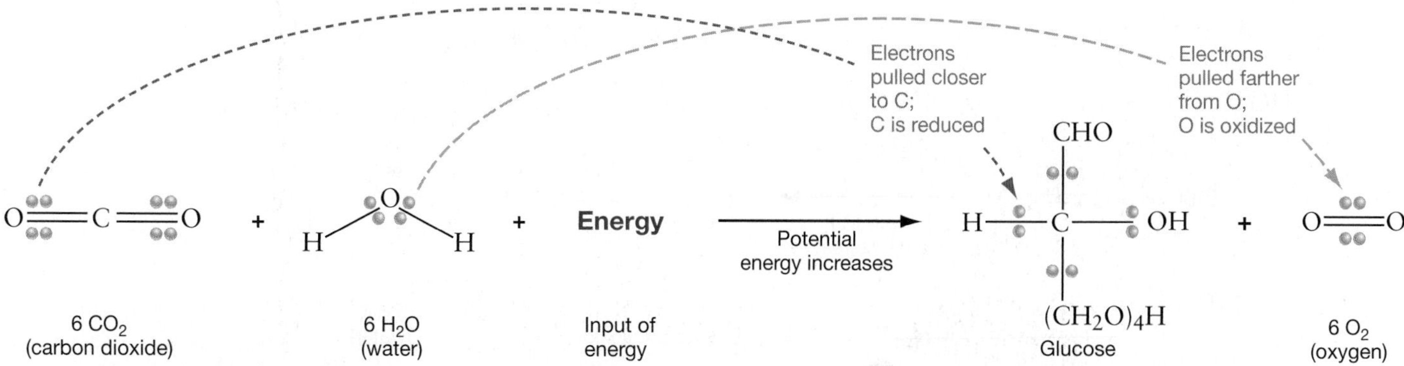

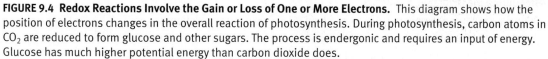

FIGURE 9.4 Redox Reactions Involve the Gain or Loss of One or More Electrons. This diagram shows how the position of electrons changes in the overall reaction of photosynthesis. During photosynthesis, carbon atoms in CO_2 are reduced to form glucose and other sugars. The process is endergonic and requires an input of energy. Glucose has much higher potential energy than carbon dioxide does.

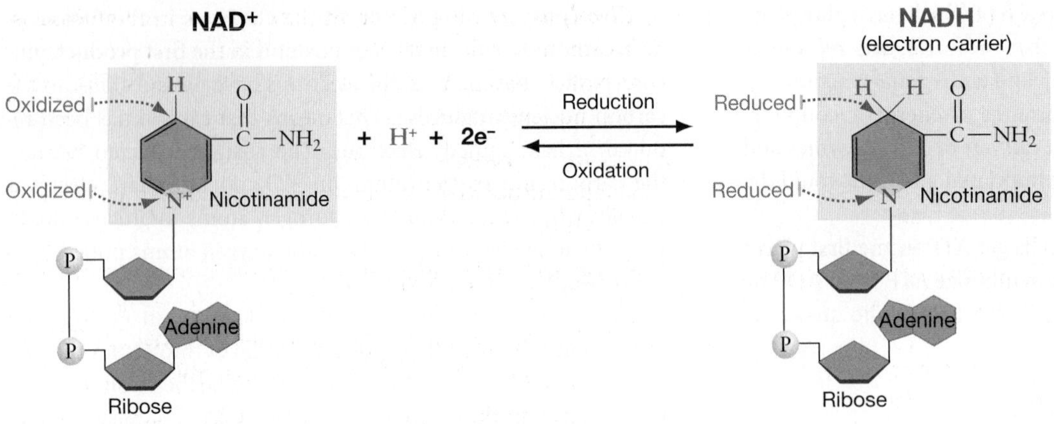

NAD+ ⇌ NADH

Oxidized → H; Oxidized → N^+ Nicotinamide; $-C(=O)-NH_2$ + H^+ + $2e^-$ (Reduction / Oxidation); Reduced → H H; Reduced → N Nicotinamide; P, Adenine, P, Ribose (electron carrier)

FIGURE 9.5 NADH is the reduced form of NAD+. NADH is an important electron carrier.

high electronegativity. Because oxygen atoms hold electrons so tightly, the electrons involved in bonds with oxygen atoms have low potential energy.

In many redox reactions in biology, understanding where oxidation and reduction have occurred becomes a matter of following hydrogen atoms—reduction means "adding H's" and oxidation means "removing H's." A good example is **nicotinamide adenine dinucleotide (NAD+)** which is reduced to form **NADH** (**Figure 9.5**). NADH readily donates electrons to other molecules. As a result, it is called an **electron carrier** and is said to have "reducing power." As you will soon see, NADH is an important electron carrier during cellular respiration.

WHAT HAPPENS WHEN GLUCOSE IS OXIDIZED? To test your understanding of redox reactions, consider what happens when glucose undergoes the uncontrolled oxidation reaction called burning:

$$\underset{\text{glucose}}{C_6H_{12}O_6} + \underset{\text{oxygen}}{6O_2} \longrightarrow \underset{\text{carbon dioxide}}{6\,CO_2} + \underset{\text{water}}{6\,H_2O} + \text{Energy}$$

The photograph at the start of the chapter shows this reaction occurring, and **Figure 9.6** provides an incomplete electron-sharing diagram.

✔If you understand the fundamental principles of reduction-oxidation, you should be able to complete Figure 9.6 (by filling in electron positions) and answer the following questions: (1) Are the carbon atoms in glucose oxidized or reduced? (2) Are the oxygen atoms in the oxygen molecule (O_2) oxidized or reduced? (3) Glucose is the molecule that acts as an electron donor in this reaction. Which molecule acts as the electron acceptor? (4) Which has higher potential energy: the reactants or the products? Based on your answer, add "Energy" to the appropriate side of Figure 9.6 with a label below indicating "Input of energy" or "Release of energy."

Before going on to the next section, you may want to visit the study area at *www.masteringbiology.com* and review the principles of redox reactions.

MB **Web Activity** Redox Reactions

When glucose burns, the change in potential energy is converted to kinetic energy in the form of heat. More specifically, a total of 686 kcal of heat is released when one mole of this sugar is oxidized.

Glucose does not burn in cells, however. Instead, the glucose in cells is oxidized through a long series of carefully controlled redox reactions. These reactions are occurring, millions of times per minute, in your cells right now. Instead of being given off as heat, much of the energy that is released is being used to make the ATP you need to read, think, move, and stay alive. In cells, the change in free energy that occurs during the oxidation of glucose is used to synthesize ATP from ADP and P_i.

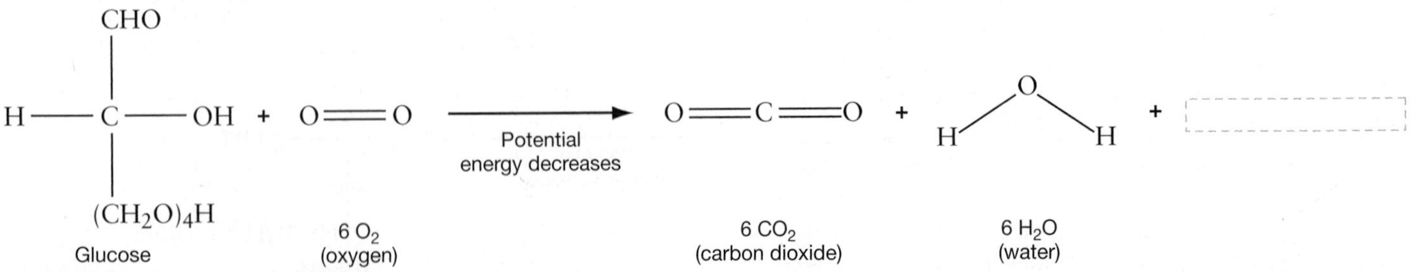

CHO | H—C—OH | $(CH_2O)_4H$ + O═O → O═C═O + H_2O (H–O–H) + [ ]

Glucose; $6\,O_2$ (oxygen); Potential energy decreases; $6\,CO_2$ (carbon dioxide); $6\,H_2O$ (water)

FIGURE 9.6 Tracking Electron Transfer during the Oxidation of Glucose.

✔**EXERCISE** Fill in the electron positions in each bond shown; then use these data to explain why the reaction is exergonic. (Check your work using the electron positions diagrammed in Figure 9.4).

CHECK YOUR UNDERSTANDING

If you understand that . . .

- ATP contains a cluster of three phosphate groups.
- When ATP or phosphate groups from ATP bind to proteins or other molecules, they gain a great deal of potential energy (and often change shape in the process).
- Many of the reactions that occur in cells are endergonic. But if a phosphate group from ATP (or ATP itself) binds to a substrate (or the enzyme that catalyzes a reaction), an exergonic reaction may occur.
- When redox reactions occur, electrons change position. Chemical energy is based on the positions of electrons, so redox reactions usually involve a change in potential energy.

✓ **You should be able to . . .**

1. Explain why ATP has such high potential energy.
2. Explain why molecules with many C—H bonds store more potential energy than molecules with many C—O bonds.

Answers are available in Appendix B.

9.2 An Overview of Cellular Respiration

In general, a cell contains only enough ATP to last from 30 seconds to a few minutes. Because it has such high potential energy, ATP is unstable and is not stored. Like many other cellular processes, the production and use of ATP is fast. Most cells are making ATP all the time.

Most of the glucose that is used to make ATP is produced by plants and other photosynthetic species. These organisms use the energy in sunlight to reduce carbon dioxide (CO_2) to glucose and other carbohydrates. While they are alive, photosynthetic species use the glucose that they produce to make ATP for themselves. When photosynthetic species decompose or are eaten, they provide glucose to animals, fungi, and many bacteria and archaea.

All organisms use glucose as a building block in the synthesis of fats, complex carbohydrates such as starch and glycogen, and other energy-storage compounds. The chemical energy stored in fats and in storage carbohydrates acts like a savings account. ATP, in contrast, is like cash. To make ATP and get cash, fats and storage carbohydrates have to be converted back to glucose. The glucose is then used to produce ATP through one of two general processes: cellular respiration or fermentation (**Figure 9.7**).

Because cellular respiration is much more efficient than fermentation, it is the primary source of ATP in most organisms. You can think of cellular respiration as a four-step process for producing ATP from a starting material with high potential energy—usually glucose. Each of the four steps consists of a series of chemical reactions, and each step has a distinctive starting molecule and a characteristic set of products.

1. ***Glycolysis*** During **glycolysis**, one molecule of glucose is broken into two molecules of the three-carbon compound pyruvate. Two ATP molecules are produced from ADP, and one molecule of NAD^+ is reduced to form NADH.
2. ***Pyruvate processing*** Pyruvate is processed to form the compound acetyl-CoA. During this step, another molecule of NADH is produced.
3. ***Citric acid cycle*** Acetyl-CoA is oxidized to two molecules of CO_2. During this sequence of reactions, more ATP and NADH are produced and **flavine adenine dinucleotide (FAD)** is reduced to form another electron carrier, **$FADH_2$**.
4. ***Electron transport and chemiosmosis*** Electrons from NADH and $FADH_2$ move through a series of proteins called an electron transport chain (ETC). The potential energy released during these redox reactions is used to create a proton gradient across a membrane; the ensuing flow of protons back across the membrane is used to make ATP.

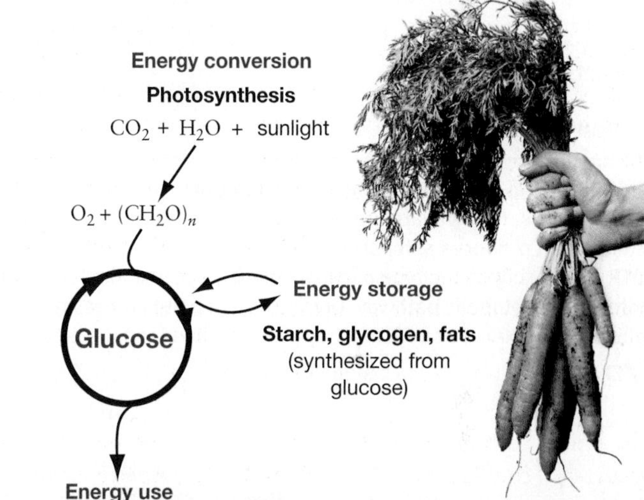

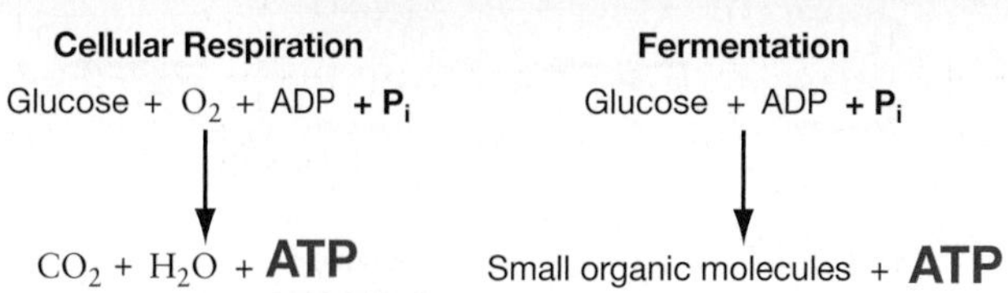

FIGURE 9.7 Glucose Is the Hub of Energy Processing in Cells. Glucose is the end-product of photosynthesis. Both plants and animals store glucose and oxidize it to provide chemical energy in the form of ATP.

Figure 9.8 summarizes the four steps in cellular respiration. Formally, **cellular respiration** is defined as any suite of reactions that produces ATP in an electron transport chain.

When you've filled in the chart at the bottom of the figure, you'll be ready to analyze each of the four steps in detail. As you delve into these details, keep asking yourself the same key questions: What goes in and what comes out? What happens to the potential energy that is released? Where does this step occur, and how is it regulated? Then take a look in the mirror. All of these processes are occurring right now, in virtually all of your cells.

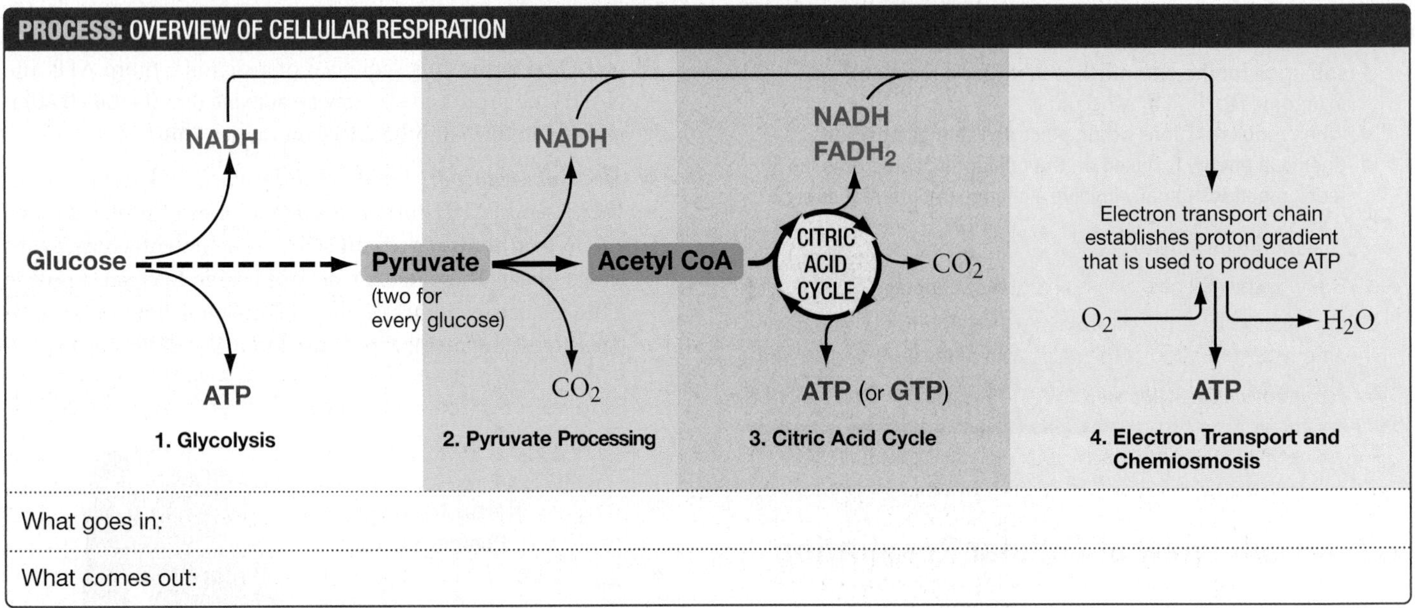

FIGURE 9.8 An Overview of Cellular Respiration. Cells produce ATP from glucose via a series of processes: (1) glycolysis, (2) pyruvate processing, (3) the citric acid cycle, and (4) electron transport and chemiosmosis. Each component produces at least some ATP or NADH (the citric acid cycle produces ATP or a related compound called GTP, depending on the type of cell involved). Because the four components are connected, glucose oxidation is an integrated metabolic pathway. Glycolysis, pyruvate processing, and the citric acid cycle complete the oxidation of glucose. The NADH and $FADH_2$ they produce then feed the electron transport chain.

✔**EXERCISE** Fill in the chart along the bottom.

FIGURE 9.9 Glycolysis Pathway. Glucose is oxidized to pyruvate through this sequence of 10 reactions. Each reaction is catalyzed by a different enzyme. The products are two net ATP (four ATP are produced, but two are invested), two molecules of NADH, and two molecules of pyruvate.

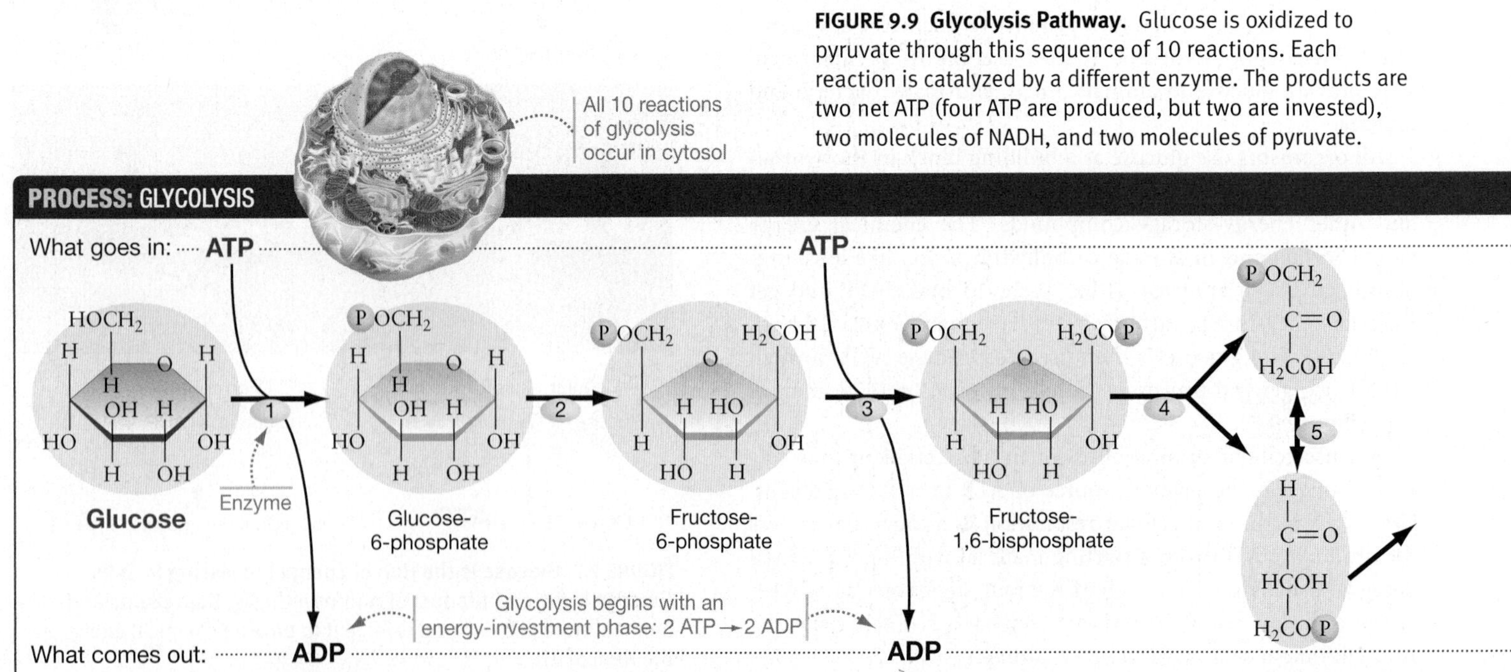

9.3 Glycolysis: Processing Glucose to Pyruvate

Because the enzymes responsible for glycolysis have been observed in nearly every bacterium, archaean, and eukaryote, it is logical to infer that the ancestor of all organisms living today made ATP by glycolysis. It's ironic, then, that the process was discovered by accident.

In the 1890s Hans and Edward Buchner were working out techniques for manufacturing extracts of baker's yeast for commercial and medicinal use. (Yeast extracts are still added to some foods as a flavor enhancer or nutritional supplement.) In one set of experiments the Buchners added sucrose, or table sugar, to their extracts. Sucrose is a disaccharide consisting of glucose and another six-carbon sugar. At the time, sucrose was commonly used as a preservative—a substance used to preserve food from decay.

Instead of preserving the yeast extracts, though, the sucrose was quickly broken down and fermented, with alcohol appearing as a by-product. This was a key finding: It showed that fermentation and other types of cellular metabolism could be studied in vitro—outside the organism. Until then, researchers thought that metabolism could take place only in intact organisms.

When researchers studied how the sugar was being processed, they found that the reactions could go on much longer than normal if inorganic phosphate was added to the mixture. This result implied that some of the compounds involved were being phosphorylated. Soon after, a molecule called fructose bisphosphate was isolated. (The prefix *bis–* means that two phosphate groups are attached to the fructose molecule at distinct locations.) Subsequent work showed that all but two of the compounds involved in glycolysis—the starting and ending molecules, glucose and pyruvate—are phosphorylated.

In 1905 researchers found that the processing of sugar by yeast extracts stopped if they boiled the reaction mix. Because enzymes were known to be inactivated by heat, this discovery suggested that enzymes were involved in at least some of the processing steps. Years later, investigators realized that each step in glycolysis is catalyzed by a different enzyme. Eventually, each of the reactions and enzymes involved was gradually worked out.

Glycolysis Is a Sequence of 10 Reactions

All 10 reactions of glycolysis occur in the cytosol (**Figure 9.9**). Note three key points about this reaction sequence:

1. Glycolysis starts by *using* ATP, not producing it. In the initial step, glucose is phosphorylated to form glucose-6-phosphate. After an enzyme rearranges this molecule to fructose-6-phosphate in the second reaction, the third reaction adds a second phosphate group, forming the fructose-1,6-bisphosphate observed by early researchers. Thus, two ATP molecules are used up before any ATP is produced.
2. Once this energy-investment phase of glycolysis is complete, the subsequent reactions represent an energy payoff phase. The sixth reaction in the sequence results in the reduction of two molecules of NAD^+; the seventh produces two molecules of ATP. This is where the energy "debt"—of two molecules of ATP invested early in glycolysis—is paid off. The final reaction in the sequence produces another two ATPs. For each molecule of glucose processed, the net yield is two molecules of NADH, two of ATP, and two of pyruvate.
3. In reactions 7 and 10 of Figure 9.9, an enzyme catalyzes the transfer of a phosphate group from a phosphorylated substrate to ADP, forming ATP. Enzyme-catalyzed reactions that result

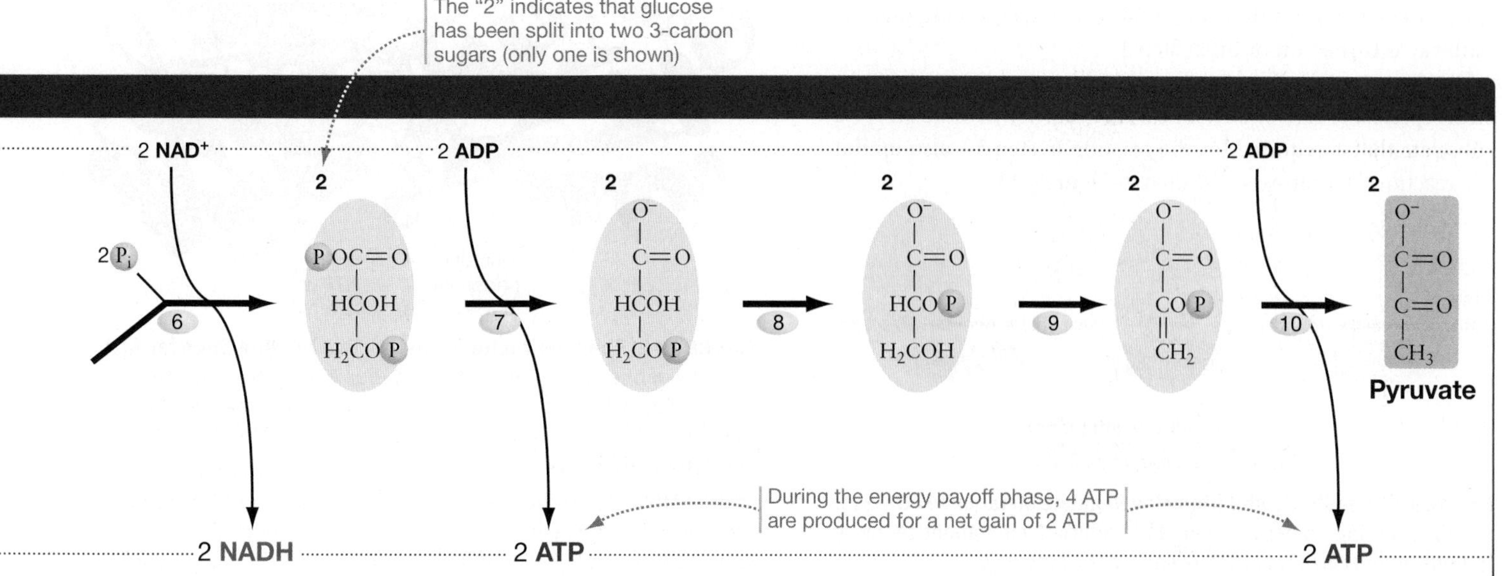

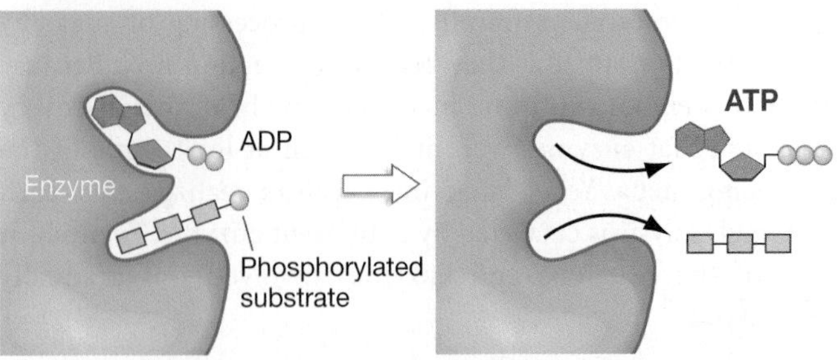

FIGURE 9.10 Substrate-Level Phosphorylation Involves an Enzyme and a Phosphorylated Substrate. Substrate-level phosphorylation occurs when an enzyme catalyzes the transfer of a phosphate group from a phosphorylated substrate to ADP, forming ATP.

in ATP production are termed **substrate-level phosphorylation** (**Figure 9.10**). The key thing to note here is that the energy to produce the ATP comes from the phosphorylated substrate—not from a proton gradient, as it does when ATP is produced by an electron transport chain.

The discovery and elucidation of the glycolytic pathway ranks as one of the great achievements in the history of biochemistry. The reactions outlined in Figure 9.9 are among the most ancient and fundamental of all life processes.

How Is Glycolysis Regulated?

An important advance occurred when biologists observed that high levels of ATP inhibit a key glycolytic enzyme called phosphofructokinase. **Phosphofructokinase** catalyzes reaction 3 in Figure 9.9—the synthesis of fructose-1,6-bisphosphate from fructose-6-phosphate. This is a crucial step in the sequence.

After reactions 1 and 2 occur, an array of enzymes can convert the products to molecules used in other metabolic pathways. Before step 3, then, the sequence can be interrupted and the intermediates used elsewhere in the cell. But once fructose-1,6-bisphosphate is synthesized, there is no point in stopping the process. Based on these observations, it makes sense that the pathway is turned on or off at step 3.

In the vast majority of cases, though, the addition of a substrate *speeds* the rate of a chemical reaction. Why would ATP—a substrate that is required for the reaction in step 3—also inhibit the reaction? The answer is shown in **Figure 9.11**.

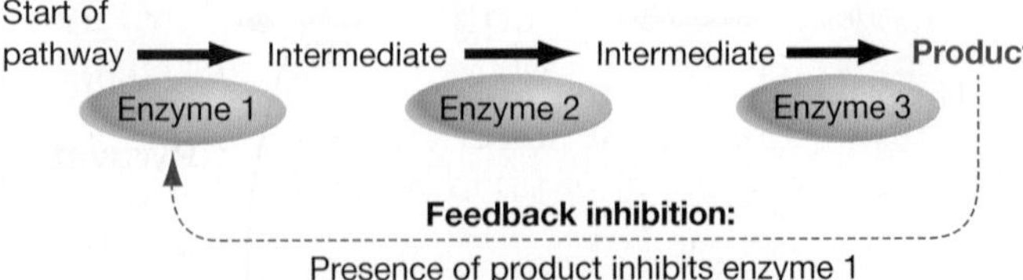

FIGURE 9.11 Feedback Inhibition May Regulate Metabolic Pathways. Feedback inhibition occurs when the product of a metabolic pathway inhibits an enzyme that is active early in the pathway.

When an enzyme in a pathway is inhibited by the product of the reaction sequence, **feedback inhibition** occurs. The product molecule "feeds back" to stop the reaction sequence when the product is abundant.

Feedback inhibition is efficient. Cells that are able to stop glycolytic reactions when ATP is abundant can conserve their stores of glucose for times when ATP is scarce. As a result, natural selection should favor individuals who have phosphofructokinase molecules that are inhibited by high concentrations of ATP.

How do high levels of the substrate inhibit the enzyme? As **Figure 9.12** shows, phosphofructokinase has two distinct binding sites for ATP. ATP can bind at the enzyme's active site or at a location that changes the enzyme's activity—a **regulatory site**.

At the active site, ATP is converted to ADP and the phosphate group is transferred to fructose-6-phosphate. This reaction results in the synthesis of fructose-1,6-bisphosphate.

If ATP concentrations are high, however, the molecule also binds at the regulatory site on phosphofructokinase. When ATP binds at this second location, the enzyme's conformation changes in a way that dramatically lowers the reaction rate at the active site. In this case, ATP acts as an allosteric regulator (see Chapter 3).

Now the question is, what happens to the pyruvate?

9.4 Processing Pyruvate to Acetyl CoA

In eukaryotes, the pyruvate produced by glycolysis is transported from the cytosol to mitochondria. As Chapter 7 noted, mitochondria are organelles found in virtually all eukaryotes.

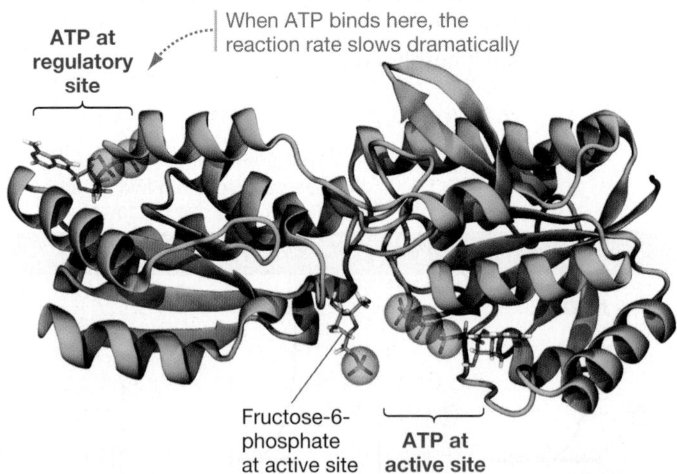

FIGURE 9.12 Phosphofructokinase Has Two Binding Sites for ATP. A model of one of the four identical subunits of phosphofructokinase. Notice the active site, where a phosphate group will be transferred from ATP to fructose-6-phosphate, and the regulatory site, where ATP binds.

✔**QUESTION** The active site has much higher affinity for ATP than the regulatory site does. What would be the consequences if the regulatory site had higher affinity for ATP than the active site did?

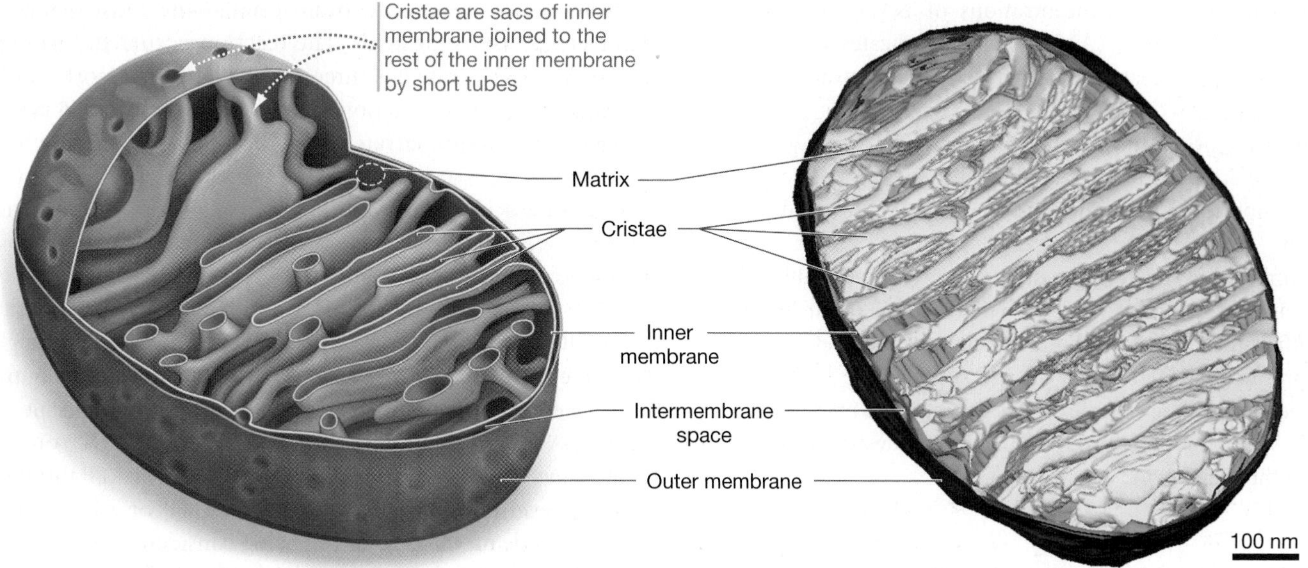

FIGURE 9.13 The Structure of the Mitochondrion. These images are based on recent research using cryo-electron tomography (the micrograph, on the right, has been colorized). Notice that the mitochondria have outer and inner membranes, and the inner membrane is connected by short tubes to sac-like cristae. Pyruvate processing occurs within the mitochondrial matrix.

Figure 9.13 shows a diagram and an image of this organelle generated with an imaging technique called cryo-electron tomography.[1] Notice that a mitochondrion has two membranes, called the inner membrane and outer membrane. The interior of the organelle is filled with layers of sac-like structures called **cristae**. Short tubes connect the cristae to the main part of the inner membrane. The region inside the inner membrane but outside the cristae is the **mitochondrial matrix**.

Pyruvate moves across the mitochondrion's outer membrane through small pores. Entry into the matrix occurs through a membrane protein called the pyruvate carrier, located in the inner membrane. Transport into the matrix is an active process. It requires ATP and represents an energy-consuming step in glucose oxidation.

Inside the mitochondrion, pyruvate reacts with a compound called **coenzyme A (CoA)**. In this and many other reactions, CoA acts as a coenzyme by accepting and then transferring an acetyl group ($-COCH_3$) to a substrate (the "A" stands for acetylation). Pyruvate reacts with CoA, through a series of steps, to produce **acetyl CoA**.

The reaction sequence occurs inside an enormous and intricate enzyme complex called **pyruvate dehydrogenase.** In eukaryotes, pyruvate dehydrogenase is located in the mitochondrial matrix. In bacteria and archaea, pyruvate dehydrogenase is located in the cytosol.

[1]Compared to images from transmission electron microscopy, cryo-electron tomography has provided a much more accurate picture of mitochondrial morphology. For a recent review, see C. Mannella, *Biochemica et Biophysica Acta* 1762 (2006):140–147.

As pyruvate is being processed, NAD^+ is reduced to NADH and one of the carbons in the pyruvate is oxidized to CO_2. The remaining two-carbon acetyl unit is transferred to CoA (**Figure 9.14**). Acetyl CoA is the final product of the pyruvate processing step in glucose oxidation. Pyruvate, NAD^+, and CoA go in; CO_2, NADH, and acetyl CoA come out.

When supplies of ATP are abundant, however, the process shuts down. Pyruvate processing stops when the pyruvate dehydrogenase complex becomes phosphorylated and changes shape. The rate of phosphorylation increases when other products—specifically acetyl CoA and NADH—are at high concentration.

These regulatory changes are more examples of feedback inhibition. Reaction products feed back to stop or slow down the pathway.

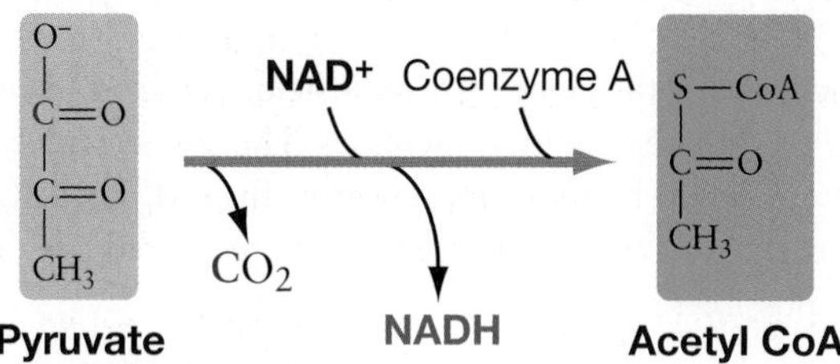

FIGURE 9.14 Pyruvate Is Oxidized to Acetyl CoA. The reaction shown here is catalyzed by pyruvate dehydrogenase.

✔**EXERCISE** Above the reaction arrow, list three molecules whose presence speeds up the reaction. Label them "Positive control." Below the reaction arrow, list three molecules whose presence slows down the reaction. Label them "Negative control by feedback inhibition."

On the contrary, high concentrations of NAD^+, CoA, or adenosine monophosphate (AMP)—which indicates low ATP supplies—*speed up* the reactions catalyzed by the pyruvate dehydrogenase complex.

Pyruvate processing is under both positive and negative control. Large supplies of products inhibit the enzyme complex; large supplies of reactants and low supplies of products stimulate it.

To summarize, pyruvate processing starts with pyruvate and ends with acetyl CoA, releasing CO_2. The reactions occur in the mitochondrial matrix, and the potential energy that is released is used to produce NADH. When energy supplies are high, the pyruvate dehydrogenase complex slows down; when energy supplies are low, the complex speeds up. Now the question is, what happens to the acetyl CoA?

9.5 The Citric Acid Cycle: Oxidizing Acetyl CoA to CO_2

While researchers were working out the sequence of reactions in glycolysis, biologists in other laboratories were focusing on a different set of redox reactions that take place in cells. These reactions involve small organic acids such as citrate, malate, and succinate. Because they have the form R-COOH, these molecules are called **carboxylic acids.**

In some cases, the redox reactions that produce carboxylic acids also produce carbon dioxide. Recall from Section 9.1 that carbon dioxide is a highly oxidized form of carbon and the endpoint of glucose metabolism. Thus, it was logical for researchers to propose that the oxidation of small carboxylic acids could be an important component of glucose oxidation.

Early researchers made three key observations about these reactions:

1. A total of eight small carboxylic acids are oxidized rapidly enough to imply that they are involved in glucose metabolism—the most rapid set of oxidation reactions predicted to occur in cells.
2. The eight carboxylic acids can react in sequence, from least to most oxidized.
3. When one of the eight carboxylic acids is added to cells, the rate of glucose oxidation increases. The added molecules do not appear to be used up, however. Instead, virtually all of the carboxylic acids added seem to be recovered later. How is this possible?

Hans Krebs solved the mystery when he realized that the reaction sequence might occur in a cycle instead of a linear pathway. Krebs had another crucial insight when he suggested that the reaction sequence was directly tied to the processing of pyruvate produced by glycolysis.

To test these hypotheses, Krebs and a colleague set out to determine whether pyruvate—the endpoint of the glycolytic pathway—could react with oxaloacetate—the most oxidized of the eight carboxylic acids. The reaction occurred, forming citrate. Because citrate has three carboxyl groups, most biologists now refer to the reaction sequence as the **citric acid cycle,** because it starts with citrate, which becomes citric acid when protonated.

The citric acid cycle is also known as the tricarboxylic acid (TCA) cycle, because it involves acids with three carboxyl groups, and the Krebs cycle, after its discoverer.

When radioactive isotopes of carbon became available in the early 1940s, researchers showed that carbon atoms cycle through the sequence of reactions just as Krebs had proposed (**Figure 9.15**). The energy released by the oxidation of one molecule of acetyl CoA is used to produce three molecules of NADH, one of $FADH_2$, and one of **guanosine triphosphate (GTP)** or ATP through substrate-level phosphorylation. Whether GTP or ATP is produced depends on the type of cell being considered.[2] For example, GTP appears to be produced in the liver cells of mammals while ATP is produced in muscle cells.

In bacteria and archaea, the enzymes responsible for the citric acid cycle are located in the cytosol. In eukaryotes, most of the enzymes responsible for the citric acid cycle are located in the mitochondrial matrix. Because glycolysis produces two molecules of pyruvate, the cycle turns twice for each molecule of glucose processed in cellular respiration.

How Is the Citric Acid Cycle Regulated?

By now, it shouldn't surprise you to learn that the citric acid cycle is carefully regulated. Reaction rates are high when ATP is scarce; reaction rates are low when ATP is abundant.

Figure 9.16 highlights the major control points. Notice that the enzyme that converts acetyl CoA to citrate is shut down when ATP binds to it. This is another example of feedback inhibition: Reaction products feed back to stop or slow down the pathway.

As Figure 9.16 indicates, feedback inhibition also occurs at two points later in the cycle. At the first of these two points, NADH binds to the enzyme's active site. This is an example of competitive inhibition (see Chapter 3). At the second point, ATP binds to an allosteric regulatory site.

The citric acid cycle can be turned off at multiple points, via several different mechanisms of feedback inhibition.

To summarize, the TCA cycle starts with acetyl and ends with CO_2. The reactions occur in the mitochondrial matrix, and the potential energy that is released is used to produce NADH, $FADH_2$, and ATP or GTP. When energy supplies are high, the cycle slows down. But a major question remains.

[2]Traditionally it was thought that the citric acid cycle produced GTP, which was later converted to ATP in the same cell. Recent work suggests that ATP is produced directly in some cell types, while GTP is produced in other cells. See C. O. Lambeth, *Biochemistry and Molecular Biology Education* 34 (2006): 21–29.

PROCESS: CITRIC ACID CYCLE

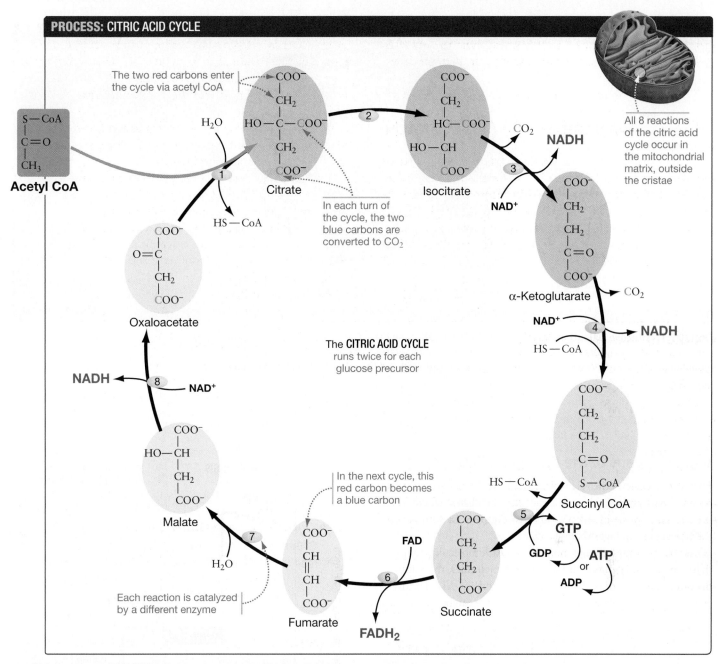

The two red carbons enter the cycle via acetyl CoA

Acetyl CoA

H_2O

1

$HS-CoA$

Citrate

2

Isocitrate

CO_2

NADH

3

NAD^+

In each turn of the cycle, the two blue carbons are converted to CO_2

All 8 reactions of the citric acid cycle occur in the mitochondrial matrix, outside the cristae

α-Ketoglutarate

CO_2

NAD^+

4

NADH

$HS-CoA$

Oxaloacetate

The **CITRIC ACID CYCLE** runs twice for each glucose precursor

NADH

8

NAD^+

Succinyl CoA

$HS-CoA$

5

GTP

GDP

ATP

or

ADP

Malate

In the next cycle, this red carbon becomes a blue carbon

7

H_2O

Fumarate

FAD

6

$FADH_2$

Succinate

Each reaction is catalyzed by a different enzyme

FIGURE 9.15 The Citric Acid Cycle Completes the Oxidation of Glucose. Acetyl CoA goes into the citric acid cycle, and carbon dioxide, NADH, $FADH_2$, and GTP come out. GTP or ATP are produced by substrate-level phosphorylation. If you follow individual carbon atoms around the cycle several times, you'll come to an important conclusion: each of the carbons in this cycle is eventually a "blue carbon" that is released as CO_2. This occurs because the "red carbons" in fumarate—which is symmetric—can flip position when malate forms.

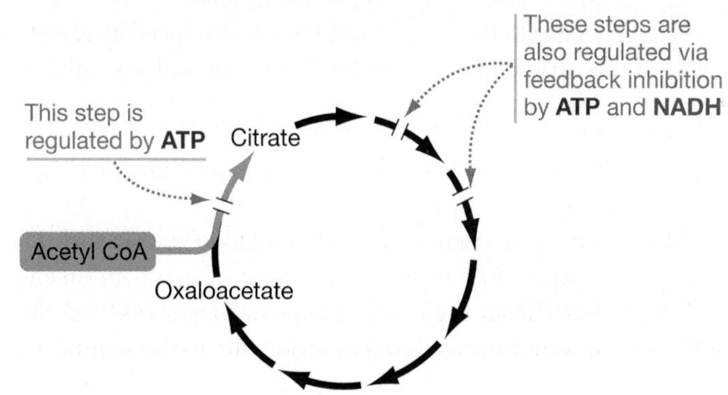

FIGURE 9.16 The Citric Acid Cycle Is Regulated by Feedback Inhibition. The citric acid cycle slows down when ATP and NADH are plentiful. ATP acts as an allosteric regulator, while NADH acts as a competitive inhibitor.

✔**QUESTION** How do allosteric regulation and competitive inhibition differ?

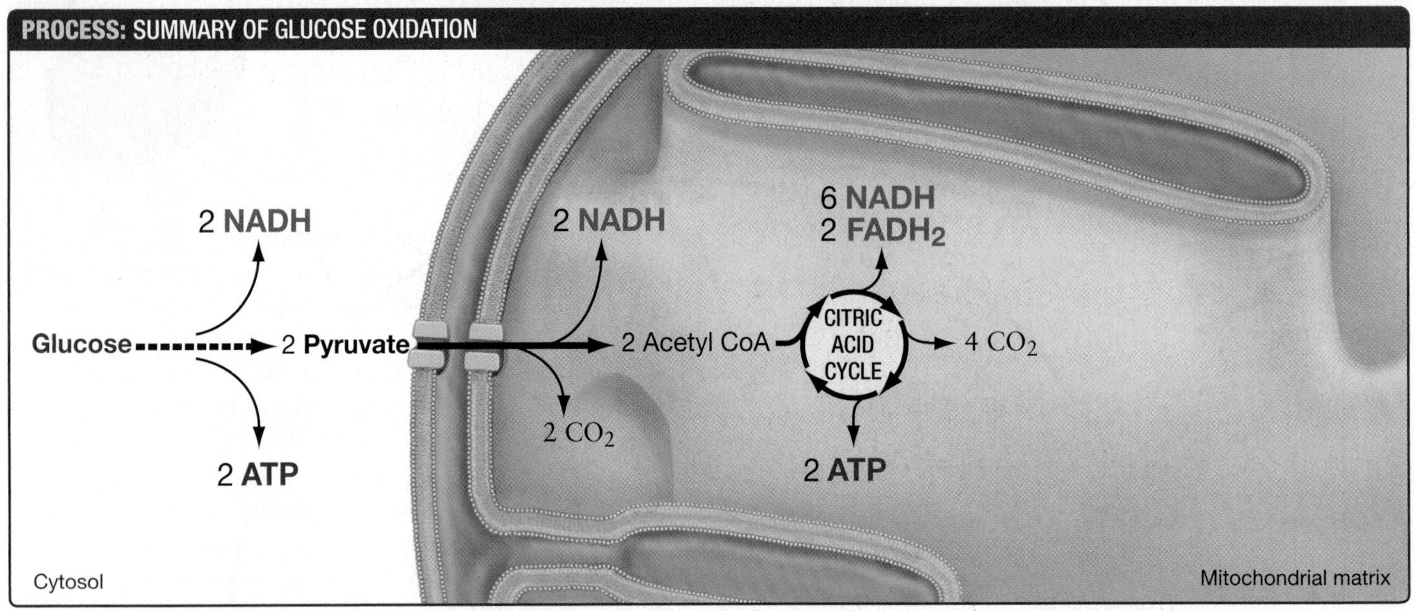

FIGURE 9.17 Glucose Oxidation Produces ATP, NADH, $FADH_2$, and CO_2. Glucose is completely oxidized to carbon dioxide via glycolysis, the subsequent oxidation of pyruvate, and then the citric acid cycle. In eukaryotes, glycolysis occurs in the cytosol; pyruvate oxidation and the citric acid cycle take place in the mitochondrial matrix.

What Happens to the NADH and $FADH_2$?

Figure 9.17 reviews the relationships of glycolysis, pyruvate processing, and the citric acid cycle and identifies where each process takes place in eukaryotic cells; **Figure 9.18** summarizes the free energy changes that take place.

As you study these figures, note that for each molecule of glucose that is fully oxidized to 6 carbon dioxide molecules, the cell produces 10 molecules of NADH, 2 of $FADH_2$, and 4 of ATP. The overall reaction for glycolysis and the citric acid cycle can be written as

$$C_6H_{12}O_6 + 10\ NAD^+ + 2\ FAD + 4\ ADP + 4\ P_i \longrightarrow 6\ CO_2 + 10\ NADH + 2\ FADH_2 + 4\ ATP$$

The ATP molecules are produced by substrate-level phosphorylation and can be used to drive endergonic reactions, power movement, or run membrane pumps. The carbon dioxide molecules are a gas that is disposed of as waste—you exhale it; plants release it or use it as a reactant in photosynthesis.

What happens to the NADH and $FADH_2$ produced by glycolysis and the citric acid cycle? Recall that the overall reaction for glucose oxidation is

$$C_6H_{12}O_6 + 6\ O_2 \longrightarrow 6\ CO_2 + 6\ H_2O + \text{Energy}$$

Glycolysis and the citric acid cycle account for the glucose, the CO_2, and—because ATP is produced—some of the chemical energy that results from the overall reaction. But the O_2 and the H_2O that appear in the overall reaction for the oxidation of

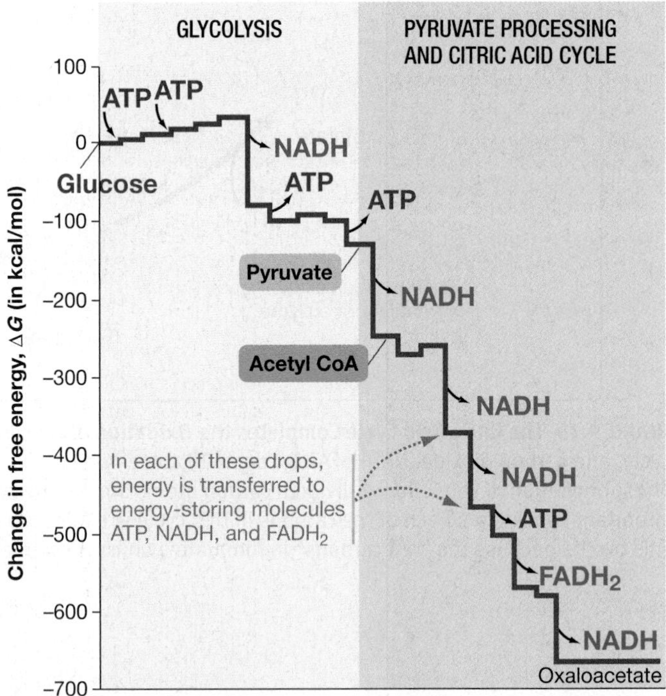

FIGURE 9.18 Free Energy Changes as Glucose Is Oxidized. If you read the vertical axis of this graph carefully, it should convince you that about 125 kcal/mol are released during glycolysis, about 125 kcal/mol during pyruvate processing, and about 430 kcal/mol during the citric acid cycle.

✔**QUESTION** Which is associated with larger changes in free energy, production of ATP or production of NADH and $FADH_2$?

glucose are still unaccounted for. As it turns out, so is much of the chemical energy. The reaction that has yet to occur is

$$NADH + FADH_2 + O_2 + ADP + P_i \longrightarrow NAD^+ + FAD + H_2O + ATP$$

In this reaction, oxygen is reduced to form water. The electrons that drive the redox reaction come from NADH and $FADH_2$. These molecules are oxidized to NAD^+ + FAD.

In effect, glycolysis, pyruvate processing, and the citric acid cycle transfer electrons from glucose to NAD^+ and FAD, creating NADH and $FADH_2$. These molecules then carry the electrons to oxygen, which serves as the final electron acceptor in eukaryotic cells. When oxygen accepts electrons, water is produced. All the components of the overall reaction for glucose oxidation are accounted for.

How does this final part of the process occur? Specifically, how is ATP generated as electrons are transferred from NADH or $FADH_2$ to O_2? In the 1960s—decades after the details of glycolysis and the citric acid cycle had been worked out—a startling answer to these questions emerged.

CHECK YOUR UNDERSTANDING

If you understand that . . .

- During glycolysis, glucose is oxidized to pyruvate, in the cytosol.
- During pyruvate processing, pyruvate is oxidized to acetyl CoA, in the mitochondrial matrix.
- In the citric acid cycle, citric acid is oxidized to carbon dioxide (CO_2), in the mitochondrial matrix.
- Glycolysis, pyruvate processing, and the citric acid cycle are all regulated processes. The cell only produces ATP when ATP is needed.

✓ You should be able to . . .

Model the following components of cellular respiration by pretending that a large piece of paper is a cell. Draw a large mitochondrion inside it. Cut out small squares of paper and label them as glucose, glycolytic reactions, citric acid cycle reactions, pyruvate dehydrogenase complex, pyruvate, acetyl CoA, CO_2, ADP $\longrightarrow$ ATP, $NAD^+ \longrightarrow$ NADH, FAD $\longrightarrow FADH_2$.

1. Put each of the squares in the appropriate location in the cell.
2. Draw arrows to connect the appropriate molecules and reactions.
3. Using dimes or paper circles for electrons, show how a total of 12 electrons from glucose are transferred to NADH or $FADH_2$ (two electrons should go to each NADH or $FADH_2$ formed) as glucose is oxidized to CO_2.
4. Label points where regulation occurs.

Once your model is working, you'll be ready to consider what happens to the NADH and $FADH_2$ you've produced.

Answers are available in Appendix B.

9.6 Electron Transport and Chemiosmosis: Building a Proton Gradient to Produce ATP

The answer to one fundamental question about the oxidation of NADH and $FADH_2$ turned out to be relatively straightforward. By isolating different components of mitochondria, researchers determined that NADH is oxidized in the inner membrane of the mitochondria and the membranes of cristae. In prokaryotes, the oxidation of NADH occurs in the plasma membrane.

Biologists who analyzed the components of these membranes made a key discovery when they isolated molecules that switch between a reduced and an oxidized state during respiration. The molecules were hypothesized to be the key to processing NADH and $FADH_2$. What are these molecules, and how do they work?

Components of the Electron Transport Chain

Collectively, the molecules responsible for the oxidation of NADH and $FADH_2$ are designated the **electron transport chain (ETC)**. As electrons are passed from one protein to another in the chain, the energy released by the redox reactions is used to pump protons across the inner membrane of mitochondria.

After this proton gradient is established, a stream of protons through the enzyme **ATP synthase** makes part of the protein spin, driving the production of ATP from ADP and P_i. Because this mode of ATP production links the phosphorylation of ADP with the oxidation of NADH and $FADH_2$, it is called **oxidative phosphorylation.**

Recall that when substrate-level phosphorylation occurs, a phosphate group is transferred from a phosphorylated substrate to ADP, forming ATP. This is not what happens when ATP synthase spins and synthesizes ATP from ADP and P_i.

Once the electrons at the bottom of the ETC are accepted by oxygen to form water, the oxidation of glucose is complete.

Several points are fundamental to understanding how the ETC works:

- Most of the molecules are proteins that contain distinctive chemical groups where the redox events take place. The active groups include ring-containing structures called flavins or iron-sulfur complexes or iron-containing heme groups. Each of these groups is readily reduced or oxidized.
- The inner membrane of the mitochondrion also contains a molecule called **ubiquinone**, which is not a protein. Ubiquinone got its name because it is nearly ubiquitous in organisms and belongs to a family of compounds called quinones. Also called **coenzyme Q** or simply **Q**, ubiquinone consists of a carbon-containing ring attached to a long tail made up of isoprene subunits. The structure of Q determines the molecule's function. The long, isoprene-rich tail is hydrophobic. As a result, Q is lipid soluble and can move throughout the mitochondrial membrane efficiently. In contrast, all but one of the proteins in the ETC are embedded in the membrane.

- The molecules involved in processing NADH and $FADH_2$ differ in electronegativity, or their tendency to hold electrons.

Because Q and the ETC proteins can cycle between a reduced state and an oxidized state, and because they differ in electronegativity, investigators realized that it should be possible to arrange these molecules into a logical sequence. The idea was that electrons would pass from a molecule with lower electronegativity to one with higher electronegativity, via a redox reaction.

As electrons moved through the chain, they would be held more and more tightly. A small amount of energy would be released in each reaction, and the potential energy in each successive bond would lessen.

Researchers worked out the sequence of compounds in the chain by experimenting with poisons that inhibit particular proteins in the inner membrane. For example, when an electron transport chain is treated with the drug antimycin A, cytochrome *b* and Q are reduced; but all of the other elements in the chain remain oxidized. This pattern only makes sense if electrons flow from NADH and $FADH_2$ to cytochrome *b* and Q before being passed on to other components.

Experiments with other poisons showed that NADH donates an electron to a flavin-containing protein at the top of the chain, while $FADH_2$ donates electrons to an iron-and sulfur-containing protein that then passes them directly to Q. After passing through each of the remaining components in the chain, the electrons are finally accepted by oxygen.

Figure 9.19 shows how electrons step down in potential energy from the electron carriers NADH and $FADH_2$ to O_2. The *x*-axis plots the sequence of redox reactions in the ETC; the *y*-axis plots the free energy changes that occur. Under standard conditions of temperature and pressure in the laboratory, the total potential energy difference from NADH to oxygen is a whopping 53 kilocalories/mole.

Once the nature of the electron transport chain became clear, biologists understood the fate of the electrons carried by NADH and $FADH_2$ and how oxygen acts as the final electron acceptor. All of the electrons that were originally present in glucose were now accounted for. This is satisfying, except for one crucial question: How is ATP produced?

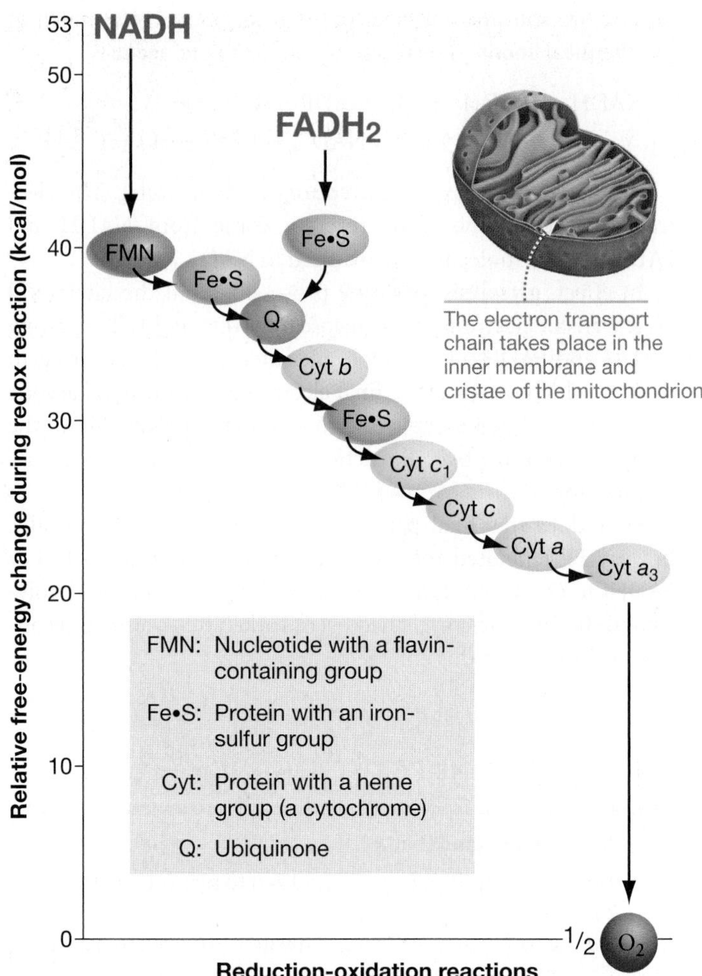

FIGURE 9.19 A Series of Reduction-Oxidation Reactions Occurs in an Electron Transport Chain. Electrons step down in potential energy from the electron carriers NADH and $FADH_2$ through an electron transport chain to a final electron acceptor. When oxygen is the final electron acceptor, water is formed. The overall free-energy change of 53 kcal/mol (from NADH to oxygen) is broken into small steps.

The Chemiosmosis Hypothesis

Throughout the 1950s most biologists working on cellular respiration assumed that electron transport chains include enzymes that catalyze substrate-level phosphorylation. Despite intense efforts, however, no one was able to find a component of the ETC that phosphorylated ADP to produce ATP.

In 1961 Peter Mitchell broke with prevailing ideas by proposing that the connection between electron transport and ATP production is indirect. Mitchell's novel hypothesis? The real job of the electron transport chain is to pump protons from the matrix of the mitochondrion through the inner membrane and out to the intermembrane space or the interior of cristae.

According to Mitchell, the pumping activity of the electron transport chain would lead to a buildup of protons in these areas. In this way, the intermembrane space and the inside of cristae would become positively charged relative to the matrix and would have a much higher concentration of protons. The result would be a strong electrochemical gradient favoring the movement of protons back into the matrix. He hypothesized that an enzyme in the inner membrane uses this **proton-motive force** to synthesize ATP.

Mitchell called the production of ATP via a proton gradient **chemiosmosis**. Although proponents of a direct link between electron transport and substrate-level phosphorylation objected vigorously to Mitchell's idea, several key experiments supported it.

Figure 9.20 illustrates how the existence of a key element in Mitchell's hypothesis was confirmed: A mitochondrial enzyme can use a proton gradient to synthesize ATP. The researchers made vesicles from artificial membranes that contained an ATP-synthesizing enzyme found in mitochondria. Along with this en-

EXPERIMENT

QUESTION: How are the electron transport chain and ATP production linked?

CHEMIOSMOTIC HYPOTHESIS: The linkage is indirect. The ETC creates a proton-motive force that drives ATP synthesis by a mitochondrial protein.

ALTERNATIVE HYPOTHESIS: The linkage is direct. The ETC is associated with enzymes that perform substrate-level phosphorylation.

EXPERIMENTAL SETUP:

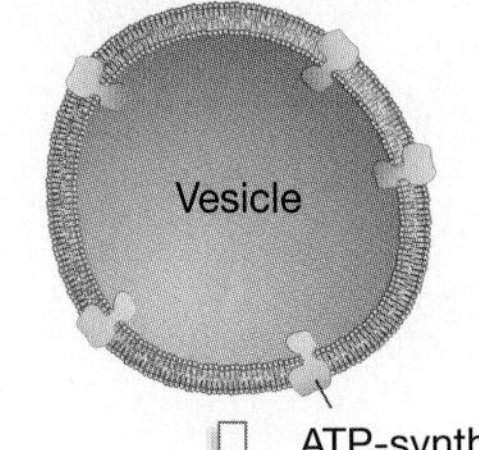

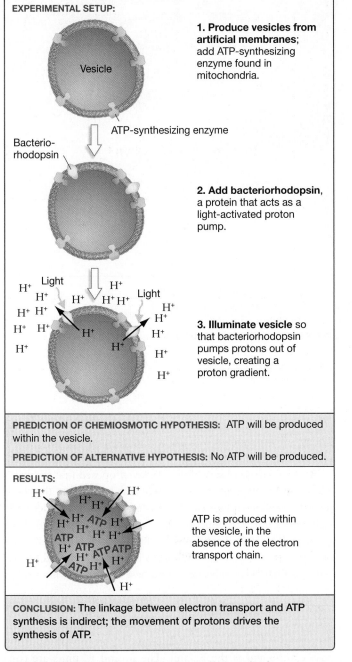

1. Produce vesicles from artificial membranes; add ATP-synthesizing enzyme found in mitochondria.

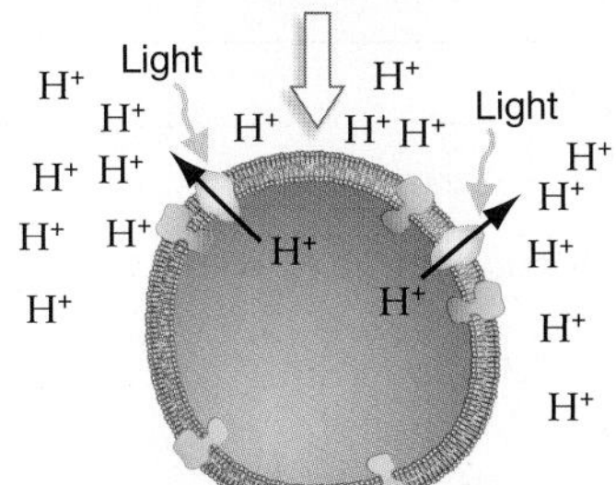

2. Add bacteriorhodopsin, a protein that acts as a light-activated proton pump.

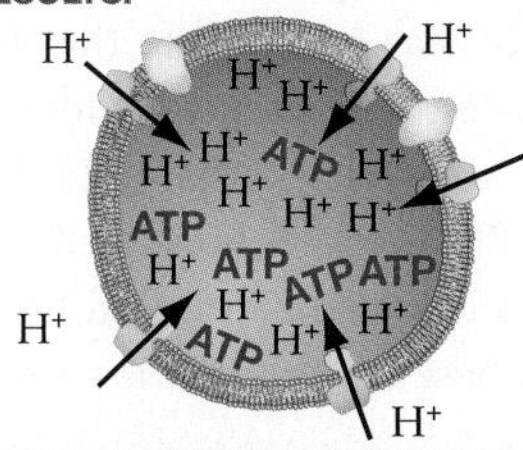

3. Illuminate vesicle so that bacteriorhodopsin pumps protons out of vesicle, creating a proton gradient.

PREDICTION OF CHEMIOSMOTIC HYPOTHESIS: ATP will be produced within the vesicle.

PREDICTION OF ALTERNATIVE HYPOTHESIS: No ATP will be produced.

RESULTS:

ATP is produced within the vesicle, in the absence of the electron transport chain.

CONCLUSION: The linkage between electron transport and ATP synthesis is indirect; the movement of protons drives the synthesis of ATP.

FIGURE 9.20 Evidence for the Chemiosmotic Hypothesis.

SOURCE: Racker, E. and W. Stoeckenius. 1974. Reconstitution of purple membrane vesicles catalyzing light-driven proton uptake and adenosine triphosphate formation. *Journal of Biological Chemistry*. 249: 662–663.

✔**QUESTION** Do you regard this as a convincing test of the chemiosmosis hypothesis? Why or why not?

zyme, they inserted bacteriorhodopsin, a well-studied membrane protein that acts as a light-activated proton pump.

When light strikes bacteriorhodopsin, it absorbs some of the light energy and changes conformation in a way that pumps protons from the interior of a membrane to the exterior. As a result, the experimental vesicles established a strong electrochemical gradient favoring proton movement to the interior. When the vesicles were illuminated to initiate proton pumping, ATP began to be produced from ADP inside the vesicles.

Mitchell's prediction was correct: In this situation, ATP production depended solely on the existence of a proton-motive force. It could occur in the *absence* of an electron transport chain. This result, and many others, have provided strong support for the hypothesis of chemiosmosis. Most ATP is produced by a flow of protons.

Chemiosmosis is like a hydroelectric dam, where the movement of water makes turbines spin and generate electricity. The electron transport chain is analogous to a series of gigantic pumps that force water up and behind the dam. The inner mitochondrial membrane functions as the dam, and ATP synthase is like the turbines inside the dam. In a mitochondrion, protons are pumped instead of water. When protons move through ATP synthase, the protein spins and generates ATP.

✔If you understand chemiosmosis, you should be able to explain why ATP production during cellular respiration is characterized as indirect. More specifically, you should be able to explain the relationship between glucose oxidation, the proton gradient, and ATP synthase.

Electron transport chains and ATP synthase occur in organisms throughout the tree of life. They are humming away in your cells now. Let's look in more detail at how they function.

How Is the Electron Transport Chain Organized?

The components of the electron transport chain are organized into four large complexes of proteins and cofactors (**Figure 9.21** on page 164). Two of the complexes pump protons. Q and the protein **cytochrome *c*** act as shuttles that transfer electrons between complexes. Q also carries a proton across the membrane along with an electron.

Research confirms that in complexes I and IV, protons actually pass directly through a sequence of electron carriers. The exact route taken by the protons is still being worked out. It is also not clear how the redox reactions taking place inside each complex—as electrons step down in potential energy—make proton movement possible.

The best-understood interaction between electron transport and proton pumping takes place in complex III. Research has shown that when Q accepts electrons from complex I or complex II, it also gains protons. The reduced form of Q then diffuses to the outer side of the inner membrane, where its electrons are used to reduce a component of complex III near the intermembrane space. The protons held by Q are released to the intermembrane space.

In this way, Q shuttles electrons and protons from one side of the membrane to the other. The electrons proceed down the

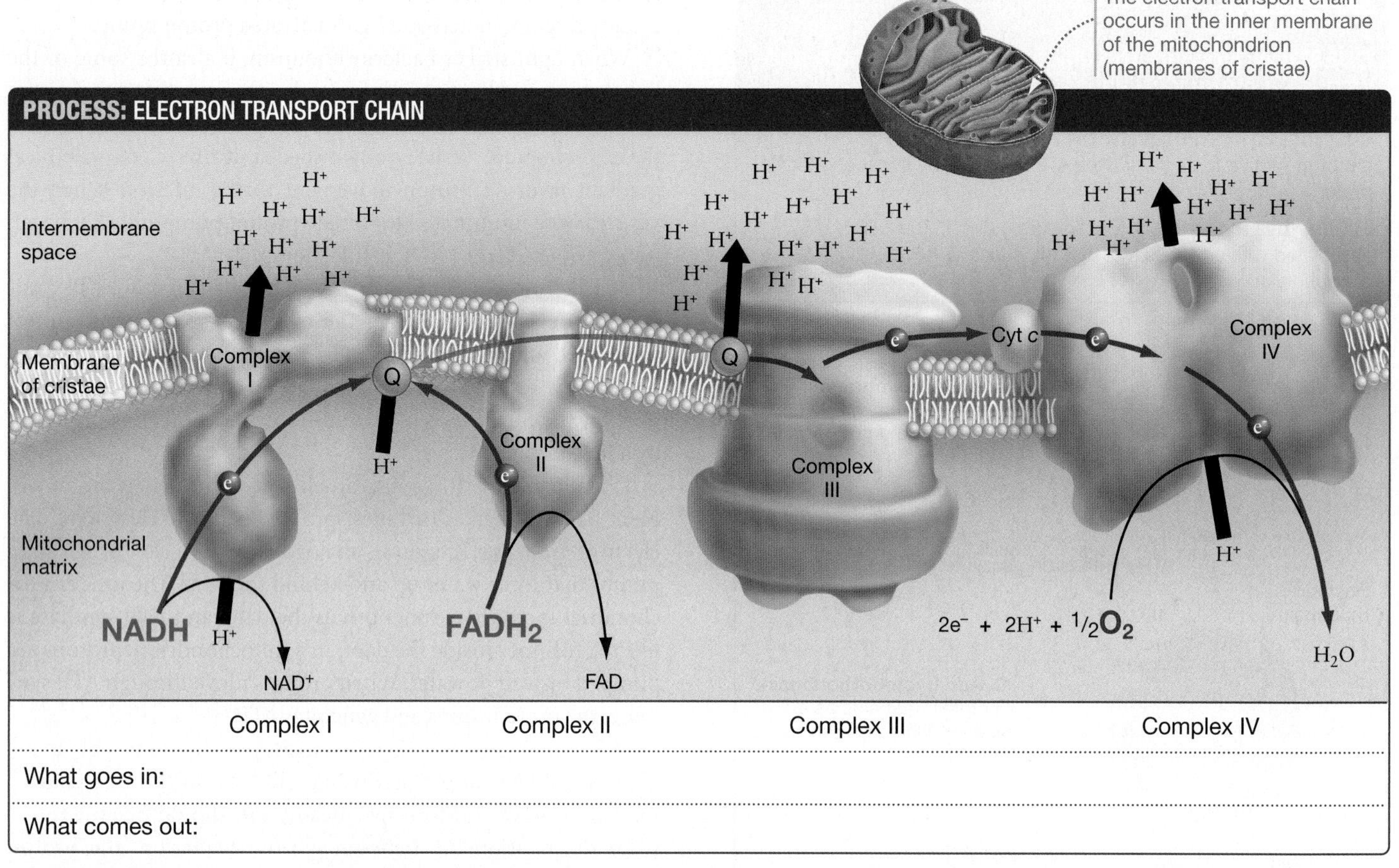

FIGURE 9.21 How Does the Electron Transport Chain Work? The individual components of the electron transport chain diagrammed in Figure 9.19 are grouped into large multiprotein complexes. Electrons are carried from one complex to another by Q and by cytochrome *c*; Q also shuttles protons across the membrane. The orange arrow indicates Q moving back and forth. Complexes I and IV use the potential energy released by the redox reactions to pump protons from the mitochondrial matrix to the intermembrane space.

✔**EXERCISE** Add an arrow across the membrane and label it "Proton gradient." In the boxes at the bottom, list "What goes in" and "What comes out" for each complex.

transport chain, and the protons released to the intermembrane space contribute to the proton-motive force.

Now, how does this proton gradient make the production of ATP possible?

The Discovery of ATP Synthase

In 1960 Efraim Racker made several key observations about how ATP is synthesized in mitochondria. When he used mitochondrial membranes to make vesicles, Racker noticed that some formed with their membrane inside out. Electron microscopy revealed that the inside-out membranes had numerous large proteins studded along their surfaces. Each protein appeared to have a base in the membrane with a lollipop-shaped stalk and a projecting knob (**Figure 9.22a**). If the solution was vibrated or treated with a compound called urea, the stalks and knobs fell off.

Racker seized on this technique to isolate the stalks and knobs and do experiments with them. For example, he found that isolated stalks and knobs could hydrolyze ATP, forming ADP and inorganic phosphate. The vesicles that contained just the base component, without the stalks and knobs, could not process ATP. The base components were, however, capable of transporting protons across the membrane.

Based on these observations, Racker proposed that the stalk-and-knob component of the protein was an enzyme that both hydrolyzes and synthesizes ATP. To test this idea, he added the stalk-and-knob components back to vesicles that had been stripped of them and confirmed that the vesicles were then capable of synthesizing ATP. Follow-up work also confirmed his hypothesis that the membrane-bound base component is a proton channel.

As **Figure 9.22b** shows, the structure of this protein complex is now well understood. The ATPase "knob" component is called the F_1 unit; the membrane-bound, proton-transporting base component is the F_o unit. The F_1 and F_o units are connected by a rotor, which spins the F_1 unit, and a stator, which interacts with the spinning F_1 unit. The entire complex is known as **ATP synthase**.

A flow of protons through the F_o unit causes the rotor connecting the two subunits to spin. By attaching long actin filaments to

(a) Vesicle formed from "inside-out" mitochondrial membrane

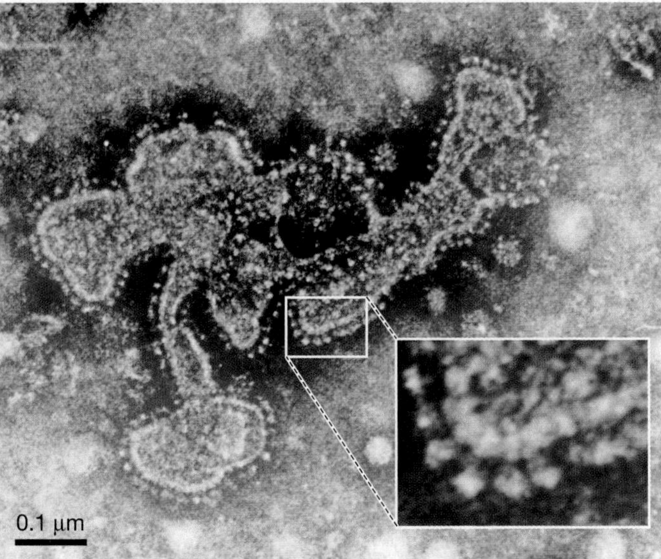

(b) The F_o unit is the base; the F_1 unit is the knob.

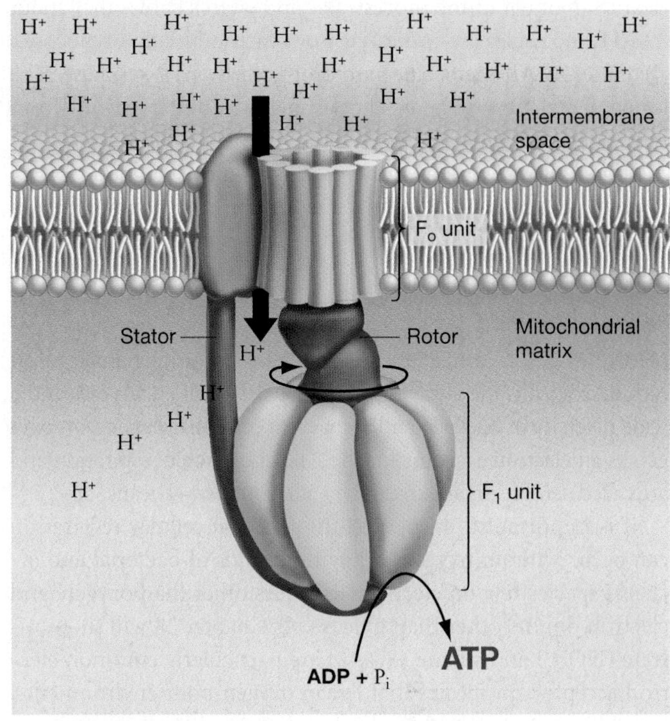

FIGURE 9.22 The Structure of ATP Synthase. (a) When patches of mitochondrial membrane turn inside out and form vesicles, proteins that have a lollipop-shaped stalk-and-knob structure face outward. Normally, the stalk and knob face inward, toward the mitochondrial matrix. **(b)** ATP synthase has two major components, designated F_o and F_1, connected by a rotor that spins.

the rotor and examining them with a videomicroscope, researchers have been able to see the rotation, which can reach speeds of 350 revolutions per second. As the F_1 unit rotates along with the rotor, its subunits are thought to change conformation in a way that catalyzes the phosphorylation of ADP to ATP. Understanding how this reaction occurs is currently the focus of intense research. ATP synthase makes most of the ATP that keeps you alive.

Organisms Use a Diversity of Electron Acceptors

Figure 9.23 summarizes glucose oxidation and cellular respiration by tracing the fate of the carbon atoms and electrons in glucose. Notice that electrons from glucose are transferred to NADH and $FADH_2$, passed through the electron transport chain, and accepted by oxygen. Proton pumping during electron transport creates the proton-motive force that drives ATP synthesis.

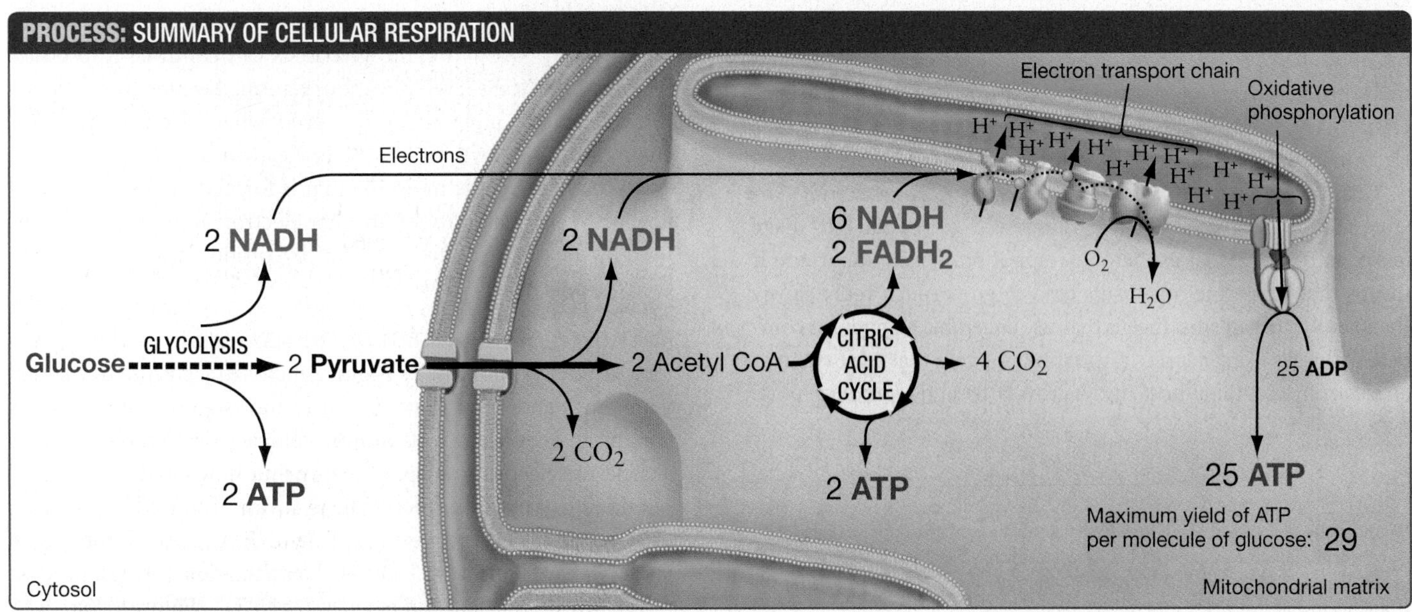

FIGURE 9.23 ATP Yield during Cellular Respiration.

The diagram also indicates the approximate yield of ATP from each component of the process. Recent research shows that about 29 ATP molecules are produced from each molecule of glucose.[3] Of these, 25 ATP molecules are produced by ATP synthase. The fundamental message here? The vast majority of the "payoff" from the oxidation of glucose occurs via oxidative phosphorylation.

If you haven't already done so, this would be a good time to review the elements of cellular respiration in the study area at *www.masteringbiology.com.*

(MB) **BioFlix™** Cellular Respiration

AEROBIC VERSUS ANAEROBIC RESPIRATION During cellular respiration, oxygen is the electron acceptor used by all eukaryotes and a wide diversity of bacteria and archaea. Species that depend on oxygen as an electron acceptor for the ETC use **aerobic** respiration and are called aerobic organisms. (The Latin root *aero–* means "air.")

It is important to recognize, though, that cellular respiration can occur without oxygen. Many thousands of bacterial and archaeal species rely on electron acceptors other than oxygen, and electron donors other than glucose. As Chapter 28 will show, nitrate (NO_3^-) and sulfate (SO_4^{2-}) are particularly common electron acceptors in species that live in oxygen-poor environments. In addition, many bacteria and archaea use H_2, H_2S, CH_4, or other inorganic compounds as electron donors—not glucose.

Cells that depend on electron acceptors other than oxygen are said to use **anaerobic** ("no air") respiration. Even though the starting and ending points of cellular respiration are different, anaerobic cells still use electron transport chains to create a proton-motive force that drives the synthesis of ATP. In bacteria and archaea, the ETC and ATP synthase are located in the plasma membrane.

AEROBIC RESPIRATION IS EFFICIENT Even though an array of compounds can serve as the final electron acceptor in cellular respiration, oxygen is the most efficient. Because oxygen holds electrons so tightly, the potential energy of electrons in a bond between an oxygen atom and a non-oxygen atom is low. As a result, there is a large difference between the potential energy of electrons in NADH and the potential energy of electrons bonded to an oxygen atom (see Figure 9.19). The large differential in potential energy means that the electron transport chain can generate a large proton-motive force.

Cells that do not use oxygen as an electron acceptor cannot generate such a large potential energy difference. As a result, they make less ATP than cells that use aerobic respiration. This is important: It means that anaerobic organisms tend to grow much more slowly than aerobic organisms. If cells that use anaerobic respiration compete with cells using aerobic respiration, the cells that use oxygen as an electron acceptor almost always grow faster and reproduce more.

[3]Traditionally, biologists thought that 36 ATP would be synthesized for every mole of glucose oxidized. More recent work has shown that actual yield is only about 29 ATP [see M. Brand, *Biochemistry and Molecular Biology Education* 31 (2003): 2–4]. Also, it's important to note that yield varies with conditions in the cell.

What happens when oxygen or other electron acceptors get used up? When this happens, the electrons carried by NADH have no place to go and the electron transport chain stops. As glycolysis, pyruvate processing, and the citric acid cycle continue, all of the NAD^+ in the cell quickly becomes NADH.

This situation is life threatening. When there is no longer any NAD^+ to supply the reactions of glycolysis, no ATP can be produced. If NAD^+ cannot be regenerated somehow, the cell will die. How do cells cope?

CHECK YOUR UNDERSTANDING

If you understand that . . .

- As electrons from NADH and $FADH_2$ move through the electron transport chain, protons are pumped into the intermembrane space of mitochondria.
- The electrochemical gradient across the inner mitochondrial membrane drives protons through ATP synthase, resulting in the production of ATP from ADP.

✓ **You should be able to . . .**

Add paper cutouts labeled ETC and ATP synthase to the model you made in Section 9.3, and then explain the steps in electron transport and chemiosmosis, using dimes to represent electrons and pennies to represent protons.

Answers are available in Appendix B.

9.7 Fermentation

Fermentation is a metabolic pathway that regenerates NAD^+ from stockpiles of NADH. It allows glycolysis to continue producing ATP in the absence of the electron acceptor required by the ETC. Fermentation occurs when pyruvate or a molecule derived from pyruvate accepts electrons from NADH (**Figure 9.24**).

When NADH gets rid of electrons in this way, NAD^+ is produced. With NAD^+ present, glycolysis can continue to produce ATP via substrate-level phosphorylation. Fermentation allows the cell to stay alive and grow, even when electron transport chains are shut down for lack of an electron acceptor.

In many cases, the molecule that is formed by the addition of an electron to pyruvate (or another electron acceptor) cannot be used by the cell. In some cases, this by-product is toxic and is excreted from the cell as waste.

MANY DIFFERENT FERMENTATION PATHWAYS EXIST If you sprint a long distance, your muscles begin metabolizing glucose so fast that your lungs and circulatory system cannot supply oxygen rapidly enough to keep electron transport chains active. When oxygen is absent, the electron transport chains shut down and NADH cannot donate its electrons there. The pyruvate produced by glycolysis then begins to accept electrons from NADH, and fermentation takes place. This process, **lactic acid fermentation**, forms a product molecule called lactate and regenerates NAD^+ (**Figure 9.25a**).

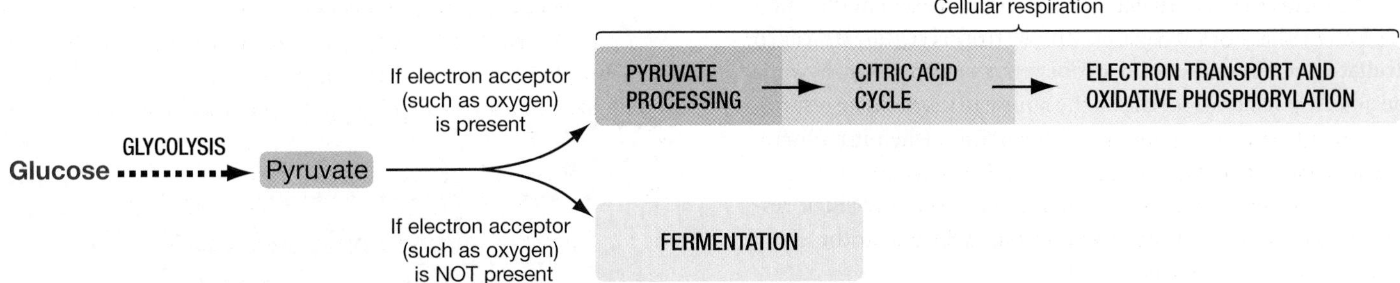

FIGURE 9.24 Cellular Respiration and Fermentation Are Alternative Pathways for Producing Energy. When oxygen or another electron acceptor used by the ETC is present in a cell, the pyruvate produced by glycolysis enters the citric acid cycle and the electron transport system is active. But if no electron acceptor is available to keep the ETC running, the pyruvate undergoes reactions known as fermentation.

Figure 9.25b illustrates a different fermentation pathway, **alcohol fermentation**, which occurs in the fungus *Saccharomyces cerevisiae*—baker's and brewer's yeast. When these fungal cells are placed in an environment such as bread dough or a bottle of grape juice and begin growing there, they quickly use up all the available oxygen. They continue to use glycolysis to metabolize sugar, however, by enzymatically converting pyruvate to the two-carbon compound acetaldehyde. This reaction gives off carbon dioxide, which causes bread to rise and produces the bubbles in champagne and beer.

(a) Lactic acid fermentation occurs in humans.

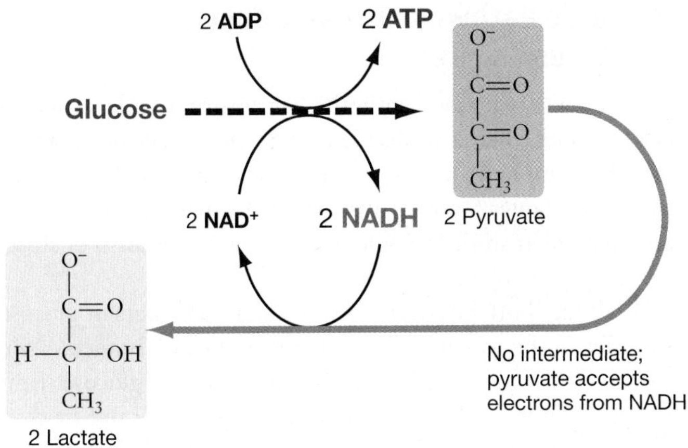

(b) Alcohol fermentation occurs in yeast.

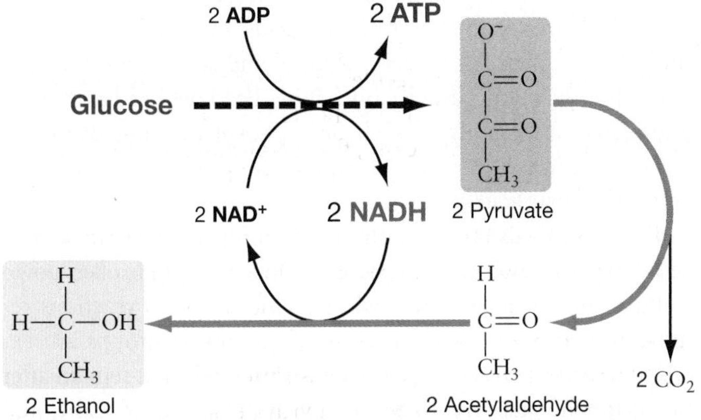

FIGURE 9.25 Fermentation Regenerates NAD$^+$ So That Glycolysis Can Continue. These are just two of the many types of fermentation that occur among the bacteria, archaea, and eukaryotes.

Acetaldehyde then accepts electrons from NADH, forming the NAD$^+$ required to keep glycolysis going. The addition of electrons to acetaldehyde forms ethanol as a waste product. The yeast cells excrete ethanol as waste. In essence, the active ingredient in alcoholic beverages is yeast urine.

Cells that employ other types of fermentation are used commercially in the production of soy sauce, tofu, yogurt, cheese, vinegar, and other products.

Bacteria and archaea that exist exclusively through fermentation are present in phenomenal numbers in the oxygen-free environment of your small intestine and in the rumen (first stomach) of cows. The rumen is a specialized digestive organ that contains over 10^{10} (10 billion) bacterial and archaeal cells per *milliliter* of fluid. The fermentations that occur in these cells produce an array of fatty acids. Cattle don't actually live off grass directly—they eat it to feed these bacteria and archaea, and use their fermentation by-products as a source of energy.

FERMENTATION AS AN ALTERNATIVE TO CELLULAR RESPIRATION Even though fermentation is a widespread type of metabolism, it is extremely inefficient compared with aerobic cellular respiration. Fermentation produces just 2 molecules of ATP for each molecule of glucose metabolized, while cellular respiration produces about 29—almost 15 times more energy per glucose molecule than fermentation. The reason for this disparity is that oxygen has much higher electronegativity than electron acceptors such as pyruvate and acetaldehyde. As a result, the potential energy drop between the start and end of fermentation is a tiny fraction of the potential energy change that occurs during cellular respiration.

Based on these observations, it is not surprising that organisms capable of both processes almost never use fermentation when an appropriate electron acceptor is available for cellular respiration. In organisms that usually use oxygen as an electron acceptor, fermentation is an alternative mode of energy production when oxygen supplies temporarily run out.

Organisms that can switch between fermentation and cellular respiration that uses oxygen as an electron acceptor are called **facultative aerobes**. The term aerobe refers to using oxygen, while the adjective facultative reflects the ability to use cellular respiration when oxygen is present and fermentation when it is absent. You are a facultative aerobe.

To review the steps in glucose oxidation and how cellular respiration interacts with fermentation pathways, go to the study area at *www.masteringbiology.com*:

 **Web Activity** Glucose Metabolism

CHECK YOUR UNDERSTANDING

If you understand that . . .

- Fermentation occurs in the absence of an ETC. It consists of reactions that oxidize NADH and regenerate the NAD^+ required for glycolysis.

✔ **You should be able to . . .**

Explain why organisms that have an ETC as well as fermentation pathways never ferment pyruvate if an electron acceptor is available.

Answers are available in Appendix B.

9.8 How Does Cellular Respiration Interact with Other Metabolic Pathways?

The enzymes, products, and intermediates involved in cellular respiration and fermentation do not exist in isolation. Instead, they are part of a huge and dynamic inventory of chemicals inside the cell.

Metabolism comprises thousands of different chemical reactions, and the amounts and identities of molecules inside cells are constantly in flux. Fermentation pathways, electron transport, and other aspects of carbohydrate metabolism may be crucial to the life of a cell, but they also have to be seen as parts of a whole (**Figure 9.26**).

This complexity can be boiled down to a simple essence, however. Cells have just two fundamental requirements: energy and carbon. They need a source of high-energy electrons for generating chemical energy in the form of ATP, and a source of carbon-containing molecules that can be used to synthesize DNA, RNA, proteins, fatty acids, and other molecules.

Reactions that break down molecules and produce ATP are called **catabolic pathways**; reactions that synthesize larger molecules from smaller components are called **anabolic pathways**.

How do glycolysis and the citric acid cycle interact with other catabolic pathways and with anabolic pathways? Let's consider how eukaryotes use molecules other than carbohydrates as fuel, and then examine how molecules involved in glycolysis and the citric acid cycle are sometimes used as building blocks to synthesize cell components.

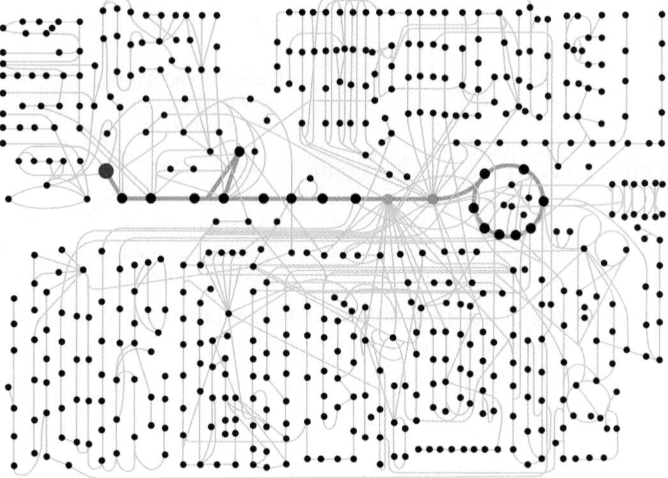

FIGURE 9.26 Metabolic Pathways Interact. A representation of a few of the thousands of chemical reactions that occur in cells. The dots represent molecules, and the lines represent enzyme-catalyzed reactions.

✔**EXERCISE** Label the large red dot, and circle the 10 reactions of glycolysis. Draw a box around the citric acid cycle.

Catabolic Pathways Break Down Molecules as Fuel

Most organisms ingest, synthesize, or absorb a wide variety of carbohydrates. These molecules range from sucrose, maltose, and other simple sugars to large polymers such as glycogen and starch. As Chapter 5 noted, glycogen is the major form of stored carbohydrate in animals, while starch is the major form of stored carbohydrate in plants.

Recall that both glycogen and starch are polymers of glucose, but differ in the way their long chains of glucose branch. Using enzyme-catalyzed reactions, cells can produce glucose from glycogen, starch, and most simple sugars. Glucose and fructose can then be processed by the enzymes of the glycolytic pathway.

Carbohydrates are not the only important source of carbon compounds used in catabolic pathways, however. As Chapter 6 pointed out, fats are highly reduced macromolecules consisting of glycerol bonded to chains of fatty acids. In cells, enzymes routinely break down fats to form glycerol and acetyl CoA. Glycerol enters the glycolytic pathway once it has been oxidized and phosphorylated to form glyceraldehyde-3-phosphate—one of the intermediates in the 10-reaction sequence of glycolysis. Acetyl CoA enters the citric acid cycle.

Proteins can also be catabolized, meaning that they can be broken down and used to produce ATP. Once they are broken down to their constituent amino acids, enzyme-catalyzed reactions remove the amino ($-NH_2$) groups. These amino groups are excreted in urine as waste. The carbon compounds that remain after this catabolic step are converted to pyruvate, acetyl CoA, and other intermediates in glycolysis and the citric acid cycle.

Figure 9.27 summarizes the catabolic pathways of carbohydrates, fats, and proteins and shows how their breakdown products feed an array of steps in glucose oxidation and cellular respi-

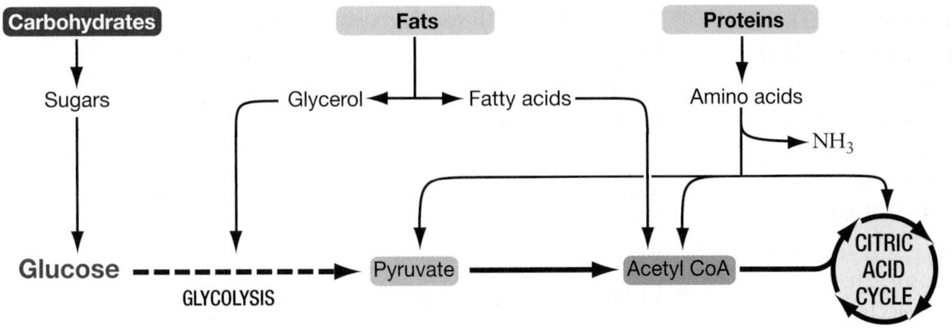

FIGURE 9.27 Catabolic Pathways: Proteins, Carbohydrates, and Fats Can All Furnish Substrates for Cellular Respiration. A variety of carbohydrates can be converted to glucose and processed by glycolysis. If carbohydrates are scarce, cells can obtain high-energy compounds from fats or even proteins for ATP production. These are catabolic reactions.

ration. When all three types of molecules are available in the cell to generate ATP, carbohydrates are used up first, then fats, and finally proteins.

Anabolic Pathways Synthesize Key Molecules

Where do cells get the precursor molecules required to synthesize amino acids, RNA, DNA, phospholipids, and other cell components? Not surprisingly, the answer often involves intermediates in carbohydrate metabolism. For example,

- In humans, about half of the 21 required amino acids can be synthesized from molecules siphoned from the citric acid cycle.
- Acetyl CoA is the starting point for anabolic pathways that result in the synthesis of fatty acids.
- The molecule that is produced by the first reaction in glycolysis can be oxidized to start the synthesis of ribose-5-phosphate—a key intermediate in the production of ribonucleotides and deoxyribonucleotides. These nucleotides, in turn, are required for manufacturing RNA and DNA.
- If ATP is abundant, pyruvate and lactate (from fermentation) can be used as a substrate in the synthesis of glucose. Excess glucose is converted to glycogen and stored.

Figure 9.28 summarizes how intermediates in carbohydrate metabolism are drawn off to synthesize macromolecules. The take-home message is that the same molecule can serve many different functions in the cell. As a result, catabolic and anabolic pathways are closely intertwined.

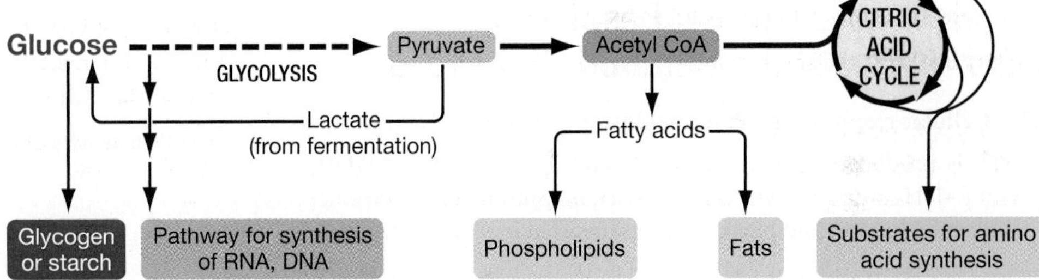

FIGURE 9.28 Anabolic Pathways: Intermediates in Carbohydrate Metabolism Can Be Drawn Off to Synthesize Cell Components. Several of the intermediates in carbohydrate metabolism act as precursor molecules in anabolic reactions leading to the synthesis of glycogen or starch, RNA, DNA, fatty acids, and amino acids.

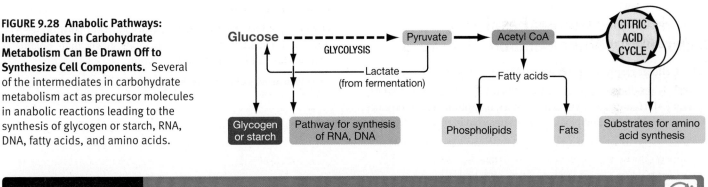

CHAPTER 9 REVIEW

For media, go to the study area at www.masteringbiology.com

Summary of Key Concepts

In cells, the endergonic reactions required for life occur in conjunction with an exergonic reaction involving ATP.

- ATP is the currency that cells use to pump ions, drive endergonic reactions, move cargo, and perform other types of work.
- Almost all of the reactions that occur inside cells are endergonic.
- When ATP or a phosphate group from ATP is added to a substrate or enzyme that participates in an endergonic reaction, the potential energy of the substrate or enzyme is raised enough to make the reaction exergonic and thus spontaneous.

✓You should be able to explain why the addition of phosphate groups raises the potential energy of proteins, and why phosphorylation often causes proteins to change shape.

Cellular respiration produces ATP from molecules with high potential energy—often glucose.

- Cells produce ATP from sugars or other compounds with high free energy by using one of two general pathways: (1) cellular respiration or (2) fermentation.
- Cellular respiration is based on redox reactions that transfer electrons from a compound with high free energy, such as glucose, to a molecule with lower free energy, such as oxygen, through an electron transport chain.
- Fermentation involves the transfer of electrons from one organic compound to another without participation by an electron transport chain.

✔You should be able to explain why cellular respiration produces so much more ATP per mole of glucose than fermentation, in terms of the total free-energy changes involved.

MB **Web Activity** Redox Reactions

Cellular respiration has four components: (1) glycolysis, (2) pyruvate processing, (3) the citric acid cycle, and (4) electron transport and chemiosmosis.

- In eukaryotes, glycolysis takes place in the cytosol. The citric acid cycle occurs in the mitochondrial matrix, and electron transport and oxidative phosphorylation proceed in the inner membranes of mitochondria.
- Glycolysis is a 10-step reaction sequence in which glucose is broken down into two molecules of pyruvate. ATP and NADH are produced.
- During pyruvate processing, a series of reactions converts pyruvate to acetyl CoA. NADH is produced and CO_2 is released.
- The citric acid cycle is an 8-step reaction cycle that begins with acetyl CoA. $FADH_2$, NADH, and GTP or ATP are produced; CO_2 is released. At the end of the citric acid cycle, glucose is completely oxidized to CO_2.
- NADH and $FADH_2$ donate electrons to an electron transport chain, which gradually steps the electrons down in potential energy until they are transferred to a final electron acceptor (often O_2). The energy that is released pumps protons across the inner mitochondrial membrane, creating an electrochemical gradient that ATP synthase uses to produce ATP.

✔You should be able to describe what would happen to NADH levels in a cell in the first few seconds after a drug has poisoned the enzyme that converts acetyl CoA to citrate.

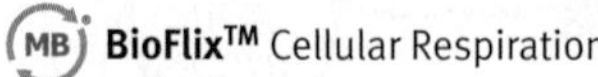
MB **BioFlix™** Cellular Respiration

Cellular respiration and fermentation are carefully regulated.

- ATP is produced only as needed. When supplies of ATP, NADH, and $FADH_2$ are high in the cell, feedback inhibition occurs: Product molecules bind to and block enzymes involved in ATP production.
- The glycolytic pathway slows when ATP binds to phosphofructokinase.
- The pyruvate dehydrogenase complex is inhibited when it is phosphorylated by ATP. It speeds up in the presence of substrates like NAD and ADP.
- The enzyme that converts acetyl CoA to citrate slows when ATP binds to it, and certain enzymes in the citric acid cycle are inhibited when NADH or ATP bind to them.
- In eukaryotes and many bacteria, fermentation only occurs if cellular respiration stops.

✔You should be able to draw a graph predicting how the rate of ATP production changes as a function of ATP concentration. (Write "ATP concentration" on the *x*-axis and "ATP production" on the *y*-axis.)

Fermentation pathways allow glycolysis to continue when the lack of an electron acceptor shuts down electron transport chains.

- Fermentation is an alternative method for processing glucose and making ATP—one that cells use when cellular respiration is not possible.
- If no electron acceptor is available, all NAD^+ is converted to NADH and glycolysis stops.
- Fermentation pathways regenerate NAD^+, so glycolysis can continue to make ATP and keep the cell alive. This happens when an organic molecule such as pyruvate accepts electrons from NADH.
- Depending on the molecule that acts as an electron acceptor, fermentation pathways produce lactate, ethanol, or other reduced organic compounds as a by-product.

✔You should be able to explain why organisms that produce ATP only via fermentation grow much more slowly than organisms that produce ATP via cellular respiration.

MB **Web Activity** Glucose Metabolism

Questions

✔TEST YOUR KNOWLEDGE

Answers are available in Appendix B

1. When does feedback inhibition occur?
   - **a.** when lack of an appropriate electron acceptor makes an electron transport chain stop
   - **b.** when an enzyme that is active early in a metabolic pathway is inhibited by a product of the pathway
   - **c.** when ATP synthase reverses and begins pumping protons out of the mitochondrial matrix
   - **d.** when cellular respiration is inhibited and fermentation begins
2. Where does the citric acid cycle occur in eukaryotes?
   - **a.** in the cytosol
   - **b.** in the matrix of mitochondria
   - **c.** in the inner membrane of mitochondria
   - **d.** in the intermembrane space of mitochondria
3. What does the chemiosmotic hypothesis claim?
   - **a.** Substrate-level phosphorylation occurs in the electron transport chain.
   - **b.** Substrate-level phosphorylation occurs in glycolysis and the citric acid cycle.
   - **c.** The electron transport chain is located in the inner membrane of mitochondria.
   - **d.** Electron transport chains generate ATP indirectly, by the creation of a proton-motive force.
4. What is the function of the reactions in a fermentation pathway?
   - **a.** to generate NADH from NAD^+, so electrons can be donated to the electron transport chain
   - **b.** to synthesize pyruvate from lactate
   - **c.** to generate NAD^+ from NADH, so glycolysis can continue
   - **d.** to synthesize electron acceptors, so that cellular respiration can continue

5. When do cells switch from cellular respiration to fermentation?
   a. when electron acceptors required by the ETC are not available
   b. when the proton-motive force runs down
   c. when NADH and $FADH_2$ supplies are low
   d. when pyruvate is not available

6. Why are NADH and $FADH_2$ said to have "reducing power"?
   a. They are the reduced forms of NAD^+ and FAD.
   b. They donate electrons to components of the ETC, reducing those components.
   c. They travel between the cytosol and the mitochondrion.
   d. They have the power to reduce carbon dioxide to glucose.

✓TEST YOUR UNDERSTANDING

Answers are available in Appendix B

1. Explain why NADH and $FADH_2$ are called electron carriers. Where do these molecules get electrons, and where do they deliver them? In eukaryotes, what molecule do these electrons reduce?
2. Compare and contrast substrate-level phosphorylation and oxidative phosphorylation.
3. Why does aerobic respiration produce much more ATP than anaerobic respiration?
4. Make a flowchart indicating the relationships among the four steps of cellular respiration. Which steps are responsible for glucose oxidation? Which produce the most ATP?
5. Explain the relationship between electron transport and oxidative phosphorylation. What does ATP synthase look like, and how does it work?
6. Describe the relationship among carbohydrate metabolism, the catabolism of proteins and fats, and anabolic pathways.

✓APPLYING CONCEPTS TO NEW SITUATIONS

Answers are available in Appendix B

1. Cyanide ($C{\equiv}N^-$) blocks complex IV of the electron transport chain. Suggest a hypothesis for what happens to the ETC when complex IV stops working. Your hypothesis should explain why cyanide poisoning in humans is fatal.
2. The presence of many sac-like cristae results in a large amount of membrane inside mitochondria. Suppose that some mitochondria had few cristae. How would their output of ATP compare with that of mitochondria with many cristae? Explain your answer.
3. When yeast cells are placed into low-oxygen environments, the mitochondria in the cells become reduced in size and number. Suggest an explanation for this observation.
4. Most agricultural societies have come up with ways to ferment the sugars in barley, wheat, rice, corn, or grapes to produce alcoholic beverages. Historians argue that this was an effective way for farmers to preserve the chemical energy in grains and fruits in a form that would not be eaten by rats or spoiled by bacteria or fungi. Why does a great deal of chemical energy remain in the products of fermentation pathways?

This chapter is part of The Big Picture. See how on pages 192–193.

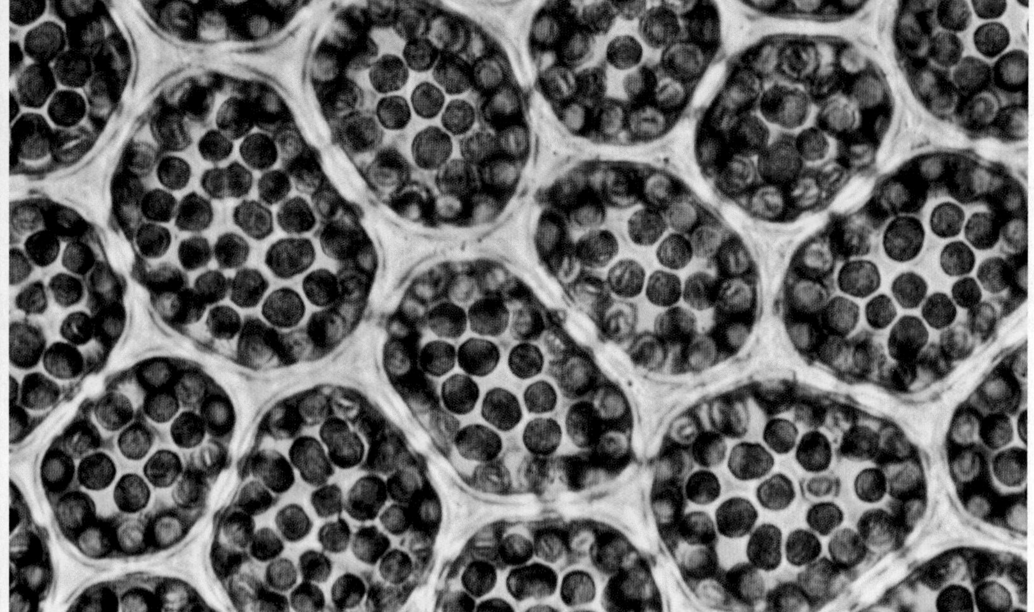

A close-up of moss cells filled with chloroplasts, where photosynthesis converts the energy in sunlight to chemical energy in the bonds of sugar. The sugar produced by photosynthetic organisms fuels cellular respiration and growth. Photosynthetic organisms, in turn, are consumed by other organisms, including you. Directly or indirectly, most organisms on Earth get their energy from photosynthesis.

10 Photosynthesis

KEY CONCEPTS

- Photosynthesis is the conversion of light energy to chemical energy stored in the bonds of carbohydrates. It consists of two linked sets of reactions.
- In the light-capturing reactions, excited electrons are used to produce the electron carrier NADPH or are donated to an electron transport chain, which results in the production of ATP via chemiosmosis.
- The reactions of the Calvin cycle start with the enzyme rubisco, which catalyzes the addition of CO_2 to a five-carbon compound. Subsequent reactions use the ATP and NADPH synthesized in the light reactions and yield a molecule required for carbohydrate production.
- In plants, CO_2 enters photosynthetic tissue through openings called stomata. The CAM and C_4 pathways increase CO_2 concentrations inside the leaves of some species and make photosynthesis more efficient.

Some three billion years ago, a novel combination of light-absorbing molecules and enzymes gave a bacterial cell the capacity to convert light energy into chemical energy in the C–C and C–H bonds of sugar. The origin of **photosynthesis**—the use of sunlight to manufacture carbohydrate—ranks as one of the great events in the history of life.

The vast majority of organisms alive today rely on photosynthesis, either directly or indirectly, to stay alive. Maples, mosses, and other photosynthetic organisms are termed **autotrophs** (literally, "self-feeders") because they make all of their own food from ions and simple molecules. Humans, houseflies, and other non-photosynthetic organisms are called **heterotrophs** ("different-feeders") because they have to obtain the sugars and many of the other macromolecules they need from other organisms.

Because there could be no heterotrophs without autotrophs, photosynthesis is fundamental to almost all life. Glycolysis may qualify as the most ancient set of energy-related chemical reactions from an evolutionary viewpoint; but ecologically—meaning, in terms of how organisms interact with each other—photosynthesis is easily the most important.

How does it happen? Let's begin with an overview, then delve into a step-by-step analysis of some of the most remarkable chemistry on Earth.

10.1 Photosynthesis Harnesses Sunlight to Make Carbohydrate

Research on photosynthesis began early in the history of biological science. Starting in the 1770s, a series of experiments showed that photosynthesis takes place only in the green parts of plants; sunlight, carbon dioxide (CO_2), and water (H_2O) are required; and oxygen (O_2) is produced as a by-product.

✔ When you see this checkmark, stop and test yourself. Answers are available in Appendix B.

By the 1840s enough was known about the process for biologists to propose that photosynthesis allows plants to convert electromagnetic energy in the form of sunlight into chemical energy in the C–C and C–H bonds of carbohydrates. When glucose is the carbohydrate that is eventually produced, the overall reaction—the sum of many independent reactions—can be written as

$$6CO_2 + 12H_2O + \text{light energy} \longrightarrow C_6H_{12}O_6 + 6O_2 + 6H_2O$$

Now read the reaction again, and note the contrast with cellular respiration. Photosynthesis is an endergonic suite of reactions that reduces carbon dioxide to glucose or other sugars. Cellular respiration is an exergonic suite of reactions that oxidizes glucose to carbon dioxide and results in the production of ATP.

Early investigators assumed that CO_2 and H_2O react directly to form the CH_2O found in carbohydrates, and that the oxygen gas (O_2) released during photosynthesis originated in the oxygen atoms of CO_2.

These researchers were wrong, though. CO_2 and H_2O participate in entirely different reactions, and the oxygen atoms in O_2 come from water.

Photosynthesis: Two Linked Sets of Reactions

During the 1930s two independent lines of research on photosynthesis converged, leading to a major advance.

The first research program, led by Cornelius van Niel, focused on photosynthesis in organisms called purple sulfur bacteria. Van Niel and his group found that these cells can grow in the laboratory on a food source that lacks sugars. Based on this observation, he concluded they must be autotrophs that manufacture their own carbohydrates. But to grow, the cells had to be exposed to sunlight and hydrogen sulfide (H_2S).

Van Niel also showed that these cells did not produce oxygen as a by-product of photosynthesis. Instead, elemental sulfur (S) accumulated in their medium. In these organisms, the overall reaction for photosynthesis was

$$CO_2 + 2H_2S + \text{light energy} \longrightarrow (CH_2O)_n + H_2O + 2S$$

Van Niel's work was crucial for two reasons:

1. It showed that CO_2 and H_2O do *not* combine directly during photosynthesis. Instead of acting as a reactant in photosynthesis in these species, H_2O is a product of the process.
2. It supported the hypothesis that the oxygen atoms in CO_2 are *not* released as oxygen gas (O_2). The purple sulfur bacteria produced no oxygen, even though carbon dioxide participated in the reaction—just as it did in plants.

Based on these findings, biologists hypothesized that the oxygen atoms released during plant photosynthesis must come from H_2O. This proposal was supported by experiments with isolated chloroplasts, which produced oxygen in the presence of sunlight even if no CO_2 was present.

The hypothesis was confirmed when heavy isotopes of oxygen—^{18}O compared with the normal isotope, ^{16}O—became available to researchers. Biologists exposed algae or plants to H_2O that contained ^{18}O, collected the oxygen gas that was given off as a by-product of photosynthesis, and confirmed that the released oxygen gas contained the heavy isotope.

As predicted, the reaction that produced this oxygen occurred only in the presence of sunlight. The light-capturing reactions of photosynthesis result in the production of oxygen from water.

A second major line of research supported these discoveries. Between 1945 and 1955, a team led by Melvin Calvin began introducing radioactively labeled carbon dioxide ($^{14}CO_2$) to algae and identifying the molecules that subsequently became labeled with the radioisotope. These experiments allowed researchers to identify the sequence of reactions involved in reducing CO_2 to sugars.

Because Calvin played an important role in this research, the reactions that reduce carbon dioxide and produce sugar came to be known as the **Calvin cycle**. Later research showed that the Calvin cycle can function only if the light-capturing reactions are occurring.

To summarize: Early research showed that photosynthesis consists of two linked sets of reactions. One set is triggered by light; the other set—the Calvin cycle—requires the products of the light-capturing reactions. The light-capturing reactions produce oxygen from water; the Calvin cycle produces sugar from carbon dioxide.

The two reactions are linked by electrons that are released when water is split to form oxygen gas. During the light-capturing reactions, these electrons are promoted to a high energy state by light and then transferred to a phosphorylated version of NAD^+, called **$NADP^+$ (nicotinamide adenine dinucleotide phosphate)**. This reaction forms **NADPH**, which functions as an electron carrier. ATP is also produced in the light-capturing reactions (see **Figure 10.1**).

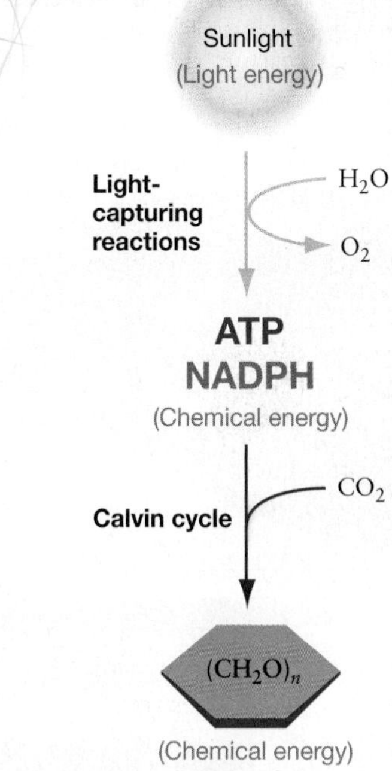

FIGURE 10.1 Photosynthesis Has Two Linked Components. In the light-capturing reactions of photosynthesis, light energy is transformed to chemical energy in the form of ATP and NADPH. During the Calvin cycle, the ATP and NADPH produced in the light-capturing reactions are used to reduce carbon dioxide to carbohydrate.

During the Calvin cycle, the electrons in NADPH and the potential energy in ATP are used to reduce CO_2 to carbohydrate. The resulting sugars are used in cellular respiration to produce ATP for the cell. Plants oxidize sugars in their mitochondria and consume O_2 in the process, just as animals and other eukaryotes do.

Where does all this activity take place?

Photosynthesis Occurs in Chloroplasts

Once experiments had established that photosynthesis takes place only in the green portions of plants, biologists focused on the bright green organelles called **chloroplasts** ("green-formed"). One leaf cell typically contains 40 to 50 chloroplasts, and a square millimeter of leaf averages about 500,000 (**Figure 10.2**).

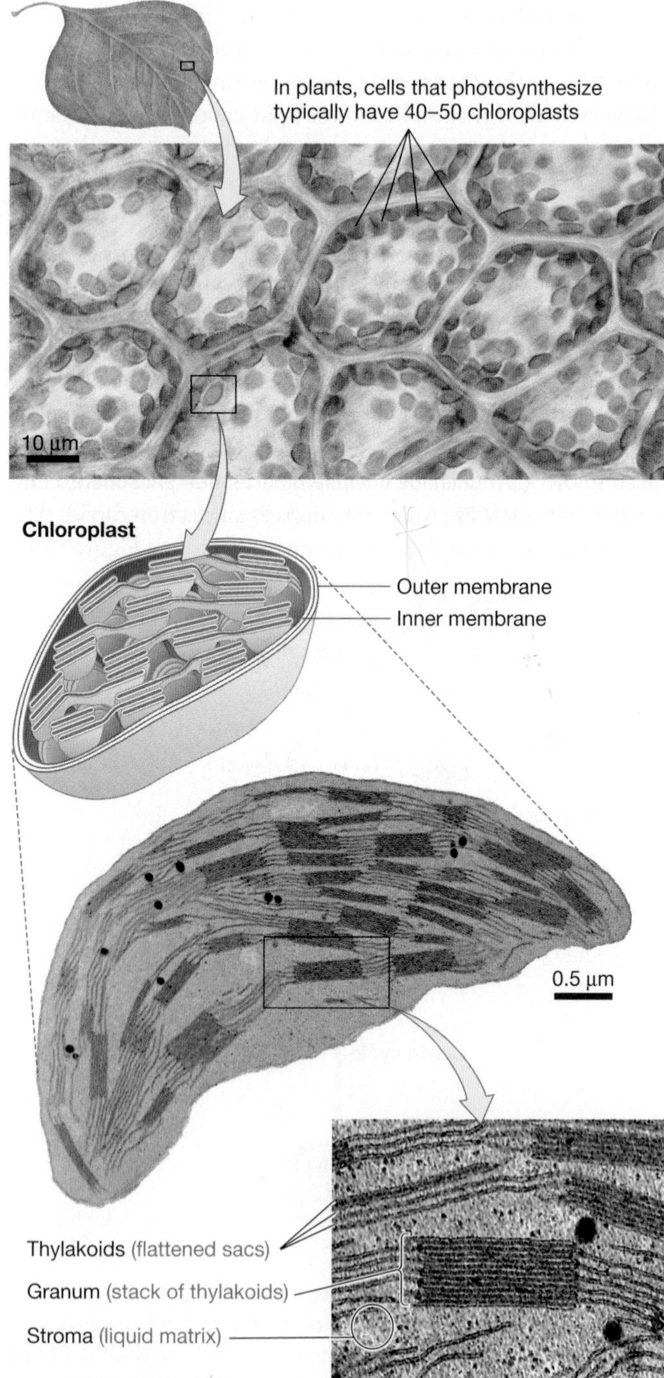

FIGURE 10.2 Photosynthesis Takes Place in Chloroplasts.

When membranes derived from chloroplasts were found to release oxygen after exposure to sunlight, the hypothesis that chloroplasts are the site of photosynthesis became widely accepted.

As Figure 10.2 shows, a chloroplast is enclosed by an outer membrane and an inner membrane. The interior is dominated by vesicle-like structures called **thylakoids**, which often occur in interconnected stacks called **grana** (singular: **granum**). The space inside a thylakoid is its **lumen**. (Recall that *lumen* is a general term for the interior of any sac-like structure. Your stomach and intestines have a lumen.) The fluid-filled space between the thylakoids and the inner membrane is the **stroma**.

When researchers analyzed the chemical composition of thylakoid membranes, they found huge quantities of pigments. **Pigments** are molecules that absorb only certain wavelengths of light—other wavelengths are either transmitted or reflected. Pigments have colors because we see the wavelengths that they do *not* absorb.

The most abundant pigment in the thylakoid membranes turned out to be chlorophyll ("green-leaf"), which reflects or transmits green light. As a result, it is responsible for the green color of plants, some algae, and many photosynthetic bacteria.

Before plunging into the details of how photosynthesis occurs, take a moment to consider just how astonishing the process is. Chemists have synthesized an amazing diversity of compounds from relatively simple starting materials, but their achievements pale in comparison to a cell that can synthesize sugar from just carbon dioxide, water, and sunlight. If photosynthesis is not *the* most sophisticated chemistry on Earth, it is certainly a contender.

10.2 How Does Chlorophyll Capture Light Energy?

The light-capturing reactions of photosynthesis begin with the simple act of sunlight striking chlorophyll. To understand the consequences of this event, it's helpful to review the nature of light.

Light is a type of electromagnetic radiation, a form of energy. Photosynthesis converts electromagnetic energy in the form of sunlight into chemical energy in the C–C and C–H bonds of sugar.

Physicists describe light's behavior as both wavelike and particle-like. Like water waves or airwaves, electromagnetic radiation is characterized by its **wavelength**—the distance between two successive wave crests (or wave troughs). The wavelength determines the type of electromagnetic radiation.

Figure 10.3 illustrates the range of wavelengths of electromagnetic radiation—the **electromagnetic spectrum**. **Visible light**, the electromagnetic radiation that humans can see, ranges in wavelength from about 400 to about 710 nanometers (nm, or 10^{-9} m). Shorter wavelengths of electromagnetic radiation contain more energy than longer wavelengths do. Thus, blue light and ultraviolet light contain much more energy than red light and infrared light do.

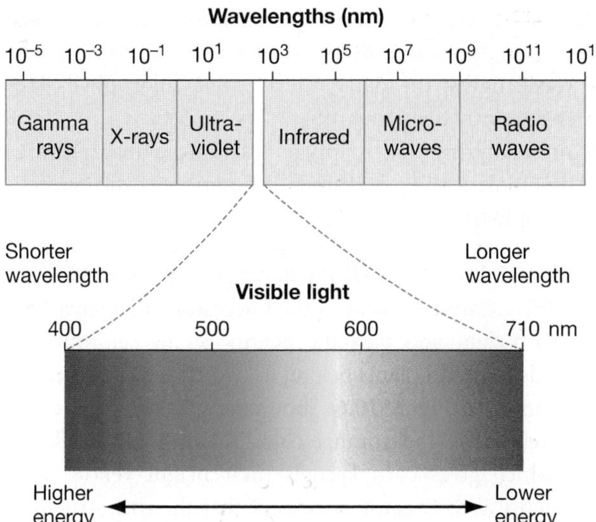

FIGURE 10.3 The Electromagnetic Spectrum. Electromagnetic energy radiates through space in the form of waves. Humans can see radiation at wavelengths between about 400 nm to 710 nm. The shorter the wavelength of electromagnetic radiation, the higher its energy.

To emphasize the particle-like nature of light, physicists point out that it exists in discrete packets called **photons**. Each photon and each wavelength of light have a characteristic amount of energy. Pigment molecules absorb this energy. How?

Photosynthetic Pigments Absorb Light

When a photon strikes an object, the photon may be absorbed, transmitted, or reflected. A pigment molecule absorbs particular wavelengths of light. Sunlight includes white light, which consists of all wavelengths in the visible portion of the electromagnetic spectrum at once.

If a pigment absorbs all of the visible wavelengths, no visible wavelength of light is reflected back to your eye, and the pigment appears black. If a pigment absorbs many or most of the wavelengths in the blue and green parts of the spectrum but transmits or reflects red wavelengths, it appears red.

What wavelengths do various plant pigments absorb? In one approach to answering this question, researchers grind up leaves and add a liquid that acts as a solvent for lipids. The solvent extracts pigment molecules from the leaf mixture. A technique called thin layer chromatography separates the pigments in the extract (**Figure 10.4a**).

To begin, spots of raw leaf extract are placed near the bottom of a stiff support that is coated with a thin layer of silica gel, cellulose, or similar porous material. The coated support is then placed in a solvent solution. As the solvent wicks upward through the coating, it carries the pigment molecules in the mixture with it. Because the pigment molecules vary in size, solubility, or both, they are carried at different rates.

Figure 10.4b shows a chromatograph from a grass-leaf extract. Notice that this leaf contains an array of pigments. To find out which wavelengths are absorbed by each of these molecules, researchers cut out a single region (color band) of the filter paper, extract the pigment, and use an instrument to record the wavelengths absorbed.

Using thin-layer chromatography, biologists have produced data like those shown in **Figure 10.5a**. This graph is an **absorption spectrum**—a graph showing the amount of light absorbed versus wavelength. Peaks indicate wavelengths where absorbance is high; troughs indicate wavelengths where absorbance is low.

Research based on these techniques has confirmed that there are two major classes of pigment in plant leaves: chlorophylls and carotenoids.

- The **chlorophylls**, designated chlorophyll *a* and chlorophyll *b*, absorb strongly in the blue and red regions of the visible spectrum. The presence of chlorophylls makes plants look green because they reflect and transmit green light, which they do not absorb.
- **Carotenoids** absorb in the blue and green parts of the visible spectrum. Thus, carotenoids appear yellow, orange, or red.

Which of these wavelengths drive photosynthesis?

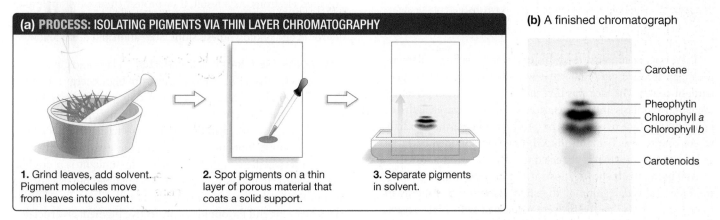

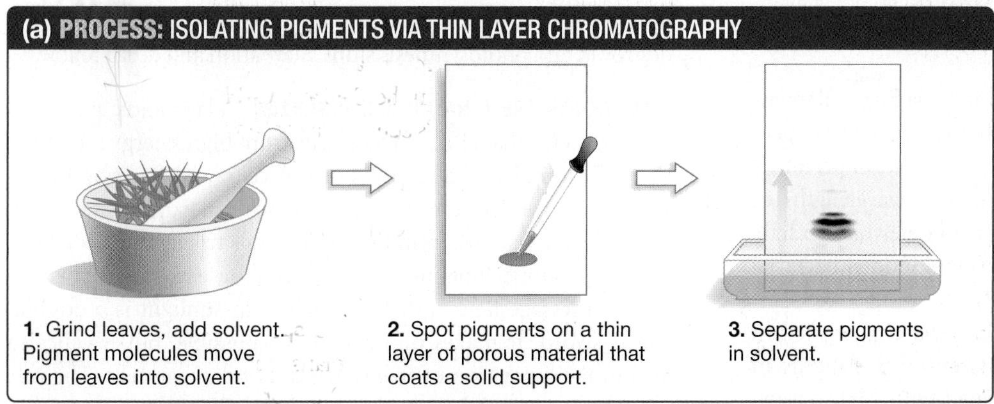

(a) PROCESS: ISOLATING PIGMENTS VIA THIN LAYER CHROMATOGRAPHY
1. Grind leaves, add solvent. Pigment molecules move from leaves into solvent.
2. Spot pigments on a thin layer of porous material that coats a solid support.
3. Separate pigments in solvent.

(b) A finished chromatograph

Carotene
Pheophytin
Chlorophyll *a*
Chlorophyll *b*
Carotenoids

FIGURE 10.4 Chromatography Is a Technique for Separating Molecules. This example shows the isolation of pigments in grass leaves. Different species of photosynthetic organisms may contain different types and quantities of pigments.

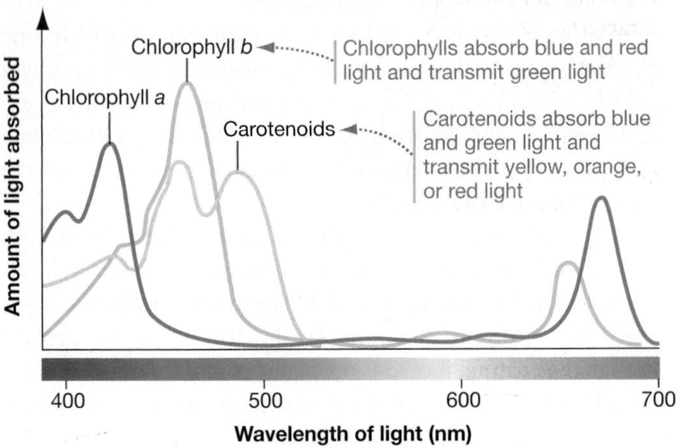

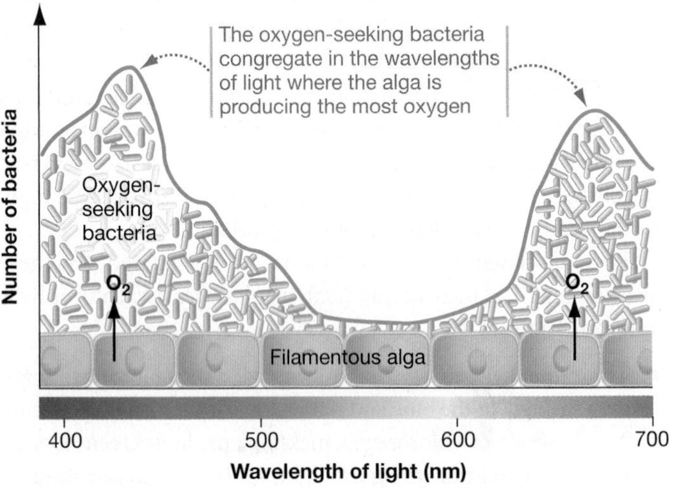

FIGURE 10.5 There Is a Strong Correlation between the Absorption Spectrum of Pigments and the Action Spectrum for Photosynthesis.

✔**EXERCISE** Some photosynthetic organisms have versions of chlorophyll pigments that absorb most strongly at 390 nm and 880 nm. Draw what the wavelength graph in part (b) would look like if these organisms were used in the experiment instead of the alga.

T. W. Engelmann answered this question by laying a filamentous alga across a glass slide that was illuminated with a spectrum of colors. The idea was that the alga would begin performing photosynthesis in response to the various wavelengths of light and produce oxygen as a by-product. To determine exactly where oxygen was being produced, Engelmann added bacterial cells from a species that is attracted to oxygen.

As **Figure 10.5b** shows, most of the bacteria congregated in the blue and red regions of the slide. Because wavelengths in these parts of the spectrum were associated with high oxygen concentrations, Engelmann concluded that they defined the **action spectrum** for photosynthesis—the wavelengths that drive the light-capturing reactions.

The data suggest that blue and red photons are the most effective at driving photosynthesis. Because the chlorophylls absorb these wavelengths, the data indicate that chlorophylls are the main photosynthetic pigments.

Before analyzing what happens during the absorption event itself, let's look at the structure and function of the carotenoids and chlorophylls.

WHAT IS THE ROLE OF CAROTENOIDS AND OTHER ACCESSORY PIGMENTS? Carotenoids are called accessory pigments because they absorb light and pass the energy on to chlorophyll. The carotenoids found in plants belong to two classes, called carotenes and xanthophylls. **Figure 10.6a** shows the structure of β-carotene, which gives carrots their orange color. A xanthophyll called zeaxanthin, which gives corn kernels their bright yellow color, is nearly identical to β-carotene, except that the ring structures on either end of the molecule contain a hydroxyl (–OH) group.

Both xanthophylls and carotenes are found in chloroplasts. In autumn, when the leaves of deciduous trees die, their chlorophyll degrades first. The wavelengths scattered by the carotenoids that remain turn forests into spectacular displays of yellow, orange, and red.

Carotenoids absorb wavelengths of light that are not absorbed by chlorophyll. As a result, they extend the range of wavelengths that can drive photosynthesis.

Researchers discovered an even more important function for carotenoids, though, by analyzing what happens to leaves when these pigments are destroyed. Many herbicides, for example, work by inhibiting enzymes that are involved in carotenoid synthesis. Plants lacking carotenoids rapidly lose their chlorophyll, turn white, and die. Based on these results, researchers have concluded that carotenoids also serve a protective function.

To understand why carotenoids are protective, recall from Chapter 2 that photons—especially the high-energy, short-wavelength photons in the ultraviolet part of the electromagnetic spectrum—contain enough energy to knock electrons out of atoms and create free radicals. Free radicals, in turn, trigger reactions that degrade molecules.

Carotenoids "quench" free radicals by accepting or stabilizing unpaired electrons. As a result, they protect chlorophyll molecules from harm. When carotenoids are absent, chlorophyll molecules are destroyed and photosynthesis stops. Starvation and death follow.

FLAVONOIDS ARE A NATURAL SUNSCREEN Flavonoids are accessory pigments that also protect plants from high-energy radiation. They are found in the vacuoles of leaf cells, and function by absorbing ultraviolet light. In their absence, chlorophyll molecules are subject to damage from the free radicals triggered by UV radiation. Flavonoids function as a sunscreen for leaves and stems.

The message here is that the energy in sunlight is a double-edged sword. It makes photosynthesis possible, but it can also lead to the formation of free radicals that damage cells. The role of carotenoids and flavonoids as protective pigments is crucial.

THE STRUCTURE OF CHLOROPHYLL As **Figure 10.6b** shows, chlorophyll *a* and chlorophyll *b* are similar in structure. Both

(a) β-carotene

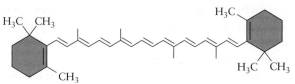

H_3C CH_3 CH_3 H_3C H_3C CH_3

(b) Chlorophylls *a* and *b*

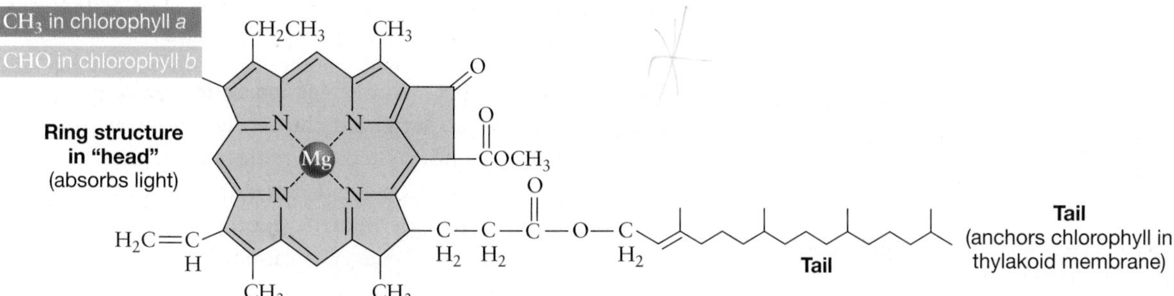

FIGURE 10.6 Photosynthetic Pigments Contain Ring Structures. **(a)** Carotene is an orange pigment found in carrot roots and other plant tissues. **(b)** Although chlorophylls *a* and *b* are very similar structurally, they have the distinct absorption spectra shown in Figure 10.5a.

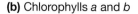

have two fundamental parts: a long tail made up of isoprene subunits (introduced in Chapter 6) and a "head" consisting of a large ring structure with a magnesium atom in the middle. The tail keeps the molecule embedded in the thylakoid membrane; the head is where light is absorbed.

Just what is "absorption"? What happens when a photon of a particular wavelength—say, red light with a wavelength of 680 nm—strikes a chlorophyll molecule?

When Light Is Absorbed, Electrons Enter an Excited State

When a photon strikes a chlorophyll molecule, the photon's energy can be transferred to an electron in the chlorophyll molecule's head region. In response, the electron is "excited," or raised to a higher energy state.

The excited electron states that are possible in a particular pigment are discrete—meaning, incremental rather than continuous—and can be represented as lines on an energy scale. These discrete energy levels are a property of the electron configurations in a particular pigment.

Figure 10.7 shows the ground state, or unexcited state, as 0 and the higher energy states as 1 and 2. If the difference between the possible energy states is the same as the energy in the photon, the photon can be absorbed and an electron is excited to that energy state.

In chlorophyll, for example, the energy difference between the ground state and state 1 is equal to the energy in a red photon, while the energy difference between state 0 and state 2 is equal to the energy in a blue photon. Thus, chlorophyll can readily absorb red photons and blue photons.

Chlorophyll does not absorb green light well, because there is no discrete step—no difference in possible energy states for its electrons—that corresponds to the amount of energy in a green photon.

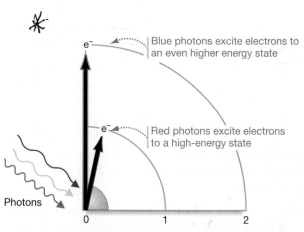

e^- Blue photons excite electrons to an even higher energy state

e^- Red photons excite electrons to a high-energy state

Photons

0 1 2

Energy state of electrons in chlorophyll

FIGURE 10.7 Electrons Are Promoted to High-Energy States When Photons Strike Chlorophyll. The discrete energy states labeled 1 and 2 are a property of chlorophyll's structure.

✔**QUESTION** Suppose a pigment had a discrete energy state that corresponded to the energy in green light. Where would you draw this energy state on this diagram?

Wavelengths in the ultraviolet part of the spectrum have so much energy that they may actually eject electrons from a pigment molecule and create a free radical. In contrast, wavelengths in the infrared regions have so little energy that in most cases they merely increase the movement of atoms in the pigment, generating heat—meaning molecular movement—rather than exciting electrons.

But if a pigment absorbs a photon with the right amount of energy, energy in the form of electromagnetic radiation is transferred to that electron. The electron now has high potential energy. What happens next?

If the excited electron simply falls back to its ground state, some of the absorbed energy is released as heat and the rest is released as electromagnetic radiation (light). This event is called

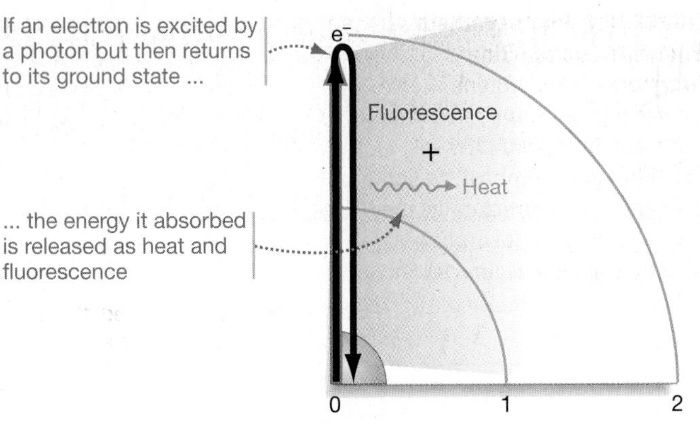
If an electron is excited by a photon but then returns to its ground state ...
... the energy it absorbed is released as heat and fluorescence
e⁻
Fluorescence
+
Heat
0
1
2
Energy state of electrons in chlorophyll

FIGURE 10.8 Fluorescence Occurs When Excited Electrons Fall Back to the Ground State.

fluorescence (**Figure 10.8**). Because some of the original photon's energy is transformed to heat, the electromagnetic radiation that is given off during fluorescence has lower energy and a longer wavelength than the original photon did.

In a chloroplast, though, only about 2 percent of the red and blue photons that chlorophyll absorbs produce fluorescence. The other 98 percent of electrons drive photosynthesis.

To understand what happens to these excited electrons, it's important to recognize that chlorophyll molecules work in groups—not individually. In the thylakoid membrane, 200–300 chlorophyll molecules and accessory pigments such as carotenoids group together with an array of proteins, forming structures called the antenna complex and the reaction center. These complexes, along with the proteins that capture and process excited electrons, comprise a **photosystem**.

THE ANTENNA COMPLEX When a red or blue photon strikes a pigment molecule in the **antenna complex**, the energy is absorbed and an electron is excited in response. This energy—but not the electron itself—is passed to a nearby chlorophyll molecule, where another electron is excited in response. This phenomenon is known as resonance. Chlorophyll molecules in the antenna complex transfer resonance energy.

Once the energy is transferred, the original excited electron falls back to its ground state. In this way, energy is transferred from one chlorophyll molecule to the next inside the antenna complex. Like a radio antenna, it receives a specific wavelength of electromagnetic radiation and transfers it to a receiver. In a photosystem, the receiver is called the reaction center.

THE REACTION CENTER At the **reaction center**, excited electrons are transferred to a specialized chlorophyll molecule that acts as an electron acceptor. When this molecule becomes reduced, the energy transformation event that started with the absorption of light becomes permanent: Electromagnetic energy is transformed to chemical energy. It cannot be reemitted as fluorescence. The redox reaction that occurs in the reaction center results in the production of chemical energy from sunlight.

Note that in the absence of light, the electron acceptor does not accept electrons. It remains in an oxidized state, because the redox reactions that transfer an electron to the electron acceptor are endergonic. But when light excites electrons in chlorophyll to a high-energy state, the reactions become exergonic. In this way, the energy in light transforms an endergonic reaction to an exergonic one.

Figure 10.9 summarizes how chlorophyll interacts with light energy by illustrating the three possible fates of electrons that are excited by photons in photosynthetic pigments. The electrons can:

1. drop back down to a low energy level and cause fluorescence, or
2. excite an electron in a nearby pigment and induce resonance, or
3. be transferred to an electron acceptor in a redox reaction.

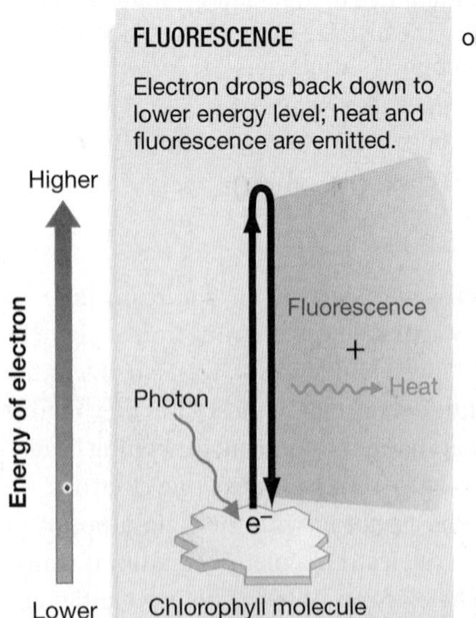
FLUORESCENCE
Electron drops back down to lower energy level; heat and fluorescence are emitted.
Higher
Energy of electron
Lower
Fluorescence
+
Heat
Photon
e⁻
Chlorophyll molecule

or

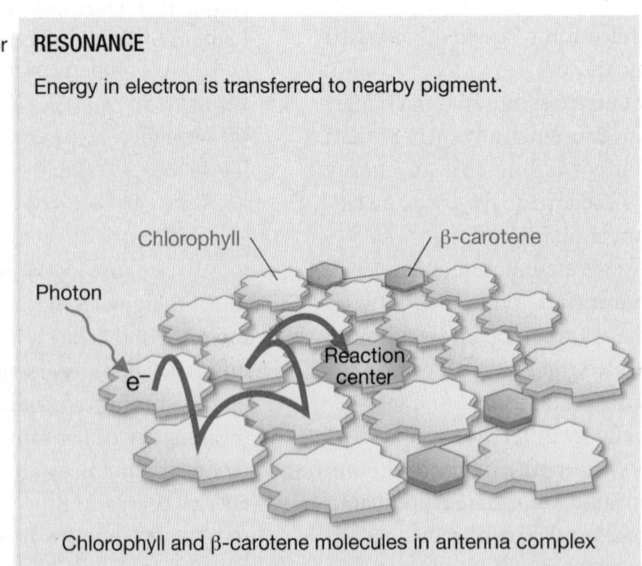
RESONANCE
Energy in electron is transferred to nearby pigment.
Chlorophyll
β-carotene
Photon
e⁻
Reaction center
Chlorophyll and β-carotene molecules in antenna complex

or

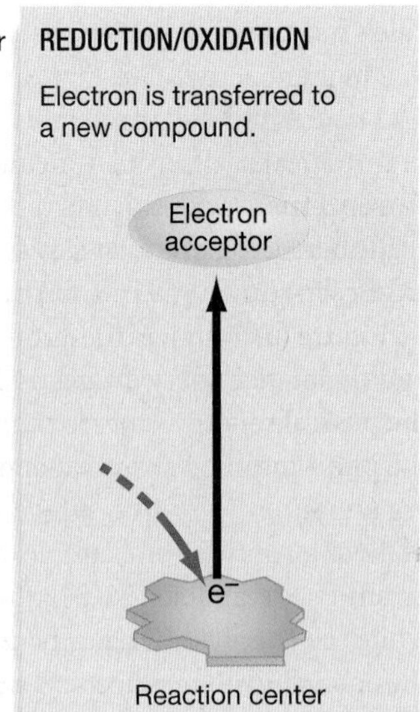
REDUCTION/OXIDATION
Electron is transferred to a new compound.
Electron acceptor
e⁻
Reaction center

FIGURE 10.9 Three Fates for Excited Electrons in Photosynthetic Pigments. When sunlight promotes electrons in pigments to a high-energy state, three things can happen: They can fluoresce, pass energy to a nearby pigment via resonance, or transfer the electron to an electron acceptor.

Fluorescence is typical of isolated pigments, resonance occurs in antenna complex pigments, and redox occurs in reaction center pigments.

Now the question is, what happens to the high-energy electrons that are transferred to the reaction center? Specifically, how are they used to manufacture sugar?

CHECK YOUR UNDERSTANDING

If you understand that . . .

- Pigments absorb specific wavelengths of light.
- When a chlorophyll molecule in the antenna complex of a chloroplast membrane absorbs red or blue light, one of its electrons is promoted to a high-energy state.
- In the antenna complex, high-energy electrons transmit their energy among chlorophyll molecules. When energy is transferred to a chlorophyll molecule in the reaction center, the electron that is excited in response reduces an electron acceptor. In this way, light energy is transformed to chemical energy.

✓ You should be able to . . .

Consider what happens when you strike a tuning fork and then touch the vibrating tuning fork against another tuning fork. Explain which event is analogous to absorbing a photon, and which event is analogous to transferring resonance energy.

Answers are available in Appendix B.

10.3 The Discovery of Photosystems I and II

During the 1950s the fate of the high-energy electrons in photosystems was the central issue facing biologists interested in photosynthesis. A key breakthrough came from a simple experiment on how green algae responded to various wavelengths of light. The algal cells being studied responded to wavelengths of 700 nm and 680 nm, which are in the far-red and red portions of the visible spectrum, respectively.

Robert Emerson found that if the cells were illuminated with either far-red light or red light, the photosynthetic response was moderate (**Figure 10.10**). But if cells were exposed to a combination of far-red and red light, the rate of photosynthesis increased dramatically—much more than the sum of the rates produced by each wavelength independently. This phenomenon was called the enhancement effect. Why it occurred was a complete mystery at the time.

The puzzle was solved by Robin Hill and Faye Bendall, who proposed that green algae and plants have two distinct types of reaction centers. One reaction center, called **photosystem II**, interacts with a different reaction center, referred to as **photosystem I** (because it was discovered first). According to the two-photosystem hypothesis, the enhancement effect occurs because photosynthesis is much more efficient when both photosystems operate together.

EXPERIMENT

QUESTION: Red and far-red light each stimulate a moderate rate of photosynthesis. How does a combination of both wavelengths affect the rate of photosynthesis?

HYPOTHESIS: When red and far-red light are combined, the rate of photosynthesis will double.

NULL HYPOTHESIS: When red and far-red light are combined, the rate of photosynthesis will be the same as for each wavelength alone.

EXPERIMENTAL SETUP:

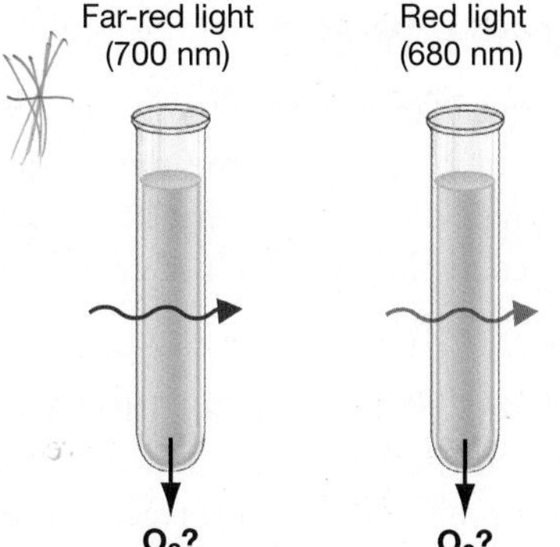

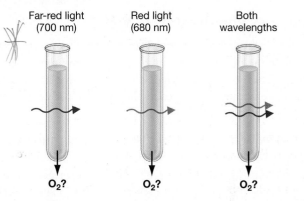

1. Expose algal cells to far-red light and then red light. Record oxygen produced as a measure of rate of photosynthesis.

2. Expose same cells to a combination of both lights.

PREDICTION: When the two wavelengths are combined, the rate of photosynthesis will double.

PREDICTION OF NULL HYPOTHESIS: When the two wavelengths are combined, the rate of photosynthesis will be the same as for each wavelength alone.

RESULTS:

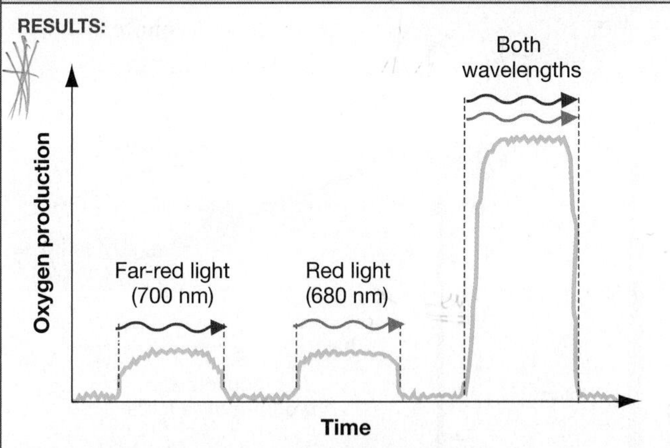

CONCLUSION: Neither hypothesis is correct. The combination of both wavelengths more than doubles the rate of photosynthesis. A new hypothesis is required to explain this enhancement effect.

FIGURE 10.10 Discovery of the "Enhancement Effect" of Red and Far-Red Light.

SOURCE: Emerson, R. and W. Arnold. 1932. The photochemical reaction in photosynthesis. *Journal of General Physiology* 16: 191–205.

✓QUESTION Was it important for the researchers to keep the density of algal cells fairly constant in each treatment? Explain why or why not.

Subsequent work has shown that the two-photosystem hypothesis is correct. In green algae and land plants, thylakoid membranes contain photosystems that differ in structure and function but complement each other.

To figure out how the two photosystems work, investigators focused on species of photosynthetic bacteria that have photosystems similar to either photosystem I or II, but not both. Once each type of photosystem was understood in isolation, researchers explored how they work in combination. Let's do the same—we'll analyze photosystem II, then photosystem I, and then how the two interact.

How Does Photosystem II Work?

To study photosystem II, researchers focused on organisms known as the purple nonsulfur bacteria and the purple sulfur bacteria. These cells have a single photosystem that has many of the same components observed in photosystem II of cyanobacteria ("blue-green bacteria"), algae, and plants.

PHEOPHYTIN ACCEPTS ELECTRONS AND TRANSFERS THEM TO AN ELECTRON TRANSPORT CHAIN In photosystem II, the action begins when the antenna complex transmits energy to the reaction center and the molecule pheophytin comes into play (**Figure 10.11**).

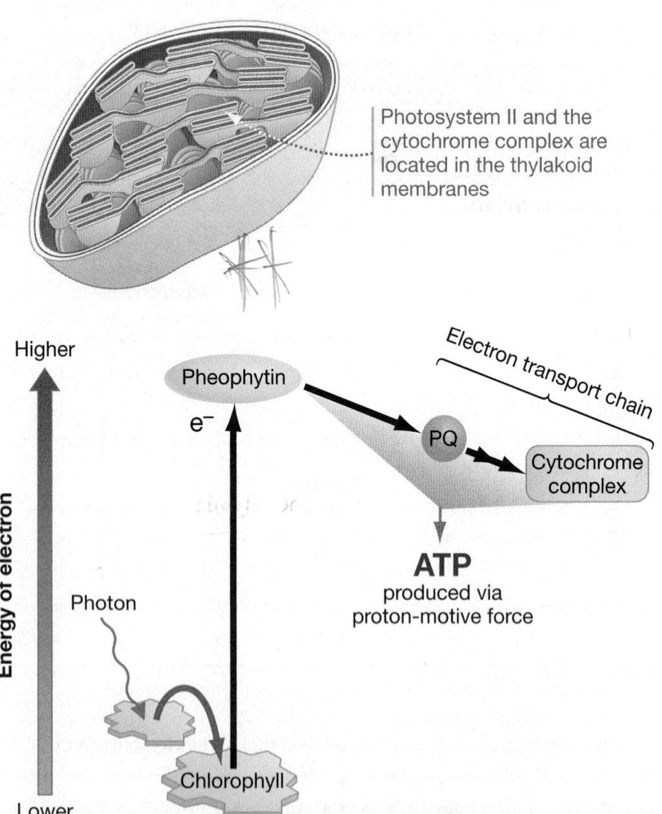

FIGURE 10.11 Photosystem II Feeds High-Energy Electrons to an Electron Transport Chain. When an excited electron leaves the chlorophyll molecule in the reaction center of photosystem II, the electron is accepted by pheophytin, transferred to plastoquinone (PQ), and then stepped down in energy along an electron transport chain.

Structurally, **pheophytin** is identical to chlorophyll except that pheophytin lacks a magnesium atom in its head region. Functionally, the two molecules are extremely different.

Instead of acting as a pigment that promotes an electron when it absorbs a photon, pheophytin acts as an electron acceptor. When an electron in the reaction center chlorophyll is excited energetically, the electron binds to pheophytin. The reduction of pheophytin (and the accompanying oxidation of the reaction center chlorophyll pigment) completes the energy transformation step that started with the absorption of light in the antenna complex.

Electrons that reach pheophytin are passed to an electron transport chain (ETC) in the thylakoid membrane. In both structure and function, this group of molecules is similar to the ETC in the inner membrane of mitochondria (Chapter 9).

- Structurally, the ETC associated with photosystem II and the ETC in the mitochondrion both contain quinones and cytochromes.
- Functionally, the redox reactions that occur in both ETCs result in protons being pumped from one side of an internal membrane to the other. The proton gradient that builds up drives ATP production via ATP synthase. Photosystem II triggers chemiosmosis and ATP synthesis in the chloroplast.

THE ELECTRON TRANSPORT CHAIN SETS UP A PROTON GRADIENT THAT DRIVES ATP SYNTHASE **Figure 10.12** explains how photosystem II works in more detail. Start by focusing on the molecule called **plastoquinone**—a quinone symbolized PQ. Note that PQ connects the excited electron in pheophytin with components of the ETC called the cytochrome complex.

Now recall from Chapter 9 that quinones are small hydrophobic molecules. Because plastoquinone is lipid soluble and not anchored to a protein, it is free to move from one side of the thylakoid membrane to the other.

When it receives electrons from pheophytin, plastoquinone carries them to the other side of the membrane and delivers them to more electronegative molecules in the chain, including a cytochrome.

In this way, plastoquinone shuttles electrons from pheophytin to the cytochrome complex. The electrons are then passed through a series of iron- and copper-containing proteins within the cytochrome complex. The potential energy released by these reactions allows protons to be added to other plastoquinone molecules, which carry them to the lumen side of the thylakoid membrane.

The protons transported by plastoquinone result in a large concentration of protons in the thylakoid lumen. When photosystem II is active, the pH of the thylakoid interior reaches 5 while the pH of the stroma hovers around 8. Because the pH scale is logarithmic, the difference of 3 units means that the concentration of H^+ is $10 \times 10 \times 10 = 1000$ times higher in the lumen than in the stroma. In addition, the stroma becomes negatively charged relative to the thylakoid lumen.

The net effect of electron transport, then, is to set up a large proton gradient that drives H^+ out of the thylakoid lumen and into the stroma. Based on your reading of Chapter 9, it should

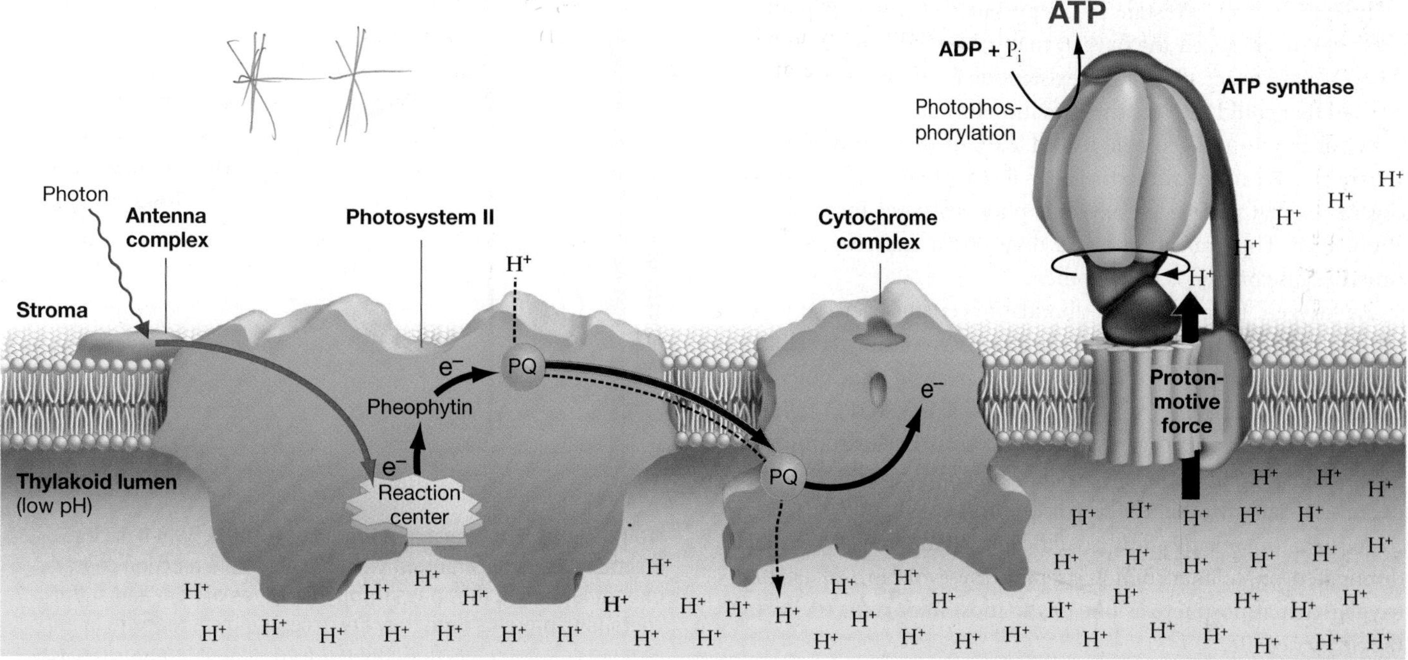

FIGURE 10.12 The ETC in PS II Pumps Protons. Plastoquinone (PQ) carries electrons from photosystem II along with protons from the stroma. The electrons are passed to the cytochrome complex, and the protons are released in the thylakoid lumen, creating a proton-motive force that drives ATP synthesis.

come as no surprise that this proton-motive force drives the production of ATP.

Specifically, proton flow is an exergonic process that drives the endergonic synthesis of ATP from ADP and P_i. The stream of protons through ATP synthase causes conformational changes that drive the phosphorylation of ADP. This process is called **photophosphorylation**.

Photophosphorylation is similar to the oxidative phosphorylation that occurs in plant and animal mitochondria. Both depend on chemiosmosis—a process that you can review in the study area at *www.masteringbiology.com.*

(MB) **Web Activity** Chemiosmosis

✔If you understand photosystem II, you should be able to compare and contrast photosystem II to the electron transfer chain of mitochondria. You also should be able to explain where high-energy electrons come from in each system and what the eventual product is.

The photosystem II story is not yet complete, however, because we haven't accounted for the electrons that flow through the system. The electron that was transferred from chlorophyll to pheophytin, in the photosystem II reaction center, needs to be replaced. In addition, the electron transport chain of photosystem II needs to donate its electrons to some final electron acceptor.

Where do the electrons required by photosystem II come from, and where do they go?

PHOTOSYSTEM II OBTAINS ELECTRONS BY OXIDIZING WATER Think back to the overall reaction for photosynthesis:

$$6CO_2 + 12H_2O + \text{light energy} \longrightarrow C_6H_{12}O_6 + 6H_2O + 6O_2$$

In the presence of sunlight, carbon dioxide and water are used to produce carbohydrate, water, and oxygen gas.

Now recall that experiments with radioisotopes of oxygen showed that the oxygen atoms in O_2 come from water, not from carbon dioxide. As it turns out, the electrons that enter photosystem II also come from water. The oxygen-generating reaction can be written as

$$2H_2O \longrightarrow 4H^+ + 4e^- + O_2$$

Because electrons are removed from water, the molecule becomes oxidized. This reaction is referred to as "splitting" water. It supplies a steady stream of electrons for photosystem II and is catalyzed by enzymes that are physically integrated into the photosystem II complex.

When excited electrons leave photosystem II and enter the ETC, the photosystem becomes so electronegative that enzymes can remove electrons from water, leaving protons and oxygen. The oxygen-generating reaction is highly endergonic. It is possible only because the energy in sunlight drives it by removing electrons from photosystem II.

OXYGENIC PHOTOSYNTHESIS AND THE OXYGEN ATMOSPHERE Among all life-forms, photosystem II is the only protein complex that can catalyze the splitting of water molecules. Organisms such as cyanobacteria, algae, and plants that have photosystem II perform **oxygenic** ("oxygen-producing") photosynthesis, because they generate oxygen as a by-product of the process. The purple sulfur and purple nonsulfur bacteria, which cannot oxidize water, perform **anoxygenic** ("no oxygen-producing") photosynthesis.

It is difficult to overstate the importance of oxygenic photosynthesis. It produced the oxygen that is keeping you alive right now. O_2 was, in fact, almost nonexistent on Earth before enzymes evolved that could catalyze the oxidation of water.

According to the geologic record, oxygen levels in the atmosphere and oceans began to rise only about 2 billion years ago, as organisms that performed oxygenic photosynthesis increased in abundance. This transition was a disaster for anaerobic organisms, because oxygen is toxic to them.

As oxygen became more abundant, certain bacterial cells evolved the ability to use it as an electron acceptor during cellular respiration. O_2 is so electronegative that it creates a huge potential energy drop for the electron transport chains involved in cellular respiration. As a result, organisms that use O_2 as an electron acceptor in cellular respiration can produce much more ATP than can organisms that use other electron acceptors (see Chapter 9).

Aerobic organisms grow so efficiently that they have long dominated our planet. Biologists rank the evolution of Earth's oxygen-rich atmosphere as one of the most important events in the history of life.

Determining exactly how photosystem II splits water and generates oxygen may be the greatest challenge currently facing researchers interested in photosynthesis. This issue has important practical applications: If human chemists could replicate the reaction, it might be possible to produce huge volumes of O_2 and hydrogen gas (H_2) from water. The resulting H_2 could provide a clean, inexpensive fuel for vehicles.

✔If you understand photosystem II, you should be able to make an energy flowchart that includes the antenna complex, ATP synthase, pheophytin, light, the proton gradient, an ETC, and a reaction center, then add notes explaining where the enzyme complex that splits water fits in.

How Does Photosystem I Work?

Recall that researchers dissected photosystem II by studying similar, but simpler, photosystems in purple nonsulfur and purple sulfur bacteria. To understand the structure and function of photosystem I, they turned to heliobacteria ("sun-bacteria").

Like purple nonsulfur and purple sulfur bacteria, heliobacteria use the energy in sunlight to promote electrons to a high-energy state. But instead of being passed to an electron transport chain that pumps protons across a membrane, the high-energy electrons in heliobacteria are used to reduce NAD^+. When NAD^+ gains two electrons and a proton, NADH is produced.

In the cyanobacteria and algae and land plants, a similar set of light-capturing reactions reduces a phosphorylated version of NAD^+, symbolized $NADP^+$, yielding NADPH. Both NADH and NADPH function as electron carriers.

Figure 10.13 explains how photosystem I works in chloroplasts—put your finger on the "2 photons" arrows and trace the steps that follow.

- Pigments in the antenna complex absorb photons and pass the energy to the reaction center.

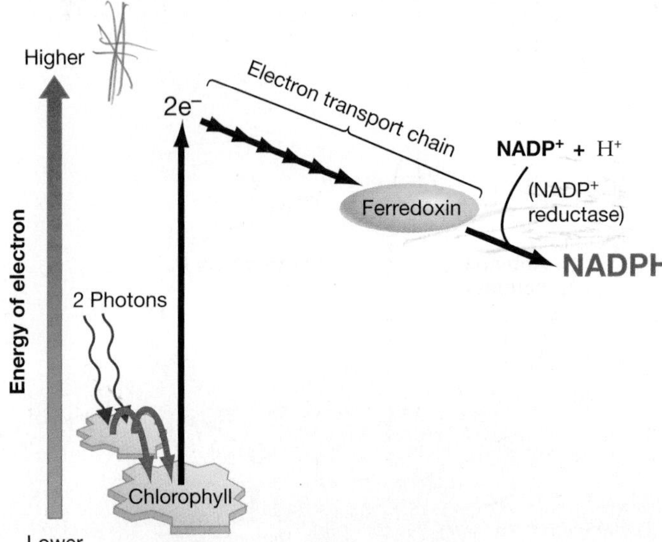

FIGURE 10.13 Photosystem I Produces NADPH. When excited electrons leave the chlorophyll molecule in the reaction center of photosystem I, they pass through a series of iron- and sulfur-containing proteins until they are accepted by ferredoxin. In an enzyme-catalyzed reaction, the reduced form of ferredoxin reacts with $NADP^+$ to produce NADPH.

- Electrons are excited in specialized chlorophyll molecules inside the reaction center.
- The high-energy electrons are passed through an electron transport chain—first to a series of iron- and sulfur-containing proteins inside the photosystem, then to a molecule called **ferredoxin**, then to the enzyme ferredoxin/$NADP^+$ oxidoreductase—also called $NADP^+$ reductase.
- $NADP^+$ reductase transfers two electrons and a proton to $NADP^+$. This reaction forms NADPH.

The photosystem itself and $NADP^+$ reductase are anchored in the thylakoid membrane; ferredoxin is closely associated with the bilayer.

To summarize: Photosystem I produces NADPH, which is similar in function to the NADH and $FADH_2$ produced by the citric acid cycle. NADPH is an electron carrier that can donate electrons to other compounds and thus reduce them. Photosystem II, in contrast, produces a proton gradient that drives the synthesis of ATP.

In combination, then, photosystems II and I produce chemical energy stored in ATP as well as reducing power in the form of NADPH. Although several groups of bacteria have just one of the two photosystems, the cyanobacteria, algae, and plants have both. In these organisms, how do the two photosystems interact?

The Z Scheme: Photosystems II and I Work Together

Figure 10.14 illustrates the **Z scheme** model for how photosystems II and I interact. The name was inspired by the proposed path of

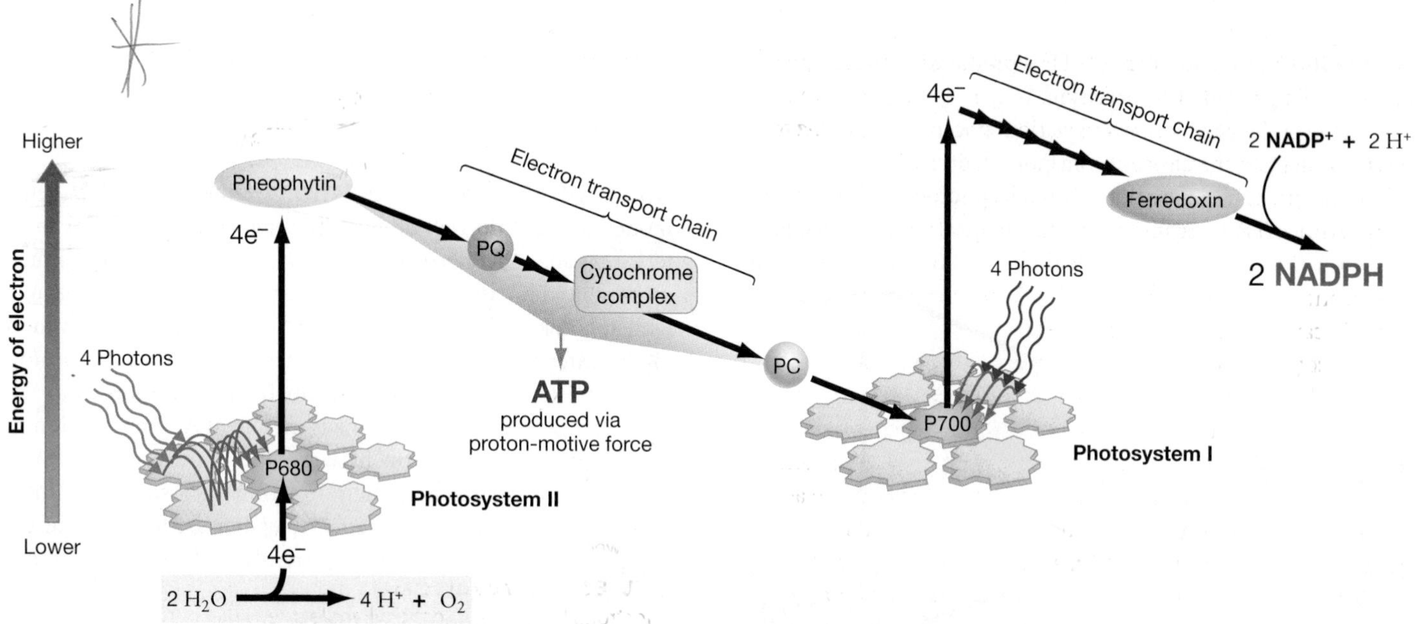

FIGURE 10.14 The Z Scheme Model Links Photosystems I and II. The Z scheme proposes that electrons from photosystem II enter photosystem I, where they are promoted to an energy state high enough to make the reduction of $NADP^+$ possible.

electrons through the two photosystems, as plotted on a vertical axis representing the changes occurring in their potential energy.

To drive home how photosynthesis works, trace the route of electrons through Figure 10.14 with your finger. Start on the lower left. The process starts when photons excite electrons in the chlorophyll molecules of photosystem II's antenna complex. When the energy in the excited electrons is transferred to the reaction center, a special pair of chlorophyll molecules named P680 passes excited electrons to pheophytin.

When pheophytin is reduced, it transfers high-energy electrons to an electron transport chain. There the electrons are gradually stepped down in potential energy through redox reactions among a series of quinones and cytochromes. Using the energy released by the redox reactions, plastoquinone (PQ) carries protons across the thylakoid membrane. ATP synthase uses the resulting proton-motive force to phosphorylate ADP, creating ATP.

When electrons reach the end of photosystem II's electron transport chain, they are passed to a small diffusible protein called **plastocyanin** (symbolized PC in Figure 10.14). Plastocyanin picks up an electron from the cytochrome complex, diffuses through the lumen of the thylakoid, and donates the electron to photosystem I.

Stop tracing for a moment, and consider the following:

- Plastocyanin is key—it forms a physical link between photosystem II and photosystem I.
- A single plastocyanin molecule can shuttle over 1000 electrons per second between photosystems.
- The flow of electrons between photosystems, by means of plastocyanin, is important because it replaces electrons that are carried away from a chlorophyll molecule called P700 in the photosystem I reaction center.

Now keep going. The electrons that flow from photosystem II to P700, via plastocyanin, are eventually transferred to the protein ferredoxin, which passes electrons to an enzyme that catalyzes the reduction of $NADP^+$ to NADPH.

Finally, direct your attention back to the lower left portion of the figure. Note that the electrons that initially left photosystem II are replaced by electrons that are stripped away from water, producing oxygen gas as a by-product.

✔You should be able to explain (1) the role of plastocyanin in linking the two photosystems, (2) where the electrons that flow through the system have their highest potential energy, and (3) why the Z scheme is sometimes referred to as **noncyclic electron flow.**

UNDERSTANDING THE ENHANCEMENT EFFECT The Z scheme helps explain the enhancement effect documented in Figure 10.10. When algal cells are illuminated with wavelengths at 680 nm, in the red portion of the spectrum, only photosystem II can run at a maximum rate. The overall rate of electron flow through the Z scheme is moderate because photosystem I's efficiency is reduced.

Similarly, when cells receive only wavelengths at 700 nm, in the far red, only photosystem I is capable of peak efficiency; photosystem II is working at a below-maximum rate, so the overall rate of electron flow is reduced.

But when both wavelengths are available at the same time, both photosystem II and photosystem I are activated and work at a maximum rate, leading to enhanced efficiency.

CYCLIC PHOTOPHOSPHORYLATION PRODUCES ADDITIONAL ATP Recent evidence indicates that an alternative electron path, called **cyclic photophosphorylation**, also occurs in green algae and plants

(**Figure 10.15**). In these species, ATP is produced in the Z-scheme and in cyclic photophosphorylation.

During cyclic photophosphorylation, photosystem I transfers electrons back to the electron transport chain in photosystem II, to augment ATP generation through photophosphorylation. This "extra" ATP is required for the chemical reactions that reduce carbon dioxide (CO_2) and produce sugars. Cyclic photophosphorylation coexists with the Z scheme and produces additional ATP.

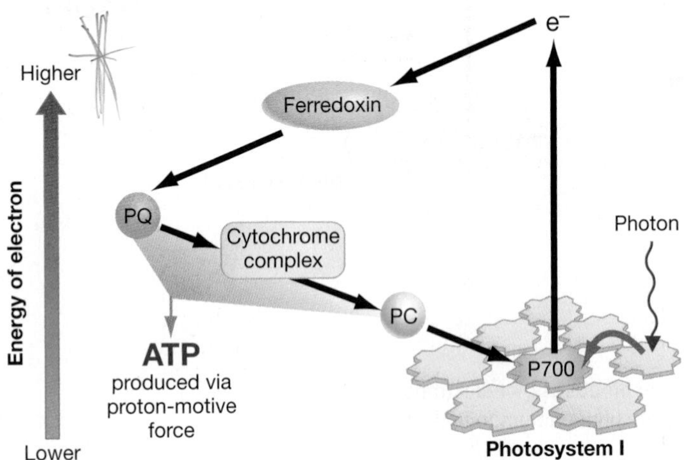

FIGURE 10.15 Cyclic Photophosphorylation Produces ATP. Cyclic electron transport is an alternative to the Z scheme. Instead of being donated to $NADP^+$, electrons cycle through the system, resulting in the production of additional ATP via photophosphorylation.

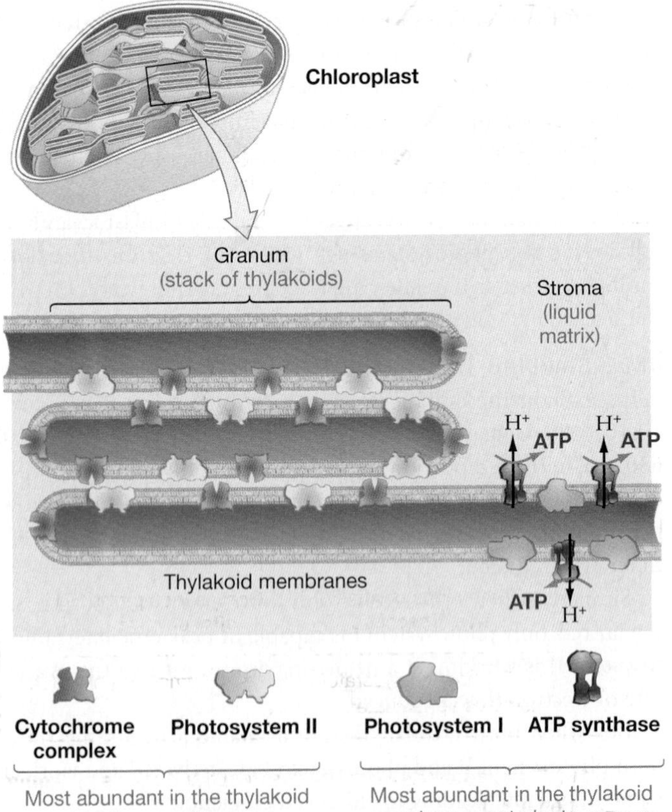

FIGURE 10.16 Photosystems I and II Occur in Separate Regions of Thylakoid Membranes within Grana. It is not yet clear how the arrangement of these components in the thylakoid relates to the paths of electrons in the Z scheme and during cyclic photophosphorylation.

WHERE ARE PS II AND PS I LOCATED? Although the Z-scheme model has held up well under experimental tests, a major mystery remains. As **Figure 10.16** shows, photosystem II is much more abundant in the interior, stacked membranes of grana, while photosystem I and ATP synthase are much more common in the exterior, unstacked membranes. Notice that the proton gradient established by photosystem II drives protons into the stroma, meaning that the stroma is the site of ATP production.

The physical separation between the photosystems is perplexing, considering that their functions are so tightly integrated according to the Z scheme. Why they are found in different parts of the thylakoid is currently the focus of intense research and debate.

The spatial relationships of PS II and PS I may be puzzling, but the fate of the ATP and NADPH produced by photosystems I and II is well documented. Chloroplasts use ATP and NADPH to reduce carbon dioxide to sugar. Your life, and the life of most organisms, depends on this process. How does it happen?

CHECK YOUR UNDERSTANDING

If you understand that . . .

- Photosystem II contributes high-energy electrons to an electron transport chain that pumps protons, creating a proton-motive force that drives ATP synthase.
- Photosystem I makes NADPH.

✓ You should be able to . . .

1. Make a model of the Z scheme using paper cutouts representing the following elements: the antenna complexes of photosystems II and I, the ETCs in photosystems I and II, pheophytin, plastoquinone, plastocyanin, cytochrome complex, ferredoxin, the reaction that splits water, and the reduction of $NADP^+$ to form NADPH.
2. Using dimes to represent electrons, explain how they flow through the photosystems.

Answers are available in Appendix B.

10.4 How Is Carbon Dioxide Reduced to Produce Glucose?

The reactions analyzed in Section 10.3 are triggered by light. This is logical, because their entire function is focused on transforming electromagnetic energy in the form of sunlight into chemical energy in the phosphate bonds of ATP and the electrons of NADPH. The reactions that produce sugar from carbon dioxide, in contrast, are not triggered directly by light. Instead, they depend on the ATP and NADPH produced by the light-capturing reactions of photosynthesis.

The Calvin Cycle Fixes Carbon

Carbon fixation is the addition of carbon dioxide to an organic compound. The word "fix" is appropriate because the process converts or fixes CO_2 to a biologically useful form. Once carbon atoms are fixed, they can be used to build the molecules found in cells.

Carbon fixation is a redox reaction—the carbon atom in CO_2 is reduced. Research on how this happens in chloroplasts gained momentum just after World War II, when radioactive isotopes of carbon became available for research purposes.

Melvin Calvin's group made great strides early in this effort, using a pulse-chase strategy. Recall from Chapter 7 that pulse-chase experiments introduce a pulse of labeled compound followed by a chase of unlabeled compound, and then follow the fate of the labeled compound over time.

In this case, the researchers fed green algae a pulse of $^{14}CO_2$ followed by a large amount of unlabeled CO_2 (**Figure 10.17**). After waiting a specified amount of time, they homogenized the cells by immersing them in hot alcohol, separated individual molecules in the extract via chromatography, and laid X-ray film over the chromatography surface.

If radioactively labeled molecules were present in the chromatograph, the energy they emitted would expose the film and create a dark spot. The labeled compounds could then be isolated and identified.

By varying the amount of time between starting the pulse of labeled $^{14}CO_2$ and analyzing the cells, Calvin and co-workers pieced together the sequence in which various intermediates formed. For example, when the team analyzed cells almost immediately after starting the $^{14}CO_2$ pulse, they found that the three-carbon compound 3-phosphoglycerate predominated. This result suggested that 3-phosphoglycerate was the initial product of carbon reduction. Stated another way, it appeared that carbon dioxide reacted with some unknown molecule to produce 3-phosphoglycerate.

This was an intriguing result, because 3-phosphoglycerate is also one of the 10 intermediates in glycolysis. The ATP and NADPH reactions manufacture carbohydrate; glycolysis breaks it down. Because the two processes are related in this way, it was logical that at least some intermediates in glycolysis and CO_2 reduction are the same.

RUBP IS THE INITIAL REACTANT WITH CO_2 Which compound reacts with CO_2 to produce 3-phosphoglycerate? This was the key, initial step. Calvin's group searched in vain for a two-carbon compound that might serve as the initial carbon dioxide acceptor and yield 3-phosphoglycerate.

Then, while Calvin was running errands one day, it occurred to him that the molecule reacting with carbon dioxide might contain five carbons, not two. Adding CO_2 to a five-carbon molecule would produce a six-carbon compound, which could then split in half to form 2 three-carbon molecules.

Experiments to test this hypothesis confirmed that the five-carbon compound **ribulose bisphosphate (RuBP)** is the initial reactant.

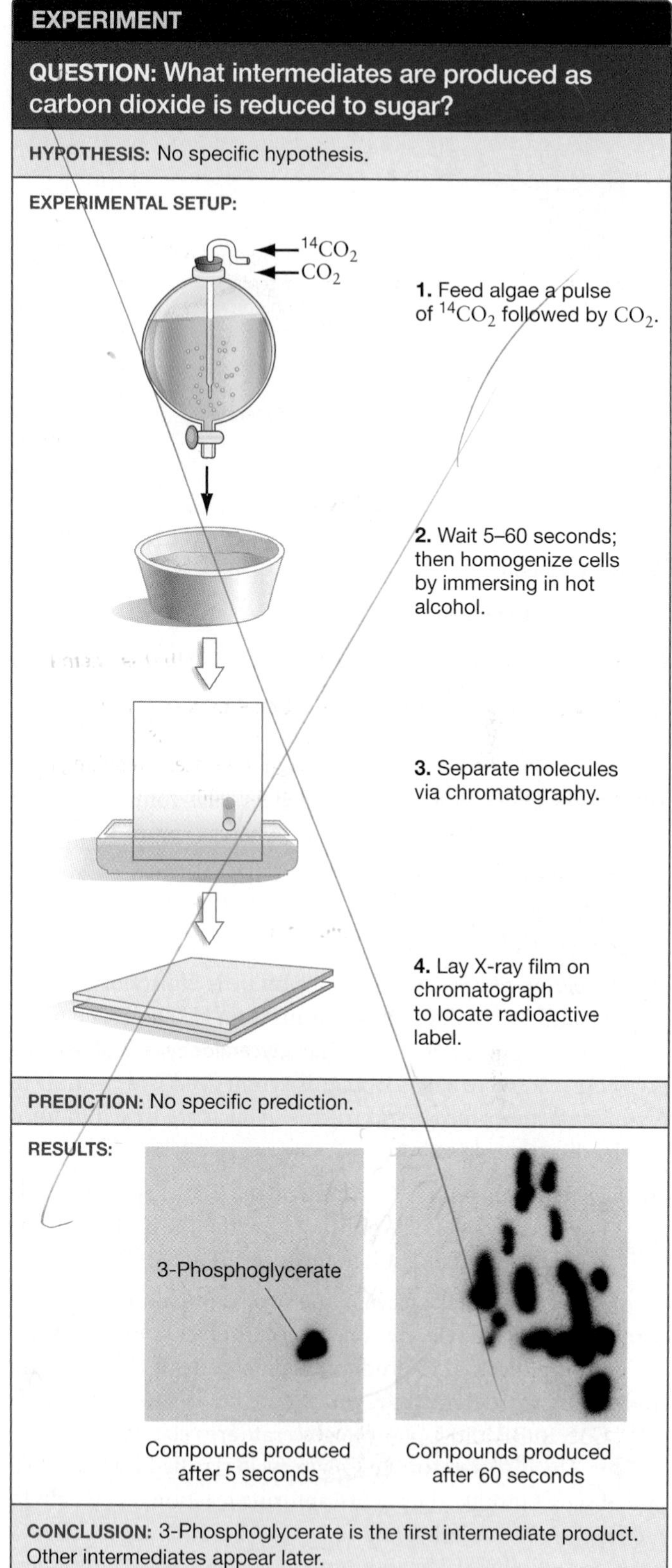

EXPERIMENT
QUESTION: What intermediates are produced as carbon dioxide is reduced to sugar?
HYPOTHESIS: No specific hypothesis.
EXPERIMENTAL SETUP:
$^{14}CO_2$
CO_2
1. Feed algae a pulse of $^{14}CO_2$ followed by CO_2.
2. Wait 5–60 seconds; then homogenize cells by immersing in hot alcohol.
3. Separate molecules via chromatography.
4. Lay X-ray film on chromatograph to locate radioactive label.
PREDICTION: No specific prediction.
RESULTS:
3-Phosphoglycerate
Compounds produced after 5 seconds
Compounds produced after 60 seconds
CONCLUSION: 3-Phosphoglycerate is the first intermediate product. Other intermediates appear later.

FIGURE 10.17 Experiments Revealed the Reaction Pathway Leading to Reduction of CO_2.

SOURCE: Benson, A. A., J. A. Bassham, et al. 1950. The path of carbon in photosynthesis. V. Paper chromatography and radioautography of the products. *Journal of the American Chemistry Society* 72: 1710–1718.

✔**QUESTION** Why wasn't this experiment based on a specific hypothesis and set of predictions?

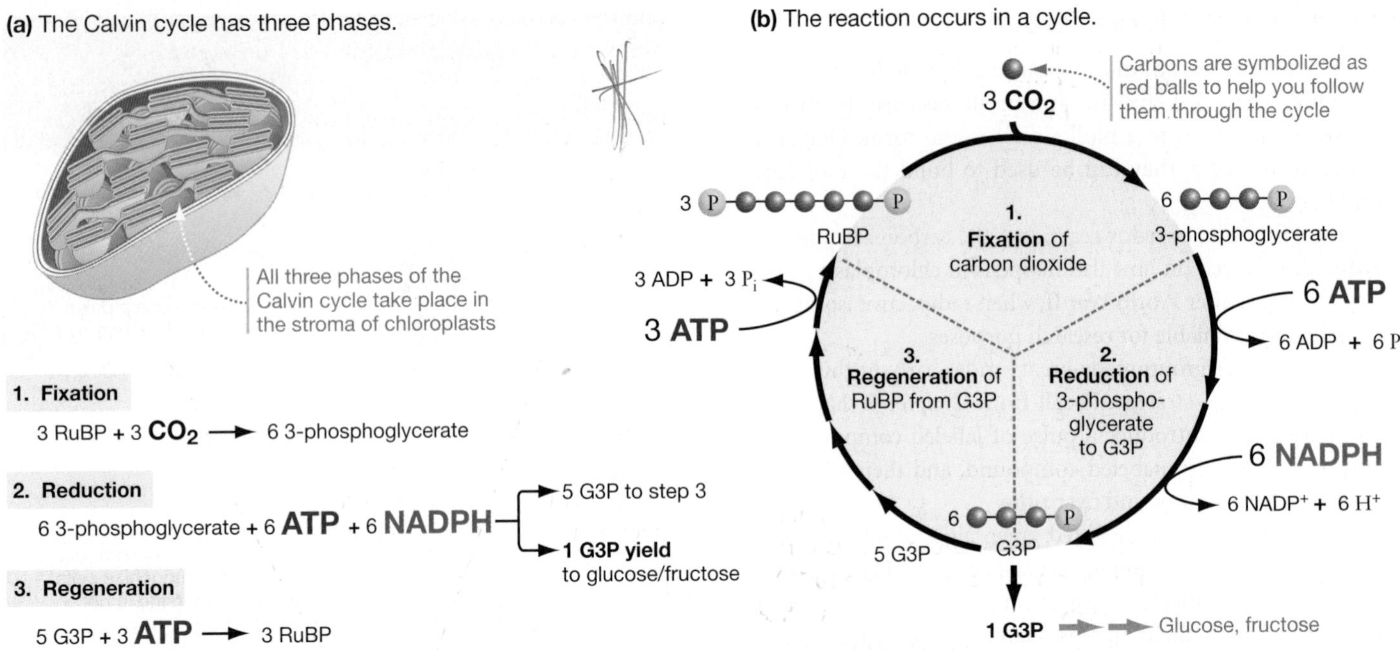

FIGURE 10.18 Carbon Dioxide Is Reduced in the Calvin Cycle. (a) Of the 6 G3Ps generated during the reduction phase, one is used in the synthesis of glucose—the other five are used to regenerate RuBP. Three molecules of CO_2 are required to make this happen. **(b)** The 3 RuBPs that are regenerated participate in fixation reactions, forming a cycle.

THE CALVIN CYCLE IS A 3-STEP PROCESS The complete Calvin cycle, as it came to be called, has three phases (**Figure 10.18**):

1. ***Fixation phase*** The Calvin cycle begins when CO_2 reacts with RuBP. This phase fixes carbon and produces two molecules of 3-phosphoglycerate.
2. ***Reduction phase*** 3-Phosphoglycerate is phosphorylated by ATP and then reduced by electrons from NADPH. The product is the phosphorylated sugar **glyceraldehyde-3-phosphate** (**G3P**). Some of the G3P that is synthesized is drawn off to manufacture glucose and fructose, which are linked to form the disaccharide sucrose.
3. ***Regeneration phase*** The rest of the G3P keeps the cycle going by serving as the substrate for the third phase in the cycle: reactions that result in the regeneration of RuBP.

All three phases take place in the stroma of chloroplasts. One turn of the Calvin cycle fixes one molecule of CO_2; three turns of the cycle are required to produce one molecule of G3P. ✔If you understand the Calvin cycle, you should be able to explain how CO_2, G3P, RuBP, and 3-phosphoglycerate are related.

The discovery of the Calvin cycle clarified how the ATP and NADPH produced by light-capturing reactions allow cells to reduce CO_2 to carbohydrate $(CH_2O)_n$. Because sugars store a great deal of potential energy, producing them takes a great deal of chemical energy. During photosynthesis, the energy required to reduce CO_2 to sugar is provided by ATP and NADPH synthesized in the light-capturing reactions.

The initial phase of the Calvin cycle—the reaction between RuBP and CO_2—is one of only two reactions that are unique to the Calvin cycle. Most reactions involved in reducing CO_2 also occur during glycolysis or other metabolic pathways.

The reaction between CO_2 and RuBP starts the transformation of carbon dioxide gas from the atmosphere into sugars. Plants use sugars to fuel cellular respiration and build leaves and other structures. Millions of non-photosynthetic organisms—from fungi to mammals—also depend on this reaction to provide the sugars they need for cellular respiration.

To review the Calvin cycle and the light-capturing reactions, go to the study area at *www.masteringbiology.com.*

MB **BioFlix™** Photosynthesis, **Web Activity** Photosynthesis

Ecologically, the addition of CO_2 to RuBP may be the most important chemical reaction on Earth. The enzyme that catalyzes it is fundamental to life. How does this protein work?

The Discovery of Rubisco

To find the enzyme that fixes CO_2, Arthur Weissbach and colleagues ground up spinach leaves, purified a large series of proteins from the resulting cell extracts, and tested each protein to see if it could catalyze the incorporation of $^{14}CO_2$ into RuBP to form 3-phosphoglycerate. Eventually they isolated the catalyst, an enzyme which is abundant in leaf tissue. The researchers' data suggested that it constituted at least 10 percent of the total protein in spinach leaves.

The CO_2-fixing enzyme, ribulose-1,5-bisphosphate carboxylase/oxygenase (commonly referred to as **rubisco**), is found in all photosynthetic organisms that use the Calvin cycle to fix carbon, and is thought to be the most abundant enzyme on Earth. The rubisco molecule is cube-shaped and has eight active sites where CO_2 is fixed.

Despite its large number of active sites, rubisco is a slow enzyme. Each active site catalyzes just three reactions per second; other enzymes typically catalyze thousands of reactions per second. Plants synthesize huge amounts of rubisco, possibly as an adaptation compensating for its lack of speed.

Besides being slow, rubisco is extremely inefficient because it catalyzes the addition of O_2 to RuBP as well as the addition of CO_2 to RuBP. This is a key point: Oxygen and carbon dioxide compete at the enzyme's active sites, which slows the rate of CO_2 reduction.

Why would an active site of rubisco accept both O_2 and CO_2? Given rubisco's importance in producing food for photosynthetic species, this trait would appear to be **maladaptive**—it reduces the fitness of individuals.

The reaction of O_2 with RuBP actually does more than just compete with the reaction of CO_2 at the same active site. One of the molecules that results from the addition of oxygen to RuBP is processed in reactions that consume ATP and O_2 and release CO_2. Part of this pathway occurs in chloroplasts, and part in peroxisomes and mitochondria. The reaction sequence resembles respiration, because it consumes oxygen and produces carbon dioxide. As a result, it is called **photorespiration** (**Figure 10.19**).

Because photorespiration consumes energy and releases fixed CO_2, it "undoes" photosynthesis. When photorespiration occurs, the overall rate of photosynthesis declines. Explaining why rubisco is so slow and why photorespiration occurs remains an unsolved problem in biological science.

Carbon Dioxide Enters Leaves through Stomata

Atmospheric carbon dioxide is a key reactant in photosynthesizing cells. It would seem straightforward, then, for CO_2 to diffuse directly into plants along a concentration gradient. But the situation is not this simple, because plants are covered with a waxy coating called a cuticle. This lipid layer prevents water from evaporating out of tissues, but it also prevents CO_2 from entering them.

How does CO_2 get into photosynthesizing tissues? The surface of a leaf is dotted with openings bordered by two distinctively shaped cells (**Figure 10.20a**). The paired cells are called **guard cells**, the opening is called a pore, and the entire structure is a **stoma** (plural: **stomata**).

If CO_2 concentrations inside the leaf are low as photosynthesis gets under way, chemical signals activate proton pumps in the membranes of guard cells. These pumps establish a charge gradient across the membrane. In response, potassium ions (K^+) move into the guard cells. When water follows along the newly created osmotic gradient, the guard cells swell and create a pore.

An open stoma allows CO_2 from the atmosphere to diffuse into air-filled spaces inside the leaf (**Figure 10.20b**). Eventually the CO_2 diffuses along a concentration gradient into the chloroplasts of photosynthesizing cells. A strong concentration gradient favoring entry of CO_2 is maintained by the Calvin cycle, which constantly uses up the CO_2 in chloroplasts.

Mechanisms for Increasing CO_2 Concentration

The oxygenation reaction that triggers photorespiration is favored when oxygen concentrations are high and CO_2 concentrations are low. But the atmosphere is 21 percent oxygen and only 0.03 percent carbon dioxide.

In addition, stomata are normally open during the day, when photosynthesis is occurring, and closed at night. But if the daytime is extremely hot and dry, leaf cells may lose a great deal of water to evaporation through their stomata. When this occurs, they must either close the openings and halt photosynthesis or risk death from dehydration. When conditions are hot and dry, then, CO_2 delivery stops—meaning that photosynthesis and growth slow. To make a bad situation worse, oxygen levels build

Reaction with carbon dioxide during photosynthesis:

RuBP + CO_2 ⟶ two 3-phosphoglycerate → used in Calvin cycle

Reaction with oxygen during photorespiration:

RuBP + O_2 ⟶ one 3-phosphoglycerate + one 2-phosphoglycolate

- one 3-phosphoglycerate → used in Calvin cycle
- one 2-phosphoglycolate → when processed, CO_2 is released and ATP is used

FIGURE 10.19 Photorespiration Competes with Photosynthesis. Rubisco catalyzes competing reactions with very different outcomes.

(a) Leaf surfaces contain stomata.

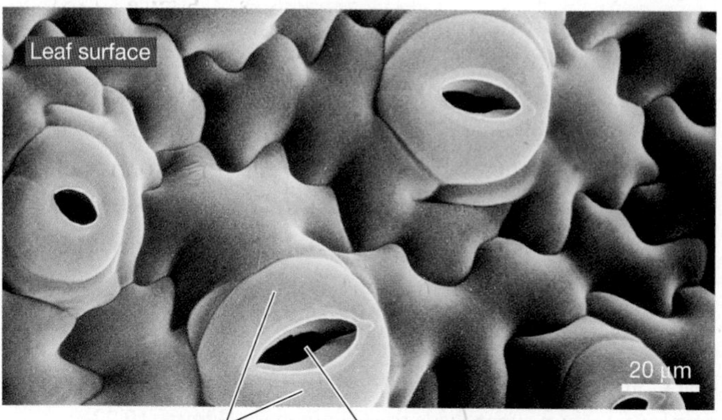

(b) Carbon dioxide diffuses into leaves through stomata.

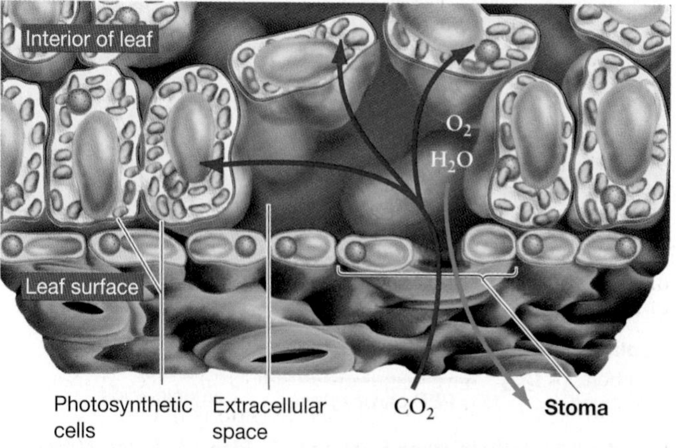

FIGURE 10.20 Leaf Cells Obtain Carbon Dioxide through Stomata.

up in leaves as cellular respiration continues, meaning that photorespiration is favored.

How do plants that live in hot, dry environments keep CO_2 supplies high enough to limit the damaging effects of photorespiration? And how can photosynthesizing cells raise CO_2 concentrations to make photosynthesis more efficient? An answer emerged in a surprising experimental result.

C_4 PHOTOSYNTHESIS After the Calvin cycle had been worked out in algae, researchers in a variety of labs used the same pulse-chase approach to investigate how carbon fixation occurs in other species. Hugo Kortschak and colleagues and Y. S. Karpilov and associates exposed leaves of sugarcane and maize (corn) to radioactive carbon dioxide ($^{14}CO_2$) and sunlight, and then isolated and identified the product molecules.

Both research teams expected to find the first of the radioactive carbon atoms in 3-phosphoglycerate—the normal product of carbon fixation by rubisco. Instead, they found that in their species, the radioactive carbon atom ended up in four-carbon compounds such as malate and aspartate.

Instead of creating a three-carbon sugar, it appeared that these species' CO_2 fixation produced four-carbon sugars. The two pathways became known as **C_3 photosynthesis** and **C_4 photosynthesis**, respectively (**Figure 10.21**).

Researchers who followed up on the initial reports found that, in some plant species, carbon dioxide can be added to RuBP by rubisco *or* to three-carbon compounds by an enzyme called **PEP carboxylase**. They also showed that the two enzymes are found in distinct cell types within the same leaf. PEP carboxylase is common in **mesophyll cells** near the surface of leaves, while rubisco is found in **bundle-sheath cells** that surround the vascular tissue in the interior of the leaf (**Figure 10.22a**). **Vascular tissue** conducts water and nutrients in plants.

This situation contrasts with C_3 plants. In C_3 species, mesophyll cells contain chloroplasts and rubisco.

Based on the observations about C_4 plants, Hal Hatch and Roger Slack proposed a three-step model to explain how CO_2 that is fixed to a four-carbon sugar feeds the Calvin cycle (**Figure 10.22b**):

Step 1 PEP carboxylase fixes CO_2 in mesophyll cells.

Step 2 The four-carbon organic acids that result travel to bundle-sheath cells through plasmodesmata (Chapter 8).

Step 3 The four-carbon organic acids release a CO_2 molecule that rubisco uses as a substrate to form 3-phosphoglycerate. This step initiates the Calvin cycle.

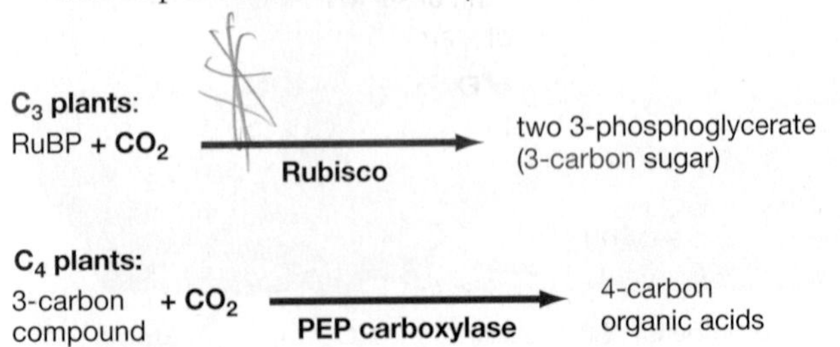

FIGURE 10.21 Initial Carbon Fixation in C_4 Plants Is Different from That in C_3 Plants.

In effect, then, the C_4 pathway acts as a CO_2 concentrator. The reactions that take place in mesophyll cells require energy in the form of ATP, but they increase CO_2 concentrations in cells where rubisco is active. Because it increases the ratio of carbon dioxide to oxygen in photosynthesizing cells, less O_2 binds to rubisco's active sites. As a result, the C_4 pathway limits the damaging effects of photorespiration.

The C_4 pathway is an adaptation that keeps CO_2 concentrations in leaves high, making photosynthesis more efficent. C_4 plants include sugarcane, maize (corn), crabgrass, and several thousand other species in 19 distinct lineages of flowering plants. This suggests that the C_4 pathway has evolved independently several times. It is not the only mechanism that plants use to continue growth under hot, dry conditions, however.

CAM PLANTS Researchers studying a group of flowering plants called the Crassulaceae discovered a second mechanism for limiting the effects of photorespiration. This photosynthetic pathway, **crassulacean acid metabolism**, or **CAM**, is a CO_2 concentrator that acts as an additional, preparatory step to the Calvin cycle.

Like the C_4 pathway, CAM increases the concentration of CO_2 inside photosynthesizing cells and involves an organic acid with

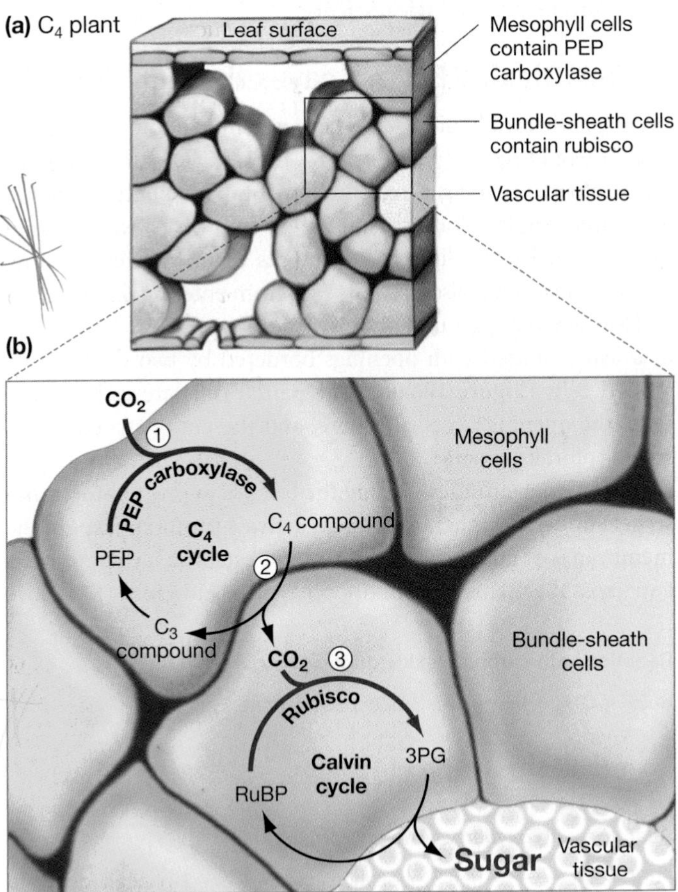

FIGURE 10.22 In C_4 plants, Carbon Fixation and the Calvin Cycle Occur in Different Cell Types. (a) The carbon-fixing enzyme PEP carboxylase is located in mesophyll cells, while rubisco is in bundle-sheath cells. **(b)** CO_2 is fixed to the three-carbon compound PEP by PEP carboxylase, forming a four-carbon organic acid. A CO_2 molecule from the four-carbon sugar then feeds the Calvin cycle.

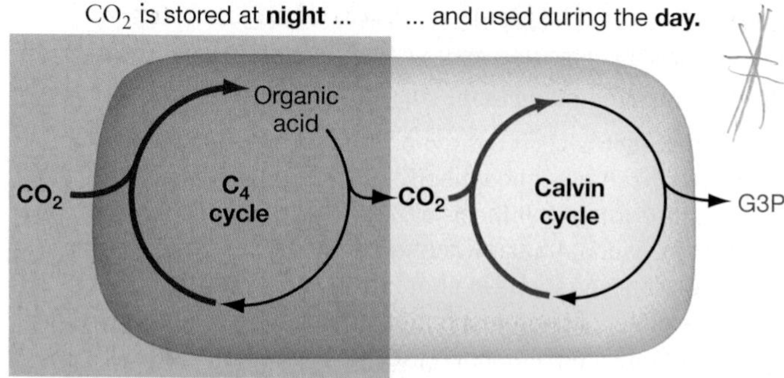

FIGURE 10.23 In CAM Plants, Carbon Fixation Occurs at Night and the Calvin Cycle Occurs during the Day.

four carbons. But unlike the C_4 pathway, CAM occurs at a different time than the Calvin cycle does—not in a different place.

CAM occurs in cacti and other species that routinely keep their stomata closed on hot, dry days. At night, when conditions are cooler and more moist, CAM plants open their stomata and take in huge quantities of CO_2. The CO_2 is temporarily fixed to organic acids and stored in the central vacuoles of photosynthesizing cells. During the day, the molecules are processed in reactions that release the CO_2 and feed the Calvin cycle (**Figure 10.23**).

 C_4 photosynthesis and CAM function as CO_2 pumps. They minimize photorespiration when stomata are closed and CO_2 cannot diffuse in directly from the atmosphere. Both are found in species that live in hot, dry environments.

But while C_4 plants stockpile CO_2 in cells where rubisco is not active, CAM plants store CO_2 when rubisco is inactive. In C_4 plants, the reactions catalyzed by PEP carboxylase and rubisco are separated in space; in CAM plants, the reactions are separated in time. You can go to the study area at *www.masteringbiology.com* to review these two processes.

(MB) **Web Activity** Strategies for Carbon Fixation

How Is Photosynthesis Regulated?

Like nuclear transport, cell-cell signaling responses, and cellular respiration, photosynthesis is regulated. Although the mechanisms responsible for turning photosynthesis on or off are still under investigation, several patterns have emerged.

- The presence of light triggers the production of proteins required for photosynthesis.
- When sugar supplies are high, the production of proteins required for photosynthesis is inhibited, but the production of proteins required to process and store sugars is stimulated.
- Rubisco is activated by regulatory molecules that are produced when light is available, but inhibited in conditions of low CO_2 availability—when photorespiration is favored.
- Inorganic phosphate is required by the light-capturing reactions, where it is incorporated into NADPH, and is released when the sugars produced by photosynthesis are processed. Low P_i levels stimulate the production of Calvin cycle proteins. When the Calvin cycle is more active and more sugar is produced and processed, inorganic phosphate is no longer tied up in NADPH and can be reused in the light-capturing reactions.

The central message here is that the rate of photosynthesis is finely tuned, to reflect changes in environmental conditions and use resources efficiently.

What Happens to the Sugar That Is Produced by Photosynthesis?

The G3P molecules that exit the Calvin cycle enter one of several reaction pathways. The most important of these reaction sequences produces the monosaccharides glucose and fructose, which combine to form the disaccharide ("two-sugar") **sucrose** (**Figure 10.24**). The process starts with G3P from the Calvin cycle, involves a series of other phosphorylated three-carbon sugars, includes the synthesis of the familiar six-carbon sugar glucose, and ends with the production of sucrose.

An alternative pathway produces glucose molecules that polymerize to form **starch**. Starch production occurs inside the chloroplast; sucrose synthesis takes place in the cytosol.

When photosynthesis is taking place slowly, almost all the glucose that is produced is used to make sucrose. Sucrose is water soluble and readily transported to other parts of the plant.

If sucrose is delivered to rapidly growing parts of the plant, it is broken down to fuel cellular respiration and growth. But if it is transported to storage cells in roots, it is converted to starch and stored for later use.

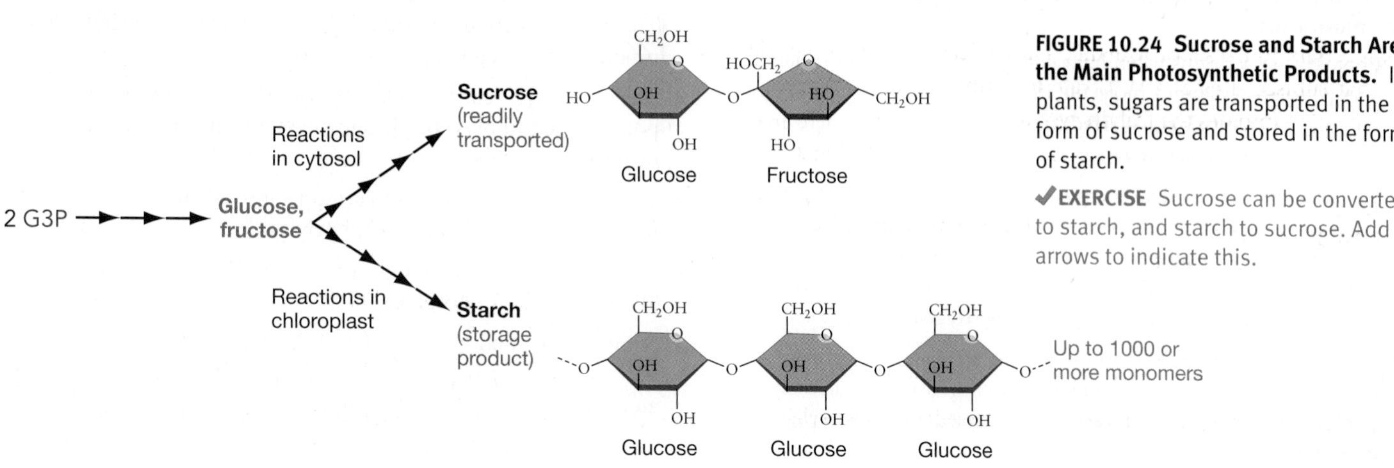

FIGURE 10.24 Sucrose and Starch Are the Main Photosynthetic Products. In plants, sugars are transported in the form of sucrose and stored in the form of starch.

✔**EXERCISE** Sucrose can be converted to starch, and starch to sucrose. Add arrows to indicate this.

When photosynthesis is proceeding rapidly and sucrose is abundant, glucose is used to synthesize starch in the chloroplasts of photosynthetic cells. Recall from Chapter 5 that starch is a polymer of glucose.

In photosynthesizing cells, starch acts as a temporary sugar-storage product. Starch is not water soluble, so it cannot be transported from photosynthetic cells to other areas of the plant. At night, the starch that is temporarily stored in leaf cells is broken down and used to manufacture sucrose molecules. The sucrose is then used by the photosynthetic cell in respiration or transported to other parts of the plant. In this way, chloroplasts provide sugars for cells throughout the plant by day and by night.

If a mouse eats the starch that is stored in a chloroplast or root cell, however, the chemical energy in the C–C and C–H bonds of the starch fuels the mouse's growth and reproduction. If an owl eats the mouse, the chemical energy in the mouse's tissues fuels the predator's growth and reproduction.

In this way, virtually all cell activity can be traced back to the chemical energy that was originally captured by photosynthesis. Photosynthesis is the staff of life.

CHECK YOUR UNDERSTANDING

If you understand that . . .

- The Calvin cycle is a three-phase process: CO_2 fixation (synthesis of 3-phosphoglycerate), production of carbohydrate (synthesis of G3P), and regeneration of RUBP.
- The C_4 and CAM pathways are mechanisms for increasing CO_2 concentrations in photosynthesizing cells. They limit the effect of photorespiration and allow photosynthesis to continue after stomata close.
- In photosynthesizing cells, sucrose is stored as starch. In all plant cells, sucrose is used to drive cellular respiration.

✓ **You should be able to . . .**

1. Describe how CO_2 is delivered to rubisco (a) via organic acids in mesophyll cells, (b) via organic acids synthesized at night and stored in vacuoles, and (c) directly.
2. Explain how the solubility of starch and sucrose affect their use by plants.

Answers are available in Appendix B.

CHAPTER 10 REVIEW

For media, go to the study area at www.masteringbiology.com

Summary of Key Concepts

Photosynthesis is the conversion of light energy to chemical energy stored in the bonds of carbohydrates. It consists of two linked sets of reactions.

- The light-capturing reactions occur in internal membranes of the chloroplast that are organized into structures called thylakoids in stacks known as grana.
- The Calvin cycle takes place in a fluid portion of the chloroplast called the stroma.

✓You should be able to explain why biologists have stopped using the once-common phrase "light-independent reactions" to describe the Calvin cycle.

In the light-capturing reactions, excited electrons are used to produce the electron carrier NADPH or are donated to an electron transport chain, which results in the production of ATP via chemiosmosis.

- Photosynthesis begins when light energy is transformed to chemical energy. After a pigment molecule in an antenna complex absorbs a photon, the energy is transferred to the reaction center, where an excited electron is transferred to an electron acceptor, reducing it.
- In photosystem II, high-energy electrons are accepted by pheophytin and passed along an electron transport chain, releasing energy that pumps protons across the thylakoid membrane. The resulting proton gradient drives the synthesis of ATP by ATP synthase. PS II takes electrons from water, releasing oxygen and protons.
- In photosystem I, high-energy electrons are accepted by iron- and sulfur-containing proteins and passed to ferredoxin. In an enzyme-catalyzed reaction, the reduced form of ferredoxin passes electrons to $NADP^+$, forming NADPH.
- The Z scheme connects photosystems II and I. Plastocyanin carries electrons from the end of PS II's ETC to PS I. They are promoted to a high-energy state in PS I's reaction center, and subsequently used to reduce $NADP^+$. Electrons from PS I may occasionally be passed back to PS II's ETC instead of being used to reduce $NADP^+$. A cyclic flow of electrons between the two photosystems boosts ATP supplies.

✓You should be able to explain why the rate of photosynthesis can be estimated by measuring the rate of oxygen production in chloroplasts.

MB **Web Activity** Chemiosmosis

The reactions of the Calvin cycle start with the enzyme rubisco, which catalyzes the addition of CO_2 to a five-carbon compound. Subsequent reactions use the ATP and NADPH synthesized in the light reactions and yield a molecule required for carbohydrate production.

- The CO_2-reduction reactions of photosynthesis depend on the products of the light-capturing reactions.
- The Calvin cycle starts when CO_2 is attached to a five-carbon compound called ribulose bisphosphate (RuBP) in a reaction catalyzed by the enzyme rubisco.
- The six-carbon compound that results immediately splits in half to form two molecules of 3-phosphoglycerate.
- Subsequently, 3-phosphoglycerate is reduced to a sugar called glyceraldehyde-3-phosphate (G3P).
- Some G3P is used to synthesize glucose and fructose, which combine to form sucrose; the rest participates in reactions that regenerate RuBP so the cycle can continue.

✔You should be able to explain why it is accurate to call the Calvin cycle a cycle.

MB BioFlix™ Photosynthesis, **Web Activity** Photosynthesis

In plants, CO_2 enters photosynthetic tissue through openings called stomata. The CAM and C_4 pathways increase CO_2 concentrations inside the leaves of some species and make photosynthesis more efficient.

- Rubisco catalyzes the addition of oxygen as well as carbon dioxide to RuBP. The reaction with oxygen leads to a loss of fixed CO_2 and ATP and is called photorespiration.
- CAM plants and C_4 plants fix CO_2 to organic acids, before it is transferred to rubisco. As a result, they can increase CO_2 levels in their tissues, reducing the effect of photorespiration and allowing photosynthesis to continue when stomata close.

✔You should be able to predict what would happen in a chloroplast containing an unusual form of rubisco—either a form that did not bind oxygen or a form that worked 10 times as fast as most forms of the enzyme.

MB **Web Activity** Strategies for Carbon Fixation

Questions

✔TEST YOUR KNOWLEDGE

Answers are available in Appendix B

1. What is resonance?
   a. when an electron is excited by a photon
   b. transformation of electromagnetic energy to chemical energy
   c. transfer of electrons among pigment molecules
   d. transfer of energy among pigment molecules
2. Why is chlorophyll green?
   a. It absorbs all wavelengths in the visible spectrum, transmitting ultraviolet and infrared light.
   b. It absorbs wavelengths only in the red and far-red portions of the spectrum (680 nm, 700 nm).
   c. It absorbs wavelengths in the blue and red parts of the visible spectrum.
   d. It absorbs wavelengths only in the blue part of the visible spectrum and transmits all other wavelengths.
3. What does it mean to say that CO_2 becomes fixed?
   a. It becomes bonded to an organic compound.
   b. It is released during cellular respiration.
   c. It acts as an electron acceptor.
   d. It acts as an electron donor.
4. What do the light-capturing reactions of photosynthesis produce?
   a. G3P
   b. RuBP
   c. ATP and NADPH
   d. sucrose or starch
5. Why do the absorption spectrum for chlorophyll and the action spectrum for photosynthesis coincide?
   a. Photosystems I and II are activated by different wavelengths of light.
   b. Wavelengths of light that are absorbed by chlorophyll trigger the light-capturing reactions.
   c. Energy from wavelengths absorbed by carotenoids is passed on to chlorophyll.
   d. The rate of photosynthesis depends on the amount of light received.
6. What happens when an excited electron is passed to an electron acceptor in a photosystem?
   a. It drops back down to its ground state, resulting in the phenomenon known as fluorescence.
   b. The chemical energy in the excited electron is released as heat.
   c. The electron acceptor is oxidized.
   d. Energy in sunlight is transformed to chemical energy.

✔TEST YOUR UNDERSTANDING

Answers are available in Appendix B

1. Explain how the energy transformation step of photosynthesis occurs. How is light energy converted to chemical energy in the form of ATP and NADPH?
2. Explain how the carbon reduction step of photosynthesis occurs. How is carbon dioxide fixed? Why are both ATP and NADPH required to produce sugar?
3. Compare and contrast photosystem II and photosystem I. What molecule connects the two photosystems?
4. When does photorespiration occur? What are its consequences for the plant?
5. Make a sketch showing how C_4 photosynthesis and CAM separate CO_2 acquisition from the Calvin cycle in space and time, respectively.
6. Why do plants need both chloroplasts and mitochondria?

✔APPLYING CONCEPTS TO NEW SITUATIONS

Answers are available in Appendix B

1. Compare and contrast mitochondria and chloroplasts. In what ways are their structures similar and different? What molecules or systems function in both types of organelles? Which enzymes or processes are unique to each organelle?
2. Some biologists claim that photorespiration is an evolutionary "holdover," because rubisco evolved over a billion years ago when O_2 levels were extremely low and CO_2 concentrations relatively high. Do you agree with this hypothesis? Why or why not?
3. In addition to their protective function, carotenoids absorb certain wavelengths of light and pass the energy on to the reaction centers of photosystem I and II. Based on their function, predict exactly where carotenoids are located in the chloroplast. Explain your rationale. How would you test your hypothesis?
4. Consider plants that occupy the top, middle, or ground layer of a forest, and algae that live near the surface of the ocean or in deeper water. Would you expect the same photosynthetic pigments to be found in species that live in these different habitats? Why or why not? How would you test your hypothesis?

This chapter is part of The Big Picture. See how on pages 192–193.

THE BIG PICTURE

It takes energy to stay alive. Use this concept map to study how the information on energy and energetics presented in this book fits together.

As you read the map, remember that chemical energy is potential energy. Potential energy is based on position of matter in space, and chemical energy is all about the position of electrons in covalent bonds. When TNT explodes, all that's happening is that electrons are moving from high-energy positions to lower-energy positions.

In essence, organisms transform energy from the Sun into chemical energy in the C–C and C–H bonds of glucose, and then into chemical energy in the P–P bonds of ATP.

The potential energy in ATP allows cells to do work: pump ions, synthesize molecules, move cargo, and send and receive signals.

Note that each box in the concept map indicates the chapter and section where you can go for review. Also, be sure to do the blue exercises in the Check Your Understanding box below.

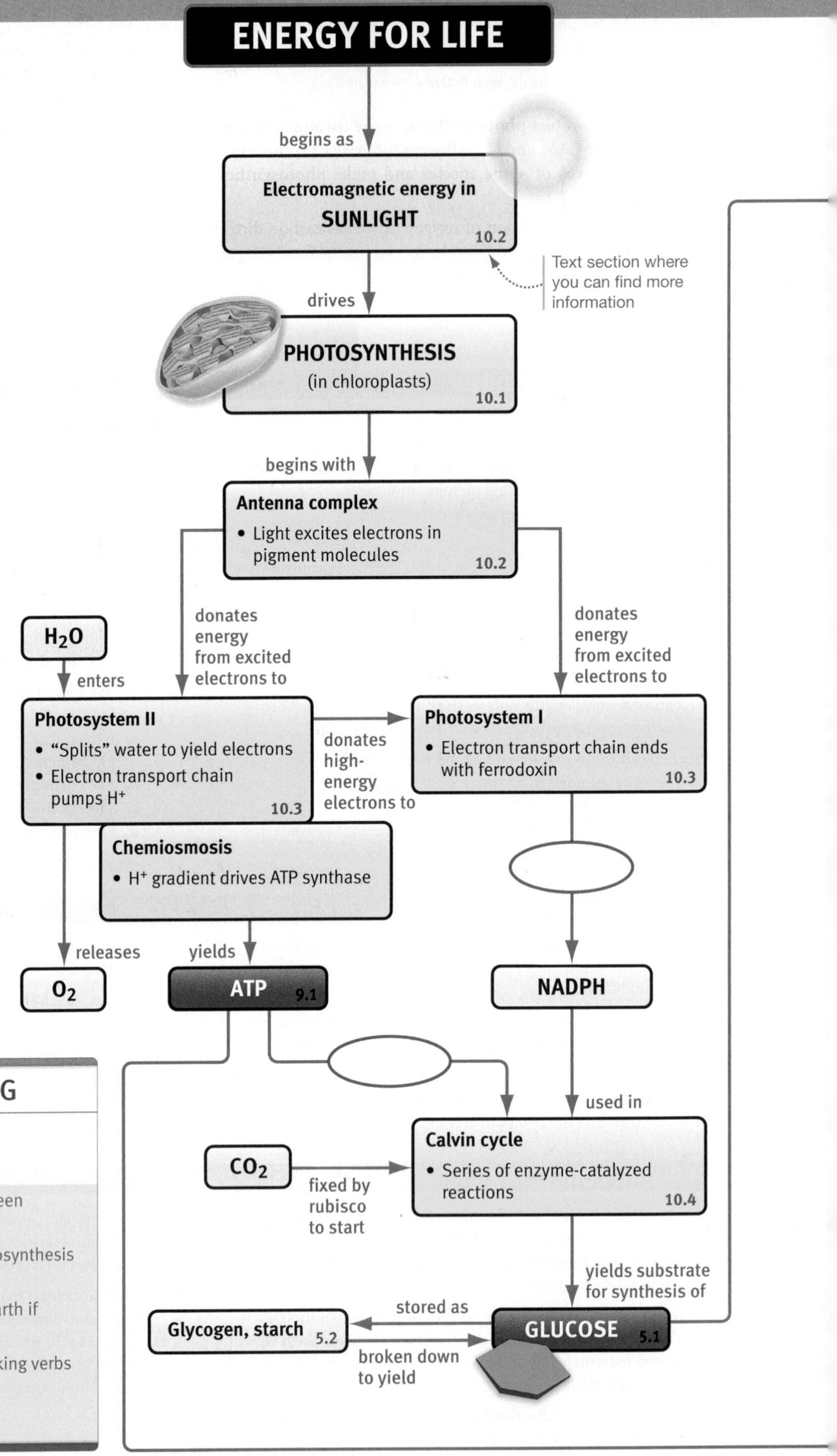

CHECK YOUR UNDERSTANDING

If you understand the big picture . . .

✔ You should be able to . . .

1. Explain how H_2O and O_2 are cycled between photosynthesis and cellular respiration.
2. Explain how CO_2 is cycled between photosynthesis and cellular respiration.
3. Describe what might happen to life on Earth if rubisco were suddenly unable to fix CO_2.
4. Fill in the blue ovals with appropriate linking verbs or phrases.

Answers are available in Appendix B.

ENERGY FOR LIFE

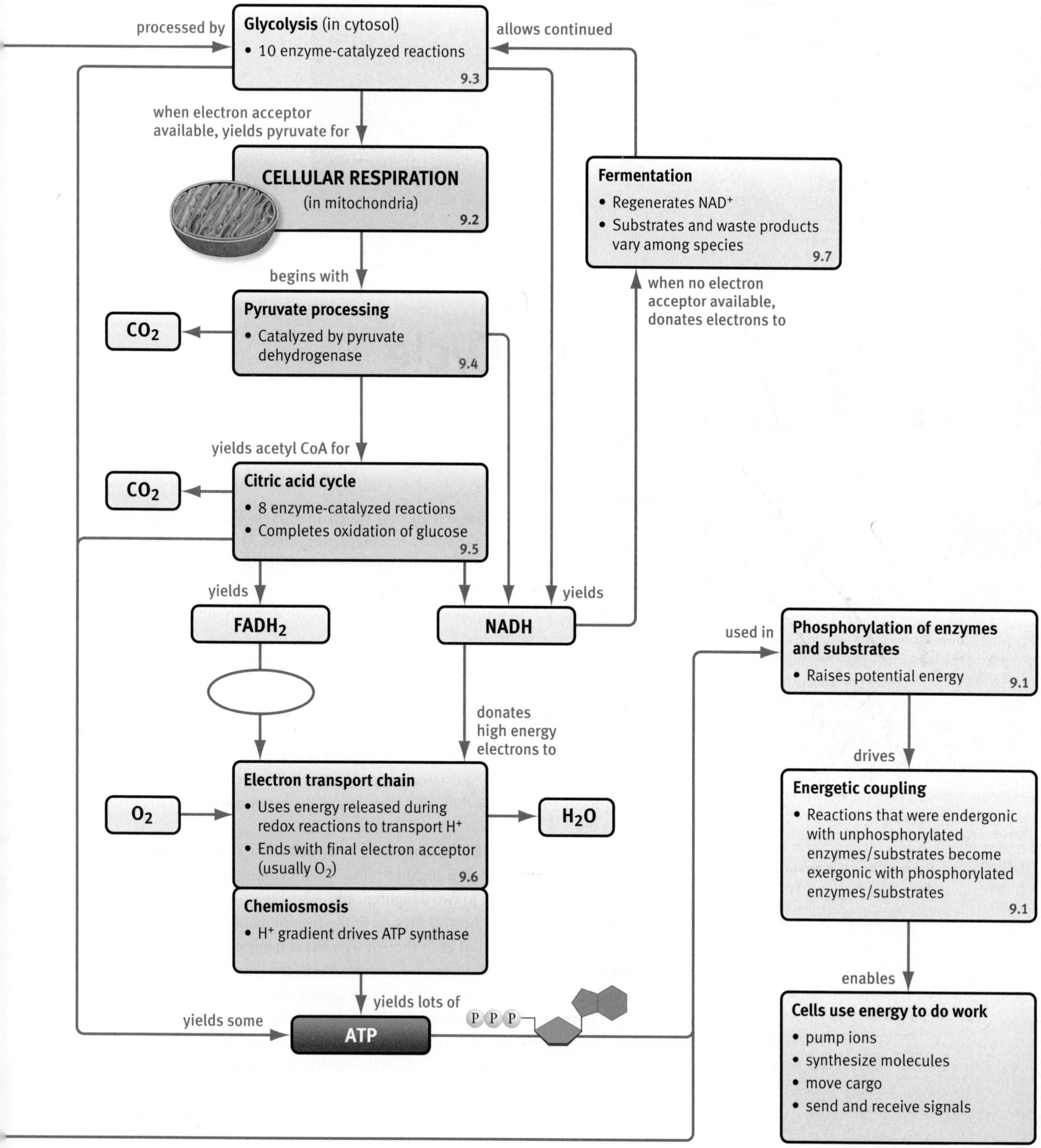

processed by

Glycolysis (in cytosol)
- 10 enzyme-catalyzed reactions

9.3

allows continued

when electron acceptor available, yields pyruvate for

CELLULAR RESPIRATION
(in mitochondria)
9.2

Fermentation
- Regenerates NAD^+
- Substrates and waste products vary among species

9.7

when no electron acceptor available, donates electrons to

begins with

Pyruvate processing
- Catalyzed by pyruvate dehydrogenase

9.4

CO_2

yields acetyl CoA for

Citric acid cycle
- 8 enzyme-catalyzed reactions
- Completes oxidation of glucose

9.5

CO_2

yields

yields

$FADH_2$

NADH

used in

Phosphorylation of enzymes and substrates
- Raises potential energy

9.1

donates high energy electrons to

drives

Electron transport chain
- Uses energy released during redox reactions to transport H^+
- Ends with final electron acceptor (usually O_2)

9.6

O_2

H_2O

Energetic coupling
- Reactions that were endergonic with unphosphorylated enzymes/substrates become exergonic with phosphorylated enzymes/substrates

9.1

Chemiosmosis
- H^+ gradient drives ATP synthase

yields lots of

yields some

ATP

P P P

enables

Cells use energy to do work
- pump ions
- synthesize molecules
- move cargo
- send and receive signals

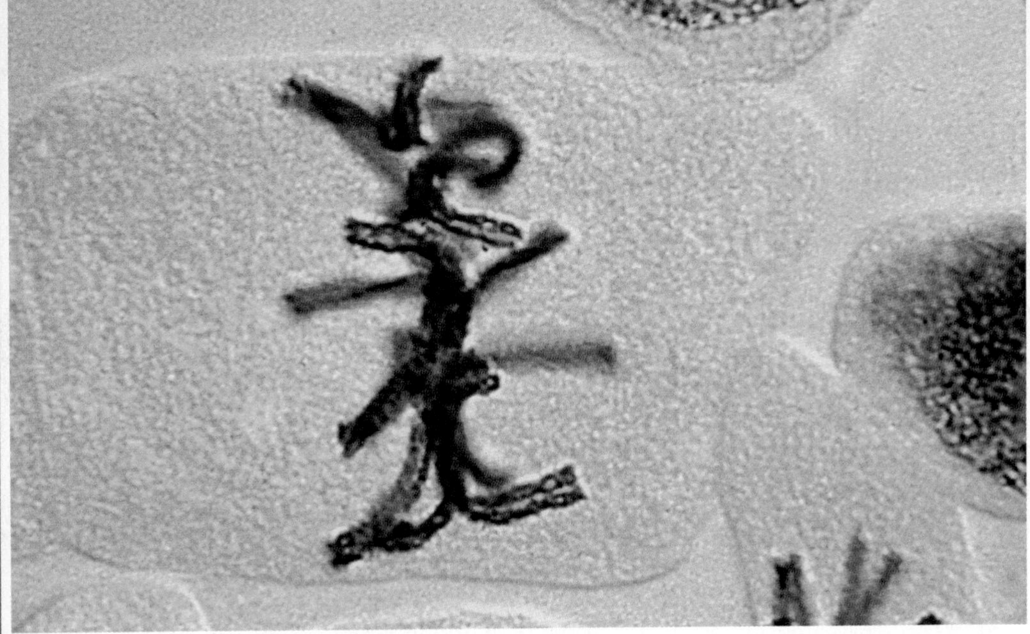

This cell, from a hyacinth, is undergoing a type of cell division called mitosis. Understanding how mitosis occurs is a major focus of this chapter.

11 The Cell Cycle

KEY CONCEPTS

- In eukaryotes, most dividing cells go through a cycle that consists of four phases.
- After chromosomes are copied during S phase, they are moved to the middle of the cell during M phase (mitosis). One chromosome copy is distributed to each of two daughter cells. Mitosis and cytokinesis produce two cells that are genetically identical to the parent cell.
- Progression through the cell cycle is carefully controlled.
- In multicellular organisms, uncontrolled cell division may lead to cancer. Different types of cancer result from different types of defects in control over the cell cycle.

Chapter 1 introduced the cell theory, which maintains that all organisms are made of cells and all cells arise from preexisting cells. Although the cell theory was widely accepted among biologists by the 1860s, most believed that new cells arose within preexisting cells by a process that resembled the growth of mineral crystals. But Rudolf Virchow proposed that new cells arise through the division of preexisting cells—that is, **cell division**.

In the late 1800s, microscopic observations of newly developing organisms, or **embryos**, confirmed Virchow's hypothesis. Multicellular eukaryotes start life as single-celled embryos and grow through a series of cell divisions.

Early studies revealed two fundamentally different ways that nuclei divide prior to cell division: meiosis and mitosis. In animals, **meiosis** leads to the production of sperm and eggs, which are the male and female reproductive cells termed **gametes**. **Mitosis** leads to the production of all other cell types, referred to as **somatic** (literally, "body-belonging") **cells**.

Mitosis and meiosis are usually accompanied by **cytokinesis** ("cell movement")—the division of the cytoplasm into two distinct cells. When cytokinesis is complete, a so-called parent cell has given rise to two daughter cells.

Mitosis and meiosis are responsible for one of the five fundamental attributes of life: reproduction (see Chapter 1). But even though they share many characteristics, mitosis and meiosis are fundamentally different. During mitosis, the genetic material is copied and then divided equally, so that daughter cells are genetically identical to the parent cell. In contrast, meiosis results in daughter cells that are genetically different from each other and that have half the amount of hereditary material as the parent cell.

This chapter focuses on mitosis; meiosis is the subject of Chapter 12. Let's begin with a look at how mitosis relates to other events in a cell's life cycle, continue with an in-depth analysis of mitosis and the cell cycle, and end by examining why uncontrolled cell division can lead to cancer.

✔ When you see this checkmark, stop and test yourself. Answers are available in Appendix B.

11.1 Mitosis and the Cell Cycle

Mitosis and cytokinesis are responsible for three key events in multicellular eukaryotes:

1. ***Growth*** The trillions of genetically identical cells that make up your body are the product of mitotic divisions that started in a single fertilized egg.
2. ***Wound repair*** When you suffer a scrape, mitosis and cytokinesis generate the cells that repair your skin.
3. ***Reproduction*** When yeast cells grow in bread dough or in a vat of beer, they are reproducing by mitosis and cytokinesis. In yeasts and other species, mitosis followed by cytokinesis is the basis of asexual reproduction. **Asexual reproduction** produces offspring that are genetically identical to the parent.

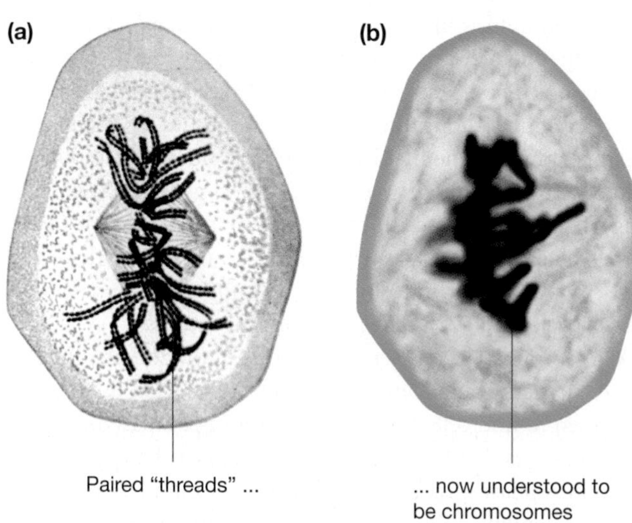

FIGURE 11.1 Chromosomes Move during Mitosis. (a) Walther Flemming's 1879 drawing of mitosis in a salamander embryo. The black threads are chromosomes. **(b)** Chromosomes can be stained with dyes and observed using the light microscope.

These processes are so basic to life that mitosis has been studied for well over a century. Like much work in biology, the research began with a simple observation.

What Is a Chromosome?

As studies of cell division in eukaryotes began, biologists found that certain chemical dyes made threadlike structures visible within nuclei. In 1879 Walther Flemming documented how the threadlike structures changed as cells in salamander embryos divided. The threads first appeared in pairs just before cell division (**Figure 11.1a**), then split to produce single, unpaired threads in the daughter cells. Flemming introduced the term mitosis, from the Greek *mitos* ("thread"), to describe this division process.

Others studied the roundworm *Ascaris*, and noted that the total number of threads in a cell was the same before and after mitosis. All of the cells in a roundworm had the same number of threads.

In 1888 Wilhelm Waldeyer coined the term **chromosome** ("colored-body") to refer to these threadlike structures (**Figure 11.1b**). A chromosome consists of a single, long DNA (deoxyribonucleic acid) double helix that is wrapped around proteins in a highly organized manner. DNA encodes the cell's hereditary information, or genetic material. A gene is a length of DNA that codes for a particular protein or ribonucleic acid (RNA) found in the cell.

Prior to mitosis, each chromosome is replicated. As mitosis starts, the chromosomes condense into compact structures that can be moved around the cell efficiently. Then one of the chromosome copies is distributed to each of two daughter cells.

Figure 11.2 illustrates unreplicated chromosomes, replicated chromosomes before they have condensed prior to mitosis, and replicated chromosomes that have condensed at the start of mitosis. Each of the DNA copies in a replicated chromosome is called a **chromatid**. The two chromatids are joined together along their entire

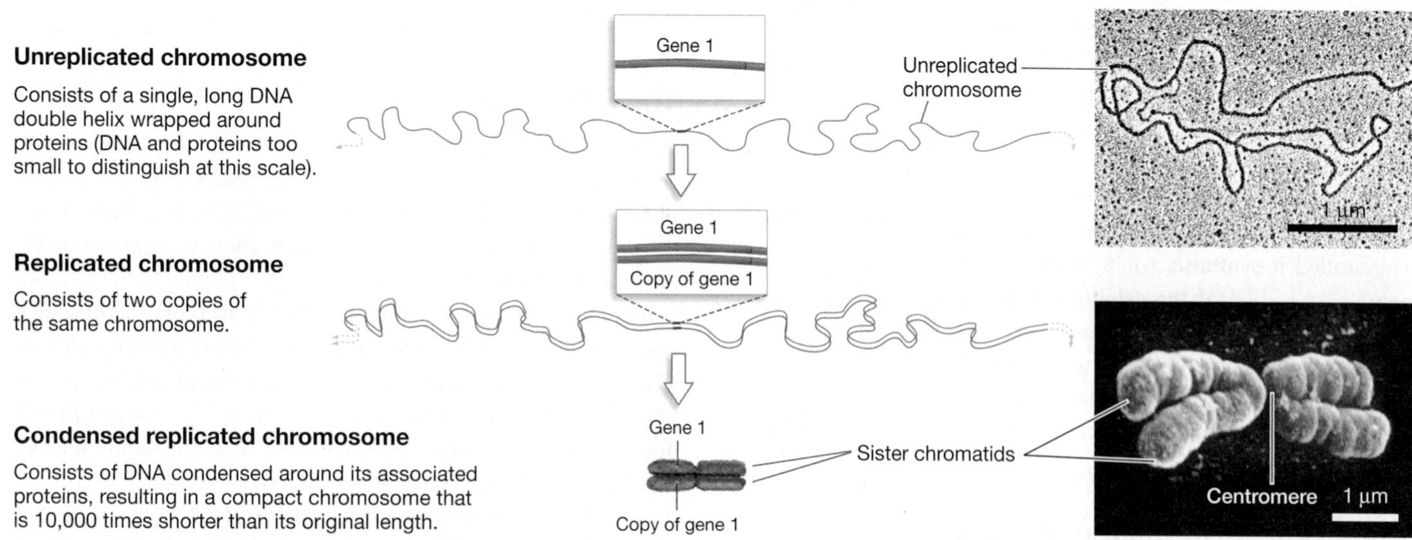

Unreplicated chromosome
Consists of a single, long DNA double helix wrapped around proteins (DNA and proteins too small to distinguish at this scale).
Gene 1
Unreplicated chromosome
1 µm
Replicated chromosome
Consists of two copies of the same chromosome.
Gene 1
Copy of gene 1
Condensed replicated chromosome
Consists of DNA condensed around its associated proteins, resulting in a compact chromosome that is 10,000 times shorter than its original length.
Gene 1
Copy of gene 1
Sister chromatids
Centromere
1 µm

FIGURE 11.2 Chromosome Morphology Changes before and during Mitosis. Chromosome replication occurs prior to mitosis. The two identical copies stay attached to each other. Early in mitosis, replicated chromosomes condense. The two copies stay attached at a region called the centromere.

length as well as at a specialized region of the chromosome called the **centromere**. Chromatids from the same chromosome are referred to as **sister chromatids**. Even though a replicated chromosome consists of two chromatids, it is still considered a single chromosome.

Cells Alternate between M Phase and Interphase

Plant and animal cells do not undergo mitosis continuously. Instead, growing cells cycle between a dividing phase called the **mitotic** (or **M**) **phase** and a nondividing phase called **interphase** ("between-phase").

Interphase is an active time: the cell is either growing and preparing to divide or fulfilling its specialized function in a multicellular individual.

With a light microscope, chromosomes can be stained and observed as discrete sructures only during M phase, when they condense into compact structures. Cells actually spend most of their time in interphase, however. No dramatic changes are observed in the nucleus during interphase, when chromosomes uncoil into the extremely long, thin structures shown in Figure 11.2a.

The Discovery of S Phase

Are chromosomes copied during M phase or interphase? To answer this question, researchers exposed dividing cells to radioactive phosphorus or radioactive thymidine. Phosphorus is a component of all deoxyribonucleotides, including thymidine, and deoxyribonucleotides are components of DNA (see Chapter 4). Both phosphorus and thymidine are incorporated into DNA as it is being synthesized.

The idea was to:

1. label DNA as chromosomes were being copied,
2. wash away any radioactive isotope that hadn't been incorporated, and
3. visualize the labeled, newly synthesized DNA by exposing the treated cells to X-ray film. Emissions from radioactive phosphorus or thymidine create a black dot in the film. This is the technique called autoradiography (see **BioSkills 9** in Appendix A).

Alma Howard and Stephen Pelc looked for black dots—indicating active DNA synthesis—inside cells right after the exposure to a radioactive isotope ended. They found black dots in interphase cells, but never in M-phase cells. This was strong evidence that DNA replication occurs during interphase.

The biologists had identified a new stage in the life of a cell. They called it **synthesis** (or **S**) **phase**. S phase is part of interphase. Replication of the genetic material is separated, in time, from the partitioning of chromosome copies during M phase.

Howard and Pelc coined the term **cell cycle** to describe the orderly sequence of events that starts with the formation of a eukaryotic cell, through the duplication of its chromosomes, to the time it undergoes division itself.

The Discovery of the Gap Phases

Howard and Pelc and researchers in other labs followed up on these early results by asking how long the S phase lasted. In most cases the experiments were done on cells that were growing in culture. Cultured cells are powerful experimental tools because they can be manipulated much more easily than cells in an intact organism (see **BioSkills 12** in Appendix A).

In one experiment, researchers exposed cultured cells to radioactive thymidine. A short time later, they stopped the labeling by flooding the solution surrounding the cultured cells with nonradioactive thymidine. This pulse-chase approach—introduced in Chapter 7—labeled cells that were in any portion of S phase.

Once the pulse ended, the biologists analyzed the labeled cells 2 hours after the end of labeling, then 4 hours after labeling, 6 hours after labeling, and so on. For each batch of cells, they recorded how many were undergoing mitosis.

One striking result emerged immediately: None of the labeled cells started mitosis immediately—even though at least some had to be at the end of S-phase when they were exposed to the pulse. Instead, it took 4 hours before the first labeled cells began mitosis.

The roughly 4-hour time lag between the end of the pulse and the appearance of the first labeled mitotic nuclei corresponds to a time lag that occurs between the end of S phase and the beginning of M phase. Stated another way, there is a gap in the cell cycle. The gap represents the period when chromosome replication is complete but mitosis has not yet begun. This interval in the cell cycle is called **G_2 phase**, for second gap.

Why second? The answer is based on another observation that the researchers made. Once some cells had started mitosis, they continued to find cells that were undergoing mitosis for another 6–8 hours. Here's a key point: All of these cells had to be somewhere in S phase—ranging from just beginning DNA synthesis to almost ending—when radioactive thymidine was available. Thus, S phase must last 6 to 8 hours.

When the times for the S, G_2, and M phases are totaled and compared with the 24 hours it takes these cells to complete one cell cycle, though, there is a discrepancy of 7 to 9 hours. This discrepancy represents the **G_1 phase**, or first gap. G_1 occurs after M phase but before S phase.

The Cell Cycle

Figure 11.3 pulls these results together into a comprehensive view of the cell cycle. There are a total of four phases in the cell cycle: M phase and an interphase consisting of the G_1, S, and G_2 phases. In the cell type diagrammed here, the G_1 phase is about twice as long as G_2.

Why do the gap phases exist? Besides copying their chromosomes during S phase, dividing cells also must replicate organelles and manufacture additional cytoplasm. Before mitosis can take place, the parent cell must grow large enough and synthesize enough organelles that its daughter cells will be normal in size and function. The two gap phases provide the time required to accomplish these tasks. They allow the cell to complete all the requirements for cell division other than chromosome replication.

Now let's turn to M phase and the process of mitosis. Once the genetic material has been copied, how do cells divide it between daughter cells?

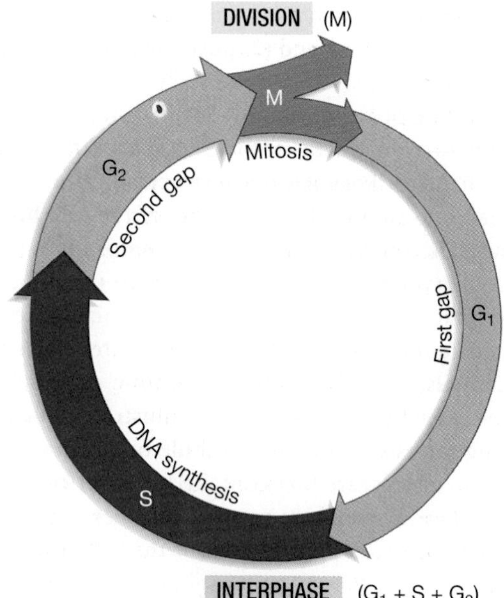

FIGURE 11.3 The Cell Cycle Has Four Phases. A representative cell cycle. The time required for the G_1 and G_2 phases varies dramatically among cells and organisms.

11.2 How Does Mitosis Take Place?

Mitosis results in the division of replicated chromosomes and the formation of two daughter nuclei with identical chromosomes and genes. Mitosis is usually accompanied by cytokinesis—cytoplasmic division and the formation of two daughter cells.

Figure 11.4 provides an overview of how chromosomes change before, during, and after mitosis and cytokinesis, beginning with a hypothetical plant cell or animal cell in G_1 phase. The first drawing shows a total of four chromosomes in the cell, but the number of chromosomes present varies widely among species—chimpanzees and potato plants have a total of 48 chromosomes in each cell; a maize (corn) plant has 20, dogs have 78, and fruit flies have 8.

Eukaryotic chromosomes consist of DNA associated with globular proteins called **histones**. In eukaryotes this DNA-protein material is called **chromatin**. During interphase, most chromatin is in a "relaxed" or uncondensed state, forming long, threadlike strands (see Figure 11.2, top).

The second drawing in Figure 11.4 shows chromosomes that have been copied prior to mitosis. Each chromosome now consists of two sister chromatids. Each chromatid contains one long DNA double helix, and sister chromatids represent exact copies of the same genetic information.

At the start of M phase, then, each chromosome consists of two sister chromatids that are attached to one another at the centromere. ✔You should be able to explain the relationship between chromosomes and (1) genes, (2) chromatin, and (3) sister chromatids.

Events in Mitosis

As the third drawing in Figure 11.4 indicates, mitosis begins when chromatin condenses to form a much more compact structure. Replicated, condensed chromosomes correspond to the paired threads observed in salamander cells by early biologists (see Figure 11.1a).

The final drawing in Figure 11.4 shows that during mitosis, the two sister chromatids separate to form independent chromosomes. One copy of each chromosome goes to each of the two daughter cells. As a result, each daughter cell receives a copy of the genetic information that is contained in each chromosome.

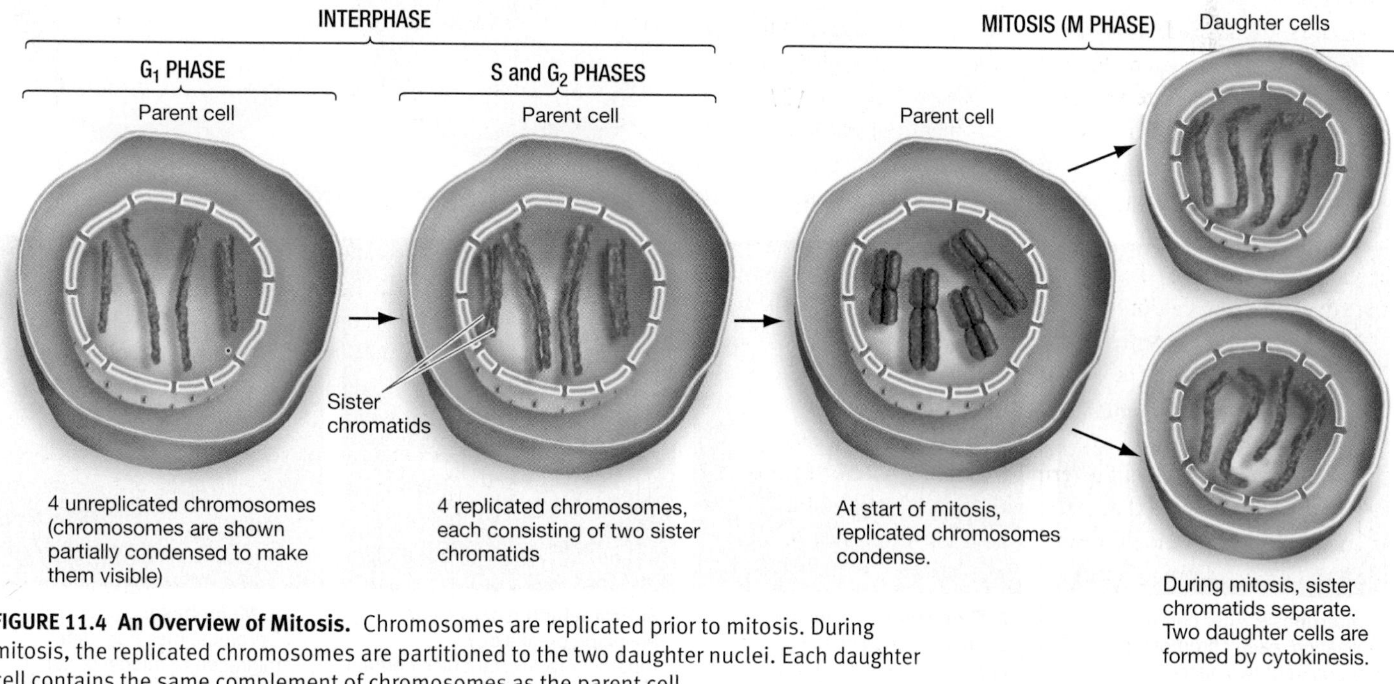

FIGURE 11.4 An Overview of Mitosis. Chromosomes are replicated prior to mitosis. During mitosis, the replicated chromosomes are partitioned to the two daughter nuclei. Each daughter cell contains the same complement of chromosomes as the parent cell.

✔**QUESTION** In the daughter cells of mitosis, are chromosomes replicated or unreplicated?

In mitosis, every daughter cell ends up with exactly the same complement of chromosomes as the parent cell had prior to replication. Thus, every daughter cell receives the same genetic information.

Although mitosis is a continuous process, biologists identify five subphases within M phase on the basis of distinctive events that occur. Some students use the mnemonic device IPPMAT to remember that interphase is followed by the mitotic subphases of prophase, prometaphase, metaphase, anaphase, and telophase.

Recall that chromosomes have already replicated during interphase (**Figure 11.5**, step 1), before mitosis begins. Now let's look at each subphase in turn.

PROPHASE Mitosis begins with the events of **prophase** ("before-phase," Figure 11.5, step 2), when chromosomes condense into compact structures. Chromosomes first become visible in the light microscope during prophase.

In the cytoplasm, prophase is marked by the formation of the spindle apparatus. The **spindle apparatus** is a structure that produces mechanical forces that **(1)** pull chromosomes to the poles of the cell during mitosis, and **(2)** push the poles of the cell away from each other.

The spindle apparatus consists of distinct populations of microtubules—components of the cytoskeleton that were introduced in Chapter 7. **Polar microtubules** extend from each spindle and overlap one another in the middle of the cell. **Kinetochore microtubules**, in contrast, attach to the chromosomes.

In all eukaryotes, the fibers that make up the spindle apparatus originate from a microtubule organizing center. Although the nature of this organizing center varies among plants, animals, fungi, and other eukaryotic groups, the spindle has the same function. Figure 11.5 illustrates an animal cell undergoing mitosis, so the microtubule organizing center is a **centrosome**—a structure that contains a pair of **centrioles** (see Chapter 7). During prophase in all eukaryotes, the spindles either begin moving to opposite sides of the cell or form on opposite sides.

PROMETAPHASE Once chromosomes have condensed, the nucleolus disappears and the nuclear envelope disintegrates. Kinetochore microtubules from each spindle apparatus attach to one

FIGURE 11.5 Mitosis and Cytokinesis. In the bottom micrographs, chromosomes are stained blue, microtubules are yellow/green, and actin filaments are red.

PROCESS: MITOSIS

Sister chromatids separate; one chromosome copy goes to each daughter nucleus.

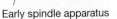

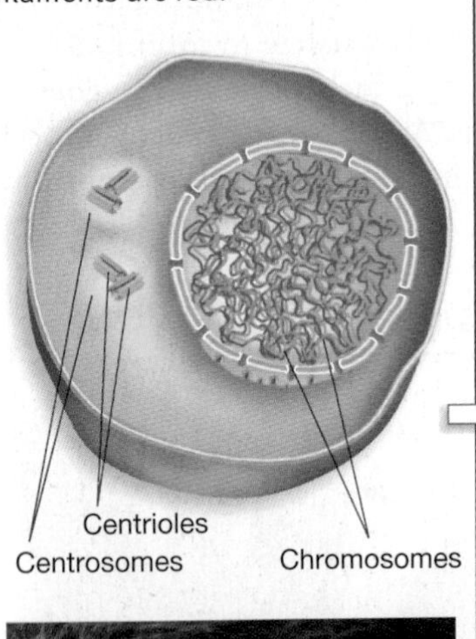

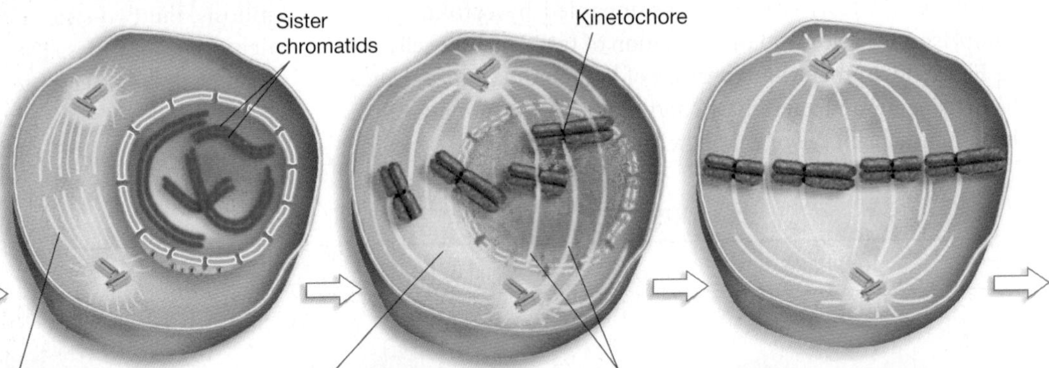

1. Interphase: After chromosome replication, each chromosome is composed of two sister chromatids. Centrosomes have replicated.

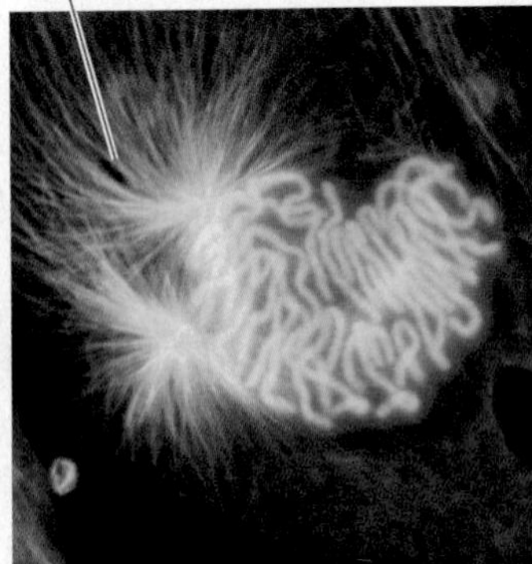

2. Prophase: Chromosomes condense, and spindle apparatus begins to form.

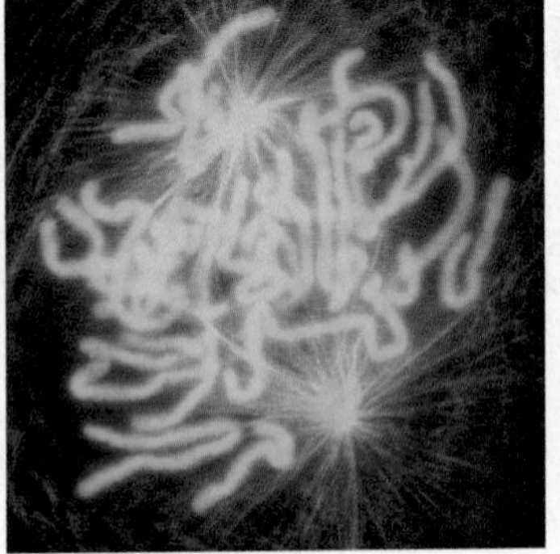

3. Prometaphase: Nuclear envelope breaks down. Kinetochore microtubules contact chromosomes at kinetochore.

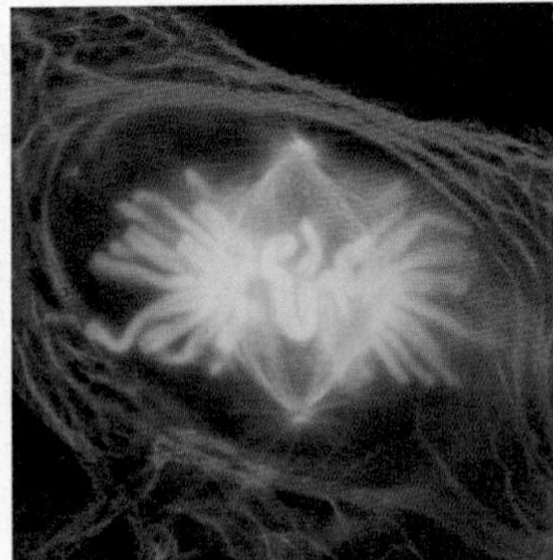

4. Metaphase: Chromosomes complete migration to middle of cell.

of the two sister chromatids of each chromosome. These events occur during **prometaphase** ("before middle-phase"); see Figure 11.5, step 3.

The attachment between the kinetochore microtubules and each chromatid is made at a structure called the **kinetochore**. Kinetochores are located at the centromere region of the chromosome. The centromere is the region where sister chromatids attach most persistently to each other during mitosis. Each chromosome has two kinetochores where microtubules attach—one on each side.

During prometaphase in animals, the centrosomes continue their movement to opposite poles of the cell. In all groups, the microtubules that are attached to the kinetochores begin moving the chromosomes to the middle of the cell.

METAPHASE During **metaphase** ("middle-phase"), animal centrosomes complete their migration to the opposite poles of the cell (Figure 11.5, step 4). In all eukaryotes, the kinetochore microtubules finish moving the chromosomes to the middle of the cell. The polar fibers that extend from each spindle overlap in the middle of the cell, forming a pole-to-pole connection.

When metaphase ends, the chromosomes are lined up along an imaginary plane called the **metaphase plate**. At this point, the formation of the spindle apparatus is complete. Each chromosome is held by kinetochore microtubules reaching to opposite poles and exerting the same amount of tension, or pull. A tug of war is occurring, with motor proteins on kinetochore microtubules pulling each chromosome in opposite directions.

ANAPHASE At the start of **anaphase** ("against-phase"), the centromeres that are holding sister chromatids together split (Figure 11.5, step 5). Because they are under tension, sister chromatids are pulled apart to create independent chromosomes as soon as they are no longer attached.

As kinetochore microtubules shorten, motor proteins pull the chromosomes to opposite poles of the cell. The two poles of the cell are also pushed away from each other by motor proteins that push the overlapping polar fibers apart.

During anaphase, then, replicated chromosomes split into two identical sets of unreplicated chromosomes. The separation of sister chromatids to opposite poles is a critical step in mitosis, because it ensures that each daughter cell receives the same complement of chromosomes.

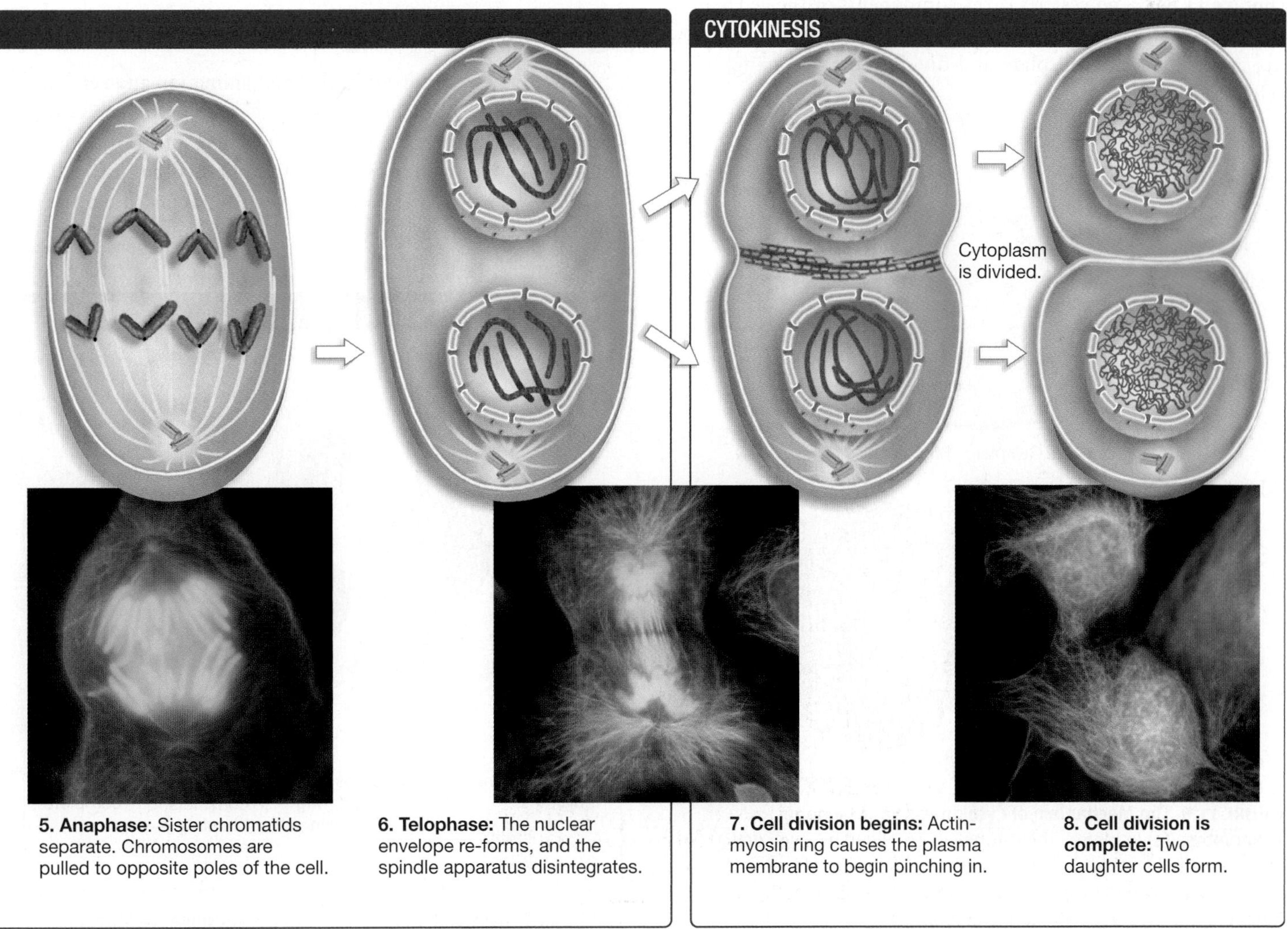
CYTOKINESIS
Cytoplasm is divided.

5. Anaphase: Sister chromatids separate. Chromosomes are pulled to opposite poles of the cell.

6. Telophase: The nuclear envelope re-forms, and the spindle apparatus disintegrates.

7. Cell division begins: Actin-myosin ring causes the plasma membrane to begin pinching in.

8. Cell division is complete: Two daughter cells form.

When anaphase is complete, each pole of the cell has an equivalent and complete collection of chromosomes that are identical to those present in the parent cell prior to chromosome replication.

TELOPHASE During **telophase** ("end-phase"), a nuclear envelope begins to form around each set of chromosomes (Figure 11.5, step 6). The spindle apparatus disintegrates, and the chromosomes begin to de-condense. Once two independent nuclei have formed, mitosis is complete.

Cytokinesis Results in Two Daughter Cells

Prior to the onset of M phase, mitochondria, lysosomes, chloroplasts, and other organelles have replicated, and the rest of the cell contents have grown. During cytokinesis (Figure 11.5, steps 7 and 8), the cytoplasm divides to form two daughter cells, each with its own nucleus and complete set of organelles. In most types of cells, mitosis is followed by cytokinesis.

In plants, cytokinesis begins when a series of microtubules and other proteins define and organize the region where the new plasma membranes and cell walls will form. Vesicles from the Golgi apparatus are transported to the middle of the dividing cell, where they form a structure called the **cell plate** (**Figure 11.6a**). The vesicles carry components of the cell wall and plasma membrane. These components gradually build up, completing the cell plate and dividing the two daughter cells.

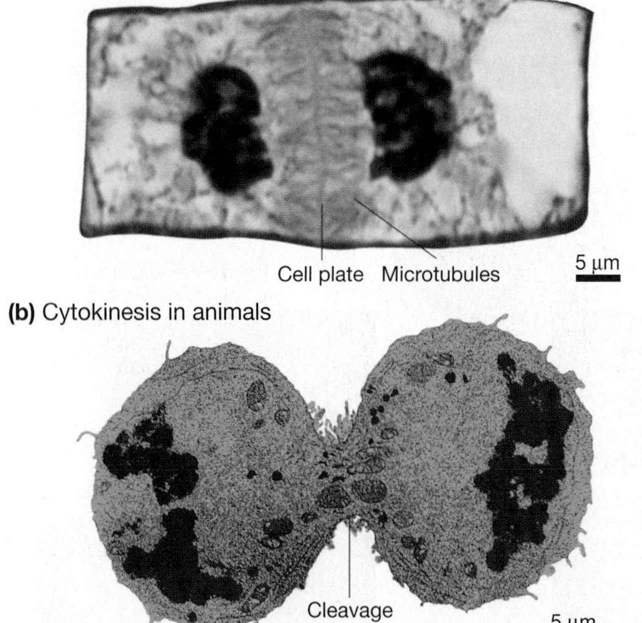
(a) Cytokinesis in plants
Cell plate
Microtubules
5 μm
(b) Cytokinesis in animals
Cleavage furrow
5 μm

FIGURE 11.6 The Mechanism of Cytokinesis Varies among Eukaryotes. (a) In plants, the cytoplasm is divided by a cell plate that forms in the middle of the parent cell. **(b)** In animals, the cytoplasm is divided by a cleavage furrow. (The cells in both micrographs have been stained or colorized.)

In animals, fungi, and slime molds, cytokinesis begins with the formation of a **cleavage furrow** (**Figure 11.6b**). The furrow appears because a ring of actin filaments forms just inside the plasma membrane, in a plane that bisects the cell. A motor protein called myosin binds to these actin filaments. When myosin binds to ATP or ADP, part of the protein moves in a way that causes actin filaments to slide (see Chapter 46).

As myosin moves the ring of actin filaments on the inside of the plasma membrane, the ring shrinks in size and tightens. Because the ring is attached to the plasma membrane, the shrinking ring pulls the membrane with it. As a result, the plasma membrane pinches inward. The actin and myosin filaments continue to slide past each other, tightening the ring further, until the original membrane pinches in two and cell division is complete.

As **Figure 11.7** shows, the mechanism of cell division in bacteria is similar to cytokinesis in animals. After the bacterial chromosome has been copied, the copies move apart and cytoskeletal elements called FtsZ fibers form a ring in the middle of the cell. The ring constricts, dividing the cell in two and producing two identical daughter cells.

Table 11.1 on page 202 summarizes the key structures involved in mitosis, and you can review the process in action in the study area at *www.masteringbiology.com.*

(MB) **BioFlix™** Mitosis, **Web Activity** The Phases of Mitosis

✔After you've studied the table and animation and reviewed Figure 11.5, you should be able to make a table with rows titled (1) spindle apparatus, (2) nuclear envelope, and (3) chromosomes, and columns titled with the five phases of mitosis. Fill in the table by summarizing what happens to each structure during each phase of mitosis.

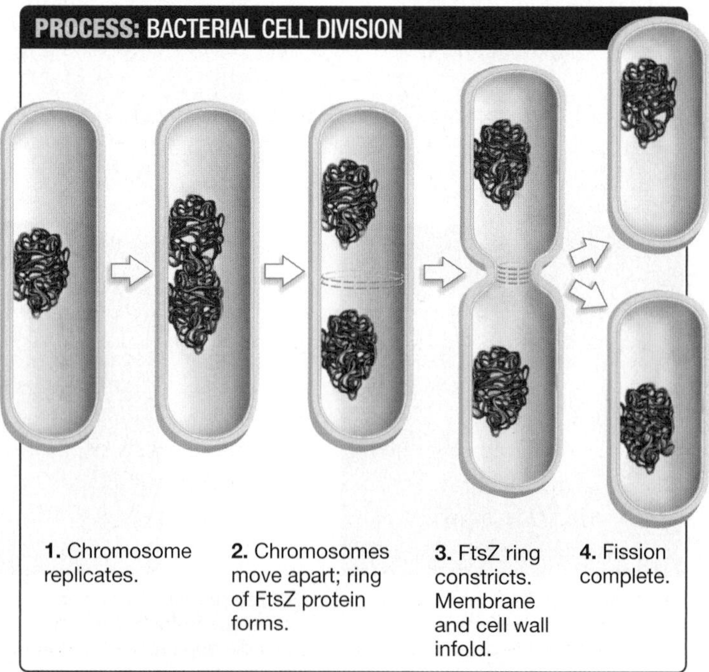
PROCESS: BACTERIAL CELL DIVISION
1. Chromosome replicates.
2. Chromosomes move apart; ring of FtsZ protein forms.
3. FtsZ ring constricts. Membrane and cell wall infold.
4. Fission complete.

FIGURE 11.7 Bacterial Cells Divide, but Do Not Undergo Mitosis.

How Do Chromosomes Move during Mitosis?

The exact and equal partitioning of genetic material to the two daughter cells is the most fundamental aspect of mitosis. How does this process occur?

To understand how sister chromatids separate and move to daughter cells, biologists have focused on understanding how the spindle apparatus functions. Do spindle microtubules act as railroad tracks, the way they do in vesicle transport? Is some sort of motor protein involved? And what is the nature of the kinetochore, where the chromosome and microtubules are joined?

MITOTIC SPINDLE FORCES The spindle apparatus is composed of microtubules. Recall from Chapter 7 that:

- microtubules are composed of α-tubulin and β-tubulin dimers,
- the number of tubulin dimers a microtubule contains determines its length, and
- microtubules are asymmetric—meaning they have a plus end and a minus end.

During mitosis, kinetochore microtubules grow from the microtubule organizing center until their plus ends attach to a kinetochore.

These observations suggest two possible mechanisms for the movement of chromosomes during anaphase. Does the kinetochore microtubule shorten due to a loss of tubulin dimers from one end? Or do intact microtubules slide past each other, like actin filaments in a contractile ring?

To test these hypotheses, biologists introduced fluorescently labeled tubulin subunits into prophase or metaphase cells. This treatment made the entire spindle apparatus visible (**Figure 11.8**, step 1). Once anaphase began, the researchers marked a region of the spindle with a bar-shaped beam of laser light. The laser bleached the fluorescence in the exposed region, darkening it—although it was still functional (Figure 11.8, step 2).

As anaphase progressed, two things happened: **(1)** The darkened region remained stationary, and **(2)** the kinetochore microtubules got shorter between the darkened region and the kinetochore.

This result suggested that the kinetochore microtubules remain stationary during anaphase, but shorten because tubulin subunits are lost from their plus ends. As they shorten at the kinetochore, the chromosomes are pulled along. How?

A KINETOCHORE MOTOR **Figure 11.9** on page 202 shows the current model for kinetochore structure and function during chromosome movement. Research is continuing, however, and it is likely that this model will be modified as additional data become available.

The kinetochore is thought to have a base that attaches to the centromere region of the chromosome and a "crown" of fibrous proteins projecting outward. Dyneins and other motor proteins are thought to be attached to the kinetochore's fibrous crown, where they can "walk" down microtubules—from the plus ends near the kinetochore toward the minus ends at the spindle.

EXPERIMENT

QUESTION: How do microtubules shorten to pull sister chromatids apart at anaphase?

HYPOTHESIS: Microtubules shorten at one end.

ALTERNATE HYPOTHESIS: Microtubules slide past each other like actin filaments.

EXPERIMENTAL SETUP:

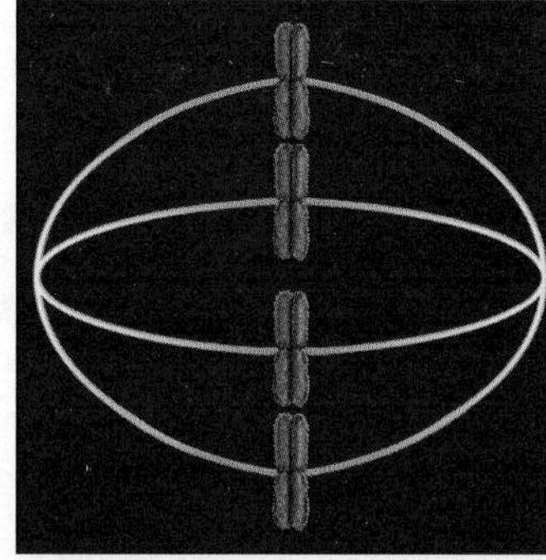

1. Use fluorescent labels to make the metaphase chromosomes fluoresce blue and the microtubules fluoresce yellow.

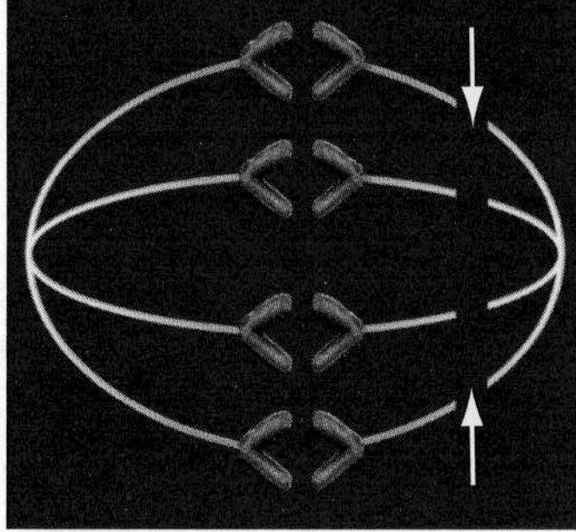

2. At the start of anaphase, darken a section of microtubules to mark them without changing their function.

PREDICTION: The darkened section will not move as chromosomes begin to move.

PREDICTION OF NULL HYPOTHESIS: The darkened section will disappear when chromosomes begin to move.

RESULTS:

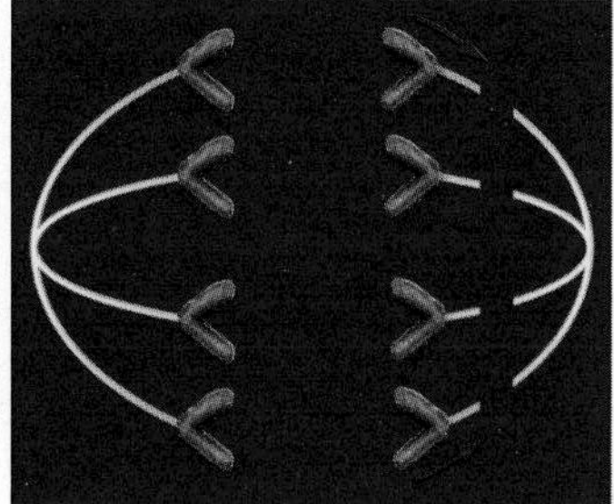

The darkened section remained visible, but the distance between chromosomes and darkened section lessened.

CONCLUSION: Microtubules shorten at one end—at the kinetochore.

FIGURE 11.8 During Anaphase, Microtubules Shorten at the Kinetochore.

SOURCE: Gorbsky, G. J., et al. 1987. Chromosomes move poleward during anaphase along stationary microtubules that coordinately disassemble from their kinetochore ends. *Journal of Cellular Biology* 104: 9–18.

✔**QUESTION** What would the outcome of the experiment be if (1) microtubules shorten at the end opposite the chromosome, or (2) microtubules slide past each other?

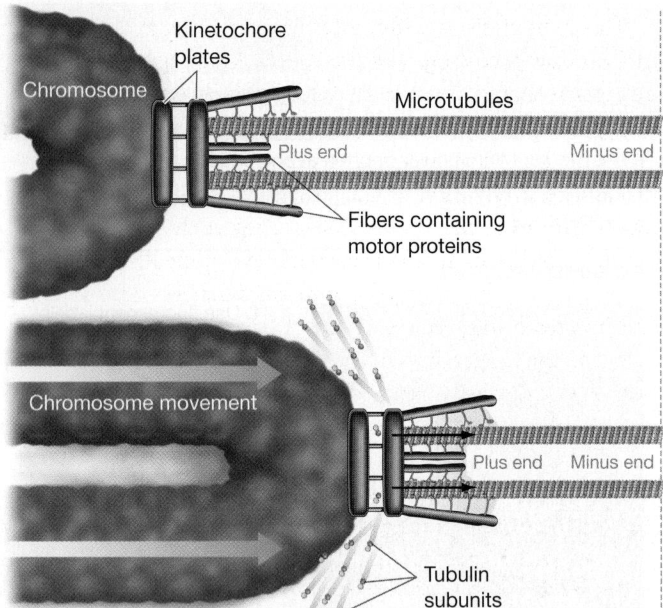

FIGURE 11.9 How Do Microtubules Move Chromosomes during Mitosis? Kinetochore microtubules shorten during anaphase, due to loss of tubulin dimers at the kinetochore. As they shorten, motor proteins walk the chromosomes down the remaining length of the kinetochore microtubules.

If these ideas are correct, the process would be similar to the way that kinesin walks down microtubules during vesicle transport or dynein walks down the microtubule doublets in flagella (see Chapter 7).

SUMMARY TABLE 11.1 **Structures Involved in Mitosis**

Structure	Definition
Chromosome	A structure composed of a DNA molecule and associated proteins
Chromatin	The material that makes up eukaryotic chromosomes; consists of a DNA molecule complexed with histone proteins
Chromatid	One strand of a replicated chromosome, with its associated proteins
Sister chromatids	The two strands of a replicated chromosome. When chromosomes are replicated, they consist of two sister chromatids. The genetic material in sister chromatids is identical. When sister chromatids separate during mitosis, they become independent chromosomes.
Centromere	The structure that joins sister chromatids
Kinetochores	The structures on sister chromatids where kinetochore microtubules attach
Microtubule organizing center	Any structure that organizes microtubules
Centrosome	The microtubule organizing center in animals
Centrioles	Cylindrical structures that comprise microtubules, located inside animal centrosomes

Biologists hypothesize that as anaphase gets under way, proteins in the kinetochore catalyze the loss of tubulin subunits at the plus end of the kinetochore microtubule, while kinetochore motor proteins walk toward the minus end. As the microtubule shortens and the motor proteins continue their walk, the chromosome is pulled to one end of the spindle apparatus.

Having explored how mitosis occurs, let's focus on how it is controlled. When does a cell divide, and when does it stop dividing?

CHECK YOUR UNDERSTANDING

If you understand that . . .

- When chromosomes replicate, each chromosome consists of two identical sister chromatids.
- After chromosomes replicate, mitosis distributes one copy of each chromosome to each daughter cell.
- Mitosis and cytokinesis produce cells with the same genetic material as that of the parent cell.

You should be able to . . .

1. Draw an unreplicated chromosome and a replicated chromosome, and label the sister chromatids and the centromere on the replicated chromosome.
2. Explain what IPPMAT stands for, and state when sister chromatids condense, move to the middle of the cell, and break apart to become independent chromosomes.

Answers are available in Appendix B.

11.3 Control of the Cell Cycle

Although the events of mitosis are virtually identical in all eukaryotes, other aspects of the cell cycle vary. In humans, for example, intestinal cells routinely divide more than twice a day to replace tissue that is lost during digestion; mature human nerve and muscle cells do not divide at all.

Most of these differences are due to variation in the length of the G_1 phase. In rapidly dividing cells, G_1 is essentially eliminated. Most nondividing cells, in contrast, are permanently stuck in G_1. Researchers refer to this arrested stage as the G_0 state, or simply "G zero." Cells that are in G_0 have effectively exited the cell cycle and are sometimes referred to as post-mitotic. Nerve cells, muscle cells, and many other cell types enter G_0 once they have matured.

A cell's division rate can also vary in response to changing conditions. For example, human liver cells normally divide about once per year. But if part of the liver is damaged or lost, the remaining cells divide every one or two days until repair is accomplished. Cells of unicellular organisms such as yeasts, bacteria, or archaea divide rapidly only if the environment is rich in nutrients; otherwise, they enter a quiescent (inactive) state.

To explain these differences, biologists hypothesized that the cell cycle must be regulated in some way. Cell cycle control is now the most prominent issue in research on cell division—partly because defects in control can lead to uncontrolled, cancerous growth.

The Discovery of Cell-Cycle Regulatory Molecules

The first solid evidence for cell-cycle control molecules came to light in 1970, when researchers found that certain chemicals, viruses, or an electric shock could fuse the membranes of two mammalian cells that were growing in culture, forming a single cell with two nuclei.

How did cell fusion experiments relate to cell-cycle regulation? When investigators fused cells that were in different stages of the cell cycle, certain nuclei changed phases. For example, when a cell in M phase was fused with one in interphase, the nucleus of the interphase cell initiated M phase (**Figure 11.10a**). The biologists hypothesized that the cytoplasm of M-phase cells contains a regulatory molecule that induces interphase cells to enter M phase.

This hypothesis was supported by experiments on the South African claw-toed frog, *Xenopus laevis*. As an egg of these frogs matures, it changes from a cell called an **oocyte**, which is arrested in a phase similar to G_2, to a mature egg that has entered M phase. The large size of these eggs—more than 1 mm in diameter—makes it relatively easy to purify their cytoplasm and use instruments with extremely fine needles to inject the eggs with cytoplasm from eggs in different stages of development.

When biologists purified cytoplasm from M-phase frog eggs and injected it into the cytoplasm of frog oocytes arrested in the G_2 phase, the immature oocytes entered M phase (**Figure 11.10b**). But when cytoplasm from interphase cells was injected into G_2 oocytes, the cells remained in the G_2 phase. The researchers

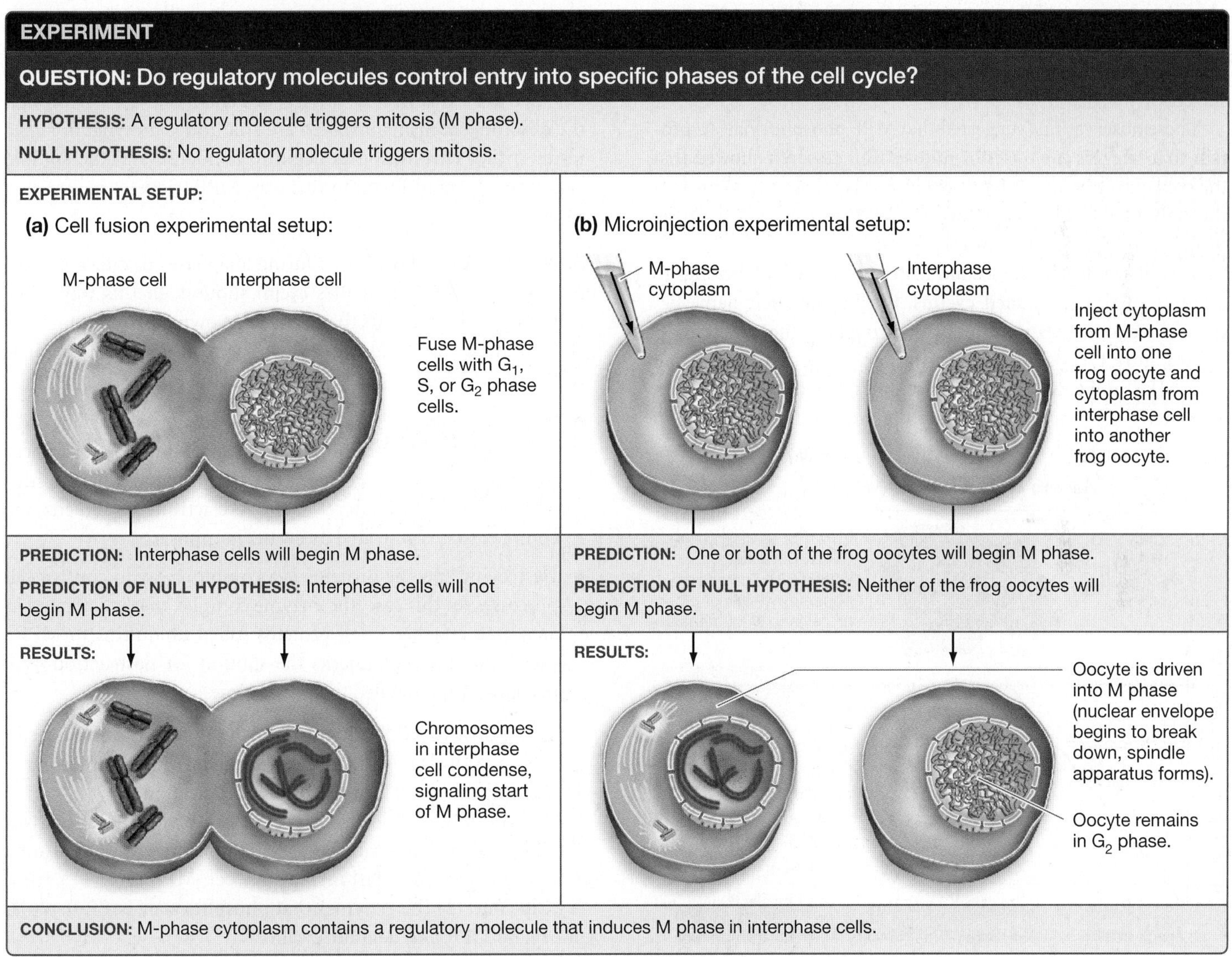

EXPERIMENT
QUESTION: Do regulatory molecules control entry into specific phases of the cell cycle?
HYPOTHESIS: A regulatory molecule triggers mitosis (M phase).
NULL HYPOTHESIS: No regulatory molecule triggers mitosis.
EXPERIMENTAL SETUP:
(a) Cell fusion experimental setup:
M-phase cell
Interphase cell
Fuse M-phase cells with G1, S, or G2 phase cells.
(b) Microinjection experimental setup:
M-phase cytoplasm
Interphase cytoplasm
Inject cytoplasm from M-phase cell into one frog oocyte and cytoplasm from interphase cell into another frog oocyte.
PREDICTION: Interphase cells will begin M phase.
PREDICTION OF NULL HYPOTHESIS: Interphase cells will not begin M phase.
PREDICTION: One or both of the frog oocytes will begin M phase.
PREDICTION OF NULL HYPOTHESIS: Neither of the frog oocytes will begin M phase.
RESULTS:
Chromosomes in interphase cell condense, signaling start of M phase.
RESULTS:
Oocyte is driven into M phase (nuclear envelope begins to break down, spindle apparatus forms).
Oocyte remains in G2 phase.
CONCLUSION: M-phase cytoplasm contains a regulatory molecule that induces M phase in interphase cells.

FIGURE 11.10 Experimental Evidence for Cell-Cycle Control Molecules. (a) When M-phase cells are fused with cells in G_1, S, or G_2 phase, the interphase chromosomes condense and begin M phase. **(b)** Microinjection experiments support the hypothesis that a regulatory molecule induces M phase.

SOURCES: Rao, P. N. and R. T. Johnson. 1970. Mammalian cell fusion studies on the regulation of DNA synthesis and mitosis. *Nature* 225: 159–164. Also Masui, Y. and C. L. Markert. 1971. Cytoplasmic control of nuclear behavior during meiotic maturation of frog oocytes. *Journal of Experimental Zoology* 177: 129–145.

✔**QUESTION** In the cell fusion experiment, how would you test the hypothesis that the fusion event itself—not something in the cytoplasm—triggered the start of M phase?

concluded that the cytoplasm of M-phase cells—but not the cytoplasm of interphase cells—contains a factor that drives immature oocytes into M phase to complete their maturation.

This factor was eventually purified and is now called **mitosis-promoting factor**, or **MPF.** Subsequent experiments showed that MPF induces mitosis in all eukaryotes. For example, injecting M-phase cytoplasm from mammalian cells into immature frog eggs results in egg maturation, and human MPF can trigger mitosis in yeast cells.

MPF appears to be a general signal that says "Start mitosis." How does it work?

MPF CONTAINS A PROTEIN KINASE AND A CYCLIN MPF is made up of two distinct polypeptide subunits. One subunit is a **protein kinase**—an enzyme that catalyzes the transfer of a phosphate group from ATP to a target protein. Recall from Chapter 9 that phosphorylation may activate or inactivate proteins. As a result, protein kinases frequently act as regulatory elements in the cell.

These observations suggested that MPF phosphorylates a protein that triggers the onset of mitosis. But research showed that the concentration of MPF protein kinase is more or less constant throughout the cell cycle. How can MPF trigger mitosis if the protein kinase subunit is always present?

The answer lies in the second MPF subunit, which belongs to a family of proteins called **cyclins**. Cyclins got their name because their concentrations fluctuate throughout the cell cycle.

As **Figure 11.11** shows, concentrations of the cyclin associated with MPF build during interphase and peak during M phase. This increase is important because the protein kinase subunit in MPF can be active only when it is bound to the cyclin subunit. As a result, the protein kinase subunit of MPF is called a **cyclin-dependent kinase**, or **Cdk**.

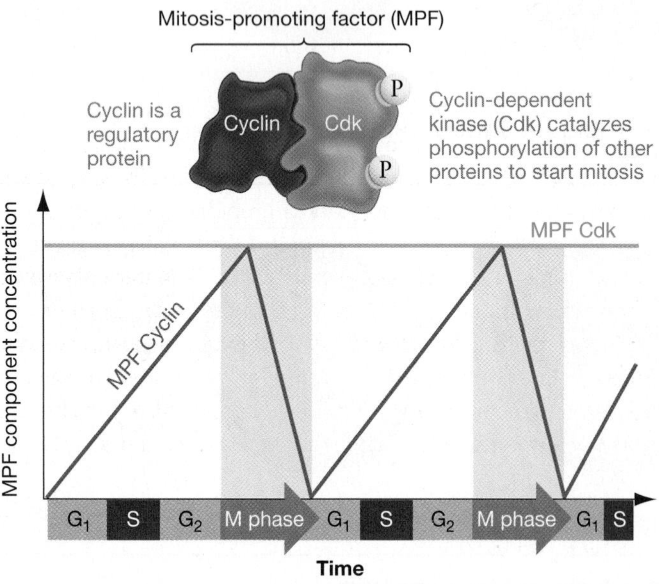

FIGURE 11.11 Cyclin Concentration Regulates MPF Concentration. Cyclin concentrations cycle in dividing cells, reaching a peak in M phase.

✔**QUESTION** Proteins that degrade cyclin are activated by events that MPF initiates. Why is this important?

To summarize, MPF is a dimer consisting of a cyclin and a cyclin-dependent kinase. The cyclin subunit functions as a regulatory protein; the kinase subunit catalyzes the phosphorylation of other proteins to start mitosis.

HOW IS MPF ACTIVATED? According to Figure 11.11, the number of complete MPF dimers builds up steadily during interphase. Why doesn't this increasing concentration of MPF trigger the onset of M phase?

The answer is that MPF's Cdk subunit becomes phosphorylated at two sites after it binds to cyclin. When Cdk is phosphorylated at both sites, MPF is inactive. Late in G_2 phase, however, enzymes cause one of the phosphate groups on the Cdk subunit to drop off. This dephosphorylation reaction changes MPF's shape in a way that activates it.

Once MPF is activated, it triggers a chain of events. Although the exact mechanisms involved are still under investigation, the result is that chromosomes begin to condense and the spindle apparatus starts to form. In this way, MPF triggers the onset of M phase.

HOW IS MPF DEACTIVATED? During anaphase, an enzyme complex begins degrading MPF's cyclin subunit. In this way, MPF triggers a chain of events that leads to its own destruction.

MPF deactivation illustrates two key concepts about regulatory systems in cells:

- **Negative feedback** occurs when a process is slowed or shut down by one of its products. Thermostats shut down furnaces when temperatures are high; phosphofructokinase is inhibited by ATP (see Chapter 9); MPF is deactivated by an enzyme complex that is activated by events in mitosis.
- Destroying specific proteins is a common way to control cell processes. In this case, the enzyme complex that is activated in anaphase attaches small proteins called ubiquitins to MPF's cyclin subunit. This marks the subunit for destruction by a protein complex called the proteasome.

In response to MPF activity, then, the concentration of cyclin declines rapidly. Slowly, it builds up again during interphase. This sets up an oscillation in cyclin concentration.

✔If you understand this aspect of cell-cycle regulation, you should be able to describe what MPF does. You should also be able to explain the relationship between MPF and (1) cyclin, (2) Cdk, and (3) the enzymes that phosphorylate MPF, dephosphorylate MPF, and degrade cyclin.

Cell-Cycle Checkpoints Can Arrest the Cell Cycle

The dramatic oscillation in cyclin concentration and activation drives the ordered events of the cell cycle. These events are occurring in your body right now. Over a 24-hour period, you swallow millions of cheek cells and lose millions of cells from your intestinal lining as waste. To replace them, cells in your cheek and intestinal tissue are making and degrading cyclin and pushing themselves through the cell cycle.

MPF is only one of many protein complexes involved in regulating the cell cycle, however. A different cyclin and protein kinase triggers the passage from G_1 phase into S phase, and several regulatory proteins maintain the G_0 state of quiescent cells. An array of regulatory molecules holds cells in particular stages or stimulates passage to the next phase.

To make sense of these observations, Leland Hartwell and Ted Weinert introduced the concept of a **cell-cycle checkpoint**. A cell-cycle checkpoint is a critical point in the cell cycle that is regulated.

Hartwell and Weinert identified checkpoints by analyzing yeast cells with defects in the cell cycle. The defective cells kept dividing under culture conditions when normal cells stopped growing, because they lacked a specific checkpoint. In the body, cells that keep dividing in this way form a mass of cells called a **tumor**.

There are three distinct checkpoints during the four phases of the cell cycle (**Figure 11.12**). In effect, interactions among regulatory molecules at each checkpoint allow a cell to "decide" whether to proceed with division. If these regulatory molecules are defective, the checkpoint may fail and cells may start dividing in an uncontrolled fashion.

G_1 CHECKPOINT The first cell-cycle checkpoint occurs late in G_1. For most cells, this checkpoint is the most important in establishing whether the cell will continue through the cycle and divide, or exit the cycle and enter G_0. What determines whether a cell passes the G_1 checkpoint?

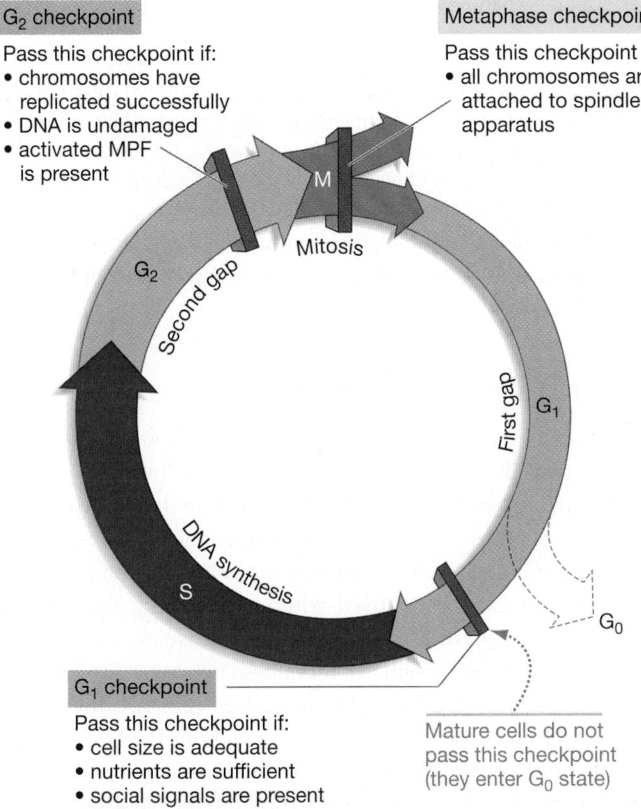

G₂ checkpoint
Pass this checkpoint if:
• chromosomes have replicated successfully
• DNA is undamaged
• activated MPF is present
Metaphase checkpoint
Pass this checkpoint if:
• all chromosomes are attached to spindle apparatus
M
Mitosis
G₂
Second gap
G₁
First gap
DNA synthesis
S
G₀
G₁ checkpoint
Pass this checkpoint if:
• cell size is adequate
• nutrients are sufficient
• social signals are present
• DNA is undamaged
Mature cells do not pass this checkpoint (they enter G₀ state)

FIGURE 11.12 The Three Cell-Cycle Checkpoints.

- *Size* Because a cell must reach a certain size before its daughter cells will be large enough to function normally, biologists hypothesize that some mechanism exists to arrest the cell cycle if the cell is too small.
- *Availability of nutrients* Unicellular organisms arrest at the G_1 checkpoint if nutrient conditions are poor.
- *Social signals* Cells in multicellular organisms pass (or do not pass) through the G_1 checkpoint in response to signaling molecules from other cells, which are termed social signals.
- *Damage to DNA* If DNA is physically damaged, the protein **p53** activates genes that either stop the cell cycle until the damage can be repaired or cause the cell's programmed, controlled destruction—a phenomenon known as **apoptosis**. In this way, p53 acts as a brake on the cell cycle.

If "brake" molecules such as p53 are defective, damaged DNA remains unrepaired. Damage in genes that regulate cell growth can lead to uncontrolled cell division. Consequently, regulatory proteins like p53 are called **tumor suppressors**.

G_2 CHECKPOINT The second checkpoint occurs after S phase, at the boundary between the G_2 and M phases. Because MPF is the key signal triggering the onset of M phase, investigators were not surprised to find that it is involved in the G_2 checkpoint.

Data suggest that if DNA is damaged or if chromosomes are not replicated correctly, the dephosphorylation and activation of MPF are blocked. When MPF is not activated, cells remain in G_2 phase. Cells at this checkpoint may also respond to signals from other cells and to internal signals relating to their size.

METAPHASE CHECKPOINT The final checkpoint occurs during mitosis. If not all chromosomes attach properly to the spindle apparatus, M phase arrests at metaphase. Specifically, anaphase is delayed until all kinetochores attach properly to the spindle apparatus. If the metaphase checkpoint did not exist, some chromosomes might not separate correctly, and daughter cells would receive too many chromosomes or not enough chromosomes.

To summarize, the three cell-cycle checkpoints have the same purpose: They prevent the division of cells that are damaged or that have other problems. The G_1 checkpoint also prevents the growth of mature cells that are in the G_0 state and should not grow any more.

You can review the cell cycle and the checkpoint concept in the study area at *www.masteringbiology.com.*

Web Activity Four Phases of the Cell Cycle

Understanding cell cycle regulation is fundamental. If one of the checkpoints fails, the affected cells may begin dividing in an uncontrolled fashion. For the organism as a whole, the consequences of uncontrolled cell division may be dire: cancer.

CHECK YOUR UNDERSTANDING

If you understand that . . .

- The cell cycle consists of four carefully controlled phases.

✓ You should be able to . . .

1. List the phases of the cell cycle, noting where checkpoints occur, and
2. Summarize how levels of MPF Cdk and MPF cyclin, and activity of MPF, change over the course of the cycle.

Answers are available in Appendix B.

11.4 Cancer: Out-of-Control Cell Division

Fifty percent of American men and 33 percent of American women will develop cancer during their lifetime. In the United States, one in four of all deaths are from cancer. It is the second leading cause of death, exceeded only by heart disease.

Cancer is a general term for disease caused by cells that divide in an uncontrolled fashion, invade nearby tissues, and spread to other sites in the body. Cancerous cells cause disease because they use nutrients and space needed by normal cells and disrupt the function of normal tissues.

Humans suffer from at least 200 types of cancer. Stated another way, cancer is not a single illness but a complex family of diseases that affect an array of organs, including the breast, colon, brain, lung, and skin. In addition, several types of cancer can affect the same organ. Skin cancers, for example, come in multiple forms.

Some cancers are relatively easy to treat; others are often fatal. **Figure 11.13** illustrates how mortality rates due to different types of cancer have changed through time in the United States.

Although cancers vary in time of onset, growth rate, seriousness, and cause, they have a unifying feature: Cancers arise from cells in which cell-cycle checkpoints have failed.

Cancerous cells have two types of defects:

1. defects that make the proteins required for cell growth active when they shouldn't be, and
2. defects that prevent tumor suppressor genes from shutting down the cell cycle.

For example, Chapter 8 introduced the protein Ras as a key component in signal transduction systems—including phosphorylation cascades that trigger cell growth. Many cancerous cells have defective forms of Ras that do not become inactivated. Instead, the defective Ras constantly sends signals that trigger mitosis and cell division.

Likewise, a large percentage of cancerous cells have defective forms of the tumor suppressor p53. Instead of being arrested or destroyed, cells with damaged DNA are allowed to continue growing.

Let's review the general characteristics of cancer and then explore why regulatory mechanisms become defective.

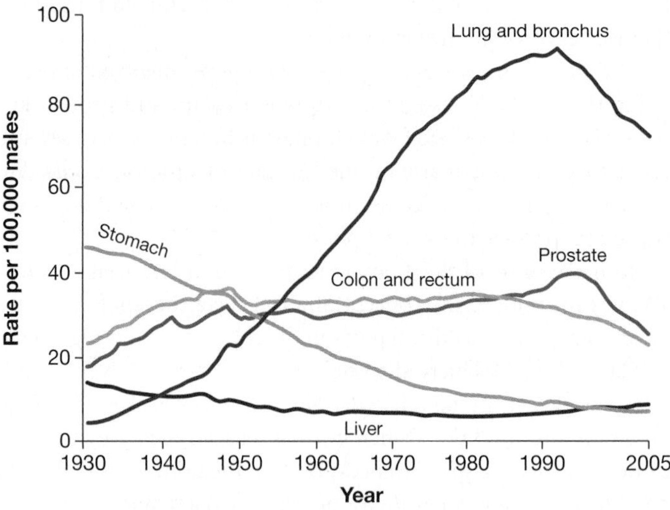

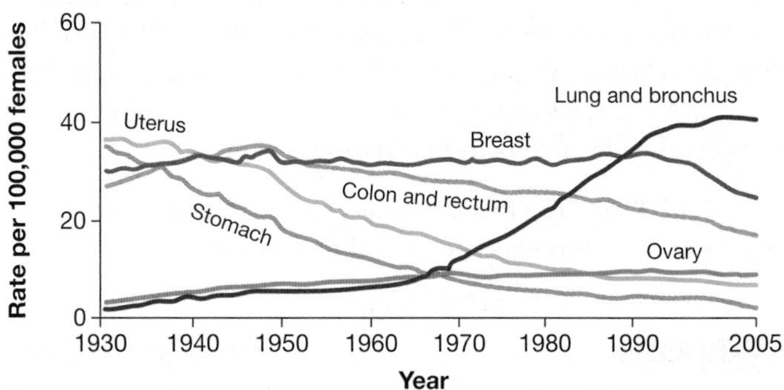

FIGURE 11.13 Cancer Death Rates in the U.S. Note that death rates could vary over time because of changes in incidence (how often people get a particular cancer) and/or treatment. (Reproduced by permission of American Cancer Society. *Cancer Facts and Figures 2009.* Atlanta: American Cancer Society, Inc.)

✓QUESTION How has the death rate due to lung cancer changed over time in males versus females? Suggest a hypothesis to explain this pattern.

Properties of Cancer Cells

When even a single cell in a multicellular organism begins to divide in an uncontrolled fashion, a mass of cells called a tumor may result. For example, most cells in the adult human brain do not divide. But if a single abnormal brain cell begins unrestrained division, the growing tumor that results can disrupt the brain's function.

If a tumor can be removed without damaging the affected organ, a cure might be achieved, so surgery is usually the first step in treatment. Often, though, surgery doesn't cure cancer. Why?

In addition to dividing quickly, cancer cells are invasive—meaning that they are able to spread throughout the body via the bloodstream or lymphatic vessels (introduced in Chapter 49), which collect excess fluid from tissues and return it to the bloodstream.

Invasiveness is a defining feature of a **malignant tumor**—one that is cancerous. Masses of noninvasive cells are noncancerous and form **benign tumors**. Some benign tumors grow slowly and are largely

(a) Benign tumor

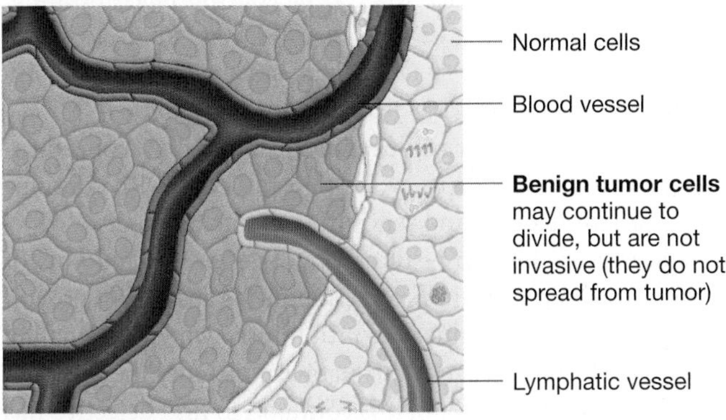

Normal cells
Blood vessel
Benign tumor cells may continue to divide, but are not invasive (they do not spread from tumor)
Lymphatic vessel

(b) Malignant tumor

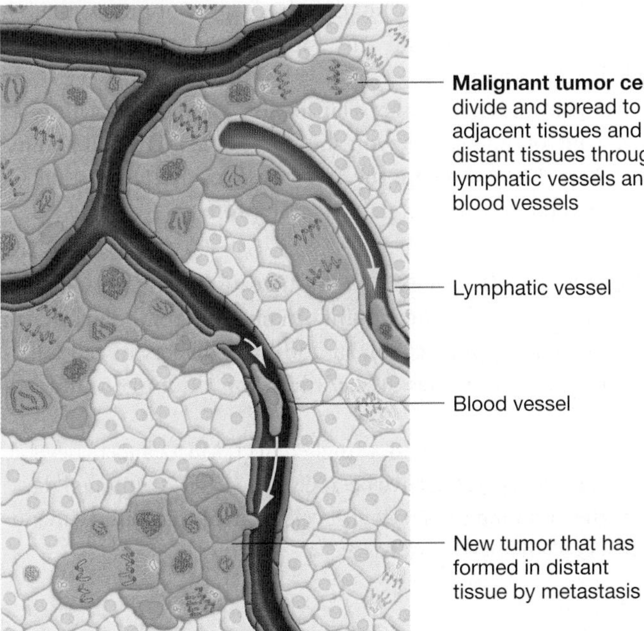

Malignant tumor cells divide and spread to adjacent tissues and to distant tissues through lymphatic vessels and blood vessels
Lymphatic vessel
Blood vessel
New tumor that has formed in distant tissue by metastasis

FIGURE 11.14 Cancers Spread to New Locations in the Body.
(a) Benign tumors grow in a single location. **(b)** Malignant tumors are metastatic—meaning that their cells can spread to distant parts of the body and initiate new tumors. Malignant tumors cause cancer.

harmless. Others grow quickly and can cause problems if they are located in the brain or other sensitive parts of the body.

Cells become malignant and cancerous if they gain the ability to detach from the original tumor and invade other tissues. By spreading from the primary tumor site, cancer cells can establish secondary tumors elsewhere in the body (**Figure 11.14**). This process is called **metastasis.**

If metastasis has occurred by the time the original tumor is detected, secondary tumors have already formed and surgical removal of the primary tumor will not lead to a cure. This is why early detection is the key to treating cancer most effectively.

Cancer Involves Loss of Cell-Cycle Control

What causes cancer at the molecular level? Recall that when many cells mature, they enter the G_0 phase—meaning their cell cycle is arrested at the G_1 checkpoint. In contrast, cells that do pass through the G_1 checkpoint are irreversibly committed to replicating their DNA and entering G_2.

Based on this observation, biologists hypothesize that many or even most types of cancer involve defects in the G_1 checkpoint. To understand the molecular nature of the disease, then, researchers have focused on understanding the normal mechanisms that operate at that checkpoint. Cancer research and research on the normal cell cycle have become two sides of the same coin.

SOCIAL CONTROL In unicellular organisms, passage through the G_1 checkpoint is thought to depend primarily on cell size and the availability of nutrients. If nutrients are plentiful, cells pass through the checkpoint and divide rapidly.

In multicellular organisms, however, cells divide in response to signals from other cells. Biologists refer to this as *social control* over cell division. The general idea is that individual cells should be allowed to divide only when their growth is in the best interests of the organism as a whole.

Social control of the cell cycle is based on **growth factors**—polypeptides or small proteins that stimulate cell division. Many growth factors were discovered by researchers who were trying to grow cells in culture. When isolated mammalian cells were placed in a culture flask and provided with adequate nutrients, they arrested in G_1 phase. The cells began to grow again only when biologists added **serum**—the liquid portion of blood that remains after blood cells and cell fragments have been removed.

Some component of serum allowed cells to pass through the G_1 checkpoint. What was it?

One of these serum components is a protein called platelet-derived growth factor (PDGF). As its name implies, PDGF is released by blood components called **platelets,** which promote blood clotting at wound sites. PDGF binds to receptor tyrosine kinases on the surface of target cells. When they receive the signal, cells divide. This promotes wound healing.

Researchers subsequently found that PDGF is produced by an array of cell types, in addition to platelets. Investigators have identified a diverse array of other growth factors as well.

For different types of cells to grow in culture, different combinations of growth factors must be supplied. Based on this result, biologists infer that different types of cells in an intact multicellular organism are controlled by different combinations of growth factors.

Cancer cells are an exception. They can often be cultured successfully without externally supplied growth factors. This observation suggests that the normal social controls on the G_1 checkpoint have broken down in cancer cells.

HOW DOES THE G_1 CHECKPOINT WORK? To understand the G_1 checkpoint, think back to how cells progress from G_2 to M phase. The key trigger is activation of MPF. Recall that MPF is a dimer of a cyclin and a Cdk, and that Cdk stands for cyclin-dependent kinase.

In G_0 cells, the arrival of growth factors stimulates the production of a key regulatory protein called E2F. E2F is analogous to MPF. When E2F is activated, it triggers the expression of genes required for S phase.

When E2F is first produced, however, it binds to a tumor suppressor protein called Rb. **Rb protein** is one of the key molecules that enforces the G_1 checkpoint. It is called Rb because it was discovered in children with retinoblastoma, a cancer that produces malignant tumors in the light-sensing tissue, or retina, of the eye.

When E2F is bound to Rb, it is in the "off" position—it can't activate the genes required for S phase. As long as Rb stays bound to E2F, the cell remains in G_0.

But as **Figure 11.15** shows, the situation changes dramatically if growth factors continue to arrive.

Step 1 Enough growth factors arrive to override the inhibitory effects of Rb.

Step 2 In addition to stimulating production of E2F, the growth factors stimulate production of cyclins that are specific to the G_1 checkpoint.

Step 3 E2F continues to bind to Rb and remain inactive. But the cyclins—which are different from those observed in MPF during G_2—begin forming G_1 cyclin-Cdk dimers. Initially, the Cdk component is phosphorylated and inactive.

Step 4 When dephosphorylation activates G_1 cyclin-Cdk complexes, they catalyze the phosphorylation of Rb.

Step 5 The phosphorylated Rb changes shape and no longer binds to E2F.

Step 6 The unbound E2F is free to activate its target genes. Production of S-phase proteins gets S phase under way.

In this way, high concentrations of growth factors function as a social signal that says, "It's OK to override Rb. Go ahead and pass the G_1 checkpoint and divide."

WHY DO SOCIAL CONTROLS AND CELL-CYCLE CHECKPOINTS FAIL? Cells can become cancerous when social controls fail—meaning, when cells begin dividing in the absence of the go-ahead signal from growth factors. One of two things can go wrong: the G_1 cyclin is overproduced, or Rb is defective.

When cyclins are overproduced and stay at high concentrations, the Cdk that binds to cyclin phosphorylates Rb continuously. This activates E2F and sends the cell into S phase.

Cyclin overproduction results from (1) excessive amounts of growth factors or (2) cyclin production in the absence of growth signals. Cyclins are produced continuously when a signaling pathway is defective. Because this pathway includes the Ras protein highlighted in Chapter 8, it is common to find defective Ras proteins in cancerous cells.

What happens if Rb is defective? When Rb is missing or does not bind normally to E2F, any E2F that is present pushes the cell through the G_1 checkpoint and into S phase, leading to uncontrolled cell division.

✔If you understand the relationship between the failure of social controls and cancer, you should be able to make diagrams like Figure 11.15 showing the consequences of (1) constant G_1 cyclin production and (2) defective Rb proteins.

CANCER IS A FAMILY OF DISEASES Because many proteins are essential to the G_1 checkpoint, many different defects can cause the checkpoint to fail. In addition, defects in cell adhesion or other properties are required for a tumor to undergo metastasis.

Cancer is seldom due to a single defect. Most cancers develop only after several genes have been damaged. The combined damage is then enough to break cell-cycle control and induce uncontrolled growth and metastasis.

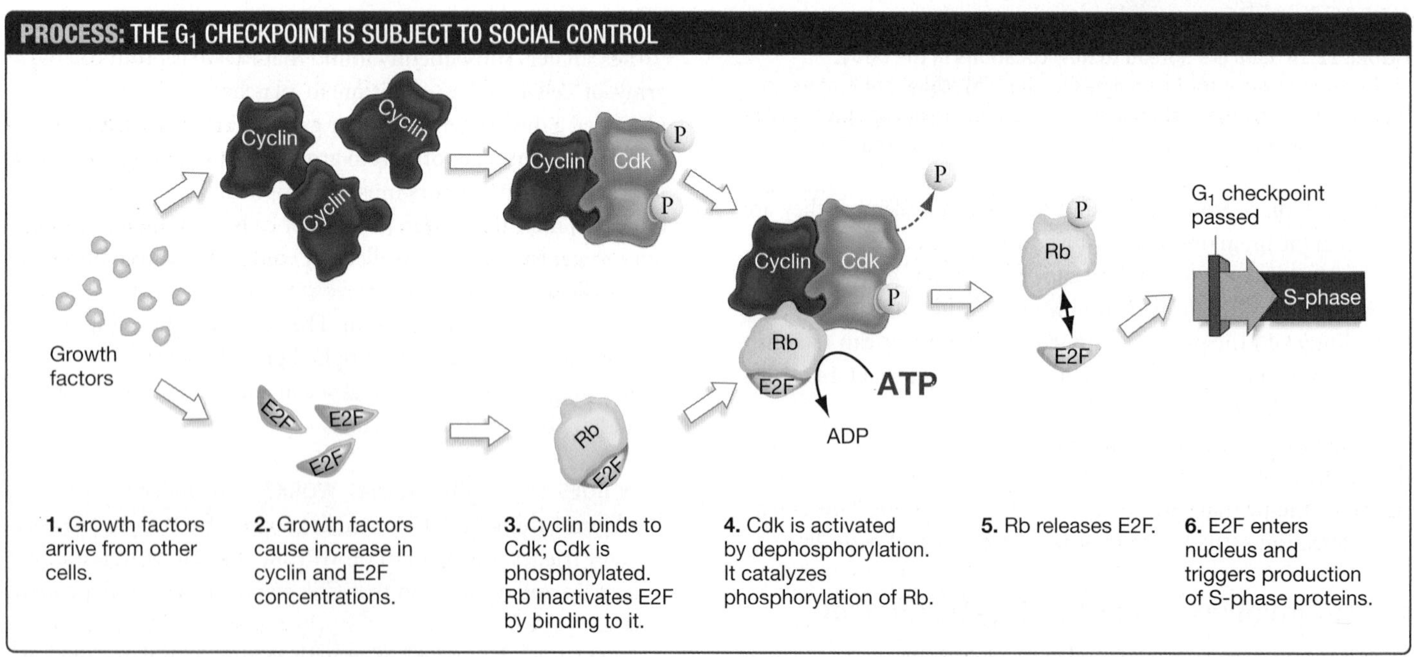
PROCESS: THE G1 CHECKPOINT IS SUBJECT TO SOCIAL CONTROL
Cyclin
Cyclin
Cyclin
Cyclin
Cdk
P
P
Growth factors
E2F
E2F
E2F
Rb
E2F
P
Cyclin
Cdk
P
Rb
E2F
ATP
ADP
P
Rb
E2F
G1 checkpoint passed
S-phase
1. Growth factors arrive from other cells.
2. Growth factors cause increase in cyclin and E2F concentrations.
3. Cyclin binds to Cdk; Cdk is phosphorylated. Rb inactivates E2F by binding to it.
4. Cdk is activated by dephosphorylation. It catalyzes phosphorylation of Rb.
5. Rb releases E2F.
6. E2F enters nucleus and triggers production of S-phase proteins.

FIGURE 11.15 Growth Factors Move Cells through the G_1 Checkpoint.

In addition, each type of cancer is due to a unique combination of errors. Hundreds if not thousands of different defects are responsible for the many different types of cancer.

Because cancer is actually a family of diseases with a complex and highly variable molecular basis, there will be no "magic bullet," or single therapy that cures all forms of the illness. Still, recent progress in understanding the cell cycle and the molecular basis of cancer has been dramatic, and cancer prevention and early detection programs are increasingly effective. The prognosis for many cancer patients is remarkably better now than it was even a few years ago. Thanks to research, almost all of us know someone who is a cancer survivor.

CHAPTER 11 REVIEW

For media, go to the study area at www.masteringbiology.com

Summary of Key Concepts

In eukaryotes, most dividing cells go through a cycle that consists of four phases.

- Dividing cells alternate between the dividing phase, called M phase, and a nondividing phase known as interphase.
- Interphase consists of S phase, when chromosomes are replicated, and gap phases called G_1 and G_2, when cell growth and replication of nonnuclear cell components occur.
- The order of cell-cycle phases is $G_1 \rightarrow S \rightarrow G_2 \rightarrow M \rightarrow G_1$ and so forth. Mature cells arrest at G_1 and enter a resting phase called G_0.

✔You should be able to explain why the gap phases exist.

After chromosomes are copied during S phase, they are moved to the middle of the cell during M phase (mitosis). One chromosome copy is distributed to each of two daughter cells. Mitosis and cytokinesis produce two cells that are genetically identical to the parent cell.

- Mitosis and cytokinesis are responsible for growth in multicellular eukaryotes and reproduction in most unicellular eukaryotes and some multicellular eukaryotes.
- Mitosis can be described as a sequence of five phases:
  (1) ***Prophase*** Chromosomes condense. The spindle apparatus begins to form.
  (2) ***Prometaphase*** The nuclear envelope disintegrates. Kinetochore microtubules make contact with chromosomes.
  (3) ***Metaphase*** Kinetochore microtubules move chromosomes to the middle of the cell.
  (4) ***Anaphase*** Kinetochore microtubules pull sister chromatids apart.
  (5) ***Telophase*** Chromosomes are pulled to opposite poles of the cell. A nuclear envelope forms around each set.
- In most cells, mitosis is followed by cytokinesis—division of all cell contents.

✔You should be able to explain why the nuclear envelope has to disintegrate prior to kinetochore microtubules contacting chromosomes.

MB **BioFlix™** Mitosis, **Web Activity** The Phases of Mitosis

Progression through the cell cycle is carefully controlled.

- Progression through the cell cycle is regulated at three checkpoints.
  (1) The G_1 checkpoint depends on sufficient nutrients, cell size, lack of DNA damage, and/or social signals.
  (2) The G_2 checkpoint delays progress until chromosome replication is complete and any damaged DNA that is present is repaired.
  (3) The M phase checkpoint delays anaphase until all chromosomes are correctly attached to the spindle apparatus.
- Cyclins and cyclin-dependent kinases (Cdks) help regulate the cell cycle. Cyclin concentrations oscillate during the cell cycle, regulating the activity of Cdks. Active Cdks phosphorylate proteins required for the next cell cycle phase.

✔You should be able to predict what happens if an influx of growth factors increases the production of cyclins.

MB **Web Activity** Four Phases of the Cell Cycle

In multicellular organisms, uncontrolled cell division may lead to cancer. Different types of cancer result from different types of defects in control over the cell cycle.

- Cancer is characterized by (1) loss of control at the G_1 checkpoint, resulting in cells that divide in an uncontrolled fashion, and (2) metastasis, or the ability of tumor cells to spread throughout the body.
- The G_1 checkpoint depends in part on Rb, which prevents progression to S phase, and G_1 cyclin-Cdk complexes that trigger progression to S phase. Defects in Rb and G_1 cyclin are common in human cancer cells.

✔You should be able to explain why there is unlikely to be a single cure for cancer.

Questions

✔TEST YOUR KNOWLEDGE

Answers are available in Appendix B

1. Which statement about the daughter cells of mitosis is correct?
   a. They differ genetically from one another and from the parent cell.
   b. They are genetically identical to one another and to the parent cell.
   c. They are genetically identical to one another but different from the parent cell.
   d. Only one of the two daughter cells is genetically identical to the parent cell.

2. Progression through the cell cycle is regulated by oscillations in the concentration of which type of molecule?
   a. p53, Rb, and other tumor suppressors
   b. receptor tyrosine kinases
   c. cyclin-dependent kinases
   d. cyclins
3. After replication, what comprises a single chromosome?
   a. the relaxed, uncondensed state
   b. a single strand of chromatin
   c. a single strand of DNA, complexed with proteins
   d. two sister chromatids
4. What major events occur during anaphase of mitosis?
   a. Chromosomes replicate, so each chromosome consists of two identical sister chromatids.
   b. Chromosomes condense and the nuclear envelope disappears.
   c. The chromosomes end up at opposite ends of the cell and two nuclear envelopes form.
   d. Sister chromatids separate, forming independent chromosomes.
5. What evidence suggests that during anaphase, kinetochore microtubules shorten at the kinetochore and not at the base of the spindle apparatus?
   a. Motor proteins are located at the kinetochore.
   b. Motor proteins are located at the kinetochore *and* at the base of the spindle apparatus.
   c. When fluorescing microtubules are darkened in the middle, the darkened segment stays stationary as the fibers shorten near the kinetochore.
   d. When fluorescing microtubules are darkened in the middle, the darkened segment moves toward the base of the spindle apparatus as the fibers shorten near the microtubule organizing center.
6. What happens if the sister chromatids of one chromosome fail to separate at anaphase?
   a. The kinetochore microtubules do not function properly.
   b. The polar microtubules do not function properly.
   c. One daughter cell receives too few chromosomes; the other receives a replicated chromosome.
   d. One daughter cell receives all the chromosomes; the other receives none.

✔TEST YOUR UNDERSTANDING

Answers are available in Appendix B

1. Sketch the phases of mitosis in a cell with four chromosomes, listing the major events of each phase. Identify at least two events that must be completed successfully for daughter cells to share an identical complement of chromosomes.
2. Make a concept map illustrating normal events at the G_1 checkpoint. Your diagram should include p53, DNA damage, Rb, E2F, social signals, G_1 Cdk, G_1 cyclin, S-phase proteins, phosphorylated (inactivated) cyclin-Cdk, dephosphorylated (activated) cyclin-Cdk, phosphorylated (inactivated) Rb.
3. Explain how cell fusion and microinjection experiments supported the hypothesis that specific molecules are involved in the transition from interphase to M phase.
4. Why are most protein kinases considered regulatory proteins?
5. Why are cyclins called cyclins? Explain their relationship to cyclin-dependent kinases.
6. Early detection is the key to surviving most cancers. Why?

✔APPLYING CONCEPTS TO NEW SITUATIONS

Answers are available in Appendix B

1. In multicellular organisms, nondividing cells stay in G_1 phase. For the cell, why is it better to be held in G_1 rather than S, G_2, or M phase?
2. When fruit fly embryos first begin to develop, mitosis occurs without cytokinesis. What is the result?
3. According to data from experiments with radioactive thymidine, the first labeled mitotic cells appear about 4 hours after the labeling period ends. From these data, researchers concluded that G_2 lasted about 4 hours. Why?
4. Cancer is primarily a disease of older people. Further, a group of individuals may share a genetic predisposition to developing certain types of cancer, yet vary a great deal in time of onset—or not get the disease at all. Discuss these observations in light of the claim that several defects usually have to occur for cancer to develop.

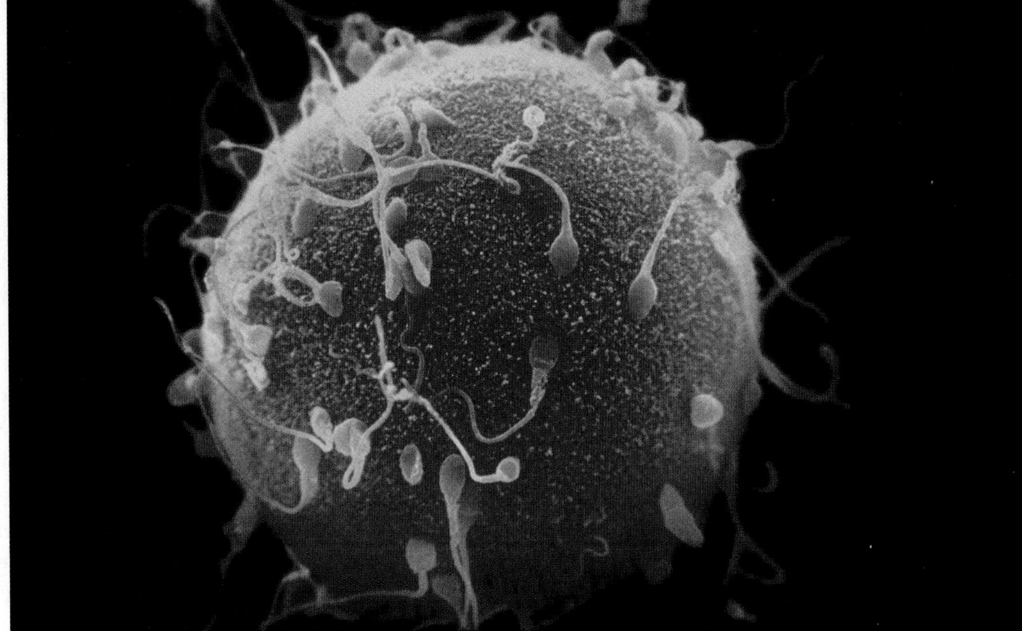

Scanning electron micrograph (with color added) showing human sperm attempting to enter a human egg. This chapter introduces the type of nuclear division called meiosis, which in animals occurs prior to the formation of sperm and eggs.

Meiosis 12

Why sex?

Simple questions—such as why sexual reproduction exists—are sometimes the best. This chapter asks what sexual reproduction is and why some organisms employ it. The focus here is on how organisms reproduce, or replicate—one of the five fundamental attributes of life introduced in Chapter 1.

For centuries people have known that during sexual reproduction, a male reproductive cell—a **sperm**—and a female reproductive cell—an **egg**—unite to form a new individual. The process of uniting sperm and egg is called **fertilization**.

The first biologists to observe fertilization studied the large, translucent eggs of sea urchins. Due to the semitransparency of the sea urchin egg cell, researchers were able to see the nuclei of a sperm and an egg fuse.

When these observations were published in 1876, they raised an important question, because biologists had already established that the number of chromosomes is constant from cell to cell within a multicellular organism. The question is, How can the chromosomes from a sperm cell and an egg cell combine, but form an offspring that has the same chromosome number as its mother and its father?

A hint at the answer came in 1883, with the observation that cells in the body of roundworms of the genus *Ascaris* have four chromosomes, while their sperm and egg nuclei have only two chromosomes apiece.

Four years later, August Weismann formally proposed a hypothesis to explain the riddle: During the formation of **gametes**—reproductive cells such as sperm and eggs—there must be a distinctive type of cell division that leads to a reduction in chromosome number. Specifically, if the sperm and egg contribute an equal number of chromosomes to the fertilized egg, Weismann reasoned, they must each contain half of the usual number of chromosomes. Then, when sperm and egg combine, the resulting cell has the same chromosome number as its mother's cells and its father's cells have.

KEY CONCEPTS

- Meiosis is a type of nuclear division resulting in cells that have half as many chromosomes as the parent cell. In animals it leads to the formation of eggs and sperm.
- Each cell produced by meiosis receives a different combination of chromosomes. Because genes are located on chromosomes, each cell produced by meiosis receives a different complement of genes. The resulting offspring are genetically distinct from each other and from their parents.
- The leading hypothesis to explain meiosis is that genetically varied offspring are more likely to thrive in environments where parasites and disease are common.
- If mistakes occur during meiosis, the resulting egg and sperm cells may contain the wrong number of chromosomes. It is rare for offspring with an incorrect number of chromosomes to develop normally.

✔ When you see this checkmark, stop and test yourself. Answers are available in Appendix B.

In the decades that followed, biologists confirmed this hypothesis by observing gamete formation in a wide variety of plant and animal species. Eventually this form of cell division came to be called meiosis (literally, "lessening-act").

Meiosis is nuclear division that leads to a halving of chromosome number. It precedes the formation of eggs and sperm in animals. To a biologist, asking "Why sex?" is equivalent to asking "Why meiosis?" Let's delve in by looking at how meiosis happens.

12.1 How Does Meiosis Occur?

To understand meiosis, it is critical to grasp some key ideas about chromosomes. For example, when cell biologists began to study the cell divisions that lead to gamete formation, they made an important observation: Each organism has a characteristic number of chromosomes.

Consider the drawings in **Figure 12.1**, based on a paper published by Walter Sutton in 1902. They show the chromosomes of the lubber grasshopper during the cell divisions leading up to the formation of a sperm. In total, there are 23 chromosomes in the cell. Your cells have 46 chromosomes; a wheat plant has 42.

Chromosomes Come in Distinct Types

Sutton realized, however, that there are just 12 distinct types of chromosomes present in a lubber grasshopper cell. The types are distinguished by size and shape. In the cells that he studied, there were two chromosomes of each type.

Sutton designated 11 of the chromosome types by the letters *a* through *k* and the twelfth by the letter X. Some years later, Nettie Stevens established that the X chromosome is associated with the sex of the individual and called it a **sex chromosome**. Non-sex chromosomes, such as *a–k* in Sutton's grasshopper cell, are known as **autosomes**.

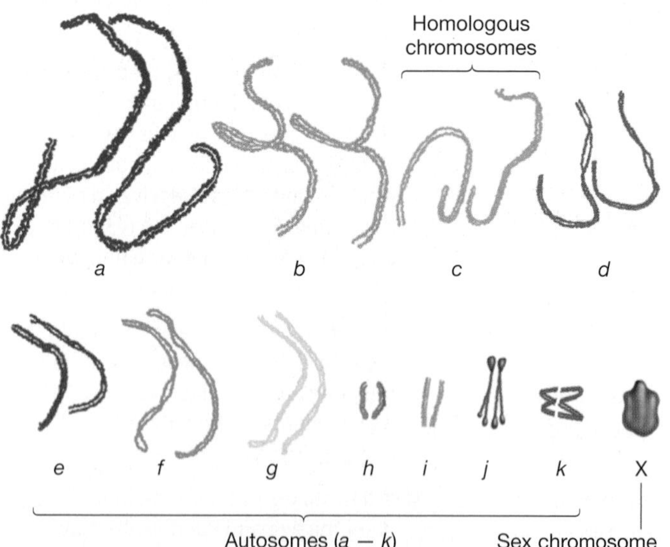

FIGURE 12.1 Cells Contain Different Types of Chromosomes, and Diploid Chromosomes Come in Pairs. Letters designate each of the 12 distinct types of chromosomes found in lubber grasshopper cells. There are two of each type of chromosome (except the X, as this is a male; a female grasshopper would have two Xs).

It turns out that lubber grasshoppers have just one type of sex chromosome. In this species, females have two sex chromosomes and are designated XX; males have just one sex chromosome and are designated XO, where the O refers to the "missing" chromosome. Two types of sex chromosomes, known as X and Y, exist in humans and other mammals. Human females have two X chromosomes, while males have one X and one Y chromosome.

Sutton also introduced an important term for discussing chromosomes, whether sex chromosomes or autosomes. He referred to the two chromosomes of each type as **homologous** ("same proportion") **chromosomes**, or simply **homologs**. The two chromosomes labeled *c*, for example, have the same size and shape and are homologous.

Later work showed that homologous chromosomes are similar not only in size and shape but also in content. Homologous chromosomes carry the same genes. A **gene** is a section of DNA that influences some hereditary trait in an individual. A trait is a characteristic. For example, each copy of chromosome *c* found in lubber grasshoppers might carry genes that influence eye formation, body size, singing behavior, and jumping ability.

The versions of a gene found on homologous chromosomes may differ, however. Biologists use the term **allele** to denote different versions of the same gene. For example, the allele for eye shape on one homolog of chromosome *c* in a lubber grasshopper might contribute roundness to the eyes, whereas the allele of the eye-shape gene on the other homolog may have the effect of producing narrower eyes (**Figure 12.2**); the particular body-size alleles present will have an influence on whether the body is larger or smaller, and so on.

Homologous chromosomes carry the same genes, but each homolog may contain different alleles.

The Concept of Ploidy

At this point in his study, Sutton had succeeded in determining the lubber grasshopper's **karyotype**—meaning the number and types of chromosomes present. As karyotyping studies became more common, cell biologists realized that, like lubber grasshop-

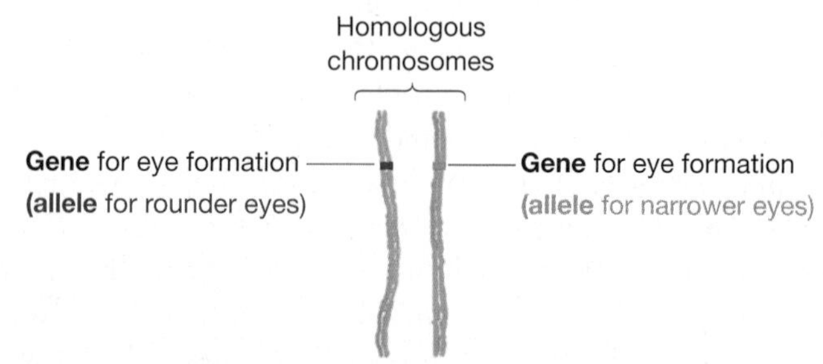

FIGURE 12.2 Homologous Chromosomes May Contain Different Alleles of the Same Gene. Only one of many possible genes is shown.

pers, the vast majority of plants and animals have more than one of each type of chromosome.

Lubber grasshoppers, humans, oak trees, and other organisms that have two versions of each type of chromosome are called **diploid** ("double-form"). Diploid organisms have two alleles of each gene—one on each of the homologous pairs of chromosomes.

Organisms whose cells contain just one of each type of chromosome—for example, bacteria, archaea, and many algae—are called **haploid** ("single-form"). Haploid organisms do not contain homologous chromosomes. They have just one allele of each gene.

Biologists use a compact notation to indicate the number of chromosomes and chromosome sets in a particular organism or type of cell:

- By convention, the letter n stands for the number of distinct types of chromosomes in a given cell and is called the **haploid number.** If sex chromosomes are present, they are counted as a single type in the haploid number. In humans, n is 23.
- To indicate the number of complete chromosome sets observed, a number is placed before the n. Thus, a cell can be n, or $2n$, or $3n$, and so on. Human autosomal cells are $2n$.

The combination of the number of sets and n is termed the cell's **ploidy**. Diploid cells or species are designated $2n$, because two chromosomes of each type are present—one from each parent. A **maternal chromosome** comes from the mother; a **paternal chromosome** comes from the father.

Humans are diploid; $2n$ is 46. Haploid cells or species are labeled simply n, because they have just one set of chromosomes—no homologs are observed. In haploid cells, the number 1 in front of n is implied and is not written out.

To summarize, the haploid number n indicates the number of distinct types of chromosomes present. In contrast, a cell's ploidy (n, $2n$, $3n$, etc.) indicates the number of each type of chromosome present. Stating a cell's ploidy is the same as stating the number of haploid chromosome sets present. ✔You should be able to state the haploid number, ploidy, and total number of chromosomes present in a female lubber grasshopper.

Later work revealed that it is common for species in some lineages—particularly certain land plants, such as ferns—to contain more than two of each type of chromosome. Instead of having two homologous chromosomes per cell, as many organisms do, **polyploid** ("many-form") species may have three or more of each type of chromosome in each cell.

Depending on the number of homologs present, polyploid species are called triploid ($3n$), tetraploid ($4n$), hexaploid ($6n$), octoploid ($8n$), and so on. Why some species are haploid versus diploid or tetraploid is currently the subject of debate and research.

Sutton and the other early cell biologists did more than just describe the karyotypes observed in their study organisms (**Table 12.1**). Through careful examination, they were able to track how chromosome numbers change during meiosis. These studies confirmed Weismann's hypothesis that a special type of cell division occurs during gamete formation.

SUMMARY TABLE 12.1 **The Number of Chromosomes Found in Some Familiar Organisms**

Organism	Haploid Chromosome Number (n)*	Diploid Chromosome Number ($2n$)
Humans	23	46
Domestic dog	36	72
Fruit fly	4	8
Chimpanzee	24	48
Bulldog ant	1	2
Garden pea	7	14
Corn (maize)	10	20

*Number of different types of chromosomes.

An Overview of Meiosis

Cells replicate each of their chromosomes before undergoing meiosis. At the start of meiosis, chromosomes are in the same state they are in prior to mitosis.

When chromosome replication is complete, each chromosome consists of two identical **sister chromatids**. Sister chromatids contain the same genetic information. They are physically joined at a portion of the chromosome called the **centromere** as well as along their entire length (**Figure 12.3**).

To understand meiosis, it is critical to understand the relationship between chromosomes and sister chromatids. An unreplicated chromosome consists of a single DNA molecule with its associated proteins, while a replicated chromosome consists of two sister chromatids. An unreplicated chromosome is a single thread; a replicated chromosome has paired threads.

The trick is to recognize that unreplicated and replicated chromosomes are both considered *single* chromosomes—even though the replicated chromosome comprises *two* sister chromatids.

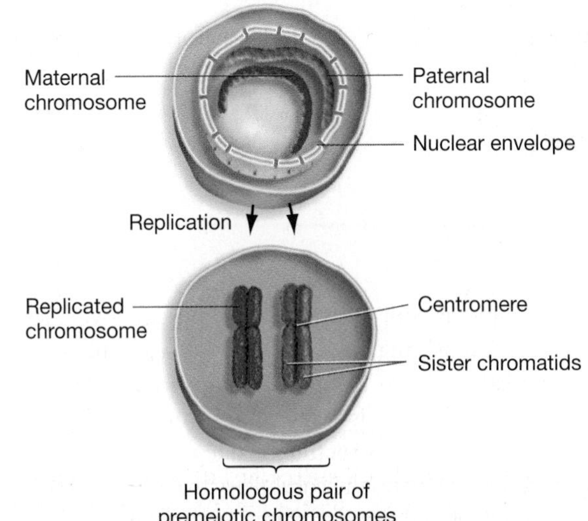

FIGURE 12.3 Each Chromosome Replicates Prior to Undergoing Meiosis.

SUMMARY TABLE 12.2 **Vocabulary for Describing the Chromosomal Makeup of a Cell**

Term	Definition	Example or Comment
Chromosome	Structure made up of DNA and proteins; carries the cell's hereditary information (genes)	Eukaryotes have linear, threadlike chromosomes; most bacteria and archaea have just one, circular, chromosome
Sex chromosome	Chromosome associated with an individual's sex	X and Y chromosomes of humans (males are XY, females XX); Z and W chromosomes of birds and butterflies (males are ZZ, females ZW)
Autosome	A non-sex chromosome	Chromosomes 1–22 in humans
Unreplicated chromosome	A chromosome that consists of a single copy (in eukaryotes, a single "thread")	
Replicated chromosome	A chromosome that has been copied; consists of two linear structures joined at the centromere	Centromere
Sister chromatids	The chromosome copies in a replicated chromosome	Sister chromatids
Homologous chromosomes (homologs)	In a diploid organism, chromosomes that are similar in size, shape, and gene content	You have a chromosome 22 from your mother (red) and a chromosome 22 from your father (blue) Homologous chromosomes
Non-sister chromatids	Chromatids belonging to homologous chromosomes	Non-sister chromatids
Tetrad	Homologous replicated chromosomes that are joined together	Tetrad
Haploid number	The number of different types of chromosomes in a cell; symbolized n	Humans have 23 different types of chromosomes ($n = 23$)
Diploid number	The number of chromosomes present in a diploid cell (see below); symbolized $2n$	In humans all cells except gametes are diploid and contain 46 chromosomes ($2n = 46$)
Ploidy	The number of each type of chromosome present	Equivalent to the number of haploid chromosome sets present
Haploid	Having one of each type of chromosome (n)	Bacteria and archaea are haploid, as are many algae; plant and animal gametes are haploid
Diploid	Having two of each type of chromosome ($2n$)	Most familiar plants and animals are diploid
Polyploid	Having more than two of each type of chromosome; cells may be triploid ($3n$), tetraploid ($4n$), hexaploid ($6n$), and so on	Seedless bananas are triploid; many ferns are tetraploid; bread wheat is hexaploid

A chromosome is still just one chromosome whether it is unreplicated—consisting of a single DNA molecule—or replicated—consisting of two identical DNA molecules. Note that an unreplicated chromosome is never called a chromatid; you can only refer to chromatids as the structures in a replicated chromosome.

Table 12.2 summarizes the vocabulary that biologists use to describe the number and types of chromosomes found in a cell and illustrates the relationship between chromosomes and chromatids. ✔If you understand this relationship, you should be able to draw the same chromosome in the replicated and unreplicated state, explain why both structures represent a single chromosome, and then label the sister chromatids in the replicated chromosome.

MEIOSIS IS TWO CELL DIVISIONS Meiosis consists of two cell divisions, called **meiosis I** and **meiosis II.** As **Figure 12.4** shows, the two divisions occur consecutively but differ sharply.

During meiosis I, the homologs in each chromosome pair separate from each other. One homolog goes to one daughter cell; the other homolog goes to the other daughter cell. The homolog that came originally from the individual's mother is colored red in

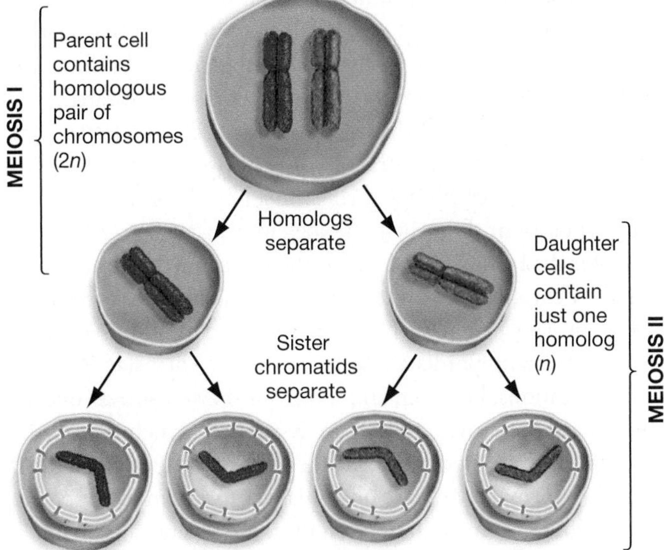

FIGURE 12.4 The Major Events in Meiosis. Meiosis reduces chromosome number by half. In diploid organisms, the products of meiosis are haploid. Maternal chromosomes are red, and paternal chromosomes are blue. Note that in this cell, $2n = 2$.

Figure 12.4; the homolog that came from the father is colored blue. It is a matter of chance which daughter cell receives which homolog.

The end result of meiosis I is that each of the two daughter cells has one of each type of chromosome instead of two, and thus half as many chromosomes as the parent cell had. During meiosis I, the diploid ($2n$) parent cell produces two haploid (n) daughter cells. Each chromosome still consists of two sister chromatids, however—meaning that chromosomes are still replicated.

During meiosis II, sister chromatids from each chromosome separate. One sister chromatid goes to one daughter cell; the other sister chromatid goes to the other daughter cell. The cell that starts meiosis II has one of each type of chromosome, but each chromosome has been replicated (meaning it still consists of two sister chromatids). The cells produced by meiosis II also have one of each type of chromosome, but now the chromosomes are unreplicated.

To reiterate, sister chromatids separate during meiosis II, just as they do during mitosis. Meiosis II is actually equivalent to mitosis occurring in a haploid cell.

As in mitosis, chromosome movements during meiosis I and II are coordinated by kinetochore microtubules that attach at the centromere of each chromosome. Movement is driven by motor proteins located at the kinetochore—the point of attachment. (Recall from Chapter 11 that the centromere is a region on the chromosome; the kinetochore is a structure in that region.)

MEIOSIS IS A REDUCTION DIVISION Sutton and a host of other early cell biologists worked out this sequence of events through careful observation of cells with the light microscope. Based on these studies, they came to a key realization: The outcome of meiosis is a reduction in chromosome number. For this reason, meiosis is known as a reduction division.

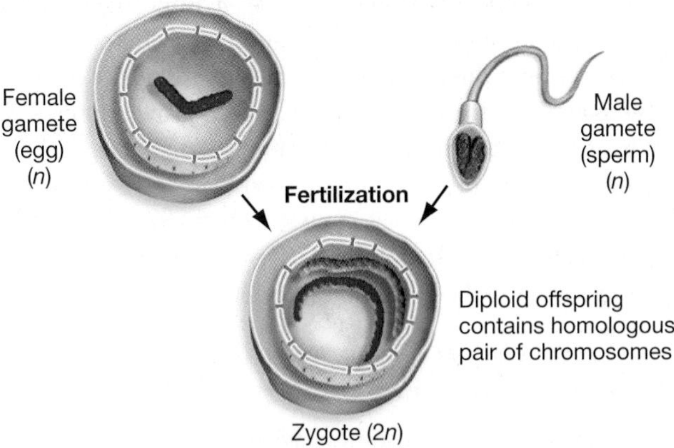

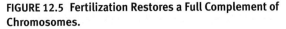

FIGURE 12.5 Fertilization Restores a Full Complement of Chromosomes.

In most plants and animals, the original cell is diploid and the four daughter cells are haploid. These four haploid daughter cells, each containing one of each homologous chromosome, eventually go on to form egg cells or sperm cells via a process called **gametogenesis** ("gamete-origin"), which is described in Chapter 48.

When two gametes fuse during fertilization, a full complement of chromosomes is restored (**Figure 12.5**). The cell that results from fertilization is diploid and is called a **zygote**. In this way, each diploid individual receives both a haploid chromosome set from its mother and a haploid set from its father.

Figure 12.6 puts these events into the context of an animal's **life cycle**—the sequence of events that occurs over the life span of an individual, from fertilization to the production of offspring. As you study the figure, note how ploidy changes as the result of meiosis and fertilization. In the case of the dog illustrated here, meiosis in a diploid adult results in the formation of haploid gametes, which combine to form a diploid zygote.

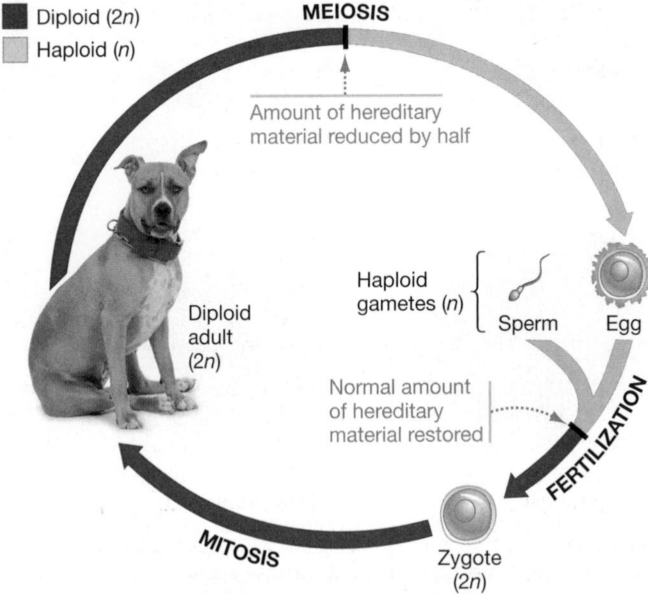

FIGURE 12.6 Ploidy Changes during the Life Cycle of a Dog. The dog is diploid throughout most of its life cycle.

✔If you understand these types of ploidy changes, you should be able to draw the "haploid-dominant" life cycle of an algal species where the only diploid phase is the zygote. Zygotes undergo meiosis to form haploid cells, which grow into haploid adults by mitosis. Adults produce haploid sperm and eggs.

Once Sutton and others had published their work on meiosis and the accompanying changes in ploidy, the mystery of fertilization was finally solved. To appreciate the consequences of meiosis fully, let's analyze the events in more detail.

The Phases of Meiosis I

Meiosis begins after chromosomes have been replicated during S phase (see Chapter 11). Prior to the start of meiosis, chromosomes are extremely long structures, just as they are during interphase of the normal cell cycle. The major steps that occur once meiosis begins are shown in **Figure 12.7**.

During early prophase I, chromosomes condense, the **spindle apparatus** forms, and the nuclear envelope begins to disappear.

Then a crucial event occurs, still during early prophase of meiosis I: Homologous chromosome pairs come together. This process is called **synapsis** and is illustrated in step 2 of Figure 12.7. Synapsis is possible because regions of homologous chromosomes that are similar at the molecular level attract one another, by means of mechanisms that are currently the subject of intense research.

The structure that results from synapsis is called a **tetrad** (*tetra* means four in Greek). A tetrad consists of two homologous chromosomes, with each homolog consisting of two sister chromatids. Chromatids from different homologs in a pair are referred to as **non-sister chromatids**. In the figure, the red-colored chromatids are non-sister chromatids with respect to the blue-colored chromatids.

During late prophase I, the non-sister chromatids begin to separate at many points along their length. They stay joined at certain locations, however, and look as if they cross over one another. Each crossover forms an X-shaped structure called a **chiasma** (plural: **chiasmata**). (In the Greek alphabet, the letter X is "chi.") Normally, at least one chiasma forms in every pair of homologous chromosomes; often there are several chiasmata.

As step 3 of Figure 12.7 shows, the chromatids that meet to form a chiasma are homologous but not sisters. Consistent with this observation, Thomas Hunt Morgan proposed that a physical exchange of paternal and maternal chromosomes occurs at chi-

FIGURE 12.7 The Phases of Meiosis. The micrographs of each phase are from a species of salamander.

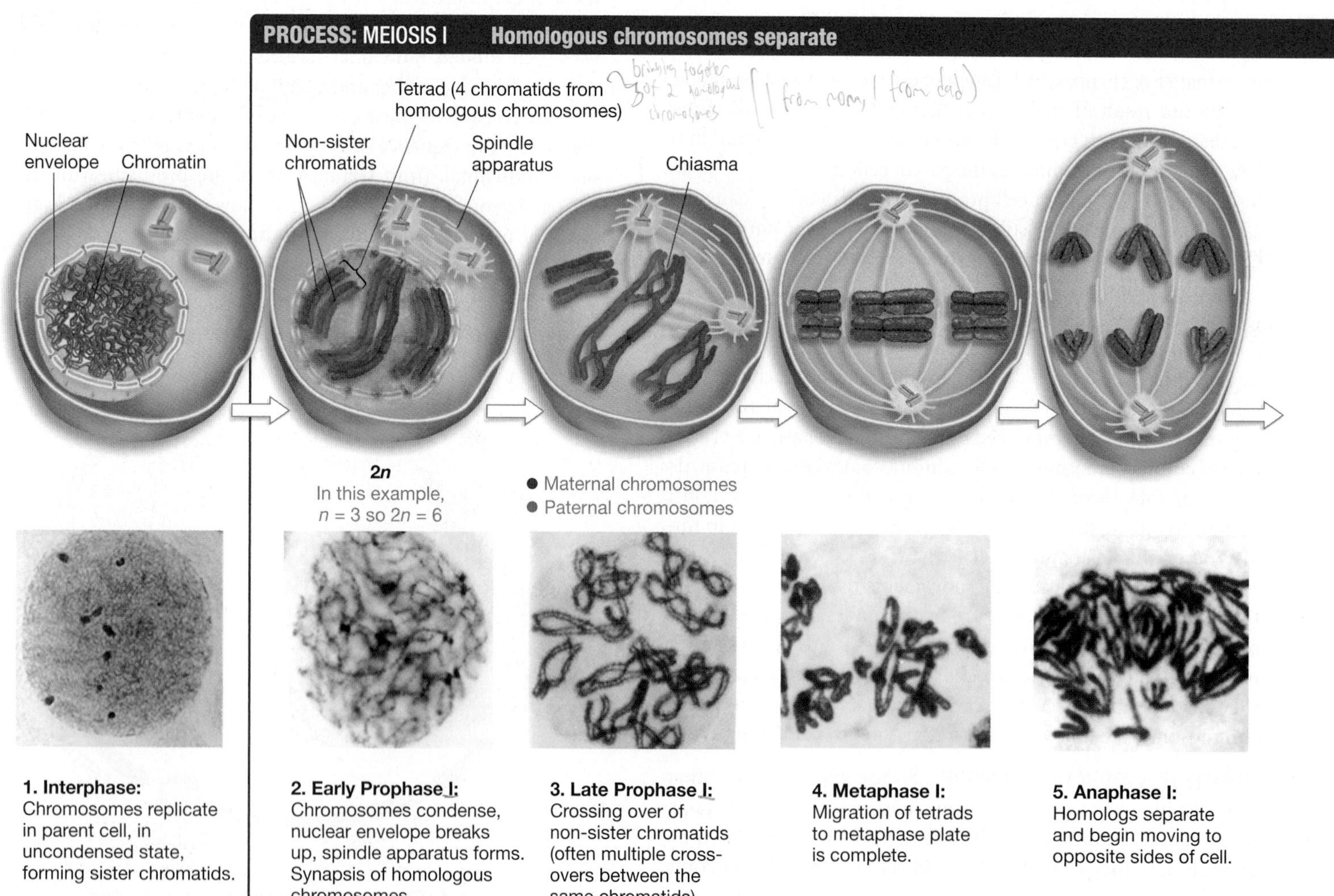
PROCESS: MEIOSIS I Homologous chromosomes separate
Tetrad (4 chromatids from homologous chromosomes)
Nuclear envelope
Chromatin
Non-sister chromatids
Spindle apparatus
Chiasma
2n
In this example, n = 3 so 2n = 6
● Maternal chromosomes
● Paternal chromosomes
1. Interphase: Chromosomes replicate in parent cell, in uncondensed state, forming sister chromatids.
2. Early Prophase I: Chromosomes condense, nuclear envelope breaks up, spindle apparatus forms. Synapsis of homologous chromosomes.
3. Late Prophase I: Crossing over of non-sister chromatids (often multiple cross-overs between the same chromatids).
4. Metaphase I: Migration of tetrads to metaphase plate is complete.
5. Anaphase I: Homologs separate and begin moving to opposite sides of cell.

asmata. According to this hypothesis, paternal and maternal chromatids break and rejoin at each chiasma, producing chromatids that have both paternal and maternal segments. Morgan called this process of chromosome exchange **crossing over**.

In step 4 of Figure 12.7, the result of crossing over is illustrated by chromosomes with a combination of red and blue segments. When crossing over occurs, the chromosomes that result have a mixture of maternal and paternal alleles.

The next major stage in meiosis I is metaphase I. This is when kinetochore microtubules move the pairs of homologous chromosomes (tetrads) to a region called the **metaphase plate**, in the middle of the cell (step 4). Two points are key here: Each tetrad moves to the metaphase plate independently of the other tetrads, and the alignment of maternal and paternal homologs from each chromosome is random.

During anaphase I, the homologous chromosomes in each tetrad separate and begin moving to opposite sides of the cell (step 5). Meiosis I concludes with telophase I, when the homologs finish moving to opposite sides of the cell (step 6). When meiosis I is complete, **cytokinesis** (division of cytoplasm) occurs and two haploid daughter cells form.

The end result of meiosis I is that one chromosome of each homologous pair is distributed to a different daughter cell. A reduction division has occurred: The daughter cells of meiosis I are haploid. The sister chromatids remain attached in each chromosome, however, meaning that the haploid daughter cells produced by meiosis I still contain replicated chromosomes.

The chromosomes in each cell are a random assortment of maternal and paternal chromosomes as a result of crossing over and the random distribution of maternal and paternal homologs during metaphase.

The preceding discussion shows that although meiosis I is a continuous process, biologists summarize the events by identifying distinct phases:

- *Early Prophase I* Replicated chromosomes condense, the spindle apparatus forms, and the nuclear envelope disappears. Synapsis of homologs forms pairs of homologous chromosomes, or tetrads. Kinetochore microtubules attach to the kinetochores at the centromeres of chromosomes.
- *Late Prophase I* Crossing over results in a mixing of chromosome segments from maternal and paternal chromosomes.

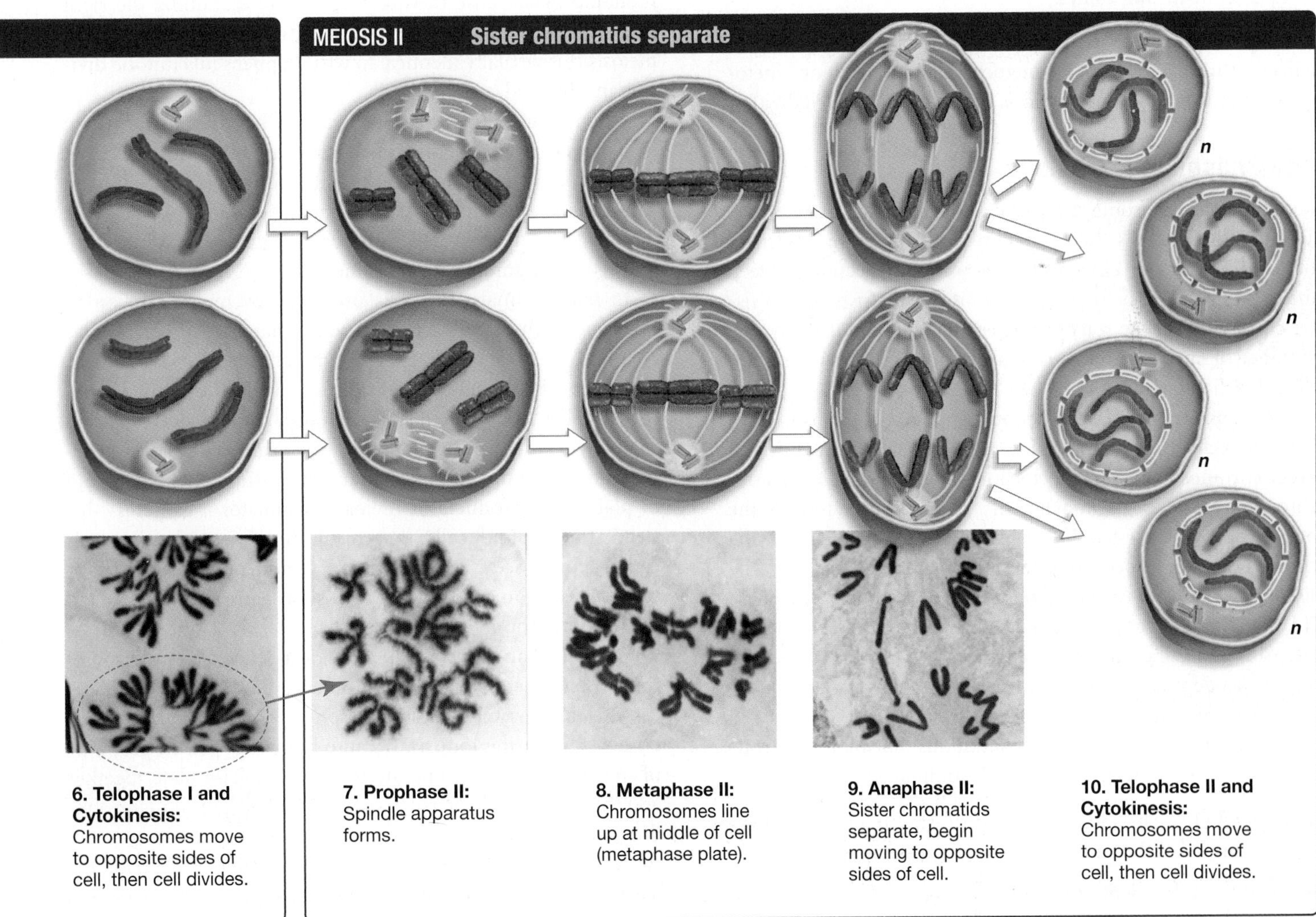

6. Telophase I and Cytokinesis: Chromosomes move to opposite sides of cell, then cell divides.

7. Prophase II: Spindle apparatus forms.

8. Metaphase II: Chromosomes line up at middle of cell (metaphase plate).

9. Anaphase II: Sister chromatids separate, begin moving to opposite sides of cell.

10. Telophase II and Cytokinesis: Chromosomes move to opposite sides of cell, then cell divides.

SUMMARY TABLE 12.3 **Key Differences between Mitosis and Meiosis**

Feature	Mitosis	Meiosis
Number of cell divisions	One	Two
Number of chromosomes in daughter cells, compared with parent cell	Same	Half
Synapsis of homologs	No	Yes
Number of crossing-over events	None	One or more per pair of homologous chromosomes
Makeup of chromosomes in daughter cells	Identical	Different—only one of each chromosome type present, paternal and maternal segments mixed within chromosomes
Role in organism life cycle	Asexual reproduction in eukaryotes; cell division for growth of multicellular organisms	Precedes production of gametes in sexually reproducing animals

- *Metaphase I* Pairs of homologous chromosomes (tetrads) move to the metaphase plate and line up.
- *Anaphase I* Homologs separate and begin moving to opposite sides of the cell.
- *Telophase I* Homologs finish moving to opposite sides of the cell. In some species, a nuclear envelope re-forms around each set of chromosomes.

Throughout, chromosome movement takes place via motor proteins that are attached to the kinetochore and walk along kinetochore microtubules. When meiosis I is complete, the cell divides.

The Phases of Meiosis II

Recall that chromosome replication occurred prior to meiosis I. Throughout meiosis I, sister chromatids remain attached. Because no further chromosome replication occurs between meiosis I and meiosis II, each chromosome consists of two sister chromatids at the start of meiosis II. And because only one member of each homologous pair of chromosomes is present, the cell is haploid.

Next, during prophase II, a spindle apparatus forms in both daughter cells. Kinetochore microtubules attach to each side of every chromosome—one kinetochore microtubule to each sister chromatid—and begin moving the chromosomes toward the middle of each cell (step 7 of Figure 12.7). In metaphase II, the chromosomes are lined up at the metaphase plate (step 8). The sister chromatids of each chromosome separate during anaphase II (step 9) and move to different daughter cells during telophase II (step 10). Once they are separated, each chromatid is considered an independent chromosome. Meiosis II results in four haploid cells, each with one chromosome of each type.

Like meiosis I, meiosis II is continuous, but biologists routinely divide it into distinct phases. To summarize,

- *Prophase II* The spindle apparatus forms. If a nuclear envelope formed at the end of meiosis I, it breaks apart.
- *Metaphase II* Replicated chromosomes, consisting of two sister chromatids, are lined up at the metaphase plate.
- *Anaphase II* Sister chromatids separate. The unreplicated chromosomes that result begin moving to opposite sides of the cell.
- *Telophase II* Chromosomes finish moving to opposite sides of the cell. A nuclear envelope forms around each haploid set of chromosomes.

It should make sense to you, after closely examining the right side of Figure 12.7, that the movement of chromosomes during meiosis II is virtually identical to what happens in a mitotic division in a haploid cell.

When meiosis II is complete, each cell divides to form two daughter cells. Because meiosis II occurs in both daughter cells of meiosis I, the process results in a total of four daughter cells from each original, parent cell. To describe meiosis in a nutshell, one diploid cell with replicated chromosomes gives rise to four haploid cells with unreplicated chromosomes.

In male animals, the cells produced by meiosis go on to form sperm through a series of events that are detailed in later chapters. In females of at least some animal species, though, meiosis begins in cells called oocytes that will eventually mature into eggs, but it then stops and restarts at several points as the oocytes mature. In mammals, for example, meiosis in females is not completed until after fertilization takes place. And in plants, the products of meiosis do not form gametes—instead, the haploid cells that result form reproductive cells called spores (see Chapter 40). For the purposes of this chapter, however, you can think of meiosis as a process that leads to gamete formation.

Table 12.3 and **Figure 12.8** provide a detailed comparison of mitosis and meiosis. The key difference between the two processes is that homologous chromosomes pair early in meiosis but do not pair during mitosis. Because homologs pair in prophase of meiosis I, they can migrate to the metaphase plate together and then separate during anaphase of meiosis I, resulting in a reduction division. ✔If you understand this key distinction between meiosis and mitosis, you should be able to describe the consequences for meiosis if homologs do not pair. Now let's delve into the details of this critical event.

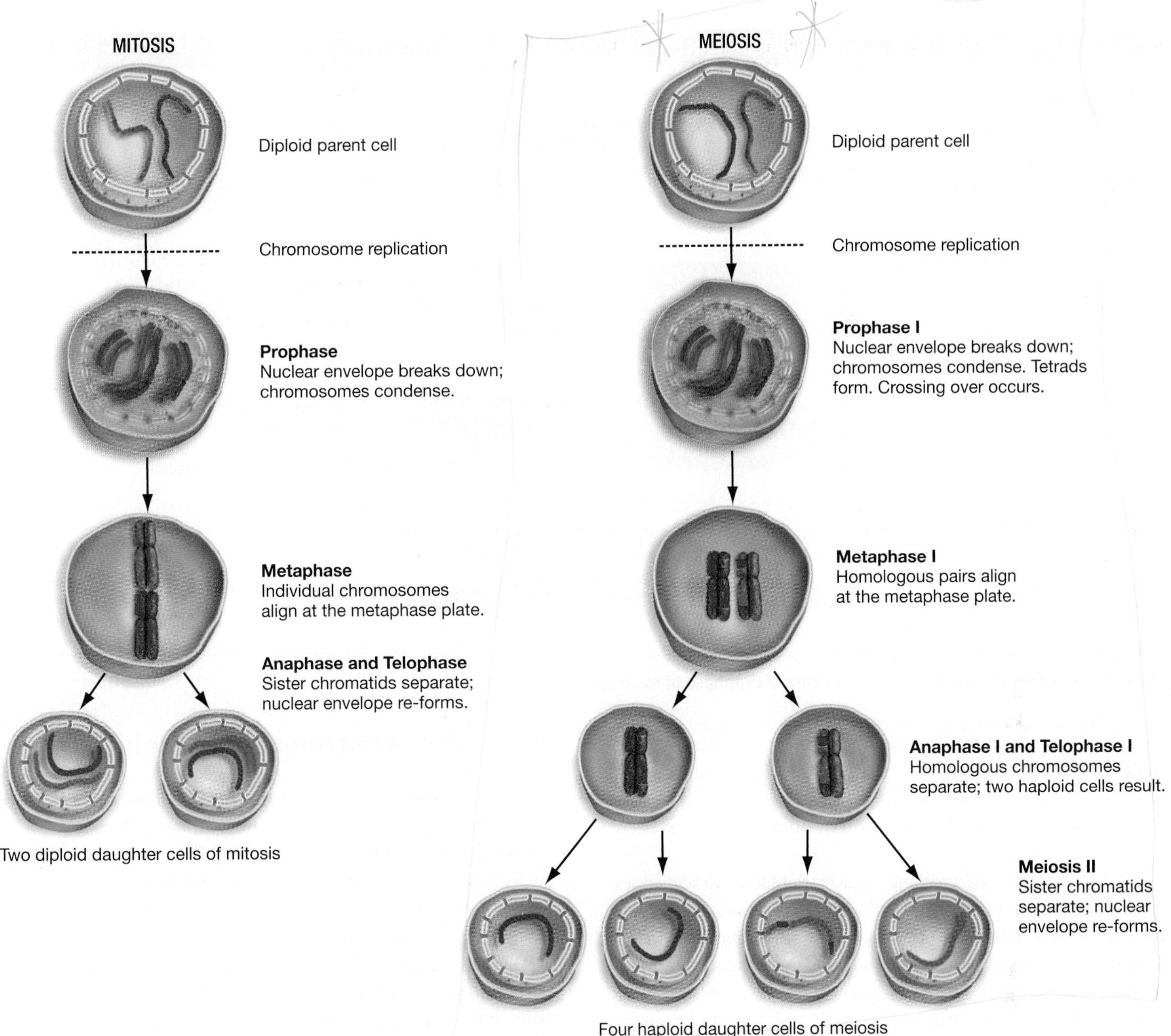

FIGURE 12.8 A Comparison of Mitosis and Meiosis. Mitosis produces two daughter cells with chromosomal complements identical to the parent cell. Meiosis produces four haploid cells with chromosomal complements unlike each other and unlike the diploid parent cell.

A Closer Look at Prophase I

Figure 12.9 on page 220 provides more detail on how several important events in prophase of meiosis I occur.

Step 1 After chromosome replication is complete, sister chromatids stay tightly joined along their entire length.

Step 2 When homologs synapse, two pairs of non-sister chromatids are brought close together and are held there by a network of proteins called the **synaptonemal complex**.

Step 3 Crossing over occurs when a complex of proteins cuts the chromosomes and then reattaches the pieces so that segments are swapped between adjacent homologs.

The key to understanding crossing over is to recognize that at each point where it occurs—as in the circles labeled "protein complex" in step 3 of Figure 12.9—the non-sister chromatids from each homolog get physically broken at the same point and *attached to each other*. As a result, segments of maternal and paternal chromosomes are swapped, as shown at the bottom of Figure 12.9.

Crossing over can occur at many locations along the length of the paired homologs, and it routinely occurs at least once between each pair of non-sister chromatids. In humans, for example, each chromosome in an egg underwent an average of 1.7 crossovers during meiosis; each chromosome in a sperm had an average of 1.1 crossovers. No one knows why crossing

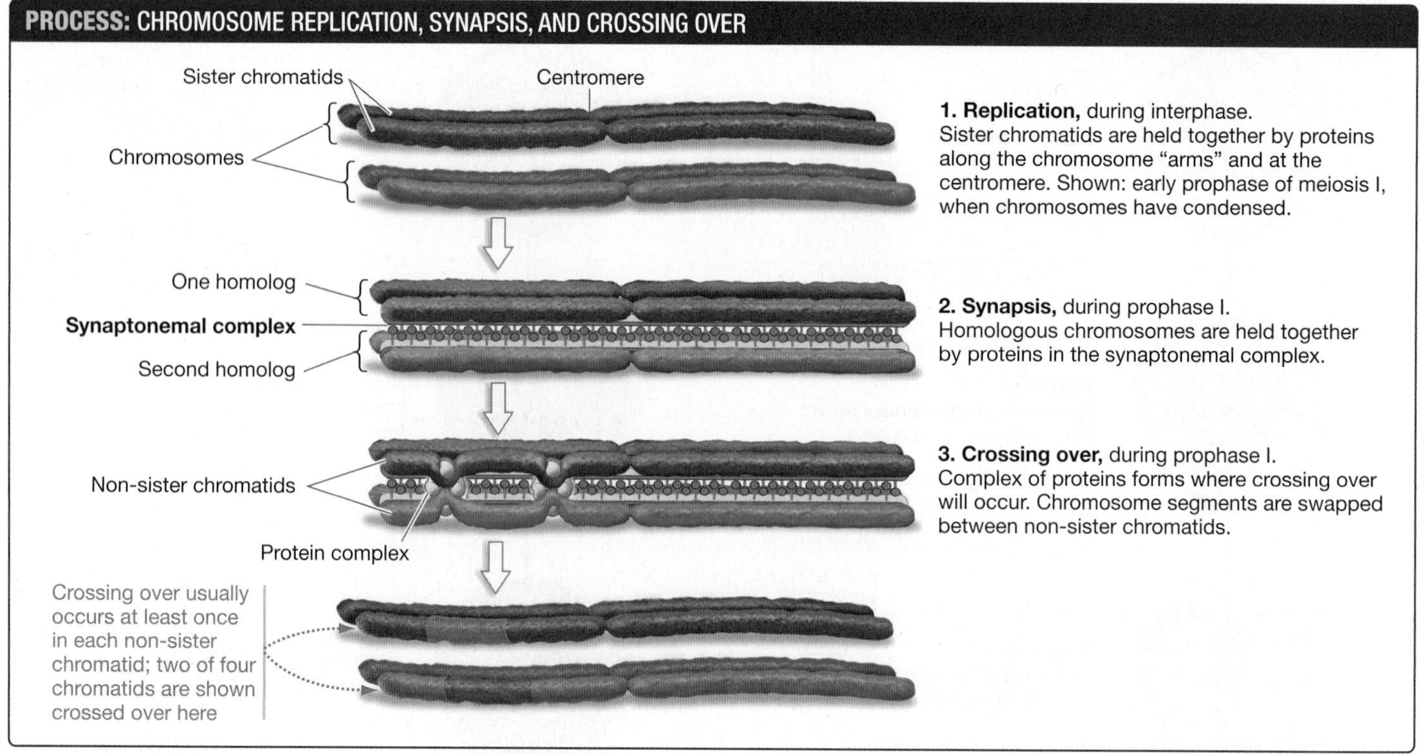

PROCESS: CHROMOSOME REPLICATION, SYNAPSIS, AND CROSSING OVER
Sister chromatids
Centromere
Chromosomes
1. Replication, during interphase.
Sister chromatids are held together by proteins along the chromosome "arms" and at the centromere. Shown: early prophase of meiosis I, when chromosomes have condensed.
One homolog
Synaptonemal complex
Second homolog
2. Synapsis, during prophase I.
Homologous chromosomes are held together by proteins in the synaptonemal complex.
Non-sister chromatids
Protein complex
3. Crossing over, during prophase I.
Complex of proteins forms where crossing over will occur. Chromosome segments are swapped between non-sister chromatids.
Crossing over usually occurs at least once in each non-sister chromatid; two of four chromatids are shown crossed over here

FIGURE 12.9 A Closer Look at Three Key Events in Prophase of Meiosis I.

over is so much more common in human females than in males. Recent work has shown that the rate of crossing over is also variable among individuals of the same sex. A gene that may be responsible for this variation in humans has even been identified.

Why is the rate of crossing over variable? And why does meiosis exist at all? These are fundamental questions in biological science. But before reading on to learn more about them, go to the study area at *www.masteringbiology.com* to review how meiosis occurs.

 **BioFlix™** Meiosis, **Web Activity** Meiosis

CHECK YOUR UNDERSTANDING

If you understand that . . .

- Meiosis is called a reduction division because the total number of chromosomes present is cut in half.
- During meiosis, a single diploid parent cell with replicated chromosomes gives rise to four haploid daughter cells.

✔ **You should be able to . . .**

1. Demonstrate the phases of meiosis I illustrated in Figure 12.7 by using pipe cleaners or pieces of cooked spaghetti.
2. Identify the event that makes meiosis a reduction division, unlike mitosis, and explain why it is responsible for reduction division.

Answers are available in Appendix B.

12.2 The Consequences of Meiosis

The cell biologists who worked out the details of meiosis in the late 1800s and early 1900s realized that the process solved the riddle of fertilization. Weismann's hypothesis—that a reduction division precedes gamete formation in animals—was confirmed.

By now, having come to appreciate that meiosis is an intricate, tightly regulated process, you shouldn't be surprised to learn that it involves dozens, if not hundreds, of different proteins. Given this complexity, it is logical to hypothesize that meiosis accomplishes something extremely important: Thanks to the independent shuffling of maternal and paternal chromosomes and crossing over during meiosis I, the chromosomes in gametes are different from the chromosomes in parental cells. Subsequently, fertilization brings haploid sets of chromosomes from a mother and father together to form a diploid offspring. The chromosome complement of this offspring is unlike that of either parent. It is a random combination of genetic material from each parent.

This change in chromosomal complement is crucial. The critical factor here is that changes in chromosome configuration occur only during sexual reproduction—*not* during asexual reproduction.

- **Asexual reproduction** refers to any mechanism of producing offspring that does not involve the fusion of gametes. Asexual reproduction in eukaryotes is usually based on mitosis. As Chapter 11 indicated, the chromosomes in the daughter cells of mitosis are identical to the chromosomes in the parental cell.
- **Sexual reproduction** refers to the production of offspring through the fusion of gametes. Sexual reproduction results in

offspring that have chromosome complements unlike their siblings' and their parents'.

Why is this difference important?

Chromosomes and Heredity

The changes in chromosomes produced by meiosis and fertilization are significant because chromosomes contain the cell's hereditary material. Stated another way, chromosomes contain the instructions for specifying what a particular trait might be in an individual. These inherited traits range from eye color and height in humans to the number or shape of the bristles on a fruit fly's leg to the color or shape of the seeds found in pea plants.

In the early 1900s, biologists began using the term gene to refer to the inherited instructions for a particular trait. Recall from Section 12.1 that the term allele refers to a particular version of a gene and that homologous chromosomes may carry different alleles.

Chromosomes are the repositories of genes, and identical copies of chromosomes are distributed to daughter cells during mitosis. Thus, cells that are produced by mitosis are genetically identical to the parent cell, and offspring produced during asexual reproduction are genetically identical to one another as well as to their parent. The offspring of asexual reproduction are **clones**—or exact copies—of their parent.

In contrast, the offspring produced by sexual reproduction are genetically different from one another and unlike either their mother or their father.

Let's begin by analyzing two aspects of meiosis that create variation among chromosomes: (1) separation and distribution of homologous chromosomes, and (2) crossing over, and then ask how these processes interact with fertilization to produce genetically variable offspring.

Independent Assortment Produces Genetic Variation

Each somatic cell in your body contains 23 homologous pairs of chromosomes and 46 chromosomes in total. Half of these chromosomes came from your mother, and half came from your father. Each chromosome is composed of genes, and genes influence particular traits. For example, one gene that affects your eye color might be located on one chromosome, while one of the genes that affects your hair color might be located on a different chromosome (**Figure 12.10a**).

Suppose that the chromosomes you inherited from your mother contain alleles that tend to produce brown eyes and black hair, but the chromosomes you inherited from your father include alleles that tend to specify green eyes and red hair. (This is a simplification for the purpose of explanation. In reality, several genes with various alleles interact in complex ways to produce human eye color and hair color.)

Will any particular gamete you produce contain the genetic instructions inherited from your mother or the instructions inherited from your father?

To answer this question, study the diagram of meiosis in **Figure 12.10b**. It shows that when pairs of homologous chromosomes line up during meiosis I and the homologs separate, a variety of combinations of maternal and paternal chromosomes can result. Each daughter cell gets a random assortment of maternal and paternal chromosomes.

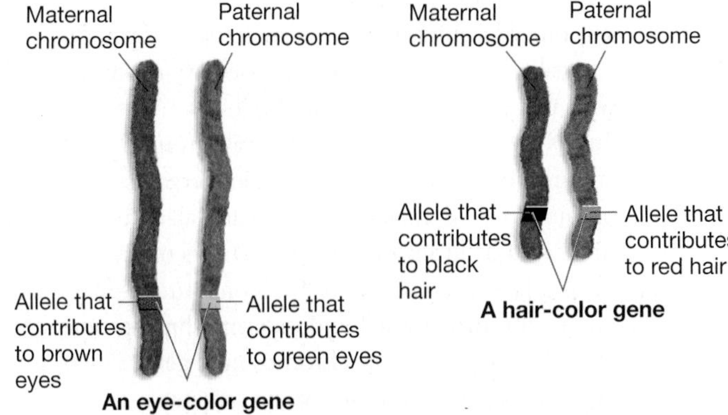

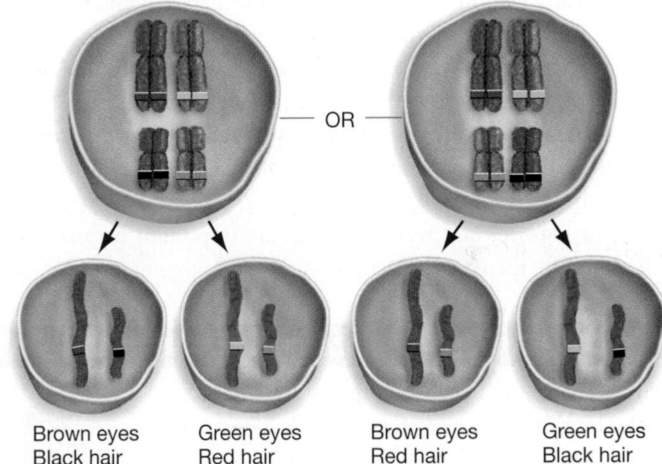

FIGURE 12.10 Independent Assortment of Homologous Chromosomes Results in Varied Combinations of Genes. **(a)** In this hypothetical example, genes that influence eye color and hair color in humans are on different chromosomes. **(b)** The cells along the bottom are products of meiosis. Notice that each cell has a different combination of genes, due to the separation of homologous chromosomes during meiosis I.

As Chapter 13 will explain in detail, this phenomenon is known as the principle of independent assortment. In the example given here, meiosis results in gametes with alleles for brown eyes and black hair, like your mother, and green eyes and red hair, like your father. But two additional combinations also occur: brown eyes and red hair, or green eyes and black hair. Four different combinations of paternal and maternal chromosomes are possible when two chromosomes are distributed to daughter cells during meiosis I.

✔If you understand how independent assortment produces genetic variation in the daughter cells of meiosis, you should be able to explain how genetic variation would be affected if maternal chromosomes always lined up together on one side of the metaphase plate during meiosis I and paternal chromosomes always lined up on the other side.

How many different combinations of maternal and paternal homologs are possible when more chromosomes are involved? In an organism with three chromosomes per haploid set ($n = 3$), eight types of gametes can be generated by randomly grouping maternal and paternal chromosomes. In general, a diploid organism can produce 2^n combinations of maternal and paternal chromosomes, where n is the haploid chromosome number. This means that you ($n = 23$) can produce 2^{23}, or about 8.4 million, gametes that differ in their combination of maternal and paternal chromosome sets. The random assortment of whole chromosomes generates an impressive amount of genetic variation among gametes.

The Role of Crossing Over

Recall from Section 12.1 that segments of paternal and maternal chromatids exchange when crossing over occurs during meiosis I. Thus, crossing over produces new combinations of alleles within a chromosome—combinations that did not exist in either parent. This phenomenon is known as recombination. **Genetic recombination** is any change in the combination of alleles on a given chromosome.

Crossing over and recombination are important because they dramatically increase the genetic variability of gametes produced by meiosis. The random assortment of homologous chromosomes during meiosis varies the combination of chromosomes present in gametes; crossing over varies the combinations of alleles within each chromosome. The number of genetically different gametes that you can produce is much more than the 8.4 million—it is virtually limitless.

How Does Fertilization Affect Genetic Variation?

Crossing over and the random mixing of maternal and paternal chromosomes ensure that each gamete is genetically unique. Even if two gametes produced by the same individual fuse to form a diploid offspring—in which case, **self-fertilization**, or "selfing," is taking place—the offspring are very likely to be genetically different from the parent (**Figure 12.11**). Selfing is common in some plant species. It also occurs in the many animal species in which single individuals contain both male and female sex organs.

Self-fertilization is rare or nonexistent in many sexually reproducing species, however. Instead, gametes from different individuals combine to form offspring. This alternative is called **outcrossing**. Outcrossing increases the genetic diversity of offspring because it combines chromosomes from different individuals, which are likely to contain different alleles.

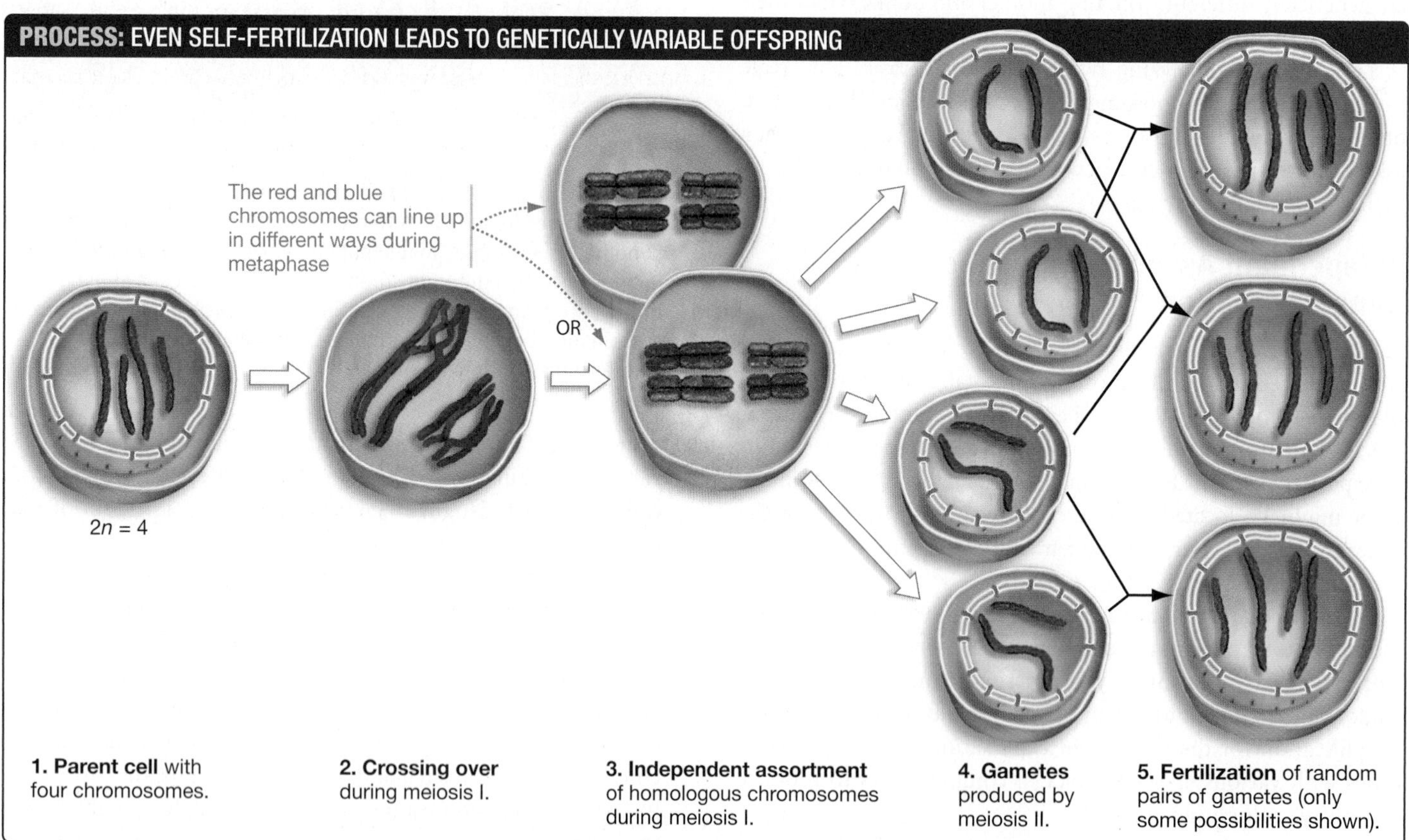
PROCESS: EVEN SELF-FERTILIZATION LEADS TO GENETICALLY VARIABLE OFFSPRING
The red and blue chromosomes can line up in different ways during metaphase
OR
2n = 4
1. Parent cell with four chromosomes.
2. Crossing over during meiosis I.
3. Independent assortment of homologous chromosomes during meiosis I.
4. Gametes produced by meiosis II.
5. Fertilization of random pairs of gametes (only some possibilities shown).

FIGURE 12.11 Crossing Over, Independent Assortment, and the Random Pairing of Gametes during Fertilization Increase Genetic Variation, Even in Offspring Produced by Self-Fertilization.

✔**EXERCISE** In step 5, only a few of the many types of offspring that could be produced are shown. Sketch two additional types that are different from those shown.

How many genetically distinct offspring can be produced when outcrossing occurs? Let's answer this question using humans as an example. Recall that a single human can produce about 8.4 million different gametes, even in the absence of crossing over. When a person mates with a member of the opposite sex, the number of possible genetic combinations that can result is equal to the product of the numbers of different gametes produced by each parent. (To understand the logic here, see **BioSkills 13** in Appendix A.) In humans this means that two parents can potentially produce 8.4 million $\times$ 8.4 million $= 70.6 \times 10^{12}$ genetically distinct offspring. This number is far greater than the total number of people who have ever lived—and the calculation does not even take into account variation generated by crossing over.

CHECK YOUR UNDERSTANDING

If you understand that . . .

- The daughter cells produced by meiosis are genetically different from the parent cell because (1) maternal and paternal homologs align randomly at metaphase of meiosis I and (2) crossing over leads to recombination within chromosomes.

✓ You should be able to . . .

1. Draw a diploid parent cell with $n = 3$ (three types of chromosomes), and then sketch six of the many genetically distinct types of daughter cells that may result when this parent cell undergoes meiosis.
2. Compare and contrast the degree of genetic variation that results from asexual reproduction, selfing, and outcrossing.

Answers are available in Appendix B.

The consequences of meiosis and sexual reproduction are simple: tremendous genetic diversity among offspring. Why is genetic diversity important?

12.3 Why Does Meiosis Exist?

Meiosis and sexual reproduction occur in only a small fraction of the lineages on the tree of life. Bacteria and archaea normally undergo only asexual reproduction; most algae, all fungi, and some animals and land plants reproduce asexually as well as sexually. Asexual reproduction is even observed in some vertebrates. Several species of guppy in the genus *Poeciliopsis*, for example, reproduce exclusively by mitosis.

Sexual reproduction is the major mode of reproduction in the insects, however, which number over 43 million species, as well as in species-rich groups like the mollusks (clams, snails, squid) and vertebrates. Yet although sexual reproduction plays an important role in the life of these organisms, scientists had no clear idea, until recently, of why it occurs. On the basis of theory, biologists had good reason to think that sexual reproduction should *not* exist.

The Paradox of Sex

In 1978 John Maynard Smith pointed out that the existence of sexual reproduction presents a paradox. Maynard Smith developed a mathematical model showing that because asexually reproducing individuals do not have to produce male offspring, their progeny can produce twice as many grand-offspring as can individuals that reproduce sexually. **Figure 12.12** diagrams this result by showing the number of females (♀) and males (♂) produced over several generations by asexual versus sexual reproduction.

In this example, each individual produces four offspring over the course of his or her lifetime. In the asexual population, each

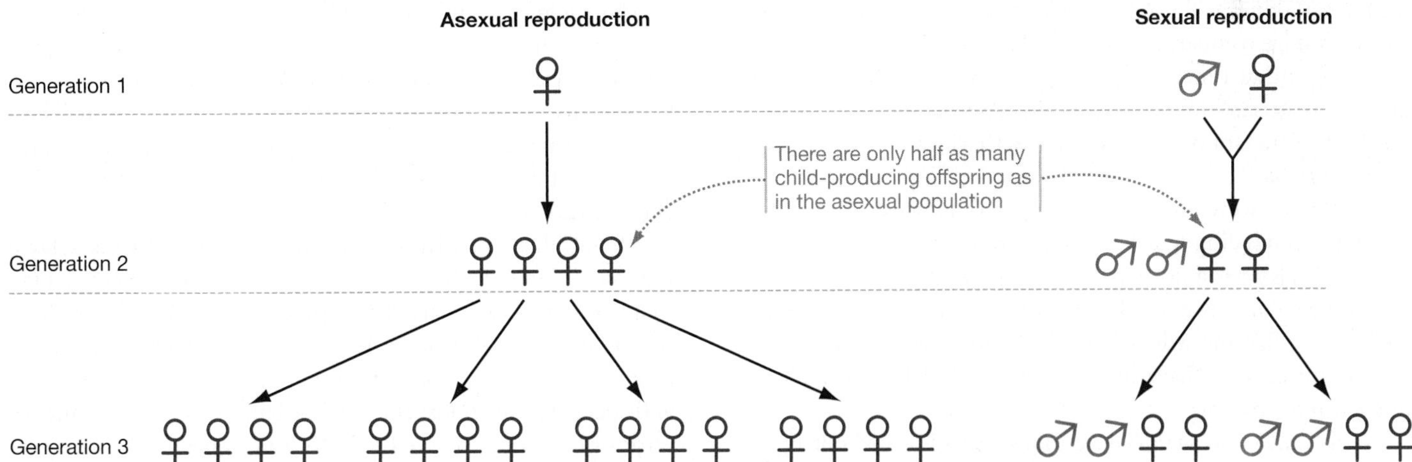

FIGURE 12.12 Asexual Reproduction Confers a Large Numerical Advantage. Each female symbol (♀) and male symbol (♂) represents an individual. This hypothetical example assumes that (1) every individual produces four offspring over the course of a lifetime, (2) sexually reproducing individuals produce half males and half females, and (3) all offspring survive to breed.

✓QUESTION How many asexually produced offspring would be present in generation 4? How many sexually produced offspring?

individual is a female that produces four offspring. But in the sexual population, it takes two individuals—one male and one female—to produce four offspring. Thus, two out of every four children that each female produces sexually—the males—cannot have children of their own. As a result, generation 2 of the sexual population has just half as many child-producing offspring as generation 2 in the asexual population. Maynard Smith referred to this as the "two-fold cost of males." Asexual reproduction is much more efficient than sexual reproduction because no males are produced.

Based on this analysis, what will happen when asexual and sexual individuals exist in the same population and compete with one another? If all other things are equal, individuals that reproduce asexually should increase in frequency in the population while individuals that reproduce sexually should decline in frequency. In fact, Maynard Smith's model predicts that sexual reproduction is so inefficient that it should be completely eliminated.

To resolve this paradox, biologists began examining the assumption "If all other things are equal." Stated another way, biologists began looking for ways that meiosis and outcrossing could lead to the production of offspring that reproduce more than asexually produced individuals do. After decades of debate and analysis, two solid answers are beginning to emerge.

The Purifying Selection Hypothesis

The first clue to unraveling the paradox of sex is a simple observation: If a gene is damaged or changed in a way that causes it to function poorly, it will be inherited by *all* of that individual's offspring when asexual reproduction occurs. Suppose the damaged gene arose in generation 1 of Figure 12.12. If the damaged gene is important enough, it might cause the four asexual females present in generation 2 to produce fewer than four offspring apiece—perhaps because the members of generation 2 die young. If so, then generation 3 will not have twice as many individuals in the asexual lineage compared to the sexual lineage.

An allele that functions poorly and lowers the fitness of an individual is said to be deleterious. Asexual individuals are doomed to transmitting all of their deleterious alleles to all of their offspring.

Suppose, however, that the same deleterious allele arose in the sexually reproducing female in generation 1 of Figure 12.12. If the female also has a normal copy of the gene, and if she mates with a male that has normal copies of the gene, then on average half of her offspring will lack the deleterious allele. Sexual individuals are likely to have offspring that lack deleterious alleles present in the parent.

Natural selection against deleterious alleles is called purifying selection. Over time, purifying selection should steadily reduce the numerical advantage of asexual reproduction.

To test this hypothesis, researchers recently compared the same genes in two closely related species of *Daphnia*, a common planktonic inhabitant of ponds and lakes. One was a species that reproduces asexually and the other reproduces sexually. As predicted, they found that individuals in the asexual species contained many more deleterious alleles than individuals in the sexual species. Results like these have convinced biologists that purifying selection is an important factor limiting the success of asexual reproduction.

The Changing-Environment Hypothesis

The second hypothesis to explain sexual reproduction also focuses on the benefits of producing genetically diverse offspring. Here's the key idea: Offspring that are genetic clones of their parents are unlikely to thrive if the environment changes.

What type of environmental change would favor genetically diverse offspring? The leading hypothesis points to changes in parasites—bacteria, viruses, fungi, and other entities that cause disease. In your own lifetime, for example, several new disease-causing agents have emerged that afflict humans. These include the SARS virus, new strains of HIV, the parasite that causes malaria, and the tuberculosis bacterium. Hundreds of genes help defend you against these types of invaders. In many cases, certain alleles help you fight off particular strains of disease-causing bacteria, eukaryotes, or viruses.

What happens if all of the offspring produced by an individual are genetically identical? If a new strain of disease-causing agent evolves, then all of the asexually produced offspring are likely to be susceptible to that new strain. But if the offspring are genetically varied, then it is likely that at least some offspring will have combinations of alleles that enable them to fight off the new disease and produce offspring of their own.

To test this idea, a research group is studying a species of snail that is native to New Zealand. This snail lives in ponds and other freshwater habitats and is susceptible to infection by over a dozen species of parasitic trematode worms. Snails that become infected cannot reproduce—the worms eat their reproductive organs. The parasites are rare in some habitats and common in others.

The biologists are interested in this snail species because some individuals reproduce only sexually while others reproduce only asexually. If the changing-environment hypothesis for the advantage of sex is correct, then the frequency of sexually reproducing individuals should be high in habitats where parasites are common, and low in habitats where parasites are rare (**Figure 12.13**).

To test these predictions, the researchers collected a large number of individuals from different habitats and calculated the frequency of individuals that reproduce sexually versus those that reproduce asexually. On the graph in Figure 12.13, the height of each bar indicates the average percentage of individuals with parasite infections; the vertical bars indicate the standard error of each average (see **BioSkills 5** in Appendix A). The data show that habitats where parasite infection rates are high have a relatively large number of sexually reproducing individuals compared with habitats that have low parasite incidence.

This result and various other studies support the changing-environment hypothesis. Although the paradox of sex remains an active area of research, more biologists are becoming convinced that sexual reproduction is helpful for two reasons: (**1**) Offspring are not doomed to inherit harmful alleles, and (**2**) at least some offspring may be able to fight off new strains of disease-causing agents.

12.4 Mistakes in Meiosis

When homologous chromosomes separate during meiosis I, a complete set of chromosomes is transmitted to each daughter cell. But what happens if there is a mistake, and the chromosomes are not properly distributed? What are the consequences for offspring if gametes contain an abnormal set of chromosomes?

In 1866 Langdon Down described a distinctive suite of co-occurring conditions observed in some humans. The syndrome was characterized by mental retardation, a high risk for heart problems and leukemia, and a degenerative brain disorder similar to Alzheimer's disease. **Down syndrome**, as the disorder came to be called, is observed in about 0.15 percent of live births (1 infant in every 666).

For over 80 years the cause of the syndrome was unknown. Then, in the late 1950s, a study of the chromosome sets of nine Down syndrome children suggested that the condition is associated with the presence of an extra copy of chromosome 21. This situation is called a **trisomy** ("three-bodies")—in this case, trisomy-21—because each cell has three copies of the chromosome. The explanation proposed for the trisomy was that a mistake had occurred during meiosis in one of the parents.

How Do Mistakes Occur?

For a gamete to get one complete set of chromosomes, two steps in meiosis must be perfectly executed.

1. The chromosomes in each homologous pair must separate from each other during the first meiotic division, so that only one homolog ends up in each daughter cell.
2. Sister chromatids must separate from each other and move to opposite poles of the dividing cell during meiosis II.

If both homologs or both sister chromatids move to the same pole of the parent cell, the products of meiosis will be abnormal. This sort of meiotic error is referred to as **nondisjunction**, because the homologs fail to separate or disjoin.

EXPERIMENT

QUESTION: Why does sexual reproduction occur?

HYPOTHESIS: In habitats where parasitism is common, sexually produced offspring have higher fitness than do asexually produced individuals.

NULL HYPOTHESIS: There is no relationship between mode of reproduction and the presence of parasites.

EXPERIMENTAL SETUP:

1. Collect snails from a wide array of habitats.

2. Document percentage of males in each population, as an index of frequency of sexual reproduction. More males means that more sexual reproduction is occurring.

3. Compare percentage of males in both populations: In one population, males are common; in the other, males are almost nonexistent. Infer that sexual reproduction is either common or almost nonexistent.

Parasitism rate in this population?

Parasitism rate in this population?

4. Assess infection rate: Document percentage of individuals infected with parasites in sexually versus asexually reproducing populations.

PREDICTION: In populations where sexual reproduction is common, parasitism rates are high. In populations with only asexual reproduction, infection rates are low.

PREDICTION OF NULL HYPOTHESIS: No difference in parasitism rate between populations that reproduce sexually versus asexually.

RESULTS:

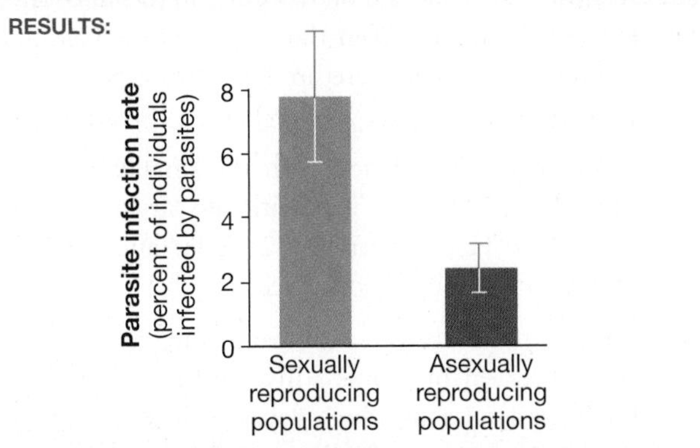

CONCLUSION: Sexual reproduction is common in habitats where parasitism is common. Asexual reproduction is common in habitats where parasitism is rare.

FIGURE 12.13 Is Sexual Reproduction Favored When Parasite Infection Rates Are High?

SOURCE: Lively, C. and J. Jokela. 2002. Temporal and spatial distributions of parasites and sex in a freshwater snail. *Evolutionary Ecology Research* 4: 219–226.

✔**QUESTION** Why was it important to collect snails from a wide array of habitats?

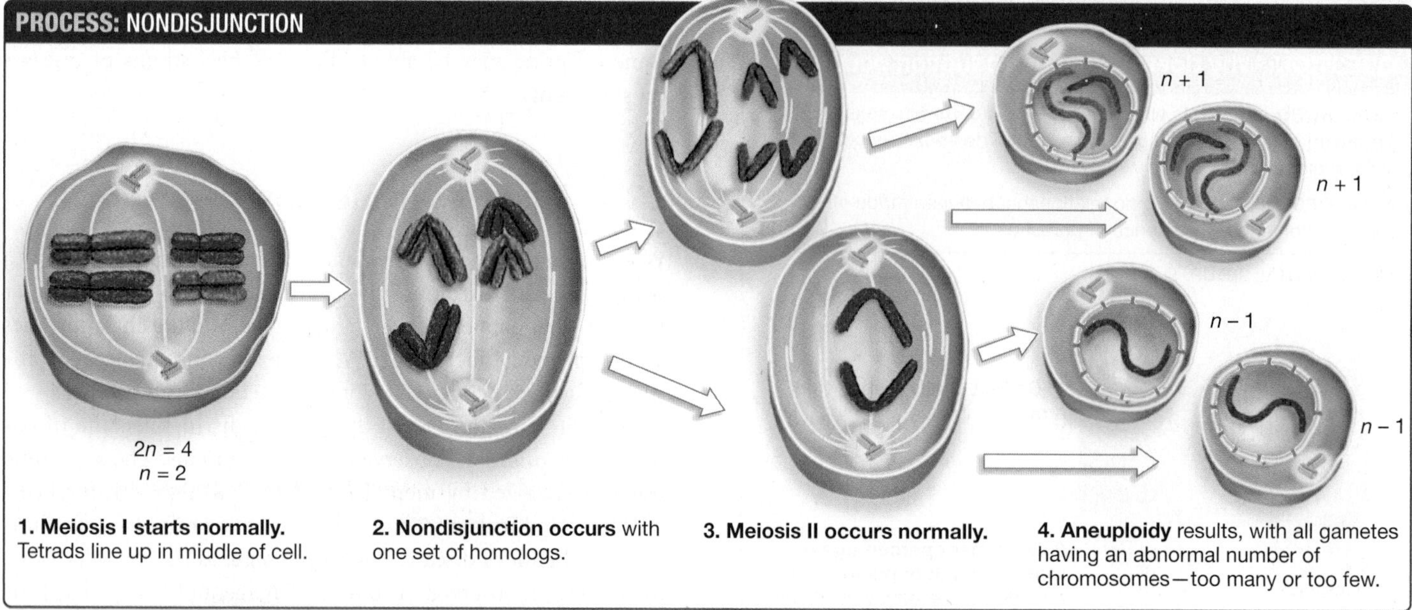
PROCESS: NONDISJUNCTION
2n = 4
n = 2
n + 1
n + 1
n − 1
n − 1
1. Meiosis I starts normally. Tetrads line up in middle of cell.
2. Nondisjunction occurs with one set of homologs.
3. Meiosis II occurs normally.
4. Aneuploidy results, with all gametes having an abnormal number of chromosomes—too many or too few.

FIGURE 12.14 Nondisjunction Leads to Gametes with Abnormal Chromosome Numbers. If homologous chromosomes fail to separate during meiosis I, the gametes that result will have an extra chromosome or will lack a chromosome.

Figure 12.14 shows what happens when homologs do not separate correctly. Notice that two daughter cells have two copies of the same chromosome—the smaller one in Figure 12.14—while the other two lack that chromosome entirely. Gametes that contain an extra chromosome are symbolized as $n + 1$; gametes that lack one chromosome are symbolized as $n - 1$.

If an $n + 1$ gamete is fertilized by a normal n gamete, the resulting zygote will be $2n + 1$. This situation is a trisomy. If the $n - 1$ gamete is fertilized by a normal n gamete, the resulting zygote will be $2n - 1$. This situation is called **monosomy**. Cells that have too many or too few chromosomes are said to be **aneuploid** ("without-form").

To review nondisjunction and its consequences, go to the study area at *www.masteringbiology.com.*

MB **Web Activity** Mistakes in Meiosis

Meiotic mistakes occur at a relatively high frequency. In humans, for example, researchers estimate that 25 percent of all conceptions produce a fertilized egg that is aneuploid. Most of these errors result from the failure of a homologous pair to separate in anaphase of meiosis I; it is relatively rare for sister chromatids to stay together during anaphase of meiosis II.

The consequences of meiotic mistakes are almost always severe when defective gametes participate in fertilization. In one study of human pregnancies that ended in early embryonic or fetal death, 38 percent of the 119 cases involved atypical chromosome complements that resulted from mistakes in meiosis. Trisomy accounted for 36 percent of the abnormal karyotypes found. It was also common to find the incorrect number of complete chromosome sets called triploidy ($3n$). Less common were abnormally sized or shaped chromosomes and monosomy ($2n - 1$). Mistakes in meiosis are the leading cause of spontaneous abortion (miscarriage) in humans.

Why Do Mistakes Occur?

The leading hypothesis to explain the incidence of trisomy and other meiotic mistakes is that they are accidents—random errors that occur during meiosis. Consistent with this proposal, there does not seem to be any genetic or inherited predisposition to trisomy or other types of meiotic dysfunction. Most cases of Down syndrome, for example, occur in families with no history of the condition. Recent research indicates that the molecular mechanism involves misattachment between microtubules and kinetochores early in meiosis I.

Table 12.4 shows data collected on trisomies in the autosomes of human fetuses and infants. Even though meiotic errors may be random, studies show that there are still strong patterns in their occurrence:

- Trisomy is much more common with the smaller chromosomes (numbers 13–22) than it is with the larger chromosomes (numbers 1–12), and trisomy-21 is far and away the most common type of trisomy observed.
- With the exception of trisomy-21, most of the trisomies and monosomies observed in humans involve the sex chromosomes. **Klinefelter syndrome**, which develops in XXY individuals, occurs in about 1 in 1000 live male births. Trisomy X (karyotype XXX) occurs in about 1 in 1000 live births and results in females who may or may not have symptoms such as

SUMMARY TABLE 12.4 **Trisomy in Human Births: Effects of Chromosome Number and Paternal versus Maternal Origin**

Trisomy (chromosome number)	Number of Cases Examined	Due to Error in Sperm	Due to Error in Egg	Maternal Errors (%)
2–12	16	3	13	81
13	7	2	5	71
14	8	2	6	75
15	11	3	8	73
16	62	0	62	100
18	73	3	70	96
21	436	29	407	93
22	11	0	11	100

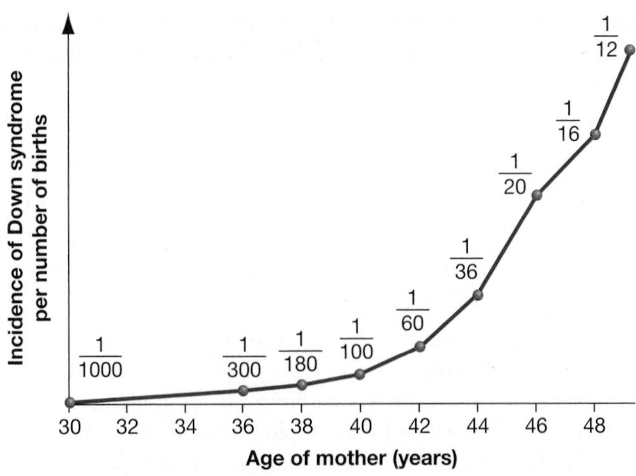

FIGURE 12.15 The Frequency of Down Syndrome Increases as a Function of a Mother's Age.

impaired mental function and sterility. **Turner syndrome** develops in XO individuals, where the "O" stands for lack of a second X, and occurs in about 1 in 5000 live births. Individuals with this syndrome are female but are sterile.

- Maternal age is an important factor in the occurrence of trisomy. In fact, maternal errors account for most incidents of trisomy. For example, over 90 percent of cases of Down syndrome are due to chromosomal defects in eggs. As **Figure 12.15** shows, the incidence of Down syndrome increases dramatically in mothers over 35 years old.

Why do these patterns occur? The frequency of nondisjunction is about equal among chromosomes, but aneuploidy tends to be lethal to embryos if it involves chromosomes that contain a large number of genes. Trisomy-21 is common because it affects the phenotype less, and is less deadly, than other trisomies. It is still a mystery, though, why such a strong correlation exists between maternal age and frequency of trisomy-21.

Studies on the mechanism of aneuploidy continue. In the meantime, one overall message is clear: Successful meiosis is critical to the health and welfare of offspring.

CHAPTER 12 REVIEW

For media, go to the study area at www.masteringbiology.com

Summary of Key Concepts

Meiosis is a type of nuclear division resulting in cells that have half as many chromosomes as the parent cell. In animals it leads to the formation of eggs and sperm.

- Chromosomes exist in sets. In diploid organisms, individuals have two versions of each type of chromosome. The two versions are called homologs. One homolog is inherited from the mother, and one from the father. Haploid organisms have just one of each type of chromosome.
- Each chromosome is replicated before meiosis begins. At the start of meiosis I, each chromosome consists of a pair of sister chromatids.
- Homologous pairs of chromosomes synapse early in meiosis I, forming a tetrad—a group of two homologous chromosomes. Non-sister chromatids undergo crossing over.
- When crossing over is complete, the pair of homologous chromosomes is moved to the metaphase plate.
- At the end of meiosis I, the homologous chromosomes are separated and distributed to two daughter cells. The daughter cells are haploid, because each receives one of each type of chromosome.
- During meiosis II, sister chromatids separate and are distributed to two daughter cells.
- From a diploid cell with replicated chromosomes, meiosis produces four haploid daughter cells with unreplicated chromosomes.

✔You should be able to explain why meiosis does not occur in bacteria—most of which have a singular, circular chromosome.

MB **BioFlix™** Meiosis, **Web Activity** Meiosis

Each cell produced by meiosis receives a different combination of chromosomes. Because genes are located on chromosomes, each cell produced by meiosis receives a different complement of genes. The resulting offspring are genetically distinct from each other and from their parents.

- When meiosis and outcrossing occur, the chromosome complements of offspring differ from one another and from their parents for three reasons:

  (1) Gametes receive a random assortment of maternal and paternal chromosomes when homologs separate in meiosis I.

  (2) Because of crossing over, each chromosome contains a random assortment of paternal and maternal alleles.

  (3) Outcrossing results in a combination of chromosome sets from different individuals.

✔You should be able to explain why monozygotic (identical) twins are much more similar than dizygotic (fraternal) twins. (Monozygotic twins arise when a fertilized egg undergoes mitosis to form two cells, which then split and develop into two separate individuals. Dizygotic twins develop from two different eggs that were fertilized by two different sperm.)

The leading hypothesis to explain meiosis is that genetically varied offspring are more likely to thrive in environments where parasites and disease are common.

- Asexual reproduction is much more efficient than sexual reproduction because males do not have to be produced and no time or energy has to be spent in courtship.
- Sexual reproduction is favored in many groups because **(1)** parents can produce offspring that lack deleterious alleles and **(2)** genetically diverse offspring are likely to include some that are better able to resist parasites.

✔You should be able to predict whether, in species that alternate between asexual and sexual reproduction, sexual reproduction occurs during seasons when environmental conditions are stable or seasons when conditions change rapidly.

If mistakes occur during meiosis, the resulting egg and sperm cells may contain the wrong number of chromosomes. It is rare for offspring with an incorrect number of chromosomes to develop normally.

- Mistakes during meiosis lead to gametes and offspring with an unbalanced set of chromosomes. Children with Down syndrome, for example, have an extra copy of chromosome 21.
- The leading hypothesis to explain meiotic mistakes is that they are random accidents resulting in a failure of homologous chromosomes or sister chromatids to separate properly during meiosis.

✔You should be able to explain what happens when none of the homologous chromosomes present separate at anaphase of meiosis I, but their sister chromatids separate normally at meiosis II.

MB **Web Activity** Mistakes in Meiosis

Questions

✔TEST YOUR KNOWLEDGE

Answers are available in Appendix B

1. In the roundworm *Ascaris*, eggs and sperm have two chromosomes, but all other cells have four. Observations such as this inspired which important hypothesis?
   a. Before gamete formation, a special type of cell division leads to a quartering of chromosome number.
   b. Before gamete formation, a special type of cell division leads to a halving of chromosome number.
   c. After gamete formation, half of the chromosomes are destroyed.
   d. After gamete formation, either the maternal or the paternal set of chromosomes disintegrates.

2. What are homologous chromosomes?
   a. chromosomes that are similar in their size, shape, and gene content
   b. similar chromosomes that are found in different individuals of the same species
   c. the two "threads" in a replicated chromosome (they are identical copies)
   d. the products of crossing over, which contain a combination of segments from maternal chromosomes and segments from paternal chromosomes

3. What is a tetrad?
   a. the X that forms when chromatids from homologous chromosomes cross over
   b. a group of four chromatids produced when homologs synapse
   c. the four points where homologous chromosomes touch as they synapse
   d. the group of four genetically identical daughter cells produced by mitosis

4. What is genetic recombination?
   a. the synapsing of homologs during prophase of meiosis I
   b. the new combination of maternal and paternal chromosome segments that results when homologs cross over
   c. the new combinations of chromosome segments that result when outcrossing occurs
   d. the combination of a haploid phase *and* a diploid phase in a life cycle

5. What is meant by a paternal chromosome?
   a. the largest chromosome in a set
   b. a chromosome that does not separate correctly during meiosis I
   c. the member of a homologous pair that was inherited from the mother
   d. the member of a homologous pair that was inherited from the father

6. Meiosis II is similar to which process?
   a. mitosis in haploid cells
   b. nondisjunction
   c. outcrossing
   d. meiosis I

TEST YOUR UNDERSTANDING

Answers are available in Appendix B

1. Explain the relationship between homologous chromosomes and the relationship between sister chromatids.
2. Lay four pens and four pencils on a tabletop, and imagine that they represent replicated chromosomes in a diploid cell with $n = 2$. Explain the phases of meiosis II by moving the pens and pencils around. (If you don't have enough pens and pencils, use strips of paper or fabric.)
3. Meiosis is called a reduction division, but all of the reduction occurs during meiosis I—no reduction occurs during meiosis II. Explain why meiosis I is a reduction division but meiosis II is not.
4. Triploid ($3n$) watermelons are produced by crossing a tetraploid ($4n$) strain with a diploid ($2n$) plant. Briefly explain why this mating produces a triploid individual. Why can mitosis proceed normally in triploid cells, but meiosis cannot?
5. Some plant breeders are concerned about the susceptibility of asexually cultivated plants, such as seedless bananas, to new strains of disease-causing bacteria, viruses, or fungi. Briefly explain their concern by discussing the differences in the genetic "outcomes" of asexual and sexual reproduction.
6. Explain why nondisjunction leads to trisomy and other types of abnormal chromosome complements. In what sense are these chromosome sets "unbalanced"?

APPLYING CONCEPTS TO NEW SITUATIONS

Answers are available in Appendix B

1. The gibbon has 44 chromosomes per diploid set, and the siamang has 50 chromosomes per diploid set. In the 1970s a chance mating between a male gibbon and a female siamang produced an offspring. Predict how many chromosomes were observed in the somatic cells of the offspring. Do you predict that this individual would be able to form viable gametes? Why or why not?
2. Meiosis results in a reassortment of maternal and paternal chromosomes. If $n = 3$ for a given organism, there are eight different combinations of paternal and maternal chromosomes. If no crossing over occurs, what is the probability that a gamete will receive *only* paternal chromosomes?
3. Some researchers predict that spontaneous abortion should be rare in older females, because they are less likely than young females to be able to have offspring in the future. How does this claim relate to Figure 12.15, which graphs mother's age versus the incidence of Down syndrome?
4. The data on snail populations that were used to test the changing-environment hypothesis have been criticized because they are observational and not experimental in nature. As a result, they do not control for factors other than parasites that might affect the frequency of sexually reproducing individuals.
   a. Design an experimental study that would provide stronger evidence that the frequency of parasite infection causes differences in the frequency of sexually versus asexually reproduced individuals in this species of snail.
   b. In defense of the existing data, comment on the value of observing patterns like this in nature, versus under controlled conditions in the laboratory.

Experiments on garden peas and sweet peas (shown here) helped launch the science of genetics.

13 Mendel and the Gene

KEY CONCEPTS

- In many species, individuals have two alleles of each gene. The principle of segregation states that prior to the formation of eggs and sperm, the alleles of each gene separate so that each egg or sperm cell receives only one of them.
- The principle of independent assortment states that alleles of different genes are transmitted to egg cells and sperm cells independently of each other.
- Genes are located on chromosomes. The principle of segregation is explained by the separation of homologous chromosomes in anaphase of meiosis I. The principle of independent assortment applies to genes found on different chromosomes and is explained by chromosomes lining up randomly in metaphase of meiosis I.
- There are important exceptions and extensions to the basic patterns of inheritance that Mendel discovered.

The science of biology is built on a series of great ideas. Two of these—the cell theory and the theory of evolution—were introduced in Chapter 1. The cell theory describes the basic structure of organisms; the theory of evolution by natural selection clarifies why species change through time. Life is cellular; populations evolve. These are two of the five fundamental attributes of life.

This chapter introduces a third great idea in biology: the chromosome theory of inheritance. The chromosome theory explained how genetic information is transmitted from one generation to the next. It shed light on a third fundamental attribute of life: Organisms process information.

An Austrian monk named Gregor Mendel laid the groundwork for the theory in 1865, when he announced that he had worked out the rules of inheritance through a series of experiments on garden peas. Another key insight emerged during the final decades of the nineteenth century, when biologists described the details of meiosis.

The chromosome theory of inheritance, formulated in 1902 by Walter Sutton and Theodor Boveri, linked these two results. This theory contends that meiosis, introduced in Chapter 12, causes the patterns of inheritance that Mendel observed. It also asserts that the hereditary factors called genes are located on chromosomes.

The chromosome theory launched the study of **genetics**, the branch of biology that focuses on the inheritance of traits. Let's start at the beginning: What are the rules of inheritance that Mendel discovered?

13.1 Mendel's Experimental System

When biological science began to emerge as an important scientific discipline, questions about **heredity**—meaning inheritance, or the transmission of traits from parents to offspring—were primarily the concern of animal breeders and horticulturists. A **trait**

✔ When you see this checkmark, stop and test yourself. Answers are available in Appendix B.

is any characteristic of an individual, ranging from overall height to the primary structure of a particular membrane protein.

In the city where Gregor Mendel lived, there was particular interest in how selective breeding could result in hardier and more productive varieties of sheep, fruit trees, and vines; and an Agricultural Society had been formed there to promote research into making selective breeding more efficient. Mendel was an active member of this society; the monastery he belonged to was also devoted to scientific teaching and research.

What Questions Was Mendel Trying to Answer?

Mendel set out to address the most fundamental of all issues concerning heredity: What are the basic patterns in the transmission of traits from parents to offspring?

At the time, two hypotheses had been formulated to answer this question:

- *Blending inheritance* claimed that the traits observed in a mother and father blend together to form the traits observed in their offspring. As a result, an offspring's traits are intermediate between the mother's and father's traits.
- *Inheritance of acquired characters* claimed that traits present in parents are modified, through use, and passed on to their offspring in the modified form.

Each of these hypotheses made predictions. Blending inheritance contended that when black sheep and white sheep mate, their hereditary determinants blend to form a new hereditary determinant for gray wool. Therefore, their offspring should be gray. Inheritance of acquired characters predicts that if giraffes extend their necks by straining to reach leaves high in the tops of trees, they subsequently produce longer-necked offspring.

These hypotheses were being promoted by the greatest scientists of Mendel's time. Are they correct?

Garden Peas Served as the First Model Organism in Genetics

After investigating and discarding several candidate species to study, Mendel chose the garden pea *Pisum sativum*. His reasons were practical: Peas are inexpensive and easy to grow from seed, have a relatively short generation time, and produce reasonably large numbers of seeds. These features made it possible for him to continue experiments over several generations and collect data from large numbers of individuals.

Peas served as a **model organism**: a species that is used for research because it is practical and because conclusions drawn from studying it turn out to apply to many other species as well. **BioSkills 14** in Appendix A introduces some of the important model organisms used in biological science today.

Two additional features of the pea made it possible for Mendel to design his experiments: He could control which parents were involved in a mating, and he could arrange matings between individuals that differed in easily recognizable traits, such as flower color or seed shape.

HOW DID MENDEL CONTROL MATINGS? **Figure 13.1a** shows the male and female reproductive organs of a garden pea flower. Sperm cells are produced in **pollen grains**, which are small sacs that mature in the male reproductive structure of the plant. Eggs are produced in the female reproductive structure.

Under normal conditions, garden peas **self-fertilize**: a flower's pollen falls on the female reproductive organ of that same flower. As **Figure 13.1b** shows, however, Mendel could circumvent this arrangement by removing the male reproductive organs from a flower before any pollen formed. Later he could transfer pollen from another pea plant to that flower's female reproductive organ with a brush. This type of mating is referred to as a **cross-fertilization** or simply a cross. Using this technique, Mendel could control the matings of his model organism.

WHAT TRAITS DID MENDEL STUDY? Mendel conducted his experiments on varieties of peas that differed in seven traits: seed shape, seed color, pod shape, pod color, flower color, flower and pod position, and stem length. Biologists refer to the observable traits of an individual, such as the shape of a pea seed or the eye color of a human, as its **phenotype** (literally, "show-type"). In the first pea populations that Mendel studied, two distinct phenotypes existed for each of the seven traits.

Mendel began his work by obtaining individuals from what breeders called pure lines or true-breeding lines. A **pure line** consists of individuals that produce offspring identical to themselves when they are self-pollinated or crossed to another member of the pure-line population. For example, earlier breeders had developed pure lines for wrinkled seeds and round seeds. During two years of trial experiments, Mendel

(a) Self-pollination

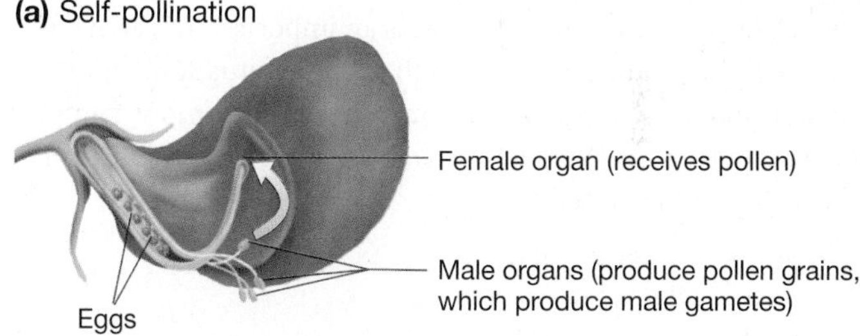

(b) Cross-pollination

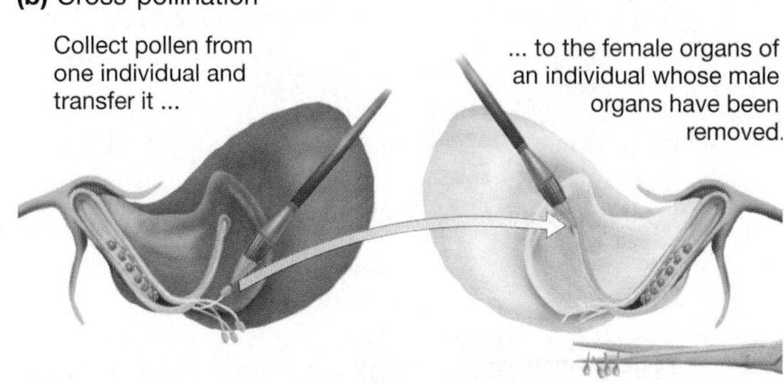

Collect pollen from one individual and transfer it ...
... to the female organs of an individual whose male organs have been removed.

FIGURE 13.1 Peas Can Be Self-Pollinated or Cross-Pollinated.
(a) Under normal conditions, garden peas pollinate themselves.
(b) Mendel developed a method of controlling the matings of his model organism.

confirmed that individuals that germinated from his wrinkled seeds produced only wrinkled-seeded offspring when they were mated to themselves or to another pure-line individual that germinated from a wrinkled seed; and that the same was true for round seeds.

Why is this important? Remember that Mendel wanted to find out how traits are transmitted from parents to offspring. Once he had confirmed that he was working with pure lines, he could predict how matings within each line would turn out; in other words, he knew what the offspring from these matings would look like. He could then compare these results with the outcomes of crosses between individuals from different pure lines.

Suppose that Mendel arranged matings between a pure-line individual with round seeds and a pure-line individual with wrinkled seeds. He knew that one parent carried a hereditary determinant for round seeds, while the other carried a hereditary determinant for wrinkled seeds. But the offspring that resulted from this mating would have both hereditary determinants. They would be **hybrids**—offspring from matings between true-breeding parents that differ in one or more traits.

Would these hybrid offspring have wrinkled seeds, round seeds, or a blended combination of wrinkled and round? What would be the seed shape in subsequent generations when hybrid individuals self-pollinated or were crossed with members of the pure lines?

13.2 Mendel's Experiments with a Single Trait

Mendel's first set of experiments consisted of crossing pure lines that differed in just one trait. This is an important research strategy in biological science: start with a simple situation. Once you understand what's going on, you can consider more complex questions, such as, What happens in crosses between individuals that differ in two traits?

Mendel began his single-trait crosses by crossing individuals from round-seeded and wrinkled-seeded pure lines. The adults used in an initial experimental cross are the **parental generation.** Their progeny (that is, offspring) are the **F_1 generation.** F_1 stands for "first filial"; the Latin roots *fili* and *filia* mean son and daughter. Subsequent generations are called the F_2 generation, F_3 generation, and so on.

The Monohybrid Cross

In his first set of crosses, Mendel took pollen from round-seeded plants and placed it on the female reproductive organs of plants from the wrinkled-seeded line. As **Figure 13.2a** shows, all of the seeds produced by progeny from this cross were round.

This was a remarkable result, for two reasons:

1. The traits did not blend together to form an intermediate phenotype. Instead, the round-seeded form appeared intact. This result was in stark contrast to the predictions of the blending-inheritance hypothesis shown in **Figure 13.2b**.
2. The genetic determinant for wrinkled seeds seemed to have disappeared. Where did it go?

DOMINANT AND RECESSIVE TRAITS To figure out what was going on, Mendel did something that turned out to be brilliant: He planted the F_1 seeds and allowed the individuals to self-pollinate when they matured.

Remember that he knew that each of these individuals had inherited a genetic determinant for round seeds and a genetic determinant for wrinkled seeds. A mating like this—between parents that each carry two different genetic determinants for the same trait—is called a **monohybrid cross.**

When he collected the seeds that were produced by many plants in the resulting F_2 generation, he observed that 5474 were round and 1850 were wrinkled (see Figure 13.2a). This observation was striking, even astonishing. The wrinkled seed shape had reappeared in the F_2 generation after disappearing completely in

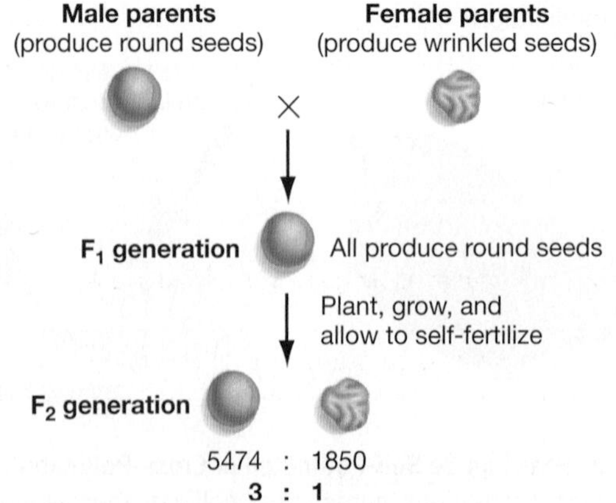

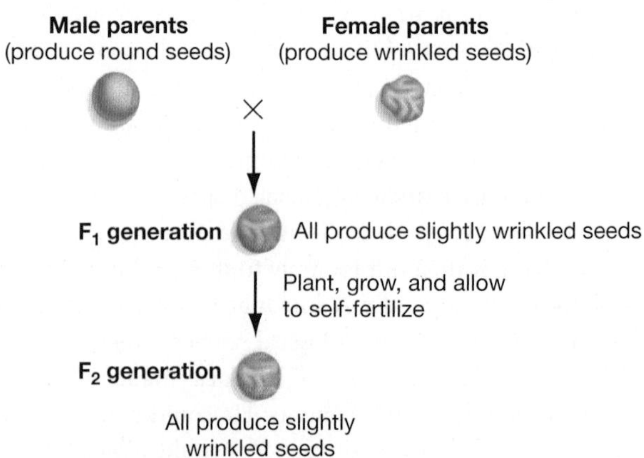

FIGURE 13.2 A Monohybrid Cross. The results of Mendel's crosses involving a single trait contrasted strongly with the predictions of the blending inheritance hypothesis.

the F_1 generation. No one had observed the phenomenon before, simply because it had been customary for biologists to stop their breeding experiments with F_1 offspring.

Mendel invented some important terms to describe this result.

- He designated wrinkled shape as a **recessive** trait relative to the round-seed trait. This was an appropriate term because none of the F_1 individuals had wrinkled seeds—meaning the wrinkled-seeds phenotype appeared to recede or temporarily become latent or hidden.
- He referred to round seeds as **dominant** to the wrinkled-seed trait. This term was apt because the round-seed phenotype appeared to dominate over the wrinkled-seed determinant when both were present.

It's important to note, though, that in genetics the term dominant has nothing to do with its everyday English usage as powerful or superior. Subsequent research has shown that individuals with the dominant phenotype do not necessarily have higher fitness than do individuals with the recessive phenotype. Nor are genetic determinants associated with a dominant phenotype necessarily more common than recessive ones. For example, a rare, dominant genetic determinant in humans causes a fatal illness—a type of brain degeneration called Huntington's disease. In genetics, the terms dominant and recessive identify *only* which phenotype is observed in individuals carrying two different genetic determinants for a given trait.

Mendel also noticed that the round and wrinkled seeds of the F_2 generation were present in a ratio of 2.96:1, or essentially 3:1. The 3:1 ratio means that for every four individuals, on average three had the dominant phenotype and one had the recessive phenotype. The results can also be stated in terms of frequencies or proportions instead of as a ratio: In this case, about 3/4 of the F_2 seeds were round and 1/4 were wrinkled.

A RECIPROCAL CROSS Mendel wanted to test the hypothesis that it mattered which parent and gamete type had a particular genetic determinant—that gender influenced the inheritance of seed shape. To do this, he performed a second set of crosses between two pure-breeding lines—this time with pollen taken from an individual from a pure line of wrinkled-seeded peas (see **Figure 13.3**).

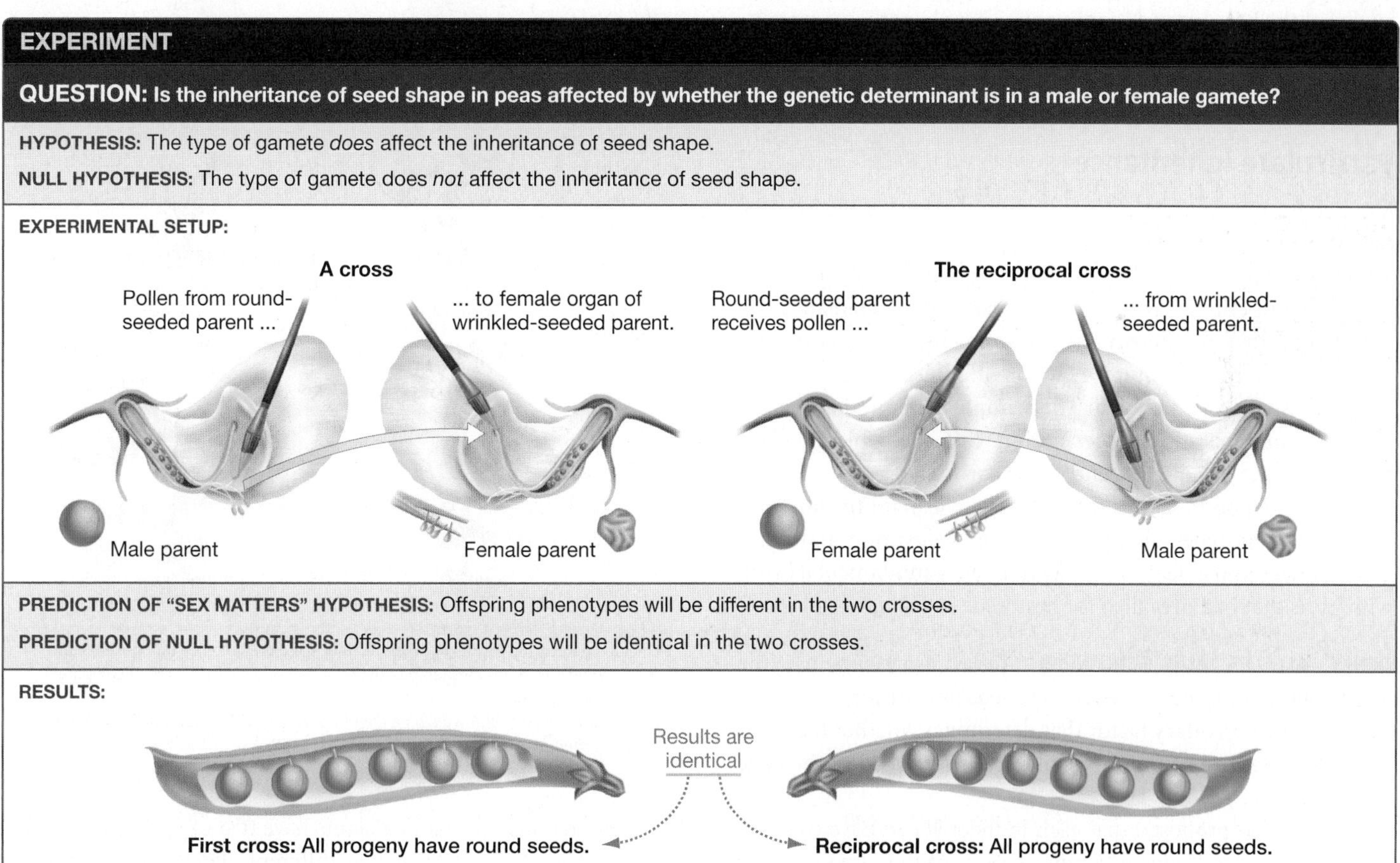

EXPERIMENT
QUESTION: Is the inheritance of seed shape in peas affected by whether the genetic determinant is in a male or female gamete?
HYPOTHESIS: The type of gamete does affect the inheritance of seed shape.
NULL HYPOTHESIS: The type of gamete does not affect the inheritance of seed shape.
EXPERIMENTAL SETUP:
A cross
Pollen from round-seeded parent ...
... to female organ of wrinkled-seeded parent.
Male parent
Female parent
The reciprocal cross
Round-seeded parent receives pollen ...
... from wrinkled-seeded parent.
Female parent
Male parent
PREDICTION OF "SEX MATTERS" HYPOTHESIS: Offspring phenotypes will be different in the two crosses.
PREDICTION OF NULL HYPOTHESIS: Offspring phenotypes will be identical in the two crosses.
RESULTS:
Results are identical
First cross: All progeny have round seeds.
Reciprocal cross: All progeny have round seeds.
CONCLUSION: It makes no difference whether the genetic determinant for seed shape comes from the male gamete or from the female gamete.

FIGURE 13.3 A Reciprocal Cross.

SOURCE: Mendel, G. 1866. Versuche über Pflanzen-hybriden. *Verhandlungen des naturforschenden Vereines in Brünn.* 4: 3–47. English translation available from ESP: Electronic Scholarly Publishing (www.esp.org).

✔**QUESTION** Some people think that experiments are failures if the hypothesis being tested is not supported. What does it mean to say that an experiment failed? Was this experiment a failure?

These experiments completed a **reciprocal cross**—a set of matings where the mother's phenotype in the initial cross is the father's phenotype in a subsequent cross, and the father's phenotype in the initial cross is the mother's phenotype in a subsequent cross.

In this case the results of the reciprocal crosses were identical: All of the F_1 progeny in the subsequent cross had round seeds, just as in the initial cross. The reciprocal cross established that it does not matter whether the genetic determinants for seed shape are located in the male or female parent.

THE RESULTS ARE GENERAL Before he tried to interpret this pattern, it was important for Mendel to establish that the results were not restricted to inheritance of seed shape. So he repeated the experiments with each of the six other traits listed earlier.

In each case, he obtained similar results (see **Table 13.1**): The products of reciprocal crosses were the same—one form of the trait was always dominant regardless of the parent it came from; the F_1 progeny showed only the dominant trait and did not exhibit an intermediate phenotype; and in the F_2 generation, the ratio of individuals with dominant and recessive phenotypes was about 3 to 1.

How could these patterns be explained? Mendel answered this question with a series of propositions about the nature and behavior of the hereditary determinants. These hypotheses are considered some of the most brilliant insights in the history of biological science.

Particulate Inheritance

Mendel's results were clearly inconsistent with either the hypothesis of blending inheritance or the hypothesis of acquired characters. To explain the patterns that he observed, Mendel proposed a competing hypothesis called **particulate inheritance**. He maintained that the hereditary determinants for traits do not blend together or acquire new or modified characteristics through use. In fact, hereditary determinants maintain their integrity from generation to generation. Instead of blending together, they act like discrete entities or particles.

Mendel's hypothesis was the only way to explain the observation that phenotypes disappeared in one generation and reappeared intact in the next. It also represented a fundamental break with ideas that had prevailed for hundreds of years.

GENES, ALLELES, AND GENOTYPES Today geneticists use the word **gene** to indicate the hereditary determinant for a trait. For example, the hereditary factor that determines whether the seeds of garden peas are round or wrinkled is referred to as the gene for seed shape.

Mendel also proposed that each individual can have two versions of any gene. Today different versions of the same gene are called **alleles**. Different alleles are responsible for the variation in the traits that Mendel studied. In the case of the gene for seed shape, one allele of this gene is responsible for the round form of the seed while another allele is responsible for the wrinkled form.

The alleles that are found in a particular individual are called the **genotype**. An individual's genotype has a profound effect on the phenotype—the physical traits.

TABLE 13.1 **F_2 Results from Mendel's Monohybrid Reciprocal Cross Experiments***

Trait	Dominant Phenotype	Recessive Phenotype	Ratio
Seed shape	5474 round	1850 wrinkled	2.96 : 1
Seed color	6022 yellow	2001 green	3.01 : 1
Pod shape	882 inflated	299 constricted	2.95 : 1
Pod color	428 green	152 yellow	2.82 : 1
Flower color	705 purple	224 white	3.15 : 1
Flower and pod position	651 axial	207 terminal	3.14 : 1
Stem length	787 tall	277 dwarf	2.96 : 1

*Mendel pooled the results from the reciprocal crosses for each trait because the results were the same whether the dominant trait originated from the male parent or the female parent.

SOURCE: Mendel, G. 1866. Versuche über Pflanzen-hybriden. *Verhandlungen des naturforschenden Vereines in Brünn.* 4: 3–47.

The hypothesis that pea plants have two of each gene—either two of the same allele or two different alleles—was important because it gave Mendel a framework for explaining dominance and recessiveness. He proposed that some alleles are dominant and others are recessive. Recall that dominance and recessiveness determine which phenotype actually appears in an individual when two different alleles are present. In garden peas, the allele for round seeds is dominant; the allele for wrinkled seeds is recessive. Therefore, as long as one allele for round seeds is present, seeds are round. When both alleles present are for wrinkled seeds (thus no allele for round seeds is present), seeds are wrinkled.

These hypotheses explain why the phenotype for wrinkled seeds disappeared in the F_1 generation and reappeared in the F_2 generation. But why did round- and wrinkled-seeded plants exist in a 3:1 ratio in the F_2 generation?

THE PRINCIPLE OF SEGREGATION To explain the 3:1 ratio of phenotypes in F_2 individuals, Mendel reasoned that the two members of each gene pair must segregate—that is, separate—into different gamete cells during the formation of eggs and sperm in the parents. As a result, each gamete contains one allele of each gene. This idea is called the **principle of segregation**.

To show how this principle works, Mendel used a letter to indicate the gene for a particular trait. For example, he used uppercase *R* to symbolize a dominant allele for seed shape and lowercase *r* to symbolize a recessive allele for seed shape. (Notice that the symbols for genes are always italicized.)

Using this notation, Mendel could describe the genotype of the individuals in the round-seed pure line as *RR* (having two of the dominant allele). The genotype of the wrinkled-seed pure line is *rr* (two of the recessive allele). Because *RR* and *rr* individuals have two copies of the same allele, they are said to be **homozygous** for the seed-shape gene (*homo* is the Greek root for "same," while *zygo* means "yoked together"). Pure-line individuals always produce offspring with the same phenotype because they are homozygous—no other allele is present.

Figure 13.4a diagrams what happened to these alleles when Mendel crossed the *RR* and *rr* pure lines. According to Mendel's analysis, *RR* parents produce eggs and sperm that all carry the *R* allele, while *rr* parents produce gametes with the *r* allele only. When two gametes—one from each parent—are fused together, they create offspring with the *Rr* genotype. Such individuals, with two different alleles for the same gene, are said to be **heterozygous** (*hetero* is the Greek root for "different"). Because the *R* allele is dominant, all of these F_1 offspring produce round seeds.

Why do the two phenotypes appear in a 3:1 ratio in the F_2 generation? Mendel proposed that during gamete formation in the F_1 (heterozygous) individuals, the paired *Rr* alleles separate into different gamete cells. As a result, about half of the gametes carry the *R* allele and half carry the *r* allele (**Figure 13.4b**). During self-fertilization, a given sperm has an equal chance of fertilizing either an *R*-bearing egg or an *r*-bearing egg. The outcome of this situation is explained in a diagram called a Punnett square.

PREDICTING OFFSPRING GENOTYPES AND PHENOTYPES WITH A PUNNETT SQUARE Years after Mendel published his work, R. C. Punnett invented a straightforward technique for predicting the genotypes and phenotypes of the offspring of different crosses. To produce a **Punnett square**, follow these steps:

1. Write each of the unique gamete genotypes produced by one parent in a horizontal row along the top of the diagram.
2. Write each of the unique gamete genotypes produced by the other parent in a vertical column down the left side of the diagram.

(a) A cross between two **homozygotes**

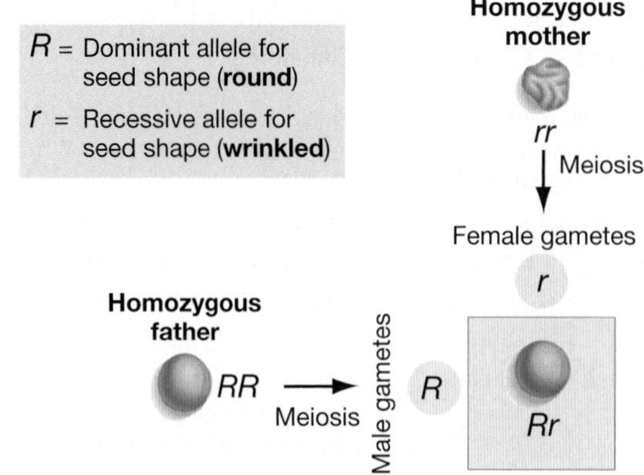

Offspring genotypes: All *Rr* (heterozygous)

Offspring phenotypes: All round seeds

(b) A cross between two **heterozygotes**

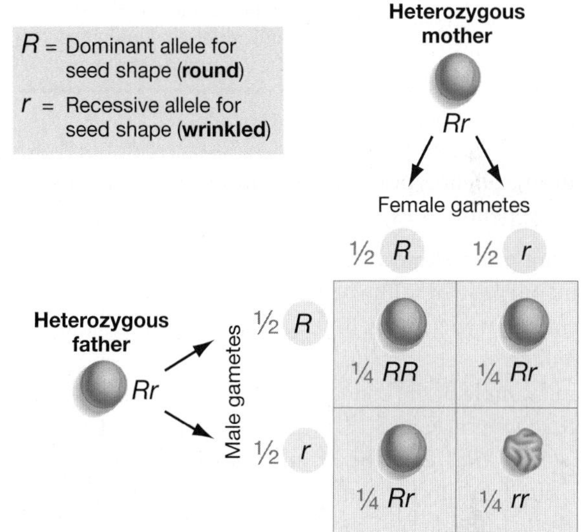

Offspring genotypes: ¼ *RR* : ½ *Rr* : ¼ *rr*

Offspring phenotypes: ¾ round : ¼ wrinkled

FIGURE 13.4 Mendel Analyzed the F_1 and F_2 Offspring of a Cross between Pure Lines. Notice that when you construct a Punnett square, you only need to list each unique type of gamete once at the head of a row or column. Therefore, for example, although the *RR* alleles segregate in the male parent of part (a), you only have to list one *R* gamete to represent the male's contribution, not two.

✔**QUESTION** In constructing a Punnett square, does it matter whether the male or female gametes go on the left or across the top? Why or why not?

3. Draw empty boxes under the horizontal row of gametes and to the right of the vertical column of gametes.
4. Fill in each box with the parental gamete genotypes written at the top of the corresponding column and at the left of the corresponding row. This step produces the offspring genotypes that result from fusion of the parental gamete genotypes.
5. Predict the proportions or ratios of each offspring genotype and phenotype by tallying the offspring genotypes and

SUMMARY TABLE 13.2 **Mendel's Model to Explain the Results of a Cross between Pure Lines**

Mendel's Claims	Comments
1. Peas have two of each gene and thus may have two different alleles of the gene.	This also turns out to be true for many other organisms.
2. Alleles do not blend together.	The hereditary determinants maintain their integrity from generation to generation.
3. Each gamete contains one of each gene (one allele).	This is due to the principle of segregation—the members of each gene pair segregate during the formation of gametes.
4. Males and females contribute equally to the genotype of their offspring.	When gametes fuse, offspring acquire a total of two of each gene—one from each parent.
5. Some alleles are dominant to other alleles.	When a dominant allele and a recessive allele for the same gene are found in the same individual, that individual has the dominant phenotype.

resulting phenotypes produced in all the boxes. **BioSkills 13** in Appendix A explains why this tallying process works.

✔If you understand these concepts, you should be able to state the purpose of a Punnett square and predict the phenotype and genotype ratios for a cross between *Yy* and *yy* pea individuals.

As an example of the concluding step in analyzing a cross, the Punnett square in Figure 13.4b predicts that 1/4 of the F_2 offspring will be *RR*, 1/2 will be *Rr*, and 1/4 will be *rr*. Because the *R* allele is dominant to the *r* allele, 3/4 of the offspring should be round seeded (the sum of the *RR* and the *Rr* offspring) and 1/4 should be wrinkled seeded (the *rr* offspring). These results are *exactly* what Mendel found in his experiments with peas. In the simplest and most elegant fashion possible, Mendel's interpretation explains the 3:1 ratio of round to wrinkled seeds observed in the F_2 offspring and the mysterious reappearance of the wrinkled seeds.

The term **genetic model** refers to a set of hypotheses that explains how a particular trait is inherited. **Table 13.2** summarizes Mendel's model for explaining the basic patterns in the transmission of traits from parents to offspring; the hypotheses it lists are sometimes referred to as Mendel's rules. They represent a radical break from the hypotheses of blending inheritance and inheritance of acquired characters that previously dominated scientific thinking about heredity.

13.3 Mendel's Experiments with Two Traits

Working with one trait at a time allowed Mendel to establish that blending inheritance does not occur. It also allowed him to infer that each pea plant had two of each gene and to recognize the principle of segregation.

Mendel's next step extended these results. The important question now was whether the principle of segregation holds true if individuals differ with respect to two traits, instead of just one. Do different genes segregate together, or independently?

The Dihybrid Cross

Mendel crossed a pure-line parent that produced round, yellow seeds with a pure-line parent that produced wrinkled, green seeds. According to his model, the F_1 offspring of this cross should be heterozygous for both genes. A mating between two such individuals—both heterozygous for two traits—is called a **dihybrid cross**.

Mendel's earlier experiments had established that the allele for yellow seeds was dominant to the allele for green seeds; these alleles were designated *Y* for yellow and *y* for green. As **Figure 13.5** indicates, two distinct possibilities existed for how the alleles of these two different genes—the gene for seed shape and the gene for seed color—would be transmitted to offspring.

- The first possibility was that the allele for seed shape and the allele for seed color present in each parent would separate from one another and be transmitted independently. This hypothesis is called independent assortment, because the two alleles would separate and be sorted into gametes independently of each other (Figure 13.5a).
- The second possibility was that the allele for seed shape and the allele for seed color would be transmitted to gametes together. This hypothesis can be called dependent assortment, because the transmission of one allele would depend on the transmission of another (Figure 13.5b).

As Figure 13.5 shows, the F_1 offspring of Mendel's mating are expected to have the dominant round and yellow phenotypes whether the different genes are transmitted together or independently. When Mendel did the cross and observed the F_1 individuals, this is exactly what he found. All of the F_1 offspring had round, yellow seeds.

The two hypotheses make radically different predictions, however, about what will be observed when the F_1 individuals are allowed to self-fertilize and produce an F_2 generation. If the different genes assort independently and combine randomly to form gametes, then each heterozygous parent should produce four different gamete genotypes, as illustrated in Figure 13.5a. A 4-row-by-4-column Punnett square results, and it predicts that there should be 9 different offspring genotypes and 4 phenotypes. Further, the yellow-round, green-round, yellow-wrinkled, and green-wrinkled phenotypes should be present in frequencies of 9/16, 3/16, 3/16, and 1/16, respectively. This is a ratio of 9:3:3:1.

On the other hand, if the alleles from each parent stay together, then a 2-row-by-2-column Punnett square would result and predict only three possible offspring genotypes and just two phenotypes, as Figure 13.5b shows. The hypothesis of dependent assortment predicts that F_2 offspring should be yellow-round or green-wrinkled, present in a ratio of 3:1.

Note that the Punnett squares make explicit predictions about the outcome of an experiment, based on a specific hypothesis

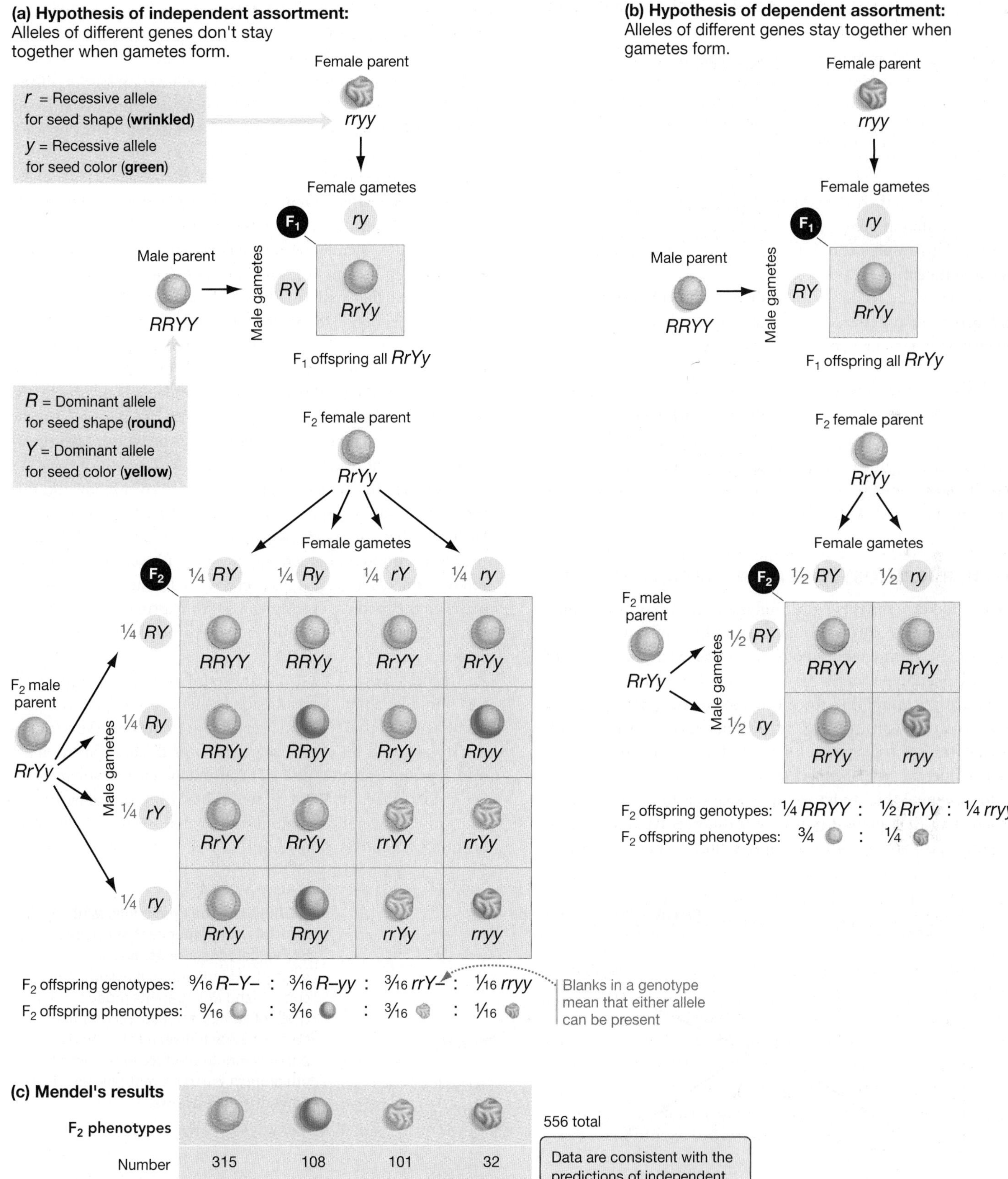

FIGURE 13.5 Mendel Analyzed the F_1 and F_2 Offspring of a Cross between Pure Lines for Two Traits. Each of two hypotheses predicted a different pattern for the outcome when alleles of different genes are transmitted to offspring: The alleles could be sorted into gametes independently of each other, or alleles from the same parent could be transmitted together, generation after generation.

about which alleles are present in each parent and how they are transmitted.

When Mendel examined the phenotypes of the F_2 offspring, he found that they conformed to the predictions of the hypothesis of independent assortment. Four phenotypes were present in frequencies that closely approximated the predicted frequencies of 9/16, 3/16, 3/16, and 1/16 and the predicted ratio of 9:3:3:1 (Figure 13.5c).

Based on these data, Mendel accepted the hypothesis that alleles of different genes are transmitted independently of one another. This result became known as the **principle of independent assortment.**

To review the logic of monohybrid and dihybrid crosses and get more practice doing them, go to the study area at *www.masteringbiology.com.*

MB **Web Activity** Mendel's Experiments

✔If you understand the principle of independent assortment, it should make sense to you that an individual with the genotype *AaBb* produces gametes with the genotypes *AB*, *Ab*, *aB*, and *ab*. You should be able to predict the genotypes of the gametes produced by individuals with the genotypes *AABb*, *PpRr*, and *AaPpRr*.

Using a Testcross to Confirm Predictions

Mendel did experiments with combinations of traits other than seed shape and color and obtained results similar to those in Figure 13.5c. Each paired set of traits produced a 9:3:3:1 ratio of progeny phenotypes in the F_2 generation. He even did a limited set of crosses examining three traits at a time. Although all of these data were consistent with the principle of independent assortment, his most powerful support for the hypothesis came from a different type of experiment.

In designing this study, Mendel's goal was to test the prediction that an *RrYy* plant produces four different types of gametes in equal proportions. To accomplish this, Mendel invented a technique called a testcross. A **testcross** uses a parent that contributes only recessive alleles to its offspring, to help determine the unknown genotype of the second parent.

Testcrosses are useful because the genetic contribution of the homozygous recessive parent is known. As a result, a testcross allows experimenters to test the genetic contribution of the other parent. If the other parent has the dominant phenotype but an unknown genotype, the results of the testcross allow researchers to infer whether that parent is homozygous or heterozygous for the dominant allele.

In this case, Mendel performed a testcross between a parent that was homozygous recessive for seed shape and color (*rryy*) and a parent that had an unknown genotype but was known from its phenotype to possess the dominant *R* and *Y* alleles (could be *RrYy* or *RRYY*). The types and proportions of offspring that could result can be predicted with the Punnett square shown in **Figure 13.6.** If the principle of independent assortment is valid, there should be four types of offspring in equal proportions if the tested parent is *RrYy*, and only one type of offspring if the tested parent is *RRYY*.

What were the actual proportions observed? Mendel did this experiment and examined the seeds produced by the progeny. He found that 1/4 (31 of 110) were round and yellow, 1/4 (26 of 110) were round and green, 1/4 (27 of 110) were wrinkled and yellow, and 1/4 (26 of 110) were wrinkled and green, which matched the predicted proportions for offspring of an *RrYy* parent. The testcross had confirmed the principle of independent assortment.

Mendel's work provided a powerful conceptual framework for thinking about transmission genetics—the patterns that occur as alleles pass from one generation to the next. This framework was based on **(1)** the segregation of discrete, paired genes into separate gametes, and **(2)** the independent assortment of genes that affect different traits. **Table 13.3** summarizes the transmission genetics vocabulary that has been introduced in Section 13.2 and Section 13.3.

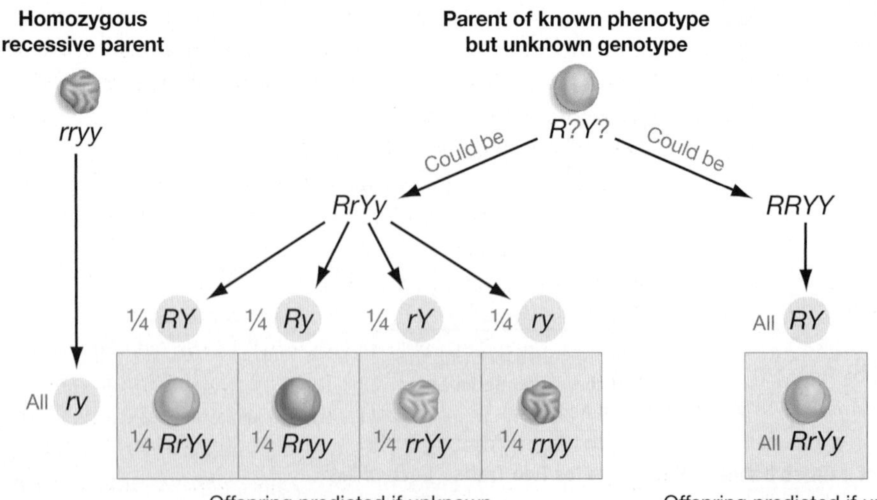

FIGURE 13.6 The Predictions Made by the Principle of Independent Assortment Can Be Evaluated in a Testcross. If the principle of independent assortment is correct, and *RrYy* parents produce four types of gametes in equal proportions, then a mating between *RrYy* and *rryy* parents should produce four types of offspring in equal proportions, as this Punnett square shows.

SUMMARY TABLE 13.3 **Vocabulary Used in Transmission (Mendelian) Genetics**

Term	Definition	Example or Comment
Gene	A hereditary factor that influences a particular trait.	This definition will change in later chapters (become more precise).
Allele	A version of a gene.	Diploid organisms have two alleles of each gene.
Genotype	A listing of the alleles in an individual.	In diploids, the genotype has two alleles of each gene; in haploids, one allele of each gene.
Phenotype	An individual's observable traits.	Influenced, but not dictated, by the genotype.
Homozygous	Having two of the same allele.	Refers to a particular gene.
Heterozygous	Having two different alleles.	Refers to a particular gene.
Dominant allele	An allele that produces the associated phenotype in heterozygotes.	Dominance does not imply high frequency or high fitness.
Recessive allele	An allele that produces the associated phenotype only in homozygotes.	Alleles appear to "recede" or disappear in heterozygotes.
Pure line	Individuals or populations that when crossed with individuals of the pure line, always produce offspring with that phenotype.	Pure-line individuals are homozygous for the gene in question.
Hybrid	Offspring from crosses between homozygous parents with different genotypes.	Offspring are heterozygous.
Reciprocal cross	A cross in which the phenotypes associated with the male and female in a prior cross are reversed.	If reciprocal crosses give identical results, the sex of the parent does not influence transmission of the trait.
Testcross	Experimental cross between a homozygous recessive individual and an individual with the dominant phenotype but an unknown genotype.	Usually used to determine whether a parent with a dominant phenotype is homozygous or heterozygous.

CHECK YOUR UNDERSTANDING

If you understand that . . .

- Mendel discovered that individuals have two alleles of each gene, and that each gamete receives one of the two alleles present in a parent. This is the principle of segregation.
- Mendel found that alleles of different genes are transmitted to gametes independently of each other. This is the principle of independent assortment.
- The alleles that Mendel analyzed were either dominant or recessive, meaning heterozygous individuals had the dominant phenotype.

✔ **You should be able to . . .**

Use the genetic problems at the end of this chapter to practice the following skills:

1. Starting with parents of known genotypes, create and analyze Punnett squares to predict the genotypes and phenotypes that will occur in their F_1 and F_2 offspring, and then calculate the expected frequency of each genotype and phenotype. (Do **Genetics Problems 2, 6, 8, 10.**)
2. Given the outcome of a cross, infer the genotypes and phenotypes of the parents. (Do **Genetics Problems 3, 4, 13.**)

Answers are available in Appendix B.

The experiments you've just reviewed were brilliant in design, execution, and interpretation. Unfortunately, they were ignored for 34 years.

13.4 The Chromosome Theory of Inheritance

Historians of science debate why Mendel's work was overlooked for so long. It is probably true that his use of ratios and proportions were difficult for biologists of that time to understand and absorb. It may also be true that the theory of blending inheritance was so well entrenched that there was a tendency to dismiss his results as peculiar or unbelievable.

Whatever the reason, Mendel's work was not appreciated until 1900, when three groups of biologists, working with a wide variety of plants and animals, independently "discovered" Mendel's work and reached the same main conclusions.

The rediscovery of Mendel's work, 16 years after his death, ignited the young field of genetics. Mendel's experiments established the basic patterns of how traits are passed from parents to offspring. But what process is responsible for these patterns? Two biologists, working independently, came up with the answer. Walter Sutton and Theodor Boveri each realized that meiosis could be responsible for Mendel's rules. When this hypothesis was published in 1902, research in genetics exploded.

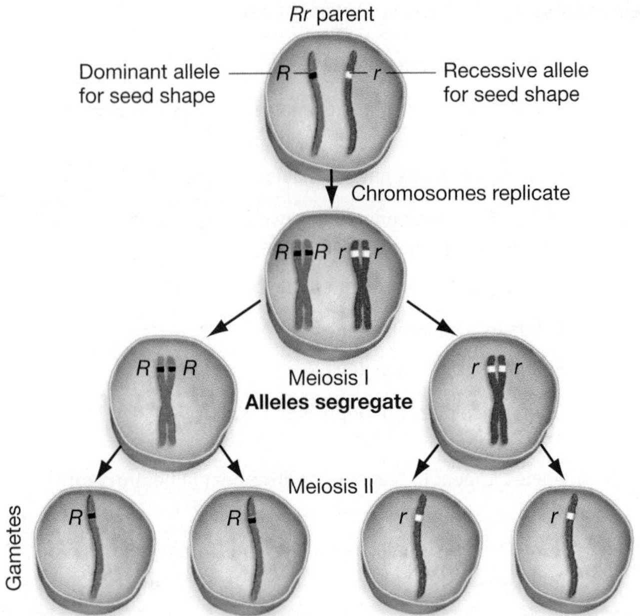

FIGURE 13.7 Meiosis Is Responsible for the Principle of Segregation. The two members of a parent's gene pair segregate into different gametes, as Mendel hypothesized, because homologous chromosomes separate during meiosis I.

Meiosis Explains Mendel's Principles

Recall from Chapter 12 that meiosis precedes gamete formation. The details of the process were worked out late in the nineteenth century. What Sutton and Boveri grasped is that meiosis not only reduces chromosome number by half but also explains the principle of segregation and the principle of independent assortment.

The cell at the top of **Figure 13.7** illustrates Sutton and Boveri's central insight: Chromosomes are composed of Mendel's hereditary determinants, or genes. In this example, the gene for seed shape is shown at a particular position along a certain chromosome. This location is known as a **locus** ("place"; plural, **loci**). A genetic locus is the physical location of a gene.

The paternal and maternal chromosomes shown in Figure 13.7 happen to possess different alleles of the gene for seed shape: One allele specifies round seeds (*R*), while the other specifies wrinkled seeds (*r*).

The subsequent steps in Figure 13.7 show how these alleles segregate into different daughter cells during meiosis I, when homologous chromosomes separate. The physical separation of alleles during anaphase of meiosis I is responsible for Mendel's principle of segregation.

Figure 13.8 follows the fates of the members of two different gene pairs—in this case, for seed shape and seed color—as meiosis proceeds. If the alleles for different genes are located on different nonhomologous chromosomes, they assort independently of

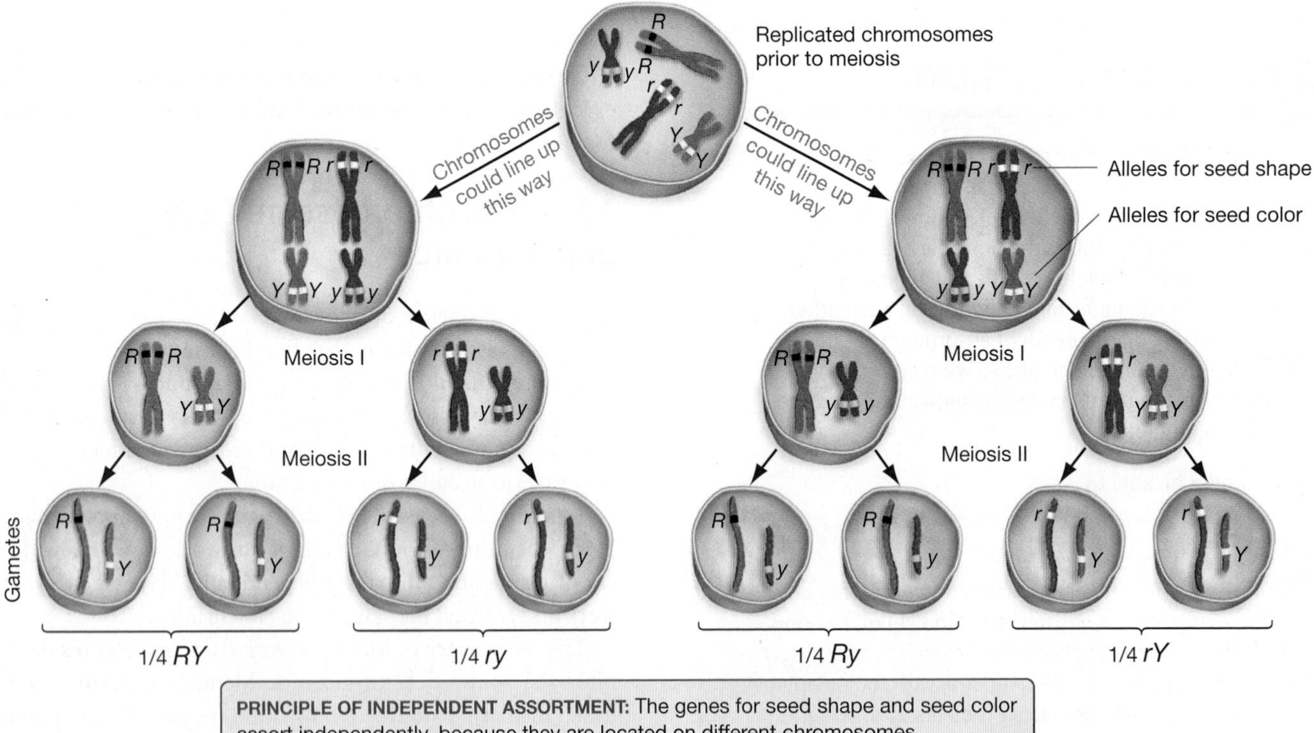

FIGURE 13.8 Meiosis Is Responsible for the Principle of Independent Assortment. The genes for different traits assort independently, as Mendel hypothesized, because nonhomologous chromosomes assort independently during meiosis I. Maternal and paternal chromosomes are shown in different colors for clarity.

one another at meiosis I. Four types of gametes, produced in equal proportions, result. This is the physical basis of Mendel's principle of independent assortment.

Sutton and Boveri formalized these observations in the **chromosome theory of inheritance**. Like other theories in biology, the chromosome theory describes a predictable pattern—a set of observations about the natural world—and a process that explains the pattern. The chromosome theory states that Mendel's rules can be explained by the independent alignment and separation of homologous chromosomes at meiosis I.

To review why meiosis explains Mendel's findings, go to the study area at *www.masteringbiology.com*.

Web Activity The Principle of Independent Assortment

When Sutton and Boveri published their ideas, however, the hypothesis that chromosomes consist of genes was untested. What experiments confirmed that chromosomes contain genes?

Testing the Chromosome Theory

During the first decade of the twentieth century, an unassuming insect rose to prominence as a model organism for testing the chromosome theory of inheritance. This organism—the fruit fly *Drosophila melanogaster*—has been at the center of genetic studies ever since (see **BioSkills 14** in Appendix A).

Drosophila melanogaster has all the attributes of a useful model organism for experimental studies in genetics: small size, ease of rearing in the lab, a short generation time (about 10 days), and abundant offspring (up to a few hundred per mating). The elaborate external anatomy of this insect also makes it possible to identify interesting phenotypic variation among individuals (**Figure 13.9a**).

Drosophila was adopted as a model organism by Thomas Hunt Morgan and his students. But because *Drosophila* is not a domesticated species like the garden pea, Morgan was not familiar with common phenotypic variants such as Mendel's round and wrinkled seeds. Consequently, an early goal of Morgan's research was simply to find and characterize individuals with different phenotypes. In these studies, the most common phenotype for each trait was referred to as **wild type**.

THE WHITE-EYE MUTANT At one point, Morgan discovered a male fly that had white eyes rather than the wild-type red eyes (**Figure 13.9b**). This individual had a discrete and easy-to-recognize phenotype different from the normal phenotype.

Morgan inferred that the white-eyed phenotype resulted from a **mutation**—a change in a gene (in this case, a gene that affects eye color). Individuals with white eyes (or other traits attributable to mutation) are referred to as **mutants**.

To explore how the white-eye trait is inherited in fruit flies, Morgan mated a red-eyed female fly with the mutant white-eyed male fly. All of the F_1 progeny had red eyes. But when Morgan did the reciprocal cross, by mating white-eyed females to red-eyed males, he got a different result: All F_1 females had red eyes, but all F_1 males had white eyes.

(a) The fruit fly *Drosophila melanogaster*

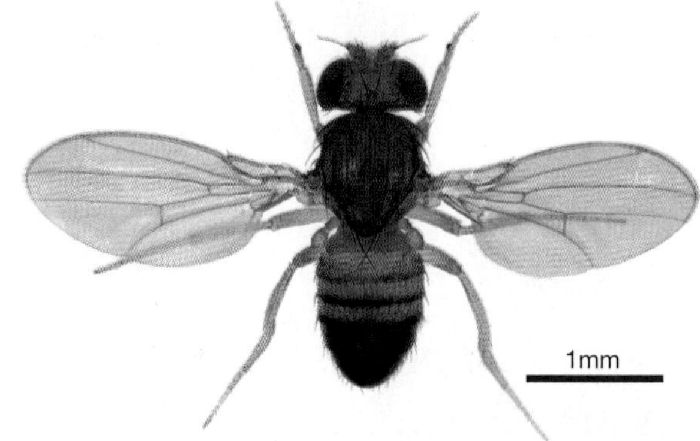

(b) Eye color is a variable trait.

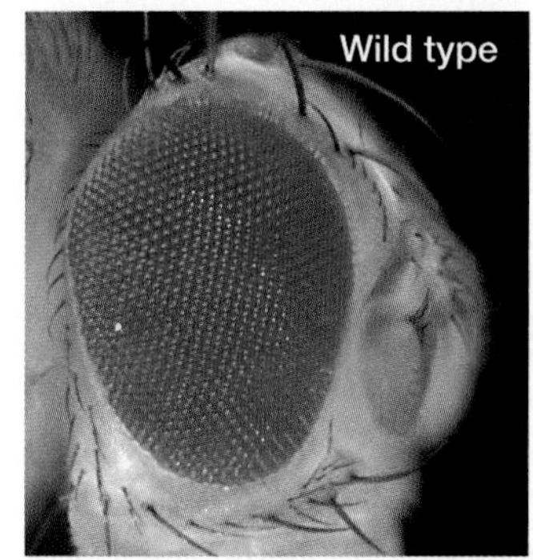

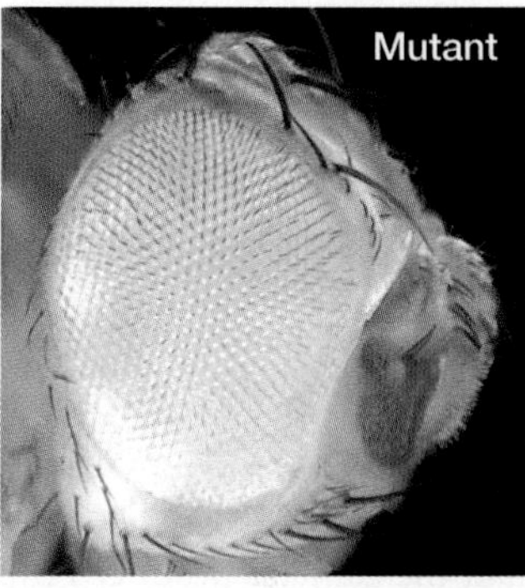

FIGURE 13.9 The Fruit Fly *Drosophila melanogaster* Is an Important Model Organism in Genetics.

Recall that Mendel's reciprocal crosses had always given results that were similar to each other. But Morgan's reciprocal crosses did not. The experiment suggested a definite relationship between the sex of the progeny and the inheritance of eye color. What was going on?

THE DISCOVERY OF SEX CHROMOSOMES Nettie Stevens began studying the karyotypes of insects about the time that Morgan began his work with *Drosophila*. In the beetle *Tenebrio molitor*, she noticed a striking difference in the chromosome complements of males and females. In females of this species, diploid cells contain 20 large chromosomes. But diploid cells in males contain 19 large chromosomes and 1 small one.

Stevens called the small chromosome the Y chromosome. This Y chromosome paired with one of the large chromosomes at meiosis I, which had already been named the X chromosome. The X and Y were different in size and shape, but they acted like homologs during meiosis. Later work showed that even though X and Y chromosomes contain different genes, they have regions that are similar enough to lead to proper pairing during prophase of meiosis I. The X and Y are now called sex chromosomes. Female beetles are XX; males are XY.

SEX LINKAGE AND THE CHROMOSOME THEORY *Drosophila* females, like *Tenebrio* females, have two X chromosomes; male fruit flies carry an X and a Y, just as *Tenebrio* males do. As a result,

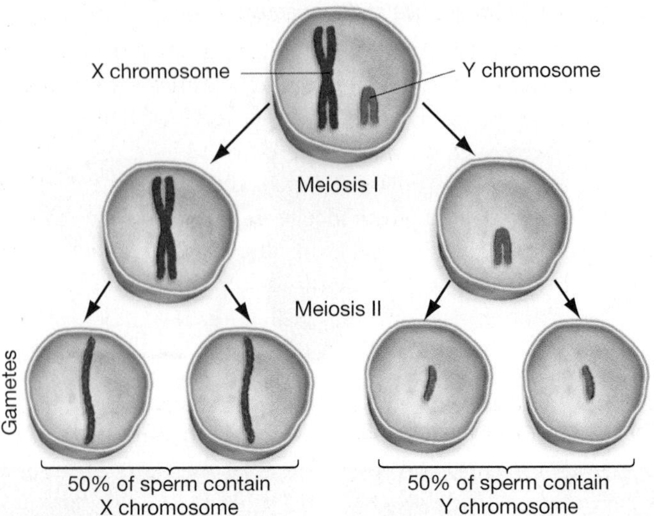

FIGURE 13.10 Sex Chromosomes Pair during Meiosis I, Then Segregate to Form X-Bearing and Y-Bearing Gametes. Sex chromosomes synapse at meiosis I in male fruit flies, even though the X and Y chromosomes differ in size and shape. No crossing over occurs between the X and Y in fruit flies. Thus, half the sperm cells that result from meiosis bear an X chromosome; half have a Y chromosome.

half the gametes produced by a male fruit fly should carry an X chromosome; the other half, a Y chromosome (**Figure 13.10**).

Morgan realized that the transmission pattern of the X chromosome in males and females explained the results of his reciprocal crosses. Specifically, he proposed that the gene for white eye color in fruit flies is located on the X chromosome and that the Y chromosome does not carry an allele of this gene.

This situation is described as **X-linked inheritance**, or simply **X-linkage**. Correspondingly, a gene residing on the Y chromosome is said to have **Y-linked inheritance**, or **Y-linkage**. The general term for such inheritance (of genes on either sex chromosome) is **sex-linked inheritance**, or **sex-linkage**.

According to the hypothesis of X-linkage, a female fruit fly has two copies of the gene that specifies eye color because she has two X chromosomes. One of these chromosomes came from her female parent, and the other from her male parent. A male, in contrast, has only one copy of the eye-color gene because he has only one X chromosome, inherited from his mother.

The Punnett squares in **Figure 13.11** show that Morgan's experimental results can be explained if the gene for eye color is located on the X chromosome, and if the allele for red color is dominant to the allele for white color. In this figure, the allele for red eyes is denoted X^W while the allele for white eyes is denoted X^w. The Y chromosome present in males is simply designated by Y. Using this notation,[1] the genotypes used in the experiment are written as X^WX^W for red-eyed females; X^wY for white-eyed males; X^wX^w for white-eyed females; and X^WY for red-eyed males. If you study the offspring genotypes, you should see that the results predicted by the hypothesis of X-linkage match the observed results.

When reciprocal crosses give different results, such as those illustrated in Figure 13.11, it is likely that the gene in question is located on a sex chromosome. Recall from Chapter 12 that non-sex chromosomes are called autosomes. Genes on non-sex chromosomes are said to show **autosomal inheritance**.

Morgan's discovery of X-linked inheritance carried an even more fundamental message. In *Drosophila*, the gene for white eye color is clearly correlated with inheritance of the X chromosome. This correlation was important evidence in support of the hypothesis that chromosomes contain genes. The discovery of X-linked inheritance convinced most biologists that the chromosome theory of inheritance was correct.

[1]Scientific papers on fruit fly genetics use a different notation. The wild-type allele is designated with a superscript +, and no X is used for X-linked traits. The red-eye allele, for example, is denoted w^+; the white-eye allele *w*. The notation used here is simplified for use in introductory courses and conforms to conventions used in human genetics.

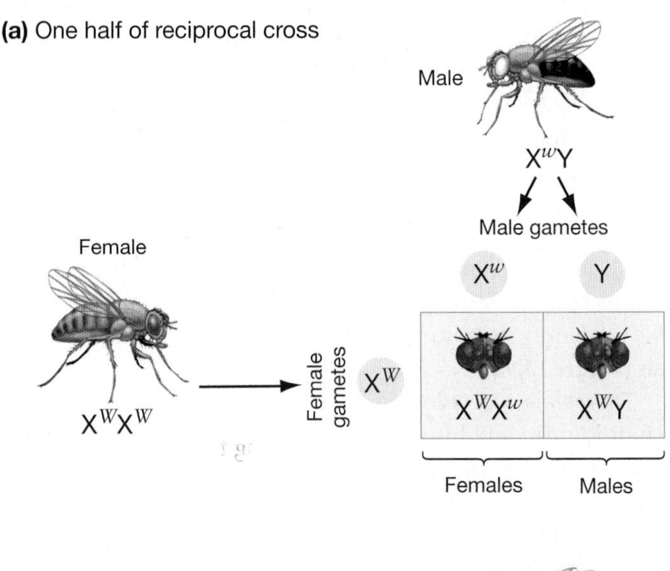

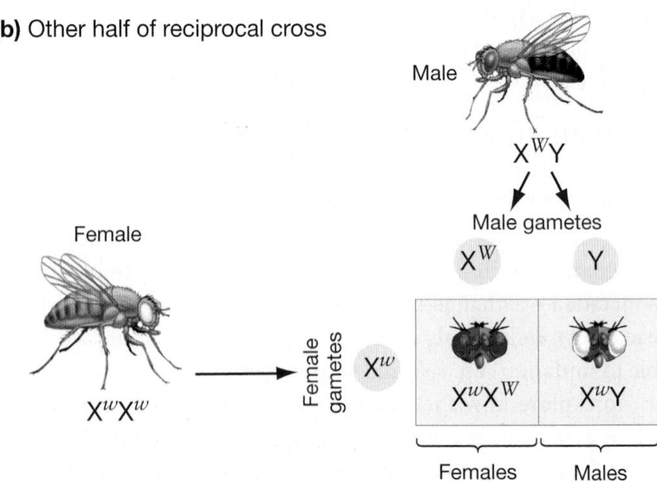

FIGURE 13.11 Reciprocal Crosses Confirm That Eye Color in *Drosophila* Is an X-Linked Trait. When Morgan crossed red-eyed females with white-eyed males and then crossed white-eyed females with red-eyed males, he observed strikingly different results.

CHECK YOUR UNDERSTANDING

If you understand that . . .

- Meiosis is the process responsible for Mendel's principle of segregation. It occurs because alleles on homologous chromosomes separate at anaphase of meiosis I.
- Meiosis is the process responsible for Mendel's principle of independent assortment. Alleles of different genes go to gametes independently because pairs of homologous chromosomes line up randomly at metaphase of meiosis I.

You should be able to . . .

1. Draw the chromosomes involved in a cross between *Pp* and *Pp* peas, and use the diagram to explain the segregation of alleles.
2. Draw the chromosomes involved in a cross between *YyRr* and *YyRr* peas, and use the diagram to explain the independent assortment of alleles.

Answers are available in Appendix B.

13.5 Extending Mendel's Rules

Biologists point out that Mendel analyzed the simplest possible genetic system. The traits that he was studying were not sex-linked. Moreover, they were influenced by just two alleles of a single gene, and each allele was completely dominant or recessive.

With these genes, Mendel was able to discover the most fundamental rules of inheritance. Investigating simple model systems is an extremely important research strategy in biological science. Researchers almost always choose to analyze the simplest situation possible before going on to explore more complicated systems. Mendel probably would have failed, as so many others had done before him, had he been trying to analyze more complex patterns of inheritance.

Once Mendel's work was rediscovered, researchers began to analyze traits and alleles whose inheritance was more complicated. If experimental crosses produced F_2 progeny that did not conform to the expected 3:1 or 9:3:3:1 ratios, researchers had a strong hint that something interesting was going on. The discovery of sex-linkage is a prominent example. How can other traits that don't appear to follow Mendel's rules contribute to a more complete understanding of heredity?

Linkage: What Happens When Genes Are Located on the Same Chromosome?

Once the chromosome theory had been tested and supported, biologists began to reevaluate Mendel's principle of independent assortment. The key issue was that genes should not undergo independent assortment if they are located on the same chromosome.

The physical association among genes on the same chromosome is called **linkage**. Notice that the terms linkage and sex-linkage are different in meaning. If two or more genes are linked, it means that they are located on the same chromosome. If a single gene is sex-linked, it means that it is located on a sex chromosome.

The first examples of linked genes were those on the X chromosome of fruit flies. After Morgan established that the white-eye gene was located on *Drosophila*'s X chromosome, he and colleagues established that one of the several genes that affects body color is also located on the X. Red eyes and gray body are the wild-type phenotypes in this species; white eyes and a yellow body occur as rare mutant phenotypes. The alleles for red eyes (X^W) and gray body (X^Y) are dominant to the alleles for white eyes (X^w) and yellow body (X^y). (Be sure not to confuse the notation for the Y chromosome in males, Y, with the yellow body allele, X^y.)

LINKED GENES AS AN EXCEPTION TO INDEPENDENT ASSORTMENT Because linked genes are physically part of the same chromosome, it is logical to predict that they should always be transmitted together during gamete formation. Stated another way, linked genes should violate the principle of independent assortment.

Recall from Section 13.4 that independent assortment is observed when genes are on different chromosomes, because the alleles of unlinked genes segregate to gametes independently of one another during meiosis I. But when genes are on the same chromosome, their alleles are carried to gametes together.

Figure 13.12 shows that a female fruit fly with one X chromosome carrying the white eye and gray body alleles, written X^{wY}, and with a second X chromosome carrying the red eye and yellow body alleles, written X^{Wy}, would be expected to generate just two classes of gametes in equal numbers during meiosis, instead of the four classes that are predicted under the principle of independent assortment. Is this what actually occurs?

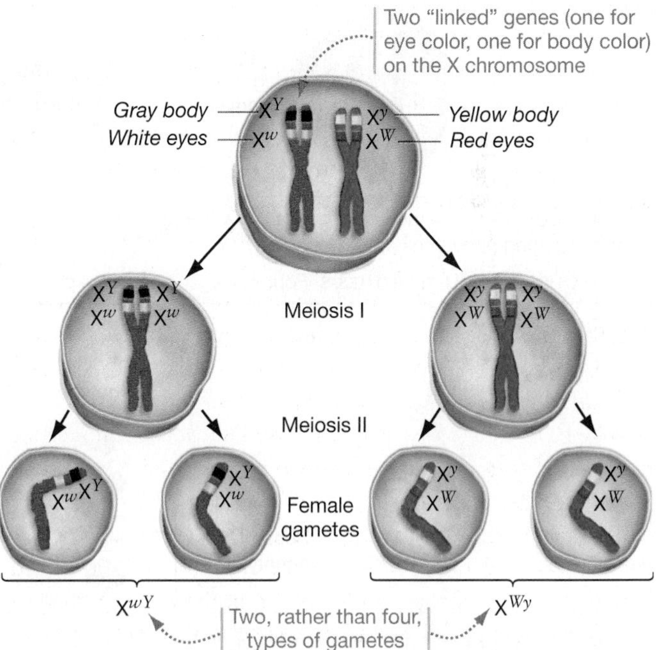

FIGURE 13.12 Linked Genes Are Inherited Together. If the *w* and *y* genes were found on different chromosomes, then this female would generate four different types of gametes instead of just two types as shown here.

EXERCISE List the four genotypes that would be generated if the white and yellow genes were not linked.

THE ROLE OF CROSSING OVER To determine whether linked traits behave as predicted, Morgan performed crosses like the one described in the "Experimental setup" section of **Figure 13.13**. In this case, $X^{wY}X^{Wy}$ females mated with $X^{wY}Y$ males.

The Results table in Figure 13.13 summarizes the phenotypes and genotypes observed in the male offspring of this experimental cross.

- Most of these males carried an X chromosome with one of the two combinations of alleles found in their mothers: X^{wY} or X^{Wy}. In these individuals, the *white* and *yellow* alleles did not segregate independently.
- A small percentage of males had novel phenotypes and genotypes: X^{wy} and X^{WY}. Morgan referred to these individuals as **recombinant** because the combination of alleles on their X chromosome was different from the combinations of alleles present in the parental generation.

To explain this result, Morgan proposed that gametes with new, recombinant genotypes were generated when crossing over occurred during prophase of meiosis I in the females.

Recall from Chapter 12 that crossing over involves a physical exchange of segments from homologous chromosomes. Crossing over occurs at least once in every synapsed pair of homologous chromosomes, and usually multiple times. (Male fruit flies are an exception to this rule. For unknown reasons, no crossing over occurs in male fruit flies.)

As **Figure 13.14** shows, a crossing-over event occurred somewhere between the *white* and *yellow* genes in the $X^{wY}X^{Wy}$ females. The recombinant chromosomes that resulted would have the genotypes X^{wy} and X^{WY} (see the middle of the bottom row of the figure). If these chromosomes ended up in a male offspring, they would produce individuals with yellow bodies and white eyes, along with individuals with gray bodies and red eyes. This is exactly what Morgan observed.

As **Box 13.1** on page 246 explains, data on the percentage of recombinant offspring that occur in crosses like the one diagrammed in Figure 13.14 can be used to estimate the location of genes, relative to one another, on the same chromosome. Data on the frequency of crossing over can be used to create a **genetic map**—a diagram showing the relative positions of genes along a particular chromosome.

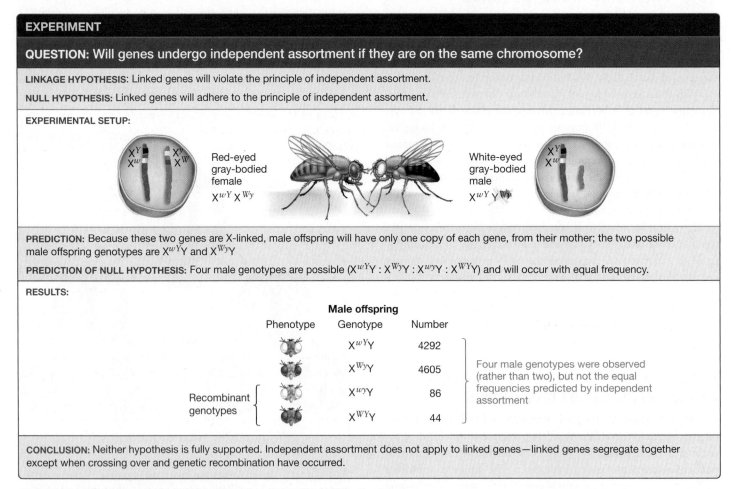

EXPERIMENT

QUESTION: Will genes undergo independent assortment if they are on the same chromosome?

LINKAGE HYPOTHESIS: Linked genes will violate the principle of independent assortment.

NULL HYPOTHESIS: Linked genes will adhere to the principle of independent assortment.

EXPERIMENTAL SETUP:

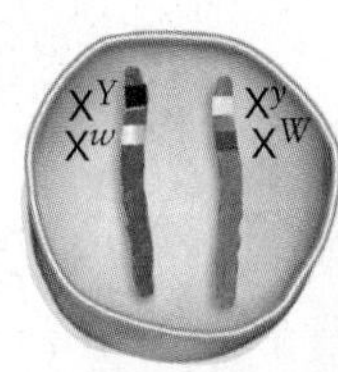

Red-eyed gray-bodied female $X^{wY}X^{Wy}$

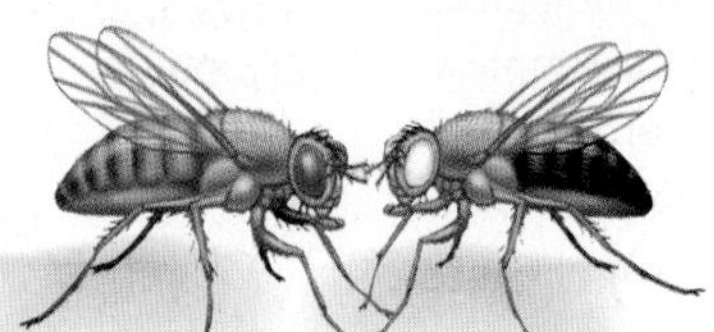

White-eyed gray-bodied male X^{wY} Y

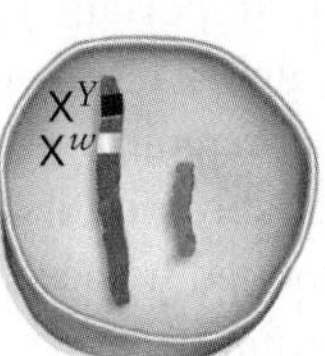

PREDICTION: Because these two genes are X-linked, male offspring will have only one copy of each gene, from their mother; the two possible male offspring genotypes are $X^{wY}Y$ and $X^{Wy}Y$

PREDICTION OF NULL HYPOTHESIS: Four male genotypes are possible ($X^{wY}Y$: $X^{Wy}Y$: $X^{wy}Y$: $X^{WY}Y$) and will occur with equal frequency.

RESULTS:

Male offspring

	Phenotype	Genotype	Number
		$X^{wY}Y$	4292
		$X^{Wy}Y$	4605
Recombinant genotypes		$X^{wy}Y$	86
Recombinant genotypes		$X^{WY}Y$	44

Four male genotypes were observed (rather than two), but not the equal frequencies predicted by independent assortment

CONCLUSION: Neither hypothesis is fully supported. Independent assortment does not apply to linked genes—linked genes segregate together except when crossing over and genetic recombination have occurred.

FIGURE 13.13 Linked Genes Are Inherited Together Unless Recombination Occurs.

SOURCE: Morgan, T. H. 1911. An attempt to analyze the constitution of the chromosomes on the basis of sex-limited inheritance in *Drosophila*. *Journal of Experimental Zoology* 11: 365–414.

✔**QUESTION** Why didn't they observe equal numbers of white-eyed, yellow-bodied males and red-eyed, gray-bodied males?

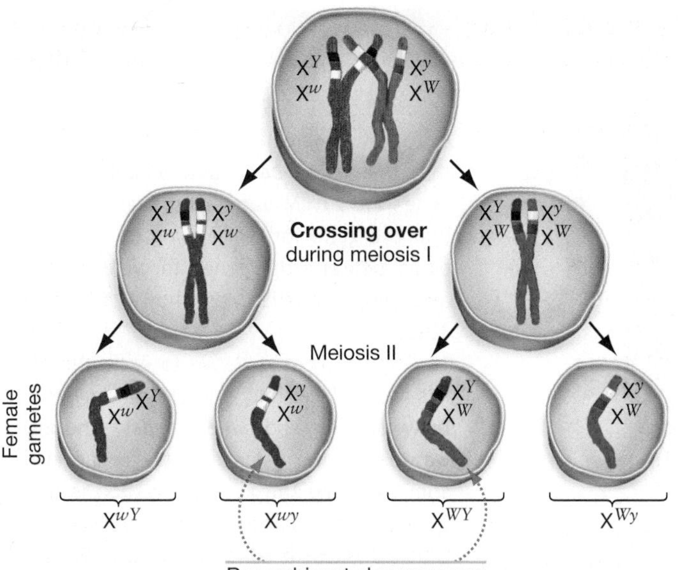

FIGURE 13.14 Genetic Recombination Results from Crossing Over. To explain the results in Figure 13.13, Morgan proposed that crossing over occurred between the *w* gene and the *y* gene in a small percentage of F_1 females during meiosis I. The recombinant chromosomes that resulted would produce the recombinant phenotypes observed in F_2 males.

The take-home message of Morgan's experiments is simple: Linked genes are inherited together unless crossing over occurs between them. When crossing over takes place, genetic recombination occurs. Linkage is an important exception to Mendel's rules.

Do Heterozygotes Always Have a Dominant or Recessive Phenotype?

The terms dominant and recessive describe which phenotype is observed when two different alleles of a gene occur in the same individual. In all seven traits that Mendel studied, only the phenotype associated with one allele—the "dominant" one—appeared in heterozygous individuals.

Not all combinations of alleles produce a completely dominant or recessive phenotype, however. In many cases, the actual phenotypes observed in heterozygous individuals conflict with the genotype and phenotype ratios that Mendel observed.

INCOMPLETE DOMINANCE Consider the flowers called four-o'-clocks, pictured in **Figure 13.15a**. In this species, biologists have developed a pure line that has purple flowers and a pure line that has white flowers. When individuals from these strains are mated, all of their offspring are lavender (**Figure 13.15b**). In Mendel's peas, crosses between purple- and white-flowered parents produced all purple-flowered offspring. Why the difference?

Biologists answered this question by examining the phenotypes of F_2 four-o'clocks. These are the progeny of self-fertilization in lavender-flowered F_1 individuals. Of the F_2 plants, 1/4 have purple flowers, 1/2 have lavender flowers, and 1/4 have white flowers. This 1:2:1 ratio of phenotypes is unlike any we have seen to date, but it exactly matches the 1:2:1 ratio of genotypes

(a) Flower color is variable in four-o'clocks.

(b) Incomplete dominance in flower color

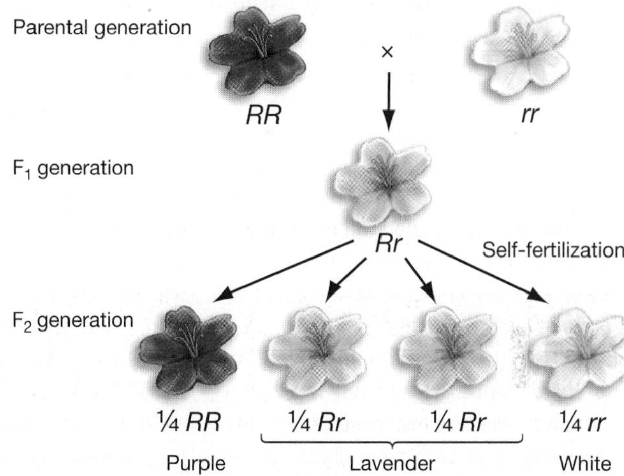

FIGURE 13.15 When Incomplete Dominance Occurs, Heterozygotes Have Intermediate Phenotypes. This cross is explained by hypothesizing that a single gene influences flower color, with alleles *R* and *r*, exhibiting incomplete dominance.

that is produced when flower color is controlled by one gene with two alleles.

To convince yourself that this explanation is sound, study the genetic model shown in Figure 13.15. According to the diagram, the inheritance of flower color genotypes in four-o'clocks and peas is identical, but the four-o'clock alleles show incomplete dominance rather than complete dominance.

When **incomplete dominance** occurs, heterozygotes have an intermediate phenotype. In the case of four-o'clocks, neither purple nor white alleles dominate. Instead, the F_1 progeny—all heterozygous—show a phenotype intermediate between the two parental strains.

Incomplete dominance illustrates an important general point: Dominance is not necessarily an all-or-none phenomenon.

CODOMINANCE Many alleles show a relationship called **codominance**. When codominance occurs, heterozygotes have the phenotype associated with each individual allele.

As an example, consider the ABO blood group in humans. The gene in question alters the polysaccharide attached to a glycoprotein (see Chapter 5) found in the plasma membranes of red blood cells. Different alleles of the *I* gene lead to the production

BOX 13.1 QUANTITATIVE METHODS: Linkage

In experiments like the one diagrammed in Figure 13.13, researchers calculate the recombination frequency as the number of offspring with recombinant phenotypes divided by the total number of offspring. With crosses involving the X-linked traits of white eyes and yellow bodies, about 1.4 percent of offspring have recombinant phenotypes and genotypes.

But in crosses with different pairs of X-linked traits, the fraction of recombinants varied. In crosses of fruit flies with X-linked genes for white eyes and a mutant phenotype called singed bristles, for example, males with recombinant chromosomes were produced about 19.6 percent of the time.

To explain these observations, Morgan proposed that genes are located in a linear array along a chromosome, and the physical distance between genes determines how frequently crossing over occurs between them. His idea was that crossing over occurs at random and can take place at locations all along the length of the chromosome. The shorter the distance between any two genes on a chromosome, the lower the probability that crossing over will take place somewhere in between (**Figure 13.16**).

A. H. Sturtevant, an undergraduate who was studying with Morgan, realized that he could figure out where genes on the same chromosome are in relation to each other based on the frequency of recombinants between various pairs. He set out to create a genetic map.

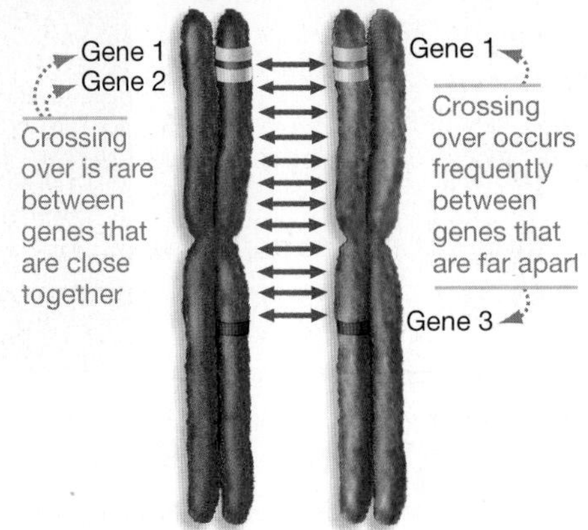

FIGURE 13.16 The Physical Distance between Genes Determines the Frequency of Crossing Over.

To define the unit of distance on his genetic map, Sturtevant used the percentage of offspring that have recombinant phenotypes with respect to two genes. He called this unit the centiMorgan (cM). One map unit, or 1 cM, represents the physical distance that produces 1 percent recombinant offspring.

The eye-color and bristle-shape genes of fruit flies are 19.6 cM apart on the X chromosome, because recombination between these genes results in 19.6 percent recombinant offspring, on average. The genes for yellow body and white eye color, in contrast, are just 1.4 cM apart.

Where is the *yellow-body* gene relative to the *singed-bristles* gene? Recombinants occurred in 21 percent of the gametes produced by females that are $X^{Ys}X^{yS}$, meaning that the *yellow-body* and *singed-bristles* genes are 21 cM apart. Sturtevant inferred that the gene for white eyes must be located *between* the genes for yellow body and singed bristles, as shown in **Figure 13.17a**.

Mapping genes relative to each other is like fitting pieces into a puzzle: placing *white* between *yellow* and *singed bristles* is the only way to make the distances between each pair sum correctly. The key observation is that 21 cM—the distance between *yellow* and *singed bristles*—is equal to 1.4 cM + 19.6 cM, or the sum of the distances between *yellow* and *white* and *white* and *singed bristles*.

Mapping is more complex when genes are farther apart. For example, crossing over occurs 50 percent of the time between genes that are 50 cM apart, making recombination indistinguishable from independent assortment. Double crossovers also become more common. To cope, researchers map genes that are relatively close, then map them to other nearby genes to gradually move down the chromosome.

Figure 13.17b provides a partial genetic map of the X chromosome in *Drosophila melanogaster*, along with the data on which the map positions are based. Using this logic and similar data, Sturtevant assembled the first genetic map.

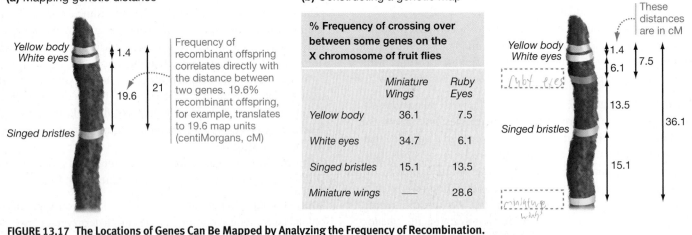

(a) Mapping genetic distance

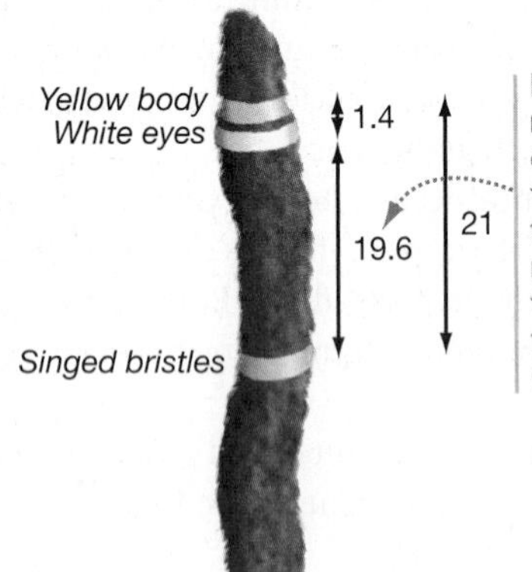

(b) Constructing a genetic map

% Frequency of crossing over between some genes on the X chromosome of fruit flies

	Miniature Wings	*Ruby Eyes*
Yellow body	36.1	7.5
White eyes	34.7	6.1
Singed bristles	15.1	13.5
Miniature wings	—	28.6

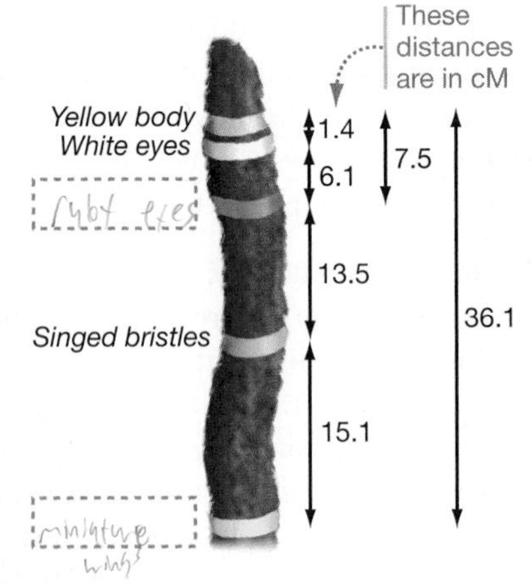

FIGURE 13.17 The Locations of Genes Can Be Mapped by Analyzing the Frequency of Recombination.
(a) The *yellow body* gene is known to be on the end of the fruit fly X chromosome. To explain the recombination frequencies observed in experimental crosses, the *yellow body, white eyes,* and *singed bristles* genes must be in the locations shown here. **(b)** A partial genetic map of the X chromosome in fruit flies.

✔**EXERCISE** In part (b), label the orange and blue genes. (Which is *ruby* and which is *miniature wings?*)

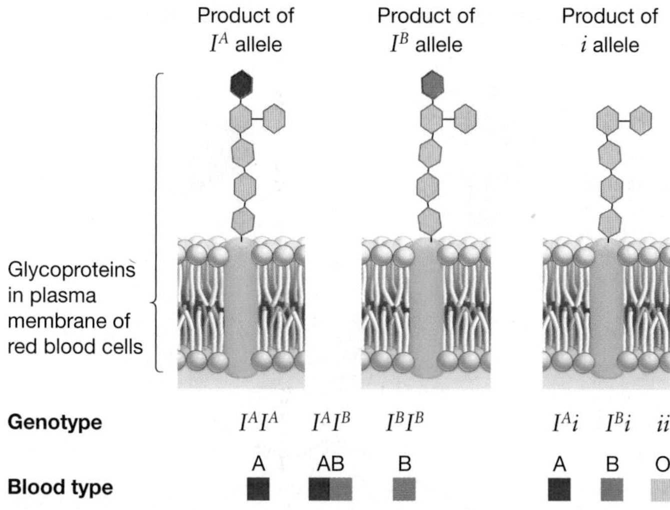

FIGURE 13.18 Phenotypes Produced by Alleles Responsible for ABO Blood Types. Alleles I^A and I^B produce a codominant phenotype when paired with each other in heterozygotes, but both produce a dominant phenotype when paired with allele *i*.

of membrane glycoproteins with different polysaccharides (**Figure 13.18**). I^AI^A individuals have blood type A—meaning that their red blood cells only have the A-type glycoprotein in their membranes. Similarly, I^BI^B individuals have blood type B. But I^AI^B individuals have blood type AB. The phenotypes associated with both alleles are present because cells have both glycoproteins on their surfaces. The alleles are codominant.

There is also an additional allele of this gene called *i*. I^Ai individuals have blood type A; I^Bi individuals have blood type B; *ii* individuals have blood type O—meaning that they lack both A and B glycoproteins on the surfaces of the red blood cells. This is striking. A and B produce a codominant phenotype when paired with each other in heterozygotes, but both produce a dominant phenotype when paired with *i*. Dominance relationships vary among alleles.

How Many Alleles and Phenotypes Exist?

Mendel worked with a total of seven traits and just 14 alleles—two for each trait. In most populations, however, it's not unusual to find dozens of alleles of a single gene. The existence of more than two alleles of the same gene is known as **multiple allelism**.

The ABO blood group in humans is multiallelic, because most populations have the I^A, I^B, and *i* alleles. As a more dramatic example, consider the gene for the β-globin protein in humans. This protein makes up part of hemoglobin, which carries oxygen from the lungs to tissues. Biologists have now identified over 500 different alleles of the β-globin gene. Many of these alleles are associated with distinctive phenotypes. Some alleles produce polypeptides with normal oxygen-carrying capacity, while others lead to reduced oxygen-carrying capacity and various types of anemia. Still other β-globin alleles are associated with adaptation to living at high altitudes, decreased stability at high temperatures, or resistance to the parasites that infect red blood cells and cause malaria.

When more than two distinct phenotypes are present in a population due to multiple allelism, the trait is **polymorphic** ("many-formed"). Oxygen-carrying capacity in humans is a highly polymorphic trait, because many alleles exist for the β-globin gene.

If you were studying the inheritance of β-globin type in humans by tracking the results of a large number of matings, multiple allelism and polymorphism would make it extremely unlikely to observe the 3:1 ratios that Mendel did. Instead, you'd observe many more than two phenotypes, in unpredictable proportions.

Does Each Gene Affect Just One Trait?

As far as is known, the alleles that Mendel analyzed affect just a single trait. The gene for seed color in garden peas, for example, does not appear to affect other aspects of the individual's phenotype. In contrast, many cases have been documented in which a single allele affects a wide variety of traits.

A gene that influences many traits, rather than just one trait, is said to be **pleiotropic** ("more-turning"). The gene responsible for **Marfan syndrome** in humans, called *FBN1*, is a good example. Although current research suggests that just a single gene is involved, individuals with Marfan syndrome exhibit a wide array of phenotypic effects: increased height, disproportionately long limbs and fingers, an abnormally shaped chest, and potentially severe heart problems. A large percentage of these individuals also suffer from problems with their backbone. The gene associated with Marfan syndrome is pleiotropic.

Pleiotropy is common. In many cases a change in a single allele affects more than one trait.

Are Phenotypes Determined by Genes?

After analyzing the results of Mendel's experiments, it would be tempting to conclude that *R* alleles dictate that seeds are round and *T* alleles dictate that individual plants are tall—that there is a strict correspondence between alleles and phenotypes.

It's important to recognize, though, that when Mendel analyzed height in his experiments, he ensured that each plant received a similar amount of sunlight and grew in similar soil. This was important because even individuals with alleles for tallness will be stunted if they are deprived of nutrients, sunlight, or water—so much so that they look similar to individuals with alleles for dwarfing. For Mendel to analyze the hereditary determinants of height, he had to control the environmental determinants of height. Let's consider how two aspects of the environment affect phenotypes: (**1**) the individual's physical surroundings and (**2**) the alleles present at other gene loci.

THE PHYSICAL ENVIRONMENT EFFECTS PHENOTYPES The phenotypes produced by most genes and alleles are strongly affected by the individual's physical environment. Consequently, an individual's phenotype is often as much a product of the physical environment as it is a product of the genotype. To capture this point, biologists refer to the combined effect of genes and environment as gene-by-environment interaction.

Gene-by-environment interactions have a profound effect on how physicians treat people with the genetic disease **phenylketonuria (PKU)**. These individuals lack an enzyme that helps convert the amino acid phenylalanine to the amino acid tyrosine. As a result, phenylalanine and a related molecule, phenylpyruvic acid, accumulate in their bodies. The molecules interfere with the development of the nervous system and produce profound mental retardation. But if PKU individuals are identified at birth and placed on a low-phenylalanine diet, then they develop normally. In many countries, newborns are routinely tested for the defect.

PKU is a genetic disease, but it is neither inevitable nor invariant. Through a simple change in their environment (their diet), individuals with a PKU genotype can have a normal phenotype.

INTERACTIONS BETWEEN GENES HAVE A PROFOUND EFFECT ON PHENOTYPES In Mendel's pea plants, a single locus influenced seed shape. Furthermore, Mendel's data showed that the seed-shape phenotype does not appear to be affected by the action of genes for seed color, seed-pod color, seed-pod shape, or other traits. The pea seeds he analyzed were round or wrinkled regardless of the types of alleles present at other loci.

In many cases, however, genes are not as independent as the gene for seed shape in peas. Consider a classic experiment published in 1905 on comb shape in chickens. The researchers, William Bateson and R. C. Punnett, crossed parents from pure-breeding lines with so-called rose and pea combs and found that the F_1 offspring had a different phenotype, called walnut combs. When these individuals bred, their offspring had walnut, rose, pea, and a fourth phenotype called single combs in a 9:3:3:1 ratio (**Figure 13.19a**). The genetic model in **Figure 13.19b** explains the data. If comb morphology results from interactions between two genes (symbolized *R* and *P*), if a dominant and a recessive allele exist for each gene, and if the four comb phenotypes are associated with the genotypes indicated at the bottom of the figure, then a cross between *RRpp* and *rrPP* parents would give the results that Bateson and Punnett observed.

When these types of gene-by-gene interactions occur, the phenotype produced by an allele depends on the action of alleles of other genes. If a chicken has an *R* allele, its phenotype depends on the allele present at the *P* gene.

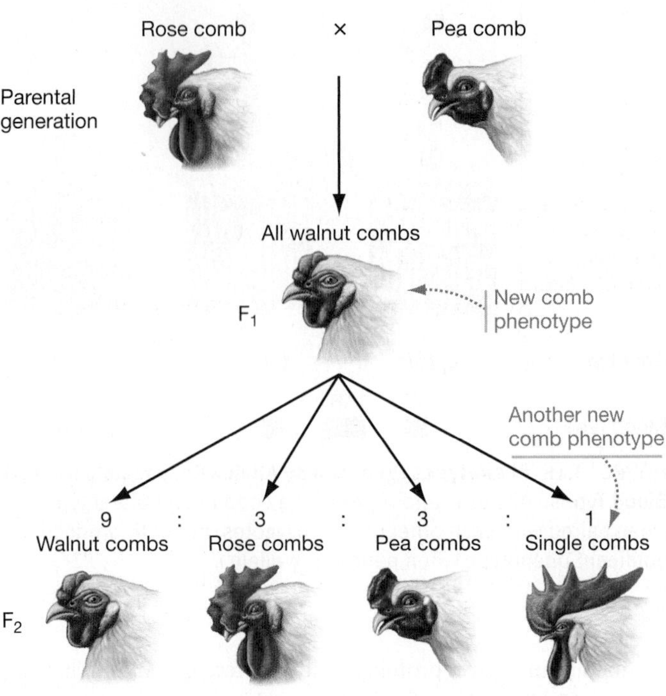

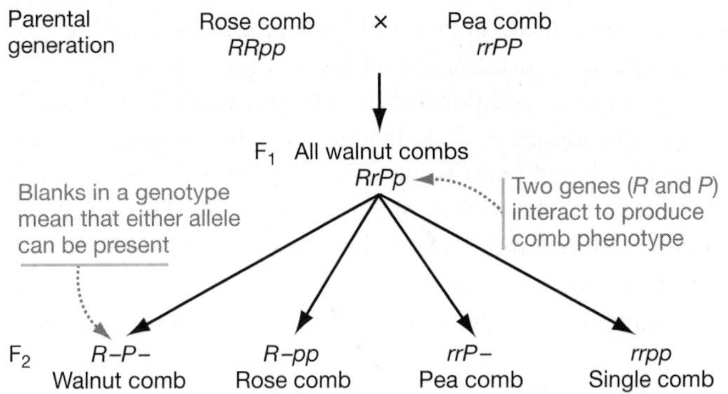

FIGURE 13.19 Genes at Different Loci Can Interact to Influence a Trait. **(a)** This cross is unusual because new phenotypes show up in the F_1 and the F_2 generation. **(b)** To explain the results, researchers hypothesized that comb shape depends on two genes that interact. The phenotype associated with any one allele depends on the genetic environment—specifically, the alleles at a second gene.

What About Traits Like Human Height and Intelligence?

Mendel worked with **discrete traits**—characteristics that are qualitatively different from each other. In garden peas, seed color is either yellow or green—no intermediate phenotypes exist. But many traits in peas and other organisms don't fall into discrete categories. In humans, for example, height, weight, and skin color fall anywhere on a continuous scale of measurement. People are not 160 cm tall or 180 cm tall, with no other heights possible.

Height and many other characteristics exhibit quantitative variation—meaning that individuals differ by degree—and are called **quantitative traits**. Like discrete traits, quantitative traits are greatly influenced by the physical environment. The effects of nutrition on human height, intelligence, and disease resistance, for example, have been well documented.

Quantitative traits share a common characteristic: When the frequencies of different values observed in a population are plot-

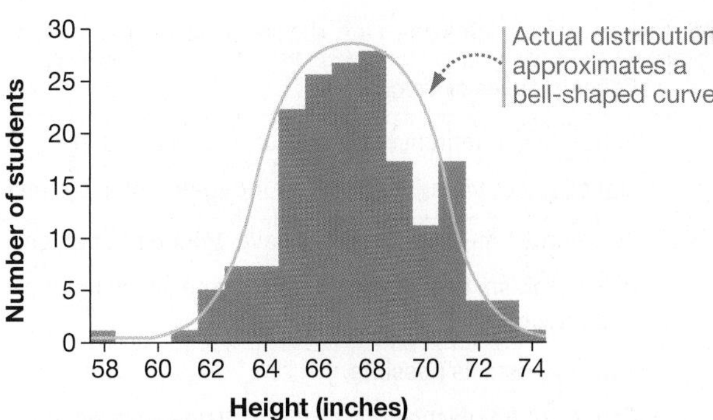

FIGURE 13.20 Quantitative Traits Have a Normal Distribution. A histogram plotting the heights of male undergraduates at Connecticut Agricultural College in 1914.

ted on a histogram, or frequency distribution (see **BioSkills 2** in Appendix A), they usually form a bell-shaped curve (**Figure 13.20**). This distribution is observed so frequently that it is often called a normal distribution. In a normal distribution, high and low values occur at low frequency; intermediate values occur at high frequency.

In 1909 Herman Nilsson-Ehle had an important insight: If many genes each contribute a small amount to the value of a quantitative trait, then a continuous, bell-shaped (normal) distribution results for the population as a whole. Nilsson-Ehle established this finding using strains of wheat that differed in kernel color. **Figure 13.21a** includes a histogram showing the distribution of F_2 phenotypes from a cross he performed between pure lines of white wheat and dark-red wheat. Notice that the frequency of colors in F_2 progeny forms a bell-shaped curve.

To explain these results, Nilsson-Ehle proposed the set of hypotheses illustrated in **Figure 13.21b**:

- The parental strains differ with respect to three genes that control kernel color: *AABBCC* produces dark-red kernels, and *aabbcc* produces white kernels.
- The three genes assort independently. When the *AaBbCc* F_1 individuals self-fertilize, white F_2 individuals would occur at a frequency of 1/4 (*aa*) × 1/4 (*bb*) × 1/4 (*cc*) = 1/64 *aabbcc*. (For help understanding the logic of this calculation, see **BioSkills 13** in Appendix A.)
- The *a*, *b*, and *c* alleles do not contribute to pigment production, but the *A*, *B*, and *C* alleles contribute to pigment production in an equal and additive way. As a result, the degree of red pigmentation is determined by the number of *A*, *B*, or *C* alleles present. Each uppercase (dominant) allele that is present makes a wheat kernel slightly darker red.

Later work showed that Nilsson-Ehle's model hypotheses were correct in virtually every detail. Quantitative traits are produced by the independent actions of many genes, although it is now clear that some genes have much greater effects on the trait in question than other genes do. As a result, the transmission of quantitative traits is said to result from polygenic ("many-genes") inheritance. In **polygenic inheritance**, each gene adds a small amount to the value of the phenotype.

In the decades immediately after the rediscovery of Mendel's work, analyses of phenomena such as sex-linkage, linkage, multiple allelism, incomplete dominance, gene-by-environment interactions, gene-by-gene interactions, and polygenic inheritance provided a fairly comprehensive answer to the question of why

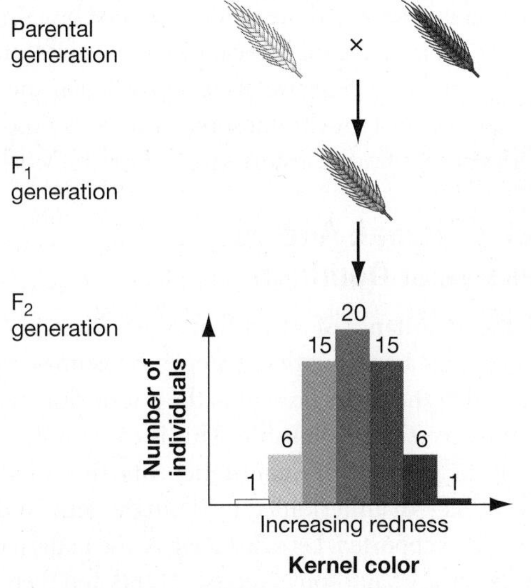

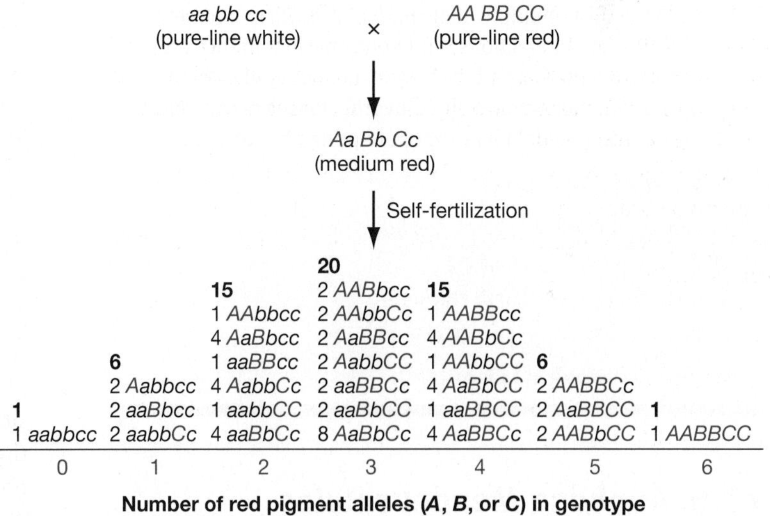

FIGURE 13.21 Quantitative Traits Result from the Action of Many Genes. (a) When wheat plants with white kernels were crossed with wheat plants with red kernels, the F_2 offspring showed a range of kernel colors. The frequency of these phenotypes approximates a normal distribution. **(b)** This model attempts to explain the results of part (a).

SUMMARY TABLE 13.4 **Some Exceptions and Extensions to Mendel's Rules**

Type of Inheritance	Definition	Consequences or Comments
Sex-linkage	Genes located on sex chromosomes.	Patterns of inheritance in males and females differ.
Linkage	Two genes found on same chromosome.	Linked genes violate principle of independent assortment.
Incomplete dominance	Heterozygotes have intermediate phenotype.	Polymorphism—heterozygotes have unique phenotype.
Codominance	Heterozygotes have phenotype of both alleles.	Polymorphism is possible—heterozygotes have unique phenotype.
Multiple allelism	In a population, more than two alleles present at a locus.	Polymorphism is possible.
Polymorphism	In a population, more than two phenotypes associated with a single gene are present.	Can result from actions of multiple alleles, incomplete dominance, and/or codominance.
Pleiotropy	A single allele affects many traits.	This is common.
Gene-by-gene interaction	In discrete traits, the phenotype associated with an allele depends on which alleles are present at another gene.	One allele can be associated with different phenotypes.
Gene-by-environment interaction	Phenotype influenced by environment experienced by individual.	Same genotypes can be associated with different phenotypes.
Polygenic inheritance of quantitative traits	Many genes are involved in specifying traits that exhibit continuous variation.	Unlike alleles that determine discrete traits, each allele adds a small amount to phenotype.

offspring resemble their parents. **Table 13.4** summarizes some of the key exceptions and extensions to Mendel's rules and gives you a chance to compare and contrast their effects on patterns of inheritance.

CHECK YOUR UNDERSTANDING

If you understand that . . .

- Genes on the same chromosome violate the principle of independent assortment. They are not transmitted to gametes independently of each other unless crossing over occurs between them.
- Sex linkage, linkage, incomplete dominance, codominance, multiple allelism, pleiotropy, environmental effects, gene interactions, and polygenic inheritance are aspects of inheritance that Mendel did not study. When they occur, monohybrid and dihybrid crosses do not result in classical Mendelian ratios of offspring phenotypes.

✔ **You should be able to . . .**

Explain why the following crosses don't produce a 3:1 phenotype ratio in F_2 offspring:

1. Rose-comb × pea-comb chickens
2. Red-kernel × white-kernel wheat plants

Answers are available in Appendix B.

13.6 Applying Mendel's Rules to Humans

When researchers set out to study how a particular gene is transmitted in wheat or fruit flies or garden peas, they begin by making a series of controlled experimental crosses. For obvious reasons, this research strategy is not possible with humans. But suppose that you are concerned about an illness that runs in your family, and you go to a genetic counselor to find out how likely your children are to have the disease. To advise you, the counselor needs to know how the trait is transmitted, including whether the gene involved is autosomal or sex-linked and what type of dominance is associated with the disease allele.

To understand the transmission of human traits, investigators have to analyze human genotypes and phenotypes that already exist. They do so by constructing a **pedigree**, or family tree, of affected individuals.

A pedigree records the genetic relationships among the individuals in a family along with each person's sex and phenotype with respect to the trait in question. If the trait is governed by a single gene, then analyzing the pedigree may reveal whether a given phenotype is due to a dominant or recessive allele and whether the gene responsible is located on a sex chromosome or on an autosome. Let's look at a series of specific case histories to see how this work is done.

Identifying Human Alleles as Recessive or Dominant

To analyze the inheritance of a trait that shows discrete variation, biologists begin by assuming that a single autosomal gene is responsible and that the alleles present in the population have a simple dominant–recessive relationship. This is the simplest possible situation. If the pattern of inheritance fits this model, then the assumptions—of inheritance by a single gene and simple dominance—are supported. Let's first analyze the pattern of inheritance that is typical of autosomal recessive traits and then examine patterns that emerge in pedigrees for autosomal dominant traits.

PATTERNS OF INHERITANCE: AUTOSOMAL RECESSIVE TRAITS In analyzing the inheritance of traits, it's helpful to distinguish conditions that *must* be met when a particular pattern of inheritance

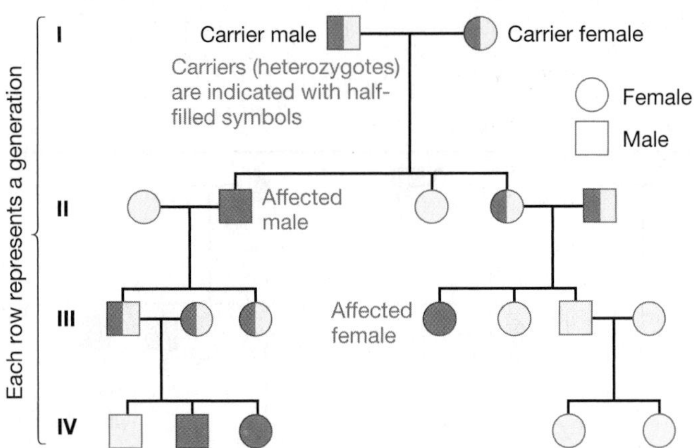

FIGURE 13.22 Pedigree of a Family with an Autosomal Recessive Disease. Diseases that are inherited as autosomal recessive traits, like sickle-cell anemia, appear in both males and females. For an individual to be affected, both parents must carry the allele responsible.

✔**QUESTION** In this pedigree, why can unaffected parents have affected offspring? If the recessive allele is rare and only one parent is affected, why are children usually unaffected?

occurs versus conditions that are *likely* to be met. For example, if a phenotype is due to an autosomal recessive allele, then

- Individuals with the trait *must* be homozygous.
- If the parents of an affected individual do not have the trait, then the parents *are likely* to be heterozygous for the trait.

Heterozygous individuals who carry a recessive allele for an inherited disease are referred to as **carriers** of the disease. These individuals carry the allele and transmit it even though they do not exhibit signs of the disease. When two carriers mate, they should produce offspring with the recessive phenotype about 25 percent of the time.

Figure 13.22 is the pedigree from a family in which an autosomal recessive trait, such as sickle-cell disease, occurs. The key feature to notice in this pedigree is that some boys and girls exhibit the trait even though their parents do not. This is the pattern you would expect to observe when the parents of an individual with the trait are heterozygous. It is also logical to observe that when an affected (homozygous) individual has children, those children do not necessarily have the trait. This pattern is predicted if affected people marry individuals who are homozygous for the wild-type allele, and is likely to occur if the recessive allele is rare in the population.

In general, a recessive phenotype should show up in offspring only when both parents have that recessive allele and pass it on to their offspring. By definition, a recessive allele produces a given phenotype only when the individual is homozygous for that allele.

PATTERNS OF INHERITANCE: AUTOSOMAL DOMINANT TRAITS By definition, when a trait is autosomal dominant, individuals who are homozygous or heterozygous for it must have the dominant phenotype. Even if one parent is heterozygous and the other is homozygous recessive, on average half of their children should show the dominant phenotype. And unless a new mutation has occurred in a gamete, any child with the trait must have a parent with this trait. The latter observation is in strong contrast to the pattern seen in autosomal recessive traits.

Figure 13.23 shows the consequences of autosomal dominant inheritance in the pedigree of a family affected by a degenerative brain disorder called **Huntington's disease**. The pedigree has two features that indicate this disease is passed to the next generation through an autosomal dominant allele. First, if a child shows the trait, then one of its parents shows the trait as well. Second, if families have a large number of children, the trait usually shows up in every generation—due to the high probability of heterozygous parents having affected children.

Identifying Human Traits as Autosomal or Sex-Linked

When it is not possible to arrange reciprocal crosses, can data in a pedigree indicate whether a trait is autosomal or sex-linked? The answer is based on a simple premise: If a trait appears about equally often in males and females, then it is likely to be autosomal.

The data in the Huntington's pedigree (Figure 13.23) indicates that the disease appears in both males and females at about equal rates. This is strong evidence that the trait is autosomal. But if males are much more likely to have the trait in question than females are, then the allele responsible is likely to be recessive and found on the X chromosome. Because so few genes occur on the Y chromosome, Y-linked inheritance is rare.

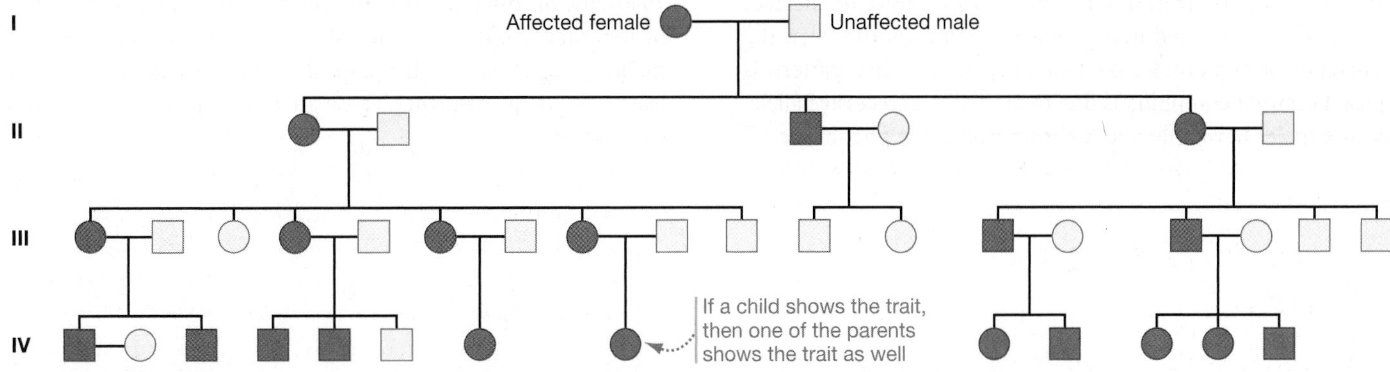

FIGURE 13.23 A Pedigree of a Family with an Autosomal Dominant Disease. Disorders like Huntington's disease, which are autosomal dominant, may appear in both males and females and tend to occur in every generation.

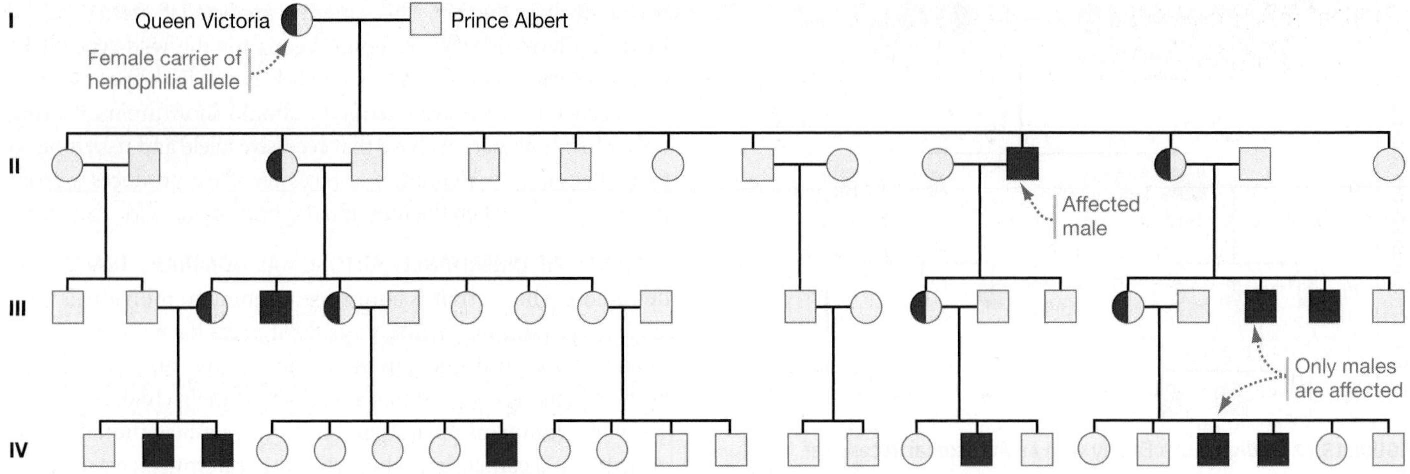

FIGURE 13.24 A Pedigree of an X-Linked Recessive Disease.

✔**QUESTION** What pattern of inheritance would you observe in a pedigree if the allele for hemophilia were X-linked dominant?

To understand why a sex bias in phenotypes implicates sex-linked inheritance, recall from Section 13.4 that sex-linked genes are located on one of the sex chromosomes. Because human males have one X chromosome and one Y chromosome, they have just one copy of each X-linked gene. But because human females have two X chromosomes, they have two copies of each X-linked gene. These simple observations are critical. In humans—just as in fruit flies and in every other species that has sex chromosomes—the pattern of inheritance in sex-linked traits is different in males and females, because the complement of sex chromosomes differs in the two sexes.

What does the pedigree of an X-linked trait look like? Let's consider the pedigree of a classic X-linked trait—the occurrence of hemophilia in the descendants of Queen Victoria, the 19th-century British monarch, and her husband Prince Albert. **Hemophilia** is caused by a defect in an important blood-clotting factor. Hemophiliacs are at a high risk of bleeding to death, because even minor injuries result in prolonged bleeding. The tip-off in the pedigree of Queen Victoria's descendants is that only males developed hemophilia (**Figure 13.24**).

Also note the affected male in generation II (the second square from the right in row II). His two sons were unaffected, but the trait reappeared in a grandson. Stated another way, the occurrence of hemophilia skipped a generation. This pattern is logical because hemophilia is due to an X-linked recessive allele. Because males have only one X chromosome, the phenotype associated with an X-linked recessive allele appears in every male that carries it. Further, the appearance of an X-linked recessive trait skips a generation in a pedigree. This pattern occurs because an affected male passes his only X chromosome on to his daughters. But because his daughters almost always received a wild-type allele from their mother, the daughters don't show the trait. They will pass the defective allele on to about half of their sons, however.

In contrast, X-linked traits that are dominant appear in every individual who has the defective allele. A good indicator of an X-linked dominant trait is a pedigree in which an affected male has all affected daughters but no affected sons. This pattern is logical because every female offspring of an affected father gets an X chromosome from him and will herself be affected, while his sons get their only X chromosome from their unaffected mother. Besides the inherited form of a bone disease called rickets, however, very few diseases are known to be due to X-linked dominant alleles.

By analyzing pedigrees in this way, biomedical researchers have been able to discover how most of the common genetic diseases in humans are inherited. As a result of such analysis, by the 1940s, the burning question in genetics was no longer the nature of inheritance but the nature of the gene itself. What are genes made of, and how are they copied so that parents pass their alleles on to their offspring? These are the questions we turn to in Chapter 14.

CHAPTER 13 REVIEW

For media, go to the study area at www.masteringbiology.com 

Summary of Key Concepts

In many species, individuals have two alleles of each gene. The principle of segregation states that prior to the formation of eggs and sperm, the alleles of each gene separate so that each egg or sperm cell receives only one of them.

- Gregor Mendel discovered that inheritance is particulate—genes do not blend together.
- The traits that Mendel studied are specified by paired hereditary determinants that separate from each other during gamete formation.

✓You should be able to predict the gamete genotypes generated by a parent with the genotype *Bb*.

The principle of independent assortment states that alleles of different genes are transmitted to egg cells and sperm cells independently of each other.

- According to the principle of independent assortment, the segregation of alleles from one gene does not affect the segregation of other gene pairs.

✓You should be able to predict the gamete genotypes generated by a parent with the genotype *BbRr*.

MB **Web Activity** Mendel's Experiments, **Web Activity** The Principle of Independent Assortment

Genes are located on chromosomes. The principle of segregation is explained by the separation of homologous chromosomes in anaphase of meiosis I. The principle of independent assortment applies to genes found on different chromosomes and is explained by chromosomes lining up randomly in metaphase of meiosis I.

- The chromosome theory of inheritance claimed that the movements of chromosomes during meiosis provide a physical basis for the principle of segregation and the principle of independent assortment.
- The chromosome theory was supported by the discovery of sex linkage. Experimental crosses with X-linked traits supported the theory's contention that genes are found on chromosomes.

✓You should be able to use the movements of chromosomes during meiosis to explain why your answers to the above two exercises are correct.

There are important exceptions and extensions to the basic patterns of inheritance that Mendel discovered.

- All genes follow the principle of independent assortment, but genes on the same chromosome do not follow the principle of independent assortment unless crossing over occurs.
- X-linked traits give different results in reciprocal crosses. Crosses with X-linked traits don't produce the 3:1 and 9:3:3:1 phenotypic ratios that Mendel observed.
- Not all heterozygotes show a dominant or recessive phenotype. Incomplete dominance occurs, and codominant phenotypes are common.
- Many genes are pleiotropic, meaning that they influence more than one trait. (There is not a strict one-to-one correspondence between genes and traits.)
- The phenotype associated with an allele is influenced by the environment that the individual experiences, and by the actions of alleles of other genes.
- Many traits are influenced by the action of many genes. In a population, these polygenic traits show quantitative instead of discrete variation. The frequency of each phenotype follows a normal distribution.
- Analyzing pedigrees has allowed researchers to deduce the basic modes of inheritance for most of the common genetic diseases in humans.

✓You should be able to explain why crossing over is said to break up linkage between alleles.

Genetics Questions

Answers are available in Appendix B

1. In studies of how traits are inherited, what makes certain species candidates for model organisms?
   a. They are the first organisms to be used in a particular type of experiment, so they are a historical "model" of what researchers expect to find.
   b. They are easy to study because a great deal is already known about them.
   c. They are the best or most fit of their type.
   d. They are easy to maintain, have a short life cycle, produce many offspring, and yield data that are relevant to many other organisms.
2. Why is the allele for wrinkled seed shape in garden peas considered recessive?
   a. It "recedes" in the F_2 generation when homozygous parents are crossed.
   b. The trait associated with the allele is not expressed in heterozygotes.
   c. Individuals with the allele have lower fitness than that of individuals with the dominant allele.
   d. The allele is less common than the dominant allele. (The wrinkled allele is a rare mutant.)
3. The alleles found in haploid organisms cannot be dominant or recessive. Why?
   a. Dominance and recessiveness describe interactions between two alleles of the same gene in the same individual.
   b. Because only one allele is present, alleles in haploid organisms are always dominant.

c. Alleles in haploid individuals are transmitted like mitochondrial DNA or chloroplast DNA.
d. Most haploid individuals are bacteria, and bacterial genetics is completely different from eukaryotic genetics.

4. Why can you infer that individuals that are "pure line" are homozygous for the gene in question?
   a. They are highly inbred.
   b. Only two alleles are present at each gene in the populations to which these individuals belong.
   c. In a pure line, phenotypes are not affected by environmental conditions or gene interactions.
   d. No other phenotype arises in a pure-line population because no other alleles are present.

5. The genes for the traits that Mendel worked with are either located on different chromosomes or so far apart on the same chromosome that crossing over almost always occurs between them. How did this circumstance help Mendel recognize the principle of independent assortment?
   a. Otherwise, his dihybrid crosses would not have produced a 9:3:3:1 ratio of F_2 phenotypes.
   b. The occurrence of individuals with unexpected phenotypes led him to the discovery of recombination.
   c. It led him to the realization that the behavior of chromosomes during meiosis explained his results.
   d. It meant that the alleles involved were either dominant or recessive, which gave 3:1 ratios in the F_1 generation.

6. What is meant by the claim that Mendel worked with the simplest possible genetic system?
   a. Discrete traits, two alleles, simple dominance and recessiveness, no sex chromosomes, and unlinked genes are the simplest situation known.
   b. The ability to self-fertilize or cross-pollinate made it simple for Mendel to set up controlled crosses.
   c. Mendel was aware of meiosis and the chromosome theory of inheritance, so it was easy to reach the conclusions he did.
   d. Mendel's experimental designs and his rules of inheritance are actually neither complex nor sophisticated.

7. Mendel's rules do not correctly predict patterns of inheritance for tightly linked genes or the inheritance of alleles that show incomplete dominance. Does this mean that his hypotheses are incorrect?
   a. Yes, because they are relevant to only a small number of organisms and traits.
   b. Yes, because not all data support his hypotheses.
   c. No, because he was not aware of meiosis or the chromosome theory of inheritance.
   d. No, it just means that his hypotheses are limited to certain conditions.

8. The artificial sweetener NutraSweet consists of a phenylalanine molecule linked to aspartic acid. The labels of diet sodas that contain NutraSweet include a warning to people with PKU. Why?
   a. NutraSweet stimulates the same taste receptors that natural sugars do.
   b. People with PKU have to avoid phenylalanine in their diet.
   c. In people with PKU, phenylalanine reacts with aspartic acid to form a toxic compound.
   d. People with PKU cannot lead normal lives, even if their environment is carefully controlled.

9. When Sutton and Boveri published the chromosome theory of inheritance, research on meiosis had not yet established that paternal and maternal homologs assort independently of each other. Then, in 1913, Elinor Carothers published a paper about a grasshopper species with an unusual karyotype: One chromosome had no homolog (meaning no pairing partner at meiosis I); another chromosome had homologs that could be distinguished under the light microscope. If chromosomes assort independently, how often should Carothers have observed each of the four products of meiosis shown in the following figure?

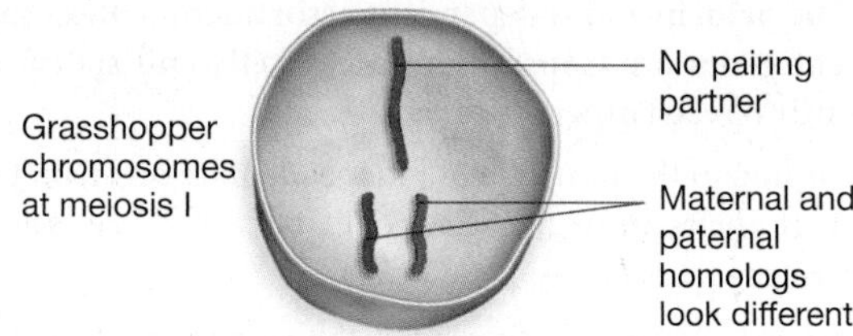

Four types of gametes possible
(each meiotic division can produce only two of the four)

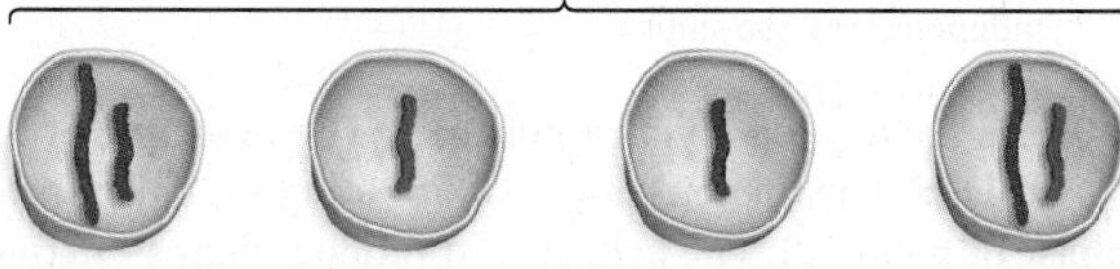

   a. Only the gametes with one of each type of chromosome would occur.
   b. The four types of gametes should be observed to occur at equal frequencies.
   c. The chromosome with no pairing partner would disintegrate, so only gametes with one copy of the other chromosome would be observed.
   d. Gametes with one of each type of chromosome would occur twice as often as gametes with just one chromosome.

10. Which of the following is the strongest evidence that a trait might be influenced by polygenic inheritance?
   a. F_1 offspring of parents with different phenotypes have an intermediate phenotype.
   b. F_1 offspring of parents with different phenotypes have the dominant phenotype.
   c. The trait shows qualitative (discrete) variation.
   d. The trait shows quantitative variation.

Solving Genetics Problems

The best way to test and extend your knowledge of transmission genetics is to work problems. Most genetics problems are set up as follows: You are given some information about the genotypes or phenotypes of one or both parents, along with data on the phenotypes of F_1 or F_2 offspring. Your task is to generate a set of hypotheses—a genetic model—to explain the results. Your hypotheses should address each of the following questions:

- Is the trait under study discrete or quantitative?

- Is the phenotype a product of one gene or many genes?
- For each gene involved, how many alleles are present—one, two, or many?
- Do the alleles involved show complete dominance, incomplete dominance, or codominance?
- Are the genes involved sex-linked or autosomal?
- If more than one gene is involved, are they linked or unlinked? If they are linked, does crossing over occur frequently?

It's also helpful to ask yourself whether gene interactions or pleiotropy might be occurring and whether it is safe to assume that the experimental design carefully controlled for effects of variation in other genes and in the environment.

In working the problem, be sure to start with the simplest possible explanation. For example, if you are dealing with a discrete trait, you might hypothesize that the cross involves a single autosomal gene with two alleles that show complete dominance. Your next step is to infer what the parental genotypes would be (according to your hypothesis), if they are not already given, and then do a Punnett square to predict what the offspring phenotypes and their frequencies should be based on your hypothesis. Next, check whether these predictions match the observed results given in the problem. If the answer is yes, you have a valid solution. But if the answer is no, you need to go back and change one of your hypotheses, redo the Punnett square, and check to see if the predictions and observations match. Keep repeating these steps until you have a model that fits the data.

Example Problems

Example 1 *Plectritis congesta* plants produce fruits that either have or do not have prominent structures called wings. The alleles involved are W^+ = winged fruit; W^- = wingless fruit. Researchers collected an array of individuals from the field and performed a series of crosses. The results are given in the following table. Complete the table by writing down the genotype of the parent or parents involved in each cross.

Parental Phenotype(s)	Number of Offspring with Winged Fruits	Number of Offspring with Wingless Fruits	Parental Genotype(s)
Wingless (self-fertilized)	0	80	
Winged (self-fertilized)	90	30	
Winged × wingless	46	0	
Winged × winged	44	0	

Solution Here you're given offspring phenotypes and you're asked to infer parental genotypes. To do this you have to propose hypothetical parental genotypes to test, make a Punnett square to predict the offspring genotypes, and then see if the predicted offspring phenotypes match the data. In this case, coming up with a hypothesis for the parental genotypes is relatively straightforward, because the problem states that the trait is due to one gene and two alleles. No information on sex is given, so assume the gene is autosomal (the simplest case). Now look at the second entry in the chart. It shows a 3:1 ratio of offspring from a winged individual that self-fertilizes. This result is consistent with the hypothesis that W^+ is dominant and W^- recessive and that this parent's genotype is W^+W^-. Now let's look at the first cross in the chart. If W^+ is dominant, then a wingless parent must be W^-W^-. When you do the Punnett square to predict offspring genotypes from selfing, you find that all the offspring will produce wingless fruits, consistent with the data. In the third cross, all the offspring make winged fruits even though one of the parents produces wingless fruits and thus is W^-W^-. This would happen only if the winged parent is W^+W^+. (If this reasoning isn't immediately clear to you, work the Punnett square.) In the fourth cross, you could get offspring that all make winged fruits if the parents were W^+W^+ and W^+W^+, or if the parents were W^+W^+ and W^+W^-. Either answer is correct. Again, you can write out the Punnett squares to see that this statement is correct.

Example 2 Two black female mice are crossed with a brown male. In several litters, female I produced 9 blacks and 7 browns; female II produced 57 blacks. What deductions can you make concerning the inheritance of black and brown coat color in mice? What are the genotypes of the parents in this case?

Solution Here you are given parental and offspring phenotypes and are asked to infer the parental genotypes. As a starting point, assume that the coat colors are due to the simplest genetic system possible: one autosomal gene with two alleles, where one allele is dominant and the other recessive. Because female II produces only black offspring, it's logical to suppose that black is dominant to brown. Let's use *B* for black and *b* for brown. Then the male parent is *bb*. To produce offspring with a 1:1 ratio of black:brown coats, female I must be *Bb*. But to produce all black offspring, female II must be *BB*. This model explains the data, so you can accept it as correct.

Genetics Problems

Answers are available in Appendix B

1. Tay-Sachs disease causes nerve cells to malfunction and results in death by age 4. Two healthy parents know from blood tests that each parent carries a recessive allele responsible for Tay-Sachs. If their first three children have the disease, what is the probability that their fourth child will not? Assuming they have not yet had a child, what is the probability that, if they have four children, all four will have the disease? If their first three children are male, what is the probability that their fourth child will be male?
2. Suppose that in garden peas the genes for seed color and seed-pod shape are linked, and that Mendel crossed *YYII* parents (which produce yellow seeds in inflated pods) with *yyii* parents (which produce green seeds in constricted pods).
   - Draw the F_1 Punnett square and predict the expected F_1 phenotype(s).
   - List the genotype(s) of gametes produced by F_1 individuals if no crossing over occurs.
   - Draw the F_2 Punnett square if no crossing over occurs. Based on this Punnett square, predict the expected phenotype(s) in the F_2 generation and the expected frequency of each phenotype.
   - If crossing over occurs during gamete formation in F_1 individuals, give the genotype of the recombinant gamete(s) that result.
   - Add the recombinant gametes to the F_2 Punnett square requested above. Will any additional phenotypes be observed at low frequency in the F_2 generation? If so, what are they?

3. The smooth feathers on the back of the neck in pigeons can be reversed by a mutation to produce a "crested" appearance in which feathers form a distinctive spike at the back of the head. A pigeon breeder examined offspring produced by a single pair of non-crested birds and recorded the following: 14 non-crested and 7 crested. She then made a series of crosses using offspring from the first cross. When she crossed two of the crested birds, all 22 of the offspring were crested. When she crossed a non-crested bird with a crested bird, 7 offspring were non-crested and 6 were crested.
   - For these three crosses, provide genotypes for parents and offspring that are consistent with these results.
   - Which allele is dominant?
4. A plant with orange-spotted flowers was grown in the greenhouse from a seed collected in the wild. The plant was self-pollinated and gave rise to the following progeny: 88 orange with spots, 34 yellow with spots, 32 orange with no spots, and 8 yellow with no spots. What can you conclude about the dominance relationships of the alleles responsible for the spotted and unspotted phenotypes? For the orange and yellow phenotypes? What can you conclude about the genotype of the original plant that had orange, spotted flowers?
5. As a genetic counselor, you routinely advise couples about the possibility of genetic disease in their offspring based on their family histories. This morning you met with an engaged couple, both of whom are phenotypically normal. The man, however, has a brother who died of Duchenne-type muscular dystrophy, an X-linked condition that results in death before the age of 20. The allele responsible for this disease is recessive. His prospective bride, whose family has no history of the disease, is worried that the couple's sons or daughters might be afflicted.
   - How would you advise this couple?
   - The sister of this man is planning to marry his fiancée's brother. How would you advise this second couple?
6. Suppose you are heterozygous for two genes that are located on different chromosomes. You carry alleles *A* and *a* for one gene and alleles *B* and *b* for the other. Draw a diagram illustrating what happens to these genes and alleles when meiosis occurs in your reproductive tissues. Label the stages of meiosis, the homologous chromosomes, sister chromatids, nonhomologous chromosomes, genes, and alleles. Be sure to list all of the genetically different gametes that could form and indicate how frequently each type should be observed. On the diagram, identify the events responsible for the principle of segregation and the principle of independent assortment.
7. Review the text's description of ABO blood types. Suppose a woman with blood type O married a man with blood type AB. What phenotypes and genotypes would you expect to observe in their offspring, and in what proportions? Answer the same question for a heterozygous mother with blood type A and a heterozygous father with blood type B.
8. An alien friend named Tukan has two sets of eyes, one set forward-looking and one set backward-looking, and smooth skin. His mate, Valco, lacks eyes but has skin covered with tiny hooks that attract all sorts of debris. Tukan and Valco have thrived on earth and have had four children, all with no eyes and smooth skin. Typical of their ways, the children interbred and produced 32 children of their own.
   - Under the models of inheritance proposed by Mendel, identify which alleles are dominant and which are recessive.
   - Provide gene symbols that would reflect the dominant–recessive allelic relationships.
   - Of the 32 children, how many would you expect to have two sets of eyes and smooth skin?
9. Phenylketonuria (PKU) is a genetic disease caused by homozygosity for a recessive mutation in the enzyme that converts the amino acid phenylalanine to tyrosine. In the absence of this enzyme, phenylalanine and some of its derivatives accumulate in the body and cause mental retardation. If individuals are identified soon enough after birth, they can be treated by a low-phenylalanine diet for the early years of their lives. As adults, though, homozygous recessive individuals are allowed to adopt a diet with normal amounts of phenylalanine. Not long after such treatments were initiated, a troubling phenomenon was observed. A high number of children born to treated mothers were mentally retarded even though the children were heterozygous for the PKU gene. Children born of treated PKU males suffered no ill effects.
   - Can you offer an explanation as to why genetically heterozygous children of treated PKU mothers might be prone to mental retardation?
   - Propose a solution to reduce the likelihood of mental retardation in children of treated PKU mothers.
10. The blending-inheritance hypothesis proposed that the genetic material from parents is unavoidably and irreversibly mixed in the offspring. As a result, offspring and later descendants should always appear intermediate in phenotype to their forebears. Mendel, in contrast, proposed that genes are discrete and that their integrity is maintained in the offspring and in subsequent generations. Suppose the year is 1890. You are a horse breeder and have just read Mendel's paper. You don't believe his results, however, because you often work with cremello (very light-colored) and chestnut (reddish-brown) horses. You know that if you cross a cremello individual from a pure-breeding line with a chestnut individual from a pure-breeding line, the offspring will be palomino—meaning they have an intermediate (golden-yellow) body color. What additional crosses would you do to test whether Mendel's model is valid in the case of genes for horse color? List the crosses and the offspring genotypes and phenotypes you'd expect to obtain. Explain why these experimental crosses would provide a test of Mendel's model.
11. Two mothers give birth to sons at the same time in a busy hospital. The son of couple 1 is afflicted with hemophilia A, which is a recessive X-linked disease. Neither parent has the disease. Couple 2 has a normal son even though the father has hemophilia A. The two couples sue the hospital in court, claiming that a careless staff member swapped their babies at birth. You appear in court as an expert witness. What do you tell the jury? Make a diagram that you can submit to the jury.
12. You have crossed two *Drosophila melanogaster* individuals that have long wings and red eyes—the wild-type phenotype. In the progeny, the mutant phenotypes called curved wings and lozenge eyes appear as follows:

Females	Males
600 long wings, red eyes	300 long wings, red eyes
200 curved wings, red eyes	100 curved wings, red eyes
	300 long wings, lozenge eyes
	100 curved wings, lozenge eyes

   - According to these data, is the curved-wing allele autosomal recessive, autosomal dominant, sex-linked recessive, or sex-linked dominant?

- Is the lozenge-eye allele autosomal recessive, autosomal dominant, sex-linked recessive, or sex-linked dominant?
- What is the genotype of the female parent?
- What is the genotype of the male parent?

13. In parakeets, two autosomal genes that are located on different chromosomes control the production of feather pigment. Gene *B* codes for an enzyme that is required for the synthesis of a blue pigment, and gene *Y* codes for an enzyme required for the synthesis of a yellow pigment. Recessive, loss-of-function mutations (resulting in no production of the affected pigment) are known for both genes. Suppose that a bird breeder has two green parakeets and mates them. The offspring are green, blue, yellow, and albino (unpigmented).
    - Based on this observation, what are the genotypes of the green parents? What is the genotype of each type of offspring? What fraction of the total progeny should exhibit each type of color?
    - Suppose that the parents were the progeny of a cross between two true-breeding strains. What two types of crosses between true-breeding strains could have produced the green parents? Indicate the genotypes and phenotypes for each cross.

14. The pedigree shown below is for the human trait called osteopetrosis, which is characterized by bone fragility and dental abscesses. Is the gene that affects bone and tooth structure autosomal or sex-linked? Is the allele for osteopetrosis dominant or recessive?

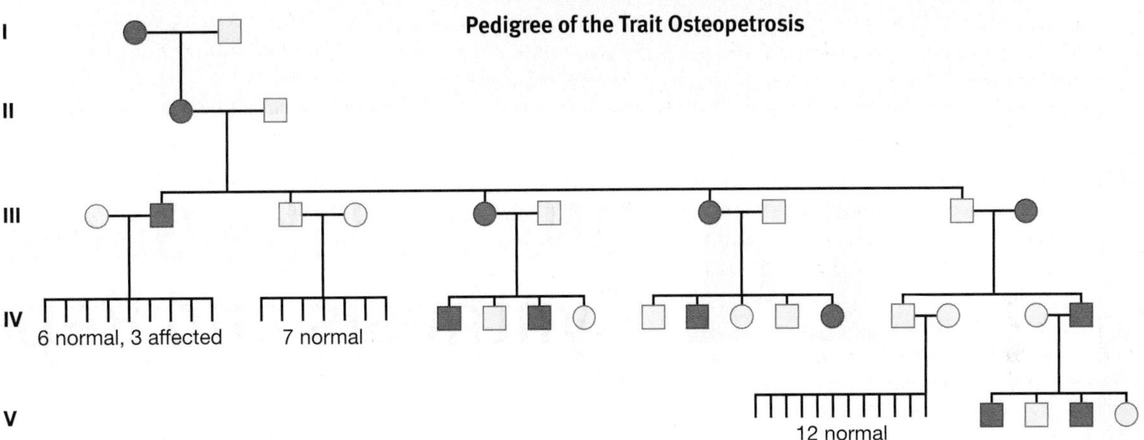

This chapter is part of The Big Picture. See how on pages 336–337.

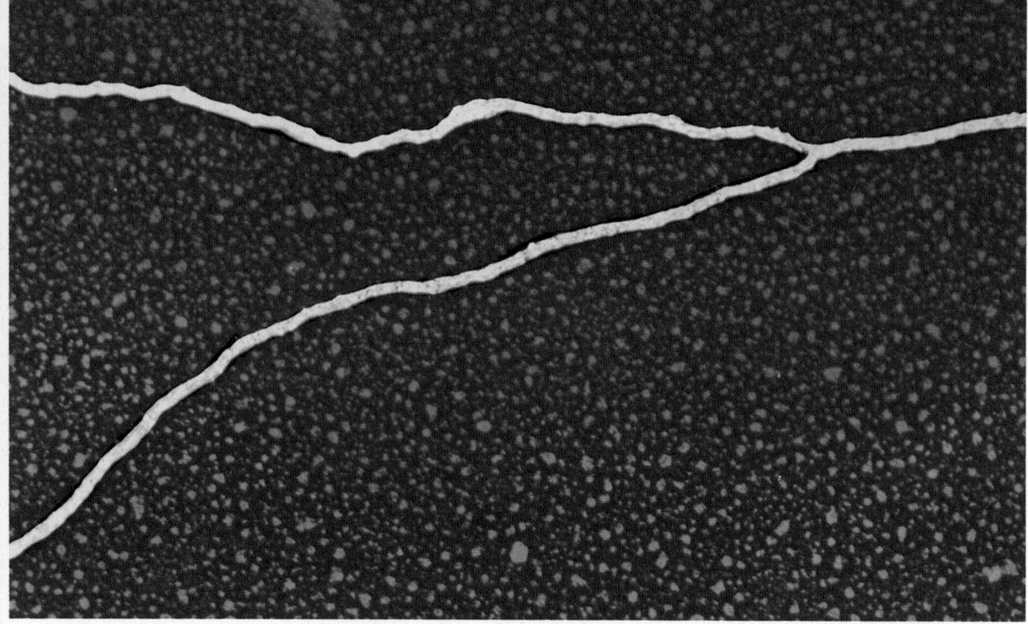

Electron micrograph (with color added) showing DNA in the process of replication. The original DNA double helix (far right) is being replicated into two DNA double helices (on the left). The point at which the two helices diverge, called the replication fork, is where DNA synthesis is taking place.

14 DNA and the Gene: Synthesis and Repair

KEY CONCEPTS

- Genes are made of DNA.
- When DNA is being copied, each strand of the double helix serves as the template for the synthesis of a complementary strand.
- DNA synthesis occurs in the $5' \rightarrow 3'$ direction only and requires a large suite of specialized enzymes. The leading strand is synthesized continuously, but the lagging strand is synthesized as a series of fragments that are then linked together.
- Specialized enzymes repair damages to DNA and fix mistakes in DNA synthesis. If these enzymes are defective, the mutation rate increases.

What are genes made of, and how are they copied so that they are faithfully passed on to offspring? These questions dominated biology during the middle of the twentieth century.

Since Mendel's time, the predominant research strategy in genetics had been to conduct a series of experimental crosses, create a genetic model to explain the types and proportions of phenotypes that resulted, and then test the model's predictions through reciprocal crosses, testcrosses, or other techniques. This strategy led to virtually all of the discoveries analyzed in Chapter 13, including Mendel's rules, sex linkage, linkage, and quantitative inheritance.

The chemical composition and molecular structure of Mendel's hereditary factors—which came to be called genes—remained a mystery, however. And even though biologists knew that genes and chromosomes were replicated during the cell cycle (see Chapter 11), with copies distributed to daughter cells during mitosis and meiosis, no one had the slightest clue about how the copying occurred.

The goal of this chapter is to explore how researchers solved these mysteries. The results provided a link between two of the five attributes of life introduced in Chapter 1: Processing genetic information and replication. How are genes copied, so they can be passed on to succeeding generations?

Let's begin with studies that identified the nature of the genetic material, explore how genes are copied during the synthesis phase of the cell cycle, and conclude by analyzing how damaged or incorrectly copied genes are repaired. Once the molecular nature of the gene was known, the nature of biological science changed forever.

✔ When you see this checkmark, stop and test yourself. Answers are available in Appendix B.

14.1 What Are Genes Made Of?

The chromosome theory of inheritance, introduced in Chapter 13, proposed that chromosomes are composed of genes. It had been known since the late 1800s that chromosomes are a complex of DNA and proteins. The question of what genes are made of, then, came down to a simple choice: DNA or protein?

Initially, most biologists backed the hypothesis that genes are made of proteins. The arguments in favor of this hypothesis were compelling. Hundreds, if not thousands, of complex and highly regulated chemical reactions occur in even the simplest living cells. The amount of information required to specify and coordinate these reactions is mind-boggling. With their almost limitless variation in structure and function, proteins are complex enough to contain this much information.

DNA, in contrast, was known to be composed of just four types of deoxyribonucleotides—the monomers introduced in Chapter 4. It was thought to be a simple molecule with some sort of repetitive and uninteresting structure. No one could imagine how such a simple compound could hold so much complex information.

DNA or protein? The experiment that settled the question is considered a classic in biological science.

The Hershey–Chase Experiment

Alfred Hershey and Martha Chase took up the question of whether genes were made of protein or DNA by studying how a virus called T2 infects the bacterium *Escherichia coli.* Hershey and Chase knew that T2 infections begin when the virus attaches to the cell wall of *E. coli* and injects its genes into the cell's interior (**Figure 14.1a**). These genes then direct the production of a new generation of virus particles inside the infected cell, which acts as a host for the parasitic virus. (For more information on viruses, see Chapter 35.)

During the infection, the exterior portion, or capsid, of the original, parent virus is left behind. The capsid remains attached to the exterior of the host cell as a "ghost" (**Figure 14.1b**). Hershey and Chase also knew that T2 is made up almost exclusively of protein and DNA. But was it protein or DNA that entered the host cell and directed the production of new viruses?

Hershey and Chase's strategy for determining the composition of the viral substance that enters the cell and acts as the hereditary material was based on two facts: **(1)** Proteins contain sulfur but not phosphorus, and **(2)** DNA contains phosphorus but not sulfur.

As **Figure 14.2** on page 260 shows, the researchers began their work by growing viruses in the presence of either a radioactive isotope of sulfur (^{35}S) or a radioactive isotope of phosphorus (^{32}P). Because these isotopes were incorporated into newly synthesized proteins and DNA, this step produced a population of viruses with radioactive proteins and a population with radioactive DNA.

Hershey and Chase allowed each set of radioactive viruses to infect *E. coli* cells. If genes consist of DNA, then radioactive protein should be found only in the ghost capsids outside the infected host cells, while radioactive DNA should be located inside the cells. But if genes consist of proteins, then only radioactive protein—and no radioactive DNA—should be found inside the cells.

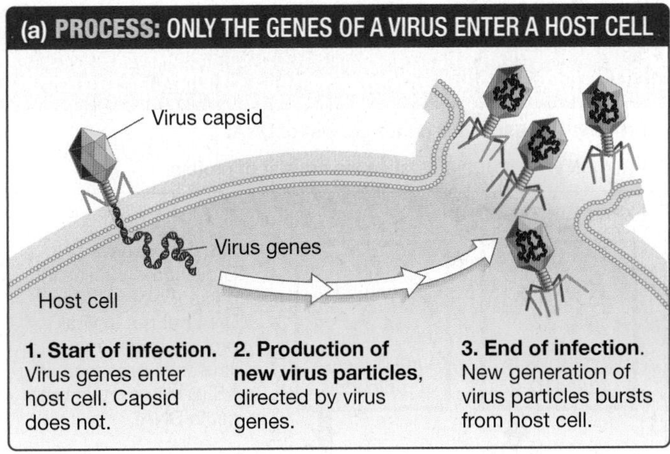

(a) PROCESS: ONLY THE GENES OF A VIRUS ENTER A HOST CELL
Virus capsid
Virus genes
Host cell
1. Start of infection. Virus genes enter host cell. Capsid does not.
2. Production of new virus particles, directed by virus genes.
3. End of infection. New generation of virus particles bursts from host cell.

(b) The virus's capsid stays outside the cell.

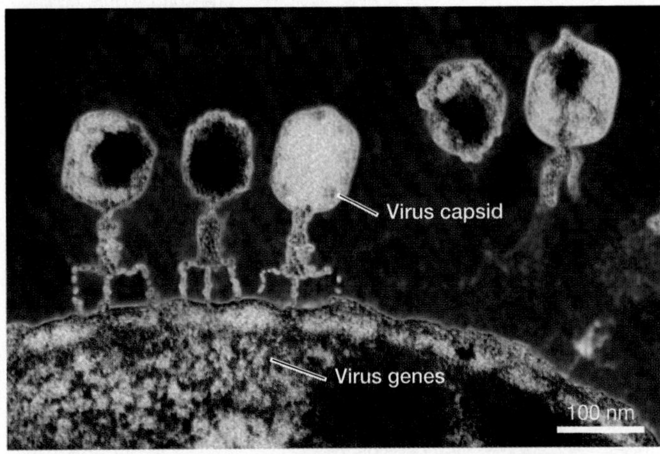

FIGURE 14.1 Viruses Inject Genes into Bacterial Cells and Leave a Capsid Behind. Color has been added to the transmission electron micrograph in (b) to make key structures more visible.

To test these predictions, Hershey and Chase sheared the ghosts off the cells by agitating each of the cultures in kitchen blenders. When the researchers spun the samples in a centrifuge, the ghosts stayed in the solution while the cells formed a pellet at the bottom of the centrifuge tube (see **BioSkills 11** in Appendix A to review how centrifugation works).

As predicted by the DNA hypothesis, the biologists found that virtually all of the radioactive protein was in the ghosts, while virtually all of the radioactive DNA was inside the host cells. Because the injected component of the virus directs the production of a new generation of virus particles, it is this component that represents the virus's genes.

After these results were published, proponents of the protein hypothesis accepted that DNA, not protein, must be the hereditary material. An astonishing claim—that DNA contained all the information for life's complexity—was correct.

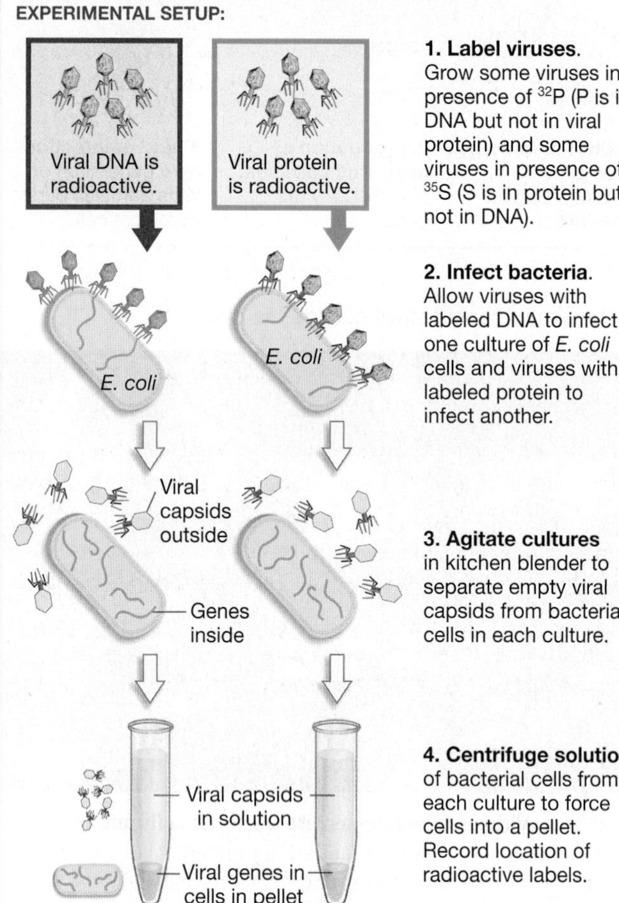

EXPERIMENT
QUESTION: Do viral genes consist of DNA or protein?
DNA HYPOTHESIS: Viral genes consist of DNA.
PROTEIN HYPOTHESIS: Viral genes consist of protein.
EXPERIMENTAL SETUP:
Viral DNA is radioactive.
Viral protein is radioactive.
1. Label viruses. Grow some viruses in presence of 32P (P is in DNA but not in viral protein) and some viruses in presence of 35S (S is in protein but not in DNA).
E. coli
E. coli
2. Infect bacteria. Allow viruses with labeled DNA to infect one culture of E. coli cells and viruses with labeled protein to infect another.
Viral capsids outside
Genes inside
3. Agitate cultures in kitchen blender to separate empty viral capsids from bacterial cells in each culture.
Viral capsids in solution
Viral genes in cells in pellet
4. Centrifuge solutions of bacterial cells from each culture to force cells into a pellet. Record location of radioactive labels.
PREDICTION OF DNA HYPOTHESIS: Radioactive DNA will be located within pellet.
PREDICTION OF PROTEIN HYPOTHESIS: Radioactive protein will be located within pellet.

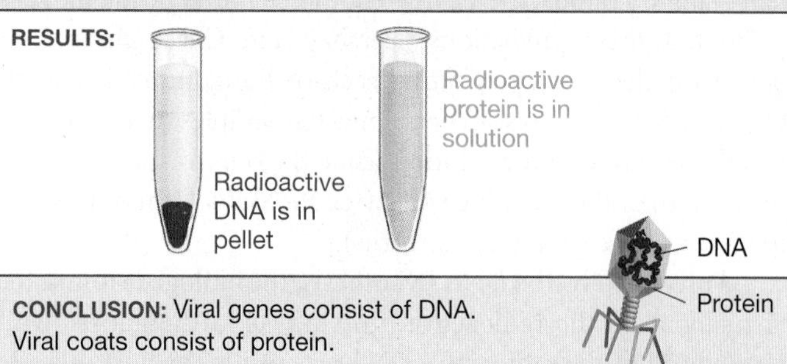

FIGURE 14.2 Experimental Evidence That DNA Is the Hereditary Material.

SOURCE: Hershey, A. D. and M. Chase. 1952. Independent functions of viral protein and nucleic acid in growth of bacteriophage. *Journal of General Physiology* 36: 39–56.

✔**QUESTION** What evidence would these investigators have to produce to convince you that the viral capsids were shaken off the bacterial cells by the agitation step?

The Secondary Structure of DNA

Chapter 4 introduced Watson and Crick's model for the secondary structure of DNA, which was proposed in 1953. Recall that DNA is a long, linear polymer made up of monomers called deoxyribonucleotides, which consist of a deoxyribose molecule (deoxyribose is a sugar), a phosphate group, and a nitrogenous base (**Figure 14.3a**). Deoxyribonucleotides link together into a polymer when a phosphodiester bond forms between a hydroxyl group on the 3′ carbon of one deoxyribose and the phosphate group attached to the 5′ carbon of another deoxyribose.

As **Figure 14.3b** shows, the primary structure of a DNA molecule has two major components: (**1**) a "backbone" made up of the sugar and phosphate groups of deoxyribonucleotides and (**2**) a series of nitrogen-containing bases that project from the backbone. A strand of DNA has a directionality, or polarity: One end has an exposed hydroxyl group on the 3′ carbon of a deoxyribose, while

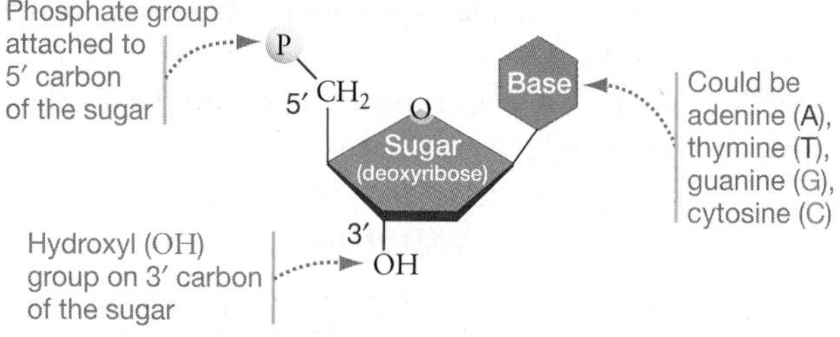

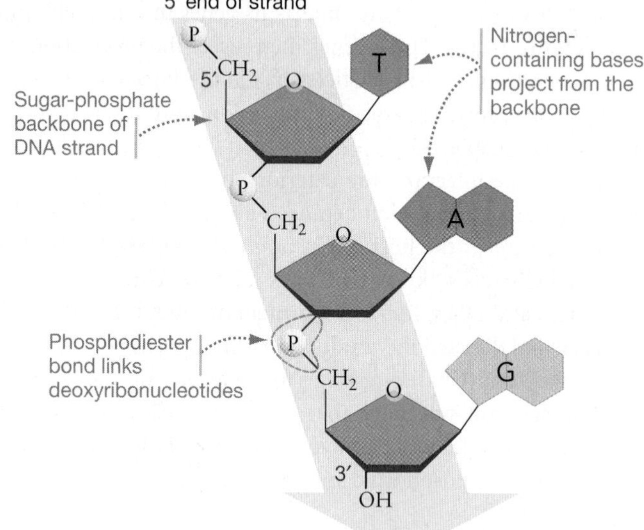

FIGURE 14.3 DNA's Primary Structure. (a) Deoxyribonucleotides are monomers that polymerize to form DNA. **(b)** DNA's primary structure is made up of a sequence of deoxyribonucleotides. Notice that the structure has a sugar-phosphate "backbone" with nitrogen-containing bases attached.

✔**EXERCISE** Write the base sequence of the DNA in part (b), in the 5′ → 3′ direction.

the other has an exposed phosphate group on a 5′ carbon. Thus, the molecule has a 3′ end and a 5′ end.

As they explored different models for the secondary structure of DNA, Watson and Crick hit on the idea of lining up two of these long strands in opposite directions, or in what is called antiparallel fashion (**Figure 14.4a**). They realized that antiparallel strands will twist around each other into a spiral or helix because certain of the nitrogen-containing bases fit together snugly in pairs inside the spiral and form hydrogen bonds (**Figure 14.4b** and **c**). The double-stranded molecule that results is called a **double helix**.

More specifically, the secondary structure is stabilized by hydrogen bonds that form between the bases adenine (A) and thymine (T) and between the bases guanine (G) and cytosine (C), along with hydrophobic interactions that the bases experience inside the helix. The pairing rules that govern this hydrogen bonding of bases are called **complementary base pairing**.

Watson and Crick realized that the A-T and G-C pairing rules suggested a way for DNA to be copied when chromosomes are replicated during S phase of the cell cycle, prior to mitosis and meiosis. They suggested that the existing strands of DNA served as a template (pattern) for the production of new strands, with bases being added to the new strands according to complementary base pairing. For example, if the template strand contained a T, then an A would be added to the new strand to pair with that T. Similarly, a G on the template strand would dictate the addition of a C on the new strand.

14.2 Testing Early Hypotheses about DNA Synthesis: The Meselson–Stahl Experiment

Complementary base pairing provided a mechanism for DNA to be copied. But many questions remained about how the copying was done. For example, biologists had three alternative hypotheses about how the old and new strands might interact during replication:

1. If the old strands of DNA separated, each could then be used as a template for the synthesis of a new, daughter strand. This hypothesis is called semiconservative replication, because each new daughter DNA molecule would consist of one old strand and one new strand.
2. If the bases temporarily turned outward so that complementary strands no longer faced each other, they could serve as a template for the synthesis of an entirely new double helix all at once. This hypothesis, called conservative replication, results in an intact parental molecule and a

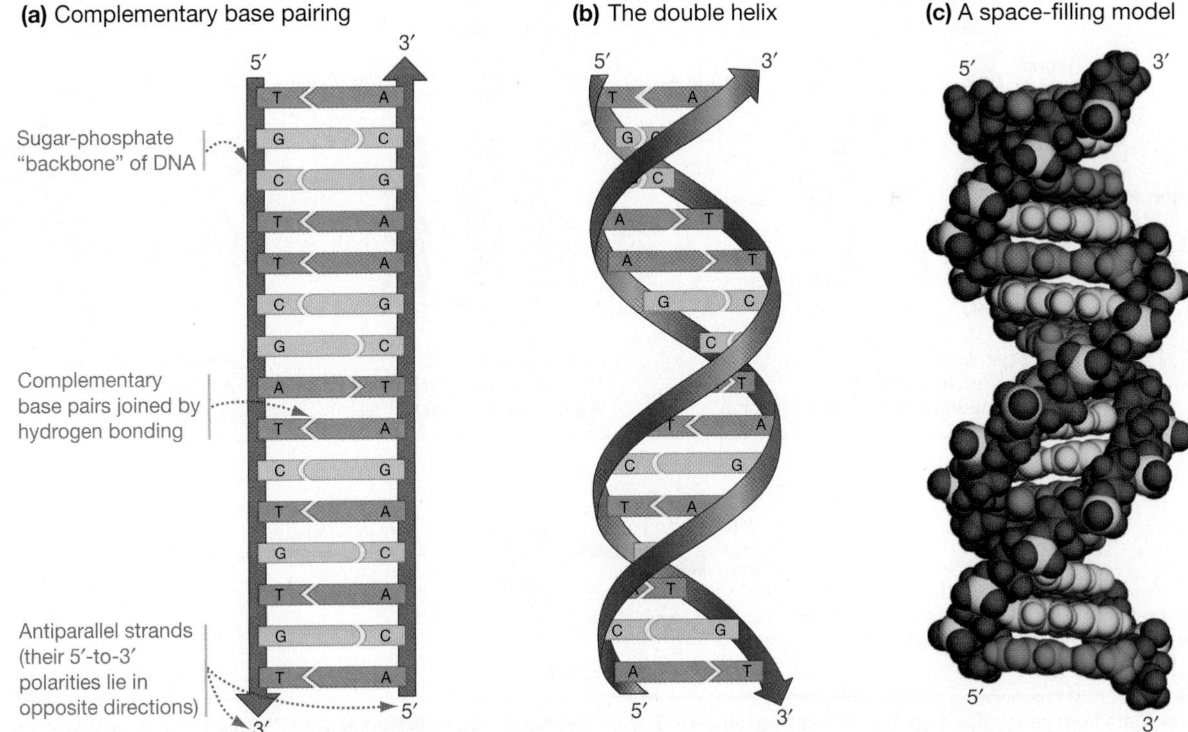

FIGURE 14.4 DNA's Secondary Structure: The Double Helix. (a) DNA normally consists of two strands, each with a sugar-phosphate backbone. Nitrogen-containing bases project from each strand and form hydrogen bonds. Only A-T and G-C pairs fit together in a way that allows hydrogen bonding to occur between the strands. **(b)** Bonding between complementary bases twists the molecule into a double helix. **(c)** This space-filling model illustrates the tight packing of atoms in the interior of the helix, where hydrogen bonding and hydrophobic interactions help stabilize the structure.

EXPERIMENT

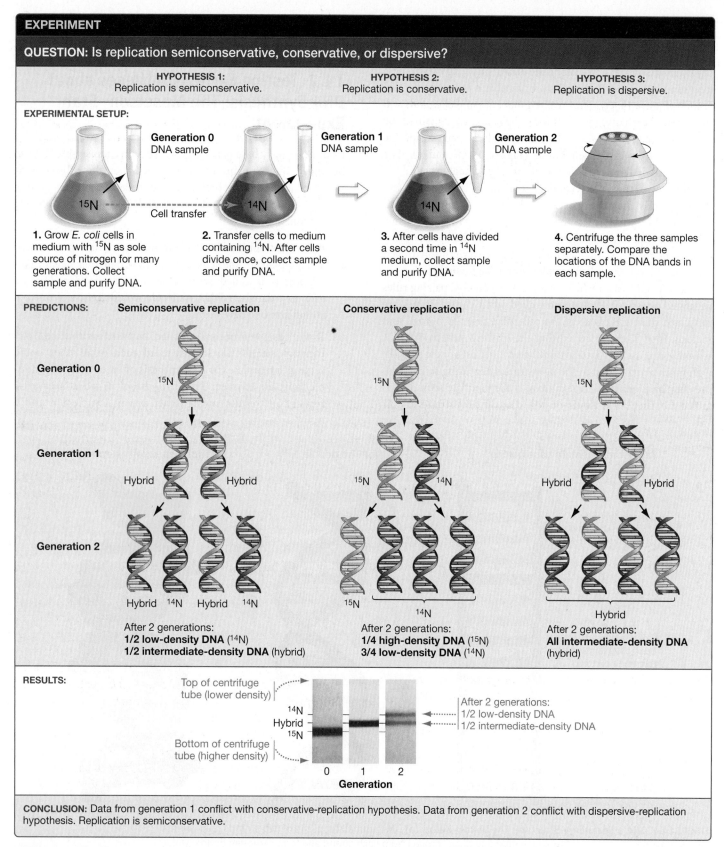

QUESTION: Is replication semiconservative, conservative, or dispersive?

HYPOTHESIS 1: Replication is semiconservative.	**HYPOTHESIS 2:** Replication is conservative.	**HYPOTHESIS 3:** Replication is dispersive.

EXPERIMENTAL SETUP:

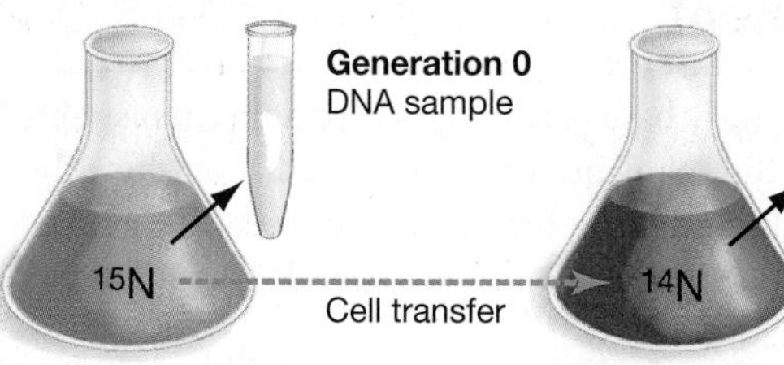
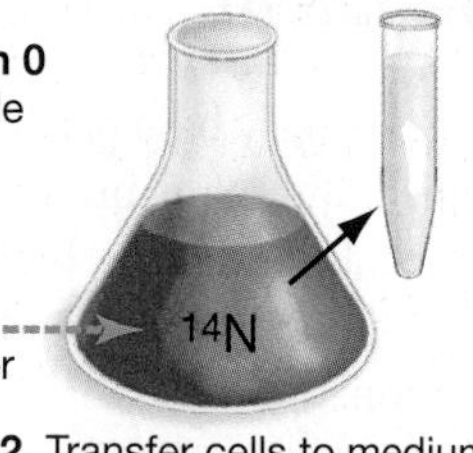
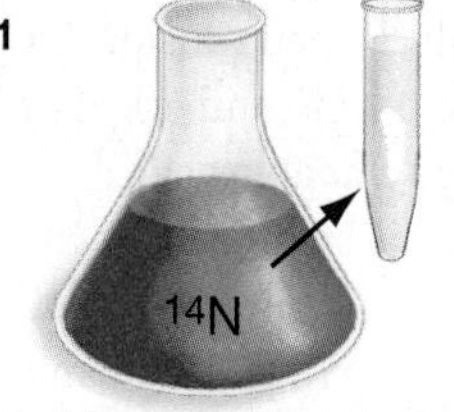
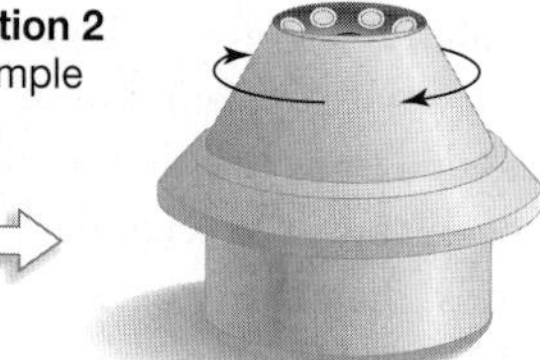

1. Grow *E. coli* cells in medium with ^{15}N as sole source of nitrogen for many generations. Collect sample and purify DNA.

2. Transfer cells to medium containing ^{14}N. After cells divide once, collect sample and purify DNA.

3. After cells have divided a second time in ^{14}N medium, collect sample and purify DNA.

4. Centrifuge the three samples separately. Compare the locations of the DNA bands in each sample.

PREDICTIONS:

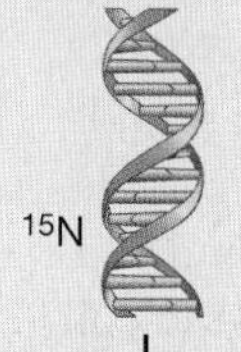
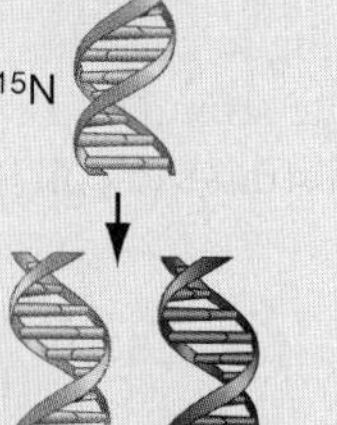
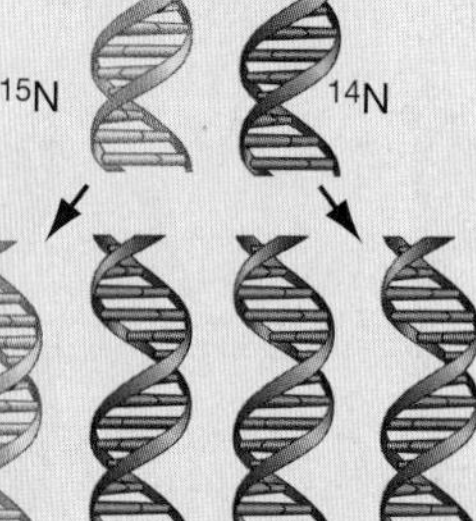
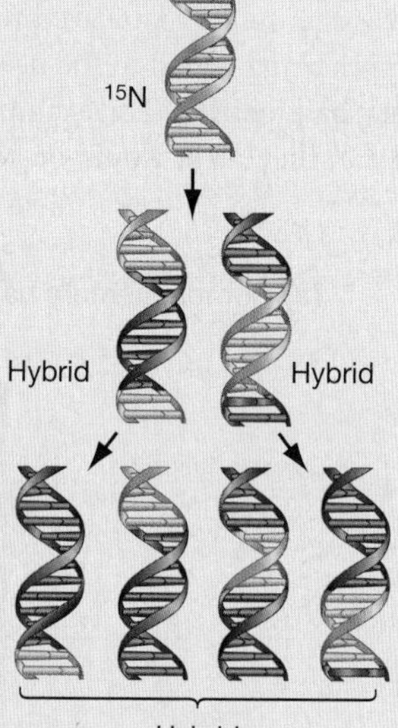

Semiconservative replication — After 2 generations:
1/2 low-density DNA (^{14}N)
1/2 intermediate-density DNA (hybrid)

Conservative replication — After 2 generations:
1/4 high-density DNA (^{15}N)
3/4 low-density DNA (^{14}N)

Dispersive replication — After 2 generations:
All intermediate-density DNA (hybrid)

RESULTS:

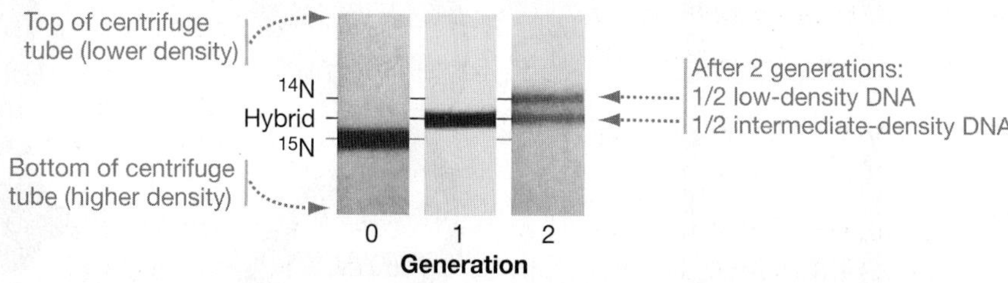

CONCLUSION: Data from generation 1 conflict with conservative-replication hypothesis. Data from generation 2 conflict with dispersive-replication hypothesis. Replication is semiconservative.

FIGURE 14.5 The Meselson–Stahl Experiment.

SOURCE: Meselson, M. and F. W. Stahl. 1958. The replication of DNA in *Escherichia coli*. *Proceedings of the National Academy of Sciences USA* 44: 671–682.

✔**EXERCISE** Meselson and Stahl actually let their experiment run for a 4th generation with cultures growing in the presence of ^{14}N. Explain what data from 3rd- and 4th-generation DNA should look like—that is, where the DNA band(s) should be.

daughter DNA molecule consisting entirely of newly synthesized strands.

3. If the parent helix was cut in short sections before being unwound, copied, and put back together, then new and old segments would alternate—stretches of old DNA would be interspersed with new DNA down the length of each daughter molecule. This possibility is called dispersive replication.

Matthew Meselson and Franklin Stahl realized that if they could tag or mark parental and daughter strands of DNA in a way that would make them distinguishable from each other, they could determine whether replication was conservative, semiconservative, or dispersive.

Before they could do any tagging, however, they needed to choose an organism to study. They decided to work with the same inhabitant of the human gastrointestinal tract, the bacterium *Escherichia coli*, that Hershey and Chase did. Because *E. coli* is small and grows quickly and readily in the laboratory, it had become a favored model organism in studies of biochemistry and molecular genetics.

Like all organisms, bacterial cells copy their entire complement of DNA, or their **genome**, before every cell division. To distinguish parental strands of DNA from daughter strands when *E. coli* replicated, Meselson and Stahl grew the cells for successive generations in the presence of different isotopes of nitrogen: first ^{15}N and later ^{14}N. Because ^{15}N contains an extra neutron, it is heavier than the normal isotope, ^{14}N.

This difference in mass, which creates a difference in density between ^{14}N-containing and ^{15}N-containing DNA, was the key to the experiment summarized in **Figure 14.5**. The logic ran as follows:

- If different nitrogen isotopes were available in the growth medium when different generations of DNA were produced, then the parental and daughter strands will have different densities.
- The technique called density gradient centrifugation separates molecules based on their density (**BioSkills 11** in Appendix A). Low-density molecules cluster in bands high in the centrifuge tube; higher-density molecules cluster in bands lower in the centrifuge tube.
- When intact, double-stranded DNA molecules are subjected to density gradient centrifugation, DNA strands that contain ^{14}N should form a band higher in the centrifuge tube; DNA strands that contain ^{15}N should form a band lower in the centrifuge tube.

In short, DNA strands containing ^{14}N and strands containing ^{15}N should form separate bands. How could this tagging system be used to test whether replication is semiconservative, conservative, or dispersive?

Meselson and Stahl began by growing *E. coli* cells with nutrients that contained only ^{15}N. They purified DNA from a sample of these cells and transferred the rest of the culture to a growth medium containing only the ^{14}N isotope. After enough time had elapsed for these cells to divide once—meaning that the DNA had been copied once—they removed a sample and isolated the DNA. After the remainder of the culture had divided again, they removed another sample and isolated its DNA.

As Figure 14.5 shows, the conservative, semiconservative, and dispersive models make distinct predictions about the makeup of the DNA molecules after replication occurs in the first and second generation.

- If replication is conservative, then the daughter cells should have double-stranded DNA with either ^{14}N or ^{15}N, but not both. As a result, two distinct DNA bands should form in the centrifuge tube—one high-density band and one low-density band.
- If replication is semiconservative or dispersive, then all of the experimental DNA should contain an equal mix of ^{14}N or ^{15}N after one generation, causing all of the DNA to form one band, intermediate in density, in the centrifuge tube.
- After two generations, the situation changes. If replication is semiconservative, then half of the daughter cells should contain only ^{14}N—meaning a second, lower-density band should appear in the centrifuge tube. But the dispersive model predicts that after two generations there will still be just one band, intermediate in density.

The photograph at the bottom of Figure 14.5 shows the experiment's results. After one generation, the density of the DNA molecules was intermediate. These data suggested that the hypothesis of conservative replication was wrong. After two generations, a lower-density band appeared in addition to the intermediate-density band. This result offered strong support for the hypothesis that DNA replication is not dispersive but semiconservative. Each newly made DNA molecule comprises one old strand and one new strand.

14.3 A Comprehensive Model for DNA Synthesis

The DNA inside a cell is like an ancient text that has been painstakingly copied and handed down, generation after generation. But while the most ancient of all human texts contain messages that are thousands of years old, the DNA in living cells has been copied and passed down for billions of years. And instead of being copied by monks or clerks, DNA is replicated by molecular scribes. What molecules are responsible for copying DNA, and how do they work?

Meselson and Stahl showed that each strand of DNA is copied in its entirety each time replication occurs, but how does DNA synthesis proceed? Does it require an input of energy in the form of ATP, or it is spontaneous? Is it catalyzed by an enzyme, or does it occur quickly on its own?

The initial breakthrough in research on the reactions that make up DNA replication came with the discovery of an enzyme called **DNA polymerase**, so named because it polymerizes deoxyribonucleotides to DNA. This protein was found to catalyze DNA synthesis. Follow-up work showed that organisms contain several types of DNA polymerases. DNA polymerase III, for example, is the enzyme that is primarily responsible for copying *E. coli*'s chromosome prior to cell division.

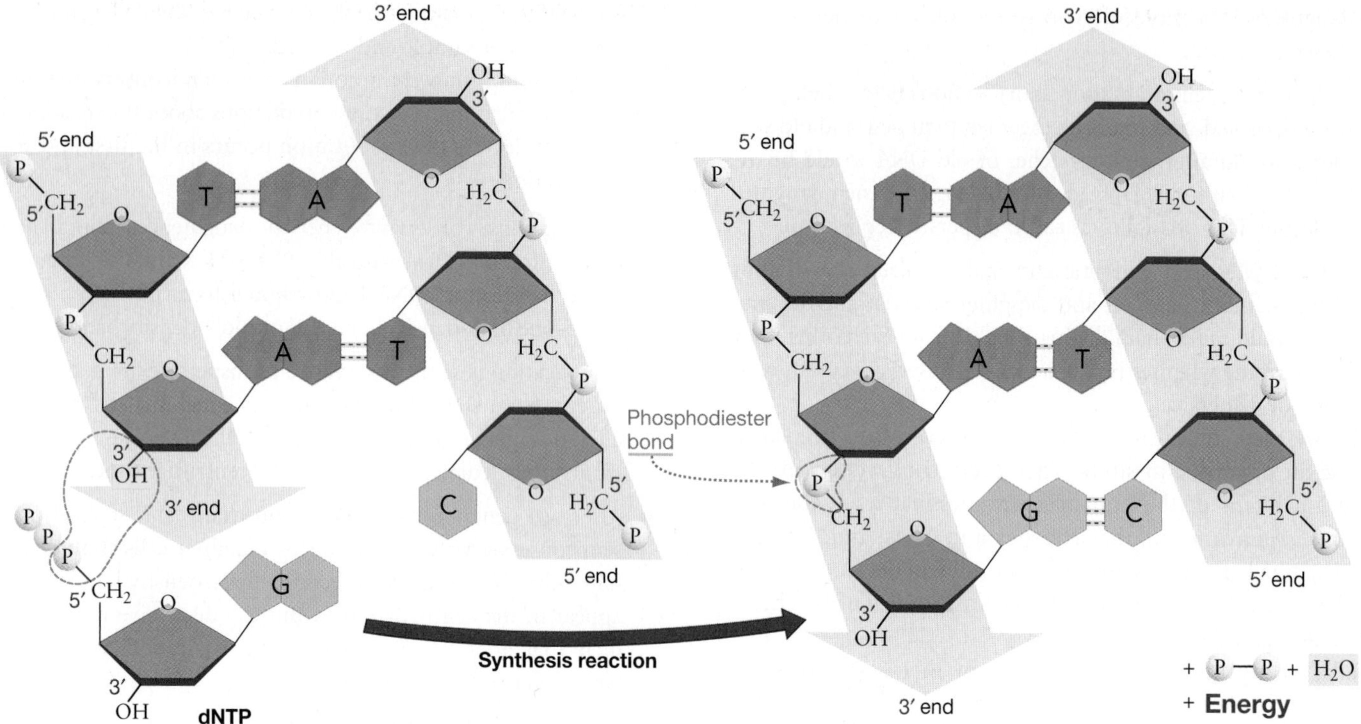

FIGURE 14.6 The DNA Synthesis Reaction. A condensation reaction results in formation of a phosphodiester bond between the 3′ carbon on the end of a DNA strand and the 5′ carbon on an incoming dNTP (deoxyribonucleoside triphosphate) monomer.

Figure 14.6 illustrates a critical characteristic of DNA polymerases: They can work in only one direction. DNA polymerases can add deoxyribonucleotides to only the 3′ end of a growing DNA chain. As a result, DNA synthesis always proceeds in the 5′ → 3′ direction. ✔**If you understand this concept, you should be able to draw two lines representing a DNA molecule, assign the 3′ to 5′ polarity of each strand, and then label the direction in which DNA synthesis will proceed for each strand.**

Figure 14.6 makes another important point about DNA synthesis. You might recall from earlier chapters that polymerization reactions are endergonic. But in cells, the reaction is exergonic because the monomers that act as substrates in the reaction are **deoxyribonucleoside triphosphates (dNTPs)**. (The *N* in dNTP stands for any of the four bases found in DNA: adenine, thymine, guanine, or cytosine). Because they have three phosphate groups close together, dNTPs have high potential energy—high enough to make the formation of phosphodiester bonds in a growing DNA strand exergonic (see Chapter 9).

How Does Replication Get Started?

Another major insight into the mechanism of DNA synthesis emerged when electron micrographs caught chromosome replication in action. As **Figure 14.7a** shows, a "bubble" forms in a chromosome when DNA is actively being synthesized. Initially, the replication bubble forms at a specific sequence of bases called the **origin of replication** (**Figure 14.7b**). Bacterial chromosomes have a single location where the replication process begins, and thus a single bubble forms. Eukaryotes have multiple sites along each chromosome where DNA synthesis begins, and thus multiple replication bubbles (**Figure 14.7c**).

Replication bubbles grow as DNA replication proceeds, because synthesis is bidirectional—that is, it occurs in both directions at the same time.

A specific suite of proteins is responsible for recognizing sites where replication begins and opening the double helix at those points. These proteins are activated by the proteins that initiate S phase in the cell cycle (see Chapter 11).

Once a replication bubble opens, a different suite of enzymes takes over and initiates replication. The action takes place in the corners of each replication bubble—at a structure called the replication fork (shown in Figure 14.7c). A **replication fork** is a Y-shaped region where the parent-DNA double helix is split into two single strands, which are then copied.

How Is the Helix Opened and Stabilized?

A battery of enzymes and specialized proteins converge on the point where the double helix opens. An enzyme called a **helicase** breaks the hydrogen bonds between deoxyribonucleotides. This reaction causes the two strands of DNA to separate. Proteins called **single-strand DNA-binding proteins** (**SSBPs**) attach to the separated strands and prevent them from snapping back into a double helix. In combination, then, the helicase and single-strand DNA-binding proteins open up the double helix and make both strands available for copying (**Figure 14.8**, step 1).

The "unzipping" process that occurs at the replication fork creates tension farther down the helix. To understand why, imag-

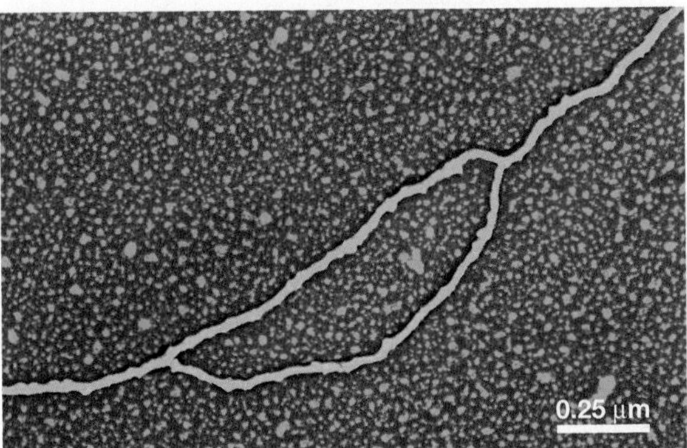

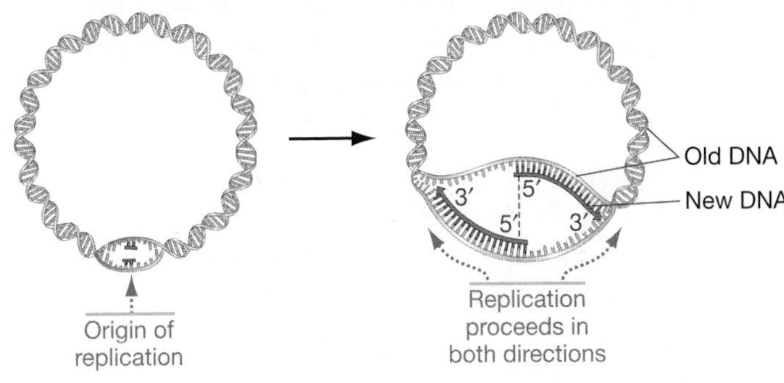

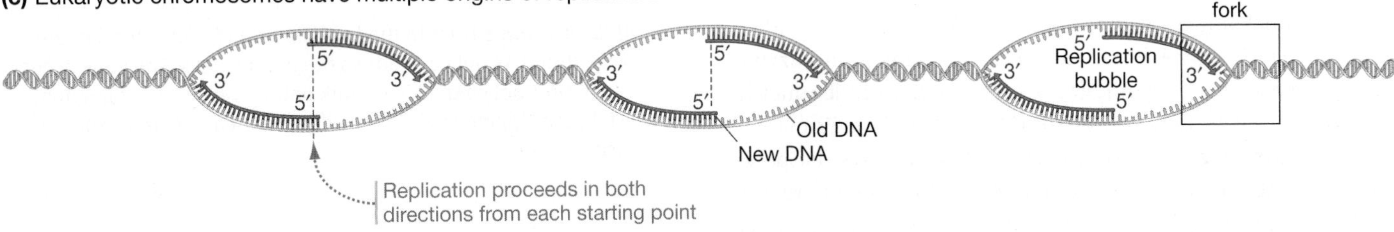

FIGURE 14.7 DNA Synthesis Proceeds in Two Directions from a Point of Origin. Color has been added to the micrograph in part (a).

ine what would happen if you started to pull apart the twisted strands of a rope. The untwisting movements at one end would force the intact section to rotate in response. If the intact end of the rope were fixed in place, though, it would eventually begin to coil on itself and kink in response to the twisting forces. This does not happen in DNA, because the twisting stress induced by helicase is relieved by the proteins called topoisomerases. A **topoisomerase** is an enzyme that cuts DNA, allows it to unwind, and rejoins it ahead of the advancing replication fork.

Now, what happens once the DNA helix is open and has stabilized?

How Is the Leading Strand Synthesized?

The keys to understanding what happens at the start of DNA synthesis are to recall that DNA polymerase works only in the $5' \rightarrow 3'$ direction and to recognize that to start synthesis, both a $3'$ end and a single-stranded template are required. The single-stranded template dictates which deoxyribonucleotide should be

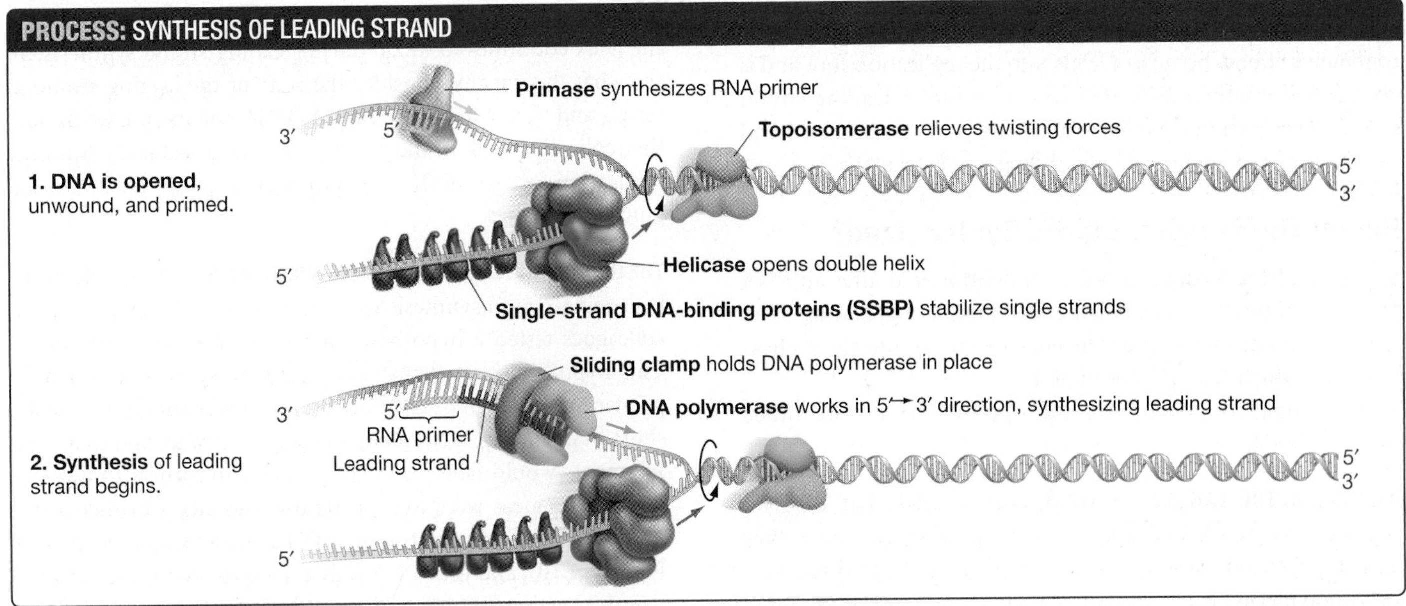

FIGURE 14.8 Leading-Strand Synthesis Is Continuous.

added next. DNA polymerase requires a **primer**—which consists of a few nucleotides bonded to the template strand—because it provides a free 3′ hydroxyl (–OH) group that can combine with an incoming dNTP to form a phosphodiester bond.

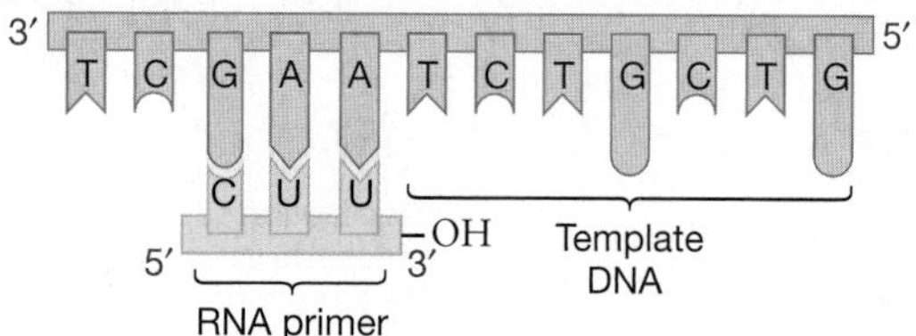

Once a primer is added to a single-stranded template, DNA polymerase begins working in the 5′ → 3′ direction and adds deoxyribonucleotides to complete the complementary strand.

Before DNA synthesis can get under way, then, an enzyme called **primase** has to synthesize a short stretch of ribonucleic acid (RNA) that acts as a primer for DNA polymerase. Primase is a type of **RNA polymerase**—an enzyme that catalyzes the polymerization of ribonucleotides into RNA (see Chapter 4 to review RNA's structure). Unlike DNA polymerases, primase and other RNA polymerases do not require a primer. These enzymes can attach ribonucleotides directly by complementary base pairing to single-stranded DNA. In this way, primase creates a primer for DNA synthesis.

Once the primer is in place, DNA polymerase begins adding deoxyribonucleotides to the 3′ end of it, producing a new strand with a sequence that is complementary to the template strand. As Figure 14.8, step 2, shows, DNA polymerase has a shape that grips the DNA strand during synthesis, a little like your hand clasping a rope. Catalysis takes place in the groove inside the enzyme, at an active site between the enzyme's "thumb" and "fingers." As DNA polymerase moves along the DNA molecule, a doughnut-shaped structure behind it, called the sliding clamp, holds the enzyme in place.

The enzyme's product is called the **leading strand**, or **continuous strand**, because it leads into the replication fork and is synthesized continuously. ✔If you understand leading-strand synthesis, you should be able to list the enzymes involved and predict the consequences if any of them are defective.

How Is the Lagging Strand Synthesized?

Synthesis of the leading strand is straightforward after an RNA primer is in place—DNA polymerase chugs along, adding bases to the 3′ end of that strand. The enzyme moves into the replication fork, which "unzips" ahead of it.

By comparison, events on the opposite strand are much more involved.

WHY DOES THE LAGGING STRAND LAG? Recall that the two strands in the DNA double helix are antiparallel—meaning they lie parallel to one another but oriented in opposite directions. DNA polymerase works in only one direction, however, so if the DNA polymerase that is synthesizing the leading strand works

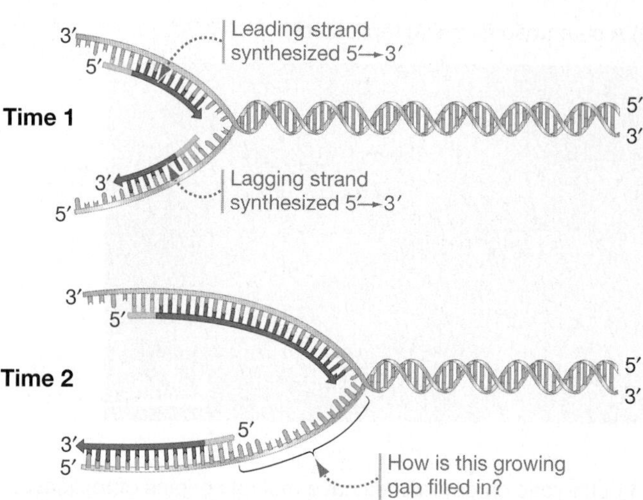

FIGURE 14.9 Gaps Form in the Lagging Strand. Gaps form in the lagging strand as helicase opens the double helix. The gaps occur because DNA polymerase can work only in the 5′ → 3′ direction, which in the lagging strand is the direction leading away from the replication fork.

into the replication fork, then a DNA polymerase must work *away* from the replication fork to synthesize the other strand in the 5′ → 3′ direction.

The strand that is synthesized in the opposite direction of the replication fork is called the **lagging strand**, because it lags behind the synthesis occurring in the fork.

The synthesis of the lagging strand starts when primase synthesizes a short stretch of RNA that acts as a primer. DNA polymerase III then adds bases to the 3′ end of the primer. The key observation is that the enzyme moves away from the replication fork. But behind it, helicase continues to open the replication fork and expose new single-stranded DNA on the lagging strand.

These events create a paradox. New single-stranded template DNA is constantly appearing behind DNA polymerase—in the direction *opposite* the direction of lagging-strand synthesis—as the helix continues to unzip during leading-strand synthesis. To see what this means, consider the state of the lagging strand at time 1 and time 2 in **Figure 14.9**. As DNA polymerase works into the replication fork in the 5′ → 3′ direction and away from the replication fork in the 5′ → 3′ direction, gaps appear in the lagging strand.

THE DISCONTINUOUS REPLICATION HYPOTHESIS The puzzle posed by lagging-strand synthesis was resolved when Reiji Okazaki and colleagues tested a hypothesis called discontinuous replication. This hypothesis stated that once primase synthesizes an RNA primer on the lagging strand, DNA polymerase might synthesize short fragments of DNA along the lagging strand, and that these fragments would later be linked together to form a continuous whole. The ideas were that primase would add a primer to the newly exposed single-stranded DNA at intervals (step 1 in **Figure 14.10**), and that DNA polymerase III would then synthesize the lagging strand until it reached the fragment produced earlier.

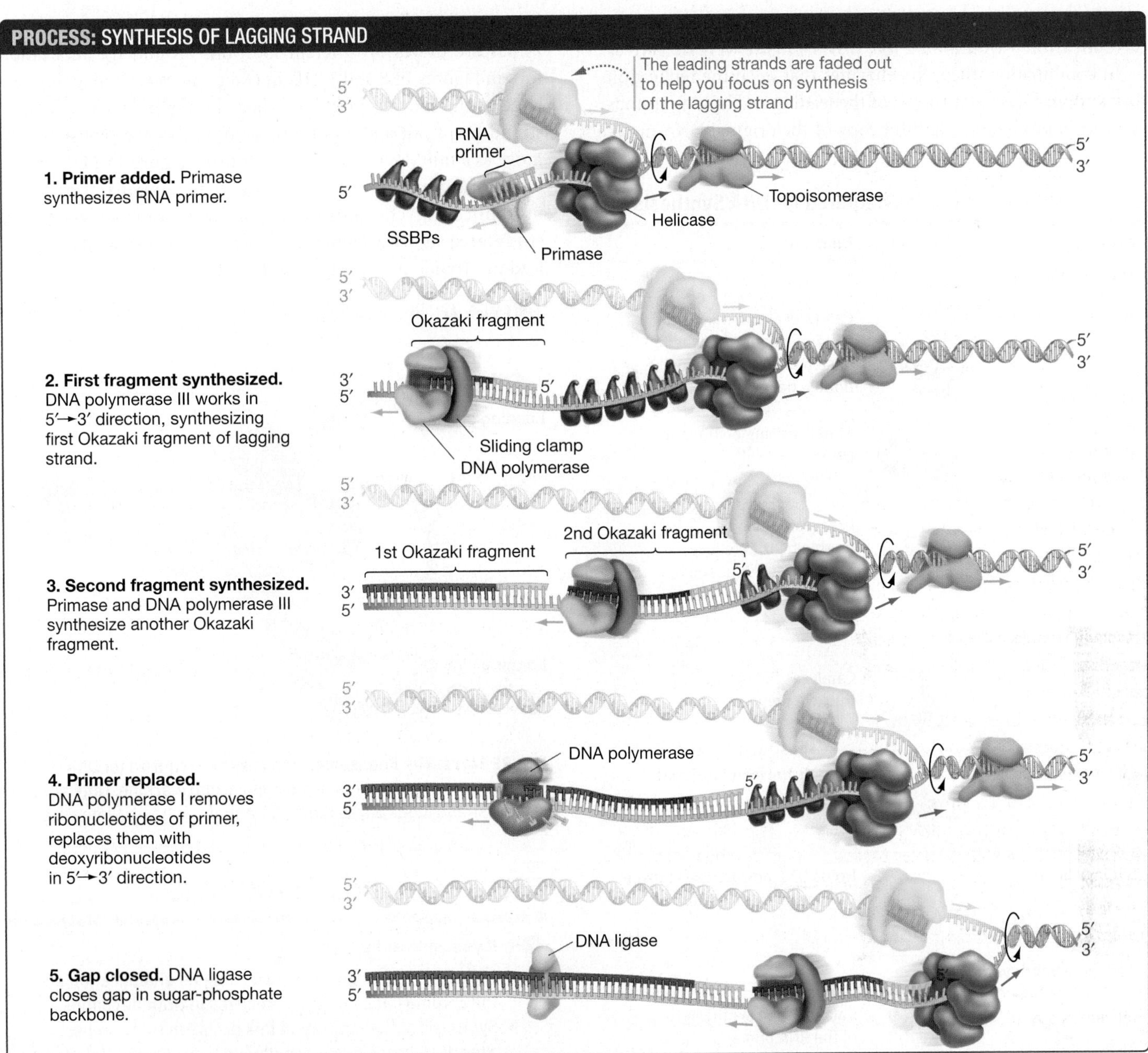
PROCESS: SYNTHESIS OF LAGGING STRAND
The leading strands are faded out to help you focus on synthesis of the lagging strand
1. Primer added. Primase synthesizes RNA primer.
RNA primer
SSBPs
Primase
Helicase
Topoisomerase
2. First fragment synthesized. DNA polymerase III works in 5′→3′ direction, synthesizing first Okazaki fragment of lagging strand.
Okazaki fragment
Sliding clamp
DNA polymerase
3. Second fragment synthesized. Primase and DNA polymerase III synthesize another Okazaki fragment.
1st Okazaki fragment
2nd Okazaki fragment
4. Primer replaced. DNA polymerase I removes ribonucleotides of primer, replaces them with deoxyribonucleotides in 5′→3′ direction.
DNA polymerase
5. Gap closed. DNA ligase closes gap in sugar-phosphate backbone.
DNA ligase
5′
3′

FIGURE 14.10 The Completion of DNA Replication.

To explore this hypothesis, Okazaki's group set out to test a key prediction: Could they document the existence of short DNA fragments produced during replication? Their critical experiment was based on the pulse-chase strategy introduced in Chapter 7. Specifically, they added a short "pulse" of a radioactive deoxyribonucleotide to *E. coli* cells, followed by a large "chase" of nonradioactive deoxyribonucleotide. According to their discontinuous replication model, some of this radioactive deoxyribonucleotide should end up in short, single-stranded fragments of DNA.

THE DISCOVERY OF OKAZAKI FRAGMENTS As predicted, the researchers succeeded in finding these fragments when they purified DNA from the experimental cells and separated the molecules by centrifugation. A small number of labeled pieces of DNA, about 1000 base pairs long, were present. These short sections came to be known as **Okazaki fragments** (see steps 2 and 3 of Figure 14.10). Subsequent work showed that Okazaki fragments in eukaryotes are smaller—just 100 to 200 base pairs long.

How are Okazaki fragments connected into a continuous whole? As step 4 of Figure 14.10 shows, a DNA polymerase removes the RNA primer at the start of each fragment and fills in the appropriate deoxyribonucleotides. Lastly, an enzyme called **DNA ligase** catalyzes the formation of a phosphodiester bond between the adjacent fragments (Figure 14.10, step 5).

Because Okazaki fragments are synthesized independently and joined together later, the lagging strand is also called the **discontinuous strand.** ✔If you understand lagging-strand synthesis, you should be able to draw what the two molecules resulting from DNA synthesis—starting from a single origin of replication—would look like if DNA ligase were defective.

In combination, then, the enzymes that open the replication fork and manage the synthesis of the leading and lagging strands succeed in producing a faithful copy of the original DNA molecule prior to mitosis or meiosis (**Table 14.1**). Although the enzymes are drawn at different locations around the replication fork in Figures 14.8 and 14.10, in reality most are joined into one large multi-enzyme machine called the **replisome.** The lagging strand loops out and around the complex so the replisome can move as a unit down the replication fork (**Figure 14.11**).

SUMMARY TABLE 14.1 **Proteins Required for DNA Synthesis**

Name	Structure	Function
Opening the helix		
Helicase		Catalyzes the breaking of hydrogen bonds between base pairs to open the double helix
Single-strand DNA-binding proteins		Stabilizes single-stranded DNA
Topoisomerase		Breaks and rejoins the DNA double helix to relieve twisting forces caused by the opening of the helix
Leading strand synthesis		
Primase		Catalyzes the synthesis of the RNA primer
DNA polymerase III		Extends the leading strand
Sliding clamp		Holds DNA polymerase in place during strand extension
Lagging strand synthesis		
Primase		Catalyzes the synthesis of the RNA primer on an Okazaki fragment
DNA polymerase III		Extends an Okazaki fragment
Sliding clamp		Holds DNA polymerase in place during strand extension
DNA polymerase I		Removes the RNA primer and replaces it with DNA
DNA ligase		Catalyzes the joining of Okazaki fragments into a continuous strand

Before moving on to consider what happens at the ends of replicating chromosomes, you may want to go to the study area at *www.masteringbiology.com* to review the steps in leading strand and lagging strand synthesis.

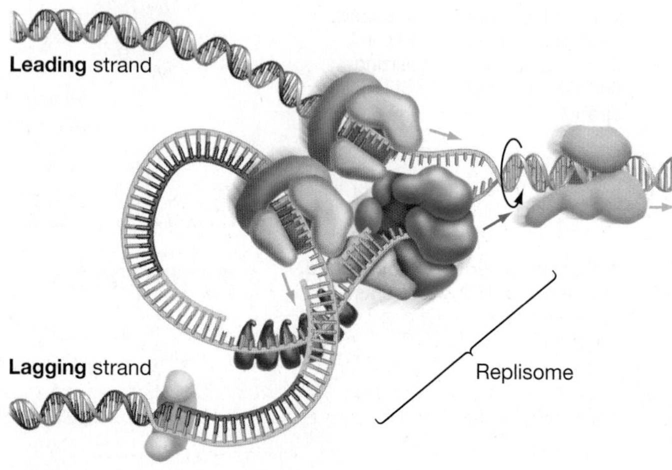

FIGURE 14.11 The Replisome. The enzymes required for DNA synthesis are organized into a multi-molecular machine. Note how the lagging strand loops out as the leading strand is being synthesized.

CHECK YOUR UNDERSTANDING

If you understand that . . .

- DNA synthesis begins at specific origins of replication on the chromosome and then proceeds in both directions.
- Synthesis at the replication fork occurs in three steps:
  Step 1 Helicase opens the double helix, SSBPs stabilize the exposed single strands, and topoisomerase prevents kinks downstream of the fork;
  Step 2 DNA polymerase synthesizes the leading strand after primase has added an RNA primer; and
  Step 3 a series of enzymes synthesizes the lagging strand.
- Lagging-strand synthesis cannot be continuous, because it moves away from the replication fork. In bacteria, enzymes called primase, DNA polymerase III, DNA polymerase I, and ligase work in sequence to synthesize Okazaki fragments and link them into a continuous whole.

✔ **You should be able to . . .**

1. Explain the function of primase.
2. Explain why there is no helicase or topoisomerase required during lagging-strand synthesis.

Answers are available in Appendix B.

14.4 Replicating the Ends of Linear Chromosomes

The circular DNA molecules in bacteria and archaea can be synthesized by the events and enzymes introduced in Section 14.3, and so can the leading and lagging strands of the linear DNA molecules found in eukaryotes. But replication at the ends of linear chromosomes is another story altogether.

The region at the end of a linear chromosome is called a **telomere** (literally, "end-part"). **Figure 14.12** illustrates the problem that arises during the replication of telomeres.

- When the replication fork reaches the end of a linear chromosome, DNA polymerase synthesizes the leading strand all the way to the end of the parent DNA template (step 1 and step 2, top strand). As a result, leading-strand synthesis results in a normal copy of the DNA molecule.
- On the lagging strand, primase adds an RNA primer close to the tip of the chromosome (see step 2, bottom strand).
- DNA polymerase synthesizes the final Okazaki fragment on the lagging strand (step 3). DNA polymerase I removes the primer.
- DNA polymerase is unable to add DNA near the tip of the chromosome because there is not enough room for primase to add a new RNA primer (step 4). As a result, the single-stranded DNA that is left must stay single stranded.

The single-stranded DNA that remains at the end of the lagging strand is eventually degraded, which results in the shortening of the chromosome. If this process were to continue unabated, every chromosome would shorten by 50 to 100 deoxyribonucleotides on average each time DNA replication occurred. Over time, linear chromosomes would be expected to disappear completely.

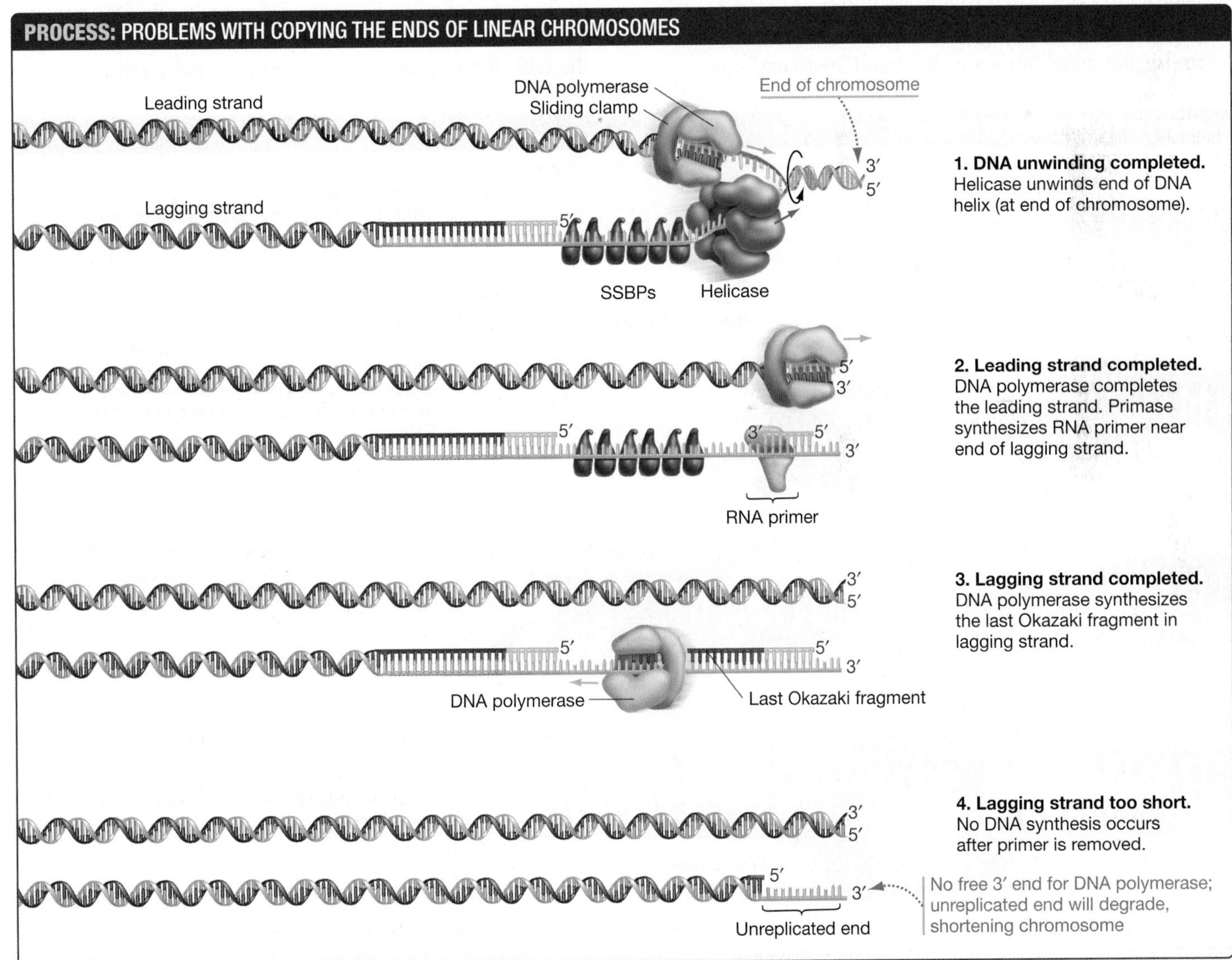
PROCESS: PROBLEMS WITH COPYING THE ENDS OF LINEAR CHROMOSOMES
Leading strand
DNA polymerase
Sliding clamp
End of chromosome
3′
5′
Lagging strand
5′
SSBPs
Helicase
1. DNA unwinding completed.
Helicase unwinds end of DNA helix (at end of chromosome).
5′
3′
5′
3′
5′
3′
RNA primer
2. Leading strand completed.
DNA polymerase completes the leading strand. Primase synthesizes RNA primer near end of lagging strand.
3′
5′
5′
5′
3′
DNA polymerase
Last Okazaki fragment
3. Lagging strand completed.
DNA polymerase synthesizes the last Okazaki fragment in lagging strand.
3′
5′
5′
3′
Unreplicated end
No free 3′ end for DNA polymerase; unreplicated end will degrade, shortening chromosome
4. Lagging strand too short.
No DNA synthesis occurs after primer is removed.

FIGURE 14.12 Telomeres Shorten during Normal DNA Replication. An RNA primer is added to the lagging strand near the end of the chromosome. Once the primer is removed, it cannot be replaced with DNA. As a result, the chromosome shortens.

How do eukaryotes maintain the integrity of their linear chromosomes? An answer emerged after two important discoveries were made:

1. Telomeres do not contain genes that code for products needed in the cell.
2. An interesting enzyme called telomerase is involved in replicating telomeres.

Instead of genes, telomeres consist of short stretches of bases that are repeated over and over. In humans, for example, the base sequence TTAGGG is repeated thousands of times. Human telomeres consist of a total of about 10,000 deoxyribonucleotides with the sequence TTAGGGTTAGGGTTAGGG....

Telomerase is remarkable because it catalyzes the synthesis of DNA from an RNA template. In fact, the enzyme carries an RNA molecule with it that acts as a built-in template, allowing telomerase to add DNA onto the end of a chromosome and prevent it from getting shorter.

Figure 14.13 shows how telomerase works.

Step 1 The unreplicated segment of the telomere at the 3′ end of the lagging strand forms a single-strand "overhang."

Step 2 Telomerase binds to the overhanging section of single-stranded parent DNA. Once the enzyme has bound, it begins catalyzing the addition of deoxyribonucleotides, in the 5′ → 3′ direction, that are complementary to its built-in RNA template.

Step 3 Telomerase moves in the 5′ → 3′ direction and continues to catalyze the addition of deoxyribonucleotides.

Step 4 Once the single-stranded "overhang" on the lagging strand is lengthened in this way, the normal machinery of DNA synthesis—primase, DNA polymerase, and ligase—resumes synthesis of the lagging strand in the 5′ → 3′ direction. The result is that the lagging strand becomes slightly longer than it was originally.

If you go to the study area at *www.masteringbiology.com*, you'll be about to review all aspects of DNA synthesis, in action:

BioFlix™ DNA Replication

It is important to recognize, though, that telomerase is not active in most types of cells. In humans, for example, active telomerase is found primarily in the cells of reproductive organs—specifically, in the cells that eventually undergo meiosis and produce gametes.

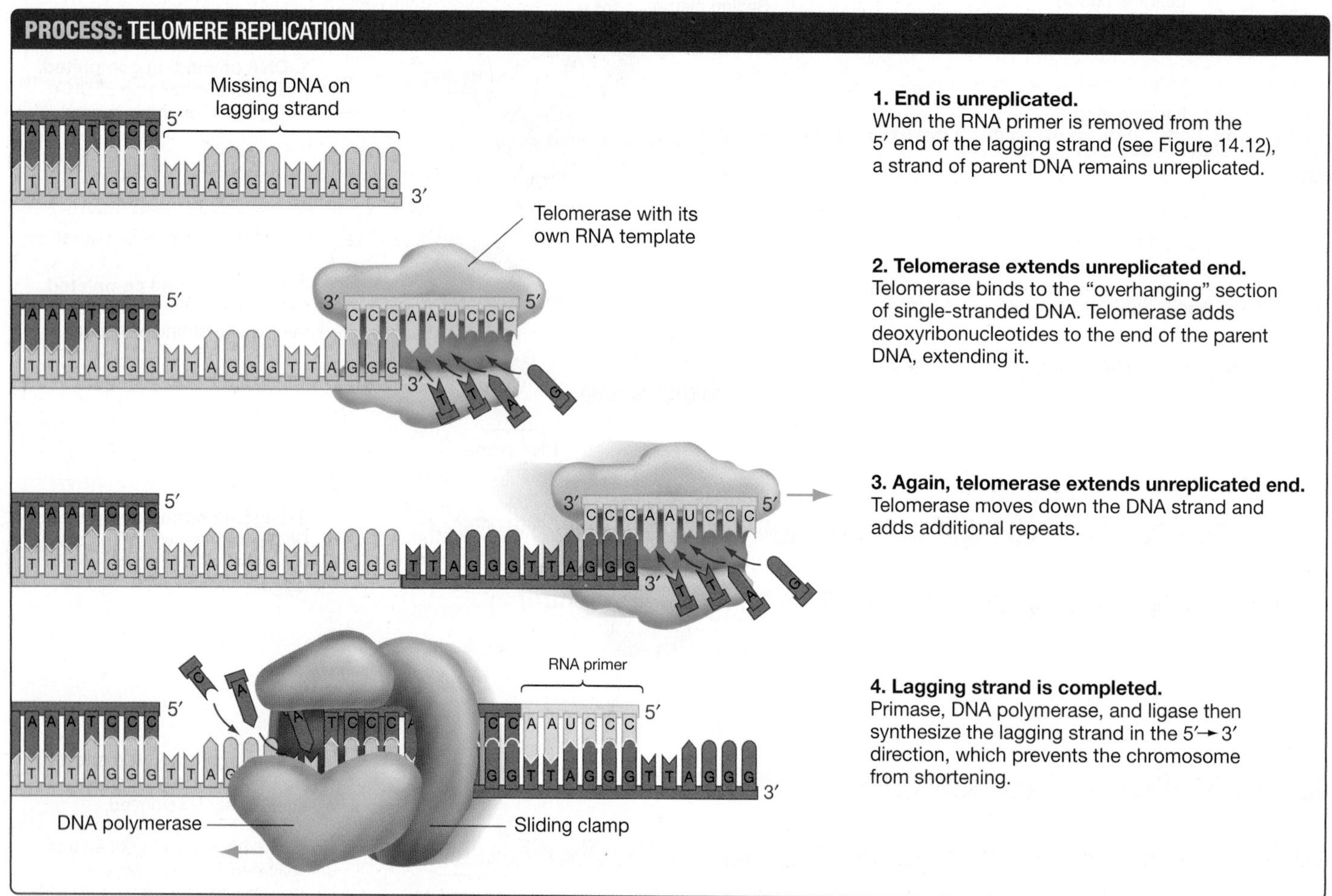

PROCESS: TELOMERE REPLICATION
Missing DNA on lagging strand
5′
3′
AAATCCC
TTTAGGGTTAGGGTTAGGG
1. End is unreplicated.
When the RNA primer is removed from the 5′ end of the lagging strand (see Figure 14.12), a strand of parent DNA remains unreplicated.
Telomerase with its own RNA template
CCCAAUCCC
2. Telomerase extends unreplicated end.
Telomerase binds to the "overhanging" section of single-stranded DNA. Telomerase adds deoxyribonucleotides to the end of the parent DNA, extending it.
3. Again, telomerase extends unreplicated end.
Telomerase moves down the DNA strand and adds additional repeats.
RNA primer
CCAAUCCC
Sliding clamp
DNA polymerase
4. Lagging strand is completed.
Primase, DNA polymerase, and ligase then synthesize the lagging strand in the 5′→3′ direction, which prevents the chromosome from shortening.

FIGURE 14.13 Telomerase Prevents Shortening of Telomeres during Replication. By extending the number of repeated sequences in the 5′ → 3′ direction, telomerase provides room for enzymes to add an RNA primer to the lagging strand. DNA polymerase can then fill in the missing section of the lagging strand.

✔**QUESTION** Would this telomerase work as well if its RNA template had a different sequence?

Somatic cells, meaning cells that are not involved in gamete formation, normally lack telomerase. As predicted, the chromosomes of somatic cells gradually shorten with each mitotic division, getting progressively smaller as an individual grows and ages.

These observations inspired a pair of important hypotheses. The first was that telomere shortening causes cells to stop dividing and enter the nondividing state called G_0 (see Chapter 11). The second was that if telomerase were mistakenly activated in a somatic cell, telomeres would fail to shorten. This would allow the cell to keep dividing and might possibly contribute to uncontrolled growth and cancer.

To test the first hypothesis, biologists added functioning telomerase to human cells growing in vitro. The treated cells continued dividing long past the age when otherwise identical cells stop growing. These results have convinced most biologists that telomere shortening has a role in limiting the amount of time cells remain in an actively growing state.

A link between continued telomerase activity and cancer formation has been harder to nail down, however. One suggestive observation is that many cancerous cells in humans and other organisms have functioning telomerase or some other mechanism for maintaining telomere length, while their noncancerous cells do not.

Could drugs that knock out telomerase be an effective way to fight cancer? So far, the data on this question are unclear. Research continues.

CHECK YOUR UNDERSTANDING

If you understand that . . .

- Chromosomes shorten during replication because the end of the lagging strand lacks a primer and cannot be synthesized.
- Shortening is prevented in certain cells—particularly those that produce sperm and egg—because telomerase adds short, repeated DNA sequences to the template strand. Primase can then add an RNA primer to the lagging strand, and DNA polymerase can fill in the missing sections.

✓ You should be able to . . .

1. Explain why telomerase is not found in bacterial cells.
2. Explain why telomerase has to have a built-in template.

Answers are available in Appendix B.

14.5 Repairing Mistakes and Damage

DNA polymerases work fast. In yeast, for example, each replication fork is estimated to move at a rate of about 50 bases per second. But the replication process is also astonishingly accurate. In organisms ranging from *E. coli* to animals, the error rate during DNA replication averages less than one mistake per *billion* deoxyribonucleotides.

This level of accuracy is critical. Humans, for example, develop from a fertilized egg that has DNA containing over 6 billion deoxyribonucleotides. The DNA inside the fertilized egg is replicated over and over to create the trillions of cells that eventually make up the adult body. If more than one or two mutations occurred during each cell division cycle as a human grew, genes would be riddled with errors by the time the individual reached maturity. Genes that contain errors are often defective.

Based on these observations, it is no exaggeration to claim that the accurate replication of DNA is a matter of life and death. Natural selection favors individuals with enzymes that copy DNA as quickly and exactly as possible.

These observations raise a key question. How can the enzymes of DNA replication be as precise as they are?

Correcting Mistakes in DNA Synthesis

As DNA polymerase marches along a parent DNA template, hydrogen bonding occurs between incoming deoxyribonucleotides and the deoxyribonucleotides on the template strand. DNA polymerases are selective about the bases they add to a growing strand because the correct base pairings (A-T and G-C) are energetically the most favorable of all possibilities for the pairing of nitrogen-containing bases. As a result, the enzyme inserts an incorrect deoxyribonucleotide only about once every 100,000 bases added (**Figure 14.14a**).

An error rate of one in 100,000 seems low, but it is much higher than the one-in-a-billion rate claimed at the start of this section. What happens when DNA polymerase makes a mistake?

DNA POLYMERASE CAN PROOFREAD Biologists were able to study why DNA synthesis is so accurate when they found cells where DNA synthesis was *in*accurate.

Specifically, researchers found mutants in *E. coli* with error rates that were 100 times greater than normal. Recall from Chapter 13 that a mutant is an individual with a novel trait caused by mutation, and that mutation is a change in the gene responsible for that trait. Many mutations change the individual's phenotype. The change may result in a trait such as white eyes in fruit flies or an elevated

(a) DNA polymerase III adds a mismatched base...

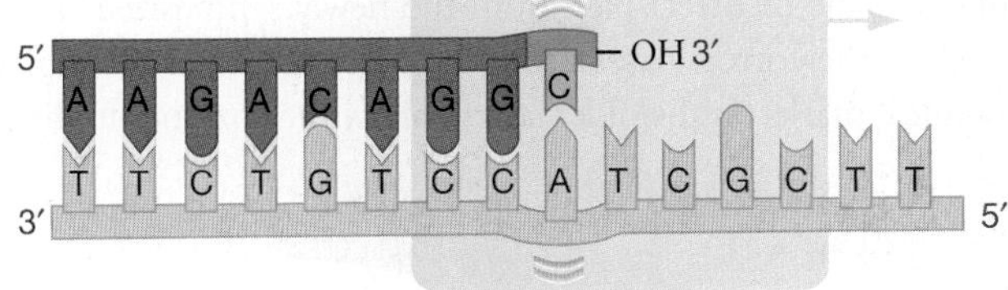

(b) ...but notices the mistake and corrects it.

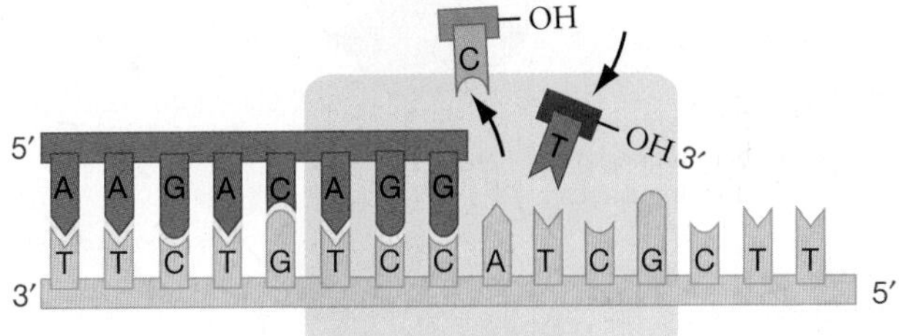

FIGURE 14.14 DNA Polymerase Can Proofread. If a mismatch such as the pairing of A with C occurs **(a)**, DNA polymerase can act as a 5′ → 3′ exonuclease, meaning that it can remove bases in that direction **(b)**. The enzyme then adds the correct base.

mutation rate in *E. coli*. At the molecular level, a mutant phenotype usually results from a change in an enzyme or other type of protein.

In the case of *E. coli* cells with high mutation rates, biologists found that the mutation responsible for the high mutation rates was a defect in a portion of the polymerase III enzyme called the ϵ (epsilon) subunit. Further analyses showed that this subunit of the enzyme acts as an exonuclease—meaning an enzyme that removes deoxyribonucleotides from the ends of DNA strands (**Figure 14.14b**). If a new deoxyribonucleotide is not correctly hydrogen bonded to a base on the complementary strand, DNA polymerase backs up and the ϵ subunit removes it.

These findings led to the conclusion that DNA polymerase III can **proofread**. If the wrong base is added during DNA synthesis, the enzyme pauses, removes the mismatched base that was just added, and then proceeds with synthesis.

Eukaryotic DNA polymerases have the same type of proofreading ability. Typically, proofreading reduces a DNA polymerase's error rate to about 1×10^{-7} (one mistake per 10 million bases). Is this accurate enough? The answer is no.

MISMATCH REPAIR If—in spite of its proofreading ability—DNA polymerase leaves a mismatched pair behind in the newly synthesized strand, a battery of enzymes springs into action to correct the problem. **Mismatch repair** occurs when mismatched bases are corrected after DNA synthesis is complete.

The proteins responsible for mismatch repair were discovered in the same way that the proofreading capability of DNA polymerase III was uncovered—by analyzing *E. coli* mutants. In this case, the mutants had normal DNA polymerase III but abnormally high mutation rates.

The first mutant gene that caused a deficiency in mismatch repair was identified in the late 1960s and was called *mutS*. (The *mut* is short for "mutator.") By the late 1980s, researchers had identified 10 proteins involved in the identification and repair of base-pair mismatches in *E. coli*.

These proteins recognize the mismatched pair, remove a section containing the incorrect base from the newly synthesized strand, and fill in the correct bases using the older strand as a template. (Chemical marks on the older strand allow the enzymes to distinguish the original strand from the newly synthesized strand.) The mismatch-repair enzymes are like a copy editor who corrects typos that the author—DNA polymerase, in this case—did not catch.

Repairing Damaged DNA

Even after DNA is synthesized and proofread and mismatches repaired, the job of ensuring accuracy doesn't end. Genes are under constant assault. Nucleotides are damaged by chemicals like the hydroxyl (OH) radicals produced during aerobic metabolism, the aflatoxin B1 found in moldy peanuts and corn, and the benzo[α]pyrene in cigarette smoke.

Radiation is another danger. Ultraviolet (UV) light, for example, can cause a covalent bond to form between adjacent pyrimidine bases. The thymine-thymine pair illustrated in **Figure 14.15** is an example. This defect, called a thymine dimer, creates kinks in the secondary structure of DNA. The kinks stall the movement of the replication fork during DNA replication and impair the enzymes responsible for using the information in genes. If the damage is not repaired, the cell may die.

To fix problems caused by chemical attack, radiation, or other events, cells have a wide array of damage-repair systems. As an example, consider the system called **nucleotide excision repair**. As step 1 in **Figure 14.16** shows, the symmetry and regularity of DNA's secondary structure makes it possible for repair proteins to recognize thymine dimers and other types of damaged bases that produce an irregularity in the molecule. Once a damaged region is recognized, enzymes remove a segment of single-stranded DNA containing the defective sequence (step 2). The presence of a DNA strand complementary to the damaged strand provides a template for synthesis of a corrected strand (steps 3 and 4).

In this way, DNA's secondary structure makes accurate repair possible, supporting the molecule's function in information storage and processing. But what happens when repair systems are defective?

Xeroderma Pigmentosum: A Case Study

Xeroderma pigmentosum (XP) is a rare autosomal recessive disease in humans. Individuals with this condition are extremely sensi-

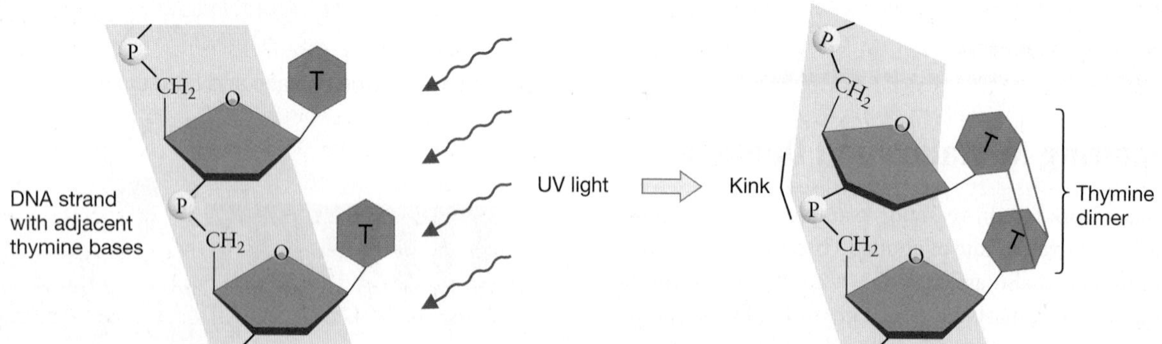

FIGURE 14.15 UV Light Damages DNA. When UV light strikes a section of DNA that has adjacent thymines, the energy can break bonds within each base and allow bonds to form *between* them. The thymine dimer that is produced causes a kink in the DNA.

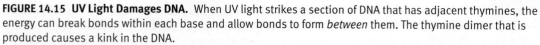

✔**QUESTION** Why are infrared wavelengths much less likely than UV to damage DNA? (Hint: see Figure 10.3).

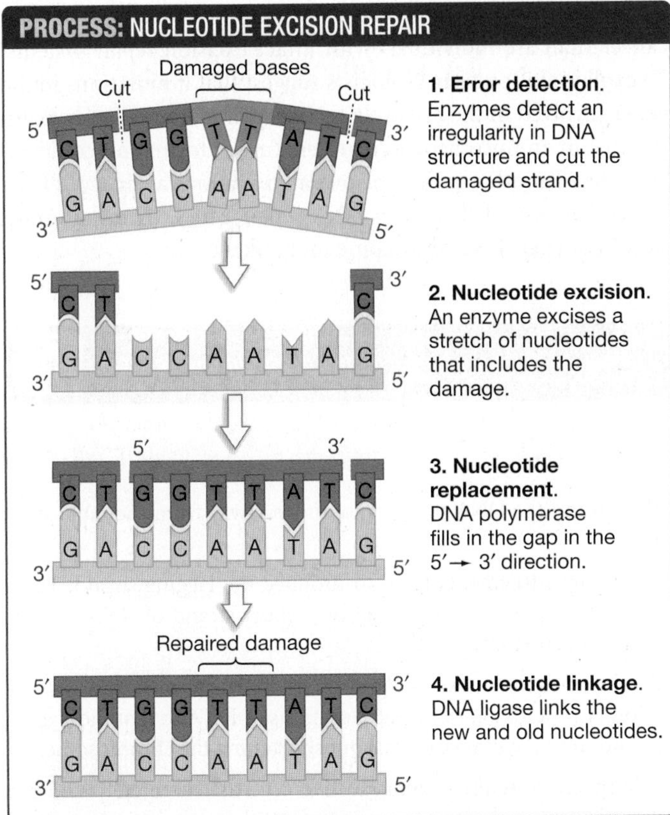

PROCESS: NUCLEOTIDE EXCISION REPAIR
Damaged bases
Cut
Cut
1. Error detection. Enzymes detect an irregularity in DNA structure and cut the damaged strand.
2. Nucleotide excision. An enzyme excises a stretch of nucleotides that includes the damage.
3. Nucleotide replacement. DNA polymerase fills in the gap in the 5′→ 3′ direction.
Repaired damage
4. Nucleotide linkage. DNA ligase links the new and old nucleotides.

FIGURE 14.16 In Excision Repair, Defective Bases Are Removed and Replaced.

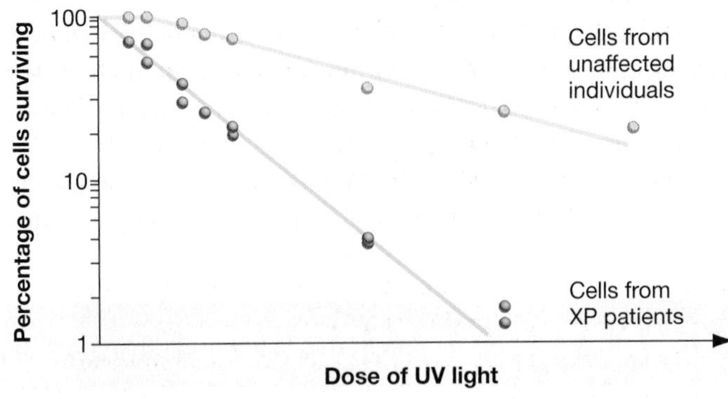

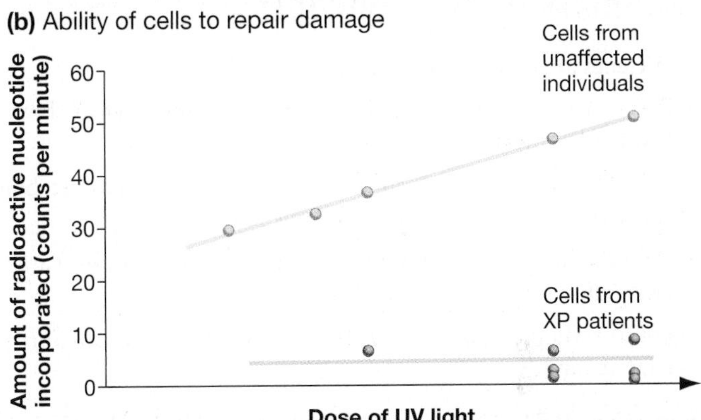

FIGURE 14.17 DNA Damage from UV Light Is Not Repaired Properly in Individuals with XP. (a) When cell cultures from unaffected individuals and from XP patients are irradiated with various doses of UV light, the percentage of cells that survive is strikingly different. **(b)** When cell cultures from unaffected individuals and from XP patients are irradiated with various doses of UV light and then fed radioactive thymidine, only unaffected individuals incorporate the labeled base into their DNA.

✔**QUESTION** Why are light-skinned people who cultivate a skin tan increasing their risk of developing cancer? (Hint: Tanning is a response to UV light.)

tive to ultraviolet (UV) light. Their skin develops lesions after even slight exposure to sunlight. In unaffected individuals, these kinds of lesions develop only after extensive exposure to UV light, X-rays, or other forms of high-energy radiation.

In 1968 James Cleaver proposed a connection between XP and DNA excision repair systems. He knew that in *E. coli*, mutations in certain genes cause DNA nucleotide excision repair proteins to fail. Cells with these mutations have an increased sensitivity to radiation. Cleaver's hypothesis was that people with XP have similar mutations. He claimed that they are extremely sensitive to sunlight because they are unable to repair the damage that occurs when the deoxyribonucleotides in DNA absorb UV light.

Cleaver and other researchers made extensive use of cell cultures to study the hypothesized connection between DNA damage, faulty nucleotide excision repair, and XP (for an introduction to cell cultures, see **BioSkills 12** in Appendix A). For example, they collected skin cells from people with XP and from people with a normal phenotype for excision repair. When these cell populations were grown in culture, the biologists exposed them to increasing amounts of ultraviolet radiation and recorded how many died.

Figure 14.17a shows the data that resulted. Note that the intensity of the radiation is graphed on the *x*-axis, with the percentage of cells surviving graphed on the *y*-axis. Note, too, that the *y*-axis is logarithmic on this graph. (For help with logarithms, see **BioSkills 7** in Appendix A; for help with reading graphs, see **BioSkills 2**). The graph has data from both healthy cells and cells from XP patients. Cell survival declined with increasing radiation dose in both types of cells, but cells from xeroderma pigmentosum individuals died off much more rapidly than cells from unaffected individuals.

The connection to excision repair systems was confirmed when Cleaver exposed cells from unaffected individuals and cells from XP individuals to various amounts of UV light, then fed the cells a radioactive deoxyribonucleotide to label DNA synthesized during the repair period. If repair is defective in XP individuals, then their cells should incorporate virtually no radioactive deoxyribonucleotide into their DNA. Cells from unaffected individuals, in contrast, should incorporate large amounts of labeled deoxyribonucleotide into their DNA.

As **Figure 14.17b** shows, this is exactly what happens. Here the amount of radioactive dNTPs incorporated into DNA is graphed against radiation dose, for both healthy and XP cells. Increasingly large amounts of radioactive dNTPs are found in the DNA of healthy cells as UV dose increases, though no such increase occurs in XP cells. These data are consistent with the hypothesis that excision repair is virtually nonexistent in XP individuals.

More recently, genetic analyses of XP patients have shown that the condition can result from mutations in any of seven genes. This discovery is not surprising in light of the large number of enzymes involved in repairing damaged DNA.

Defects in the genes required for DNA repair are frequently associated with cancer. Individuals with xeroderma pigmentosum, for example, are 1000 to 2000 times more likely to get skin cancer than are individuals with intact excision repair systems. To explain this pattern, biologists suggest that if mutations in the genes involved in the cell cycle go unrepaired, the cell may begin to grow in an uncontrolled manner. Tumor formation could result. Stated another way, if the overall mutation rate in a cell is elevated because of defects in DNA repair genes, then the mutations that trigger cancer become more likely.

CHAPTER 14 REVIEW

For media, go to the study area at www.masteringbiology.com 

Summary of Key Concepts

Genes are made of DNA.

- Experiments on viruses that had labeled proteins or DNA showed that DNA is the hereditary material.
- DNA's primary structure consists of a sugar-phosphate backbone and a sequence of nitrogen-containing bases.
- DNA's secondary structure consists of two strands in an antiparallel orientation. The strands twist into a helix.

✔You should be able to explain why the secondary structure of DNA is stable.

When DNA is being copied, each strand of the double helix serves as the template for the synthesis of a complementary strand.

- By labeling DNA with ^{15}N or ^{14}N, researchers were able to validate the hypothesis that DNA replication is semiconservative, meaning that each strand is copied in its entirety.
- Each strand of a parent DNA molecule provides a template for the synthesis of a daughter strand, resulting in two complete DNA double helices.

✔You should be able to write a sequence of double-stranded DNA that is 10 base-pairs long, separate the strands, and, without comparing them, write in the bases that are added during DNA replication.

DNA synthesis occurs in the $5' \rightarrow 3'$ direction only and requires a large suite of specialized enzymes. The leading strand is synthesized continuously, but the lagging strand is synthesized as a series of fragments that are then linked together.

- DNA synthesis is an enzyme-catalyzed reaction that takes place in one direction.
- DNA synthesis requires both a template and a primer sequence, and it takes place at the replication fork, where the double helix is being opened.
- Synthesis of the leading strand in the $5' \rightarrow 3'$ direction is continuous, but synthesis of the lagging strand is more complex because on that strand the DNA polymerase moves away from the replication fork.
- Short DNA fragments form to produce the lagging strand. These Okazaki fragments are primed by a short strand of RNA and are linked together after synthesis.
- At the ends of linear chromosomes in eukaryotes, the enzyme telomerase adds short, repeated sections of DNA so that the lagging strand can be synthesized without shortening the chromosome.
- Telomerase is active in reproductive cells that eventually undergo meiosis. As a result, gametes contain chromosomes of normal length.

✔You should be able to compare and contrast the functions of the three polymerases introduced in this chapter: DNA polymerase, primase, and telomerase.

(MB) **Web Activity** DNA Synthesis, **BioFlix™** DNA Replication

Specialized enzymes repair damages to DNA and fix mistakes in DNA synthesis. If these enzymes are defective, the mutation rate increases.

- DNA replication is remarkably accurate because DNA polymerase proofreads and because mismatch-repair enzymes remove incorrect bases once synthesis is complete and replace them with the correct sequence.
- DNA repair occurs after bases have been damaged by chemicals or radiation.
- Nucleotide excision repair systems cut out damaged portions of genes and replace them with correct sequences.
- Several types of human cancers are associated with defects in the genes responsible for DNA repair.

✔You should be able to explain the logical connections between failure of repair systems, increases in mutation rate, and high likelihood of cancer developing.

Questions

✔TEST YOUR KNOWLEDGE

Answers are available in Appendix B

1. What does it mean to say that strands in a double helix are antiparallel?
   a. Their primary sequences consist of a sequence of *complementary* bases.
   b. They each have a sugar-phosphate backbone.
   c. They each have a $5' \rightarrow 3'$ directionality.
   d. They have opposite directionality, or polarity.

2. Which of the following is *not* a property of DNA polymerase?
   a. It catalyzes the addition of dNTPs only in the $5' \rightarrow 3'$ direction.
   b. It requires a primer to work.
   c. It is associated with a sliding clamp only on the leading strand.
   d. It can proofread because it has an exonuclease activity.
3. What is the function of topoisomerase?
   a. holding DNA polymerase steady as it moves down the leading or lagging strand
   b. opening the DNA helix at the replication fork
   c. stabilizing single strands of DNA, once the replication fork is open
   d. preventing kinks in DNA as the replication fork opens and unwinds
4. What is the function of primase?
   a. synthesis of the short section of double-stranded DNA required by DNA polymerase
   b. synthesis of a short section of RNA, complementary to single-stranded DNA
   c. closing the gap at the $3'$ end of DNA after excision repair
   d. removing primers and synthesizing a short section of DNA to replace them
5. Where and how are Okazaki fragments synthesized?
   a. on the leading strand, in a $5' \rightarrow 3'$ direction
   b. on the leading strand, in a $3' \rightarrow 5'$ direction
   c. on the lagging strand, in a $5' \rightarrow 3'$ direction
   d. on the lagging strand, in a $3' \rightarrow 5'$ direction
6. What does telomerase do?
   a. It adds a protein primer to the ends of linear chromosomes.
   b. It adds double-stranded DNA to the "blunt end" of a linear chromosome.
   c. It adds double-stranded DNA to the lagging strand at the end of a linear chromosome.
   d. It adds single-stranded DNA to the lagging strand at the end of a linear chromosome.

TEST YOUR UNDERSTANDING

Answers are available in Appendix B

1. Researchers design experiments so that only one thing is different between the treatments that are being compared. In the Hershey-Chase experiment, what was this single difference?
2. What does it mean to say that DNA replication is "bidirectional" from the point of origin?
3. Why is the synthesis of the lagging strand of DNA discontinuous? How is it possible for the synthesis of the leading strand to be continuous?
4. Explain how telomerase prevents linear chromosomes from shortening during replication.
5. List the three events that increase the accuracy of DNA replication, in chronological order. Indicate which events involve DNA polymerase III and which involve specialized repair enzymes.
6. Explain how the structure of DNA makes it relatively easy for proteins to recognize base-pair mismatches or damaged bases. How does DNA's secondary structure make it possible for damaged sections or incorrect bases to be removed and repaired?

APPLYING CONCEPTS TO NEW SITUATIONS

Answers are available in Appendix B

1. If DNA polymerase III did not require a primer, which steps in DNA synthesis would differ from what is observed? Would any special enzymes be required to replicate telomeres? Explain your answers.
2. In the late 1950s Herbert Taylor grew bean root-tip cells in a solution of radioactive thymidine and allowed them to undergo one round of DNA replication. He then transferred the cells to a solution without the radioactive deoxyribonucleotide, allowed them to replicate again, and examined their chromosomes for the presence of radioactivity. His results are shown in the following figure, where red indicates a radioactive chromatid.

Nonradioactive chromatid

Radioactive chromatids

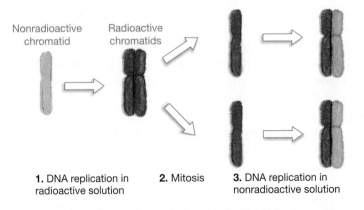

1. DNA replication in radioactive solution

2. Mitosis

3. DNA replication in nonradioactive solution

   a. Draw diagrams explaining the pattern of radioactivity observed in the sister chromatids after the first and second rounds of replication.
   b. What would the results of Taylor's experiment be if eukaryotes used a conservative mode of DNA replication?

3. The graph that follows shows the survival of four different *E. coli* strains after exposure to increasing doses of ultraviolet light. The wild-type strain is normal, but the other strains have a mutation in either a gene called *uvrA*, a gene called *recA*, or both.

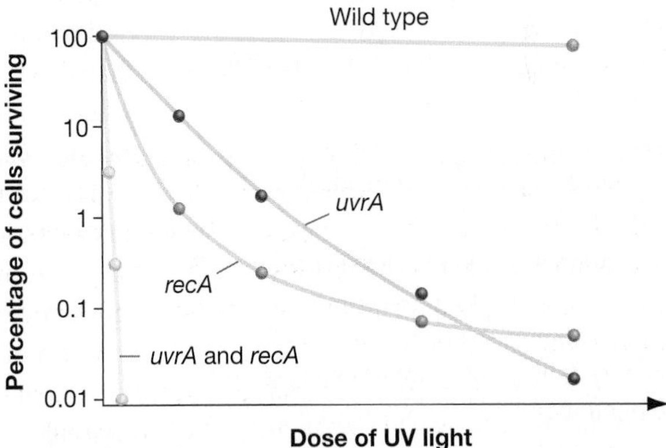

   a. Which strains are most sensitive to UV light? Which strains are least sensitive?
   b. What are the relative contributions of these genes to the repair of UV damage?

4. One widely used test to identify whether certain chemicals, such as pesticides or herbicides, might be carcinogenic (cancer causing) consists of exposing bacterial cells to the chemical and recording whether the exposure leads to an increased mutation rate. In effect, this test equates cancer-causing chemicals with mutation-causing chemicals. Why is this an informative test?

This chapter is part of The Big Picture. See how on pages 336–337.

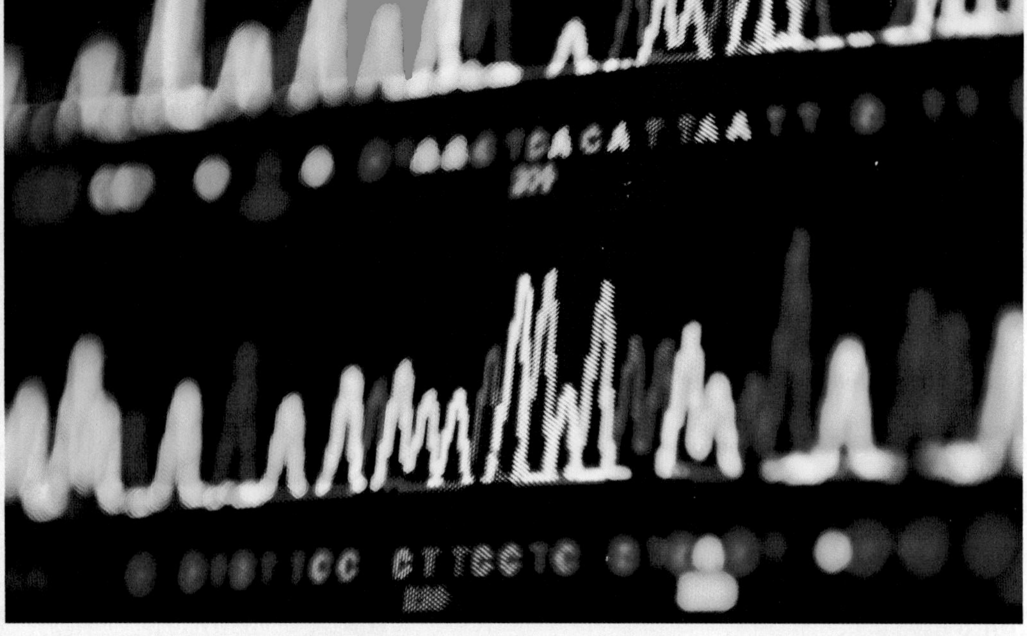

This computer screen is showing output from a DNA sequencer. Each peak corresponds to a different base in DNA—an A, T, G, or C. This chapter explores how DNA sequences in organisms are related to their phenotypes.

15 How Genes Work

KEY CONCEPTS

- Most genes code for proteins.
- DNA is transcribed to messenger RNA by RNA polymerase, and then messenger RNA is translated to proteins by ribosomes. In this way, genetic information is converted from DNA to RNA to proteins.
- Each amino acid in a protein is specified by a group of three bases in messenger RNA.
- Mutations are random changes in DNA, ranging in extent from single bases to large chromosome regions, that may or may not produce changes in the phenotype.

DNA has been called the blueprint of life. If an organism's DNA is like a set of blueprints, then its cells are like construction sites, and the enzymes inside a cell are like construction workers. But how does the DNA inside each cell assemble this team of skilled laborers and specify the construction materials needed to build and maintain the cell, and perhaps remodel it when conditions change?

Mendel provided insights that made the study of these questions possible. He discovered that particular alleles are associated with certain phenotypes and that alleles do not change when transmitted from parent to offspring. The resulting chromosome theory of inheritance established that genes are found in chromosomes, whose movement during meiosis explains Mendel's results.

The science of molecular biology began with the discovery that DNA is the hereditary material, and that DNA is a double-helical structure containing sequences of four bases. From these early advances, it was clear that genes are made of DNA and that genes carry the instructions for making and maintaining an individual.

But biologists still didn't know how the information in DNA is translated into action. How does **gene expression**—the process of converting archived information into molecules that actually do things in the cell—occur?

This chapter introduces some of the most pivotal ideas in all of biology—ideas that connect genotypes to phenotypes by revealing how genes work at the molecular level. They also speak to the heart of a key attribute of life: processing genetic information to produce a living organism.

Understanding how genes work triggered a major transition in biological science. Instead of thinking about genes solely in relation to their effects on eye color in fruit flies or on seed shape in garden peas, biologists could begin analyzing the molecular composition of genes and their products. The molecular revolution in biology took flight.

✔ When you see this checkmark, stop and test yourself. Answers are available in Appendix B.

15.1 What Do Genes Do?

Although biologists of the early twentieth century made tremendous progress in understanding how genes are inherited, an explicit hypothesis explaining what genes do did not appear until 1941. That year George Beadle and Edward Tatum published a series of breakthrough experiments on a bread mold called *Neurospora crassa.*

Beadle and Tatum's research was inspired by an idea that was brilliant in its simplicity. As Beadle said: "One ought to be able to discover what genes do by making them defective." The idea was to knock out a gene by damaging it and then infer what the gene does by observing the phenotype of the mutant individual.

Today, alleles that do not function at all are called **knock-out, null**, or **loss-of-function alleles**. Creating knock-out mutant alleles and analyzing their effects is still one of the most common research strategies in studies of gene function. But Beadle and Tatum were the pioneers.

The One-Gene, One-Enzyme Hypothesis

To start their work, Beadle and Tatum exposed a large number of *N. crassa* individuals to radiation. As Chapter 14 indicated, high-energy radiation damages the double-helical structure of DNA—often in a way that makes the affected gene nonfunctional.

Their next step was to examine the mutant individuals. Eventually they succeeded in finding mutant *N. crassa* individuals that could not make specific compounds. For example, one of the mutants could not make pyridoxine, also called vitamin B6, even though normal individuals can. Further, Beadle and Tatum showed that the inability to synthesize pyridoxine was due to a defect in a single gene, and that the inability to synthesize other molecules was due to defects in other genes.

These results inspired their **one-gene, one-enzyme hypothesis.** Beadle and Tatum proposed that the mutant *N. crassa* individual could not make pyridoxine because it lacked an enzyme required to synthesize the compound and that the lack of the enzyme was due to a genetic defect. Based on analyses of knock-out mutants, the one-gene, one-enzyme hypothesis claimed that each gene contains the information needed to make an enzyme.

An Experimental Test of the Hypothesis

Three years later, Adrian Srb and Norman Horowitz published a rigorous test of the one-gene, one-enzyme hypothesis. These biologists focused on the ability of *N. crassa* individuals to synthesize the amino acid arginine. In the lab, normal cells of this bread mold grow well on a laboratory culture medium that lacks arginine. This is possible because *N. crassa* cells are able to synthesize their own arginine.

Previous work had shown that organisms synthesize arginine in a series of steps called a **metabolic pathway.** As **Figure 15.1** shows, compounds called ornithine and citrulline are intermediate products in the metabolic pathway leading to arginine. Specific enzymes are required to synthesize ornithine, convert ornithine to citrulline, and change citrulline to arginine. Srb and Horowitz hypothesized that specific genes in *N. crassa* cells are responsible for producing each of the three enzymes involved.

To test this idea, Srb and Horowitz used radiation to create a large number of mutant individuals. High-energy radiation is equally likely to damage DNA in any part of the organism's genome, however, and most organisms have thousands or tens of thousands of genes. Of the many mutants the biologists created, how could they find the handful that specifically knocked out a step in the pathway for arginine synthesis?

To find the mutants they were looking for, the researchers performed what is now known as a genetic screen. A **genetic screen** is any technique for picking certain types of mutants out of many thousands of randomly generated mutants.

Srb and Horowitz began their screen by raising colonies of irradiated cells on a medium that included arginine. Then they transferred a sample of each colony to a medium that *lacked* arginine. If an individual could grow in the presence of arginine but failed to grow without arginine, they concluded that it couldn't make its own arginine.

The biologists followed up by confirming that the offspring of these cells also had this defect. Based on these data, they were confident that they had isolated individuals with mutations in one or more of the genes for the enzymes shown in Figure 15.1.

To test the one-gene, one-enzyme hypothesis, the biologists grew the mutants on normal media that lacked arginine and were supplemented in each case either with nothing, ornithine, citrulline, or arginine.

As **Figure 15.2** on page 278 shows, the results from these growth experiments were dramatic. Some of the mutant cells were able to grow on some of these media but not on others. More specifically, the mutants fell into three distinct classes, which the researchers called *arg1*, *arg2*, and *arg3*.

As the Interpretation section of the figure shows, the data make sense if each type of mutant lacked a different, specific step in a metabolic pathway because of a defect in a particular gene. In short, Srb and Horowitz had documented a correlation

FIGURE 15.1 Different Enzymes Catalyze Each Step in the Metabolic Pathway for Arginine.

✔**QUESTION** If a cell lacked enzyme 2 but received ornithine in its diet, could it still grow? Could it still grow if it received citrulline instead?

EXPERIMENT

QUESTION: What do genes do?

HYPOTHESIS: Each gene contains the information required to make a different enzyme.

NULL HYPOTHESIS: Genes have nothing to do with making enzymes.

EXPERIMENTAL STRATEGY: Knock out specific genes. Test to see if the enzymes required for different steps in the pathway for synthesizing arginine are missing.

EXPERIMENTAL SETUP: Isolate mutant *N. crassa* that cannot synthesize arginine. Grow each type of mutant on normal medium that is:

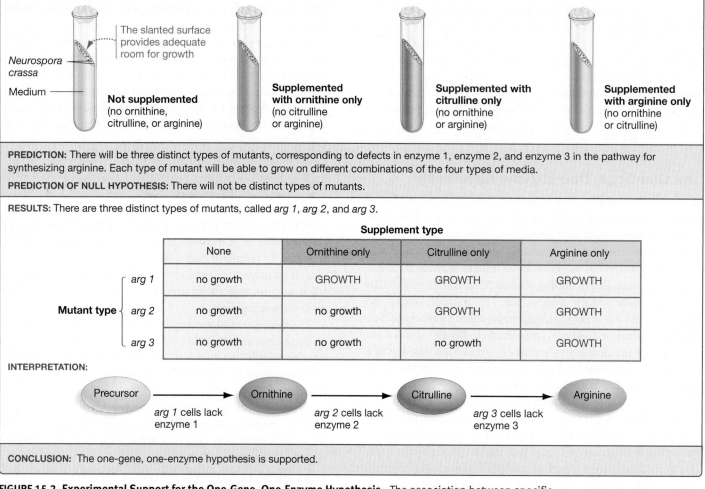

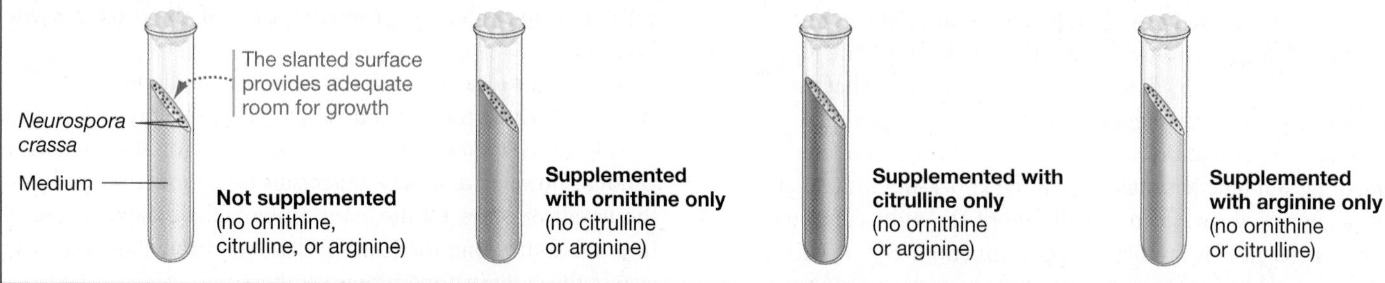

PREDICTION: There will be three distinct types of mutants, corresponding to defects in enzyme 1, enzyme 2, and enzyme 3 in the pathway for synthesizing arginine. Each type of mutant will be able to grow on different combinations of the four types of media.

PREDICTION OF NULL HYPOTHESIS: There will not be distinct types of mutants.

RESULTS: There are three distinct types of mutants, called *arg 1*, *arg 2*, and *arg 3*.

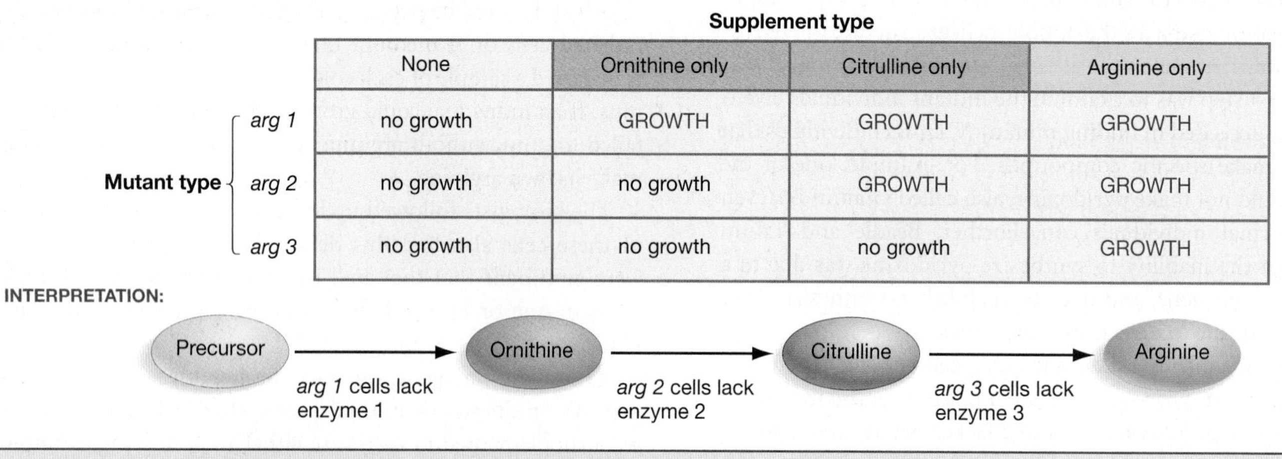

Supplement type

Mutant type	None	Ornithine only	Citrulline only	Arginine only
arg 1	no growth	GROWTH	GROWTH	GROWTH
arg 2	no growth	no growth	GROWTH	GROWTH
arg 3	no growth	no growth	no growth	GROWTH

INTERPRETATION:

CONCLUSION: The one-gene, one-enzyme hypothesis is supported.

FIGURE 15.2 Experimental Support for the One-Gene, One-Enzyme Hypothesis. The association between specific genetic defects in *N. crassa* and specific deficits in the metabolic pathway for arginine synthesis provided evidence that the one-gene, one-enzyme hypothesis was correct.

SOURCE: Srb, A. M. and N. H. Horowitz. 1944. The ornithine cycle in *Neurospora* and its genetic control. *Journal of Biological Chemistry* 154: 129–139.

✔**QUESTION** Experimental designs must be repeatable, so that other investigators can try the experiment themselves to check the results. Name three things that these researchers would need to describe, so that others could repeat this experiment.

between a specific genetic defect and a defect at a specific point in a metabolic pathway.

This experiment convinced most investigators that the one-gene, one-enzyme hypothesis was correct. To review its design, results, and interpretation, go to the study area at *www.masteringbiology.com*.

 **Web Activity** The One-Gene, One-Enzyme Hypothesis

Follow-up work showed that genes dictate the structures of all the proteins produced by an organism—not just the structures of enzymes. Biologists finally understood what most genes do: They contain the instructions for making proteins.

In many cases, though, proteins are made up of different polypeptides, each of which is a product of a different gene. Consequently, for greater accuracy, the one-gene, one-enzyme hypothesis is best called the one-gene, one-polypeptide hypothesis.

15.2 The Central Dogma of Molecular Biology

How does a gene specify the production of a protein? As soon as Beadle and Tatum's hypothesis had been supported in *N. crassa* and a variety of other organisms, this question became a central one.

Part of the answer lay in the molecular structure of the gene. Biochemists knew that the primary components of DNA were four nitrogen-containing bases: the pyrimidines thymine (abbreviated T) and cytosine (C), and the purines adenine (A) and guanine (G). They also knew that these bases were connected in a linear sequence by a sugar-phosphate backbone. Watson and Crick's model for the secondary structure of the DNA molecule, introduced in Chapter 4 and reviewed in Chapter 14, revealed that two strands of DNA are wound into a double helix, held together by hydrogen bonds between the complementary base pairs A-T and G-C.

Given DNA's structure, it appeared extremely unlikely that DNA directly catalyzed the reactions that produce proteins. Its shape was too regular to suggest that it could bind a wide variety of substrate molecules and lower the activation energy for chemical reactions. So how did information translate into action?

The Genetic Code Hypothesis

Crick proposed that the sequence of bases in DNA might act as a code. His idea was that DNA was *only* an information-storage molecule. The instructions it contained would have to be read and then translated into proteins.

Crick offered Morse code as an analogy. Morse code is a message-transmission system using dots and dashes to represent the letters of the alphabet and in that way convey all the complex information of human language. Crick was proposing that different combinations of bases could specify the 20 amino acids, just as different combinations of dots and dashes specify the 26 letters of the alphabet. A particular stretch of DNA, then, could contain the information needed to produce the amino acid sequence of a particular enzyme.

In code form, the tremendous quantity of information required to build and operate a cell could be stored compactly. This information could also be copied through complementary base pairing and transmitted efficiently from one generation to the next.

It soon became apparent, however, that the information encoded in the base sequence of DNA is not translated into the amino acid sequence of proteins directly. Instead, the link between DNA as information repository and proteins as cellular machines is indirect.

RNA as the Intermediary between Genes and Proteins

The first clue that the biological information in DNA must go through an intermediary in order to produce proteins came from data on the structure of cells. In eukaryotic cells, DNA is enclosed within a membrane-bound organelle called the nucleus (see Chapter 7). But the cells' ribosomes, where protein synthesis takes place, are outside the nucleus, in the cytoplasm.

To make sense of this observation, François Jacob and Jacques Monod suggested that RNA molecules act as a link between genes and the protein-manufacturing centers. Jacob and Monod's hypothesis is illustrated in **Figure 15.3**. They predicted that short-lived molecules of RNA, which they called **messenger RNA**, or **mRNA**, carry information from DNA to the site of protein synthesis. Messenger RNA is one of several distinct types of RNA in cells.

Follow-up research confirmed that the messenger RNA hypothesis is correct. One particularly important piece of evidence was the discovery of an enzyme that catalyzes the synthesis of RNA. This protein is called **RNA polymerase**, because it polymerizes ribonucleotides into strands of RNA.

RNA polymerase synthesizes RNA molecules according to the information provided by the sequence of bases in a particular stretch of DNA. Unlike DNA polymerase, RNA polymerase does not require a primer to begin connecting ribonucleotides together to produce a strand of RNA.

To test the mRNA hypothesis more directly, researchers created a reaction mix containing three critical elements: **(1)** the enzyme RNA polymerase; **(2)** ribonucleotides containing the bases adenine (A), uracil (U), guanine (G), and cytosine (C); and **(3)** copies of a strand of synthetic DNA that contained deoxyribonucleotides in which the only base was thymine (T).

Why uracil? Recall from Chapter 4 that RNA contains the base uracil instead of thymine. If an RNA molecule binds to a DNA or RNA molecule, uracil forms complementary base pairs with adenine.

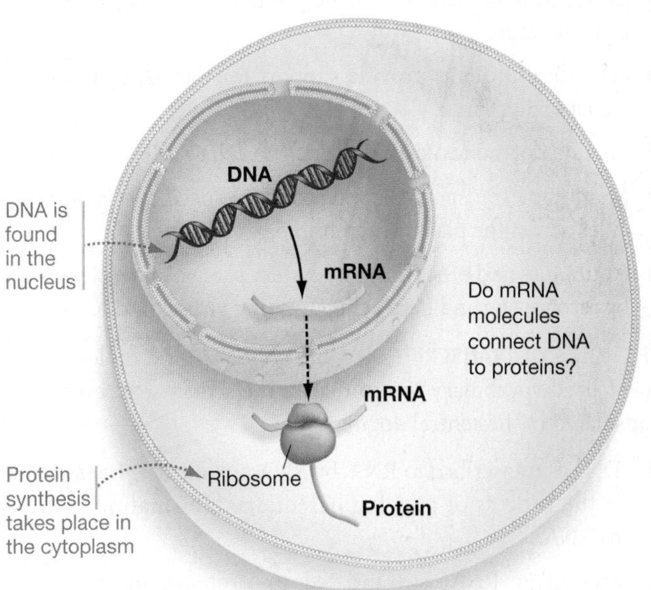

FIGURE 15.3 The Messenger RNA Hypothesis. In the cells of plants, animals, fungi, and other eukaryotes, most DNA is found only in the nucleus, but proteins are manufactured outside the nucleus, at ribosomes. Biologists proposed that the information coded in DNA is carried from inside the nucleus out to the ribosomes by messenger RNA (mRNA).

The DNA strand in the experiment had the sequence TTTTTT..., however. After allowing the polymerization reaction to proceed, the biologists isolated RNA molecules that contained only the base adenine.

This result provided strong support for the hypothesis that RNA polymerase synthesizes RNA according to the rules of complementary base pairing introduced in Chapter 4, because thymine pairs with adenine. Similar experiments showed that synthetic DNAs containing no bases other than cytosine result in the production of RNA molecules containing no bases other than guanine.

Dissecting the Central Dogma

Once the mRNA hypothesis was accepted, Francis Crick articulated what became known as the central dogma of molecular biology. The **central dogma** summarizes the flow of information in cells. It simply states that DNA codes for RNA, which codes for proteins:

$$\text{DNA} \longrightarrow \text{RNA} \longrightarrow \text{proteins}$$

Crick's simple statement encapsulates much of the research reviewed in this chapter and the preceding one. DNA is the hereditary material. Genes consist of specific stretches of DNA that code for products used in the cell. The sequence of bases in DNA specifies the sequence of bases in an RNA molecule, which specifies the sequence of amino acids in a protein. In this way, genes ultimately code for proteins.

Many proteins function as enzymes that catalyze chemical reactions in the cell. Other proteins perform the types of roles introduced in earlier chapters:

- Motor proteins and contractile proteins move the cell itself or cellular cargo.
- Structural proteins provide support for the cell or tracks for transporting cargo.
- Peptide hormones carry signals from cell to cell.
- Membrane transport proteins conduct specific ions or molecules across the plasma membrane.
- Antibodies and other immune system proteins provide defense by recognizing and destroying invading viruses and bacteria.

THE ROLES OF TRANSCRIPTION AND TRANSLATION Biologists use specialized vocabulary to summarize the sequence of events encapsulated in the central dogma.

1. DNA is transcribed to RNA, by RNA polymerase. **Transcription** is the process of copying hereditary information in DNA to RNA.
2. Messenger RNA is translated to proteins, in ribosomes. **Translation** is the process of using the information in nucleic acids to synthesize proteins.

The term transcription is appropriate. In everyday English, transcription simply means making a copy of information. The scientific use is similar because it conveys the idea that DNA acts as a permanent record—an information archive or blueprint. This permanent record is copied, during transcription, to produce the short-lived form called mRNA.

Translation is also an appropriate term. In everyday English, translation refers to transferring information from one language to another. In biology, translation is the transfer of information from one type of molecule to another—from the "language" of nucleic acids to the "language" of proteins. Translation is also referred to simply as protein synthesis.

The following diagram summarizes the relationship between transcription and translation, as well as the relationships among DNA, RNA, and proteins:

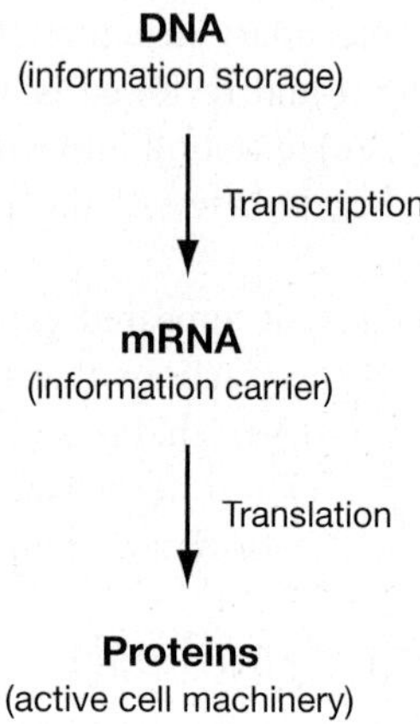

Gene expression occurs via transcription and translation.

LINKING GENOTYPES AND PHENOTYPES According to the central dogma, an organism's genotype is determined by the sequence of bases in its DNA, while its phenotype is a product of the proteins it produces.

To appreciate this point, consider that the enzymes and other proteins encoded by genes are what make the "stuff" of the cell and dictate which chemical reactions occur inside. In populations of the Oldfield mouse native to southeastern North America, for example, individuals have a gene for a protein called the melanocortin receptor. This receptor influences how much dark pigment is deposited in fur. An important aspect of a mouse's phenotype—its coat color—is determined in part by the DNA sequence at the gene for this receptor-producing enzyme (**Figure 15.4a**).

Later work revealed that a gene's alleles differ in their DNA sequence. As a result, the proteins produced by different alleles of the gene may differ in their amino acid sequence. If the primary structures of proteins vary, their functions are likely to vary as well.

To drive this point home, look at the DNA sequence in the portion of the melanocortin receptor gene shown in **Figure 15.4b**, and compare it with the sequence in Figure 15.4a. The sequences differ—meaning that they are different alleles. Now look at the protein products of each allele, and note that one of the amino acids in the protein's primary structure differs—one allele specifies an arginine residue; the other specifies a cysteine residue.

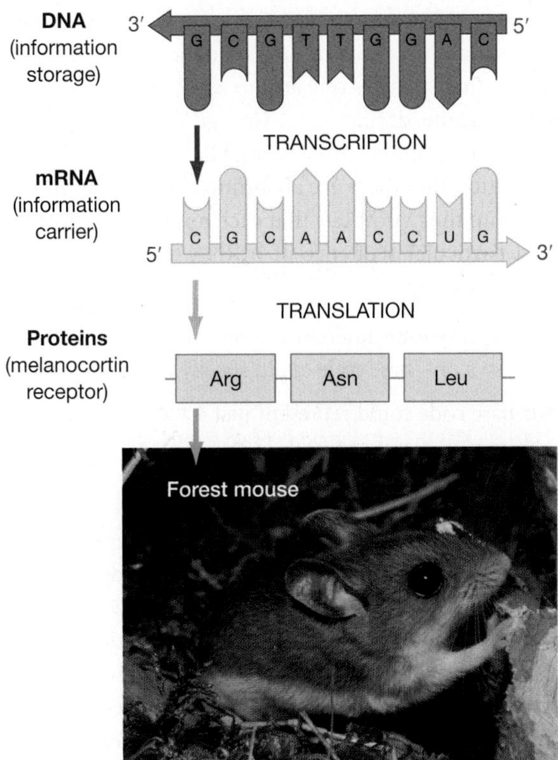

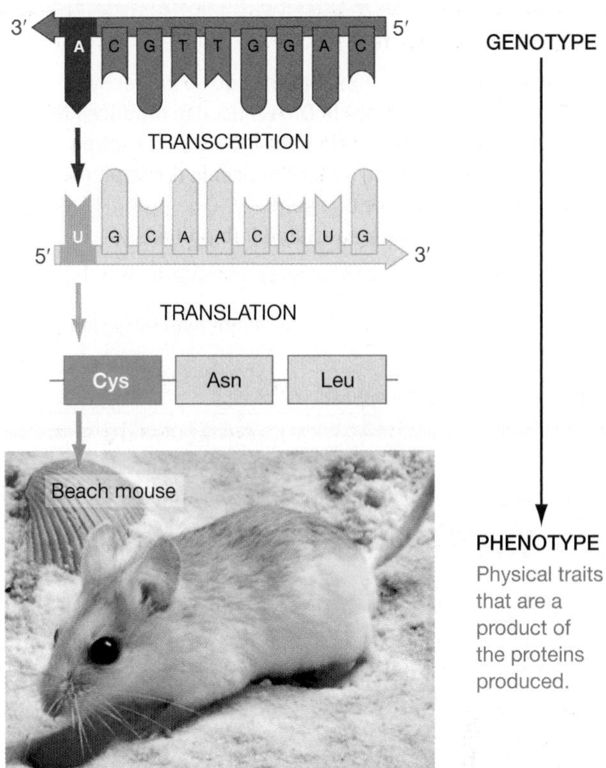

FIGURE 15.4 The Flow of Information in the Cell. The central dogma revealed the connection between genotype and phenotype. The DNA sequences given in parts **(a)** and **(b)** are from different alleles (genotypes) that influence coat color (phenotypes) in Oldfield mice. Forest-dwelling mice are dark, which camouflages them in their forested habitats. Beach-dwelling mice are light, which camouflages them in their sandy habitat.

At the protein level, the phenotypes associated with these alleles differ. The consequences for the entire mouse are striking: Melanocortin receptors that have arginine in this location tend to deposit a large amount of pigment, but receptors that have cysteine in this location tend to deposit small amounts of pigment. Whether a mouse is dark or light depends, in large part, on a single base change in its DNA sequence. In this case, a tiny difference in genotype produces a large change in phenotype. The central dogma links genotypes to phenotypes.

EXCEPTIONS TO THE CENTRAL DOGMA The central dogma provided an important conceptual framework for the burgeoning field called molecular genetics and inspired a series of fundamental questions about how genes and cells work. But important modifications to the central dogma have occurred in the decades since Frances Crick first proposed it:

- Many genes code for RNA molecules that do not function as mRNAs—they are not translated into proteins.
- In some cases, information flows from RNA back to DNA.

The discovery of a wide array of distinct RNA types ranks among the most profound advances in the past decade of biological science. Messenger RNA is just one of seven major types currently recognized. Several types of RNA in addition to mRNA are involved in protein synthesis, while other RNAs help regulate which genes are transcribed and which proteins are active in a cell (see Chapter 18). For the genes coding for these types of RNA, information flow would be diagrammed as simply DNA → RNA.

In the early 1970s, the discovery of "reverse" information flow created the kind of excitement now being generated by the discovery of so many different kinds of RNA. Some viruses, for example, have genes comprised of RNA. When RNA viruses infect a cell, a specialized viral polymerase called **reverse transcriptase** synthesizes a DNA version of the RNA genes. In these viruses, information flows from RNA to DNA.

The human immunodeficiency virus (HIV), which causes AIDS, is an RNA virus. Several of the most-commonly prescribed drugs for AIDS patients fight the infection by poisoning the virus's reverse transcriptase. The drugs prevent viruses from replicating efficiently by disrupting reverse information flow.

The punchline? Crick's hypothesis is a central concept in biology, but cells, viruses, and researchers aren't dogmatic about it.

CHECK YOUR UNDERSTANDING

If you understand that . . .

- Genes code for proteins, but they do so indirectly.
- The sequence of bases in DNA is used to produce RNA, including messenger RNA (mRNA), via transcription. The sequence of bases in an RNA molecule is complementary to the DNA sequence of a gene.
- Messenger RNAs are translated into proteins.

You should be able to . . .

Explain why errors in transcription might lead to changes in a cell's phenotype.

Answers are available in Appendix B.

15.3 The Genetic Code

Once biologists understood the general pattern of information flow in the cell, the next challenge was to understand the final link between DNA and proteins. Exactly how does the sequence of bases in a strand of mRNA code for the sequence of amino acids in a protein?

If this question could be answered, biologists would have cracked the **genetic code**—the rules that specify the relationship between a sequence of nucleotides in DNA or RNA and the sequence of amino acids in a protein. Researchers from all over the world took up the challenge. A race was on.

How Long Is a Word in the Genetic Code?

The first step in cracking the genetic code was to determine how many bases make up a "word." In a sequence of mRNA, how long is a message that specifies one amino acid?

Based on some simple logic, George Gamow suggested that each code word contains three bases. His reasoning derived from the observation that there are 20 amino acids commonly used in cells and from the hypothesis that each amino acid must be specified by a particular sequence of mRNA. **Figure 15.5** illustrates Gamow's reasoning:

- There are only four different bases in ribonucleotides (A, U, G, and C), so a one-base code could specify only four amino acids.
- A two-base code could represent just 4×4, or 16, amino acids.
- A three-base code could specify $4 \times 4 \times 4$, or 64, different amino acids.

A three-base code provides more than enough messages to code for all 20 amino acids. A three-base code is known as a **triplet code**.

Gamow's hypothesis suggested that the genetic code could be redundant. That is, more than one triplet of bases might specify the same amino acid. As a result, different three-base sequences in an mRNA—say, AAA and AAG—might code for the same amino acid—say, lysine.

The group of three bases that specifies a particular amino acid is called a **codon**. According to the triplet code hypothesis, many of the 64 codons that are possible might actually specify the same amino acids.

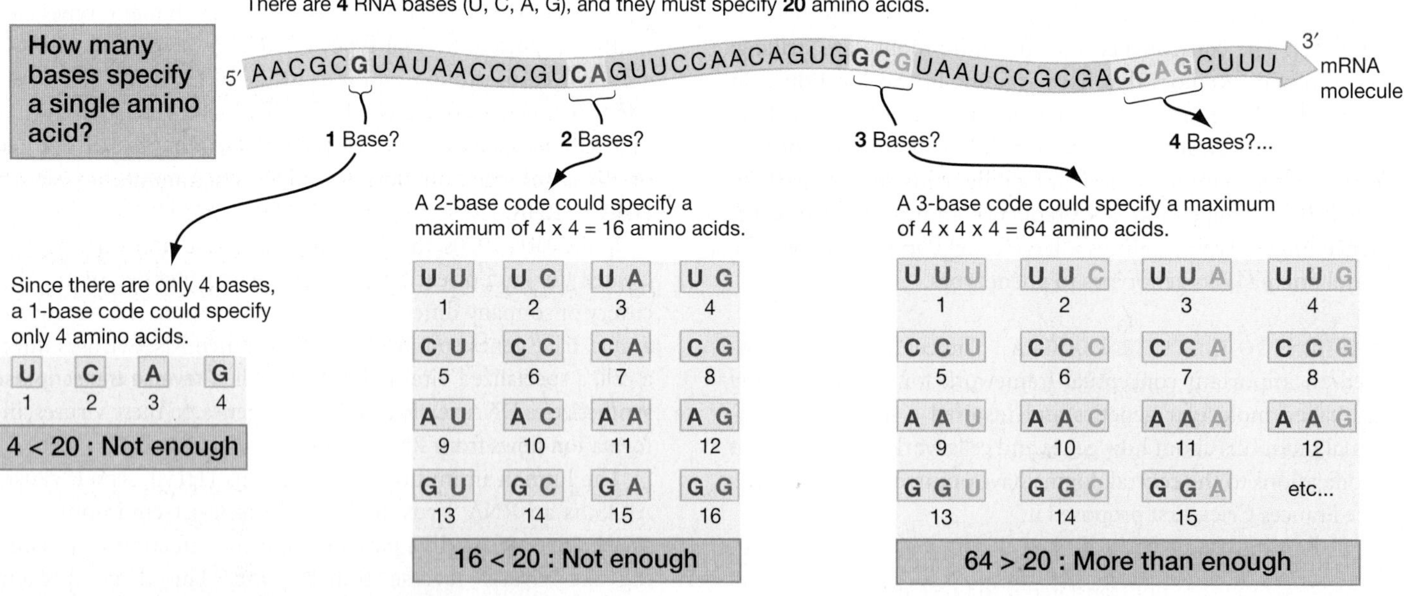

FIGURE 15.5 In the Genetic Code, How Many Bases Form a "Word"?

QUESTION How many amino acids could be specified by a four-base code? Why did biologists conclude that a four-base code was extremely unlikely? Your answer should include an explanation for why a four-base code is unlikely to evolve.

Work by Francis Crick and Sydney Brenner confirmed that codons are three bases long. Their experiments used chemicals that caused an occasional addition or deletion of a base in DNA. As predicted for a triplet code, a one-base addition or deletion in the base sequence led to a loss of function in the gene being studied. This is because a single addition or deletion mutation throws the sequence of codons, or the **reading frame**, out of register.

To understand how a reading frame works, consider the sentence

"The fat cat ate the rat."

The reading frame of this sentence is a three-letter word and a space. If the fourth letter in this sentence—the *f* in *fat*—were deleted, the reading frame would transform the sentence into

"The atc ata tet her at."

This is gibberish.

When the reading frame in a DNA sequence is thrown out of register by the addition or deletion of a base, the composition of each codon changes just like the letters in each word of the example sentence above. The protein produced from the altered DNA sequence has a completely different sequence of amino acids. In terms of its normal function, this protein is gibberish.

Crick and Brenner were also able to produce DNA sequences that had deletions or additions of two base pairs or three base pairs. The only time functional proteins were produced was when three bases were removed. In the sentence

"The fat cat ate the rat."

the combination of removing one letter from each of the first three words might result in

"Tha tca ate the rat."

Just as the altered sentence still conveys some meaning, genes with three deletion mutations were able to produce a functional protein.

The researchers interpreted these results as strong evidence in favor of the triplet code hypothesis. Most other biologists agreed. For a much more in-depth look at the experiments that confirmed the triplet code, go to the study area at *www.masteringbiology.com.*

Web Activity The Triplet Nature of the Genetic Code

The confirmation of the triplet code launched a long, laborious effort to determine which amino acid is specified by each of the 64 codons. Ultimately, it was successful.

How Did Researchers Crack the Code?

The initial advance in deciphering the genetic code came in 1961, when Marshall Nirenberg and Heinrich Matthaei created a method for synthesizing RNAs of known sequence. They began by creating a long polymer of uracil-containing ribonucleotides. These synthetic RNAs were added to an in vitro system for synthesizing proteins. The researchers analyzed the resulting amino acid chain and determined that it was polyphenylalanine—a polymer consisting of the amino acid phenylalanine.

This result was strong evidence that the RNA triplet UUU codes for the amino acid phenylalanine. By complementary base pairing, it was clear that the corresponding DNA sequence would be AAA. This initial work was followed by experiments using RNAs consisting of only A or C. RNAs consisting of only AAAAA... produced polypeptides consisting of only lysine; poly-C RNAs (RNAs consisting of only CCCCC...) produced polypeptides composed entirely of proline.

Nirenberg and Philip Leder later devised a system for synthesizing specific codons. With these they performed a series of experiments in which they added each codon to a cell extract containing the 20 different amino acids, ribosomes, and other molecules required for protein synthesis. As Chapter 7 noted, ribosomes are the multimolecular machines at which proteins are synthesized. Then the researchers determined which amino acid became bound to the ribosomes when a particular codon was present. For example, when the codon CAC was in the reaction mix, the amino acid histidine would bind to the ribosomes. This result confirmed that CAC codes for histidine.

These ribosome-binding experiments allowed Nirenberg and Leder to determine which of the 64 codons coded for each of the 20 amino acids.

In addition, researchers discovered that certain codons are punctuation marks signaling "start of message" or "end of message." These codons indicate that protein synthesis should start at a given codon or that the protein chain is complete.

- There is one **start codon** (AUG), which signals that protein synthesis should begin at that point on the mRNA molecule. The AUG codon codes for the amino acid methionine.
- There are three **stop codons**, also called termination codons (UAA, UAG, and UGA). The stop codons signal that the protein is complete, and end the translation process.

The complete genetic code is given in **Figure 15.6** on page 284. Deciphering it was a tremendous achievement. It represents more than five years of work by several teams of researchers.

ANALYZING THE CODE Once biologists had cracked the genetic code, they realized that it has a series of important properties.

- *It is redundant.* All amino acids except methionine and tryptophan are coded by more than one codon.
- *It is unambiguous.* A single codon never codes for more than one amino acid.
- *It is nearly universal.* With a few minor exceptions, all codons specify the same amino acids in all organisms.
- *It is conservative.* When several codons specify the same amino acid, the first two bases in those codons are almost always identical.

The last point is subtle, but important. Here's the key: If a mutation in DNA or an error in transcription or translation affects

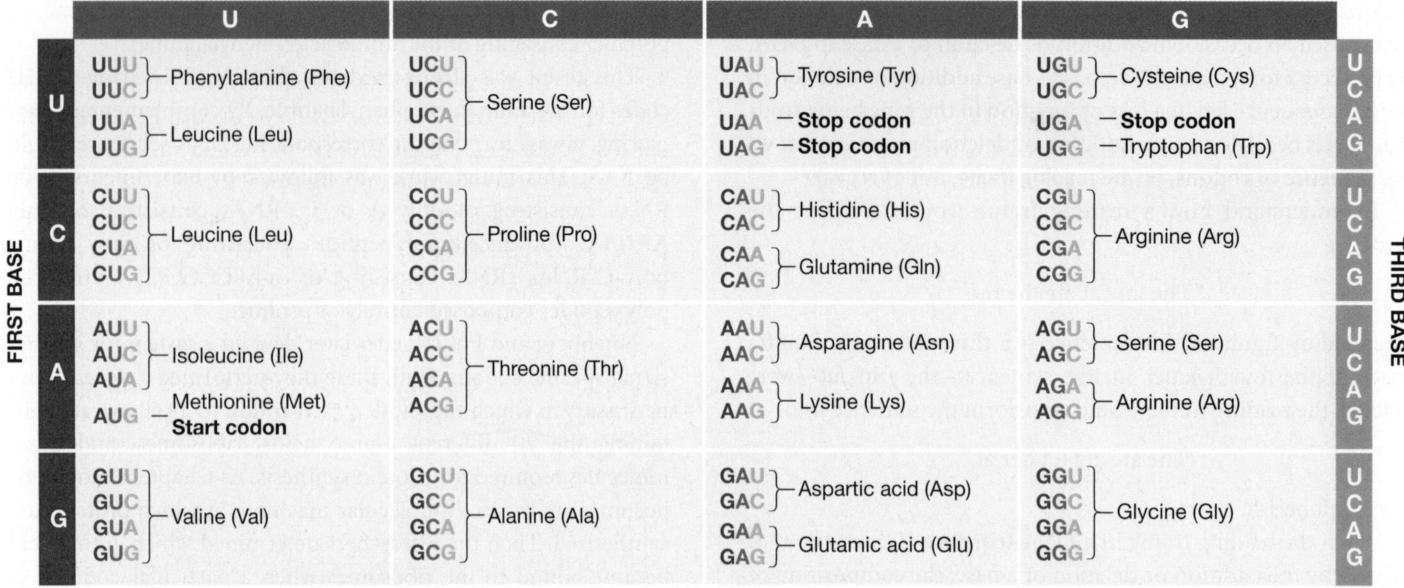

SECOND BASE
U
C
A
G
FIRST BASE
THIRD BASE
UUU
UUC
Phenylalanine (Phe)
UUA
UUG
Leucine (Leu)
UCU
UCC
UCA
UCG
Serine (Ser)
UAU
UAC
Tyrosine (Tyr)
UAA – Stop codon
UAG – Stop codon
UGU
UGC
Cysteine (Cys)
UGA – Stop codon
UGG – Tryptophan (Trp)
CUU
CUC
CUA
CUG
Leucine (Leu)
CCU
CCC
CCA
CCG
Proline (Pro)
CAU
CAC
Histidine (His)
CAA
CAG
Glutamine (Gln)
CGU
CGC
CGA
CGG
Arginine (Arg)
AUU
AUC
AUA
Isoleucine (Ile)
AUG – Methionine (Met) Start codon
ACU
ACC
ACA
ACG
Threonine (Thr)
AAU
AAC
Asparagine (Asn)
AAA
AAG
Lysine (Lys)
AGU
AGC
Serine (Ser)
AGA
AGG
Arginine (Arg)
GUU
GUC
GUA
GUG
Valine (Val)
GCU
GCC
GCA
GCG
Alanine (Ala)
GAU
GAC
Aspartic acid (Asp)
GAA
GAG
Glutamic acid (Glu)
GGU
GGC
GGA
GGG
Glycine (Gly)

FIGURE 15.6 The Genetic Code. To read an mRNA codon, locate its first base in the red band on the left side; from there, move rightward to the box under the codon's second base in the blue band along the top; and lastly, move up and down within that box to the level of the codon's third base in the green band on the right side. The 64 codons, along with the amino acid or other signal that they specify, are displayed in the boxes. By convention, codons are always written in the 5′ → 3′ direction.

the third position in a codon, it is unlikely to change the amino acid in the final protein. This feature makes individuals less vulnerable to small, random changes or errors in their DNA sequences. Compared with randomly generated codes, the existing genetic code efficiently minimizes the phenotypic effects of small changes in DNA and errors during translation. Stated another way, the genetic code does not represent a random assemblage of bases, like letters drawn from a hat. It has been honed by natural selection, and is remarkably efficient.

USING THE CODE Using the genetic code, biologists can work forwards or backwards in the central dogma to:

1. Predict the codons and amino acid sequence encoded by a particular DNA sequence (see **Figure 15.7a**).

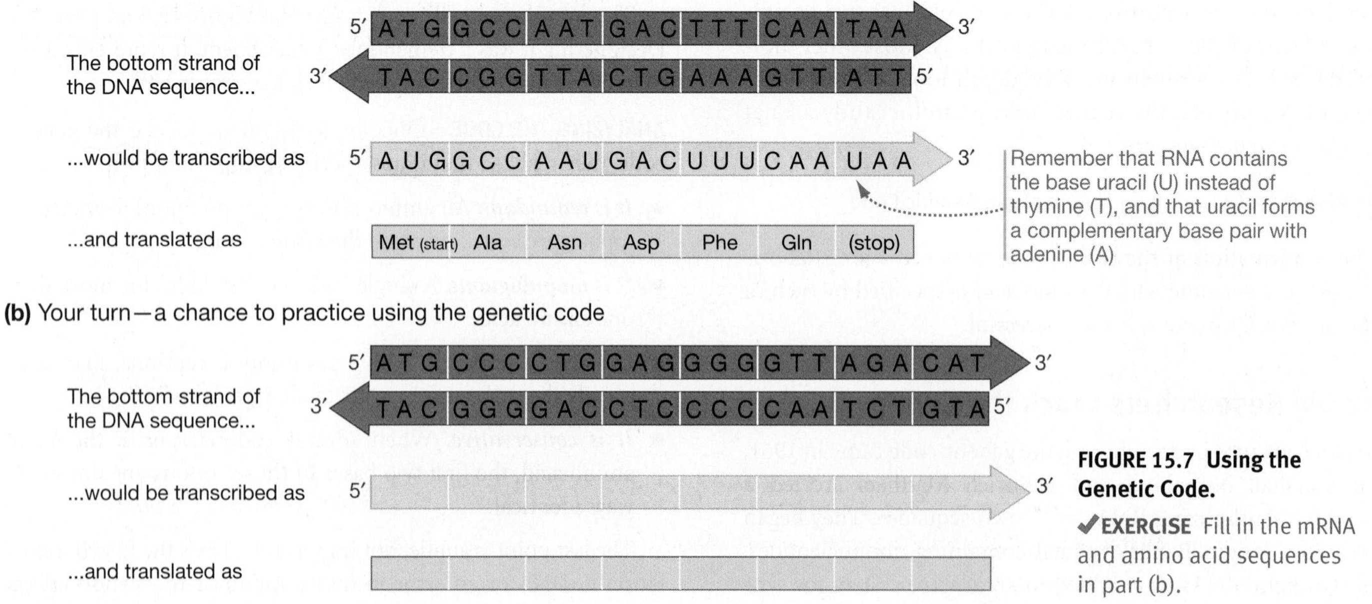

FIGURE 15.7 Using the Genetic Code.

✔**EXERCISE** Fill in the mRNA and amino acid sequences in part (b).

2. Approximate the mRNA and DNA sequence that would code for a particular sequence of amino acids.

Why can you only *approximate* an mRNA or DNA sequence from a given amino acid sequence? The answer lies in the code's redundancy. If a polypeptide contains phenylalanine, you don't know if the codon responsible is UUU or UUC.

✔If you understand how to read the genetic code, you should be able to do the following tasks: (1) Identify the codons in Figure 15.4 and decide whether they are translated correctly. (2) Complete the Exercise for **Figure 15.7b**. (3) Write an mRNA that codes for the amino acid sequence Ala-Asn-Asp-Phe-Gln and yet is different from the one given in Figure 15.7a. Indicate the mRNA's 5′ → 3′ polarity. Then write the double-stranded DNA that corresponds to this mRNA. Indicate the 5′ → 3′ polarity of both DNA strands.

Once the central dogma and genetic code were understood, biologists were finally able to explore and eventually understand the molecular basis of mutation. How do novel traits—such as dwarfing in garden peas and white eye color in fuit flies—come to be?

CHECK YOUR UNDERSTANDING

If you understand that . . .

- The sequence of bases in mRNA constitutes a code. Particular combinations of three bases lead to the addition of specific amino acids to the protein encoded by the gene.
- The genetic code is redundant. It consists of 64 combinations of bases, but only 20 amino acids and a stop "punctuation mark" need to be specified.

✔ **You should be able to . . .**

Consider the consequences of a mutation in a DNA sequence ATA to one of the following sequences: GTA, TTA, or GCA.

1. In each case, specify the resulting change in the mRNA codon.
2. In each case, describe the effect on the resulting protein.

Answers are available in Appendix B.

15.4 What Is the Molecular Basis of Mutation?

This chapter has explored how the information archived in DNA is put into action in the form of working RNAs and proteins. Now the question is, what happens if the information in DNA changes? What are the consequences for the cell?

A **mutation** is any permanent change in an organism's DNA. It is a modification in a cell's information archive—a change in its genotype. Mutations create new alleles.

Because changes in the genotype can change the mRNA codons that are transcribed and the amino acids that are translated, mutations may lead to changes that affect the primary structure of proteins and thus the organism's phenotype.

Point Mutation

Figure 15.8 shows how a common type of mutation occurs. If DNA polymerase mistakenly inserts the wrong base as it synthesizes a new strand of DNA, and if proofreading by DNA polymerase and the mismatch repair system fail to correct the mismatched base before another round of DNA replication occurs, a change in the sequence of bases in DNA results. A single base change such as this is called a **point mutation**.

What happens when point mutations are transcribed and translated? To answer this question, look back at Figure 15.4 and recall that a change in a single base in DNA is associated with a difference in coat color in certain populations of mice. The DNA sequence in Figure 15.4a is found in dark-colored mice that live in forest habitats; the sequence in Figure 15.4b is found in light-colored mice that live in beach habitats.

Because beach-dwelling populations are younger than the nearby forest-dwelling populations, researchers hypothesize that the following sequence of events occurred:

1. Forest mice colonized beach habitats.
2. Either before or after the colonization event, a point mutation occurred that altered the melanocortin receptor gene and resulted in some offspring having light coats.

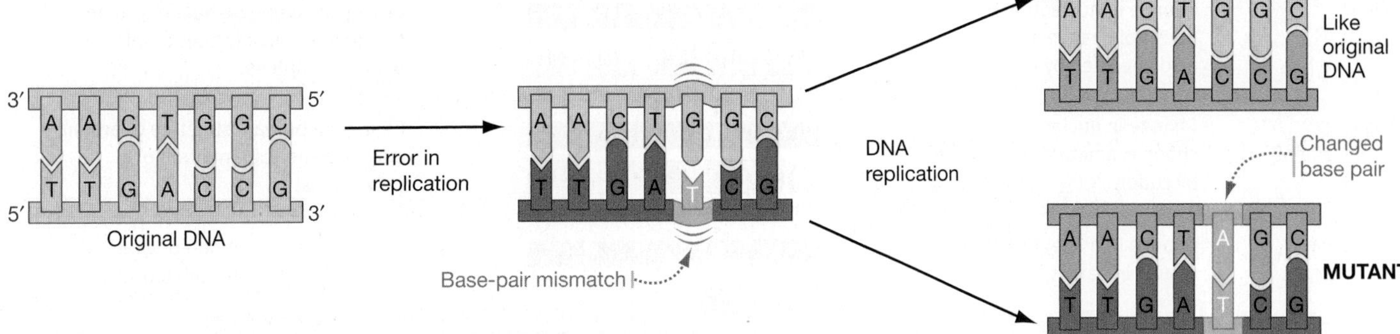

FIGURE 15.8 Unrepaired Mistakes in DNA Synthesis Lead to Point Mutations.

✔**QUESTION** Why is it logical that the type of mutation illustrated here is termed a point mutation?

3. Light-colored mice are camouflaged in beach habitats; in sandy environments, they suffer lower predation than dark-colored mice.
4. Over time, the allele created by the point mutation increased in frequency in beach-dwelling populations.

Point mutations that cause these types of changes in the amino acid sequence of proteins are called **missense mutations** or **replacement mutations**. But note that if the same G to A change had occurred in the third position of the same DNA codon, instead of the first position, there would have been no change in the mRNA or protein produced. The mRNA codons CGC and CGU both code for arginine. A point mutation that does not change the amino acid sequence of the gene product is called a **silent mutation**.

In terms of the impact on organisms, biologists divide mutations into three categories:

1. **Beneficial** Some mutations increase the fitness of the organism—meaning, its ability to survive and reproduce—in certain environments. The G-to-A mutation is beneficial in beach habitats because it camouflages mice.
2. **Neutral** If a mutation has no affect on fitness, it is termed neutral. Silent mutations are usually neutral.
3. **Deleterious** Because organisms tend to be well-adapted to their current habitat, and because mutations are random changes in the genotype, most mutations lower fitness. The G-to-A mutation would be deleterious in the forest habitat.

Recent studies indicate that the vast majority of point mutations are neutral or slightly deleterious. **Table 15.1** summarizes the types of point mutations that have been documented and reviews their consequences for the amino acid sequences of proteins and for fitness.

Chromosome-Level Mutations

Besides documenting various types of point mutations, biologists study mutations consisting of larger-scale changes in the composition of chromosomes. Chapter 12 introduced **polyploidy**, which is a change in the number of each type of chromosome present, and **aneuploidy**, the addition or deletion of a chromosome. Polyploidy, aneuploidy, and other changes in chromosome number result from chance mistakes in the partitioning of chromosomes during meiosis or mitosis.

But in addition to changes in overall chromosome number, the composition of individual chromosomes can change in important ways. For example, chromosome segments can become detached when accidental breaks in chromosomes occur. The segments may become flipped and rejoin—a phenomenon known as a chromosome **inversion**—or become attached to a different chromosome, an event called chromosome **translocation**.

Like point mutations, chromosome-level mutations can be beneficial, neutral, or deleterious. But because they represent massive changes in the genotype, chromosome alterations are usually deleterious. It is common, for example, to find that the chromosomes of cancerous cells exhibit aneuploidy, inversions, translocations, or combinations of such mutations. **Figure 15.9** drives this home by comparing the **karyotype**—the complete set of chromosomes in a cell—of a normal versus cancerous human cell.

To summarize, point mutations and chromosome-level mutations are random changes in DNA that produce new genes, new alleles, and new traits. At the level of individuals, mutations can cause disease or death or lead to increases in fitness. At the level of populations, mutations furnish the heritable variation that Mendel and Morgan analyzed and that makes evolution possible. These phenomena are explored more fully in Unit 5.

SUMMARY TABLE 15.1 **Known Types of Point Mutations**

Name	Definition	Example	Consequence
	Original DNA sequence of non-template (coding) strand	TAT TGG CTA GTA CAT Tyr–Trp–Leu–Val–His — Original polypeptide	
Silent	Change in nucleotide that does not change amino acid specified by codon	TAC TGG CTA GTA CAT Tyr–Trp–Leu–Val–His	Change in genotype but no change in phenotype. Usually neutral with respect to fitness.
Missense (replacement)	Change in nucleotide that changes amino acid specified by codon	TAT TGT CTA GTA CAT Tyr–**Cys**–Leu–Val–His	Change in primary structure of protein may be beneficial, neutral, or deleterious.
Nonsense	Change in nucleotide that results in early stop codon	TAT TGA CTA GTA CAT Tyr **STOP**	Premature termination—polypeptide is truncated. Usually deleterious.
Frameshift	Addition or deletion of a nucleotide	TAT TCG GCT AGT ACA T Tyr–**Ser**–**Ala**–**Ser**–**Thr**	Reading frame is shifted—massive missense. Usually deleterious.

(a) Normal human karyotype

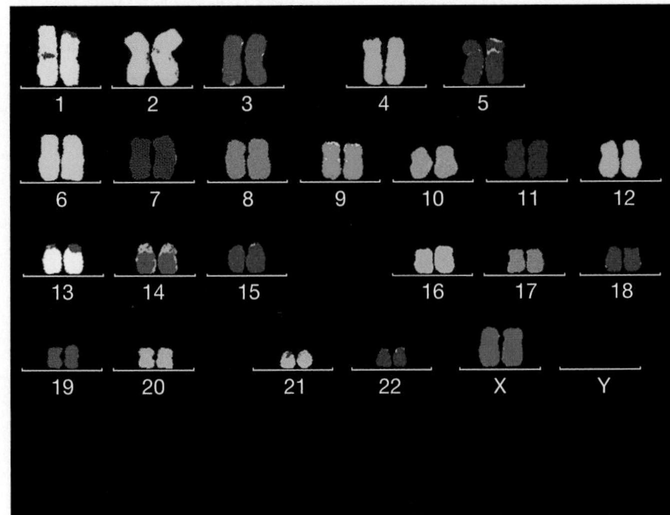

(b) Karyotype from a human cancerous cell

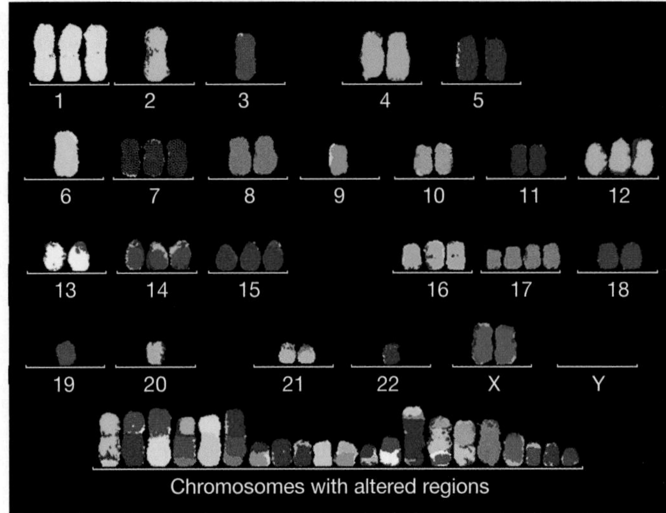

FIGURE 15.9 Chromosome-Level Mutations. Karyotypes are micrographs of all the chromosomes present in a cell. A normal human karyotype **(a)** includes pairs of 23 standard chromosomes, whereas a karyotype from a cancerous cell **(b)** exhibits many irregularities. Some whole chromosomes are missing, extra copies of others are present, and many individual chromosomes contain translocated or inverted fragments.

CHAPTER 15 REVIEW

For media, go to the study area at www.masteringbiology.com

Summary of Key Concepts

Most genes code for proteins.

- Because proteins serve so many functions in the cell, the one-gene, one-enzyme hypothesis was restated as one-gene, one-protein, or more accurately, one-gene, one-polypeptide.

✔You should be able to explain why the one-gene, one-polypeptide hypothesis is now considered too narrow, and use the form "one gene, . . ." to state a modern, more-accurate equivalent.

MB **Web Activity** The One-Gene, One-Enzyme Hypothesis

DNA is transcribed to messenger RNA by RNA polymerase, and then messenger RNA is translated to proteins by ribosomes. In this way, genetic information is converted from DNA to RNA to proteins.

- DNA does not code for proteins directly. Instead, mRNA molecules are transcribed from DNA and then translated into proteins.
- The flow of information, from DNA to RNA to proteins, is called the central dogma of molecular biology.
- Many RNAs do not act as messengers for protein synthesis. They perform other functions in the cell.
- In some viruses, information flows from RNA to DNA.

✔You should be able to explain why transcription and translation are appropriate terms, in terms of how information is processed in a cell.

Each amino acid in a protein is specified by a group of three bases in messenger RNA.

- By synthesizing RNAs of known base composition and then observing the results of translation, researchers were able to unravel the genetic code.
- The genetic code is read in triplets.
- The code is redundant—meaning that most of the 20 amino acids are specified by more than one codon.
- Certain codons signal when translation starts and stops.

✔You should be able to explain why some changes in DNA sequences do not change the corresponding protein—meaning, why certain changes in an organism's genotype do not change its phenotype.

MB **Web Activity** The Triplet Nature of the Genetic Code

Mutations are random changes in DNA, ranging in extent from single bases to large chromosome regions, that may or may not produce changes in the phenotype.

- Depending on the location and type of alteration in DNA and its impact on the resulting RNA or protein product, a mutation can be beneficial, deleterious, or neutral with respect to fitness.
- Mutations may produce novel proteins and RNAs. They are the source of the heritable variation that makes evolution possible.

✔You should be able to explain why a novel mutation was important during the evolution of beach-dwelling populations of Oldfield mice, and explain what would happen if the same mutation occurred in a population that lives in forest habitats.

Questions

✓ TEST YOUR KNOWLEDGE

Answers are available in Appendix B

1. What does the one-gene, one-enzyme hypothesis state?
   a. Genes are composed of stretches of DNA.
   b. Genes are made of protein.
   c. Genes code for ribozymes.
   d. A single gene codes for a single protein.
2. Which of the following is an important exception to the central dogma of molecular biology?
   a. Many genes code for RNAs that function directly in the cell.
   b. DNA is the repository of genetic information in all organisms (though not all viruses).
   c. Messenger RNA is a short-lived "information carrier."
   d. Proteins are responsible for most aspects of the phenotype.
3. DNA's primary structure is made up of just four different bases, and its secondary structure is regular and highly stable. How can a molecule with these characteristics hold all of the information required to build and maintain a cell?
   a. The information is first transcribed, then translated.
   b. The messenger RNA produced from DNA has much more complex secondary structures, and thus holds much more information.
   c. A protein produced (indirectly) from DNA has much more complex primary and secondary structures, and thus holds much more information.
   d. The information in DNA is in code form.
4. Why did researchers suspect that DNA does not code for proteins directly?
   a. In eukaryotes, DNA is found inside the nucleus, but proteins are produced outside the nucleus.
   b. In prokaryotes, DNA and proteins are never found together.
   c. When DNA was damaged by ultraviolet radiation or other sources of energy, the proteins in the cell did not change.
   d. There are several distinct types of RNA, of which only one functions as messenger RNA.
5. Which of the following describes an important experimental strategy in deciphering the genetic code?
   a. comparing the amino acid sequences of proteins with the base sequence of their genes
   b. analyzing the sequence of RNAs produced from known DNA sequences
   c. analyzing mutants that changed the code
   d. examining the polypeptides produced when RNAs of known sequence were translated
6. What is a stop codon?
   a. The place where transcription ends.
   b. The place where translation ends.
   c. The end of a chromosome.
   d. Any codon that does not code for an amino acid.

✓ TEST YOUR UNDERSTANDING

Answers are available in Appendix B

1. Explain why Morse code is an appropriate analogy for the genetic code.
2. Draw a hypothetical metabolic pathway composed of five substrates, five enzymes, and a product called Biological Sciazine. Number the substrates 1–5, and label the enzymes A–E, in order. (For instance, enzyme A catalyzes the reaction between substrates 1 and 2.)
   - Suppose a mutation made the gene for enzyme C nonfunctional. What molecule would accumulate in the affected cells?
   - Suppose some individuals can survive if given substrate 5 in the diet. But they die if given substrates 1, 2, 3, and 4. Which enzyme in the pathway is affected by this mutation?
3. Why did experiments with *Neurospora crassa* mutants support the one-gene, one-enzyme hypothesis?
4. Explain how a single-base deletion disrupts the reading frame of a gene. Include an example.
5. When researchers discovered that a combination of three deletion mutations or three addition mutations would restore the function of a gene, most biologists were convinced that the genetic code was read in triplets. Explain the logic behind this conclusion.
6. Explain why all point mutations change the genotype, but why only some point mutations change the phenotype.

✓ APPLYING CONCEPTS TO NEW SITUATIONS

Answers are available in Appendix B

1. Recall that DNA and RNA are synthesized only in the $5' \rightarrow 3'$ direction and that DNA and RNA sequences are always written in the $5' \rightarrow 3'$ direction. Consider the following DNA sequence:

   ```
   5' TTGAAATGCCCGTTTGGAGATCGGGTTACAGCTAGTCAAAG 3'
   3' AACTTTACGGGCAAACCACTAGCCCAATGTCGATCAGTTTC 5'
   ```

   - Identify bases in the bottom strand that encode start and stop codons.
   - Write the mRNA sequence that would be transcribed between start and stop codons if the bottom strand served as the template.
   - Write the amino acid sequence that would be translated from the mRNA sequence you just wrote.
2. What problems would arise if the genetic code contained only 22 codons—one for each amino acid, a start signal, and a stop signal? (Hint: When DNA is copied prior to mitosis or meiosis, random errors occur that change its primary base sequence.)
3. Scientists say that a phenomenon is a "black box" if they can describe it and study its effects but don't yet know the underlying mechanism that causes it. In what sense was genetics—meaning the transmission of heritable traits—a black box before the central dogma of molecular biology was understood?
4. One of the possibilities that researchers interested in the genetic code had to consider was that the code was overlapping, meaning that a single base could be part of more than one codon. Make a diagram showing how an overlapping code would work, assuming that each codon is three bases long. As an example, use the sequence 5′ AUGUUACGGAAUUGA 3′.

This chapter is part of The Big Picture. See how on pages 336–337.

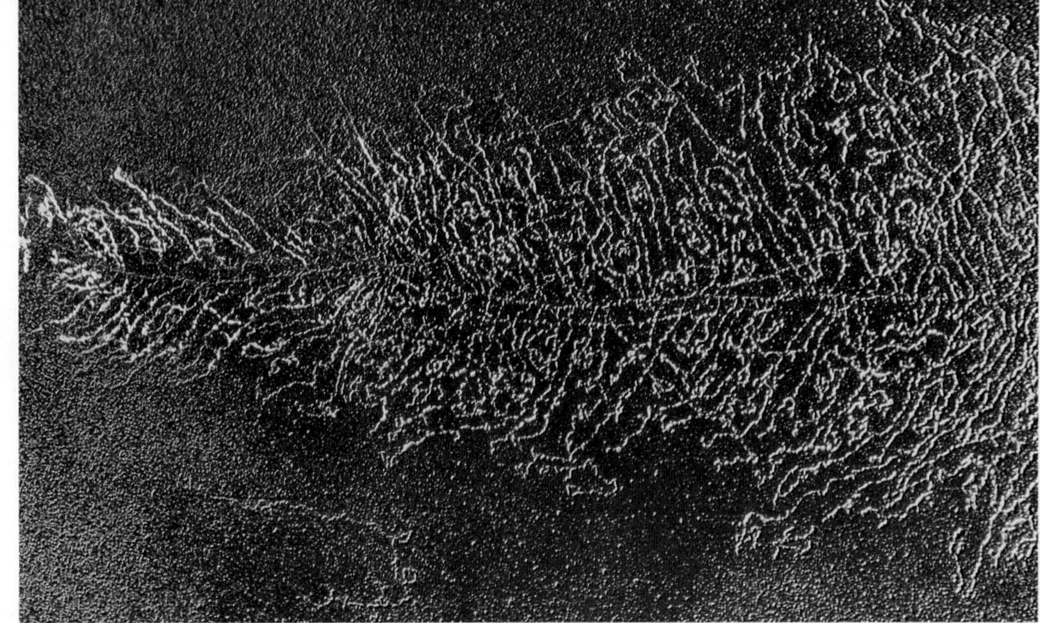

In this segment of a chromosome inside a frog egg, extensive transcription is taking place. The horizontal strand in the middle of this micrograph is DNA; the strands that have been colored yellow and red, and that are coming off on either side, are RNA molecules.

Transcription, RNA Processing, and Translation 16

KEY CONCEPTS

- After RNA polymerase binds DNA with the help of other proteins, it catalyzes the production of an RNA molecule whose base sequence is complementary to the base sequence of the DNA template strand.
- Eukaryotic genes contain regions called exons and regions called introns; during RNA processing, the regions coded by introns are removed, and the ends of the RNA receive a cap and tail.
- Ribosomes translate mRNAs into proteins with the help of intermediary molecules called transfer RNAs (tRNAs).
- Each transfer RNA carries an amino acid corresponding to the tRNA's three-base-long anticodon.
- In the ribosome, the tRNA anticodon binds to a three-base-long mRNA codon, causing the amino acid carried by the transfer RNA to be added to the growing protein.

Proteins are the stuff of life. They give shape to our cells, control the chemical reactions that go on inside them, and regulate how materials move into, out of, and through them.

No one knows exactly how many different proteins can be made in your body's cells, but 100,000 is a reasonable estimate. Some of these proteins may not be produced at all in some types of cells; others may be present in quantities ranging from millions of copies to fewer than a dozen.

A cell builds the proteins it needs from instructions encoded in its DNA. Chapter 15 introduced the central dogma of molecular biology—the flow of information from DNA to mRNA to protein. Once this pattern of information flow had been established, biologists puzzled over how cells actually accomplish the two major steps in the process: transcription and translation. Specifically, how does RNA polymerase know where to start transcribing a gene, and where to end? And once an RNA message is produced, how is the linear sequence of ribonucleotides translated into the linear sequence of amino acids in a protein?

As Chapter 1 pointed out, processing genetic information is one of five key attributes of life. This chapter delves into the molecular mechanisms of gene expression—the blood and guts of the central dogma. It starts with the monomers that build an RNA and ends with a finished protein.

16.1 An Overview of Transcription

The first step in converting genetic information into proteins is to synthesize a messenger RNA version of the instructions archived in DNA. Enzymes called RNA polymerases, introduced in Chapter 15, are responsible for synthesizing mRNA.

✔ When you see this checkmark, stop and test yourself. Answers are available in Appendix B.

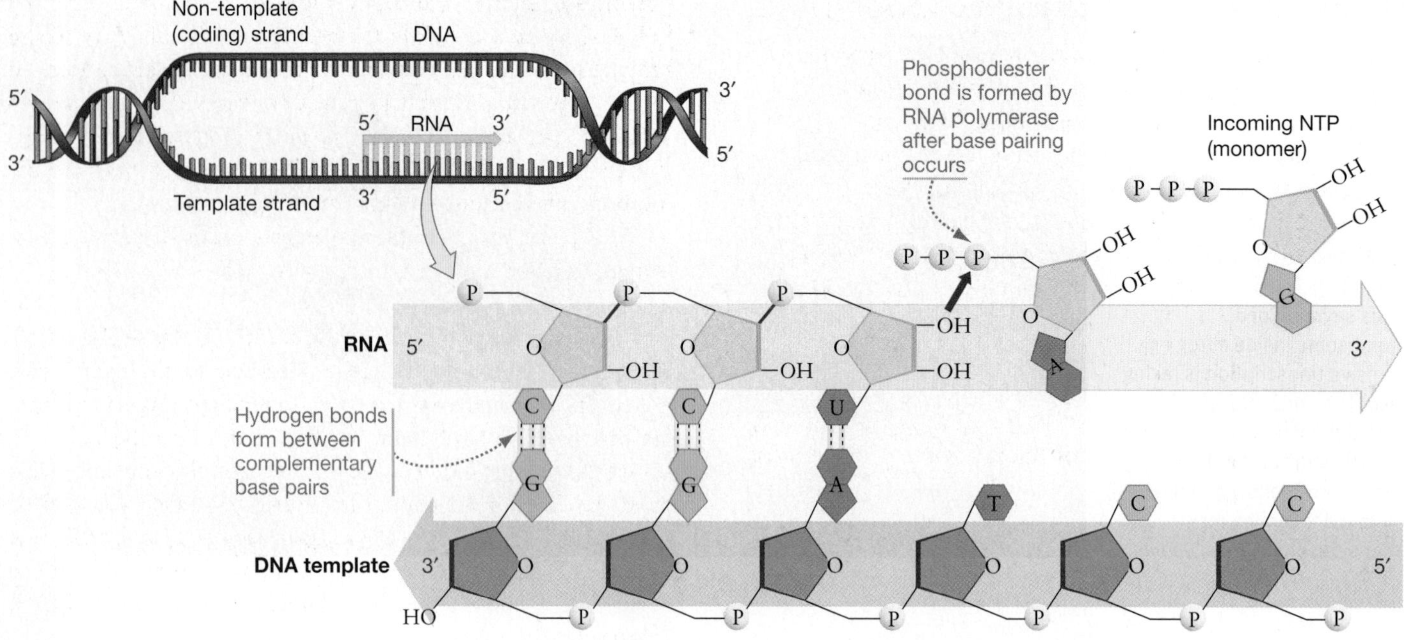

FIGURE 16.1 Transcription Is the Synthesis of RNA from a DNA Template. The reaction catalyzed by RNA polymerase (not shown) results in the formation of a phosphodiester bond between ribonucleotides. RNA polymerase produces an RNA strand whose sequence is complementary to the bases in the DNA template.

✔**QUESTION** In which direction is RNA synthesized, 5′→3′ or 3′→5′? In which direction is the DNA template "read"?

Figure 16.1 shows how the polymerization reaction occurs. Note the incoming monomer—a ribonucleotide triphosphate, or NTP—at the far right of the diagram. NTPs are like the dNTPs introduced in Chapter 14's discussion of DNA synthesis, except that they have a hydroxyl (−OH) group on the 2′ carbon. Once an NTP that matches a base on the DNA template is in place, RNA polymerase catalyzes the formation of a phosphodiester bond between the 3′ end of the growing mRNA chain and the new ribonucleotide. As this 5′→3′ matching-and-catalysis process continues, an RNA that is complementary to the gene is synthesized. This is transcription.

Notice that only one of the two DNA strands is used as a template and transcribed, or "read," by RNA polymerase.

- The strand that is read by the enzyme is called the **template strand**.
- The other strand is called the **non-template strand** or **coding strand**. Coding strand is a particularly appropriate name, because its sequence matches the sequence of the RNA that is transcribed from the template strand and codes for a polypeptide.

The coding strand and the RNA don't match exactly, however, because RNA has uracil (U) rather than the thymine (T) found in the coding strand. For the same reason, an adenine (A) in the DNA template strand specifies a U in the complementary RNA strand.

Once the basic role of RNA polymerase was understood, biologists turned to questions about how the enzyme works.

Characteristics of RNA Polymerase

Like the DNA polymerases introduced in Chapter 14, an RNA polymerase performs a template-directed synthesis in the 5′ to 3′ direction. But unlike DNA polymerases, RNA polymerases do not require a primer to begin transcription.

Bacteria have a single RNA polymerase. Eukaryotes, in contrast, have three distinct types. As **Table 16.1** shows, RNA polymerase I, II, and III—often referred to as pol I, pol II, and pol III—each transcribe only certain types of RNA in eukaryotes. RNA pol II is the only polymerase that transcribes the genes that code for proteins and produces mRNA.

TABLE 16.1 **Eukaryotic RNA Polymerases**

Name of Enzyme	Type of Gene Transcribed
RNA polymerase I (RNA pol I)	Genes that code for most of the large RNA molecules (rRNAs) found in ribosomes
RNA polymerase II (RNA pol II)	Protein-coding genes (produce mRNAs); also, genes that code for RNAs that function in ribosome assembly, and in processing and regulation of mRNAs
RNA polymerase III (RNA pol III)	Genes that code for transfer RNAs (tRNAs), for one of the small rRNAs found in ribosomes, and for noncoding RNAs (ncRNAs); also, genes that code for RNAs that function in ribosome assembly, and in processing and regulation of mRNAs

Initiation: How Does Transcription Begin?

How does RNA polymerase know where to start transcription on the DNA template? The answer to this question defined what biologists now call the **initiation** phase of transcription.

Soon after the discovery of bacterial RNA polymerase, researchers realized that the enzyme cannot initiate transcription on its own. Instead, a detachable protein subunit called sigma must bind to the polymerase before transcription can begin.

Bacterial RNA polymerase and sigma form what biologists call a **holoenzyme** (literally, "whole enzyme"; **Figure 16.2a**). A holoenzyme consists of a **core enzyme**, which contains the active site for catalysis, and other required proteins.

If bacterial RNA polymerase is the core enzyme of this holoenzyme, what does sigma do? When researchers mixed the polymerase, sigma, and DNA together, they found that the holoenzyme bound tightly to specific sections of DNA. These binding sites were named **promoters**, because they are sections of DNA where transcription begins. The discovery of promoters suggested that sigma's function is regulatory in nature. Sigma appeared to be responsible for guiding RNA polymerase to specific locations where transcription should begin.

What is the nature of these specific locations? What do promoters look like, and what do they do?

(a) RNA polymerase and sigma form a holoenzyme.

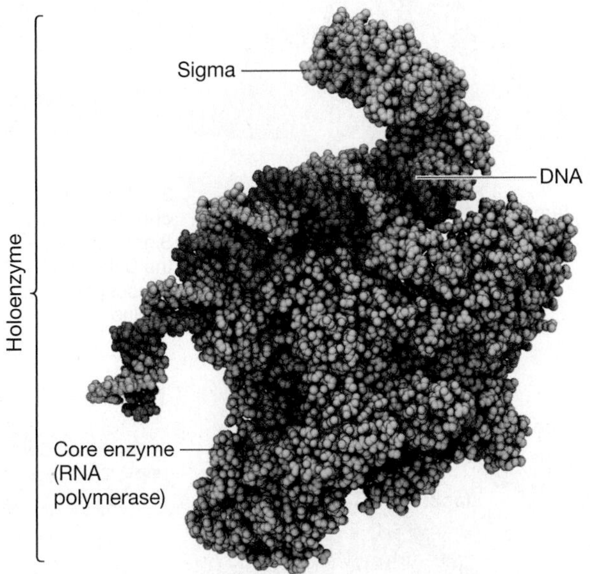

(b) Sigma recognizes and binds to the promoter.

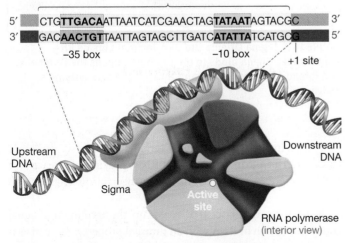

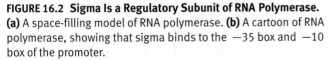

FIGURE 16.2 Sigma Is a Regulatory Subunit of RNA Polymerase. **(a)** A space-filling model of RNA polymerase. **(b)** A cartoon of RNA polymerase, showing that sigma binds to the −35 box and −10 box of the promoter.

BACTERIAL AND EUKARYOTIC PROMOTERS David Pribnow offered an initial answer to these questions in the mid-1970s. When Pribnow analyzed the base sequence of promoters from various bacteria and from viruses that infect bacteria, he found that the promoters were 40–50 base pairs long and had a particular section in common: a series of bases identical or similar to TATAAT. This six-base-pair sequence is now known as the −10 box, because it is centered about 10 bases from the point where bacterial RNA polymerase starts transcription (**Figure 16.2b**).

DNA that is located in the direction RNA polymerase moves during transcription is said to be **downstream** from the point of reference; DNA located in the opposite direction is said to be **upstream**. Thus, the −10 box is centered about 10 bases upstream from the transcription start site. The place where transcription begins is called the +1 site.

Soon after the discovery of the −10 box, researchers recognized that the sequence TTGACA occurred in these same promoters and was centered about 35 bases upstream from the +1 site. This second key sequence is called the −35 box. Although all promoters have a −10 box and a −35 box, the sequences outside these boxes (but within the promoter) vary.

Eukaryotic genes have promoters that signal where transcription should begin, just as bacteria do. Promoters in eukaryotic DNA are much more diverse and complex than bacterial promoters, however. Many of the eukaryotic promoters include a unique sequence called the **TATA box**, centered about 30 base pairs upstream of the transcription start site.

THE ROLE OF SIGMA SUBUNITS AND BASAL TRANSCRIPTION FACTORS In bacteria, transcription begins when sigma, as part of the holoenzyme complex, binds to the −35 and −10 boxes. Sigma, and not RNA polymerase, makes the initial contact with DNA that starts transcription. This key observation supports the hypothesis that sigma is a regulatory protein. Sigma tells RNA polymerase where and when to start synthesizing RNA.

Recent work has shown that most bacteria have several types of sigma proteins, each with a distinct structure and function. *Escherichia coli* has seven different sigma proteins, for example, while *Streptomyces coelicolor* has more than 60. Each of these proteins binds to promoters with slightly different DNA base sequences outside of the −10 and −35 boxes. Thus, each type of sigma protein allows RNA polymerase to bind to a different type of promoter and therefore a different kind of gene.

In some cases, certain sigma proteins bind to promoters for genes that have similar functions. For example, one type of sigma initiates the transcription of genes that help the cell cope with high temperatures.

The type of sigma protein in the RNA polymerase holoenzyme determines which type of genes will be transcribed. Controlling which sigma proteins are active is one of the ways that bacterial cells control which genes are expressed.

Transcription initiation in eukaryotes is similar to the process in bacteria in that RNA polymerase does not bind directly to promoter sequences by itself. Instead, proteins that came to be called **basal transcription factors** initiate eukaryotic transcription by binding to the appropriate promoter region in DNA. In eukaryotes, the function of basal transcription factors is analogous to the function of the sigma proteins in bacteria, except that many proteins are involved, instead of one, and basal transcription factors are not part of a holoenzyme. The basal transcription factors assemble at the promoter first, and RNA polymerase follows.

EVENTS INSIDE THE HOLOENZYME Once sigma binds to a promoter for a bacterial gene, the DNA helix opens, creating two separated strands of DNA, as shown in **Figure 16.3**, steps 1 and 2. As step 2 shows, the template strand is threaded through a channel that leads to the active site inside RNA polymerase. Ribonucleoside triphosphates—the monomers that will be strung together into RNA—enter a channel at the bottom of the enzyme and diffuse to the active site.

When an incoming NTP pairs with a complementary base on the template strand of DNA, RNA polymerization begins. The reaction catalyzed by RNA polymerase is exergonic and spontaneous because NTPs have so much potential energy, owing to their three phosphate groups. As step 3 of Figure 16.3 shows, sigma is released once RNA synthesis is under way. The initiation phase of transcription is complete.

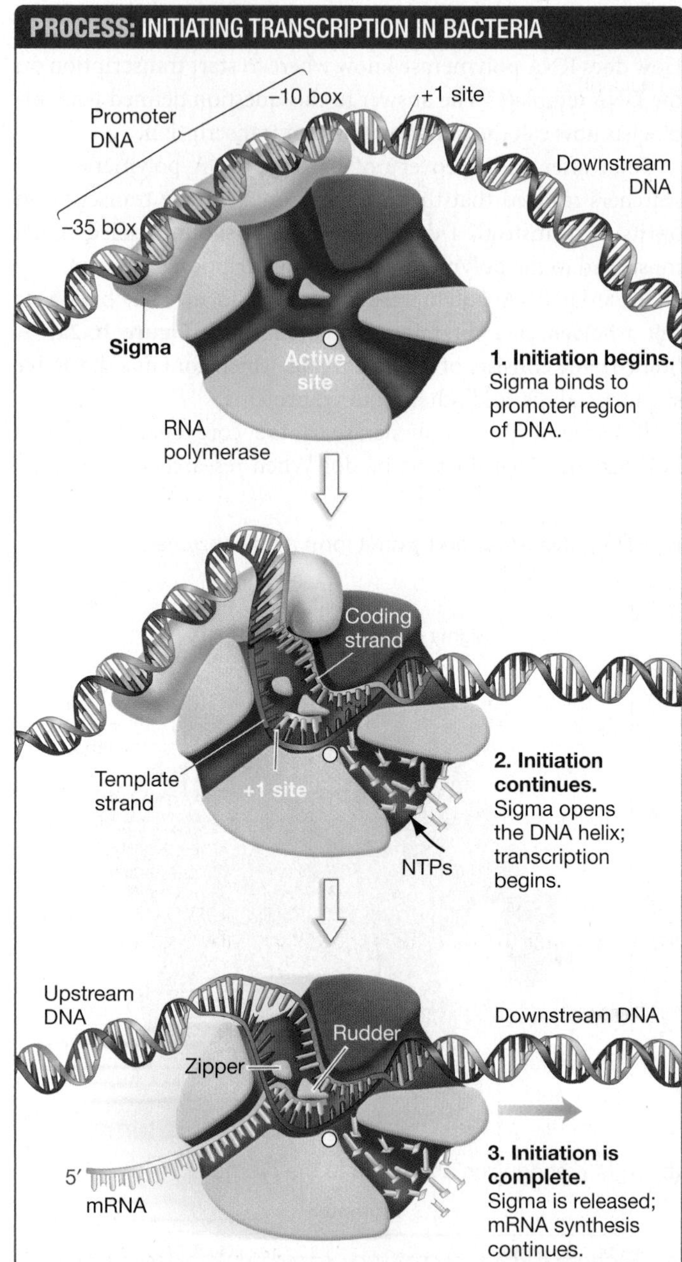

FIGURE 16.3 Sigma Orients the DNA Template inside RNA Polymerase. Sigma binds to the promoter, opens the DNA helix, and threads the template strand through the core enzyme's active site.

Elongation and Termination

Once RNA polymerase begins moving along the DNA template in the 3′→5′ direction, synthesizing RNA in the 5′→3′ direction, the **elongation** phase of transcription is under way. In the interior of the enzyme, a group of amino acids called the rudder helps steer the template and non-template strands through channels inside the enzyme (see Figure 16.3, step 3). Meanwhile, the enzyme's active site catalyzes the addition of nucleotides to the 3′ end of the growing RNA molecule at the rate of about 50 nucleotides per second. A group of projecting amino acids called the enzyme's zipper then helps separate the newly synthesized RNA from the DNA template.

Note that during the elongation phase of transcription, all of the prominent channels and grooves in the enzyme are filled (Figure 16.3, step 3). Double-stranded DNA goes into and out of one groove; ribonucleoside triphosphates enter another; and the growing RNA strand exits to the rear. In this way, the enzyme's structure correlates closely with its function.

Transcription ends with a **termination** phase. In most cases, transcription stops when RNA polymerase reaches a DNA sequence that functions as a transcription-termination signal.

In bacteria, the bases that make up the termination signal code for a stretch of RNA with an unusual property: As soon as it is synthesized, the RNA sequence folds back on itself and forms a short double helix that is held together by complementary base pairing. The secondary structure that results is called a hairpin (**Figure 16.4**). The formation of the hairpin structure is thought

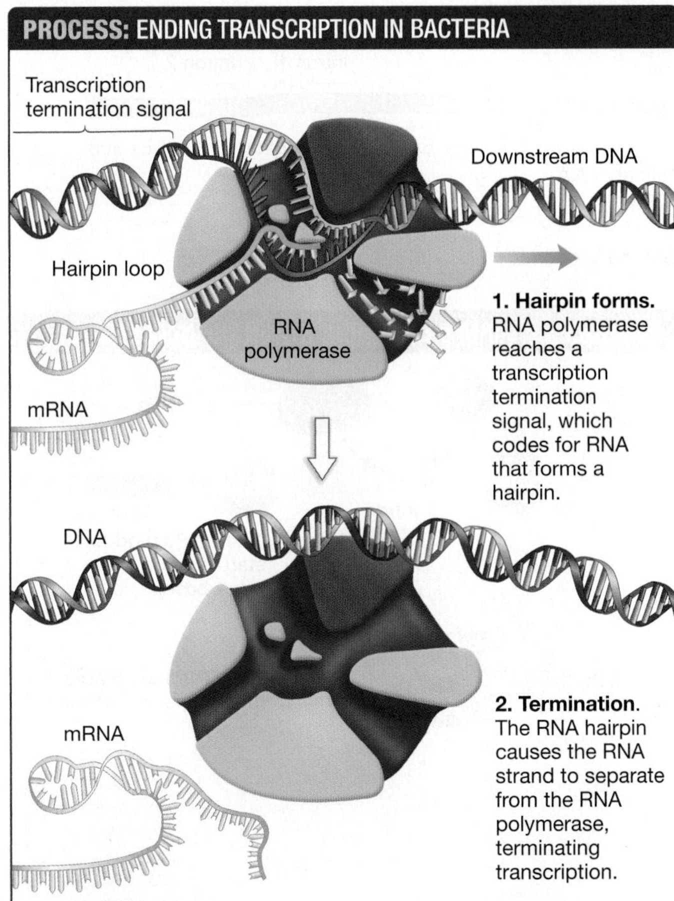
PROCESS: ENDING TRANSCRIPTION IN BACTERIA
Transcription termination signal
Downstream DNA
Hairpin loop
RNA polymerase
mRNA
1. Hairpin forms. RNA polymerase reaches a transcription termination signal, which codes for RNA that forms a hairpin.
DNA
mRNA
2. Termination. The RNA hairpin causes the RNA strand to separate from the RNA polymerase, terminating transcription.

FIGURE 16.4 Transcription Terminates When a Hairpin Forms.

CHECK YOUR UNDERSTANDING

If you understand that . . .

- Transcription initiation depends on interactions between proteins associated with RNA polymerase and a promoter sequence in DNA.
- In bacteria, sigma binds to RNA polymerase and contacts the promoter. In eukaryotes, basal transcription factors bind to the promoter and recruit RNA polymerase.
- During transcription elongation, ribonucleotide triphosphates are the substrate for a polymerization reaction catalyzed by RNA polymerase. The enzyme adds ribonucleotides that are complementary to the template strand in DNA.
- Transcription ends when a termination signal at the end of the gene leads to the disassociation of RNA polymerase and the DNA template.

✔ **You should be able to . . .**

1. Explain why different sigma proteins are able to bind to different promoters.
2. Explain why ribonucleotide triphosphates, rather than ribonucleotides (which have one phosphate group), are the monomers required for RNA synthesis.

Answers are available in Appendix B.

to disrupt the interaction between RNA polymerase and the RNA transcript, resulting in the physical separation of the enzyme and its product.

To review key concepts about transcription and see the initiation, elongation, and termination phases in action, go to the study area at *www.masteringbiology.com.*

 **Web Activity** RNA Synthesis

16.2 RNA Processing in Eukaryotes

The molecular machinery required for transcription is much more complex in eukaryotes than in bacteria. But the contrast between single sigma proteins and multiple basal transcription factors, or between bacterial and eukaryotic promoters, is mild compared to what may be the most striking contrast in gene expression between bacteria and eukaryotes: In bacteria, the information in DNA is converted to mRNA directly. In eukaryotes, it isn't.

When transcription terminates in bacteria, the result is a mature mRNA that is ready to be translated into a protein. But when a eukaryotic gene is transcribed, the product is an immature **primary transcript**, or pre-mRNA. Before primary transcripts can be translated, they have to be processed in a complex series of steps.

Why? And how? And what consequences does this RNA processing requirement have for gene expression in eukaryotes?

The Startling Discovery of Eukaryotic Genes in Pieces

Eukaryotic genes do not consist of one continuous DNA sequence that codes for a product, as do bacterial genes. Instead, the regions in a eukaryotic gene that code for proteins are intermittently interrupted by stretches of hundreds or many thousands of intervening bases.

Although these intervening bases are part of the gene, they do not code for a product. To make a functional mRNA, then, eukaryotic cells must dispose of certain sequences inside the primary transcipt and then combine the separated coding sections into an integrated whole.

What sort of data would provoke such a startling claim? The first evidence came from work that Phillip Sharp and colleagues carried out in the late 1970s to determine how DNA templates are transcribed.

They began one of their experiments by heating DNA molecules sufficiently to break the hydrogen bonds between complementary bases. This treatment separated the two strands. The single-stranded DNA was then incubated with the mRNA encoded by the sequence. The team's intention was to promote base pairing between the mRNA and the single-stranded DNA.

The researchers expected that the mRNA would form base pairs with the DNA sequence that acted as the template for its synthesis—that the mRNA and DNA would match up exactly. But when the team examined the DNA–RNA hybrid molecules with the electron microscope, they observed the structure shown

(a) Micrograph of DNA-RNA hybrid

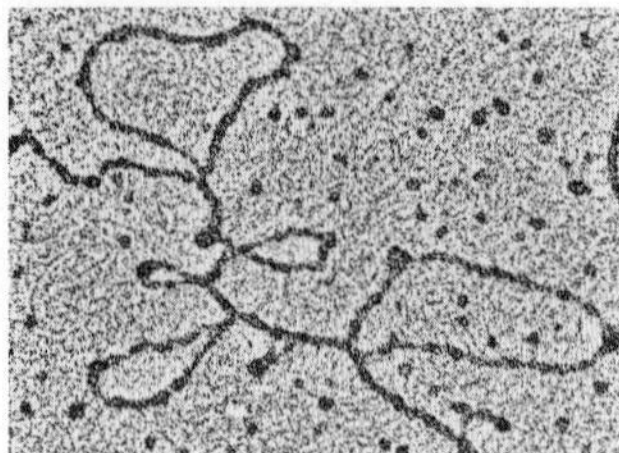

(b) Interpretation of micrograph

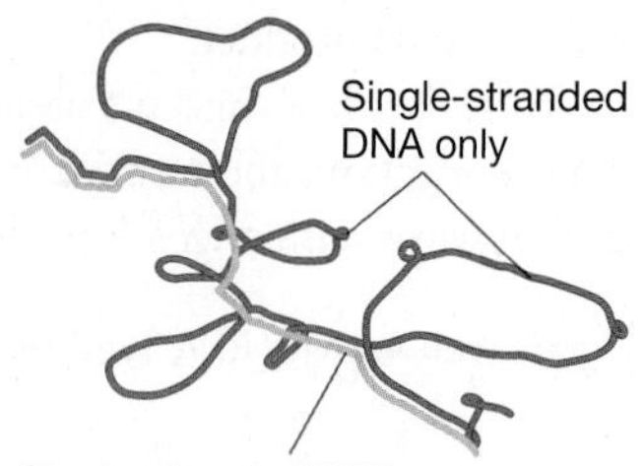

FIGURE 16.5 The Discovery of Noncoding Regions of DNA. The loops in the micrograph and drawing represent regions of DNA that do not have an equivalent sequence in the mRNA. These intervening regions are "extra" DNA compared with the sequences in the mRNA.

✔**QUESTION** If the noncoding regions of the gene did not exist, what would the micrograph in part (a) look like?

in **Figure 16.5a**. Instead of matching up exactly, parts of the DNA formed loops.

What was going on? As **Figure 16.5b** shows, Sharp's group interpreted these loops as stretches of nucleotides that are present in the DNA template strand but are *not* in the corresponding mRNA.

Sharp's group and a team headed by Richard Roberts went on to propose that there is not a one-to-one correspondence between the nucleotide sequence of a eukaryotic gene and its mRNA. As an analogy, it could be said that eukaryotic genes do not carry messages such as "Biology is my favorite course of all time." Instead, eukaryotic genes carry messages that read something like "BIOL τηεπροτεινχοδινγρεγιονσοφγενεσOGY IS MY FAVORαρειντ ερρθπτεϑβυνονψοϑινγϑITE COURSE OF ανϑηαωετοβεσ πλιχεϑτογετηερ ALL TIME." Here the sections of noncoding sequence are represented with Greek letters. They must be removed from the mRNA before it can carry an intelligible message to the translation machinery.

When it became clear that the genes-in-pieces hypothesis was correct, Walter Gilbert suggested that regions of eukaryotic genes that are part of the final mRNA be referred to as **exons** (because they are *ex*pressed) and the sections of primary transcript not in mRNA be referred to as **introns** (because they are *int*ervening). Exons code for segments of functional proteins or RNAs; introns do not. Introns are sections of genes that are not represented in the final mRNA product. As a result, eukaryotic genes are much larger than their corresponding mature RNA transcripts.

RNA Splicing

The transcription of eukaryotic genes by RNA polymerase generates a primary RNA transcript that contains both the exon and intron regions (**Figure 16.6a**). As transcription proceeds, the introns are removed from the growing RNA strand by a process known as **splicing**. In this phase of information processing, pieces of the primary transcript are removed and the remaining segments are joined together. Splicing occurs while transcription is still under way and results in an RNA that contains an uninterrupted genetic message.

(a) Introns must be removed from eukaryotic RNA transcripts.

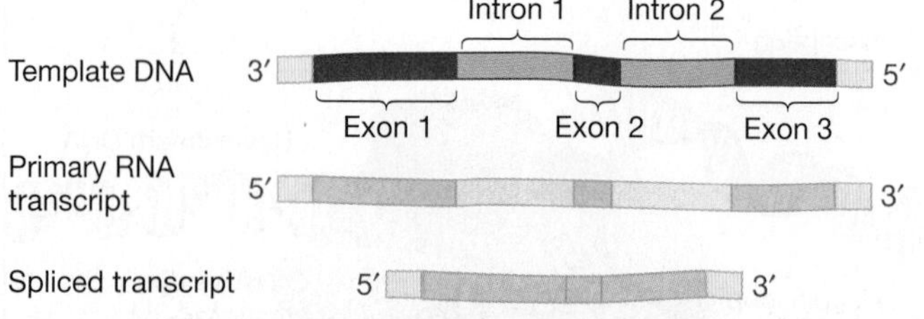

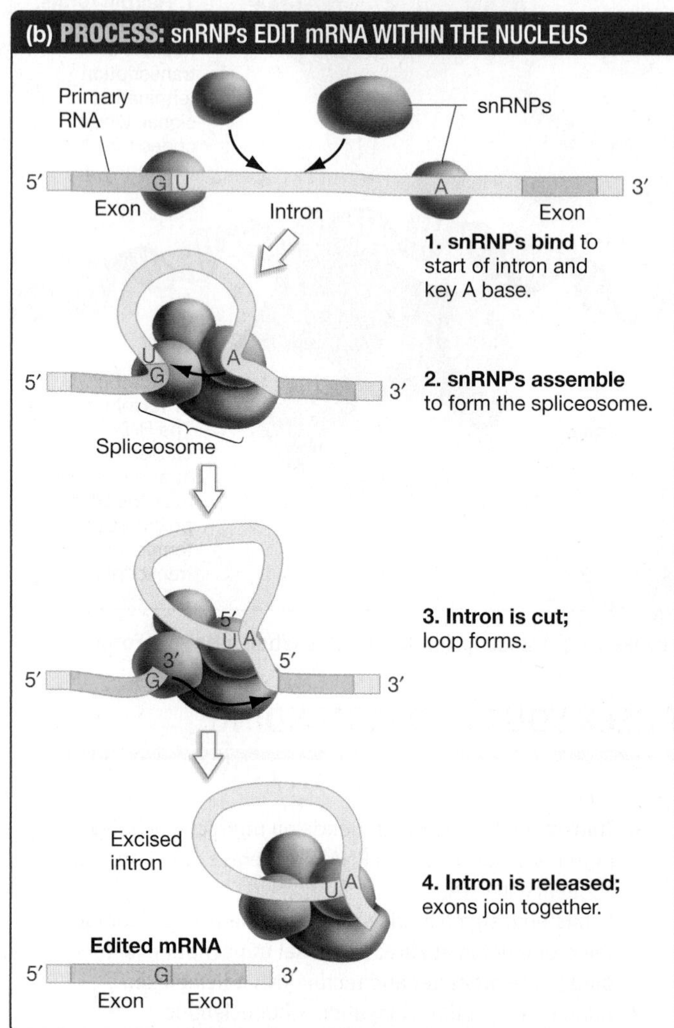

FIGURE 16.6 Introns Are Spliced Out of the Original mRNA.

Figure 16.6b provides more detail about how introns are removed from genes. In most cases, splicing is catalyzed by complexes of proteins and specialized RNAs, called small nuclear RNAs (snRNAs). These protein-plus-RNA complexes are known as **small nuclear ribonucleoproteins**, or **snRNPs** (pronounced "snurps"). The process can be broken into four steps:

1. The process begins when snRNPs bind to the 5′ exon–intron boundary, which is marked by the bases GU and a key adenine ribonucleotide near the end of the intron.
2. Once the initial snRNPs are in place, other snRNPs arrive to form a multipart complex called a **spliceosome**. The

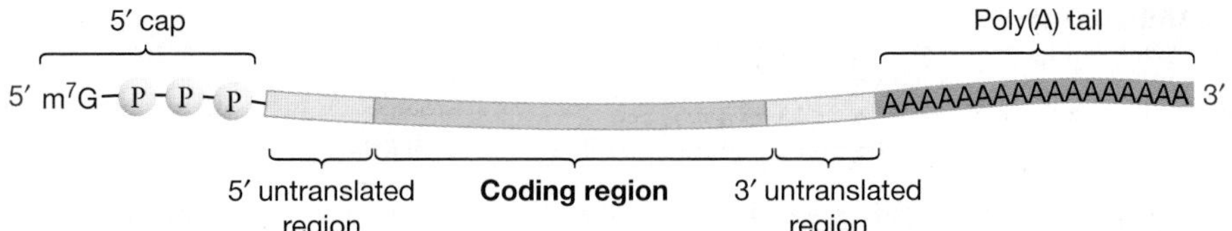

FIGURE 16.7 In Eukaryotes, Mature mRNAs Have a Cap and a Tail. Eukaryotic mRNAs have a cap consisting of a molecule called 7-methylguanylate (symbolized as m^7G) bonded to three phosphate groups; the tail is made up of a long series of adenine residues.

spliceosomes found in human cells contain about 145 different proteins and RNAs, making them the most complex molecular machines known.

3. The intron forms a loop with the key adenine at its connecting point.
4. The loop is cut out, and a phosphodiester bond links the exons on either side, producing a contiguous coding sequence.

Splicing is now complete. In most cases, the excised intron is degraded to ribonucleotide monophosphates.

Current data suggest that both the cutting and rejoining reactions that occur during splicing are catalyzed by the snRNA molecules in the spliceosome—meaning that the reactions are catalyzed by a ribozyme. Section 16.5 will demonstrate that ribozymes also play a key role in translation. As the RNA world hypothesis introduced in Chapter 4 predicts, proteins are not the only important catalysts in cells.

Adding Caps and Tails to Transcripts

Intron splicing of primary RNA transcripts in eukaryotes is followed by other important processing steps.

- As soon as the 5′ end of a eukaryotic RNA emerges from RNA polymerase, enzymes add a structure called the 5′ **cap** (**Figure 16.7**). The cap consists of the molecule 7-methylguanylate and three phosphate groups.
- An enzyme cleaves the 3′ end of most RNAs once transcription is complete, and another enzyme adds a long row of 100–250 adenine nucleotides, not encoded on the DNA template strand, known as the **poly(A) tail**.

With the addition of the cap and tail and completion of splicing, processing of the primary RNA transcript is complete. The product is a mature mRNA.

Figure 16.7 also shows that in the mature RNA molecule, the coding sequence for the polypeptide is flanked by sequences that are not destined to be translated. These 5′ and 3′ untranslated regions (or UTRs) help stabilize the mature RNA and regulate its translation. The mRNAs in bacteria also possess 5′ and 3′ UTRs.

Not long after the caps and tails on eukaryotic mRNAs were discovered, evidence began to accumulate that they protect mRNAs from degradation by ribonucleases and enhance the efficiency of translation. For example:

- Experimental mRNAs that have a cap and a tail last longer when they are introduced into cells than do experimental mRNAs that lack a cap, a tail, or both a cap and a tail.
- Experimental mRNAs with caps and tails produce more proteins than do experimental mRNAs without caps and tails.

Follow-up work has shown that the 5′ cap and the poly(A) tail serve as recognition signals for the translation machinery. It has also confirmed that they extend the life span of an mRNA by protecting the message from degradation by ribonucleases in the cytosol.

RNA processing is the general term for any of the modifications, such as splicing or poly(A) tail addition, needed to convert a primary transcript into a mature RNA. It is summarized in **Table 16.2** on page 296 along with other important differences in how mRNAs are produced in eukaryotes as compared to bacteria.

CHECK YOUR UNDERSTANDING

If you understand that . . .

- Eukaryotic genes consist of exons, which are parts of the primary transcript that remain in mRNA, and introns, which are regions of the primary transcript that are removed in forming mRNA.
- Introns are spliced out of primary RNA transcripts by multimolecular machines called spliceosomes.
- Enzymes add a 5′ cap and a poly(A) tail to spliced transcripts, producing a mature mRNA that is ready to be translated.

✔ **You should be able to . . .**

1. Explain why "ribonucleoprotein" is an appropriate name for the subunits of the spliceosome.
2. Explain the function of the 5′ cap and the poly(A) tail.

Answers are available in Appendix B.

16.3 An Introduction to Translation

To synthesize a protein, the sequence of bases in a messenger RNA molecule is translated into a sequence of amino acids in a polypeptide. The genetic code presented in Chapter 15 specifies the correspondence between each triplet codon in mRNA and the amino acid it codes for. But how are the amino acids assembled into a polypeptide according to the information in messenger RNA?

Studies of translation in cell-free systems proved extremely effective at answering this question. Once in vitro translation systems had been developed from human cells, *E. coli*, and a variety of other organisms, biologists could see that the sequence of events is similar in bacteria, archaea, and eukaryotes.

Ribosomes Are the Site of Protein Synthesis

The first question that biologists answered about translation concerned where it occurs. The answer grew from a simple observation: There is a strong positive correlation between the presence

TABLE 16.2 **Comparing Transcription in Bacteria and Eukaryotes**

Point of Comparison	Bacteria	Eukaryotes
RNA polymerase(s)	One	Three; each produces a different class of RNA
Promoter structure	Typically contains a −35 box and a −10 box	Complex and variable; often includes a TATA box about −30 from the transcription start site
Protein(s) involved in contacting promoter	Sigma; different versions of sigma bind to different promoters	Many basal transcription factors
RNA-processing	None	Extensive; several processing steps occur in the nucleus before RNA is exported to the cytosol for translation: **1.** Enzyme-catalyzed addition of 5′ cap **2.** Splicing (intron removal) by spliceosome **3.** Enzyme-catalyzed addition of 3′ poly(A) tail

of small structures known as **ribosomes** in a given type of cell and the rate at which that cell synthesizes proteins. For example:

- Immature human red blood cells divide rapidly, synthesize millions of copies of the protein hemoglobin, and contain large numbers of ribosomes.
- The same cells at maturity have low rates of protein synthesis and very few ribosomes.

Based on this correlation, investigators proposed that ribosomes are the site of protein synthesis in the cell.

To test this hypothesis, Roy Britten and collaborators did a pulse-chase experiment similar in design to experiments introduced in Chapter 7. Recall that a pulse-chase experiment labels a population of molecules as they are being produced. The location of the tagged molecules is then followed over time.

In this case, the tagging was done by supplying a pulse of radioactive sulfur atoms that would be incorporated into the amino acids methionine and cysteine, followed by a chase of unlabeled sulfur atoms. If the ribosome hypothesis were correct, the radioactive signal should be associated with ribosomes for a short period of time—while the amino acids were being polymerized into proteins. Later, when translation was complete, all of the radioactivity should be found in proteins, not in the ribosomes.

This is exactly what the researchers found. Based on these data, biologists concluded that proteins are synthesized at ribosomes and then released.

Comparing Translation in Bacteria and Eukaryotes

About a decade after the ribosome hypothesis was confirmed, electron micrographs showed bacterial ribosomes in action (**Figure 16.8a**). The images showed that in bacteria, ribosomes attach to mRNAs and begin synthesizing proteins even before transcription is complete. In fact, multiple ribosomes attach to each

(a) Bacterial ribosomes during translation

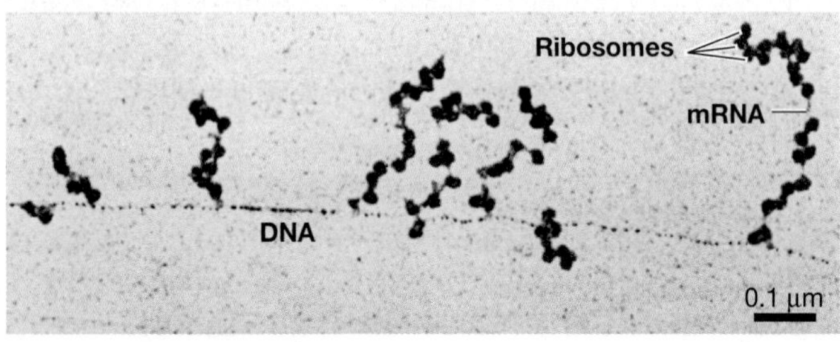

(b) In bacteria, transcription and translation are tightly coupled.

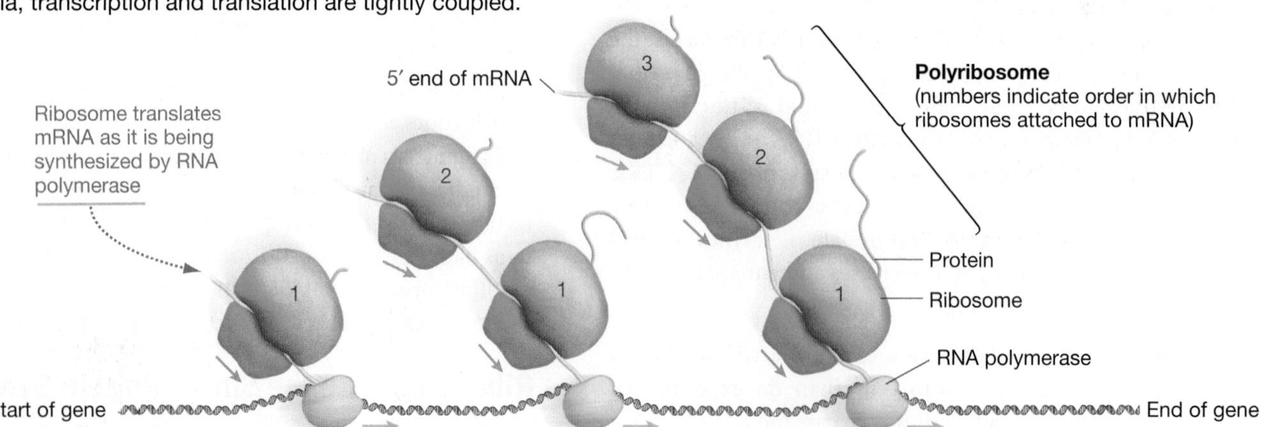

FIGURE 16.8 Transcription and Translation Occur Simultaneously in Bacteria. In bacteria, ribosomes attach to mRNA transcripts and begin translation while RNA polymerase is still transcribing the DNA template strand.

mRNA, forming a **polyribosome** (**Figure 16.8b**). In this way, many copies of a protein can be produced from a single mRNA.

Transcription and translation can occur concurrently in bacteria because there is no nuclear envelope to separate the two processes. Thus, transcription and translation are physically connected.

The situation is different in eukaryotes. In these organisms, primary transcripts are processed in the nucleus to produce a mature mRNA, which is then exported to the cytosol (**Figure 16.9a**). Once mRNAs are outside the nucleus, ribosomes attach to them and begin translation. As in bacteria, polyribosomes form (**Figure 16.9b**). But in eukaryotes, transcription and translation are separated in time and space.

(a) mRNAs are exported to the cytosol.

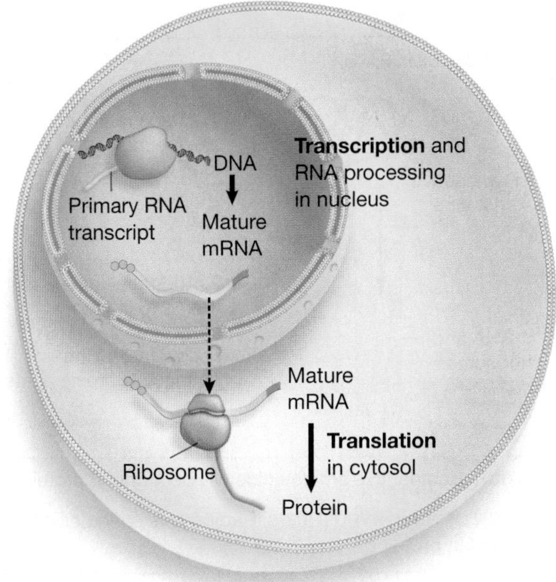

(b) Polyribosomes form in the cytosol.

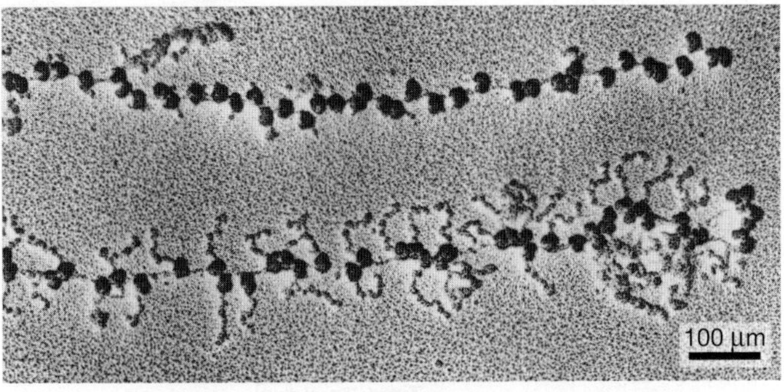

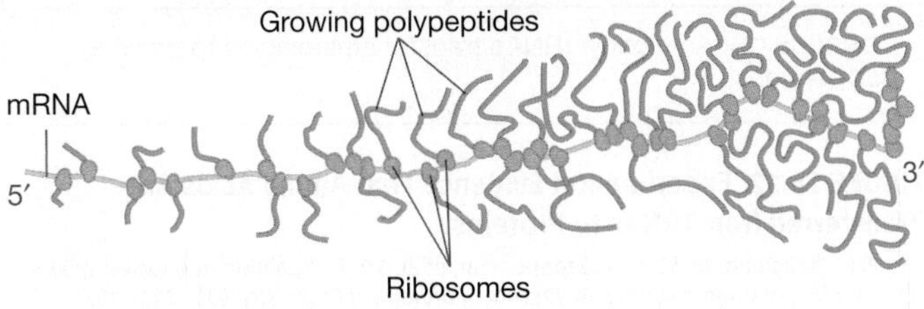

FIGURE 16.9 Transcription and Translation Are Separated in Space and Time in Eukaryotes.

How Does an mRNA Triplet Specify an Amino Acid?

When an mRNA interacts with a ribosome, hereditary instructions encoded in nucleic acids are translated into a different chemical language—the amino acid sequences found in proteins. The discovery of the genetic code revealed that triplet codons in mRNA specify particular amino acids in a protein. How does this conversion happen?

One early hypothesis was that mRNA codons and amino acids interact directly. This hypothesis proposed that the bases in a particular codon were complementary in shape or charge to the side group of a particular amino acid (**Figure 16.10a**). But Francis Crick pointed out that the idea didn't make chemical sense. For example, how could the nucleic acid bases interact with a hydrophobic amino acid side group, which does not form hydrogen bonds?

Crick proposed an alternative hypothesis. As **Figure 16.10b** shows, he suggested that some sort of adapter molecule holds amino acids in place while interacting directly and specifically with a codon in mRNA by hydrogen bonding. In essence, Crick predicted the existence of a chemical go-between that produced a physical connection between the two types of molecules. As it turns out, Crick was right.

16.4 The Structure and Function of Transfer RNA

Crick's adapter molecule was discovered by accident. Biologists were trying to work out an in vitro protein synthesis system and discovered that ribosomes, mRNA, amino acids, ATP, and a molecule called guanosine triphosphate, or GTP, had to be present for translation to occur. (GTP is similar to ATP but contains guanosine instead of adenosine.)

(a) Hypothesis 1: Amino acids interact directly with mRNA codons.

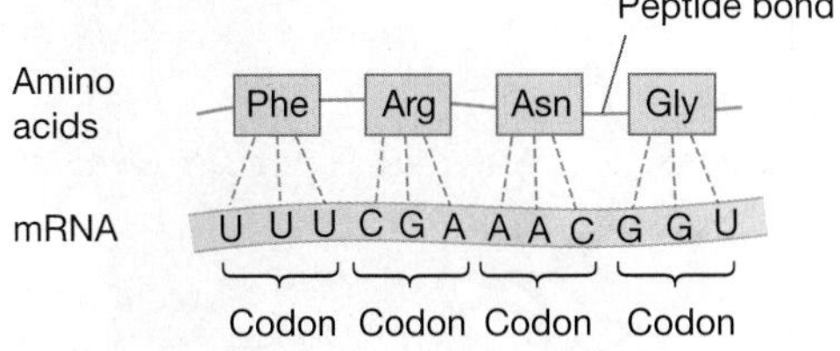

(b) Hypothesis 2: Adapter molecules hold amino acids and interact with mRNA codons.

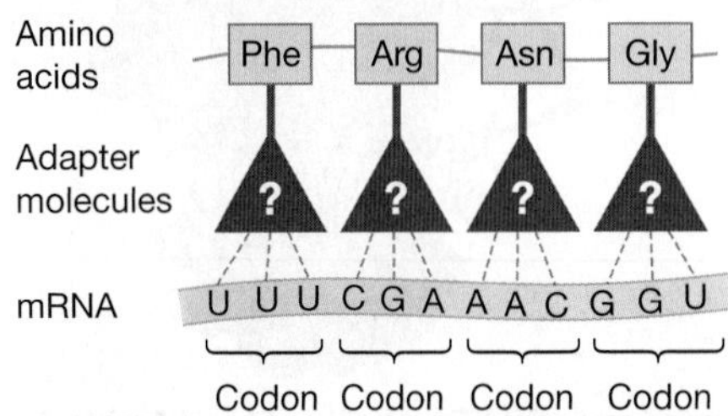

FIGURE 16.10 How Do mRNA Codons Interact with Amino Acids?

These results were logical: ribosomes provide the catalytic machinery, mRNAs contribute the message to be translated, amino acids are the building blocks of proteins, and ATP and GTP supply potential energy to drive the endergonic polymerization reactions responsible for forming proteins.

But in addition, a cellular fraction that contained a previously unknown type of RNA turned out to be indispensable. If this type of RNA is missing, protein synthesis does not occur. What is this mysterious RNA, and why is it essential to translation?

The novel class of RNAs eventually became known as **transfer RNA (tRNA)**. The role of tRNA in translation was a mystery until some researchers happened to add a radioactive amino acid—leucine—to an in vitro protein synthesis system. The treatment was actually done as a control for an unrelated experiment. To the researchers' amazement, some of the radioactive leucine attached to tRNA molecules.

Follow-up experiments revealed several key facts about this process:

- An input of energy, in the form of ATP, is required to attach an amino acid to a tRNA.
- Enzymes called **aminoacyl tRNA synthetases** catalyze the addition of amino acids to tRNAs—what biologists call "charging" a tRNA.
- For each of the 20 major amino acids, there is a different aminoacyl tRNA synthetase and one or more tRNAs.

The combination of a tRNA molecule covalently linked to an amino acid is called an **aminoacyl tRNA**. **Figure 16.11** shows an aminoacyl tRNA synthetase still bound to a tRNA that has just been charged with an amino acid. Note how tightly the two structures fit together—making it possible for the enzyme and substrate to interact in an extremely precise way.

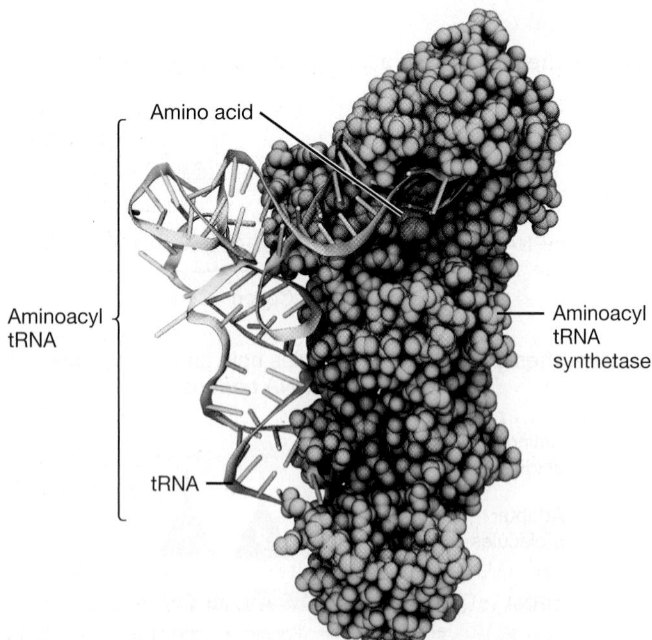

FIGURE 16.11 Aminoacyl tRNA Synthetases Load Amino Acids onto the Appropriate tRNA.

What happens to the amino acids bound to tRNAs? To answer this question, biologists tracked the fate of radioactive leucine molecules that were attached to tRNAs. They found that the amino acids are transferred from aminoacyl tRNAs to proteins.

The data supporting this conclusion are shown in **Figure 16.12**. The graph in the "Results" section of this figure shows that ra-

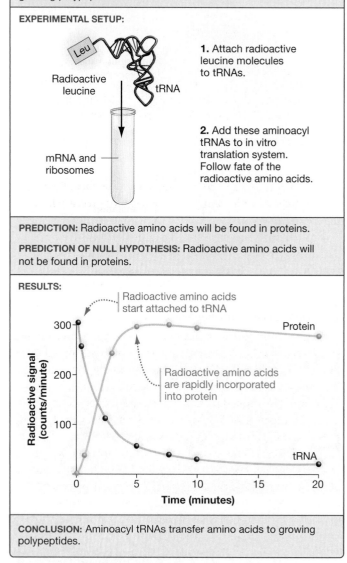

EXPERIMENT

QUESTION: What happens to the amino acids attached to tRNAs?

HYPOTHESIS: Aminoacyl tRNAs transfer amino acids to growing polypeptides.

NULL HYPOTHESIS: Aminoacyl tRNAs do not transfer amino acids to growing polypeptides.

EXPERIMENTAL SETUP:

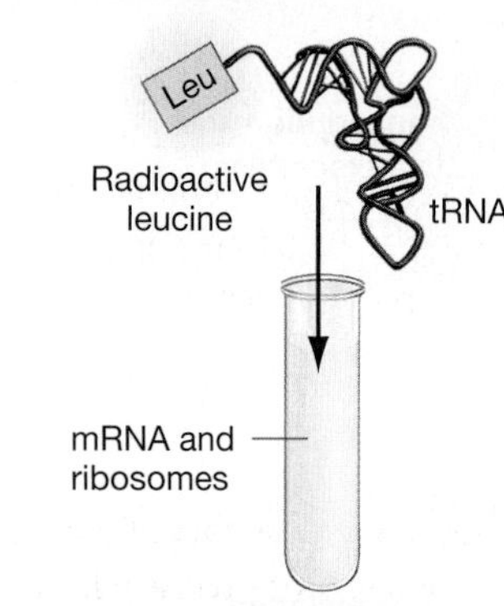

1. Attach radioactive leucine molecules to tRNAs.

2. Add these aminoacyl tRNAs to in vitro translation system. Follow fate of the radioactive amino acids.

PREDICTION: Radioactive amino acids will be found in proteins.

PREDICTION OF NULL HYPOTHESIS: Radioactive amino acids will not be found in proteins.

RESULTS:

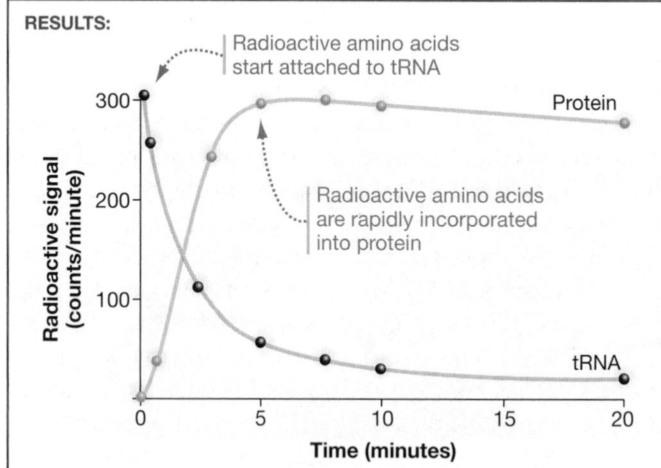

CONCLUSION: Aminoacyl tRNAs transfer amino acids to growing polypeptides.

FIGURE 16.12 Experimental Evidence That Amino Acids Are Transferred from tRNAs to Proteins.

SOURCE: Hoagland, M. B., M. L. Stephenson, et al. 1958. A soluble ribonucleic acid intermediate in protein synthesis. *Journal of Biological Chemistry* 231: 241–257.

✔**QUESTION** What would the graphed results look like if the null hypothesis were correct?

dioactive amino acids are lost from tRNAs and incorporated into polypeptides synthesized in ribosomes. To understand this conclusion, do the following:

1. Put your finger on the point on the *x*-axis that indicates that one minute has passed since the start of the experiment.
2. Read up until you hit the green line and the gray line. The green line represents data from proteins; the gray line represents data from tRNAs.
3. Check the *y*-axis—which indicates the amount of radioactive leucine present—at each point.
4. It should be clear that early in the experiment, almost all of the radioactive leucine is attached to tRNA, not protein.

Next, do the same four steps at the point on the *x*-axis labeled 10 minutes (since the start of the experiment). Your conclusion now should be that late in the experiment, almost all of the radioactive leucine is attached to proteins, not tRNA.

These results inspired the use of "transfer" in tRNA's name, because amino acids are transferred from the RNA to the growing end of a new polypeptide. The experiment also confirmed that aminoacyl tRNAs act as the interpreter in the translation process: tRNAs are Crick's adapter molecules.

What Do tRNAs Look Like?

Transfer RNAs serve as chemical go-betweens that allow amino acids to interact with an mRNA template. But precisely how does the connection occur?

This question was answered by research on tRNA's molecular structure. The initial studies established the sequence of nucleotides in various tRNAs, or what is termed their primary structure. Transfer RNA sequences are relatively short, ranging from 75 to 85 nucleotides in length.

When biologists studied the primary sequence closely, they noticed that certain parts of the molecules can form secondary structures. Specifically, some sequences of bases in the tRNA molecule can form hydrogen bonds with complementary base sequences elsewhere in the same molecule. As a result, portions of the molecule should form the stem-and-loop structures introduced in Chapter 4. The stems are short stretches of double-stranded RNA; the loops are single-stranded.

Two aspects of this secondary structure proved especially interesting. A CCA sequence at the 3′ end of each tRNA molecule offered a binding site for amino acids, while a triplet on the loop at the far end of the structure could serve as an anticodon. An **anticodon** is a set of three ribonucleotides that forms base pairs with the mRNA codon.

Later, X-ray crystallography studies revealed that tRNAs also have tertiary structure. Recall from Chapter 3 that the tertiary structure of a molecule is the three-dimensional arrangement of its atoms and is usually a product of folding. As **Figure 16.13** shows, tRNAs fold into an L-shaped molecule. The anticodon is at one end of the structure; the CCA sequence and attached amino acid is at the other end.

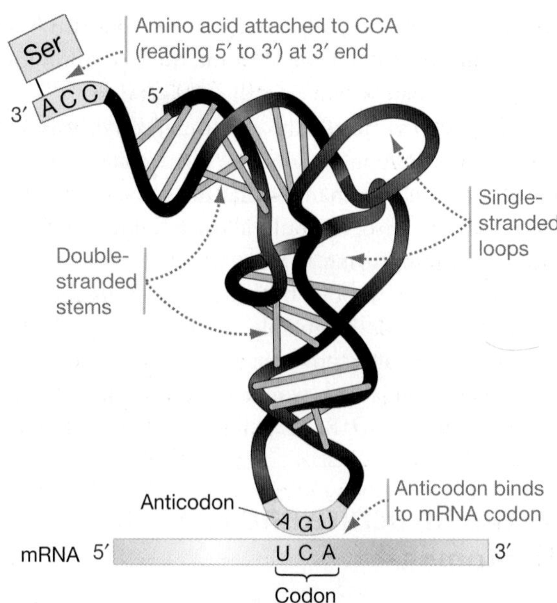

FIGURE 16.13 The Structure of Transfer RNA. The anticodon of an aminoacyl tRNA forms complementary base pairs with an mRNA codon.

All of the tRNAs in a cell have the same structure, shaped like an upside-down L. They vary at the anticodon and attached amino acid. The tertiary structure of tRNAs is important because it maintains a precise physical distance between the anticodon and amino acid. As it turns out, this separation is key to the positioning of the amino acid and the anticodon in the ribosome.

✔If you understand the structure of tRNAs, you should be able to (1) describe where on the upside-down, L-shaped structure the amino acid attaches, and (2) explain the relationship between the anticodon of a tRNA and a codon in an mRNA.

How Many tRNAs Are There?

After characterizing all of the different types of tRNAs available in cells, biologists encountered a paradox. According to the genetic code introduced in Chapter 15, the 20 most common amino acids found in proteins are specified by 61 different mRNA codons. Instead of containing 61 different tRNAs with 61 different anticodons, though, most cells contain only about 40. How can all 61 mRNA codons in the genome be translated with only two-thirds that number of tRNAs?

To resolve this paradox, Francis Crick proposed what is known as the **wobble hypothesis**. Recall from Chapter 15 that:

1. Many amino acids are specified by more than one codon.
2. Codons for the same amino acid tend to have the same nucleotides at the first and second positions but a different nucleotide at the third position.

For example, both of the codons CAA and CAG code for the amino acid glutamine. (Codons are always written in the $5' \rightarrow 3'$ direction.) Surprisingly, experimental data have shown that a tRNA with an anticodon of GUU can base pair with both CAA and CAG in mRNA. (Anticodons are written in the $3' \rightarrow 5'$

direction.) The GUU anticodon matches the first two bases (C and A) in both codons, but the U in the third position forms a nonstandard base pair with a G in the CAG codon.

Crick proposed that inside the ribosome, certain bases in the third position of tRNA anticodons can bind to bases in the third position of a codon in a manner that does not match Watson-Crick base pairing. If so, it would allow a limited flexibility, or "wobble," in the base pairing.

According to the wobble hypothesis, a nonstandard base pair—such as G-U—is acceptable in the third position of a codon as long as it does not change the amino acid that the codon specifies. In this way, wobble in the third position of a codon allows just 40 or so tRNAs to bind to all 61 mRNA codons.

16.5 The Structure and Function of Ribosomes

Recall that protein synthesis occurs when the sequence of bases in an RNA message is translated into a sequence of amino acids in a polypeptide. The conversion of each mRNA codon begins when the anticodon of an aminoacyl tRNA binds to the codon. The conversion is complete when a peptide bond forms between the tRNA's amino acid and the growing polypeptide chain.

Both of these events take place inside a ribosome. Biologists have known since the 1930s that ribosomes contain a considerable amount of protein along with a great deal of **ribosomal RNA (rRNA)**. Later work showed that ribosomes can be separated into two major substructures, called the large subunit and small subunit. Each ribosome subunit consists of a complex of RNA molecules and proteins. The small subunit holds the mRNA in place during translation; the large subunit is where peptide-bond formation takes place.

Figure 16.14 shows how all of the molecules required for translation fit together. Note that during protein synthesis, three distinct tRNAs are lined up inside the ribosome. All three are bound to their corresponding mRNA codon at the base of the structure.

- The tRNA that is on the right in the figure, and colored red, carries an amino acid. This tRNA's position in the ribosome is called the A site—"A" for acceptor or aminoacyl.
- The tRNA that is in the middle (green) holds the growing polypeptide chain and occupies the P site, for peptidyl, inside the ribosome. (Think of "P" for peptide-bond formation.)
- The left-hand (blue) tRNA no longer has an amino acid attached and is about to leave the ribosome. It occupies the ribosome's E site—"E" for exit.

Because all tRNAs have similar secondary and tertiary structure, they all fit equally well in the A, P, and E sites.

The ribosome is a molecular machine that synthesizes proteins in a three-step sequence:

1. An aminoacyl tRNA diffuses into the A site; its anticodon binds to a codon in mRNA.
2. A peptide bond forms between the amino acid held by the aminoacyl tRNA in the A site and the growing polypeptide, which was held by a tRNA in the P site.
3. The ribosome moves ahead, and all three tRNAs move one position down the line. The tRNA in the E site exits; the tRNA in the P site moves to the E site; and the tRNA in the A site switches to the P site.

The protein that is being synthesized grows by one amino acid each time this three-step sequence repeats. The process occurs up to 20 times per second in bacterial ribosomes and about

(a) Diagram of ribosome during translation (interior view)

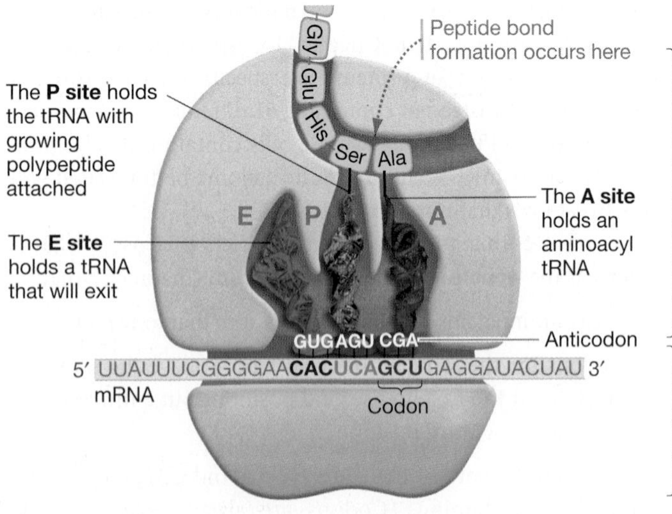

(b) Model of ribosome during translation (exterior view)

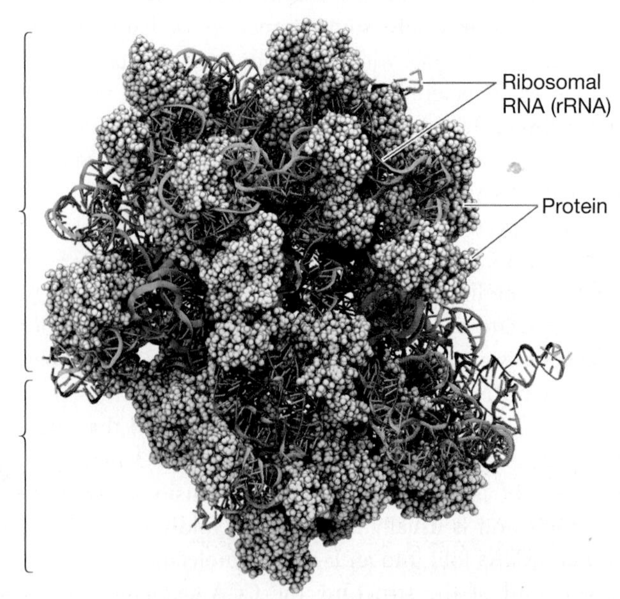

FIGURE 16.14 The Structure of the Ribosome. Ribosomes have three distinct sites in their interior where tRNAs are found.

2 times per second in eukaryotic ribosomes. Protein synthesis starts at the amino end (N-terminus) of a polypeptide and proceeds to the carboxy end (C-terminus; see Chapter 3).

This introduction to how tRNAs, mRNAs, and ribosomes interact during protein synthesis leaves several key questions unanswered, however. How do mRNAs and ribosomes get together to start the process? Once protein synthesis is under way, how is peptide-bond formation catalyzed inside the ribosome? And how does protein synthesis conclude when the ribosome reaches the end of the message? Let's consider each question in turn.

Initiating Translation

To translate an mRNA properly, a ribosome must begin at a specific point in the message, translate the mRNA up to the message's termination codon, and then stop. Using the same terminology that they apply to transcription, biologists call these three phases of protein synthesis initiation, elongation, and termination, respectively.

One key to understanding translation initiation is to recall, from Chapter 15, that a start codon (usually AUG) is found near the 5′ end of all mRNAs and that it codes for the amino acid methionine. The presence of this start codon is an aspect of initiation that is common to both bacteria and eukaryotes.

Figure 16.15 shows how translation gets under way in bacteria. The process begins when a section of rRNA in a small ribosomal subunit binds to a complementary sequence on an mRNA. The mRNA region is called the **ribosome binding site**, or **Shine-Dalgarno sequence**, after the biologists who discovered it. The site is about six nucleotides upstream from the AUG start codon. It consists of all or part of the sequence 5′-AGGAGGU-3. The complementary sequence in the rRNA of the small subunit reads 3′-UCCUCCA-5′.

This initial interaction between the small subunit and the message is mediated by proteins called **initiation factors** (Figure 16.15, step 1). In eukaryotes, initiation factors bind to the 5′ cap on mRNAs and guide it to the ribosome.

Once the Shine-Dalgarno sequence has attached to the small ribosomal subunit, an aminoacyl tRNA bearing a modified form of methionine called *N*-formylmethionine (abbreviated *f*-met) binds to the AUG start codon (Figure 16.15, step 2). In eukaryotes, this initial amino acid is normal methionine.

Initiation is complete when the large subunit joins the complex (Figure 16.15, step 3). When the ribosome is completely assembled, the tRNA bearing *f*-met occupies the P site.

To summarize, translation initiation is a three-step process in bacteria: **(1)** The mRNA binds to a small ribosomal subunit, **(2)** the initiator aminoacyl tRNA bearing *f*-met binds to the start codon, and **(3)** the large ribosomal subunit binds, completing the complex.

Elongation: Extending the Polypeptide

At the start of elongation, the E and A sites in the ribosome are empty of tRNAs. As a result, an mRNA codon is exposed at the base of the A site. As step 1 in **Figure 16.16** on page 302 illustrates, elongation proceeds when an aminoacyl tRNA binds to the codon in the A site by complementary base pairing between anticodon and codon.

When both the P site and A site are occupied by tRNAs, the amino acids on the tRNAs are in the ribosome's active site. This is where peptide bond formation—the essence of protein synthesis—occurs.

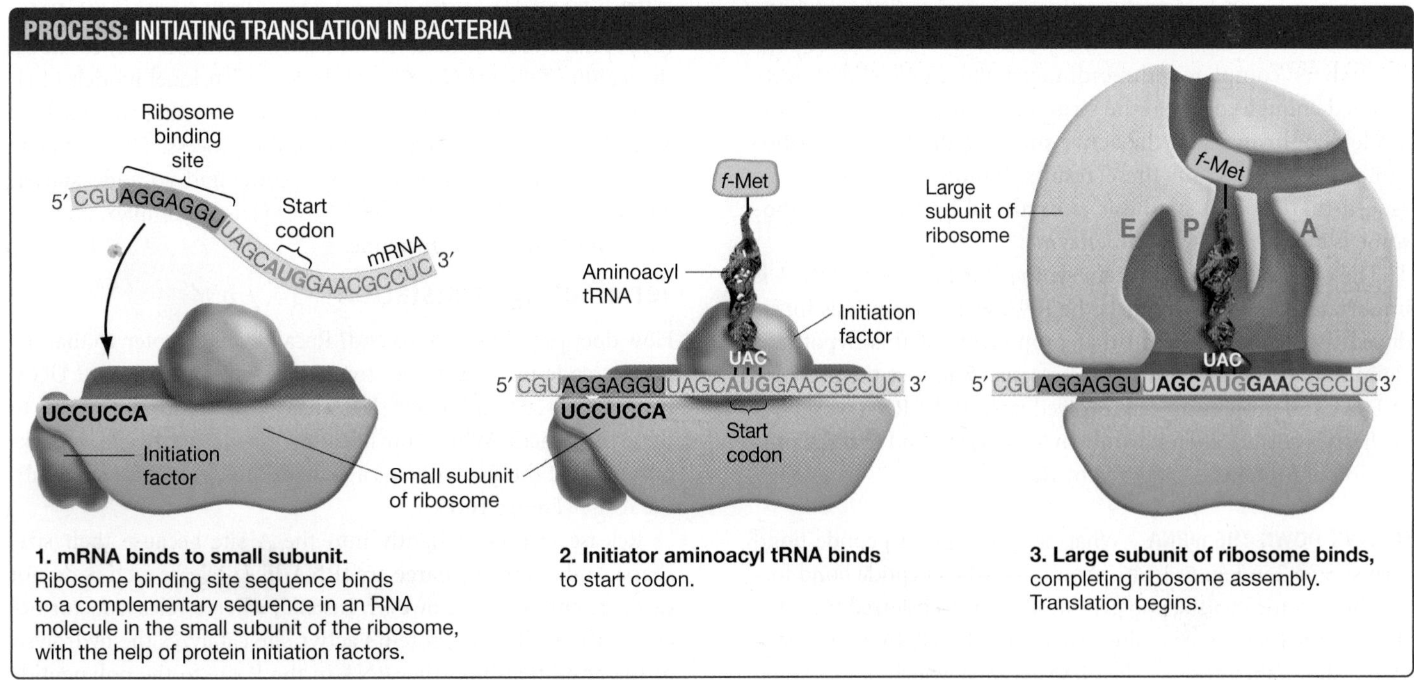
PROCESS: INITIATING TRANSLATION IN BACTERIA
Ribosome binding site
5′ CGUAGGAGGUUAGCAUGGAACGCCUC 3′
Start codon
mRNA
UCCUCCA
Initiation factor
Small subunit of ribosome
f-Met
Aminoacyl tRNA
Initiation factor
UAC
5′ CGUAGGAGGUUAGCAUGGAACGCCUC 3′
UCCUCCA
Start codon
Large subunit of ribosome
f-Met
E
P
A
UAC
5′ CGUAGGAGGUUAGCAUGGAACGCCUC 3′
1. mRNA binds to small subunit.
Ribosome binding site sequence binds to a complementary sequence in an RNA molecule in the small subunit of the ribosome, with the help of protein initiation factors.
2. Initiator aminoacyl tRNA binds to start codon.
3. Large subunit of ribosome binds, completing ribosome assembly. Translation begins.

FIGURE 16.15 Initiation Proteins Manage Assembly of Ribosomes and mRNAs.

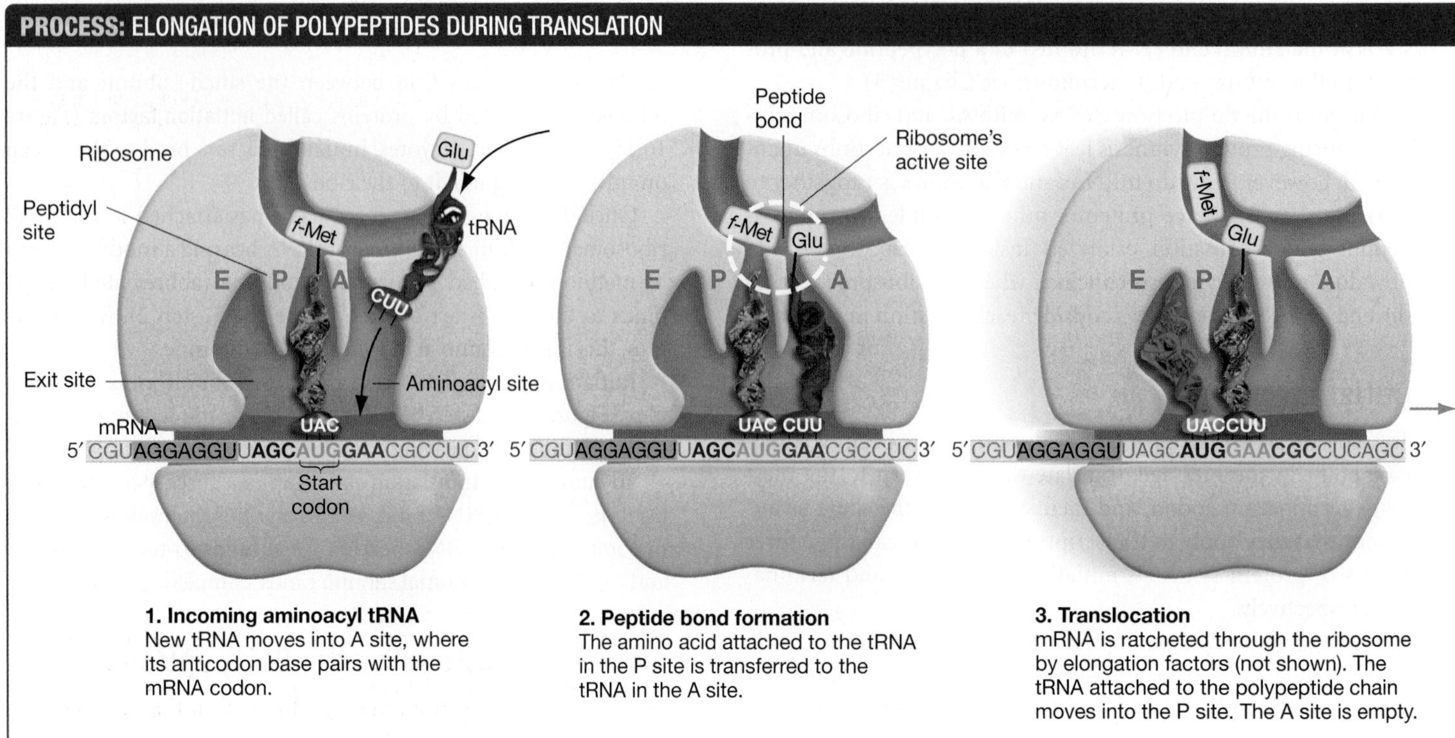
PROCESS: ELONGATION OF POLYPEPTIDES DURING TRANSLATION
Ribosome
Peptidyl site
Glu
f-Met
tRNA
E
P
A
CUU
Exit site
Aminoacyl site
mRNA
UAC
5′ CGUAGGAGGUUAGCAUGGAACGCCUC 3′
Start codon
Peptide bond
Ribosome's active site
f-Met
Glu
E
P
A
UAC CUU
5′ CGUAGGAGGUUAGCAUGGAACGCCUC 3′
f-Met
Glu
E
P
A
UACCUU
5′ CGUAGGAGGUUAGCAUGGAACGCCUCAGC 3′
1. Incoming aminoacyl tRNA
New tRNA moves into A site, where its anticodon base pairs with the mRNA codon.
2. Peptide bond formation
The amino acid attached to the tRNA in the P site is transferred to the tRNA in the A site.
3. Translocation
mRNA is ratcheted through the ribosome by elongation factors (not shown). The tRNA attached to the polypeptide chain moves into the P site. The A site is empty.

FIGURE 16.16 During Elongation, Peptide Bond Formation Occurs inside the Ribosome.

Peptide bond formation is considered one of the most important reactions that takes place in cells, because manufacturing proteins is among the most fundamental of all cell processes. The question is, what makes it happen?

IS THE RIBOSOME AN ENZYME OR A RIBOZYME? Because ribosomes contain both proteins and RNA, researchers had argued for decades over whether the active site consisted of protein or RNA. The debate was not resolved until the year 2000, when researchers completed three-dimensional models that were detailed enough to reveal the structure of the active site. These models confirmed that the active site consists entirely of ribosomal RNA. Based on these results, biologists are now convinced that protein synthesis is catalyzed by RNA. The ribosome is a ribozyme—not an enzyme.

The observation that protein synthesis is catalyzed by RNA is important because it supports the RNA world hypothesis introduced in Chapter 4. Recall that proponents of this hypothesis claim that life began with RNA molecules and that the presence of DNA and proteins in cells evolved later. If the RNA world hypothesis is correct, then it would make sense to find that the production of proteins is catalyzed by RNA.

MOVING DOWN THE mRNA What happens after a peptide bond forms? Step 2 in Figure 16.16 shows that when peptide bond formation is complete, the polypeptide chain is transferred from the tRNA in the P site to the amino acid held by the tRNA in the A site. Step 3 shows the process called **translocation**, which occurs when proteins called **elongation factors** move the mRNA so that it ratchets through the ribosome in the $5' \rightarrow 3'$ direction. Translocation is an energy-demanding event that requires GTP.

Translocation does several things: It moves the empty tRNA into the E site; it moves the tRNA containing the growing polypeptide into the P site; and it opens the A site and exposes a new mRNA codon. If the E site is occupied when translocation occurs, the tRNA there is ejected into the cytosol.

The three steps in elongation—(**1**) arrival of aminoacyl tRNA, (**2**) peptide bond formation, and (**3**) translocation—repeat down the length of the mRNA. Recent three-dimensional models of ribosomes in various stages of translation show that the machine as a whole is highly dynamic during the process. The ribosome constantly changes shape as tRNAs come and go and catalysis and translocation occur. The ribosome is a complex and dynamic multimolecular machine.

Terminating Translation

How does protein synthesis end? Recall from Chapter 15 that the genetic code includes three stop codons: UAA, UAG, and UGA. In most cells, no aminoacyl tRNA has an anticodon that binds to these sequences. When translocation opens the A site and exposes one of the stop codons, a protein called a **release factor** fills the A site (**Figure 16.17**).

Release factors fit tightly into the A site because their size, shape, and electrical charge are tRNA-like. Release factors do not carry an amino acid, however. Instead, when a release factor occupies the A site, the protein's active site catalyzes the hydrolysis of the bond that links the tRNA in the P site to the polypeptide chain. This reaction frees the polypeptide.

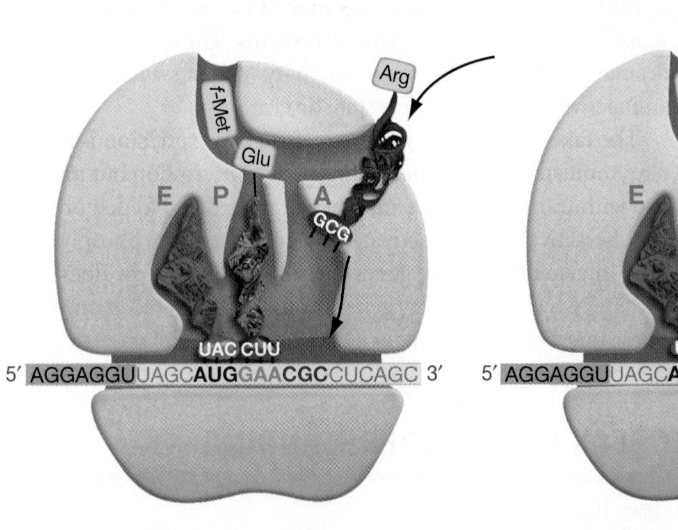

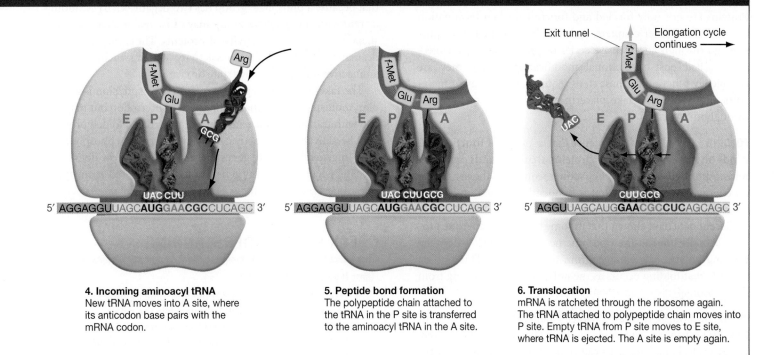

4. **Incoming aminoacyl tRNA**
New tRNA moves into A site, where its anticodon base pairs with the mRNA codon.

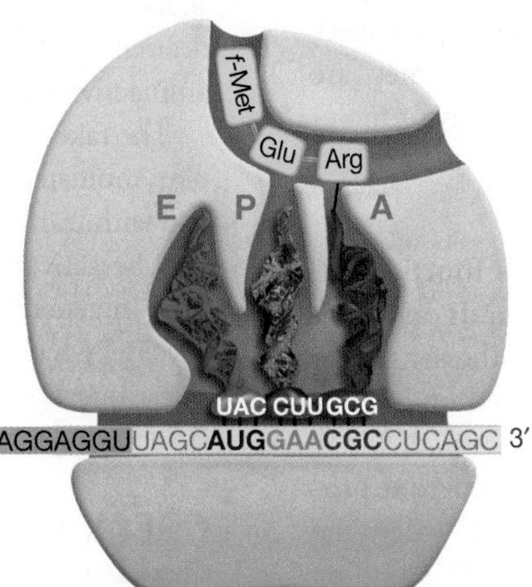

5. **Peptide bond formation**
The polypeptide chain attached to the tRNA in the P site is transferred to the aminoacyl tRNA in the A site.

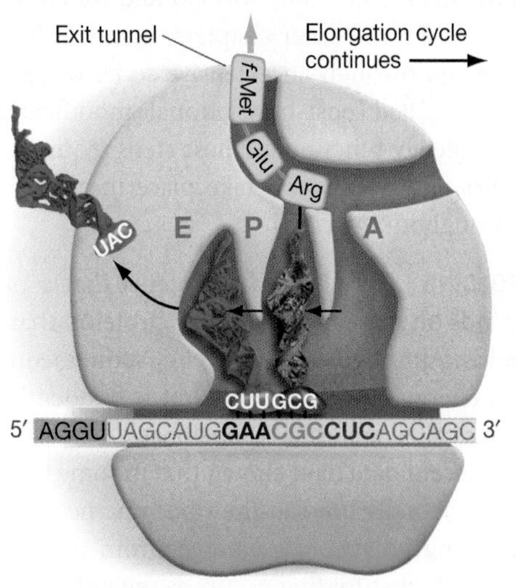

6. **Translocation**
mRNA is ratcheted through the ribosome again. The tRNA attached to polypeptide chain moves into P site. Empty tRNA from P site moves to E site, where tRNA is ejected. The A site is empty again.

The newly synthesized polypeptide is released from the ribosome, the ribosome separates from the mRNA, and the two ribosomal subunits dissociate. The subunits are ready to attach to the start codon of another message and start translation anew.

To review the initiation, elongation, and termination phases of translation, go to the study area at *www.masteringbiology.com.*

MB **Web Activity** Synthesizing Proteins

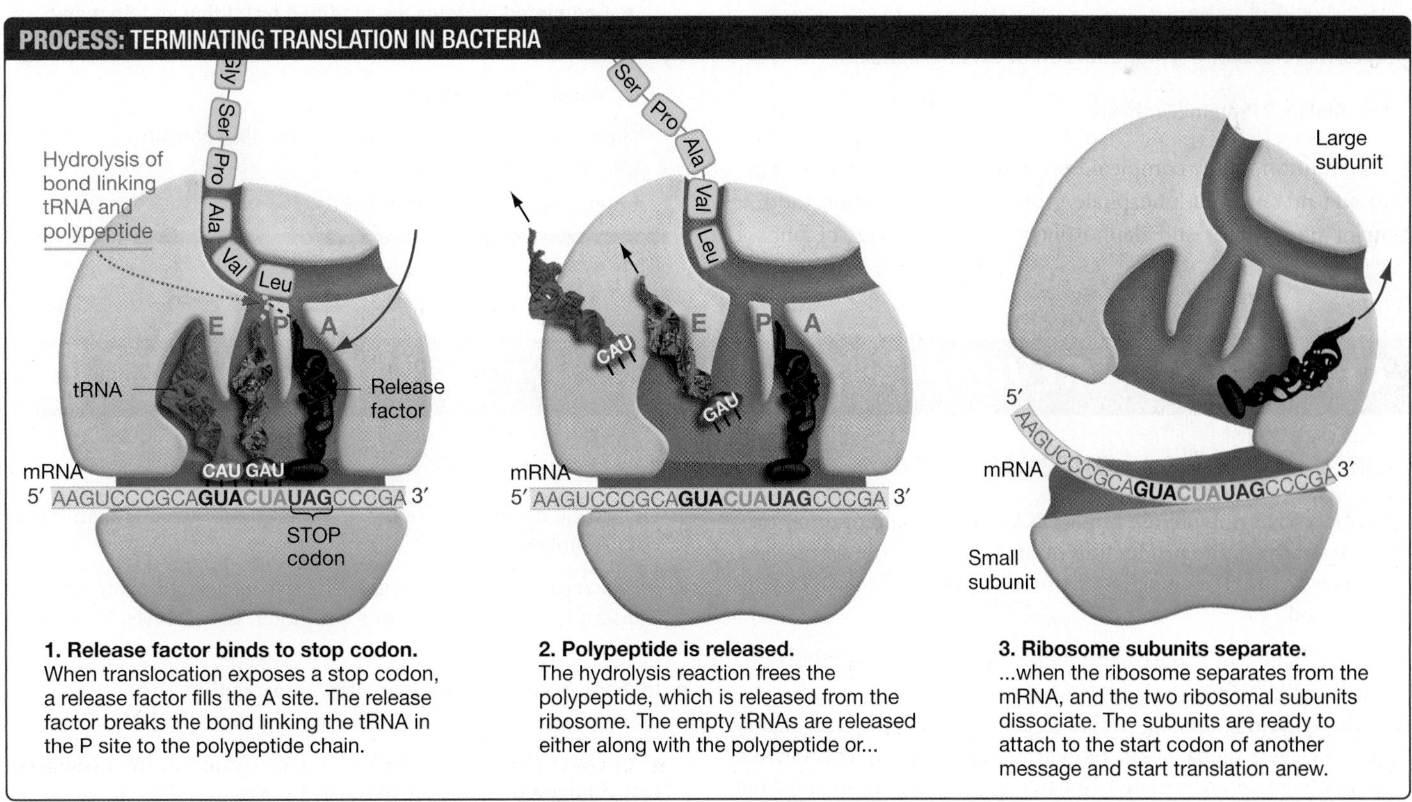

1. **Release factor binds to stop codon.**
When translocation exposes a stop codon, a release factor fills the A site. The release factor breaks the bond linking the tRNA in the P site to the polypeptide chain.

2. **Polypeptide is released.**
The hydrolysis reaction frees the polypeptide, which is released from the ribosome. The empty tRNAs are released either along with the polypeptide or...

3. **Ribosome subunits separate.**
...when the ribosome separates from the mRNA, and the two ribosomal subunits dissociate. The subunits are ready to attach to the start codon of another message and start translation anew.

FIGURE 16.17 Release Factors Manage Translation Termination.

Post-Translational Modifications

Proteins are not fully formed and functional when termination occurs. From earlier chapters, it should be clear that most proteins go through an extensive series of processing steps, collectively called post-translational modification, before they are completely functional. These steps require a wide array of molecules and events and take place in a wide variety of locations throughout the cell.

FOLDING Recall from Chapter 3 that a protein's function depends on its shape, and that a protein's shape depends on how it folds. Although folding occurs spontaneously, in the sense that no energy input is required, it is frequently speeded up by proteins called **molecular chaperones.**

Recent data have shown that in some bacteria, chaperone proteins actually bind to the ribosome near the "tunnel" where the growing polypeptide emerges from the ribosome. This finding suggests that folding occurs as the polypeptide is emerging from the ribosome.

CHEMICAL MODIFICATIONS Chapter 7 pointed out that many eukaryotic proteins are extensively modified after they are synthesized. For example, in the organelles called the rough endoplasmic reticulum and the Golgi apparatus, small chemical groups may be added to proteins—often, sugar or lipid groups that are critical for normal functioning. In some cases, the proteins receive a carbohydrate-based sorting signal that serves as an address label and ensures that the molecule will be carried to the correct location in the cell. (Other proteins have a sorting signal built into their primary structure.)

In the study area at *www.masteringbiology.com*, you'll find animations and tutorials to help you review transcription, RNA processing, translation, and addition of sorting signals.

In addition, many completed proteins are altered by enzymes that add or remove a phosphate group. Phosphorylation (addition of phosphate) and dephosphorylation (removal of phosphate) of proteins were introduced in Chapters 9 and 11. Recall that because a phosphate group has two negative charges, adding or removing a phosphate group may cause major changes in the shape and chemical reactivity of proteins. These changes have a dramatic effect on the protein's activity—often switching it from an inactive state to an active state or vice versa.

The take-home message here is that gene expression is a complex, multistep process that begins with transcription but may not end with translation. Instead, even completed and folded proteins may be activated or deactivated by events such as phosphorylation. In general, how are genes turned on or off? How does a cell "decide" which of its many genes should be expressed at any time? These questions are the focus of the next two chapters.

CHECK YOUR UNDERSTANDING

If you understand that . . .

- Translation begins when (1) the ribosome binding site on an mRNA binds to an rRNA sequence in the small ribosomal subunit, (2) the initiator aminoacyl tRNA binds to the start codon in the mRNA, and (3) the large subunit of the ribosome attaches to the small subunit to complete the ribosome.
- Translation elongation occurs when (1) an appropriate aminoacyl tRNA enters the A site, (2) a peptide bond forms between the amino acid held by that tRNA in the A site and the polypeptide held by the tRNA in the P site, and (3) the ribosome moves down the mRNA one codon.
- Translation ends when the ribosome reaches a stop codon.
- Completed proteins are modified by folding and, in many cases, addition of sugar, lipid, or phosphate groups.

✔ **You should be able to . . .**

Explain why the E, P, and A sites in the ribosome are appropriately named.

Answers are available in Appendix B.

CHAPTER 16 REVIEW

For media, go to the study area at www.masteringbiology.com

Summary of Key Concepts

After RNA polymerase binds DNA with the help of other proteins, it catalyzes the production of an RNA molecule whose base sequence is complementary to the base sequence of the DNA template strand.

- RNA polymerase begins transcription by binding to promoter sequences in DNA.
- In bacteria, this binding occurs in conjunction with a regulatory protein called sigma. Sigma recognizes particular sequences within promoters that are centered 10 bases and 35 bases upstream from the start of the actual genetic message.
- Eukaryotic promoters are more complex and variable than bacterial promoters are.
- In eukaryotes, transcription begins when a large array of basal transcription factors bind to a promoter. In response, RNA polymerase binds to the site.
- In both bacteria and eukaryotes, transcription ends when RNA polymerase encounters a termination signal on the DNA template.

✔You should be able to predict the consequences of a mutation that changes the sequence of nucleotides in a promoter.

Eukaryotic genes contain regions called exons and regions called introns; during RNA processing, the regions coded by introns are removed, and the ends of the RNA receive a cap and tail.

- Stretches of noncoding RNA called introns are spliced out by complex molecular machines called spliceosomes.
- A "cap" is added to the 5′ end of a primary transcript, and a poly(A) tail is added to the 3′ end.
- The cap and tail serve as recognition signals for the translation machinery and protect the message from degradation by ribonucleases.

✔You should be able to explain why RNA processing does not occur in bacteria.

Ribosomes translate mRNAs into proteins with the help of intermediary molecules called transfer RNAs.

- Experiments with radioactively labeled amino acids confirmed that ribosomes are the site of protein synthesis.
- Experiments with radioactively labeled amino acids confirmed that transfer RNAs (tRNAs) serve as the chemical bridge between the RNA message and the polypeptide product.

✔You should be able to explain why the name transfer RNA is appropriate.

Each transfer RNA carries an amino acid corresponding to the tRNA's three-base-long anticodon.

- tRNAs have an L-shaped tertiary structure. One leg of the L contains the anticodon, which forms complementary base pairs with the mRNA codon. The other leg holds the amino acid appropriate for that codon.
- Imprecise pairing—or "wobbling"—is allowed in the third position of a codon–anticodon pairing, so only about 40 different tRNAs are required to translate the 61 codons that code for amino acids.

✔You should be able to explain the relationships between an aminoacyl tRNA, a tRNA, an amino acid, and aminoacyl tRNA synthetase.

In the ribosome, the tRNA anticodon binds to a three-base-long mRNA codon, causing the amino acid carried by the transfer RNA to be added to the growing protein.

- Protein synthesis occurs in three steps: (1) an incoming aminoacyl tRNA occupies the A site; (2) the growing polypeptide chain is transferred from a peptidyl tRNA in the ribosome's P site to the amino acid bound to the tRNA in the A site, and a peptide bond is formed; and (3) the ribosome is translocated to the next codon on the mRNA, accompanied by ejection of the empty tRNA from the E site.
- Peptide bond formation is catalyzed by a ribozyme (RNA), not an enzyme (protein).
- Proteins fold into their three-dimensional conformation (tertiary structure), sometimes with the aid of chaperone proteins.
- Proteins may be targeted to specific locations in the cell by post-translational modification that adds signal sequences. Some proteins remain inactive until modified by phosphorylation.

✔You should be able to create a concept map (see **BioSkills 8** in Appendix A) that describes the relationships among the following concepts and structures: translation, initiation, elongation, termination, protein folding, initial amino acid, chemical modification, mRNA, charged tRNAs in A site, growing polypeptide in P site, empty tRNA in E site, start codon, ribosome binding site, initiation factors, ribosome subunits.

(MB) **Web Activity** Synthesizing Proteins, **BioFlix™** Protein Synthesis

Questions

✔TEST YOUR KNOWLEDGE

Answers are available in Appendix B

1. How did the A site of the ribosome get its name?
   a. It is where amino acids are affixed to tRNAs, producing aminoacyl tRNAs.
   b. It is where the amino group on the growing polypeptide chain is available for peptide bond formation.
   c. It is the site occupied by incoming aminoacyl tRNAs.
   d. It is surrounded by α-helices of ribosomal proteins.
2. How did the P site of the ribosome get its name?
   a. It is where the promoter resides.
   b. It is made up of protein.
   c. It is where peptidyl tRNAs reside.
   d. It is where a growing polypeptide chain is phosphorylated.
3. What is a molecular chaperone?
   a. a protein that recognizes the promoter and guides the binding of RNA polymerase
   b. a protein that activates or deactivates another protein by adding or removing a phosphate group
   c. a protein that is a component of the large ribosomal subunit and that assists with peptide bond formation
   d. a protein that helps newly translated proteins fold into their proper three-dimensional configuration
4. The three types of RNA polymerase found in eukaryotic cells transcribe different types of genes. What does RNA polymerase II produce?
   a. rRNAs
   b. tRNAs
   c. mRNAs
   d. spliceosomes
5. What is an anticodon?
   a. the part of an mRNA that signals translation termination
   b. the part of an mRNA that signals the start of translation
   c. the part of a tRNA that binds to a codon in mRNA
   d. the part of a tRNA that accepts an amino acid, through a reaction catalyzed by tRNA synthetase
6. What do researchers observe when mRNAs that lack a cap and tail are added to a cell—compared to identical mRNAs that have a cap and tail?
   a. Elongation factors cannot bind to the experimental mRNAs.
   b. Basal transcription factors cannot bind to the promoter.
   c. Introns cannot be spliced out (the spliceosome is inhibited).
   d. Translation rate decreases; mRNA life span decreases.

✔ TEST YOUR UNDERSTANDING

Answers are available in Appendix B

1. Explain the relationship among eukaryotic promoter sequences, basal transcription factors, and RNA polymerase. Explain the relationship among bacterial promoter sequences, sigma, and RNA polymerase.
2. According to the wobble rules, the correct amino acid can be added to a growing polypeptide chain even if the third base in the mRNA codon is not complementary to the third base in the tRNA anticodon. How do the wobble rules relate to the redundancy of the genetic code?
3. Why does splicing occur in eukaryotic mRNAs? Where does it occur, and how are snRNPs involved?
4. Describe the sequence of events that occurs during translation as a protein elongates by one amino acid and the ribosome moves down the mRNA. Your answer should specify what is happening in the ribosome's A site, P site, and E site.
5. What evidence supports the hypothesis that peptide bond formation is catalyzed by a ribozyme?
6. In an aminoacyl tRNA, why is the observed distance between the amino acid and the anticodon important?

✔ APPLYING CONCEPTS TO NEW SITUATIONS

Answers are available in Appendix B

1. The 5′ cap and poly(A) tail on eukaryotic mRNAs protect the message from degradation by ribonucleases. But why do ribonucleases exist? What function would an enzyme that destroys messages serve? Answer this question using the example of an mRNA for a hormone that causes human heart rate to increase.
2. The nucleotide shown below is called cordycepin.

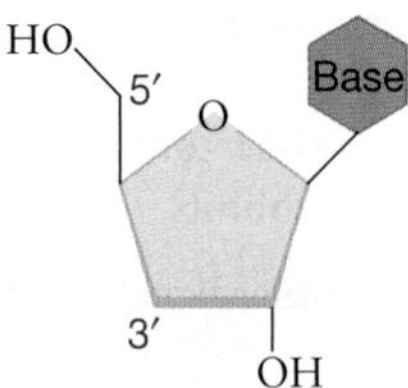

If cordycepin triphosphate, which has three phosphate groups bonded to the 5′ hydroxyl group in the figure, is added to a cell-free transcription reaction, the nucleotide is added onto the growing RNA chain. This observation confirms that synthesis occurs in the 5′→3′ direction. Explain why cordycepin triphosphate cannot be added to the 5′ end of an RNA.

3. Certain portions of the rRNAs in the large subunit of the ribosome are very similar in all organisms. To make sense of this finding, Carl Woese suggests that the conserved sequences have an important functional role. His logic is that these conserved sequences are so important to cell function that any changes in the sequences cause death. Which specific portions of the ribosome would you expect to be identical or nearly identical in all organisms, and which would you expect to be more variable? Explain your logic.
4. Recent structural models show that a poison called α-amanitin inhibits transcription by binding to a site inside RNA pol II but not to the active site itself. Based on the model of bacterial RNA polymerase in Figure 16.2, predict where α-amanitin binds and why it inhibits transcription.

This chapter is part of The Big Picture. See how on pages 336–337.

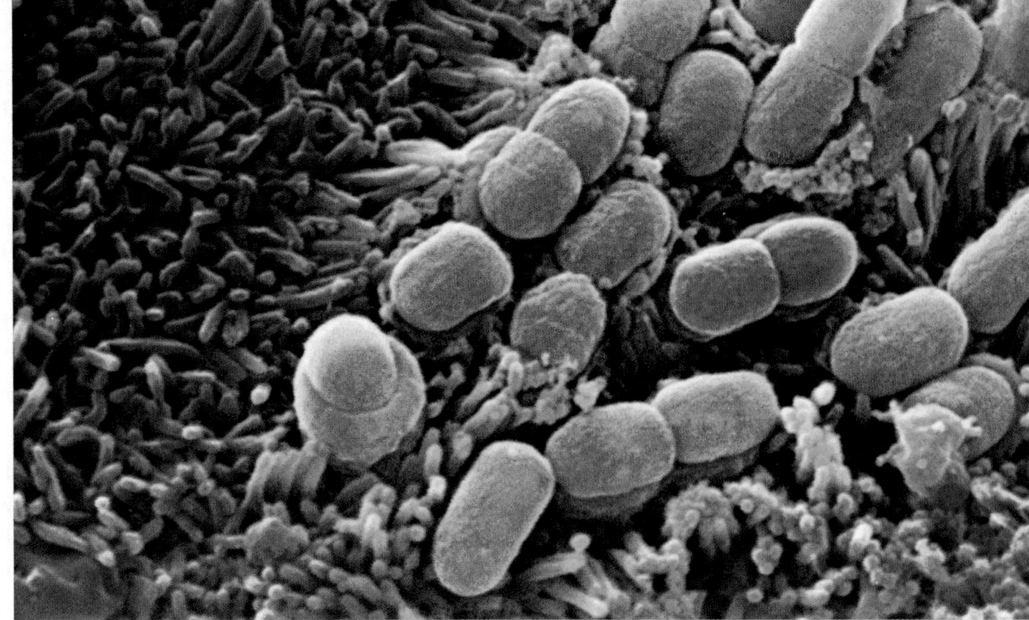

The structures that have been colored blue in this scanning electron micrograph are projections from human intestinal cells; the structures colored yellow are the bacterium *Escherichia coli*. In the intestine, the nutrients available to bacteria constantly change. This chapter explores how changes in gene expression help bacteria respond to environmental changes.

Control of Gene Expression in Bacteria 17

KEY CONCEPTS

- In bacteria, gene expression can be controlled at three levels: transcription, translation, or post-translation (protein activation).
- Changes in gene expression allow bacterial cells to respond to environmental changes.
- Transcriptional control can be negative or positive. Negative control occurs when a regulatory protein prevents transcription. Positive control occurs when a regulatory protein increases the transcription rate.

Imagine waiting eagerly to hear the opening lines of a wonderfully melodic symphony played by a renowned orchestra. The crowd applauds as the celebrated conductor comes onstage, then hushes as he takes the podium. He cocks the baton; the musicians raise their instruments. As the baton comes down, every instrument begins blaring a different tune at full volume. A tuba plays "Dixie"; a violinist renders "In-A-Gadda-Da-Vida"; and a cellist begins Mexico's national anthem. A snare drum lays down beats for OutKast's "Hey Ya," while the bass drum simulates the cannons in the "1812 Overture." Instead of music, there is pandemonium. The conductor staggers off stage, clutching his heart.

A cacophony like this would result if a bacterial cell "played" all its genes at full volume all the time. The *Escherichia coli* cells living in your gut right now have over 4300 genes. If all of those genes were expressed at the fastest possible rate at all times, the *E. coli* cells would stagger off the stage, too. But this does not happen. Cells are extremely selective about which genes are expressed, in what amounts, and when.

This chapter explores how bacterial cells control the activity, or expression, of their genes. Gene expression is the process of converting information that is archived in DNA into molecules that actually do things in the cell. It occurs when a protein or other gene product is synthesized and active.

Previous chapters detailed how genetic information is processed in cells; this chapter focuses on *when* genetic information is used. Let's begin by reviewing some of the environmental challenges that bacterial cells face, and then explore how these organisms meet them.

17.1 Gene Regulation and Information Flow

The bacteria that live in and on your body vastly outnumber your own cells. Consider just one of the species present: the gut-dwelling *Escherichia coli*. These cells can use a wide array of carbohydrates to supply the carbon and energy they need. But as your diet changes from day to day, the availability of different sugars in your intestines varies. Each type of nutrient requires a different membrane transport protein to bring the molecule into the cell and a different suite of enzymes to process it. Precise control of gene expression gives *E. coli* the ability to use the available sugars efficiently.

To understand why precise control over gene expression is so important, you have to realize that bacterial cells from an array of species can be packed 2 cm thick along your intestinal walls. All of these organisms are competing for space and nutrients. In an environment like this, a cell has to use resources efficiently if it's going to be able to survive and reproduce. An individual that synthesizes proteins it doesn't need has fewer resources to devote to making the proteins it does need. Such cells are losers—they compete less successfully for the resources that are required to produce offspring.

Realizing this, biologists predicted that most gene expression is triggered by specific signals from the environment, such as the presence of specific sugars. Did you drink milk at your last meal, or eat French fries and a candy bar? Each type of food contains different sugars. Each sugar should induce a different response from the *E. coli* cells in your intestine. Just as a conductor needs to regulate the orchestra's musicians, cells need to regulate which proteins they produce.

Mechanisms of Regulation—An Overview

The flow of information from DNA to RNA to activation of the final gene product occurs in three steps, represented by arrows in the following diagram:

DNA ⟶ mRNA ⟶ protein ⟶ activated protein

Gene expression can be controlled at any of these arrows. The arrow from DNA to RNA represents transcription—the making of messenger RNA (mRNA). The arrow from RNA to protein represents translation, in which ribosomes read the information in mRNA and use that information to synthesize a protein. The arrow from protein to activated protein represents post-translational modifications—including folding, addition of carbohydrate or lipid groups, or perhaps phosphorylation.

How can a bacterial cell avoid producing proteins that are not needed at a particular time, and thus use resources efficiently? A look at the flow of information from DNA to protein suggests three possible mechanisms:

1. The cell could avoid making the mRNAs for particular enzymes. If there is no mRNA, then ribosomes cannot make the gene product. **Transcriptional control** occurs when regulatory proteins affect RNA polymerase's ability to bind to a promoter and initiate transcription:

   DNA –✕→ mRNA ⟶ protein ⟶ activated protein

2. If the mRNA for an enzyme has been transcribed, the cell might have a way to prevent the mRNA from being translated into protein. **Translational control** occurs when regulatory molecules alter the length of time an mRNA survives before it is degraded by ribonucleases, or affect translation initiation, or affect elongation factors and other proteins during the translation process:

   DNA ⟶ mRNA –✕→ protein ⟶ activated protein

3. Chapter 16 pointed out that some proteins are manufactured in an inactive form and have to be activated by chemical modification, such as the addition of a phosphate group. This type of regulation is **post-translational control**:

   DNA ⟶ mRNA ⟶ protein –✕→ activated protein

Which of these three forms of control occur in bacteria? The short answer to this question is "all of the above." As **Figure 17.1** shows, many factors affect how much active protein is produced from a particular gene.

- Transcriptional control is particularly important due to its efficiency—it saves the most energy for the cell, because it stops the process at the earliest possible point.
- Translational control is advantageous because it allows a cell to make rapid changes in the relative amounts of different proteins.

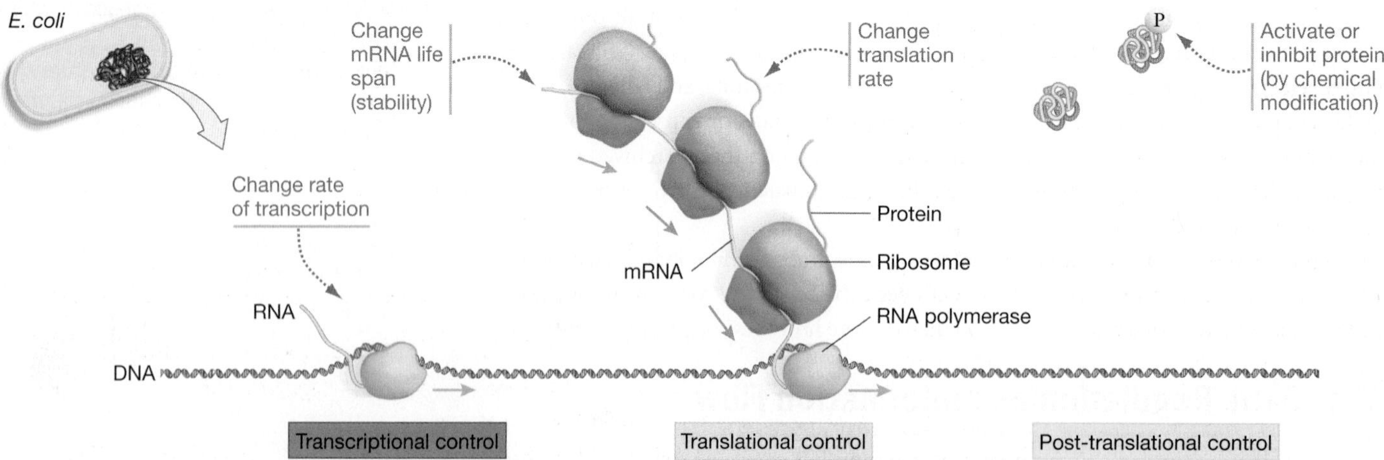

FIGURE 17.1 Gene Expression in Bacteria Can Be Regulated at Three Levels.

✔**EXERCISE** Label the mode of regulation that is the slowest in response time, and that which is fastest. Label the most efficient and least efficient in resource use.

- Post-translational control is significant as well, because it provides the most rapid response of all three mechanisms.

Among these mechanisms of gene regulation, there is a clear trade-off between the speed of response and the conservation of ATP, amino acids, and other resources. Transcriptional control is slow but efficient in resource use; post-translational control is fast but energetically expensive.

Although this chapter focuses almost exclusively on mechanisms of transcriptional control, it is important to keep in mind that bacteria possess translational and post-translational controls as well, and that some genes—such as those that code for the enzymes required for glycolysis—are transcribed all the time, or **constitutively**. Finally, it is critical to realize that gene expression is not an all-or-none proposition. Genes are not just "on" or "off"—instead, the level of expression is highly variable.

Variation in gene expression allows cells to respond to changes in their environment.

Metabolizing Lactose—A Model System

As Chapters 13 through 16 have shown, many of the great advances in genetics have been achieved through the analysis of model systems. Mendel studied garden peas and discovered the fundamental patterns of gene transmission; Morgan studied fruit flies and confirmed the chromosome theory of inheritance; an array of researchers used viruses and *E. coli* to work out the mechanisms of DNA synthesis, transcription, and translation. In studies of gene regulation, the key model system has been the metabolism of the sugar lactose in *E. coli.*

Jacques Monod, François Jacob, and many colleagues introduced lactose metabolism in *E. coli* as a model system during the 1950s and 1960s. Although they worked with a single species of bacteria, their results had a profound effect on thinking about gene regulation in all organisms. Some details turned out to be specific to the *E. coli* genes responsible for lactose metabolism, but many of the results are universal.

Escherichia coli can use a wide variety of sugars for ATP production, via cellular respiration or fermentation. These sugars also serve as raw material in the synthesis of amino acids, vitamins, and other complex compounds. Glucose is *E. coli's* preferred carbon source, however—meaning that it is the source of energy and carbon atoms that the organism uses most efficiently.

A preference for glucose makes sense, because glycolysis begins with glucose and is the main pathway for the production of ATP. Lactose, the sugar found in milk, is also used by *E. coli*, but only when glucose supplies are depleted. Recall from Chapter 5 that lactose is a disaccharide made up of one molecule of glucose and one molecule of galactose.

To use lactose, *E. coli* must first transport the sugar into the cell. Once lactose is inside the cell, the enzyme β-galactosidase catalyzes a reaction that breaks the sugar down into glucose and galactose. The glucose released by this reaction goes directly into the glycolytic pathway; other enzymes convert the galactose to a substance that can also be processed in the glycolytic pathway.

In the early 1950s, biologists discovered that *E. coli* produces high levels of β-galactosidase only when lactose is present in the environment. Based on this observation, researchers proposed that lactose itself regulates the gene for β-galactosidase—meaning that lactose acts as an inducer. An **inducer** is a substrate in a reaction, and it stimulates the expression of a specific gene or genes.

In the late 1950s Jacques Monod investigated how the presence of glucose affects the regulation of the β-galactosidase gene. Would *E. coli* produce high levels of β-galactosidase when both glucose and lactose were present in the surrounding environment? As the experiment summarized in **Figure 17.2** shows, the

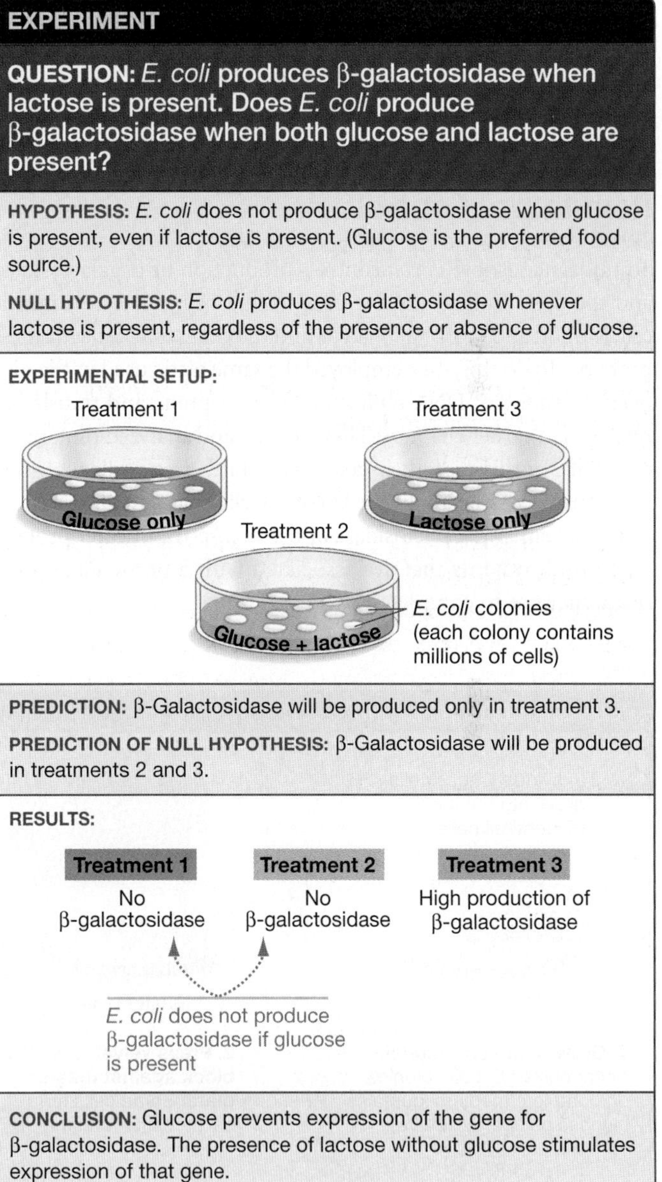

EXPERIMENT
QUESTION: E. coli produces β-galactosidase when lactose is present. Does E. coli produce β-galactosidase when both glucose and lactose are present?
HYPOTHESIS: E. coli does not produce β-galactosidase when glucose is present, even if lactose is present. (Glucose is the preferred food source.)
NULL HYPOTHESIS: E. coli produces β-galactosidase whenever lactose is present, regardless of the presence or absence of glucose.
EXPERIMENTAL SETUP:
Treatment 1
Treatment 2
Treatment 3
Glucose only
Glucose + lactose
Lactose only
E. coli colonies (each colony contains millions of cells)
PREDICTION: β-Galactosidase will be produced only in treatment 3.
PREDICTION OF NULL HYPOTHESIS: β-Galactosidase will be produced in treatments 2 and 3.
RESULTS:
Treatment 1
No β-galactosidase
Treatment 2
No β-galactosidase
Treatment 3
High production of β-galactosidase
E. coli does not produce β-galactosidase if glucose is present
CONCLUSION: Glucose prevents expression of the gene for β-galactosidase. The presence of lactose without glucose stimulates expression of that gene.

FIGURE 17.2 Glucose Affects the Regulation of the β-Galactosidase Gene.

SOURCE: Pardee, A. B., F. Jacob, and J. Monod. 1959. The genetic control and cytoplasmic expression of "inducibility" in the synthesis of β-glactosidase by *E. coli.* *Journal of Molecular Biology* 1: 165–178.

✔**QUESTION** How would you control growing conditions in the three treatments, so that the results of this experiment are valid?

answer was no. The enzyme β-galactosidase is produced only when lactose is present and glucose is not present.

Monod teamed up with François Jacob to investigate exactly how lactose and glucose regulate the genes responsible for lactose metabolism—the gene for the membrane protein that imports lactose and the gene for β-galactosidase. Discoveries about how these genes are regulated shed light on how genes in all organisms are controlled. Research on this system is still going strong, over 50 years later.

✔You should be able to make a chart summarizing the molecules involved in regulating lactose use in *E. coli.* There should be 10 rows and 2 columns. Title the first column "Name" and the second column "Function." The rows are *lacZ, lacY,* operator, promoter, CAP site, repressor, CAP, cAMP, lactose, and glucose. As you read this chapter, fill in the "Function" column.

17.2 Identifying Genes under Regulatory Control

To understand how *E. coli* controls production of β-galactosidase and the membrane transport protein that brings lactose into the cell, Monod and Jacob first had to find the genes that code for these proteins. To do this, they employed the same tactic used in the pioneering studies of DNA replication, transcription, and translation reviewed in earlier chapters: They isolated and analyzed mutant individuals. In this case, their goal was to find *E. coli* cells that were not capable of metabolizing lactose. Cells that can't use lactose must lack either β-galactosidase or the lactose-transporter protein.

To find mutants that are associated with a particular trait, a researcher has to complete two steps:

1. Generate a large number of individuals with mutations at random locations in their genomes. Monod and colleagues accomplished this step by exposing *E. coli* populations to X-rays, UV light, or **mutagens**—chemicals that damage DNA and increase mutation rates.
2. Screen the mutants to find individuals with defects in the process or biochemical pathway in question—in this case, defects in lactose metabolism. Recall from Chapter 15 that a genetic screen is any technique for selecting individuals with certain types of mutations out of a large population.

The researchers were looking for cells that cannot grow in an environment that contains only lactose as an energy source. Normal cells grow well in this environment. How could the researchers select cells on the basis of *lack* of growth?

Replica Plating to Find Mutant Genes

Replica plating and growth on indicator plates were key techniques in the search for mutants with defects in lactose metabolism. **Figure 17.3** shows how **replica plating** works.

Step 1 When mutants with defects in lactose metabolism are desired, mutagenized bacteria are spread on a "master plate" filled with gelatinous agar containing glucose. It is important that the mutant cells you want to identify are capable of growing on the master plate. The bacteria are then allowed to grow, so that each cell produces a colony—a large number of identical cells descended from a single cell.

Step 2 A block covered with a piece of sterilized velvet is pressed onto the master plate. Because of the contact, cells from each colony on the master plate are transferred to the velvet.

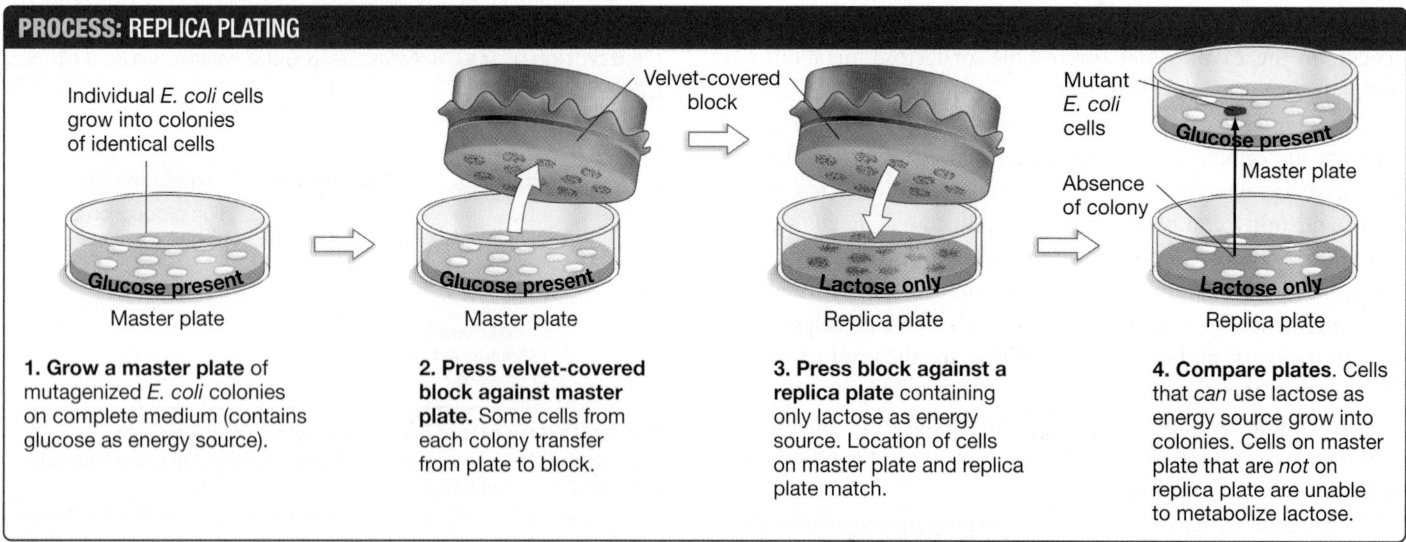

PROCESS: REPLICA PLATING
Individual E. coli cells grow into colonies of identical cells
Glucose present
Master plate
1. Grow a master plate of mutagenized E. coli colonies on complete medium (contains glucose as energy source).
Velvet-covered block
Glucose present
Master plate
2. Press velvet-covered block against master plate. Some cells from each colony transfer from plate to block.
Lactose only
Replica plate
3. Press block against a replica plate containing only lactose as energy source. Location of cells on master plate and replica plate match.
Mutant E. coli cells
Glucose present
Master plate
Absence of colony
Lactose only
Replica plate
4. Compare plates. Cells that can use lactose as energy source grow into colonies. Cells on master plate that are not on replica plate are unable to metabolize lactose.

FIGURE 17.3 Replica Plating Is a Technique for Identifying Mutant Cells. Here, replica plating is used to isolate mutant *E. coli* cells with a deficiency in lactose metabolism. There are two requirements for replica plating: (1) The location of colonies transferred from the master plate to the replica plate must match exactly on both plates, and (2) the media in the two plates must differ by just one component.

✔**QUESTION** How would you alter this protocol to isolate mutant cells with a deficiency in the enzymes required to synthesize tryptophan?

Step 3 The velvet is pressed onto a plate containing a **medium**—a liquid or solid that supports growth—that differs from the master plate by a single component. In this case, the second medium has only lactose—no glucose—as the source of carbon and energy. Cells from the velvet stick to the plate's surface, producing an exact copy of the locations of the colonies on the master plate. This copy is called the replica plate.

Step 4 After these cells grow, compare the colonies that thrive on the replica plate's medium with those on the master plate. In this example, colonies that grow on the master plate but are missing on the replica plate represent mutants deficient in lactose metabolism. By picking these particular colonies from the master plate, researchers build a collection of lactose mutants.

Monod also used an alternative strategy based on **indicator plates**, where mutants with metabolic deficiencies are observed directly. In this case, he added a compound that is acted on by β-galactosidase. The compound acts as an indicator molecule for the presence of functioning β-galactosidase because one of the molecules produced by the reaction is yellow. Colonies with normal β-galactosidase turn yellow; colonies that have a defect in the β-galactosidase enzyme or its production stay white.

Different Classes of Lactose Metabolism Mutants

The initial mutant screen yielded the three types of mutants summarized in **Table 17.1**. In one class, the mutant cells were unable to cleave the indicator molecule—even if lactose was present inside the cells to induce production of the β-galactosidase protein. The investigators concluded that these mutants must lack a functioning version of the β-galactosidase protein—meaning the gene that encodes β-galactosidase is defective. This gene was designated *lacZ*, and the mutant allele *lacZ*$^-$.

TABLE 17.1 **Three Types of Lactose Metabolism Mutants in *E. coli***

Observed Phenotype	Interpretation	Inferred Genotype
1. Cells cannot cleave indicator molecule even if lactose is present as an inducer.	No β-galactosidase; gene for β-galactosidase is defective. Call this gene *lacZ*.	*lacZ*$^-$
2. Cells cannot accumulate lactose.	No membrane protein (galactoside permease) to import lactose; gene for galactoside permease is defective. Call this gene *lacY*.	*lacY*$^-$
3. Cells cleave indicator molecule even if lactose is absent as an inducer.	Constitutive expression of *lacZ* and *lacY*; gene for regulatory protein that shuts down *lacZ* and *lacY* is defective—it does not need to be induced by lactose. Call this gene *lacI*.	*lacI*$^-$

In the second class of mutants, the cells failed to accumulate lactose inside the cell. In normal cells the concentration of lactose is about 100 times that of lactose in the surrounding environment, but in the mutant cells lactose concentrations were much lower. To explain this result, Jacob and Monod hypothesized that the mutant cells had defective copies of the membrane protein responsible for transporting lactose into the cell. This protein was identified and named galactoside permease; the gene that encodes it was designated *lacY*. **Figure 17.4** summarizes the functions of β-galactosidase and galactoside permease.

The third and most surprising class of mutants did not show normal regulation of the expression of β-galactosidase and galactoside permease. Instead, these mutants made the proteins all the time—even if no lactose was present.

Cells that are abnormal because they produce a product at all times are called **constitutive mutants**. The gene that mutated to produce constitutive β-galactosidase expression was named *lacI*. The letter "I" signified that these mutants did not need an inducer—lactose—to express β-galactosidase or galactoside permease.

To understand the reasoning here, recall that in normal cells, the expression of these genes is induced by the presence of lactose. But in cells with a mutant form of *lacI* (*lacI*$^-$ mutants), gene expression occurs with or without lactose. This means that *lacI*$^-$ mutants have a defect in gene regulation. In these mutants, the gene remains "on" when it should be turned off.

To pull these observations together, the researchers hypothesized that the normal product of the *lacI* gene prevents the transcription of *lacZ* and *lacY* when lactose is absent. Because lactose triggers production of β-galactosidase, it was reasonable to expect that the *lacI* gene or gene product interacts with lactose in some way. (Later work showed that the inducer is actually a derivative of lactose called *allolactose*. For the sake of historical accuracy and simplicity, however, this discussion refers to lactose itself as the inducer.)

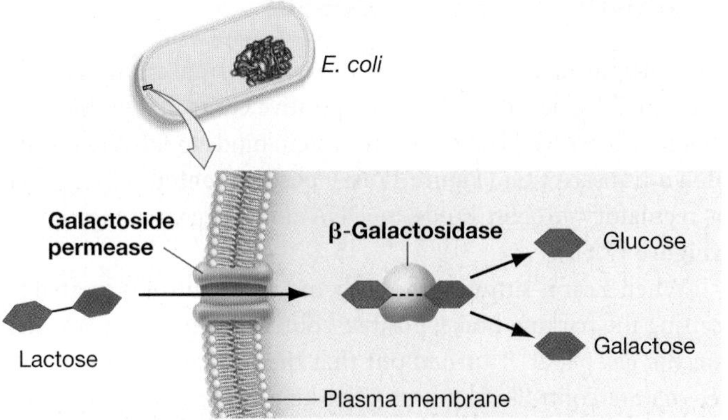

FIGURE 17.4 Two Proteins *E. coli* Needs for Using Lactose. For *E. coli* cells to use lactose, the membrane protein galactoside permease must be present to bring the sugar into the cell. Then the enzyme β-galactosidase must be present to break lactose into its glucose and galactose subunits.

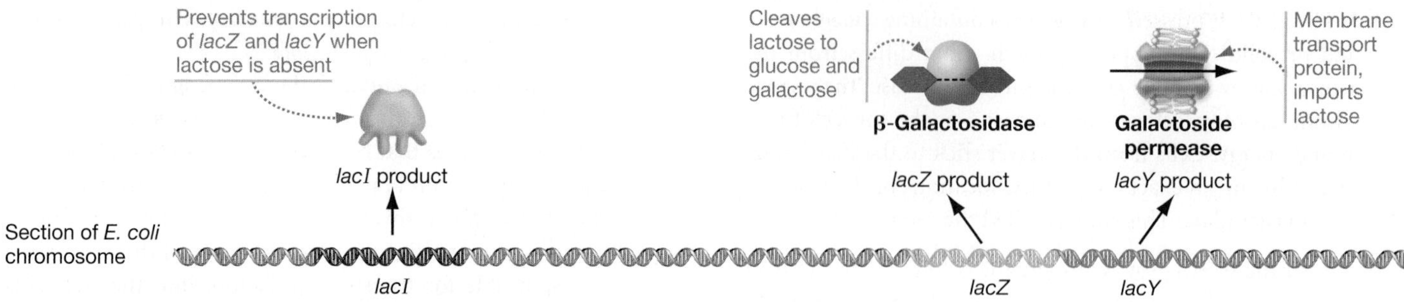

FIGURE 17.5 The *lac* Genes Are in Close Physical Proximity.

Several Genes Are Involved in Lactose Metabolism

Jacob and Monod had succeeded in identifying three genes involved in lactose metabolism: *lacZ*, *lacY*, and *lacI*. They had concluded that *lacZ* and *lacY* code for proteins required for the metabolism and import of lactose, while *lacI* is responsible for some sort of regulatory function. When lactose is absent, the *lacI* gene or gene product shuts down the expression of *lacZ* and *lacY*. But when lactose is present, the opposite occurs—transcription of *lacZ* and *lacY* is induced.

✔If you understand the genes involved in lactose metabolism, you should be able to describe the specific function of *lacZ* and *lacY*. You should also be able to describe the effect of the *lacI* gene product when lactose is present versus absent and explain why these effects are logical.

Jacob and Monod followed up on these experiments by mapping the physical location of the three genes on *E. coli*'s circular chromosome (**Figure 17.5**). Their data showed that the genes are close together. This was a crucial finding, because it suggested that both *lacZ* and *lacY* might be controlled by *lacI*. Could one regulatory gene govern more than one protein-encoding gene? If so, how does *lacI* actually work? And why do lactose and glucose have opposite effects on it?

17.3 Mechanisms of Negative Control: Discovery of the Repressor

In principle, there are two general ways that transcription can be regulated: by negative control or positive control. **Negative control** occurs when a regulatory protein binds to DNA and shuts down transcription (**Figure 17.6a**); **positive control** occurs when a regulatory protein binds to DNA and triggers transcription (**Figure 17.6b**).

When a car is sitting at the curb, negative control is exerted by setting the parking brake; positive control occurs when you step on the gas pedal. It turned out that the *lacZ* and *lacY* genes in *E. coli* are controlled by both a parking brake and a gas pedal. These genes are under both negative control and positive control.

The hypothesis that the *lacZ* and *lacY* genes might be under negative control originated with Leo Szilard in the late 1950s. Szilard suggested to Monod that the *lacI* gene codes for a product that represses transcription of the *lacZ* and *lacY* genes.

Stated another way, the *lacI* gene produces an inhibitor that exerts negative control over the *lacZ* and *lacY* genes. This transcription inhibitor was called a **repressor**. It was thought to bind directly to DNA near or on the promoter for the *lacZ* and *lacY* genes (**Figure 17.7a**).

To explain how the presence of lactose triggers transcription in normal cells, Szilard and Monod proposed that lactose interacts with the repressor in a way that makes the repressor release from its binding site (**Figure 17.7b**). The idea was that lactose induces transcription by removing negative control. The repressor is the parking brake; lactose releases the parking brake.

(a) Negative control: Regulatory protein *shuts down* transcription.

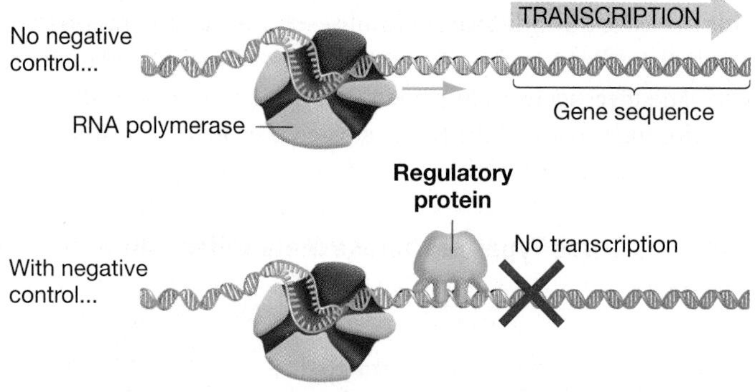

(b) Positive control: Regulatory protein *triggers* transcription.

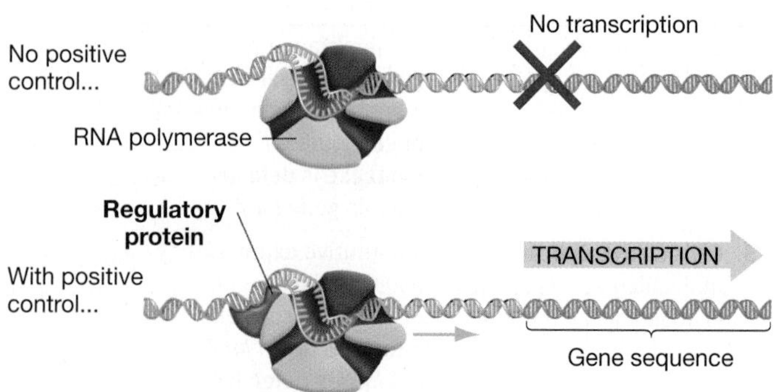

FIGURE 17.6 Genes Are Regulated by Negative Control, Positive Control, or Both. Section 17.4 explores positive control and its interaction with negative control.

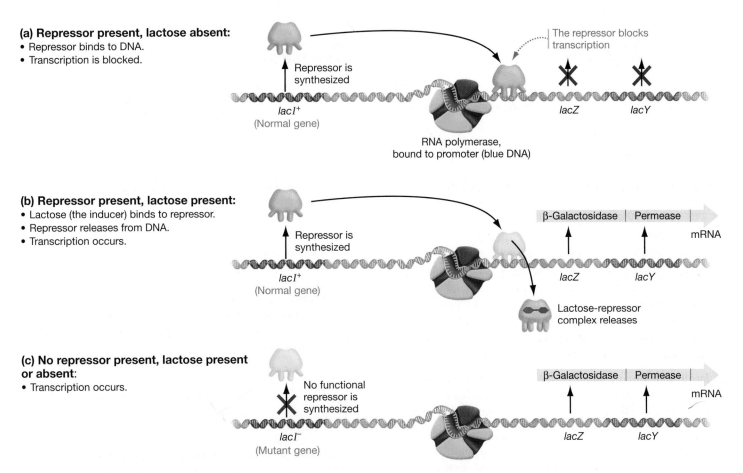

(a) Repressor present, lactose absent:
- Repressor binds to DNA.
- Transcription is blocked.

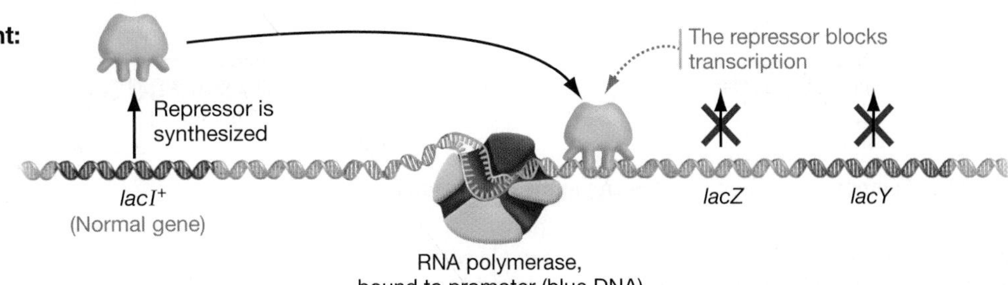

(b) Repressor present, lactose present:
- Lactose (the inducer) binds to repressor.
- Repressor releases from DNA.
- Transcription occurs.

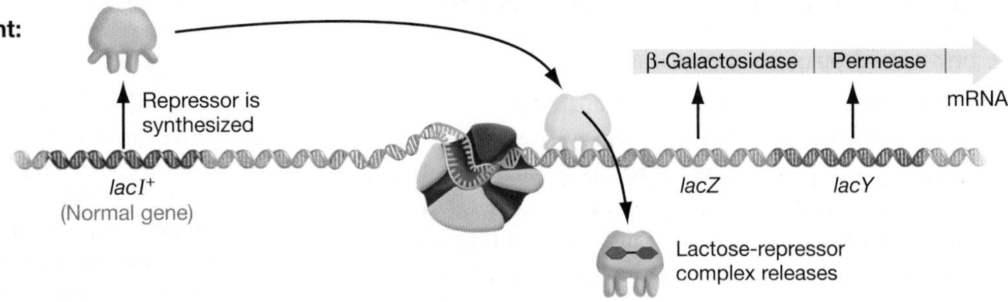

(c) No repressor present, lactose present or absent:
- Transcription occurs.

FIGURE 17.7 The Hypothesis of Negative Control. The negative-control hypothesis maintains that **(a)** transcription of genes involved in lactose use is normally blocked by a repressor molecule that binds to DNA on or near the promoter for *lacZ* and *lacY*, **(b)** lactose induces transcription of *lacZ* and *lacY* by interacting with the repressor, and **(c)** when a functional repressor is absent, transcription proceeds.

What about the constitutive mutants? **Figure 17.7c** shows that constitutive transcription is observed in *lacI*$^-$ mutants because a functional repressor is absent—the parking brake is broken.

To test these hypotheses, Jacob and Monod and co-workers created *E. coli* cells that had functioning copies of the genes for β-galactosidase and galactoside permease but lacked a functional gene for the repressor. As predicted, these cells made β-galactosidase all the time. But when these cells received a functioning copy of the repressor gene, β-galactosidase production declined and then stopped.

This result supported the hypothesis that the repressor codes for a protein that shuts down transcription—that it is indeed the "parking brake" on transcription. However, if an inducer such as lactose was then added to the experimental cells, β-galactosidase activity resumed. This result supported the hypothesis that lactose removes the repressor—it releases the parking brake.

What's the take-home message? The *lacI* gene codes for a repressor protein that exerts negative control on *lacZ* and *lacY*. Lactose acts as an inducer by removing the repressor and ending negative control.

The *lac* Operon

Jacob and Monod summarized the results of their experiments with a comprehensive model of negative control that was published in 1961. One of their key conclusions was that the genes for β-galactosidase and galactoside permease are controlled together. To encapsulate this idea, they coined the term **operon** for a set of coordinately regulated bacterial genes that are transcribed together into one mRNA. Logically enough, the group of genes involved in lactose metabolism was termed the *lac* operon.

Later, a gene called *lacA* was found to be tightly linked to *lacY* and *lacZ* and part of the same operon. The *lacA* gene codes for the enzyme transacetylase. The enzyme's function is protective in nature. It catalyzes reactions that allow certain types of sugars to be exported from the cell when they are too abundant.

Three hypotheses are central to the Jacob–Monod model of *lac* operon regulation:

1. The *lacZ*, *lacY*, and *lacA* genes are adjacent and are transcribed into one mRNA initiated from the single promoter of the *lac* operon. As a result, the expression of the three genes is coordinated.
2. The repressor is a protein encoded by *lacI* that binds to DNA and prevents transcription of *lacZ*, *lacY*, and *lacA*. Jacob and Monod proposed that lacI is expressed constitutively, and that the repressor binds to a section of DNA in the *lac* operon called the **operator**.

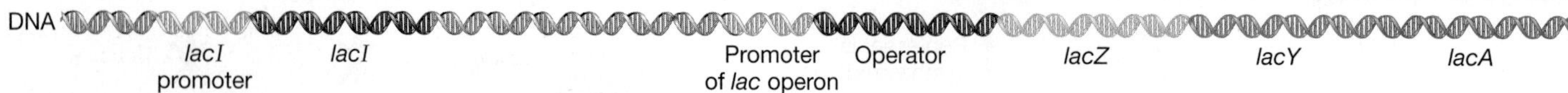

FIGURE 17.8 Components of Negative Control in the *lac* Operon.

✔**EXERCISE** Using small, colored bits of candy or paper, add the repressor protein to the figure. Then add RNA polymerase, and then add lactose. At each step, explain what happens after the molecule is added.

3. The inducer (lactose) binds to the repressor. When it does, the repressor changes shape. The shape change causes the repressor to drop off the DNA strand.

These hypotheses are summarized in **Figure 17.8**.

Recall from Chapter 3 that this form of control over protein function is called **allosteric regulation**. In allosteric regulation, a small molecule binds directly to a protein and causes it to change its shape and activity. When the inducer binds to the repressor, negative control ends and transcription can proceed.

Years after their model of negative control was published, experiments confirmed the existence of the operator:

- Jacob and Monod found *E. coli* mutants that had normal forms of the repressor but still expressed *lacZ* constitutively. In these cells, the repressor protein couldn't function because the operator was abnormal.
- Walter Gilbert and Benno Müller-Hill tagged copies of the repressor protein with a radioactive atom and showed that it physically binds to the DNA sequences of the operator.

Why Has the *lac* Operon Model Been So Important?

The *lac* operon has been an immensely important model system in genetics, for two reasons. First, follow-up work showed that numerous bacterial genes and operons are under negative control by repressor proteins—meaning that Jacob and Monod's findings were general. Second, the *lac* operon model introduced a fundamentally important idea: Gene expression is regulated by physical contact between regulatory proteins and specific regulatory sites in DNA. Publication of the *lac* operon model was a watershed event in the history of biological science.

Besides confirming the existence of negative control, regulatory proteins, and regulatory sites in DNA, work on the *lac* operon offered an important example of post-translational control over gene expression. To understand why, you have to realize that the repressor protein is always present—because it is transcribed and translated constitutively at low levels. When a rapid change in *lac* operon activity is required, it does not occur by means of changes in the transcription or translation of new repressor proteins. Instead, the activity of *existing* repressor proteins is altered.

This turns out to be a common type of control. In virtually all cases, the activity of key regulatory proteins is controlled by post-translational modifications.

CHECK YOUR UNDERSTANDING

If you understand that . . .

- Negative control occurs when something must be taken away for transcription to occur.
- The *lac* operon repressor exerts negative control over three protein-coding genes by binding to the operator site in DNA, near the promoter.
- For transcription to occur in the *lac* operon, an inducer molecule (a derivative of lactose) must bind to the repressor, causing it to release from the operator.

✔ **You should be able to . . .**

1. Explain why lactose should induce transcription of the *lac* operon.
2. Diagram the *lac* operon, showing the relative positions of the operator, the promoter, and the three protein-coding genes, and indicate what is happening at the operon in the absence of lactose and in the presence of lactose.

Answers are available in Appendix B.

17.4 Mechanisms of Positive Control: Catabolite Repression

The model of negative control over the *lac* operon, summarized in Figure 17.7, is elegant and successful in explaining experimental results. But it is not complete. After studying the model, you may think of an important question that it fails to answer: Where does glucose fit in?

Transcription of the *lac* operon is drastically reduced when glucose is present in the environment—even when lactose is available to induce β-galactosidase expression. This makes sense, given that glucose is *E. coli*'s preferred carbon source. When glucose is already present, the cell doesn't need to cleave lactose as its way of acquiring glucose.

The hydrolysis of lactose into its glucose and galactose subunits, by the enzyme β-galactosidase, is an example of catabolism. In many cases, operons that encode catabolic enzymes (like β-galactosidase) are inhibited when the end product of the reaction, the catabolite, is abundant (**Figure 17.9a**). Biologists use the term **catabolite repression** to describe a situation like this. Catabolite repression is a form of feedback inhibition (see Chapter 9), sometimes called end-point or end-product inhibition.

Figure 17.9b shows how catabolite repression affects the lactose metabolism. In catabolite repression of the *lac* operon, glu-

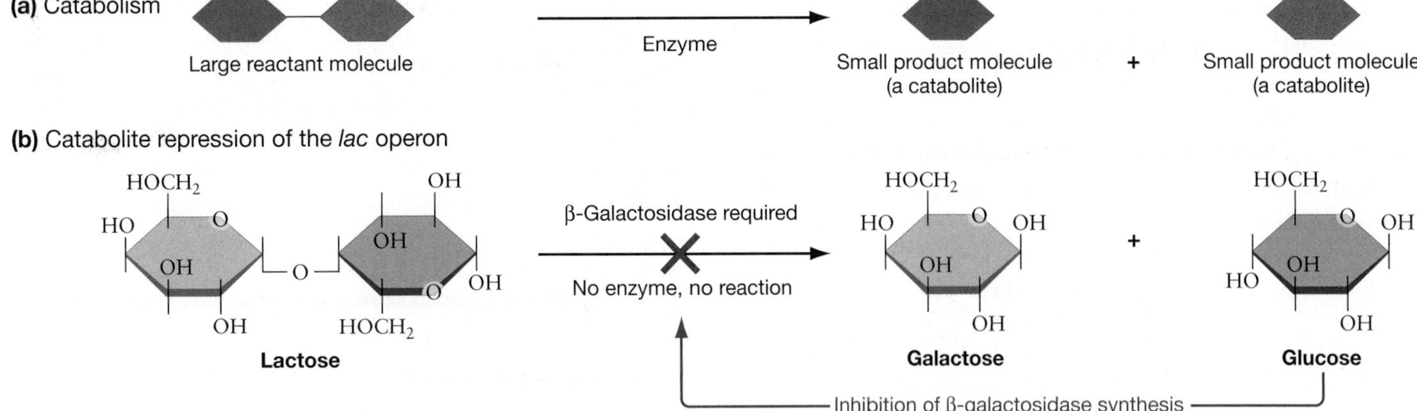

FIGURE 17.9 Catabolite Repression Is a Mechanism of Gene Regulation. (a) A generalized example of catabolism. **(b)** Catabolite repression occurs when one of the small product molecules of catabolism represses the production of the enzyme responsible for the reaction. In the case of lactose metabolism, the production of β-galactosidase is suppressed when glucose is present.

cose is the catabolite. When glucose is abundant in the cell, transcription of the *lac* operon is decreased.

The CAP Protein and Binding Site

How does glucose prevent expression of the *lac* operon? An answer to this question began to emerge when researchers discovered a second major control element in the *lac* operon—in this case, an example of positive control.

Positive control of the *lac* operon depends on a regulatory protein called **catabolite activator protein (CAP)** and a DNA sequence known as the **CAP binding site**, which is located just upstream of the *lac* promoter. The CAP protein binds to the CAP binding site and triggers transcription of the *lac* operon.

To understand how CAP works, it's important to realize that not all promoters are created equal. Strong promoters allow efficient initiation of transcription by RNA polymerase; weak ones support much less efficient initiation of transcription.

The *lac* promoter is weak. But when the CAP regulatory protein is bound to the CAP site just upstream of the *lac* promoter, the protein interacts with RNA polymerase in a way that allows transcription to begin much more frequently.

Because CAP binding greatly strengthens the *lac* promoter, CAP exerts positive control of the *lac* operon. When CAP is active, transcription increases. After the inducer removes the parking brake, CAP can push the gas pedal to the floor.

Researchers also discovered that CAP, like the repressor protein, is allosterically regulated. CAP can't bind to the CAP binding site unless the regulatory molecule **cyclic AMP (cAMP)** binds to it (**Figure 17.10**). Cyclic AMP is the green light that tells CAP to floor it.

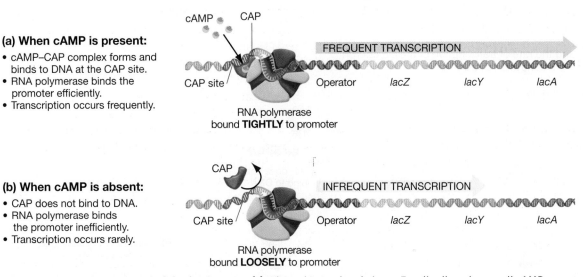

(a) When cAMP is present:
- cAMP–CAP complex forms and binds to DNA at the CAP site.
- RNA polymerase binds the promoter efficiently.
- Transcription occurs frequently.

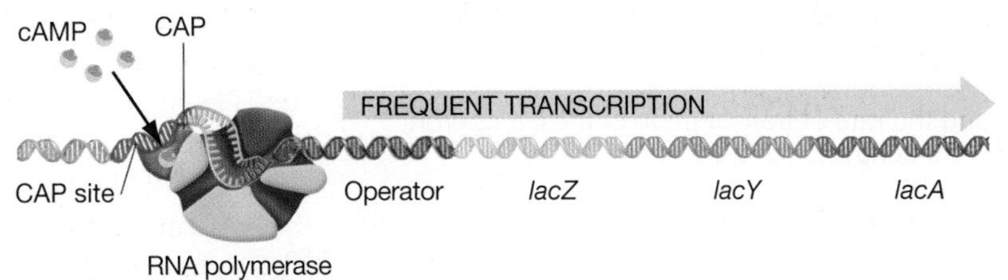

(b) When cAMP is absent:
- CAP does not bind to DNA.
- RNA polymerase binds the promoter inefficiently.
- Transcription occurs rarely.

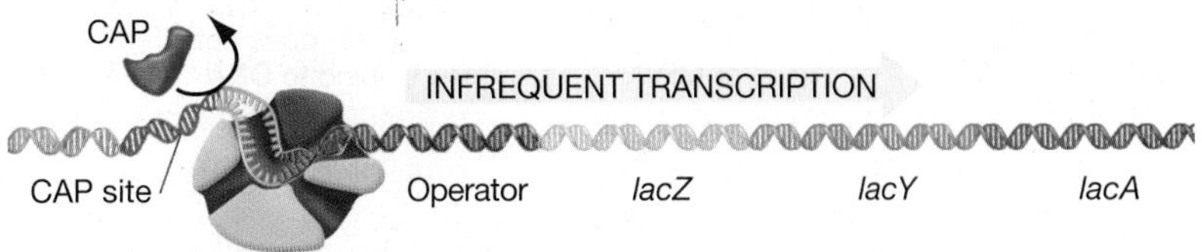

FIGURE 17.10 Positive Control of the *lac* Operon. (a) When glucose levels in an *E. coli* cell are low, cyclic AMP (cAMP) is produced. cAMP then interacts with CAP to increase transcription of the *lac* operon. **(b)** When glucose is abundant, cAMP is rare in the cell, and positive control does not occur.

✔**EXERCISE** Above each part, write "Glucose high" or "Glucose low" as appropriate.

How Does Glucose Influence Formation of the CAP–cAMP Complex?

Where does glucose fit into all this? Its role is indirect: Glucose regulates cAMP levels. When extracellular glucose concentrations are high, intracellular cAMP concentrations are low; when extracellular glucose concentrations are low, intracellular cAMP concentrations are high.

This seesaw is driven by the enzyme **adenylyl cyclase**, which produces cAMP from ATP (**Figure 17.11a**). Adenylyl cyclase's activity is inhibited by extracellular glucose.

Imagine a situation in which glucose is abundant outside the cell (**Figure 17.11b, top**). In this state, adenylyl cyclase activity is low. Therefore, cAMP levels inside the cell are low. The CAP–cAMP complex is unable to form, so CAP does not have the conformation that allows it to bind to the CAP site and stimulate *lac* operon transcription.

Conversely, when the extracellular concentration of glucose is low, the intracellular concentration of cAMP increases (**Figure 17.11b, bottom**). The CAP–cAMP complex forms, binds to the CAP site, and allows RNA polymerase to initiate transcription efficiently.

Figure 17.12 summarizes how positive control and negative control combine to regulate the *lac* operon. For even more review, go to the study area at *www.masteringbiology.com.*

(MB) **Web Activity** The *lac* Operon

The general message of this chapter is that interactions among regulatory elements produce finely tuned control over gene expression. Because positive and negative control elements are superimposed, *E. coli* fully activates the genes for lactose metabolism only when lactose is available and when glucose is scarce or absent. With these controls over gene expression, the bacteria are able to compete, grow, and reproduce efficiently.

CHECK YOUR UNDERSTANDING

If you understand that . . .

- Positive control occurs when something must be added for transcription to occur.
- CAP exerts positive control over the *lac* operon by binding to the CAP site and increasing the transcription rate.
- CAP can bind only when it is complexed with cAMP—a signaling molecule whose presence indicates that glucose levels are low.

✓ You should be able to . . .

1. Diagram what happens at the operon when cAMP levels are low and when cAMP levels are high.
2. Explain how positive control and negative control work together to control the *lac* operon in the presence or absence of glucose and lactose.

Answers are available in Appendix B.

(a) The enzyme adenylyl cyclase catalyzes production of cAMP from ATP.

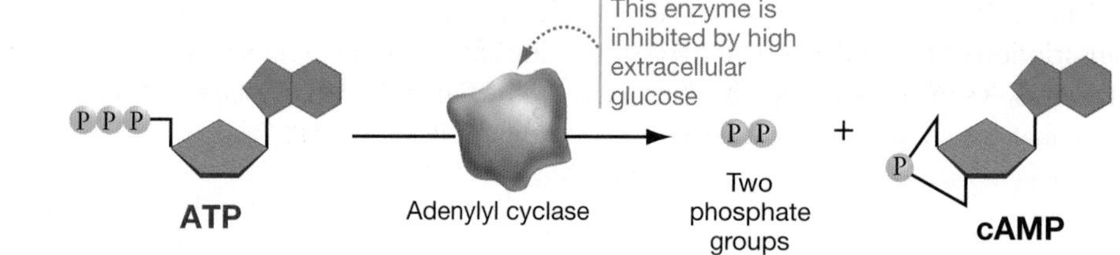

(b) The amount of cAMP and the rate of transcription of the *lac* operon are inversely related to the concentration of glucose.

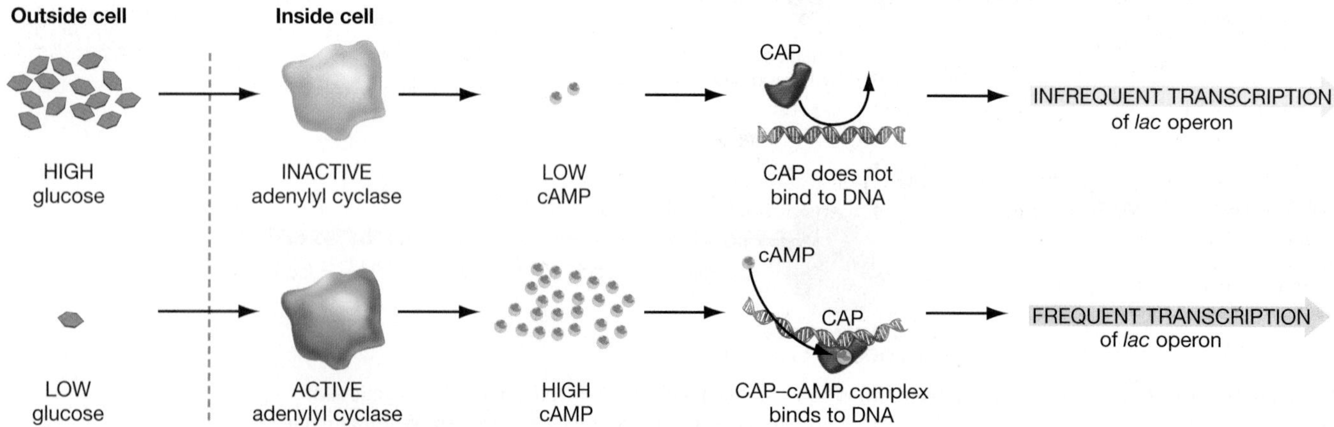

FIGURE 17.11 Cyclic AMP (cAMP) Is Synthesized When Glucose Levels Are Low.

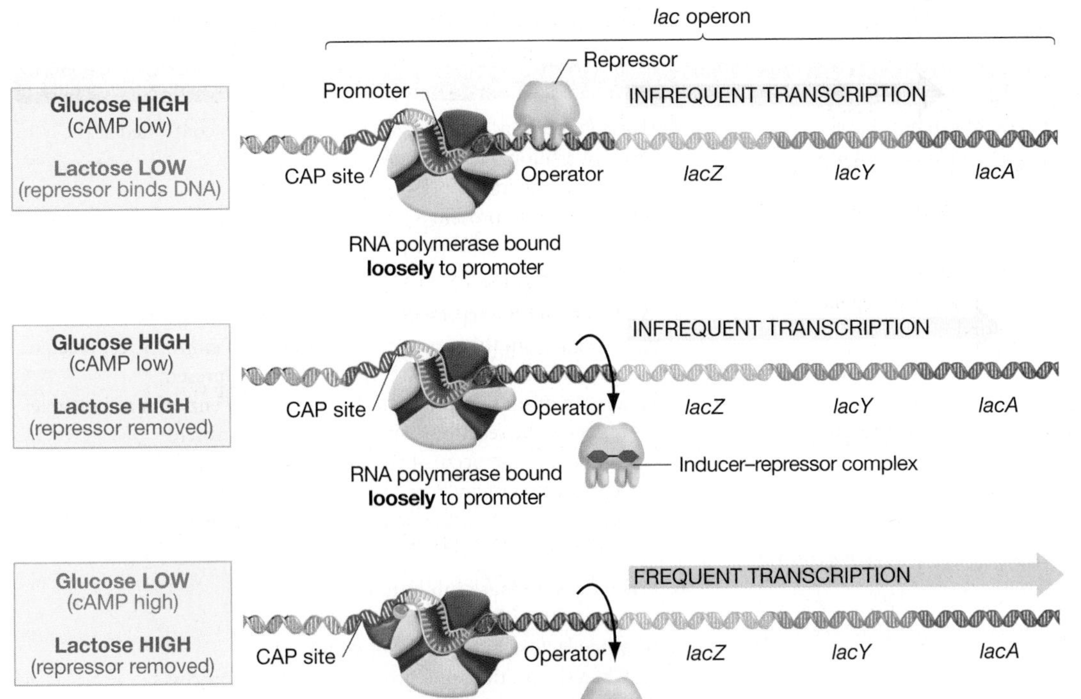

FIGURE 17.12 An Overview of *lac* Operon Regulation.

✔**EXERCISE** Create a flowchart called "Release of negative control" that contains the following terms: Operator, Lactose, Transcription of *lac* operon, RNA polymerase at promoter, and Repressor. Put the terms in order and connect them with labels indicating how the elements relate to each other. Do a similar flowchart called "Elements of positive control" with the terms cAMP, RNA polymerase at promoter, CAP–cAMP, Adenylyl cyclase, CAP, CAP site, Glucose, and Transcription of *lac* operon.

CHAPTER 17 REVIEW

For media, go to the study area at www.masteringbiology.com

Summary of Key Concepts

In bacteria, gene expression can be controlled at three levels: transcription, translation, or post-translation (protein activation).

- Among the levels of gene regulation, there is a trade-off between the speed of response to changed conditions and the efficient use of resources.

✔You should be able to describe one element of the *lac* operon that is under transcriptional control and one element that is under post-translational control.

Changes in gene expression allow bacterial cells to respond to environmental changes.

- Transcription may be constitutive or inducible.
- Constitutive expression occurs in genes whose products are required at all times, such as genes that encode glycolytic enzymes.
- Expression of most genes is induced by environmental signals—meaning that gene products are produced or activated only when they are needed.

✔You should be able to predict how gene expression changes in your gut-dwelling bacteria after you eat a dessert versus drink a glass of milk.

Transcriptional control can be negative or positive. Negative control occurs when a regulatory protein prevents transcription. Positive control occurs when a regulatory protein increases the transcription rate.

- The *lac* operon is under both negative and positive control.
- Negative control occurs because a repressor protein binds to a DNA operator sequence near the promoter of the protein-encoding genes.
- When lactose is present, it binds to the repressor and causes it to fall off the operator, allowing transcription to occur.
- Positive control occurs because low glucose levels induce production of cAMP. After cAMP binds to CAP, the complex binds to the CAP site and triggers rapid transcription.

✔You should be able to extend this chapter's use of a car analogy for negative and positive control by identifying (1) how the operator, repressor, and inducer relate to a parking brake; and (2) how CAP, cAMP, and the CAP binding site relate to a gas pedal.

MB **Web Activity** The *lac* Operon

Questions

TEST YOUR KNOWLEDGE

Answers are available in Appendix B

1. After completing a genetic screen, what do researchers have?
   a. a master plate and a replica plate—each of which contains identical copies of bacterial colonies
   b. individuals with mutations that disable genes required for a particular process or metabolic pathway
   c. a large collection of mutagenized cells
   d. mutants with defects in genes for enzymes—*not* in regulatory sites in DNA (e.g., promoters or operators)

2. Why are the genes involved in lactose metabolism considered to be an operon?
   a. They occupy adjacent locations on the *E. coli* chromosome.
   b. They have a similar function.
   c. They are all required for normal cell function.
   d. They are under the control of the same promoter.

3. In the *lac* operon, which regulatory molecule exerts negative control by inhibiting transcription?
   a. lactose (the inducer)
   b. the repressor protein
   c. the catabolite activator protein (CAP)
   d. cAMP

4. In the *lac* operon, which regulatory molecule is controlled at the post-translational level, by cAMP?
   a. lactose (the inducer)
   b. the repressor protein
   c. the catabolite activator protein (CAP)
   d. cAMP

5. What is catabolite repression?
   a. a mechanism that turns off the synthesis of enzymes responsible for catabolic reactions when the product is present
   b. a mechanism that turns off the synthesis of enzymes responsible for catabolic reactions when the product is absent
   c. repression that occurs because of allosteric changes in a regulatory protein
   d. repression that occurs because of allosteric changes in a DNA sequence

6. When does feedback inhibition occur?
   a. when allosteric regulation occurs
   b. when lactose binds to the operator
   c. when the product of a process inhibits the process
   d. when the product of a process triggers the process

TEST YOUR UNDERSTANDING

Answers are available in Appendix B

1. *E. coli* expresses genes for glycolytic enzymes constitutively. Why?
2. Explain the difference between positive and negative control over transcription.
3. Why is it advantageous for the *lac* operon in *E. coli* to be under both positive control and negative control? What would happen if only negative control occurred? What would happen if only positive control occurred?
4. In *E. coli*, rising levels of cAMP can be considered a starvation signal. Explain.
5. CAP is also known as the cAMP-receptor protein. Why?
6. The galactose released when β-galactosidase cleaves lactose enters the glycolytic pathway in *E. coli* after a series of enzyme-catalyzed, ATP-requiring reactions has converted the galactose to glucose-6-phosphate. Why, then, is glucose, not lactose, the preferred sugar in *E. coli*?

APPLYING CONCEPTS TO NEW SITUATIONS

Answers are available in Appendix B

1. You are interested in using bacteria to metabolize wastes at an old chemical plant and convert them into harmless compounds. You find bacteria that are able to tolerate high levels of the toxic compounds toluene and benzene, and you suspect that it is due to the ability of the bacteria to break these compounds into less-toxic products. If that is true, these toluene- and benzene-resistant strains will be valuable for cleaning up toxic sites. How could you find out whether these bacteria are metabolizing toluene as a source of carbon compounds?
2. Assuming that the bacteria you examined in Level 3, Problem 1 do have an enzymatic pathway to break down toluene, would you predict that the genes involved are constitutively expressed, under positive control, or under negative control? Why?
3. The *lacI* gene mutants produce β-galactosidase constitutively because no repressors are present to bind to the operator. Other repressor mutants have been isolated that are called $LacI^S$ mutants. These repressor proteins continue to bind to the operator, even in the presence of the inducer. How would this mutation affect the function of the *lac* operon? Specifically, how well would $LacI^S$ mutants do in an environment that has lactose as its sole sugar?
4. X-gal is a colorless, lactose-like molecule that can be split into two fragments by β-galactosidase. One of these product molecules is blue. The following photograph shows *E. coli* colonies growing in a medium that contains X-gal.

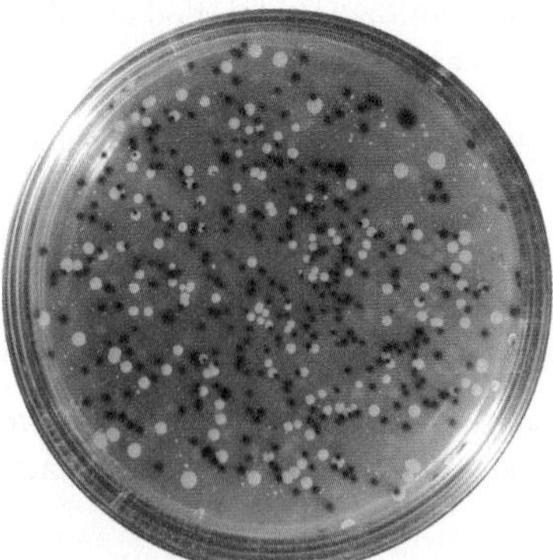

Find three colonies whose cells have functioning copies of β-galactosidase. Find three colonies whose cells have mutations in the *lacZ* locus or in one of the genes involved in regulation of *lacZ*. Suppose you could analyze the sequence of the β-galactosidase gene from each of the mutant colonies. How would these data help you distinguish which cells are structural mutants and which are regulatory mutants?

This chapter is part of The Big Picture. See how on pages 336–337.

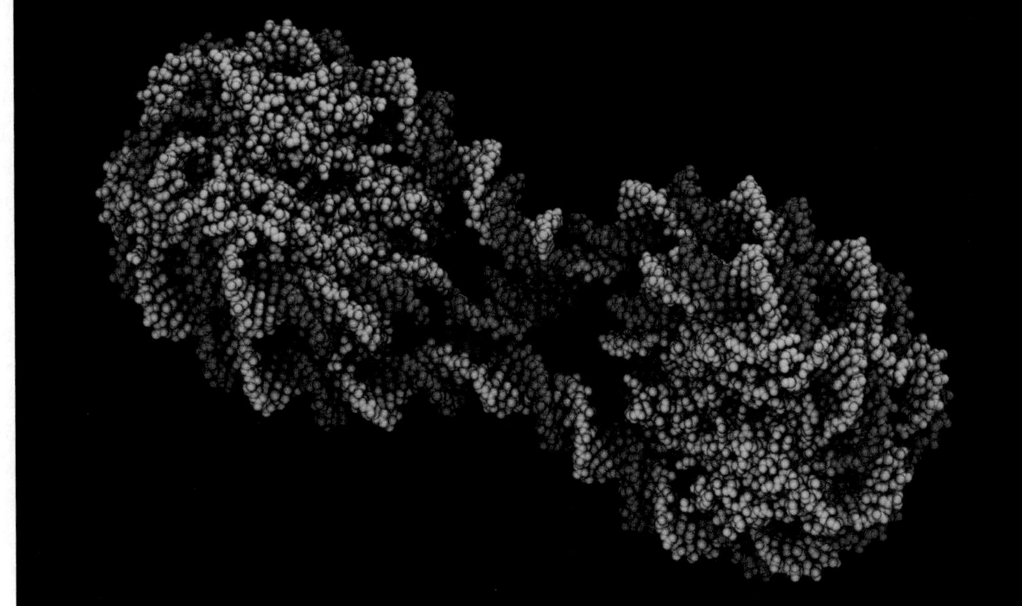

A model of eukaryotic DNA in the condensed state. The DNA (shown in red and pink) is wrapped around proteins (in green). The DNA has to be uncoiled before transcription can take place.

Control of Gene Expression in Eukaryotes 18

KEY CONCEPTS

- Changes in gene expression allow eukaryotic cells to respond to changes in the environment and cause distinct cell types to develop.
- Eukaryotic DNA is packaged with proteins into structures that must be opened before transcription can occur.
- In eukaryotes, transcription is triggered by regulatory proteins that bind to the promoter and to sequences close to and far from the promoter.
- Once transcription is complete, gene expression is controlled by (1) alternative splicing, which allows a single gene to code for several different products; (2) molecules that regulate the life span of mRNAs; and (3) activation or inactivation of protein products.
- Cancer can develop when mutations disable genes that regulate cell-cycle control genes.

Bacteria regulate gene expression to respond to changes in their environment. As Chapter 17 indicated, *Escherichia coli* thrive best if the genes that are required to import and cleave lactose are expressed only when the cells are relying on lactose as a source of energy—when glucose is absent and lactose is present.

Unicellular eukaryotes face similar challenges. Consider the yeast *Saccharomyces cerevisiae*, which is used extensively in the production of beer, wine, and bread. In nature this species lives on the skins of grapes and other fruits, where temperature and humidity can change dramatically. In addition, the sugars that the cells use as food vary in type and concentration as the fruit ripens, falls, and rots. For yeast cells to grow and reproduce efficiently, gene expression has to be modified in response to these changes.

The cells that make up multicellular eukaryotes face additional challenges. Consider your body, which contains trillions of cells, each with a specialized structure and function. You have heart muscle cells, lung cells, nerve cells, skin cells, and so on. Even though these cells look different, they contain the same genes. Your bone cells and blood cells aren't different because of a difference in their genes but because they *express* different genes. Your bone cells have blood cell genes—they just don't transcribe them.

Why? The answer is that your cells respond to their environment, just as bacteria and unicellular eukaryotes do. But there's a key difference. The cells in a multicellular eukaryote express different genes in response to changes in the internal environment—specifically, to signals from other cells. As a human being or an oak tree develops, cells that are located in different parts of the organism are exposed to different cell-cell signals. As a result, they express different genes. **Differential gene expression** is responsible for creating different cell types, arranging them into tissues, and coordinating their activity to form the multicellular society we call an individual.

How does all of this regulation and differentiation happen? Unit 4 introduces the signals that trigger the formation of muscle, bone, leaf, and flower cells. In contrast, this

✓ When you see this checkmark, stop and test yourself. Answers are available in Appendix B.

chapter focuses on what happens after a eukaryotic cell receives such a signal. Let's start with an overview of how gene expression can be controlled, and close with a look at how defects in the process can help trigger cancer.

18.1 Mechanisms of Gene Regulation in Eukaryotes—An Overview

Like bacteria, eukaryotes can control gene expression at the levels of transcription, translation, and post-translation (in the last, by protein activation or inactivation). But as **Figure 18.1** shows, three additional levels of control occur in eukaryotes as genetic information flows from DNA to proteins.

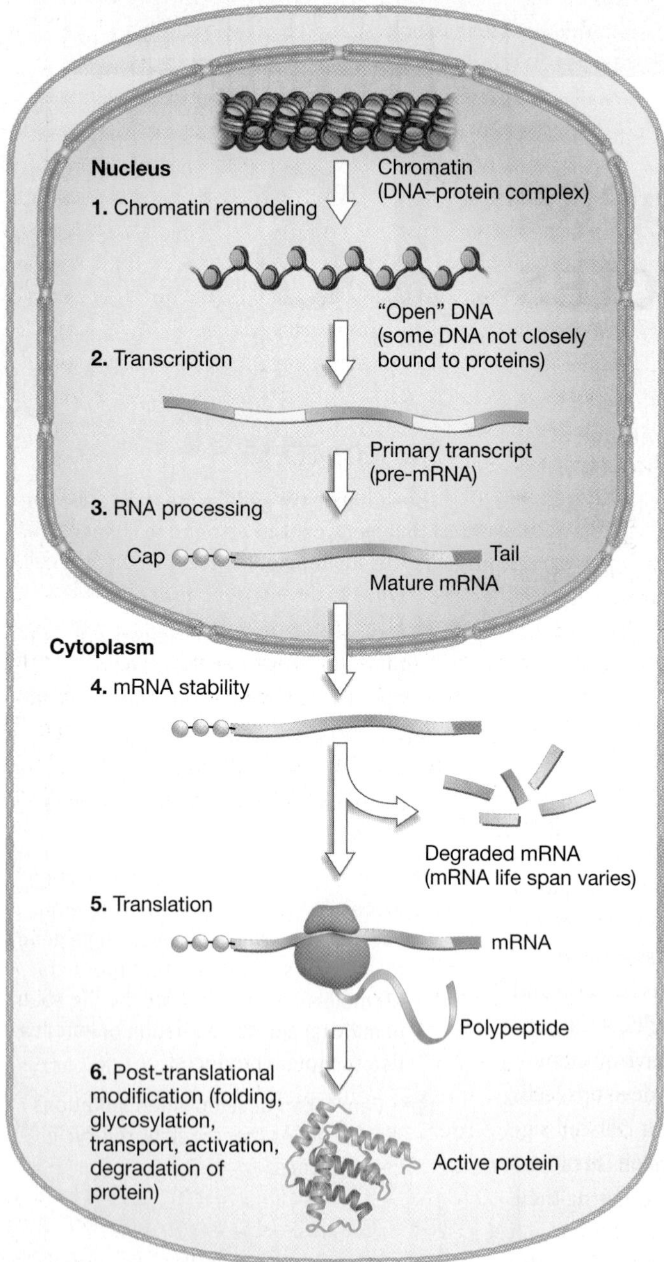

FIGURE 18.1 In Eukaryotes, Gene Expression Can Be Controlled at Many Different Levels.

The first additional level of control involves the DNA–protein complex at the top of the figure. In eukaryotes, DNA is wrapped around proteins to create a DNA–protein complex called **chromatin**. Eukaryotic genes have promoters, just as bacterial genes do, but before transcription can begin in eukaryotes, the stretch of DNA containing the promoter must be released from tight interactions with proteins, so that RNA polymerase can make contact with the promoter. To capture this idea, biologists say that **chromatin remodeling** must occur prior to transcription.

The second level of regulation that is unique to eukaryotes is **RNA processing**. These are the steps required to produce a mature, processed mRNA from a primary RNA transcript. Recall from Chapter 16 that introns have to be spliced out of primary transcripts. In many cases, carefully orchestrated alternative splicing patterns occur—meaning that different combinations of exons are included in the mRNA. If a cell switches from one splicing pattern to another, a different gene product results.

Third, mRNA life span is regulated in eukaryotes. mRNAs that are active in the cell for a long time tend to be translated more than mRNAs that have a short life span.

Each of the six potential control points shown in Figure 18.1 is employed at certain times in a typical eukaryotic cell. This chapter explores all six—chromatin remodeling, transcription, RNA processing, mRNA stability, translation, and post-translational modification of proteins.

To appreciate the breadth and complexity of gene regulation in eukaryotes, let's follow the series of events that occur as an embryonic cell responds to a developmental signal. Suppose a molecule arrives that specifies the production of a muscle-specific protein. What happens next?

18.2 Chromatin Remodeling

For the arrival of a signaling molecule from outside the cell to result in the transcription of a particular gene, the chromatin around the target gene must be drastically remodeled. To appreciate why, consider that a typical cell in your body contains about 6 billion base pairs of DNA. Lined up end to end, these nucleotide pairs would form a double helix about 2 m (6.5 feet) long. But the nucleus that holds this DNA is only about 5 μm in diameter—less than the thickness of this page.

In eukaryotes, DNA is packed inside the nucleus so tightly that RNA polymerase can't access it. Part of this packing is done by supercoiling—meaning the DNA double helix is twisted on itself many times, just as in bacterial chromosomes (see Chapter 7). But supercoiling is just part of the packing system found in eukaryotes.

What Is Chromatin's Basic Structure?

The first data on the chemical composition of eukaryotic DNA were published in the early 1900s, when researchers established that eukaryotic DNA is intimately associated with proteins. Later work documented that the most abundant DNA-associated proteins belong to a group called the **histones**. Chromatin consists of DNA complexed with histones and other proteins.

In the 1970s electron micrographs like the one in **Figure 18.2a** revealed that chromatin has a regular structure. In some preparations for electron microscopy, chromatin actually looked like beads on a string. The "beads" came to be called **nucleosomes**.

More details emerged in 1984 when researchers determined the three-dimensional structure of eukaryotic DNA by using X-ray crystallography (see **BioSkills 10** in Appendix A). The X-ray crystallographic data indicated that each nucleosome consists of DNA wrapped almost twice around a core of eight histone proteins. As **Figure 18.2b** indicates, a histone called H1 "seals" DNA to each set of eight nucleosomal histones. Between each pair of nucleosomes there is a "linker" stretch of DNA.

(a) Nucleosomes in chromatin

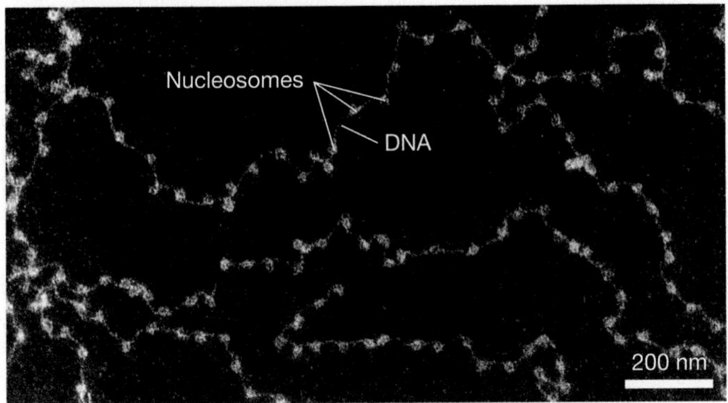

(b) Nucleosome structure

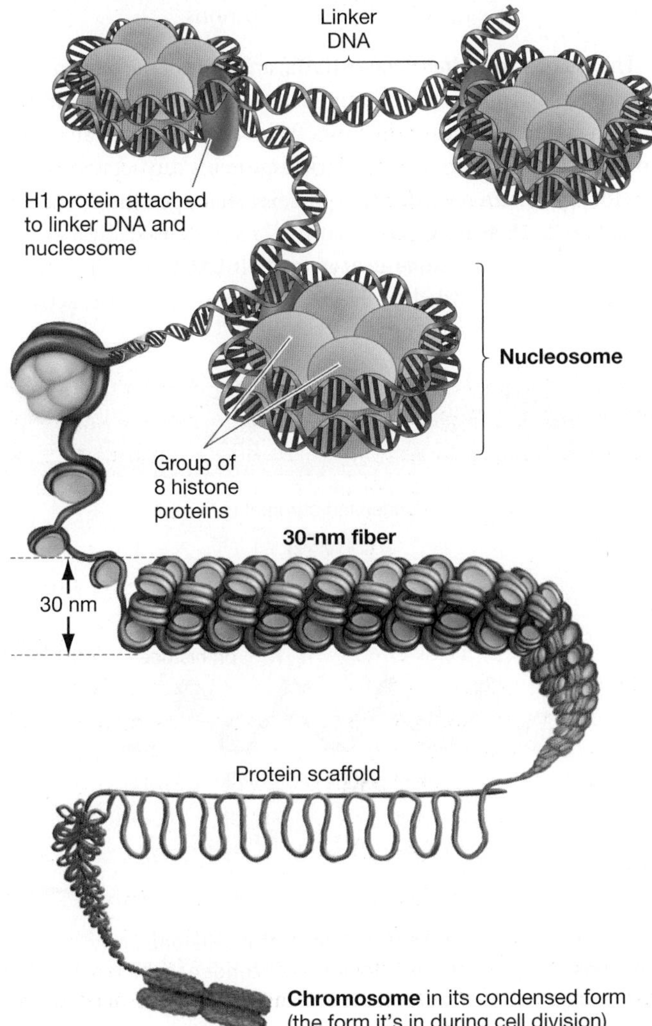

FIGURE 18.2 Chromatin Has Several Levels of Organization.

The intimate association between DNA and histones occurs in part because DNA is negatively charged and histones are positively charged. DNA has a negative charge because of its phosphate groups; histones are positively charged because they contain many lysine or arginine residues or both (see Chapter 3).

More recent work has shown that there is another layer of complexity in eukaryotic DNA. H1 histones interact with each other and with histones in other nucleosomes to produce a tightly packed structure like that shown in Figure 18.2b. Based on its width, this structure is called the 30-nanometer fiber. (Recall that a nanometer is one-billionth of a meter and is abbreviated nm.)

Finally, the 30-nm fibers are attached, at intervals along their length, to proteins that form a scaffold or framework inside the nucleus. In this way, the entire chromosome is organized and held in place. When chromosomes condense prior to mitosis or meiosis, the scaffold proteins and 30-nm fibers are folded and packed into still larger, more-compact structures.

A eukaryotic chromosome, then, is made up of chromatin that has several layers of organization: The DNA is wrapped around histones to form nucleosomes; nucleosomes are packed into 30-nm fibers, 30-nm fibers are attached to scaffold proteins, and the entire assembly can be folded into the highly condensed structure observed during cell division.

Although recent studies have confirmed that bacterial DNA interacts with proteins that may be organized in nucleosome-like structures, nothing like the 30-nm fibers or higher-order arrangements have been observed in bacterial chromosomes.

The elaborate structure of eukaryotic chromatin does more than just package DNA so that it fits into the nucleus. Chromatin structure also has profound implications for the control of gene expression. To appreciate this point, consider the 30-nm fiber illustrated in Figure 18.2b. If this tightly packed stretch of DNA contains a promoter, how can RNA polymerase bind to it and initiate transcription?

Evidence That Chromatin Structure Is Altered in Active Genes

Once the nucleosome-based structure of chromatin was established, biologists hypothesized that the close physical interaction between DNA and histones must be altered for RNA polymerase to make contact with DNA. More specifically, biologists hypothesized that a gene could not be transcribed until the chromatin near its promoter was remodeled.

The central idea is that chromatin must be relaxed, or decondensed, for RNA polymerase to bind to the promoter. If so, then chromatin remodeling would represent the first step in the control of eukaryotic gene expression. Two types of studies have provided strong support for this hypothesis.

"CLOSED" DNA IS PROTECTED FROM DNASE DNase is an enzyme that cuts DNA at random locations. The enzyme cannot cut DNA efficiently if the molecule is tightly complexed with histones, however. As **Figure 18.3** shows, the enzyme works effectively only if DNA is in the "open" configuration.

Harold Weintraub and Mark Groudine used this observation to test the hypothesis that the DNA of actively transcribed genes is in an open configuration. They compared chromatin structure in two genes in blood cells of chickens: the β-globin and ovalbumin genes. β-Globin is a protein that is part of the hemoglobin found in red blood cells; ovalbumin is a major protein of egg white. In blood cells, the β-globin gene is transcribed at high levels, but the ovalbumin gene is not transcribed at all.

After treating blood cells with DNase and then analyzing the state of the β-globin and ovalbumin genes, the researchers found that DNase cut up the β-globin gene much more readily than the ovalbumin gene. They interpreted this finding as evidence that chromatin in blood cells was in an open configuration at the β-globin gene but closed at the ovalbumin gene. Analogous studies using DNase on different genes and in different cell types yielded similar results.

HISTONE MUTANTS The second type of evidence in support of the chromatin-remodeling hypothesis comes from studies of mutant brewer's yeast cells that do not produce the usual complement of histones. In these mutant cells, many yeast genes that are normally never transcribed are instead transcribed at high levels at all times.

To interpret this finding, biologists hypothesized that the lack of histone proteins prevented the assembly of normal chromatin. If the absence of normal histone–DNA interactions promotes transcription, then the presence of normal histone–DNA interactions must prevent it.

Taken together, the data suggest that in their normal, or default, state, eukaryotic genes are turned off. This is a new mechanism of negative control—different from the repressor proteins introduced in Chapter 17. When DNA is wrapped into a 30-nm fiber, the parking brake is on. If so, then gene expression depends on chromatin being opened up in the promoter region.

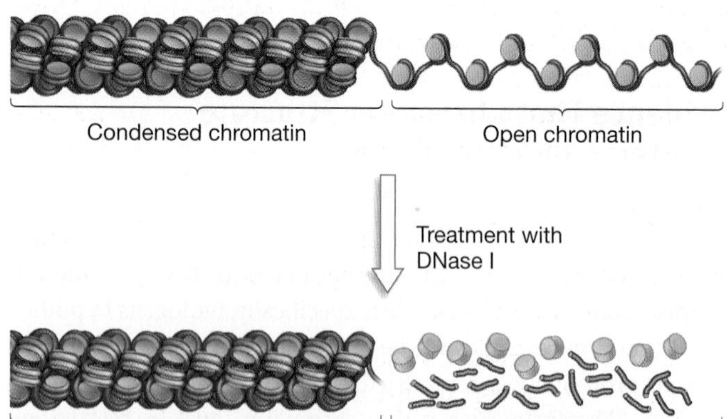

FIGURE 18.3 DNase Assay for Chromatin Structure. DNase is an enzyme that cuts DNA at random locations. It cannot cut intact condensed chromatin.

How Is Chromatin Altered?

Research on chromatin remodeling has been proceeding at a furious pace, and has recently succeeded in identifying some of the key players. Multi-protein machines called chromatin-remodeling complexes, which reshape chromatin through a series of ATP-dependent reactions, are particularly important. In essence, the remodeling complexes make contact with DNA on the surface of the nucleosome, then twist it in a way that allows the DNA to loop out and away from the protein core, where it can be transcribed.

Other major players in chromatin-remodeling work by adding small molecules to histone proteins. To date, eight distinct types of chemical modifications have been discovered. Two of the best-studied involve adding acetyl (CH_3COOH) or methyl (CH_3) groups. These processes are known as **acetylation** and **methylation**, respectively.

- Acetylation of histones is usually associated with positive control—meaning, activation of genes through binding of a regulatory protein.
- Methylation can be correlated with either activation or inactivation, depending on which histones are altered and where the methyl groups occur. Methylation can set the parking brake or release it, depending on conditions.

The **histone acetyl transferases (HATs)** are some of the best-studied enzymes that modify chromatin. HATs acetylate the positively charged lysine residues in histones. When a HAT adds an acetyl group to selected histones, the marked proteins can act as binding sites for chromatin remodeling complexes that open the DNA.

As **Figure 18.4** shows, chromatin is "recondensed" by a group of enzymes called **histone deacetylases (HDACs)**, which remove the acetyl groups added by HATs. Histone deacetylase activity

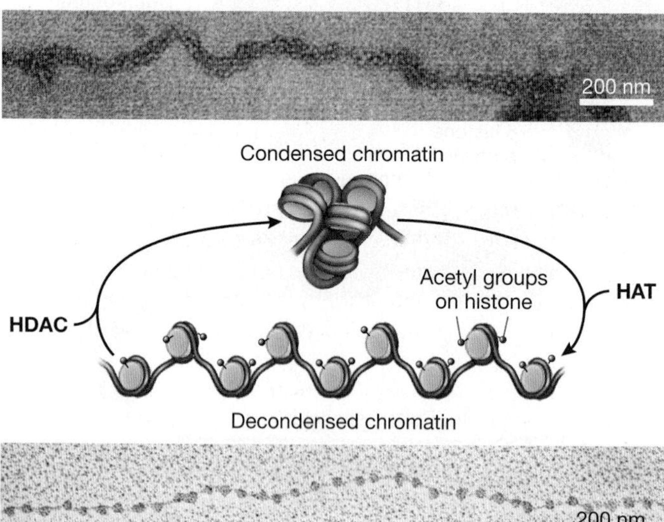

FIGURE 18.4 Acetyl Groups Open Chromatin. Histone acetyl transferases (HATs) cause chromatin to decondense; histone deacetylases (HDACs) cause it to condense.

✔**QUESTION** Are HATs and HDACs elements in positive control or negative control? Explain your reasoning.

reverses the effects of acetylation, returning chromatin to its default condensed state. If HATs are an on switch for transcription, HDACs are the off switch.

The take-home message from work on chromatin remodeling is simple: The state of the histone proteins that are complexed with DNA is a critical determinant of whether transcription can occur.

Chromatin Modifications Can Be Inherited

Within a single individual, the pattern of chromatin modifications varies from one cell type to another. For example, suppose you analyzed the same gene in a cell that was destined to give rise to muscle and in a cell that is a precursor to part of the brain. This and other genes would likely have a completely different pattern of acetylation and methylation in the two cell types.

To capture this point, biologists propose that a **histone code** exists. The histone code hypothesis contends that precise patterns of chemical modifications of histones contain information, analogous to the way the genetic code stores information. But instead of dictating the amino acid sequence in a gene product, the histone code influences whether or not a particular gene is expressed.

The histone code hypothesis is attracting interest among biologists because of the now well-established evidence that some or most of the chemical modifications that distinguish a muscle- from a brain-associated cell are passed on to daughter cells during mitosis. This is a key observation, because it means that daughter cells inherit patterns of gene expression from their parent cells.

Histone modifications are an example of **epigenetic inheritance**, the collective term for patterns of inheritance that are not due to differences in DNA sequences. If a cell received a "become a muscle cell" signal early in development, it would not only modify its chromatin in distinctive ways but pass those modifications on to its descendants. Muscle cells are different from brain cells in part because they inherited differently modified histones—not different types of genes.

Now the question is, what happens once a section of DNA is opened up by chromatin remodeling and exposed to RNA polymerase?

CHECK YOUR UNDERSTANDING

If you understand that . . .

- Eukaryotic DNA is wrapped tightly around histones, forming nucleosomes, which are then coiled into dense structures called 30-nm fibers.
- Before transcription can begin, the DNA–protein complex must be relaxed so that RNA polymerase can contact the promoter.
- Specific chemical modifications to histones play a key role in determining whether or not chromatin is opened and a gene is expressed.
- In many if not most cases, the histone modifications in a cell are passed on to its daughter cells.

✓ You should be able to . . .

1. Compare and contrast the structure of DNA in bacteria versus that in eukaryotes.
2. Explain why certain patterns of acetylation or methylation could influence whether a cell became a muscle cell or a brain cell.

Answers are available in Appendix B.

18.3 Initiating Transcription: Regulatory Sequences and Regulatory Proteins

Chapter 16 introduced the **promoter**—the site in DNA where RNA polymerase binds to initiate transcription. In position and function, eukaryotic promoters are similar to bacterial promoters.

Most eukaryotic promoters are located close to the point where RNA polymerase begins transcription, and all contain highly conserved sequences analogous to the −35 box and −10 box in bacterial promoters. To date, three such conserved sequences have been observed; each eukaryotic promoter appears to have two of the three. One of the most common of these conserved sequences is known as the **TATA box**.

Once a promoter has been exposed by chromatin remodeling, the first step in transcription is an interaction with the **TATA-binding protein (TBP)**. But binding by TBP does not guarantee that a gene will be expressed. In eukaryotes, a wide array of other DNA sequences and proteins are involved.

Some Regulatory Sequences Are Near the Promoter

Regulatory sequences are sections of DNA that, like the prokaryotic CAP-binding site and operators introduced in Chapter 17, are involved in controlling the activity of genes. The first regulatory sequences in eukaryotic DNA were discovered in the late 1970s, when Yasuji Oshima and co-workers set out to understand how yeast cells control the metabolism of the sugar galactose.

When galactose is absent, *S. cerevisiae* cells produce only tiny quantities of the enzymes required to metabolize it. But when galactose is present, transcription of the genes encoding these enzymes increases by a factor of 1000.

The team's first major result was the discovery of mutant cells that failed to produce any of the five enzymes required for galactose metabolism, even if galactose was present. To interpret this observation, they hypothesized that

1. The five genes are regulated together, even though they are not on the same chromosome;
2. Normal cells have a CAP-like regulatory protein that exerts positive control over the five genes;
3. The mutant cells have a loss-of-function mutation that completely disables the regulatory protein.

Other researchers were able to isolate the regulatory protein and confirm that it binds to a short stretch of DNA located just

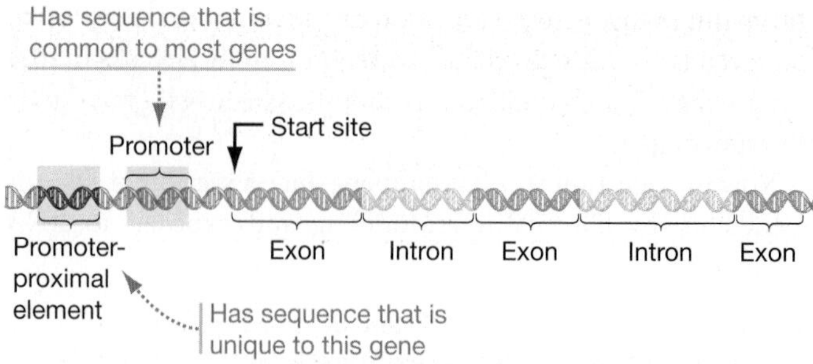

FIGURE 18.5 Promoter-Proximal Elements Regulate the Expression of Some Eukaryotic Genes. Note that exons and introns are not drawn to scale throughout this chapter. They are typically very large compared with promoters and promoter-proximal (regulatory) elements.

upstream from the promoter for the five genes required for galactose use. The location and structure of this regulatory sequence are comparable to those of the CAP-binding site in the *lac* operon of *E. coli.*

Similar regulatory sequences have now been found in a wide array of eukaryotic genes and species. Because such sequences are located close to the promoter and bind regulatory proteins, they are termed **promoter-proximal elements** (**Figure 18.5**).

Unlike the promoter itself, promoter-proximal elements have sequences that are unique to specific genes. In this way, they furnish a mechanism for eukaryotic cells to express certain genes but not others.

The discovery of promoter-proximal elements and a mechanism of positive control suggested a satisfying parallel between gene regulation in bacteria and in eukaryotes. This picture changed, however, when researchers discovered a new class of eukaryotic DNA regulatory sequences—sequences unlike anything in bacteria.

Some Regulatory Sequences Are Far from the Promoter

Susumu Tonegawa and colleagues made a discovery that may rank as the most startling in the history of research on gene expression. Tonegawa's group was exploring how human immune system cells regulate the genes involved in the production of antibodies. **Antibodies** are proteins that bind to specific sites on other molecules (see **BioSkills 9** in Appendix A). Produced by your immune system, antibodies bind to viruses and bacteria and mark them for destruction by other cells.

The antibody gene in question is broken into many introns and exons. Introns are DNA sequences spliced out of the primary mRNA transcript; exons are regions of eukaryotic genes that are included in the mature RNA once splicing is complete. The biologists used techniques that will be introduced in Chapter 19 to place copies of an intron in new locations and found that, when they placed the intron close to a gene, the gene's transcription rate increased. Based on this observation, Tonegawa and co-workers hypothesized that the intron contained some sort of regulatory sequence.

To test this hypothesis, the biologists performed what is now considered a classic experiment (**Figure 18.6**). The protocol was simple in concept:

Step 1 Start with a human antibody-producing gene that includes an intron flanked by two exons.

Step 2 Use enzymes to remove several different specific pieces of the intron from different samples of the gene.

Step 3 Use enzymes to ligate (link) the remaining sections of the gene back together. Each of the modified genes produced in this way is missing a different section of the intron.

Step 4 Insert the various versions of the gene into mouse cells, which do not normally contain the antibody-producing gene. Some mouse cells receive normal copies of the gene; others receive copies of the gene missing different portions of the intron; still others receive no gene.

To test their hypothesis, they analyzed the mRNAs produced by each of the experimental cells and compared them with mRNA from a normal human cell and from mouse cells that received the normal gene. If a regulatory sequence is located inside the intron, then some of the genes with modified introns should lack that sequence and fail to transcribe the antibody gene. As the "Results" section of Figure 18.6 shows, some of the modified copies of the gene were not transcribed at all. Based on these results, Tonegawa and co-workers proposed that the intron contains a regulatory sequence that is required for transcription to occur.

This result was remarkable for two reasons: (**1**) The regulatory sequence was thousands of bases away from the promoter, and (**2**) it was downstream of the promoter instead of upstream. Regulatory elements that are far from the promoter are termed **enhancers**.

Follow-up work has shown that enhancers occur in all eukaryotes and that they have several key characteristics:

- Enhancers can be more than 100,000 bases away from the promoter. They can be located in introns or in untranscribed sequences on either the 5′ or 3′ side of the gene (see **Figure 18.7**). Researchers have yet to find enhancers located in exons.
- Like promoter-proximal elements, many types of enhancers exist.
- Enhancers can work even if their normal $5' \rightarrow 3'$ orientation is flipped.
- Most genes have more than one enhancer.
- Enhancers can work even if they are moved to a new location in the vicinity of the gene, on the same chromosome.

Enhancers are regulatory sequences unique to eukaryotes. When regulatory proteins bind to enhancers, transcription begins. Thus, enhancers are a gas pedal—an element in positive control.

In addition, eukaryotic genomes contain regulatory sequences that are similar in structure to enhancers but opposite in function. These sequences are **silencers**. When regulatory proteins

EXPERIMENT

QUESTION: Can a regulatory sequence in DNA be located far from the promoter?

HYPOTHESIS: A regulatory sequence exists far from the promoter—in fact, in the intron—of an antibody-producing gene.

NULL HYPOTHESIS: Regulatory sequences are located close to the promoter—not in introns or in other distant sequences.

EXPERIMENTAL SETUP:

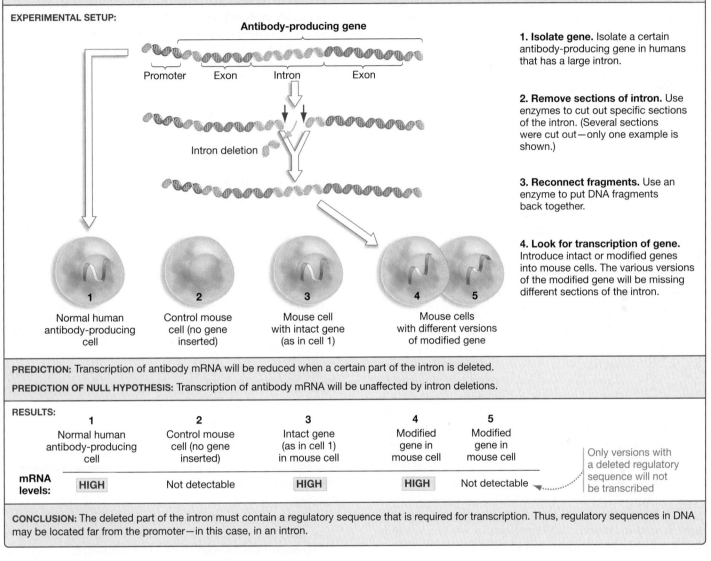

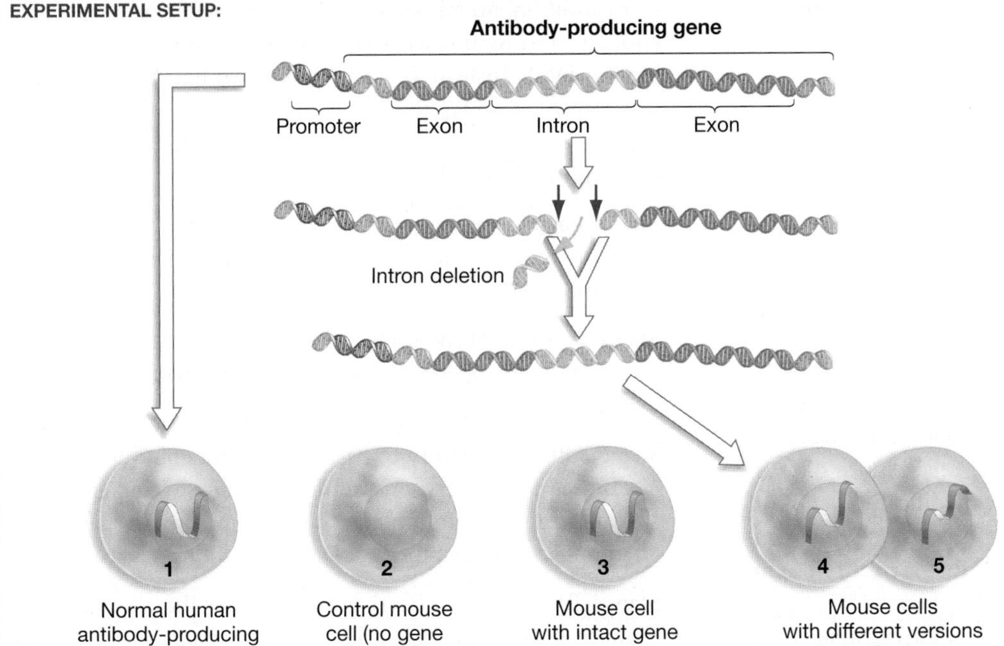

1. Isolate gene. Isolate a certain antibody-producing gene in humans that has a large intron.

2. Remove sections of intron. Use enzymes to cut out specific sections of the intron. (Several sections were cut out—only one example is shown.)

3. Reconnect fragments. Use an enzyme to put DNA fragments back together.

4. Look for transcription of gene. Introduce intact or modified genes into mouse cells. The various versions of the modified gene will be missing different sections of the intron.

PREDICTION: Transcription of antibody mRNA will be reduced when a certain part of the intron is deleted.

PREDICTION OF NULL HYPOTHESIS: Transcription of antibody mRNA will be unaffected by intron deletions.

RESULTS:

	1 Normal human antibody-producing cell	2 Control mouse cell (no gene inserted)	3 Intact gene (as in cell 1) in mouse cell	4 Modified gene in mouse cell	5 Modified gene in mouse cell
mRNA levels:	HIGH	Not detectable	HIGH	HIGH	Not detectable

Only versions with a deleted regulatory sequence will not be transcribed

CONCLUSION: The deleted part of the intron must contain a regulatory sequence that is required for transcription. Thus, regulatory sequences in DNA may be located far from the promoter—in this case, in an intron.

FIGURE 18.6 Evidence That Enhancers Are Required for Transcription.

SOURCE: Hozumi, N., and S. Tonegawa. 1976. Evidence for somatic rearrangement of immunoglobulin genes coding for variable and constant regions. *Proceedings of the National Academy of Sciences, USA* 73: 3628–3632.

✔**QUESTION** Why did the researchers assess mRNA production in human cells that did *not* receive a modified or intact gene?

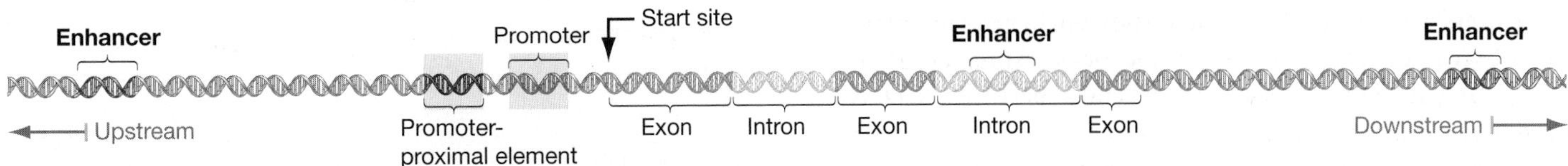

FIGURE 18.7 Enhancers Are Far from the Genes They Regulate. All of the colored sections of the DNA strand shown here (including the enhancers) are considered part of the same gene.

✔**EXERCISE** Compare and contrast the structure of this typical eukaryotic gene and the structure of a bacterial operon.

bind to silencers, transcription is shut down. Silencers are a brake—an element in negative control.

The discovery of enhancers and silencers expanded the catalog of regulatory sites known in organisms and inspired researchers to reconsider the nature of the gene. Biologists began defining the **gene** as the DNA that codes for a functional polypeptide or RNA molecule *and* the regulatory sequences required for expression.

The Role of Regulatory Proteins in Differential Gene Expression

Experiments that followed up on Tonegawa's work supported the hypothesis that enhancers are binding sites for proteins that regulate transcription. By analyzing mutant yeast, fruit flies, and roundworms that have defects in the expression of particular genes, biologists have identified a large number of regulatory proteins that bind to enhancers and silencers.

These results support one of the most general statements researchers are able to make about gene regulation in eukaryotes: In multicellular species, different types of cells express different genes because they have different histone modifications and contain different regulatory proteins. The regulatory proteins, in turn, are produced in response to signals that arrive from other cells early in embryonic development.

If a "become a muscle cell" signal arrives, for example, it triggers the production of regulatory proteins that are specific to muscle cells. Because the regulatory proteins bind to specific enhancers and silencers and promoter-proximal elements, they trigger the production of muscle-specific proteins. But if no "become a muscle cell" signal arrives, then no muscle-specific regulatory proteins are produced and no muscle-specific gene expression takes place.

Differential gene expression is a result of the production or activation of specific regulatory proteins. Eukaryotic genes are turned on when specific regulatory proteins bind to enhancers and promoter-proximal elements; the genes are turned off when regulatory proteins bind to silencers or when chromatin remains condensed. Distinctive regulatory proteins are what make a muscle cell a muscle cell and a bone cell a bone cell.

The Initiation Complex

Many questions remain about how transcription is initiated in eukaryotes. What is clear is that two broad classes of regulatory proteins interact with regulatory sequences at the start of transcription:

- **Regulatory transcription factors** are proteins that bind to enhancers, silencers, or promoter-proximal elements. These transcription factors are responsible for the expression of particular genes in particular cell types and at particular stages of development.
- **Basal transcription factors** interact with the promoter and are not restricted to particular cell types. Basal transcription factors must be present for transcription to occur, but they do not provide much in the way of regulation.

TBP, for example, is a basal transcription factor that is common to all genes. Other basal transcription factors are specific to promoters recognized by RNA polymerase I, II, or III. ✓If you understand this concept, you should be able to compare and contrast the regulatory and basal transcription factors found in muscle cells versus nerve cells.

In addition, proteins that make up the **mediator complex** have a role in starting transcription. The mediator complex does not bind to DNA. Instead, it creates a physical link between regulatory transcription factors and basal transcription factors.

Figure 18.8 summarizes a current model for how transcription is initiated in eukaryotes.

Step 1 Regulatory transcription factors bind to DNA and recruit chromatin-remodeling complexes and histone acetyl transferases (HATs).

Step 2 Once the chromatin-remodeling complexes and HATs are in place, they open a broad swath of chromatin that includes the promoter region.

Step 3 Other regulatory transcription factors bind to the newly exposed enhancers or promoter-proximal elements; basal transcription factors bind to the promoter. When mediator complexes connect the two, DNA has to loop.

Step 4 RNA polymerase II is recruited to the site, forming a multi-protein machine called the **basal transcription complex**. Transcription can begin.

✓If you understand this model, you should be able to explain why DNA forms loops near the promoter in order for transcription to begin.

CHECK YOUR UNDERSTANDING

If you understand that . . .

- Eukaryotic genes have regulatory sequences called promoter-proximal elements close to their promoters.
- Eukaryotic genes also have regulatory sequences called enhancers or silencers far from their promoters.
- Transcription initiation is a multistep process that begins when regulatory transcription factors bind to DNA and recruit proteins that open chromatin.
- Interactions between regulatory transcription factors and basal transcription factors result in the formation of the basal transcription complex and the arrival of RNA polymerase at the gene's start site.

✓ **You should be able to . . .**

1. Compare and contrast the nature of regulatory sequences and regulatory proteins in bacteria versus eukaryotes.
2. Explain why the presence of certain regulatory proteins could influence whether a cell became a muscle cell or a brain cell.

Answers are available in Appendix B.

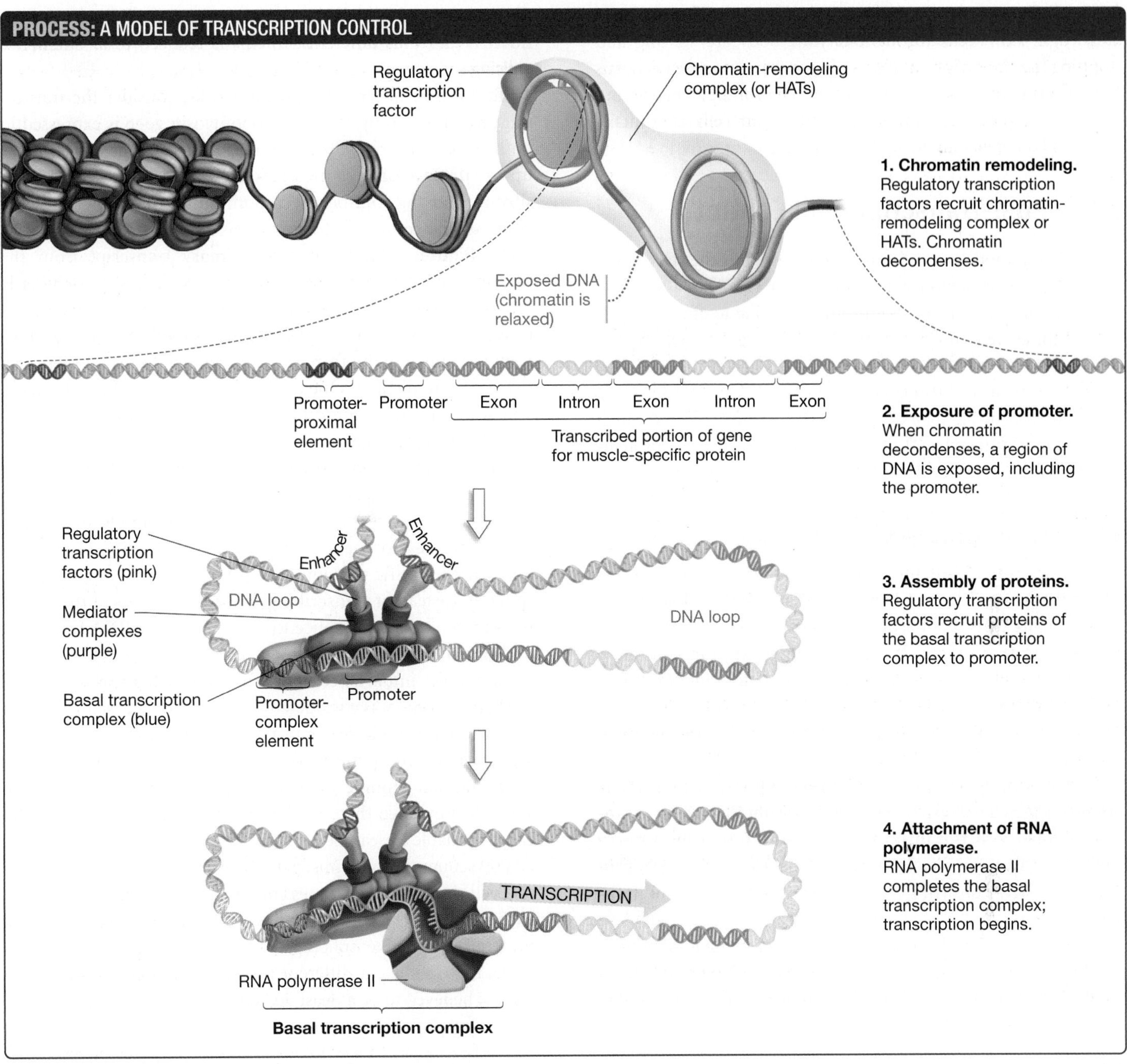

FIGURE 18.8 The Elements of Transcription Control in Eukaryotes. According to the currently accepted model, transcription is initiated through a series of steps that remodel chromatin and assemble the basal transcription complex that recruits RNA polymerase.

The assembly of the basal transcription complex depends on interactions among regulatory transcription factors that are bound to enhancers, silencers, and promoter-proximal elements. The result is a large, multimolecular machine that is positioned at the start site and able to start transcription. The completed basal transcription complex can contain as many as 60 proteins, including RNA polymerase.

Compared with what happens in bacteria, where just 3 to 5 proteins may interact at the promoter to initiate transcription, the process in eukaryotes is remarkably complicated. To review the many players involved and their interactions, go to the study area at *www.masteringbiology.com.*

(MB) **Web Activity** Transcription Initiation in Eukaryotes

Currently, biologists are working to understand exactly how regulatory and basal transcription factors interact to control formation of the basal transcription complex. This research is important, for transcription initiation lies at the heart of gene expression. Precise regulation of transcription is critical not only to the development of embryos but also to the daily life of eukaryotes. Right now, cells throughout your body are starting and stopping the transcription of specific genes in response to signals from nearby and distant cells. As the environment inside and outside your body continually changes, your cells continually change which genes are being transcribed.

18.4 Post-Transcriptional Control

Chromatin remodeling and transcription are just the start of the story of gene regulation. Once an mRNA is made, a series of events has to occur if the final product is going to affect the cell. Each of these events offers an opportunity to regulate gene expression, and each is used in some cells at least some of the time. The control mechanisms include (1) splicing mRNAs in various ways, (2) modifying the life span of mRNAs or altering the rate at which translation is initiated, and (3) activating or inactivating proteins after translation has occurred. Let's consider each in turn.

Alternative Splicing of mRNAs

Introns are spliced out in the nucleus as the primary RNA is transcribed. Recall from Chapter 16 that the RNA that results from splicing consists of sequences encoded by exons, it is protected by a cap on the 5′ end and a long poly(A) tail on the 3′ end, and the splicing is accomplished by the molecular machines called **spliceosomes**. What Chapter 16 did not mention, however, is that splicing provides an opportunity for the regulation of gene expression.

During splicing, changes in gene expression are possible because selected exons may be removed from the primary transcript along with the introns. As a result, the same primary RNA transcript can yield more than one kind of mature, processed mRNA, consisting of different combinations of transcribed exons.

This is important. If these mature mRNAs contain differences in their ribonucleotide sequence, then the polypeptides translated from them will likewise differ. Splicing the same primary RNA transcript in different ways to produce different mature mRNAs and thus different proteins is referred to as **alternative splicing**.

To see how alternative splicing works, consider the muscle-cell protein tropomyosin. The tropomyosin gene is expressed in both skeletal muscle cells and smooth muscle cells, which make up two distinct kinds of muscle tissue. Skeletal muscle is responsible for moving your bones; smooth muscle lines many parts of your gut and certain blood vessels.

As **Figure 18.9a** shows, the primary transcript from the tropomyosin gene contains 14 exons. In each type of muscle cell, a different subset of the 14 exons are spliced together to produce two different mRNAs (**Figure 18.9b**). As a result of alternative splicing, the tropomyosin proteins found in these two cell types are distinct. One of the many reasons skeletal muscle and smooth muscle are different is that they contain different types of tropomyosin.

Alternative splicing is controlled by proteins that bind to RNAs in the nucleus and interact with spliceosomes. When cells that are destined to become skeletal muscle or smooth muscle are developing, they receive signals leading to the production of specific proteins that are active in the regulation of splicing. Instead of transcribing different versions of the tropomyosin gene, the cells splice the same primary RNA transcript in different ways.

Before the importance of alternative splicing was widely appreciated, a gene was considered to be a nucleotide sequence that encodes one specific protein or RNA, along with its regulatory sequences. Based on this view, estimates for the number of genes in the human genome were typically in the range of 60,000 to 100,000. But once the complete human genome sequence became available, researchers realized that we may have as few as 20,000 sequences for primary mRNA transcripts.

Even though our genomes contain a relatively low number of such sequences, recent data indicate that over 90 percent of them undergo alternative splicing and produce multiple products. Thus, the number of different proteins that your cells can produce is believed to be at least 50,000.

(a) Tropomyosin gene

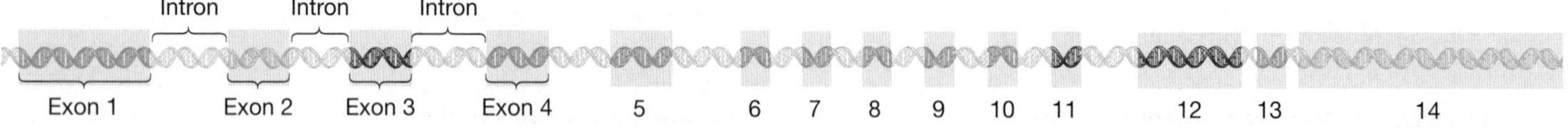

(b) Alternative splicing produces more than one mature mRNA.

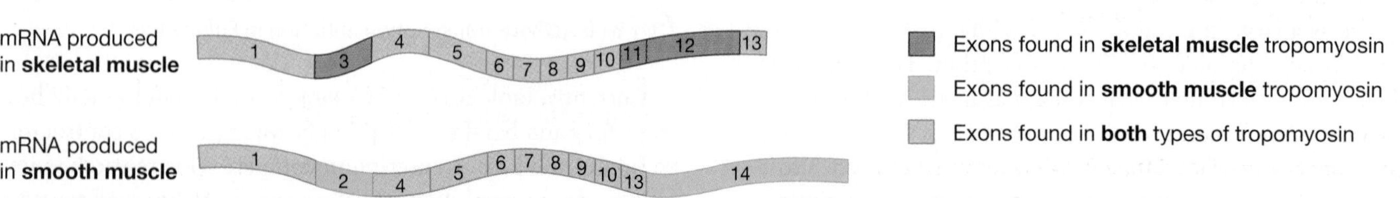

FIGURE 18.9 Alternative Splicing Produces More than One Mature mRNA from the Same Gene.

Thanks to results like these, the definition of the gene is changing once again: Genes now have to be thought of as the coding and regulatory sequences that direct the production of one or more related polypeptides or RNAs.

The current record holder for the number of distinct mRNA sequences derived from one gene is the *Dscam* gene in the fruit fly *Drosophila melanogaster*. The products of this gene help to guide growing nerve cells within the embryo. Because the primary transcript is spliced into about 38,000 distinct forms of mRNA, the *Dscam* gene can produce thousands of different products.

Alternative splicing ranks as a major mechanism in the control of gene expression in multicellular eukaryotes. ✔If you understand alternative splicing, you should be able to explain why it does not occur in bacteria and describe where it occurs in Figure 18.1.

mRNA Stability and RNA Interference

Once splicing is complete and processed mRNAs are exported to the cytoplasm, new regulatory mechanisms come into play. For example, it has long been known that the life span of an mRNA in the cell can vary. The mRNA for casein—the major protein in milk—is produced in the mammary gland tissue of female mammals. Normally, many of these mRNAs persist in the cell for just an hour, and little casein protein is produced. But when a female mouse is lactating, regulatory molecules help the mRNAs persist almost 30 times longer—leading to a huge increase in the production of casein. In this instance, mRNA stability is associated with changes in the length of the poly(A) tail.

In many cases, the life span of an mRNA is controlled by tiny, single-stranded RNA molecules that bind to complementary sequences in the mRNA. Once part of an mRNA becomes double stranded in this way, specific proteins degrade the mRNA or prevent it from being translated into a polypeptide. This phenomenon is known as **RNA interference**. How does it work?

Figure 18.10 walks through the sequence of events.

Step 1 RNA interference begins when RNA polymerase transcribes DNA sequences that code for an unusual product—a small RNA molecule that doubles back on itself to form a hairpin. Hairpin formation occurs because pairs of sequences within the RNA transcript are complementary.

Step 2 Some of the RNA is trimmed by enzymes in the nucleus; then the double-stranded segment that remains is exported to the cytoplasm.

Step 3 In the cytoplasm, the double-stranded RNA sequence is cut by another enzyme into molecules that are typically about 22 nucleotides long.

Step 4 One of the strands from this short RNA is taken up by a group of proteins called the RNA-induced silencing complex, or RISC. The RNA strand held by the RISC is a **microRNA (miRNA)**, or a small interfering RNA (siRNA)

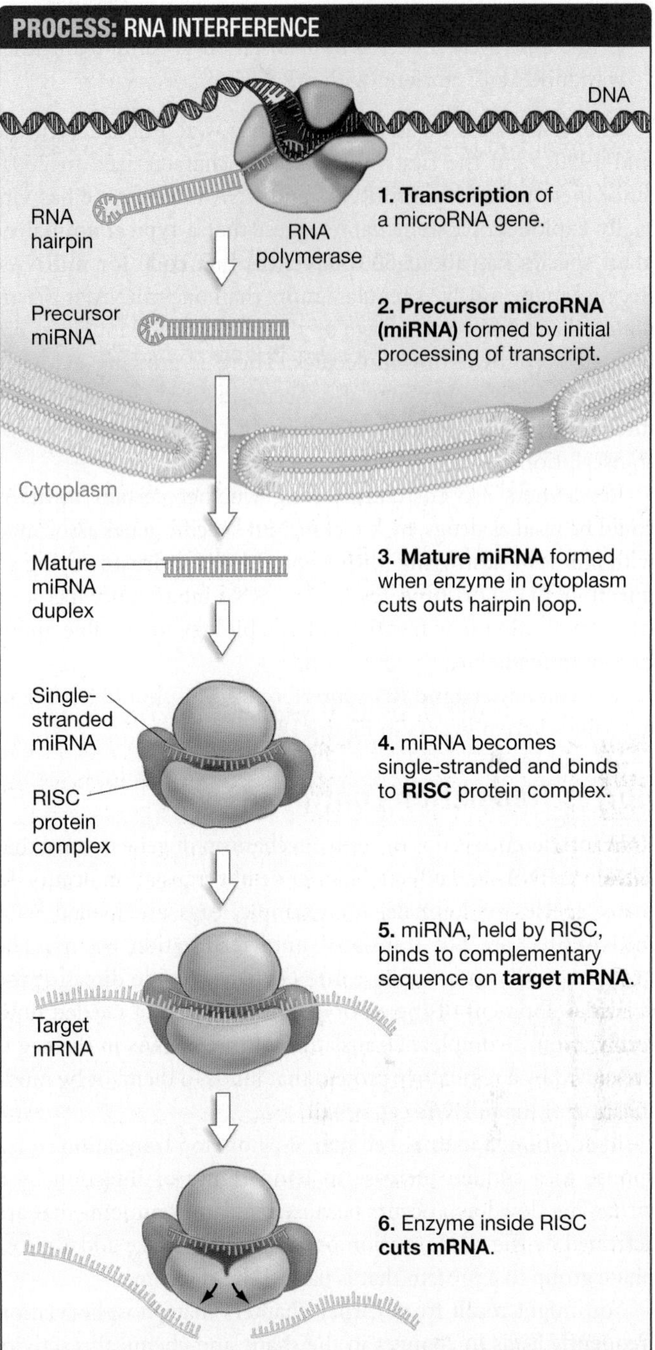

FIGURE 18.10 MicroRNAs Target Certain mRNAs for Destruction. In essence, miRNAs bind to mRNAs by complementary base pairing and target them for destruction by the RISC protein complex. The steps required to process and activate an miRNA are important because they allow the production of miRNAs to be carefully controlled.

Step 5 Once it is part of a RISC, the miRNA binds to its complementary sequences in a target mRNA.

Step 6 If the match between an miRNA and an mRNA is perfect, an enzyme in the RISC destroys the mRNA by cutting it in two. In effect, tight binding by an miRNA is a "kiss of death" for the mRNA. If the match isn't perfect, however, the mRNA is not destroyed. Instead, its translation is inhibited. Either way, miRNAs "interfere" with mRNAs.

The first papers on RNA interference were published in the mid-1990s, and the first miRNAs were characterized in 2001. Since then, research on miRNAs and RNA interference has virtually exploded. Current data suggest that a typical animal or plant species has about 500 sequences that code for miRNAs. Because many miRNAs regulate more than one mRNA, it is estimated that a large percentage of all animal and plant genes are regulated by these tiny molecules. There is growing evidence that miRNA-like elements also occur in bacterial cells. RNA interference is increasingly recognized as a key aspect of post-transcriptional control.

Researchers are currently testing whether certain miRNAs could be used as drugs, by knocking out specific genes associated with illness or destroying mRNAs produced by viruses during an infection. In a short time, research on RNA interference has gone from an exciting new frontier in basic biology to possible applications in medicine.

✔If you understand RNA interference, you should be able to describe where it occurs in Figure 18.1.

How Is Translation Controlled?

RNA interference is not the only mechanism of gene control that acts on mRNAs and affects whether or not translation occurs. In many species of animals, for example, eggs are loaded with mRNAs that are not translated until fertilization occurs. The proteins produced from these mRNAs play a part in directing the early development of the embryo, but they are not needed until fertilization is complete. Translation of the mRNAs in the egg is prevented by a regulatory protein that binds to them, or by modification of the mRNA's cap or tail.

In addition, a mature cell may slow or stop translation in response to a sudden increase in temperature or infection by a virus. The slowdown occurs because regulatory proteins that are activated by the viral invasion or temperature spike add a phosphate group to a protein that is part of the ribosome.

You might recall from earlier chapters that phosphorylation frequently leads to changes in the shape and chemical reactivity of proteins. In the case of the phosphorylated ribosomal protein, the shape change slows or prevents translation.

For the cell, this dramatic change in gene expression can mean the difference between life and death. High temperatures disrupt protein folding, so shutting down translation prevents the production of improperly folded polypeptides. If the insult is an invading virus, the cell stops the infection because it avoids manufacturing viral proteins.

Mechanisms like these are a reminder that gene expression can be regulated at multiple points: at the level of chromatin structure, transcription initiation, RNA processing, mRNA availability, and translation rate. But that's not all. Let's look now at the last level possible: Altering protein activity, after translation is complete.

Post-Translational Control

Chapter 17 explained that in bacteria, mechanisms of post-translational regulation are important because they allow cells to respond to new conditions rapidly. The same is true for eukaryotes. Instead of waiting for transcription, RNA processing, and translation to occur, the cell can respond to altered conditions by quickly activating or inactivating existing proteins.

There is a trade-off, however: speed is gained at the expense of efficiency. Transcription, RNA processing, and translation use up energy and materials; it is wasteful to produce proteins that won't be used.

You have already encountered several important mechanisms of post-translational control over gene expression.

- Proteins are folded into their final, active conformation by chaperone proteins (see Chapter 16). Folding is required for proteins to function normally, and folding is regulated by the presence of chaperones.
- Enzymes may modify proteins by adding carbohydrate groups (see Chapter 7) or cleaving off certain amino acids.
- Phosphorylation is an extremely common mechanism for activating or deactivating proteins. You might recall that Chapter 11 featured the activation of cyclin-Cdk complexes by phosphorylation and the subsequent entry into M phase of the cell cycle.

There is yet another key mechanism of post-translational control, however: the targeted destruction of proteins. Chapter 11 introduced this phenomenon by describing the short life span of cyclin proteins. Once a cell is well into M phase, an enzyme begins adding a small polypeptide called ubiquitin to the cyclin proteins. Ubiquitin got its name because it is ubiquitous in cells. A multi-molecular machine called the **proteasome** recognizes proteins that have a ubiquitin tag, and cuts them into short segments. In this way, ubiquitinization is a mechanism for controlling the life span of proteins in a cell.

CHECK YOUR UNDERSTANDING

If you understand that . . .

- Alternative splicing allows a single gene to code for many products.
- RNA interference is one of several mechanisms for controlling an mRNA's life span and translation rate.
- Ubiquitin-tagging and destruction by proteasomes is one of several mechanisms for controlling a protein's life span or activity.

✔ **You should be able to . . .**

1. Explain why the discovery of alternative splicing forced biologists to change their definition of the gene.
2. Explain why "RNA interference" is aptly named.

Answers are available in Appendix B.

18.5 How Does Gene Expression in Bacteria Compare with That in Eukaryotes?

Biologists have been studying the control of gene expression for over 50 years. Almost as soon as they knew that information in DNA is transcribed into RNA and then translated into proteins, researchers began asking questions about how that flow of information is regulated. **Table 18.1** summarizes what biologists have learned over the past half century about how bacterial and eukaryotic gene expression is controlled—organized by the six steps in gene expression introduced in Figure 18.1.

How does the regulation of gene expression differ in bacteria and eukaryotes? Biologists point to four fundamental contrasts, two of which involve levels of control that exist in eukaryotes but not in bacteria:

1. ***Packaging*** The chromatin structure of eukaryotic DNA must be opened in order for TBP, the basal transcription complex, and RNA polymerase to gain access to genes and initiate transcription. A key insight here is that, because eukaryotic DNA is packaged so tightly, the default state of transcription in eukaryotes is the "off" state. In contrast, the default state of transcription in bacteria, which lack histone proteins and have freely accessible promoters, is "on." Chromatin structure provides a mechanism of negative control that does not exist in bacteria.
2. ***Alternative splicing*** Prior to translation, primary transcripts in eukaryotes must be spliced—an occurrence that is extremely rare in bacteria. The one-to-one correspondence between the number of genes and the number of gene products observed in bacteria is not seen in eukaryotes. Instead, each eukaryotic gene may code for one to thousands of distinct products.
3. ***Complexity*** Transcriptional control is much more complex in eukaryotes than in bacteria. The function of sigma proteins in bacteria is analogous to the role of the basal transcription complex in eukaryotes. Likewise, the function of CAP, the repressor, and other regulatory proteins is analogous to the role of regulatory transcription factors in eukaryotes. But the sheer number of eukaryotic proteins involved in regulating transcription dwarfs that in bacteria, as does the complexity of their interactions.
4. ***Coordinated expression*** In bacteria, genes that take part in the same cellular response are organized into operons controlled by a single promoter. Because their mRNAs are translated together, several proteins are produced in a coordinated fashion. In contrast, operons are rare in eukaryotes. Eukaryotic genes that are physically scattered can be expressed at the same time because a single set of regulatory transcription factors can trigger the transcription of several genes. For example, muscle-specific genes found on several different chromosomes can be transcribed in response to the same muscle-specific regulatory transcription factor. Recent data also suggest that in some cases, genes on different chromosomes may be physically associated inside the nucleus, and share regulatory elements. These are the means by which eukaryotes coordinate the expression of functionally related genes.

SUMMARY TABLE 18.1 **Regulating Gene Expression in Bacteria and Eukaryotes**

Level of Regulation	Bacteria	Eukaryotes
Chromatin remodeling	• Limited packaging of DNA • Remodeling not a major issue in regulating gene expression.	• Extensive packaging of DNA • Chromatin must be opened for transcription to begin.
Transcription	• Positive and negative control by regulatory proteins that act at sites close to the promoter • Sigma interacts with promoter.	• Positive and negative control by regulatory proteins that act at sites close to and far from promoter • Large basal transcription complex interacts with promoter. • Mediator complex required.
RNA processing	• None documented	• Extensive processing: alternative splicing of introns addition of 5′ cap and 3′ tail
mRNA stability	• Some RNA interference documented	• For many genes, RNA interference limits life span or translation rate.
Translation	• Regulatory proteins bind to mRNAs and/or ribosome and affect translation rate.	• Regulatory proteins bind to mRNAs and/or ribosome and affect translation rate.
Post-translational modification	• Folding by chaperone proteins • Chemical modification (e.g., phosphorylation) may change activity.	• Folding by chaperone proteins • Chemical modification (glycosylation, phosphorylation) • Ubiquination targets proteins for destruction by proteasome.

To date, biologists do not have a good explanation for why gene expression is so much more complex in unicellular eukaryotes than it is in bacteria. All unicellular organisms have to respond to environmental changes in an appropriate way. So why do unicellular algae, yeasts, and other eukaryotes have ways to regulate gene expression that bacteria lack? After decades of research, the answer is still not clear.

It is easier to generate a hypothesis to explain why gene expression is complex in multicellular eukaryotes. In these organisms, cells have to differentiate as an individual develops. Changes in gene expression are responsible for the differentiation of muscle cells, bone cells, leaf cells, flower cells, and so on in response to signals from other cells. The need for each cell type to have a unique pattern of gene expression may explain why control of gene expression is so much more complex in multicellular eukaryotes than in bacteria.

The effort to understand how developmental signals produce cell-specific gene expression in multicellular organisms represents one of two great frontiers in gene-expression research. The other major frontier is the quest to understand how certain defects in gene regulation result in uncontrolled cell growth and the suite of diseases called cancer.

18.6 Linking Cancer with Defects in Gene Regulation

Normal regulation of gene expression results in the orderly development of an embryo and, in juveniles and adults, appropriate responses to environmental change. Abnormal regulation of gene expression, in contrast, can lead to developmental abnormalities and diseases such as cancer.

Hundreds of distinct forms of cancer exist. These diseases are enormously varied in terms of their initial cause, the tissues they affect, their rate of progression, and their outcome. Because the underlying defects, symptoms, and consequences are so diverse, cancer is not a single disease but a family of related diseases.

All cancers are characterized by uncontrolled cell growth. But for most cancers to become dangerous, two other events are required: The rapidly growing cells must metastasize, meaning that some cells leave their point of origin and invade other tissues (see Chapter 11), and the cancer cells must stimulate the growth of blood vessels that supply them with nutrients. Here let's focus on the first step in cancer formation, and the question of what causes uncontrolled cell growth.

Causes of Uncontrolled Cell Growth

Each type of cancer is caused by a different set of genetic defects that lead to uncontrolled cell growth. Many cancers are associated with mutations in regulatory transcription factors. These mutations lead to cancer when they affect one of two classes of genes: (1) genes that stop or slow the cell cycle, and (2) genes that trigger cell growth and division by initiating specific phases in the cell cycle.

Genes that stop or slow the cell cycle are **tumor suppressor** genes. The products of these genes prevent the cell cycle from progressing unless specific signals indicate that conditions are right for moving forward with mitosis and cell division. If a mutation disrupts normal function of a tumor suppressor gene, then a key brake on the cell cycle is eliminated.

Genes that encourage cell growth by triggering specific phases in the cell cycle are called **proto-oncogenes** (literally, "first-cancer-genes"). In normal cells, proto-oncogenes are required to initiate each phase in the cell cycle. They are active only when conditions are appropriate for growth, however. In cancerous cells, defects in the regulation of proto-oncogenes cause these genes to stimulate growth at all times. In such cases, a mutation has converted the proto-oncogene into an **oncogene**—an allele that promotes cancer development.

p53: A Case Study

To gain a deeper understanding of how defects in gene expression can lead to cancer, consider research on the gene that is most often defective in human cancers. The gene is called ***p53*** because the protein it codes for has a molecular weight of approximately 53 kilodaltons. Sequencing studies have revealed that mutant, nonfunctional forms of *p53* are found in over half of all human cancers. The *p53* gene codes for a regulatory transcription factor.

Researchers began to understand what *p53* does when they exposed normal, noncancerous human cells to UV radiation and noticed that levels of p53 protein increased markedly. Recall from Chapter 14 that UV radiation damages DNA. Follow-up studies confirmed that there is a close correlation between DNA damage and the amount of p53 in a cell. In addition, analyses of the protein's primary structure suggested that it might contain a region that binds to DNA.

These observations inspired the hypothesis that p53 is a transcription factor that serves as the master brake on the cell cycle. In this model, p53 is activated after DNA damage occurs. The activated protein binds to the enhancers of genes that arrest the cell cycle. Once these genes are activated, the cell has time to repair its DNA before continuing to grow and divide.

Research has shown that this model of p53 function is correct in almost every detail.

- Three-dimensional models generated by X-ray crystallography studies confirmed that p53 binds directly to DNA.
- Virtually all of the p53 mutations associated with cancer are located in the protein's DNA-binding site (see **Figure 18.11** on page 333)—suggesting that defective forms of the protein can't bind to enhancers.
- One of the genes that p53 regulates codes for a protein that prevents cell-cycle regulatory proteins from triggering M (mitosis) phase.

Experiments have also shown that when a cell's DNA is extensively damaged and cannot be repaired, p53 activates the transcription of genes that cause the cell to take its own life by

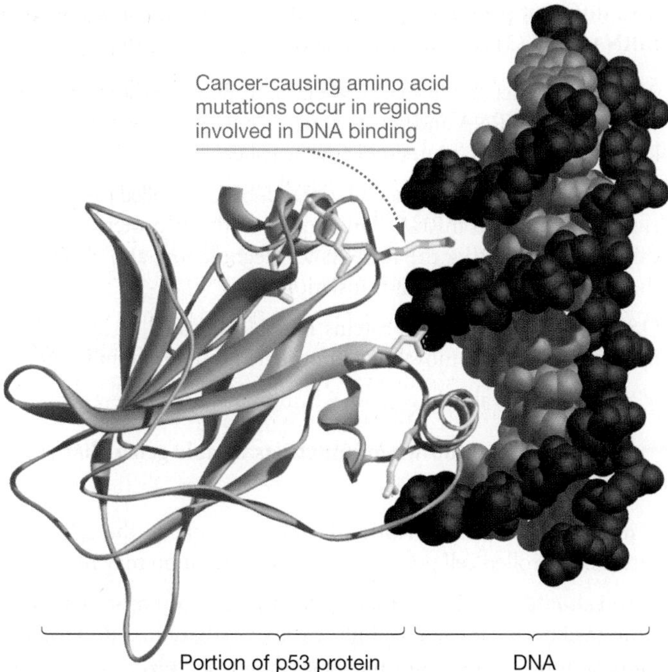
Cancer-causing amino acid mutations occur in regions involved in DNA binding
Portion of p53 protein
DNA

FIGURE 18.11 p53 Is a Transcription Factor that Serves as a Master Brake on the Cell Cycle When DNA Is Damaged. In cancer patients, the amino acids highlighted in yellow often differ from those in the normal protein, preventing p53 from binding to DNA.

apoptosis (see Chapter 11). If mutations in the *p53* gene make the protein product inactive, then damaged cells are not shut down or killed but instead continue to move through the cell cycle. They are likely to contain many mutations, however, because of the damage they have sustained to their DNA. If these mutations create oncogenes, the cells have taken a key step on the road to cancer. The p53 protein is like a quality control officer. If it is missing, things can go downhill.

The role of *p53* in preventing cancer is so fundamental that biologists call this gene "the guardian of the genome." Currently, research is forging ahead on two fronts: Biologists are striving to (1) identify more of the genes that are regulated by the p53 protein, and (2) find molecules that could act as anticancer drugs by mimicking p53's shape and activity.

CHECK YOUR UNDERSTANDING

If you understand that . . .

- Cancer is associated with mutations that lead to loss of control over the cell cycle.
- Uncontrolled cell growth may result when a mutation in a regulatory gene creates a protein that activates the cell cycle constitutively.
- Uncontrolled cell growth may result when a mutation prevents a tumor suppressor gene product from shutting down the cell cycle in damaged cells.

✓ **You should be able to . . .**

1. Explain why cancer has a common pattern (uncontrolled cell growth), but not a common cause. Your answer should refer to the six levels of gene regulation outlined in Figure 18.1.
2. Explain why loss-of-function mutations in *p53* are observed in so many cancers.

Answers are available in Appendix B.

CHAPTER 18 REVIEW

For media, go to the study area at www.masteringbiology.com 

Summary of Key Concepts

Changes in gene expression allow eukaryotic cells to respond to changes in the environment and cause distinct cell types to develop.

- In a multicellular eukaryote, cells are different not because they have different genes but because they express different genes.
- Gene expression is regulated at six distinct levels: Chromatin has to be remodeled, the transcription of specific genes may be initiated or repressed, mRNAs may be spliced in different ways to produce a different product, the life span of specific mRNAs may be extended or shortened, translation rate may be increased or decreased, and the life span or activity of particular proteins may be altered.

✓You should be able to describe how the presence of the nuclear envelope, and the physical separation of transcription and translation, influence the levels of gene regulation observed in eukaryotes versus bacteria and archaea.

Eukaryotic DNA is packaged with proteins into structures that must be opened before transcription can occur.

- Eukaryotic DNA is wrapped around histone proteins to form bead-like nucleosomes that are then coiled into 30-nm fibers and higher-order chromatin structures.
- Transcription cannot be initiated until the interaction between DNA and histones in chromatin is relaxed.
- The state of chromatin depends on the acetylation or methylation of histones and the action of molecular machines called chromatin-remodeling complexes.

✓You should be able to explain why chromatin remodeling has to be the first step in gene activation.

In eukaryotes, transcription is triggered by regulatory proteins that bind to the promoter and to sequences close to and far from the promoter.

- Regulatory transcription factors are proteins that bind to regulatory sequences called (1) enhancers and silencers, which are often located at a distance from the gene in question, or (2) promoter-proximal sequences, near the start of the coding sequence.
- The first regulatory transcription factors that bind to DNA recruit proteins that loosen the histones' grip on the gene, making the promoter accessible to basal transcription factors.
- Interactions between regulatory and basal transcription factors lead to the formation of the basal transcription complex.
- Once the basal transcription complex is assembled, RNA polymerase is recruited to the site and transcription begins.

✔You should be able to draw a model of a eukaryotic gene undergoing transcription. Label enhancers, promoter-proximal elements, the promoter, regulatory transcription factors, basal transcription factors, and RNA polymerase.

MB **Web Activity** Transcription Initiation in Eukaryotes

Once transcription is complete, gene expression is controlled by (1) alternative splicing, which allows a single gene to code for several different products; (2) molecules that regulate the life span of mRNAs; and (3) activation or inactivation of protein products.

- Alternative splicing allows a single gene to produce more than one version of an mRNA and more than one kind of protein. It is regulated by proteins that interact with the spliceosome.
- RNA interference occurs when tiny strands of RNA, called microRNAs (miRNAs), bind to mRNAs in company with the protein complex called RISC, marking the mRNAs for degradation, and also when the miRNAs merely inhibit translation.
- Once translation occurs, proteins may be activated or inactivated by the addition or removal of a phosphate group or other events.

✔You should be able to explain why humans have so few genes.

Cancer can develop when mutations disable genes that regulate cell-cycle control genes.

- If mutations alter transcription factors that control the cell cycle, then uncontrolled cell growth and tumor formation may result.

✔You should be able to explain why cancer is common in people who have been exposed to high levels of radiation, and why it is more common in older people than younger people.

Questions

✔TEST YOUR KNOWLEDGE

Answers are available in Appendix B

1. What is chromatin?
   a. the protein core of the nucleosome, which consists of histones
   b. the 30-nm fiber
   c. the DNA-protein complex found in eukaryotes
   d. the histone *and* non-histone proteins in eukaryotic nuclei
2. What is a tumor suppressor?
   a. a gene associated with tumor formation when its product does not function
   b. a gene associated with tumor formation when its product functions normally
   c. a gene that accelerates the cell cycle and leads to uncontrolled cell growth
   d. a gene that codes for a transcription factor involved in tumor formation
3. Which of the following statements about enhancers is correct?
   a. They contain a unique base sequence called a TATA box.
   b. They are located only in 5'-flanking regions.
   c. They are located only in introns.
   d. They are found in a variety of locations and are functional in any orientation.
4. In eukaryotes, why are certain genes expressed only in certain types of cells?
   a. Different cell types contain different genes.
   b. Different cell types have the same genes but different promoters.
   c. Different cell types have the same genes but different enhancers.
   d. Different cell types have different regulatory transcription factors.
5. What is alternative splicing?
   a. the phosphorylation events that lead to different types of post-translational regulation
   b. mRNA processing events that lead to different combinations of exons being spliced together
   c. folding events that lead to proteins with alternative conformations
   d. action by regulatory proteins that leads to changes in the life span of an mRNA
6. What types of proteins bind to promoter-proximal elements?
   a. the basal transcription complex
   b. the basal transcription complex plus RNA polymerase
   c. basal transcription factors
   d. regulatory transcription factors

✓TEST YOUR UNDERSTANDING

Answers are available in Appendix B

1. Compare and contrast (a) enhancers and the CAP site; (b) promoter-proximal elements and the *lac* operon operator; and (c) basal transcription factors and sigma.
2. Explain how alternative splicing could play a role in changing eukaryotic gene expression in response to changes in the environment.
3. Compare and contrast (a) enhancers and silencers; (b) promoter-proximal elements and enhancers; and (c) transcription factors and the mediator complex.
4. Explain the relationship between complementary base pairing and RNA interference.
5. Explain the concept of the histone code. Your answer should compare and contrast the structure of chromatin in muscle cells versus nerve cells.
6. Explain why mutations in *p53* can lead to loss of control over the cell cycle and to the development of cancer.

✓APPLYING CONCEPTS TO NEW SITUATIONS

Answers are available in Appendix B

1. Histone proteins have been extremely highly conserved during evolution. The histones found in fruit flies and humans, for example, are nearly identical in amino acid sequence. Offer an explanation for this observation. (Hint: What are the consequences of a mutation in a histone?)
2. Cancers are most common in tissues where cell division is common, such as blood cells and cells in the lining of the lungs or gut. Why is this observation logical?
3. Levels of p53 protein in the cytoplasm increase after DNA damage. Design an experiment to determine whether this increase is due to increased transcription of the *p53* gene or to activation of preexisting p53 proteins by a post-translational mechanism such as phosphorylation.
4. Suggest a way that miRNAs that are complementary to viral RNAs could be useful as drugs.

This chapter is part of The Big Picture. See how on pages 336–337.

The Big Picture

Copying, using, and transmitting genetic information is fundamental to life. Cells use the genetic information archived in their DNA to respond to changes in the environment and, in multicellular organisms, to develop into specific cell types.

Hereditary information is transmitted to offspring with random changes called mutation.

Thus, genetic information is dynamic—both within generations and between generations.

Note that each box in the concept map indicates the chapter and section where you can go for review. Also, be sure to do the blue exercises in the Check Your Understanding box below.

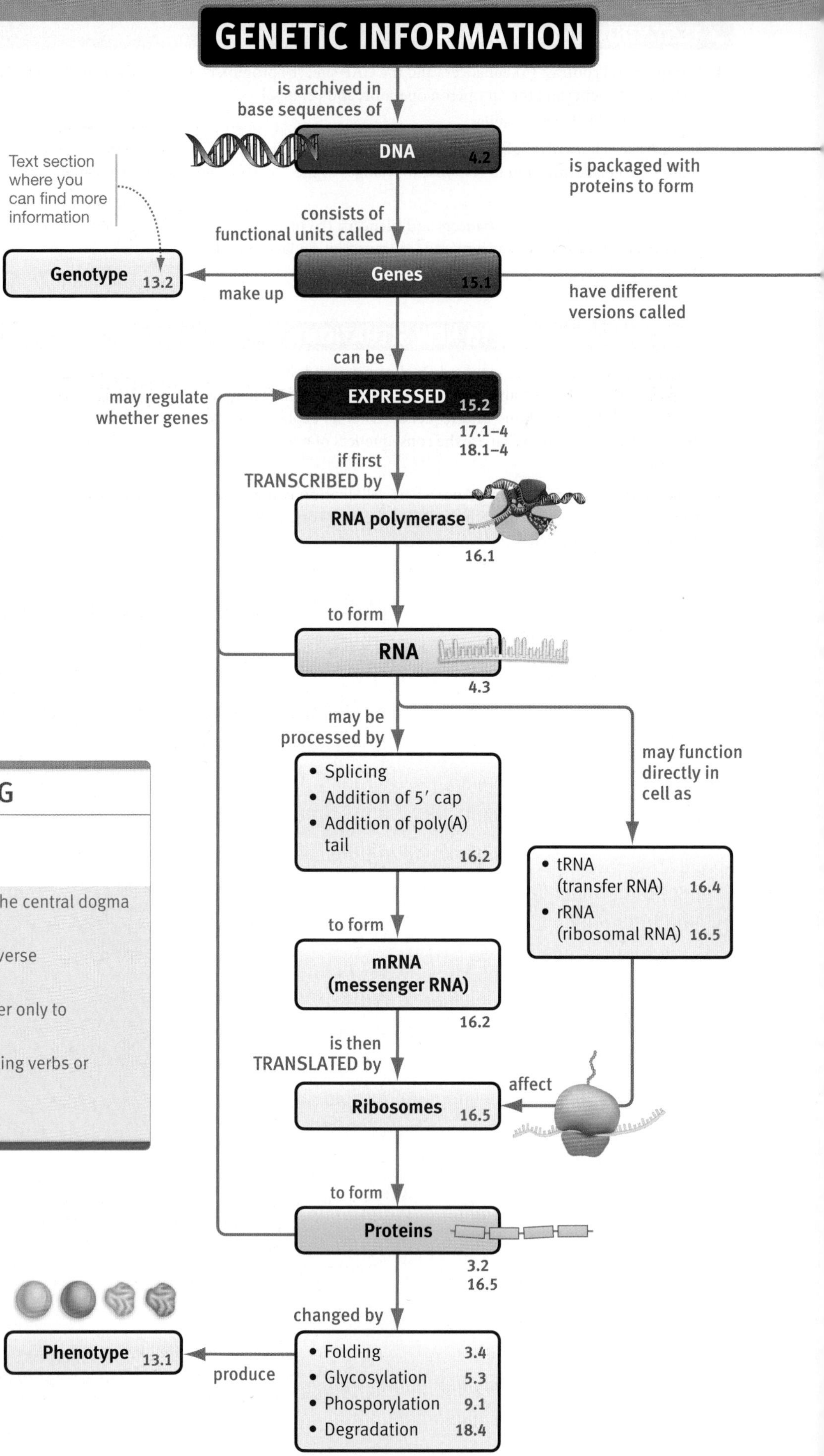

CHECK YOUR UNDERSTANDING

If you understand the big picture . . .

✔ You should be able to . . .

1. Draw stars next to the three elements of the central dogma of molecular biology.
2. Add arrows and labels indicating what reverse transcriptase does.
3. Draw an E in the corners of boxes that refer only to eukaryotes, not prokaryotes.
4. Fill in the blue ovals with appropriate linking verbs or phrases.

Answers are available in Appendix B.

More Big Picture activities are available in the study area at www.masteringbiology.com  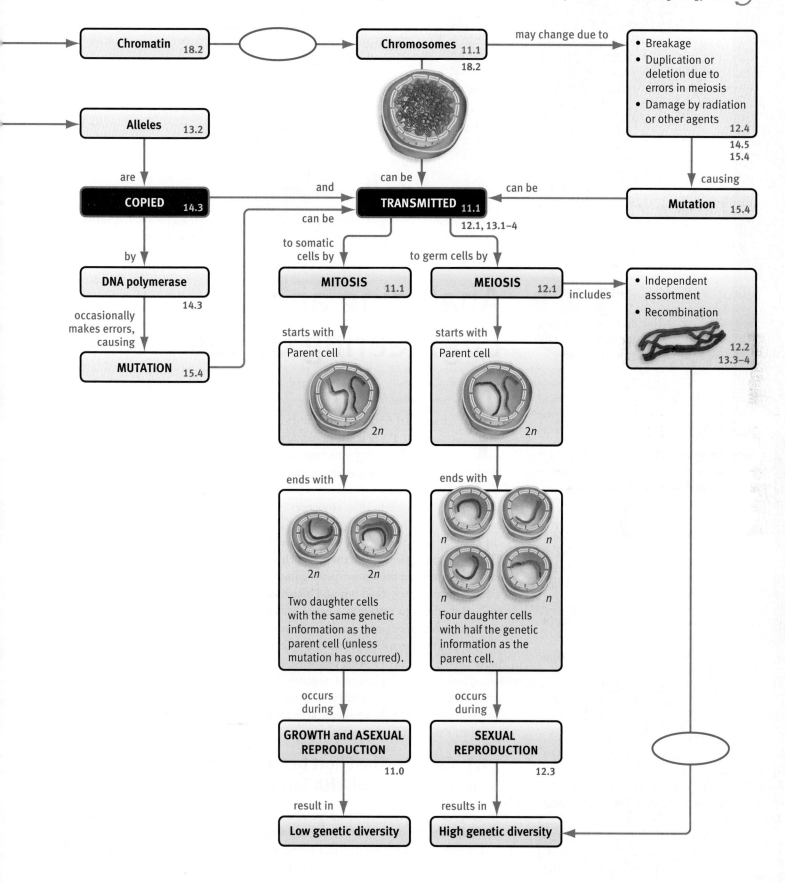

Chromatin 18.2 → [ ] → Chromosomes 11.1, 18.2

Chromosomes may change due to:
- Breakage
- Duplication or deletion due to errors in meiosis
- Damage by radiation or other agents

12.4, 14.5, 15.4

causing → Mutation 15.4

Alleles 13.2 are → **COPIED** 14.3 by → DNA polymerase 14.3 occasionally makes errors, causing → MUTATION 15.4

COPIED and → **TRANSMITTED** 11.1; Chromosomes can be → TRANSMITTED; Mutation can be → TRANSMITTED; MUTATION can be → TRANSMITTED

TRANSMITTED 12.1, 13.1–4:

to somatic cells by → MITOSIS 11.1

to germ cells by → MEIOSIS 12.1

MEIOSIS includes:
- Independent assortment
- Recombination

12.2, 13.3–4

MITOSIS starts with: Parent cell (2n)

MITOSIS ends with: 2n, 2n. Two daughter cells with the same genetic information as the parent cell (unless mutation has occurred).

occurs during → GROWTH and ASEXUAL REPRODUCTION 11.0 result in → Low genetic diversity

MEIOSIS starts with: Parent cell (2n)

MEIOSIS ends with: n, n, n, n. Four daughter cells with half the genetic information as the parent cell.

occurs during → SEXUAL REPRODUCTION 12.3 results in → High genetic diversity

Independent assortment / Recombination → [ ] → High genetic diversity

The rice plants in these bottles have been genetically engineered—using techniques introduced in this chapter—to produce a molecule needed for a key vitamin.

19 Analyzing and Engineering Genes

KEY CONCEPTS

- Enzymes that cut DNA at specific locations and other enzymes that piece DNA segments back together allow biologists to move genes from one place to another.
- Biologists can obtain many identical copies of a gene by (1) inserting it into a bacterial cell that copies the gene each time the cell divides or (2) by conducting a polymerase chain reaction.
- The sequence of bases in a gene can be determined by dideoxy sequencing.
- If individuals with a certain phenotype also tend to share a genetic marker (a known site in DNA that is unrelated to the phenotype), the gene responsible for the phenotype is likely to be near that marker.
- Researchers are attempting to insert genes into humans to cure genetic diseases. Efforts to insert genes into plants have been much more successful.

The molecular revolution in biological science got its start when researchers confirmed that DNA is the hereditary material and succeeded in describing the molecule's secondary structure. But when biologists discovered how to remove DNA sequences from an organism, manipulate them, and insert them into different individuals, the molecular revolution really took off.

Efforts to manipulate DNA sequences in organisms are often referred to as genetic engineering. Genetic engineering became possible with the discovery of enzymes that cut DNA at specific sites and of other enzymes that paste DNA sequences together. These new molecular tools were extremely powerful. Biologists no longer had to rely solely on controlled breeding experiments to change the genetic characteristics of individuals. Instead, they could mix and match specific DNA sequences in the lab. Because successful efforts to manipulate genes usually result in novel combinations of DNA, techniques used to engineer genes are often referred to as **recombinant DNA technology**.

This chapter uses a series of case histories to introduce basic molecular biology techniques in the context of solving problems. It also considers the ethical and economic issues raised by efforts to manipulate genes. What are the potential perils and benefits of introducing recombinant genes into human beings, food plants, and other organisms? This question, one of the great ethical challenges of the twenty-first century, is a recurrent theme in the following pages.

19.1 Case 1—The Effort to Cure Pituitary Dwarfism: Basic Recombinant DNA Technologies

To understand the basic techniques and tools of genetic engineering, let's consider the role they played in developing a treatment for pituitary dwarfism in humans.

The pituitary gland is a structure at the base of the mammalian brain that produces several important biomolecules, including a protein that stimulates growth. This

✔ When you see this checkmark, stop and test yourself. Answers are available in Appendix B.

protein, which was found to be just 191 amino acids long, was named human growth hormone (HGH). In humans, the gene that codes for it is called *GH1*.

The discovery of growth hormone led researchers immediately to suspect that at least some forms of inherited dwarfism might be caused by a defect in the *GH1* gene. This hypothesis was confirmed by studies showing that people with certain types of dwarfism produce little growth hormone or none at all. These people have defective copies of *GH1* and exhibit pituitary dwarfism, type I (**Figure 19.1a**).

By studying the pedigrees of families in which dwarfism was common, several teams of researchers established that pituitary dwarfism, type I, is an autosomal recessive trait (see Chapter 13). In other words, affected individuals have two copies of the defective allele. Individuals who are affected by pituitary dwarfism have normal body proportions but grow more slowly than average people, reach puberty from two to ten years later than average, and are short in stature as adults—typically no more than 120 cm (4 feet) tall (**Figure 19.1b**).

Why Did Early Efforts to Treat the Disease Fail?

Once the molecular basis of pituitary dwarfism was understood, physicians began treating the disease with injections of naturally produced growth hormone. This approach was inspired by the spectacular success that had been achieved in treating type I diabetes mellitus. Diabetes mellitus is caused by a deficiency of the peptide hormone insulin, and clinicians had been able to alleviate the disease's symptoms by injecting patients with insulin from pigs.

Early trials showed that people with pituitary dwarfism could be treated successfully with growth hormone therapy, but only if the protein came from humans. Growth hormones isolated from pigs, cows, or other animals were ineffective. Until the 1980s, however, the only source of human growth hormone was pituitary glands dissected from human cadavers. As a result, the drug was extremely scarce and expensive.

Meeting demand turned out to be the least of the problems with growth hormone therapy, however. To understand why, recall from Chapter 3 that infectious proteins called prions can cause degenerative brain disorders in mammals. When some of the children treated with human growth hormone developed a prion disease in their teens and twenties, physicians realized that the supply of growth hormone was contaminated with a prion protein from the brains of the cadavers supplying the hormone. In 1984, the use of growth hormone isolated from cadavers was banned.

Steps in Engineering a Safe Supply of Growth Hormone

To replace natural sources of growth hormone, researchers turned to genetic engineering. Their plan was to insert fully functional copies of human *GH1* into the bacterium *Escherichia coli*, which they hoped would then produce huge quantities of recombinant progeny. If the plan worked, the recombinant cells would produce uncontaminated growth hormone in sufficient quantities to meet demand at an affordable price.

The plan required investigators to find *GH1*, obtain many copies of the gene, and insert them into *E. coli* cells. Their ability to do these things hinged on three of the most basic tools in molecular biology. Let's consider each in turn.

USING REVERSE TRANSCRIPTASE TO PRODUCE cDNAs Chapter 15 mentioned that an enzyme called reverse transcriptase is responsible for a major exception to the central dogma of molecular biology: It allows information to flow from RNA to DNA. More specifically, reverse transcriptase catalyzes the synthesis of DNA from an RNA template.

DNA that is produced from RNA is called **complementary DNA**, or **cDNA**. Although reverse transcriptase initially produces a

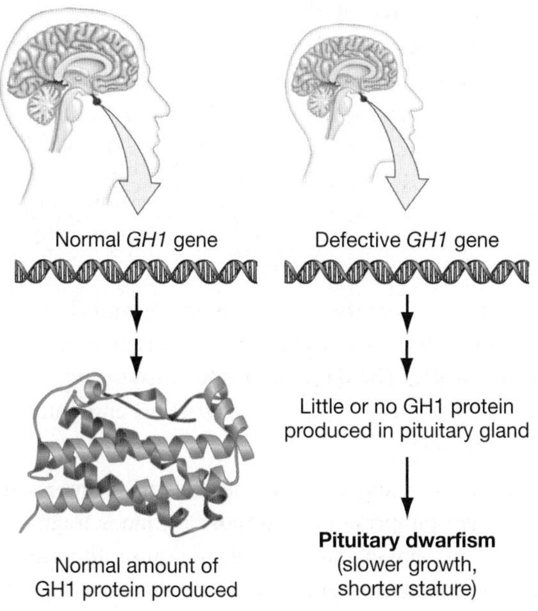

FIGURE 19.1 Pituitary dwarfism is a genetic disease. (a) If mutations in the human *GH1* sequence are severe enough to knock out the gene, pituitary dwarfism may result. **(b)** William Harrison and Charles Stratton, in a photo taken about 1860. Harrison and Stratton were both celebrated comedians and performers. Stratton, whose stage name was Tom Thumb, enjoyed audiences in the White House with Abraham Lincoln and Buckingham Palace with Queen Victoria. Stratton had pituitary dwarfism; Harrison had normal height.

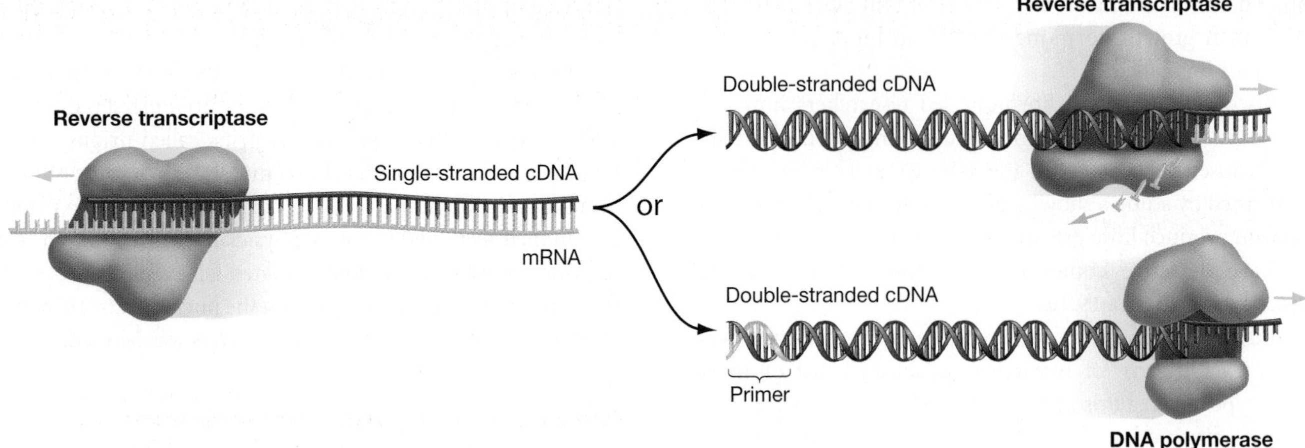

FIGURE 19.2 Reverse Transcriptase Catalyzes the Synthesis of DNA from RNA. The single-stranded DNA produced by reverse transcriptase is complementary to the RNA template. The cDNA can be made double stranded by reverse transcriptase or DNA polymerase. DNA polymerase requires a primer.

single-stranded cDNA, it is also capable of synthesizing the complementary strand to yield a double-stranded DNA. In many cases, however, researchers add a primer to single-stranded cDNAs and use DNA polymerase to synthesize the second strand (**Figure 19.2**).

Reverse transcriptase played a key role in the search for the growth hormone gene. Knowing that *GH1* is actively transcribed in cells from the pituitary gland, researchers isolated mRNAs from pituitary gland cells and used reverse transcriptase to reverse-transcribe those mRNAs to cDNAs. The reaction products could be expected to contain double-stranded cDNAs corresponding to each gene that is actively expressed in pituitary cells.

The next move? Isolating each of the cDNAs and making many identical copies of them.

USING PLASMIDS IN CLONING Efforts to produce many copies of a gene are referred to as **DNA cloning**. If a researcher says that she has cloned a gene, it means that she has isolated it and then produced many identical copies.

In many cases, researchers can clone a gene by inserting it into a small, circular DNA molecule called a **plasmid**. You might recall from Chapter 7 that plasmids are common in bacterial cells. They are physically separate from the bacterial chromosome, and are not required by the cell for normal growth and reproduction. Most replicate independently of the chromosome. Some plasmids carry genes for antibiotic resistance or other traits that increase the cell's ability to grow in a particular environment.

Researchers realized that if they could splice a loose piece of DNA into a plasmid and then insert the modified plasmid into a bacterial cell, the engineered plasmid would be replicated and passed on to daughter cells as the bacterium grew and divided. If this recombinant bacterium were then placed in a nutrient broth and allowed to grow and reproduce overnight, billions of copies of the original cell, each containing identical modified plasmid DNA, would result. When a plasmid is used in this way—to make copies of a foreign DNA sequence—it is called a **cloning vector**, or simply a **vector**.

Biologists harvest the recombinant genes by breaking the bacteria open, isolating all of the DNA, and then separating the plasmids from the main chromosomes. But how do they insert a gene into a plasmid in the first place?

USING RESTRICTION ENDONUCLEASES AND DNA LIGASE TO CUT AND PASTE DNA To cut a gene out for later insertion into a cloning vector, researchers use enzymes called restriction endonucleases. A **restriction endonuclease** is a bacterial enzyme that cuts DNA molecules at specific base sequences. In bacterial cells, these enzymes cut up DNA from invading viruses and prevent the cell from becoming fatally infected.

Most of the 400 known restriction endonucleases cut DNA only at sites that form palindromes. In English, a word or sentence is a palindrome if it reads the same way backward as it does forward. "Madam, I'm Adam" is an example. In biology, a stretch of double-stranded DNA forms a palindrome if the $5' \rightarrow 3'$ sequence of one strand is identical to the $5' \rightarrow 3'$ sequence on the antiparallel, complementary strand.

To insert the pituitary gland cDNAs into plasmids, researchers performed the sequence of steps outlined in **Figure 19.3**.

Step 1 The left side of the figure shows a plasmid containing a palindromic sequence that is cut by a specific restriction endonuclease. As the right side of the figure shows, the researchers attached the same palindromic sequence to the ends of each cDNA in their sample.

Step 2 They cut the recognition sites in each plasmid (left) and at the ends of each cDNA (right) with a restriction endonuclease called EcoRI. (The name stands for *Escherichia coli* restriction I, because it was the first restriction endonuclease discovered in *E. coli*.)

Step 3 Like most restriction endonucleases, EcoRI makes a staggered cut in the palindrome. The resulting DNA fragments are described as having **sticky ends**, because the single-stranded bases on one fragment are complementary to the

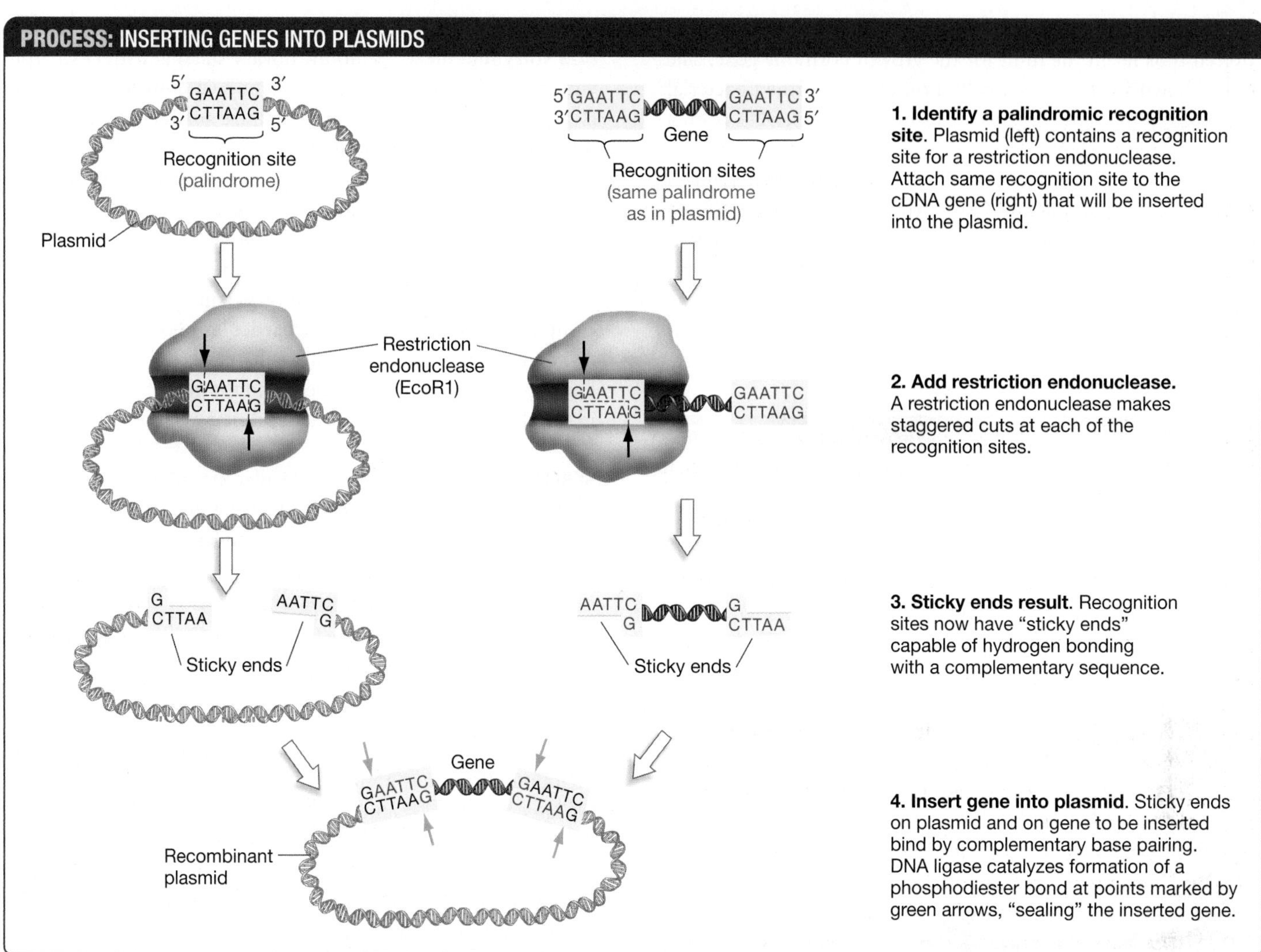
PROCESS: INSERTING GENES INTO PLASMIDS
5′ GAATTC 3′
3′ CTTAAG 5′
Recognition site
(palindrome)
Plasmid
5′GAATTC GAATTC 3′
3′CTTAAG CTTAAG 5′
Gene
Recognition sites
(same palindrome
as in plasmid)
1. Identify a palindromic recognition site. Plasmid (left) contains a recognition site for a restriction endonuclease. Attach same recognition site to the cDNA gene (right) that will be inserted into the plasmid.
Restriction
endonuclease
(EcoR1)
GAATTC
CTTAAG
GAATTC
CTTAAG
GAATTC
CTTAAG
2. Add restriction endonuclease. A restriction endonuclease makes staggered cuts at each of the recognition sites.
G
CTTAA
AATTC
G
Sticky ends
AATTC
G
G
CTTAA
Sticky ends
3. Sticky ends result. Recognition sites now have "sticky ends" capable of hydrogen bonding with a complementary sequence.
Gene
GAATTC
CTTAAG
GAATTC
CTTAAG
Recombinant
plasmid
4. Insert gene into plasmid. Sticky ends on plasmid and on gene to be inserted bind by complementary base pairing. DNA ligase catalyzes formation of a phosphodiester bond at points marked by green arrows, "sealing" the inserted gene.

FIGURE 19.3 Genes Can Be Inserted into Plasmids in Preparation for Cloning. Once a gene has been inserted into a plasmid, the recombinant plasmid can be introduced into bacterial cells that grow and divide to produce many identical copies of the gene.

single-stranded bases on the other fragment. As a result, the two ends can pair up and hydrogen bond to one another: the complementary sequences in the sticky ends of the plasmid (in black) will bind to the sticky ends in the cDNA (in red) by complementary base pairing.

Step 4 Finally, researchers used **DNA ligase**—introduced in Chapter 14 as the enzyme that connects Okazaki fragments during DNA replication—to seal the recombinant pieces of DNA together at the arrows marked in green.

The importance of the creation of sticky ends in DNA cannot be overstated. If restriction sites in different DNA sequences are cut with the same restriction endonuclease, the presence of the same sticky ends in both samples of DNA allows the resulting fragments to be spliced together by complementary base pairing. This is the essence of recombinant DNA technology—the ability to create novel combinations of DNA sequences by cutting specific sequences and pasting them into new locations.

After performing this procedure, the researchers who were hunting for the growth hormone gene had a set of recombinant plasmids. Each contained a cDNA made from one of the many human pituitary gland mRNAs.

TRANSFORMATION: INTRODUCING RECOMBINANT PLASMIDS INTO BACTERIAL CELLS If a recombinant plasmid can be inserted into a bacterial or yeast cell, the foreign DNA will be copied and transmitted to new cells as the host cell grows and divides. In this way, researchers can obtain millions or billions of copies of specific genes. How is the insertion brought about?

Cells that take up DNA from the environment and incorporate it into their genomes are said to undergo **transformation**. To transform bacterial cells with a plasmid, researchers increase the permeability of the cell's plasma membranes using a specific chemical treatment or an electrical shock.

Typically, just a single plasmid enters the cell during this treatment. The cells are then spread out on plates at a low enough density to ensure that each cell is physically isolated. The individual cells grow into colonies containing millions of identical cells.

PRODUCING A cDNA LIBRARY **Figure 19.4** summarizes the steps covered thus far in the hunt for the growth hormone gene. The result, shown in step 5, is a collection of transformed bacterial cells. Each of the cells contains a plasmid with one cDNA from a pituitary gland mRNA.

A collection of DNA sequences, each of which is inserted into a vector, is called a **DNA library**. If the sequences are cDNAs from a particular cell type or tissue, the library is called a **cDNA library**. If the sequences are fragments of DNA that collectively represent the entire genome of an individual, the library is called a **genomic library**.

DNA libraries are made up of cloned genes. Each gene present can be produced in large quantity and isolated in pure form.

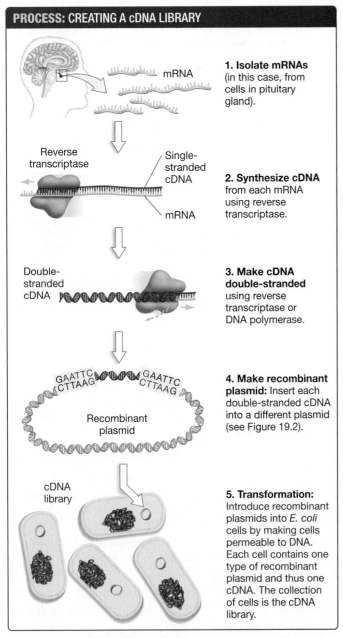

PROCESS: CREATING A cDNA LIBRARY

mRNA

1. Isolate mRNAs (in this case, from cells in pituitary gland).

Reverse transcriptase

Single-stranded cDNA

mRNA

2. Synthesize cDNA from each mRNA using reverse transcriptase.

Double-stranded cDNA

3. Make cDNA double-stranded using reverse transcriptase or DNA polymerase.

GAATTC CTTAAG

GAATTC CTTAAG

Recombinant plasmid

4. Make recombinant plasmid: Insert each double-stranded cDNA into a different plasmid (see Figure 19.2).

cDNA library

5. Transformation: Introduce recombinant plasmids into *E. coli* cells by making cells permeable to DNA. Each cell contains one type of recombinant plasmid and thus one cDNA. The collection of cells is the cDNA library.

FIGURE 19.4 Complementary DNA Libraries Represent a Collection of the mRNAs in a Cell.

✔**QUESTION** Would each type of cDNA in the library be represented just once? Why or why not?

✔If you understand this concept, you should be able to describe how you could make a genomic library starting with DNA from your own cells and using the restriction endonuclease EcoR1 to cut the genome into fragments that are small enough to insert into plasmids or other vectors.

DNA libraries are important because they give researchers a way to store information from a particular cell type or genome in a form that is accessible. But like a college library, a DNA library isn't very useful unless there is a way to retrieve specific pieces of information. At your school's library, you use call numbers or computer searches to retrieve a particular book or article. How do you go about retrieving a particular gene from a DNA library? For example, how did researchers find the growth hormone gene in the cDNA library of the human pituitary gland?

SCREENING A DNA LIBRARY Molecular biologists are often faced with the task of finding one specific gene in a large collection of DNA fragments. To do this requires a **probe**—a marked molecule that binds to the molecule the biologist is looking for.

A DNA probe is a single-stranded fragment that will bind to a single-stranded complementary sequence in the sample of DNA being analyzed. By binding to the target sequence, the probe marks the fragment containing that sequence, distinguishing it from all the other DNA fragments in the sample. As **Figure 19.5** shows, a DNA probe must be labeled in some way so that it can be found after it has bound to the complementary sequence in the large sample of fragments.

✔If you understand the concept of a DNA probe, you should be able to explain why the probe must be single stranded and labeled in order to work, and why it binds to just one specific fragment. You should also be able to indicate where a probe with the

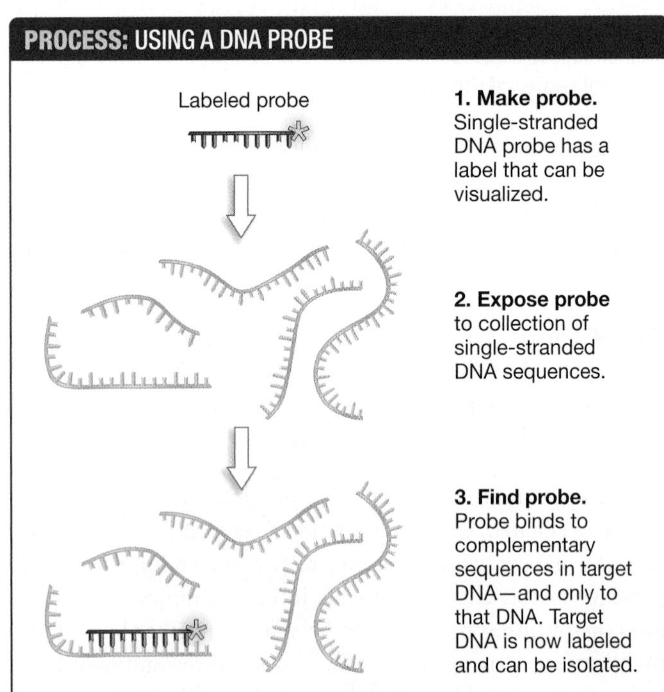

PROCESS: USING A DNA PROBE
Labeled probe
1. Make probe.
Single-stranded DNA probe has a label that can be visualized.
2. Expose probe to collection of single-stranded DNA sequences.
3. Find probe.
Probe binds to complementary sequences in target DNA—and only to that DNA. Target DNA is now labeled and can be isolated.

FIGURE 19.5 DNA Probes Bind to Specific Target Sequences among Many Different Sequences.

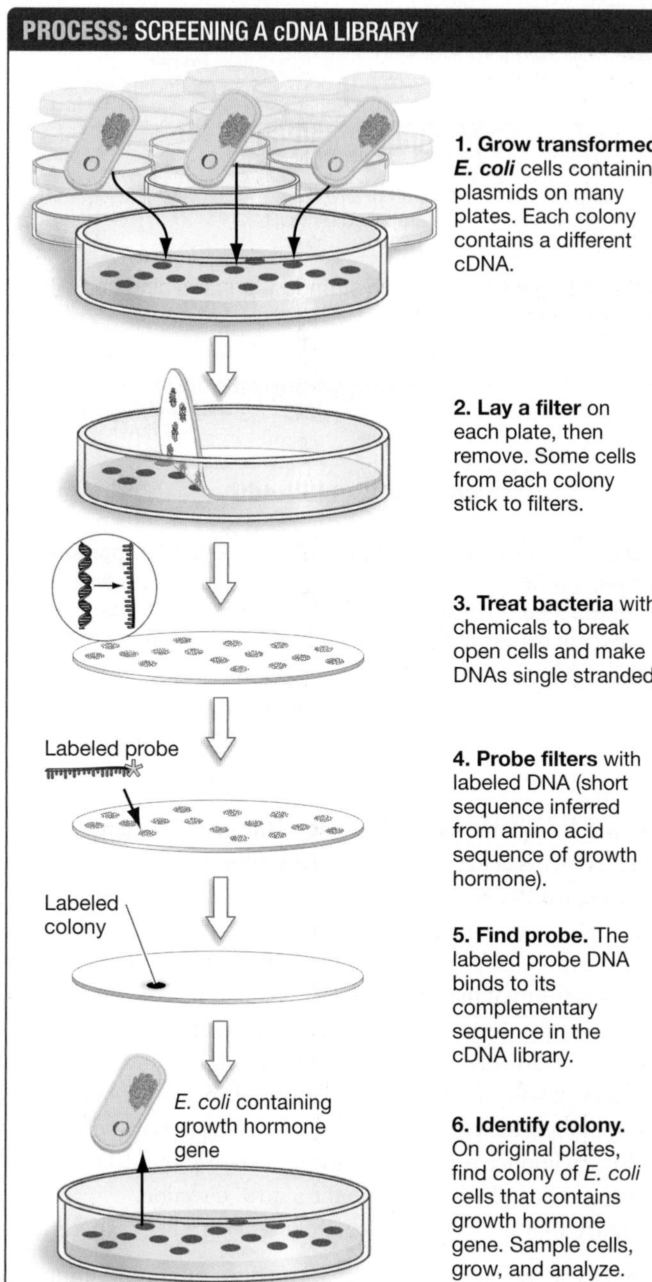

PROCESS: SCREENING A cDNA LIBRARY
1. Grow transformed E. coli cells containing plasmids on many plates. Each colony contains a different cDNA.
2. Lay a filter on each plate, then remove. Some cells from each colony stick to filters.
3. Treat bacteria with chemicals to break open cells and make DNAs single stranded.
Labeled probe
4. Probe filters with labeled DNA (short sequence inferred from amino acid sequence of growth hormone).
Labeled colony
5. Find probe. The labeled probe DNA binds to its complementary sequence in the cDNA library.
E. coli containing growth hormone gene
6. Identify colony. On original plates, find colony of E. coli cells that contains growth hormone gene. Sample cells, grow, and analyze.

FIGURE 19.6 Finding Specific Genes by Probing a cDNA Library.

sequence AATCG (recall that sequences are always written 5′ to 3′) will bind to a target DNA with the sequence TTTTACCCA TTTACGATTGGCCT (again written 5′ to 3′).

To find an appropriate probe for the human growth hormone gene, researchers began by using the genetic code to predict the approximate DNA sequence of *GH1*. This was possible because the sequence of amino acids in the polypeptide was known. Thus, the researchers could roughly infer the mRNA codon and DNA sequence that coded for each amino acid. You made similar inferences in some of the exercises in Chapter 15. But recall from Chapter 15 that the genetic code is redundant, with more than one codon for most amino acids. As a result, the sequence inferred for the growth hormone gene was actually a set of related sequences.

The next step was to synthesize many copies of a short, single-stranded stretch of DNA that was complementary to the inferred sequence. Because these molecules would bind to single-stranded fragments from the actual gene by complementary base pairing, they could act as a probe. In this case, the label the researchers attached to the probe was a radioactive atom.

Figure 19.6 shows how researchers used this probe to find the plasmid in the cDNA library that contained *GH1*. (For more information on how to use probes, see **BioSkills 9** in Appendix A.) As predicted, the labeled probe bound to its complementary sequence in the cDNA library—identifying the recombinant cell that contained human growth hormone.

MASS-PRODUCING GROWTH HORMONE To accomplish their goal of producing large quantities of the human growth hormone, the investigators used recombinant DNA techniques to transfer the growth hormone cDNA to a new plasmid. The plasmid in question contained a promoter sequence recognized by *E. coli*'s RNA polymerase holoenzyme (see Chapter 16). The recombinant plasmids were then introduced into *E. coli* cells.

The transformed *E. coli* cells that resulted contained a gene for human growth hormone attached to an *E. coli* promoter. These cells began to transcribe and translate the human growth hormone gene. Human growth hormone accumulated in the cells and was subsequently isolated and purified.

Today, bacterial cells containing the human growth hormone gene are grown in huge quantities. These cells have proved to be a safe and reliable source of the human growth hormone protein. The effort to cure pituitary dwarfism using recombinant DNA technology was a spectacular success—a triumph of applied biology, or **biotechnology**. To review this work, go to the study area at *www.masteringbiology.com*.

Web Activity Producing Human Growth Hormone

Ethical Concerns over Recombinant Growth Hormone

As supplies of growth hormone increased, physicians used it in treating not only people with pituitary dwarfism but also children of short stature who had no actual growth hormone deficiency. Even though the treatment requires several injections per week until adult stature is reached, growth hormone therapy was popular because it often increased the height of these children by a few centimeters.

In essence, growth hormone was being used as a cosmetic—a way to improve appearance in cultures where height is deemed attractive. But if short people are discriminated against in a culture, is a medical treatment a better solution than education and changes in attitudes? And what if parents wanted a tall child to be even taller, to enhance her potential success as, say, a basketball player?

Currently, the U.S. Food and Drug Administration has approved the use of human growth hormone for only the shortest 1.2 percent of children. These individuals are projected to reach adult heights of less than 160 cm (5′3″) in males and 150 cm (4′11″) in women.

Growth hormone has also become a popular performance-enhancing drug for athletes, because it improves the maintenance of bone density and muscle mass. Part of its popularity stems from the fact that it is virtually undetectable in the drug tests currently administered by governing bodies.

Should athletes be able to enhance their physical skills by taking hormones or other types of drugs? Is the drug safe at the dosages athletes are using? These questions are being debated by physicians, researchers, agencies that govern sports, and legislative bodies.

In the meantime, it is clear that while solving one important problem, recombinant DNA technology created others. One of this chapter's recurring themes is that genetic engineering has costs that must be carefully weighed against its benefits.

CHECK YOUR UNDERSTANDING

If you understand that . . .

- The essence of recombinant DNA technology is to cut DNA into fragments with a restriction endonuclease, paste specific sequences together by complementary base pairing of sticky ends and the action of DNA ligase, and insert the resulting recombinant genes into a bacterial (or yeast) cell so that the genes are expressed.
- A DNA library consists of cloned sequences that have been inserted into plasmids or other vectors. A probe can be used to find specific sequences in the library.

You should be able to . . .

1. Explain why restriction endonucleases create DNA fragments with sticky ends.
2. Explain why "probe" is an appropriate term for a labeled sequence that is used to find a particular gene in a DNA library.

Answers are available in Appendix B.

19.2 Case 2—Amplification of Fossil DNA: The Polymerase Chain Reaction

Inserting a gene into a bacterial plasmid is one method for cloning DNA. The polymerase chain reaction is another.

The **polymerase chain reaction (PCR)** is an in vitro DNA synthesis reaction in which a specific section of DNA is replicated over and over, by DNA polymerase, to amplify the number of copies of that sequence. It's a technique for generating many identical copies of a particular section of DNA.

Requirements of PCR

Although PCR is much faster and technologically easier than cloning genes into a DNA library, there is a catch: PCR is possible only when a researcher already has some information about DNA sequences near the gene in question. Sequence information is required because to do a polymerase chain reaction, you have to start by synthesizing short lengths of single-stranded DNA

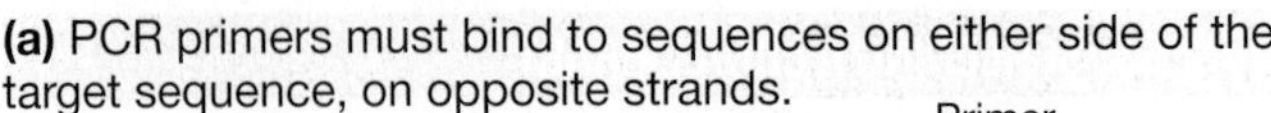

(a) PCR primers must bind to sequences on either side of the target sequence, on opposite strands.

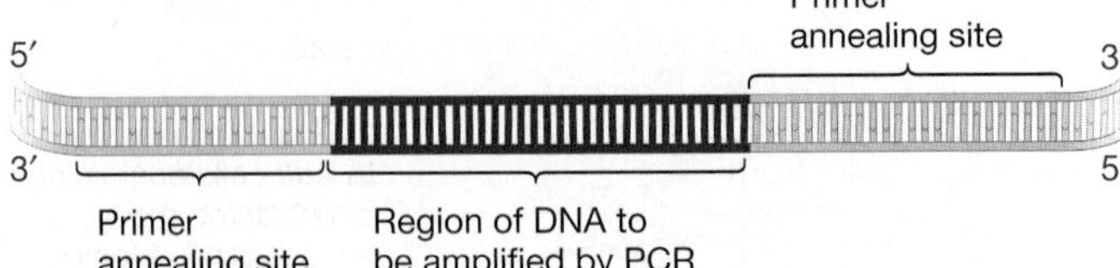

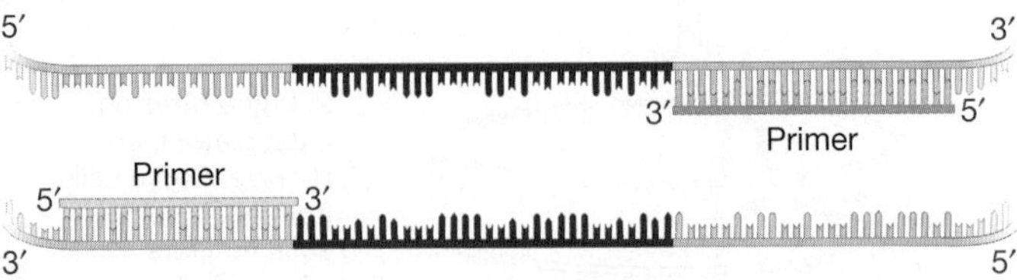

FIGURE 19.7 The Polymerase Chain Reaction Requires Appropriate Primers. (a) The "primer-annealing sites" are sequences where a primer will bind. To design an appropriate primer, the base sequence at these annealing sites must be known. **(b)** The primers bind to single-stranded target DNA, as shown.

EXERCISE Indicate where DNA polymerase would begin to work on each strand; add an arrow indicating the direction of DNA synthesis.

that match sequences on either side of the gene of interest. These short segments act as primers for the synthesis reaction.

As **Figure 19.7a** shows, the primer sequences must be complementary to bases on either side of the target gene—the DNA you wish to copy. One primer is complementary to a sequence on one strand upstream of the target DNA; the other primer is complementary to a sequence on the other strand, downstream of the target DNA. If the target DNA molecule is made single stranded, then the primers will bond, or anneal, to their complementary sequences, as shown in **Figure 19.7b**. You might recall that DNA polymerase cannot work without a primer. Once the primers are bound, DNA polymerase can extend each strand in the 5′ to 3′ direction.

Figure 19.8 shows the sequence of events in a PCR experiment.

Step 1 The researcher creates a reaction mix containing an abundant supply of the four deoxyribonucleoside triphosphates (dNTPs; see Chapter 14), a DNA sample that includes the gene of interest, many copies of the two primers, and an enzyme called *Taq* polymerase (see below).

Step 2 The reaction mix is heated to 95°C. At this temperature, the double-stranded template DNA denatures. This means that the two DNA strands separate, forming single-stranded templates.

Step 3 The mixture is allowed to cool to 50–60°C. In this temperature range, the primers bond, or anneal, to complementary portions of the single-stranded template DNA. This step is called primer annealing.

Step 4 The reaction mix is heated to 72°C. At this temperature *Taq* polymerase synthesizes the complementary DNA strand from the dNTPs, starting at the primer. This step is called *extension.*

Step 5 Repeat steps 2 through 4.

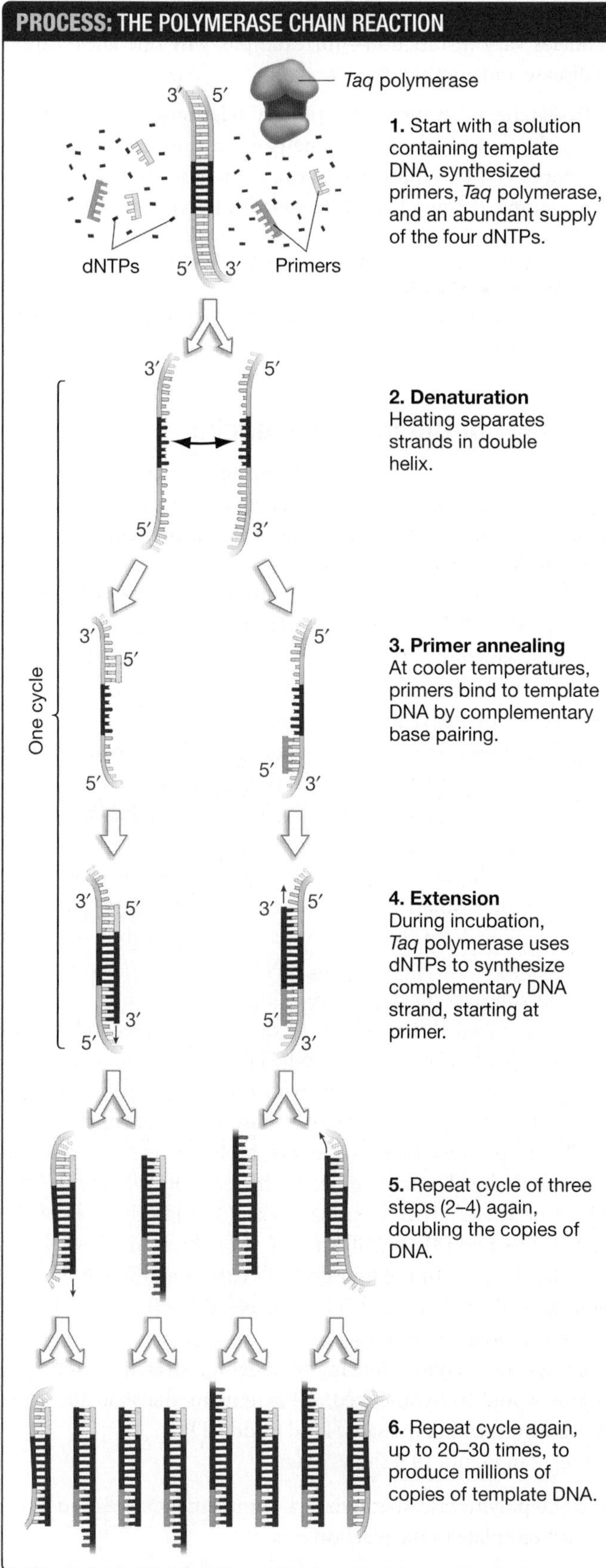

PROCESS: THE POLYMERASE CHAIN REACTION
Taq polymerase
3′
5′
dNTPs
5′
3′
Primers
1. Start with a solution containing template DNA, synthesized primers, Taq polymerase, and an abundant supply of the four dNTPs.
2. Denaturation
Heating separates strands in double helix.
One cycle
3. Primer annealing
At cooler temperatures, primers bind to template DNA by complementary base pairing.
4. Extension
During incubation, Taq polymerase uses dNTPs to synthesize complementary DNA strand, starting at primer.
5. Repeat cycle of three steps (2–4) again, doubling the copies of DNA.
6. Repeat cycle again, up to 20–30 times, to produce millions of copies of template DNA.

FIGURE 19.8 The Polymerase Chain Reaction Is a Method for Producing Many Copies of a Specific Sequence. Each PCR cycle (denaturation, primer annealing, and extension) results in a doubling of the number of target and primer sequences.

Step 6 Continue repeating steps 2 through 4 until the necessary number of copies is obtained.

Today, the temperature changes required in each step are controlled by automated PCR machines.

Taq polymerase is a DNA polymerase found in the bacterium *Thermus aquaticus*, which was originally discovered in a hot spring inside Yellowstone National Park, Wyoming. Researchers use *Taq* polymerase because the PCR reaction mix has to be heated, and *Taq* polymerase is heat stable. Most DNA polymerases are destroyed at high temperature, but *Taq* polymerase continues to function normally even when heated to 95°C.

The denaturation, primer annealing, and extension steps constitute a single PCR cycle. If one copy of the template sequence existed in the original sample, then two copies are present at the end of the first cycle (see step 4 in Figure 19.8). These two copies then act as templates for the second cycle—another round of denaturation, primer annealing, and extension—after which four copies of the target gene are present (see step 5).

Each time the cycle repeats, the amount of template sequence in the reaction mixture doubles (step 6). Doubling occurs because each newly synthesized segment of DNA serves as a template in the subsequent cycle, along with the previously synthesized segments. Starting with a single copy, successive cycles result in the production of 2, 4, 8, 16, 32, 64, 128, 256 copies, and so on. A total of n cycles can generate 2^n copies; so in just 20 cycles, one sequence can be amplified to over a million copies. By performing up to 30 cycles, researchers obtain enormous numbers of copies of the template sequence.

Before reading further, be sure to review how PCR works in the study area at *www.masteringbiology.com*.

 **Web Activity** The Polymerase Chain Reaction

PCR in Action

To understand why PCR is so valuable, consider a study by biologist Svante Pääbo and colleagues, who wanted to analyze DNA recovered from the 30,000-year-old bones of a fossilized human of the species *Homo neanderthalensis*. Their goal was to determine the sequence of bases in the ancient DNA and compare it with DNA from modern humans (*Homo sapiens*).

If modern humans have sequences that are identical or almost identical to the sequences found in Neanderthals, it would suggest that some of us inherited DNA directly from a Neanderthal ancestor. That could happen only if *H. sapiens* and *H. neanderthalensis* interbred while they coexisted in Europe.

The Neanderthal bone was so old, however, that most of the DNA in it had degraded into tiny fragments. The biologists could recover only a minute amount of DNA that was still in moderate-sized pieces. Fortunately, the Neanderthal DNA sample included a few fragments of the gene region that Pääbo's team wanted to study. The researchers were able to design primers that bracketed this gene region, based on the sequence of highly conserved sections of the same gene from *H. sapiens*.

Using PCR, the researchers produced millions of copies of the Neanderthal DNA fragment. After analyzing these sequences, the team found that they differ from the same gene segment found in modern humans. Subsequent work with DNA from 14 other Neanderthal fossils, from locations throughout Europe, gave the same result. These data support the hypothesis that Neanderthals never interbred with modern humans—even though the two species lived in the same areas of Europe at the same time.

PCR has been used to study other fossil DNAs, as well. The current record holder for oldest DNA to be amplified by PCR came from 17-million-year-old magnolia trees. PCR is useful any time a researcher needs a large number of copies of a particular gene. For example,

- Forensic biologists, who use biological analyses to help solve crimes, clone DNA from tiny drops of blood or hair gathered at crime scenes. The copied DNA can then be analyzed to identify victims or implicate perpetrators.
- Genetic counselors, who advise pregnant couples on how likely their offspring are to suffer from inherited diseases, can use PCR to find out if an embryo being carried by a client has alleles associated with deadly illness.

Because the complete genomes of a wide array of organisms have now been sequenced, researchers can find appropriate primer sequences to use in cloning almost any target gene by PCR. The polymerase chain reaction is now one of the most basic and widely used techniques in molecular biology.

CHECK YOUR UNDERSTANDING

If you understand that . . .

- PCR is a technique (summarized in Figure 19.8) for amplifying a specific region of DNA into millions of copies, which can then be sequenced or used for other types of analyses.

✓ You should be able to . . .

1. Explain the purpose of the denaturation, annealing, and extension steps in a PCR cycle, and why "chain reaction" is an appropriate part of PCR's name.
2. Write down the sequence of a DNA strand 50 base pairs long, then design 20-base-pair-long primers that would allow you to amplify the segment by PCR.

Answers are available in Appendix B.

19.3 Case 3—Sanger's Breakthrough Innovation: Dideoxy DNA Sequencing

Once researchers have cloned a gene from a DNA library or by PCR, determining the gene's base sequence is usually one of the first things they want to do. Understanding a gene's sequence is valuable for a variety of reasons. For example,

- Once a gene's sequence is known, the amino acid sequence of its product can be inferred from the genetic code. Knowing a protein's primary structure often provides clues to its function.
- Comparing sequences is fundamental to understanding why alleles vary in function—for example, why one allele causes disease and another doesn't.
- Researchers can infer evolutionary relationships by comparing the sequences of the same gene in different species (see Chapter 1). This information can be used to study an array of questions, ranging from how new traits evolve to where new diseases come from.

How do researchers sequence DNA? In 1977 Frederick Sanger published a technique called dideoxy sequencing that is still in use today.

The Logic of Dideoxy Sequencing

As **Figure 19.9** shows, **dideoxy sequencing** is a clever variation on the basic in vitro DNA synthesis reaction. But saying "clever" is an understatement. Sanger had to link three important insights to make his sequencing strategy work.

The first insight was to use monomers called dideoxyribonucleoside triphosphates (ddNTPs) along with deoxyribonucleoside triphosphates (dNTPs) in the reaction mix. As the top of Figure 19.9 shows, ddNTPs are identical to dNTPs except that they lack a hydroxyl group at their 3′ carbon. Four types of ddNTPs are used in dideoxy sequencing, each named according to whether it contains adenine (ddATP), thymine (ddTTP), cytosine (ddCTP), or guanine (ddGTP). The use of ddNTPs inspired the name "dideoxy" sequencing.

Sanger realized that if a ddNTP were added to a growing DNA strand, it would terminate synthesis. Why? After a ddNTP is added, no hydroxyl group is available on a 3′ carbon to link to the 5′ carbon on an incoming dNTP monomer. As a result, DNA polymerization stops once a ddNTP is added.

Sanger linked this property of ddNTPs to a second fundamental insight: Every time a ddNTP is added to a growing strand, the result is a fragment with a length corresponding to the position in the template of a base complementary to the ddNTP. To produce these fragments, biologists create a reaction mix containing **(1)** many copies of the template DNA with **(2)** a primer attached, **(3)** DNA polymerase, **(4)** a large supply of the four dNTPs, and **(5)** a small amount of the four ddNTPs (Figure 19.9, step 1). Each of the four ddNTPs carries a different fluorescent tag. (In the figure, ddGTP is purple, ddCTP is blue, ddATP is green, and ddTTP is orange.)

Under these conditions, many daughter strands of different lengths would be synthesized. All fragments that had the same length would end in the same kind of ddNTP.

Step 2 in Figure 19.9 shows why:

- DNA polymerase synthesizes a complementary strand from each template in the reaction mix.
- The synthesis of each one of these complementary strands starts at the same point—the primer.
- Because there are many dNTPs and relatively few ddNTPs in the reaction mix, dNTPs are usually incorporated opposite

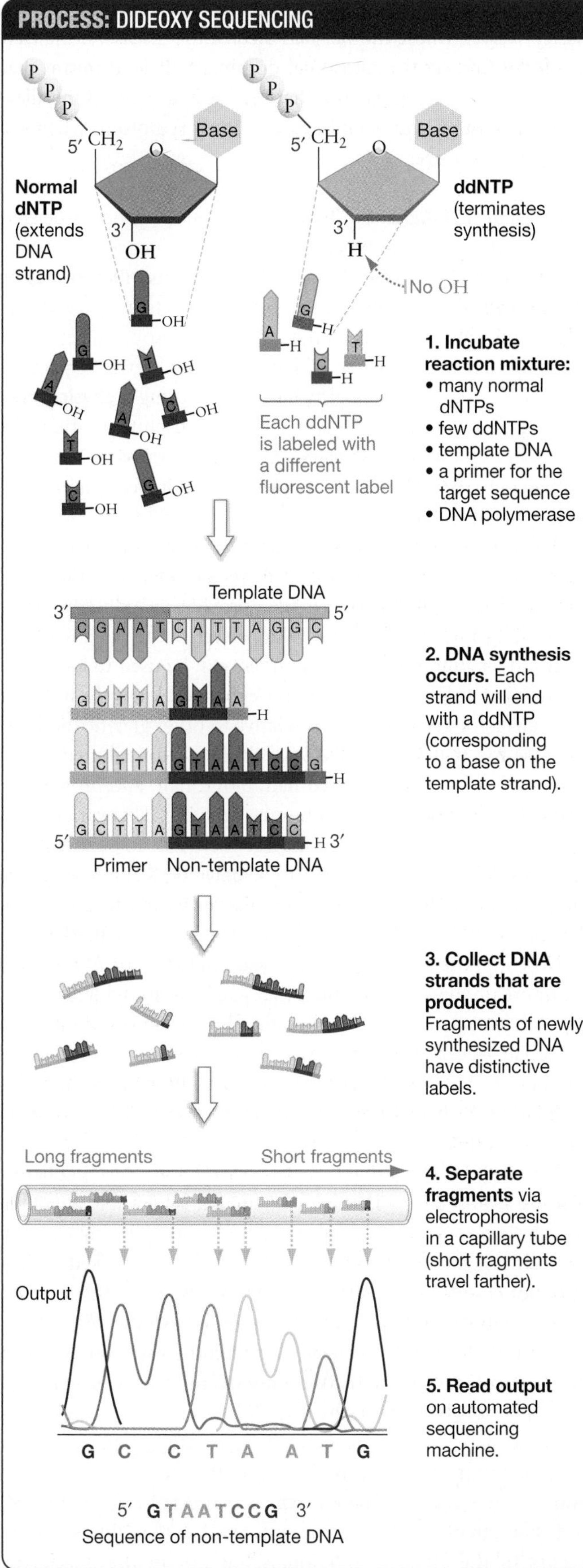
PROCESS: DIDEOXY SEQUENCING
Normal dNTP (extends DNA strand)
ddNTP (terminates synthesis)
Base
No OH
Each ddNTP is labeled with a different fluorescent label
1. Incubate reaction mixture:
• many normal dNTPs
• few ddNTPs
• template DNA
• a primer for the target sequence
• DNA polymerase
Template DNA
2. DNA synthesis occurs. Each strand will end with a ddNTP (corresponding to a base on the template strand).
Primer
Non-template DNA
3. Collect DNA strands that are produced. Fragments of newly synthesized DNA have distinctive labels.
Long fragments
Short fragments
4. Separate fragments via electrophoresis in a capillary tube (short fragments travel farther).
Output
G C C T A A T G
5. Read output on automated sequencing machine.
5′ GTAATCCG 3′
Sequence of non-template DNA

FIGURE 19.9 Dideoxy Sequencing Is a Method for Determining the Base Sequence of DNA.

each complementary base on the template strand as DNA works its way along the template strand.

- Occasionally, one of the few ddNTPs present is incorporated into the growing strand, opposite the corresponding base in the template. Which complementary base in the template strand gets paired with a ddNTP is random.
- The addition of the ddNTP at this location stops further elongation.
- "Stops" of this kind happen for each base in the template strand. As a result, the overall reaction produces a collection of newly synthesized strands whose various lengths correspond to the location of each base in the template strand (see step 3 in Figure 19.9). Each fragment will fluoresce the color of its ddNTP.

Sanger's third insight? When the fragments produced by the synthesis reactions are lined up by size, the dideoxy monomers on the successive fragments reveal the sequence of bases in the template DNA. To line up fragments in order of size, biologists separate them using gel electrophoresis (step 4 in Figure 19.9). As step 5 shows, a machine can read the pattern of fluorescence, indicating the sequence of bases in the newly synthesized strand.

Dideoxy sequencing ranks among the greatest of all technological advances in the history of biological science. Its impact is comparable to the development of the light microscope, the electron microscope, microelectrodes for recording membrane potentials, and recombinant gene technology.

"Next Generation" Sequencing

Sequencing technology is advancing at a blindingly fast pace. Dideoxy sequencing remains the method of choice the first time a genome is sequenced from a particular species. But new approaches now make it possible to sequence other individuals of that species—to compare with the original—much faster and more cheaply than the dideoxy method.

Most of the newer methods are based on detecting the pyrophosphate molecule (two phosphate groups bonded together; see the "P–P" in Figure 14.6) that is released after a DNA polymerase adds a dNTP to a growing DNA strand. The pyrophosphates that are released during the synthesis reaction drive a set of other reactions that result in luminescence, which can be detected. Sequencing strategies like this don't involve any ddNTPs, and are sometimes called pyrosequencing or sequencing-by-synthesis.

Because pyrosequencing can be done with tiny quantities of template DNA and reactants, the process can be miniaturized. Hundreds of thousands of template DNA fragments can be attached to the same glass slide and sequenced at the same time—resulting in dramatic savings of time and money.

Sequencing the human genome for the first time took four years and cost $300 million U.S. dollars. Now researchers are contemplating the possibility of sequencing your genome in 4 minutes for less than $10,000.

CHECK YOUR UNDERSTANDING

If you understand that . . .

- Dideoxy DNA sequencing, which makes use of monomers that truncate the synthesis of DNA in an identifiable way, allows researchers to determine the sequence of bases present in a length of DNA.

✔ **You should be able to . . .**

Explain why labeled ddNTPs have to be present in small numbers relative to the number of unlabeled dNTPs.

Answers are available in Appendix B.

19.4 Case 4—The Huntington's Disease Story: Finding Genes by Mapping

Mendel had no idea what a "hereditary determinant" actually was. But now we know. Biology's molecular revolution has allowed researchers to find and characterize individual genes—to explore the connection between genotype and phenotype as explicitly and directly as possible. The question is, How do researchers find the genes associated with certain traits in the first place? How do you find the gene responsible for seed shape in peas, or white eyes in flies, or DNA polymerase III in *E. coli*?

Conceptually, the answer is simple: You begin with a map of known sites in the genome, then look for an association between one of those known sites and the phenotype you're interested in. The gene that affects the phenotype is probably close to the known site.

In practice, the process is not so simple. As an example of how this type of gene hunt is done, let's consider one of the first successful searches ever conducted for a human gene—the gene associated with Huntington's disease.

How Was the Huntington's Disease Gene Found?

Huntington's disease is a rare but devastating illness. Typically, affected individuals first show symptoms between the ages of 35 and 45. At onset, an individual appears to be clumsier than normal and tends to develop small tics and abnormal movements. As the disease progresses, uncontrollable movements become more pronounced. Eventually the affected individual twists and writhes involuntarily. Personality and intelligence are also compromised—to the extent that the early stage of this disease is sometimes misdiagnosed as the brain disorder schizophrenia. The illness may continue to progress for 10 to 20 years and is eventually fatal.

Because Huntington's disease appeared to run in families, physicians suspected that it was a genetic disease. More specifically, an analysis of pedigrees from families affected by Huntington's disease suggested that the trait was due to a single, autosomal dominant allele (see Chapter 13). To test this hypothesis, researchers set out to locate and identify the gene or genes involved. It took over 10 years of intensive effort to reach this goal.

The search for the Huntington's disease gene was led by Nancy Wexler, whose mother had died of the disease. If the trait was indeed due to an autosomal dominant allele, it meant that there was a 50 percent chance that Wexler had received the allele from her mother and would begin to show symptoms when she reached middle age.

USING GENETIC MARKERS To locate the gene or genes associated with a particular phenotype, such as a disease, researchers traditionally started with a **genetic map**, also known as a **linkage map** or **meiotic map** (see Chapter 13). Recall that a genetic map shows the relative positions of genes on the same chromosome, determined by analyzing the frequency of recombination between pairs of genes. More recently, biologists have begun using a **physical map** of the genome. A physical map records the absolute position of a gene—in numbers of base pairs—along a chromosome. Genome sequencing (see Chapter 20) is producing physical maps for a wide array of species.

A genetic map is valuable in gene hunts because it contains **genetic markers**—easily identified genes or sequences that have known locations. Each genetic marker provides a landmark—a position along a chromosome that is known relative to other markers.

To understand how genetic markers can be used to locate the positions of unknown genes, suppose that you knew the position of a hair-color gene in humans relative to other genetic markers. Suppose too that various alleles of this gene contributed to the development of black hair, red hair, blond hair, and brown hair in the group of people that you were studying. This variation in phenotype associated with the marker is crucial. To be useful in a gene hunt, a genetic marker has to be **polymorphic**, meaning that the phenotype associated with the marker varies. In our hypothetical example, hair color is a polymorphic genetic marker.

Now suppose that the genetic disease called cystic fibrosis was common among the individuals that you were studying. Further, suppose that people who had cystic fibrosis almost always had black hair—even though they were just as likely as unaffected individuals to have any other inherited trait observed in the study population, such as the presence or absence of a widow's peak or detached earlobes.

If you observe that a certain marker and a certain phenotype are almost always inherited together, it is logical to conclude that the genes involved are physically close to each other on the same chromosome—meaning that they are closely linked. If they were not closely linked, then crossing over between them would be common and they would *not* be inherited together. As a result, you could infer that the gene for cystic fibrosis was very close to the hair-color gene. ✔If you understand this concept, you should be able to explain why it's helpful to hunt for genes using a genetic map with many genetic markers rather than only a few.

Gene hunts in humans boil down to this: Researchers have to find a large number of people who are affected and unaffected, and then attempt to locate a genetic marker that almost always occurs in the affected individuals but not in the unaffected people. If such a marker is found, the disease gene is almost guaranteed to be nearby.

The types of genetic markers used in gene mapping have changed over time. Today, researchers often have a large catalog of genetic markers available, including the particularly abundant markers known as **single nucleotide polymorphisms** (**SNPs**, pronounced "snips"). A SNP is a site in DNA where some individuals in the population have a different base. For example, some might have an A at a certain site while others have a C. That site would qualify as a SNP.

In the late 1970s and early 1980s, when biologists were searching for the Huntington's gene, SNPs were unknown. The best genetic markers available were restriction sites—short stretches of DNA where restriction endonucleases cut the double helix. These sequences are also known as restriction endonuclease recognition sites.

The restriction sites that Wexler's team used were polymorphic: Some individuals had sites where cuts occurred, but in other individuals, the DNA sequence at the same site was different, and no cuts occurred. Thus, each restriction site was either present in an individual or not present—just as an individual might have an A instead of a C at a certain SNP. Wexler's team was looking for restriction sites that were almost always present in diseased individuals but absent in healthy individuals, or vice versa (**Figure 19.10**).

AN ASSOCIATION STUDY Once a genetic map containing many genetic markers has been assembled, biologists need help from individuals affected by an inherited disease to find the gene in question. Recall that the fundamental goal is to find a genetic marker that is almost always inherited along with the disease-causing allele. Biologists call this an association study. Gene hunts based on association studies are more likely to be successful if large groups are involved. Large sample sizes minimize the probability that researchers will observe an association between one or more markers and the disease just by chance, rather than because they are closely linked.

Huntington's disease is rare, so obtaining a large sample size is problematic. But Wexler's Huntington's disease team was fortunate to find a large, extended family affected with the disease living along the shores of Lake Maracaibo, Venezuela.

From historical records, the researchers deduced that the Huntington's disease allele was introduced to this family by a European sailor or trader who visited the area in the early 1800s. When family members agreed to participate in the study, there were over 3000 of his descendants living in the area. One hundred of these people had been diagnosed with Huntington's disease. To help in the search for the gene, family members agreed to donate skin or blood samples for DNA analysis.

When Wexler's team looked for associations between the presence or absence of the disease phenotype and the genetic markers observed in each family member, markers called A, B, C, and D turned out to be especially important. These four combinations of restriction sites represent alleles. The key finding was that the C allele was almost always found in diseased individuals. Apparently, the English sailor who introduced the Huntington's disease allele also had the C allele—this particular combination of restriction sites—in his DNA. The C allele and the Huntington's disease gene are so close together that recombination between them—which would put an A or B or D allele next to the Huntington's disease allele in his descendants—has been extremely rare.

From the genetic map of humans that was available at the time, Wexler's team knew that the C restriction sites were on chromosome 4. Eventually the team succeeded in narrowing down the location of the C sites, and thus the Huntington's disease gene, to a region about 500,000 base pairs long. Because the haploid human genome contains over 3 billion base pairs, this was a huge step in focusing the search for the gene.

PINPOINTING THE DEFECT Once the general location of the Huntington's disease gene was known, biologists looked in that region for exons that encode a functional mRNA. Then they sequenced exons from diseased and normal individuals, compared the data, and pinpointed specific bases that differed between the two groups of individuals. In this way, dideoxy sequencing played a key role in the gene hunt.

When this analysis was complete, the research team found that individuals with Huntington's disease have an unusual number of CAG codons near the 5′ end of a particular gene. CAG codes for glutamine. Healthy individuals have 11–25 copies of the CAG codon at that location, while affected individuals have 42 or more copies.

When the Huntington's disease research team confirmed that the increase in the CAG codon was always observed in affected individuals, the team concluded that the long search for the Huntington's disease gene was over. They named the newly discovered gene *IT15* and its protein product huntingtin. In both affected and normal individuals, the huntingtin protein is involved in the early development of nerve cells. It is only later in life that mutant forms of the protein cause disease.

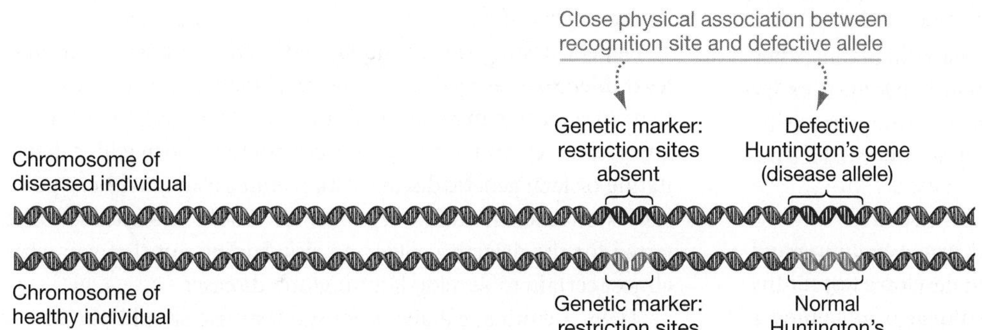

FIGURE 19.10 Genetic Markers and Disease Alleles Are Inherited Together if They Are Closely Linked. Because of genetic recombination, genetic markers that are far from the gene of interest are equally likely to be found in both affected and unaffected individuals—there will be no association between the marker and either the normal or the disease-causing allele.

What Are the Benefits of Finding a Disease Gene?

How have efforts to find disease genes improved human health and welfare? Has the effort to locate the Huntington's disease gene helped researchers and physicians understand and treat the illness? Biomedical researchers point to three major benefits of successful disease-gene hunts.

IMPROVED UNDERSTANDING OF THE PHENOTYPE Once a disease gene is found and its sequence is known, researchers can usually figure out why its product causes disease. In the case of *IT15*, autopsies of Huntington's patients had shown that their brains decrease in size, and that the brain tissue contains clumps, or aggregates, of the huntingtin protein.

Huntingtin aggregates are a direct consequence of changes in the number of CAG repeats in the *IT15* gene. Long stretches of polyglutamine (a polymer of the amino acid glutamine) are known to result in the formation of protein aggregates. The leading hypothesis to explain Huntington's disease proposes that a gradual buildup of polyglutamine aggregates triggers neurons to undergo apoptosis, or programmed cell death.

These results explained why Huntington's disease is pleiotropic (see Chapter 13). Patients suffer from abnormal movements *and* personality changes because neurons from throughout the brain are being killed. The results also help explain why the disease takes so long to appear, and why it is progressive: The defective huntingtin proteins take time to build up to deadly levels, but then continue to increase over time. Finally, understanding the molecular mechanism responsible for the illness explained why the disease allele is dominant. One copy of the defective gene is enough to produce fatal concentrations of aggregates.

THERAPY Once *IT15* was found, biologists began a search for new therapies for Huntington's disease by introducing the defective allele into mice, using the types of genetic engineering techniques discussed in Section 19.5. These mice are called **transgenic** (literally, "across-genes"), because they have alleles that have been modified by genetic engineering.

Transgenic mice that produce defective versions of the huntingtin protein develop a version of Huntington's disease, exhibiting tremors and abnormal movements, higher-than-normal levels of aggression toward litter and cage mates, and a loss of neurons in the brain. Laboratory animals with disease symptoms that parallel those of a human disease are said to provide an **animal model** of the disease.

Animal models are valuable in disease research, because they can be used to test potential treatments before investigators try them on human patients. For example, research groups are using transgenic mice to test drugs that appear to prevent or reduce the aggregation of the huntingtin protein.

GENETIC TESTING When the Huntington's gene was found and sequenced, biologists used the knowledge to develop a test for the presence of the defective allele. The test consists of obtaining a DNA sample from an individual and using the polymerase chain reaction to amplify the chromosome region that contains the CAG repeats responsible for the disease. If the number of CAG repeats is 35 or less, the individual is not considered at risk. Forty or more repeats results in a positive diagnosis for Huntington's.

Thanks to genetic maps based on SNPs, gene hunts are increasingly successful. Biologists have recently documented alleles associated with a predisposition to developing type I diabetes, type II diabetes, breast cancer, obesity, coronary heart disease, bipolar disorder, Crohn's disease, and rheumatoid arthritis. Testing for these alleles is now available.

Three types of genetic testing are currently done for genetic diseases:

1. *Carrier testing* can determine whether an individual is a carrier for a genetic disease. Before starting their own family, people from families affected by a genetic disease frequently want to know whether they carry the allele responsible. This is especially true for diseases, such as cystic fibrosis (CF), that are due to recessive alleles. If only one of the prospective parents has the allele, then none of the children they have together should develop CF. But if both prospective parents carry the allele, then each child they produce has a 25 percent chance of having CF.
2. *Prenatal testing* analyzes DNA from fetal cells obtained early in gestation. Alleles for genetic diseases can then be amplified by PCR and sequenced. Suppose that two parents, both carrying the CF allele, decide to have children but do not want to have a child that will suffer from the disease. They can use the results of prenatal testing as a basis for deciding either to continue or terminate a pregnancy.
3. *Adult testing* is particularly important for diseases that have a strong environmental component. For example, five to ten percent of the women who develop breast cancer have inherited a faulty gene. If testing reveals that a woman has a susceptibility allele, she can be checked for the onset of cancer more frequently than normal. She may also be able to make dietary and other changes that make her less likely to actually develop the illness.

Ethical Concerns over Genetic Testing

Tables 19.1 and **19.2** on pages 352–353 summarize the tools and techniques discussed in this chapter. Although these advances have already cured serious diseases and promise additional benefits for human health and welfare, they also raise difficult ethical issues.

Genetic testing, for example, can create serious moral and legal dilemmas, as well as harrowing personal choices. Consider that some people maintain that it is morally wrong to terminate any pregnancy, even if the fetus is certain to be born with a debilitating or fatal genetic disease. Think about Nancy Wexler's position soon after the discovery of *IT15:* Would you choose to be tested for the defective allele and risk finding out that you were almost certain to develop Huntington's disease?

There are other, equally serious, questions. Should physicians agree to test people for genetic diseases that have no cure? Should it be legal for health insurance companies to test clients? If so, can companies refuse to insure people at risk for diseases that require expensive treatments? What about life insurance companies? Or employers?

These questions are being debated by political and religious leaders, health-care workers, philosophers, and the public at large. In many cases, we've yet to find answers.

CHECK YOUR UNDERSTANDING

If you understand that . . .

- Genes for particular traits can be located if they are closely linked to a known genetic marker and are thus inherited along with the marker.

✔ You should be able to . . .

Describe how you would design a study with the goal of identifying alleles associated with the development of alcoholism in humans.

Answers are available in Appendix B.

19.5 Case 5—Severe Immune Disorders: The Potential of Gene Therapy

For physicians who treat inherited diseases such as Huntington's, sickle-cell anemia, and cystic fibrosis, the ultimate goal is to replace or augment defective copies of the gene with normal alleles. This approach to treatment is called **gene therapy**.

For gene therapy to succeed, two crucial requirements must be met. First, the allele associated with the healthy phenotype must be sequenced and its regulatory sites must be understood. Second, a method must be available for introducing this allele into affected individuals. The DNA has to be introduced in a way that ensures expression of the gene in the correct tissues, in the correct amount, and at the correct time. If the defective allele is dominant, then the introduction step may be even more complicated: In at least some cases, the introduced allele must physically replace or block the expression of the undesirable dominant allele.

How Can Novel Alleles Be Introduced into Human Cells?

Section 19.1 reviewed how recombinant DNA sequences are packaged into plasmids and taken up by *E. coli* cells. Humans and other mammals lack plasmids, however, and their cells do not take up foreign DNA in response to chemical or electric treatments as efficiently as bacterial cells do. How can foreign genes be introduced efficiently into human cells?

To date, researchers have focused on packaging foreign DNA into viruses for transport into human cells. As Chapter 35 will detail, viral infection begins when a virus particle enters or attaches to a host cell and inserts its genome into that host cell. In some cases the viral DNA becomes integrated into a host-cell chromosome. This tendency enables viruses that infect human cells to be used as vectors to carry engineered alleles into the chromosomes of target cells. Potentially, the alleles delivered by the virus could be expressed and produce a product capable of curing a genetic disease.

Currently, the vectors of choice in gene therapy are retroviruses. When a **retrovirus** infects a human cell, reverse transcriptase catalyzes the production of a DNA copy of the virus's RNA genome. Other viral enzymes catalyze the insertion of the viral DNA into a host-cell chromosome. If human genes can be packaged into a retrovirus, then the virus is capable of inserting the human alleles into a chromosome in a target cell (**Figure 19.11**).

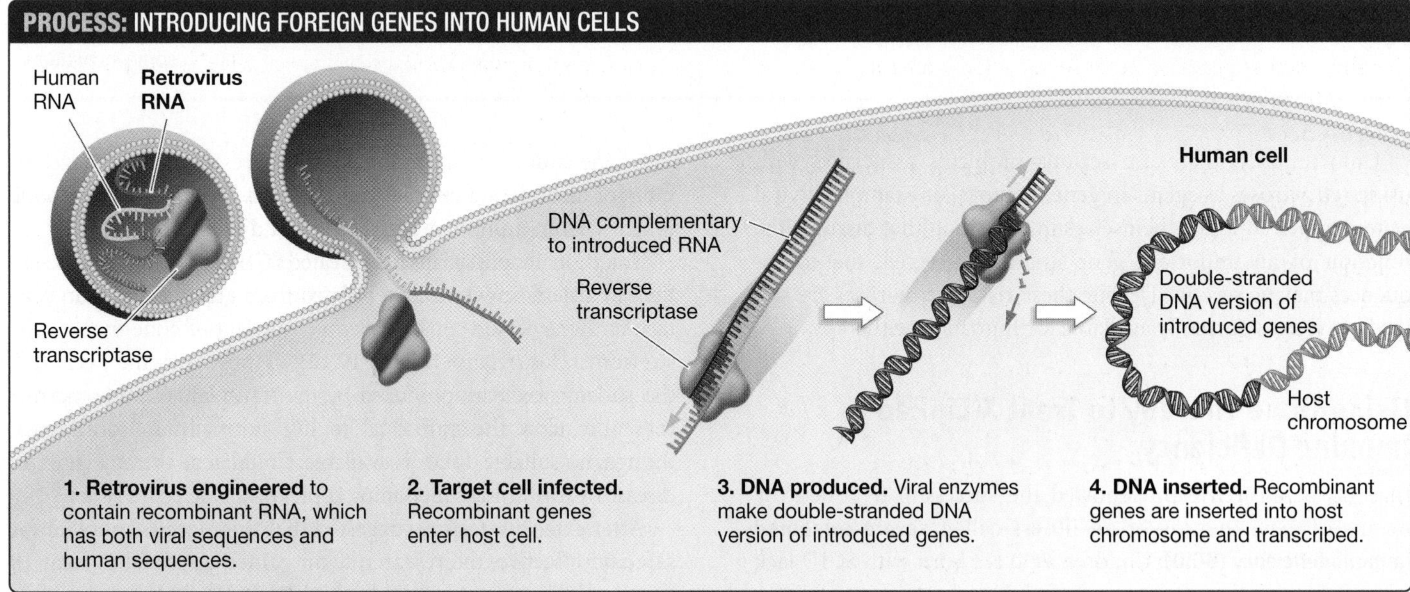
PROCESS: INTRODUCING FOREIGN GENES INTO HUMAN CELLS
Human RNA
Retrovirus RNA
Reverse transcriptase
Human cell
DNA complementary to introduced RNA
Reverse transcriptase
Double-stranded DNA version of introduced genes
Host chromosome
1. Retrovirus engineered to contain recombinant RNA, which has both viral sequences and human sequences.
2. Target cell infected. Recombinant genes enter host cell.
3. DNA produced. Viral enzymes make double-stranded DNA version of introduced genes.
4. DNA inserted. Recombinant genes are inserted into host chromosome and transcribed.

FIGURE 19.11 Retroviruses Insert Their Genes into Host-Cell Chromosomes.

✔QUESTION What happens if the recombinant DNA is inserted in the middle of a gene that is critical to normal cell function?

SUMMARY TABLE 19.1 **Some Common Tools Used in Genetic Engineering**

Tool	Description	How Used	Illustration
Reverse transcriptase	Enzyme that catalyzes synthesis of a complementary DNA (cDNA) from an RNA template.	Many applications, including making cDNAs used in constructing a genetic library.	cDNA RNA
Restriction endonucleases	Enzymes that cut DNA at a specific sequence—often a palindromic sequence that is six base pairs long.	Allows researchers to cut DNA at specific locations. Cuts in palindromic sites create "sticky ends."	GAATTC CTTAAG
DNA ligase	An enzyme that catalyzes the formation of a phosphodiester bond between nucleotides on the same DNA strand.	Ligates (joins) sequences that were cut with a restriction endonuclease. Gives researchers the ability to splice fragments of DNA together.	GAATTC CTTAAG
Plasmids	Small, extrachromosomal loops of DNA found in many bacteria and in some yeast.	After a target gene is inserted into a plasmid, the recombinant plasmid serves as a vector for transferring the gene into a bacterial or yeast cell, so the gene can be cloned.	GAATTC CTTAAG GAATTC CTTAAG
***Taq* polymerase**	DNA polymerase from the bacterium *Thermus aquaticus*. Catalyzes synthesis of DNA from a primed DNA template; remains stable at 95°C.	Responsible for the "primer extension" step in the polymerase chain reaction. Heat stability allows enzyme to be active even after denaturation step of PCR cycle at 95°C.	
Single nucleotide polymorphisms (SNPs)	Sites in DNA where the identity of the base varies among individuals in a population.	One of many polymorphic types of DNA sequences that are useful in creating the genetic maps required for gene hunts.	Sequence of some individuals ... GATTCTTATTAACTACGC ... ... GATTCTTCTTAACTACGC ... Sequence of some individuals

Unfortunately, there are serious problems associated with using retroviruses as agents in gene therapy. For example, if viral genes happen to insert themselves in a position that disrupts the function of an important gene in the target cell, the consequences may be serious. Despite these risks, retroviruses are still the best vectors currently available for human gene therapy.

Using Gene Therapy to Treat X-Linked Immune Deficiency

In 2000, a research team reported the successful treatment by means of gene therapy of an illness called **severe combined immunodeficiency (SCID)**. Children who are born with SCID lack a normal immune system and are unable to fight off infections.

The type of SCID the team treated is designated SCID-X1, because it is caused by mutations in a gene on the X chromosome. The gene codes for a receptor protein necessary for the development of immune system cells called T cells. T cells develop in bone marrow, from undifferentiated cells that divide continuously.

Traditionally, physicians have treated SCID-X1 by keeping the patient in a sterile environment, isolated from any direct human contact, until the person could receive a transplant of bone-marrow tissue from a close relative (**Figure 19.12**). In most cases, the T cells that the patient needs are produced by the transplanted bone-marrow cells and allow the individual to live normally. In some cases, though, no suitable donor is available. Could gene therapy cure this disease by furnishing functioning copies of the defective gene?

After extensive testing suggested that their treatment plan was safe and effective, the research team gained approval to treat 10 boys, each less than 1 year old, who had SCID-X1 but no suitable bone-marrow donor. The researchers removed bone marrow from each patient, collected the stem cells that produce mature

SUMMARY TABLE 19.2 **Some Common Techniques Used in Genetic Engineering**

Technique	Description	How Used	Illustration
Recombinant DNA technology ("genetic engineering")	Taking a copy of a gene from one individual and placing it in the genome of a different individual (often of a different species).	Many applications, including DNA cloning, gene therapy (see Section 19.5), and biotechnology (see Sections 19.1 and 19.6).	Inserted gene
Genetic libraries	A collection of all DNA sequences present in a particular source. The library consists of individual DNA fragments that are isolated and inserted into a plasmid or other vector, so they can be cloned.	cDNA libraries allow researchers to catalog the genes being expressed in a particular cell type. Genomic libraries allow researchers to archive all the DNA sequences present in a genome. Libraries can be screened to find a particular target gene.	Stored cDNA
DNA probing/screening a genetic library	Use of a labeled, known DNA fragment to hybridize (by complementary base pairing) with a collection of unlabeled, unknown fragments.	Allows a researcher to find a particular DNA sequence in a large collection of sequences.	Labeled probe
Polymerase chain reaction (PCR)	A DNA synthesis reaction that uses known primer sequences on either side of a target gene. Reaction is based on many cycles of primer annealing, primer extension, and DNA denaturation.	Produces many identical copies of a target sequence. A shortcut method for DNA cloning.	
Dideoxy sequencing	In vitro DNA synthesis reaction that includes dideoxyribonucleotide triphosphates (ddNTPs) as monomers.	Determining the base sequence of a gene or other section of DNA.	G C C T A A T G
Genetic mapping	Creation of a map showing the relative positions of genes or specific DNA sequences on chromosomes. Done by analyzing the frequency of recombination between sequences (see Chapter 13).	Many applications, including use of mapped genetic markers to find unknown genes associated with diseases or other distinctive phenotypes.	Yellow body, White eyes, Ruby eyes; 1.4, 6.1

FIGURE 19.12 A "Bubble Child."
Children with SCID cannot fight off bacterial or viral infections. As a result, such children must live in a sterile environment.

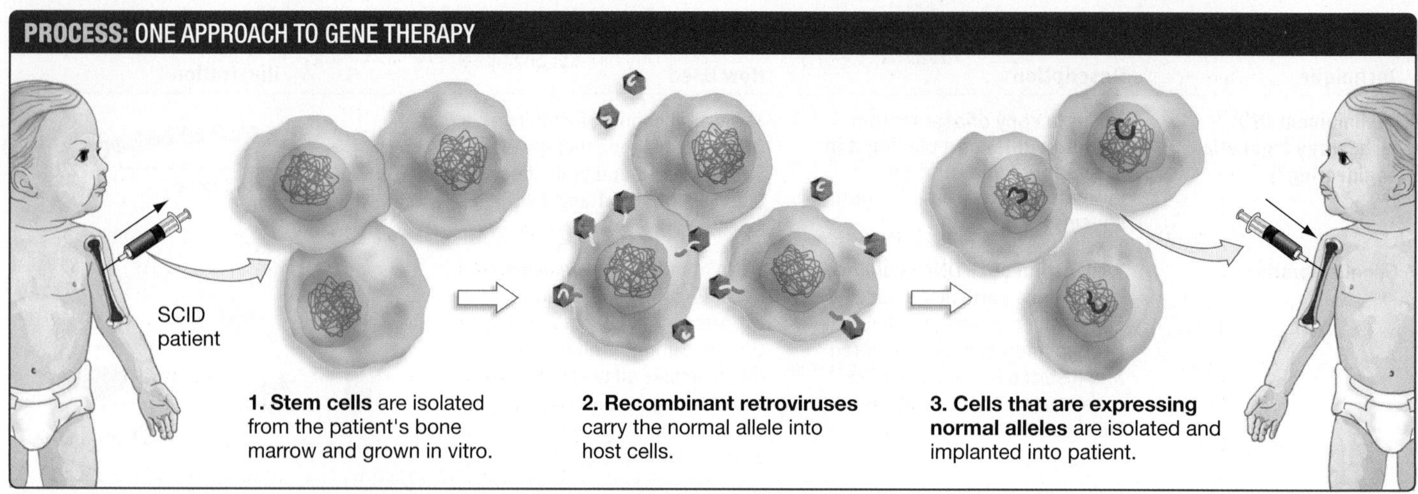
PROCESS: ONE APPROACH TO GENE THERAPY
SCID patient
1. Stem cells are isolated from the patient's bone marrow and grown in vitro.
2. Recombinant retroviruses carry the normal allele into host cells.
3. Cells that are expressing normal alleles are isolated and implanted into patient.

FIGURE 19.13 Curing a Genetic Disorder Caused by a Loss-of-Function Allele. For gene therapy to work, copies of a normal allele have to be introduced into a patient's cells and be expressed.

T cells, and infected those cells with an engineered retrovirus that carried the normal receptor gene. Cells that began to produce normal receptor protein were then isolated and transferred back into the patients (**Figure 19.13**).

Within four months after reinsertion of the transformed marrow cells, nine of the boys had normal levels of functioning T cells. These patients were removed from germ-free isolation rooms and began residing at home, where they grew and developed normally.

Subsequently, however, four of the boys developed a cancer characterized by unchecked growth of T cells. Follow-up analyses of their bone-marrow cells showed that a viral-borne receptor gene had been inserted either near a gene for a regulatory transcription factor that triggers the growth of T cells, or near a gene for a cyclin that drives the cell cycle (see Chapter 11). The viral sequences apparently acted as an enhancer and led to constitutive expression of the transcription factor or cyclin.

As this book goes to press, three of the four boys have responded to cancer chemotherapy and are healthy. The fourth did not respond to treatment and died of cancer.

The tenth boy to receive gene therapy never succeeded in producing T cells at all. For unknown reasons, his recombinant stem cells failed to function normally when they were transplanted back into his bone marrow. Fortunately, physicians were later able to find a bone marrow donor whose cells were a close enough match to his to make a successful transplant possible.

Ethical Concerns over Gene Therapy

Throughout the history of medicine, efforts to test new drugs, vaccines, and surgical protocols have always carried a risk for the patients involved. Gene therapy experiments are no different. The researchers who run gene therapy trials must explain the risks clearly and make every effort to minimize them.

The initial report on the development of cancer in the boys who received gene therapy for SCID-X1 concluded with the following statement: "We have proposed . . . a halt to our trial until further evaluation of the causes of this adverse event and a careful reassessment of the risks and benefits of continuing our study of gene therapy."

When recombinant DNA technology first became possible, many researchers thought they would live to see most or all of the serious inherited diseases in humans cured by gene therapy. Several decades later, that optimism is tempered. In humans, gene therapy is still highly experimental and extremely expensive. Researchers are beginning to doubt whether gene therapy will ever be safe, effective, and affordable enough to be considered an important medical advance.

19.6 Case 6—The Development of Golden Rice: Biotechnology in Agriculture

Progress in human gene therapy has been slow, but progress in transforming crop plants with recombinant genes has been breathtakingly rapid. Worldwide, about 114 million hectares (282 million acres) of transgenic crops were grown in 2007. You have almost certainly eaten food from a genetically modified plant at some time, if not today.

Recent efforts to develop transgenic plants have focused on three general objectives:

1. ***Reducing losses from herbivore damage*** For instance, researchers have transferred a gene from the bacterium *Bacillis thuringiensis* into corn; the presence of the "Bt toxin" encoded by this gene protects the plant from corn borers and other caterpillar pests.
2. ***Reducing competition with weeds*** An example is the genetic engineering of soybeans for resistance to an herbicide—a molecule that kills plants—called glyphosate. Soybean fields with the engineered strain can be sprayed with glyphosate to kill weeds without harming the soybeans.

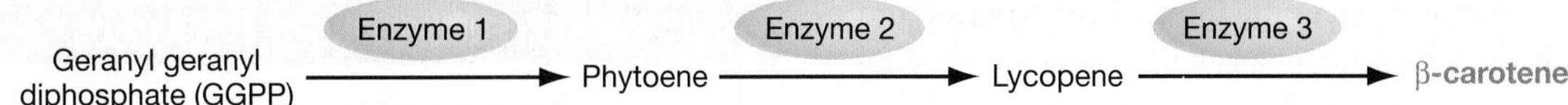
Enzyme 1
Enzyme 2
Enzyme 3
Geranyl geranyl diphosphate (GGPP)
Phytoene
Lycopene
β-carotene

FIGURE 19.14 Synthetic Pathway for β-Carotene. GGPP is a molecule found in rice seeds. Three enzymes are required to produce β-carotene from GGPP.

3. ***Improving the quality of the product consumed by people*** This objective is exemplified by the crop plants, including soybeans and canola, that have been engineered to produce a higher percentage of unsaturated fatty acids relative to saturated fatty acids (see Chapter 6). Saturated fatty acids can contribute to heart disease, so crops with less of them are healthier to eat.

How is this work done?

Rice as a Target Crop

Almost half the world's population depends on rice as its staple food. Unfortunately, rice is a poor source of certain vitamins and essential nutrients—including vitamin A. Vitamin A deficiency causes blindness in 250,000 Southeast Asian children each year. It also increases susceptibility to diarrhea, respiratory infections, and childhood diseases such as measles.

Humans and other mammals synthesize vitamin A from a precursor molecule known as β-carotene (beta-carotene). β-carotene belongs to a family of plant pigments called the carotenoids (see Chapter 10). Carotenoids are orange, yellow, and red and are especially abundant in carrots.

Rice plants synthesize β-carotene in their chloroplasts but not in the part of the seed that is eaten by humans. Could genetic engineering produce a strain of rice that synthesized β-carotene in the carbohydrate-rich seed tissue called endosperm, which humans do eat?

Synthesizing β-Carotene in Rice

To explore the possibility of genetically engineering rice, a research team searched for compounds in rice endosperm that could serve as precursors for the synthesis of β-carotene. They found that maturing rice endosperm contains a molecule called geranyl geranyl diphosphate (GGPP), which is an intermediate in the synthetic pathway that leads to the production of carotenoids.

As **Figure 19.14** shows, three enzymes are required to produce β-carotene from GGPP. If genes that encode these enzymes could be introduced into rice plants along with regulatory sequences that would trigger their synthesis in endosperm, the researchers could produce a transgenic strain of rice that would contain β-carotene.

Fortunately, genes that encode two of the required enzymes had already been isolated from daffodils, and the gene for the third enzyme had been purified from a bacterium. Because the sequences had been inserted into plasmids and grown in bacteria, many copies were available for manipulation. To each of the coding sequences in the plasmids, biologists added the promoter region from an endosperm-specific protein. This segment would promote transcription of the recombinant sequences in endosperm cells.

Next, the three sets of sequences had to be inserted into rice plants. How are foreign genes introduced into plants?

The *Agrobacterium* Transformation System

Agrobacterium tumefaciens is a bacterium that infects plant tissues and triggers formation of tumorlike growths called galls. When researchers looked into how these infections occur, they found that a plasmid carried by the *Agrobacterium* cells, called a **Ti (tumor-inducing) plasmid**, plays a key role (**Figure 19.15**).

Ti plasmids contain several functionally distinct sets of genes. One set encodes products that allow the bacterium to bind to the cell walls of a host. Another set, referred to as the virulence genes, encodes the proteins required to transfer part of the Ti DNA, called T-DNA (transferred DNA), into the interior of the plant cell. The T-DNA then travels to the plant cell's nucleus and integrates into the plant's chromosomal DNA (Figure 19.15, step 1).

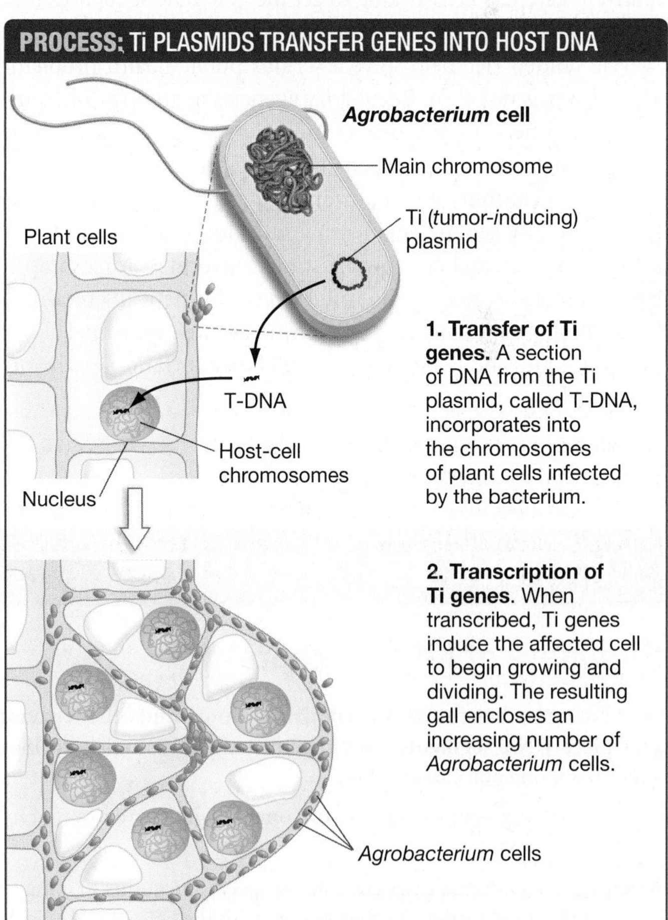
PROCESS: Ti PLASMIDS TRANSFER GENES INTO HOST DNA
Agrobacterium cell
Main chromosome
Ti (tumor-inducing) plasmid
Plant cells
1. Transfer of Ti genes. A section of DNA from the Ti plasmid, called T-DNA, incorporates into the chromosomes of plant cells infected by the bacterium.
T-DNA
Host-cell chromosomes
Nucleus
2. Transcription of Ti genes. When transcribed, Ti genes induce the affected cell to begin growing and dividing. The resulting gall encloses an increasing number of Agrobacterium cells.
Agrobacterium cells

FIGURE 19.15 Agrobacterium Infections Introduce Genes into a Host-Cell Chromosome. Ti plasmids of *Agrobacterium* cells induce gall formation—a tumorlike growth.

When transcribed, T-DNA induces the infected cell to grow and divide. The result is the formation of a gall that houses a growing population of *Agrobacterium* cells (Figure 19.15, step 2).

Researchers soon realized that the Ti plasmid offers an efficient way to introduce recombinant genes into plant cells. Follow-up experiments confirmed that recombinant genes could be added to the T-DNA that integrates into the host chromosome, that the gall-inducing genes could be removed from the T-DNA, and that the resulting sequence is efficiently transferred and expressed in its new host plant.

Using the Ti Plasmid to Produce Golden Rice

To generate a strain of rice that produces all three enzymes needed to synthesize β-carotene in endosperm, the researchers exposed embryos to *Agrobacterium* cells containing genetically modified Ti plasmids (**Figure 19.16**). The transformed plants were then grown in the greenhouse. When the transgenic individuals had matured and produced seeds, the researchers found that some rice grains contained so much β-carotene that they appeared yellow. The biologists called the engineered plants "golden rice."

Follow-up experiments used gene sequences from maize, rather than daffodil, and resulted in individuals that produce 23 times as much β-carotene in their seeds as the original transformants. Currently, researchers are working to get the new gene sequences into the rice strains that are most commonly planted in Southeast Asia.

Will golden rice help solve a serious public health problem? The answer is not clear. Regulatory agencies in an array of countries would need to approve its use, and seed would have to be made available to farmers at an affordable price.

In the meantime, it's important to recognize that each solution offered by genetic engineering introduces new issues to resolve. Some researchers and consumer advocates, for example, have expressed concerns about the safety of genetically modified foods. Biology students and others who are well informed about the techniques and issues involved will be important participants in this debate.

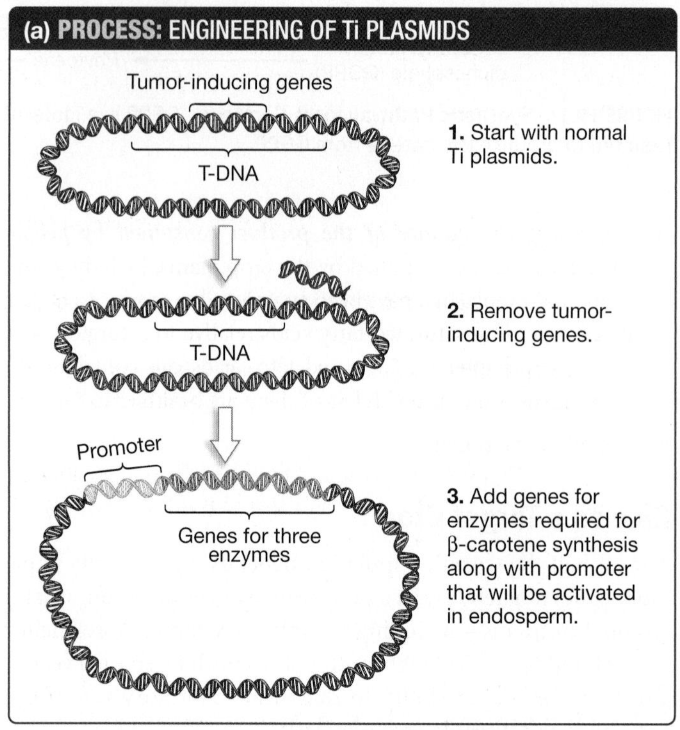
(a) PROCESS: ENGINEERING OF Ti PLASMIDS
Tumor-inducing genes
T-DNA
1. Start with normal Ti plasmids.
T-DNA
2. Remove tumor-inducing genes.
Promoter
Genes for three enzymes
3. Add genes for enzymes required for β-carotene synthesis along with promoter that will be activated in endosperm.

(b) Golden rice (right) is engineered to synthesize β-carotene.

FIGURE 19.16 Constructing a Ti plasmid that will produce "Golden Rice." Golden rice is a transgenic strain capable of synthesizing β-carotene in the endosperm of its seeds.

CHAPTER 19 REVIEW

For media, go to the study area at www.masteringbiology.com

MB

Summary of Key Concepts

Enzymes that cut DNA at specific locations and other enzymes that piece DNA segments back together allow biologists to move genes from one place to another.

- In genetic engineering, alleles from one individual are inserted into another individual.
- Restriction endonucleases cut DNA at specific locations so it may be inserted into plasmids or other vectors with the help of DNA ligase.
- In many cases, engineered alleles are modified by the addition of certain types of promoters or other regulatory sequences.

✔You should be able to explain why the discovery of DNA ligase was crucial to genetic engineering. What problem would arise for genetic engineers if the enzyme did not exist?

Biologists can obtain many identical copies of a gene by (1) inserting it into a bacterial cell that copies the gene each time the cell divides or (2) by conducting a polymerase chain reaction.

- To clone the gene for the human growth hormone, researchers isolated the gene, introduced it into a plasmid that was taken up by *E. coli* cells, and cultured the *E. coli* cells.

- The polymerase chain reaction (PCR) depends on having primers that bracket a target stretch of DNA. These allow *Taq* polymerase, a heat-stable form of DNA polymerase, to amplify a single target DNA sequence to millions of identical copies.

✔You should be able to list the advantages and disadvantages of plasmids versus PCR for cloning genes.

(MB) **Web Activity** Producing Human Growth Hormone, **Web Activity** The Polymerase Chain Reaction

The sequence of bases in a gene can be determined by dideoxy sequencing.

- Dideoxy sequencing is based on an in vitro synthesis reaction in which dideoxyribonucleotides stop DNA replication at each base in the sequence.
- When the DNA fragments generated by a dideoxy sequencing reaction are separated via gel electrophoresis, the sequence of nucleotides in the gene can be determined.

✔You should be able to explain why the newly synthesized DNA fragments—when they are lined up by size—correspond to the sequence of bases in the template DNA.

If individuals with a certain phenotype also tend to share a genetic marker (a known site in DNA that is unrelated to the phenotype), the gene responsible for the phenotype is likely to be near that marker.

- To find the gene associated with Huntington's disease, investigators analyzed a large number of polymorphic genetic markers in affected and unaffected individuals. Their goal was to find a marker that was inherited along with the allele responsible for the disease.
- Once an association study pinpointed the general area where the gene was located, biologists could sequence DNA from the region to determine exactly where the gene of interest was located.

✔You should be able to explain why genetic markers that lack polymorphism are not useful in gene hunts.

Researchers are attempting to insert genes into humans to cure genetic diseases. Efforts to insert genes into plants have been much more successful.

- Once genes are located and characterized, they can be introduced into other individuals or species in an effort to change their traits.
- Genetic transformation can occur in several ways, depending on the species involved.
- In humans, recombinant DNA must be introduced by viruses. Because this is technically difficult and possibly dangerous, progress in human gene therapy has been slow.
- Certain bacteria that infect plants have plasmids that integrate their genes into the host-plant genome. By adding recombinant alleles to these plasmids, researchers have been able to introduce alleles that improve crops.

✔You should be able to explain how genes are inserted into plants.

Questions

✔TEST YOUR KNOWLEDGE

Answers are available in Appendix B

1. What do restriction endonucleases do?
   a. They cleave bacterial cell walls and allow viruses to enter the cells.
   b. They join pieces of DNA by catalyzing the formation of phosphodiester bonds between them.
   c. They cut stretches of DNA at specific sites known as recognition sequences.
   d. They act as genetic markers in the chromosome maps used in gene hunts.
2. What is a plasmid?
   a. an organelle found in many bacteria and certain eukaryotes
   b. a circular DNA molecule that in some cases replicates independently of the main chromosome(s)
   c. a type of virus that has a DNA genome and that infects certain types of human cells, including lung and respiratory tract tissue
   d. a type of virus that has an RNA genome, codes for reverse transcriptase, and inserts a cDNA copy of its genome into host cells
3. When present in a DNA synthesis reaction mixture, a ddNTP molecule is added to the growing chain of DNA. No further nucleotides can be added afterward. Why?
   a. There are not enough dNTPs available.
   b. A ddNTP can be inserted at various locations in the sequence, so fragments of different length form—each ending with a ddNTP.
   c. The 5′ carbon on the ddNTP lacks a hydroxyl group, so no phosphodiester bond can form.
   d. The 3′ carbon on the ddNTP lacks a hydroxyl group, so no phosphodiester bond can form.
4. Once the gene that causes Huntington's disease was found, researchers introduced the defective allele into mice to create an animal model of the disease. Why was this model valuable?
   a. It allowed them to test potential drug therapies without endangering human patients.
   b. It allowed them to study how the gene is regulated.
   c. It allowed them to make large quantities of the huntingtin protein.
   d. It allowed them to study how the gene was transmitted from parents to offspring.
5. To begin the hunt for the human growth hormone gene, researchers created a cDNA library from cells in the pituitary gland. What did this library contain?
   a. only the sequence encoding growth hormone
   b. DNA versions of all the mRNAs in the pituitary-gland cells
   c. all of the coding sequences in the human genome, but no introns
   d. all of the coding sequences in the human genome, including introns
6. What does it mean to say that a genetic marker and a disease gene are closely linked?
   a. The marker lies within the coding region for the disease gene.
   b. The sequence of the marker and the sequence of the disease gene are extremely similar.
   c. The marker and the disease gene are on different chromosomes.
   d. The marker and the disease gene are in close physical proximity and tend to be inherited together.

TEST YOUR UNDERSTANDING

Answers are available in Appendix B

1. Explain how restriction endonucleases and DNA ligase are used to insert foreign genes into plasmids and create recombinant DNA. Make a drawing that shows why sticky ends are sticky, and that identifies the exact location where DNA ligase catalyzes a key reaction.
2. Explain the function of a vector in genetic engineering. List one attribute of a "perfect" vector.
3. What is a cDNA library? Would you expect the cDNA library from a human muscle cell to be different from the cDNA library from a human nerve cell in the same individual? Explain why or why not.
4. What are genetic markers, and how are they used to create a genetic map?
5. Researchers added the promoter sequence from an endosperm-specific gene to the Ti plasmids used in creating golden rice. Why was this step important? Comment on the roles of promoter and enhancer sequences in genetic engineering in eukaryotes.
6. Compare and contrast PCR with the DNA synthesis that occurs in cells (see Chapter 14).

APPLYING CONCEPTS TO NEW SITUATIONS

Answers are available in Appendix B

1. Suppose you had a large amount of sequence data, similar to the data that Nancy Wexler's team had in the region of the Huntington's disease gene, and that you knew that genes of the species being studied typically contain about 1500 bases. How would you use the genetic code (see Chapter 15) and information on the structure of promoters (see Chapters 17 and 18) to find the precise location of one or more genes in your sequence?
2. Is it possible for human gene therapy to alter germ-line cells so that individuals could pass altered alleles on to their offspring? Explain why or why not.
3. Describe similarities between how researchers screen a DNA library and how they perform a genetic screen—for example, for mutant *E. coli* cells that cannot metabolize lactose (see Chapter 17).
4. A friend of yours is doing a series of PCR reactions and comes to you for advice. She purchased two sets of primers, hoping that one set would amplify the template sequence shown here. (The dashed lines in the template sequence stand for a long sequence of bases.) Neither of the primer pairs produced any product DNA, however.

	Primer a		Primer b
Primer Pair 1:	5' CAAGTCC 3'	&	5' GCTGGAC 3'
Primer Pair 2:	5' GGACTTG 3'	&	5' GTCCAGC 3'
Template:	5' ATTCGGACTTG---GTCCAGCTAGAGG 3'		
	3' TAAGCCTGAAC---CAGGTCGATCTCC 5'		

   a. Explain why each primer pair didn't work. Indicate whether both primers are at fault or just one primer is the problem.
   b. Your friend doesn't want to buy new primers. She asks you whether she can salvage this experiment. What do you tell your friend to do?

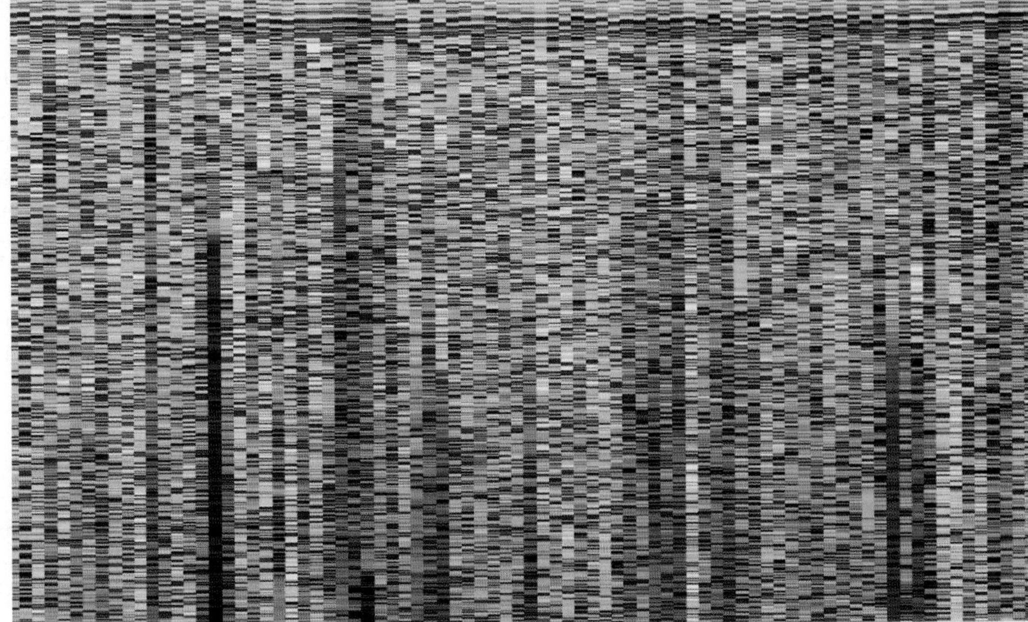

Output from an automated genome-sequencing machine, representing about 48,000 bases from the human genome. Each vertical stripe represents the sequence of a stretch of DNA.

Genomics 20

The first data sets describing the complete DNA sequence, or **genome**, of humans were published in February 2001. These papers were immediately hailed as a landmark in the history of science. In just 50 years, biologists had gone from not understanding the molecular nature of the gene to knowing the molecular makeup of every gene present in our species.

Years later, the multinational effort called the **Human Genome Project** is still producing a wide array of data on the locations and functions of genes and other types of DNA sequences found in humans. It's important to recognize, though, that research on *Homo sapiens* is part of a much larger, ongoing effort to sequence genomes from an array of other eukaryotes, hundreds of bacteria, and dozens of archaea. The effort to sequence, interpret, and compare whole genomes is referred to as **genomics**. The pace of progress in this field is nothing short of explosive.

Progress in genomics has triggered the development of a related field called functional genomics. Genomics supplies a list of the genes present in an organism; **functional genomics** answers questions about the functioning of that genome, such as when genes are expressed and how gene products interact.

As an introductory biology student, you are part of the first generation trained in the genome era. Genomics is revolutionizing biological science and will almost certainly be an important part of your personal and professional life. Let's delve in.

KEY CONCEPTS

- Once a genome has been completely sequenced, researchers use a variety of techniques to identify which sequences code for products and which act as regulatory sites.
- Bacterial and archaeal genomes are relatively small. Among species, there is a positive correlation between total gene number and metabolic capabilities. Gene transfer between species is also common.
- Eukaryotic genomes are large and complex. They include many sequences that have little to no effect on the fitness of the organism, and many transcribed sequences whose function is not known.
- Data and techniques derived from genome sequencing projects are being used to analyze cancer cells.

20.1 Whole-Genome Sequencing

Genomics has moved to the cutting edge of research in biology, largely because technological advances—including the development of the dideoxy sequencing and pyrosequencing techniques introduced in Chapter 19—have made it possible to obtain large quantities of high-quality sequence data in a reasonable amount of time, with a reasonable amount of money.

✓ When you see this checkmark, stop and test yourself. Answers are available in Appendix B.

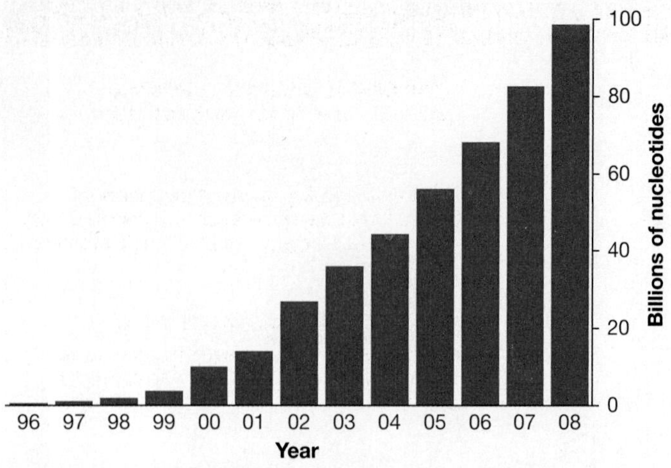

FIGURE 20.1 The Total Number of Bases Sequenced Is Growing Rapidly. Data from the DNA Data Bank of Japan.

As data become less expensive and faster to acquire, the pace of genome sequencing accelerates. The result is that an almost mind-boggling number of sequences are now being generated. As this book goes to press, the primary international repository for DNA sequence data contains over 106 *billion* nucleotides. By way of comparison, a haploid human genome contains about 3 billion bases.

Figure 20.1 gives a visual sense of the growth in sequence data by plotting time versus the number of nucleotides, in billions. There are three large international online repositories for sequence data. The numbers plotted here were compiled by one of them: the DNA Data Bank of Japan. The height of each bar on the graph represents the size of the database in that year. If you compare the increase from one year to the next to the total in the previous year, and do that for several years, you will see that the database is growing at an average rate of about 30 percent per year.

How Are Complete Genomes Sequenced?

Genomes range in size from about a half million base pairs to several billion. But even under the best conditions, a single sequencing reaction can analyze only about 1000 base pairs. How do investigators break a genome into sequencing-sized pieces and then figure out how the thousands or millions of pieces go back together?

SHOTGUN SEQUENCING When researchers set out to sequence the genome of a certain species for the first time, they usually rely on an approach known as **shotgun sequencing**. In shotgun sequencing, a genome is broken up into a set of overlapping fragments that are small enough to be sequenced. The regions of overlap are then used as guides for putting the sequenced fragments back into the correct order (**Figure 20.2**).

Step 1 High-frequency sound waves, or sonication, are used to randomly break a genome into pieces about 160 kilobases (kb) long (1 kb = 1000 bases).

Step 2 Each 160-kb piece is inserted into a plasmid called a **bacterial artificial chromosome (BAC)**. BACs are able to replicate large segments of DNA. Using techniques introduced in Chapter 19, each BAC is then inserted into a different *Escherichia coli* cell, creating a **BAC library**. A BAC library is a genomic library: a set of all the DNA sequences in a particular genome, split into small segments and inserted into cloning vectors (see Chapter 19). By separating the cells in a BAC library and allowing each cell to grow into a large colony, researchers can isolate large numbers of each 160-kb fragment.

Step 3 After many copies of each 160-kb fragment have been produced, the DNA is again broken into fragments—but this time, the segments are about 1 kb long.

Step 4 These small fragments are then inserted into plasmids and placed inside bacterial cells. (Note that by this point the genome has been broken down twice, into increasingly manageable pieces: 160-kb fragments in BACs and 1-kb segments in plasmids.) The plasmids are copied many times as the bacterial cells grow into a large population. Large numbers of each 1-kb fragment are then available for sequencing reactions.

Step 5 Next, the 1-kb fragments from each 160-kb BAC clone are sequenced, and computer programs analyze regions where the ends of the various 1-kb segments overlap. Overlaps occur because there were many copies made of each 160-kb segment, and these copies were fragmented randomly by sonication.

Step 6 The computer mixes and matches segments from a single BAC clone until an alignment consistent with all available data is obtained and the BACs have been reconstructed.

Step 7 The ends of the reconstructed BACs are analyzed in a similar way. The goal is to arrange each 160-kb segment in its correct position along the chromosome, based on regions of overlap.

In essence, the shotgun strategy consists of breaking a genome into tiny fragments, sequencing the fragments, and then putting the sequence data back into the correct order. To review the logic of the shotgun sequencing approach, go to the study area at *www.masteringbiology.com.*

MB **Web Activity** Human Genome Sequencing Strategies

✔If you understand shotgun sequencing, you should be able to explain why it is essential for fragments to have regions of overlap.

THE IMPACT OF NEXT-GENERATION SEQUENCING STRATEGIES Chapter 19 introduced pyrosequencing approaches as faster and cheaper alternatives to traditional dideoxy sequencing. The speed and efficiency are due to:

- Miniaturization that makes it possible to sequence many DNA fragments at the same time.
- Avoiding slow and expensive steps like cloning fragments into BACs and plasmids to produce many copies of each fragment.

The key difference is that pyrosequencing reactions take place on a single DNA fragment—rather than many copies of the same fragment as required by dideoxy sequencing. With pyrosequencing, a genome is sheared into many tiny fragments which are then separated and sequenced directly.

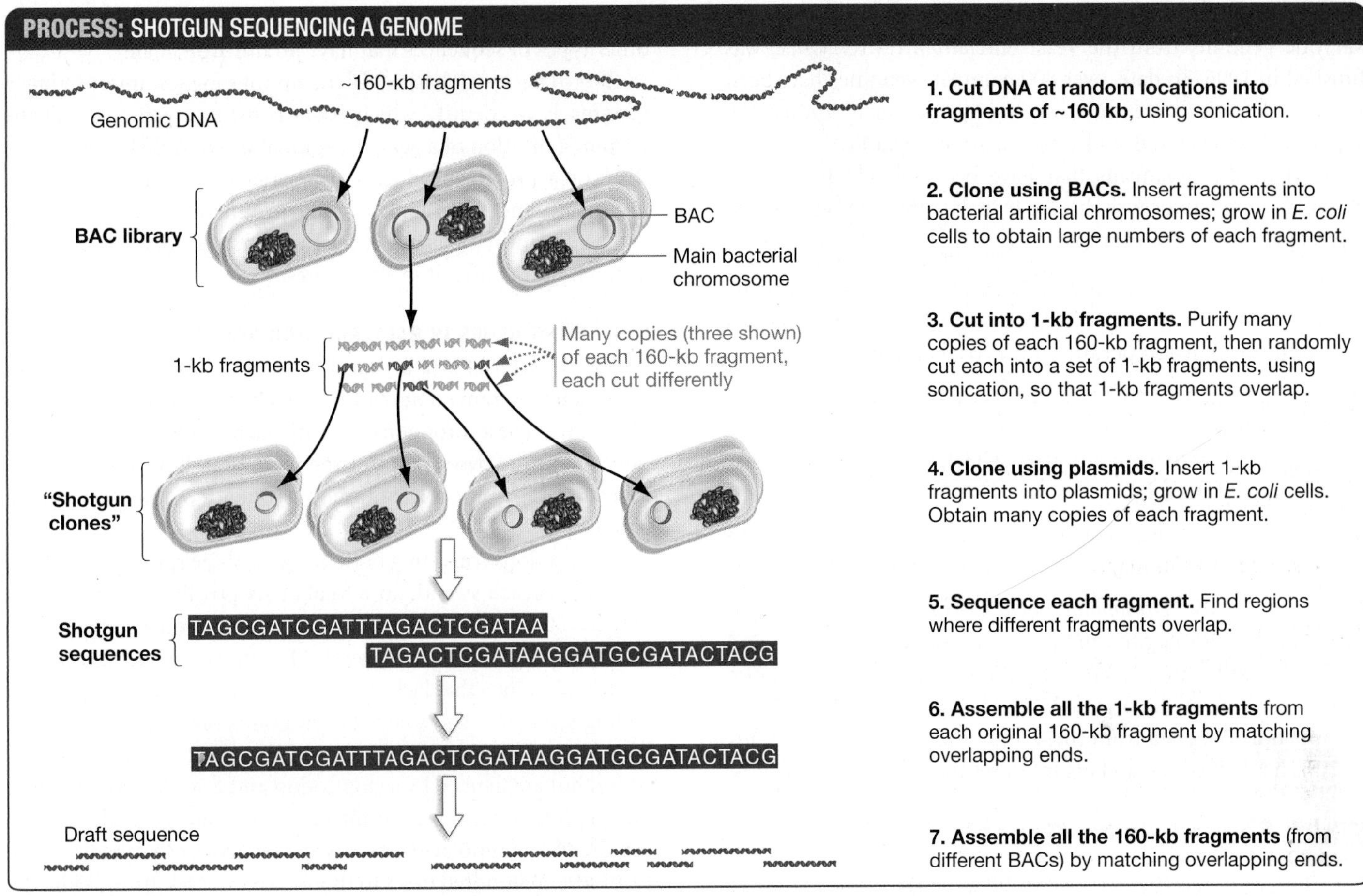

PROCESS: SHOTGUN SEQUENCING A GENOME
~160-kb fragments
Genomic DNA
1. Cut DNA at random locations into fragments of ~160 kb, using sonication.
BAC library
BAC
Main bacterial chromosome
2. Clone using BACs. Insert fragments into bacterial artificial chromosomes; grow in E. coli cells to obtain large numbers of each fragment.
1-kb fragments
Many copies (three shown) of each 160-kb fragment, each cut differently
3. Cut into 1-kb fragments. Purify many copies of each 160-kb fragment, then randomly cut each into a set of 1-kb fragments, using sonication, so that 1-kb fragments overlap.
"Shotgun clones"
4. Clone using plasmids. Insert 1-kb fragments into plasmids; grow in E. coli cells. Obtain many copies of each fragment.
Shotgun sequences
TAGCGATCGATTTAGACTCGATAA
TAGACTCGATAAGGATGCGATACTACG
5. Sequence each fragment. Find regions where different fragments overlap.
TAGCGATCGATTTAGACTCGATAAGGATGCGATACTACG
6. Assemble all the 1-kb fragments from each original 160-kb fragment by matching overlapping ends.
Draft sequence
7. Assemble all the 160-kb fragments (from different BACs) by matching overlapping ends.

FIGURE 20.2 Shotgun Sequencing Breaks Large Genomes into Many Short Segments.

✔**QUESTION** A shotgun blast produces many small, scattered pieces of shot. Why is "shotgun" an appropriate way to describe this sequencing strategy?

The downside of pyrosequencing is that it only works with fragments of about 100 base pairs in length. These fragments are too small to be pieced back together to reconstruct a complete genome accurately. But if a complete genome is already available for the organism, pyrosequencing offers a relatively quick and inexpensive way to sequence the entire genome from a particular individual—with all the tiny fragments being arranged in the correct order by being compared to the "master genome."

BIOINFORMATICS How do researchers piece together the millions of fragments produced by a shotgun sequencing or pyrosequencing study? And once a complete genome is assembled, how are the raw sequence data and any annotations—meaning information about the position, potential product, and function of a particular sequence—made available to the international community of researchers?

The answer is **bioinformatics**—a field that fuses computer science and biology in an effort to manage, analyze, and interpret biological information. For example, researchers in bioinformatics have created searchable databases that hold annotated sequence information, so that investigators can evaluate the similarities between newly discovered genes and genes that had been studied previously in the same or other species.

Because the amount of data is so large, the computational challenges in genomics are formidable. Thus far, sophisticated algorithms and continually improving computer hardware have allowed researchers to keep pace with the rate of data acquisition. The vast quantity of data generated by genome sequencing centers has made bioinformatics a key to continued progress.

Which Genomes Are Being Sequenced, and Why?

The first genome to be sequenced from an organism—not a virus—came from a bacterium that lives in the human upper respiratory tract. *Haemophilus influenzae* has one circular chromosome and a total of 1,830,138 base pairs of DNA. Its genome was small enough to sequence completely in a reasonable amount of time and within a reasonable budget, given the technology available in the early 1990s. *H. influenzae* was an important research subject because it causes earaches and respiratory tract infections in children. One strain is also capable of infecting the membranes surrounding the brain and spinal cord, causing meningitis.

Publication of the *H. influenzae* genome in 1995 was quickly followed by publication of complete genomes sequenced from an assortment of bacteria and archaea. Sequencing of the first eukaryotic genome, from the yeast *Saccharomyces cerevisiae*, was finished in 1996. To date, over 800 complete genomes have been sequenced; projects are underway to sequence an additional 100 archaeal, 2079 bacterial, and 1004 eukaryotic genomes.

Most of the organisms that have been selected for whole-genome sequencing cause disease or have other interesting biological properties. For example,

- Genomes of bacteria and archaea from hot environments have been sequenced in the hopes of discovering enzymes useful for high-temperature industrial applications.
- Certain bacterial and archaeal species have been chosen because they perform interesting chemical reactions, such as the synthesis of methane (natural gas; CH_4). Researchers hope that these organisms might act as a source of commercial products.
- The rice genome was sequenced because rice is the main food source for most humans.
- The fruit fly *Drosophila melanogaster*, the roundworm *Caenorhabditis elegans*, the house mouse *Mus musculus*, and the mustard plant *Arabidopsis thaliana* were analyzed because they serve as model organisms in biology (see **BioSkills 14** in Appendix A). Data from these and other well-studied organisms has helped researchers interpret the human genome.

Which Sequences Are Genes?

Obtaining raw sequence data is just the beginning of the effort to understand a genome. As researchers point out, raw sequence data are analogous to the parts list for a house. The list would read something like "windowwabeborogovestaircasedoorjubjub . . . ," however, because it has no punctuation and contains portions that appear to have no meaning.

Where do the genes for "window," "staircase," and "door" start and end? Are the segments that read "wabeborogove" and "jubjub" important in gene regulation, or are they simply spacers or other types of sequences that have no function at all?

The most basic task in annotating or interpreting a genome is to identify which bases constitute genes. Recall the current definition of a gene: A segment of DNA that codes for an RNA or a protein product—or a series of alternatively spliced products—and that regulates their production. In bacteria and archaea, identifying genes is relatively straightforward. The task is much more difficult in eukaryotes.

IDENTIFYING GENES IN BACTERIAL AND ARCHAEAL GENOMES To interpret bacterial and archaeal genomes, biologists begin with computer programs that scan the sequence of a genome in both directions. These programs identify each reading frame that is possible on the two strands of the DNA. Recall from Chapter 15 that a reading frame is a continuous sequence of non-overlapping codons.

With codons consisting of three bases, three reading frames are possible on each strand, for a total of six possible reading frames (**Figure 20.3**). Because randomly generated sequences contain a stop codon about one in every 20 codons on average, a long stretch of codons that lacks a stop codon is a good indication of a coding sequence. The computer programs draw attention to any "gene-sized" stretches of sequence that lack an internal stop codon but are flanked by a stop codon and a start codon. Because polypeptides range in size from a few dozen amino acids to many hundreds of amino acids, gene-sized stretches of sequence range from several hundred bases to thousands of bases. In addition, the computer programs look for sequences typical of promoters, operators, or other regulatory sites. DNA segments that are identified in this way are called **open reading frames**, or **ORFs**.

Once an ORF is found, a computer program compares its sequence with the sequences of known genes from well-studied species. If the ORF is unlike any gene that has so far been described in any species, further research is required before it can

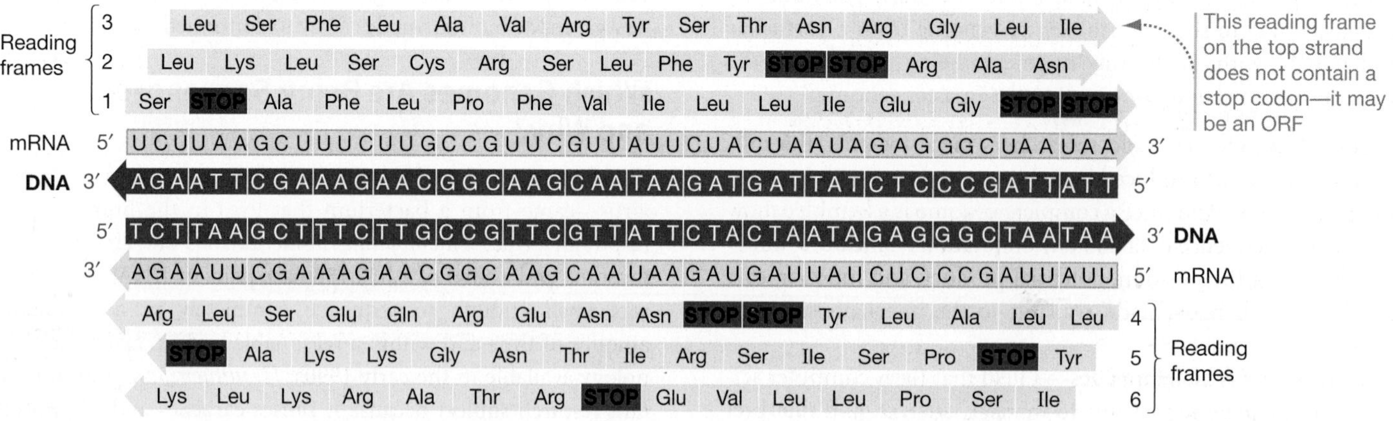

FIGURE 20.3 Open Reading Frames May Be the Locations of Genes. Computer programs scan the three possible reading frames on each strand of DNA and use the genetic code to translate each codon. A long stretch of codons that lacks a stop codon may be an open reading frame (ORF)—a possible gene.

✔**QUESTION** To predict the mRNA codons that would be produced by a particular reading frame, a computer analyzes the DNA in the 3′ to 5′ direction. Why?

actually be considered a gene. A "hit," in contrast, means that the ORF shares a significant amount of sequence with a known gene from another species.

Similarities between genes in different species are usually due to **homology**. If genes are homologous, it means they are similar because they are related by descent from a common ancestor. Homologous genes have similar base sequences and frequently the same or a similar function. For example, consider the genes introduced in Chapter 14 that code for enzymes involved in repairing mismatches in DNA. The mismatch-repair genes in *E. coli*, yeast, and humans are similar in structure, DNA sequence, and function. To explain this similarity, biologists hypothesize that the common ancestor of all cells living today had mismatch-repair genes—thus, the descendants of this ancestral species also have versions of these genes.

Researchers can confirm that an ORF is actually a gene by finding that it is homologous to a known gene. They can also analyze the product that would be produced by an ORF and see if it conforms to a known gene. It has taken much more complex analyses, though, to find eukaryotic genes.

IDENTIFYING GENES IN EUKARYOTIC GENOMES Mining eukaryotic sequence data for genes is complicated. Because coding regions are broken up by introns, for example, it is not possible to scan for long ORFs. Instead, researchers combine an array of approaches.

Perhaps the most productive gene-finding strategy has been to isolate mRNAs from cells, use reverse transcriptase to produce a cDNA version of each mRNA, and sequence a portion of the resulting molecule to produce an **expressed sequence tag**, or **EST**. ESTs represent protein-coding genes. To locate the gene responsible, researchers use the EST to find the matching sequence in genomic DNA.

Although ESTs and other gene-finding strategies have been productive, it will probably be many years before biologists are convinced they have identified all of the coding regions in even a single eukaryotic genome. As that effort continues, though, researchers are analyzing the data and making some remarkable observations. Let's first consider what genome sequencing has revealed about the nature of bacterial and archaeal genomes and then move on to eukaryotes. Is the effort to sequence whole genomes paying off?

20.2 Bacterial and Archaeal Genomes

By the time you read this paragraph, biologists will have obtained extensive, if not complete, sequence data from close to 2300 distinct bacterial and archaeal species or strains. For example, researchers have sequenced the genome of a laboratory population of *Escherichia coli*—derived from the harmless strain that lives in your gut—as well as the genome of a form that causes severe disease in humans. As a result, researchers can now compare the genomes of closely related cells that have different ways of life.

This section focuses on a simple question: Based on data published between 1995 and 2009, what general observations have biologists been able to make about the nature of all these bacterial and archaeal genomes?

The Natural History of Prokaryotic Genomes

In a sense, biologists who are working in genomics can be compared to the naturalists of the eighteenth and nineteenth centuries. These early biologists explored the globe, collecting the plants and animals they encountered. Their goal was to describe what existed. Similarly, the first task of a genome sequencer is to catalog what is in a genome—specifically, the number, type, and organization of genes. Several interesting conclusions can be drawn from relatively straightforward observations about the data obtained thus far.

- In bacteria, there is a general correlation between the size of a genome and the metabolic capabilities of the organism. Species that live in a wide array of habitats and use a wide array of molecules for food have large genomes; parasites have small genomes (**Figure 20.4**). (A parasite lives off a host—making use of much of the host's biochemical machinery—rather than synthesizing its own molecules.)
- The function of many bacterial genes is still unknown. *E. coli* may be the most intensively studied of all organisms, but the function of over 30 percent of its genes is unknown.
- There is tremendous genetic diversity among bacteria and archaea. About 15 percent of the genes in each species' genome appear to be unique, based on the genomes sequenced to date. That is, about one in six genes in one of these species has yet to be found in a different type of prokaryote.
- Redundancy among genes within a genome is common. *E. coli* has 86 genes whose DNA sequences are nearly identical—meaning that the proteins they produce are nearly alike in structure and presumably in function. Biologists hypothesize that slightly different forms of the same protein are produced in response to slight changes in environmental conditions.

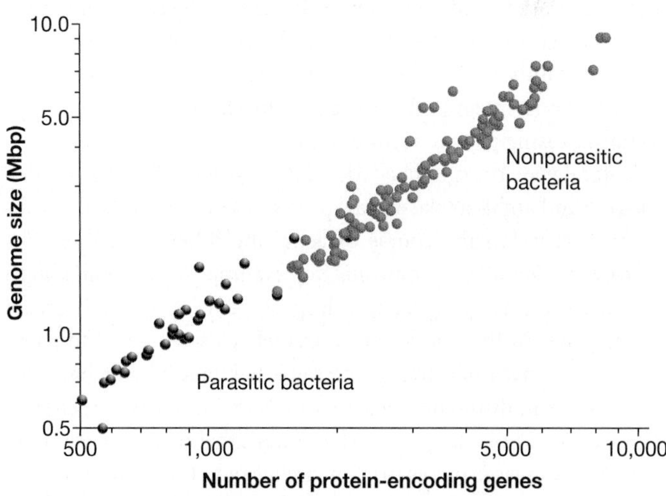

FIGURE 20.4 Parasites Have Small Genomes. When reading this graph, keep in mind that both axes are plotted on logarithmic scales (see **BioSkills 7** in Appendix A). Gene number and genome size are positively correlated—meaning that they increase together. But note that almost all of the parasitic species have genomes that are smaller than the nonparasitic species.

- Multiple chromosomes are more common than anticipated. Several species of bacteria and archaea have two different circular chromosomes instead of one. And at least some bacteria have linear chromosomes.
- Many species contain the small, extrachromosomal DNA molecules called plasmids. Recall from Chapter 19 that plasmids contain a small number of genes, though not genes that are absolutely essential for growth. Plasmids may be exchanged between cells of the same species or even of different species.

Perhaps the most surprising observation of all, however, is that in many bacterial and archaeal species, a significant proportion of the genome appears to have been acquired from other, often distantly related, species. How could this happen?

Lateral Gene Transfer

The movement of DNA from one species to another is called **lateral gene transfer**. Recent estimates suggest that over 50 percent of archaean species and 30–50% of bacterial species have at least one gene acquired by lateral gene transfer. Over the course of evolution, lateral gene transfer has been an important source of new genes and allelic diversity in bacteria and archaea.

Biologists use two general criteria to identify genes that were acquired via lateral gene transfer: (**1**) a gene is much more similar to genes in distantly related species than to those in closely related species and (**2**) the proportion of G-C base pairs to A-T base pairs in a particular gene or series of genes is markedly different from the base composition of the rest of the genome. In many cases, the proportion of G-C bases in a genome is characteristic of the particular genus or species.

How can genes move from one species to another? In at least some cases, plasmids appear to be responsible. For example, most of the genes that are responsible for conferring resistance to antibiotics are found on plasmids. Researchers have documented the transfer of plasmid-borne antibiotic-resistance genes between distantly related bacteria. In some cases, genes from plasmids become integrated into the main chromosome of a bacterium, causing genetic recombination.

Lateral gene transfer may also occur by transformation—when bacteria and archaea take up raw pieces of DNA from the environment, perhaps in the course of acquiring other molecules. The bacterium *Thermotoga maritima*, for example, occupies the high-temperature environments near deep-sea vents. Almost 25 percent of the genes in this species are extremely closely related to genes found in archaea that live in the same habitats. The archaea-like genes occur in distinctive clusters within the *T. maritima* genome. These observations support the hypothesis that the sequences were transferred in large pieces from an archaean to the bacterium.

Similar types of direct gene transfer are hypothesized to have occurred in the bacterium *Chlamydia trachomatis*. This organism is a major cause of blindness in humans from Africa and Asia; it also causes chlamydia, the most common sexually transmitted bacterial disease in the United States. The *C. trachomatis* genome contains 35 genes that resemble eukaryotic genes in structure. Because *C. trachomatis* lives inside the cells that it parasitizes, the most logical explanation for this observation is that the bacterium occasionally takes up DNA directly from its host cell, resulting in a eukaryote-to-bacterium transfer.

Environmental Sequencing

Biologists continue to sequence the genomes of individual species and strains. But recently some research groups have taken a different approach: Cataloging all of the genes present in a community of bacteria and archaea. This type of research program is called **environmental sequencing** or **metagenomics**. The subject of these studies is genes—not organisms.

The first environmental sequencing study was conducted in the Sargasso Sea, near the islands of the Bahamas. Researchers chose the spot because it is extremely nutrient-poor and species-poor—a desert in the ocean. This is a common strategy in biological science: Start by studying a simple system, then go on to more complex situations. To inventory the complete array of bacterial genes present, the research group collected cells from different water depths and locations, isolated DNA from the samples, and sequenced the DNA.

After analyzing over 1 billion base pairs, the team concluded that at least 1800 bacterial species were present, of which 148 were previously undiscovered. They also identified more than 1.2 million alleles that had never before been characterized. These alleles included over 780 sequences that code for proteins similar to the rhodopsin found in the cells of your retina—a molecule that is absorbing the light entering your eye right now. Follow-up work suggests that most of the Sargasso Sea rhodopsin-like molecules are also absorbing light, and that bacterial cells use the energy of the light to pump protons across their plasma membranes—creating a chemiosmotic gradient that can synthesize ATP (see Chapter 9).

CHECK YOUR UNDERSTANDING

If you understand that . . .

- The size of bacterial and archaeal genomes correlates with the cell's metabolic capabilities.
- Lateral gene transfer—movement of DNA from one species to another—is extensive in prokaryotes.
- Environmental sequencing catalogs all of the genes found in a particular habitat.

✔ **You should be able to . . .**

1. Explain why it is logical to observe that parasitic bacteria have small genomes.
2. Explain the logic behind claiming that a gene's similarity to a gene in a distantly related species, and dissimilarity to the same gene in closely related species, is evidence for lateral gene transfer.

Answers are available in Appendix B.

A similar study of bacterial and archaeal genes in the human gut found genes for enzymes that digest the complex polysaccharides of plant cell walls. Environmental sequencing is providing new insights about the living world, from how rhodopsin-like proteins help bacteria thrive in a desert to how your lunch is being digested.

20.3 Eukaryotic Genomes

Sequencing eukaryotic genomes presents two daunting challenges. The first is size. The genomes of bacteria and archaea range from 580,070 base pairs in *Mycoplasma genitalium* to over 6.3 million base pairs in *Pseudomonas aeruginosa*, but eukaryotic genomes are even larger. The haploid genome of *Saccharomyces cerevisiae* (baker's yeast), a unicellular eukaryote, contains over 12 million base pairs. The roundworm *Caenorhabditis elegans* has a genome of 97 million base pairs; the fruit-fly genome contains 180 million base pairs; the mustard plant *Arabidopsis thaliana*'s genome has 130 million base pairs; and those of humans, rats, mice, and cattle contain roughly 3 billion base pairs each.

The second great challenge in sequencing eukaryotic genes is coping with noncoding sequences that are repeated many times. Many eukaryotic genomes are dominated by repeated DNA sequences that occur between genes or inside introns and do not code for products used by the organism. These repeated sequences greatly complicate the work of aligning and interpreting sequence data. If such sequences don't code for a product, why do they exist?

Parasitic and Repeated Sequences

In many eukaryotes, the exons and regulatory sequences associated with genes make up a relatively small percentage of the genome. Over 90 percent of a bacterial or archaeal genome consists of genes, but about 50 percent of an average eukaryotic genome consists of repeated sequences that do not code for a product used by the cell.

When noncoding and repeated sequences were discovered, they were initially considered "junk DNA" that was nonfunctional and probably unimportant and uninteresting. But subsequent work has shown that many of the repeated sequences observed in eukaryotes are actually derived from sequences known as transposable elements.

Transposable elements are segments of DNA that are capable of being inserted into new locations, or transposing, in a genome. They are similar to viruses, except that viruses leave a host cell that they have infected and find a new cell to infect. In contrast, transposable elements never leave their host cell—they simply make copies of themselves that become inserted in new locations. Transposable elements are passed from parents to offspring, generation after generation, because they are part of the genome.

A transposable element is an example of what biologists call a selfish gene: a DNA sequence that survives and reproduces but does not increase the fitness of the host genome. Transposable elements and viruses are classified as parasitic because it takes time and resources to copy them along with the rest of the genome, and because they can disrupt gene function when they insert in a new location. As a result, they decrease their host's fitness.

HOW DO TRANSPOSABLE ELEMENTS WORK? Transposable elements come in a wide variety of types and spread through genomes in a variety of ways. Different species—*E. coli*, fruit flies, yeast, and humans, for example—contain distinct types of transposable elements.

As an example of how these selfish genes work, consider a well-studied type called a **long interspersed nuclear element (LINE)** that is found in humans and other eukaryotes. Because LINEs are so similar to the retroviruses, introduced in Chapter 19, biologists hypothesize that they are derived from them evolutionarily. Your genome contains tens of thousands of LINEs, each between 1000 and 5000 bases long. **Figure 20.5** on page 366 illustrates the steps that allow an active LINE to transpose.

Most of the LINEs observed in the human genome do not actually function, however, because they don't contain a promoter or the genes for either reverse transcriptase or integrase. To make sense of this observation, researchers hypothesize that the insertion process illustrated in steps 6 and 7 of Figure 20.5 is usually disrupted in some way, leaving the inserted replica of the original line incomplete.

Virtually every prokaryotic and eukaryotic genome examined to date contains at least some transposable elements. They vary widely in type and number, however, and bacterial and archaeal genomes have relatively few transposable elements compared to most eukaryotes studied thus far. This observation has inspired the hypothesis that bacteria and archaea either have efficient means of removing parasitic sequences or can somehow thwart insertion events. To date, however, this hypothesis has yet to be tested rigorously.

Research on transposable elements and lateral gene transfer has revolutionized how biologists view the genome. Many genomes are riddled with parasitic sequences, and others have undergone radical change in response to lateral gene transfer events. In other words, genomes are much more dynamic and complex than previously thought. Their size and composition can change dramatically over time.

REPEATED SEQUENCES AND DNA FINGERPRINTING In addition to containing repeated sequences from transposable elements, eukaryotic genomes have several thousand loci called **short tandem repeats (STRs)**. These are small sequences repeated one after another contiguously along part of a chromosome.

There are two major classes of STRs:

1. Repeating units that are just 1 to 5 bases long are known as **microsatellites** or **simple sequence repeats**. The most common type of microsatellite is a repeated stretch of the dinucleotide AC, giving the sequence ACACACAC....
2. Repeating units that are 6 to 500 bases long are known as **minisatellites** or **variable number tandem repeats (VNTRs)**.

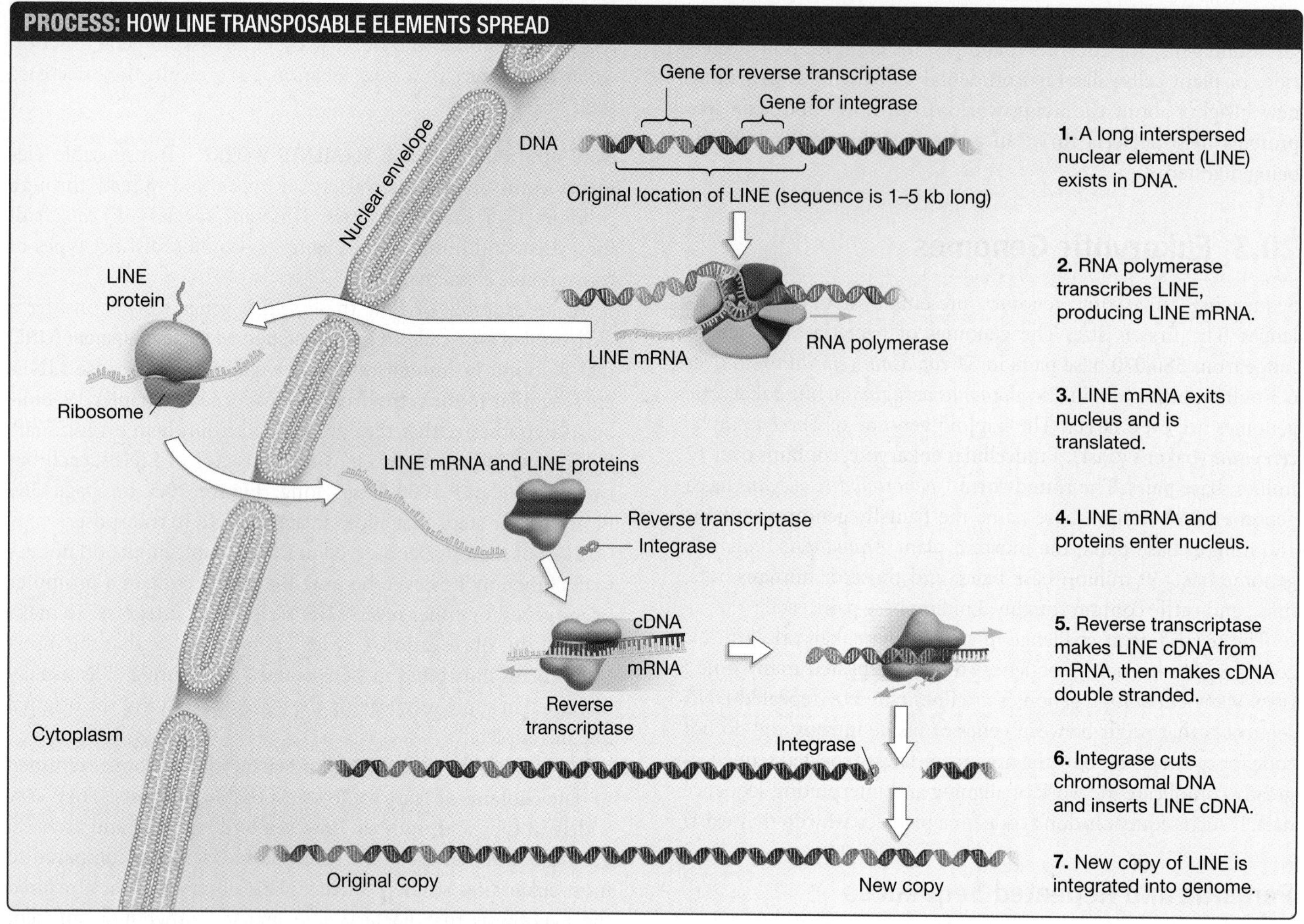

FIGURE 20.5 Transposable Elements Spread within a Genome.

Microsatellite sequences are thought to originate when DNA polymerase skips or mistakenly adds extra bases during replication; the origin of minisatellites is still unclear. Together, the two types of repeated sequences make up 3 percent of the human genome.

Soon after these sequences were first characterized, Alec Jeffreys and co-workers established that microsatellite and minisatellite loci are "hypervariable," meaning that they vary among individuals much more than any other type of sequence does.

Figure 20.6 illustrates one mechanism to explain why microsatellites and minisatellites have so many different alleles: a process called **unequal crossover**. Here's how it works: Homologous chromosomes sometimes align incorrectly during prophase of meiosis I. Instead of lining up in exactly the same location, the two chromosomes pair in a way that matches up bases in different DNA repeats. When crossover occurs, the resulting chromosomes have different numbers of repeats.

Repeated sequences are particularly prone to unequal crossover, because their homologs are so similar that they are likely to misalign. If the region in question has a unique number of repeats, it represents a unique allele. Like any other alleles, microsatellite and minisatellite alleles are transmitted from parents to offspring.

Misalignment and errors by DNA polymerase are so common in these sequences that, in most eukaryotes, the genome of virtually every individual has at least one new allele. This variation in repeat number among individuals is the basis of most DNA fingerprinting. **DNA fingerprinting** refers to any technique for identifying individuals based on the unique features of their genomes. Because microsatellite and minisatellite sequences vary so much among individuals, they are now the sequences of choice for DNA fingerprinting.

To fingerprint an individual, researchers obtain a DNA sample and perform the polymerase chain reaction (PCR), using primers that flank a region containing an STR (**Figure 20.7a**). Once the region has been cloned, it can be analyzed to determine the number of repeats present. Primers are now available for many different STR loci, so researchers can efficiently analyze the alleles present at many STRs.

These advances have important practical implications. For example, DNA fingerprinting of blood or semen found at crime scenes has been used to show that people who were accused of crimes were actually innocent. DNA fingerprinting has also been used as evidence to convict criminals or assign paternity in birds, humans, and other species that have well-characterized microsatellite or minisatellite sequences (**Figure 20.7b**).

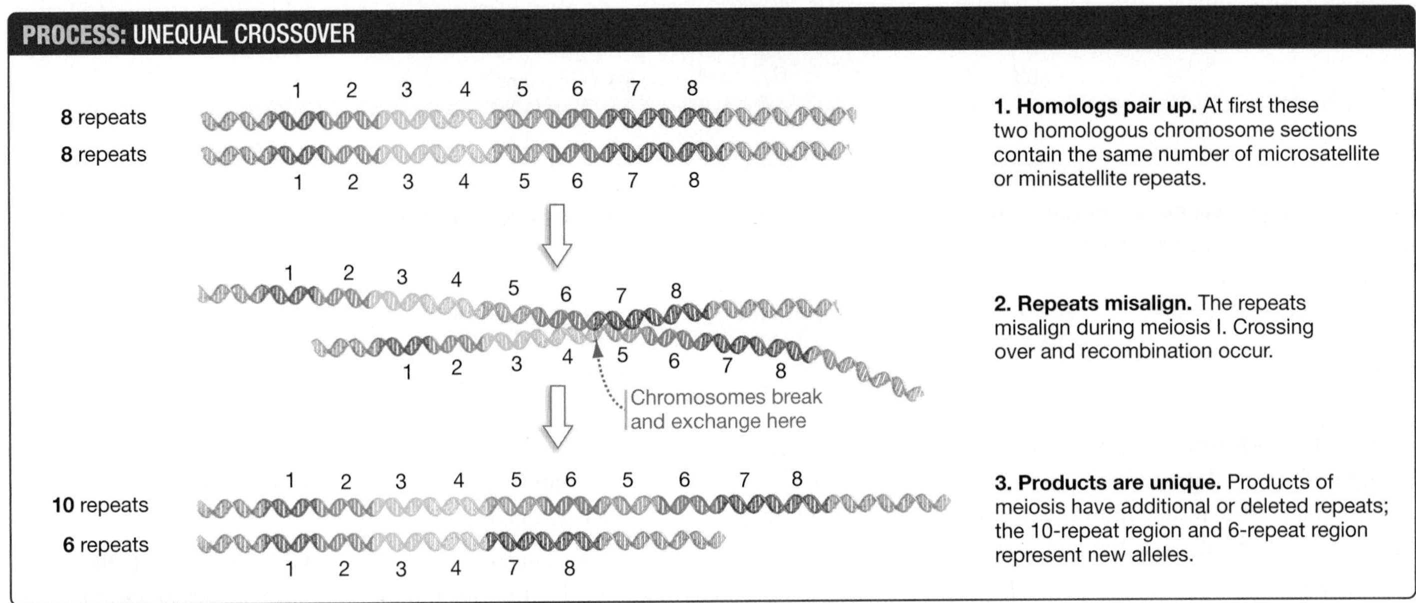
PROCESS: UNEQUAL CROSSOVER
8 repeats
8 repeats
1 2 3 4 5 6 7 8
1. Homologs pair up. At first these two homologous chromosome sections contain the same number of microsatellite or minisatellite repeats.
2. Repeats misalign. The repeats misalign during meiosis I. Crossing over and recombination occur.
Chromosomes break and exchange here
10 repeats
6 repeats
1 2 3 4 5 6 5 6 7 8
1 2 3 4 7 8
3. Products are unique. Products of meiosis have additional or deleted repeats; the 10-repeat region and 6-repeat region represent new alleles.

FIGURE 20.6 Unequal Crossover Adds or Deletes DNA Repeats.

Now that we've reviewed the characteristics of some particularly prominent types of noncoding sequences in eukaryotes, let's consider the characteristics of coding sequences in these genomes. We start with the most basic question of all: Where do eukaryotic genes come from?

Gene Families

In eukaryotes, the major source of new genes is the duplication of existing genes. Biologists infer that genes have been duplicated recently when they find groups of similar genes clustered along the same chromosome. The genes are usually similar in general structural features, such as the arrangement of exons and introns, and in their base sequence. Within a species, genes that are extremely similar to each other in structure and function are considered to be part of the same **gene family.**

The degree of sequence similarity among members of a gene family varies. In the genes that code for ribosomal RNAs (rRNAs) in vertebrates, the sequences are virtually identical—meaning that each individual has many exact copies of the same gene. In other cases, though, the proportion of bases that are identical is 50 percent or less.

HOW DO GENE FAMILIES ARISE? Genes that make up gene families are hypothesized to have arisen from a common ancestral sequence through gene duplication. When **gene duplication** occurs, an extra copy of a gene is added to the genome.

The most common type of gene duplication results from unequal crossover during meiosis—the same process that resulted in extra microsatellite and minisatellite repeats in Figure 20.6. Gene-sized segments of chromosomes can be deleted or duplicated if homologous chromosomes misalign during prophase of meiosis I and an unequal crossover occurs. Like microsatellites or minisatellites, the duplicated segments are arranged in tandem—one after the other.

(a) Using PCR to amplify minisatellite and microsatellite loci.

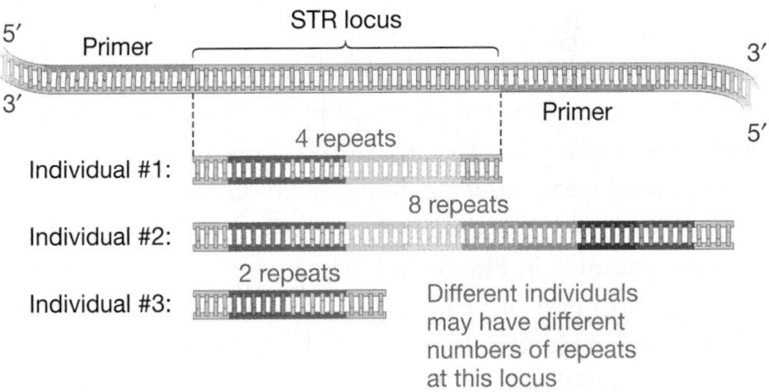
STR locus
5′
Primer
3′
3′
Primer
5′
4 repeats
Individual #1:
8 repeats
Individual #2:
2 repeats
Individual #3:
Different individuals may have different numbers of repeats at this locus

(b) Compare number of STR repeats in alleles to test paternity.

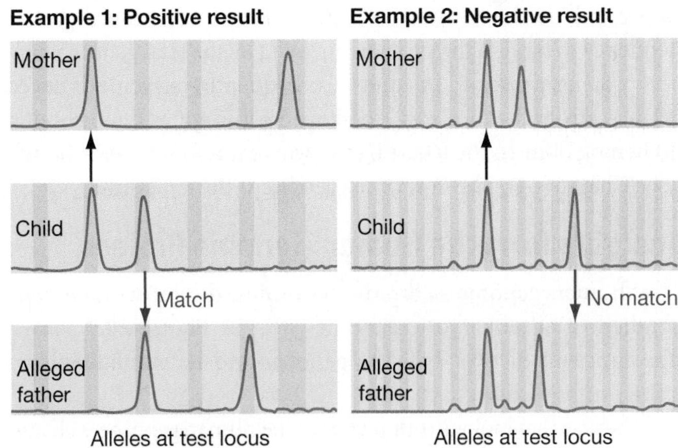
Example 1: Positive result
Example 2: Negative result
Mother
Mother
Child
Child
Match
No match
Alleged father
Alleged father
Alleles at test locus
Alleles at test locus

FIGURE 20.7 DNA Fingerprinting Can Be Used to Identify Parents. **(a)** The lengths of minisatelite and microsatelite loci vary among individuals. **(b)** Here, the position of each peak indicates the number of repeats at a particular locus. Each individual has two alleles at each locus, and thus two peaks. The two peaks from a child should line up with one peak from each parent. Typically 6 to 16 loci are tested to determine paternity.

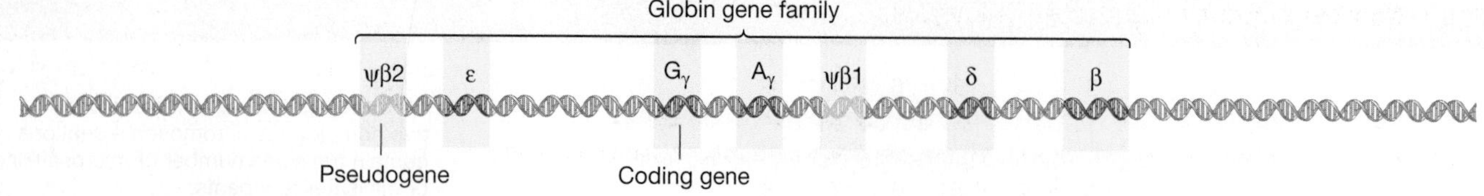

FIGURE 20.8 Gene Families Are Closely Related Genes. Most of the genes illustrated here are expressed at different times during development.

✔**EXERCISE** Suppose that during prophase of meiosis I, the β locus on one chromosome aligned with the ψβ2 locus on another chromosome, and then crossing over occurred in the noncoding sequences just to the left (as oriented in the figure) of this β-ψβ2 pairing. List the order of the globin-family genes that would result on each chromosome.

NEW GENES—NEW FUNCTIONS? Gene duplication is important because the original gene is still functional and produces a normal product. As a result, the new, duplicated stretches of sequence are redundant. In some cases the duplicated genes retain their original function and provide additional quantities of the same product. But if mutations in the duplicated sequence alter the protein product, and if the altered protein product performs a valuable new function in the cell, then an important new gene has been created.

Alternatively, mutations in the duplicated region may make its expression impossible. For example, a mutation could produce a stop codon in the middle of an exon. A member of a gene family that resembles a working gene but does not code for a functional product, due to early stop codons or other defects, is called a **pseudogene**. Pseudogenes have no function.

As an example of a gene family, consider the human globin genes diagrammed in **Figure 20.8**. These sequences code for proteins that form part of hemoglobin—the oxygen-carrying molecule in your red blood cells. Note that the globin gene family contains several pseudogenes, along with several genes that code for oxygen-transporting proteins. Each coding gene in the family serves a slightly different function. For example, some genes are active only in the fetus or the adult. Oxygen has a much greater tendency to bind to the proteins encoded by the fetal genes than to the proteins expressed in adults. Consequently, oxygen is induced to move from the mother's blood, where it is not as tightly bound to hemoglobin, to the fetus's blood, where it is more tightly bound (see Chapter 44). The flow of oxygen keeps the fetus alive.

Insights from the Human Genome Project

The human genome is rapidly becoming the most intensively studied of all eukaryotic genomes. But as **Figure 20.9** shows, the function of over half the genes found in humans is currently unknown.

The way that biologists think about the human genome is changing rapidly, however, due to two recent and dramatic discoveries:

1. Genes for miRNAs are much more common than previously thought. As Chapter 18 noted, miRNAs are small molecules involved in regulating the life span of mRNAs.
2. A much larger proportion of the genome is transcribed than previously thought. Many of these sequences produce RNAs that never leave the nucleus. Because their role in the cell is unknown, researchers call them transcripts of unknown function (TUFs).

Clearly, a great deal remains to be learned about the human genome. In the meantime, two important questions have emerged from early studies. Let's consider each of them in turn.

WHY DO HUMANS HAVE SO FEW GENES? Of all observations about the nature of eukaryotic genomes, perhaps the most striking is that organisms with complex morphology and behavior do not appear to have particularly large numbers of genes. **Table 20.1** indicates the estimated number of genes found in selected eukaryotes. Notice that the total number of genes in *Homo sapiens*, which is considered a particularly complex organism, is about the same as in roundworms, dogs, chickens, and rats, and substantially lower than the number of genes in mice, rice, and the weedy mustard plant *Arabidopsis thaliana*.

Before the human genome was sequenced, many biologists expected that humans would have at least 100,000 genes. But the most recent estimates suggest that we have only a fifth of that number.

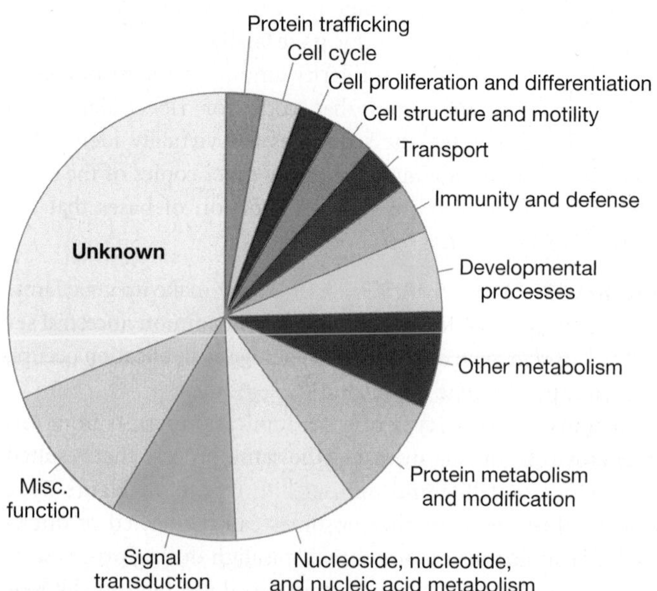

FIGURE 20.9 The Function of Many Human Protein-Coding Genes Is Unknown.

TABLE 20.1 **Number of Genes in Selected Eukaryotes**

Species	Description	Genome Size (millions of base pairs)	Estimated Number of Genes
Saccharomyces cerevisiae	Baker's and brewer's yeast; a unicellular fungus; an important model organism in biochemistry and genetics	12.1	6000
Plasmodium falciparum	Single-celled, parasitic eukaryote; causes malaria in humans	22.8	5268
Drosophila melanogaster	Fruit fly; an important model organism in genetics and developmental biology	120	13,600
Caenorhabditis elegans	A roundworm; an important model organism in developmental biology	97	19,000
Canis familiarus	Domestic dog	2410	19,300
Gallus gallus	Chicken	1050	21,500
Homo sapiens	Human	3000	20,500
Rattus norvegicus	Norway rat; an important model organism in physiology and behavior	2750	21,000
Mus musculus	House mouse; an important model organism in genetics and developmental biology	2500	37,000
Arabidopsis thaliana	A mustard plant; an important model organism in genetics and developmental biology	140	27,500
Oryza sativa	Rice	389	41,000

How can this be? In prokaryotes there is a correlation between genome size, gene number, a cell's metabolic capabilities, and the cell's ability to live in a variety of habitats. Similarly, it makes sense that plants have exceptionally large numbers of genes, because they synthesize so many different and complex molecules from just carbon dioxide, nitrate ions, phosphate ions, and other simple nutrients. The large numbers of genes enable plants to produce large numbers of enzymes. But why isn't there a stronger correlation between gene number and morphological and behavioral complexity in animals?

The leading hypothesis to explain this observation is based on **alternative splicing**. Recall from Chapter 18 that the exons of a particular gene can be spliced in ways that produce distinct mature mRNAs. As a result, a single eukaryotic gene can code for multiple transcripts and thus multiple proteins. The alternative-splicing hypothesis claims that multicellular eukaryotes do not need enormous numbers of distinct genes. Instead, alternative splicing creates different proteins from the same gene.

In support of the alternative-splicing hypothesis, researchers have analyzed the mRNAs produced by human genes and have estimated that each gene produces an average of more than three distinct transcripts. If this result is valid, the actual number of different proteins that can be produced is more than triple the gene number. Humans may have fewer than 20,000 genes according to current estimates, but these genes may have the ability to produce 100,000 different transcripts.

HOW CAN THE HUMAN AND CHIMP GENOMES BE SO SIMILAR? Comparing the numbers of genes found in humans and in mice created a paradox that may be resolved by the alternative-splicing hypothesis. But comparing the base sequences of genes in humans and chimps has created another paradox.

Here is the issue: At the level of base sequences, human beings and chimpanzees are 98.8 percent identical on average. Of the homologous genes analyzed in humans and chimps, 29 percent are identical in amino acid sequence; the average difference between homologous proteins is just two amino acids. If humans and chimps are so similar genetically, why do they appear to be so different in their morphology and behavior?

The leading hypothesis to resolve this paradox cites the importance of regulatory genes and regulatory sequences. Recall from Chapter 18 that a **regulatory sequence** is a section of DNA involved in controlling the activity of other genes; it may be a promoter, a promoter-proximal element, an enhancer, or a silencer. The term **structural gene**, in contrast, refers to a sequence that codes for a tRNA, rRNA, protein, or other type of product. **Regulatory genes** code for regulatory transcription factors that alter the expression of specific genes.

To resolve the sequence-similarity paradox, biologists propose that even though many structural genes in closely related species, such as humans and chimps, are identical or nearly identical, regulatory sequences and regulatory genes in the two species might contain important differences. Suppose the structural gene for human growth hormone is identical in base sequence to that for chimp growth hormone. Even so, if differences in transcription factors, enhancers, or promoters change the pattern of expression of that gene—perhaps turning it on later and longer in humans than in chimps—then height and other characteristics will differ even though the structural gene is the same.

Based on current analyses, biologists suggest that the human genome contains about 3000 different regulatory transcription factors. Subtle mutations in these proteins and the regulatory sites that they bind to could have a significant effect on gene expression and thus on the phenotype. Subtle differences in TUFs could also play a role, if it turns out that these nucleus-restricted RNAs play a role in regulating gene expression.

These observations suggest that most of the genetic changes responsible for the rapid evolution of humans over the past 5 million years have been due to changes in regulatory genes and sequences and alternative splicing, rather than to changes in structural genes. To date, however, there are no specific examples of changes in the regulatory sequences responsible for the phenotypic differences observed between humans and chimps or other closely related species. The regulatory hypothesis still needs to be tested rigorously.

CHECK YOUR UNDERSTANDING

If you understand that . . .

- Eukaryotic genomes are riddled with parasitic sequences that do not contribute to the fitness of the organism.
- Simple repeated sequences are common in eukaryotic genomes.
- In eukaryotes, many of the coding sequences are organized into families of genes with related functions.
- The recent discovery of many miRNA genes and TUFs suggests that the regulation of eukaryotic gene expression may be much more complex than previously thought.

✓ You should be able to . . .

1. Explain why simple tandem repeats make DNA fingerprinting possible.
2. Explain how unequal crossover leads to duplicated sequences.

Answers are available in Appendix B.

20.4 Functional Genomics and Proteomics

Eric Lander has compared the sequencing of the human genome to the establishment of the periodic table of the elements in chemistry. Once the periodic table was validated, chemists focused on understanding how the elements combine to form molecules. Similarly, biologists now want to understand how the elements of the human genome combine to produce an individual.

Remember that a genome sequence is essentially a parts list. Once that list is assembled, researchers delve deeper to understand how genes interact to produce an organism.

What Is Functional Genomics?

For decades, biologists have worked at understanding how and when individual genes are expressed. Research on the *lac* operon, reviewed in Chapter 17, is typical. But now researchers can ask how and when *all* of the genes in an organism are expressed.

Large-scale analysis of gene expression is part of functional genomics—research on how genes work together to produce a phenotype. The effort is motivated by the realization that gene products do not exist in a vacuum. Instead, groups of RNAs and proteins act together to respond to environmental challenges such as extreme heat or drought. Similarly, distinct groups of genes are transcribed at different stages as a multicellular eukaryote grows and develops.

One of the most basic tools used in functional genomics is called a microarray. A **DNA microarray** consists of a large number of single-stranded DNA segments that are permanently affixed to a glass slide. For example, the slide pictured in **Figure 20.10** contains thousands of spots, each of which contains single-stranded DNA from a unique exon found in the human genome.

A typical experiment done with a DNA microarray would follow the protocol outlined in **Figure 20.11**. For example, if the researchers' goal were to compare genes that are expressed during normal cell activity with those expressed under heat stress, the first step would be to isolate the mRNAs being produced in the cells in question under each of the two conditions: in control cells functioning at normal temperature and in cells of the same kind exposed to high temperatures.

Once they had purified mRNAs from the two populations of cells (step 1), investigators would use reverse transcriptase to make a single-stranded cDNA version of each RNA in each of the two samples (see Chapter 19). In addition to the four standard dNTPs, one of the DNA building blocks used in synthesizing the cDNA would carry a fluorescent label (step 2). The label used for the cDNA of the control cells would glow one color (let's say green), while the label chosen for the cDNA of the heat-stressed cells would glow another color (let's say red).

The labeled cDNAs of both colors would then be used to probe the microarray (step 3). As Chapter 19 noted, a probe allows an investigator to find a particular molecule in a sample containing many different molecules. In this case, the labeled cDNAs will bind to the single-stranded DNAs on the plate by complementary base pairing.

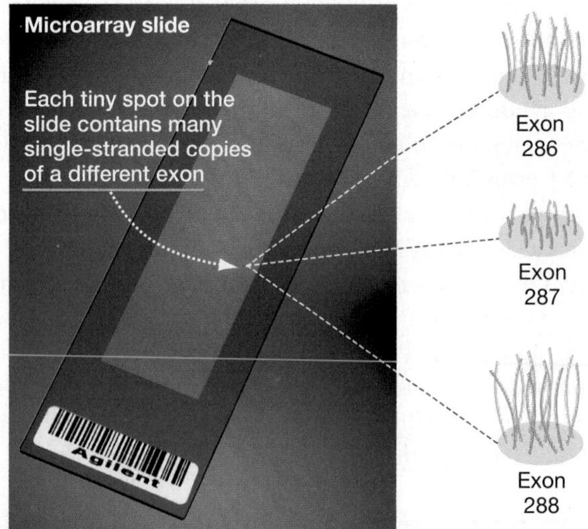

FIGURE 20.10 DNA Microarrays Represent Every Gene in a Genome. To create a DNA microarray, investigators arrange thousands of short, single-stranded DNA sequences onto spots on a glass plate. The DNAs are exons from the genome of a particular species.

Out of all the exons in the genome, then, only the exons that are being expressed by the two populations of cells will be labeled. In our example, genes that are expressed by the control cells under normal conditions will be labeled green, while those expressed by the same kind of cells during heat stress will be labeled red. If one of the exons in the microarray is expressed under both sets of conditions, then both green- and red-labeled cDNAs will bind to that spot and make it appear yellow (step 4).

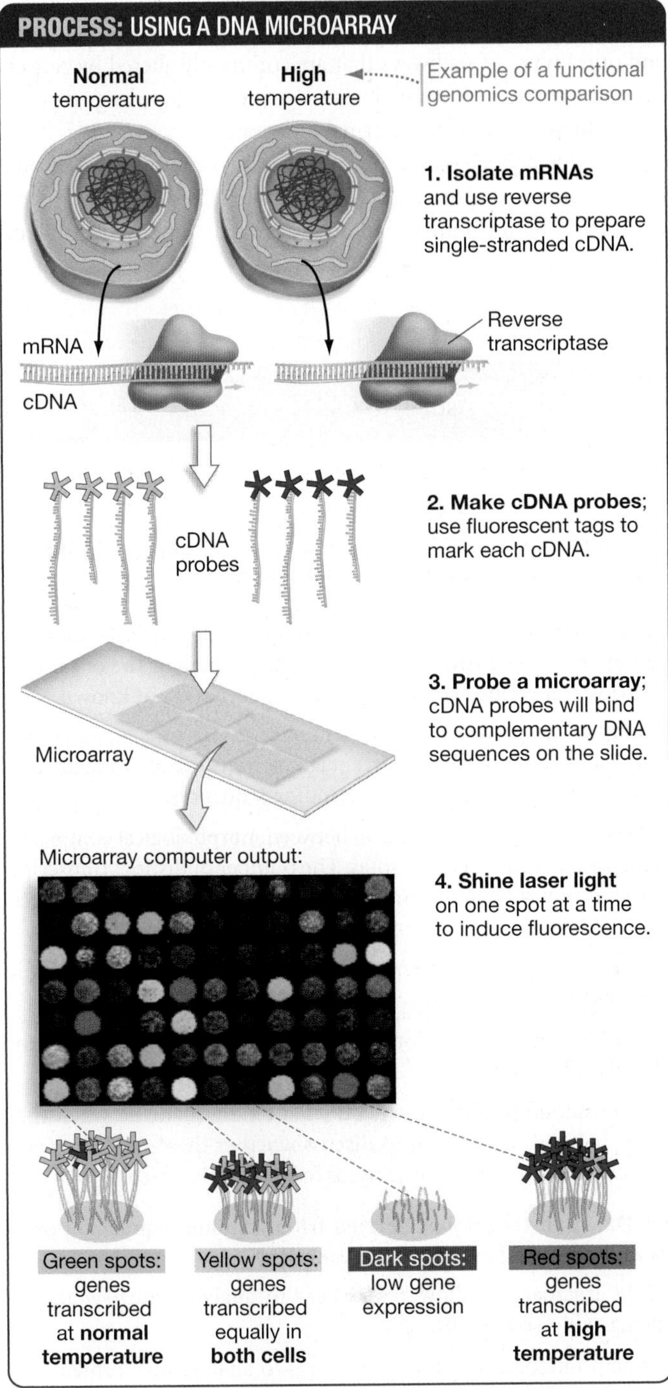

PROCESS: USING A DNA MICROARRAY
Normal temperature
High temperature
Example of a functional genomics comparison
1. Isolate mRNAs and use reverse transcriptase to prepare single-stranded cDNA.
mRNA
cDNA
Reverse transcriptase
2. Make cDNA probes; use fluorescent tags to mark each cDNA.
cDNA probes
3. Probe a microarray; cDNA probes will bind to complementary DNA sequences on the slide.
Microarray
Microarray computer output:
4. Shine laser light on one spot at a time to induce fluorescence.
Green spots: genes transcribed at normal temperature
Yellow spots: genes transcribed equally in both cells
Dark spots: low gene expression
Red spots: genes transcribed at high temperature

FIGURE 20.11 DNA Microarrays Are Used to Study Changes in Gene Expression. By probing a microarray with labeled cDNAs synthesized from mRNAs, researchers can identify which coding sequences are being transcribed. Here mRNAs from cells growing at normal temperature are green, while mRNAs from cells growing at high temperature are red.

A microarray lets researchers study the expression of thousands of genes at a time. As a result, they can identify which sets of genes are expressed in concert under specific sets of conditions.

Once a microarray has been used, the bound cDNA probes can be removed. The original DNAs remain in place, so the slide may then be reused to assess gene expression in a different type of cell, or in the same cell type under different conditions. Researchers can use microarrays to establish which genes are transcribed in different organs and tissues, during cancerous growth, or in response to changes in environmental conditions—such as starvation, the presence of a toxin, or a viral infection. ✔If you understand the concept of how microarrays are used, you should be able to explain how you would use a DNA microarray to compare the genes expressed in brain cells versus liver cells of an adult human.

What Is Proteomics?

The Greek root *–ome*, meaning "all," inspired the term genome. Similarly, biologists use the term **transcriptome** in referring to the complete set of genes that are transcribed in a particular cell, and **proteome** in referring to the complete set of proteins that are produced. **Proteomics**, it follows, is the large-scale study of all the proteins in a cell or organism.

Proteomic studies begin by identifying the proteins present in a cell or organelle. Then researchers attempt to determine the locations of the proteins and how they interact. They may also try to document how the proteins that are present change through time or compare with those in other cells. Instead of studying individual proteins or how two proteins might interact, proteomics is based on studying all of the proteins present at once.

One approach to studying protein–protein interactions is similar to the use of DNA microarrays, except that large numbers of proteins, rather than DNA sequences, are affixed on a glass plate. This microarray of proteins is then treated with an assortment of proteins produced by the same organism. These proteins are labeled with a fluorescent or radioactive tag. If any labeled proteins bind to the proteins in the microarray, the two molecules may also interact in the cell. In this way, researchers hope to identify proteins that physically bind to one another—like the G proteins and associated enzymes introduced in Chapter 8, or the cyclin and Cdk molecules introduced in Chapter 11. Microarray technology is allowing biologists to study protein–protein interactions on a massive scale.

Applied Genomics in Action: Understanding Cancer

Chapter 11 and Chapter 18 introduced the family of diseases called cancer. Recall that cancer develops when an array of genes—starting with genes involved in the control of cell growth—no longer function properly, due to mutation.

Biomedical researchers have been increasingly successful at developing drugs that fight cancers effectively. Many of these drugs work by killing any rapidly dividing cell. But a variety of cells in the body have to divide rapidly in order to function normally. Cancer chemotherapy also kills these cells. This is why chemotherapy patients suffer side effects like hair loss, skin and digestive problems, and weakened immune systems.

Could safer and more-effective therapies be devised, if researchers knew exactly which genes were mutated in cancer cells? Researchers are using tools created by advances in genomics to deepen our understanding of cancer. For example,

- Investigators are using microarrays to compare which genes are expressed in cancer cells versus normal cells. A recent analysis of 40 such experiments has identified a common suite of 69 genes that are mis-expressed in the majority of cancers in the dataset.
- Using data from the human genome project, researchers have sequenced tens of thousands of genes from cancerous cells. Comparisons with normal cells have identified common sets of genes that are mutated in cancerous cells. Some of these analyses suggest that as many as 120 distinct mutations may play a role in driving the development of different cancers.
- The complete genome sequences of cancerous and noncancerous cells from the same person identified over 600 mutations in the cancerous cells.

Studies like these raise the possibility of identifying cell-signaling and other types of pathways that are commonly altered by cancer-causing mutations. If so, biologists hope to develop drugs that can restore these pathways to a normal function.

The general hope is that "pure research" in genomics—motivated by the simple desire to understand ourselves and our fellow organisms better—may lead to improved human health and welfare.

CHAPTER 20 REVIEW

For media, go to the study area at www.masteringbiology.com

Summary of Key Concepts

Once a genome has been completely sequenced, researchers use a variety of techniques to identify which sequences code for products and which act as regulatory sites.

- Recent advances in dideoxy sequencing and pyrosequencing have allowed investigators to sequence DNA more rapidly and cheaply, resulting in a flood of genome data.
- Researchers annotate genome sequences by finding genes and determining their function.
- To identify genes in bacteria and archaea, researchers use computers to scan the genome for start and stop codons that are in the same reading frame and that are separated by gene-sized stretches of sequence.
- To identify genes in eukaryotes, researchers study RNAs as a way of characterizing actively transcribed sequences.
- Among species, homologous genes are identified on the basis of similarities in sequence and structure, and are inferred to have similar function.

✔You should be able to describe how a research group that discovered a gene for coat color in mice would determine whether a homologous gene exists in the human genome.

Bacterial and archaeal genomes are relatively small. Among species, there is a positive correlation between total gene number and metabolic capabilities. Gene transfer between species is also common.

- Parasitic bacteria tend to have small genomes; bacteria and archaea that live in a broad array of habitats or that use a wide variety of nutrients tend to have larger genomes.
- The function of many of the genes identified in bacteria and archaea is still unknown. Redundancy among genes is also high.
- Lateral gene transfer is common in bacteria and archaea. It is an important source of new genes in many species.

✔You should be able to describe two mechanisms responsible for lateral gene transfer in bacteria and archaea.

Eukaryotic genomes are large and complex. They include many sequences that have little to no effect on the fitness of the organism, and many transcribed sequences whose function is not known.

- Compared with prokaryotic genomes, eukaryotic genomes are large and contain a high percentage of transposable elements, repeated sequences, and other noncoding sequences.
- There is no obvious correlation between morphological complexity and gene number in eukaryotes. The number of distinct transcripts produced may be much larger than the actual gene number in certain species, however, as a result of alternative splicing.
- Gene duplication has been an important source of new genes in eukaryotes.
- Changes in gene regulation appear to have been important during human evolution.

✔You should be able to explain what biologists mean when they refer to "junk DNA," and to discuss whether these sequences lack function and are uninteresting, as originally proposed.

Data and techniques derived from genome sequencing projects are being used to analyze cancer cells.

- DNA microarrays allow researchers to analyze which genes are being expressed in cells.
- Protein microarrays allow researchers to analyze how proteins interact in cells.
- DNA microarrays and DNA sequencing are being used to identify mutations that are common to many different types of cancer.

✔You should be able to explain what you would conclude if microarray experiments indicated that a particular gene was consistently expressed in cancer cells but not in normal cells from the same tissue.

Questions

TEST YOUR KNOWLEDGE

Answers are available in Appendix B

1. What is an open reading frame in bacteria?
   a. a gene whose function is already known
   b. a DNA section that is thought to code for a protein because it is similar to a complementary DNA (cDNA)
   c. a DNA section that is thought to code for a protein because it has a start codon and a stop codon flanking hundreds of base pairs
   d. any member of a gene family
2. What best describes the logic behind shotgun sequencing?
   a. Break the genome into tiny pieces. Sequence each piece. Use overlapping ends to assemble the pieces in the correct order.
   b. Start with one end of each chromosome. Sequence straight through to the other end of the chromosome.
   c. Use a variety of techniques to identify genes and ORFs. Sequence these segments—not the noncoding and repeated sequences.
   d. Break the genome into pieces. Map the location of each piece. Then sequence each piece.
3. What are minisatellites and microsatellites?
   a. small, extrachromosomal loops of DNA that are similar to plasmids
   b. parts of viruses that have become integrated into the genome of an organism
   c. incomplete or "dead" remains of transposable elements in a host cell
   d. short and simple repeated sequences in DNA
4. What is the leading hypothesis to explain the paradox that large, morphologically complex eukaryotes such as humans have relatively small numbers of genes?
   a. lateral transfer of genes from other species
   b. alternative splicing of mRNAs
   c. polyploidy, or the doubling of the genome's entire chromosome complement
   d. expansion of gene families through gene duplication
5. What evidence do biologists use to infer that a gene is part of a gene family?
   a. Its sequence is exactly identical to that of another gene.
   b. Its structure—meaning its pattern of exons and introns—is identical to that of a gene found in another species.
   c. Its composition, in terms of percentage of A-T and G-C pairs, is unique.
   d. Its sequence, structure, and composition are similar to those of another gene in the same genome.
6. What is a pseudogene?
   a. a coding sequence that originated in a lateral gene transfer
   b. a gene whose function has not yet been established
   c. a polymorphic gene—meaning that more than one allele is present in a population
   d. a gene whose sequence is similar to that of functioning genes but does not produce a functioning product

TEST YOUR UNDERSTANDING

Answers are available in Appendix B

1. Explain how open reading frames are identified in the genomes of bacteria and archaea. Why is it more difficult to find open reading frames in eukaryotes?
2. Why is it logical to observe that bacterial species found in a wide array of habitats have relatively large genomes?
3. Why are LINEs and other repeated sequences referred to as "genomic parasites"?
4. Why can DNA fingerprinting help identify an individual's relatives?
5. Researchers can create microarrays of short, single-stranded DNAs that represent many or all of the exons in a genome. Explain how these microarrays are used to document changes in the transcription of genes over time or in response to environmental challenges.
6. Explain the concept of homology and how identifying homologous genes helps researchers identify the function of unknown genes. Are duplicated sequences that form gene families homologous? Explain.

APPLYING CONCEPTS TO NEW SITUATIONS

Answers are available in Appendix B

1. Parasites lack genes for many of the enzymes found in their hosts. Most parasites, however, have evolved from free-living ancestors that had larger genomes. Based on these observations, W. Ford Doolittle claims that the loss of genes in parasites represents an evolutionary trend. He summarizes his hypothesis with the quip "use it or lose it." What does he mean?
2. According to eyewitness accounts, communist revolutionaries executed Nicholas II, the last czar of Russia, along with his wife and five children, the family physician, and several servants. Many decades after this event, a grave purported to hold the remains of the royal family was discovered. Biologists were asked to analyze DNA from each adult and juvenile skeleton and determine whether the bodies were indeed those of several young siblings, two parents, and several unrelated adults. If the grave was authentic, describe how similar the DNA fingerprints of each skeleton would be relative to the fingerprints of other individuals in the grave.
3. The human genome contains a gene that encodes a protein called syncytin. This gene is expressed in placental cells during pregnancy. The syncytin gene is nearly identical in DNA sequence to a gene in a virus that infects humans. In this virus, the syncytin-like gene codes for a protein found in the virus's outer envelope. State a hypothesis to explain the similarity between the two genes.
4. A recent study used microarrays to compare the patterns of expression of genes that are active in the brain, liver, and blood of chimpanzees and humans. Although the overall patterns of gene expression were similar in the liver and blood of the two species, expression patterns were strikingly different in the brain. How does this study relate to the hypothesis that most differences between humans and chimps involve changes in gene regulation?

A young fish, still attached to the nutrient-filled yolk in the egg. This chapter introduces the processes responsible for transforming a fertilized egg into an individual that has specialized cells, tissues, and organs.

21 Principles of Development

KEY CONCEPTS

- During development, cells divide, die, move, or expand in a directed manner; specialize; and interact with other cells.
- Cells become specialized because they express different genes, not because they contain different genes.
- Cells interact continuously by means of cell-cell signals during development.
- When development begins, early cell-cell signals trigger a cascade of effects that cause increasing specialization as development proceeds.
- Mutations in genes responsible for development may lead to the evolution of new body sizes, shapes, and structures.

What question qualifies as the greatest current challenge in biological science? Although there are many candidates, one of the most compelling is the question addressed in Unit 4: How does a multicellular individual develop from a single cell—a fertilized egg?

It's important to pause for a moment and think about the magnitude of this problem. For example, at one time you consisted of a single cell. If you had been able to watch your own development, you would have seen that cell divide rapidly and form a ball of tiny, identical-looking cells. At that point, the fertilized egg had given rise to an **embryo**—a young, developing organism. After continued cell division, large groups of cells suddenly began moving into the interior of the embryo. Cell division continued at a rapid pace. After a week or two the embryo elongated, and a recognizable head and tail portion appeared. Tiny precursors of vertebrae became visible, along with rudimentary eyes. Eventually buds emerged and went on to form your limbs. As development continued, the embryo eventually became you.

Biologists who have watched this process in humans or other organisms never cease to marvel at it. How does all the growth and formation of distinctive body parts happen?

To understand how researchers are answering this question, you'll need to draw on what you've already learned about cell-cell interactions, the regulation of gene expression, and a host of other topics described in previous units. Part of the excitement surrounding developmental biology is that it draws on insights from biochemistry, cell biology, genetics, and evolution. It is one of the most interdisciplinary fields in all of biological science. Let's delve in.

✔ When you see this checkmark, stop and test yourself. Answers are available in Appendix B.

21.1 Shared Developmental Processes

Over one hundred years of research has culminated in one of the great insights of contemporary biology: A few fundamental principles are common to all developmental sequences observed in multicellular organisms. Their discovery has brought a unified understanding to the variation observed in how the embryos of fruit flies, oak trees, and humans grow. And it has given biologists a framework for explaining how a single cell can give rise to a complex, multicellular individual.

An individual develops as cells divide, move, or expand in a directed way; begin to express certain genes rather than others; and signal to each other about where they are, what they are doing, and what type of cell they are becoming. In addition, selected cells die in a regulated manner during development (**Table 21.1**). We'll consider each of these processes, and then go on to consider more specific questions about how cells interact and specialize.

Cell Proliferation

For an embryo to grow and develop, its cells have to proliferate—they have to divide and make more cells. This statement may strike you as obvious. Less obvious, but equally important, is the following point: The location, timing, and extent of cell division have to be tightly controlled.

Chapter 11 introduced mitosis and cytokinesis, which are responsible for cell proliferation in eukaryotes. That chapter also introduced the stages of the cell cycle and how they are controlled. You might recall that cells initiate mitosis in response to a regulatory protein complex called mitosis-promoting factor (MPF), that each stage of the cell cycle has checkpoints that are carefully regulated, and that cells continue to grow or stop growing in response to what biologists call "social controls"—meaning, signals from other cells.

In both plants and animals, most cells stop growing when they mature. But both plants and animals have specific populations of undifferentiated cells that keep proliferating throughout the individual's life.

- In plants these cells are grouped into **meristems**. The meristematic tissues described in Chapter 23 are present in the same locations in embryonic and adult plants and perform the same function—giving rise to the stems, roots, leaves, flowers, and other structures that develop throughout life.
- In animals, these cells are called **stem cells**. Embryonic stem cells can give rise to almost any differentiated cell type in the body. In juveniles and adults, stem cells are found in specific locations in the body, where they proliferate to replace skin and blood and gut cells that die, repair wounds, and create a constant supply of disease-fighting cells in the immune system.

SUMMARY TABLE 21.1 **Five Essential Developmental Processes**

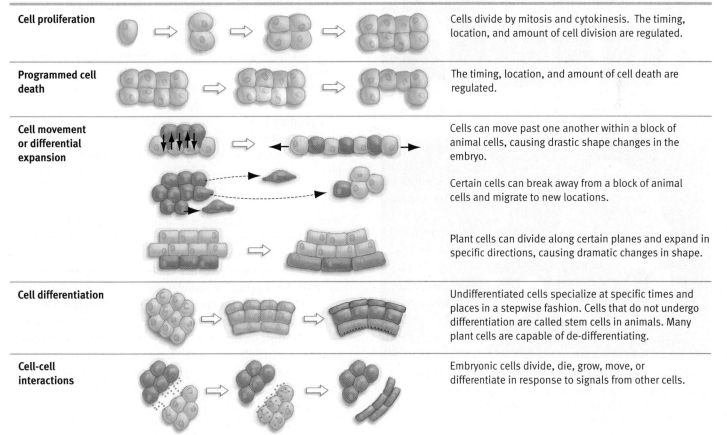

Cell proliferation	Cells divide by mitosis and cytokinesis. The timing, location, and amount of cell division are regulated.
Programmed cell death	The timing, location, and amount of cell death are regulated.
Cell movement or differential expansion	Cells can move past one another within a block of animal cells, causing drastic shape changes in the embryo. Certain cells can break away from a block of animal cells and migrate to new locations. Plant cells can divide along certain planes and expand in specific directions, causing dramatic changes in shape.
Cell differentiation	Undifferentiated cells specialize at specific times and places in a stepwise fashion. Cells that do not undergo differentiation are called stem cells in animals. Many plant cells are capable of de-differentiating.
Cell-cell interactions	Embryonic cells divide, die, grow, move, or differentiate in response to signals from other cells.

Programmed Cell Death

Cells grow or stop growing in response to signals from other cells. But in some cases, they also commit suicide in response to signals from other cells.

Cell death is a highly regulated aspect of plant and animal development. **Programmed cell death** is called **apoptosis** (literally, "falling away") and occurs as certain tissues and organs take shape. As the feet of a chicken embryo develop, for example, cells that are initially present between the toes must die in order for separate toes to form (**Figure 21.1a**). In plants, programmed cell death allows flower petals to fall after pollination has occurred, and causes leaves to be lost in autumn.

Some of the key initial studies on apoptosis were done on the roundworm *Caenorhabditis* (pronounced *see-nor-ab-DIE-tis*) *elegans*. This species is a popular model organism in developmental biology because its entire array of organs and tissues consists of only about a thousand cells. And because they are transparent, biologists can follow individual cells throughout development. (For more on *C. elegans*, see **BioSkills 14** in Appendix A.)

As a *C. elegans* individual matures, 131 of its cells undergo apoptosis. To explore how this happens, Hillary Ellis and Robert Horvitz found embryos that do *not* exhibit the normal pattern of cell deaths, and then used techniques introduced in Chapter 19 to locate the mutant genes responsible for the defect. Their initial work uncovered two genes that are essential for apoptosis. The researchers proposed that the genes are part of a genetic program for apoptosis—in short, that they are cell-suicide genes.

Follow-up work, in which researchers searched databases of known DNA sequences, confirmed that mice have similar genes in terms of their sequence and structure. When a research team used genetic engineering techniques to produce mice in which both copies of one of their cell-suicide genes were disrupted, the embryos that resulted had severe malformation of the brain (**Figure 21.1b**). The defect occurred because cells that would normally die early in development survived.

The same genes have now been found in humans and other mammals. Normal apoptosis is important in the development of human embryos, and abnormal apoptosis—either too much or too little—has been implicated in certain diseases of adults. For example, inappropriate activation of programmed cell death is involved in some neurodegenerative diseases, including ALS (Lou Gehrig's disease).

(a) Chicken embryo with normal (left) and defective (right) cell-suicide genes.

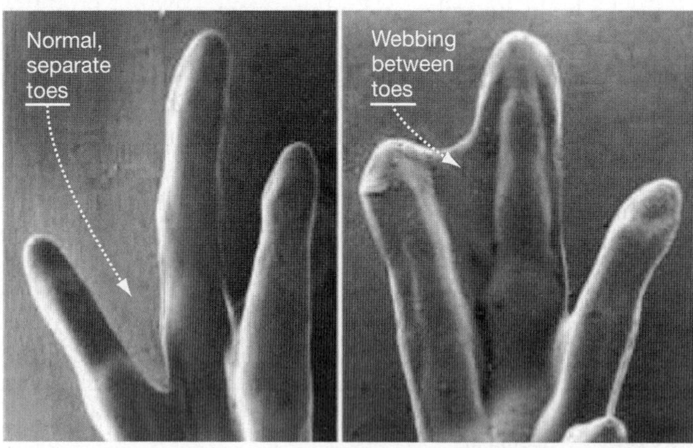

(b) Mouse embryo with normal (left) and defective (right) cell-suicide genes.

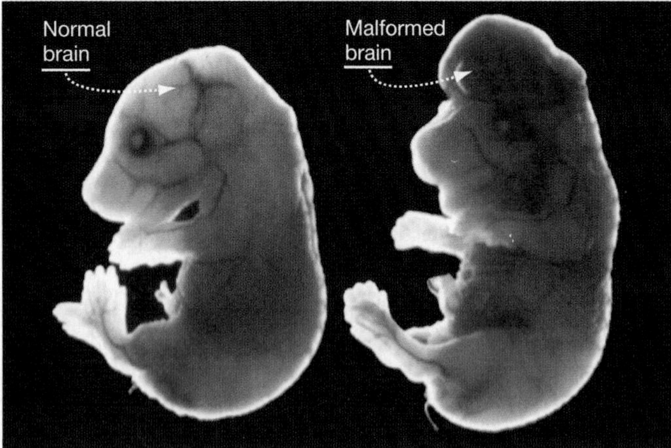

FIGURE 21.1 Defects Occur When Programmed Cell Death Fails.

✔**QUESTION** If the defective version of the cell-suicide gene is recessive, what would the embryo on the right side of part (b) look like if it were heterozygous?

Cell Movement or Cell Growth

Besides proliferating or dying, many animal cells have to move to a new location in order for normal development to occur. In plants, the cells do not move, but changes in the orientation of cell division control the direction of subsequent cell proliferation and cell expansion.

Some of the most dramatic cell movements during animal development occur early in the process, after rapid cell divisions have produced a mass of similar-looking cells. During a sequence of events called **gastrulation**, cells in different parts of the mass rearrange themselves into three distinctive layers, which then give rise to the skin, gut, and other basic parts of the body (gastrulation is described in Chapter 22).

Later in development, certain animal cells break away from their original sites and migrate to new locations in the embryo. There they give rise to germ cells (sperm or eggs), pigment-containing cells, precursors of blood cells, or certain nerve cells. If any of these cell movements is inhibited, or if migrating cells end up in the wrong place or move at the wrong time, the embryo is likely to be deformed or die before development is complete.

Plant cells, in contrast, are encased in stiff cell walls and do not move. Instead, the directionality of cell division and cell expansion is carefully regulated. During development, changes in the direction of cell growth result in the proper formation of straight and bent stems, leaf veins, and other structures.

In plants, changes in cell size can make an individual "move." For example, Chapter 39 will detail how plants grow toward light. This response is not due to changes in the rate of cell proliferation in meristems but to changes in the stem. In response to signals from cells that receive light, cells on one side of the stem expand—causing the entire shoot to bend the other way.

Differential cell growth is a key part of plant development, just as regulated cell movement is a key part of animal development.

Cell Differentiation

As development progresses, most cells undergo **differentiation**—the process of becoming a specialized type of cell. As a result, a fertilized egg may give rise to hundreds of distinctive cell types.

In the case of your own development, some of your embryonic cells differentiated to form muscle cells that contract and relax, while others became nerve cells that conduct electrical signals throughout the body. As an oak tree develops, some cells secrete thickened walls and transport water as part of a woody stem, while others become flattened and secrete the waxy coating found on the surface of leaves.

Differentiation is a progressive, step-by-step process. Initially, cells have the capacity to differentiate into any cell type. Meristematic cells in plants and stem cells in animals remain in this state, but most cells become committed to a certain cell fate early in development and later become differentiated—meaning that they begin to look and behave like a specific cell type.

Some plant cells are also capable of "de-differentiating." They can change their structure and function, even after they have specialized. For example, de-differentiation occurs when a branch of a western redcedar tree droops down low enough to make contact with the soil. Cells in the branch de-differentiate and then redifferentiate to form root cells, resulting in the growth of a fully formed root where the branch initially rested on the ground.

These types of plant cells are said to be totipotent ("all-powerful"). Totipotent cells highlight an important difference between plant and animal development. Once differentiated, animal cells cannot de-differentiate and redifferentiate. They are what they are.

Cell-Cell Interactions

Chapter 8 introduced the topic of cell-cell interactions by examining the extracellular matrix found between cells. That chapter also explored how adjacent cells are attached and exchange materials and how cells respond to signals from other cells. During development, the most important cell-cell interactions involve sending and receiving signals.

You might recall from Section 8.3 that when a signal arrives at the surface of a cell, its message is received and processed. In most cases, cells change their activity in response to these signals. Embryonic cells grow, move, or differentiate in response to signals from other cells.

Some of the most exciting research in developmental biology is focused on how developing cells send cell-cell signals and respond to them. In many cases, the signal transduction pathways introduced in Chapter 8 trigger the production of the transcription factors introduced in Chapter 18. As a result, the arrival of cell-cell signals changes patterns of gene expression and thus the embryonic cell's structure and behavior. In this way, the fate of a cell inside an embryo hinges on the signals it receives from other cells.

21.2 The Role of Differential Gene Expression in Development

The differentiation of a cell occurs through differential gene expression. The muscle cells in your body are different from your nerve cells because they express different genes and therefore produce different proteins. The water-transport cells in an oak tree are different from its leaf-surface cells for the same reason.

If you think about these statements, you'll realize that they have to be true. The only way that cells can have different structures and functions is if they contain different molecules. What is less obvious is whether cells express different genes because they contain different genes, or whether all the cells in a body contain the same genes but express only a specialized subset.

Evidence That Differentiated Plant Cells Are Genetically Equivalent

If cells from the stem of a cedar tree can de-differentiate to form roots, the cells involved must contain the genes required by root cells. Gardeners and farmers have known for centuries that in many plant species, complete new individuals can be produced from a small section of a root or shoot.

From observations like this, researchers strongly suspected that all plant cells contain the same genes—meaning that they are genetically equivalent. This suspicion was confirmed in the 1950s when biologists were able to grow entire tobacco plants or carrots from a single, differentiated cell taken from an adult. (For more information on how plant cells are grown in culture, see **BioSkills 12** in Appendix A.) These experiments confirmed that differentiated plant cells are genetically equivalent.

How is it possible for plant cells to de-differentiate and redifferentiate so readily? This question has been much more difficult to answer. Somehow, the processes that control gene expression—changes in chromatin structure, regulatory transcription factors, RNA processing, miRNA activity, and so on—get reprogrammed. How this happens remains a mystery.

Evidence That Differentiated Animal Cells Are Genetically Equivalent

In contrast to plants, the issue of genetic equivalence was extremely difficult to resolve for animals. Serious experimental work began in the 1950s, but the question wasn't settled until the late 1990s.

Early experiments were based on transferring nuclei from differentiated frog cells into unfertilized eggs whose nuclei had been removed. Some of these transplanted nuclei were able to direct the development of tadpoles successfully. These results provided strong evidence that all cells in the same individual are genetically equivalent.

This conclusion was confirmed in 1997, when Ian Wilmut and colleagues reported the results of nuclear transfer experiments in sheep. As **Figure 21.2** shows, the researchers removed mammary-gland cells from a 6-year-old pregnant female, grew them in culture, and fused them with eggs whose nuclei had been removed. As the upper drawings show, the eggs came from a black-faced breed of sheep, while the donor nuclei came from a white-faced breed. After developing in culture, the resulting embryos were implanted in the uteri of surrogate mothers. In one of several hundred such transfer attempts, a white-faced lamb named Dolly was born.

Genetic tests showed that Dolly was a **clone**—a copy—of the white-faced donor of the mammary-gland cell. Dolly grew into a fertile adult and, by normal mating, produced her own lamb named Bonnie. Soon after, other research groups reported similar results in mice and cows. More recently, horses, monkeys, and dogs have been cloned. The procedure remains technically difficult, however, and the vast majority of nuclear transplant experiments fail.

Taken together, research on cloning plants and animals has shown that in most cases, cellular differentiation does not involve changes in the genetic makeup of cells. Instead, it results from differential gene expression. There are some important exceptions to this rule, however. For example, small stretches of DNA are rearranged in certain immune system cells in humans and other mammals, late in development. As a result, many immune cells are genetically unique. Chapter 49 explains this phenomenon.

How Does Differential Gene Expression Occur?

Chapter 18 emphasized that eukaryotic cells control gene expression at several different levels: chromatin remodeling, chromatin modification, transcription regulation, alternative splicing of mRNAs, selective destruction of mRNAs, translation rate, and activation and deactivation of proteins after they are translated. All of these processes occur during development. Which is responsible for differentiation?

The answer is transcriptional control. To understand why, ask yourself whether a muscle cell should produce mRNAs or proteins that are specifically required by nerve cells. The answer is no. If it did, it would also have to produce microRNAs that disable nerve-cell mRNAs or regulatory proteins that keep nerve-cell proteins inactivated. It is much more logical to expect that muscle cells transcribe only genes required by muscle cells. This is exactly what researchers have found.

Transcription is the fundamental level of control in differential gene expression during development. In eukaryotes, transcription is controlled primarily by the presence of **regulatory transcription factors** that influence chromatin remodeling and bind to promoter-proximal elements, enhancers, silencers, or other regulatory sites in DNA.

This simple insight is extremely important. To understand differentiation, researchers have to understand how and why regulatory transcription factors vary among cells.

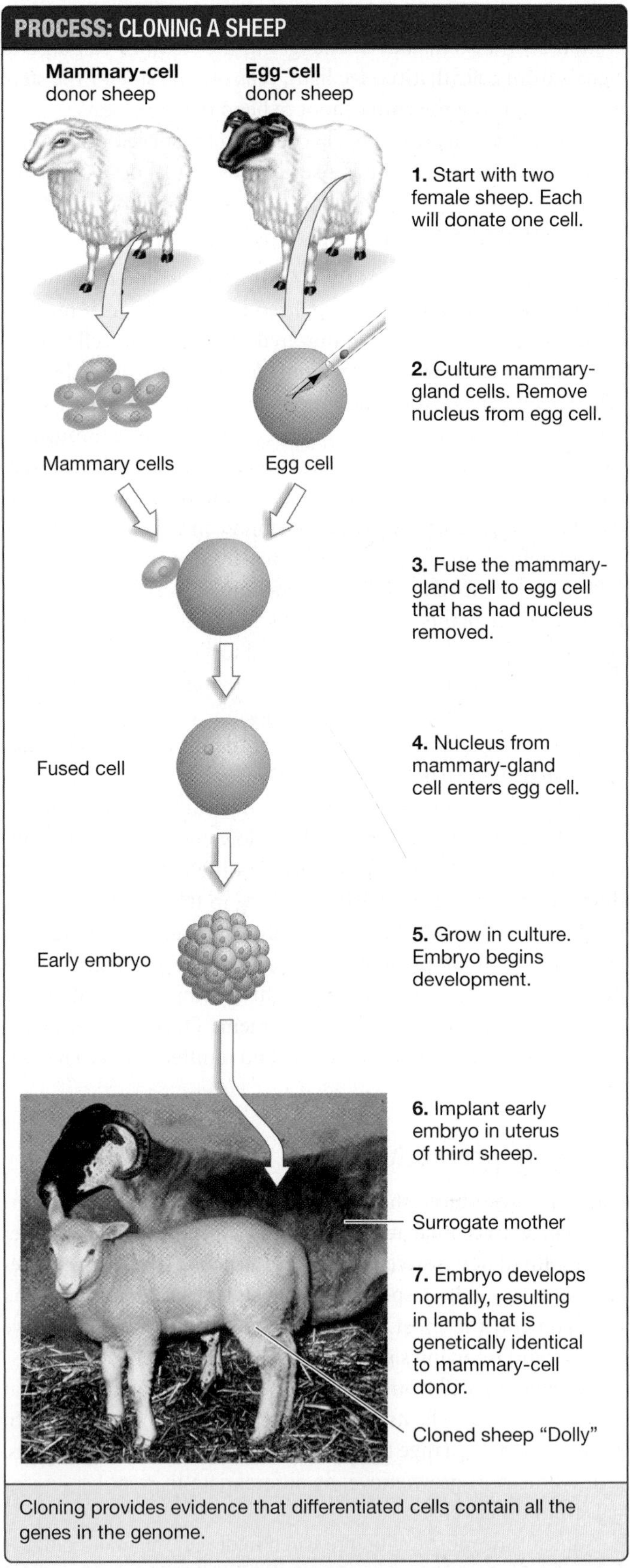
PROCESS: CLONING A SHEEP
Mammary-cell donor sheep
Egg-cell donor sheep
1. Start with two female sheep. Each will donate one cell.
2. Culture mammary-gland cells. Remove nucleus from egg cell.
Mammary cells
Egg cell
3. Fuse the mammary-gland cell to egg cell that has had nucleus removed.
Fused cell
4. Nucleus from mammary-gland cell enters egg cell.
Early embryo
5. Grow in culture. Embryo begins development.
6. Implant early embryo in uterus of third sheep.
Surrogate mother
7. Embryo develops normally, resulting in lamb that is genetically identical to mammary-cell donor.
Cloned sheep "Dolly"
Cloning provides evidence that differentiated cells contain all the genes in the genome.

FIGURE 21.2 Mammals Can Be Cloned by Transplanting Nuclei from Mature Cells. The lamb that resulted from this experiment was identical to the white-faced individual that donated the nucleus, not the black-faced egg donor or surrogate mother.

CHECK YOUR UNDERSTANDING

If you understand that . . .

- Differentiation occurs because embryonic cells express distinctive subsets of genes, not because they contain different genes.
- Differential gene expression is predominantly based on transcriptional control. Different types of cells have different types of transcription factors.

✓ You should be able to . . .

Explain the evidence for genetic equivalence in both plant and animal cells.

Answers are available in Appendix B.

21.3 Cell-Cell Signals Trigger Differential Gene Expression

To understand development, you have to think like a cell. Suppose that you were one of the hundreds or thousands of cells in a developing animal embryo. Your fate—whether you ended up as part of an arm or a kidney, and whether you differentiated into a nerve cell or a blood-vessel cell—would depend on your location along four axes: time (meaning the organism's current stage of development) plus three spatial dimensions—the three body axes illustrated in **Figure 21.3**.

1. One axis runs **anterior**, toward the head, to **posterior**, toward the tail.
2. One axis runs **ventral**, toward the belly, to **dorsal**, toward the back.
3. One axis runs left to right.

Cell-cell signals tell cells where they are in time and space. This information activates transcription factors that turn specific genes on or off, resulting in differentiation. As development proceeds, the distinctive suites of genes that are activated at successive stages determine the fate of each cell.

Let's consider how this process happens, beginning with one of the first developmental signals ever discovered. Although you'll be analyzing what happens as a fruit-fly embryo develops, keep an important point in mind: Principles that were discovered in fruit flies are relevant to virtually all multicellular organisms studied to date—from mustard plants to humans.

Master Regulators Set Up the Major Body Axes

Biologists use the term **pattern formation** to describe the events that determine the spatial organization of an embryo. If a molecule signals that a target cell is in the embryo's head, or tail, or dorsal side, or ventral side, that molecule is involved in pattern formation.

Pattern formation is progressive. Early signals act as master regulators that set up the general anterior–posterior, dorsal–ventral, and left–right axes of an embryo. Genes activated by

(a) The three body axes observed in humans and other animals...

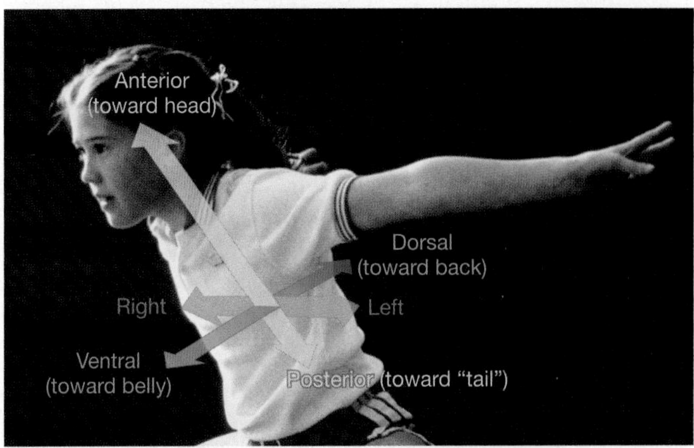

(b) ...are initially established in embryos.

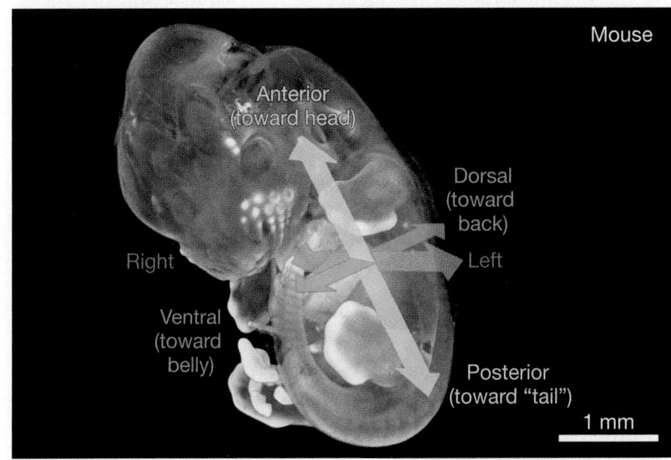

FIGURE 21.3 Most Animals Have Three Major Body Axes.

these master regulators send signals with more specific information about the cells' physical location. As growth continues, the process repeats: New signals arrive and activate genes that specify finer and finer control over what a cell becomes.

THE DISCOVERY OF *BICOID* The insights that led to the discovery of master regulators emerged from work on the fruit fly *Drosophila melanogaster*—a key model organism in genetics and developmental studies (see **BioSkills 14** in Appendix A).

Christiane Nüsslein-Volhard and Eric Wieschaus started this work in the 1970s. They began by exposing adult flies to treatments that cause mutations and examining their offspring for defects in development. One of the most dramatic mutations they found affected the anterior–posterior axis of the embryos. As **Figure 21.4** on page 380 shows, the mutant embryos were missing all of the structures normally found in the anterior end. Instead, the anterior end contained some structures normally found in the posterior.

The gene responsible for this phenotype was dubbed *bicoid*, meaning "two tailed." Based on its phenotype, Nüsslein-Volhard and Wieschaus suspected that the *bicoid* gene's product must provide positional information. In other words, they hypothesized

(a) A normal fruit-fly embryo

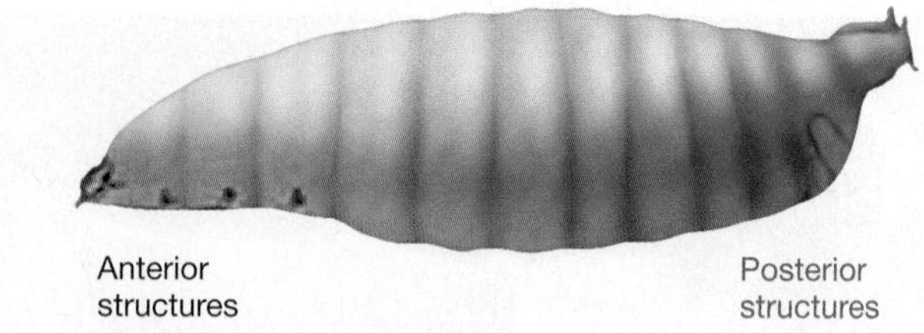

(b) A *bicoid* mutant

Posterior structures
Posterior structures

FIGURE 21.4 Pattern-Formation Mutants Have Misshapen Bodies.

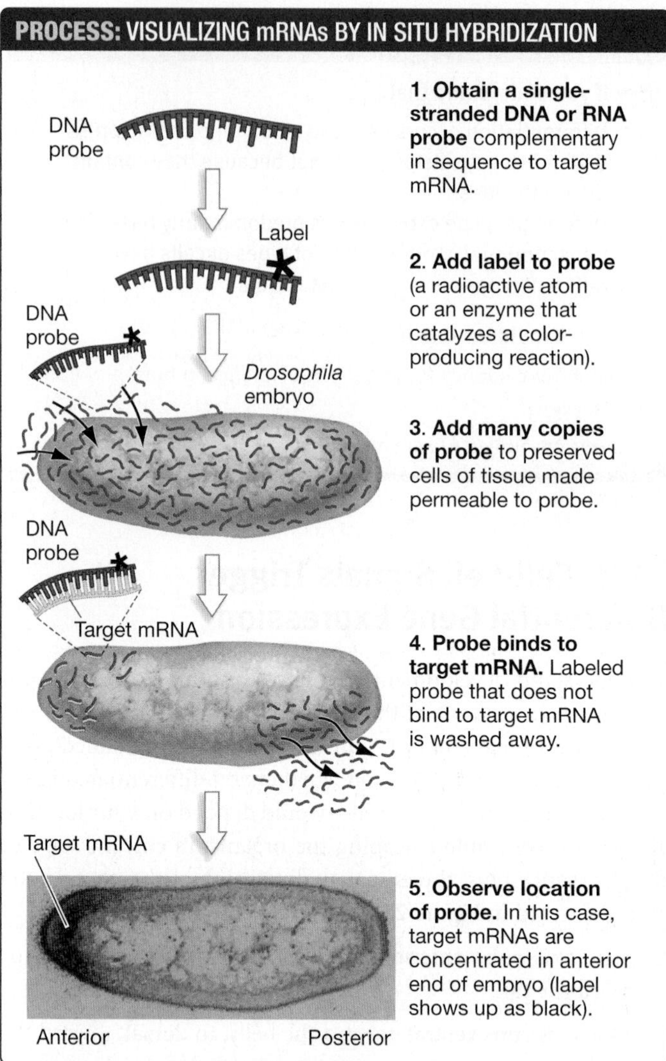

3. Add many copies of probe to preserved cells of tissue made permeable to probe.

4. Probe binds to target mRNA. Labeled probe that does not bind to target mRNA is washed away.

5. Observe location of probe. In this case, target mRNAs are concentrated in anterior end of embryo (label shows up as black).

FIGURE 21.5 In Situ Hybridization Allows Researchers to Pinpoint the Location of Specific mRNAs. The micrograph in the last step shows the location of mRNA from the *bicoid* gene in a fruit-fly embryo.

✔**QUESTION** In situ hybridization is typically used to identify cells that are expressing a particular gene. Why is it valid to claim that labeled cells are expressing the gene in question?

that the *bicoid* gene coded for a signal that tells cells where they are located along the anterior–posterior body axis.

THE IMPORTANCE OF CONCENTRATION GRADIENTS To test their hypothesis, Nüsslein-Volhard and colleagues cloned and sequenced the gene using techniques introduced in Chapter 19. Then they used **in situ** (literally, "in place") **hybridization** to find where *bicoid* mRNAs are located in embryos. As **Figure 21.5** shows, in situ hybridization works by adding a label to single-stranded copies of DNA or RNA molecules—specifically, to molecules that are complementary in sequence to the mRNA of interest. In this case, the probes were designed to bind to *bicoid* mRNA inside the embryo. As a result, the labeled probes marked the location of the mRNAs.

When Nüsslein-Volhard's group treated eggs and early embryos with labeled copies of a probe that bound to *bicoid* mRNA, they found the mRNA to be highly localized at the anterior end (see step 5 in Figure 21.5). Follow-up work showed that when these mRNAs are translated, the protein product forms a steep concentration gradient: Bicoid protein is abundant in the anterior end but declines to progressively lower concentrations in the posterior end.

Later work showed that the Bicoid protein is a regulatory transcription factor (see Chapter 18). It binds to DNA and activates genes required for the formation of anterior structures. In effect, a high concentration of Bicoid is a signal that says, "You're in the head region." A medium concentration of Bicoid means, "You're in the middle of the body." A low concentration indicates, "You're in the posterior" (**Figure 21.6**). When Bicoid is lacking, cells throughout the embryo get the "you're in the posterior" message—leading to the mutant phenotype you saw in Figure 21.4.

To review how the *bicoid* gene works, go to the study area at *www.masteringbiology.com.*

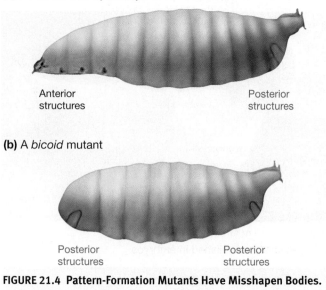 **Web Activity** Early Pattern Formation in *Drosophila*

AUXIN'S ROLE IN PLANT DEVELOPMENT To capture Bicoid's role in the development of a fruit-fly embryo, biologists refer to it as a "master regulator." The idea is that Bicoid gets the cell differentiation process underway, by providing information on where cells are in the body, extremely early in development.

Plants also have a master regulator. But unlike Bicoid, the plant master regulator is not a transcription factor. Instead, it is a **hormone**: a signaling molecule that travels through the body and acts on distant target cells. In plant embryos, the cell-cell signal called auxin enters cells and triggers the production of transcription factors that affect differentiation.

Auxin is produced in meristematic cells at the tip, or apex, of the growing embryo—what will become the top of the stem—and is transported toward the base—what will become the root. In the process, a concentration gradient forms. A high concen-

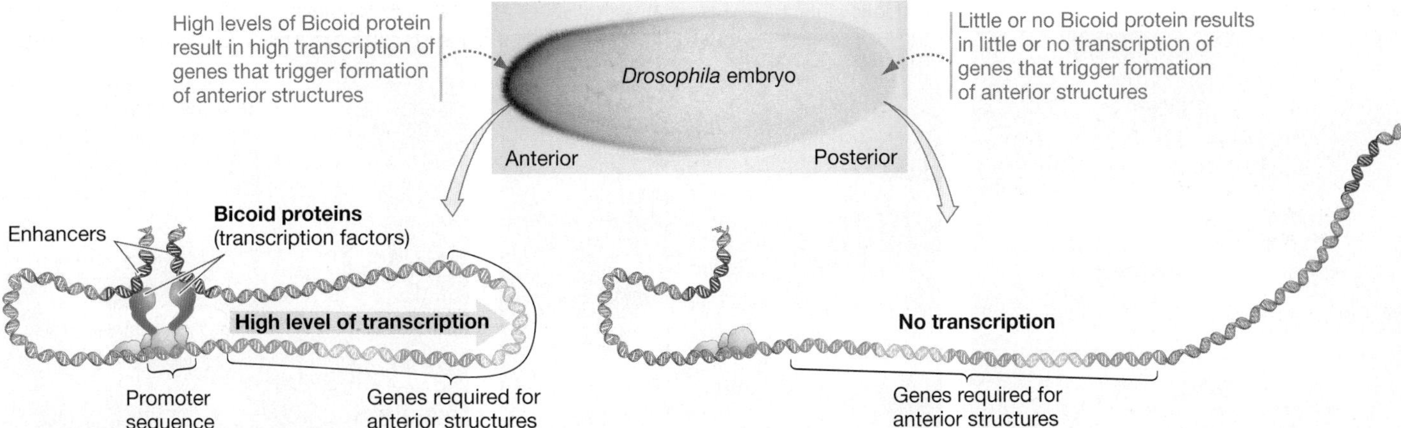
High levels of Bicoid protein result in high transcription of genes that trigger formation of anterior structures
Drosophila embryo
Little or no Bicoid protein results in little or no transcription of genes that trigger formation of anterior structures
Anterior
Posterior
Enhancers
Bicoid proteins (transcription factors)
High level of transcription
No transcription
Promoter sequence
Genes required for anterior structures
Genes required for anterior structures

FIGURE 21.6 Differential Gene Expression Occurs through the Presence or Absence of Regulatory Transcription Factors. Bicoid protein is a regulatory transcription factor that triggers the formation of anterior structures. Because the concentration of Bicoid decreases toward the embryo's posterior, different genes are expressed in the anterior of the embryo than in the posterior.

tration of auxin is a signal that means, "You're near the top of the shoot"; when auxin accumulates at the root it signals, "You're near the base of the root."

In both plants and animals, molecules that provide spatial information during early embryonic development, via a concentration gradient, are called **morphogens** ("form-source"). Bicoid and auxin are morphogens that have a fundamental impact on early development. Because they are present in a concentration gradient, they provide cells with information about their position along the anterior–posterior or the apical–basal body axis. Other initial signals are found in concentration gradients that tell cells where they are along the other body axes.

Regulatory Genes Provide Increasingly Specific Positional Information

The initial work on *bicoid* illustrated the importance of cell-cell signals and interactions as a general theme in animal and plant development. Follow-up work was instrumental in focusing attention on a second fundamental developmental principle common to both plants and animals: Differentiation is a progressive, step-by-step process.

Along with the "two-posteriored monsters" they named *bicoid* mutants, Nüsslein-Volhard and Wieschaus found an array of embryos that had normal anterior–posterior patterning but defects in how their body segments became organized later in development. A **segment** is a region of an animal body that contains a distinct set of structures and is repeated along its length. The mutants had defective **segmentation genes**.

Researchers have now identified three classes of segmentation genes:

1. Sequences called gap genes are expressed first, in broad regions along the head-to-tail axis (**Figure 21.7a**). They define the general position of segments (what part of the body the segments are in).

Anterior — Posterior

(a) Gap genes

Early in development, gap genes define the *general position of head, thorax, and abdominal regions.*

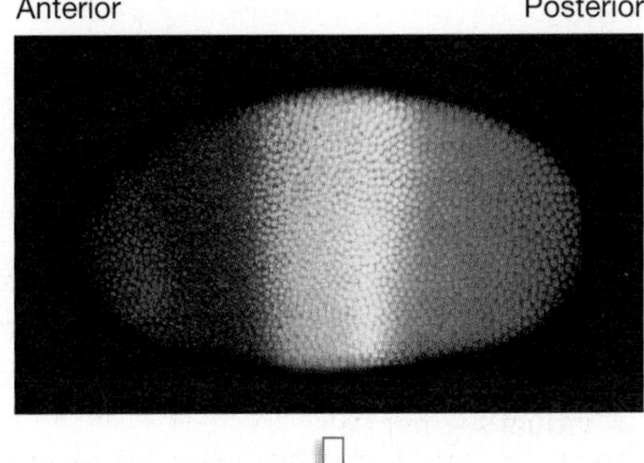

(b) Pair-rule genes

Later in development, pair-rule genes demarcate the *edges of individual segments*.

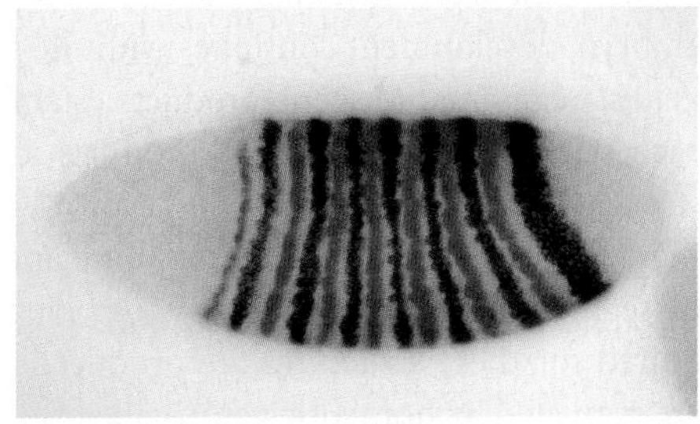

(c) Segment polarity genes

Still later, segment polarity genes delineate boundaries *within individual segments*.

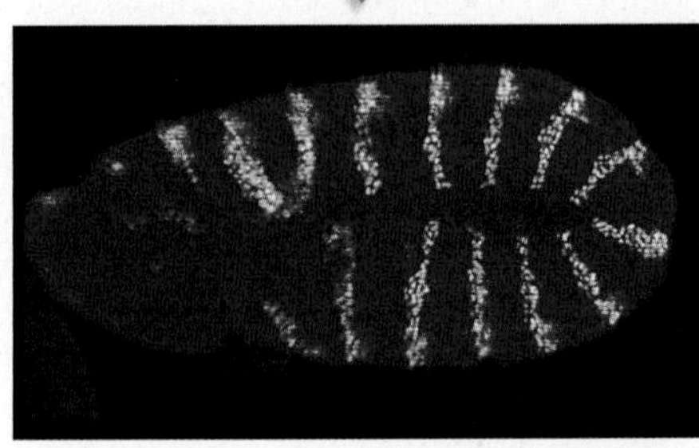

FIGURE 21.7 Sequences of Genes Demarcate Body Segments in Fruit Flies. The embryos in **(a)** and **(b)** were stained for two different gene products from segmentation genes. The embryo in **(c)** was stained for one gene product.

Normal fruit fly

Antennae

Haltere

Homeotic mutant

Wings in place of halteres

Homeotic mutant

Legs in place of antennae

FIGURE 21.8 Homeotic Mutants in *Drosophila* Have Structures in the Wrong Locations. As these colorized, scanning electron micrographs show, homeotic mutants in fruit flies include individuals with wings growing where small, stabilizing structures called halteres should be, or legs growing where antennae should be.

2. Pair-rule genes are expressed next, in alternating bands (**Figure 21.7b**). This pattern and order of expression suggest that pair-rule genes demarcate the edges of individual segments.
3. Segment polarity genes are expressed later, in more restricted bands (**Figure 21.7c**). This pattern and order of expression imply that they delineate boundaries within individual segments.

Once segmentation gene products have established the identity of each segment along the anterior–posterior axis of a fly embryo, development continues with the activation of **homeotic genes**. Segmentation gene products establish the boundaries of each segment; homeotic gene products identify each segment's structural role. More specifically, homeotic gene products trigger the development of structures that are appropriate to each type of segment, such as antennae, wings, or legs. The proteins required for these structures are produced by effector genes that are regulated by homeotic genes called the **Hox genes**.

The *Hox* genes were discovered when researchers found adult fruit flies with body parts in the wrong place. For example, a series of mutations in *Hox* genes can transform a segment in the middle part of the body to the segment just anterior. Instead of bearing the pair of small stabilizer structures called halteres, the transformed segment bears a pair of wings. The mutant has four wings instead of two (**Figure 21.8**).

This type of replacement of one structure by another is termed **homeosis** ("like-condition"). Homeosis occurs when cells get incorrect information about where they are in the body.

To put all this information into perspective, biologists recognize that the genes involved in early development form a regulatory cascade (**Figure 21.9**). Master regulators trigger the production of other regulatory signals and transcription factors, which trigger production of another set of signals and regulatory proteins, and so on.

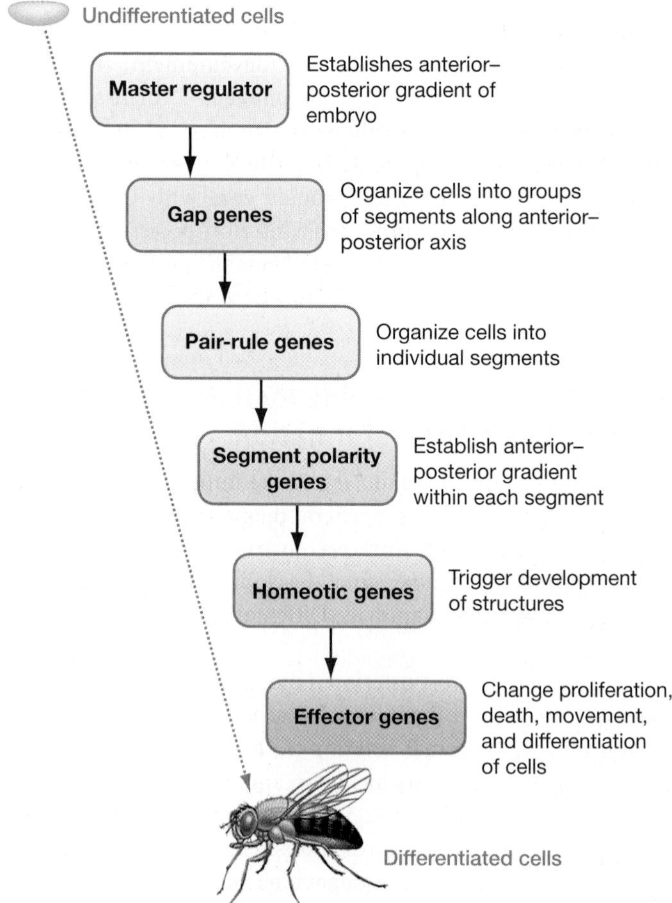

Undifferentiated cells
Master regulator
Establishes anterior–posterior gradient of embryo
Gap genes
Organize cells into groups of segments along anterior–posterior axis
Pair-rule genes
Organize cells into individual segments
Segment polarity genes
Establish anterior–posterior gradient within each segment
Homeotic genes
Trigger development of structures
Effector genes
Change proliferation, death, movement, and differentiation of cells
Differentiated cells

FIGURE 21.9 A Regulatory Gene Cascade in Fruit Flies. Thus far, researchers have described 9 gap genes, 8 pair-rule genes, 9 segment polarity genes, and 11 homeotic (*Hox*) genes in *Drosophila*. Researchers are currently trying to identify the effector genes responsible for producing key structures in this species.

Similar types of regulatory cascades are being found in the other body axes of fruit flies, and in other animal and plant species. In plants, for example, sequences called *MADS-box* genes—detailed in Chapter 23—are analogous to the homeotic genes of animals.

Regulatory genes act in a sequence, triggering gene cascades that provide progressively detailed information about where cells are located in time and space. This positional information, in turn, causes changes in cell proliferation, death, movement, differentiation, and interaction.

- Because the signals and transcription factors vary in identity and concentration along the three major body axes, cells in different locations receive unique positional information.
- Each level in a regulatory cascade provides a more specific level of information about where a cell is.
- As regulatory cascades proceed over time, a cell's fate becomes more and more finely determined.

Cell-Cell Signals and Regulatory Genes Are Evolutionarily Conserved

After homeotic genes were discovered and characterized in fruit flies, researchers began looking for similar genes in other animals. The results were striking.

Investigators have found that clusters of *Hox* genes occur in virtually every animal examined to date, including frogs, crustaceans (crabs and their relatives), birds, various types of worms, mice, and humans. Although the number of *Hox* genes varies widely among species, their chromosomal organization is similar to that of the fly homeotic genes (**Figure 21.10a**).

Recent studies of *Hox* genes have shown that they are expressed along the head-to-tail axis of the mouse embryo in the same sequence as in fruit flies (**Figure 21.10b**). In addition, experiments have shown that when mouse *Hox* genes are altered by mutation, defects in pattern formation result. Based on these data, biologists conclude that in flies, mice, humans, and most other animals, *Hox* genes play a key role in identifying the position of cells along the head-to-tail axis of the body.

This conclusion was supported in dramatic fashion when researchers in William McGinnis's lab introduced the *Hoxb6* gene from mice into fruit-fly eggs. The *Hoxb6* gene in mice is similar in structure and sequence to the *Antp* gene of flies. Because it was introduced without its normal regulatory sequences, the *Hoxb6* gene was expressed throughout the treated fly embryos. The resulting larvae had defects identical to those observed in naturally occurring fly mutants in which the *Antp* gene is mistakenly expressed throughout the embryo. This is a stunning result: A mouse allele not only affected the development of a fly but also mimicked the effect of a specific fly allele.

To interpret these observations, biologists hypothesize that the genes in *Hox* complexes of animals are homologous—meaning that they are similar because they are descended from genes in a common ancestor. If this hypothesis is correct, it means that the first *Hox* genes arose before the origin of animals. For the past billion years, *Hox* gene products have been helping to direct the development of animals.

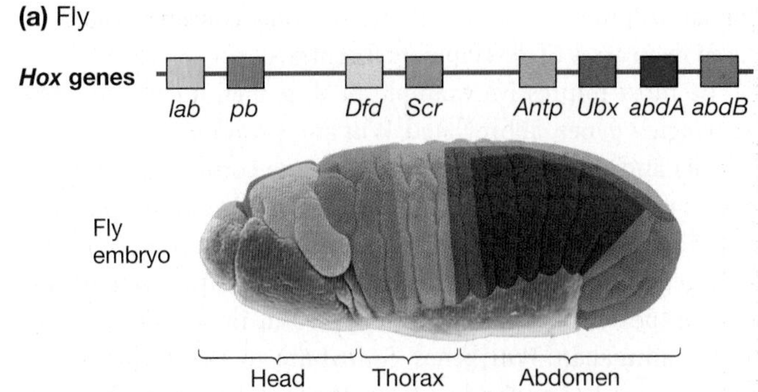

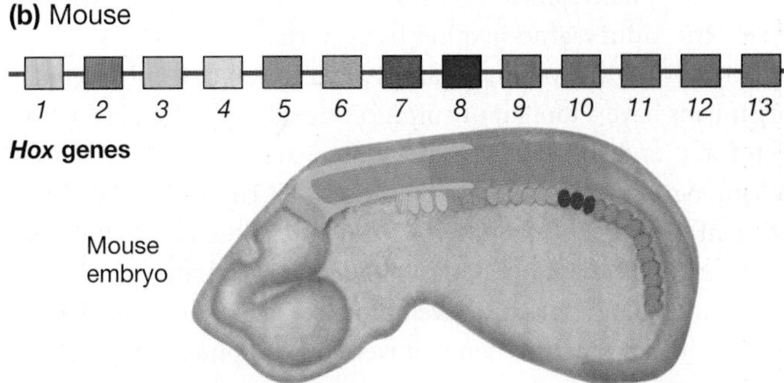

FIGURE 21.10 Organization and Expression of *Hox* Genes. The location of *Hox* genes on the chromosome correlates with their pattern of expression in the embryos of flies and mice. The genes represented by same-color boxes in parts **(a)** and **(b)** are considered homologous.

✔**QUESTION** What evidence would support the claim that these genes are homologous?

The take-home message from these studies is that at least some of the molecular mechanisms of pattern formation have been highly conserved during animal evolution. The discovery of these shared mechanisms is one of the most significant results to have emerged from animal development studies to date.

Although animal bodies are spectacularly diverse in size and shape, the underlying mechanisms responsible for their development are similar. Regulatory gene cascades occur in all animals, and all plants, too. Within each of these lineages, many of the elements in gene cascades are shared. What varies among species is less a matter of which genes are present and more a matter of when, where, and in what quantity similar genes are expressed.

Common Signaling Pathways Are Active in Many Contexts

Regulatory gene cascades and the evolutionary conservation of key developmental genes are general features of animal and plant development. But biologists have articulated another organizing principle as well: During development, the same regulatory transcription factors and cell-cell signals are active in a variety of contexts.

Regulatory proteins and signals are not only conserved; they are reused in an array of developmental contexts.

The most impressive example of this phenomenon may be the *wingless* genes, abbreviated *Wnt* and pronounced "wint." In humans and other mammals, the genome contains a family of 15 *Wnt* genes. Each of these genes codes for proteins responsible for cell-cell signaling during development. More specifically, these genes are part of the regulatory cascade that sets up the anterior–posterior axis in the embryo. But in addition, signals from mammalian *Wnt* genes contribute to the formation of back muscles, the midbrain region, the limbs, the gonads (testes or ovaries), hair follicles, parts of the intestine, and structures inside the kidney, among other body parts.

To capture this point, biologists like to say that multicellular organisms have a tool kit of common signals, signal-transduction pathways, and regulatory proteins that are used over and over during development. The common tool kit can direct the development of dramatically different structures because the tools are deployed at different times and in different locations.

As an analogy for this developmental principle, consider the signal called a pinch. When you were little, a pinch on the cheek from your grandmother indicated something very different from a pinch on the arm from an older sibling. It also means something different from a pinch you might receive now from someone you are romantically attracted to. In development, as in human communication, the context in which a signal is sent and received—its location, timing, and intensity—has a major effect on the signal's meaning and consequences.

Also by analogy, the elements of a regulatory gene cascade or a particular signal transduction pathway are like the tools that a carpenter uses. The same hammer and circular saw can be used to build a shack or a palace. The key is the timing and extent of hammer and saw "expression"—when and how the tools are used during the development of a structure.

CHECK YOUR UNDERSTANDING

If you understand that . . .

- Cell-cell signals trigger the production of specific sets of transcription factors that change gene expression in receiving cells.
- Different cells express different genes because they receive different sets of signals, and because they produce different sets of transcription factors in response.

✓ You should be able to . . .

1. Explain why the gradient of Bicoid protein delivers information about where cells are along the anterior–posterior axis of a fly embryo.
2. Describe how the cascade of transcription factors triggered by Bicoid leads to the gradual differentiation of segments along the anterior–posterior axis of a fly embryo.

Answers are available in Appendix B.

21.4 Changes in Developmental Pathways Underlie Evolutionary Change

Sections 21.1 through 21.3 have shown that for an embryo to develop, cells have to proliferate, die, move or expand, differentiate, and interact in specific ways. Differentiation is caused by differential gene expression. It results from signals that tell cells where they are in time and space, triggering a complex cascade of regulatory transcription factors.

If any of these processes is disrupted, the embryo is likely to die. But if one of these processes is modified in some slight way, the effect may be a structure with a different size or shape or activity. As a result, the embryo will develop new features, and the adult will have a novel phenotype.

Once biologists began working out the regulatory signals and cascades introduced earlier in the chapter, they realized that the genetic changes altering these developmental processes must be the foundation of evolutionary change. The increase in body size that has occurred during human evolution, for example, must have resulted from mutations that altered the signals, regulatory sequences in DNA, or transcription factors that are involved in the amount and timing of cell proliferation throughout the body.

A research field called evolutionary-developmental biology, or **evo-devo**, focuses on understanding how changes in developmentally important genes have led to the evolution of new phenotypes such as the flower, the leaf, the limbs found in tetrapods (amphibians, reptiles, mammals), and the limbs of arthropods (crabs, insects, millipedes). As an example of how this work is done, let's consider an instance of limb *loss*—in snakes.

Although some snakes do not develop any sort of forelimb or hind limb at all, boas and pythons have tiny pelvic (hip) bones and a rudimentary femur (thigh bone) and claw (**Figure 21.11**). The fossil record shows that the ancestor of all snakes had four functional legs along with feet and toes. Their closest living relatives, the lizards, also have limbs.

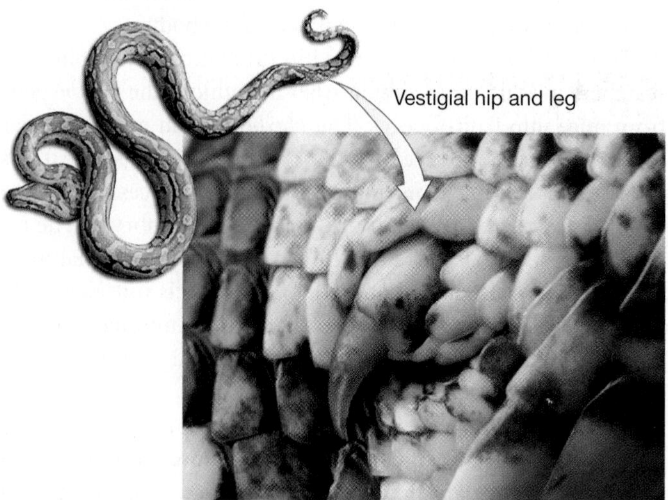

FIGURE 21.11 Some Snakes Have Rudimentary Hind Limbs. This claw, of a South African python, is attached to a vestigial femur and hip.

(a) Pattern of gene expression in tetrapods with forelimbs.

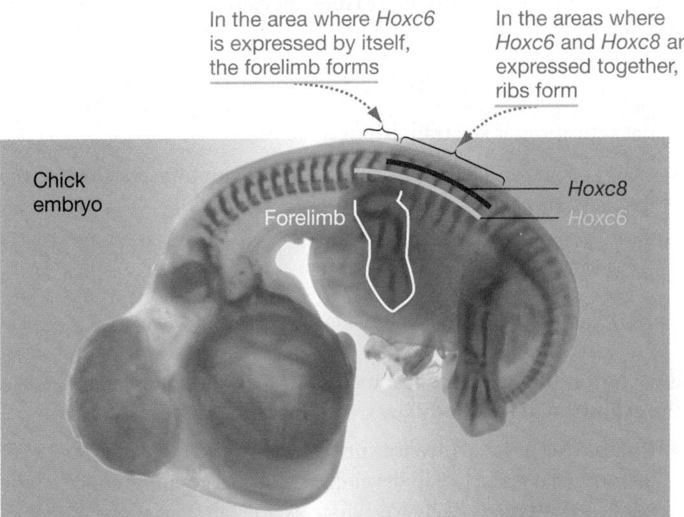

In the area where *Hoxc6* is expressed by itself, the forelimb forms
In the areas where *Hoxc6* and *Hoxc8* are expressed together, ribs form
Chick embryo
Forelimb
Hoxc8
Hoxc6

(b) Pattern of gene expression in snakes (no forelimbs).

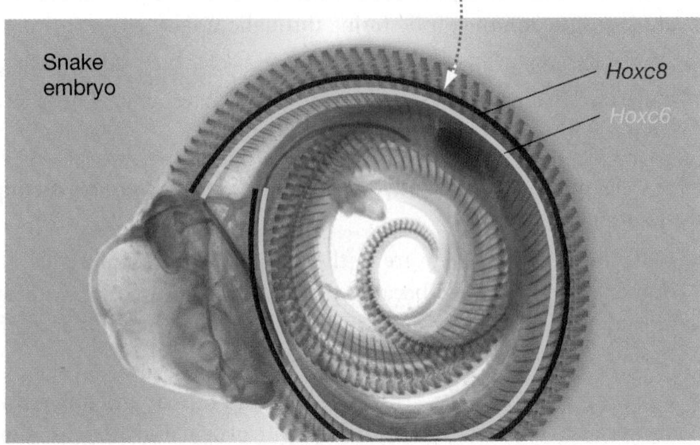

Hoxc6 and *Hoxc8* are always expressed together, so no forelimbs form
Snake embryo
Hoxc8
Hoxc6

FIGURE 21.12 Changes in Homeotic Gene Expression Led to Limb Loss in Snakes. *Hoxc6* and *Hoxc8* code for transcription factors. If both transcription factors are present in a population of cells along the anterior–posterior axis of a tetrapod, genes that lead to the formation of vertebrae and ribs are activated. But if only *Hoxc6* is expressed, then genes that lead to the formation of forelimbs are activated.

How did snakes lose their legs? What developmental changes were responsible for this dramatic evolutionary event?

Researchers were able to answer this question because the cell-cell signals and regulatory cascades responsible for limb development are well understood. In chicken embryos, for example, *Hox* genes called *Hoxc6* and *Hoxc8* are expressed together in cells where ribs form, but *Hoxc6* is expressed without *Hoxc8* in the region that gives rise to the forelimbs (**Figure 21.12a**). The situation is different in snakes. As **Figure 21.12b** shows, *Hoxc6* and *Hoxc8* are expressed together throughout the snake embryo—including in the region where forelimbs should form. These data suggest that a change in the regulation of a homeotic gene—specifically, where *Hoxc8* is expressed—led to the evolutionary loss of the forelimb. Snakes make ribs instead of forelimbs.

Hind limb loss, in contrast, is due to defects in a signaling molecule encoded by the gene *Sonic hedgehog*. This gene is homologous to one of the segment polarity genes found in fruit flies. These observations suggest that a mutation in a gene responsible for setting up the anterior–posterior axis led to the evolutionary loss of the hind limb. Recent work has shown that defects in *Sonic hedgehog* signaling were also responsible for the loss of limbs in whales.

Both in forelimb and hind limb loss, biologists can point to alterations in a specific transcription factor or cell-cell signal that explains why an important evolutionary change occurred. Changes in a regulatory cascade that is triggered early in development led to changes in the adult body of a lizard-like ancestor. Snakes have been legless ever since.

CHAPTER 21 REVIEW

For media, go to the study area at www.masteringbiology.com 

Summary of Key Concepts

During development, cells divide, die, move, or expand in a directed manner; specialize; and interact with other cells.

- Cells have to proliferate in a regulated manner as the body develops.
- Some cells have to undergo programmed cell death as part of normal development.
- In animals, masses of cells move in a coordinated way early in development; later, certain cells move to specific locations in the body. Plant cells do not move, but normal plant development depends on precise control over the plane of cell division and directed expansion of cells.
- Cells undergo a step-by-step process that leads to differentiation, or the production of specialized cell types.
- Cell-cell signals provide a constant flow of information about where cells are in space and time.

✔ You should be able to predict how the development of an oak seedling would be affected by a drug that cuts the normal rate of cell proliferation by half.

Cells become specialized because they express different genes, not because they contain different genes.

- In animals and plants, cells differentiate due to differential gene expression, not differential loss of genes.
- Meristematic cells in plants and stem cells in animals do not differentiate, but continue proliferating throughout life.

✔ You should be able to explain why it's possible to propagate banana trees from cuttings—specifically, why parenchyma cells in a stem can give rise to vascular tissue in a root.

Cells interact continuously by means of cell-cell signals during development.

- Cells "know" where they are in the body and how far along development has progressed because they produce and receive a steady stream of signals. These signals are produced by other cells and diffuse or are transported throughout the body.
- Cells begin expressing a distinctive suite of proteins and differentiate because they receive a distinctive suite of signals. Differentiation depends on both the types and quantities of signals received.

✔ You should be able to explain why concentration gradients in transcription factors or cell-cell signals convey information.

When development begins, early cell-cell signals trigger a cascade of effects that cause increasing specialization as development proceeds.

- Differentiation occurs in a step-by-step manner because cell-cell signals trigger a cascade of effects that over time lead to the sequential activation of key transcription factors.
- Cells become specialized because they contain transcription factors that activate certain genes and not others.

✔ You should be able to explain the concept of a "master regulator" in early development.

MB **Web Activity** Early Pattern Formation in *Drosophila*

Mutations in genes responsible for development may lead to the evolution of new body sizes, shapes, and structures.

- If the cell-cell signals, transcription factors, and regulatory DNA sequences that are active in development undergo mutation, then the adult phenotype is likely to change as a result. Variation in phenotypes is the basis of evolutionary change.

✔ You should be able to explain the evidence behind the claim that changes in cell-cell signals and transcription factors were responsible for the loss of limbs in snakes.

Questions

✔ TEST YOUR KNOWLEDGE

Answers are available in Appendix B

1. What is apoptosis?
   - **a.** an experimental technique used to kill specific cells
   - **b.** programmed cell death that is required for normal development
   - **c.** a pathological condition observed only in damaged or diseased organisms
   - **d.** a developmental mechanism unique to the roundworm *C. elegans*
2. What is the function of stem cells in adult mammals?
   - **a.** Some of their daughter cells remain as stem cells and continue to divide throughout life.
   - **b.** They give rise to hair, fingernails, and other structures that grow throughout life.
   - **c.** They produce compounds that stem blood loss from wounds.
   - **d.** They produce cells that differentiate to replace dead or damaged cells.
3. Why are in situ hybridizations such a valuable tool for studying development?
   - **a.** They identify the location of specific mRNAs, and so provide a picture of differential gene expression.
   - **b.** They allow researchers to understand how cell-cell signals and regulatory transcription factors interact.
   - **c.** They provide data on homology—the presence of similar genes in different species.
   - **d.** They can be done with RNA or DNA probes.
4. What does it mean to say that differentiation is "progressive?"
   - **a.** Differentiation gets more efficient over time.
   - **b.** Differentiation gets more complex over time.
   - **c.** Cells become increasingly more specialized over time.
   - **d.** Differentiation is triggered by master regulators.
5. What is a homeotic mutant?
   - **a.** an individual with a structure located in the wrong place
   - **b.** an individual with an abnormal head-to-tail axis
   - **c.** in flies, an individual that is missing segments; in *Arabidopsis*, an individual that is missing a hypocotyl or other embryonic structure
   - **d.** an individual with double the normal number of structures or segments
6. What is "homeosis"?
   - **a.** replacement of one structure by another structure (a structure develops in the wrong location)
   - **b.** the observation that the same signaling pathway is used during the development of many different structures in the same species (in different locations or at different times)
   - **c.** the observation that the same regulatory cascades are conserved among animal or plant species
   - **d.** similarity between species due to descent from a common ancestor

✔ TEST YOUR UNDERSTANDING

Answers are available in Appendix B

1. What does it mean to say that cell proliferation, death, movement, or expansion; differentiation; and interaction are shared developmental processes?
2. Explain the logic behind using nuclear transplant experiments to test the hypothesis that all animal cells in the body are genetically equivalent.

3. How did researchers go about looking for genes that have important roles in establishing the anterior–posterior body axis and body segments in *Drosophila*?
4. Why is it significant that many of the genes involved in development encode regulatory transcription factors?
5. Explain the connection between the existence of regulatory cascades and the observation that differentiation is a step-by-step process.
6. What evidence suggests that at least some of the molecular mechanisms responsible for pattern formation have been highly conserved over the course of animal evolution?

✓ APPLYING CONCEPTS TO NEW SITUATIONS

Answers are available in Appendix B

1. Recent research has shown that the products of two different *Drosophila* genes are required to keep *bicoid* mRNA concentrated at the anterior end of the egg. In individuals with mutant forms of these proteins, *bicoid* mRNA diffuses farther toward the posterior pole than it normally does. Predict what effect these mutations will have on segmentation of the larva.
2. In 1992 David Vaux and colleagues used genetic engineering technology to introduce an active human gene for a protein that inhibits apoptosis into embryos of the roundworm *C. elegans*. When the team examined the embryos, they found that cells that normally undergo programmed cell death survived. What is the significance of this observation?
3. Suppose that physicians implanted human stem cells into the brains of patients with Parkinson's disease, in the area of the brain where the patients had lost many nerve cells. What would have to happen for the stem cells to differentiate into functioning nerve cells?
4. Fruit flies have 6 legs; spiders have 8 legs; centipedes have many; earthworms have none. All of these species are segmented and have *Hox* genes. Based on data in this chapter, generate a hypothesis to explain how variation in leg number evolved in these species.

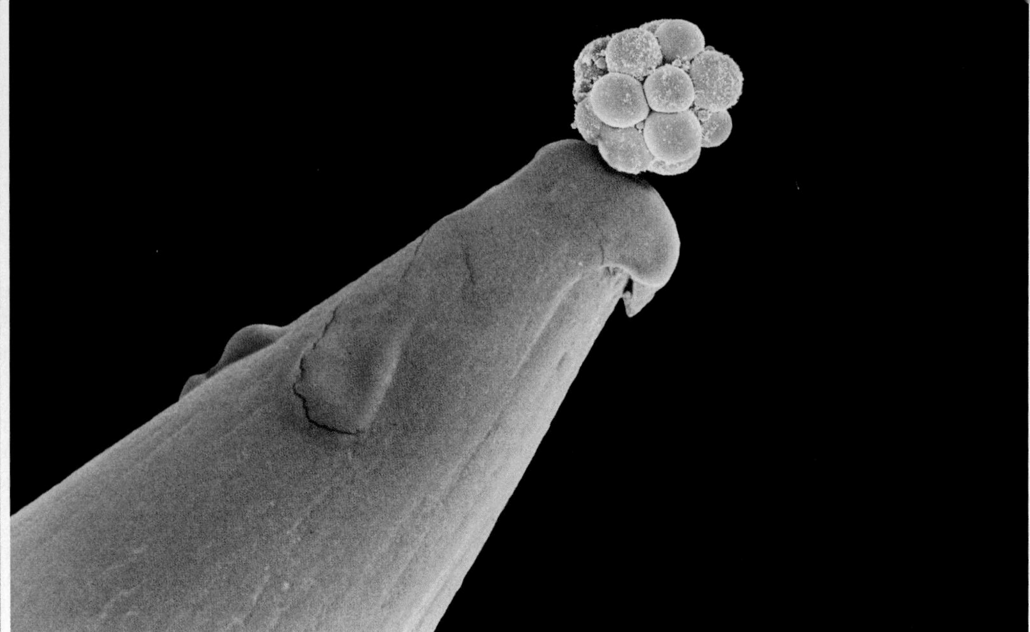

A human embryo, about 3 days old, on the tip of a pin.

22 An Introduction to Animal Development

KEY CONCEPTS

- Fertilization begins with specific interactions between proteins on the plasma membranes of sperm and egg.
- The earliest cell divisions divide the fertilized egg into a mass of cells whose individual fates depend on key regulatory molecules they contain and the signals they receive.
- Early in development, cells engage in collective, coordinated movements to form distinct tissue layers. Each layer gives rise to a different set of tissues and organs in the adult.
- As development proceeds, specialized organs and other structures form through the interacting effects of cell-cell signals, cell proliferation, cell movements, and differentiation. Differentiation is complete when cells express tissue-specific proteins.

When the physician, researcher, and writer Lewis Thomas considered how a human being develops from a fertilized egg, he could only marvel: "You start out as a single cell derived from the coupling of a sperm and an egg, this divides into two, then four, then eight, and so on, and at a certain stage there emerges a single cell which will have as all its progeny the human brain. The mere existence of that cell should be one of the great astonishments of the earth. People ought to be walking around all day, all through their waking hours, calling to each other in endless wonderment, talking of nothing except that cell. It is an unbelievable thing, and yet there it is, popping neatly into its place amid the jumbled cells of every one of the several billion human embryos around the planet, just as if it were the easiest thing in the world to do."[1]

What molecular mechanisms are responsible for the development of the brain and other tissues and organs? Chapter 21 introduced the basic developmental processes of cell proliferation, death, movement, or expansion; differentiation; and interactions, and went on to explore differentiation in depth. That chapter's central message was that cell-cell signals cause specific sets of transcription factors to be produced in various cells throughout the embryo, resulting in differential gene expression and differentiation. Brain cells "know" that they are brain cells because they have received specific types and concentrations of cell-cell signals; in response, they produce brain-specific proteins.

This chapter's goal is to use the general developmental principles presented in Chapter 21 to understand the overall sequence of events responsible for animal development. We begin with sperm and egg and end with formation of a differentiated cell—in this case, a muscle cell. At each step, remember that it is cell-cell signals and regulatory proteins that allow developing cells to divide, move, interact, and differentiate—to do the things required for them to develop correctly and contribute to the formation of a new individual.

[1]Lewis Thomas, *The Medusa and the Snail* (New York: Viking Press, 1979), 156. Thomas was exercising some poetic license here. The brain actually arises from a group of cells in the embryo rather than from a single cell.

✔ When you see this checkmark, stop and test yourself. Answers are available in Appendix B.

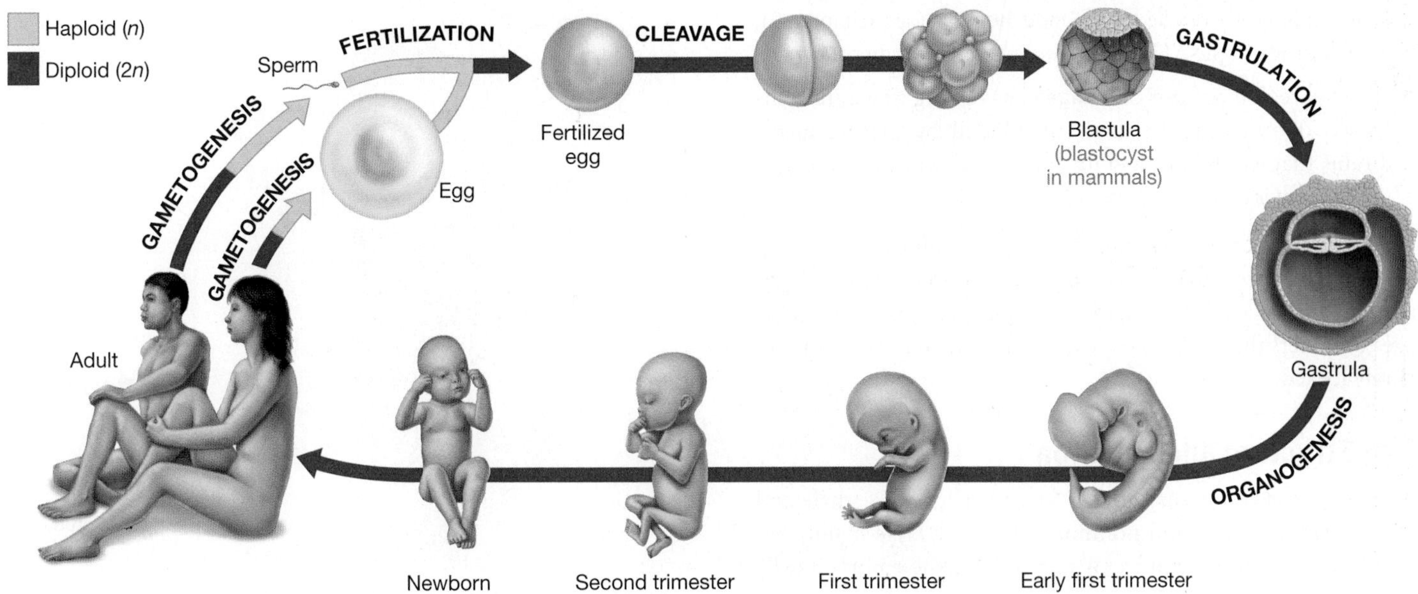

FIGURE 22.1 Development Proceeds in Ordered Phases. In animals, the development of a new individual starts with the formation of gametes (sperm and eggs) and continues with fertilization and early cell divisions (cleavage) that result in a blastula. Gastrulation then rearranges the blastula into a gastrula. Organogenesis follows and leads to the formation of adult tissues and structures.

22.1 Gamete Structure and Function

Recall from Chapter 12 that gametes are haploid reproductive cells. In animals and many other organisms, male gametes are called sperm and female gametes are called eggs. The sequence of mitotic and meiotic cell divisions leading up to the production of animal sperm and eggs—a process called gametogenesis—is outlined in Chapter 48.

As **Figure 22.1** shows, gametogenesis in animals is followed by **fertilization**, or the union of a sperm and an egg. Fertilization is the subject of Section 22.2; the subsequent steps in development—called cleavage, gastrulation, and organogenesis—are introduced in Sections 22.3 through 22.5.

Here let's focus on the gamete itself. Sperm and eggs form in the reproductive organs of adult organisms. The DNA and cytoplasm in these reproductive cells are the initial components of the new individual. Both sperm and egg contribute chromosomes—usually a haploid genome containing one allele of each gene—to the offspring. But the similarity ends there. Egg cells contribute much more than chromosomes, and are routinely hundreds or thousands of times larger than sperm cells.

Sperm Structure and Function

Sperm cells begin to develop after meiosis has resulted in the production of a haploid nucleus. As a mammalian sperm cell matures, it acquires the four main compartments shown in **Figure 22.2**: the head, neck, midpiece, and tail.

- The head region contains the nucleus and an enzyme-filled structure called the **acrosome**. The enzymes stored in the acrosome allow the sperm to penetrate the barriers surrounding the egg.
- The neck encloses a **centriole** that will combine with a centriole contributed by the egg to form a centrosome (see Chapter 7). The centrosome is required for spindle formation during mitosis (see Chapter 11).

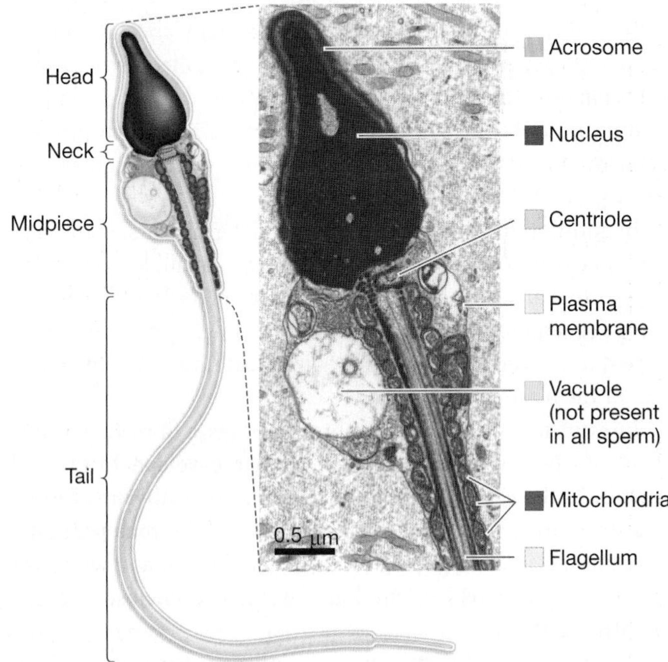

FIGURE 22.2 Sperm Structure. The morphology of human sperm is typical of many mammal species. The cell is specialized for motility and fusing with an egg cell.

- The midpiece is packed with **mitochondria** (see Chapter 9), which produce the ATP required to power movement.
- The tail region consists of a **flagellum**—a long structure composed of microtubules, and surrounded by plasma membrane, that whips back and forth to make swimming possible (see Chapter 7).

Sperm are race cars—stripped down, streamlined, souped-up cells that are specialized for racing other sperm to the egg. Eggs, in comparison, are like semitrailers—bulky, far less mobile storage containers that are packed with valuable merchandise and securely locked.

Egg Structure and Function

Animal gametes are easy to tell apart: Sperm are small and motile; eggs are large and nonmotile. Eggs are large mainly because they contain the nutrients required for the embryo's early development. The quantity of nutrients present in the egg varies widely among species, however.

- In mammals, embryos start to obtain nutrition through a maternal organ called the **placenta** within a week or two after fertilization. Thus, the egg only has to supply nutrients for early development and is relatively small.
- In species where females lay eggs directly into the environment, the stores in the egg are the *only* source of nutrients until organs have formed and a larva or juvenile hatches and begins to feed. In these species, the nutrients required for early development are provided by **yolk**—a fat- and protein-rich cytoplasm that is loaded into egg cells as they mature. Yolk may be present as one large mass or as many small granules.

In addition to nutrients, the eggs of many species contain key developmental regulatory molecules called **cytoplasmic determinants** that control the early events of development. The *bicoid* mRNA introduced in Chapter 21 is a cytoplasmic determinant. Its protein product acts as a master regulator in development.

In addition to yolk and cytoplasmic determinants, many eggs contain organelles called **cortical granules**—small, enzyme-filled vesicles that are activated during fertilization. As the egg matures, cortical granules are synthesized in the Golgi apparatus, transported to the cell surface, and localized just under the inner surface of the plasma membrane.

Just outside the plasma membrane of eggs, a fibrous, mat-like sheet of glycoproteins called the **vitelline envelope** forms and surrounds the egg. In the eggs of humans and other mammals, this structure is unusually thick and is called the **zona pellucida**. In some species, a large gelatinous matrix known as a jelly layer surrounds the vitelline envelope to further enclose the egg (**Figure 22.3**).

To summarize, the egg is loaded with rich stores of nutrients and guarded by one or more layers of protective material. How does a sperm cell penetrate these coatings to reach the egg's plasma membrane and fertilize the egg?

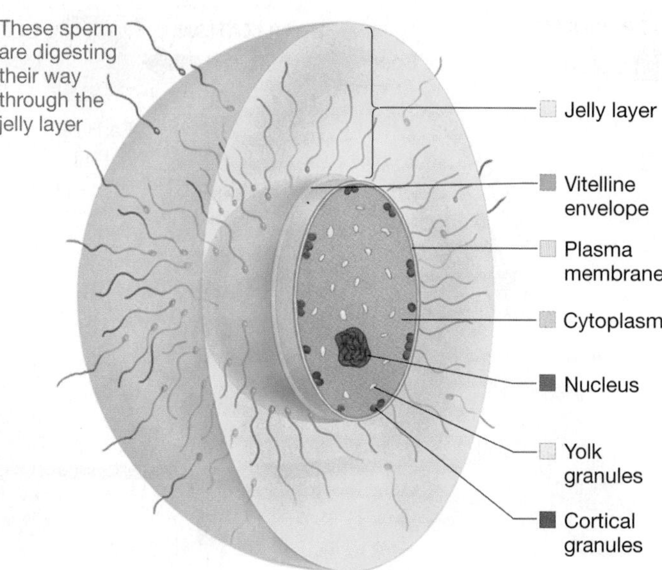

FIGURE 22.3 Sea Urchin Eggs Are Covered by a Jelly Layer. In many types of animals that lay eggs in water, the egg is surrounded with a protective jelly layer. The other features shown here are also typical of many animal eggs.

22.2 Fertilization

Fertilization may seem like a simple process: A sperm cell fuses with an egg cell to form a diploid cell known as a **zygote**—a fertilized egg. The process is actually extraordinarily complex, however.

For fertilization to take place, sperm and egg cells must be in the same place at the same time. Then they must recognize and bind to each other. Next they have to fuse—even though most of the other cells in the body do not fuse with cells they contact. In most species, fusion must also be limited to a single sperm so that the zygote does not receive extra chromosomes. Finally, the fusion of the two gametes has to trigger the onset of development. This complexity has made fertilization a fascinating research topic. The contact that takes place between sperm and egg at fertilization qualifies as the best studied of all cell-to-cell interactions.

Research on fertilization began in earnest early in the twentieth century, when biologists started to study the sperm–egg interaction in sea urchins. Like most aquatic animals, sea urchins shed their gametes into the surrounding water and fertilization occurs externally.

When the sperm and egg of sea urchins meet, the head of the sperm initially encounters the jelly layer of the egg cell. To reach the plasma membrane of the egg cell, the sperm head digests its way through the egg's jelly layer and vitelline envelope using enzymes released from the acrosome—the structure at the tip of the sperm's head. Once the sperm head contacts the egg-cell surface, the plasma membranes of the egg and sperm fuse. The sperm nucleus, mitochondria, and centriole enter the egg—though the sperm mitochondria later disintegrate—and the sperm and egg nuclei fuse to form the zygote nucleus. Fertilization is then complete.

But what prevents cross-species fertilization, and the production of dysfunctional hybrid offspring? After all, in many habitats sperm and eggs from a particular sea urchin species float in seawater along with eggs and sperm from other sea urchin species and many other organisms.

How Do Gametes from the Same Species Recognize Each Other?

In the 1970s, Victor Vacquier and co-workers succeeded in identifying a protein on the head of sea urchin sperm that binds to the surface of sea urchin eggs in a species-specific manner. They called this protein bindin. Follow-up work showed that the bindin proteins from even very closely related species are distinct. As a result, bindin should ensure that a sperm binds only to eggs from the same species.

But what does bindin bind to? If bindin acts as a key, what acts as the lock? Kathleen Foltz and William Lennarz hypothesized that egg-cell membranes contain a receptor for bindin. To test this hypothesis, they set out to isolate the bindin receptor. More specifically, they set out to isolate the part of the bindin receptor that is exposed on the outside of the egg. They predicted that this region of the protein interacts with the bindin on sperm.

Figure 22.4 illustrates Foltz and Lennarz's experimental approach. They began by treating the surface of sea urchin eggs with a **protease**—an enzyme that cleaves peptide bonds. When the investigators isolated the protein fragments that were released from the egg surface, they found one that bound to sperm and to isolated bindin molecules. Further, this binding occurred in a species-specific manner. A protein fragment from the eggs of one species bound to sperm of its own species, but did not bind to sperm of different species. Based on these observations, the biologists claimed that they had found the outward-facing portion of the egg-cell receptor protein for sperm.

During sea urchin fertilization, species-specific bindin molecules on sperm interact with species-specific receptors on the surface of the egg. This interaction is required for the plasma membranes of sperm and egg to fuse. Strong evidence exists for the same types of protein–protein interactions between the sperm and egg cells of mammals, although the exact proteins involved have yet to be identified.

✔If you understand the importance of protein–protein interactions for fusion of sperm and egg-cell membranes, you should be able to explain why a molecule that bound to a bindin-like protein on the human sperm head would function as an effective contraceptive.

Why Does Only One Sperm Enter the Egg?

In early studies on sea urchin fertilization, researchers noticed that only one sperm succeeded in fertilizing the egg, even when dozens or even hundreds of sperm were clustered around the vitelline envelope. This observation makes sense: Multiple fertilization, or **polyspermy**, would result in a zygote that had more than two copies of each chromosome. Sea urchin embryos with more than two copies of each chromosome die. How do animals avoid polyspermy?

In sea urchins, fertilization results in the erection of a physical barrier to sperm entry. The process begins when sperm entry causes calcium ions (Ca^{2+}) to be released from storage areas inside the egg. As **Figure 22.5a** on page 392 shows, a wave of ion release starts at the point of sperm entry and propagates throughout the egg.

EXPERIMENT

QUESTION: If bindin is the protein on the surface of sperm that acts like a key, what is the "lock"?

HYPOTHESIS: The lock is a receptor protein on the surface of sea urchin eggs. Bindin binds to this receptor.

NULL HYPOTHESIS: The lock is not a receptor protein on the surface of sea urchin eggs.

EXPERIMENTAL SETUP:

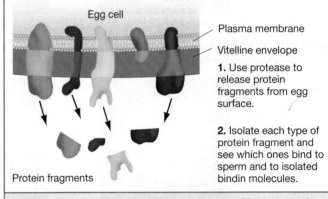

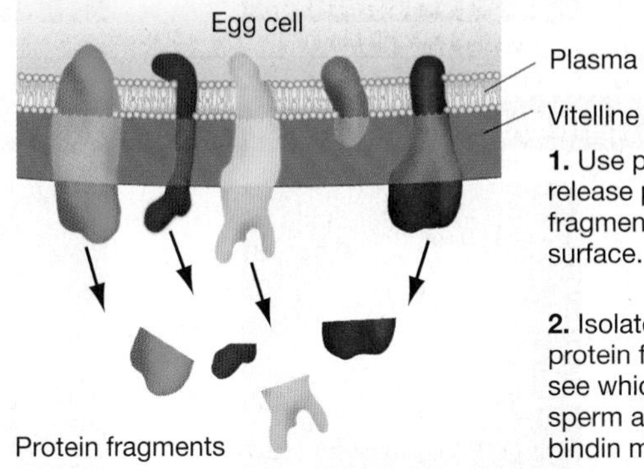

Egg cell
Plasma membrane
Vitelline envelope
1. Use protease to release protein fragments from egg surface.
2. Isolate each type of protein fragment and see which ones bind to sperm and to isolated bindin molecules.
Protein fragments

PREDICTION: One of the protein fragments isolated from egg surface will bind to sperm and to isolated bindin molecules.

PREDICTION OF NULL HYPOTHESIS: The protein fragments isolated from egg surface will not bind to sperm or to isolated bindin molecules.

RESULTS:

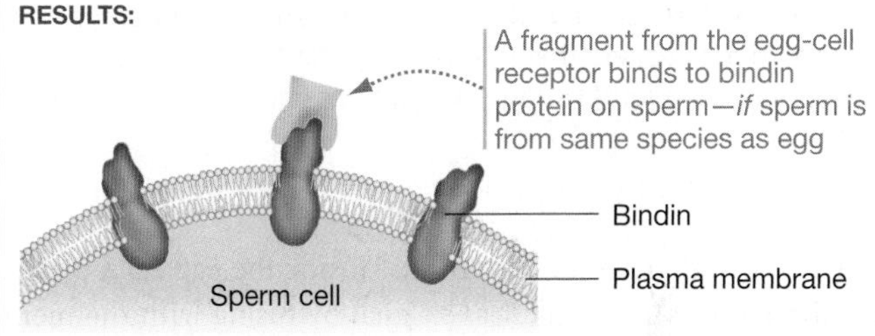

A fragment from the egg-cell receptor binds to bindin protein on sperm—*if* sperm is from same species as egg
Bindin
Plasma membrane
Sperm cell

CONCLUSION: The "lock" is a receptor protein on the surface of eggs that binds to bindin (the "key") in a species-specific manner.

FIGURE 22.4 The Egg-Cell Receptor for Sperm Was Isolated and Characterized.

SOURCE: Foltz, K. R., and W. J. Lennarz. 1990. Purification and characterization of an extracellular fragment of the sea urchin egg receptor for sperm. *Journal of Cellular Biology* 111: 2951–2959.

✔**QUESTION** Experimental designs that depend on an exhaustive search of all possibilities are sometimes called "brute force" strategies. In what sense was this experiment a "brute force" approach?

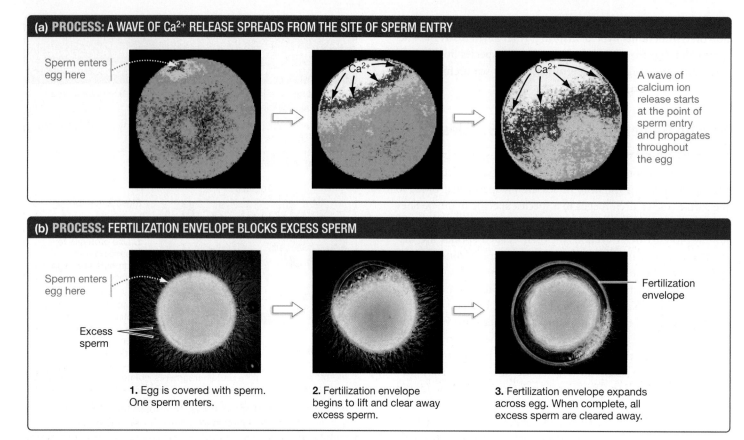

(a) PROCESS: A WAVE OF Ca^{2+} RELEASE SPREADS FROM THE SITE OF SPERM ENTRY

Sperm enters egg here

Ca^{2+}

Ca^{2+}

A wave of calcium ion release starts at the point of sperm entry and propagates throughout the egg

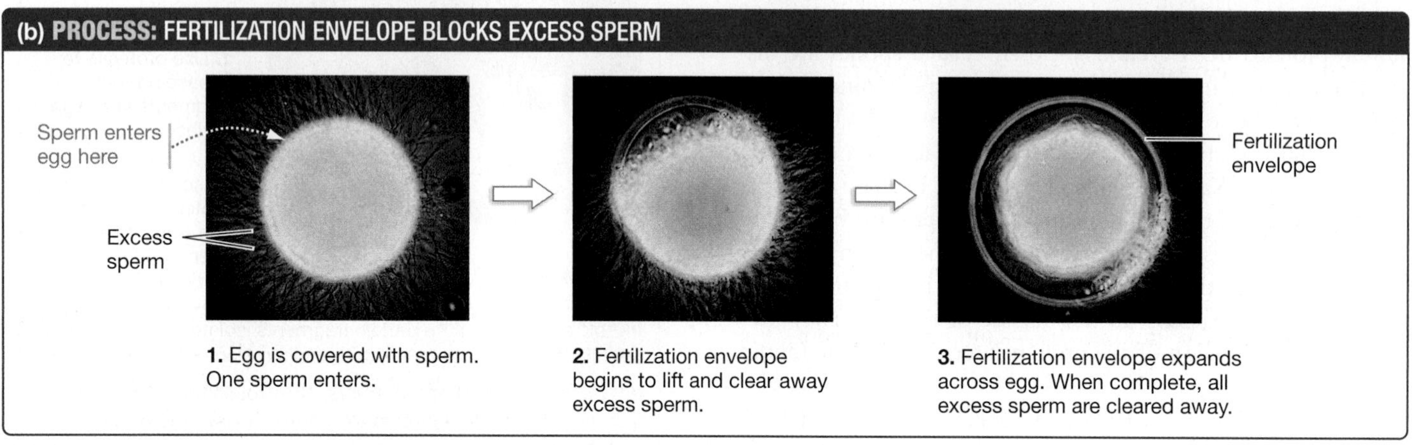
(b) PROCESS: FERTILIZATION ENVELOPE BLOCKS EXCESS SPERM
Sperm enters egg here
Excess sperm
Fertilization envelope
1. Egg is covered with sperm. One sperm enters.
2. Fertilization envelope begins to lift and clear away excess sperm.
3. Fertilization envelope expands across egg. When complete, all excess sperm are cleared away.

FIGURE 22.5 A Physical Barrier Erected after Fertilization Prevents Polyspermy. (a) During fertilization, a wave of Ca^{2+} begins at the point of sperm entry and spreads under the egg membrane in about 30 seconds. The white dots are from a reagent that fluoresces in the presence of calcium ions. **(b)** In response to increased Ca^{2+} concentrations, a fertilization envelope rises in about 40 seconds and clears away excess sperm.

The cortical granules located just inside the egg cell's plasma membrane respond to the Ca^{2+} signal by fusing with the membrane and releasing their contents to the exterior. These contents include proteases that digest the exterior-facing fragment of the egg-cell receptor for sperm, which prevents any new sperm from binding to the egg surface.

In addition, ions and other compounds released by the cortical granules accumulate between the egg cell's plasma membrane and the vitelline envelope. Because these solutes are highly concentrated, they cause water to flow into the space between the plasma membrane and vitelline envelope by osmosis (see Chapter 6). The influx of water causes the envelope matrix to lift away from the cell, forming a **fertilization envelope** (**Figure 22.5b**) that keeps additional sperm from contacting the egg's plasma membrane.

Mammal eggs do not produce a fertilization envelope, but enzymes that are released from cortical granules perform a function similar to the sea urchin enzymes that destroy the egg-cell receptor for sperm. More specifically, the mammalian enzymes modify proteins on the egg-cell surface in a way that prevents binding by additional sperm.

Once a sperm nucleus does enter a sea urchin or human egg, the two haploid nuclei fuse to form a diploid zygote. Development can begin.

CHECK YOUR UNDERSTANDING

If you understand that . . .

- Fertilization depends on specific interactions between proteins on the plasma membranes of animal sperm and egg cells.
- Enzymes that digest the egg-cell receptor for sperm help prevent polyspermy.

✓ You should be able to . . .

1. Predict the consequences of mutations that change the structure of bindin or the egg-cell receptor for sperm.
2. Describe the role of calcium signaling in processes that block polyspermy.

Answers are available in Appendix B.

22.3 Cleavage

Cleavage refers to the rapid cell divisions that take place in a zygote immediately after fertilization. Cleavage is the first step in **embryogenesis**—the process by which a single-celled zygote becomes a multicellular embryo.

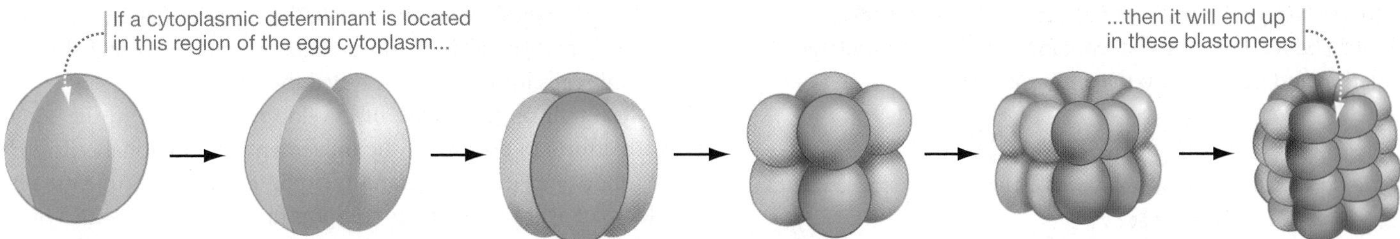

FIGURE 22.6 Cytoplasmic Determinants Are Sequestered into Certain Blastomeres during Cleavage.

In most animals, cleavage partitions the egg cytoplasm without additional cell growth taking place. The zygote simply divides into two cells, then four cells, then eight, and so on, without concurrent growth overall. The key feature of cleavage is that the cytoplasm present in the egg is divided into a larger and larger number of smaller and smaller daughter cells.

Cleavage is fast—the fastest cell divisions recorded over an individual's lifetime. In frogs, cleavage produces 37,000 cells in under two days. When cleavage occurs in fruit flies, mitosis can occur every 10 minutes, producing about 50,000 cells in half a day.

The cells that are created by cleavage divisions are called **blastomeres** (literally, "bud-part"). When cleavage is complete, the embryo consists of a mass of blastomere cells called a **blastula** ("little-sprout, bud").

Partitioning Cytoplasmic Determinants

The key to understanding cleavage is to analyze what is happening to the egg cytoplasm as it is divided up into many blastomeres. For example, think back to the Bicoid protein introduced in Chapter 21. Bicoid is a transcription factor whose distribution in the fruit-fly egg displays an anterior–posterior concentration gradient. When a fly egg undergoes cleavage, then, only nuclei in the anterior end of the egg are exposed to a high concentration of Bicoid.

As Section 22.1 indicated, a molecule that exists in eggs and helps direct early development is called a cytoplasmic determinant. It follows, then, that *bicoid* mRNA is a cytoplasmic determinant.

Cytoplasmic determinants are found in specific locations within the egg cytoplasm, so they end up in specific populations of blastomeres. As a result, certain regulatory gene cascades are triggered only in certain blastomeres (**Figure 22.6**). By dividing the egg cytoplasm to precisely distribute cytoplasmic determinants to certain cells, cleavage initiates the step-by-step process that, in combination with signals received from other cells, results in cell differentiation.

Cleavage in Mammals

In humans and other mammals, cleavage occurs in the **fallopian tube**, or **oviduct**: a structure that connects reproductive organs called the ovary and the uterus (**Figure 22.7**). The **ovary** is the organ in which the egg matures, and the **uterus** is the organ in which the embryo develops. Fertilization occurs near the ovary, and cleavage occurs as the embryo travels down the length of the fallopian tube toward the uterus.

Cleavage results in a specialized type of blastula called a **blastocyst** ("sprout-bag"), which has two major populations of cells. The exterior of the blastocyst is a thin-walled, hollow structure called the **trophoblast** ("feeding-sprout"). Inside the trophoblast there is a cluster of cells called the **inner cell mass (ICM)**.

After arriving at the uterus, the blastocyst implants into the uterine wall, and an organ called the placenta begins to form. The **placenta** is derived from a mixture of maternal cells and trophoblast cells. It has a key function: it allows nutrients and wastes

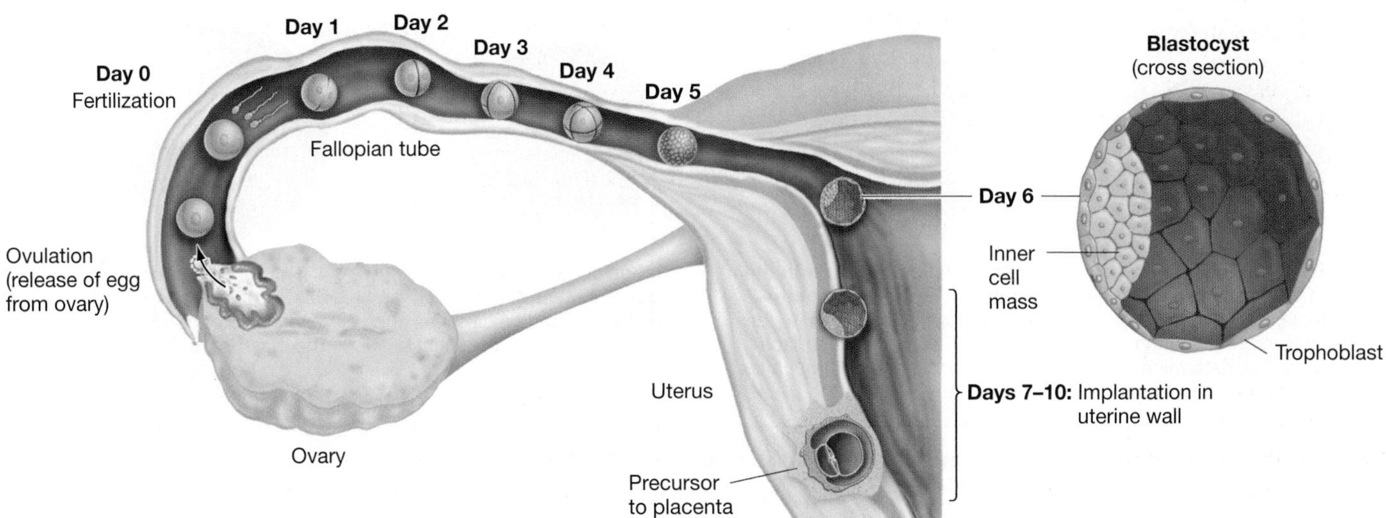

FIGURE 22.7 In Humans, Cleavage Occurs before Implantation into the Uterus.

to be exchanged between the mother's blood and the embryo's blood. The ICM, in contrast, contains the cells that undergo gastrulation and develop into the embryo.

CHECK YOUR UNDERSTANDING

If you understand that . . .

- Cleavage is a series of rapid cell divisions that result in formation of a blastula.
- During cleavage, the egg cytoplasm is divided.
- Because certain regulatory signals and transcription factors are found at specific locations in the egg cytoplasm of most species, different blastomeres end up with different sets of cytoplasmic determinants.

✓ You should be able to . . .

Explain why cytoplasmic determinants have to be localized in specific parts of the egg, in order for them to end up in a defined subset of blastomeres.

Answers are available in Appendix B.

22.4 Gastrulation

As cleavage nears completion, cell division slows. During the next phase of the developmental sequence, cell proliferation stops being the most important developmental process. Instead, cell movement becomes primary. During **gastrulation**, extensive and highly organized cell movements radically rearrange the embryonic cells into a structure called the gastrula.

Research on this phase of development started in the 1920s with efforts to document the movement of individual cells during newt and frog gastrulation. In these early experiments, tiny blocks of agar (a gelatinous compound) were soaked with a nontoxic dye. The dyed blocks were then pressed against the surface of blastula-stage embryos so that a small number of blastomeres became marked with dye. By allowing marked embryos to develop and then examining them at intervals during gastrulation, researchers were able to follow the movement of cells.

Formation of Germ Layers

The pattern of gastrulation varies widely among species of animals, but the general outcome is the same: Gastrulation results in the formation of embryonic tissue layers. A **tissue** is an integrated set of cells that function as a unit.

Most early animal embryos have just three primary tissue layers: (1) ectoderm ("outside skin"), (2) mesoderm ("middle skin"), and (3) endoderm ("inner skin"). These embryonic tissues are called **germ layers** because they give rise to the organs and tissues of the adult. Adults of most animal species have a wide array of tissues, including muscle tissue, connective tissue, and nerve tissue.

Figure 22.8 shows how the cell movements of gastrulation result in the formation of the three embryonic tissue layers in a

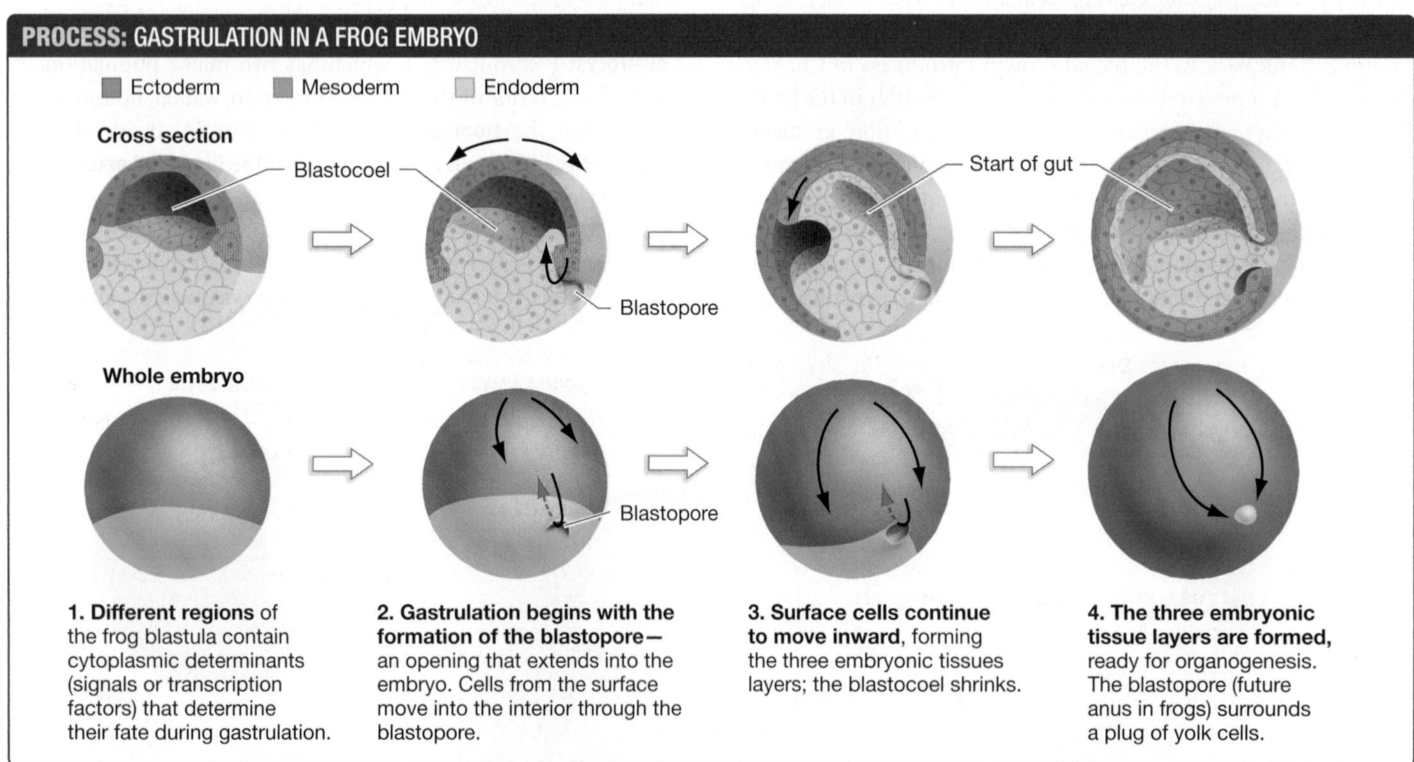

PROCESS: GASTRULATION IN A FROG EMBRYO
Ectoderm
Mesoderm
Endoderm
Cross section
Blastocoel
Blastopore
Start of gut
Whole embryo
Blastopore
1. Different regions of the frog blastula contain cytoplasmic determinants (signals or transcription factors) that determine their fate during gastrulation.
2. Gastrulation begins with the formation of the blastopore— an opening that extends into the embryo. Cells from the surface move into the interior through the blastopore.
3. Surface cells continue to move inward, forming the three embryonic tissues layers; the blastocoel shrinks.
4. The three embryonic tissue layers are formed, ready for organogenesis. The blastopore (future anus in frogs) surrounds a plug of yolk cells.

FIGURE 22.8 Precise Cell Movements during Gastrulation Form Distinct Layers of Tissues.

✓EXERCISE In frogs, the initial opening of the blastopore is located on what will become the dorsal side of the larva and adult. The blastopore becomes the anus of the larva and adult. On the last drawing in each row, mark where the head, tail, back, and belly will develop.

frog embryo. The drawings are based on experiments with dyed blastomeres, and they distinguish between blastomeres according to the type of embryonic tissue they will become. Cells that will become ectoderm are shown here in blue; cells destined to form mesoderm are shown in pink, and cells that will form endoderm are colored yellow. These cell fates are determined by cell-cell signals, transcription factors, and cytoplasmic determinants in the blastomeres.

Examine the steps in the figure carefully.

Step 1 The frog blastula contains a fluid-filled interior space called the **blastocoel**. The blastula of many animal species is hollow or partially hollow, as it is here.

Step 2 As gastrulation begins, cells begin moving into the blastocoel through an invagination—similar to the indentation that forms when you push a finger into a balloon. In frogs this invagination is slit-like and eventually forms the **blastopore**.

Step 3 Cells from the periphery continue to move to the interior of the embryo through the blastopore, forming a tube-like structure that will become the gut or digestive tract.

Step 4 When gastrulation is complete, the ectoderm, mesoderm, and endoderm cells are arranged in three distinct layers.

To review these steps and other features of early development, go to the study area at *www.masteringbiology.com.*

MB **Web Activity** Early Stages of Animal Development

Figure 22.9 shows the fate of the embryonic tissues as development proceeds: **Ectoderm** forms the outer covering of the adult body and the nervous system; **mesoderm** gives rise to muscle, most internal organs, and connective tissues such as bone and cartilage; and **endoderm** produces the lining of the digestive tract or gut, along with some of the associated organs.

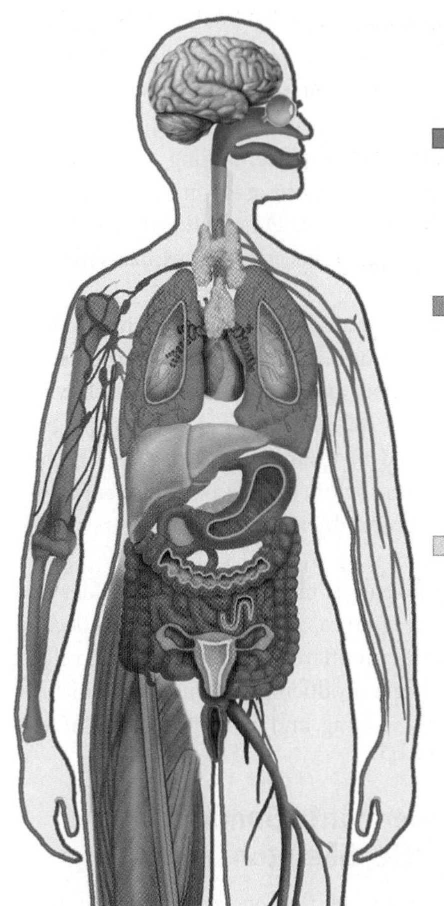

FIGURE 22.9 The Three Embryonic Tissues Give Rise to Different Adult Tissues and Organs.

✔**QUESTION** In a gastrula, mesoderm is sandwiched between ectoderm, which is on the outside of the embryo, and endoderm, which is on the inside. Relate this observation to the positions of adult tissues and organs derived from mesoderm, ectoderm, and endoderm.

Definition of Body Axes

In addition to establishing the embryonic tissue layers, gastrulation has another major outcome: The major body axes become visible. In frogs, for example, the blastopore becomes the anus, and the region just above the blastopore (as drawn in Figure 22.8) becomes the dorsal, or back, side of the embryo. In this way, the anterior–posterior and dorsal–ventral axes of the body become apparent.

In frogs, the major body axes were partially determined early in development by Bicoid-like cytoplasmic determinants that were stored in the egg and partitioned into blastomeres during cleavage. These master regulators start regulatory gene cascades that begin the long, step-by-step process of differentiation introduced in Chapter 21.

Current research on gastrulation is focused on understanding how cells move in such an organized way—that is, on the mechanism responsible for the actual movement and on the mechanism by which the cells navigate. But like the embryo itself, we'll move along to the next phase in the developmental sequence—the formation of tissues, organs, and other basic structures as the body takes shape.

CHECK YOUR UNDERSTANDING

If you understand that . . .

- Gastrulation consists of coordinated cell movements that reorganize the embryonic cells and result in the formation of embryonic germ layers and a tube that will eventually become the gut. The embryonic tissue layers are ectoderm, mesoderm, and endoderm.
- Once gastrulation is complete, the major body axes are visible and organs and other structures can begin to form.

✔ **You should be able to . . .**

1. Name the three embryonic germ layers and describe their relative positions in the gastrula.
2. Describe the relationship between the cells that invaginate early in gastrulation, the formation of the gut, and the formation of the anterior–posterior axis of the body.

Answers are available in Appendix B.

22.5 Organogenesis

Organogenesis ("organ-origin") is the process of tissue and organ formation that begins once gastrulation is complete and the embryonic germ layers are in place. During organogenesis, cells proliferate and become differentiated—meaning they become specialized for different jobs. Differentiated cells have a distinctive structure and function because they express a distinctive suite of genes.

As Chapter 21 noted, differentiation is a progressive, stepwise process. An irreversible commitment to become a particular cell type is the end point of a long and complex sequence of events, mediated by the cascades of cell-cell signals and regulatory transcription factors introduced in Chapter 21. Thus, cells become destined to a specific fate long before they become differentiated—meaning, before they begin producing products that are specific to a certain cell type.

To explore how organogenesis takes place, let's consider how muscle tissue forms in the embryo, and how muscle cells become differentiated. What causes undifferentiated mesodermal cells to form muscle tissue and begin producing the muscle-specific proteins that make breathing, walking, and turning the pages of textbooks possible?

Organizing Mesoderm into Somites: Precursors of Muscle, Skeleton, and Skin

Figure 22.10 illustrates some of the key events that occur as organogenesis begins in frogs, humans, and other vertebrates. In this cross section of a frog embryo, the dorsal (back) side of the individual is at the top and the ventral (belly) side is at the bottom.

You should already be familiar with the ectoderm, mesoderm, and endoderm layers shown in step 1 of the figure, along with the early gut cavity. But you should also note that a new structure has appeared in the dorsal mesoderm—a rod-like element called the **notochord**. This structure is shown in cross section in the figure and runs the length of the anterior–posterior axis of the embryo, just under the dorsal surface.

The notochord is unique to the group of animals called the **chordates**, which includes humans and other vertebrates. In some species of chordates, the notochord is a long-lasting structure that functions as a simple internal skeleton—it stiffens the body and makes efficient swimming movements possible. But in species like frogs and humans, the notochord is transient. It appears only in embryos. As organogenesis proceeds, many of the cells in the notochord undergo apoptosis—programmed cell death.

NEURAL TUBE FORMATION The short-lived notochord observed in many chordate embryos functions as a key organizing element during organogenesis. As step 2 in Figure 22.10 shows, cells in or near the notochord produce signaling molecules that induce the ectoderm on the dorsal (back) side of the embryo to fold. Folding results from changes in cell shape—analogous to the differential cell expansion that occurs in plant cell development (see Chapter 21).

The folding of the ectoderm begins when signals from the notochord trigger massive changes in cytoskeletal elements inside each of the dorsal ectodermal cells. As the cytoskeleton is reorganized, the ectodermal cells extend in length and then constrict at their dorsal end and expand at their ventral end. The constriction above and expansion below makes the sheet of cells fold upward.

As folding continues, a structure called the **neural tube** forms (step 3). The neural tube is the precursor of the brain and spinal cord. Besides producing the signals that direct formation of the neural tube, the notochord furnishes physical support as the ectodermal cells fold into their final configuration.

SOMITE FORMATION As organogenesis continues, mesodermal cells near the neural tube become organized into blocks of tissue called **somites**. Somites form on both sides of the neural tube down the length of the body (**Figure 22.11**). Somite formation is

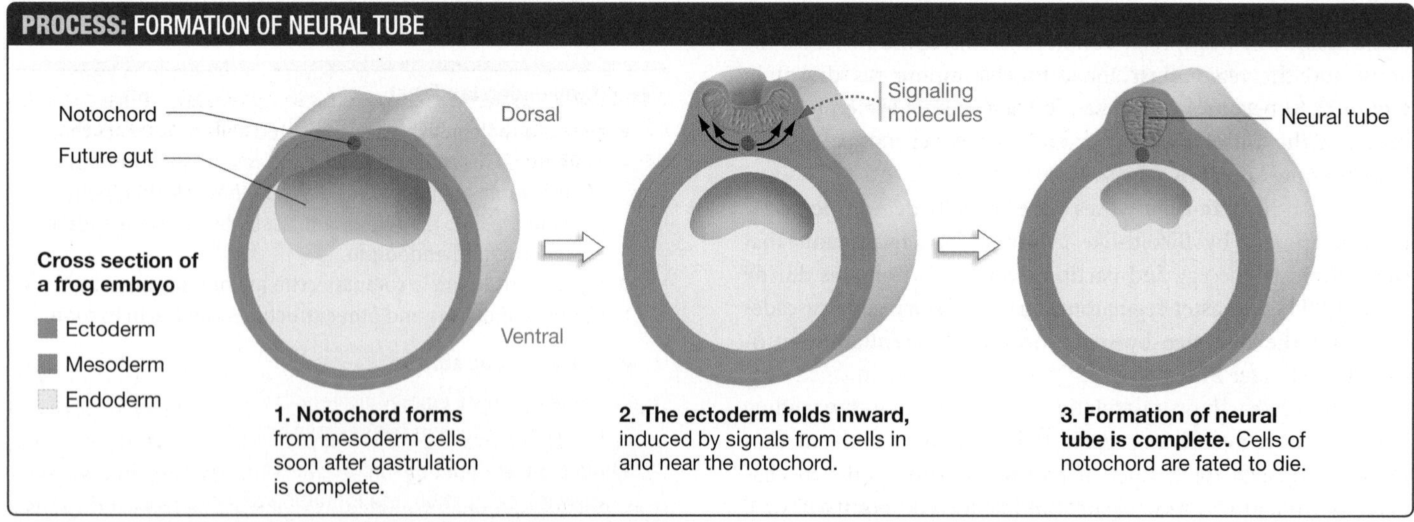
PROCESS: FORMATION OF NEURAL TUBE
Notochord
Future gut
Dorsal
Ventral
Cross section of a frog embryo
Ectoderm
Mesoderm
Endoderm
1. Notochord forms from mesoderm cells soon after gastrulation is complete.
Signaling molecules
2. The ectoderm folds inward, induced by signals from cells in and near the notochord.
Neural tube
3. Formation of neural tube is complete. Cells of notochord are fated to die.

FIGURE 22.10 The Notochord and Neural Tube Form Early in Organogenesis. In vertebrates, the notochord forms from mesoderm cells soon after gastrulation is complete. Molecules produced in or near the notochord induce the formation of the neural tube and other structures along the dorsal (back) side of the embryo.

(a) Surface view of somites

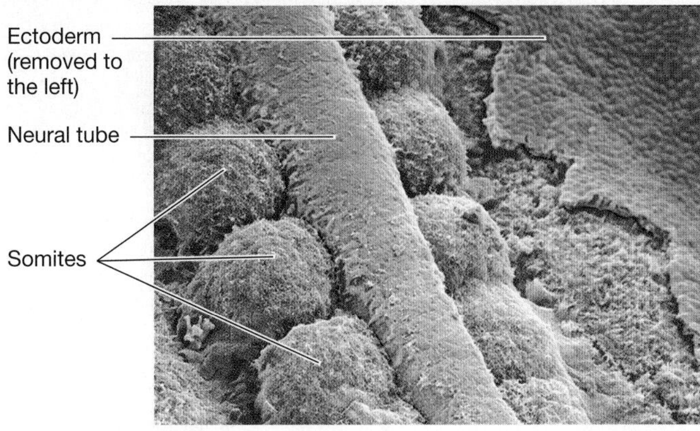

(b) Cross section of somites

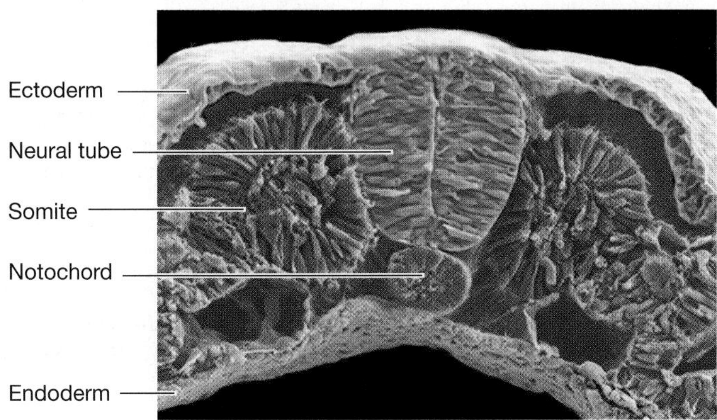

FIGURE 22.11 Somites Form on Both Sides of the Neural Tube. Somites are made of mesodermal cells (color-coded pink in these chick embryo micrographs).

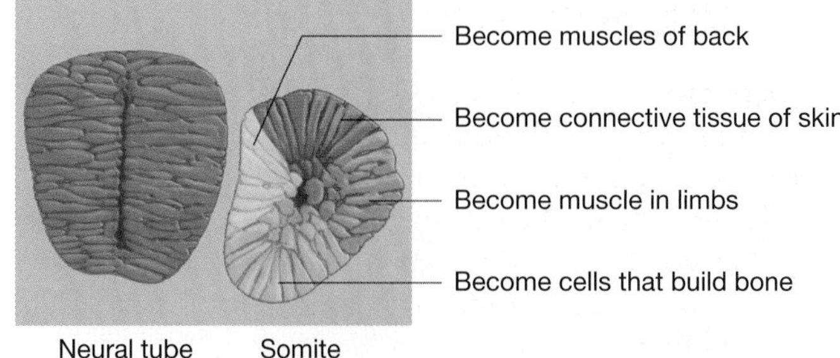

FIGURE 22.12 Within a Somite, a Cell's Position Determines Its Fate. Each somite eventually breaks up into distinct populations of cells, each of which gives rise to distinct tissue.

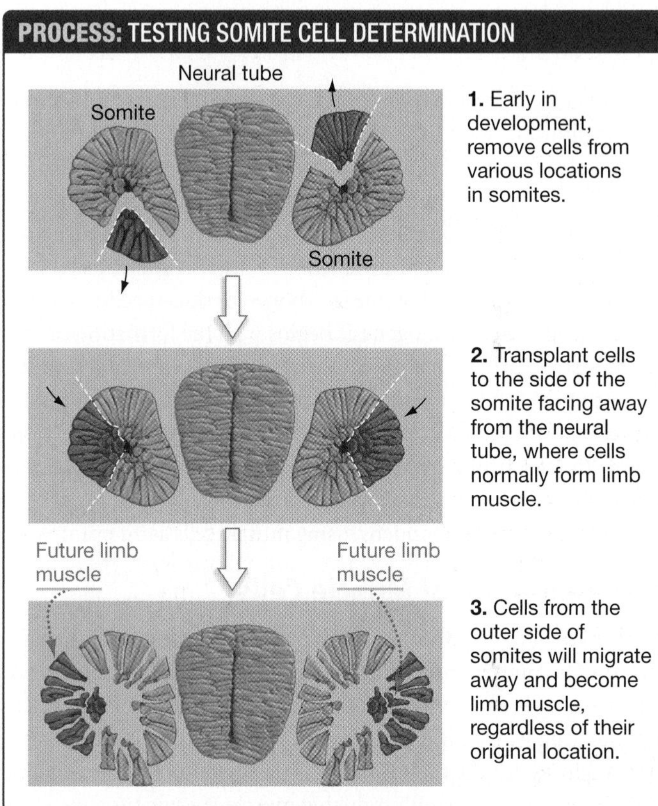

PROCESS: TESTING SOMITE CELL DETERMINATION
Neural tube
Somite
Somite
1. Early in development, remove cells from various locations in somites.
2. Transplant cells to the side of the somite facing away from the neural tube, where cells normally form limb muscle.
Future limb muscle
Future limb muscle
3. Cells from the outer side of somites will migrate away and become limb muscle, regardless of their original location.

FIGURE 22.13 Experimental Evidence for Progressive Differentiation of Mesodermal Cells in Somites. By transplanting somite cells early in somite formation to new locations within the structure, researchers showed that the cells' fates are not determined initially.

a response to changes in the cell adhesion molecules that keep mesodermal cells attached to each other (see Chapter 8).

Somites are transient structures, just as the notochord is. But by marking cells and following them over time, researchers have found that somite cells form a variety of structures.

As development proceeds, somite cells break away in distinct groups that migrate to their final location in the developing embryo, and there continue to proliferate. These cell movements are critical to organogenesis. Cells from somites build the vertebrae and ribs, the deeper layers of the skin that covers the back, and the muscles of the back, body wall, and limbs. As **Figure 22.12** shows, the destiny of a somite cell depends on its position within the somite at a particular time during organogenesis.

By transplanting cells from one location to another within the somite, researchers found that initially any of those cells can become any of the body's somite-derived elements. For example, biologists transplanted cells from various parts of the somite to the sector farthest from the neural tube. All the transplants eventually became committed to form limb muscles (**Figure 22.13**). Cells transplanted later in development, however, failed to become the cell type associated with their new position. Instead, the transplanted cells differentiated into the cell types they would normally have formed in their original position, even though they were now in a new and inappropriate location.

CELL DETERMINATION As the somite matures, its cells become irreversibly determined—meaning that the cell type they eventually become has been decided—based on their location within the somite. Note that a cell's fate (1) becomes fixed in a step-by-step process—guided by regulatory gene cascades—and (2) is a function of its location in the embryo.

Recent studies of this process, called **determination**, have shown that distinct populations of cells in a somite differentiate in response

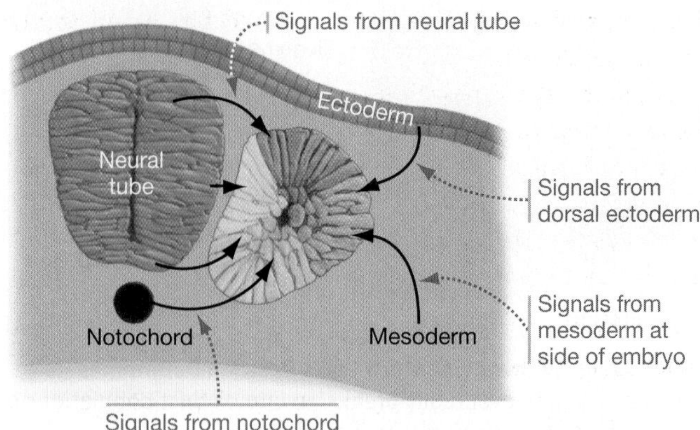

FIGURE 22.14 Somite Cells Differentiate in Response to Signals from Nearby Tissues. Somite cells become determined in response to signals from the neural tube, notochord, and nearby ectoderm and mesoderm.

to distinct local combinations of cell-cell signals. **Figure 22.14** shows signals diffusing away from cells in the notochord, the neural tube, and nearby ectoderm and mesoderm to act on specific populations of target cells in the somite. These different combinations of cell-cell signals direct the movement of somite cells when the structure breaks up and trigger the production of transcription factors required for the expression of muscle-, bone-, or skin-specific proteins.

To summarize, organogenesis begins with the formation of embryonic structures such as the notochord, neural tube, and somites. It continues with the formation of bones, muscles, skin, and other organs and tissues found in a larval or juvenile individual. At this point, the cells that make up these structures begin expressing tissue-specific proteins. Let's take a closer look at how this crucial step in organogenesis happens, using muscle cells as an example.

Differentiation of Muscle Cells

Why do cells in the portion of somite located farthest from the neural tube become committed to producing muscle? Harold Weintraub and colleagues answered this question in the late 1980s by experimenting with cells called myoblasts. A myoblast is a cell that is derived from cells in the somite. It is destined to become a muscle cell but has not yet begun producing muscle-specific proteins.

Weintraub and co-workers hypothesized that myoblasts contain at least one regulatory protein that commits them to their fate. Their idea was that myoblasts begin producing this muscle-determining protein after they receive an appropriate set of cell-cell signals from nearby tissues. In effect, they were looking for a regulatory transcription factor in myoblasts that dictates, "I will become a muscle cell."

Figure 22.15 outlines how the biologists went about searching for this hypothetical protein.

EXPERIMENT

QUESTION: Why do cells in a certain part of somites become committed to producing muscle in response to the signals they receive?

HYPOTHESIS: Cell-cell signals trigger production of a regulatory protein (or proteins) that commits myoblasts to their fate.

NULL HYPOTHESIS: Myoblasts do not contain a regulatory protein that commits them to their fate.

EXPERIMENTAL SETUP:

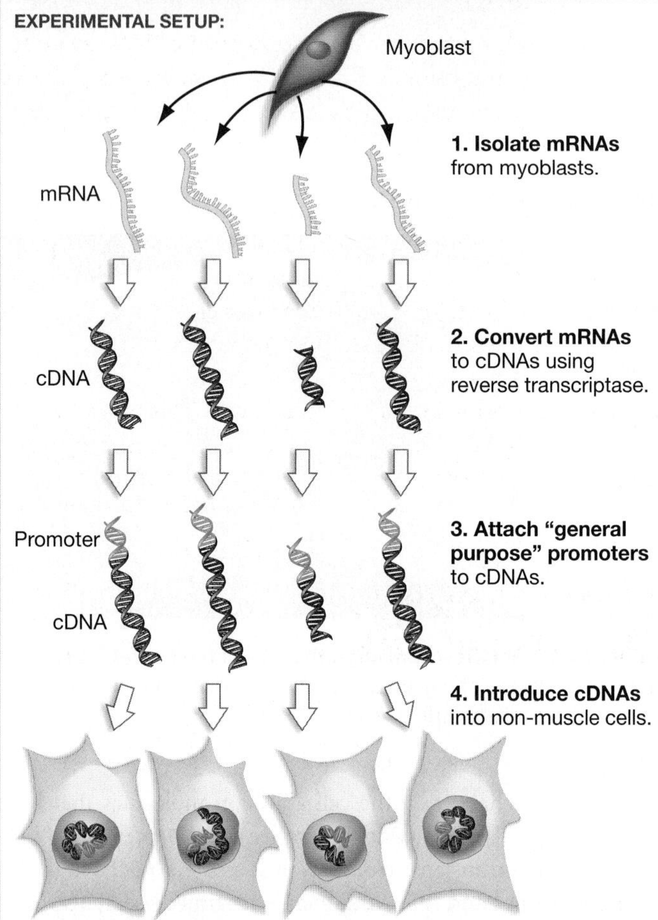

PREDICTION: One of the myoblast-derived cDNAs will convert non-muscle cells into cells that produce muscle-specific proteins.

PREDICTION OF NULL HYPOTHESIS: None of the myoblast-derived cDNAs will convert non-muscle cells into cells that produce muscle-specific proteins.

RESULTS:

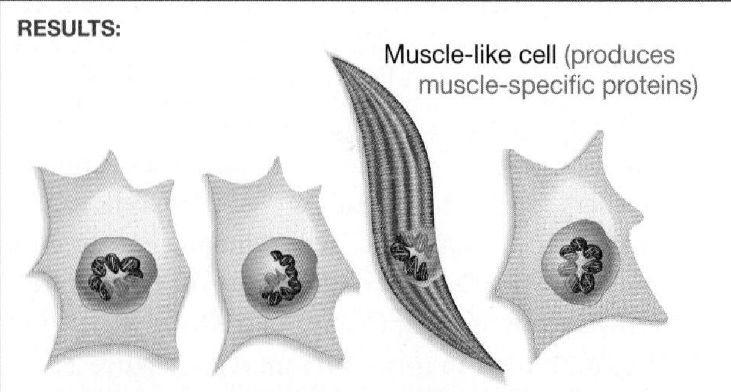

CONCLUSION: Certain somite cells contain a regulatory protein (later called MyoD) that commits them to differentiate into muscle.

FIGURE 22.15 The Search for a Gene That Causes Muscle-Cell Differentiation.

SOURCE: Weintraub, H., et al. 1989. Activation of muscle-specific genes in pigment, nerve, fat, liver, and fibroblast cell lines by forced expression of *MyoD*. *Proceedings of the National Academy of Sciences, USA* 86: 5434–5438.

✔**QUESTION** Why did the researchers have to attach a "general purpose" promoter to the cDNAs?

1. They began by isolating mRNAs from myoblasts.
2. They used reverse transcriptase to convert the mRNAs to cDNAs. Because myoblasts transcribe genes required for a muscle cell to function, the cDNAs included muscle-specific genes.
3. They attached a type of promoter to the cDNAs that would ensure the genes' expression in any type of cell.
4. They introduced the recombinant genes into non-muscle cells called fibroblasts and monitored the development of the transformed cells.

As predicted, one of the myoblast-derived cDNAs converted fibroblasts into muscle-like cells. Follow-up experiments showed that the same gene product could convert pigment cells, nerve cells, fat cells, and liver cells into cells that produced muscle-specific proteins.

Weintraub's group called the protein product of this gene **MyoD**, for *myo*blast *d*etermination. Subsequent work showed that the *MyoD* gene encodes a regulatory transcription factor and that the MyoD protein binds to enhancer elements located upstream of muscle-specific genes (see Chapter 18).

In addition, researchers found that the MyoD protein activates further expression of the *MyoD* gene. This was a key observation because it meant that once *MyoD* is turned on, it triggers its own expression—meaning that the gene continues to be transcribed. Other researchers have found that genes closely related to *MyoD* are also required for the differentiation of muscle cells.

To put this specific example into context, think back to the sequence of events occurring in early development and to the principles of development introduced in Chapter 21. Differentiation is a step-by-step process that is complete when cells begin producing proteins that are specific to a particular cell type. By what path did the cells in your bicep become muscle cells?

- Fertilization of the egg inside your mother triggered the onset of cleavage, resulting in a blastocyst.
- Certain cells in the blastocyst began producing signals that triggered regulatory gene cascades—changes in gene expression that in turn led specific cells, during the positional changes of gastrulation, to become mesoderm in your back.
- Early in organogenesis, signals from the notochord and nearby cells induced the production of MyoD and other muscle-determining proteins in certain populations of cells from somites. In response, these target cells were committed to becoming muscle and moved into your upper arm as it formed.
- Later, the MyoD-containing cells began expressing muscle-specific proteins.

All of these steps made it possible for you to move your arms after birth. The rest, as they say, is history.

CHAPTER 22 REVIEW

For media, go to the study area at www.masteringbiology.com 

Summary of Key Concepts

Fertilization begins with specific interactions between proteins on the plasma membranes of sperm and egg.

- Sperm cells only contribute a haploid genome and a centriole to the embryo.
- Eggs contribute a large amount of cytoplasm—usually containing cytoplasmic determinants and a large supply of nutrients—to the embryo in addition to a haploid genome and a centriole.
- Fertilization only occurs after a protein on the sperm head binds to a specific receptor on the egg-cell membrane.

✔You should be able to explain why bindin and the egg-cell receptor for sperm are said to interact like a lock and key.

The earliest cell divisions divide the fertilized egg into a mass of cells whose individual fates depend on key regulatory molecules they contain and the signals they receive.

- Early embryonic development begins with cleavage—a series of cell divisions that divide the egg cytoplasm into a large number of cells.
- During cleavage, an array of signals and regulatory transcription factors are apportioned to different blastomeres. As a result, different blastomeres have different fates.
- Once cleavage is complete, the embryo consists of a mass of cells.

✔You should be able to describe how a Bicoid-like master regulator might be partitioned into cells during cleavage in a human embryo.

Early in development, cells engage in collective, coordinated movements to form distinct tissue layers. Each layer gives rise to a different set of tissues and organs in the adult.

- Embryonic cells undergo dramatic movements during gastrulation.
- Gastrulation has two major consequences: (1) embryonic tissues are arranged into endoderm, mesoderm, and ectoderm layers; and (2) the back-to-belly and head-to-tail axes of the body become visible.

✔You should be able to compare and contrast a mesodermal cell in the anterior end of an embryo and a mesodermal cell in the posterior end, in terms of the cell-cell signals they receive and the regulatory transcription factors they contain.

MB **Web Activity** Early Stages of Animal Development

As development proceeds, specialized organs and other structures form through the interacting effects of cell-cell signals, cell proliferation, cell movements, and differentiation. Differentiation is complete when cells express tissue-specific proteins.

- Tissues, organs, and other structures form during the developmental phase called organogenesis.
- Early in vertebrate organogenesis, cells in the notochord release cell-cell signals that induce the formation of a neural tube—precursor to the brain and spinal cord.

- Blocks of mesodermal cells, called somites, form next to the neural tube.
- In response to cell-cell signals from the notochord, neural tube, and other nearby tissues, cells in the somites move to new positions, proliferate, and begin expressing tissue-specific proteins.

✔You should be able to give examples of cell interaction (via cell-cell signals), expansion, movement, proliferation, and differentiation during neural tube, somite, and muscle development in mammals.

Questions

✔TEST YOUR KNOWLEDGE

Answers are available in Appendix B

1. What is bindin, and what does it bind to?
   a. a protein on mammal embryos that allows them to attach to the uterine wall
   b. a protein on the plasma membrane of frog blastomeres that allows them to bind to each other and cohere
   c. a protein on the sea urchin sperm-cell nucleus that binds to the egg-cell nucleus
   d. a protein on the sea urchin sperm head that binds to an egg-cell receptor
2. How is calcium signaling involved in blocking polyspermy?
   a. It triggers a regulatory gene cascade.
   b. Accumulation of calcium ions causes an outward flow of water, raising the fertilization envelope.
   c. It triggers fusion of cortical granules with the plasma membrane.
   d. Calcium ions bind to egg-cell receptors for sperm, blocking binding sites.
3. What happens during cleavage?
   a. The neural tube—precursor of the spinal cord and brain—forms.
   b. Basal and apical cells—precursors of the suspensor and embryo, respectively—form.
   c. The fertilized egg divides without growth occurring, forming a mass of cells.
   d. Massive movements of cells make the primary body axes visible and organize the three embryonic tissues.
4. What happens during gastrulation?
   a. The neural tube—precursor of the spinal cord and brain—forms.
   b. Basal and apical cells—precursors of the suspensor and embryo, respectively—form.
   c. The fertilized egg divides without growth occurring, forming a ball of cells.
   d. Massive movements of cells make the primary body axes visible and organize the three embryonic tissues.
5. In animals, which adult tissues and organs are derived from ectoderm?
   a. lining of the digestive tract and associated organs
   b. blood, heart, kidney, bone, and muscle
   c. nerve cells and skin
   d. blastopore and blastocoel
6. During organogenesis in vertebrates, which of the following does *not* occur?
   a. establishment of the anterior–posterior axis
   b. differentiation of cells
   c. movement of cells into new positions
   d. extensive cell-cell signaling

✔TEST YOUR UNDERSTANDING

Answers are available in Appendix B

1. Explain why eggs are so much larger than sperm.
2. Why do elaborate mechanisms exist to prevent polyspermy?
3. Blastomeres look identical. Explain why they are not.
4. Why are ectoderm, mesoderm, and endoderm called germ layers?
5. Explain how cell transplantation experiments provided evidence that cells in a somite are determined in a step-by-step fashion.
6. What evidence supports the hypothesis that MyoD triggers differentiation of muscle cells?

✔APPLYING CONCEPTS TO NEW SITUATIONS

Answers are available in Appendix B

1. At the molecular level, explain why bindin on the sperm of the sea urchin *Strongylocentrotus purpuratus* attaches to the egg-cell receptor for sperm on eggs of *S. purpuratus*, but why bindin from other species of *Strongylocentrotus* cannot.
2. A molecule called α-amanitin inhibits translation. When researchers treat fruit-fly eggs with this compound, the early stages of cleavage proceed normally after fertilization. What can you conclude from this experiment?
3. In the marine organisms called sea squirts, eggs contain a yellow pigment. Blastomeres that contain this yellow pigment become muscle cells. State a hypothesis to explain this observation.
4. After organogenesis, frog embryos are about the same size as a fertilized egg. Why?

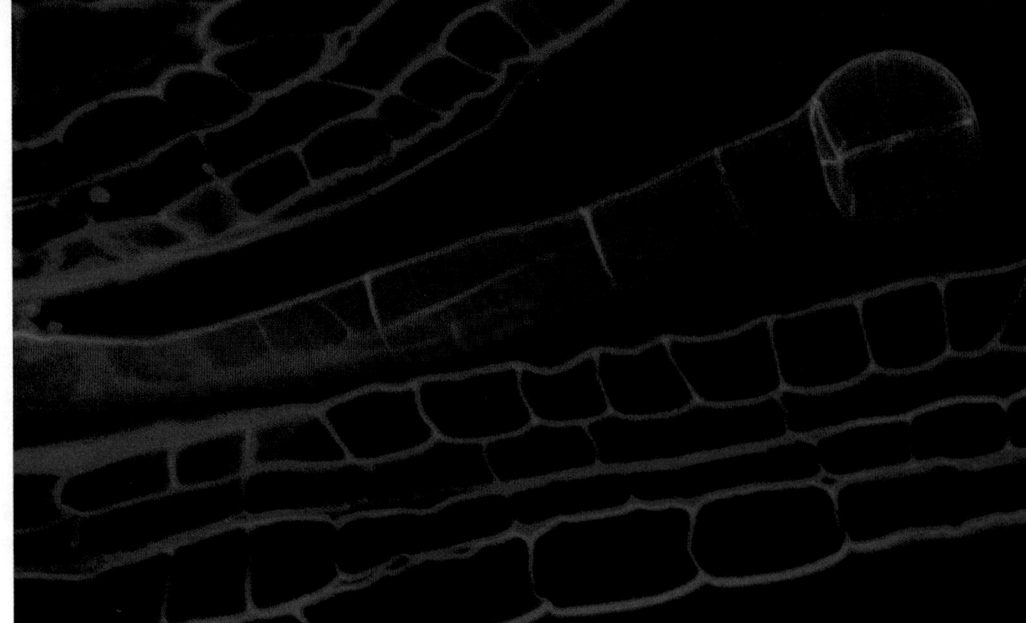

The ball-like cells on the far right of this image, colored red here, are an *Arabidopsis thaliana* embryo at the 8-cell stage. This chapter explores how plant embryos develop into adults that continue to grow throughout life and reproduce.

An Introduction to Plant Development 23

About the year 2750 B.C., perhaps more than a century before the first pyramids were constructed in ancient Egypt, a blessed event occurred on a mountainside in southwestern North America. A sperm cell and an egg cell from a bristlecone pine tree fused to form a zygote. The offspring produced by that encounter has been growing and developing ever since—for over 4700 years.

Unlike many animals, plants continue to grow and develop throughout their lives, whether that life lasts two weeks or thousands of years. In many plant species, for example, the entire aboveground shoot system dies each fall and grows back the next spring. Continuous development is a hallmark of plant biology.

In addition, many plant cells retain the ability to de-differentiate—to stop producing a certain suite of tissue-specific proteins and begin producing proteins typical of a different type of tissue. Under the right circumstances, for example, a photosynthetic cell in a plant stem can de-differentiate and give rise to cells that function as storage cells in a root. This extraordinary flexibility is another hallmark of plant development. Animal development is rigid and inflexible in comparison with that of plants.

The striking differences between plant and animal development are not surprising given their very different ways of life. Most animals move around to find food and mates, while plants stand their ground, make their own food, and use water, wind, or insects to transport their gametes.

Despite numerous differences, though, animal and plant development are both governed by the common principles described in Chapter 21. As plant cells proliferate, they divide in precise orientations and expand in specific directions. Like the cells in animal embryos, developing plant cells communicate constantly by cell-cell signals, differentiate due to specific combinations of cell-cell signals and regulatory transcription factors, and may undergo programmed cell death. This unity of underlying principles is one of the great discoveries emerging from research on plant developmental biology.

KEY CONCEPTS

- In sharp contrast to animals, plants develop continuously, do not commit cells to gamete production until late in development, and produce gametes by mitosis in haploid cells.
- In flowering plants, double fertilization results in the production of a zygote and a nutritive tissue that supports embryogenesis.
- Embryogenesis results in the formation of the major body axes and three types of embryonic tissue.
- Vegetative development is the function of meristems, where cell division continues throughout life—producing cells that go on to differentiate.
- When the function of a meristem shifts from vegetative to reproductive development, key regulatory transcription factors are activated and control the position and identity of floral organs.

✓ When you see this checkmark, stop and test yourself. Answers are available in Appendix B.

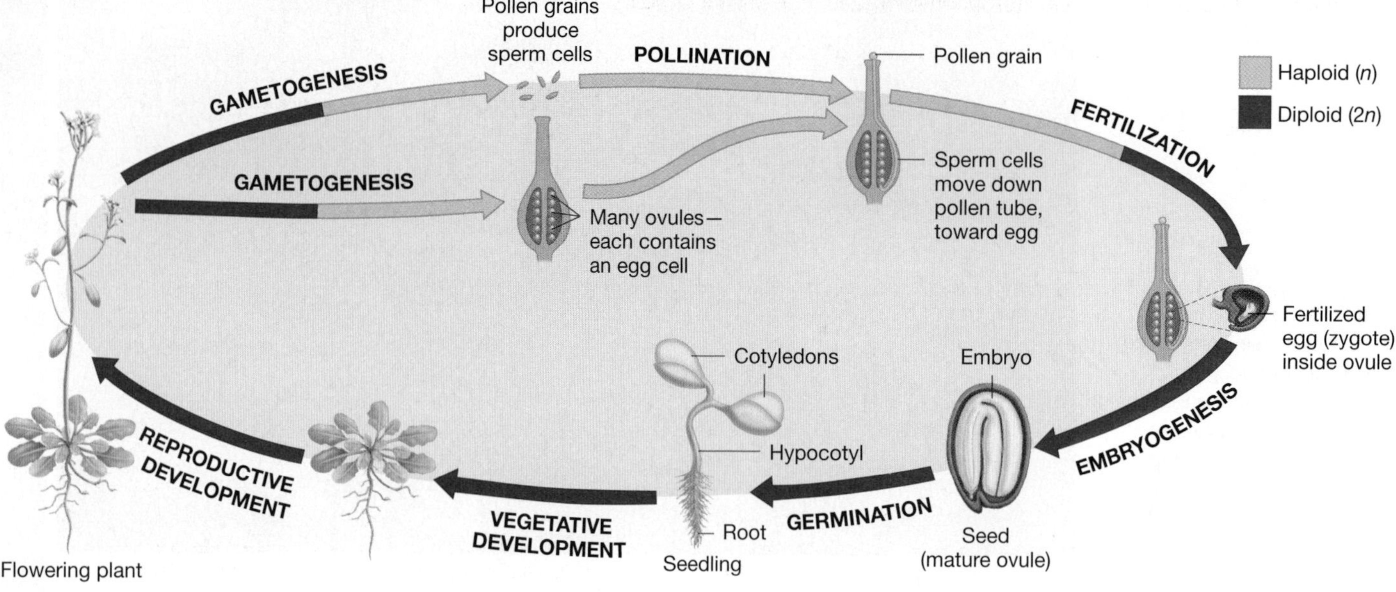

FIGURE 23.1 An Overview of Development in *Arabidopsis*. This chapter traces the events in the life of a plant, from fertilization and embryogenesis to vegetative and reproductive development.

23.1 Gametogenesis, Pollination, and Fertilization

Figure 23.1 provides an overview of development in a flowering plant—the most species-rich and abundant group of plants (see Chapter 30). In this figure and in many other places in this chapter, a small species called *Arabidopsis thaliana*—the best-studied of all land plants—will serve as a model organism (see **BioSkills 14** in Appendix A).

Note that the sequence of events, or life cycle, diagrammed in Figure 23.1 begins with gametogenesis, or gamete formation. In *Arabidopsis* and other flowering plants, an egg is fertilized inside a protective, womb-like structure called the **ovule**. Development continues inside the ovule with **embryogenesis**—literally, "embryo-origin." In many plant species, embryogenesis ends with the maturation of the ovule into a **seed**—a structure that contains the embryo and a supply of nutrients surrounded by a protective coat.

An embryo may remain in a nongrowing state inside the seed for months, years, or in some cases, centuries. When conditions are favorable, however, the seed undergoes **germination**—meaning that it resumes growth—to form a seedling. Organogenesis, which continues throughout life, then forms the three vegetative organs: roots, leaves, and stems. **Vegetative organs** are the nonreproductive portions of the plant body. Later, cells in the stem will become converted to reproductive structures, initiating gamete production and the sexual phase of the plant life cycle.

How Are Sperm and Egg Produced?

Of all the differences between plant and animal development, one of the most dramatic is seen during gametogenesis. In animals, gametes are produced by differentiation that occurs in the products of meiosis. In plants, gametes are produced by mitosis that occurs in haploid cells produced by meiosis.

Figure 23.2a introduces how the process works in the male reproductive organs of flowering plants such as *Arabidopsis*. Note that diploid cells undergo meiosis to form haploid cells, which then divide by mitosis to give rise to tiny, multicellular structures called **pollen grains**. One of the haploid cells inside the pollen grain will later divide by mitosis, giving rise to two sperm cells.

Figure 23.2b illustrates the general sequence of events leading to egg formation in *Arabidopsis* and other flowering plants. A diploid cell inside the ovule divides by meiosis, producing four daughter cells. Only one of these cells survives; the other three undergo a programmed death. The surviving cell divides by mitosis several times to produce a tiny, multicellular structure called the embryo sac. Inside the embryo sac, a haploid cell differentiates into an egg.

Because the haploid, multicellular structures called pollen grains and embryo sacs alternate with a diploid, multicellular plant as one generation gives rise to the next, this type of life cycle is called **alternation of generations**. Alternation of generations is explained more thoroughly in Chapter 29 and Chapter 30; Chapter 40 explains the steps involved in gametogenesis in plants in much more detail. Now the question is, Once a pollen grain lands on the structure that holds the egg, what happens?

Pollen–Stigma Interactions

Pollen grains are carried to a mature flower by wind, water, insects, bats, birds, or some other agent. If the pollen grains land

(a) Production of sperm in the male reproductive organs of a flowering plant.

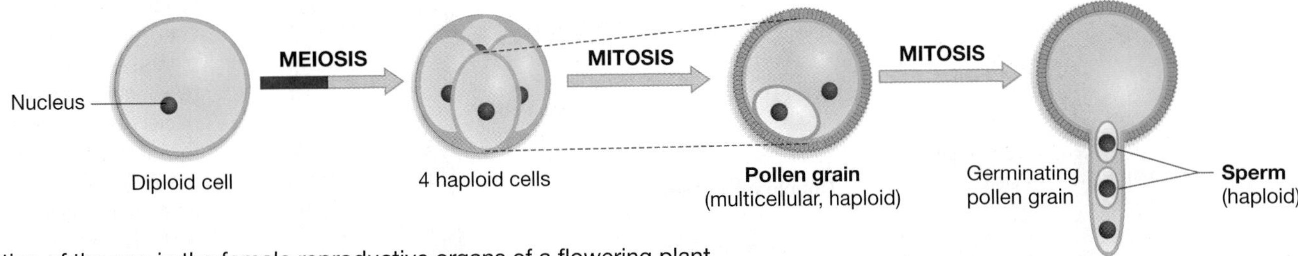

(b) Production of the egg in the female reproductive organs of a flowering plant.

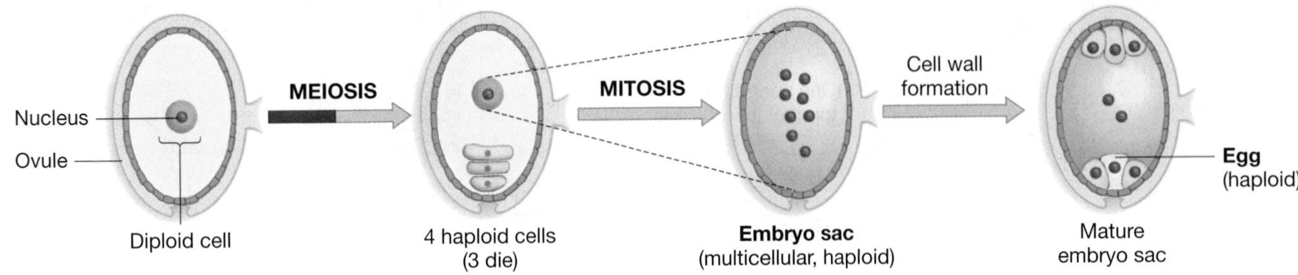

FIGURE 23.2 An Overview of Gametogenesis in Flowering Plants. Unlike the situation in animals, plants produce haploid multicellular structures. A haploid cell inside the pollen grain produces sperm by mitosis; a haploid cell inside the embryo sac produces an egg by mitosis.

(a) Pollen grains interact with the stigma.

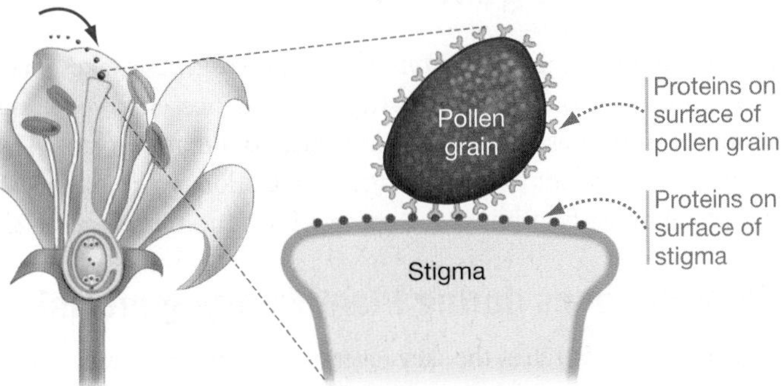

(b) Sperm move to the egg through a pollen tube.

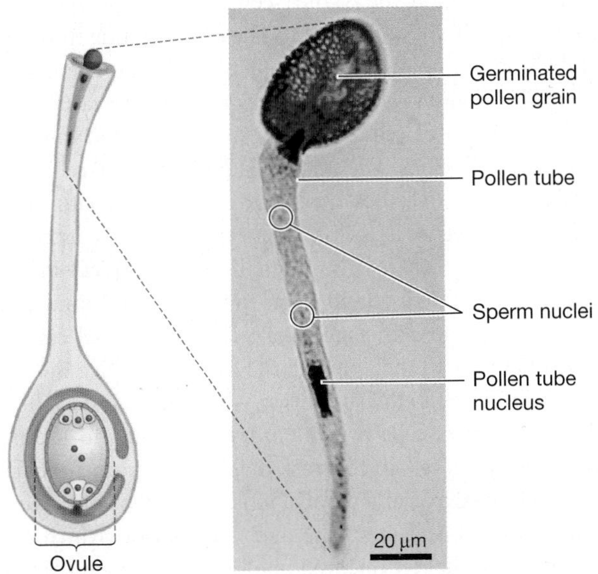

on a floral structure called the stigma, one of the best-studied cell-cell interactions in all of plant biology takes place.

For the pollination and fertilization process to continue successfully, proteins on the surface of the pollen grain have to interact with proteins on the surface of the stigma (**Figure 23.3a**). Like bindin attaching to the egg-cell receptor for sperm in sea urchins (see Chapter 22), the specificity of these interactions prevents fertilization from occurring between members of different species. In many cases, it also prevents self-fertilization—the union of a sperm and egg from the same individual (see Chapter 40 for details on these cell-cell interactions).

If the protein–protein interactions on the surface of the stigma are successful, a **pollen tube** begins to extend from the pollen grain, down the stigma, toward the egg (**Figure 23.3b**). Recent research suggests that this growth is guided by cell-cell signals released from the egg at the base of the female reproductive structure, or **carpel.**

Double Fertilization

Another dramatic contrast with animal development is seen when the pollen tube reaches the base of the carpel. The two sperm cells exit the pollen tube, pass through the wall of the ovule, and enter the embryo sac. The embryo sac contains the egg

FIGURE 23.3 The Fertilization Process Begins When Pollen and Stigma Interact.

✔**QUESTION** How does the interaction between pollen and stigma in flowering plants compare to the interaction between bindin and the egg-cell receptor for sperm in sea urchins?

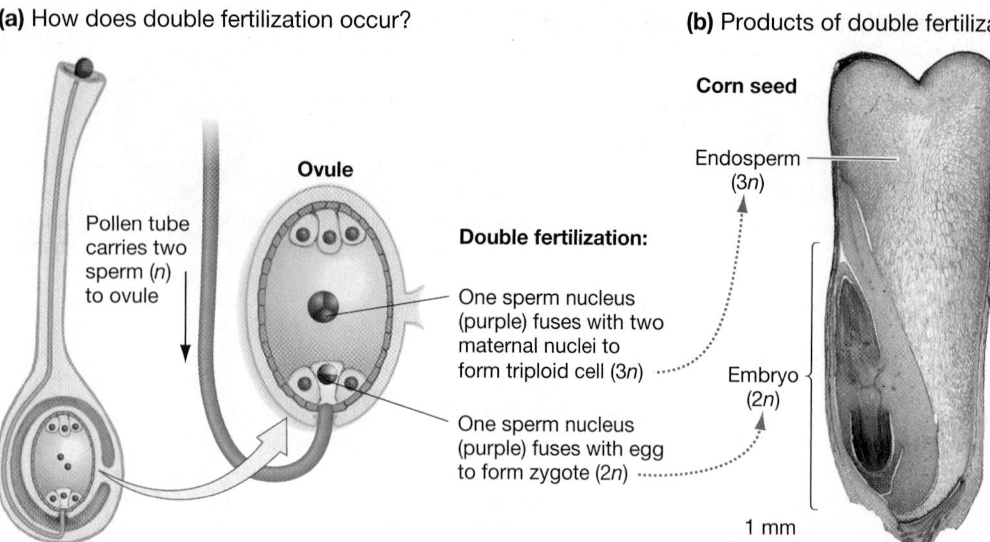
(a) How does double fertilization occur?
Ovule
Pollen tube carries two sperm (n) to ovule
Double fertilization:
One sperm nucleus (purple) fuses with two maternal nuclei to form triploid cell (3n)
One sperm nucleus (purple) fuses with egg to form zygote (2n)
(b) Products of double fertilization
Corn seed
Endosperm (3n)
Embryo (2n)
1 mm

FIGURE 23.4 Double Fertilization Leads to the Formation of Endosperm. (a) In flowering plants, a sperm (*n*) fuses with two or more haploid maternal nuclei near the egg to form a cell that gives rise to endosperm tissue. In many species, endosperm cells are $3n$. **(b)** Endosperm is a nutritive tissue packed with proteins, carbohydrates, and fats or oils.

cell and a maternal cell that in many species contains two haploid nuclei. One sperm nucleus fuses with the egg to form the diploid zygote, while the other sperm nucleus fuses with the cell that has two maternal haploid nuclei to form a triploid ($3n$) cell. This event is known as **double fertilization**. The process is aptly named, as two cell-fusion events occur (**Figure 23.4a**).

The triploid cell divides repeatedly by mitosis to form a triploid nutritive tissue called endosperm. **Endosperm** ("inside-seed") provides the proteins, carbohydrates, and fats or oils required for embryonic development, seed germination, and early seedling growth.

In species with large seeds, the endosperm grows into a sizeable nutrient reservoir as the ovule matures (**Figure 23.4b**). When you eat wheat, rice, corn, or other grains, you are eating primarily endosperm. Functionally, endosperm is analogous to the yolk found in most animal eggs.

Once double fertilization is complete, the stage is set for embryogenesis—the early development of a new individual.

CHECK YOUR UNDERSTANDING

If you understand that . . .

- Gametes arise by mitosis in haploid cells that were produced by meiosis.
- Fertilization is preceded by specific interactions between proteins on the surfaces of plant pollen grains and stigmas.
- In flowering plants, double fertilization results in the production of a zygote and a (usually) triploid cell that grows into a nutritive tissue.

✔ **You should be able to . . .**

1. Describe when meiosis and mitosis occur during the development of a sperm cell in *Arabidopsis*.
2. Suggest a hypothesis to explain why protein–protein interactions between pollen grains and stigmas are important to the individual's ability to reproduce. (Hint: consider what might happen if the protein–protein interactions did *not* occur, and any pollen grain could germinate.)

Answers are available in Appendix B.

23.2 Embryogenesis

In flowering plants such as *Arabidopsis*, embryogenesis takes place inside the ovule as the seed is maturing. In essence, embryogenesis produces a tiny, less-developed precursor to a mature plant. The process is equivalent to the cleavage, gastrulation, and organogenesis phases of early animal development introduced in Chapter 22.

What Happens during Plant Embryogenesis?

Figure 23.5 illustrates the key events in embryogenesis, using *Arabidopsis* as a model organism. This sequence of events was worked out through careful observation of different-aged embryos.

After fertilization, the zygote (seen in step 1) undergoes a highly asymmetric cell division (step 2). The cells resulting from this initial division are unlike in size, content, and fate. The bottom, or basal, cell is large and is dominated by an extensive vacuole. It gives rise to a column of cells called the suspensor, which anchors the embryo as it develops. The small cell above the basal cell, called the apical cell, is rich in cytoplasm and gives rise to the mature embryo.

The asymmetries in the basal and apical cell help establish one of the primary axes of the plant body: the **apical–basal axis. Apical** refers to the tip; **basal** refers to the base, or foundation.

As steps 3 and 4 in Figure 23.5 show, the basal cell divides perpendicularly to the apical–basal axis to produce the suspensor. Only one cell in the suspensor—the one closest to the apical cell—contributes cells to the embryo and thus to the mature adult. The apical cell, in contrast, divides both perpendicularly to the apical–basal axis and parallel to it. These divisions give rise to a simple ball of cells at the tip of the suspensor (which can also be seen in the chapter opening photograph on page 401). At this point, the embryo is said to be in the globular stage.

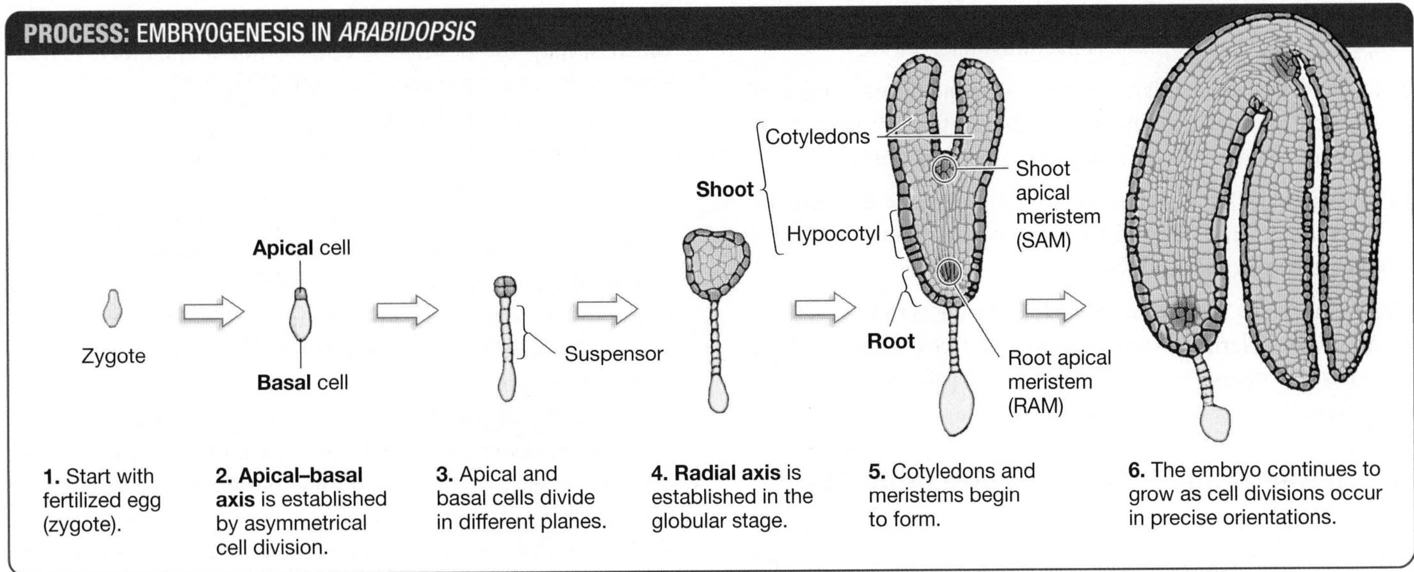

FIGURE 23.5 The Stages of Embryogenesis. Embryogenesis takes place inside the developing seed.

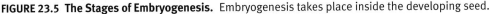

✔**EXERCISE** Label the apical–basal and radial axes on the globular-stage embryo.

Because of the distinctive-looking cells covering its exterior, there is now a visible difference between cells in the interior of the embryo and cells on the surface. This change creates the second major body axis, the radial axis. The radial axis extends from the interior of the body out to the exterior.

The initial events in embryogenesis illustrate a general point about plant development: The fate of a plant cell can be summed up in the old quip about the three keys to success in real estate—"location, location, location." Starting with the initial division that creates the apical and basal cells, plant cells differentiate based on where they are in the body.

As the ball of cells continues to grow and develop, the embryonic leaves, or **cotyledons**, begin to take shape (step 5). The cotyledons are connected to the developing root by a stem-like structure called the **hypocotyl**. Together, the cotyledons and hypocotyl make up the **shoot**, which will become the aboveground portion of the body. The shoot system functions in photosynthesis and reproduction. The **root**, in contrast, forms the belowground portion of the body. The root system anchors the individual and functions as a water- and nutrient-gathering structure.

Once the apical–basal and radial axes are established, and as the cotyledons, hypocotyl, and root begin to take shape, groups of cells called the **shoot apical meristem (SAM)** and **root apical meristem (RAM)** form. As Chapter 21 noted, **meristem** is a tissue consisting of undifferentiated cells that divide repeatedly into daughter cells, some of which become specialized.

Meristem cells are analogous to the stem cells found in animals, except that cells derived from plant meristems differentiate into a much wider array of types. The root meristem can form all the underground portions of the plant, and the shoot meristem can form all the aerial portions, including reproductive structures. Throughout a plant's life, meristematic tissues continue to produce cells that can differentiate into adult tissues and structures.

Note that all of this growth and development takes place without the aid of the cell migration seen in gastrulation in animals. Because plant cells have stiff cell walls, they do not move. Consequently, for the cotyledons and other embryonic structures to take shape, cell divisions must occur in precise orientations (step 6 of Figure 23.5), and the resulting cells must exhibit differential growth—meaning that some cells grow larger than others.

In addition to establishing the two body axes, early development in *Arabidopsis* produces three embryonic tissues (**Figure 23.6**).

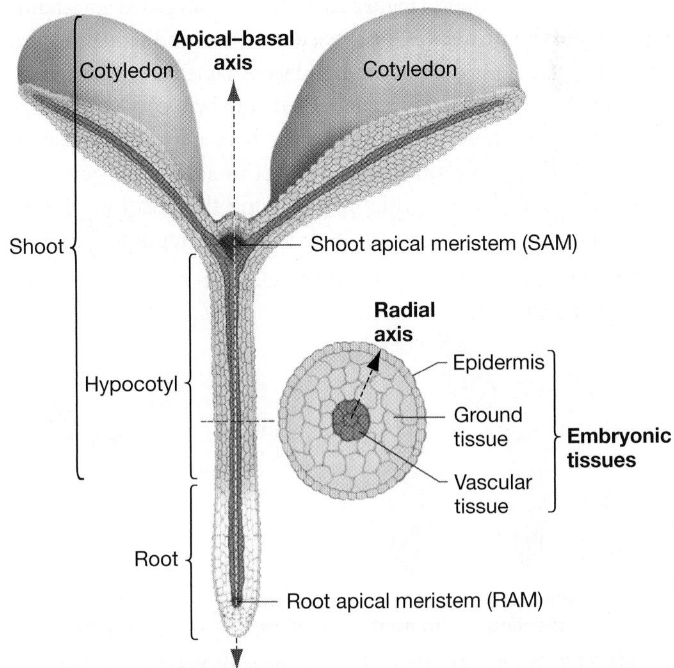

FIGURE 23.6 Embryogenesis Establishes the Two Body Axes and Three Embryonic Tissues.

- The **epidermis** (literally, "over-skin") is an outer covering of specialized cells that protects the individual.
- Inside the epidermal layer of cells is **ground tissue**, a mass of cells that may later differentiate into cells that are specialized for photosynthesis, food storage, or other functions.
- The **vascular tissue** in the center of the plant will eventually differentiate into specialized cells that transport food and water between root and shoot.

These embryonic tissues are analogous to the ectoderm, mesoderm, and endoderm of developing animals (see Chapter 22). The three tissue systems are arranged in a radial pattern, echoing the radial axis.

Which Genes and Proteins Set Up Body Axes?

The genetic approach to development that was pioneered with research on *Drosophila melanogaster* has also proven to be a powerful way of studying plant embryogenesis. Although the specific genes involved are different in plants than in animals, the basic mechanism by which genes direct the earliest events in development is similar.

Research on the genetics of early development in plants was pioneered by Gerd Jürgens and colleagues in the 1990s. This research group set out to identify genes that are transcribed in the zygote or embryo of *Arabidopsis* and that are responsible for establishing the apical–basal axis of the plant body. It's no coincidence that this effort was similar to the work on anterior–posterior pattern formation in *Drosophila* introduced in Chapter 21. Jürgens had participated in the work with flies.

The biologists' initial goal was to identify individuals with developmental defects at the seedling stage. More specifically, they were looking for mutants that lacked particular regions along the apical–basal axis of the body. The team succeeded in finding several bizarre-looking mutants (**Figure 23.7**). Individuals called apical mutants lacked the first leaves, or cotyledons. Some individuals lacked the embryonic stem, or hypocotyl, and were named central mutants. Individuals dubbed basal mutants lacked both hypocotyls and roots.

To interpret these results, the researchers suggested that each type of *Arabidopsis* mutant had a defect in a different gene and that each gene played a role in specifying the position of cells along the apical–basal axis of the body. They hypothesized that these genes are analogous to the segmentation genes of fruit flies, which specify the destiny of cells within well-defined regions along the anterior–posterior axis of insects.

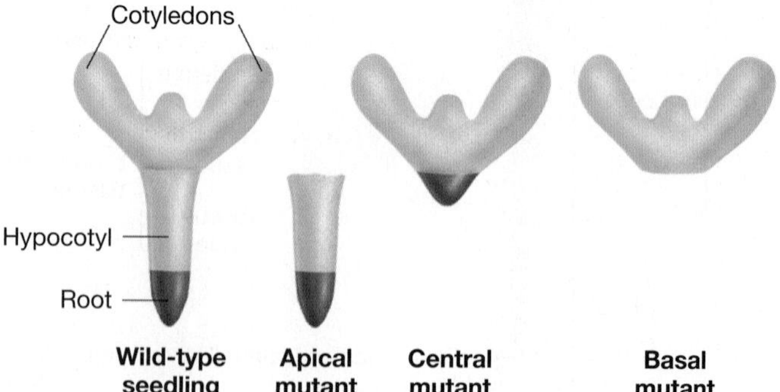

FIGURE 23.7 Mutant *Arabidopsis* Embryos with Misshapen Bodies. Researchers have identified *Arabidopsis* mutant individuals missing specific sections of the body along the apical–basal axis.

What are these *Arabidopsis* genes, and how do they exert their effects? To answer these questions, consider the gene responsible for the mutants lacking hypocotyls and roots. This gene has been cloned and sequenced and named *MONOPTEROS*. Because its DNA sequence indicates that its protein product has a DNA-binding domain, *MONOPTEROS* is hypothesized to encode a transcription factor that regulates the activity of target genes.

The MONOPTEROS protein, in turn, is activated in response to signals from auxin. Auxin is a cell-to-cell signal molecule that was introduced in Chapter 21. It is produced in the shoot apical meristem and transported toward the basal parts of the individual. (For more detail on auxin's prominent role as a long-distance signal, or hormone, see Chapter 39.) Recall that the concentration of auxin along the apical–basal axis of a plant forms a concentration gradient that provides positional information, not unlike the Bicoid concentration gradient found in fruit-fly embryos (see Chapter 21).

The take-home message from these results is that the auxin signal is part of a regulatory cascade that triggers the activation of MONOPTEROS and other regulatory transcription factors specific to cells in the developing hypocotyl and roots. Although the genes and proteins involved in forming the cotyledons, hypocotyl, and root of *Arabidopsis* are not yet understood in as much detail as the segmentation genes of *Drosophila*, several important similarities are clear. Both developmental pathways are based on cell-to-cell signals and regulatory cascades that result in the step-by-step specification of a cell's position and fate.

Many questions remain about embryogenesis in *Arabidopsis*. How is auxin production turned on as the shoot apical meristem first begins to form in embryos? Once production of the *MONOPTEROS* gene product begins, what target genes are affected? What genes other than *MONOPTEROS* are found in the regulatory cascade responsible for development along the apical–basal axis? These questions present a host of interesting challenges to current researchers.

CHECK YOUR UNDERSTANDING

If you understand that . . .

- Early embryonic development results in the formation of the apical–basal and radial axes of the plant body and three embryonic tissues.
- The early structures of the shoot and root systems form along the body's apical–basal axis.
- The genes responsible for setting up the body axes are currently the focus of intense research.

✓ You should be able to . . .

1. Relate the "location, location, location" quip to the differentiation of epidermal, ground, and vascular tissue.
2. Predict the effect on the *MONOPTEROS* gene of adding auxin to embryonic root cells.

Answers are available in Appendix B.

23.3 Vegetative Development

For a plant to thrive, it has to adjust to constantly changing conditions. Consider just one such condition—the availability of light.

You might recall from Chapter 10 that plants use wavelengths in the blue and red portions of the spectrum to drive photosynthesis. Now suppose that you are an oak tree with a life expectancy of 300 years. The quality and quantity of light that your leaves receive depends on where you happen to germinate. Are you growing on flat ground? With a southern exposure in full sun? With a northern exposure rarely exposed to full sun?

In addition, from the time you emerge from an acorn to the time of your death, the light you receive will depend on changes in climate and weather as well as on your size relative to the size and proximity of plants that compete with you for light and shade you. Finally, the leaves in your bottommost branches experience a different light regime from the leaves at your apex.

How do plants cope with all this variation in their living conditions? Unlike most animals, they don't move around to find a place that suits their requirements. Instead, they adjust to their immediate surroundings, in large part through the continuous growth and development of roots, stems, and leaves. If an oak tree is heavily shaded on one side, it stops growing in that direction and extends branches on the other side. If it is heavily shaded on all sides, its growth is directed upward. This constant adjustment to changing environmental conditions is possible because of the meristems that are located at the tips of shoots and roots.

Meristems Provide Lifelong Growth and Development

When embryonic development is complete, the basic body axes are established and the initial structures in the root and shoot systems have formed. For the rest of the individual's life, further development is driven by the meristems (**Figure 23.8a**). Meristematic tissue is located at each tip in the shoot and at the tips of root systems. As a result, the individual is capable of growing in any direction aboveground or belowground, depending on conditions.

Figure 23.8b provides a close-up view of a shoot apical meristem, or SAM. The cells within the meristem are small and undifferentiated. Within each meristem, the rate and direction of cell growth are dictated by cell-cell signals produced in response to environmental cues, such as the arrival of spring, the presence of abundant water, or the amount of light striking the plant. Just below the meristem, daughter cells produced by mitosis and cytokinesis in the meristem initially differentiate into epidermal, ground, or vascular tissue. Eventually these cells will differentiate into more specialized cell types.

Careful microscopy allowed biologists to tease out the sequence of events that occur as meristems grow, and intense research continues to explore how interactions between auxin and other cell-cell signals influence the fate of cells produced by meristems.

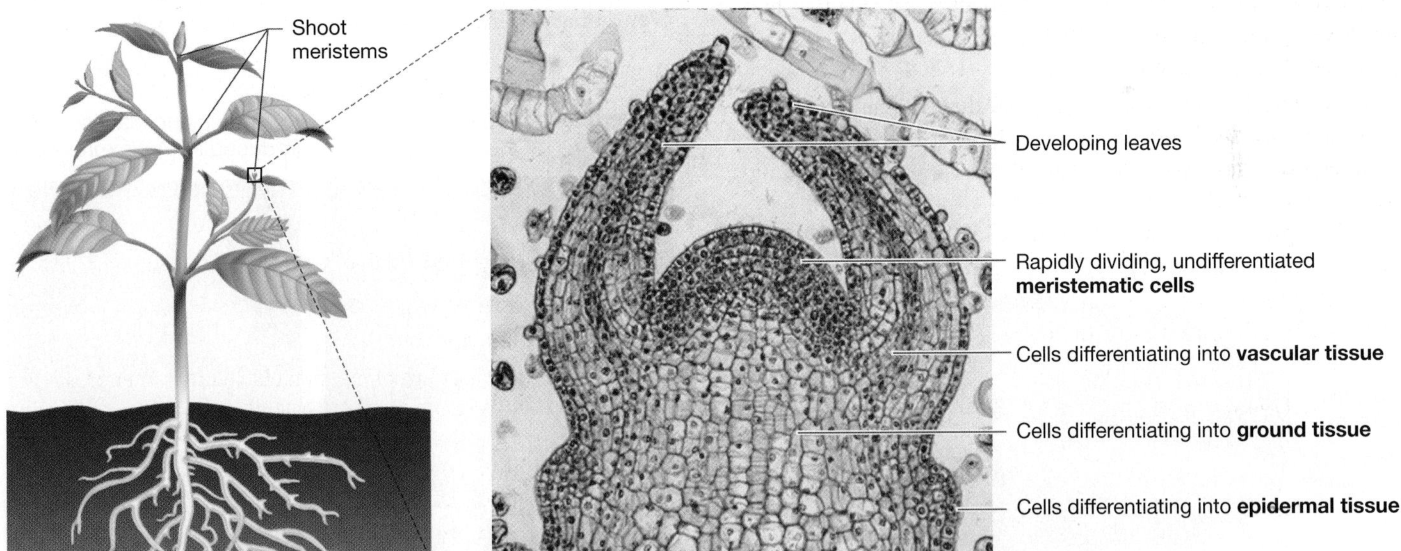

FIGURE 23.8 Meristems Are Where Development Takes Place. (a) Each tip in the root and shoot system contains a meristem. The individual can grow in any direction to which a meristem is oriented. **(b)** When meristem cells divide, the daughter cells either remain undifferentiated and continue to function as meristem cells or differentiate into new epidermal, ground, or vascular cells.

Recently researchers have also taken up the question of which genes respond to these cell-cell signals. In particular, which genes and gene products direct the formation of specific structures during vegetative development? Let's consider one especially well studied example—the genetic control of shape in developing leaves.

Which Genes and Proteins Determine Leaf Shape?

Applying tiny amounts of auxin to cells within a SAM induces leaves to grow there. This observation suggests that the initiation of a leaf depends on the concentration of auxin in parts of a SAM, although other types of cell-cell signals are undoubtedly involved as well. Once a leaf begins to grow, the next key event in its development is the formation of three axes: the proximal–distal, lateral, and adaxial–abaxial (upper–lower) axes shown in **Figure 23.9**. Proximal is toward the main body, while distal is away from the main body; the lateral axis runs from the middle of a leaf toward its margin. The amount and direction of growth along these axes determines the shape of the leaf.

Recent research has begun to identify the genes responsible for specifying these three leaf axes. For example, analyses of mutant snapdragons—a flowering plant you may have seen growing in a garden near your home—and other species has shown that a gene called *PHANTASTICA* (abbreviated *PHAN*) is critical in setting up the adaxial–abaxial axis of leaves.

The protein product of *PHAN* has a DNA-binding domain and acts as a regulatory transcription factor. *PHAN* triggers the expression of genes that cause cells to form the upper surface of leaves and suppresses transcription of genes required for forming the lower leaf surface. It is part of a regulatory cascade that begins with auxin and other cell-cell signals and ends with the growth of a normal-shaped leaf.

Recent research on tomatoes suggests that changes in *PHAN* expression may also underlie at least some of the evolutionary changes observed in leaf shape. Leaf shape varies widely among species (see Chapter 36). Simple leaves consist of a single blade, but as **Figure 23.10a** shows, tomatoes have compound leaves—each leaf blade is divided into smaller units called leaflets. Other species have palmately compound leaves, meaning that leaflets radiate from a single point.

To explore whether changes in *PHAN* might have a role in the evolution of various leaf shapes, a team of biologists used techniques introduced in Chapter 19 to create transgenic tomato plants. In these individuals, the *PHAN* gene product was blocked to a moderate or large extent.

As **Figure 23.10b** shows, leaf shape in the transgenic individuals was dramatically different. Some of the individuals had simple leaves that were cup-shaped, while others had several leaflets emerging from the same point. Although it is still uncertain why changing the specification of the upper–lower leaf axis affects overall shape, it appears clear that *PHAN* expression plays a role.

These results have inspired the hypothesis that at least some evolutionary changes in leaf size and shape are due to mutations that created new alleles of genes that regulate *PHAN* expression. Alleles that result in lowered *PHAN* expression might lead to simple leaves with a single blade; alleles that increase the extent of *PHAN* expression might result in compound leaves like those of normal tomatoes.

Mutations that alter development alter phenotypes. Variation in phenotype is the raw material of evolution (see Chapter 1 and Chapter 24).

By experimentally altering the genes that regulate development, researchers are beginning to understand the genetic changes leading to novel types of leaves. If this research continues to be productive, biologists will have another example of how changes in regulatory pathways that direct development underlie evolutionary change (see Chapter 21).

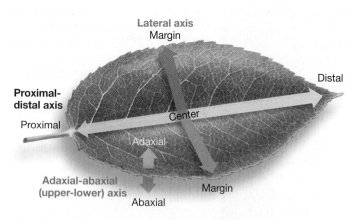

FIGURE 23.9 Leaves Have Three Axes. Overall, the plant body has just two axes: apical–basal and radial. Individually, however, every leaf has the three axes shown here. (To remember the difference between adaxial and abaxial, pretend that the *b* in a*b*axial stands for "below.")

✔**QUESTION** How do the body axes in plants compare and contrast with the body axes in animals?

(a) Normal tomato leaf

(b) Transgenic leaves (reduced *PHAN* expression)

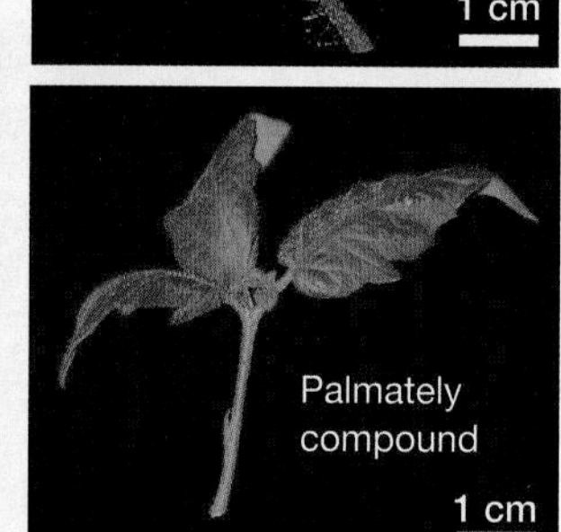

FIGURE 23.10 Changes in Genes That Establish Leaf Axes Can Change Leaf Shape.

SOURCE: Reprinted by permission from *Nature* (Vol 424, July 24, 2003, p. 439, by Kim, M., et al.). © 2004 Macmillan Publishers Ltd. Photographs supplied by Dr. Neelima Sinah.

CHECK YOUR UNDERSTANDING

If you understand that . . .

- Vegetative development is continuous, and vegetative growth can occur in all directions.
- Vegetative growth and development occur in ways that increase an individual's ability to survive and produce offspring in the face of changing environmental conditions.
- Continuous development is possible in plants because meristematic tissue is present at the tips of each component of the root and shoot system.

✓ You should be able to . . .

Make a sketch of a flowering plant and label the locations of SAMs and RAMs. Make another sketch of what the same plant would look like after extensive growth, given that a cow defecated near its left side.

Answers are available in Appendix B.

23.4 Reproductive Development

Among the many startling contrasts observed between animal and plant development, one of the most important concerns the development of reproductive tissues and organs. In animals, the cells that give rise to sperm and egg cells are set aside, or sequestered, early in development. These **germ cells** migrate to the ovaries or testes once the reproductive organs have developed. As a result, the cells that give rise to animal gametes undergo relatively few rounds of mitosis—perhaps 20 to 50—prior to meiosis and gametogenesis.

In contrast, plants do not have germ cells that are set aside early in development. Instead, flowering and gametogenesis occur when a SAM converts from vegetative development to reproductive development.

As a result, meristematic cells that have divided hundreds of times can give rise to the reproductive organs of plants, and eventually to sperm and eggs. Because mutations occur during each cell cycle, plants generate much more genetic variation by mutation than animals do.

Although biologists have only begun to explore the consequences of this fact, research on the mechanisms responsible for the formation of reproductive structures has been intense. To introduce this huge body of work, let's consider some highlights of research on *Arabidopsis.*

The Floral Meristem and the Flower

When specialized proteins in *Arabidopsis* sense that nights are getting shorter and the temperature is favorable, they trigger the production of signals that convert SAMs from vegetative to reproductive development. A SAM converted in this way is called a **floral meristem**; instead of vegetative structures, it produces flowers, which contain the plant's reproductive organs. The genes that take part in the regulatory cascade responsible for the maturation of a floral meristem are now well characterized.

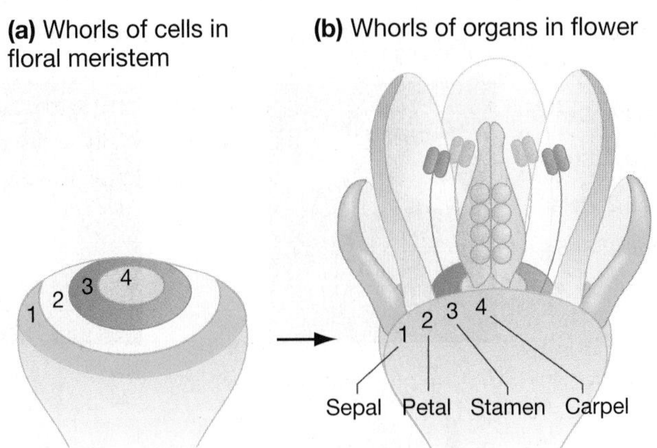

FIGURE 23.11 Flowers Are Composed of Four Organs.

As an *Arabidopsis* flower develops, the floral meristem produces four kinds of organs: (1) sepals, (2) petals, (3) stamens, and (4) carpels. Each of these organs is a modified leaf (**Figure 23.11**).

- **Sepals** are located around the outside of the flower and provide protection as the organ develops. In some species, sepals are also colorful and function in attracting pollinators.
- Inside the sepals is a whorl, or circular arrangement, of **petals**, which enclose the male and female reproductive organs. If insects or other animals have to be attracted to pollinate the species in question, the petals may be colored to help advertise the reproductive structures.
- The pollen-producing organs, or **stamens**, are located in a whorl inside the petals.
- In the center of the entire structure are egg-producing reproductive organs, or **carpels**. (Ovules are located at the base of carpels.)

The question is, How does the floral meristem produce these four organs in the characteristic pattern of whorls within whorls?

The first hint of an answer came in the late 1800s, when researchers realized that several types of mutant flowering plants—including some popular garden plants—were homeotic. In the mutant individuals, one kind of floral organ was replaced by another. For example, one homeotic mutant had flowers with sepals, petals, another ring of petals, and carpels instead of having sepals, petals, stamens, and carpels. These mutants were similar to the *Drosophila* homeotic mutants described in Chapter 21, where individuals have legs or antennae growing in the wrong location—in place of the appropriate structure.

Just as an analysis of homeotic mutants in fruit flies triggered a breakthrough in understanding the genetic control of body axis formation and segmentation in animals, an analysis of homeotic floral mutants in *Arabidopsis* triggered a breakthrough in understanding the genetic control of flower structure in plants.

The Genetic Control of Flower Structures

Over 100 years after floral homeotic mutants were first described, Elliot Meyerowitz and colleagues assembled a large collection of *Arabidopsis* individuals with homeotic mutations in flower

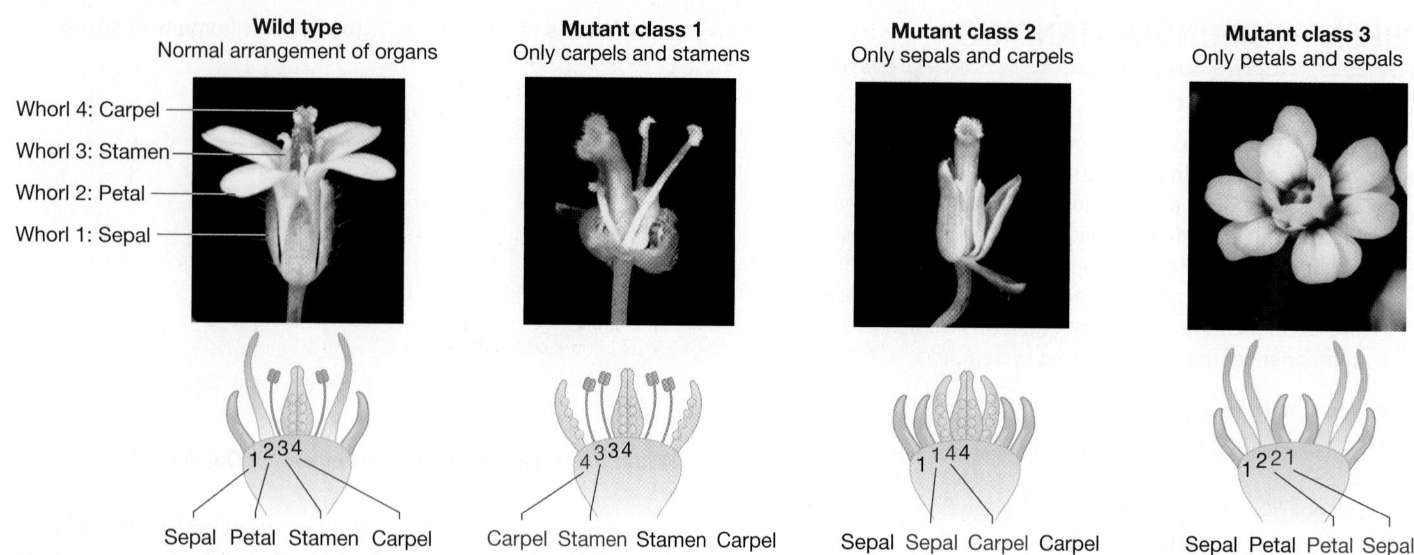

FIGURE 23.12 Homeotic Mutants in *Arabidopsis* Flowers Have Organs in the Wrong Locations.

structure. The researchers' goal was to identify and characterize the genes responsible for specifying the four floral organs.

The group found that the mutants could be sorted into the three general classes shown in **Figure 23.12**. The classes were distinguished by the type of homeotic transformation that occurred. Some mutants had only carpels and stamens; others had only sepals and carpels; still others had only petals and sepals. The key observation was that each type of mutant lacked the elements found in *two* of the four whorls.

What was going on? To begin searching for an answer, the biologists hypothesized that each of the three classes of homeotic mutants was caused by a defect in a single gene. They reasoned that if three genes are responsible for setting up the pattern of a flower, the mutants suggested a hypothesis for how the three gene products interact. Because they referred to the three hypothetical genes as *A*, *B*, and *C*, the hypothesis is called the ABC model.

THE ABC MODEL Three basic ideas underlie the ABC model (**Figure 23.13a**):

- Each of the three genes involved is expressed in two adjacent whorls.
- Because each gene is expressed in two adjacent whorls, a total of four different combinations of gene products can occur.
- Each of these four combinations of gene products triggers the development of a different floral organ.

Specifically, the Meyerowitz group proposed that (1) the A protein alone causes cells to form sepals, (2) a combination of A and B proteins sets up the formation of petals, (3) B and C combined specify stamens, and (4) the C protein alone designates cells as the precursors of carpels.

Does this model explain how the three classes of homeotic mutants occur? The answer is yes, if two additional elements are added to the model:

- The A protein inhibits production of the C protein.
- The C protein inhibits production of the A protein.

Then the patterns of gene expression correspond to the mutant phenotypes, as shown in **Figure 23.13b**.

For example, if the *A* gene is disabled by mutation, then it no longer inhibits the expression of the *C* gene and all cells produce the C protein. As a result, cells in the outermost whorl express only C protein and develop into carpels, while cells in the whorl just to the inside produce B and C proteins and develop into stamens.

✔If you understand the ABC model, you should be able to explain why biologists did not hypothesize that four genes are involved—one that specifies each of the four floral organs—and why they proposed that each gene involved is expressed in two adjacent whorls.

TESTING THE MODEL Although the ABC model is plausible and appeared to explain the data, it needed to be tested. To accomplish this, Meyerowitz and co-workers mapped the genes responsible for the mutant phenotypes and cloned the appropriate DNA sequences, using techniques introduced in Chapter 19. Once they had isolated the genes, they were able to obtain and use single-stranded DNAs to perform in situ hybridizations (see Chapter 21). The goal was to document the pattern of expression of the *A*, *B*, and *C* genes and see if that pattern corresponded to the model's predictions.

As anticipated, the mRNAs for each of the three genes showed up in the sets of whorls predicted by the model. The *A* gene is expressed in the outer two whorls, the *B* gene is expressed in the middle two whorls, and the *C* gene is expressed in the inner two whorls.

This result strongly supported the validity of the ABC model. Just as different combinations of *Hox* gene products specify the identity of fly segments, different combinations of floral identity genes specify the parts of a flower.

***MADS-BOX* GENES** Similarities to principles of animal development did not end there. When Meyerowitz and others analyzed the

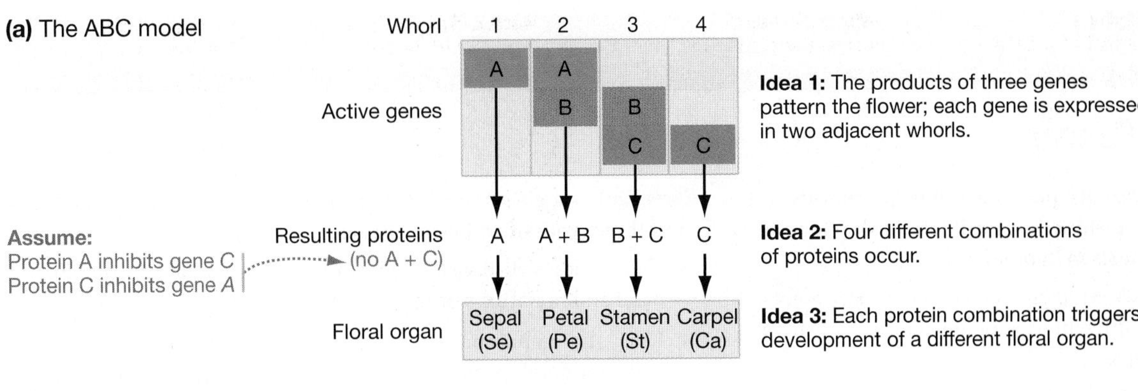

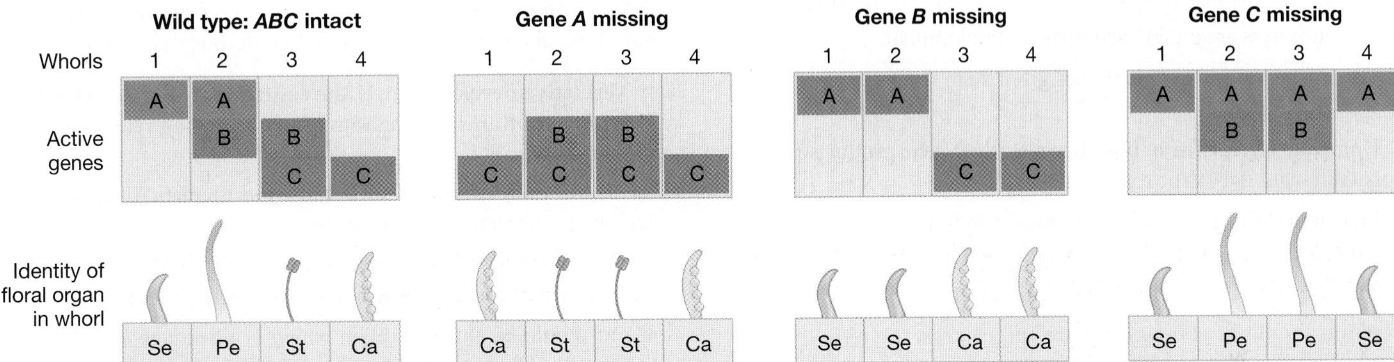

FIGURE 23.13 The ABC Model for Genetic Control of Flower Development. The ABC model is a hypothesis to explain why three types of homeotic mutants exist in *Arabidopsis* flowers.

DNA sequences of the floral organ identity genes, they discovered that all three genes contained a segment that coded for a long sequence of 58 amino acids, called the **MADS box**, that binds to DNA. Based on this observation, the researchers hypothesized that the floral genes encode regulatory transcription factors that bind to enhancers or other regulatory sequences and trigger the expression of genes required for sepal, petal, carpel, and stamen formation.

The researchers found many remarkable similarities between the *MADS-box* genes and the homeotic genes found in *Drosophila* and other animals.

- The *Hox* genes of animals have a region called the homeobox that is functionally similar to the MADS box: It encodes a DNA-binding domain in Hox proteins.
- Both *Hox* and *MADS-box* gene products are parts of regulatory cascades that lead to the specification of important structures. *Hox* genes regulate the expression of genes responsible for forming limbs or other structures in specific parts of the body; *MADS-box* genes regulate the expression of genes responsible for forming flowers.
- Both sets of genes tell cells where they are in the body and can produce homeotic mutants if they do not work properly.

Although different genes are involved, the underlying genetic principles for putting a multicellular body together are similar in plants and animals.

Biologists are currently working to identify the genes targeted by the ABC proteins, just as they are working to identify the genes controlled by the *Hox* products in animals. Eventually, they hope to understand the complete regulatory cascade from initiation of the floral meristem to the expression of proteins that are specific to petals or other floral organs.

Ultimately, the goal is to explore how changes in this cascade could have led to the evolution of the spectacular diversity of flowers observed today. Were mutations in *MADS-box* genes, or the effector genes that they control, responsible for the flamoyant petals of the lady-slipper orchid, the broad, red sepals of the pointsettia, and the elaborate stamens of a sugar maple tree? Stay tuned.

CHECK YOUR UNDERSTANDING

If you understand that . . .

- Reproductive development in plants begins when SAMs are converted to floral meristems.
- In *Arabidopsis*, development of the four floral organs depends on the expression of regulatory transcription factors encoded by the *A*, *B*, and *C* genes.

You should be able to . . .

1. Describe a mechanism whereby an A protein could inhibit expression of a *C* gene.
2. Compare and contrast the *MADS-box* genes of plants with the *Hox* genes of animals.

Answers are available in Appendix B.

CHAPTER 23 REVIEW

For media, go to the study area at www.masteringbiology.com 

Summary of Key Concepts

In sharp contrast to animals, plants develop continuously, do not commit cells to gamete production until late in development, and produce gametes by mitosis in haploid cells.

- The striking differences between plant and animal development are not surprising given their different ways of life. But many principles are shared in the two groups.
- In both plants and animals, fertilization is mediated by protein–protein interactions between cells produced by each parent.
- The body axes are established early in development.
- Three types of embryonic tissues give rise to the array of adult tissues and organs.
- Homeotic genes exist in both lineages; their gene products participate in regulatory gene cascades.
- Meristem cells and stem cells are similar in function.
- Differentiation is based on cell-cell signals that trigger and modulate cascades of regulatory transcription factors.

✔You should be able to compare and contrast the roles of auxin and Bicoid.

In flowering plants, double fertilization results in the production of a zygote and a nutritive tissue that supports embryogenesis.

- The pollination and fertilization process begins with interactions between proteins on the surface of pollen grains and proteins on the surface of the stigma.
- Pollen tubes grow toward the egg and form a conduit for cells.
- In flowering plants, one sperm nucleus fertilizes the egg to form a zygote while a second sperm nucleus fuses with a (usually) diploid cell to form a triploid cell.
- The triploid cell goes on to produce the nutritive tissue called endosperm.

✔You should be able to explain what's "double" about double fertilization.

Embryogenesis results in the formation of the major body axes and three types of embryonic tissue.

- The earliest cell divisions in embryogenesis establish the apical–basal and radial axes of the individual.
- Later, the embryonic structures called the cotyledons, hypocotyl, and root develop, and the embryonic epidermal, ground, and vascular tissues form.

✔You should be able to sketch a seedling and label the apical–basal and radial axes, cotyledons, hypocotyl, and root.

Vegetative development is the function of meristems, where cell division continues throughout life—producing cells that go on to differentiate.

- Plants can grow continuously because meristematic tissue exists at each tip in the root and shoot system.
- Continuous growth and development allows plants to respond to environmental conditions that change throughout their life.

✔You should be able to explain why it is advantageous for plants to have more than one SAM and one RAM, in terms of their ability to grow and reproduce.

When the function of a meristem shifts from vegetative to reproductive development, key regulatory transcription factors are activated and control the position and identity of floral organs.

- Meristems that carry out vegetative development can convert to reproductive development instead, as a response to changes in day length or other environmental cues.
- Once a floral meristem is established in *Arabidopsis*, combinations of regulatory transcription factors encoded by *A*, *B*, and *C* genes interact to produce the flower's sepals, petals, stamens, and carpels.

✔You should be able to explain why individuals with defective alleles at the *B* gene are considered homeotic mutants.

Questions

✔TEST YOUR KNOWLEDGE

Answers are available in Appendix B

1. What is the fate of the two cells found inside pollen grains prior to germination?
   a. One cell directs development of the pollen tube; the other gives rise to sperm cells by mitosis.
   b. One cell directs development of the pollen tube; the other gives rise to sperm cells by meiosis.
   c. One cell fertilizes the egg; the other fuses with a diploid cell to form triploid endosperm.
   d. One cell initiates germination by interacting with proteins on the surface of the stigma; the other gives rise to sperm cells.

2. Which of the following does *not* represent a contrast between plant and animal development?
   a. Under certain conditions, plant cells can "de-differentiate" readily.
   b. The fate of a cell is determined in part by its location in the embryo.
   c. Germ cells are set aside early in development.
   d. Plant cells do not move.

3. What evidence suggests that auxin concentrations help determine where leaves form near SAMs?
   a. Auxin is produced in SAMs and transported from there toward the root.

b. Auxin is present in a concentration gradient, with higher concentrations apically and lower concentrations basally.
c. Auxin concentrations are relatively constant along the radial axis of the body, and leaves form along the radial axis.
d. Addition of small quantities of auxin to a SAM can induce leaf development.

4. Which of the following does *not* occur during embryogenesis?
   a. formation of the radial axis
   b. production of the suspensor
   c. formation of the cotyledons and hypocotyl
   d. formation of the lateral and proximal–distal axes

5. When does the apical–basal axis first become apparent?
   a. when the epidermal, ground, and vascular tissues form
   b. when the cotyledons, hypocotyl, and root form
   c. when the first cell division produces the apical cell and basal cell
   d. during the globular stage, when the suspensor is complete

6. What evidence suggests that changes in the way that the *PHAN* gene is expressed could be partly responsible for evolutionary changes in leaf shape?
   a. If *PHAN* gene expression is manipulated experimentally, individuals produce leaf types found in different species.
   b. Experiments have shown that *PHAN* plays a role in establishing the upper–lower surface axis in leaves.
   c. Sequencing studies and other data have shown that *PHAN* encodes a regulatory transcription factor.
   d. All plant species surveyed to date have a gene homologous to *PHAN*.

✔TEST YOUR UNDERSTANDING

Answers are available in Appendix B

1. When do important protein–protein interactions occur as a pollen grain interacts with a stigma? Describe the major type of cell-cell communication that occurs.
2. In plants, reproductive tissues may develop late in life, from cells that have undergone mitosis hundreds or thousands of times. How does this differ from animals, and what are the consequences?
3. Compare and contrast the stem cells of animals with the meristems of plants. How are these cells similar? How are they different?
4. In what sense are the epidermal, ground, and vascular tissues produced in the SAMs and RAMs of a 300-year-old oak tree "embryonic"?
5. When in situ hybridization experiments documented where *A*, *B*, and *C* genes were expressed in developing *Arabidopsis* flowers, it was considered strong support for the ABC model. Explain why.
6. Give an example of how each of the fundamental developmental processes—cell proliferation, death, expansion, interaction, and differentiation—plays a role in plant development.

✔APPLYING CONCEPTS TO NEW SITUATIONS

Answers are available in Appendix B

1. When growing conditions are extremely poor, many plant species stop vegetative growth and put all of their energy into reproductive growth. Propose a hypothesis to explain why.
2. When growth occurs in an unlimited or unrestricted way, it is said to be indeterminate. But when growth is of limited duration and then stops, it is said to be determinant. Which process is observed in vegetative development and which in reproductive development? Explain your logic.
3. Leaves that grow at the top of a tree are typically smaller than leaves that grow at the bottom of the same tree. Small leaves lose less water in bright sunlight; large leaves capture more light in shade. Explain how this size difference might develop, in terms of changes in gene expression.
4. Make a sketch showing the locations of SAM and RAM in a young oak tree. Suppose that a concrete sidewalk gets added on the left side of the tree, preventing water penetration below. A billboard goes up on the right side, cutting off light but not water penetration from that direction. But after a few years, an underground water pipe under the billboard begins to leak, providing abundant water year round. Draw the expected locations of SAM and RAM in the same tree 50 years after your initial sketch. Explain your logic.

APPENDIX A **BioSkills**

BIOSKILLS 1

THE METRIC SYSTEM

The metric system is the system of units of measure used in every country of the world but three (Liberia, Myanmar, and the United States). It is also the basis of the SI system used in scientific publications.

The popularity of the metric system is based on its consistency and ease of use. These attributes, in turn, arise from the system's use of the base 10. For example, each unit of length in the system is related to all other measures of length in the system by a multiple of 10. There are 10 millimeters in a centimeter; 100 centimeters in a meter; 1000 meters in a kilometer.

Measures of length in the English system, in contrast, do not relate to each other in a regular way. Inches are routinely divided into 16ths; there are 12 inches in a foot; 3 feet in a yard; 5280 feet (or 1760 yards) in a mile.

TABLE B1.2 **Prefixes Used in the Metric System**

Prefix	Abbreviation	Definition
nano-	n	$0.000\ 000\ 001 = 10^{-9}$
micro-	μ	$0.000\ 001 = 10^{-6}$
milli-	m	$0.001 = 10^{-3}$
centi-	c	$0.01 = 10^{-2}$
deci-	d	$0.1 = 10^{-1}$
—	—	$1 = 10^{0}$
kilo-	k	$1000 = 10^{3}$
mega-	M	$1\ 000\ 000 = 10^{6}$
giga-	G	$1\ 000\ 000\ 000 = 10^{9}$

TABLE B1.1 **Metric System Units and Conversions**

Measurement	Unit of Measurement and Abbreviation	Metric System Equivalent	Converting Metric Units to English Units
Length	kilometer (km)	$1\ \text{km} = 1000\ \text{m} = 10^{3}\ \text{m}$	1 km = 0.62 mile
	meter (m)	1 m = 100 cm	1 m = 1.09 yards = 3.28 feet = 39.37 inches
	centimeter (cm)	$1\ \text{cm} = 0.01\ \text{m} = 10^{-2}\ \text{m}$	1 cm = 0.3937 inch
	millimeter (mm)	$1\ \text{mm} = 0.001\ \text{m} = 10^{-3}\ \text{m}$	1 mm = 0.039 inch
	micrometer (μm)	$1\ \mu\text{m} = 10^{-6}\ \text{m} = 10^{-3}\ \text{mm}$	
	nanometer (nm)	$1\ \text{nm} = 10^{-9}\ \text{m} = 10^{-3}\ \mu\text{m}$	
Area	hectare (ha)	$1\ \text{ha} = 10{,}000\ \text{m}^2$	1 ha = 2.47 acres
	square meter (m^2)	$1\ \text{m}^2 = 10{,}000\ \text{cm}^2$	$1\ \text{m}^2 = 1.196$ square yards
	square centimeter (cm^2)	$1\ \text{cm}^2 = 100\ \text{mm}^2 = 10^{-4}\ \text{m}^2$	$1\ \text{cm}^2 = 0.155$ square inch
Volume	liter (L)	1 L = 1000 mL	1 L = 1.06 quarts
	milliliter (mL)	$1\ \text{mL} = 1000\ \mu\text{L} = 10^{-3}\ \text{L}$	1 mL = 0.034 fluid ounce
	microliter (μL)	$1\ \mu\text{L} = 10^{-6}\ \text{L}$	
Mass	kilogram (kg)	1 kg = 1000 g	1 kg = 2.20 pounds
	gram (g)	1 g = 1000 mg	1 g = 0.035 ounce
	milligram (mg)	$1\ \text{mg} = 1000\ \mu\text{g} = 10^{-3}\ \text{g}$	
	microgram (μg)	$1\ \mu\text{g} = 10^{-6}\ \text{g}$	
Temperature	Kelvin (K)*		K = °C + 273.15
	degrees Celsius (°C)		$°\text{C} = \frac{5}{9}(°\text{F} - 32)$
	degrees Fahrenheit (°F)		$°\text{F} = \frac{9}{5}°\text{C} + 32$

*Absolute zero is −273.15°C = 0 K

If you have grown up in the United States and are accustomed to using the English system, it is extremely important to begin developing a working familiarity with metric units and values. The tables and questions below should help you get started with this process.

✔ Questions

1. Some friends of yours just competed in a 5-kilometer run. How many miles did they run?
2. An American football field is 120 yards long, while rugby fields are 144 meters long. In yards, how much longer is a rugby field than an American football field?
3. What is your normal body temperature in degrees Celsius? (Normal body temperature is 98.6°F.)
4. What is your current weight in kilograms?
5. A friend asks you to buy a gallon of milk. How many liters would you buy to get approximately the same volume?

BIOSKILLS 2

READING GRAPHS

Graphs are the most common way to report data, for a simple reason. Compared to reading raw numerical values in a table or list, a graph makes it much easier to understand what the data mean.

Learning how to read and interpret graphs is one of the most basic skills you'll need to acquire as a biology student. As when learning piano or soccer or anything else, you need to understand a few key ideas to get started and then have a chance to practice—a *lot*—with some guidance and feedback.

Getting Started

To start reading a graph, you need to do three things: read the axes, figure out what the data points represent—that is, where they came from—and think about the overall message of the data. Let's consider each in turn.

What Do the Axes Represent?

Graphs have two axes: one horizontal and one vertical. The horizontal axis of a graph is also called the *x*-axis or the abscissa. The vertical axis of a graph is also called the *y*-axis or the ordinate. Each axis represents a variable that takes on a range of values. These values are indicated by the ticks and labels on the axis. Note that each axis should *always* be clearly labeled with the unit or treatment it represents.

Figure B2.1 shows a scatterplot—a type of graph where continuous data are graphed on each axis. Continuous data can take an array of values over a range. In contrast, discrete data can take only a restricted set of values. If you were graphing the average height of men and women in your class, height is a continuous variable but gender is a discrete variable.

In the example in the figure, the *x*-axis represents time in units of generations of maize; the *y*-axis represents the average percentage of the dry weight of a maize kernel that is protein.

To create a graph, researchers plot the independent variable on the *x*-axis and the dependent variable on the *y*-axis (Figure B2.1a). The terms independent and dependent are used because the values on the *y*-axis depend on the *x*-axis values. In our example, the researchers wanted to show how the protein content of maize kernels in a study population changed over time. Thus, the protein concentration plotted on the *y*-axis depended on the year (generation) plotted on the *x*-axis. The value on the *y*-axis always depends on the value on the *x*-axis, but not vice versa.

In many graphs in biology, the independent variable is either time or the various treatments used in an experiment. In these cases, the *y*-axis records how some quantity changes as a function of time or as the outcome of the treatments applied to the experimental cells or organisms.

What Do the Data Points Represent?

Once you've read the axes, you need to figure out what each data point is. In our maize kernel example, the data point in Figure B2.1b represents the average percentage of protein found in a sample of kernels from a study population in a particular generation.

If it's difficult to figure out what the data points are, ask yourself where they came from—meaning, how the researchers got them. You can do this by understanding how the study was done and by understanding what is being plotted on each axis. The *y*-axis will tell you what they measured; the *x*-axis will usually tell you when they measured it or what group was measured. In some cases—for example, in a plot of average body size versus average brain size in primates—the *x*-axis will report a second variable that was measured.

What Is the Overall Trend or Message?

Look at the data as a whole, and figure out what they mean. Figure B2.1c suggests an interpretation of the maize kernel example. If the graph shows how some quantity changes over time, ask yourself if that quantity is increasing, decreasing, fluctuating up and down, or staying the same. Then ask whether the pattern is the same over time or whether it changes over time.

When you're interpreting a graph, it's extremely important to limit your conclusions to the data presented. Don't extrapolate beyond the data, unless you are explicitly making a prediction based on the assumption that present trends will continue. For example, you can't say that average % protein content was increasing in the population before the experiment started, or that it will continue to increase in the future. You can only say what the data tell you.

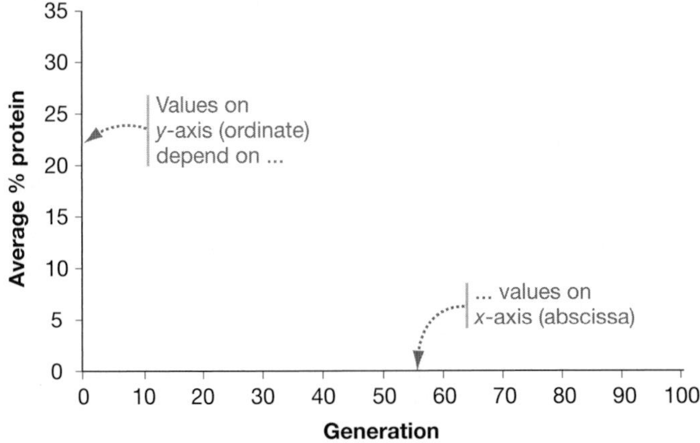

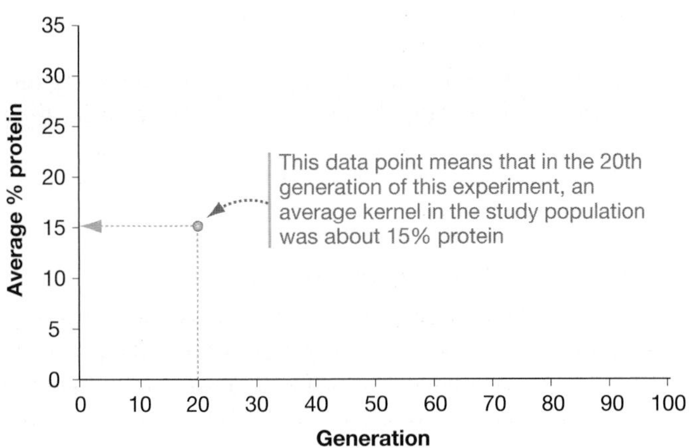

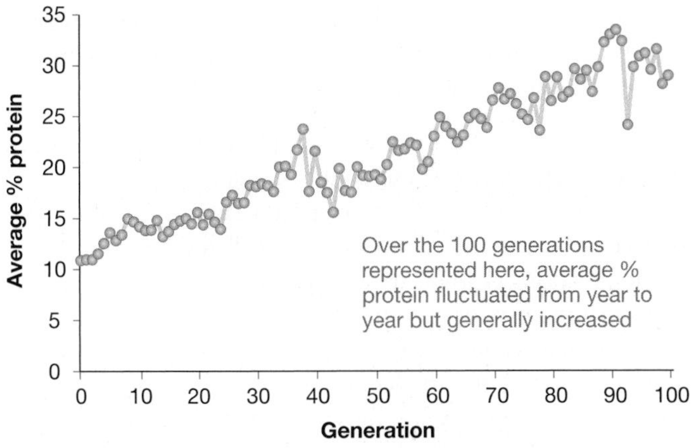

FIGURE B2.1 Scatterplots Are Used to Graph Continuous Data.

Types of Graphs

Most of the graphs in this text are scatterplots like the one shown here, where individual data points are plotted.

Sometimes the data points in a scatterplot will be by themselves, sometimes they will be connected by dot-to-dot lines to help make the overall trend clearer, as in this figure, and sometimes they will have a smooth line through them. A smooth line through data points—sometimes straight, sometimes curved—is a mathematical "line of best fit." A line of best fit represents a mathematical function that summarizes the relationship between the *x* and *y* variables. It is "best" in the sense of fitting the data points most precisely.

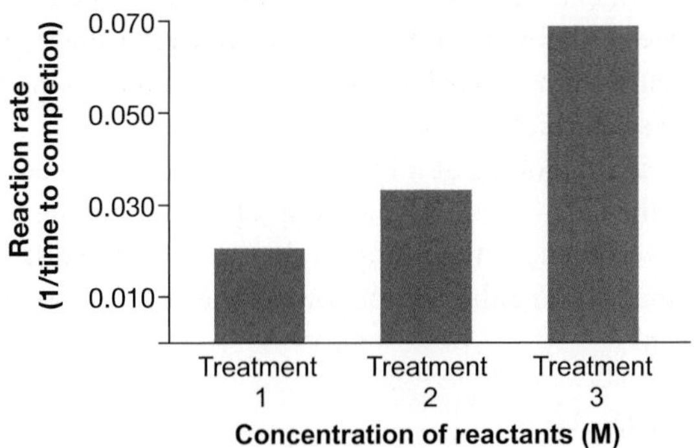

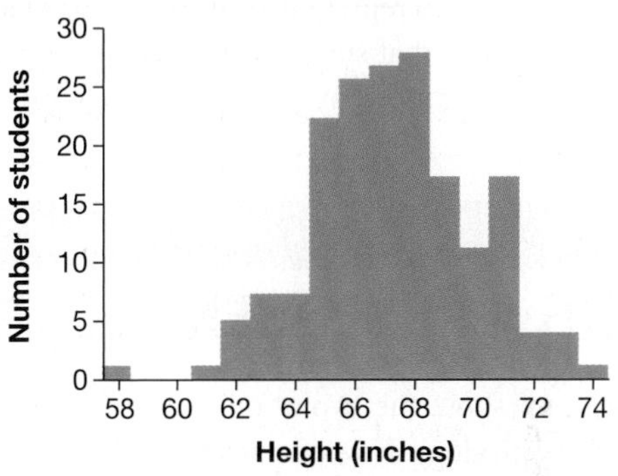

FIGURE B2.2 Bar Charts and Histograms. (a) Bar charts are used to graph data that are discontinuous or categorical. **(b)** Histograms show the distribution of frequencies or values in a population.

Scatterplots are the most appropriate type of graph when the data have a continuous range of values and you want to show individual data points. But you will also come across two other major types of graphs in this text:

- *Bar charts* plot data that have discrete or categorical values instead of a continuous range of values. In many cases the bars might represent different treatment groups in an experiment, as in **Figure B2.2a**. In this graph, the height of the bar indicates the average value.
- *Histograms* illustrate frequency data and can be plotted as numbers or percentages. **Figure B2.2b** shows an example where height is plotted on the *x*-axis and the number of students in a population is plotted on the *y*-axis. Each rectangle indicates the number of individuals in each interval of height, which reflects the relative frequency, in this population, of people whose heights are in that interval. The measurements could also be recalculated so that the *y*-axis would report the proportion of people in each interval. Then the sum of all the bars would equal 100 percent.

When you are looking at a bar chart that plots values from different treatments in an experiment, ask yourself if these values are the same or different. If the bar chart reports averages over discrete ranges of values, ask what trend is implied—as you would for a scatterplot.

When you are looking at a histogram, ask whether there is a "hump" in the data—indicating a group of values that are more frequent than others. Is the hump in the center of the distribution of values, toward the left, or toward the right? If so, what does it mean?

Getting Practice

Working with this text will give you lots of practice with reading graphs—they appear in almost every chapter. In many cases we've inserted an arrow to represent your instructor's hand at the whiteboard, with a label that suggests an interpretation or draws your attention to an important point on the graph. In other cases, you should be able to figure out what the data mean on your own or with the help of other students or your instructor.

✔Questions

1. What is the total change in average percent protein in maize kernels, from the start of the experiment until the end?
2. What was the trend in average percent protein in maize kernels between generation 37 and generation 42?
3. Would the conclusions from the bar chart in Figure B2.2a be different if the data and label for Treatment 3 were put on the far left and the data and label for Treatment 1 on the far right?
4. In Figure B2.2b, about how many students in this class are 70 inches tall?
5. What is the most common height in the class graphed in Figure B2.2b?

BIOSKILLS 3

READING A PHYLOGENETIC TREE

Phylogenetic trees show the evolutionary relationships among species, just as a genealogy shows the relationships among people in your family. They are unusual diagrams, however, and it can take practice to interpret them correctly.

To understand how evolutionary trees work, consider **Figure B3.1**. Notice that a phylogenetic tree consists of branches, nodes, and tips.

- Branches represent populations through time. In this text, branches are drawn as horizontal lines. In most cases the length of the branch is arbitrary and has no meaning, but in some cases branch lengths are proportional to time (if so, there will be a scale at the bottom of the tree). The vertical lines on the tree represent splitting events, where one group broke into two independent groups. Their length is arbitrary—chosen simply to make the tree more readable.
- Nodes (also called forks) occur where an ancestral group splits into two or more descendant groups (see point B in Figure B3.1). Thus, each node represents the most recent common ancestor of the two or more descendant populations that emerge from it. If more than two descendant groups emerge from a node, the node is called a polytomy (see node C).
- Tips (also called terminal nodes) are the tree's endpoints, which represent groups living today or a dead end—a branch ending in extinction. The names at the tips can represent species or larger groups such as mammals or conifers.

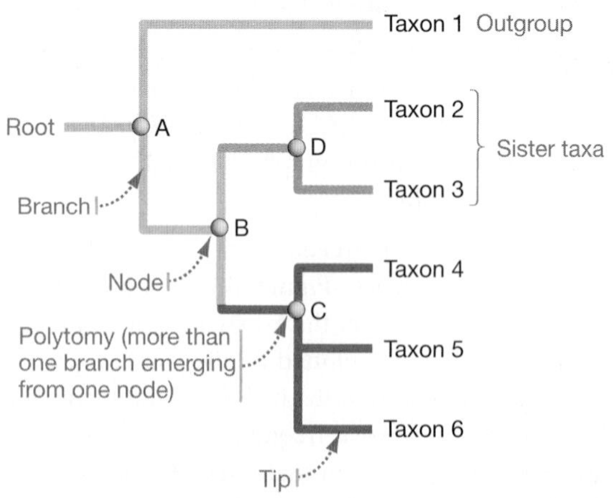

FIGURE B3.1 Phylogenetic Trees Have Roots, Branches, Nodes, and Tips.

✔**EXERCISE** Circle all four monophyletic groups present.

Recall from Chapter 1 that a taxon (plural: taxa) is any named group of organisms. A taxon could be a single species, such as *Homo sapiens*, or a large group of species, such as Primates. Tips connected by a single node on a tree are called sister taxa.

The phylogenetic trees used in this text are all rooted. This means that the first, or most basal, node on the tree—the one on the far left in this book—is the most ancient. To determine where the root on a tree occurs, biologists include one or more outgroup species when they are collecting data to estimate a particular phylogeny. An outgroup is a taxonomic group that is known to have diverged prior to the rest of the taxa in the study.

In Figure B3.1, "Taxon 1" is an outgroup to the monophyletic group consisting of taxa 2–6. A monophyletic group consists of an ancestral species and all of its descendants. The root of a tree is placed between the outgroup and the monophyletic group being studied. This position in Figure B3.1 is node A.

Understanding monophyletic groups is fundamental to reading and estimating phylogenetic trees. Monophyletic groups may also be called lineages or clades and can be identified using the "one-snip test": If you cut any branch on a phylogenetic tree, all of the branches and tips that fall off represent a monophyletic group. Using the one-snip test, you should be able to convince yourself that the monophyletic groups on a tree are nested. In Figure B3.1, for example, the monophyletic group comprising node A and taxa 1–6 contains a monophyletic group consisting of node B and taxa 2–6, which includes the monophyletic group represented by node C and taxa 4–6.

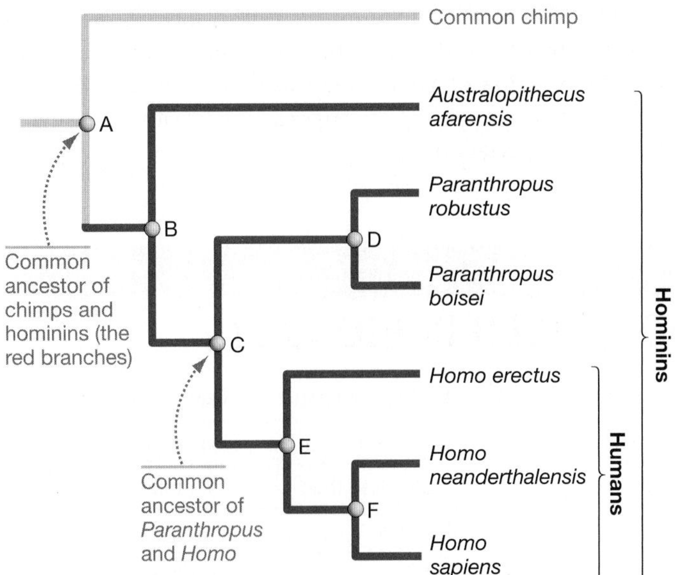

FIGURE B3.2 An Example of a Phylogenetic Tree. A phylogenetic tree showing the relationships of species in the monophyletic group called hominins.

✔**EXERCISE** All of the hominins walked on two legs—unlike chimps and all of the other primates. Add a mark on the phylogeny to show where upright posture evolved, and label it "origin of walking on two legs." Circle and label a pair of sister species. Label an outgroup to the monophyletic group called humans (species in the genus *Homo*).

To put all these new terms and concepts to work, consider the phylogenetic tree in **Figure B3.2**, which shows the relationships between common chimpanzees and six human and humanlike species that lived over the past 5–6 million years. Chimps functioned as an outgroup in the analysis that led to this tree, so the root was placed at node A. The branches marked in red identify a monophyletic group called the hominins.

To practice how to read a tree, put your finger at the tree's root, at the far left, and work your way to the right. At node A, the ancestral population split into two descendant populations. One of these populations eventually evolved into today's chimps; the other gave rise to the six species of hominins pictured. Now continue moving your finger toward the tips of the tree until you hit node C. It should make sense to you that at this splitting event, one descendant population eventually gave rise to two *Paranthropus* species, while the other became the ancestor of humans—species in the genus *Homo*. If multiple branches emerge from a node, creating a polytomy, it means that the populations split from one another so quickly that it is not possible to tell which split off earlier or later.

As you study Figure B3.2, you should consider a couple of important points.

1. There are many equivalent ways of drawing this tree. For example, this version shows *Homo sapiens* on the bottom. But the tree would be identical if the two branches emerging from node E were rotated 180°, so that the species appeared in the

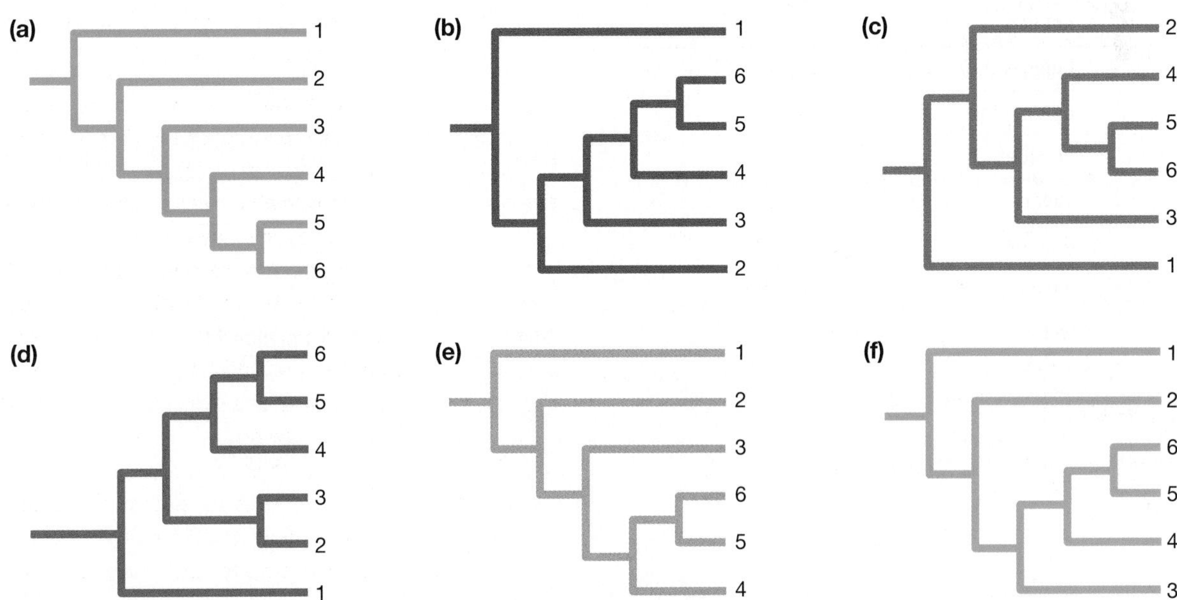

FIGURE B3.3 Alternative Ways of Drawing the Same Tree.

✔**QUESTION** Five of these six trees describe exactly the same relationships among taxa 1 through 6. Identify the tree that is different from the other five.

order *Homo sapiens*, *Homo neanderthalensis*, *Homo erectus*. Trees are read from root to tips, not from top to bottom or bottom to top.

2. No species on any tree is any higher or lower than any other. Chimps and *Homo sapiens* have been evolving exactly the same amount of time since their divergence from a common ancestor—neither species is higher or lower than the other. It is legitimate to say that more-ancient groups like *Australopithecus afarensis* have traits that are ancestral or more basal—meaning, that appeared earlier in evolution—compared to traits that appear in *Homo sapiens*, which are referred to as more derived.

Figure B3.3 on page B:5 presents a chance to test your tree-reading ability. Five of the six trees shown in this diagram are identical in terms of the evolutionary relationships they represent. One differs. The key to understanding the difference is to recognize that the ordering of tips does not matter in a tree—only the ordering of nodes (branch points) matters. You can think of a tree like a mobile: the tips can rotate without changing the underlying relationships.

BIOSKILLS 4

SOME COMMON LATIN AND GREEK ROOTS USED IN BIOLOGY

Greek or Latin Root	English Translation	Example Term
a, an	not	anaerobic
aero	air	aerobic
allo	other	allopatric
amphi	on both sides	amphipathic
anti	against	antibody
auto	self	autotroph
bi	two	bilateral symmetry
bio	life, living	bioinformatics
blast	bud, sprout	blastula
co	with	cofactor
cyto	cell	cytoplasm
di	two	diploid
ecto	outer	ectoparasite
endo	inner, within	endoparasite
epi	outer, upon	epidermis
exo	outside	exothermic
glyco	sugary	glycolysis
hetero	different	heterozygous
homo	alike	homozygous
hydro	water	hydrolysis
hyper	over, more than	hypertonic
hypo	under, less than	hypotonic
inter	between	interspecific
intra	within	intraspecific
iso	same	isotonic
logo, logy	study of	morphology
lyse, lysis	loosen, burst	glycolysis
macro	large	macromolecule
meta	change, turning point	metamorphosis
micro	small	microfilament
morph	form	morphology
oligo	few	oligopeptide
para	beside	parathyroid gland
photo	light	photosynthesis
poly	many	polymer
soma	body	somatic cells
sym, syn	together	symbioticm, synapsis
trans	across	translation
tri	three	trisomy
zygo	yoked together	zygote

✔ Questions

Provide literal translations of the following terms:

1. heterozygote
2. glycolysis
3. morphology
4. trisomy

BIOSKILLS 5

USING STATISTICAL TESTS AND INTERPRETING STANDARD ERROR BARS

When biologists do an experiment, they collect data on individuals in a treatment group and a control group, or several such comparison groups. Then they want to know whether the individuals in the two (or more) groups are different. For example, Chapter 2 introduces an experiment in which student researchers measured how fast a product formed when they set up a reaction with three different concentrations of reactants. Each treatment—meaning, each combination of reactant concentrations—was replicated many times.

Figure B5.1 graphs the average reaction rate for each of the three treatments in the experiment. Note that Treatments 1, 2, and 3 represent increasing concentrations of reactants. The thin "I-beams" on each bar indicate the standard error of each average. The standard error is a quantity that indicates the uncertainty in the calculation of an average.

For example, if two trials with the same concentration of reactants had a reaction rate of 0.075 and two trials had a reaction rate of 0.025, then the average reaction rate would be 0.50. In this case, the standard error would be large. But if two trials had a reaction rate of 0.051 and two had a reaction rate of 0.049, the average would still be 0.050, but the standard error would be small.

In effect, the standard error quantifies how confident you are that the average you've calculated is the average you'd observe if you did the experiment under the same conditions an extremely large number of times. It is a measure of precision.

Once they had calculated these averages and standard errors, the students wanted to answer a question: Does reaction rate increase when reactant concentration increases?

After looking at the data, you might conclude that the answer is yes. But how could you come to a conclusion like this objectively, instead of subjectively?

The answer is to use a statistical test. This can be thought of as a three-step process.

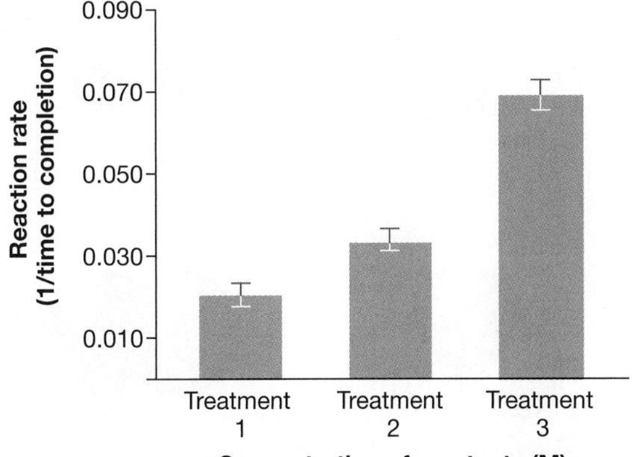

FIGURE B5.1 Standard Error Bars Indicate the Uncertainty in an Average.

1. Specify the null hypothesis, which is that reactant concentration has no effect on reaction rate.
2. Calculate a test statistic, which is a number that characterizes the size of the difference among the treatments. In this case, the test statistic compares the actual differences in reaction rates among treatments to the difference predicted by the null hypothesis. The null hypothesis predicts that there should be no difference.
3. The third step is to determine the probability of getting a test statistic as large as the one calculated just by chance. The answer comes from a reference distribution—a mathematical function that specifies the probability of getting various values of the test statistic if the null hypothesis is correct. (If you take a statistics course, you'll learn which test statistics and reference distributions are relevant to different types of data.)

You are very likely to see small differences among treatment groups just by chance—even if no differences actually exist. If you flipped a coin 10 times, for example, you are unlikely to get exactly five heads and five tails, even if the coin is fair. A reference distribution tells you how likely you are to get each of the possible outcomes of the 10 flips if the coin is fair, just by chance.

In this case, the reference distribution indicated that if the null hypothesis of no actual difference in reaction rates is correct, you would see differences as large as those observed only 0.01 percent of the time just by chance. By convention, biologists consider a difference among treatment groups to be statistically significant if you have less than a 5 percent probability of observing it just by chance. Based on this convention, the student researchers were able to claim that the null hypothesis is not correct for reactant concentration. According to their data, the reaction they studied really does happen faster when reactant concentration increases.

It is likely that you'll be doing actual statistical tests early in your undergraduate career. To use this text, though, you only need to be aware of what statistical testing does. And you should take care to inspect the standard error bars on graphs in this book. As a *very* rough rule of thumb, averages often turn out to be significantly different, according to an appropriate statistical test, if there is no overlap between two times the standard errors.

✔Question

Suppose you estimated the average height of students in your class by sampling two individuals at random. Then suppose you estimated the same quantity by sampling every individual that showed up for class on a particular day. Which estimate of the average is likely to have the smallest standard error, and why?

READING CHEMICAL STRUCTURES

If you haven't had much chemistry yet, learning basic biological chemistry can be a challenge. One of the stumbling blocks is simply being able to read chemical structures efficiently and understand what they mean. This skill will come much easier once you have a little notation under your belt and you understand some basic symbols.

Atoms are the basic building blocks of everything in the universe, just as cells are the basic building blocks of your body. Every atom has a 1- or 2-letter symbol. **Table B6.1** shows the symbols for most of the atoms you'll encounter in this book. You should memorize these. The table also offers details on how the atoms form bonds as well as how they are represented in visual models.

When atoms attach to each other by covalent bonding, a molecule forms. Biologists have a couple of different ways of representing molecules—you'll see each of these in the book and in class.

- A molecular formula like those in **Figure B6.1a** simply lists the atoms present in a molecule, with subscripts indicating how many of each atom are present. If the formula has no subscript, only one of that type of atom is present. A methane (natural gas) molecule, for example, can be written as CH_4. It consists of one carbon atom and four hydrogen atoms.
- Structural formulas like those in **Figure B6.1b** show which atoms in the molecule are bonded to each other, with each bond indicated by a dash. The structural formula for methane indicates that each of the four hydrogen atoms forms one covalent bond with carbon, and that carbon makes a total of four covalent bonds. Single covalent bonds are symbolized by a single dash; double bonds are indicated by two dashes.

Even simple molecules have distinctive shapes, because different atoms make covalent bonds at different angles. Ball-and-stick and space-filling models show the geometry of the bonds accurately.

TABLE B6.1 **Some Attributes of Atoms Found in Organisms**

Atom	Symbol	Number of Bonds It Can Form	Standard Color Code*
Hydrogen	H	1	white
Carbon	C	4	black
Nitrogen	N	3	blue
Oxygen	O	2	red
Sodium	Na	1	—
Magnesium	Mg	2	—
Phosphorus	P	5	orange or purple
Sulfur	S	2	yellow
Chlorine	Cl	1	—
Potassium	K	1	—
Calcium	Ca	2	—

*In ball-and-stick or space-filling models.

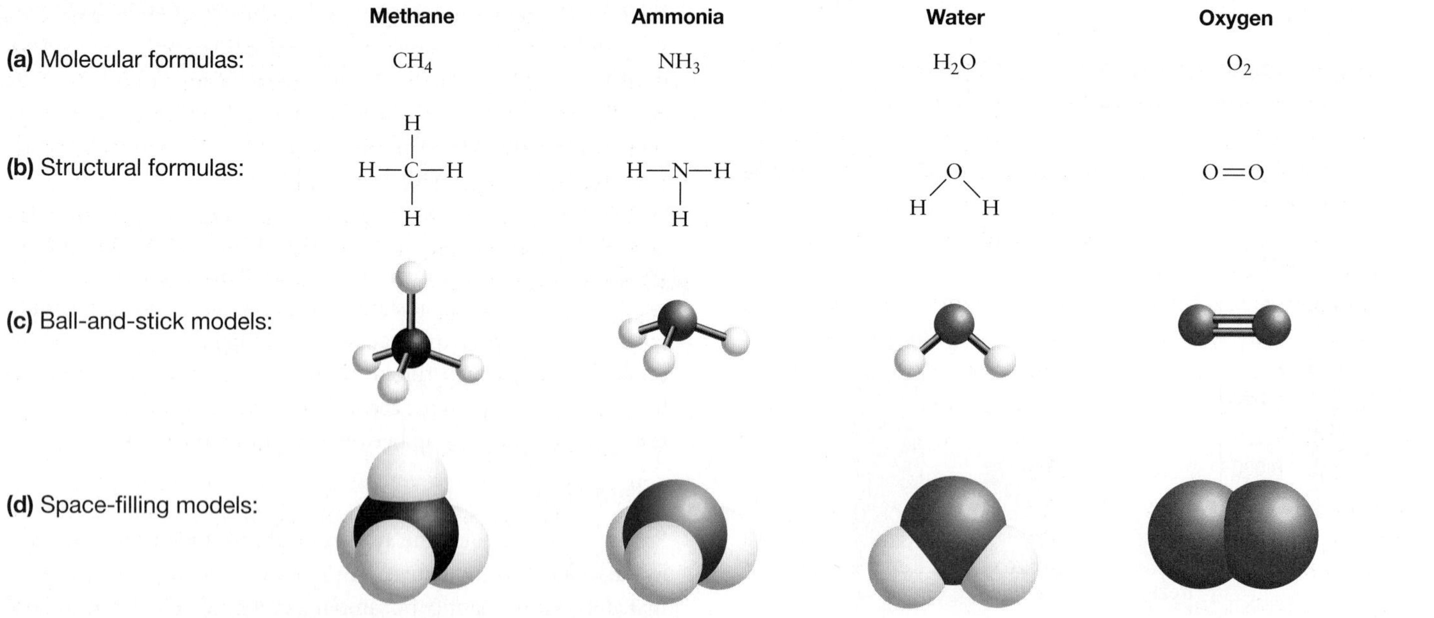

	Methane	Ammonia	Water	Oxygen
(a) Molecular formulas:	CH_4	NH_3	H_2O	O_2
(b) Structural formulas:	H—C—H (with H above and below C)	H—N—H (with H below N)	H—O—H (bent)	O=O
(c) Ball-and-stick models:				
(d) Space-filling models:				

FIGURE B6.1 Molecules Can Be Represented in Several Different Ways.

✔**EXERCISE** Carbon dioxide consists of a carbon atom that forms a double bond with each of two oxygen atoms, for a total of four bonds. It is a linear molecule. Write carbon dioxide's molecular formula, then draw its structural formula, a ball-and-stick model, and a space-filling model.

- In a ball-and-stick model, a stick is used to represent each covalent bond (see **Figure B6.1c**).
- In space-filling models, the atoms are simply stuck onto each other in their proper places (see **Figure B6.1d**).

To learn more about a molecule when you look at a chemical structure, ask yourself three questions:

1. ***Is the molecule polar—meaning that some parts are more negatively or positively charged than others?*** Molecules that contain nitrogen or oxygen atoms are often polar, because these atoms have such high electronegativity (see Chapter 2). This trait is important because polar molecules dissolve in water.
2. ***Does the structural formula show atoms that might participate in chemical reactions?*** For example, are there charged atoms or amino or carboxyl (–COOH) groups that might act as a base or an acid?
3. ***In ball-and-stick and especially space-filling models of large molecules, are there interesting aspects of overall shape?*** For example, is there a groove where a protein might bind to DNA, or a cleft where a substrate might undergo a reaction in an enzyme?

BIOSKILLS 7

USING LOGARITHMS

You have probably been introduced to logarithms and logarithmic notation in algebra courses, and you will encounter logarithms at several points in this course. Logarithms are a way of working with powers—meaning, numbers that are multiplied by themselves one or more times.

Scientists use exponential notation to represent powers. For example,

$$a^x = y$$

means that if you multiply a by itself x times, you get y. In exponential notation, a is called the base and x is called the exponent. The entire expression is called an exponential function.

What if you know y and a, and you want to know x? This is where logarithms come in. You can solve for exponents using logarithms. For example,

$$x = \log_a y$$

This equation reads, x is equal to the logarithm of y to the base a. Logarithms are a way of working with exponential functions. They are important because so many processes in biology (and chemistry and physics, for that matter) are exponential in nature. To understand what's going on, you have to describe the process with an exponential function and then use logarithms to work with that function.

Although a base can be any number, most scientists use just two bases when they employ logarithmic notation: 10 and e. Logarithms to the base 10 are so common that they are usually symbolized in the form $\log y$ instead of $\log_{10} y$. A logarithm to the base e is called a natural logarithm and is symbolized ln (pronounced *EL-EN*) instead of log. You write "the natural logarithm of y" as $\ln y$. The base e is an irrational number (like π) that is approximately equal to 2.718. Like 10, e is just a number. But both 10 and e have qualities that make them convenient to use in biology (and chemistry, and physics).

Most scientific calculators have keys that allow you to solve problems involving base 10 and base e. For example, if you know y, they'll tell you what $\log y$ or $\ln y$ are—meaning that they'll solve for x in our example above. They'll also allow you to find a number when you know its logarithm to base 10 or base e. Stated another way, they'll tell you what y is if you know x, and y is equal to e^x or 10^x. This is called taking an antilog. In most cases, you'll use the inverse or second function button on your calculator to find an antilog (above the log or ln key).

To get some practice with your calculator, consider the equation

$$10^2 = 100$$

If you enter 100 in your calculator and then press the log key, the screen should say 2. The logarithm tells you what the exponent is. Now press the antilog key while 2 is on the screen. The calculator screen should return to 100. The antilog solves the exponential function, given the base and the exponent.

If your background in algebra isn't strong, you'll want to get more practice working with logarithms—you'll see them frequently during your undergraduate career. Remember that once you understand the basic notation, there's nothing mysterious about logarithms. They are simply a way of working with exponential functions, which describe what happens when something is multiplied by itself a number of times—like cells that divide and then divide again and then again.

Using logarithms will also come up when you are studying something that can have a large range of values, like the concentration of hydrogen ions in a solution or the intensity of sound that the human ear can detect. In cases like this, it's convenient to express the numbers involved as exponents. Using exponents makes a large range of numbers smaller and more tractable. For example, instead of saying that hydrogen ion concentration in a solution can range from 1 to 10^{-14}, the pH scale allows you to simply say that it ranges from 1 to 14. Instead of giving the actual value, you're expressing it as an exponent. It just simplifies things.

✓Questions

In Chapter 52, you'll use the equation $N_t = N_0 e^{rt}$.

1. What type of function does this equation describe?
2. After taking the natural logarithm of both sides, how would you write the equation?

BIOSKILLS 8

MAKING CONCEPT MAPS

A concept map is a graphical device for organizing and expressing what you know about a topic. It has two main elements: (**1**) concepts that are identified by words or short phrases and placed in a box or circle, and (**2**) labeled arrows that physically link two concepts and explain the relationship between them. The concepts are arranged hierarchically on a page, with the most general concepts at the top and the most specific ideas at the bottom.

The combination of a concept, a linking word, and a second concept is called a proposition. Good concept maps also have cross-links—meaning, labeled arrows that connect different elements in the hierarchy, as you read down the page.

Concept maps were initially developed by Joseph Novak in the early 1970s and have proven to be an effective studying and learning tool. They can be particularly valuable if constructed by a group, or when different individuals exchange and critique concept maps they have created independently. Although concept maps vary widely in quality and can be graded using objective criteria, there are many equally valid ways of making a high-quality concept map on a particular topic.

When you are asked to make a concept map in this text, you will usually be given at least a partial list of concepts to use. As an example, suppose you were asked to create a concept map on experimental design and were given the following concepts: results, predictions, control treatment, experimental treatment, controlled (identical) conditions, conclusions, experiment, hypothesis to be tested, null hypothesis. One possible concept map is shown in **Figure B8.1**.

Good concept maps have four qualities:

- They exhibit an organized hierarchy, indicating how each concept on the map relates to larger and smaller concepts.
- The concept words are specific—not vague.
- The propositions are accurate.
- There is cross-linking between different elements in the hierarchy of concepts.

As you practice making concept maps, go through these criteria and use them to evaluate your own work, as well as the work of fellow students.

✔Exercises

1. In many cases, investigators contrast a hypothesis being tested with an alternative hypothesis that does not qualify as a null hypothesis. Add an "Alternative hypothesis" concept to the map in Figure B8.1, along with other concepts and labeled linking arrows needed to indicate its relationship to other information on the map.
2. Add a box for the concept "Statistical testing" (see **BioSkills 5**) along with appropriately labeled linking arrows.

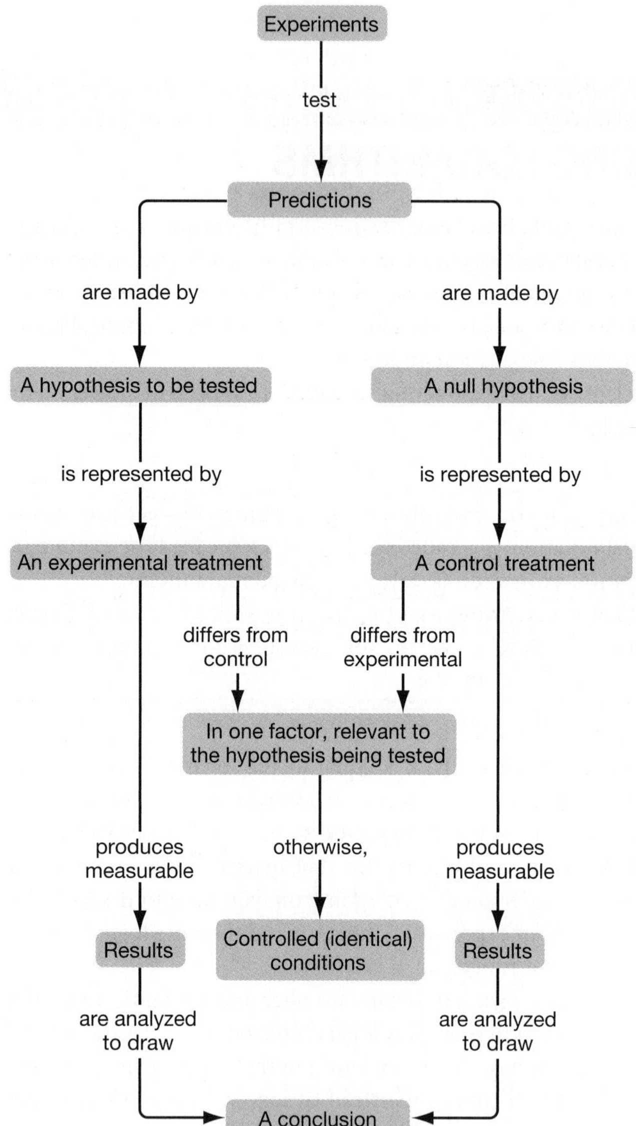

FIGURE B8.1 A Concept Map on Principles of Experimental Design.

BIOSKILLS 9

SEPARATING AND VISUALIZING MOLECULES

To study a molecule, you have to be able to isolate it. This is a two-step process: the molecule has to be separated from other molecules in a mixture, and then physically picked out or located in a purified form. This BioSkill focuses on the techniques that biologists use to separate nucleic acids and proteins and then find the particular one they are interested in.

Using Electrophoresis to Separate Molecules

In molecular biology, the standard technique for separating proteins and nucleic acids is called gel electrophoresis or, simply, electrophoresis (literally "electricity-moving"). You may be using electrophoresis in a lab for this course, and you will certainly be analyzing data derived from electrophoresis in this text.

The principle behind electrophoresis is simple. Both proteins and nucleic acids carry a charge. As a result, these molecules move when placed in an electric field. Negatively charged molecules move toward the positive electrode; positively charged molecules move toward the negative electrode.

To separate a mixture of macromolecules so that each can be isolated and analyzed, researchers place the sample in a gelatinous substance. More specifically, the sample is placed in a "well"—a slot in a sheet or slab of the gelatinous substance. The "gel" itself consists of long molecules that form a matrix of fibers. The presence of the fibers keeps molecules in the sample from moving around randomly, but the gelatinous matrix also has pores through which the molecules can pass.

When an electrical field is applied across the gel, the molecules in the well move through the gel toward an electrode. Molecules that are smaller or more highly charged for their size move faster than do larger or less highly charged molecules. As they move, then, the molecules separate by size and by charge. Small and highly charged molecules end up at the bottom of the gel; large, less-charged molecules remain near the top.

An Example "Run"

Figure B9.1 shows the electrophoresis setup used in an experiment investigating how RNA molecules polymerize In this case, the investigators wanted to document how long RNA molecules became over time, when ribonucleotide triphosphates were present in a particular type of solution.

Step 1 shows how they loaded samples of macromolecules, taken on different days during the experiment, into wells at the top of the gel slab. This is a general observation: Each well holds a different sample. In this and many other cases, the researchers also filled a well with a sample containing fragments of known size, called a size standard or "ladder."

In step 2 the researchers immersed the gel in a solution that conducts electricity and applied a voltage across the gel. The molecules in each well started to run down the gel, forming a lane. After several hours of allowing the molecules to move, they removed the electric field (step 3). By then, molecules of different size and charge had separated from one another. In this case,

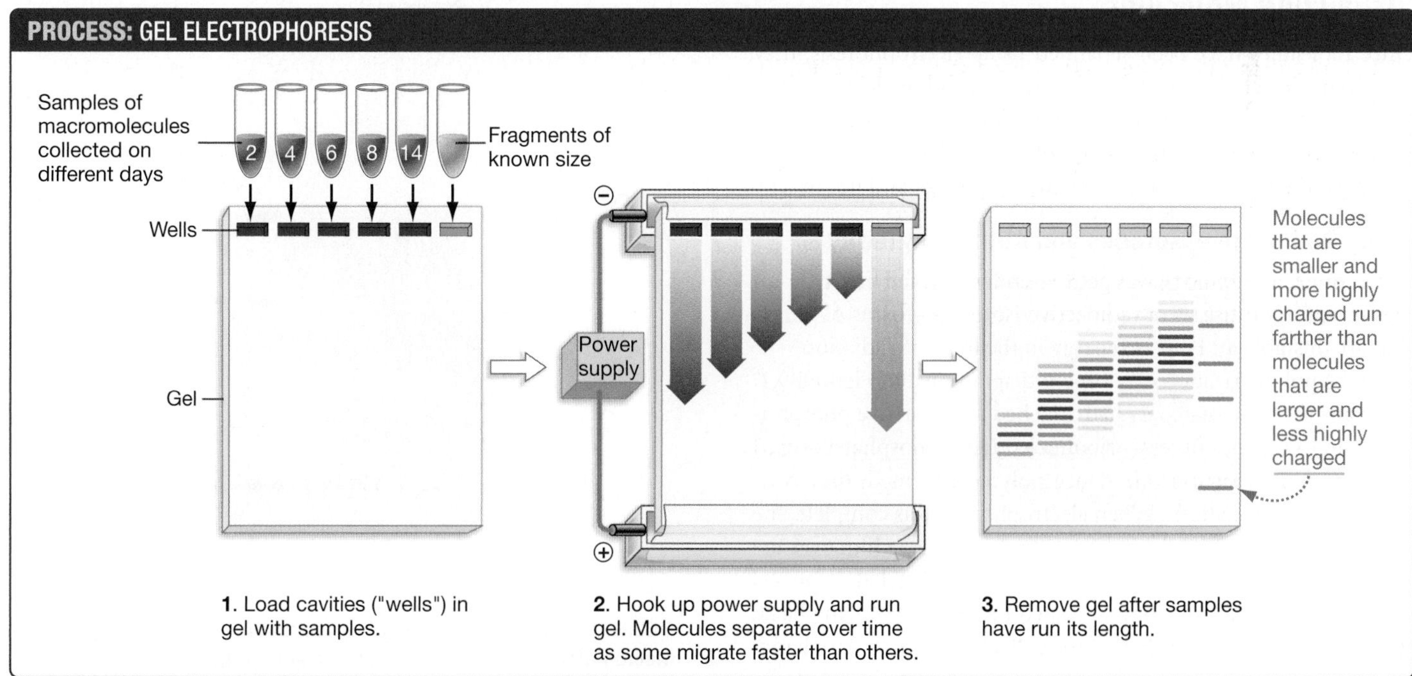
PROCESS: GEL ELECTROPHORESIS
Samples of macromolecules collected on different days
2
4
6
8
14
Fragments of known size
Wells
Gel
Power supply
Molecules that are smaller and more highly charged run farther than molecules that are larger and less highly charged
1. Load cavities ("wells") in gel with samples.
2. Hook up power supply and run gel. Molecules separate over time as some migrate faster than others.
3. Remove gel after samples have run its length.

FIGURE B9.1 Macromolecules Can Be Separated via Gel Electrophoresis.

✔ **QUESTION** DNA and RNA run toward the positive electrode. Why are these molecules negatively charged?

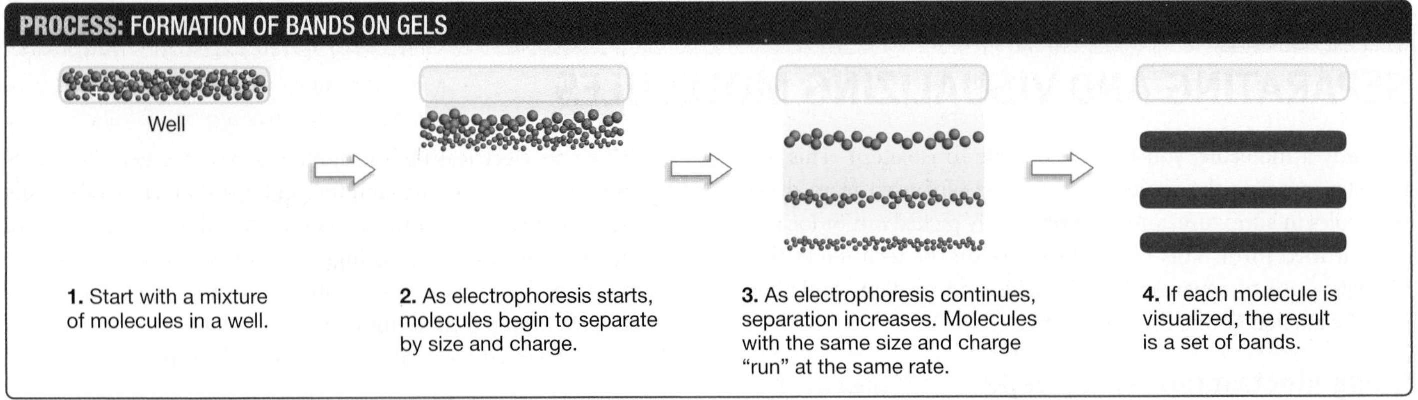

PROCESS: FORMATION OF BANDS ON GELS
Well
1. Start with a mixture of molecules in a well.
2. As electrophoresis starts, molecules begin to separate by size and charge.
3. As electrophoresis continues, separation increases. Molecules with the same size and charge "run" at the same rate.
4. If each molecule is visualized, the result is a set of bands.

FIGURE B9.2 On a Gel, Alike Molecules Form Bands.

small RNA molecules had reached the bottom of the gel. Above them were larger RNA molecules, which had run more slowly.

Why Do Separated Molecules Form Bands?

When researchers visualize a particular molecule on a gel, using techniques described below, the image that results consists of bands: shallow lines that are as wide as a lane in the gel. Why?

To understand the answer, study **Figure B9.2**. The top panel shows the original mixture of molecules. In this cartoon, the size of each dot represents the size of each molecule. The key is to realize that the original sample contains many copies of each specific molecule, and that these copies run down the length of the gel together—meaning, at the same rate—because they have the same size and charge.

It's that simple: Alike molecules form a band because they stay together.

Visualizing Molecules

Once molecules have been separated using electrophoresis, they have to be detected. Unfortunately, proteins and nucleic acids are invisible unless they are tagged in some way. Let's consider two of the most common tagging systems, then consider how researchers can tag and visualize specific molecules of interest and not others.

Using Radioactive Isotopes and Autoradiography

When molecular biology was getting under way, the first types of tags in common use were radioactive isotopes—forms of atoms that are unstable and release energy in the form of radiation.

In the polymerization experiment diagrammed in Figure B9.1, for example, the researchers had attached a radioactive phosphorus atom to the monomers—ribonucleotide triphosphates—used in the original reaction mix. Once polymers formed, they contained radioactive atoms. When electrophoresis was complete, the investigators visualized the polymers by laying X-ray film over the gel. Because radioactive emissions expose film, a black dot appears wherever a radioactive atom is located in the gel. So many black dots occur so close together that the collection forms a dark band.

This technique for visualizing macromolecules is called autoradiography. The autoradiograph that resulted from the polymerization experiment is shown in **Figure B9.3**. The samples, taken on days 2, 4, 6, 8, and 14 of the experiment, are labeled along the bottom. The far right lane contains macromolecules of known size; this lane is used to estimate the size of the molecules in the experimental samples. The bands that appear in each sample lane represent the different polymers that had formed.

Reading a Gel

One of the keys to interpreting or "reading" a gel is to realize that darker bands contain more radioactive markers, indicating the presence of many radioactive molecules. Lighter bands contain fewer molecules.

To read a gel, then, you look for (**1**) the presence or absence of bands in some lanes—meaning, some experimental samples—versus others, and (**2**) contrasts in the darkness of the bands—meaning, differences in the number of molecules present.

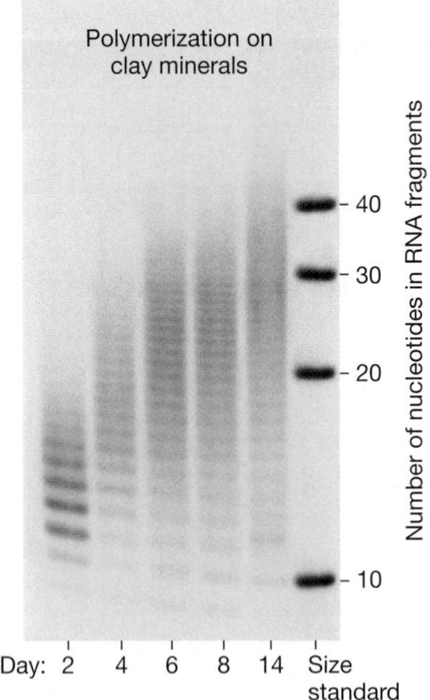

FIGURE B9.3 Autoradiography Is a Technique for Visualizing Macromolecules. The molecules in a gel can be visualized in a number of ways. In this case, the RNA molecules in the gel exposed an X-ray film because they had radioactive atoms attached. When developed, the film is called an autoradiograph.

For example, several conclusions can be drawn from the data in Figure B9.3. First, a variety of polymers formed at each stage. After the second day, for example, polymers from 12 to 18 monomers long had formed on the clay particles used in this experiment. Second, the overall length of polymers produced increased with time. At the end of the fourteenth day, most of the RNA molecules were between 20 and 40 monomers long.

Starting in the late 1990s and early 2000s, it became much more common to tag nucleic acids with fluorescent tags. Once electrophoresis is complete, the presence of fluorescence can be detected by exposing the gel to an appropriate wavelength of light; the fluorescent tag fluoresces or glows in response (fluorescence is explained in Chapter 10).

Fluorescent tags have important advantages over radioactive isotopes: (**1**) They are safer to handle. (**2**) They are faster—you don't have to wait hours or days for the radioactive isotope to expose a film. (**3**) They come in multiple colors, so you can tag several different molecules in the same experiment and detect them independently.

Using Nucleic Acid Probes

In many cases, researchers want to find one specific molecule—a certain DNA sequence, for example—in the collection of molecules on a gel. How is this possible? The answer hinges on using a particular molecule as a probe.

Chapter 19 explains how probes work in detail. Here it's enough to get the general idea: A probe is a marked molecule that binds specifically to your molecule of interest. The "mark" is often a radioactive atom or a fluorescent tag.

If you are looking for a particular DNA or RNA sequence on a gel, for example, you can expose the gel to a single-stranded probe that binds to the target sequence by complementary base pairing. Once it has bound, you can detect the band through autoradiography or fluorescence.

- Southern blotting is a technique for making DNA fragments that have been run out on a gel single-stranded, transferring them from the gel to a nylon filter, and then probing them to identify segments of interest. The technique was named after its inventor, Edwin Southern.
- Northern blotting is a technique for transferring RNA fragments from a gel to a nylon filter, and then probing them to detect target segments. The name is a lighthearted play on Southern blotting—the protocol from which it was derived.

Using Antibody Probes

How can researchers find a particular protein out of a large collection of different proteins? The answer is to use an antibody. An antibody is a protein that binds specifically to a section of a different protein. The structure of antibodies and their function in the immune system are explored in detail in Chapter 49.

To use an antibody as a probe, investigators attach a tag molecule—often an enzyme that catalyzes a color-forming reaction—to the antibody and allow it to react with proteins in a mixture. The antibody will stick to the specific protein that it binds to, and then can be visualized thanks to the tag it carries.

If the proteins in question have been separated by gel electrophoresis, the result is called a Western blot. The name Western is an extension of the Southern and Northern pattern.

✔Question

Suppose you've been given a gel that has been stained for "RNA X." One lane contains no bands. Two lanes have a band in the same location, even though one of the bands is barely visible and the other is extremely dark. The fourth lane has a light band located above the bands in the other lanes. Interpret the gel.

BIOSKILLS 10

BIOLOGICAL IMAGING: MICROSCOPY AND X-RAY CRYSTALLOGRAPHY

A lot of biology happens at levels that can't be detected with the naked eye. Biologists use an array of microscopes to study small multicellular organisms, individual cells, and the contents of cells. And to understand what individual macromolecules or multimolecular machines like ribosomes look like, researchers use data from a technique called X-ray crystallography.

You'll probably use dissecting microscopes and compound light microscopes to view specimens during your labs for this course, and throughout this text you'll be seeing images generated from other types of microscopy and from X-ray crystallographic data. One of the fundamental skills you'll be acquiring as an introductory student, then, is a basic understanding of how these techniques work. The key is to recognize that each approach for visualizing microscopic structures has strengths and weaknesses. As a result, each technique is appropriate for studying certain types or aspects of cells or molecules.

Light Microscopy

If you use a dissecting microscope during labs, you'll recognize that it works by magnifying light that bounces off a whole specimen—often a live organism. You'll be able to view the specimen in three dimensions, which is why these instruments are sometimes called stereomicroscopes, but the maximum magnification possible is only about 20 to 40 times normal size (20× to 40×).

To view smaller objects, you'll probably use a compound microscope. Compound microscopes magnify light that is passed *through* a specimen. The instruments used in introductory labs are usually capable of 400× magnifications; the most sophisticated compound microscopes available can achieve magnifications of about 2000×. This is enough to view individual bacterial or eukaryotic cells and see large structures inside cells, like condensed chromosomes (see Chapter 11). To prepare a specimen for viewing under a compound light microscope, the tissues or cells are usually sliced to create a section thin enough for light to pass through efficiently. The section is then dyed to increase contrast and make structures visible. In many cases, different types of dyes are used to highlight different types of structures.

Electron Microscopy

Until the 1950s, the compound microscope was the biologist's only tool for viewing cells directly. But the invention of the electron microscope provided a new way to view specimens. Two basic types of electron microscopy are now available: one that allows researchers to examine cross sections of cells at extremely high magnification, and one that offers a view of surfaces at somewhat lower magnification.

Transmission Electron Microscopy (TEM)

The transmission electron microscope is an extraordinarily effective tool for viewing cell structure at high magnification. TEM forms an image from electrons that pass through a specimen, just as a light microscope forms an image from light rays that pass through a specimen.

Biologists who want to view a cell under a transmission electron microscope begin by "fixing" the cell, meaning that they treat it with a chemical agent that stabilizes the cell's structure and contents while disturbing them as little as possible. Then the researcher permeates the cell with an epoxy plastic that stiffens the structure. Once this epoxy hardens, the cell can be cut into extremely thin sections with a glass or diamond knife. Finally, the sectioned specimens are impregnated with a metal—often lead. (The reason for this last step is explained shortly.)

Figure B10.1a outlines how the transmission electron microscope works. A beam of electrons is produced by a tungsten filament at the top of a column and directed downward. (All of the air is pumped out of the column, so that the electron beam isn't scattered by collisions with air molecules.) The electron beam passes through a series of lenses and through the specimen. The lenses are actually electromagnets, which alter the path of the beam much like a glass lens in a dissecting or compound microscope bends light. The lenses magnify and focus the image on a screen at the bottom of the column. There the electrons strike a coating of fluorescent crystals, which emit visible light in response—just like a television screen. When the microscopist moves the screen out of the way and allows the electrons to expose a sheet of black-and-white film, the result is a micrograph—a photograph of an image produced by microscopy.

The image itself is created by electrons that pass through the specimen. If no specimen were in place, all the electrons would pass through and the screen (and micrograph) would be uniformly bright. Unfortunately, cell materials by themselves would also appear fairly uniform and bright. This is because an atom's ability to deflect an electron depends on its mass. In turn, an atom's mass is a function of its atomic number. The hydrogen, carbon, oxygen, and nitrogen atoms that dominate biological molecules have low atomic numbers. This is why cell biologists must saturate cell sections with lead solutions. Lead has a high atomic number and scatters electrons effectively. Different macromolecules take up lead atoms in different amounts, so the

(a) Transmission electron microscopy: High magnification of cross sections

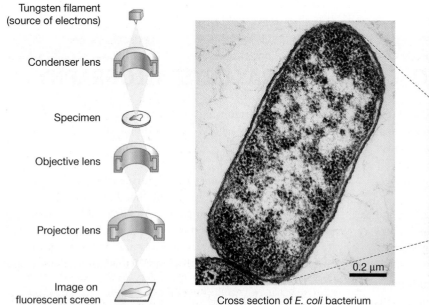

Tungsten filament (source of electrons)

Condenser lens

Specimen

Objective lens

Projector lens

Image on fluorescent screen

0.2 µm

Cross section of *E. coli* bacterium

(b) Scanning electron microscopy: Lower magnification of surfaces

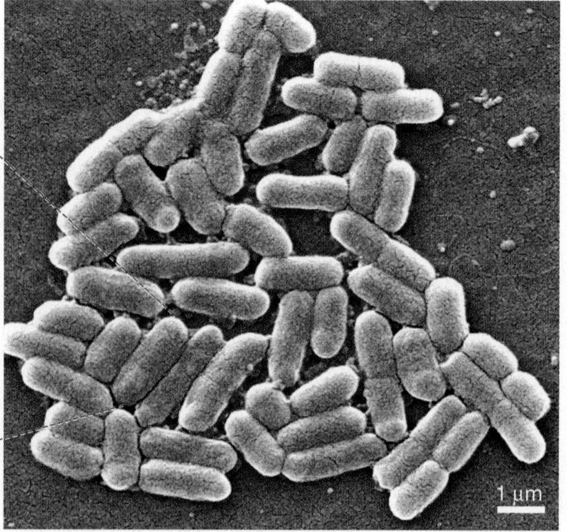

Surface view of *E. coli* bacteria

FIGURE B10.1 There Are Two Basic Types of Electron Microscopy.

metal acts as a "stain" that produces contrast. With TEM, areas of dense metal scatter the electron beam most, producing dark areas in micrographs.

The advantage of TEM is that it can magnify objects up to 250,000×—meaning that intracellular structures are clearly visible. The downsides are that researchers are restricted to observing dead, sectioned material, and they must take care that the preparation process does not distort the specimen.

Scanning Electron Microscopy (SEM)

The scanning electron microscope is the most useful tool biologists have for looking at the surfaces of structures. Materials are prepared for scanning electron microscopy by coating their surfaces with a layer of metal atoms. To create an image of this surface, the microscope scans the surface with a narrow beam of electrons. Electrons that are reflected back from the surface or that are emitted by the metal atoms in response to the beam then strike a detector. The signal from the detector controls a second electron beam, which scans a TV-like screen and forms an image magnified up to 50,000 times the object's size.

Because SEM records shadows and highlights, it provides images with a three-dimensional appearance (**Figure B10.1b**). It cannot magnify objects nearly as much as TEM can, however.

Studying Live Cells and Real-Time Processes

Until the 1960s, it was not possible for biologists to get clear, high-magnification images of living cells. But a series of innovations over the past 50 years has made it possible to observe organelles and subcellular structures in action.

The development of video microscopy, where the image from a light microscope is captured by a video camera instead of by an eye or a film camera, proved revolutionary. It allowed specimens to be viewed at higher magnification, because video cameras are more sensitive to small differences in contrast than are the human eye or still cameras. It also made it easier to keep live specimens functioning normally, because the increased light sensitivity of video cameras allows them to be used with low illumination, so specimens don't overheat. And when it became possible to digitize video images, researchers began using computers to remove out-of-focus background material and increase image clarity.

A more recent innovation was the use of a fluorescent molecule called green fluorescent protein, or GFP, which allows researchers to tag specific molecules or structures and follow their movement over time. GFP is naturally synthesized in jellyfish that fluoresce, or emit light. By affixing GFP molecules to another protein and then inserting it into a cell, investigators can follow the protein's fate over time and even videotape its movement. For example, researchers have videotaped GFP-tagged proteins being transported from the rough ER through the Golgi apparatus and out to the plasma membrane. This is cell biology: the movie.

GFP's influence has been so profound that the researchers who developed its use in microscopy were awarded the 2008 Nobel prize in Chemistry.

Visualizing Structures in 3-D

The world is three-dimensional. To understand how microscopic structures and macromolecules work, it is essential to understand their shape and spatial relationships. Consider three techniques currently being used to reconstruct the 3-D structure of cells, organelles, and macromolecules.

- *Confocal microscopy* is carried out by mounting cells that have been treated with one or more fluorescing tags on a microscope slide and then focusing a beam of ultraviolet light at a specific depth within the specimen. The fluorescing tag emits visible light in response. A detector for this light is then set up at exactly the position where the emitted light comes into focus. The result is a sharp image of a precise plane in the cell being studied (**Figure B10.2**). By altering the focal plane, a researcher can record images from an array of depths in the specimen; a computer can then be used to generate a 3-D image of the cell.
- *Electron tomography* uses a transmission electron microscope to generate a 3-D image of an organelle or other subcellular structure. The specimen is rotated around a single axis, with the researcher taking many "snapshots." The individual images are then pieced together with a computer. This technique has provided a much more accurate view of mitochondrial structure than was possible using traditional TEM (see Chapter 7).
- *X-ray crystallography, or X-ray diffraction analysis,* is the most widely used technique for reconstructing the 3-D structure of molecules. As its name implies, the procedure is based on bombarding crystals of a molecule with X-rays. X-rays are scattered in precise ways when they interact with the electrons surrounding the atoms in a crystal, producing a diffraction pattern that can be recorded on X-ray film or other types of

(a) Conventional fluorescence image of single cell

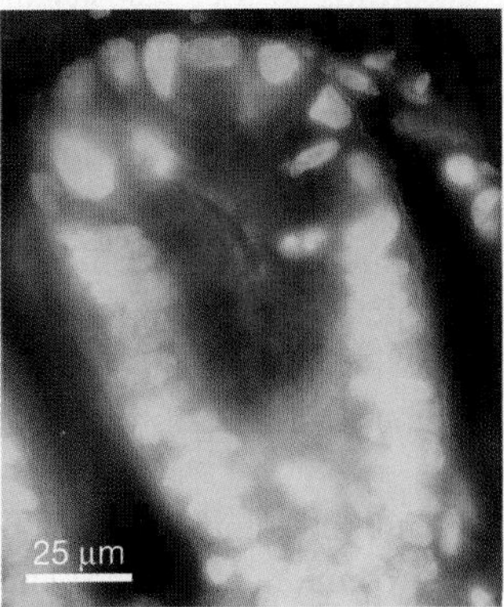

(b) Confocal fluorescence image of same cell

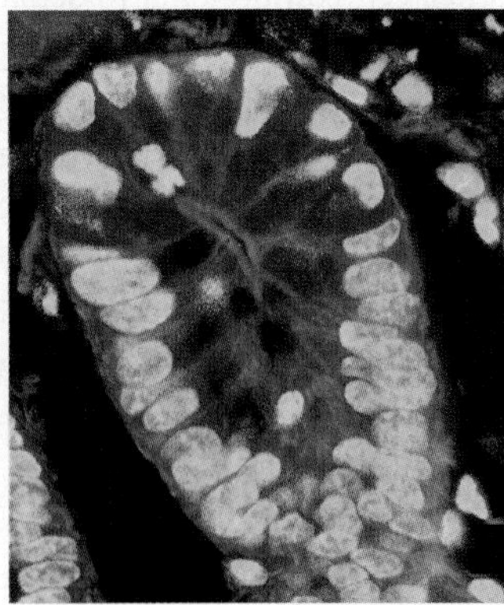

FIGURE B10.2 Confocal Microscopy Provides Sharp Images of Living Cells. **(a)** The conventional image of this mouse intestinal cell is blurred, because it results from light emitted by the entire cell. **(b)** The confocal image is sharp, because it results from light emitted at a single plane inside the cell.

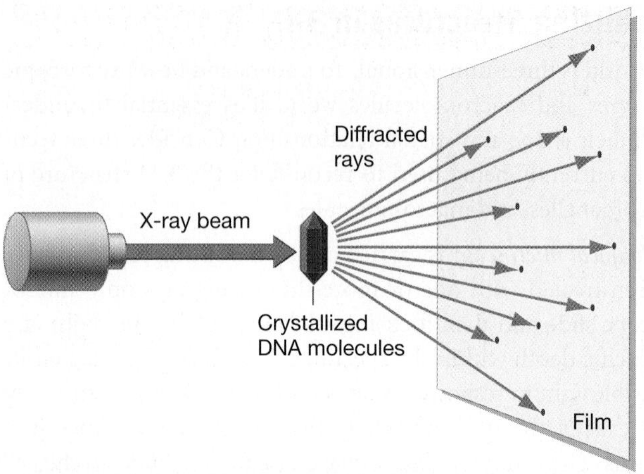

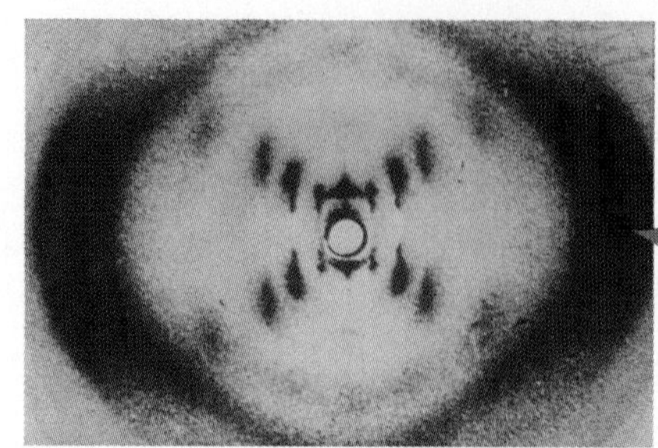

FIGURE B10.3 X-Ray Crystallography. When crystallized molecules are bombarded with X-rays, the radiation is scattered in distinctive patterns. The photograph at the right shows an X-ray film that recorded the pattern of scattered radiation from DNA molecules.

detectors (**Figure B10.3**). By varying the orientation of the X-ray beam as it strikes a crystal and documenting the diffraction patterns that result, researchers can construct a map representing the density of electrons in the crystal. By relating these electron-density maps to information about the primary structure of the nucleic acid or protein, a 3-D model of the molecule can be built. Virtually all of the molecular models used in this book were built from X-ray crystallographic data.

✓Questions

1. Suppose you are looking at a transmission electron micrograph of a cancerous cell from the human liver. No mitochondria are present. Does this mean that the cell lacks mitochondria?
2. X-ray crystallography is time consuming and technically difficult. Why is the effort to understand the structure of biological molecules worthwhile? What's the payoff?

BIOSKILLS 11

SEPARATING CELL COMPONENTS BY CENTRIFUGATION

Biologists use a technique called differential centrifugation to isolate specific cell components. Differential centrifugation is based on breaking cells apart to create a complex mixture and then separating components in a centrifuge. A centrifuge accomplishes this by spinning cells in a solution that allows molecules and other cell components to separate according to their density or size and shape. The individual parts of the cell can then be purified and studied in detail, in isolation from other parts of the cell.

The first step in preparing a cell sample for centrifugation is to release the organelles and cell components by breaking the cells apart. This can be done by putting them in a hypotonic solution, by exposing them to high-frequency vibration, by treating cells with a detergent, or by grinding them up. Each of these methods breaks apart plasma membranes and releases the contents of the cells.

The resulting pieces of plasma membrane quickly reseal to form small vesicles, often trapping cell components inside. The solution that results from the homogenization step is a mixture of these vesicles, free-floating macromolecules released from the cells, and organelles. A solution such as this is called a cell extract or cell homogenate.

When a cell homogenate is placed in a centrifuge tube and spun at high speed, the components that are in solution tend to move outward, along the red arrow in **Figure B11.1a**. The effect is similar to a merry-go-round, which seems to push you outward in a straight line away from the spinning platform. In response to this outward-directed force, the solution containing the cell homogenate exerts a centripetal (literally, "center-seeking") force that pushes the homogenate away from the bottom of the tube. Larger, denser molecules or particles resist this inward force more readily than do smaller, less dense ones and so reach the bottom of the centrifuge tube faster.

To separate the components of a cell extract, researchers often perform a series of centrifuge runs. Steps 1 and 2 of **Figure B11.1b** illustrate how an initial treatment at low speed causes larger, heavier parts of the homogenate to move below smaller, lighter parts. The material that collects at the bottom of the tube is called the pellet, and the solution and solutes left behind form the supernatant ("above swimming"). The supernatant is placed in a fresh tube and centrifuged at increasingly higher speeds and longer durations. Each centrifuge run continues to separate cell components based on their size and density.

To accomplish even finer separation of macromolecules or organelles, researchers frequently follow up with centrifugation at extremely high speeds. One strategy is based on filling the centrifuge tube with a series of sucrose solutions of increasing

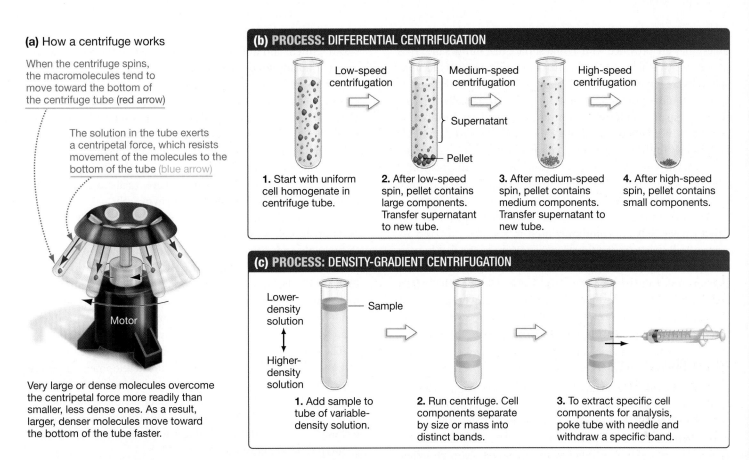

(a) How a centrifuge works

When the centrifuge spins, the macromolecules tend to move toward the bottom of the centrifuge tube (red arrow)

The solution in the tube exerts a centripetal force, which resists movement of the molecules to the bottom of the tube (blue arrow)

Motor

Very large or dense molecules overcome the centripetal force more readily than smaller, less dense ones. As a result, larger, denser molecules move toward the bottom of the tube faster.

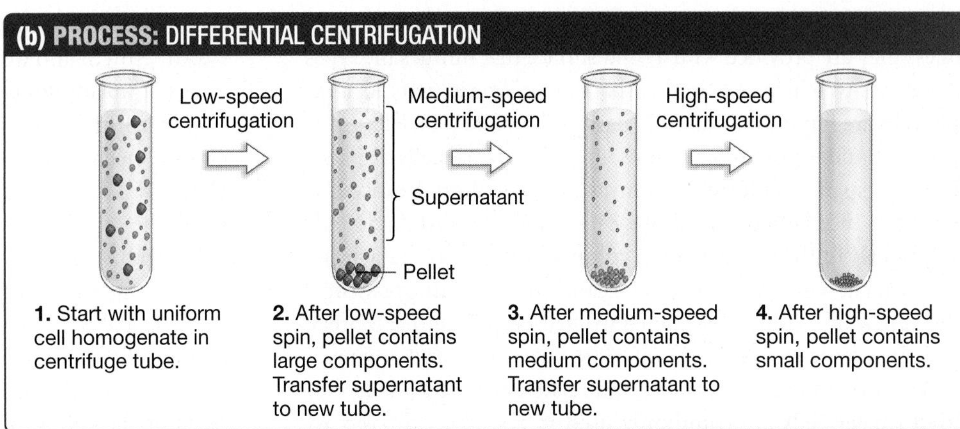

(b) PROCESS: DIFFERENTIAL CENTRIFUGATION
Low-speed centrifugation
Medium-speed centrifugation
High-speed centrifugation
Supernatant
Pellet
1. Start with uniform cell homogenate in centrifuge tube.
2. After low-speed spin, pellet contains large components. Transfer supernatant to new tube.
3. After medium-speed spin, pellet contains medium components. Transfer supernatant to new tube.
4. After high-speed spin, pellet contains small components.

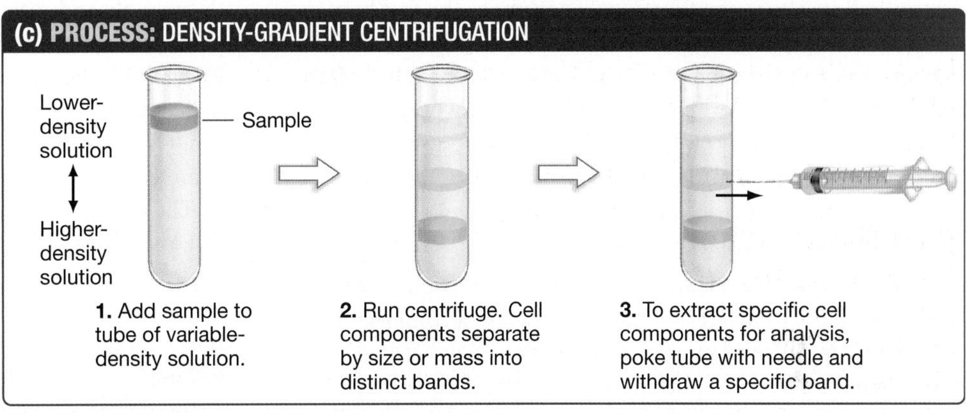

(c) PROCESS: DENSITY-GRADIENT CENTRIFUGATION
Lower-density solution
Sample
Higher-density solution
1. Add sample to tube of variable-density solution.
2. Run centrifuge. Cell components separate by size or mass into distinct bands.
3. To extract specific cell components for analysis, poke tube with needle and withdraw a specific band.

FIGURE B11.1 Cell Components Can Be Separated by Centrifugation. (a) The forces inside a centrifuge tube allow cell components to be separated. **(b)** Through a series of centrifuge runs made at increasingly higher speeds, an investigator can separate fractions of a cell homogenate by size via differential centrifugation. **(c)** A high-speed centrifuge run can achieve extremely fine separation among cell components by density-gradient centrifugation.

density (**Figure B11.1c**). The density gradient allows cell components to separate on the basis of small differences in size, shape, and density. When the centrifuge run is complete, each cell component occupies a distinct band of material in the tube, based on how quickly each moves through the increasingly dense gradient of sucrose solution during the centrifuge run. A researcher can then collect the material in each band for further study.

✓ Questions

1. Electrophoresis separates molecules by charge and size. What is the physical basis for separating molecules or cell components via centrifugation?
2. Compared to other types of centrifugation, why would extremely high-speed centrifugation allow similar molecules to separate?

BIOSKILLS 12

CELL AND TISSUE CULTURE METHODS

For researchers, there are important advantages to growing plant and animal cells and tissues outside the organism itself. Cell and tissue cultures provide large populations of a single type of cell or tissue and the opportunity to control experimental conditions precisely.

Animal Cell Culture

The first successful attempt to culture animal cells occurred in 1907, when a researcher cultivated amphibian nerve cells in a drop of fluid from the spinal cord. But it was not until the 1950s and 1960s that biologists could routinely culture plant and animal cells in the laboratory. The long lag time was due to the difficulty of re-creating conditions that exist in the intact organism precisely enough for cells to grow normally.

To grow in culture, animal cells must be provided with a liquid mixture containing the nutrients, vitamins, and hormones that stimulate growth. Initially, this mixture was serum, the liquid portion of blood; now serum-free media are available for certain cell types. Serum-free media are preferred because they are much more precisely defined chemically than serum.

In addition, many types of animal cells will not grow in culture unless they are provided with a solid surface that mimics the types of surfaces to which cells in the intact organisms adhere. As a result, cells are typically cultured in flasks (**Figure B12.1a**, left).

Even under optimal conditions, though, normal cells display a finite life span in culture. In contrast, many cultured cancerous cells grow indefinitely. In culture, cancerous cells also do not adhere tightly to the surface of the culture flask. These characteristics correlate with two features of cancerous cells in organisms: They grow in a continuous, uncontrolled fashion and can break away from the original site to infiltrate new tissues and organs.

Because of their immortality and relative ease of growth, cultured cancer cells are commonly used in research on basic aspects of cell structure and function. For example, the first human cell type to be grown in culture was isolated in 1951 from a malignant tumor of the uterine cervix. These cells are called HeLa cells in honor of their donor, Henrietta Lacks, who died soon thereafter from her cancer. HeLa cells continue to grow in laboratories around the world (Figure B12.1a, right).

Plant Tissue Culture

Certain cells found in plants are totipotent—meaning that they retain the ability to divide and differentiate into a complete, mature plant, including new types of tissue. These cells, called parenchyma cells, are important in wound healing and asexual reproduction. But they also allow researchers to grow complete adult plants in the laboratory, starting with a small number of parenchyma cells.

Biologists who grow plants in tissue culture begin by placing parenchyma cells in a liquid or solid medium containing all the nutrients required for cell maintenance and growth. In the early days of plant tissue culture, investigators found not only that specific growth signals called hormones were required for successful growth and differentiation but also that the relative abundance of hormones present was critical to success.

The earliest experiments on hormone interactions in tissue cultures were done with tobacco cells in the 1950s by Folke Skoog and co-workers. These researchers found that when the hormone called auxin was added to the culture by itself, the cells enlarged but did not divide. But if the team added roughly equal amounts of auxin and another growth signal called cytokinin to the cells, the cells began to divide and eventually formed a callus, or an undifferentiated mass of parenchyma cells.

By varying the proportion of auxin to cytokinins in different parts of the callus and through time, the team could stimulate the growth and differentiation of root and shoot systems and produce whole new plants (**Figure B12.1b**). A high ratio of auxin to cytokinin led to the differentiation of a root system, while a high ratio of cytokinin to auxin led to the development of a shoot system. Eventually Skoog's team was able to produce a complete plant from just one parenchyma cell.

The ability to grow whole new plants in tissue culture from just one cell has been instrumental in the development of genetic engineering (see Chapter 19). Researchers insert recombinant genes into target cells, test the cells to identify those that successfully express the recombinant genes, and then use tissue culture techniques to grow those cells into adult individuals with a novel genotype and phenotype.

✔Questions

1. What is a limitation of how experiments on HeLa cells are interpreted?
2. State a disadvantage of doing experiments on plants that have been propagated from single cells growing in tissue culture.

(a) Animal cell culture: immortal HeLa cancer cells

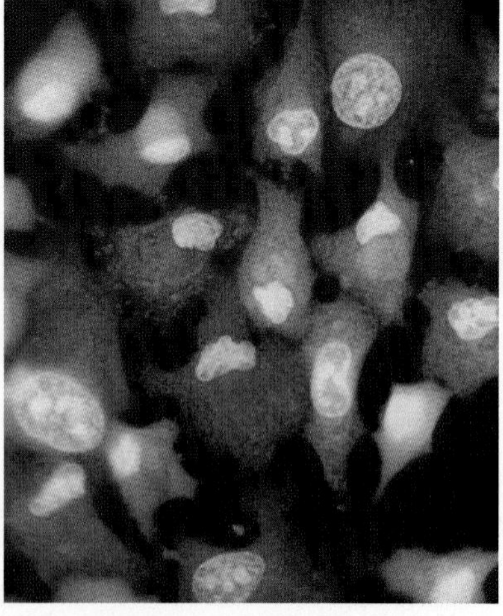

(b) Plant tissue culture: tobacco callus

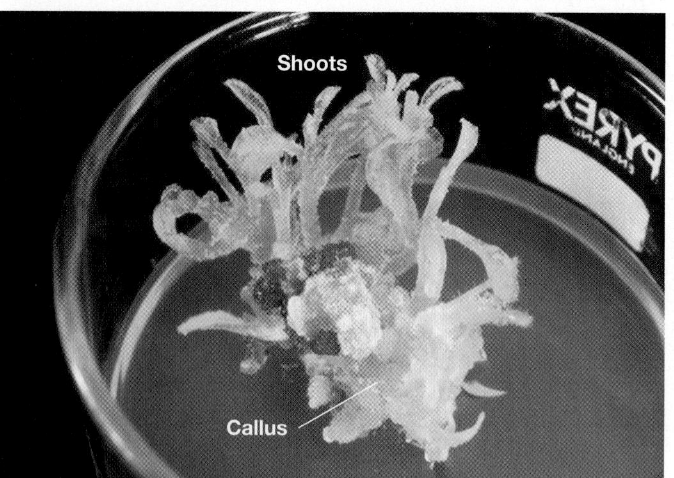

FIGURE B12.1 Animal and Plant Cells Can Be Grown in the Lab.

BIOSKILLS 13

COMBINING PROBABILITIES

In several cases in this text, you'll need to combine probabilities from different events in order to solve a problem. One of the most common applications is in genetics problems. For example, Punnett squares work because they are based on two fundamental rules of probability. Each rule pertains to a distinct situation.

The Both-And Rule

The both-and rule—also known as the product rule or multiplication rule—applies when you want to know the probability that two or more independent events occur together. Let's use the rolling of two dice as an example. What is the probability of rolling two sixes? These two events are independent, because the probability of rolling a six on one die has no effect on the probability of rolling a six on the other die. (In the same way, the probability of getting a gamete with allele *R* from one parent has no effect on the probability of getting a gamete with allele *R* from the other parent. Gametes fuse randomly.)

The probability of rolling a six on the first die is 1/6. The probability of rolling a six on the second die is also 1/6. The probability of rolling a six on *both* dice, then, is $1/6 \times 1/6 = 1/36$. In other words, if you rolled two dice 36 times, on average you would expect to roll two sixes once.

In the case of a cross between two parents heterozygous at the *R* gene, the probability of getting allele *R* from the father is 1/2 and the probability of getting *R* from the mother is 1/2. Thus, the probability of getting both alleles and creating an offspring with genotype *RR* is $1/2 \times 1/2 = 1/4$.

The Either-Or Rule

The either-or rule—also known as the sum rule or addition rule—applies when you want to know the probability of an event happening when there are several different ways for the same event or outcome to occur. In this case, the probability that the event will occur is the sum of the probabilities of each way that it can occur.

For example, suppose you wanted to know the probability of rolling either a one or a six when you toss a die. The probability of drawing each is 1/6, so the probability of getting one or the other is $1/6 + 1/6 = 1/3$. If you rolled a die three times, on average you'd expect to get a one or a six once.

In the case of a cross between two parents heterozygous at the *R* gene, the probability of getting an *R* allele from the father and an *r* allele from the mother is $1/2 \times 1/2 = 1/4$. Similarly, the probability of getting an *r* allele from the father and an *R* allele from the mother is $1/2 \times 1/2 = 1/4$. Thus, the combined probability of getting the *Rr* genotype in either of the two ways is $1/4 + 1/4 = 1/2$.

✓Questions

1. Suppose that four students each toss a coin. What is the probability of four "tails"?
2. After a single roll of a die, what is the probability of getting either a two, a three, or a six?

BIOSKILLS 14

MODEL ORGANISMS

Research in biological science starts with a question. In most cases, the question is inspired by an observation about a cell or an organism. To answer it, biologists have to study a particular species. Study organisms are often called model organisms, because investigators hope that they serve as a model for what is going on in a wide array of species.

Model organisms are chosen because they are convenient to study and because they have attributes that make them appropriate for the particular research proposed. They tend to have some common characteristics:

- ***Short generation time and rapid reproduction*** This is important because it makes it possible to produce offspring quickly and perform many experiments in a short amount of time—you don't have to wait long for individuals to grow.
- ***Large numbers of offspring*** This is particularly important in genetics, where many offspring phenotypes and genotypes need to be assessed to get a large sample size.
- ***Small size, simple feeding and habitat requirements*** These attributes make it relatively cheap and easy to maintain individuals in the lab.

The following notes highlight just a few model organisms supporting current work in biological science.

Escherichia coli

Of all model organisms in biology, perhaps none has been more important than the bacterium *Escherichia coli*—a common inhabitant of the human gut. The strain that is most commonly worked on today, called K-12 (**Figure B14.1a** on page B:20), was originally isolated from a hospital patient in 1922.

During the last half of the twentieth century, key results in molecular biology originated in studies of *E. coli*. These results include the discovery of enzymes such as DNA polymerase, RNA polymerase, DNA repair enzymes, and restriction endonucleases; the elucidation of ribosome structure and function; and the initial characterization of promoters, regulatory transcription factors, regulatory sites in DNA, and operons. In many cases, initial discoveries made in *E. coli* allowed researchers to confirm that homologous enzymes and processes existed in an array of organisms, often ranging from other bacteria to yeast, mice, and humans.

The success of *E. coli* as a model for other species inspired Jacques Monod's claim that "Once we understand the biology of *Escherichia coli*, we will understand the biology of an elephant." The genome of *E. coli* K-12 was sequenced in 1997, and the strain continues to be a workhorse in studies of gene function, biochemistry, and particularly biotechnology. Much remains to be learned, however. Despite over 60 years of intensive study, the function of about a third of the *E. coli* genome is still unknown.

In the lab, *E. coli* is usually grown in suspension culture, where cells are introduced to a liquid nutrient medium, or on plates containing agar—a gelatinous mix of polysaccharides. Under optimal growing conditions—meaning before cells begin to get crowded and compete for space and nutrients—a cell takes just 30 minutes on average to grow and divide. At this rate, a single cell can produce a population of over a million descendants in just 10 hours. Except for new mutations, all of the descendant cells are genetically identical.

Dictyostelium discoideum

The cellular slime mold *Dictyostelium discoideum* is not always slimy, and it is not a mold—meaning a type of fungus. Instead, it is an amoeba. Amoeba is a general term that biologists use to characterize a unicellular eukaryote that lacks a cell wall and is extremely flexible in shape. *Dictyostelium* has long fascinated biologists because it is a social organism. Independent cells sometimes aggregate to form a multicellular structure.

Under most conditions, *Dictyostelium* cells are haploid (n) and move about in decaying vegetation on forest floors or other habitats. They feed on bacteria by engulfing them whole. When these cells reproduce, they can do so sexually by fusing with another cell then undergoing meiosis, or asexually by mitosis, which is more common. If food begins to run out, the cells begin to aggregate. In many cases, tens of thousands of cells cohere to form a 2-mm-long mass called a slug (**Figure B14.1b**). (This is not the slug that is related to snails.)

After migrating to a sunlit location, the slug stops and individual cells differentiate according to their position in the slug. Some form a stalk; others form a mass of spores at the tip of the stalk. (A spore is a single cell that develops into an adult organism, but it is not formed from gamete fusion like a zygote.) The entire structure, stalk plus mass of spores, is called a fruiting body. Cells that form spores secrete a tough coat and represent a durable resting stage. The fruiting body eventually dries out, and the wind disperses the spores to new locations, where more food might be available.

Dictyostelium has been an important model organism for investigating questions about eukaryotes:

- Cells in a slug are initially identical in morphology but then differentiate into distinctive stalk cells and spores. Studying this process helped biologists better understand how cells in plant and animal embryos differentiate into distinct cell types.
- The process of slug formation has helped biologists study how animal cells move and how they aggregate as they form specific types of tissues.
- When *Dictyostelium* cells aggregate to form a slug, they stick to each other. The discovery of membrane proteins responsible for cell-cell adhesion helped biologists understand some of the general principles of multicellular life highlighted in Chapter 8.

Arabidopsis thaliana

In the early days of biology, the best-studied plants were agricultural varieties such as maize (corn), rice, and garden peas. When biologists began to unravel the mechanisms responsible for oxygenic photosynthesis in the early to mid-1900s, they relied on green algae that were relatively easy to grow and manipulate in the lab—often the unicellular species *Chlamydomonas reinhardii*—as an experimental subject.

Although crop plants and green algae continue to be the subject of considerable research, a new model organism emerged in the 1980s and now serves as the preeminent experimental subject in plant biology. That organism is *Arabidopsis thaliana*, commonly known as thale cress or wall cress (**Figure B14.1c**).

Arabidopsis is a member of the mustard family, or Brassicaceae, so it is closely related to radishes and broccoli. In nature it is a weed—meaning a species that is adapted to thrive in habitats where soils have been disturbed.

One of the most attractive aspects of working with *Arabidopsis* is that individuals can grow from a seed into a mature, seed-producing plant in just four to six weeks. Several other attributes make it an effective subject for study: It has just five chromosomes, has a relatively small genome with limited numbers of repetitive sequences, can self-fertilize as well as undergo cross-fertilization, can be grown in a relatively small amount of space and with a minimum of care in the greenhouse, and produces up to 10,000 seeds per individual per generation.

Arabidopsis has been instrumental in a variety of studies in plant molecular genetics and development, and it is increasingly popular in ecological and evolutionary studies. In addition, the entire genome of the species has now been sequenced, and studies have benefited from the development of an international "*Arabidopsis* community"—a combination of informal and formal associations of investigators who work on *Arabidopsis* and use regular meetings, e-mail, and the Internet to share data, techniques, and seed stocks.

Saccharomyces cerevisiae

When biologists want to answer basic questions about how eukaryotic cells work, they often turn to the yeast *Saccharomyces cerevisiae.*

S. cerevisiae is unicellular and relatively easy to culture and manipulate in the lab (**Figure B14.1d**). In good conditions, yeast cells grow and divide almost as rapidly as bacteria. As a result, the

(a) Bacterium *Escherichia coli* (strain K-12)

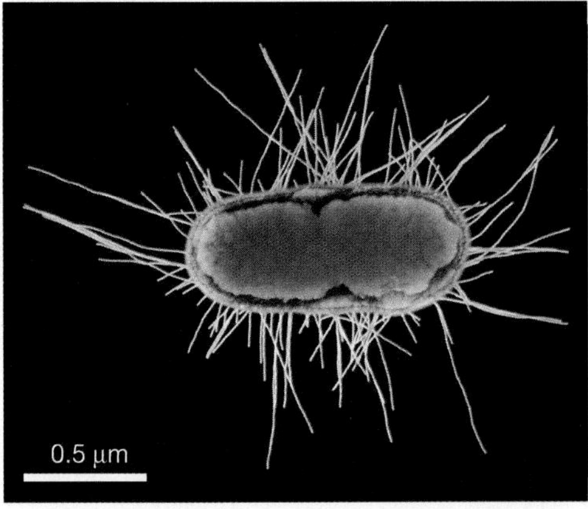

(b) Slime mold *Dictyostelium discoideum*

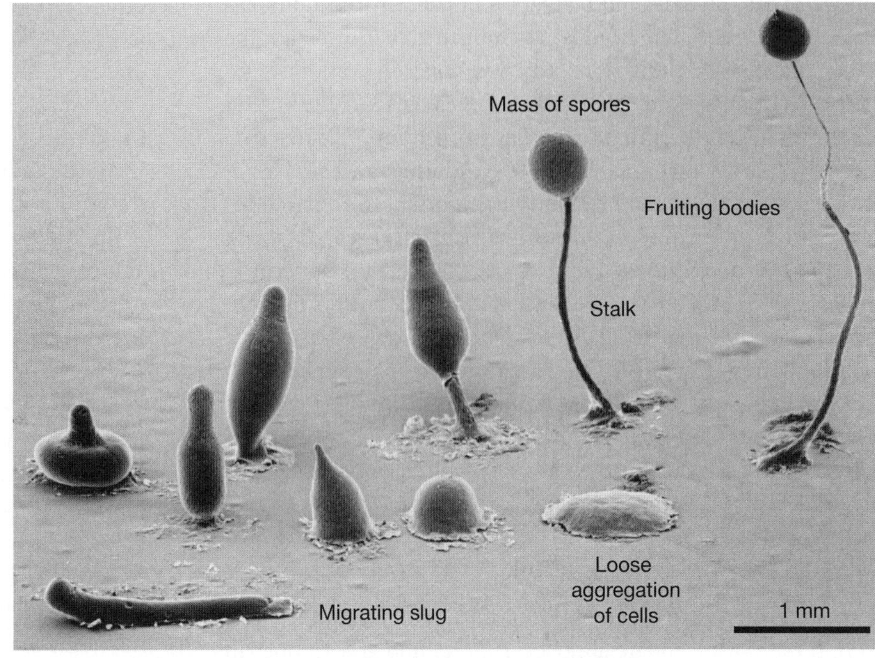

(c) Thale cress *Arabidopsis thaliana*

(e) Fruit fly *Drosophila melanogaster*

(f) Roundworm *Caenorhabditis elegans*

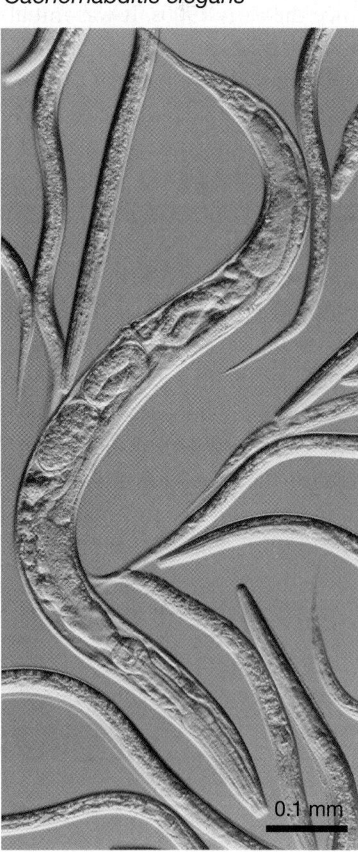

(d) Yeast *Saccharomyces cerevisiae*

(g) Mouse *Mus musculus*

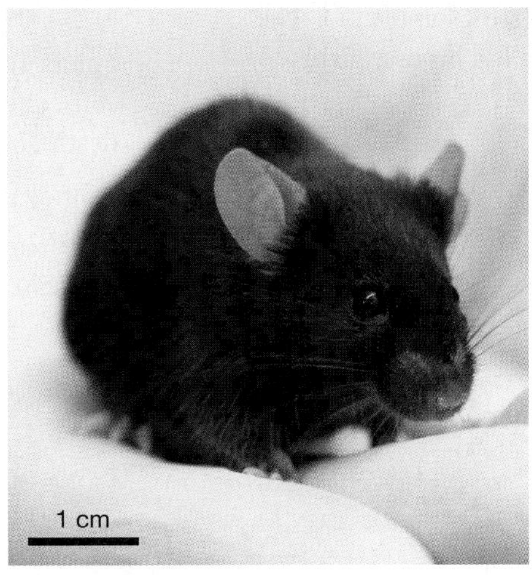

FIGURE B14.1 Model Organisms.

✔**QUESTION** *E. coli* is grown at a temperature of 37°C. Why?

species has become the organism of choice for experiments on control of the cell cycle and regulation of gene expression in eukaryotes. For example, research has confirmed that several of the genes controlling cell division and DNA repair in yeast have homologs in humans; and when mutated, these genes contribute to cancer. Strains of yeast that carry these mutations are now being used to test drugs that might be effective against cancer.

S. cerevisiae has become even more important in efforts to interpret the genomes of organisms like rice, mice, zebrafish, and humans. It is much easier to investigate the function of particular genes in *S. cerevisiae* by creating mutants or transferring specific alleles among individuals than it is to do the same experiments in mice or zebrafish. Once the function of a gene has been established in yeast, biologists can look for the homologous gene in other eukaryotes. If such a gene exists, they can usually infer that it has a function similar to its role in *S. cerevisiae.* It was also the first eukaryote with a completely sequenced genome.

Drosophila melanogaster

If you walk into a biology building on any university campus around the world, you are almost certain to find at least one lab where the fruit fly *Drosophila melanogaster* is being studied (**Figure B14.1e**).

Drosophila has been a key experimental subject in genetics since the early 1900s. It was initially chosen as a focus for study by T. H. Morgan because it can be reared in the laboratory easily and inexpensively, matings can be arranged, the life cycle is completed in less than two weeks, and females lay a large number of eggs. These traits made fruit flies valuable subjects for breeding experiments designed to test hypotheses about how traits are transmitted from parents to offspring (see Chapter 13).

More recently, *Drosophila* has also become a key model organism in the field of developmental biology. The use of flies in developmental studies was inspired in large part by the work of Christianne Nüsslein-Volhard and Eric Wieschaus, who in the 1980s isolated flies with genetic defects in early embryonic development. By investigating the nature of these defects, researchers have gained valuable insights into how various gene products influence the development of eukaryotes (see Chapter 21). The complete genome sequence of *Drosophila* has been available to investigators since the year 2000.

Caenorhabditis elegans

The roundworm *Caenorhabditis elegans* emerged as a model organism in developmental biology in the 1970s, due largely to work by Sydney Brenner and colleagues. (*Caenorhabditis* is pronounced *see-no-rab-DIE-tiss.*)

C. elegans was chosen for three reasons: **(1)** Its cuticle (soft outer layer) is transparent, making individual cells relatively easy to observe (**Figure B14.1f**); **(2)** adults have exactly 959 nonreproductive cells; and, most important, **(3)** the fate of each cell in an embryo can be predicted because cell fates are invariant among individuals. For example, when researchers examine a 33-cell *C. elegans* embryo, they know exactly which of the 959 cells in the adult will be derived from each of those 33 embryonic cells.

In addition, *C. elegans* are small (less than 1 mm long), are able to self-fertilize or cross-fertilize, and undergo early development in just 16 hours. The entire genome of *C. elegans* has now been sequenced.

Mus musculus

Because the house mouse *Mus musculus* is the most important model organism among mammals, it is especially prominent in biomedical research—where researchers need to work on individuals with strong genetic and developmental similarities to humans.

The house mouse was an intelligent choice of model organism in mammals because it is small and thus relatively inexpensive to maintain in captivity, and because it breeds rapidly. A litter can contain 10 offspring and generation time is only 12 weeks—meaning that several generations can be produced in a year. Descendants of wild house mice have been selected for docility and other traits that make them easy to handle and rear; these populations are referred to as laboratory mice (**Figure B14.1g**).

Some of the most valuable laboratory mice are strains with distinctive, well-characterized genotypes. Inbred strains are virtually homogenous genetically (see Chapter 25) and are useful in experiments where gene-by-gene or gene-by-environment interactions have to be controlled. Other populations carry mutations that knock out genes and cause diseases similar to those observed in humans. These individuals are useful for identifying the cause of genetic diseases and testing drugs or other types of therapies.

✔Questions

Which model organisms described here would be the best choice for the following studies? In each case, explain your reasoning.

1. A study of why specific cells in an embryo die at certain points in normal development. One goal is to understand the consequences for the individual when programmed cell death does not occur when it is supposed to.
2. A study of proteins that are required for cell-cell adhesion.
3. Research on a gene suspected to be involved in the formation of breast cancer in humans.

APPENDIX B Answers

CHAPTER 1

Check Your Understanding (CYU)

CYU p. 5 The data points would all be about 11 percent, indicating no change in average kernel protein content over time. **CYU p. 8** From the sequence data provided, Species A and B differ only in one nucleotide of the rRNA sequence (position 10 from left). Species C differs from Species A and B in four nucleotides (positions 1, 2, 9, and 10). A correctly drawn phylogenetic tree would indicate that Species A and B appear to be closely related, with Species C being more distantly related [see Figure A1.1]. **CYU p. 12** The key here is to test predation rates during the hottest part of the day (when desert ants actually feed) versus other parts of the day. The experiment would best be done in the field, where natural predators are present. One approach would be to capture a large number of ants, divide the group in two, and measure predation rates (number of ants killed per hour) when they are placed in normal habitat during the hottest part of the day versus an hour before (or after). (1) The control group here is the normal condition—ants out during the hottest part of the day. If you didn't include a control, a critic could argue that predation did or did not occur because of your experimental setup or manipulation, not because of differences in temperature. (2) You would need to make sure that there is no difference in body size, walking speed, how they were captured and maintained, or other traits that might make the ants in the two groups more or less susceptible to predators. They should also be put out in the same habitat, so the presence of predators is the same in the two treatments.

You Should Be Able To (YSBAT)

YSBAT p. 5 Average kernel protein content declined, from 11 percent to a much lower value. (The experiment is still continuing; to date the low-selection line is about 5 percent.) **YSBAT p. 12** (1) A critic could argue that the ants weren't navigating normally, because they had been caught and released and transferred to a new channel. (2) A critic could argue that the ants can't navigate normally on their manipulated legs.

Caption Questions and Exercises

Figure 1.2 If Pasteur had done any of the things listed, he would have had more than one variable in his experiment. This would allow critics to claim that he got different results because of the differences in broth types, heating, or flask types—not the difference in exposure to preexisting cells. The results would not be definitive. **Figure 1.4** Molds and other fungi are more closely related to green algae because they differ from plants at two positions (5 and 8 from left), but differ from green algae at only one position (8). **Figure 1.6** The eukaryotic cell is roughly 10 times (more exactly, 8.5 times) the size of the prokaryotic cell. **Figure 1.8** You should be skeptical, because the results could be due to the single individual being sick or injured or unusual in some other way.

Summary of Key Concepts

KC 1.1 Dead cells cannot maintain a difference in conditions inside the cell versus outside, replicate, use energy, or process information. **KC 1.2** Observations on thousands of diverse species supported the claim that all organisms consist of cells. The hypothesis that all cells come from preexisting cells was supported when Pasteur showed that new cells do not arise and grow in a boiled liquid unless they are introduced from the air. **KC 1.3** If seeds with higher protein content leave the most offspring, then individuals with low protein in their seeds will become rare over time. **KC 1.4** A newly discovered species can be classified as a member of the Bacteria if the sequence of its rRNA contains some features found only in Bacteria. The same logic applies to classifying a new species in the Archaea or Eukarya. **KC 1.5** (1) A hypothesis is an explanation of how the world works; a prediction is an outcome you should observe if the hypothesis is correct. (2) Experiments are convincing because they measure predictions from two opposing hypotheses. Both predicted actions cannot occur, so one hypothesis will be supported while the other will not.

Test Your Knowledge

1. d; **2.** d; **3.** c; **4.** b; **5.** d; **6.** c

Test Your Understanding

1. That the entity they discovered replicates, processes information, acquires and uses energy, is cellular, and that its populations evolve. **2.** [See Figure A1.2] **3.** The rule ensured that two different organisms would never end up with the same name. **4.** Over time, traits that increased the fitness of individuals in this habitat became increasingly more frequent in the population. **5.** Individuals with certain traits selected, in the sense that they produce the most offspring. **6.** Yes. If evolution is defined as "change in the characteristics of a population over time," then those organisms that are most closely related should have experienced less change over time. On a phylogenetic tree, species with substantially similar rRNA sequences would be diagrammed with a closer common ancestor—one that had the sequences they inherited—than the ancestors shared between species with dissimilar rRNA sequences.

Applying Concepts to New Situations

1. A scientific theory is not a guess—it is an idea whose validity can be tested with data. Both the cell theory and the theory of evolution have been validated by large bodies of observational and experimental data. **2.** If all eukaryotes living today have a nucleus, then it is logical to conclude that the nucleus arose in a common ancestor of all eukaryotes, indicated by the arrow you should have added to the figure on page 14 [see Figure A1.3]. If it had arisen in a common ancestor of Bacteria or Archaea, then species in those groups would have had to lose the trait—an unlikely event. **3.** The data set was so large and diverse that it was no longer reasonable to argue that noncellular life-forms would be discovered. **4.** Yes. The heritable traits that confer resistance to HIV should increase over time.

CHAPTER 2

Check Your Understanding (CYU)

CYU p. 21 [See Figure A2.1] **CYU p. 33** (1) Gibbs equation: $\Delta G = \Delta H - T\Delta S$. ΔG symbolizes the change in the Gibbs free energy. ΔH represents the difference in potential energy between the products and the reactants. T represents the temperature (in degrees Kelvin) at which the reaction is taking place. ΔS symbolizes the change in entropy (disorder). (2) When ΔH is negative—meaning that the reactants have lower potential energy than the products—and when ΔS is positive, meaning that the products have higher entropy (are more disordered) than the reactants.

You Should Be Able To (YSBAT)

YSBAT p. 22 (1) $^{\delta+}H{-}O^{\delta-}{-}H^{\delta+}$. (2) If water were linear, the partial negative charge on oxygen would have partial positive charges on either side. Compared to the actual, bent molecule, the partial negative charge would be much less exposed and less able to participate in hydrogen bonding. **YSBAT p. 26** The pH 8 solution has 100 times fewer protons than the pH 6 solution. **YSBAT p. 30** (1) As T increases, ΔG is more likely to be negative, making the reaction spontaneous. (2) Exothermic reactions may be nonspontaneous if they result in a decrease in entropy—meaning that the products are more ordered than the reactants (ΔS is negative).

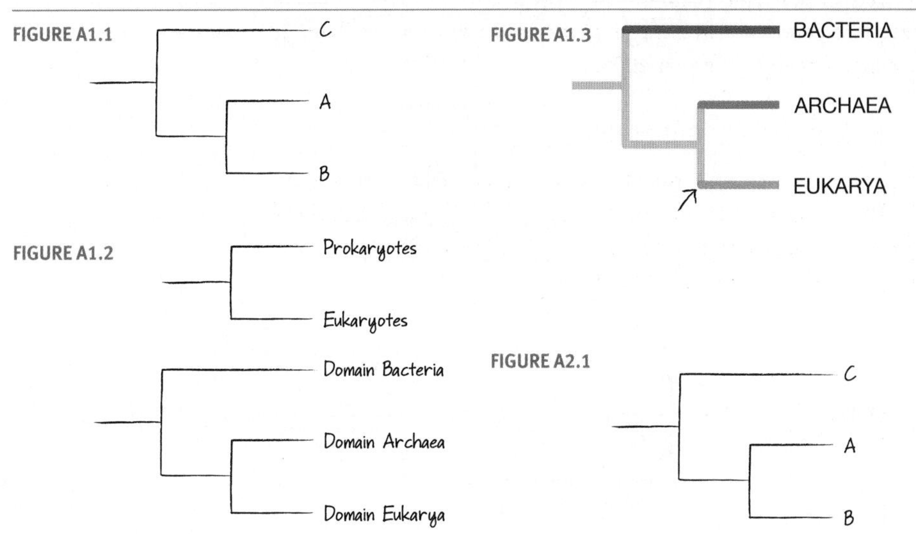

Caption Questions and Exercises

Figure 2.7 Oxygen and nitrogen have high electronegativity. They hold shared electrons more tightly than C, H, and many other atoms, resulting in polar bonds. **Figure 2.13** Oils are nonpolar. They have long chains of carbon atoms bonded to hydrogen atoms, which share electrons evenly because their electronegativities are similar. When an oil and water are mixed, the polar water molecules interact with each other via hydrogen bonding much more strongly than they interact with the nonpolar oil molecules, which interact with themselves instead. **Figure 2.16** The coffee becomes less acidic because milk is more alkaline (pH 6.5) than black coffee (pH 5). **Table 2.2** (Row 1) "Cause": Electrostatic attraction between partial charges on water molecules and opposite charges on ions; hydrogen bonds between water and other polar molecules. (Row 2) "Biological Consequences": Ice floats on denser liquid water. If it were denser and sank, oceans and lakes would fill with ice. (Row 4) "Cause": Liquid water must absorb lots of heat energy to break hydrogen bonds and change to a gas. One possible concept map relating the structure of water to its properties is shown below [see Figure A2.2]. **Figure 2.19** [See Figure A2.3] **Figure 2.22** The reaction rate at a specific set of reactant concentrations, averaged over many replicates. **Table 2.3** All the functional groups in Table 2.3, except the sulfhydryl group (−SH), are highly polar. The sulfhydryl group is only very slightly polar.

Summary of Key Concepts

KC 2.1 [See Figure A2.4] **KC 2.2** Because these functional groups are polar, water forms hydrogen bonds with them. **KC 2.3** Chemical energy—meaning, potential energy stored in chemical bonds—is transformed to heat. **KC 2.4** Spontaneous chemical reactions lead toward disorder and a release of energy. Cells are full of highly ordered molecules that contain high-energy bonds. Energy must be added to make these molecules and maintain life. **KC 2.5** Electrons in carbon-carbon bonds are held loosely and equally between the carbon atoms, whereas electrons in a carbon-oxygen bond are held tightly by the oxygen. Because of this difference, molecules with carbon-carbon bonds have more potential energy than carbon dioxide with its two carbon-oxygen bonds. Carbon-carbon bonds are often present in large, highly ordered molecules. The entropy of a group of such molecules, a measure of their disorder, is much less than the entropy of a group of simple, less-ordered molecules such as carbon dioxide.

Test Your Knowledge

1. b; **2.** a; **3.** a; **4.** d; **5.** d; **6.** d

Test Your Understanding

1. The reaction lowers the pH of the solution by releasing extra H^+ into the solution. If additional CO_2 is added, the sequence of reactions would be driven to the right, which would make the ocean more acidic. **2.** No. Shells that are farther from the protons (positive charges) in the nucleus house electrons that have greater potential energy than shells closer to the nucleus. **3.** [See Figure A2.5] The electron orbitals surrounding oxygen are oriented in the shape of a tetrahedron. Two of these orbitals are occupied by unshared electrons and the other two are shared with hydrogen atoms, resulting in the bent shape of the water molecule. Due to its greater electronegativity, oxygen attracts the shared electrons more strongly than does hydrogen, so it has a partial negative charge while the hydrogen atoms have a partial positive charge. **4.** In a covalent bond, the shared electrons are extremely close to two nuclei. In a hydrogen bond, the (+) and (−) charges are only partial, and the attraction is at a much greater distance. **5.** Water forms large numbers of hydrogen bonds, which must be broken before the molecules can start moving faster—meaning that their temperature has increased. **6.** The overall shape of an organic molecule is determined by its carbon framework. The functional groups attached to the carbons determine the molecule's chemical behavior, because these groups are likely to interact with other molecules.

Applying Concepts to New Situations

1. In CO_2, the double bonds between the carbon and each of the oxygen atoms lock the molecule into a linear shape that is nonpolar. In water, the partial charges on

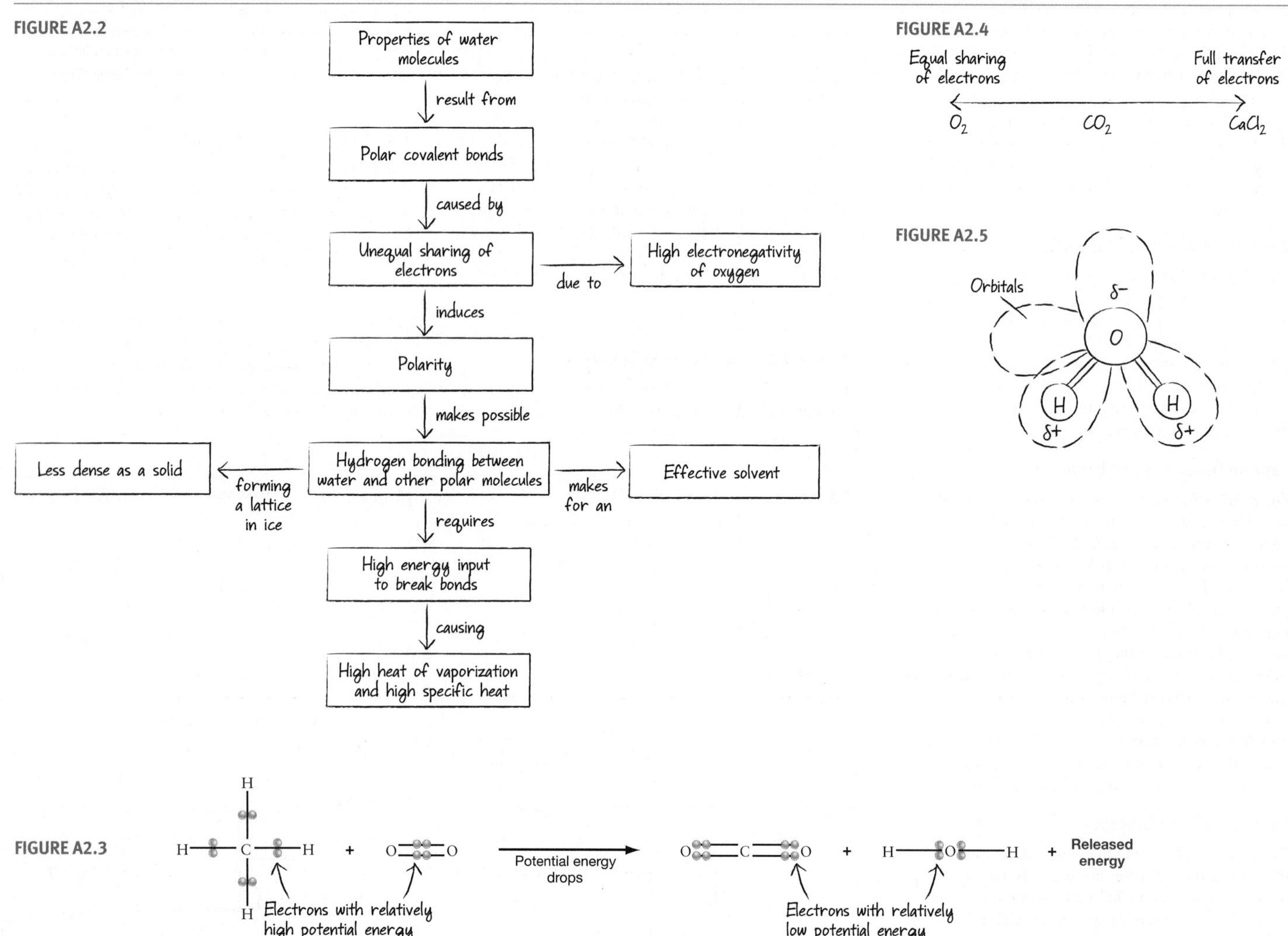

oxygen and hydrogen allow water to form hydrogen bonds with other polar or charged substances. In addition, the single covalent bonds in water are much easier to break than the double covalent bonds in carbon dioxide, making it more reactive. **2.** When oxygen is covalently bonded to carbon or hydrogen, the bond electrons spend more time near the oxygen atom. In contrast, carbon and hydrogen have roughly similar electronegativities, and they tend to share the electrons of a covalent bond more equally. **3.** The Sun will burn out when the mass of all of its component atoms is finally converted to energy. **4.** In hot weather, water absorbs large amounts of heat due to its high specific heat and high heat of vaporization. In cold weather, water releases the large amount of heat that it has absorbed.

CHAPTER 3

Check Your Understanding (CYU)

CYU p. 44 [See Figure A3.1] **CYU p. 51** Secondary, tertiary, and quaternary structure all depend on bonds and other interactions between amino acids that are linked in a chain (primary structure). **CYU p. 56** (1) Substrates can react only if they meet in a precise orientation. (2) The shape change alters the shape of the active site so that substrates no longer fit or can't be oriented correctly for a reaction to occur.

You Should Be Able To (YSBAT)

YSBAT p. 50 The sequence of amino acids in the two polypeptide chains constitutes the primary structure. Each chain contains α-helices and a β-pleated sheet (secondary structures), and is folded into a specific three-dimensional shape (tertiary structure). The overall three-dimensional shape of the dimeric Cro protein, formed by association of two identical polypeptides, constitutes its quaternary structure. **YSBAT p. 53** More—because catalysts lower the activation energy required for the reaction to proceed, more molecules at any given temperature have enough kinetic energy to supply the activation energy. **YSBAT p. 54** (1) orienting substrates, (2) transition state, (3) R-groups, (4) structure.

Caption Questions and Exercises

Figure 3.1 The water-filled flask is the ocean; the gas-filled flask is the atmosphere; the condensed water droplets are rain; the electrical sparks are lightning. **Figure 3.3** The green R-groups contain mostly C and H, which have roughly equal electronegativities. Electrons are evenly shared in C—H bonds and C—S bonds, so the groups are nonpolar. Most of the pink R-groups have a highly electronegative oxygen atom with a partial negative charge, making them polar. Cysteine has a sulfur that is slightly more electronegative than hydrogen, so it will be less polar than the other pink groups. **Figure 3.6** A polar covalent bond. The electrons are shared in the C—N bond, but because N is more electronegative than C, the bond is polar. **Figure 3.16** In order for a chemical reaction to proceed, the kinetic energy of the reactant molecules and atoms must be greater than the activation energy. Only the molecules to the right of the line you drew will be able to react. **Figure 3.18** No—a catalyst only changes the activation energy, not the overall free energy change. **Figure 3.22** [See Figure A3.2]

Summary of Key Concepts

KC 3.1 Many possible correct answers, including: the presence of an active site in an enzyme that is precisely shaped to fit a substrate or substrates in the correct orientation for a reaction to occur; the "donut" shape of porin allowing certain substances to pass through it; the butterfly shape of TATA-box binding protein being precisely the right size for a DNA molecule to fit. **KC 3.2** R-groups containing partial charges or full charges can form hydrogen bonds with water, make the amino acid soluble. Nonpolar R-groups do not interact with water and make the amino acid insoluble. **KC 3.3** [See Figure A3.3] **KC 3.4** When a catalyst is present, the reactants and products don't change—only the transition state changes. Thus, the overall free energy change in the reaction is the same—only the activation energy changes.

Test Your Knowledge

1. d; **2.** c; **3.** a; **4.** b; **5.** a; **6.** b

Test Your Understanding

1. The shape of reactant molecules (the key) fits into the active site of an enzyme (the lock). The model assumed that the enzyme is rigid; in fact it is flexible and dynamic. **2.** It will fold in on itself, away from water, when placed in an aqueous solution. **3.** Both are mechanisms that regulate enzymes; the difference is whether the regulatory molecule binds at the active site (competitive inhibition) or away from the active site (allosteric regulation). **4.** Energy is required. Polymerization is a nonspontaneous reaction because the product molecules have lower entropy and greater potential energy than the reactants. **5.** Proteins are highly variable in overall shape and chemical properties due to variation in the composition of R-groups and the array of secondary through quaternary structures that are possible. This variation allows them to fulfill many different roles in the cell. Diversity in the shape and reactivity of active sites makes them effective catalysts. **6.** The function of an enzyme depends on the shape and chemical reactivity of its active site. Temperature and pH affect enzyme function because they can change the shape and chemical reactivity of the active site—either by disrupting bonds or interfering with acid-base reactions.

Applying Concepts to New Situations

1. Yes—within limits. As Figure 3.23 shows, some bacteria thrive in very hot or highly acidic environments on Earth. But extremely low pHs or high temperatures are likely to denature proteins. **2.** The data suggest that the enzyme and substrate form a transition state that requires a change in the shape of the active site, with each movement corresponding to one reaction. **3.** Without the coenzyme, the free radical–containing transition state would not be stabilized and the reaction rate would drop dramatically. **4.** Residues in the active site changed in a way that made the drugs less likely to bind and prevent the normal substrate from binding.

FIGURE A3.1

FIGURE A3.2

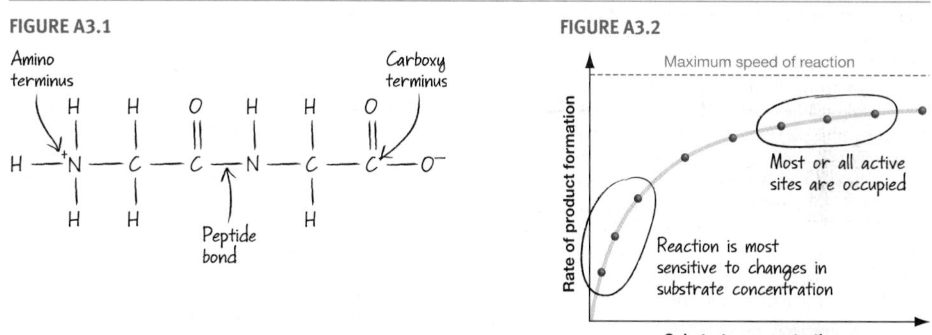

FIGURE A3.3

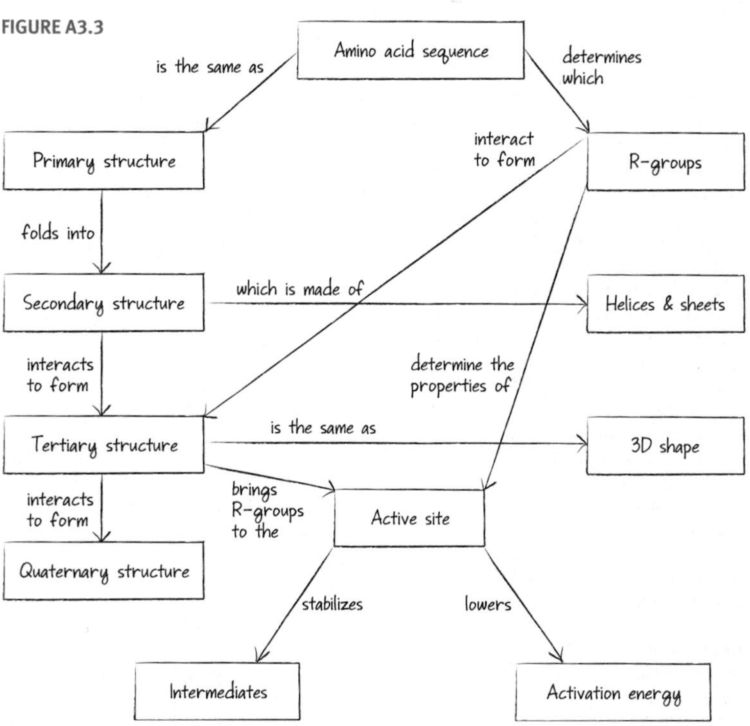

CHAPTER 4

Check Your Understanding (CYU)

CYU p. 62 [See Figure A4.1] **CYU p. 66** [See Figure A4.2]

You Should Be Able To (YSBAT)

YSBAT p. 60 [See Figure A4.3] **YSBAT p. 64** If the two strands are parallel, the nitrogenous bases are not aligned in a way that allows hydrogen bond formation. No helix would form.

Caption Questions and Exercises

Figure 4.3 5′ UAGC 3′. **Figure 4.9** Endergonic—energy must be added (as heat) for the reaction to occur.

Summary of Key Concepts

KC 4.1 (1) Both polymerize via energy-demanding condensation reactions that add a monomer to a growing chain. But proteins polymerize via formation of peptide bonds while nucleic acids polymerize via formation of phosphodiester linkages. (2) Proteins have an amino terminus at one end and a carboxyl terminus at the other end, while nucleic acids have a 5′ end (with an unlinked phosphate group) and a 3′ end (with an unlinked hydroxyl group). (3) Both have a "backbone"; in proteins it consists of amino and carbonyl groups, while in nucleic acids it consists of phosphate groups and sugars. **KC 4.2** C-G pairs involve three hydrogen bonds, so they are more stable than A-T pairs with just two bonds. **KC 4.3** A single-stranded RNA molecule has unpaired bases that can pair up with other bases on the same RNA strand, thereby folding the molecule into stem-and-loop configurations with a particular three-dimensional shape. The unpaired bases also can bind to another RNA molecule, linking the two together in a quaternary structure. Because DNA molecules are double stranded, with no unpaired bases, internal folding and interaction with other DNA molecules is not possible. **KC 4.4** If the copied ribozymes are not exactly like the template molecule, then some are likely to have changes that make them more efficient as catalysts. If there were no variation—meaning, if copying were exact—there could be no improvement in efficiency or other change over time.

Test Your Knowledge

1. c; **2.** d; **3.** a; **4.** b; **5.** a; **6.** c

Test Your Understanding

1. [See Figure A4.4] **2.** The addition of the phosphate groups raises the potential energy of the resulting monomers (nucleoside triphosphates) enough that the polymerization reaction is exergonic. **3.** Nucleic acids are directional because the two ends of the polymer are different. One end has a free phosphate group on a 5′ carbon; the other end has a free hydroxyl group bonded to a 3′ carbon. **4.** DNA is a more stable molecule than RNA because it lacks a hydroxyl group on the 2′ carbon and is therefore resistant to degradation, and because the two sugar-phosphate backbones are linked by many hydrogen bonds between nitrogenous bases. DNA is more stable than proteins because it is symmetrical and has few exposed chemical groups that could participate in chemical reactions. **5.** Hairpin structures form when folding of the sugar-phosphate backbone allows the bases in one part of an RNA strand to align with bases in another segment of the same RNA strand in an antiparallel fashion, so they can hydrogen-bond and form a double helix. **6.** DNA has limited catalytic ability because it (1) lacks functional groups that can stabilize transition states and (2) has a regular structure that is not conducive to forming active sites. RNA molecules can catalyze some reactions because they (1) have exposed hydroxyl functional groups that can stabilize transition states and (2) can fold into shapes that create active sites. Proteins are the most effective catalysts because (1) amino acids have a wide range of chemical reactivity, and (2) they fold into an enormous diversity of shapes that can create active sites.

Applying Concepts to New Situations

1. Yes—if the complementary bases lined up over the entire length of the two strands, they would twist into a double helix analogous to a DNA molecule. The same types of hydrogen bonds and hydrophobic interactions would occur as observed in the "stem" portions of hairpins in single-stranded RNA. **2.** An RNA replicase would undergo replication and be able to evolve. It would process information in the sense of copying itself, and would use energy to drive endergonic polymerization reactions. It would not be bound by a membrane, however, and would not be able to acquire energy. It would best be considered as an intermediate step between nonlife and "true life," as outlined in Chapter 1. **3.** It is unlikely that bases would align properly for hydrogen bonding to occur, so hydrophobic interactions would probably be more important. **4.** No single right answer (these are opinion questions). Note that when biologists design studies about the origin of life, they base their experimental conditions on the best available current knowledge about early Earth conditions. New information about these conditions might require a revision of the experiment that leads to different results and conclusions.

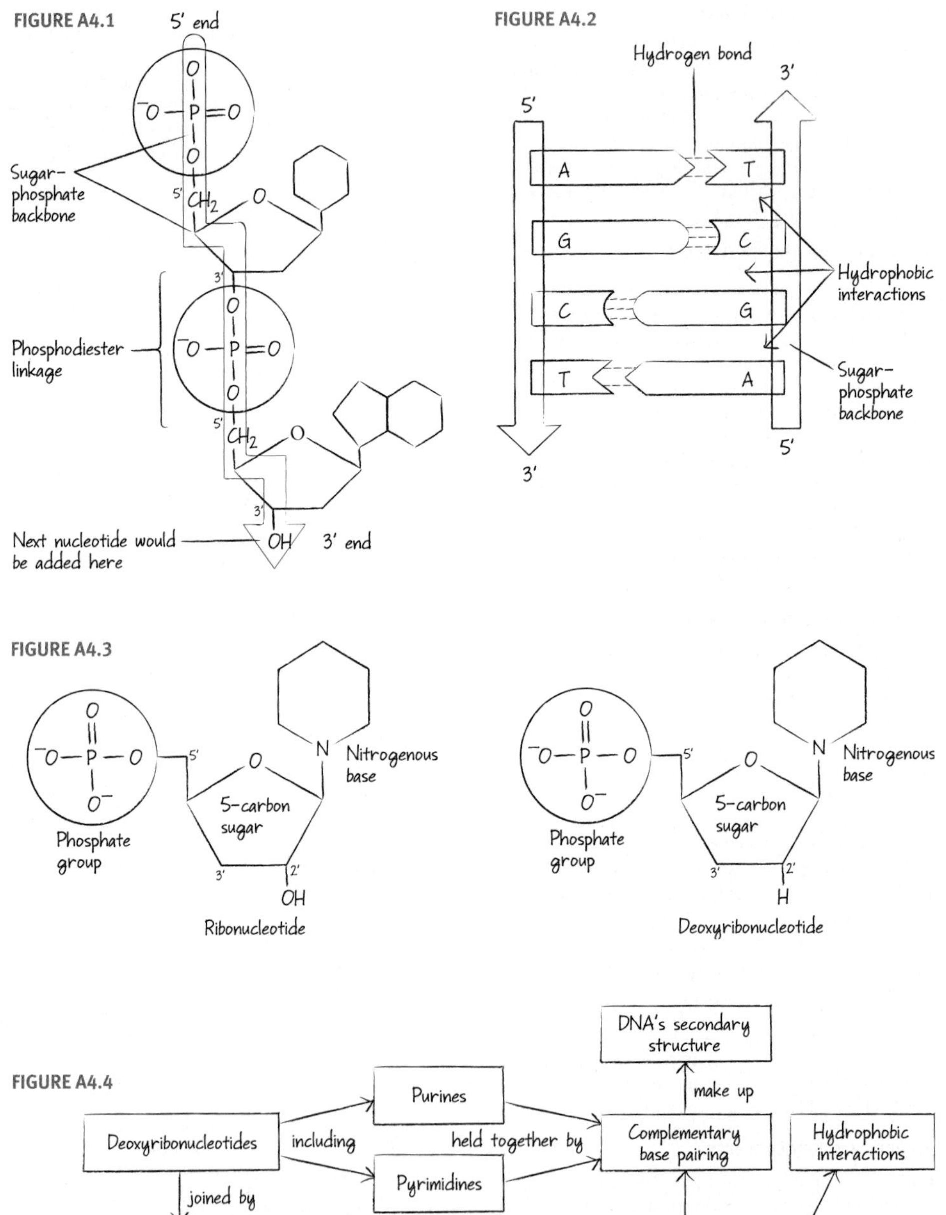

CHAPTER 5

Check Your Understanding (CYU)

CYU p. 73 [See Figure A5.1] **CYU p. 76** They could differ in (1) location of linkages (e.g., 1,4 or 1,6); (2) types of linkages (e.g., α or β); (3) the sequence of the monomers (e.g., two galactose then two glucose, versus alternating galactose and glucose); and/or (4) whether the four monomers are linked in a line or whether they branch. **CYU p. 79** (1) Aspect 1: The β-1,4-glycosidic linkages in these molecules are difficult to degrade. Aspect 2: When individual molecules of these carbohydrates align, bonds form between them and produce tough fibers or sheets. (2) Most are probably being broken down into glucose, which in turn is being broken down in reactions that lead to the synthesis of ATP. In short, the carbohydrates are providing you with chemical energy.

Caption Questions and Exercises

Figure 5.2 [See Figure A5.2] **Figure 5.7** All of the C—C and C—H bonds should be circled.

Summary of Key Concepts

KC 5.1 Molecules have to interact in an extremely specific orientation in order for a reaction to occur. Changing the location of a functional group by even one carbon can mean that the molecule will undergo completely different types of reactions. **KC 5.2** The orientations of the α- versus β-glycosidic linkages are different, so the molecules in a chain end up in a spiral versus linear arrangement. **KC 5.3** (1) Polysaccharides used for energy storage are formed entirely from glucose monomers joined by α-glycosidic linkages; structural polysaccharides are comprised of glucose or other sugars joined by β-glycosidic linkages. (2) The monomers in energy-storage polysaccharides are linked in a spiral arrangement; the monomers in structural polysaccharides are linked in a linear arrangement. (3) Energy-storage polysaccharides may branch; structural polysaccharides do not. (4) Individual chains of energy-storage polysaccharides do not associate with each other; adjacent chains of structural polysaccharides are linked by hydrogen bonds or covalent bonds.

Test Your Knowledge

1. d; **2.** a; **3.** c; **4.** a; **5.** d; **6.** b

Test Your Understanding

1. Carbohydrates are ideal for signaling the identity of the cell because they are so diverse structurally. This diversity enables them to serve as very specific identity tags for cells. **2.** When you compare the glucose monomers in an α-1,4-glycosidic linkage versus a β-1,4-glycosidic linkage, the linkages are located on opposite sides of the plane of the glucose rings (e.g., "above" versus "below" the plane), and the glucose monomers are linked in the same orientation versus having every other glucose flipped in orientation. β-1,4-glycosidic linkages are much more difficult for enzymes to break, so they resist degradation. **3.** Starch and glucose both consist of glucose monomers joined by α-1,4-glycosidic linkages, and both function as storage carbohydrates. Starch is composed of an unbranched amylose and branched amylopectin, while glycogen is even more highly branched. **4.** The electrons in the C=O bonds of carbon dioxide molecules are held tightly by the highly electronegative oxygen atoms, so have low potential energy. The electrons in the C—C and C—H bonds of carbohydrates are shared equally and have much higher potential energy. **5.** They have β-1,4-glycosidic linkages, which are resistant to degradation and put glucose monomers in positions where hydrogen bonds can form between adjacent strands, forming strong fibers. **6.** Glycogen has α-1,4-glycosidic linkages with α-1,6-glycosidic linkages at branch points; cellulose has β-1,4-glycosidic linkages and is not branched—instead, it forms hydrogen bonds with adjacent cellulose molecules. Glycogen is an energy-storage molecule; cellulose is a structural polysaccharide.

Applying Concepts to New Situations

1. Carbohydrates are energy-storage molecules, so minimizing their consumption may reduce total energy intake. Lack of available carbohydrate also forces the body to use fats for energy, reducing the amount of fat that is stored. **2.** Lactose is made up of glucose and galactose. **3.** Amylase breaks down the starch in the cracker into glucose monomers, which stimulate the sweet receptors in your tongue. **4.** When bacteria contact lysozyme, their cell walls, which contain peptidoglycan, begin to degrade, leading to the death of the bacteria.

CHAPTER 6

Check Your Understanding (CYU)

CYU p. 85 (1) Fats consist of three fatty acids linked to glycerol; steroids have a distinctive four-ring structure with a side group attached; phospholipids have a hydrophilic, phosphate-containing "head" region and a hydrocarbon tail. (2) In cholesterol, the hydrocarbon steroid rings are hydrophobic; the hydroxyl group is hydrophilic. In phospholipids, the phosphate-containing head group is hydrophilic; the fatty acid chains are hydrophobic. **CYU p. 89** [See Table A6.1] **CYU p. 92** [See Figure A6.1] **CYU p. 99** Passive transport does not require an

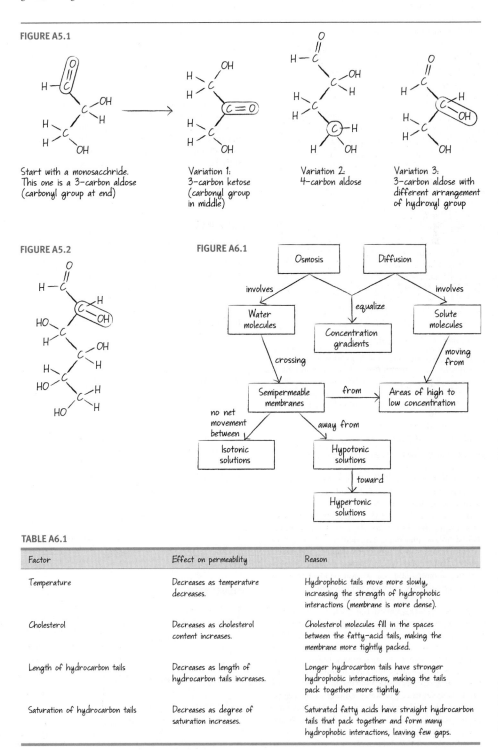

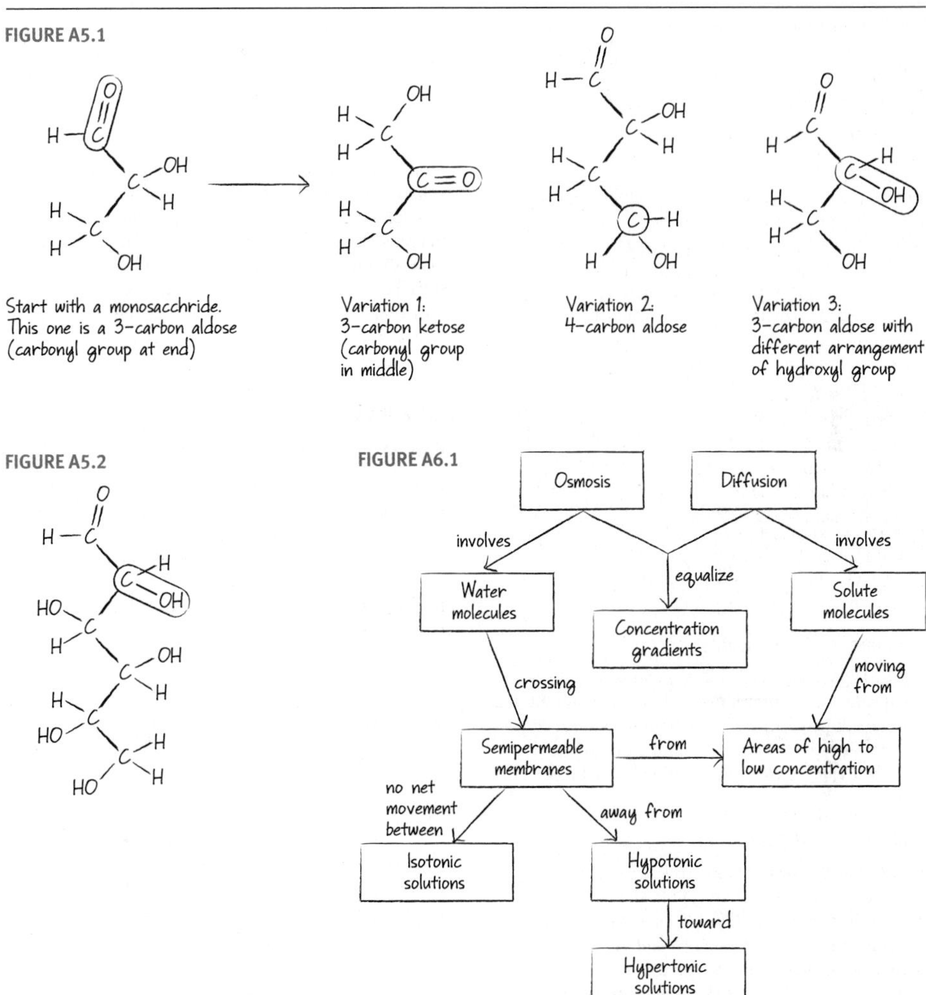

TABLE A6.1

Factor	Effect on permeability	Reason
Temperature	Decreases as temperature decreases.	Hydrophobic tails move more slowly, increasing the strength of hydrophobic interactions (membrane is more dense).
Cholesterol	Decreases as cholesterol content increases.	Cholesterol molecules fill in the spaces between the fatty-acid tails, making the membrane more tightly packed.
Length of hydrocarbon tails	Decreases as length of hydrocarbon tails increases.	Longer hydrocarbon tails have stronger hydrophobic interactions, making the tails pack together more tightly.
Saturation of hydrocarbon tails	Decreases as degree of saturation increases.	Saturated fatty acids have straight hydrocarbon tails that pack together and form many hydrophobic interactions, leaving few gaps.

expenditure of energy—it happens as a result of existing concentration or electrical gradients. Active transport is active in the sense of requiring energy. In cotransport, a second ion or molecule is transported against its electrochemical gradient with ("co") an ion that is transported along its electrochemical gradient.

You Should Be Able To (YSBAT)

YSBAT p. 84 Fats have many C—C and C—H bonds, which have high free energy. They are hydrophobic because they contain long hydrocarbon chains and lack any highly polar groups. **YSBAT p. 86** No, because both are relatively large molecules that are polar or carry a charge. **YSBAT p. 91** (1) Add a large number of green dots to the inside of the liposome. (2) Add a large number of green dots to the solution outside the liposome. **YSBAT p. 94** Your arrow should point out of the cell. There is no concentration gradient, but the outside has a net positive charge, which favors outward movement of negative ions. **YSBAT p. 96** [See Figure A6.2]

Caption Questions and Exercises

Figure 6.3 At the polar hydroxyl group in cholesterol and the polar head group in phospholipids. **Figure 6.14** Higher, because less water would have to move to the right side to achieve equilibrium. **Figure 6.21** The electrical gradient would reverse, because the inside of the membrane would be more positive than the outside. The concentration gradient for sodium would also reverse; sodium would diffuse to the exterior. **Figure 6.22** No—the 10 replicates where no current was recorded probably represent instances where the CFTR protein was damaged and not functioning properly. (In general, no experimental method works "perfectly.") **Figure 6.27** "Diffusion": description as given; no proteins involved. "Facilitated diffusion": Passive movement of ions or molecules that cannot cross a phospholipid bilayer readily, along a concentration gradient. Facilitated by channels or transporters. "Active transport": Active (energy-demanding) movement of ions or molecules against an electrochemical gradient.

Summary of Key Concepts

KC 6.1 Highly permeable bilayers consist of phospholipids with short, unsaturated hydrocarbon tails. **KC 6.2** (1) The solute will diffuse until both sides are at equal concentrations. (2) Water will diffuse toward the side with the higher solute concentration. **KC 6.3** [See Figure A6.3]

Test Your Knowledge

1. b; **2.** a; **3.** b; **4.** d; **5.** c; **6.** d

Test Your Understanding

1. No, because they have no polar end to interact with water. Instead, these lipids would collect and float on the surface of water, or collect in droplets suspended in water, reducing their interaction with water to a minimum. **2.** Hydrophilic, phosphate-containing head groups interact with water; hydrophobic, fatty-acid tails associate with each other. A bilayer has a lower potential energy and thus is more stable than are independent phospholipids in solution. **3.** Ethanol's polarity reduces the speed at which it can cross a membrane, but its small size and lack of charge increase the speed at which it crosses membranes. Ethanol probably crosses a membrane more slowly than water (which is also polar, but smaller) but faster than larger polar molecules, and much faster than charged molecules. **4.** If no membrane separates the two solutions, then the constant, random motion of solute and water molecules causes even mixing of the two solutions. No net diffusion of water molecules will occur. **5.** Only hydrophobic amino acids interact with the nonpolar lipid tails in the interior of a phospholipid bilayer. Figure 3.3 lists the following nonpolar, hydrophobic amino acids: glycine, alanine, valine, leucine, isoleucine, methionine, phenylalanine, tryptophan, and proline. According to Table 3.1, tyrosine and cysteine are also relatively hydrophobic, suggesting that they might also be found in the interior of transmembrane proteins. **6.** CO_2 will diffuse into the cell; water will move out via osmosis. Na^+ and K^+ will enter the cell through gramicidin by facilitated diffusion. Cl^- will not move.

Applying Concepts to New Situations

1. Flip-flops should be rare, because they require polar head group to pass through the hydrophobic portion of the lipid bilayer. To test this prediction, you could monitor the number of dyed phospholipids that transfer from one side of the membrane to the other in a given period of time. **2.** The kinks in unsaturated hydrocarbon tails keep membranes fluid, even at low temperature, because they prevent extensive hydrophobic interactions. In hot environments, it would be advantageous for phospholipids to have saturated tails, to prevent membranes from being too fluid. **3.** Adding a methyl group makes a drug more hydrophobic and thus more likely to pass through a lipid bilayer. Adding a charged group make it hydrophilic and reduces its ability to pass through the lipid bilayer. **4.** Detergents are amphipathic. Their hydrophobic ends interact with grease while their hydrophilic ends interact with water. This forms a bridge between the grease and the water, effectively making the grease dissolve in water and allowing it to be washed away.

CHAPTER 7

Check Your Understanding (CYU)

CYU p. 105 (1) Photosynthetic membranes increase food production by providing a large surface area to hold the pigments and enzymes required for photosynthesis. (2) The presence of a magnetic mineral, which changes position as the cell moves through a magnetic field, allows the cell to move in a directed way. (3) The layer of thick, strong material stiffens the cell and provides protection from mechanical damage. **CYU p. 115** (1) Both organelles contain specific sets of enzymes, and the interior of lysosomes is acidic. Peroxisomes contain catalase and other enzymes that process fatty acids and toxins via oxidation reactions. Lysosomal enzymes digest macromolecules, releasing monomers that can be recycled into new macromolecules. (2) From top to bottom, the cells in the last column should read as follows: administrative/information hub, protein factory, large molecule manufacturing and shipping (protein synthesis and folding center, protein finishing and shipping line, fat factory, waste processing and recycling center), fatty-acid processing and detox center, warehouse, power station, food-manufacturing facility, support beams, perimeter fencing with secured gates, and leave blank. **CYU p. 122** (1) Proteins that lack a signal sequence will not be delivered to the ER. They are released into the cytosol. (2) The proteins will not be secreted—they will be found in lysosomes. **CYU p. 128** The products will not be able to move through the system normally, because their microtubule "roads" are absent.

You Should Be Able To (YSBAT)

YSBAT p. 128 (1) The microtubules at the bottom of the axoneme would slide to the right, but the axoneme would not bend. (2) Nothing would happen (dynein wouldn't move).

Caption Questions and Exercises

Figure 7.1 [See Figure A7.1] **Figure 7.15** The vesicles deliver digestive enzymes that are required for the mature lysosome to function. **Figure 7.16** Storing the toxins in vacuoles prevents the toxins from damaging the plant's own organelles and cells. **Figure 7.19** The cell wall is outside the plasma membrane. **Figure 7.21** [See Figure A7.2] **Figure 7.23** "Prediction": The labeled tail region fragments or the labeled core region fragments of the nucleoplasmin protein will be found in the cell nucleus. "Prediction of null hypothesis": No labeled fragments of the nucleoplasmin protein will be found in the nucleus of the cell. "Conclusion": The "Send to nucleus" zip code is in the tail region of the nucleoplasmin protein.

FIGURE A6.2

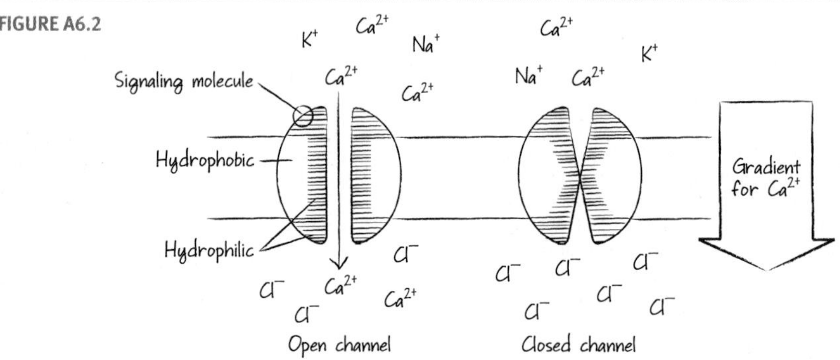

FIGURE A6.3

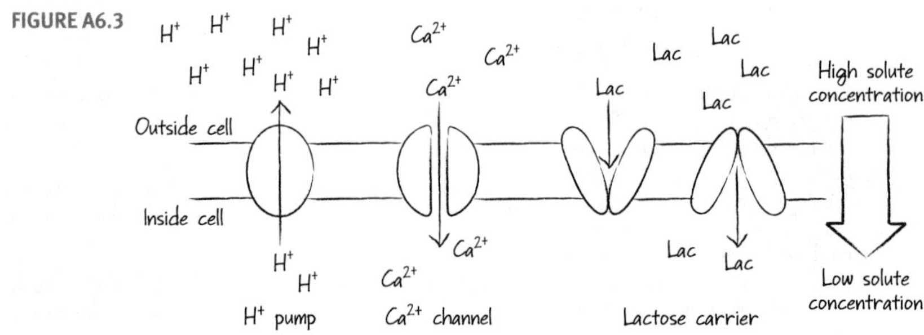

Summary of Key Concepts

KC 7.1 (1) They will be unable to make ATP and will die. (2) They will be limp and less able to resist attack by predators. In many environments they will be unable to resist the osmotic pressure of water entering the cytoplasm and will burst. (3) They will be unable to synthesize new proteins and will die. **KC 7.2** As a group, proteins have complex and highly diverse shapes and chemical properties. Individual proteins, then, can recognize and bind to molecular zip codes in a very specific way (like a key fitting into a lock). **KC 7.3** Actin filaments are made up of two strands of actin monomers, microtubules are made up of tubulin protein dimers that form a tube, intermediate filaments are made up of a variety of different protein monomers. Actin filaments and microtubules exhibit polarity (or directionality), with new subunits constantly being added or subtracted at either end (but added faster to the plus end). All three elements provide structural support, but only actin filaments and microtubules are involved in movement and cell division. **KC 7.4** Micrographs of cells at increasing time intervals after the pulse treatment represent time-lapse photography—a set of still images that show change through time.

Test Your Knowledge

1. a; **2.** c; **3.** c; **4.** b; **5.** a; **6.** d

Test Your Understanding

1. All cells are bound by a plasma membrane, are filled with cytoplasm, carry their genetic information (DNA) in chromosomes, and contain ribosomes (the sites of protein synthesis). Some prokaryotes have organelles not found in plants or animals, such as a magnetite-containing structure. Plant cells have chloroplasts, vacuoles, and a cell wall. Animal cells contain lysosomes and lack a cell wall. **2.** Ribosome in cytoplasm (protein is synthesized) → Rough ER (protein is folded and initially modified) → Transport vesicle → Golgi apparatus (protein is glycosylated; has molecular tag indicating destination) → Transport vesicle → Plasma membrane → Extracellular space. **3.** When a globular "head" section of kinesin binds and releases ATP, it undergoes a conformational change that swings it forward and binds it to the microtubule. The two globular head regions alternate between swinging and binding movements, causing the protein to "walk" down the microtubule. **4.** Changes in the length of microfilaments and microtubules contribute to changes in the size and shape of the cytoskeleton. **5.** Plasma membrane proteins are produced by the endomembrane system. Cells that function in membrane transport processes should have an extensive endomembrane system, including RER. **6.** All microtubules have the same basic structure, consisting of tubulin dimers that form a tube with plus and minus ends. The microtubules involved in structural support have no associated motor proteins; those involved in vesicle movement and flagella do. The microtubules that make up flagella are grouped in a 9 doublets + 2 single-microtubules arrangement. Structural and vesicle transport microtubules are single structures.

Applying Concepts to New Situations

1. Since mannose-6-phosphate serves as a lysosome tag in the endomembrane system, material that is ingested by phagocytic cells of the immune system might be targeted to lysosomes in the same way. To test this hypothesis, you could expose immune system cells to a population of bacteria that have labeled proteins, isolate the proteins after ingestion, and test to see whether they now have a mannose-6-phosphate tag attached. **2.** The proteins must receive a molecular zip code that binds to a receptor on the surface of peroxisomes. They could diffuse randomly to peroxisomes, or be transported in a directed way by vesicles or specialized motor proteins. **3.** (a) "Housekeeping"—destruction of macromolecules

FIGURE A7.1

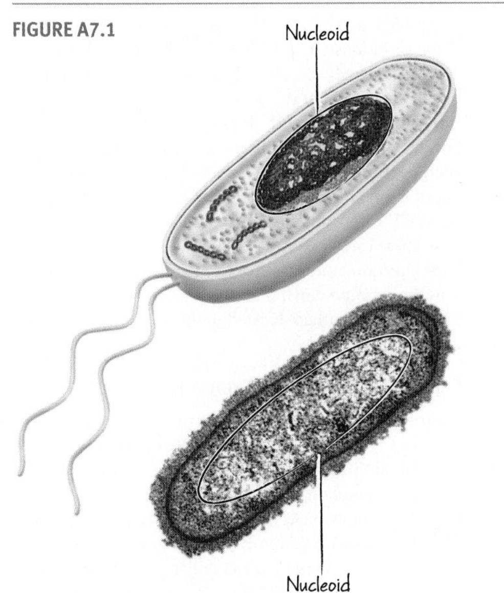

FIGURE A7.2

(a) Animal pancreatic cell: Exports digestive enzymes.

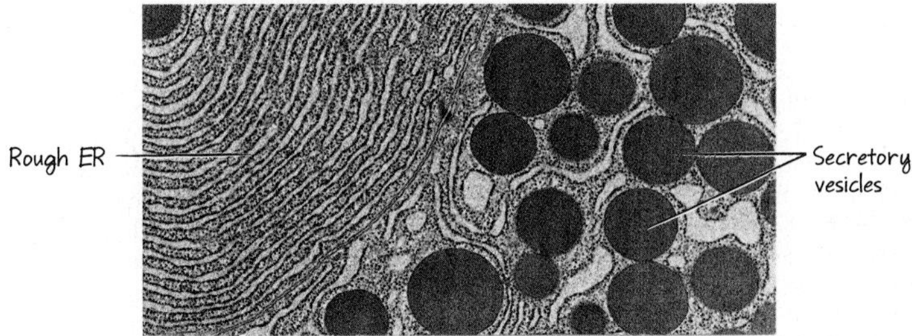

(b) Animal testis cell: Exports lipid-soluble signals.

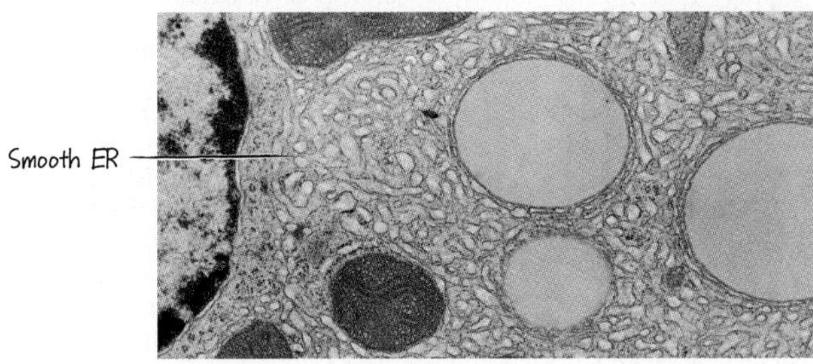

(c) Plant leaf cell: Manufactures ATP and sugar.

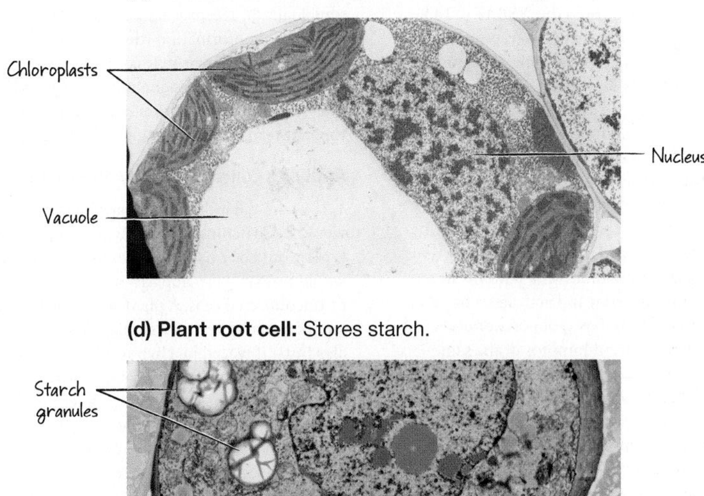

(d) Plant root cell: Stores starch.

that are dangerous or no longer needed. (b) Structural support (e.g., wood is made up of cells with extensive cell walls). (c) Detoxification of dangerous compounds via oxidation reactions, or processing of stored fatty acids. **4.** The cell wall prevents the cells from bursting in hypotonic (aqueous) environments, or protects the cell from damage. Expose mutant and normal individuals to (1) a hypotonic environment, and (2) a eukaryote or virus that routinely attacks and kills this archaeal species. If the first hypothesis is correct, the mutant cells will burst but the normal cells will live. If the second hypothesis is correct, more mutant cells will die than normal cells.

CHAPTER 8

Check Your Understanding (CYU)

CYU p. 134 Plant cell walls and animal ECMs are both fiber composites. In plant cell walls the fiber component consists of cross-linked cellulose fibers and the ground substance is pectin. In animal ECMs the fiber component consists of collagen fibrils and the ground substance is gel-forming polysaccharides. **CYU p. 139** (1) The three structures differ in composition, but their function is similar. The middle lamella in plants is composed of pectins that glue adjacent cells together. Tight junctions are made up of membrane proteins that line up and "stitch" adjacent cells together. Desmosomes are "rivet-like" structures composed of proteins that link the cytoskeletons of adjacent cells. (2) The plasma membranes of adjacent plant cells are continuous at plasmodesmata, so cytoplasm, smooth endoplasmic reticulum, and other cell components are shared. Gap junctions connect adjacent animal cells by forming protein-lined pores. The openings allow ions and small molecules to be shared between cells. **CYU p. 146** (1) Each cell-cell signal binds to a specific receptor protein. A cell can respond to a signal only if it has the appropriate receptor. Different types of cells have different types of receptors. (2) Signals are amplified if one or more steps in a signal transduction pathway, involving either second messengers or a phosphorylation cascade, results in the activation of multiple downstream molecules.

You Should Be Able To (YSBAT)

YSBAT p. 138 (1) Yes—molecules can pass through a middle lamella because it is made up of gelatinous material that is not watertight. (2) Developing muscle cells could not adhere normally and muscle tissue would not form properly. The embryo would die. **YSBAT p. 143** The spy is the signal that arrives at the receptor (the castle gate). The guard is the G protein–linked receptor in the plasma membrane, and the queen is the G protein. The commander of the guard is the enzyme that catalyzes production of a second messenger (the soldiers). **YSBAT p. 144** (1) The red dominos (RTK components) would be the first dominos in the chain and the first to be tipped to initiate the domino cascade. The black domino (Ras) would be the second domino in line. Then there would be green, blue, pink, and yellow dominos (other kinases) set up in branching patterns to represent how Ras phosphorylates and activates a number of different kinases. Each of those colored dominos would be followed by a line of dominos of the same color representing the target proteins of those kinases. (2) Tipping the red domino would knock down the black domino (Ras) and would initiate a branching domino cascade that is dependent on the black domino's falling. The red domino represents how RTK activates Ras and how Ras initiates a cascade of phosphorylation.

Caption Questions and Exercises

Figure 8.9 "Prediction of null hypothesis": Cells will not adhere, or will adhere randomly with respect to cell type and species. **Figure 8.10** "Prediction": Cells treated with an antibody that blocks membrane proteins involved in adhesion will not adhere. "Prediction of null hypothesis": All cells will adhere normally. **Figure 8.12** Top: Tight junctions; Middle: Desmosome; Bottom: Gap junctions.

Summary of Key Concepts

KC 8.1 (1) Adjacent cells fall apart from each other. (2) Nearby cells fall apart from each other; both cells and tissues are weaker and more susceptible to damage. **KC 8.2** Epithelium separates compartments, creating an inside and outside. If tight junctions degrade, materials from the outside will diffuse in and materials from the inside will diffuse out. **KC 8.3** Adrenalin binds to both heart and liver cells, but the activated receptors trigger different signal transduction pathways and lead to different cell responses. **KC 8.4** If only one step existed in a signal transduction pathway, it would be difficult for the process to be regulated by many different pathways. Multiple steps provide points for different pathways to regulate the response.

Test Your Knowledge

1. b; **2.** b; **3.** a; **4.** b; **5.** d; **6.** c

Test Your Understanding

1. The cross-linked fiber components withstand tension; the ground substance withstands compression. **2.** If each enzyme in the cascade phosphorylates many enzymes in the next step of the cascade, the initial signal will be amplified many times over. **3.** Although both are made up of membrane-spanning proteins, tight junctions seal adjacent animal cells together while gap junctions allow a flow of material between them. Middle lamellae are pectin-rich layers that glue adjacent plant cells together; plasmodesmata are gaps in the cell walls of adjacent cells where the plasma membrane is continuous, allowing exchange of materials between them. **4.** When dissociated cells from two sponge species were mixed, the cells sorted themselves into distinct aggregates that contained only cells of the same species. By blocking membrane proteins with antibodies and isolating cells that would *not* adhere, researchers found that specialized groups of proteins, including cadherins, are responsible for selective adhesion. **5.** Signal crosses plasma membrane and binds to intracellular receptor (reception) $\rightarrow$ receptor changes conformation, and signal-receptor complex moves to target site (processing) $\rightarrow$ signal-receptor complex binds to a target molecule (e.g., a gene or membrane pump), which changes its activity (response) $\rightarrow$ signal falls off receptor or is destroyed; receptor changes to inactive conformation (deactivation). **6.** Information from different signals may conflict or be reinforcing. "Cross-talk" between signaling pathways allows cells to integrate information from many signals at the same time, instead of responding to each signal in isolation.

Applying Concepts to New Situations

1. If an animal lacked an extracellular matrix, its cells would be structurally weak. Aggregations of similar cell types would be unable to form tissues, so individuals would probably develop as a mass of poorly connected or unconnected cells. A plant species with a similar disability would experience similar problems, but would also burst if placed in a hypotonic environment. **2.** (a) The response would have to be extremely local—the activated signal-receptor complex would have to affect nearby proteins. (b) Little or no amplification could occur, because there could never be more than one molecule that triggers the response. (c) The only way to regulate the response would be to block the receptor or make it more responsive to the signal. **3.** Chitin forms chains that can cross-link with one another. Therefore, the fungal cell wall likely has a structure similar to that of the plant cell wall (see Figure 8.2, left) except that chitin, rather than cellulose, is the major fiber component. **4.** Antibody binding would prevent activation of the associated G protein. Even if the appropriate signal were present, no second messenger could be produced and no cell response could occur. Permanent binding by a drug would either activate the G protein constantly or block activation completely.

CHAPTER 9

Check Your Understanding (CYU)

CYU p. 153 (1) In part, because its three phosphate groups have four negative charges in close proximity. The electrons repulse each other, raising their potential energy. (2) Electrons in C—O bonds are held more tightly than electrons in C—H bonds, so they have lower potential energy. **CYU p. 161** (1) and (2) are combined with the answer to CYU p. 166 below. (3) Start with 12 dimes on glucose. (These dimes represent the 12 electrons that will be moved to electron carriers during redox reactions throughout glycolysis and the citric acid cycle.) Move two dimes to the NADH box generated by glycolysis and the other 10 dimes to the pyruvate box. Then move these 10 dimes through the pyruvate dehydrogenase box, placing two of them in the NADH box next to pyruvate dehydrogenase and the remaining eight dimes in the acetyl CoA box. Next move the eight dimes in the acetyl CoA box through the citric acid cycle, placing six of them in the NADH box and two in the $FADH_2$ box generated during the citric acid cycle. (4) These boxes are marked with stars in the diagram. **CYU p. 166** [See Figure A9.1] To illustrate the chemiosmotic mechanism, take the dimes (electrons) piled on the NADH and $FADH_2$ boxes and move them through the ETC. While moving these dimes, also move pennies from the mitochondrial matrix to the intermembrane space. As the dimes exit the ETC, add them to oxygen to generate water. Once all of the pennies have been pumped by the ETC into the intermembrane space, move them through ATP synthase back into the mitochondrial matrix to fuel the formation of ATP. **CYU p. 168** Electron acceptors such as oxygen have a much lower electronegativity than pyruvate. Donating an electron to O_2 causes a greater drop in potential energy, making it possible to generate much more ATP per molecule of glucose.

You Should Be Able To (YSBAT)

YSBAT p. 150 (1) Ribulose and carbon dioxide will not react because the reaction is endergonic and cannot proceed without the input of energy. (2) It is "activated" because it now has high potential energy. (3) The endergonic reaction (ribulose + CO_2) is coupled to the exergonic reaction (ATP hydrolysis) via the phosphorylation of ribulose. **YSBAT p. 152** (1) They are oxidized. (2) They are reduced. (3) Oxygen. (4) The reactants have higher energy. "Energy" should be added to the product side of the equation with "Release of energy" written below it. **YSBAT p. 163** "Indirect" is accurate because the energy released during glucose oxidation is stored in reduced electron carriers (instead of being used to produce ATP directly). The energy released as electrons from these carriers flow through the ETC is used to generate a proton gradient across a membrane. These protons then diffuse down their concentration gradient through ATP synthase located in the membrane, which drives ATP synthesis.

Caption Questions and Exercises

Figure 9.6 In glucose, put two dots in the middle of each C—C or C—H bond, and two dots next to the O in the C—O bond. In O_2, put four electrons in the middle of the double bond. In CO_2, put four electrons next to each O. In H_2O, put two electrons near O in each O—H bond. The electrons are held more tightly in the products than in the reactants, so potential energy has decreased. The difference in potential energy is released, so the reaction is exergonic. **Figure 9.8** *Glycolysis:* "What goes in" = glu-

FIGURE A9.1

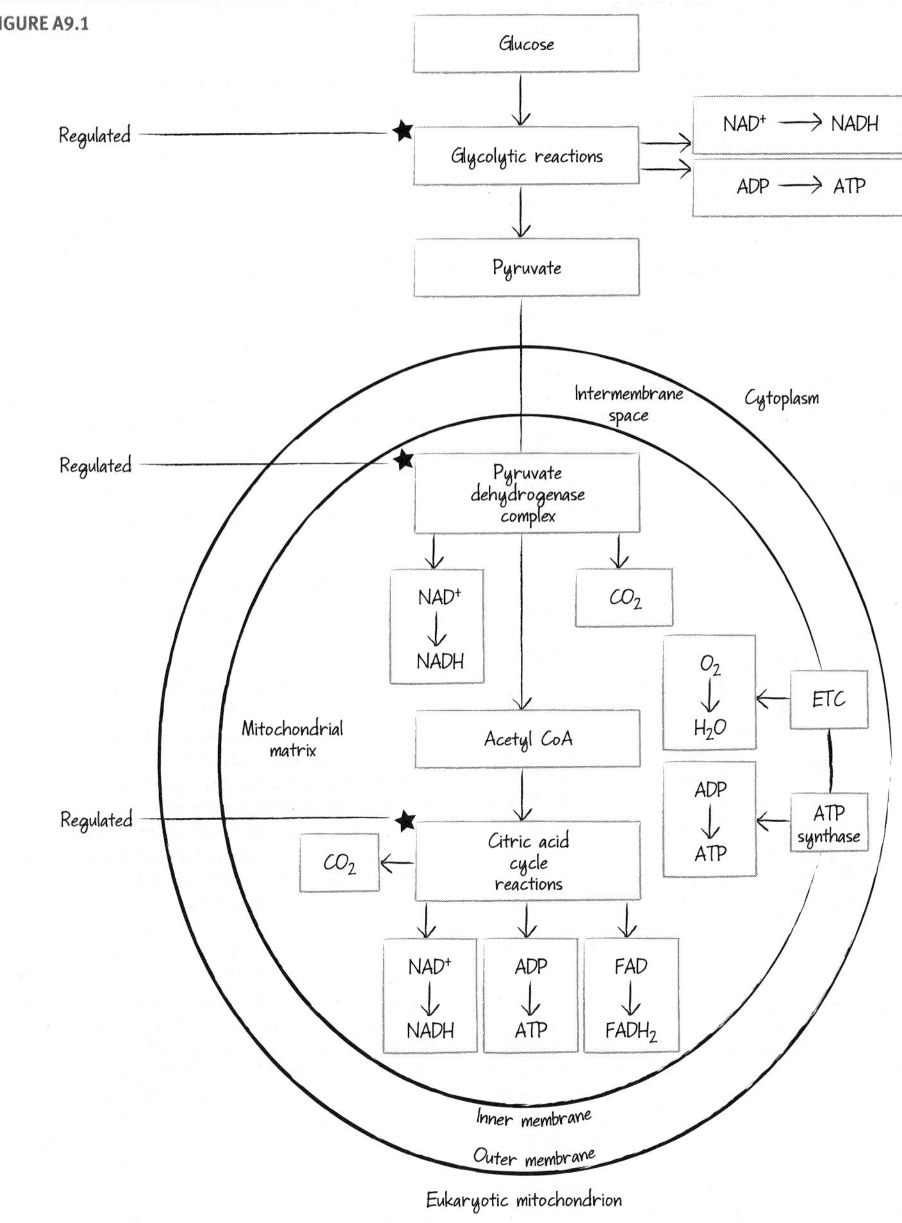

FIGURE A9.2

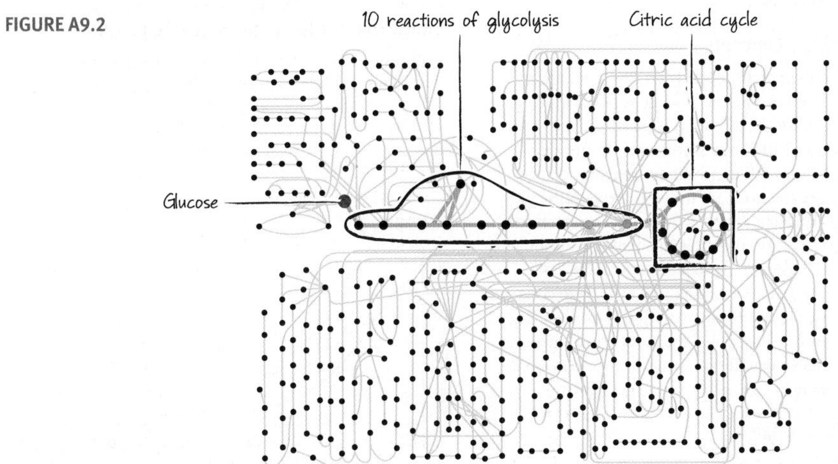

cose, NAD^+, ADP, inorganic phosphate; "What comes out" = pyruvate, NADH, ATP. *Pyruvate processing:* "What goes in" = pyruvate, NAD^+; "What comes out" = NADH, CO_2, acetyl CoA. *Citric acid cycle:* "What goes in" = acetyl CoA, NAD^+, FADH, GDP or ADP, inorganic phosphate; "What comes out" = NADH, $FADH_2$, ATP or GTP, CO_2. *Electron transport and chemiosmosis:* "What goes in" = NADH, $FADH_2$, O_2; "What comes out" = ATP, H_2O, NAD^+, FAD. **Figure 9.12** If the regulatory site had a higher affinity for ATP than the active site, then ATP would always be bound at the regulatory site, and glycolysis would always proceed at a very slow rate. **Figure 9.14** "Positive control": AMP, NAD^+, CoA (reaction substrates). "Negative control by feedback inhibition": acetyl CoA, NADH, ATP (reaction products). **Figure 9.16** An allosteric regulator binds somewhere other than the active site and causes shape changes that slow the reaction or speed it up. A competitive inhibitor binds at the active site and keeps the substrate from binding. **Figure 9.18** Production of NADH and $FADH_2$. (In the line representing free energy, there are greater drops at each occurrence of NADH and $FADH_2$ versus ATP formation.) **Figure 9.20** Yes, because the artificial vesicles include only an ATP-synthesizing enzyme and a proton pump that creates a proton-motive force. The only energy source in this experiment is the proton gradient, so the result strongly supports the chemiosmotic hypothesis. (Note: This experiment does not test whether all ATP production in the ETC is through chemiosmosis. It does not preclude the possibility that the ETC also performs some substrate-level phosphorylation.) **Figure 9.21** The proton gradient arrow should start above in the inner membrane space and point down across the membrane into the mitochondrial matrix. *Complex I:* "What goes in" = NADH and H^+; "What comes out" = NAD^+, e^-, H^+. *Complex II:* "What goes in" = $FADH_2$; "What comes out" = FAD, e^-. *Complex III:* "What goes in" = e^-; "What comes out" = e^-. *Complex IV:* "What goes in" = e^-, H^+, O_2; "What comes out" = H_2O, H^+. **Figure 9.26** [See Figure A9.2]

Summary of Key Concepts

KC 9.1 Phosphorylation adds two negative charges in a small area. The electrical repulsion that results raises the protein's potential energy and its tertiary structure. **KC 9.2** The free-energy drop from glucose to oxygen (or another electron acceptor, during cellular respiration) is much greater than the free energy drop from glucose to pyruvate (during fermentation). Thus, there is more free energy available to use in synthesizing ATP. **KC 9.3** NADH would decrease if a drug poisoned the acetyl CoA-to-citrate enzyme, but increase if a drug poisoned ATP synthase. **KC 9.4** [See Figure A9.3] **KC 9.5** Organisms that produce ATP by fermentation grow more slowly than those that produce ATP via cellular respiration

FIGURE A9.3

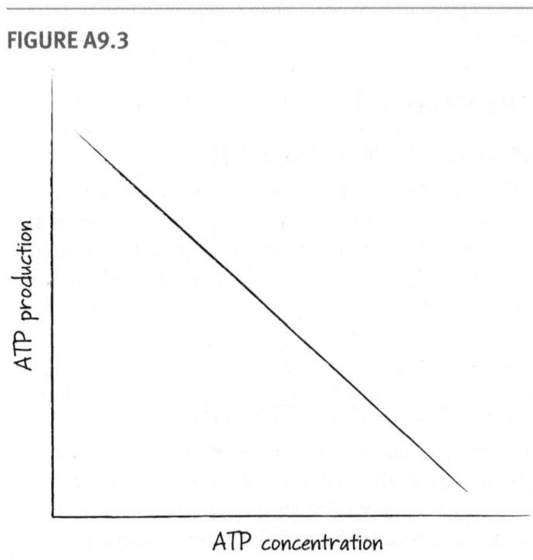

simply because fermentation produces fewer ATP molecules per glucose molecule than cellular respiration does.

Test Your Knowledge

1. b; **2.** b; **3.** d; **4.** c; **5.** a; **6.** b

Test Your Understanding

1. NADH and $FADH_2$ get their electrons from the intermediates of glycolysis, pyruvate processing, and the TCA cycle and deliver them to the ETC, where they reduce O_2. **2.** Both processes produce ATP from ADP and P_i, but substrate-level phosphorylation occurs when enzymes remove a "high-energy" phosphate from a substrate and directly transfer it to ADP, while oxidative phosphorylation is based on electrons moving through an ETC and production of a proton-motive force that drives ATP synthase. **3.** Aerobic respiration is much more productive because oxygen has extremely high electronegativity compared to other electron acceptors, resulting in a greater release of energy during electron transport and more proton pumping. **4.** Glycolysis → Pyruvate processing → TCA cycle → ETC and chemiosmosis. The first three steps are responsible for glucose oxidation; the final step produces the most ATP. **5.** Electron transport makes oxidative phosphorylation possible (oxidative phosphorylation occurs when electrons have been transported through the ETC and a proton gradient has been established). ATP synthase consists of an F_o unit and F_1 unit joined by a stalk. When protons flow through the F_o unit, the stalk and F_1 unit spin. The motion drives the synthesis of ATP from ADP and P_i. **6.** Stored carbohydrates can be broken down into glucose that enters the glycolytic pathway. If carbohydrates are absent, products from fat and protein catabolism can be used to fuel cellular respiration or fermentation. If ATP is plentiful, anabolic reactions use intermediates of the glycolytic pathway and the TCA cycle to synthesize carbohydrates, fats, and proteins.

Applying Concepts to New Situations

1. When complex IV is blocked, electrons can no longer be transferred to oxygen, the final acceptor, and cellular respiration stops. Fermentation could keep glycolysis going, but it is inefficient and unlikely to fuel a cell's energy needs over the long term. Cells that lack the enzymes required for fermentation would die first. **2.** Because mitochondria with few cristae would have fewer electron transport chains and ATP synthase molecules, they would produce much less ATP than mitochondria with numerous cristae. **3.** When oxygen is unavailable for cellular respiration, yeast cells switch to fermentation, which occurs in the cytosol. They are unlikely to expend large amounts of energy and materials to maintain mitochondria. **4.** The potential energy drop between glucose and pyruvate is relatively small. When pyruvate acts as the electron acceptor during fermentation, a great deal of potential energy remains in the alcohol or other fermentation products.

CHAPTER 10

Check Your Understanding (CYU)

CYU p. 179 When a pigment absorbs energy and an electron is excited in response, the event is analogous to striking the tuning fork. When an excited electron transmits resonance energy to another pigment molecule, the event is analogous to touching a vibrating tuning fork to another and making it vibrate. **CYU p. 184** (1) Your model should conform to the relationships shown in Figure 10.13. (2) Put four dimes on the water-splitting reaction ($2\ H_2O \rightarrow 4\ H^+ + O_2$). When four photons hit the PSII antenna complex, move the dimes to pheophytin, then to the PSII ETC (plastoquinone and the cytochrome complex). The dimes go to plastocyanin, then to the antenna complex of PSI. When four photons strike, move the dimes to the PSI ETC and ferrodoxin, then combine them with $NADP^+ + 2\ H^+$ to form NADPH. **CYU p.190** (1) (a) C_4 plants use PEP carboxylase to fix CO_2 into organic acids in mesophyll cells. These organic acids are then pumped into bundle-sheath cells, where they release carbon dioxide to rubisco. (b) CAM plants take in CO_2 at night, and have enzymes that fix it into organic acids stored in the central vacuoles of photosynthesizing cells. During the day, the organic acids are processed to release CO_2 to rubisco. (c) By diffusion through a plant's stomata when they are open. (2) Sucrose can be transported to other parts of the plant because it is water soluble. Starch is not water soluble, so it cannot be transported. It acts as a storage product, instead.

You Should Be Able To (YSBAT)

YSBAT p. 181 Photosystem II is very similar to the ETC in mitochondria. Both use the energy released from electron transfer to pump protons, generating a proton gradient that is used to generate ATP through ATP synthase. The high-energy electrons for PSII come from chlorophyll via water, however, while they come from NADH and $FADH_2$ for the ETC. The eventual products of PSII are ATP and oxygen; the eventual products of the mitochondrial ETC are ATP and water. **YSBAT p. 182** Light → Antenna complex → Reaction center → Pheophytin → ETC → Proton gradient → ATP synthase. Electrons from water are donated to the reaction center. When they absorb resonance energy and reduce pheophytin, they convert electromagnetic energy to chemical energy. **YSBAT p. 183** (1) Plastocyanin transfers electrons that move through PSII to the reaction center of PSI. (2) After they are excited by a photon and donated to the initial electron acceptor. (3) The electrons that originate from PSII's reaction center end up in NADPH—they are not cycled back to form water. **YSBAT p. 186** Rubisco joins CO_2 with RuBP to form a six-carbon intermediate that is immediately hydrolyzed to yield two molecules of 3-phosphoglycerate, which participate in reactions that yield G3P.

Caption Questions and Exercises

Figure 10.5 [See Figure A10.1] **Figure 10.7** The energy state corresponding to a photon of green light would be located between the energy states corresponding to red and blue photons. **Figure 10.10** Yes—otherwise, changes in the rate of photosynthesis could be due to changes in the density of photosynthetic cells, not differences in wavelengths received. **Figure 10.17** The researchers didn't have any basis on which to predict these intermediates. They needed to perform the experiment to identify them. **Figure 10.24** One arrow should point from sucrose to starch, another from starch to sucrose. Note that the interconversion requires transport between the cytosol and chloroplast.

Summary of Key Concepts

KC 10.1 The Calvin cycle depends on the ATP and NADPH produced by PSI and PSII, so it is not independent of light. **KC 10.2** Oxygen is produced by a critical step in photosynthesis: splitting water to provide electrons to PSII. If oxygen production increases, it means that more electrons are moving through the photosystems. **KC 10.3** The "final" product of the reaction sequence reacts to form the molecule that is the substrate for the initial reaction in the sequence. **KC 10.4** Both forms of rubisco would increase carbohydrate production: Oxygen would not compete with CO_2 for binding to the active site of rubisco and the competing reactions of photorespiration would not occur, or rubisco would fix 10 times as much CO_2 in a given period of time.

Test Your Knowledge

1. d; **2.** c; **3.** a; **4.** c; **5.** b; **6.** d

Test Your Understanding

1. In PSII, it occurs when excited electrons from the reaction center are accepted by pheophytin, reducing it. In PSI, it occurs when excited electrons from the reaction center are accepted by ferredoxin. **2.** CO_2 is fixed when rubisco catalyzes the reaction between carbon dioxide and ribulose bisphosphate to form 3-phosphoglycerate. Subsequent reactions that produce sugar require phosphorylation by ATP and high-energy electrons from NADPH. **3.** PSII and PSI both have reaction centers, an energy transformation step, and an ETC. PSII produces ATP while PSI produces NADPH, and only PSII splits water to obtain electrons. Plastocyanin transfers electrons between the two photosystems, connecting them. **4.** Photorespiration occurs when levels of CO_2 are low and O_2 are high. Less sugar is produced because (1) CO_2 doesn't participate in the initial reaction catalyzed by rubisco, and (2) when rubisco catalyzes the reaction with O_2 instead, one of the products is eventually broken down to CO_2 in a process that uses ATP. **5.** See Figure 10.22b and Figure 10.23. **6.** Photosynthesis in chloroplasts produces glucose, which is processed in mitochondria to produce ATP.

Applying Concepts to New Situations

1. Both organelles have a double membrane, extensive internal membranes, ETCs that include cytochromes and quinones, and ATP synthase. Only chloroplasts have chlorophyll and other pigment molecules that absorb light. Chloroplasts use NADPH as an electron carrier; mitochondria use NADH and $FADH_2$. The Calvin cycle occurs only in chloroplasts; the TCA cycle occurs only in mitochondria. **2.** Because CO_2 and O_2 levels minimized the impact of photorespiration when rubisco evolved, the hypothesis is credible. But once O_2 levels increased, any change in rubisco that minimized photorespiration would give individuals a huge advantage over organisms with "old" forms of rubisco. There has been plenty of time for such changes to occur, making the "holdover" hypothesis less credible. **3.** Carotenoids should be located close to the chlorophyll molecules in the reaction center, so they can pass energy along and neutralize free radicals that could damage chlorophyll. One way to test this hypothesis is to isolate thylakoid membranes and test for the presence of carotenoid and chlorophyll molecules. **4.** No—they are unlikely to have the same complement of photosynthetic pigments. Different wavelengths of light are available in various layers of a forest and water depths. It is logical to predict that plants and algae have pigments that absorb the available wavelengths efficiently. One way to test this hypothesis would be to isolate pigments from species in different locations, and test the absorbance spectra of each.

THE BIG PICTURE: ENERGY

Check Your Understanding (CYU), p. 192

1. Photosynthesis uses H_2O as a substrate and releases O_2 as a by-product; cellular respiration uses O_2 as a sub-

FIGURE A10.1

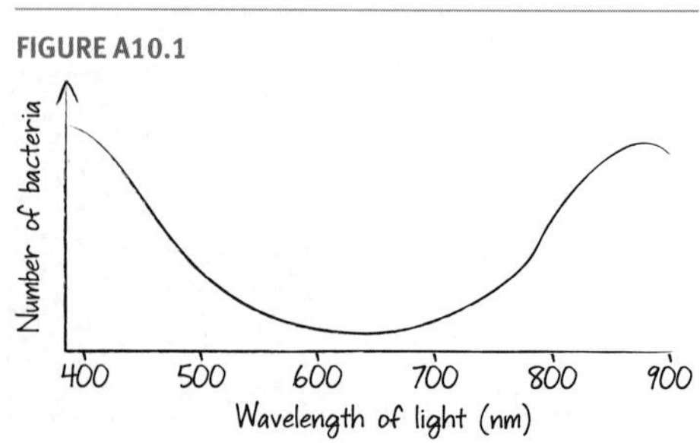

strate and releases H_2O as a by-product. **2.** Photosynthesis uses CO_2 as a substrate; cellular respiration releases CO_2 as a by-product. **3.** CO_2 fixation would essentially stop; CO_2 would continue to be released by cellular respiration. CO_2 levels in the atmosphere would increase rapidly, and production of new plant tissue would cease—meaning that most animals would quickly starve to death. **4.** ATP "is used by" the Calvin cycle; photosystem I "yields" NADPH.

CHAPTER 11

Check Your Understanding (CYU)

CYU p. 202 (1) [See Figure A11.1] (2) Interphase, prophase, prometaphase, metaphase, anaphase, telophase. Sister chromatids condense in prophase, move to the middle of the cell in metaphase, break apart in anaphase. **CYU p. 206** (1) $G_1 \rightarrow S \rightarrow G_2 \rightarrow M \rightarrow G_1$, etc. Checkpoints occur at the end of G_1 and G_2 and during M phase. (2) MPF Cdk levels are fairly constant throughout the cycle, but MPF cyclin increases. MPF activity increases at the end of G_2, initiating M phase, and declines at the end of M phase.

You Should Be Able To (YSBAT)

YSBAT p. 197 (1) Genes are segments of chromosomes that code for proteins. (2) Chromosomes are made of chromatin. (3) Sister chromatids are copies of the same chromosome, joined together. **YSBAT p. 200** [See Table A11.1] **YSBAT p. 204** MPF activates proteins that get mitosis under way. MPF consists of cyclin and a Cdk, and is inactivated by phosphorylation (at two sites) and activated by dephosphorylation (at one site). Enzymes that degrade cyclin reduce MPF levels. **YSBAT p. 208** [See Figure A11.2]

Caption Questions and Exercises

Figure 11.6 Unreplicated. **Figure 11.10** Fuse two interphase cells. If the regulatory molecule hypothesis is correct, neither cell should start M phase. But if the fusion event itself is the trigger, then at least one cell should start M phase. **Figure 11.13** In effect, MPF turns itself off after it is activated. If this didn't happen, the cell would undergo mitosis again right away. **Figure 11.15** After 1990, the death rate from lung cancer decreased in men but not in women. One hypothesis to explain the data is that rates of cigarette smoking declined in men but increased in women.

Summary of Key Concepts

KC 11.1 The gap phases give the cell time to replicate organelles and grow prior to division, as well as perform the normal functions required to stay alive. **KC 11.2** Kinetochore microtubules originate in the cytoplasm but chromosomes are in the nucleus. If the nuclear envelope did not disintegrate, the microtubules could not reach the chromosomes. **KC 11.3** The cell cycle will move forward, from G_1 to S phase, triggering cell division. **KC 11.4** A wide variety of defects, and multiple defects, cause a cell to become cancerous. No one drug or therapy can cure all of the defects involved.

Test Your Knowledge

1. b; **2.** d; **3.** d; **4.** d; **5.** c; **6.** c

Test Your Understanding

1. [See Figure A11.3] For daughter cells to have identical complements of chromosomes, all the chromosomes must be replicated during the S phase, the spindle apparatus must connect with the kinetochores of each replicated chromosome in prometaphase, and the sister chromatids of each replicated chromosome must separate in anaphase. **2.** One possible concept map is shown

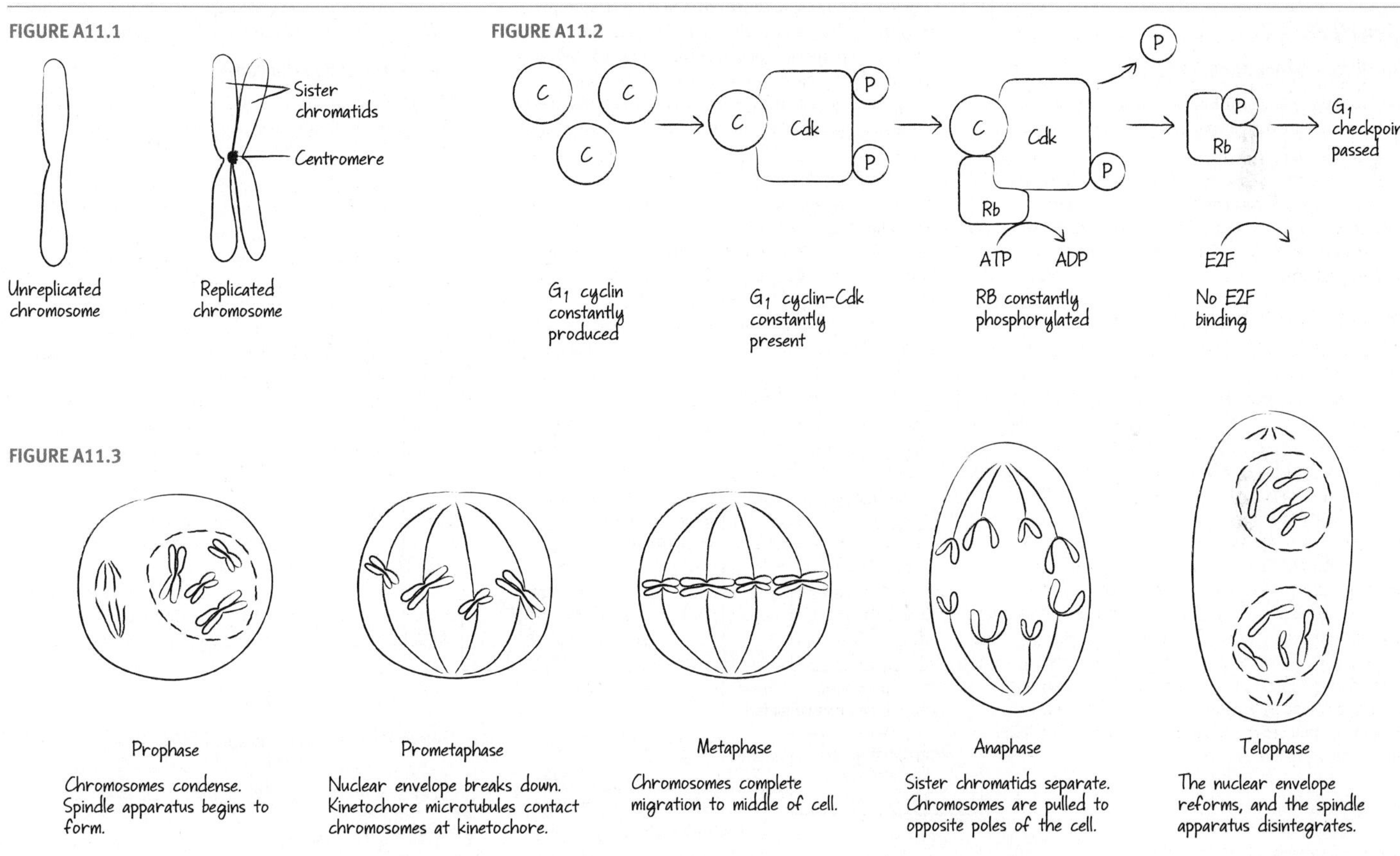

TABLE A11.1

	Prophase	Prometaphase	Metaphase	Anaphase	Telophase
Mitotic spindles	Grow	Contact chromosomes	Move chromosomes	Shorten	Break down
Nuclear envelope	Disintegrates	Nonexistent	Nonexistent	Nonexistent	Re-forms
Chromosomes	Condense	Attach to kinetochore microtubules	Move to metaphase plate	Sister chromatids separate	Collect at opposite poles

below [see Figure A11.4]. **3.** Cell fusion experiments suggested that something in mitotic cells initiated mitosis in other cells. Microinjection experiments suggested that this "something" is in the cytoplasm of M-phase cells. **4.** Protein kinases phosphorylate other proteins. Phosphorylation changes a protein's shape, altering its function (activating or inactivating it). As a result, protein kinases regulate the function of other proteins. **5.** Cyclin concentrations cycle during the cell cycle. At high concentration, cyclins bind to a specific cyclin-dependent kinase (or Cdk), forming an active protein kinase, such as MPF. **6.** If a cancer has not yet metastasized, the tumor can be completely removed by surgery.

Applying Concepts to New Situations

1. It is efficient: If the cell is not going to divide, there is no reason to invest energy in replication. **2.** A single cell with many identical nuclei in it. **3.** Labeled cells are in S phase, so 4 hours passed between the end of S phase and the start of M phase. **4.** Cancer requires many defects. Older cells have had more time to accumulate defects. Individuals with a genetic predisposition to cancer start out with some cancer-related defects, but this does not mean that the additional defects required for cancer to occur will develop.

CHAPTER 12

Check Your Understanding (CYU)

CYU p. 220 (1) Use four long and four short pipe cleaners (or pieces of cooked spaghetti) to represent the chromatids of two replicated homologous chromosomes (four total chromosomes). Mark two long and two short ones with a colored marker pen to distinguish maternal and paternal copies of these chromosomes. Twist identical pipe cleaners (e.g., the two long colored ones) together to simulate replicated chromosomes. Arrange the pipe cleaners to depict the different phases of meiosis I as follows: *Early prophase I:* Align sister chromatids of each homologous pair to form two tetrads. *Late prophase I:* Form one or more crossovers between non-sister chromatids in each tetrad. (This is hard to simulate with pipe cleaners—you'll have to imagine that each chromatid now contains both maternal and paternal segments.) *Metaphase I:* Line up homologous pairs (the two pairs of short pipe cleaners and the two pairs of long pipe cleaners) at the metaphase plate. *Anaphase I:* Separate homologs. Each homolog still consists of sister chromatids joined at the centromere. *Telophase I and cytokinesis:* Move homologs apart to depict formation of two haploid cells, each containing a single replicated copy of two different chromosomes. (2) During anaphase I, homologs (not sister chromatids, as in mitosis) are separated, making the cell products of meiosis I haploid. **CYU p. 223** (1) [See Figure A12.1] Maternal chromosomes are white and paternal chromosomes are black. Daughter cells with many other possible combinations of chromosomes than shown could result from meiosis of this parent cell. (2) Asexual reproduction generates no appreciable genetic diversity. Self-fertilization is preceded by meiosis so it generates gametes, through crossing over and independent assortment, that have combinations of alleles not present in the parent. Outcrossing generates the most genetic diversity among offspring because it produces new combinations of alleles from two different individuals.

You Should Be Able To (YSBAT)

YSBAT p. 213 $n = 12$; the organism is diploid with $2n = 24$ (in females; 23 in males). **YSBAT p. 214** [See Figure A12.2] Because the two sister chromatids are identical and attached, it is sensible to consider them as parts of a single chromosome. **YSBAT p. 216** [See Figure A12.3] **YSBAT p. 218** Crossing over would not occur and the daughter cells produced by meiosis would be diploid, not haploid. There would be no reduction division. **YSBAT p. 221** Each gamete would inherit either all maternal or all paternal chromosomes. This would limit genetic variation in the offspring by precluding the many possible gametes containing various combinations of maternal and paternal chromosomes.

Caption Questions and Exercises

Figure 12.11 [See Figure A12.4] **Figure 12.12** Asexually: 64 (16 individuals from generation three produce 4 offspring per individual). Sexually: 16 (8 individuals from generation three form 4 couples; each couple produces 4 offspring). **Figure 12.13** Sampling widely reduces the possibility that the results are due to an environmental factor—other than the presence of parasites—that differs between sexually and asexually reproducing populations.

Summary of Key Concepts

KC 12.1 Haploid cells cannot undergo meiosis, because there is no homolog to synapse with—haploid cells can-

FIGURE A11.4

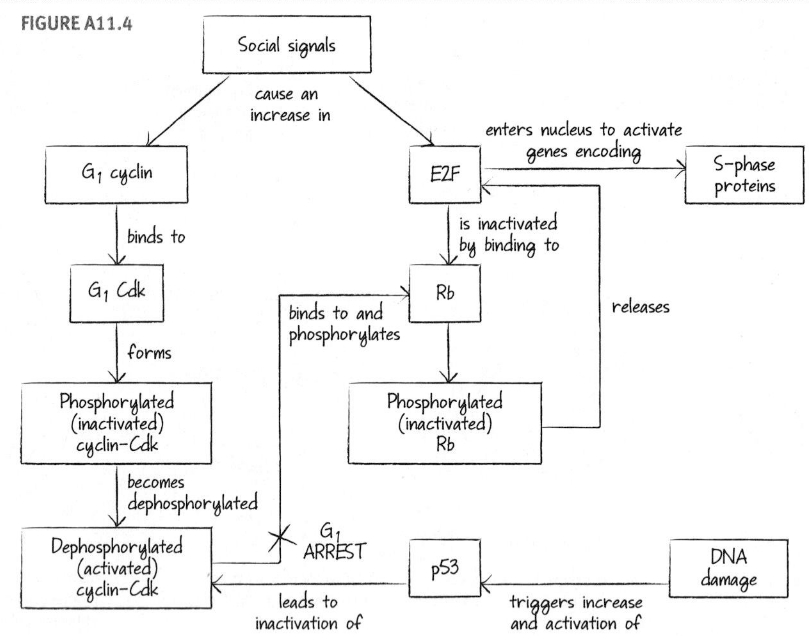

FIGURE A12.1

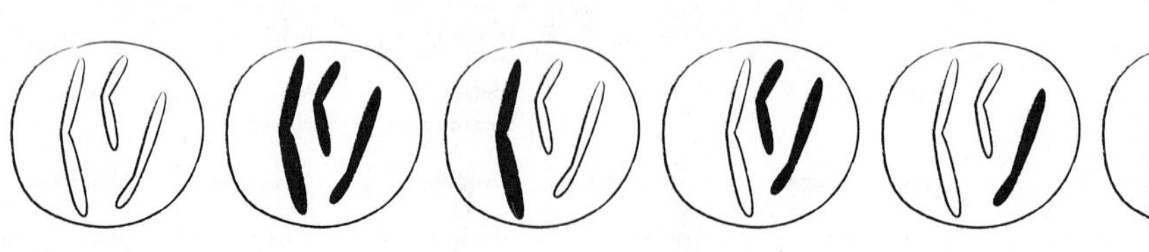

not undergo a reduction in ploidy. **KC 12.2** Monozygotic twins are genetically identical—the cells that originally gave rise to the two individuals were the product of mitosis. Dizygotic twins each arise from gametes that are the product of meiosis. Genetically, they are no more alike than any pair of siblings. **KC 12.3** Sexual reproduction will likely occur during seasons when conditions are changing rapidly, because genetically diverse offspring may have an advantage in the new conditions. **KC 12.4** The resulting gametes will be diploid instead of haploid.

Test Your Knowledge

1. b; **2.** a; **3.** b; **4.** b; **5.** d; **6.** a

Test Your Understanding

1. Homologous chromosomes are similar in size, shape, and gene content, and originate from different parents. Sister chromatids are exact copies of a chromosome that are generated when chromosomes are replicated (S phase of the cell cycle). **2.** Refer to Figure 12.6 as a guide for this exercise. The four pens represent the chromatids in one replicated homologous pair; the four pencils, the chromatids in a different homologous pair. To simulate meiosis II, make two "haploid cells"—each with a pair of pens and a pair of pencils representing two replicated chromosomes (one of each type in this species). Line them up in the middle of the cell, then separate the two pens and the two pencils in each cell such that one pen and one pencil goes to each of four daughter cells. **3.** Meiosis I is a reduction division because homologs separate—daughter cells have just one of each type of chromosome instead of two. Meiosis II is not a reduction division because sister chromatids separate—daughter cells have unreplicated chromosomes instead of replicated chromosomes, but still just one of each type. **4.** Tetraploids produce diploid gametes, which combine with a haploid gamete from a diploid individual to form a triploid offspring. Mitosis proceeds normally in triploid cells because replicated chromosomes align independently before sister chromatids separate. But during meiosis in a triploid, homologous chromosomes can't pair up correctly. The third set of chromosomes does not have a homologous partner to pair with. **5.** Asexually produced individuals are genetically identical, so if one is susceptible to a new disease, all are. Sexually produced individuals are genetically unique, so if a new disease strain evolves, at least some plants are likely to be resistant to it. **6.** If homologs or sister chromatids do not separate normally, the daughter cells will receive one too many chromosomes or one too few. The resulting chromosome sets are considered unbalanced because they do not have the normal number of copies of each gene.

Applying Concepts to New Situations

1. The gibbon would have 22 chromosomes in each gamete, and the siamang would have 25. Each somatic cell of the offspring would have 47 chromosomes. The offspring should be sterile because it has an unbalanced set of chromosomes, so not all of the chromosomes would have a homolog to pair with in meiosis prophase I. **2.** One in eight. **3.** Aneuploidy is the major cause of spontaneous abortion. If spontaneous abortion is rare in older women, it would result in a higher incidence of aneuploid conditions such as Down syndrome in older women, as recorded in the figure. **4.** (a) Such a study might be done in the laboratory, controlling conditions in identical tanks. A population of snails can be established in each tank. A parasite could be added to one tank, and then changes in the frequency of sexual reproduction in the snail populations can be monitored and observed over time. (b) In nature, all of the forces that can act to produce change in a population are present. A laboratory test is only an approximation of what occurs in the wild. For example, sexually reproduced offspring might only have an advantage when the parasite-ridden population is also stressed by food shortages or high or low temperatures. If so, an experiment in the lab with ample food and preferred temperatures would give an incorrect result.

CHAPTER 13

Check Your Understanding (CYU)

CYU p. 239 See answers to Genetics Problems on p. A:14. **CYU p. 243** (1) [See Figure A13.1] Segregation of alleles occurs when homologs that carry those alleles are separated during anaphase I. One allele ends up in each

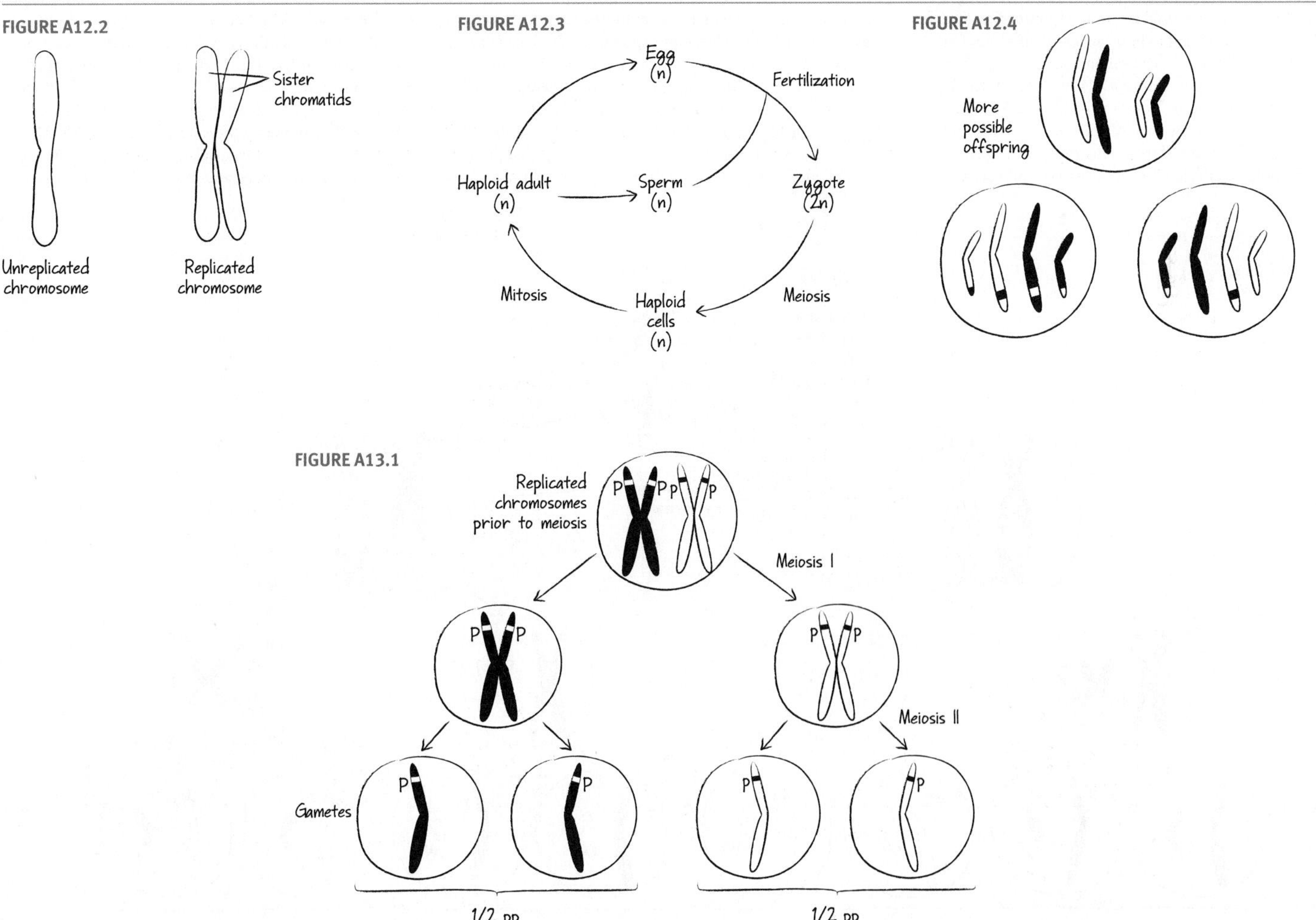

daughter cell. (2) [See Figure A13.2] Independent assortment occurs because homologous pairs line up randomly at the metaphase plate during metaphase I. The figure shows two alternative arrangements of homologs in metaphase I. As a result, it is equally possible for a gamete to receive the following four combinations of alleles: *YR*, *Yr*, *yR*, *yr*. **CYU p. 250** (1) The comb phenotype results from interactions between alleles at two different genes, not a single gene. Matings between rose and pea comb chickens produce F_2 offspring that may have a new combination of alleles, and thus new phenotypes.
(2) Kernel color in wheat is influenced by alleles at many different genes, not a single gene. F_2 offspring have a normal distribution of phenotypes, not a 3:1 ratio.

You Should Be Able To (YSBAT)

YSBAT p. 236 Punnett squares are an efficient way to predict offspring genotypes from parents of known genotype. The phenotype ratios are 1:1 round:wrinkled; the genotype ratios are 1:1 *Yy*:*yy*. **YSBAT p. 238** *AABb* → *AB* and *Ab*. *PpRr* → *PR*, *Pr*, *pR*, and *pr*. *AaPpRr* → *APR*, *APr*, *ApR*, *Apr*, *aPR*, *apR*, *apr*, and *aPr*.

Caption Questions and Exercises

Figure 13.3 An experiment is a failure if you didn't learn anything from it. That is not the case here. **Figure 13.4** No—the outcome (the expected offspring genotypes that the Punnett square generates) will be the same.
Figure 13.12 X^{WY}, X^{Wy}, X^{wY}, X^{wy}. **Figure 13.13** Random chance (or perhaps red-eyed, gray-bodied males don't survive well). **Figure 13.17** The gene colored orange is *ruby*; the gene colored blue is *miniature wings*.
Figure 13.22 If both parents are heterozygous (carriers), on average 1/4 of their children will be affected. If the allele is rare, then it is unlikely for the spouse of an affected (homozygous) parent to be a carrier—hence, children will not be affected—they will only get one allele from their affected parent. **Figure 13.24** All affected females would have all affected sons and daughters, no matter what the father's genotype. Affected fathers would transmit the allele to half of their daughters, so on average half would be affected if the mother was unaffected.

Summary of Key Concepts

KC 13.1 *B* and *b*. **KC 13.2** *BR*, *Br*, *bR*, and *br*, in equal proportions. **KC 13.3** The *B* and *b* alleles are located on different but homologous chromosomes, which separate into different daughter cells during meiosis I. The *BbRr* notation indicates that the *B* and *R* genes are on different chromosomes. As a result, the chromosomes line up independently of each other in metaphase of meiosis I. The *B* allele is equally likely to go to a daughter cell with *R* as with *r*; likewise the *b* allele is equally likely to go to a daughter cell with *R* as with *r*. **KC 13.4** Instead of two alleles on the same chromosome being transmitted together, crossing over separates them so they are transmitted independently of each other. They are no longer physically linked.

Genetics Questions

1. d; **2.** b; **3.** a; **4.** d; **5.** a; **6.** a; **7.** d; **8.** b; **9.** b; **10.** d

Genetics Problems

1. 3/4; 1/256 (see **BioSkills 13** in Appendix A); 1/2 (the probabilities of transmitting the alleles or having sons does not change over time). **2.** Your answer to the first three parts should conform to the F_1 and F_2 crosses diagrammed in Figure 13.5b, except that different alleles and traits are being analyzed. The recombinant gametes would be *Yi* and *yI*. Yes—there would be some individuals with round, green seeds and with wrinkled, yellow seeds. **3.** Cross 1: non-crested (*Cc*) × non-crested (*Cc*) = 14 non-crested (*C_*); 7 crested (*cc*). Cross 2: crested (*cc*) × crested (*cc*) = 22 crested (*cc*). Cross 3: non-crested (*Cc*) × crested (*cc*) = 7 non-crested (*Cc*); 6 crested (*cc*). Non-crested (*C*) is the dominant allele. **4.** This is a dihybrid cross that yields progeny phenotypes in a 9:3:3:1 ratio. Let *O* stand for the allele for orange petals and *o* the allele for yellow petals; let *S* stand for the allele for spotted petals and *s* the allele for unspotted petals. Start with the hypothesis that *O* is dominant to *o*, that *S* is dominant to *s*, that the two genes are found on different chromosomes so they assort independently, and that the parent individual's genotype is *OoSs*. If you do a Punnett square for the *OoSs* × *OoSs* mating, you'll find that progeny phenotypes should be in the observed 9:3:3:1 proportions. **5.** Let *D* stand for the normal allele and *d* for the allele responsible for Duchenne-type muscular dystrophy. The woman's family has no history of the disease, so her genotype is almost certainly *DD*. The man is not afflicted, so he must be *DY*. (The trait is X-linked, so he has only one allele; the "*Y*" stands for the Y chromosome.) Their children are not at risk. The man's sister could be a carrier, however—meaning she has the genotype *Dd*. If so, then half of the second couple's male children are likely to be affected. **6.** Your stages of meiosis should look like Figure 12.6, except with $2n = 4$ instead of $2n = 6$. The *A* and *a* alleles could be on the red and blue versions of the longest chromosome, and the *B* and *b* alleles could be on the red and blue versions of the smallest chromosomes. The places you draw them are the locations of the *A* and *B* genes, but each chromosome has only one allele. Each pair of red and blue chromosomes is a homologous pair. Sister chromatids bear the same allele (e.g., both sister chromatids of the long blue chromosomes might bear the *a* allele). Chromatids from the longest and shortest chromosomes are not homologous. To identify the events that result in the principles of segregation and independent assortment, see Figures 13.7 and 13.8 and substitute *A*, *a*, and *B*, *b* for *R*, *r* and *Y*, *y*.
7. Half of their offspring should have the genotype iI^A and the type A blood phenotype. The other half of their offspring should have the genotype iI^B and the type B blood phenotype. Second case: the genotype and phenotype ratios would be 1:1:1:1 I^AI^B (type AB) : I^Ai (type A) : I^Bi (type B) : *ii* (type O). **8.** Because the children of Tukan and Valco had no eyes and smooth skin, you can conclude that the allele for eyelessness is dominant to eyes and the allele for smooth skin is dominant to hooked skin. *E* = eyeless, *e* = two eye sets, *S* = smooth skin, *s* = hooked skin. Tukan is *eeSS*; Valco is *EEss*. The children are all *EeSs*. Grandchildren with eyes and smooth skin are *eeS_*. Assuming that the genes are on different chromosomes, one-fourth of the children's gametes are *ee* and three-fourths are *S_*. So ¼ *ee* × ¾ *S_* × 32 = 6 children would be expected to have two sets of eyes and smooth skin. **9.** Although the mothers were treated as children by a reduction of dietary phenylalanine, they would have accumulated phenylalanine and its derivatives once they

FIGURE A13.2

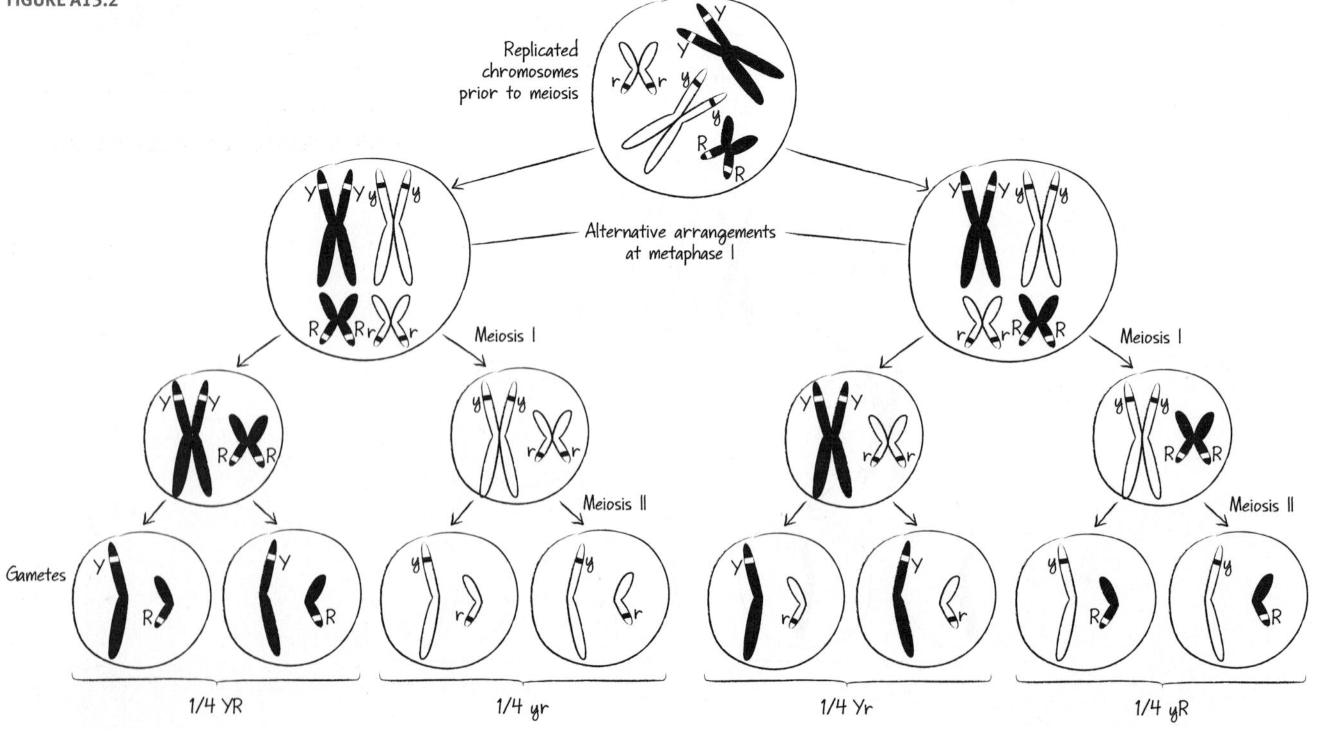

went off the low-phenylalanine diet as young adults. Children born of such mothers were therefore exposed to high levels of phenylalanine during pregnancy. For this reason, a low-phenylalanine diet is recommended for such mothers throughout the pregnancy. **10.** According to Mendel's model, palomino individuals should be heterozygous at the locus for coat color. If you mated palomino individuals, you would expect to see a combination of chestnut, palomino, and cremello offspring. If blending inheritance occurred, however, all of the offspring should be palomino. **11.** Because this is an X-linked trait, the father who has hemophilia could not have passed the trait on to his son. Thus, the mother in couple 1 must be a carrier and have passed the recessive allele on to her son, who is XY and affected. To educate a jury about the situation, you should draw what happens to the X and Y during meiosis, then make a drawing showing the chromosomes in couple 1 and couple 2, with a Punnett square showing how these chromosomes get passed to the affected and unaffected children. **12.** The curved-wing allele is autosomal recessive; the lozenge-eye allele is sex-linked (specifically, X-linked) recessive. Let *L* be the allele for long wings and *l* be the allele for curved wings; let X^R be the allele for red eyes and X^r the allele for lozenge eyes. The female parent is LlX^RX^r; the male parent is LlX^RY. **13.** Albinism indicates the absence of pigment, so let *b* stand for an allele that gives the absence of blue and *y* for an allele that gives the absence of yellow pigment. If blue and yellow pigment blend to give green, then both green parents are *BbYy*. The green phenotype is found in *BBYY, BBYy, BbYY,* and *BbYy* offspring. The blue phenotype is found in *BByy* or *Bbyy* offspring. The yellow phenotype is observed in *bbYY* or *bbYy* offspring. Albino offspring are *bbyy.* The phenotypes of the offspring should be in the ratio 9:3:3:1 as green:blue:yellow:albino. Two types of crosses yield *BbYy* F_1 offspring: *BByy* × *bbYY* (blue × yellow) and *BBYY* × *bbyy* (green × albino). **14.** Autosomal dominant.

CHAPTER 14

Check Your Understanding (CYU)

CYU p. 268 (1) DNA polymerase adds only nucleotides to the free 3′-OH on a strand. Primase synthesizes a short RNA sequence that provides the free 3′ end necessary for DNA polymerase to start working. (2) The helix is already opened and the downstream portions relaxed. **CYU p. 271** (1) Telomerase is not needed—bacterial chromosomes don't shorten after replication because they are circular. DNA polymerase can synthesize a complementary strand all the way around. (2) Otherwise there would be no template to synthesize the extension of a single DNA strand.

You Should Be Able To (YSBAT)

YSBAT p. 264 [See Figure A14.1] The new strands grow in opposite directions, each in 5′ → 3′ direction. **YSBAT p. 266** Helicase, topoisomerase, single-strand DNA-binding proteins, primase, and DNA polymerase are all required for leading-strand synthesis. If any one of these proteins is nonfunctional, then DNA replication will not occur. **YSBAT p. 267** [See Figure A14.2] If DNA ligase was defective, then the leading strand would be continuous, and the lagging strand would have gaps in it where the Okazaki fragments had not been joined.

Caption Questions and Exercises

Figure 14.2 The lack of radioactive protein in the pellet (after centrifugation) is strong evidence; they could also make micrographs of infected bacterial cells before and after agitation. **Figure 14.3** 5′ TAG 3′. **Figure 14.5** The same two bands should appear, but the upper band (DNA containing only ^{14}N) should get bigger and darker and the lower band (hybrid DNA) should get smaller and lighter in color, as there is less and less heavy DNA in each succeeding generation. **Figure 14.13** As long as the RNA template could bind to the "overhanging" section of single-stranded DNA, any sequence could produce a longer strand. For example, 5′ CCCAUUCCC 3′ would work just as well. **Figure 14.15** They are much lower in energy. **Figure 14.17** Exposure to UV radiation can cause formation of thymine dimers. If thymine dimers are not repaired, they represent mutations. If such mutations occur in genes controlling the cell cycle, cells can grow abnormally, resulting in cancers.

Summary of Key Concepts

KC 14.1 Complementary bases hydrogen bond, and hydrophobic interactions occur between bases stacked inside the double helix. **KC 14.2** The bases added during DNA replication are shown in color type.

Original DNA:	CAATTACGGA GTTAATGCCT
Replicated DNA:	CAATTACGGA GTTAATGCCT CAATTACGGA GTTAATGCCT

KC 14.3 DNA polymerase synthesizes DNA by adding deoxyribonucleotides to the free 3′-OH group of an existing nucleotide sequence (primer), using single-stranded DNA as a template. Primase synthesizes short RNA sequences, using single-stranded DNA as a template. In contrast to DNA polymerase, primase can begin synthesis de novo, that is, without a primer. Telomerase adds deoxyribonucleotides to the unreplicated 3′ end of the lagging strand, using as the template a short RNA molecule that is associated with the enzyme. **KC 14.4** If errors in DNA aren't corrected, they represent mutations. When DNA repair systems fail, the mutation rate increases. As the mutation rate increases, the chance that one or more cell cycle genes will be mutated increases. Mutations in these genes often result in uncontrolled cell division, ultimately leading to cancer.

Test Your Knowledge

1. d; **2.** c; **3.** d; **4.** b; **5.** c; **6.** d

Test Your Understanding

1. Whether DNA or proteins were labeled. **2.** There are two replication forks at each point of origin—replication proceeds in both directions at the same time. **3.** On the lagging strand, DNA polymerase moves away from the replication fork. When helicase unwinds a new section of DNA, primase must build a new primer on the lagging strand (closer to the fork) and another polymerase molecule must begin synthesis at this point. On the leading strand, DNA polymerase moves in the same direction as helicase, so synthesis can continue without interruption from one primer (at the origin of replication). **4.** Telomerase binds to the 3′ overhang at the end of a chromosome. Once bound, it begins catalyzing the addition of deoxyribonucleotides to the overhang in the 5′ → 3′ direction, lengthening the overhang. This allows primase, DNA polymerase, and ligase to catalyze the addition of deoxyribonucleotides to the lagging strand in the 5′ → 3′ direction, restoring the lagging strand to its original length. **5.** First, DNA polymerases are highly selective in matching complementary bases. Second, DNA polymerase III proofreads by recognizing mismatched bases, cutting out the incorrect base, and replacing it with the correct base. Third, proteins of the mismatch repair system excise incorrect bases and insert the proper ones. **6.** DNA is a symmetrical double helix. A mismatch (e.g., A-G or C-T) or damage to a base pair (e.g., thymine dimer) will cause a distortion or kink in the molecule that is recognizable by repair enzymes. Because the bases on the two strands are complementary, repair enzymes are able to replace the damaged bases easily and accurately.

Applying Concepts to New Situations

1. [See Figure A14.3] Primase would not have to build primers on the leading and lagging strands, because DNA polymerase III would be able to use a "raw" single-stranded template. It is likely that telomerase would still

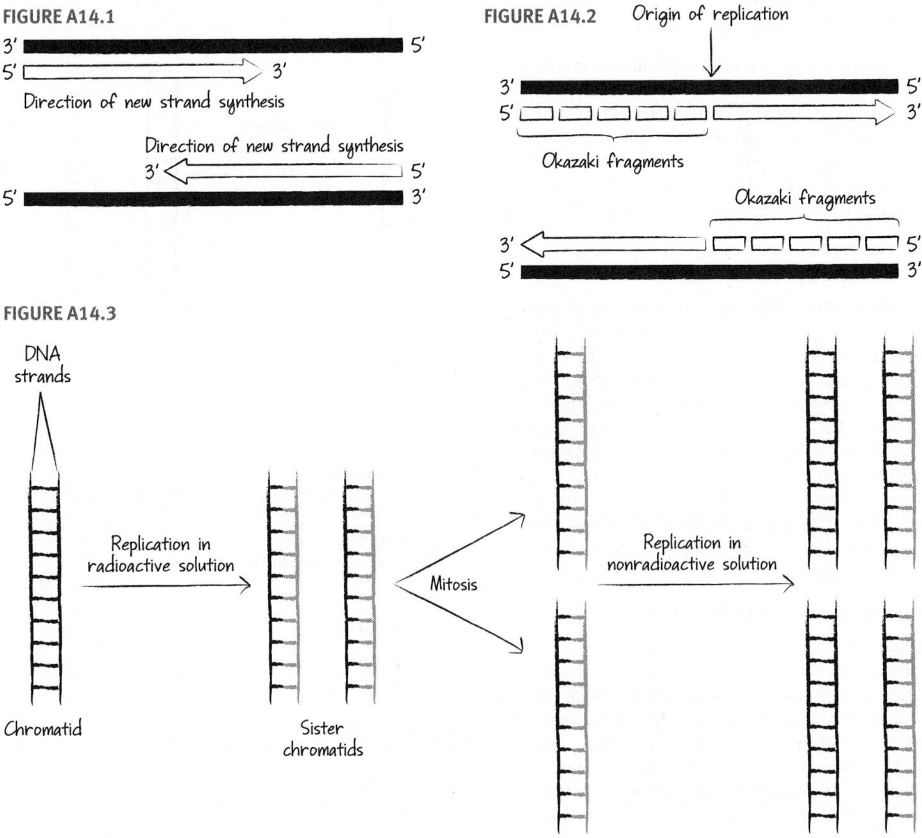

be required because, like primase, DNA polymerase would probably not have room to attach to the last end of the lagging strand, leaving a single-stranded overhang. **2.** (a) In Figure A.14 below, the gray lines represent DNA strands containing radioactivity. Following replication in a radioactive solution, one DNA strand of each sister chromatid is radioactive. (b) [See Figure A14.4] **3.** (a) The double mutant of both *uvrA* and *recA* is most sensitive to UV light; the single mutants are in between; and the wild type is least sensitive. (b) The *recA* gene contributes more to UV repair through most of the UV dose levels. But at very high UV doses, the *uvrA* gene is somewhat more important than the *recA* gene. **4.** Many types of cancer arise because of mutations, especially those in genes involved in controlling cell growth. If chemicals that increase mutation rates in bacteria also increase mutation rates in humans, then the probability of mutations in cell control genes, and hence of cancer formation, also increases.

CHAPTER 15

Check Your Understanding (CYU)

CYU p. 282 If an error in transcription changes the sequence of bases in an mRNA, it may change the protein that is subsequently produced—and thus change the cell's phenotype. Changes in the sequence of other types of RNAs might also affect the phenotype. **CYU p. 285** (1) Remember that DNA and mRNA sequences are written in the $5' \rightarrow 3'$ direction unless indicated otherwise. Thus, to convert DNA triplets (codons) to mRNA codons, the DNA sequence is read "backwards" ($3' \rightarrow 5'$) and U (rather than T) is the base transcribed from A.

DNA codon	mRNA codon	Amino acid
ATA	UAU	Tyrosine
GTA	UAC	Tyrosine
TTA	UAA	Stop
GCA	CGU	Arginine

(2) The ATA $\rightarrow$ GTA mutation would have no effect on the protein. The ATA $\rightarrow$ TTA mutation introduces a stop codon, so the resulting polypeptide would be shortened. This would result in synthesis of a mutant protein much shorter than the original protein. The ATA $\rightarrow$ GCA mutation might have a profound effect on the protein's conformation because arginine's structure is different from tyrosine's.

You Should Be Able To (YSBAT)

YSBAT p. 285 (1) The codons in Figure 15.5 are translated correctly. (2) [See Figure A15.1] (3) There are many possibilities (just pick alternate codons for one or more of the amino acids); one is an mRNA sequence of (running $5' \rightarrow 3'$) GCG-AAC-GAU-UUC-CAG. To get the corresponding DNA sequence, write this sequence but substitute Ts for Us: GCG-AAC-GAT-TTC-CAG. This strand also runs in the $5' \rightarrow 3'$ direction. Now write the complementary bases, which will be in the $3' \rightarrow 5'$ direction: CGC-TTC-CTA-AAG-GTC. When this second strand is transcribed by RNA polymerase, it will produce the mRNA given with the proper $5' \rightarrow 3'$ orientation.

Caption Questions and Exercises

Figure 15.1 No, it could not make citrulline from ornithine without enzyme 2. Yes, it would no longer need enzyme 2 to make citrulline. **Figure 15.2** Many possibilities: strain of fungi used, exact method for creating mutants and harvesting spores to plant, exact growing conditions (temperature, light, recipe for growth medium—including concentration of supplemented amino acids), objective criteria for determining growth or no-growth. **Figure 15.5** $4 \times 4 \times 4 \times 4 = 256$ amino acids. A 4-base code is unlikely to evolve because only 20 amino acids are commonly used in the synthesis of proteins—there is no selective advantage for a code to be more complicated than necessary. **Figure 15.7** [See Figure A15.1] **Figure 15.8** The DNA sequence changes only at a single location, or point.

Summary of Key Concepts

KC 15.1 Not all genes code for polypeptides. A better statement might be something like, "one-gene, one RNA or polypeptide product." **KC 15.2** Transcription means to copy, and mRNA is a short-lived copy of the information in DNA. Translation means to change languages, and proteins are a different "chemical language" than nucleic acids (DNA and RNA). **KC 15.3** If a change in a DNA sequence doesn't change the amino acid specified by the corresponding mRNA codon, then the change in genotype doesn't change the phenotype. Many changes in the third positions of codons do not change the phenotype, because the genetic code is redundant. **KC 15.4** The new mutation produced light coat colors, which allowed beach-dwelling mice to be camouflaged and survive better. If the same mutation occurred in a woodland-dwelling population, the light-colored individuals would probably be spotted by predators and eaten.

Test Your Knowledge

1. d; **2.** a; **3.** d; **4.** a; **5.** d; **6.** b

Test Your Understanding

1. In Morse code, different combinations of dots and dashes code for the complexity of the English language. In the genetic code, different combinations of bases code for the complexity of proteins in the cell.

2.

Substrate 1 $\xrightarrow{A}$ Substrate 2 $\xrightarrow{B}$ Substrate 3 $\xrightarrow{C}$

Substrate 4 $\xrightarrow{D}$ Substrate 5 $\xrightarrow{E}$ BiologicalSciazine

Substrate 3 would accumulate. Hypothesis: The individuals have a mutation in the gene for enzyme D. **3.** They supported an important prediction of the hypothesis: Losing a gene (via mutation) resulted in loss of an enzyme. **4.** It puts the three-nucleotide codons out of register. If the first base is deleted from the sequence ATG-CGA-GAC-TTA, then the reading frame becomes TGC-GAG-ACT-TA. **5.** In a triplet code, addition or deletion of 1–2 bases disrupts the reading frame "downstream" of the mutation site(s), resulting in a dysfunctional protein. But addition or deletion of 3 bases restores the reading frame—the normal sequence is disrupted only between the first and third mutation. The resulting protein is altered, but may still be able to function normally. Only a triplet code would show these patterns. **6.** A point mutation changes the nucleotide sequence of an existing allele, creating a new one, so it always changes the genotype. But because the genetic code is redundant, some point mutations do not change the resulting mRNA codons and thus do not change the protein product.

Applying Concepts to New Situations

1. [See Figure A15.2] **2.** Every copying error would result in a mutation that would change the amino acid sequence of the protein and would likely affect its function. **3.** Before the central dogma was understood, DNA was known to be the hereditary material, but no one knew how particular sequences of bases resulted in the production of RNA and protein products. The central dogma clarified

FIGURE A14.4

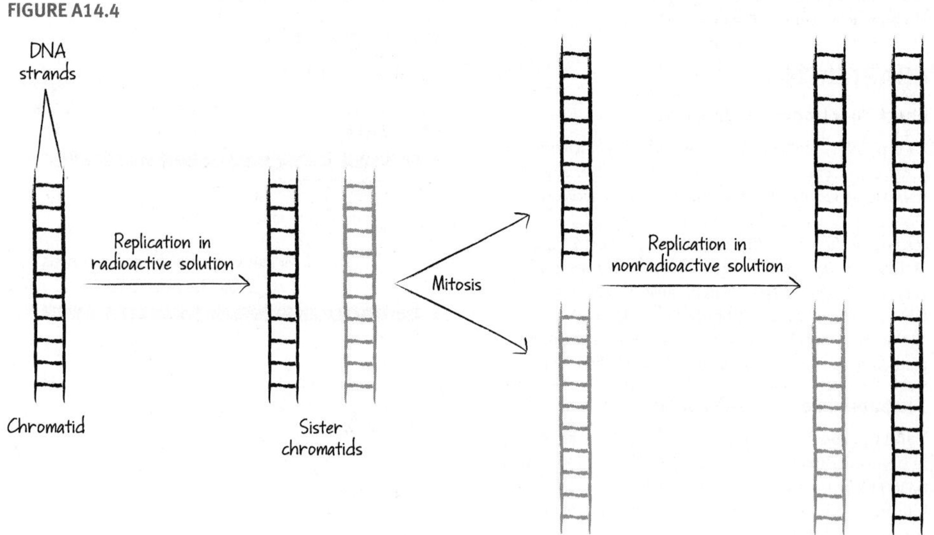

FIGURE A15.1

mRNA sequence:
5′ AUG-CCC-CUG-GAG-GGG-GUU-AGA-CAU 3′

Amino acid sequence:
Met-Pro-Leu-Glu-Gly-Val-Arg-His

FIGURE A15.2

Bottom DNA strand:
5′ AACTT-TAC(start)-GGG-CAA-ACC-ACT-AGC-CCA-ATG-TCG-ATC(stop)-AGTTTC 3′

mRNA sequence:
5′ AUG-CCC-GUU-UGG-AGA-UCG-GGU-UAC-AGC-UAG 3′

Amino acid sequence:
Met-Pro-Val-Trp-Arg-Ser-Gly-Tyr-Ser

how genotypes produce phenotypes. **4.** A triplet code that overlapped by one base might work as follows:

```
5′ AUGUUACGGAAUUGA 3′
    GUU
     AUC
      CGG
       GAA
        AUU
         UGA
```

CHAPTER 16

Check Your Understanding (CYU)

CYU p. 293 (1) Different types of sigma proteins have different amino acid sequences and overall structures, so they are able to bind to stretches of DNA that differ in their base sequence. (2) NTPs are required because the three phosphate groups raise the monomer's potential energy enough to make the polymerization reaction exergonic. **CYU p. 295** (1) The subunits contain both RNA (the "ribonucleo" in the name) and proteins. (2) The cap and tail protect mRNAs from degradation and facilitate translation. **CYU p. 304** E is for Exit—the site where empty tRNAs are ejected; P is for Peptidyl (or peptide bond)—the site where peptide bond formation takes place; A is for Aminoacyl—the site where aminoacyl tRNAs enter.

You Should Be Able To (YSBAT)

YSBAT p. 299 (1) The amino acid attaches on the top left of the L-shaped structure. (2) The anticodon is antiparallel in orientation to the mRNA codon, and contains the complementary bases.

Caption Questions and Exercises

Figure 16.1 RNA is synthesized in the $5' \rightarrow 3'$ direction; the DNA template is "read" $3' \rightarrow 5'$. **Figure 16.5** There would be no loops—the molecules would match up exactly. **Figure 16.12** If the amino acids stayed attached to the tRNAs, the gray line in the graph would stay high and the green line low. If the amino acids were transferred to some other cell component, the gray line would decline but the green line would be low.

Summary of Key Concepts

KC 16.1 The new sequence would probably not bind sigma or basal transcription factors as effectively as the original sequence. If so, transcription rate at that gene would probably decrease. **KC 16.2** Bacterial genes lack introns, so their RNA products do not require splicing. And because bacterial mRNAs are translated immediately, they do not need a cap and tail. **KC 16.3** These RNAs transfer amino acids to growing polypeptides; they transfer information in nucleic acids (an mRNA codon) to a protein product. **KC 16.4** An aminoacyl tRNA synthetase catalyzes the addition of an amino acid to a tRNA, forming an aminoacyl tRNA. **KC 16.5** One possible concept map is shown below [see Figure A16.1].

Test Your Knowledge

1. c; **2.** c; **3.** d; **4.** c; **5.** c; **6.** d

Test Your Understanding

1. Basal transcription factors bind to promoter sequences in eukaryotic DNA and facilitate the positioning of RNA polymerase. As part of the RNA polymerase holoenzyme, sigma binds to a promoter sequence in bacterial DNA and initiates transcription by the core enzyme. **2.** If the wobble rules did not exist, it would take one tRNA for each amino-acid-specifying codon in the genetic code. This is inefficient. The wobble rules allow a single tRNA to match several mRNA codons, reducing the inefficiency caused by redundancy in the genetic code. **3.** Eukaryotic mRNAs contain introns which must be removed by splicing. Splicing takes place in the nucleus and is accomplished by a large complex called a spliceosome, composed of many snRNPs. **4.** After a peptide bond forms between the polypeptide and the amino acid held by the tRNA in the A site, the ribosome moves down the mRNA. As it does, an empty tRNA leaves the E site. The now-empty tRNA that was in the P site enters the E site; the tRNA holding the polypeptide chain moves from the A site to the P site, and a new aminoacyl tRNA enters the A site. **5.** The ribosome's active site is made up of RNA, not protein. **6.** The separation allows the aminoacyl tRNA to fit into the ribosome correctly, such that the anticodon binds to the mRNA and the amino acid fills the active site.

Applying Concepts to New Situations

1. Ribonucleases degrade mRNAs that are no longer needed by the cell. If an mRNA for a hormone that increased heart rate were never degraded, the hormone would be produced continuously and heart rate would stay elevated—a dangerous situation. **2.** Cordycepin triphosphate has a triphosphate group on the $5'$ carbon, but no $3'$-OH group. If RNA were built $3' \rightarrow 5'$, then the incoming nucleoside triphosphates would have to have an OH group on the $3'$ carbon in order for a phosphodiester bond to be formed. Since cordycepin lacks a $3'$-OH, it could not be added to the $5'$ end of an RNA chain growing in the $3' \rightarrow 5'$ direction. **3.** The regions most crucial to the ribosome's function should be the most highly conserved: the active site, the E-, P-, and A-sites, and the site where mRNAs initially bind. Other regions should be more variable among species. **4.** The most likely locations are one of the grooves or channels where RNA, DNA, and ribonucleotides move through the enzyme—plugging one of them would prevent transcription.

CHAPTER 17

Check Your Understanding (CYU)

CYU p. 314 (1) The genes for metabolizing lactose should be expressed only when lactose is available. (2) [See Figure A17.1] **CYU p. 316** (1)

FIGURE A16.1

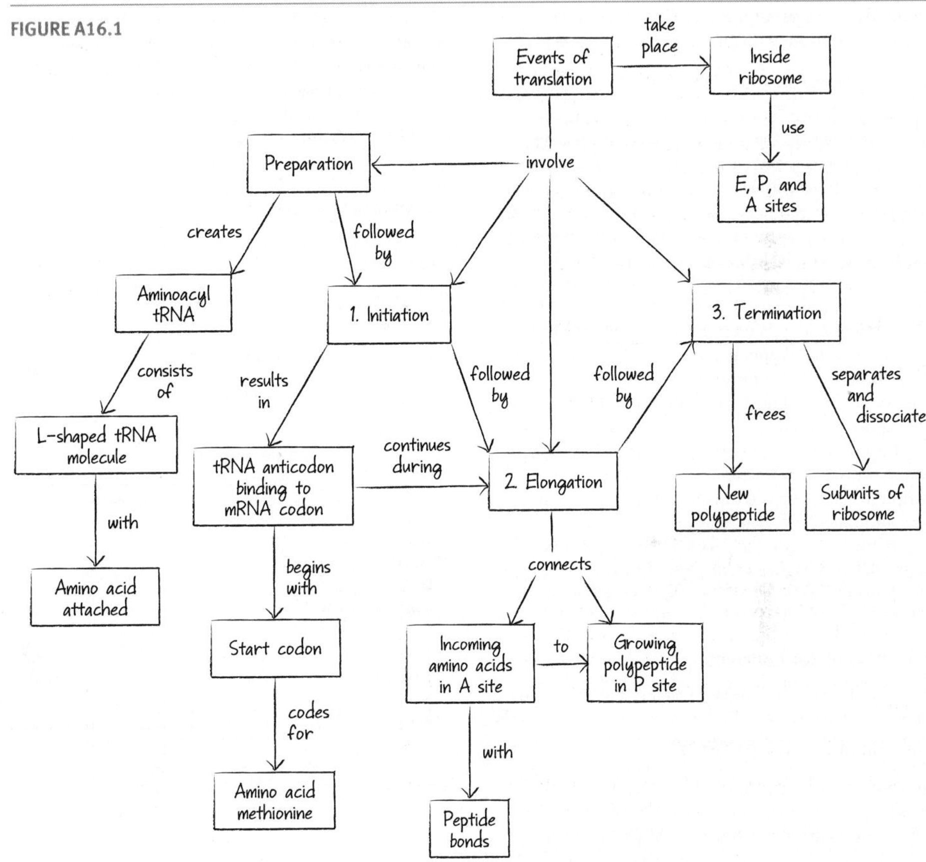

FIGURE A17.1

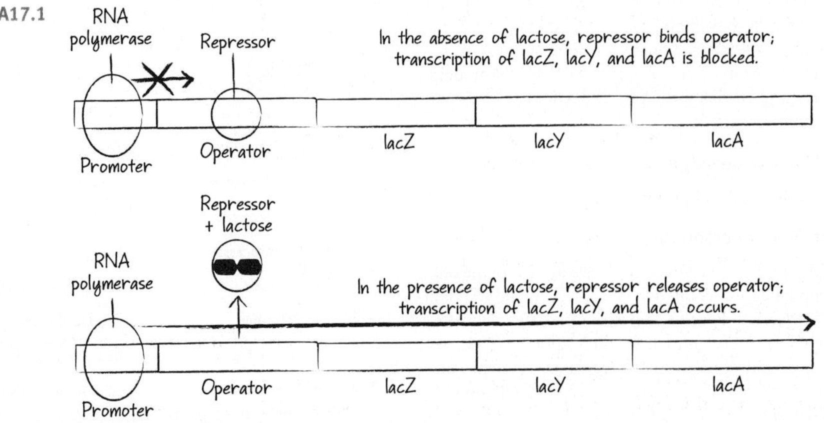

[See Figure A17.2] (2) Two conditions must be met for the *lac* operon to be expressed at a high rate. First, glucose levels must be low. Absence of glucose exerts positive control, because the cAMP-CAP complex binds to the CAP site and increases transcription rate. Second, lactose must be present. Lack of lactose exerts negative control, because the repressor remains attached to the operator and reduces transcription rate.

You Should Be Able To (YSBAT)

YSBAT p. 310 [See Table A17.1] **YSBAT p. 312** *lacZ* codes for the β-galactosidase enzyme, which breaks the disaccharide lactose into glucose and galactose. *lacY* codes for the lactose permease enzyme, which transports lactose into the bacterial cell. *lacI* codes for a protein that shuts down production of the other *lac* products. When lactose is absent, the *lacI* product prevents transcription. This is logical because there is no reason for the cell to make β-galactosidase and lactose permease if there is no lactose to metabolize. But when lactose is present, it interacts with *lacI* in some way so that *lacZ* and *lacY* are induced (their transcription can occur). When lactose is present, the enzymes that metabolize it are expressed.

Caption Questions and Exercises

Figure 17.1 Write "Slowest response, most efficient resource use" next to the transcriptional control label. Write "Fastest response, least efficient resource use" next to the post-translational control label. **Figure 17.2** Plates from all three treatments must be identical and contain identical growth medium, except for the presence of the sugars labeled in the figure. Also, all plates must be grown under the same physical conditions (temperature, light) for the same time. **Figure 17.3** Use a medium with all 20 amino acids when producing a master plate of mutagenized *E. coli* colonies, then use a replica plate that contains all of the amino acids except tryptophan. Choose cells from the master plate that did *not* grow on the replica plate. **Figure 17.8** Put the "Repressor protein" on the operator. No transcription will take place. Then put the "RNA polymerase" on the promoter. No transcription will take place. Finally, put "lactose" on the repressor protein and then remove the resulting lactose-repressor complex from the operon. Transcription will begin. **Figure 17.10** Write "Glucose low" above the drawing in part (a). Write "Glucose high" above the drawing in part (b).

Summary of Key Concepts

KC 17.1 Production of β-galactosidase and galactosidase permease are under transcriptional control—transcription depends on the action of regulatory proteins. The activity of the repressor and CAP—the regulatory proteins—are under post-translational control. **KC 17.2** After a dessert, glucose concentrations in your gut should be high, causing the *lac* operon in *E. coli* to shut down. After drinking milk, glucose concentrations should be low and lactose concentrations high, causing the *lac* operon in *E. coli* cells to be highly expressed. **KC 17.3** (1) The operator is the parking brake; the repressor locks it in place, and the inducer releases it. (2) The CAP-binding site is the gas pedal; the CAP-cAMP complex is a heavy foot on it.

Test Your Knowledge

1. b; **2.** d; **3.** b; **4.** c; **5.** a; **6.** c

Test Your Understanding

1. The glycolytic enzymes are always needed in the cell, because they are required to produce ATP, and ATP is always needed. **2.** Positive control means that a regulatory protein, when present, causes transcription to increase. Negative control means that a regulatory protein, when present, prevents transcription. **3.** The combination of positive and negative control allows *E. coli* to shut down the *lac* operon when glucose is present or lactose absent, but activate it when glucose is absent and lactose present. If only negative control occurred, the operon would be activated when glucose is present. If only positive control occurred, the operon would be activated when lactose is absent. **4.** cAMP levels rise when glucose is low. If glucose is low, the cell is in danger of starving unless it can switch to another food source. **5.** CAP acts like a receptor for cAMP—when cAMP binds to it, its activity changes. **6.** Unlike lactose, glucose can enter the glycolytic pathway directly—without being converted to another molecule. Less ATP and fewer enzymes are needed to acquire and use glucose as an energy source.

Applying Concepts to New Situations

1. Set up cultures with individuals that all come from the same colony of toluene-tolerating bacteria. Half of the cultures should have toluene as the only source of carbon; half should have glucose or another common source of carbon as well as toluene. Cells will grow in both cultures if they are able to use toluene as a source of carbon. **2.** Because toluene is not common in the environment, it is most likely that the toluene genes are under negative control. That is, the presence of toluene—by removing some brake on expression of the enzyme's genes—would trigger production of whatever enzymes are needed to break down toluene. **3.** The *LacI*^{S} mutants would not respond to the inducer, so the repressor would remain on the operator and block transcription even in the presence of the inducer, lactose. These mutants would soon die in an environment where lactose was the only sugar present. **4.** Cells with functioning β-galactosidase are blue; cells that are not producing normal β-galactosidase are white. The sequence of the β-galactosidase gene in structural mutants would differ from the sequence of the wild-type gene, whereas the sequence of the β-galactosidase gene in regulatory mutants would be the same as the sequence of the wild-type gene.

CHAPTER 18

Check Your Understanding (CYU)

CYU p. 323 (1) The basic double helical structure of DNA is the same, but the DNA in eukaryotic cells is complexed with histones to form chromatin, which is tightly packed into 30-nm fibers attached to scaffolding proteins. In bacteria, DNA is associated with histone-like proteins but is not organized into complex structures. (2) Addition of acetyl or methyl groups to histones can cause chromatin to open or close. Different patterns of acetylation or methylation will determine which genes in muscle cells versus brain cells can be transcribed and which are not available for transcription. **CYU p. 326** (1) Bacterial regulatory sequences are found close to the promoter; eukaryotic regulatory sequences can be close to the promoter or far from it. Bacterial regulatory proteins interact directly with RNA polymerase to initiate or prevent transcription; eukaryotic regulatory proteins influence transcription by altering chromatin structure or binding to the basal transcription complex through mediator proteins. (2) Certain regulatory proteins open

FIGURE A17.2

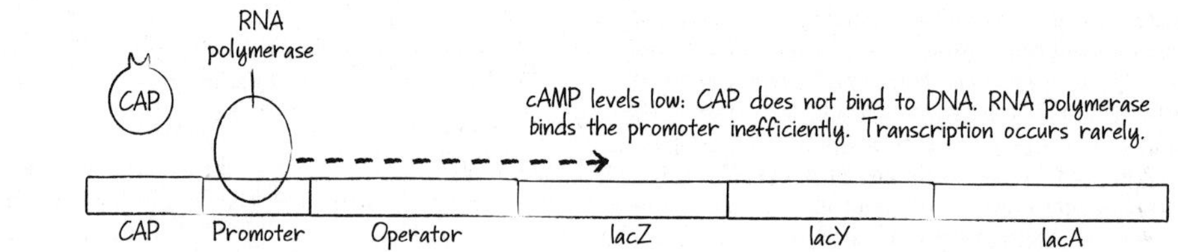

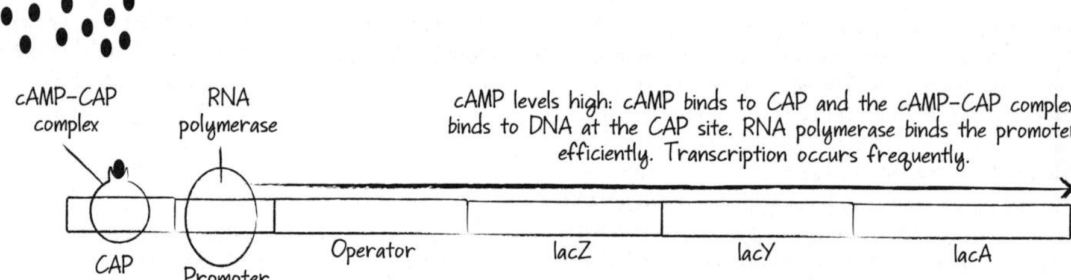

TABLE A17.1

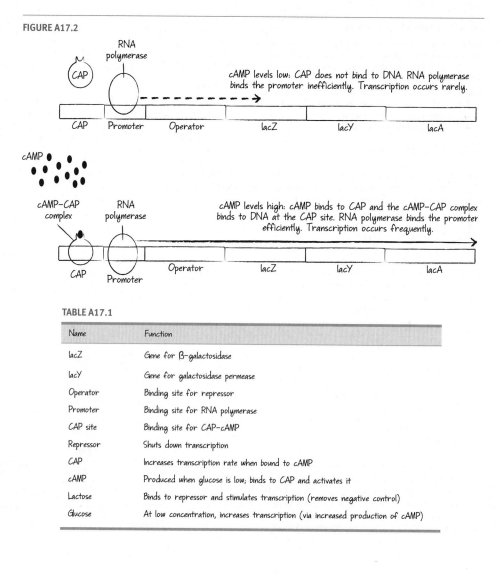

Name	Function
lacZ	Gene for β-galactosidase
lacY	Gene for galactosidase permease
Operator	Binding site for repressor
Promoter	Binding site for RNA polymerase
CAP site	Binding site for CAP-cAMP
Repressor	Shuts down transcription
CAP	Increases transcription rate when bound to cAMP
cAMP	Produced when glucose is low; binds to CAP and activates it
Lactose	Binds to repressor and stimulates transcription (removes negative control)
Glucose	At low concentration, increases transcription (via increased production of cAMP)

chromatin at muscle- or brain-specific genes, and then activate or repress the transcription of cell-type-specific genes. Muscle-specific genes are expressed only if muscle-specific regulatory proteins are produced and activated. **CYU p. 330** (1) It became clear that a single gene can code for multiple products instead of a single one. (2) miRNAs interfere with mRNAs by targeting them for destruction or preventing them from being translated. **CYU p. 333** (1) Many different types of mutations can disrupt control of the cell cycle and initiate cancer. These mutations can affect any of the six levels of control over gene regulation outlined in Figure 18.1. (2) The p53 protein is responsible for shutting down the cell cycle in cells with damaged DNA. If the protein does not function, then cells with damaged DNA—and thus many mutations—continue to divide. If these cells have mutations in genes that regulate the cell cycle, then they may continue to divide in an uncontrolled fashion.

You Should Be Able To (YSBAT)

YSBAT p. 326a The basal transcription factors found in muscle and nerve cells are similar or identical; the regulatory transcription factors found in each cell type are different. **YSBAT p. 326b** DNA forms loops when distant regulatory regions, such as silencers and enhancers, are brought close to the promoter through binding of regulatory transcription factors to the mediator complex. **YSBAT p. 329** Alternative splicing does not occur in bacteria because bacterial genes do not contain introns—in bacteria, each gene codes for a single product. Alternative splicing is part of step 3 in Figure 18.1. **YSBAT p. 330** Step 4 and step 5. RNA interference either (1) decreases the life span of mRNAs or (2) inhibits translation.

Caption Questions and Exercises

Figure 18.4 Acetylation of histones decondenses chromatin and allows transcription to begin, so HATs are elements in positive control. Deacetylation condenses the chromatin and inactivates transcription, so HDACs are elements in negative control. **Figure 18.6** This treatment allowed them to measure the normal amount of antibody mRNA produced, for comparison with the other treatments. **Figure 18.7** A typical eukaryotic gene usually contains introns and is regulated by multiple enhancers. Bacterial operons lack introns and enhancers. The promoter-proximal element found in some eukaryotic genes is comparable to the CAP binding site in the *lac* operon of bacteria. Bacterial operons have a single promoter but code for more than one protein; eukaryotic genes code for a single product.

Summary of Key Concepts

KC 18.1 Because eukaryotic RNAs are not translated as soon as transcription occurs, it is possible for RNA processing to occur, which creates variation in the mRNAs produced from a primary RNA transcript and their life span. **KC 18.2** The default state of eukaryotic genes is "off," because the highly condensed state of the chromatin makes DNA unavailable to RNA polymerase. **KC 18.3** [See Figure A18.1] **KC 18.4** Alternative splicing makes it possible for a single gene to code for multiple products. **KC 18.5** Cancer is a disease that results from accumulated mutations. Radiation damages DNA and increases the rate of mutation. Mutations accumulate as cells undergo repeated divisions, so are more common in older people.

Test Your Knowledge

1. c; **2.** a; **3.** d; **4.** d; **5.** b; **6.** d

Test Your Understanding

1. (a) Enhancers and the CAP site are similar because both are sites in DNA where regulatory proteins bind. They are different because enhancers generally are located at great distances from the promoter, whereas the CAP site is located near the promoter. (b) Promoter-proximal elements and the *lac* operon operator are both regulatory sites in DNA located close to the promoter. (c) Basal transcription factors and sigma are proteins that must bind to the promoter before RNA polymerase can initiate transcription. They differ because sigma is part of the RNA polymerase holoenzyme, while the basal transcription complex recruits RNA polymerase to the promoter. **2.** If changes in the environment cause changes in how spliceosomes function, then the RNAs and proteins produced from a particular gene could change in a way that helps the cell cope with the new environmental conditions. **3.** (a) Enhancers and silencers are both regulatory sequences located at a distance from the promoter. Enhancers bind regulatory transcription factors that activate transcription; silencers bind regulatory proteins that shut down transcription. (b) Promoter-proximal elements and enhancers are both regulatory sequences that bind positive regulatory transcription factors. Promoter-proximal elements are located close to the promoter; enhancers are far from the promoter. (c) Transcription factors bind to regulatory sites in DNA; the mediator complex does not bind to DNA but instead forms a bridge between regulatory transcription factors and basal transcription factors. **4.** RNA interference happens when complementary base pairing occurs between a small single-stranded miRNA and a single-stranded mRNA. **5.** Certain chemical modifications of histones mark sections of chromatin for transcription activation or repression. These modifications are passed to daughter cells when cells divide. A certain array of chromatin modifications or "histone codes" is associated with the production of muscle- or nerve-specific proteins. **6.** Lack of functional p53 protein prevents damaged cells from being destroyed or shut down (which would prevent their continuing to divide). As a result, cells with damaged DNA and thus many mutations continue to divide, possibly initiating cancerous growth.

Applying Concepts to New Situations

1. In eukaryotic cells, histones are involved in the basic folding of DNA molecules in chromatin. Mutations in histones are likely to interfere with this folding, and thus the expression of large numbers of genes. Individuals with mutant forms of histones are likely to die as a result—meaning that histone structure should be highly conserved over time. **2.** Tissues that divide more often must replicate their DNA more often; thus they are more susceptible to mutation. Because the cell cycle is accelerated in these cell types, it is logical to expect that control over the cell cycle could easily be lost. **3.** You could treat a culture of DNA-damaged cells with a drug that stops transcription and then compare them with untreated DNA-damaged cells. If transcriptional control regulates p53 levels, then the p53 level would be lower in the treated cells versus control cells. If control of p53 levels is post-translational, then in both cultures the p53 level would be the same. *Other approaches:* add labeled NTPs to damaged cells and see if they are incorporated into mRNAs for p53; or add labeled amino acids to damaged cells and see if they are incorporated into completed p53 proteins; or add labeled phosphate groups and see if they are added to p53 proteins. **4.** The miRNA would bind to viral RNA inside infected cells and prevent the viral RNA from working, thus stopping the infection.

THE BIG PICTURE: GENETIC INFORMATION

Check Your Understanding (CYU), p. 336

1. Star = DNA, mRNA, proteins. **2.** RNA "is reverse transcribed by" reverse transcriptase "to form" DNA. **3.** E = splicing, etc.; E = meiosis and sexual reproduction (along with their links). **4.** Chromatin "makes up" chromosomes; independent assortment and recombination "contribute to" high genetic diversity.

CHAPTER 19

Check Your Understanding (CYU)

CYU p. 344 (1) When the endonuclease makes a staggered cut in a palindromic sequence and the strands separate, the single-stranded bases that are left will bind ("stick") to the single-stranded bases left where the endonuclease cut the same palindrome at a different location. (2) The word probe means to examine thoroughly. A DNA probe "examines" a large set of sequences thoroughly, and binds to one—the one that has a complementary base sequence. **CYU p. 346** (1) Denaturation makes DNA single stranded so the primer can bind to the sequence during the annealing step. Once the primer is in place, *Taq* polymerase can synthesize the rest of the strand during the extension step. It is a "chain reaction" because the products of each reaction cycle are used in the next reaction cycle—this is why the number of copies doubles in each cycle. (2) One of many possible

FIGURE A18.1

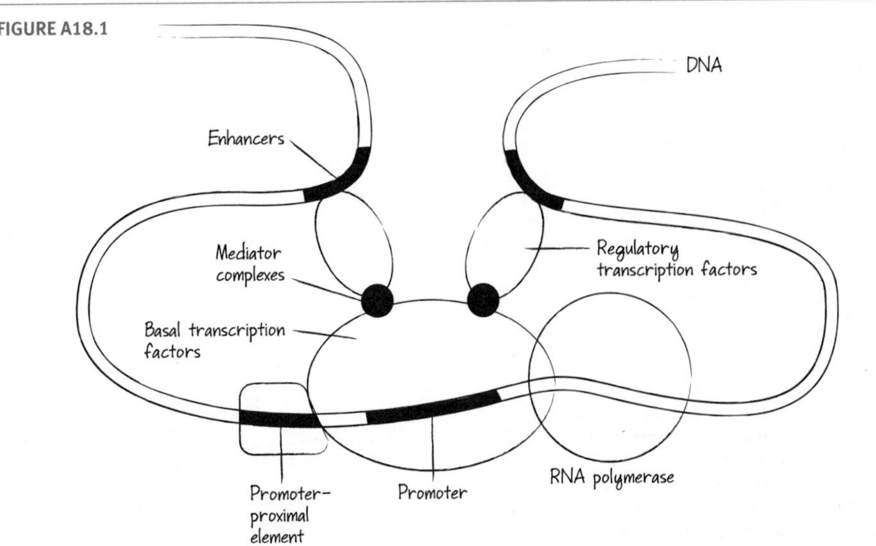

answers is shown below [see Figure A19.1]. **CYU p. 348** If ddNTPs were present at high concentration, they would almost always be incorporated—meaning that only fragments from the first complementary base in the sequence would be produced. **CYU p. 351** Start with a genetic map with as many polymorphic markers as possible. Get the genotype at these markers for a large number of individuals who have the same type of alcoholism—one that is thought to have a genetic component—as well as a large number of unaffected individuals. Look for particular versions of a marker that is almost always found in affected individuals. Genes that contribute to a predisposition to alcoholism will be near that marker.

You Should Be Able To (YSBAT)

YSBAT p. 342a Isolate the DNA and cut it into small fragments with EcoR1, which leaves sticky ends. Cut copies of a plasmid or other vector with EcoR1. Mix the fragments and plasmids under conditions that promote complementary base pairing by sticky ends of fragments and plasmids. Use DNA ligase to catalyze formation of phosphodiester bonds and seal the sequences. **YSBAT p. 342b** The probe must be single stranded so that it will bind by complementary base pairing to the target DNA, and it must be labeled so that it can be detected. The probe will only base pair with fragments that include a sequence complementary to the probe's sequence. A probe with the sequence 5′ AATCG 3′ will bind to the region of the target DNA that has the sequence 5′ AATCG 3′ as shown below:

```
       5′ AATCG 3′
3′ TCCGGTTAGCATTACCATTTT 5′
```

YSBAT p. 348 Using a map with many markers makes it more likely that there will be one very tightly linked to the gene of interest—meaning that a version of the marker will almost always be associated with the phenotype you are tracking.

Caption Questions and Exercises

Figure 19.4 No—many times, because there were many copies of each type of mRNA present in the pituitary cells and many pituitary cells were used to prepare the library. **Figure 19.7** The polymerase will begin at the 3′ end of each primer. On the top strand in part (b), it will move to the left; on the bottom strand it will move to the right. As always, synthesis is in the 5′ to 3′ direction. **Figure 19.10** The insertion will probably disrupt the gene and have serious consequences for the cell and potentially the individual.

Summary of Key Concepts

KC 19.1 When "sticky ends" anneal, the recombinant product would not be stable—it would tend to break apart because no phosphodiester bond would form between the DNA fragments. **KC 19.2** PCR advantages: It is relatively fast and easy, and can amplify a DNA sequence that is rare in the sample. PCR disadvantage: It requires knowledge of sequences on either side of the target gene, so primers can be designed. Plasmid advantages: No knowledge of the sequence is required. Plasmid disadvantages: It is slower and technically more difficult than PCR. **KC 19.3** The length of each fragment is dictated by where a ddNTP was incorporated into the growing strand, and each ddNTP corresponds to a base on the template strand. Thus, the sequence of fragment sizes corresponds to the sequence on the template DNA. **KC 19.4** Both affected and unaffected individuals have the same marker, so you have no way of knowing that it is close to the gene you are interested in. **KC 19.5** Genes can be inserted into the Ti plasmids, and the recombinant plasmids transferred to plant cells.

Test Your Knowledge

1. c; **2.** b; **3.** d; **4.** a; **5.** b; **6.** d

Test Your Understanding

1. When a restriction endonuclease cuts a "foreign gene" sequence and a plasmid, the same sticky ends are created on the excised foreign gene and the cut plasmid. After the sticky ends on the foreign gene and the plasmid anneal, DNA ligase catalyzes closure of the DNA backbone, sealing the foreign gene into the plasmid DNA. [See Figure A19.2] **2.** Vectors hold a piece of foreign DNA and carry it into an intact host cell. A "perfect" vector (1) has well-characterized restriction endonuclease sites, (2) can be inserted into host cells easily, (3) expresses recombinant genes in the host cell in a controlled way, and (4) doesn't damage the host cell. **3.** A cDNA library is a collection of complementary DNAs made from all the mRNAs present in a certain group of cells. A cDNA library from a human nerve cell would be different from one made from a human muscle cell, because nerve cells and muscle cells express many different, cell type–specific genes. **4.** Genetic markers are genes or other loci that have known locations in the genome. When these locations are diagrammed, they represent the physical relationships among landmarks—in other words, they form a map. **5.** The genes for making β-carotene need to be expressed in the endosperm. Adding an endosperm-specific promoter adjacent to the genes ensures that they are expressed in the endosperm. In general, including specific promoter and enhancer sequences with recombinant genes helps researchers control where and when the recombinant genes are expressed. **6.** PCR and cellular DNA synthesis are similar in the sense of producing copies of a template DNA. Both rely on primers and DNA polymerase. The major difference between the two is that PCR copies only a specific target sequence, while the entire genome is copied during cellular DNA synthesis.

Applying Concepts to New Situations

1. You could use a computer program to identify likely promoter sequences in the sequence data, and then look for sequences just downstream that have an AUG start codon and a stop codon about 1500 base pairs apart. **2.** Yes—you would transform cells that are the precursors to sperm or eggs, so that the recombinant DNA is found in mature sperm or eggs and passed on to offspring. **3.** In both techniques, researchers use an indicator to identify either a gene of interest or a colony of bacteria with a particular trait. The problem is the same—picking one particular thing (a certain gene or a cell with a particular mutation) out of a large collection. **4.** (a) Primer 1b binds to the top right strand and would allow DNA polymerase to synthesize the top strand across the target gene. Primer 1a, however, binds to the upper left strand and would allow DNA polymerase to synthesize the upper strand *away* from the target gene. Primer 2a binds to the bottom left strand and would allow DNA polymerase to synthesize the bottom strand across the target gene. Primer 2b, however, binds to the bottom right strand and would allow DNA polymerase to synthesize away from the target gene. (b) She could use primer 1b with primer 2a.

CHAPTER 20

Check Your Understanding (CYU)

CYU p. 364 (1) Parasites don't need genes that code for enzymes required to synthesize molecules they acquire from their hosts. (2) If two closely related species inherited the same gene from a common ancestor, the genes should be similar. But if one species acquired the same gene from a distantly related species via lateral gene transfer, then the genes should be much less similar. **CYU p. 370** (1) Unequal crossover and mistakes in DNA synthesis are so common in simple tandem repeats that new alleles—consisting of the same chromosome region with different repeat numbers—are created almost every generation, producing a molecular fingerprint. (2) Unequal crossover occurs when homologous chromosomes

FIGURE A19.1

5′ CATGACTATTACGTATCGGGTACTATGCTATCGATCTAGCTACGCTAGCT 3′
3′ GTACTGATAATGCATAGCCCATGATACGATAGCTAGATCGATGCGATCGA 5′

Probe #1, which will anneal to the 3′ end of the top strand:

5′ AGCTAGCGTAGCTAGATCGAT 3′

Probe #2, which will anneal to the 3′ end of the bottom strand:

5′ CATGACTATTACGTATCGGGG 3′

The primers will bind to the separated strands of the parent DNA sequence as follows:

5′ CATGACTATTACGTATCGGGTACTATGCTATCGATCTAGCTACGCTAGCT 3′
3′ TAGCTAGATCGATGCGATCGA 5′

5′ CATGACTATTACGTATCGGG 3′
3′ GTACTGATAATGCATAGCCCATGATACGATAGCTAGATCGATGCGATCGA 5′

FIGURE A19.2

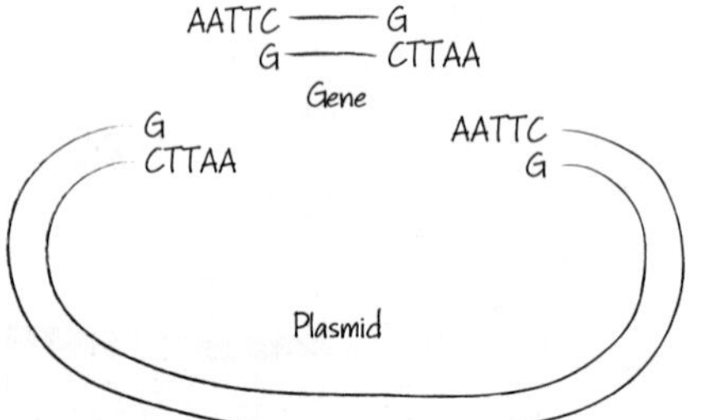

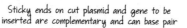

Sticky ends on cut plasmid and gene to be inserted are complementary and can base pair

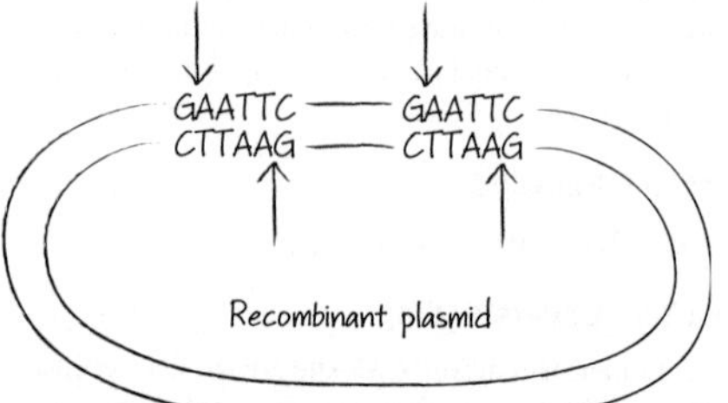

DNA ligase forms four phosphodiester bonds between gene and plasmid DNA at the points indicated by arrows

align incorrectly during synapsis. When crossing over occurs within a misaligned section of chromosome, one homolog ends up with less DNA and the other ends up with more DNA. The extra DNA is made up of sequences from the other chromosome that are duplicates of the DNA on the original strand.

You Should Be Able To (YSBAT)

YSBAT p. 360 If no overlap occurred, there would be no way of ordering the fragments correctly. You would only be able to put fragments in random order—not the correct order. **YSBAT p. 371** Start with a microarray containing exons from a large number of human genes. Isolate mRNAs from brain tissue and liver tissue, and make labeled cDNAs from each. Probe the microarray with each type of probe, and record where binding occurs. Binding events identify exons that are transcribed in each type of tissue. Compare the results to identify genes that are expressed in brain but not liver, or liver but not brain.

Caption Questions and Exercises

Figure 20.2 Shotgun sequencing is based on fragmenting the genome into many small pieces. **Figure 20.3** Because the mRNA codons that match up with each strand are oriented in the 5′ to 3′ direction. **Figure 20.8** Chromosome 1:
ψβ2-ϵ-Gγ-Aγ-ψβ1-δ-ψβ2-ϵ-Gγ-Aγ-ψβ1-δ-β
Chromosome 2:
β

Summary of Key Concepts

KC 20.1 If a search of human gene sequence databases revealed a gene that was similar in base sequence, and if follow-up work confirmed that the mouse and human genes were similar in their pattern of exons and introns and regulatory sequences, then the researchers could claim that they are homologous. **KC 20.2** DNA can move from one prokaryotic species to another packaged in plasmids or viruses, or by the direct uptake of DNA fragments from the environment (transformation). **KC 20.3** Junk DNA refers to noncoding regions of DNA—sequences that are not transcribed into RNA. The function of repeated sequences is not yet clear, but many noncoding sequences are interesting because they are relicts from transposable elements. **KC 20.4** Expression of the gene may be involved in the development of cancer.

Test Your Knowledge

1. c; **2.** a; **3.** d; **4.** b; **5.** d; **6.** d

Test Your Understanding

1. Computer programs are used to scan sequences in both directions to find ATG start codons, a gene-sized logical sequence with recognizable codons, and then a stop codon. One can also look for characteristic promoter, operator, and other regulatory sites. It is more difficult to identify open reading frames in eukaryotes because their genomes are so much larger and because of the presence of introns and repeated sequences. **2.** To have the potential to thrive in different environments, a cell requires a large array of enzymes and proteins—to use different sources of food, cope with different temperature or pH conditions and so on—and thus a large array of genes. **3.** They use the transcription and translation machinery of the host cell to copy themselves and insert themselves into the host's genome, and this may have a negative effect on the host cell. **4.** Each individual has a DNA fingerprint—a unique collection of microsatellite and minisatellite alleles (length variants). Because alleles are inherited, closely related individuals share more alleles than more distantly related individuals. **5.** A DNA microarray experiment identifies which genes are being expressed in a particular cell at a particular time. If a series of experiments shows that different genes are expressed in cells at different times or under different conditions, it implies that expression was turned on or off in response to changes in age or changes in conditions. **6.** Homology is a similarity among different species that is due to their inheritance from a common ancestor. If a newly sequenced gene is found to be homologous with a known gene of a different species, it is assumed that the gene products have similar function. Duplicated genes are homologous because they share a common ancestor gene.

Applying Concepts to New Situations

1. If "gene A" is not necessary for existence, it can be lost by an event like unequal crossing over (on the chromosome with deleted segments) with no ill effects on the organism. In fact, individuals who have lost unnecessary genes are probably at a competitive advantage, because they no longer have to spend time and energy copying and repairing unused genes. **2.** If the grave was authentic, it might include two very different parental patterns along with three children whose patterns each represented a mix between the two parents. The other unrelated individuals would have patterns not shared by anyone else in the grave. **3.** The genes may have similar DNA sequences because the viral gene was incorporated into the genome of a human host in the past, and subsequently passed on to future generations. **4.** You would expect that the livers and blood of chimps and humans would function similarly, but that strong differences occur in brain function. The microarray data support this prediction and suggest that even though the brain proteins might have similar sequences, chimp and human brains are different because certain genes are turned on or off at different times and expressed in different amounts.

CHAPTER 21

Check Your Understanding (CYU)

CYU p. 379 Biologists have been able to grow entire plants from a single differentiated cell taken from an adult. They have also succeeded at producing animals by transferring the nucleus of a fully differentiated cell to an egg whose nucleus has been removed. **CYU p. 384** (1) Bicoid is a transcription factor, so cells that experience a high versus medium versus low concentration transcribe different amounts of Bicoid-regulated genes. (2) Bicoid proteins control the expression of gap genes, which—in effect—tell cells which third of the embryo they are in along the anterior-posterior axis. Gap genes control the expression of pair-rule genes, which organize the embryo into individual segments. Pair-rule genes control expression of segment polarity genes, which establish an anterior-posterior polarity to each individual segment. Segment polarity genes turn on homeotic genes, which then trigger genes for producing segment-specific structures like wings or legs.

Caption Questions and Exercises

Figure 21.2 It would look like the embryo of the left side of part (b)—a normal phenotype. **Figure 21.5** In cells that are expressing a particular gene, the gene is transcribed into RNA. In situ hybridization locates these copies of RNA, identifying which cells are expressing the gene. **Figure 21.10** Genes are usually considered homologous if they have similar structure—meaning similar DNA sequence and exon-intron structure.

Summary of Key Concepts

KC 21.1 The seedling would grow very slowly and be small, but development should proceed normally otherwise. **KC 21.2** Parenchyma cells in the stem still contain a complete set of genes, so if they can be de-differentiated and received appropriate signals from other cells, they could start producing vascular-tissue-specific proteins. **KC 21.3** Different concentrations of transcription factors or cell-cell signals will cause differences in the amount of expression from certain genes—so a high concentration means something different than a low concentration. **KC 21.4** The idea is that a single molecule can set up the anterior-posterior or apical-basal axis of the embryo. Once the axis is established, a sequence of events follows in which cells get increasingly more precise information about where they are located along that axis—based on the original concentration of the master regulator. **KC 21.5** When the patterns of expression of two key genes involved in limb versus rib formation are compared in chick and snake embryos, they differ. In snakes, the "form ribs here" pattern occurs instead of the "form legs here" pattern.

Test Your Knowledge

1. b; **2.** d; **3.** a; **4.** c; **5.** a; **6.** a

Test Your Understanding

1. They occur in both plants and animals, and are responsible for the changes that occur as an embryo develops. **2.** If the transplanted nucleus has undergone some permanent change or loss of genetic information during development, it should not be able to direct the development of a viable adult. But if the transplanted nucleus is genetically equivalent to the nucleus of a fertilized egg, then it should be capable of directing the development of a new individual. **3.** The researchers exposed adult flies to treatments that induce mutations, then looked for embryos with defects in the anterior-posterior body axis or body segmentation. The embryos had mutations in genes required for body axis formation and segmentation. **4.** Development—specifically differentiation—depends on changes in gene expression. Changes in gene expression depend on differences in regulatory transcription factors. **5.** Differentiation is triggered by the presence or absence of external signals. These signals trigger the production of transcription factors, which induce other transcription factors, and so on—a sequence that constitutes a regulatory cascade—as development progresses. At each step in the cascade, a new subset of genes is activated—resulting in a step-by-step progression from undifferentiated to fully differentiated cells. **6.** *Hox* genes were first discovered in *Drosophila* and have been identified in nearly every animal. The number of *Hox* genes varies across species, but their arrangement on the chromosome and pattern of expression in the embryo are strikingly similar across animal species.

Applying Concepts to New Situations

1. There would be more Bicoid protein farther toward the posterior and less in the anterior—meaning that the anterior segments would be "less anterior" in their characteristics and the posterior segments would be "more anterior." **2.** The result means that the human gene can be transcribed in worms and that its protein product has performed the same function in worms as in humans. This is strong evidence that the genes are homologous, and that their function has been evolutionarily conserved. **3.** The stem cells would have to be stimulated with nerve-specific signals to trigger their differentiation into nerve cells. **4.** Homeotic genes—such as *Hox* genes—are responsible for triggering the production of structures like wings or legs in particular segments. One hypothesis to explain the variation is that changes in gene expression led to different numbers of segments that express leg-forming *Hox* genes.

CHAPTER 22

Check Your Understanding (CYU)

CYU p. 392 (1) If a mutation in either bindin or the egg-cell receptor for sperm prevented the proteins from binding to each other, then fusion of sperm and egg-cell membranes could not take place and fertilization would not occur. (2) The entry of a sperm triggers a wave of calcium ion release around the egg-cell membrane. Thus, calcium ions act as a signal that indicates "A sperm has entered the egg." In response to the increase in calcium ion concentration, cortical granules fuse with the egg-cell membrane and release their contents to the exterior, causing the fertilization envelope to develop. **CYU p. 394** If cytoplasmic determinants were distributed in equal concentration throughout the egg, all blastomeres would end up with the same types and concentrations of cytoplasmic determinants. **CYU p. 395** (1) Endoderm is in the interior, ectoderm is on the outside, mesoderm is in between. (2) These cells become endoderm that forms the gut lining. The direction of the gut helps define the anterior-posterior axis of the body—it connects mouth and anus.

You Should Be Able To (YSBAT)

YSBAT p. 391 If the bindin-like protein were blocked, it would not be able to bind to the egg-cell receptor for sperm. Fertilization would not take place.

Caption Questions and Exercises

Figure 22.4 The researchers didn't start out with a hypothesis about which egg-cell membrane protein bound to bindin. Instead, they tested every type of protein projecting from the egg-cell surface to see which one, if any, bound to bindin. **Figure 22.8** [See Figure A22.1] **Figure 22.9** Ectoderm forms most of the structures on the outside of the adult. Endoderm forms mostly interior structures—for example, the inner lining of the respiratory and digestive tracts. Mesoderm forms structures located between the ectoderm-derived and endoderm-derived structures, including bones and muscles and the bulk of most of the organ systems. **Figure 22.15** It made gene expression possible in non-muscle cells. A "general purpose" promoter can lead to transcription of an inserted cDNA in any type of cell, including fibroblasts.

Summary of Key Concepts

KC 22.1 Bindin and the egg-cell receptor for sperm bind to each other. For this to happen, their tertiary structures must fit together somewhat like the structures of a lock and key. **KC 22.2** If the master regulator were localized to a certain part of the egg cytoplasm or formed a Bicoid-like concentration gradient throughout the eggs, blastomeres would contain or not contain the regulator or contain a distinct concentration of it. **KC 22.3** Both cells would contain cell-cell signals and transcription factors specific to mesodermal cells, but the anterior cell would contain anterior-specific signals and regulatory transcription factors while the posterior cell would contain posterior-specific signals and regulators. **KC 22.4** Cell-cell signals from the notochord direct the formation of the neural tube; subsequently, cell-cell signals from the notochord, neural tube, and ectoderm direct the differentiation of somite cells. Notochord cells die later in development. Cells expand during neural tube formation. Somite cells move to new positions, proliferate, and differentiate.

Test Your Knowledge

1. d; **2.** c; **3.** c; **4.** d; **5.** c; **6.** a

Test Your Understanding

1. Eggs contain stores of nutrients. **2.** If more than one sperm nucleus entered the egg, the resulting cell would contain more than two copies of each chromosome. Mitosis could not occur correctly and the embryo would be deformed or die. **3.** They contain different types and/or concentrations of cytoplasmic determinants. **4.** They give rise to all of the tissue and organs of the adult (from the Latin *germen*, meaning shoot or sprout). **5.** When transplanted early in development, somite cells become the cell type associated with their new location. But when transplanted later in development, somite cells become the cell type associated with the original location. These observations indicate that the same cell is not committed to a particular fate until later in development. **6.** If the gene for MyoD is expressed in non-muscle cells, the cells begin producing muscle-specific proteins. It triggers differential gene expression typical of muscle cells.

Applying Concepts to New Situations

1. The interactions that allow proteins to bind to each other are extremely specific—certain amino acids have to line up in certain positions with specific charges and/or chemical groups exposed. Sperm-egg binding is species specific because each form of bindin binds to a different egg cell receptor for sperm. **2.** Translation is not needed for cleavage to occur normally—meaning that all the proteins needed for early cleavage are present in the egg. **3.** The yellow pigment is a cytoplasmic determinant that triggers a gene regulatory cascade leading to differentiation of muscle cells. **4.** The embryo cannot yet feed, so all of its activity is fueled by the nutrient stores in the egg. It is not taking in a lot of new molecules that would allow it to get bigger.

FIGURE A22.1

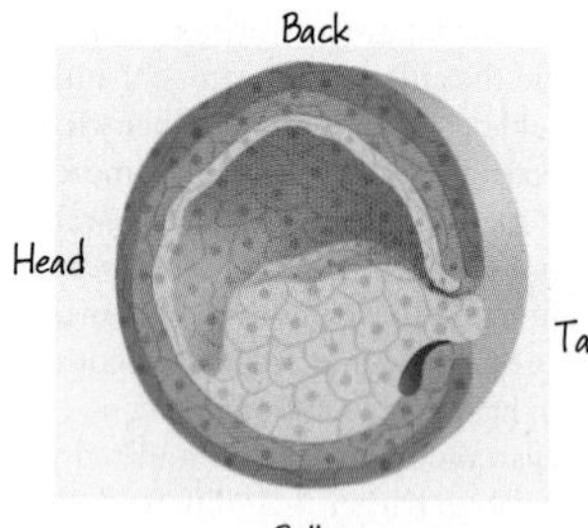

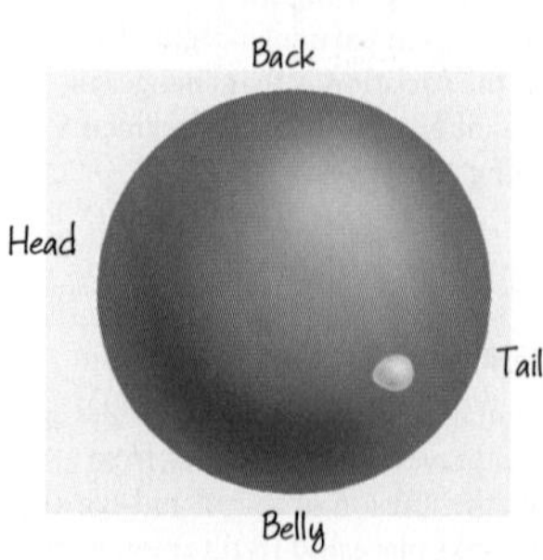

CHAPTER 23

Check Your Understanding (CYU)

CYU p. 404 (1) In the male reproductive organs of a mature flower, cells undergo meiosis to produce haploid cells. Each of these haploid cells divides by mitosis to produce a pollen grain containing haploid cells. One of the cells in the pollen grain divides again by mitosis to form two sperm cells. (2) They ensure that the plant is fertilized by a pollen grain from the same species. If any pollen could germinate on any stigma, pollen from a pine tree could grow on the stigma of an orchid. **CYU p. 406** (1) Cells located on the outside of the embryo become epidermal tissue; cells in the interior become vascular tissue; cells in between become ground tissue. (2) The *MONOPTEROS* gene product would be overactivated. The embryonic cells would get information indicating that they are in the shoot apex, where MONOPTEROS levels are high. As a result, the root would not develop normally. **CYU p. 409** [See Figure A23.1] If a new supply of fertilizer (cow dung) appeared near the plant's left side, the roots on that side would extend toward the nutrients. The shoot system should grow in response to the new nutrients but, all other things being equal (e.g., no shading of plant from sunlight), shoot growth should be symmetrical. **CYU p. 411** (1) The A protein might function as a transcriptional repressor by binding to a regulatory site near the *C* gene, thereby preventing transcription of the *C* gene. (2) Like *Hox* genes, *MADS-box* genes code for DNA-binding proteins that function as transcriptional regulators. Both Hox proteins and MADS-box proteins are involved in regulatory transcription cascades that lead to development of specific structures in specific locations. Although they perform similar functions, *Hox* genes and *MADS-box* genes evolved independently and differ in their base sequences. Thus, these genes and their protein products are not homologous.

You Should Be Able To (YSBAT)

YSBAT p. 410 If four genes coded for flower structure and each gene specified one of the four organs, then each mutant should lack an element in only one of the whorls. If each of three genes is expressed in two adjacent whorls, then four different gene expression patterns are possible—one for each of the four different whorls.

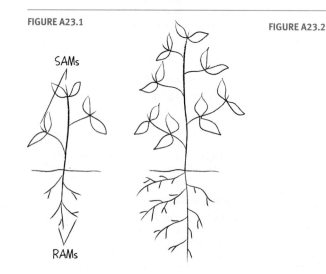

FIGURE A23.1

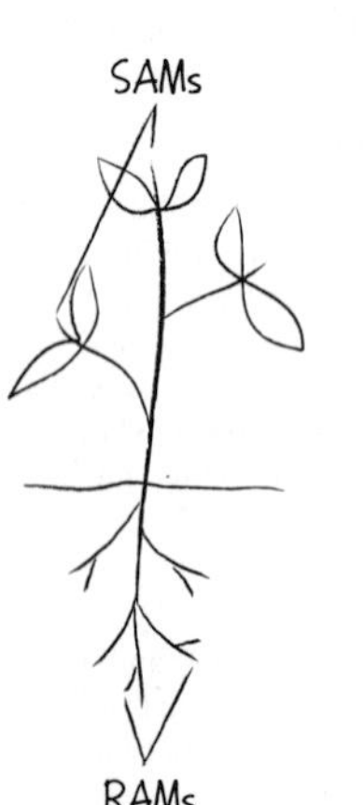

FIGURE A23.2

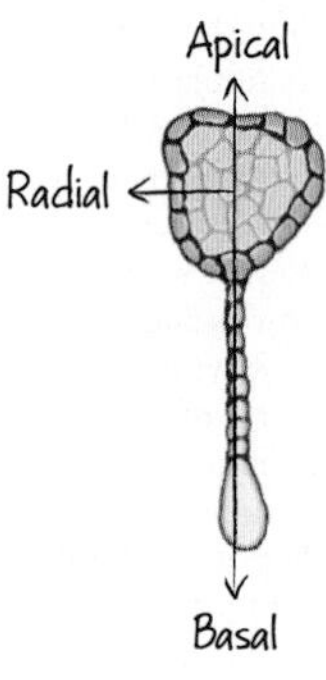

Caption Questions and Exercises

Figure 23.3 In both cases, protein-protein interactions between male and female structures in two different individuals prevent fertilization between members of different species. **Figure 23.5** [See Figure A23.2] **Figure 23.9** Many animals have three body axes analogous to the three axes of leaves: anterior-posterior (like the leaf's proximal-distal), dorsal-ventral (like the leaf's adaxial-abaxial), and a left-right axis (like the leaf's lateral axis). The main plant body, in contrast, has only two axes: apical-basal and radial.

Summary of Key Concepts

KC 23.1 Bicoid defines the anterior-posterior axis of the fly embryo; auxin defines the apical-basal axis of the plant embryo. Both work via concentration gradients. Bicoid is a regulatory transcription factor while auxin is a cell-cell signal. **KC 23.2** There are two fusion events between sperm and haploid cells in the female reproductive structure: one sperm fuses with the egg and leads to the development of an embryo; the other sperm fuses with two (or more) nuclei and leads to the development of endosperm. **KC 23.3** [See Figure A23.3] **KC 23.4** The presence of multiple SAMs and RAMs provides plants with flexibility in adapting to environmental changes and acquiring sunlight and nutrients, thereby increasing their ability to grow and reproduce. For example, multiple SAMs allow a plant to extend shoots (branches) in multiple directions to receive optimal sunlight, and multiple RAMs allow it to extend roots in multiple directions to acquire optimal water and nutrients. Also, a plant would be in trouble if its only SAM or RAM got eaten. **KC 23.5** Homeotic mutants are defined by a phenotype in which one body structure is replaced by another. *Arabidopsis* plants with a defective *B* gene have flowers in which petals are replaced by sepals and stamens are replaced by carpels.

Test Your Knowledge

1. a; **2.** b; **3.** d; **4.** d; **5.** c; **6.** a

Test Your Understanding

1. Proteins on the surface of a pollen grain interact with proteins on the surface of the stigma. Signals from the egg direct the growth of the pollen tube. **2.** Animal cells that lead to sperm or egg formation are "sequestered" early in development, so that they are involved only in reproduction and undergo few rounds of mitosis. Because mutations occur in every round of mitosis, plant eggs and sperm should contain many more mutations than animal eggs and sperm. **3.** Both meristems and stem cells can produce differentiated cells in an adult. Meristems are much more flexible than stem cells, as they can form all parts of the plant, and they increase the size of the body. Animal stem cells can usually develop into only a few differentiated cell types, and only replace lost or damaged cells. **4.** Just like the three tissue layers in plant embryos, the tissues produced in the SAMs and RAMs of a 300-year-old oak tree can differentiate into all of the specialized cell types found in a mature plant. **5.** The ABC model predicted that the *A*, *B*, and *C* genes would each be expressed in specific whorls in the developing flower. The experimental data showed that the prediction was correct. **6.** Cell proliferation is responsible for expansion of meristems and growth in size. Cell death occurs when leaves drop, and when three or four meiotic products die during gametogenesis in female tissues. Asymmetrical cell proliferation along with cell expansion is responsible for giving the adult plant a particular shape. Differentiation is responsible for generating functional tissues and organs in the adult plant. Cell-cell signals are responsible for setting up the apical-basal axis of the embryo and for guiding the pollen tube to the egg.

Applying Concepts to New Situations

1. If an individual is likely to die, it should throw all of its remaining resources into reproduction. **2.** Because vegetative growth is continuous and occurs in all directions (unrestricted), it could be considered indeterminate. However, reproductive growth produces mature reproductive organs and stops when those organs are complete. Because its duration is limited, it could be considered determinant. **3.** High light conditions could trigger changes in gene expression that reduced the rate of cell proliferation during leaf development. In contrast, low light conditions might trigger changes in gene expression that increase the rate of cell proliferation. **4.** [See Figure A23.4] In a young oak tree, SAMs would be evenly distributed in the upper parts of the plant and the RAMs would be evenly distributed in the tips of the roots. But after 50 years in the situation described above, the RAMs would be concentrated on the right side of the plant's roots, enabling the roots to grow into the area where water is available from the leaky pipe next to the building. The SAMs would be concentrated on the left side of the plant, allowing the upper parts of the tree to grow away from billboard, where light is available.

CHAPTER 24

Check Your Understanding (CYU)

CYU p. 429 (1) *Postulate 1:* Traits vary within a population. *Postulate 2:* Some of the trait variation is heritable. *Postulate 3:* There is variation in reproductive success (some individuals produce more offspring than others). *Postulate 4:* Individuals with certain heritable traits produce the most offspring. The first two postulates describe heritable variation; the second two describe differential reproductive success. (2) Beak size and shape and body size vary among individual finches, in part because of differences in their genotypes (some alleles lead to larger or narrower beaks, for example). When a drought hit, individuals with deep beaks survived better and produced more offspring than individuals with shallow beaks. **CYU p. 432** (1) When certain individuals are selected, their traits do not change—they simply produce more offspring than other individuals. (2) Adaptations are not optimal solutions to challenges posed by a particular environment because they are compromised by (1) the necessity of meeting many challenges at the same time (an adaptive "solution" to one problem—such as flying faster—may make another problem worse—such as being maneuverable in flight); (2) lack of "optimal" alleles or presence of alleles that affect more than one trait; and (3) the necessity of selecting only preexisting variation in traits.

You Should Be Able To (YSBAT)

YSBAT p. 425 (1) Relapse occurred because the few bacteria remaining after drug therapy were not eliminated by the patient's weakened immune system and began to reproduce quickly. (2) No—almost all of the cells present at the start of the infection would have been resistant to the drug. **YSBAT p. 430** In biology, an adaptation is any heritable trait that increases an individual's ability to produce offspring in a particular environment. In everyday English, adaptation is often used to refer to an individual's nonheritable adjustment to meet an environmental challenge, a phenomenon that biologists call acclimation. The phenotypic changes resulting from acclimation are not passed on to offspring.

Caption Questions and Exercises

Figure 24.4 The theory of special creation would claim that the fossil and extant sloths were both created 6000 years ago, but that the fossil species became extinct during the flood in Noah's time, described in the Bible. It is not clear how the theory of special creation would explain transitional features observed in the fossil record. **Figure 24.5** If vestigial traits result from inheritance of acquired characteristics, some individuals must have lost the traits during their own lifetimes and passed the reduced traits on to their offspring. For example, a certain monkey's long tail might have been bitten off by a predator, or an ape's hair might have been pulled out of its skin by a rival during a fight. The new traits would then somehow have passed to the individuals' eggs and sperm, resulting in shorter-tailed and less-hairy offspring, until humans with a coccyx and goose bumps resulted. **Figure 24.16** "Prediction": Beak measurements were different before and after the drought. "Prediction of null hypothesis": No difference in beak measurements before and after the drought.

Summary of Key Concepts

KC 24.1 Under the theory of special creation, changes in *Mycobacterium* populations would be explained as individual creative events governed by an intelligent creator. Under the theory of evolution by inheritance of acquired characters, changes in *Mycobacterium* populations would be explained by the cells trying to transcribe genes in the presence of the drug, and their *rpoB* gene becoming altered as a result. **KC 24.2** In biology, fitness is the ability of an individual to produce offspring, relative to that ability in other individuals in the population.

FIGURE A23.3

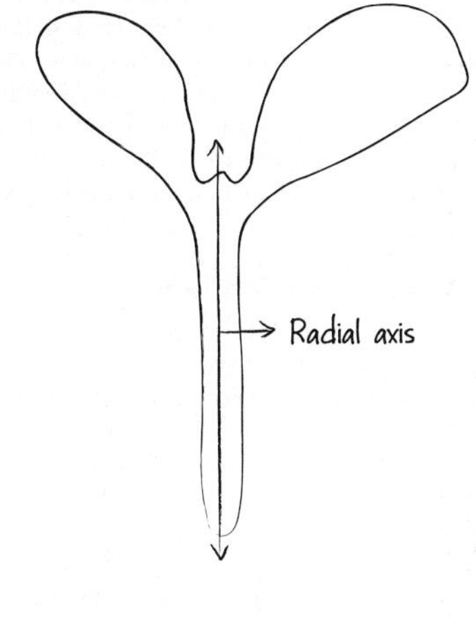

FIGURE A23.4

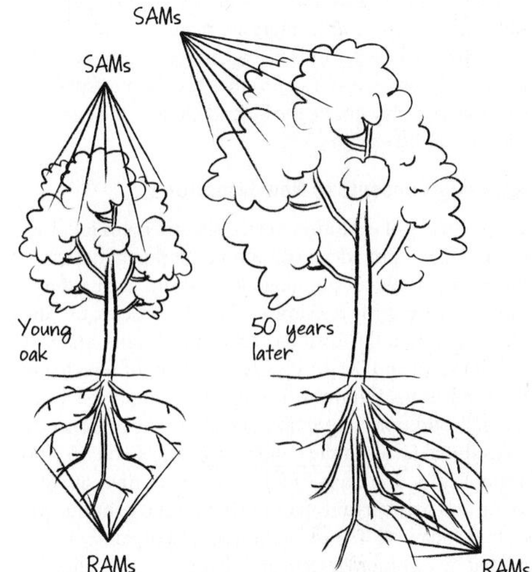

In everyday English, fitness is a physical attribute that is acquired as a result of practice or exercise. **KC 24.3** Brain size in *H. sapiens* might be constrained by the need for babies to pass through the mother's birth canal, by the energy required to maintain a large brain as an adult, or by lack of genetic variation for even larger brain size. Flying speed might be constrained by loss of maneuverability (and thus less success in hunting), the energy demands of extremely rapid flight, or lack of genetic variation for even faster flight.

Test Your Knowledge

1. b; **2.** d; **3.** a; **4.** d; **5.** a; **6.** c

Test Your Understanding

1. The theory of evolution by natural selection predicts that when a population's environment changes, individuals with certain traits will produce the most offspring, and the frequency of those traits will increase in the population. These changes should accumulate over long periods of time, eventually resulting in the formation of a new species that is geographically close to its ancestral species, and a fossil record should reflect changes in traits over time. The theory of special creation predicts that species do not change over time (or do so only due to divine intervention). Finally, the theory of acquired characters predicts that changes that occur in individuals in response to the environment are passed on to the next generation. **2.** Mutation produces new genetic variations, at random, without any forethought as to which variations might prove adaptive in the future. Individuals with mutations that are disadvantageous won't produce many offspring, but individuals with beneficial mutations will produce many offspring. **3.** The evidence for within-patient evolution is that DNA sequences at the start and end of treatment were identical except for a single nucleotide change in the *rpo* gene. If rifampin were banned, it is likely that *rpoB* mutant strains would have had lower fitness in the drug-free environment and would not continue to increase in frequency in *M. tuberculosis* populations. **4.** Typological thinking is based on the idea that species are unchanging and that any differences between individuals within a species are unimportant. Population thinking treats variation among individuals as critically important. That variation is what natural selection acts on, or selects. Population thinking was a radical break with typological thinking. **5.** Yes—the predictions made by a theory can be tested with indirect evidence in convincing ways. For example, if evolutionary theory predicts that certain transitional fossils should be found in rocks of a certain age, then researchers can search in those rocks and see if those fossils are indeed found. New groups of species found on island chains should be each others' closest relatives. And so on. **6.** Organisms do not necessarily become more complex over time; there are many examples of traits that have been lost. The biggest and strongest organisms do not necessarily produce the most offspring because they may not be the best adapted to the environment at that time.

Applying Concepts to New Situations

1. The theory of evolution fits the six criteria as follows: (1) and (2): It provides a common underlying mechanism responsible for puzzling observations such as homology, geographic proximity of similar species, the law of succession in the fossil record, vestigial traits, and extinctions. (3) and (4): It suggests new lines of research to test predictions about the outcome of changing environmental conditions in populations, about the presence of transitional forms in the fossil record, and so on. (5) It is a simple idea that explains the tremendous diversity of living and fossil organisms and why species continue to change today. (6) The realization that all organisms are related by common descent and that none are higher or lower than others was a surprise. **2.** It is well documented that people who are healthy and well fed grow taller than people who are sick and malnourished. In the absence of data indicating that alleles associated with increased height have increased in frequency recently, it is more logical to hypothesize that the observed change is due to changes in the environment—not evolutionary (genetic) changes. **3.** Compare the sequences of the 20 genes from many samples of preserved human tissue with the sequences of the same genes from a large sample of currently living humans. If evolution has occurred, the frequency of alleles correlated with "tallness" should be significantly greater in living humans than in those who lived a hundred years ago. **4.** The ability to tan is an adaptation because it is passed on from one generation to the next. The tan itself is not passed on, but the ability to tan is passed on; therefore it is heritable and can be classified as an adaptation.

CHAPTER 25

Check Your Understanding (CYU)

CYU p. 440 Given the observed genotype frequencies, the observed allele frequencies are freq(A_1) = 0.574 + ½(0.339) = 0.744; freq(A_2) = ½(0.339) + 0.087 = 0.256. Given these allele frequencies, the genotype frequencies expected under the Hardy-Weinberg principle are A_1A_1: $0.744^2 = 0.554$; A_1A_2: $2(0.744 \times 0.256) = 0.381$; A_2A_2: $0.256^2 = 0.066$. There are 4 percent too few heterozygotes observed, relative to the expected proportion. One of the assumptions of the Hardy-Weinberg principle is not met at this gene in this population, at this time. **CYU p. 446** (1) When allele frequencies fluctuate randomly up and down, sooner or later the frequency of an allele will hit 0. That allele thus is lost from the population, and the other allele at that locus is fixed. (2) In small populations, sampling error is large. For example, the accidental death of a few individuals would have a large impact on allele frequencies. **CYU p. 455** (1) Sperm are small and hence relatively cheap to produce, whereas eggs are large and require a large investment of resources to produce. (2) Sperm are inexpensive to produce, so reproductive success for males depends on their ability to find mates—not on their ability to find resources to produce sperm. The opposite pattern holds for females. Sexual selection is based on variation in ability to find mates, so is more intense in males—leading to more exaggerated traits.

You Should Be Able To (YSBAT)

YSBAT p. 444 If allele frequencies are changing due to drift, the populations in Table 25.1 would behave like the simulated populations in Figure 25.6—frequencies would drift up and down over time, and diverge. **YSBAT p. 451** The proportions of homozygotes should increase, and the proportions of heterozygotes should decrease. **YSBAT p. 452** (1) More recessive deleterious alleles are found in homozygotes and eliminated by natural selection. (2) If there are few or no deleterious recessives in a population, there is less inbreeding depression.

Caption Questions and Exercises

Table 25.1 The observed allele frequencies, calculated from the observed genotype frequencies, are 0.43 for *M* and 0.57 for *N*. The expected genotypes, calculated from the observed allele frequencies under the Hardy-Weinberg principle, are 0.185 for *MM*; 0.49 for *MN*; 0.325 for *NN*. **Figure 25.6** [See Figure A25.1] **Figure 25.8** Original population: freq(A_1) = (9 + 9 + 11) / 54 = 0.54; New population: freq(A_1) = (2 + 2 + 1) / 6 = 0.83. The frequency of A_1 has increased dramatically. **Figure 25.11** Yes—the frozen cells are traces of organisms that lived in the past. **Figure 25.13** The difference between number of flowers in individuals from the two types of matings represents inbreeding depression. This difference increases with age, which means that inbreeding depression increases with age. **Figure 25.16** An allele in males, because a successful male can have as many as 100 offspring—10 times more than a successful female.

Summary of Key Concepts

KC 25.1 There will be an excess of observed genotypes containing the favored allele compared to the proportion expected under Hardy-Weinberg proportions. **KC 25.2** Genetic drift will rapidly reduce genetic variation in small populations. Because captive individuals are usually housed under optimal conditions, selection will not be as intense as it would be in the wild. Alleles that allow individuals to thrive in captivity will increase; alleles that might lower fitness may not be eliminated. **KC 25.3** It increased homozygosity of recessive alleles associated with diseases such as hemophilia. As a result, the royal families were plagued by genetic diseases. **KC 25.4** In this case, reproductive success in females will depend on the number of male mates they can attract. It is reasonable to predict that female red-necked phalaropes are under intense sexual selection for traits that can help them attract mates and are more brightly colored than males.

Test Your Knowledge

1. b; **2.** a; **3.** a; **4.** b; **5.** d; **6.** c

Test Your Understanding

1. Selection may decrease genetic variation or maintain it (e.g., if heterozygotes are favored). Drift reduces it. Gene flow may increase or decrease it, depending on whether immigrants bring new alleles or emigrants remove alleles. Mutation increases it. **2.** Selective pressures often change over time or in different areas occupied by the same species. Even if selective pressures do not vary, mutation continually introduces new alleles. **3.** Sexual selection is most intense on the sex that makes the least investment in the offspring, so that sex tends to have exaggerated traits that make individuals successful in competition for mates. In this way, sexual dimorphism evolves. **4.** The prediction of a null hypothesis states what you should observe if the hypothesis you are testing is not correct. If you are testing the hypothesis that allele frequencies are changing or that mating is nonrandom with respect to a certain gene, the Hardy-Weinberg principle furnishes the appropriate null. **5.** Even if individuals in a small population mate at random or attempt to avoid inbreeding, over time all individuals in a small population are closely related. There are simply not enough nonrelatives to mate with. **6.** Alleles associated with extreme phenotypes are eliminated; alleles associated with intermediate phenotypes increase in frequency.

FIGURE A25.1

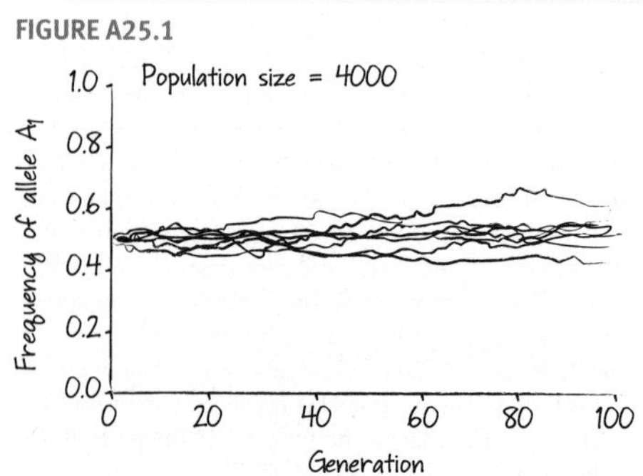

Applying Concepts to New Situations

1. The frequency of heterozygotes is $2p_1p_2$, or $2 \times 0.01 \times 0.99$, which is 0.0198. **2.** Inbreeding causes a decrease in heterozygotes, and a loss of fitness due to the increased numbers of homozygotes for deleterious alleles. Introducing new genetic variants through gene flow will increase allelic diversity in the population and result in more heterozygotes. **3.** If males never lift a finger to help females raise children, the fundamental asymmetry of sex is pronounced and sexual dimorphism should be high. If males invest a great deal in raising offspring, then the fundamental asymmetry of sex is small and sexual dimorphism should be low. **4.** Captive-bred young, transferred adults, and habitat corridors could be introduced to counteract the damaging effects of drift and inbreeding in the small, isolated populations. It is important, though, to introduce individuals only from similar habitats and connect patches of similar habitat with corridors—to avoid introducing alleles that lower fitness.

CHAPTER 26

Check Your Understanding (CYU)

CYU p. 462 (1) Biological species concept: Reproductive isolation means that no gene flow is occurring. Morphological species concept: If populations are evolving independently, they may have evolved morphological differences. Phylogenetic species concept: If populations are evolving independently, they should have synapomorphies that identify them as independent twigs on the tree of life. (2) Biological species concept: It cannot be used to evaluate fossils, species that reproduce asexually, or species that do not occur in the same area and therefore never have the opportunity to mate. Morphological species concept: It cannot identify species that differ in traits other than morphology and is subjective by nature—it can lead to differences of opinion that cannot be resolved by data. Phylogenetic species concept: Reliable phylogenetic information exists only for a small number of organisms. **CYU p. 468** Use one of the cases given in the chapter to illustrate colonization (Galápagos finches), vicariance (shrimp), or habitat specialization (apple maggot flies). In each case, drift will cause allele frequencies to change randomly. Differences in the habitats occupied by the isolated populations will cause allele frequencies to change under natural selection.

You Should Be Able To (YSBAT)

YSBAT p. 465 Selection will favor individuals that (1) breed on apples and have enzymes that are good at digesting apple fruit and that function well as larvae develop in warm temperatures and spend a relatively long time overwintering, and (2) breed on hawthorns and have enzymes that are good at digesting hawthorn fruit and that function well as larvae develop in cool temperatures and spend a relatively short time overwintering. **YSBAT p. 467a** A diploid individual experiences a defect in meiosis resulting in the formation of diploid gametes. The individual self-fertilizes, producing tetraploid offspring. The tetraploid individuals self-fertilize or mate with other tetraploid individuals, producing a tetraploid population. **YSBAT p. 467b** A tetraploid ($4n$) species gives rise to diploid ($2n$) gametes, and a diploid species gives rise to haploid (n) gametes. When a haploid gamete (n) fertilizes a diploid gamete ($2n$), a triploid offspring results. If this individual self-fertilizes, then a population of hexaploid wheat is formed.

Caption Questions and Exercises

Figure 26.3 Save one from the Atlantic Coast and one from the Gulf Coast. Their DNA sequences differ more than do subspecies within a geographic region, so you would be preserving more genetic diversity. **Figure 26.7** It creates genetic isolation—the precondition for divergence. **Figure 26.11** If the same processes were at work in the past as are at work in an experiment with living organisms, then the outcome of the experiment should be a valid replication of the outcome that occurred in the past.

Summary of Key Concepts

KC 26.1 (Many possibilities) Catch a group of fruit flies that are living in a lab, separate the individuals into different cages, and expose the two new populations to dramatically different environmental conditions—heat, lighting, food sources. In essence: create genetic isolation and then conditions for divergence due to drift and selection. **KC 26.2** Human populations would not be considered separate species under the biological concept because human populations can successfully interbreed. They would not be considered separate species under the morphological species concept because all human populations have the same basic morphology. (Although human races differ in minor superficial attributes such as skin color and hair texture, they have virtually identical anatomy and physiology in all other regards.) Nor would human populations be separate species under the phylogenetic species concept because they all arose from a very recent common ancestor. (DNA comparisons have revealed that human races are remarkably similar genetically and do not differ enough genetically to qualify even for subspecies status.) **KC 26.3** The sample experiment described above is an example of vicariance, because the experimenter fragmented the habitat into isolated cages. (If your own experiment differed from the one described above, then a different mechanism of speciation may have been involved.) **KC 26.4** Hybrid individuals will increase in frequency, and natural selection will favor parents that hybridize. The two parent populations may go extinct if hybrids live in the same environment. If the hybrids occupy a different environment, a new species may form.

Test Your Knowledge

1. a; **2.** b; **3.** d; **4.** a; **5.** a; **6.** a

Test Your Understanding

1. Direct observation: Galápagos finch colonization, apple maggot flies, *Tragopogon* allopolyploids, maidenhair ferns. Indirect evidence: snapping shrimp. **2.** On the basis of biological and morphospecies criteria (breeding range and morphological traits), six different species/subspecies of seaside sparrows were recognized. However, based on phylogenetic criteria (comparison of gene sequences), biologists concluded that seaside sparrows represent only two monophyletic groups. **3.** The colonizing population is typically small, and genetic drift has a large effect on smaller populations. If the newly colonized habitat differs from the original one, natural selection will favor individuals with alleles that increase fitness in the new environment. **4.** Some flies breed on apple fruits, others breed on hawthorn fruits. This reduces gene flow. Because apple fruits and hawthorn fruits have different scents and other characteristics, selection is causing the populations to diverge. **5.** Cells that go through many rounds of mitosis often accumulate errors, including mistakes that produce a tetraploid cell. If such a cell becomes part of a developing flower, and goes through meiosis, it will produce diploid gametes that can fuse to form polyploidy offspring. This is less likely to happen in animals because the future reproductive cells undergo fewer rounds of mitosis, so they have fewer chances to become polyploid. **6.** If the species have lived in the same area for a long time, there has been opportunity for selection to favor traits that prevent hybridization. This is reinforcement. If the species do not live in the same area, then there has been no opportunity for selection to favor traits that prevent hybridization.

Applying Concepts to New Situations

1. Decreasing. Gene flow tends to equalize allele frequencies among populations. **2.** These data cast doubt on the hypothesis. If founder events trigger speciation, then at least some speciation events due to introductions should be in progress. **3.** Two things: (1) Some flies happen to have alleles that allow them to respond to apple scents; others happen to have alleles that allow them to respond to hawthorn scents. There is no "need" involved. (2) The alleles for scent response exist; they are not acquired by spending time on the fruit. **4.** If the populations and habitat fragments are small enough, the species is likely to dwindle to extinction due to inbreeding and loss of genetic variation or catastrophes like a severe storm or a disease outbreak. If the populations survive, they are likely to diverge into new species because they are genetically isolated and because the habitats may differ.

CHAPTER 27

Check Your Understanding (CYU)

CYU p. 479 Hair and limb structures in humans and whales are examples of homology because they are traits that can be traced to a common ancestor. All mammals have hair and similar limb bone structure. However, extensive hair loss and advanced social behavior in whales and humans are examples of homoplasy. These traits are not common to all mammalian species and likely arose independently during the evolution of specific mammalian lineages. **CYU p. 488** (1) Doushantuo fossils are microscopic and include sponges and embryos from unknown species. Ediacaran fossils include sponges, jellyfish, and comb jellies along with traces of many other unidentified animals. Most were filter feeders. Burgess Shale fossils include every major animal lineage. The species present are larger, have much more complex morphology, and lived in a wide array of niches. (2) There were many resources available because no other animals (or other types of organisms) existed to exploit them. Also, the evolution of new species in new niches made new niches available for predators. Morphological innovations like limbs and complex mouthparts were important because they made it possible for animals to live in habitats other than the benthic area. **CYU p. 492** (1) Evidence includes the high content of iridium in sedimentary rock formed at the K-T boundary, shocked quartz and microtektite in rock layers that date 65 mya, and a large crater that was found off the coast of Mexico's Yucatán peninsula with microtektite in the crater's walls. Any of these three could be considered convincing, as they are known to form only at impact sites. Taken together, they are particularly convincing. (2) Mass extinctions are caused by such rapid and unusual environmental changes that any species is "lucky" to survive—it is not possible to adapt to such fast changes and such unusual environments.

You Should Be Able To (YSBAT)

YSBAT p. 479 Because whales and hippos share SINEs 4–7 and no other species has these four, they are synapomorphies that define them as a monophyletic group. The similarity is unlikely to be due to homoplasy because the chance of four SINEs inserting in exactly the same place in two different species, independently, is almost astronomically small. **YSBAT p. 485** In habitats on the mainland, there are more competitors and so species are already using the niches (resources) that are available to silverswords in Hawaii and *Anolis* lizards on Caribbean islands.

Caption Questions and Exercises

Figure 27.5 [See Figure A27.1] **Figure 27.16** The dinosaurs went extinct during the end-Cretaceous.

Summary of Key Concepts

KC 27.1 Humans are the only mammal that walks upright on two legs. **KC 27.2** They are the hardest structures in vertebrates and plants, so fossilize most readily. **KC 27.3** They could eat food that was available off the substrate. And once animals lived off the bottom, it created an opportunity for other animals who could eat them to evolve. **KC 27.4** Human-induced extinctions today are not due to poor adaptation to the normal environment. Humans are changing habitats rapidly and in unusual ways, so it is not possible for most species to adapt to these changes rapidly enough to avoid going extinct.

Test Your Knowledge

1. c; **2.** a; **3.** b; **4.** d; **5.** d; **6.** d

Test Your Understanding

1. The fossil record is biased because recent, abundant organisms with hard parts that live underground or in environments where sediments are being deposited are most likely to fossilize. Even so, it is the only data available on what organisms that lived in the past looked like, and where they lived. **2.** Diverse animal forms are found in the Cambrian that are not present in earlier strata. An enormous amount of diversity appeared in a time frame that was short compared to the sweep of earlier Earth history. **3.** Homoplasy is rare relative to homology, so phylogenetic trees that minimize the total change required are usually more accurate. In the case of the artiodactyl astralagus, parsimony was misleading because the astralagus was lost when whales evolved limblessness—creating two changes (a loss following a gain) instead of one (a gain). **4.** (Many answers are possible.) Adaptive radiation of *Anolis* lizards after they colonized new islands in the Caribbean. *Hypothesis:* After lizards arrived on each new island, where there were no predators or competitors, they rapidly diversified to occupy four distinct types of habitats on each island. (Many answers are possible.) Adaptive radiation during the Cambrian period following a morphological innovation. *Hypothesis:* Additional *Hox* genes made it possible to organize a large, complex body; the evolution of complex mouthparts and limbs made it possible for animals to move and find food in new ways. **5.** Environments all over the world deteriorated rapidly—large parts of the ocean lacked oxygen, many coastal habitats disappeared due to change in sea level, and little oxygen was available in the atmosphere. The underlying cause of these changes is not known. **6.** If a trait evolved in a common ancestor, then all of its descendants should share that trait, unless it is lost.

Applying Concepts to New Situations

1. Place the corpse in an environment in which decomposition is slow. One possibility is a swamp or bog; another is a beach or other coastal environment where mud or sand are being deposited. **2.** The fossil record and phylogenetic trees of extinct species indicate that whales evolved from a semiaquatic ancestor. Phylogenetic trees of living species indicate that hippos and whales share a recent common ancestor—meaning that one of the semiaquatic organisms that gave rise to today's whales also gave rise to the semiaquatic species that are today's hippos. **3.** It would be helpful to have (1) a large crater or other physical evidence from a major impact dated at 251 Mya, (2) shocked quartz and microtektites in rock layers dating to the end-Permian era, and (3) high levels of iridium or other elements common in space rocks and rare on Earth, dated to the time of the extinctions. **4.** Oxygen is an effective final electron acceptor in cellular respiration because of its high electronegativity. Organisms that use it as a final electron acceptor can produce more usable energy than organisms that do not use oxygen, but only if it is available. With more available energy, aerobic organisms can grow larger and move faster.

THE BIG PICTURE: EVOLUTION

Check Your Understanding (CYU), p. 494

1. Circle = inbreeding, sexual selection, natural selection, genetic drift, mutation, and gene flow. **2.** Adaptation "increases" fitness; synapomorphies "identify branches on" the tree of life. **3.** Several answers possible: e.g., the flower in angiosperms, the pharyngeal jaw in cichlid fishes, feathers and flight in birds. **4.** Genetic drift, mutation, and gene flow "are random with respect to" fitness.

CHAPTER 28

Check Your Understanding (CYU)

CYU p. 504 (1) Conditions should mimic a spill—sand or stones with a layer of crude oil or seawater with oil floating on top. Add samples, from sites contaminated with oil, that might contain cells capable of using molecules in oil as electron donors or electron acceptors. Other conditions (temperature, pH, etc.) should be realistic. (2) Use direct sequencing. Choose a well-studied gene, such as 16S RNA, with reliable PCR primers. After isolating DNA from a soil sample, do a PCR reaction and clone the resulting genes into plasmids grown in *E. coli* cells, so that each culture contains a different gene. Once many copies are available, sequence the genes and use the data to place the original organisms on the tree of life. **CYU p. 512** Eukaryotes can only (1) fix carbon via the Calvin-Benson pathway, (2) use aerobic respiration with organic compounds as electron donors, and (3) perform oxygenic photosynthesis. Among bacteria and archaea, there is much more diversity in pathways for carbon fixation, respiration, and photosynthesis, along with many more fermentation pathways.

You Should Be Able To (YSBAT)

YSBAT p. 506 cyanobacteria—photoautotrophs; *Clostridium aceticum*—chemoorganoautotroph; *Nitrosomonas* sp.—chemolithotrophs; heliobacteria—photoheterotrophs; *Escherichia coli*—chemoorganoheterotroph; *Beggitoa*—chemolithotrophic heterotrophs **YSBAT p. 513a, YSBAT p. 513b, YSBAT p. 514a, YSBAT p. 514b, YSBAT p. 515** [See Figure A28.1]

Caption Questions and Exercises

Figure 28.1 Prokaryotic—as it would require just one evolutionary change, the origin of the nuclear envelope in Eukarya. If it were eukaryotic, it would require that the nuclear envelope was lost in both Bacteria and Archaea—two changes and less parsimonious (see Chapter 27). **Table 28.1** [See Figure A28.2] **Figure 28.2** Improved nutrition made people better able to fight off disease, and improved sanitation lowered transmission of disease-causing bacteria. **Figure 28.4** Weak—different culture conditions may have revealed different species. **Figure 28.9** Table 28.5 contains the answers. For example, for organisms called sulfate reducers you would have H_2 as the electron donor, SO_4^{2-} as the electron acceptor, and H_2S as the reduced by-product. For humans the electron donor is glucose ($C_6H_{12}O_6$), the electron acceptor is O_2, and the reduced by-product is water (H_2O). **Table 28.5** Organotrophs use sugars, which are organic compounds, as their electron donor—getting energy by "feeding" on them. Sulfate reducers use sulfate ions as their electron acceptors, thus reducing the sulfate ion. Methanogens generate methane as a by-product. **Figure 28.11** Aerobic respiration. More free energy is released when oxygen is the final electron acceptor than when any other molecule is used, so more ATP can be produced and used for growth.

FIGURE A27.1

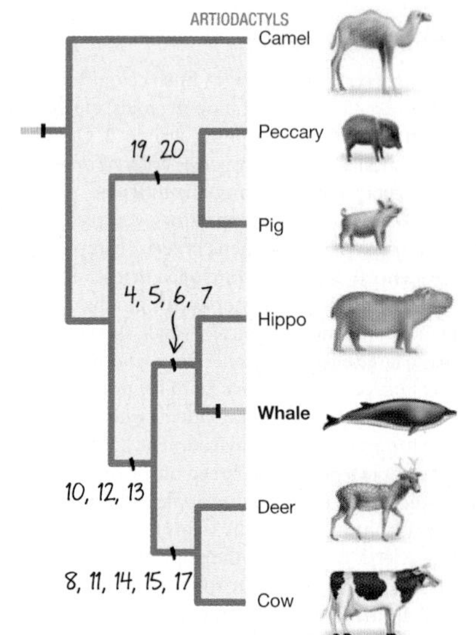

FIGURE A28.1

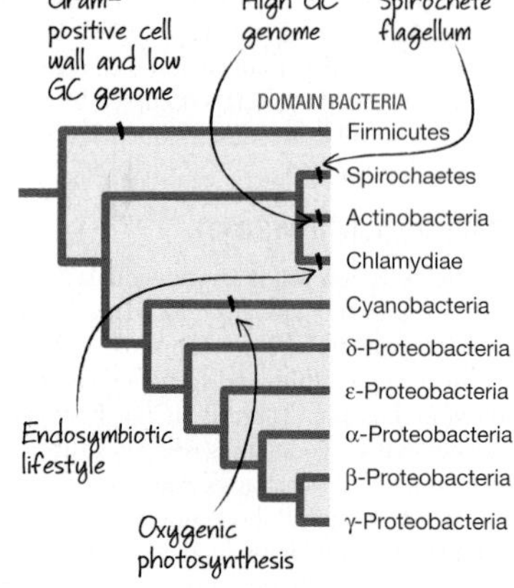

FIGURE A28.2

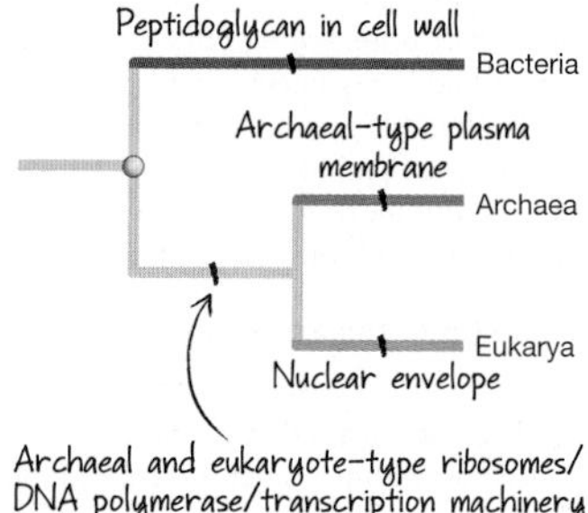

Summary of Key Concepts

KC 28.1 Eukaryotes have a nuclear envelope that encloses their chromosomes; bacteria and archaea do not. Bacteria have cell walls that contain peptidoglycan, and archaea have phospholipids containing isoprene subunits in their plasma membranes. Thus, the exteriors of a bacterium and archaeon are radically different. Archaea and eukaryotes also have similar machinery for processing genetic information. **KC 28.2** They are compatible, and thrive in each others' presence—one species' waste product is the other species' food. **KC 28.3** If bacteria and archaea did not exist, then (1) the atmosphere would have little or no oxygen, and (2) almost all nitrogen would exist in molecular form (the gas N_2).

Test Your Knowledge

1. c; **2.** b; **3.** d; **4.** d; **5.** b; **6.** c

Test Your Understanding

1. An electron donor provides the potential energy required to produce ATP. **2.** Yes. The array of substances that bacteria and archaea can use as electron donors, electron acceptors, and fermentation substrates, along with the diversity of ways that they can fix carbon and perform photosynthesis, allows them to live just about anywhere. **3.** You can sample and amplify genes—getting DNA sequence data that allow you to place unseen species on the tree of life. **4.** Large amounts of potential energy are released and ATP produced when oxygen is the electron acceptor, because oxygen is so electronegative. Large body size and high growth rates are not possible without large amounts of ATP. **5.** [See Figure A28.3] The following can serve as both electron donors and by-products: CH_4, NO_2^-, H_2S. The following can serve as both electron acceptors and by-products: SO_4^{2-}, CO_2, and NO_3^-. **6.** They are paraphyletic because the prokaryotes include some (bacteria and archaea) but not all (eukaryotes) groups derived from the common ancestor of all organisms living today.

Applying Concepts to New Situations

1. This result supports their hypothesis, because the drug poisons the enzymes of the electron transport chain and prevents electron transfer to Fe^{3+}, which is required to drive magnetite synthesis. If magnetite had still formed, another explanation would have been needed. **2.** Hypothesis: A high rate of tooth cavities in Western children is due to an excess of sucrose in the diet, which is absent from the diets of East African children. To test this hypothesis, have East African children switched to a diet that contains sucrose and monitor the presence of *S. mutans*. Have Western children switch to a diet lacking sucrose and monitor the presence of *S. mutans*. **3.** Look in waters or soil polluted with benzene-containing compounds. Put samples from these environments in culture tubes where benzene is the only source of carbon. Monitor the cultures and study the cells that grow efficiently. **4.** They should get energy from reduced organic compounds "stolen" from their hosts.

CHAPTER 29

Check Your Understanding (CYU)

CYU p. 523 (1) *Plasmodium* species are transmitted to humans by mosquitoes. If mosquitoes can be prevented from biting people, they cannot spread the disease. (2) Iron added → primary producers (photosynthetic protists and bacteria) bloom → more carbon dioxide taken up from atmosphere during photosynthesis → consumers bloom, eat primary producers → bodies of primary producers and consumers fall to bottom of ocean → large deposits of carbon-containing compounds form on ocean floor. **CYU p. 526** (1) Opisthokonts have a flagellum at the base or back of the cell; alveolate cells contain unique support structures called alveoli; stramenopiles have straw-like hairs on their flagella. (2) In direct sequencing, DNA is isolated directly from the environment and analyzed to place species on the tree of life. It is not necessary to actually see the species being studied.

You Should Be Able To (YSBAT)

YSBAT p. 526a [See Figure A29.1 on p. A:28] **YSBAT p. 526b** Membrane infoldings observed in bacterial species today support the hypothesis's plausibility—they confirm that the initial steps could have actually happened. The continuity of the nuclear envelope and ER are consistent with the hypothesis, which predicts that the two structures are derived from the same source (infolded membranes). **YSBAT p. 528** A photosynthetic bacterium (e.g., a cyanobacterium) could have been engulfed by a larger eukaryotic cell. If it was not digested, it could continue to photosynthesize and supply sugars to the host cell. **YSBAT p. 531** The chloroplast genes label should come off of the branch that leads to cyanobacteria. (If you had a phylogeny just of the cyanobacteria, the chloroplast branch would be located somewhere inside.) **YSBAT p. 532** The acquisition of the mitochondrion and the chloroplast represent the transfer of entire genomes, and not just single genes, to a new organism. **YSBAT p. 533** Yes—when food is scarce or population density is high, the environment is changing rapidly (deteriorating). Offspring that are genetically unlike their parents may be better able to cope with the new and challenging environment. **YSBAT p. 536a** (1) Alternation of generations refers to a life cycle in which there are multicellular haploid phases and multicellular diploid phases. A gametophyte is the multicellular haploid phase; the sporophyte is the multicellular diploid phase. A spore is a cell that grows into a multicellular individual, but is not produced by fusion of two cells. A zygote is a cell that grows into a multicellular individual, but *is* produced by fusion of two cells (gametes). Gametes are haploid cells that fuse to form a zygote. (2) [See Figure A29.2 on p. A:28] **YSBAT p. 536b, YSBAT p. 536c, YSBAT p. 536d, YSBAT p. 537a, YSBAT p. 537b, YSBAT p. 537c** [See Figure A29.1 on p. A:28] **YSBAT p. 537d** It is most likely that the two types of amoebae evolved independently. The alternative hypothesis is that the common ancestor of alveolates, stramenopiles, rhizarians, plants, opisthokonts, and amoebozoa were amoeboid, and that this growth form was lost many times. [See Figure A29.1 on p. A:28] **YSBAT p. 538a, YSBAT p. 538b** [See Figure A29.1 on p. A:28] **YSBAT p. 539a** If euglenids could take in food via phagocytosis (ingestive feeding), then it would have provided a mechanism by which a smaller photosynthetic protist could have been engulfed and incorporated into the cell via secondary endosymbiosis. **YSBAT p. 539b, YSBAT p. 540a, YSBAT p. 540b, YSBAT p. 541a, YSBAT p. 541b, YSBAT p. 542a, YSBAT p. 542b, YSBAT p. 543** [See Figure A29.1 on p. A:28]

Caption Questions and Exercises

Figure 29.9 Two—one derived from the original bacterium and one derived from the eukaryotic cell that engulfed the bacterium. **Table 29.3** Yes—green algae, euglenids, and chlorarachniophytes all have chlorophyll *a* and *b*, as predicted by the hypothesis that a green algal chloroplast was transferred to the ancestor of euglenids and of chlorarachniophytes. Red algae have chlorophyll *a*, and chromalveolates (which include the brown algae, diatoms, and dinoflagellates) have chlorophyll *a* and *c*. If the hypothesis is correct, chlorophyll *c* must have evolved in an ancestor of the chromalveolates independently of the acquisition of chlorophyll *a* from red algae.

Summary of Key Concepts

KC 29.1 Photosynthetic protists use CO_2 and light to produce sugars and other organic compounds, so they furnish the first or primary source of organic material in an ecosystem. **KC 29.2** (1) Outside membrane was from host eukaryote; inside from engulfed cyanobacterium. (2) From the outside in, the four membranes are derived from the eukaryote that engulfed a chloroplast-containing eukaryote, the plasma membrane of the eukaryote that was engulfed, the outer membrane of the engulfed cell's chloroplast, and the inner membrane of its chloroplast. **KC 29.3** Each set of pigments absorbs most strongly in a certain part of the electromagnetic spectrum. If different species have different pigments, they can live in close proximity and perform photosynthesis without competing for the same wavelengths of light. **KC 29.4** A gametophyte is haploid, and a sporophyte is diploid.

Test Your Knowledge

1. b; **2.** b; **3.** a; **4.** b; **5.** d; **6.** b

FIGURE A28.3

TABLE 28.5 **Some Electron Donors and Acceptors Used by Bacteria and Archaea**

Electron Donor	Electron Acceptor	By-Products: From Electron Donor	By-Products: From Electron Acceptor	Category*
Sugars	O_2	CO_2	H_2O	Organotrophs
H_2 or organic compounds	SO_4^{2-}	H_2O or CO	H_2S or S^{2-}	Sulfate reducers
H_2	CO_2	H_2O	CH_4	Methanogens
CH_4	O_2	CO_2	H_2O	Methanotrophs
S^{2-} or H_2S	O_2	SO_4^{2-}	H_2O	Sulfur bacteria
Organic compounds	Fe^{3+}	CO_2	Fe^{2+}	Iron reducers
NH_3	O_2	NO_2^-	H_2O	Nitrifiers
Organic compounds	NO_3^-	CO_2	N_2O, NO, or N_2	Denitrifiers (or nitrate reducers)
NO_2^-	O_2	NO_3^-	H_2O	Nitrosifiers

*The name biologists use to identify species that use a particular metabolic strategy.

Test Your Understanding

1. An extensive cytoskeleton is required to shape the cell membrane so that the pseudopod can form and surround the food item. A cell wall is not flexible enough to wrap around the food item as the plasma membrane does. **2.** Because all eukaryotes living today have cells with a nuclear envelope, it is valid to infer that their common ancestor also had a nuclear envelope. Because bacteria and archaea do not have a nuclear envelope, it is valid to infer that the trait arose in the common ancestor of eukaryotes. **3.** Meiosis is required for alternation of generations because it converts a diploid phase of the life cycle to a haploid phase. Without it, the generations wouldn't "alternate." **4.** The host cell provided a protected environment and carbon compounds for the endosymbiont; the endosymbiont provided increased ATP from the carbon compounds. **5.** It confirmed a fundamental prediction made by the hypothesis, and could not be explained by any alternative hypothesis. **6.** All alveolates have alveoli, which are unique structures that function in supporting the cell. Among alveolates there are species that are (1) ingestive feeders, photosynthetic, or parasitic, and (2) move using cilia, flagella, or a type of amoeboid movement.

Applying Concepts to New Situations

1. The observation suggests that eukaryotes did not acquire mitochondria until there was enough oxygen present to make aerobic respiration efficient. (Oxygen is also poisonous to cells at high concentration, so mitochondria gave the early eukaryotes a way to "detoxify" oxygen.) **2.** If the apicoplast that is found in *Plasmodium* (the organism that causes malaria) is genetically similar to chloroplasts, and if glyphosate poisons chloroplasts, it is reasonable to hypothesize that glyphosate will poison the apicoplast and potentially kill the *Plasmodium.* This would be a good treatment strategy for malaria because humans have no chloroplasts, provided that the glyphosate produces no other effects that would be detrimental to humans. **3.** Primary producers usually grow faster when CO_2 concentration increases, but to date, they have not grown fast enough to make CO_2 levels drop—CO_2 levels have been increasing steadily over decades. **4.** Given that lateral gene transfer can occur at different points in a phylogenetic history, specific genes can become part of a lineage by a different route from that taken by other genes of the organism. In the case of chlorophyll *a,* its history traces back to a bacterium being engulfed by a protist and forming a chloroplast.

CHAPTER 30

Check Your Understanding (CYU)

CYU p. 552 (1) Green algae and land plants share an array of morphological traits that are synapomorphies, including the chlorophylls they contain, (2) green algae appear before land plants in the fossil record, and (3) on phylogenetic trees estimated from DNA sequence data, green algae and land plants share a most recent common ancestor, with green algae being the initial groups to diverge and land plants diverging subsequently. **CYU p. 566** (1) Cuticle prevents water loss from the plant; vascular tissue moves water up from the soil and moves photosynthetic products down to the roots. (2) [See Figure A30.1]

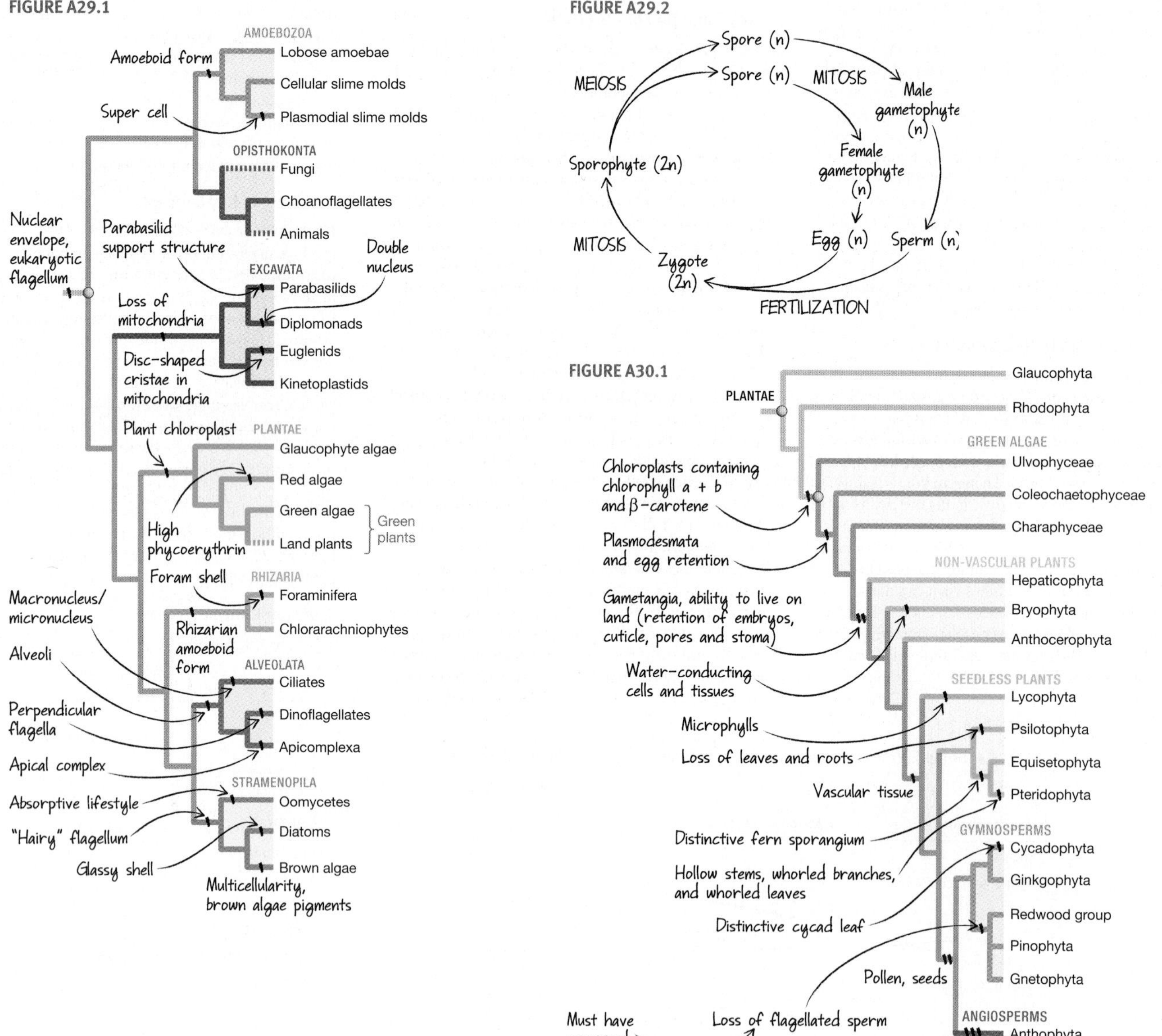

You Should Be Able To (YSBAT)

YSBAT p. 558 Alternation of generations occurs when there are multicellular haploid individuals and multicellular diploid individuals in a life cycle. A sporophyte is a multicellular diploid individual that produces spores by meiosis. A spore is a single cell that is not formed by fusion of two cells, and that grows by mitosis into a multicellular adult. **YSBAT p. 559** In the hornwort photo, the sporophyte is the spike-like green and brown structure; the gametophyte is the leafy-looking structure underneath. The horsetail gametophyte is the microscopic individual on the left; the sporophyte is the much larger individual on the right. **YSBAT p. 569a, YSBAT p. 569b, YSBAT p. 570, YSBAT p. 571, YSBAT p. 572a, YSBAT p. 572b, YSBAT, p. 573, YSBAT p. 574** [See Figure A30.1]

Caption Questions and Exercises

Figure 30.7 If you find the common ancestor of all the green algae (at the base of the Ulvophyceae), the lineages that are collectively called green algae don't include that common ancestor and all of its descendants—only some of its descendants. The same is true for non-vascular plants (the common ancestor here is at the base of the Hepaticophyta) and seedless vascular plants (the common ancestor here is at the base of the Lycophyta). **Figure 30.9** Water flows more easily through a short, wide pipe than through a long, skinny one because there is less resistance from the walls of the pipe. **Figure 30.10** To map an innovation on a phylogenetic tree, biologists determine the location(s) on the tree that is consistent with all descendants from that point having the innovation (unless, in rare cases, the innovation was lost in a few descendants). **Figure 30.15** There are multicellular haploid stages and multicellular diploid stages in these plants. **Figure 30.19** Gymnosperm gametophytes are microscopic, so they are even smaller than fern gametophytes. The gymnosperm gametophyte is completely dependent on the sporophyte for nutrition, while fern gametophytes are not. Fern gametophytes are photosynthetic and even supply nutrition to the young sporophytes. **Figure 30.20** Consistent—the fossil data suggest that gymnosperms evolved earlier, and gymnosperms have larger gametophytes than angiosperms. **Figure 30.22** It tests the hypothesis that the presence of yarn on the spur changes pollinator behavior and thus reproductive success. **Figure 30.24** [See Figure A30.1]

Summary of Key Concepts

KC 30.1 Rates of soil formation will drop; rates of soil loss will increase. **KC 30.2** Because so many adaptations are required—starting with cuticle and pores or stomata—it is unlikely that the transition from water to land would happen more than once. Stated another way, the aquatic and terrestrial environments are so different for plants that the transition was difficult and thus unlikely. **KC 30.3** Both spores and seeds have a tough, protective coat, so can survive while being dispersed to a new location. Seeds have the advantage of carrying a store of nutrients with them—when a spore germinates, it has to make its own food via photosynthesis right away.

Test Your Knowledge

1. c; **2.** b; **3.** d; **4.** c; **5.** c; **6.** a

Test Your Understanding

1. Plants build and hold soils required for human agriculture and forestry, and increase water supplies that humans can use for drinking, irrigation, or industrial use. Plants release oxygen that we breathe. **2.** Cuticle prevents water loss from leaves but also prevents entry of CO_2 required for photosynthesis. Stomata allow CO_2 to diffuse but can close to minimize water loss. Liverwort pores allow gas exchange but cannot be closed if conditions become dry. Liverworts that lack pores have a cuticle that is thin enough to allow some gas exchange. **3.** They provided the support needed for plants to grow upright and not fall over in response to wind or gravity. Erect growth allowed plants to compete for light. **4.** (1) Gametangia are found in all land plant groups except angiosperms; (2) transfer cells are found in all land plants; (3) pollen is found in gymnosperms and angiosperms; (4) seeds are found in gymnosperms and angiosperms, (5) fruit is found in angiosperms. **5.** In a gametophyte-dominant life cycle, the gametophyte is larger and longer lived than the sporophyte and produces most of the nutrition. In a sporophyte-dominant life cycle, the sporophyte generation is the larger, longer-lived, and photosynthetic phase of the life cycle. **6.** Homosporous plants produce a single type of spore that develops into a gametophyte that produces both egg and sperm. Heterosporous plants produce two different types of spores that develop into two different gametophytes that produce either egg or sperm. In a tulip, the microsporangium is found within the stamen, and the megasporangium is found within the ovule. Microspores divide by mitosis to form male gametophytes (pollen grains); megaspores divide by mitosis to form the female gametophyte.

Applying Concepts to New Situations

1. Homosporous plants produce a single type of spore that develops into a gametophyte that produces both egg and sperm. Heterosporous plants produce two different types of spores that develop into two different gametophytes that produce either egg or sperm. In a tulip, the microsporangium is found within the stamen, and the megasporangium is found within the ovule. Microspores divide by mitosis to form male gametophytes (pollen grains); megaspores divide by mitosis to form the female gametophyte. **2.** The combination represents a compromise between efficiency and safety. Tracheids can still transport water if vessels become blocked by air bubbles. **3.** A "reversion" to wind pollination might be favored by natural selection because it is costly to produce a flower that can attract animal pollinators. Because wind-pollinated species grow in dense clusters, they can maximize the chance that the wind will carry pollen from one individual to another (less likely if the individuals are far apart). Wind-pollinated deciduous trees flower in early spring before their developing leaves begin to block the wind. **4.** Alter one characteristic of a flower and present the flower to the normal pollinator. As a control, present the normal (unaltered) flower to the normal pollinator. Record the amount of time the pollinator spends in the flower, the amount of pollen removed, or some other measure of pollination success. Repeat for other altered characteristics. Analyze the data to determine which altered characteristic affects pollination success the most.

CHAPTER 31

Check Your Understanding (CYU)

CYU p. 586 (1) Because they are made up of a network of thin, branching hyphae, mycelia have a large surface area, which makes absorption efficient. (2) Swimming spores and gametes, zygosporangia, basidia, asci. **CYU p. 594** (1) When birch tree seedlings are grown in the presence and absence of EMF, individuals denied their normal EMF are not able to acquire sufficient nitrogen and phosphorus. Isotope-tracing experiments also show that sugars are transferred from host plant to EMF and that nitrogen and phosphorus are transferred from EMF to host plant. (2) Meiosis and production of haploid spores.

You Should Be Able To (YSBAT)

YSBAT p. 591 Human sperm and egg undergo plasmogamy followed by karyogamy during fertilization, but heterokaryosis does not occur in humans or other eukaryotes besides fungi. **YSBAT p. 594, YSBAT p. 596a, YSBAT p. 596b, YSBAT p. 596c, YSBAT p. 597, YSBAT p. 598a, YSBAT p. 598b** [See Figure A31.1]

Caption Questions and Exercises

Figure 31.10 Labeled-nutrient experiments identify which nutrients are exchanged and in which direction. They explain *why* plants do better in the presence of mycorrhizae, and why the fungus also benefits.

FIGURE A31.1

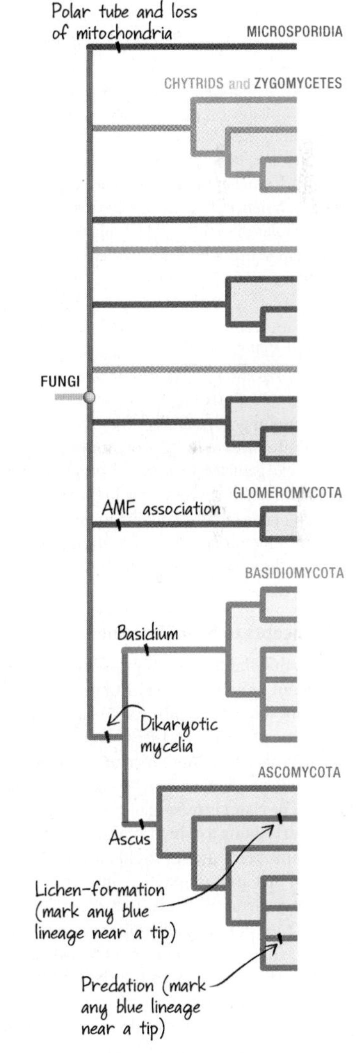

Figure 31.11 [See Figure A31.2] **Figure 31.13** The haploid mycelium. **Figure 31.14** Mycelia produced by asexual reproduction are genetically identical to their parent; mycelia produced by sexual reproduction are genetically different from both parents (each spore has a unique genotype).

Summary of Key Concepts

KC 31.1 Loss of mycorrhizal fungi would decrease nutrient delivery to plants and reduce their growth inside the experimental plots, compared to control plots with intact fungi. Also, lack of fungi would slow decay of dead plant material, causing a dramatic buildup of dead organic material. **KC 31.2** Absorption has to occur across a surface (usually via transport proteins in the plasma membrane). Other things being equal, more absorption will occur across a large surface area than a small surface area. **KC 31.3** Haploid hyphae fuse, forming a heterokaryotic mycelium. When karyogamy occurs, a diploid nucleus forms—just as when gametes fuse.

Test Your Knowledge

1. a; **2.** d; **3.** c; **4.** b; **5.** b; **6.** d

Test Your Understanding

1. Along with a few bacteria, fungi are the only organisms that can digest wood completely. If the wood is not digested, carbon remains trapped in wood. Without fungi, CO_2 would be tied up and unavailable for photosynthesis, and the presence of undecayed organic matter would reduce the space available for plants to grow. **2.** Fungi produce enzymes that degrade cellulose and lignin. **3.** Plant roots have much smaller surface area than EMF or AMF. Hyphae are much smaller than the smallest portions of plant roots so can penetrate dead material more efficiently. Extracellular digestion—which plant roots cannot do—allows fungi to break large molecules into small compounds that can be absorbed. **4.** DNA, RNA, and ATP all contain phosphorus. Without adequate amounts of phosphorus, plants cannot grow by synthesizing more DNA, RNA, and ATP. **5.** Both compounds are processed via extracellular digestion. Different enzymes are involved, however. The degradation of lignin is uncontrolled and does not yield useful products; digestion of cellulose is controlled and produces useful glucose molecules. **6.** Most fungi do not make gametes. Instead, the products of specific alleles identify mating types—probably to encourage fusion between hyphae that are dissimilar genetically, resulting in the production of genetically diverse offspring. It is possible for thousands of different mating-type alleles to exist in a population.

Applying Concepts to New Situations

1. (1) Confirm that the chytrid fungus is found only in sick frogs and not healthy frogs. (2) Isolate the chytrid fungus and grow it in a pure culture. (3) Expose healthy frogs to the cultured fungus and see if they become sick. (4) Isolate the fungus from the experimental frogs, grow it in culture, and test whether it is the same as the original fungus. **2.** The claim is reasonable because fungi have so many ways of making a living from plant tissues. For example, the same plant species could have several species of fungi that are endophytic, mycorrhizal, or parasitic, as well as an array of species that break down its tissues when it dies. **3.** Each of the different cellulase enzymes attacks cellulose in a different way, so producing all the enzymes together increases the efficiency of the fungus in breaking down cellulose completely. It is likely that lignin peroxidase is produced along with cellulases, so they can act in concert to degrade wood. To test this idea, you could harvest enzymes secreted from a mycelium before and after it contacts wood in a culture dish, and see if the cellulases and lignin peroxidase appear together once the mycelium begins growing on the wood. **4.** Collect a large array of colorful mushrooms that are poisonous and capture mushroom-eating animals, such as squirrels. Present a hungry squirrel with a choice of mushrooms that have been dyed or painted a drab color versus treated with a solution that is identical to the dye or paint used but uncolored. Record which mushrooms the squirrel eats. Repeat the test with many squirrels and many mushrooms.

CHAPTER 32

Check Your Understanding (CYU)

CYU p. 609 (1) Bilateral symmetry led to the evolution of long, slender body plans. Triploblasty gave rise to an inner tube (gut), an outer tube (body lining), and muscles and organs in between. The coelom functions as a hydrostatic skeleton to facilitate movement in soft-bodied animals that lack limbs. (2) When an animal moves through an environment in one direction, its ability to acquire food and perceive and respond to threats is greater if its feeding, sensing, and information-processing structures are at the leading end. **CYU p. 617** (1) The mouthparts of deposit feeders are relatively simple, as they simply gulp relatively soft material. The mouthparts of mass feeders are more complex, because they have to tear off and process chunks of relatively hard material. (2) Gametes that are shed into aquatic environments can float or swim. This cannot happen on land, so internal fertilization is more common.

You Should Be Able To (YSBAT)

YSBAT p. 606 If you do the exercise correctly, the balloon should wriggle. The wriggling movements would produce movement if the balloon (or animal) were surrounded by water or soil. **YSBAT p. 618** The origin of epithelial tissue should be marked on the same branch as multicellularity. **YSBAT p. 619** The origin of cnidocytes should be marked on the Cnidaria branch. **YSBAT p. 620** The origin of cilia-powered swimming should be marked on the Ctenophora branch.

Caption Questions and Exercises

Figure 32.6 The animal would get shorter and fatter. **Figure 32.9** The label and bar should go on the branch to the left of the clam shell and "Mollusk" label. **Figure 32.19** Stained *Dll* gene products would be located in the legs of the insect but would not be concentrated anywhere in the onychophoran and segmented worm. **Figure 32.22** No—there is no multicellular haploid form.

Summary of Key Concepts

KC 32.1 It should increase dramatically—animals would no longer be consuming plant material. **KC 32.2** Your drawing should show the tube-within-a-tube design of a worm, with one end labeled head and containing the mouth and brain. From the brain, one or two major nerve tracts should run the length of the body. The gut should go from the mouth to the other end, ending in the anus. In between the gut and outer body wall, there should be blocks or layers of muscle. **KC 32.3** During gastrulation, the initial pore becomes the mouth in protostomes, whereas in deuterostomes the pore becomes the anus. The protostome coelom is formed from blocks of mesodermal tissue that form cavities. The deuterostome coelom forms when mesodermal tissue around the gut pinches off. **KC 32.4** Natural selection has produced eye structures that function well in a particular species' habitat, and mollusks live in a wide array of habitats. Some mollusks live buried in sand and have no eyes—eyes would have no function in this habitat. Mollusks like squid and octopuses live in open water and hunt prey. They have complex eyes that form images and help them find food. **KC32.5** Juveniles look like miniature adults and feed on the same foods as adults. Larvae look much different from adults, live in different habitats, and eat different foods. The difference is important because larvae do not compete with adults for food.

FIGURE A31.2

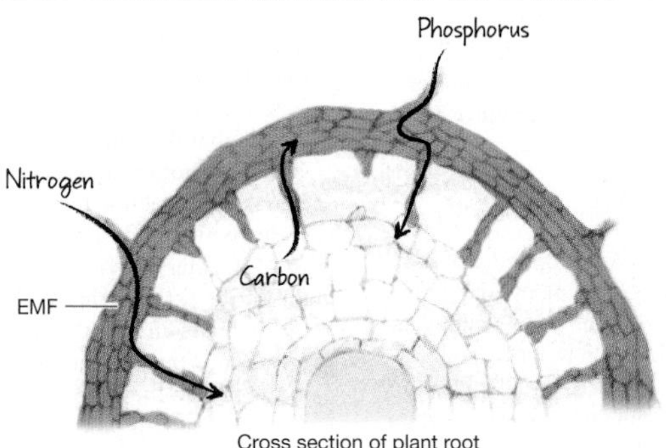

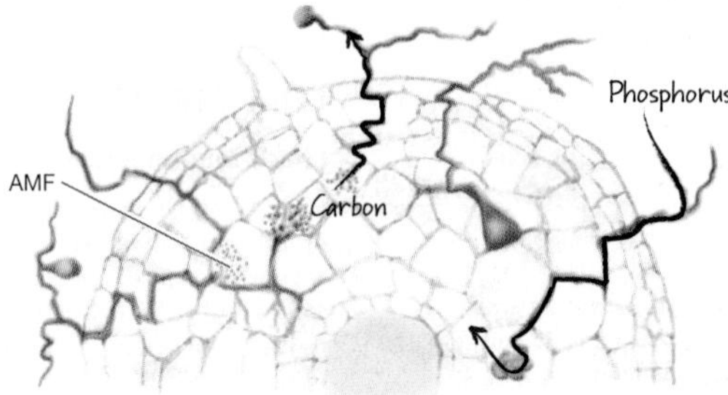

Test Your Knowledge

1. a; **2.** d; **3.** b; **4.** a; **5.** d; **6.** c

Test Your Understanding

1. Diploblasts have two types of embryonic tissue; triploblasts have three. Mesoderm made it possible for an enclosed, muscle-lined cavity to develop, creating a coelom. **2.** A hydrostatic skeleton is a pressurized, fluid-filled chamber (the coelom) that is surrounded by muscles. Contractions on either side of the body change the pressure of the fluid in the hydrostatic skeleton and thus the shape of the coelom. When muscle contractions are coordinated throughout the length of the animal, the shape of the hydrostatic skeleton changes and results in movement. **3.** There are many unicellular organisms that are heterotrophic, but they can consume only small packets of food. Animals are multicellular, so they are larger and can consume larger packets of food. **4.** A long, hollow structure that is sharp enough to pierce the wall of the stem; a simple, muscular mouth opening that can gulp soft, decaying material; sharp, stiff structures that can move in a scissoring action; a long, hollow structure that can be inserted into the flower. **5.** In oviparous species, the mother adds nutrient-rich yolk to the egg that nourishes the developing embryo. In viviparous species, the mother transfers nutrients directly from her body to the growing embryo. **6.** Radially symmetric organisms encounter the environment in many directions, so it is advantageous to have an equal number of neurons located throughout the body. Bilaterally symmetric organisms encounter the environment in one direction, so it is advantageous to have most neurons clustered in the head region.

Applying Concepts to New Situations

1. Yes—if the same gene is found in nematodes and humans, it was likely found in the common ancestor of protostomes and deuterostomes. If so, then fruit flies should also have this gene. **2.** Mosquitoes, because larval and adult mosquitoes feed on different food sources in different habitats and do not compete with one another. You could test this prediction by collecting data on the total number of mosquito versus tick species, their abundance, and geographic distribution. **3.** It should resemble a nerve net, because echinoderms need to take in and process information from multiple directions—not just one. **4.** Web-building spiders are some of the only terrestrial animals that suspension feed.

CHAPTER 33

Check Your Understanding (CYU)

CYU p. 630 (1) Arthropods have a tube-within-a-tube body plan with a drastically reduced coelom. They have a hemocoel body cavity that holds internal organs and body fluids. The body is segmented, with segments grouped into tagmata. They have jointed limbs and an exoskeleton made of chitin. Compared to unjointed limbs, jointed limbs allow animals to move faster and with more precision. (2) Supporting the body (and moving) without support from water. Preventing the body from drying out. Facilitating gas exchange.

You Should Be Able To (YSBAT)

YSBAT p. 631a, YSBAT p. 631b, YSBAT p. 632, YSBAT p. 633, YSBAT p. 637, YSBAT p. 638 [See Figure A33.1]

Caption Questions and Exercises

Figure 33.3 The tuft is the cluster of ciliated tentacles in part (a); the wheel is the ring of cilia in part (b). **Figure 33.6** Different phyla of worms have different feeding strategies, different mouthparts, and eat different foods, so they are not in direct competition for food. **Figure 33.22** [See Figure A33.2]

Summary of Key Concepts

KC 33.1 All of the phyla that are considered protostomes share a pattern of early development found in no other animals. Based on this observation, it is logical to claim that this pattern of development arose in the common ancestor of these phyla—meaning that it is a synapomorphy. **KC 33.2** Flatworms likely lost the adaptation of a coelom because as they evolved a thin, flattened body plan, the coelom was no longer needed for internal gas circulation or movement. Arthropods have a drastically reduced coelom because they have evolved limbs and complex musculature for movement and no longer require the coelom to provide a hydrostatic skeleton. **KC 33.3** (1) Because larvae and adults live in a different habitat, young and adults do not compete for food. (2) Because larvae can swim, they can disperse to new locations.

Test Your Knowledge

1. d; **2.** a; **3.** c; **4.** c; **5.** a; **6.** b

Test Your Understanding

1. Ecdysozoans grow by molting; lophotrochozoans grow by incremental additions to their bodies. In addition, some lophotrochozoan phyla have lophophores and/or trochophore larvae. Both groups are bilaterally symmetric triploblasts with the protostome pattern of development. **2.** Multiple times, because phylogenies show that segmented groups evolved from unsegmented ancestors in both lophotrochozoans (annelids) and ecdysozoans (arthropods). **3.** The ability to live on land offered the opportunity of exploiting new habitats and food sources, with minimal competition from other animals. **4.** An exoskeleton provided a stiff surface for muscle attachment—facilitating rapid, precise movement—as well as protection from enemies. **5.** The ability to fly allowed insects to disperse to new habitats and find new food sources efficiently. **6.** The leading hypothesis is variation in mouthpart structure and feeding strategies.

Applying Concepts to New Situations

1. [See Figure A33.3] **2.** If the ancestors of brachiopods and mollusks lived in similar habitats and experienced natural selection that favored similar traits, then they would have evolved to have similar forms and habitats. This is called convergent evolution (see Chapter 27). **3.** Aplacophora are probably basal—meaning that they would branch off between the ancestral form at the base of the tree and the rest of the groups. Because they have reduced forms of key molluscan synapomorphies, it is likely that they are the least-derived of the Mollusca. Because they lack a shell for muscle attachment and a well-developed foot, they probably move slowly via undulating movements and live in benthic habitats. **4.** Agree—it is likely that the common ancestor of arthropods was marine, and that terrestrial forms evolved from aquatic ancestors independently in crabs, isopods, and insects.

CHAPTER 34

Check Your Understanding (CYU)

CYU p. 650 (1) If it's an echinoderm, it should have five-part radial symmetry, a calcium carbonate endoskeleton just underneath the skin, and a water vascular system (e.g., visible podia). **CYU p. 659** (1) Jaws allow animals to capture food efficiently and process it by crushing or tearing. (2) The increased cushioning from enclosed fluids provided mechanical support, and the increased surface area made transport of gases and other materials more efficient.

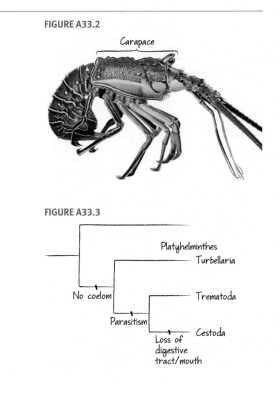

FIGURE A33.1

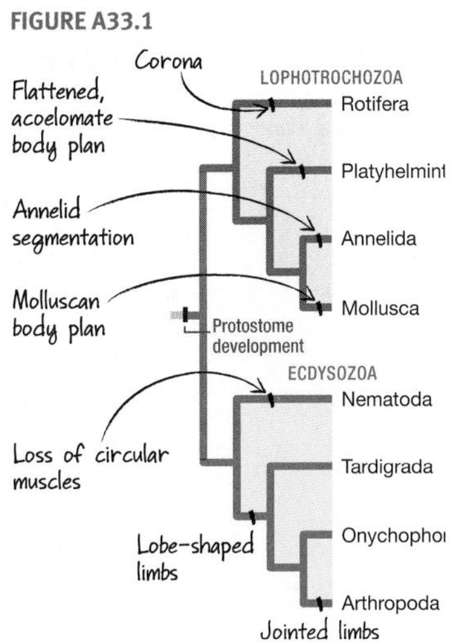

FIGURE A33.2

FIGURE A33.3

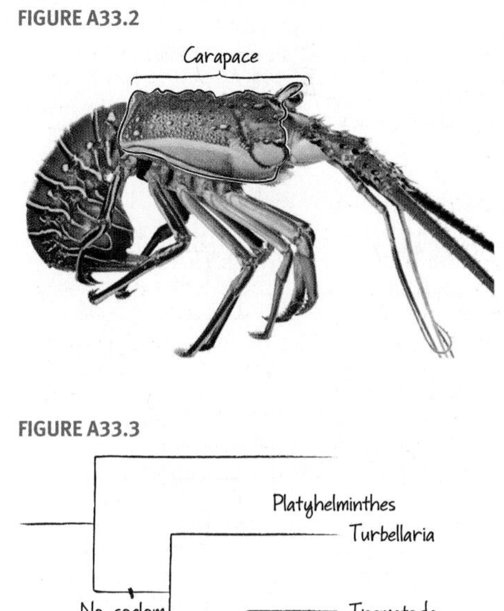

You Should Be Able To (YSBAT)

YSBAT p. 649a, YSBAT p. 649b, YSBAT p. 650 [See Figure A34.1] **YSBAT p. 655** [See Figure A34.2] **YSBAT p. 660** [See Figure A34.3] **YSBAT p. 662a, YSBAT p. 662b, YSBAT p. 663, YSBAT p. 664** [See Figure A34.4] **YSBAT p. 665, YSBAT p. 666a, YSBAT p. 666b, YSBAT p. 667a, YSBAT p. 667b, YSBAT p. 668** [See Figure A34.3]

Caption Questions and Exercises

Figure 34.1 Invertebrates are a paraphyletic group. The group does not include all the descendants of the common ancestor because vertebrates are excluded. **Figure 34.12** Mammals and birds are equally related to amphibians, because birds and mammals share a common ancestor, and this ancestor shares a common ancestor with amphibians. **Figure 34.14** Yes—even a rudimentary jaw could have been useful to better grasp prey or to change the size of the mouth. **Figure 34.16** If this phylogeny were estimated on the basis of limb traits, it would be based on the *assumption* that the limb evolved from fish fins in a series of steps. But because the phylogeny was estimated from other types of data, it is legitimate to use it to analyze the evolution of the limb without any assumptions about how limbs evolved. **Figure 34.20** They are smaller—their function in an amniotic egg has been taken over by the placenta. **Figure 34.37** Four hominin species existed 2.2 mya, five existed 1.8 mya, and four existed 100,000 years ago. **Figure 34.38** The forehead became much larger with the face becoming "flatter"; the brow ridges are less prominent in later skulls than in earlier skulls.

Summary of Key Concepts

KC 34.1 (1) Mussels and clams will increase dramatically; (2) kelp density will increase. **KC 34.2** Like today's lungfish, the earliest tetrapods could have used their limbs for pulling themselves along the substrate in shallow-water habitats. **KC 34.3** The earliest fossils of *H. sapiens* are found in Africa, and phylogenetic analyses of living human populations indicate that the most basal groups are all African.

Test Your Knowledge

1. a; **2.** a; **3.** d; **4.** a; **5.** d; **6.** b

Test Your Understanding

1. It is an enclosed, fluid-filled structure surrounded by muscle. Muscular contraction forces water into the tube feet, resulting in extension of the podia. **2.** Pharyngeal gill slits function in suspension feeding. The notochord furnishes a simple endoskeleton that stiffens the body; electrical signals that coordinate movement are carried by the dorsal hollow nerve cord to the muscles in the tail, which beats back and forth to make swimming possible. Cephalochordates and ascidians are chordates but they are not vertebrates. **3.** If jaws are derived forms of gill arches, then the same genes and the same cells should be involved in the development of the jaw and the gill arches. **4.** Homologous genes are involved in the formation of the fins of ray-finned fish and the limbs of tetrapods. This observation supports a prediction of the fins-to-limbs hypothesis. **5.** The hominins fulfill the criteria for an adaptive radiation: Over a short time interval, many species that occupy an array of foods and habitats evolved. Changes in tooth and jaw structure, tool use, and body size suggest that different hominin species exploited different types of food. **6.** Increased parental care allows offspring to be better developed, and thus have increased chances of survival, before they have to live on their own.

Applying Concepts to New Situations

1. If confirmed, it means that pharyngeal gill slits were present in the earliest echinoderms and lost later. **2.** No—most of the feathered dinosaurs known from fossils did not fly. (The first feathers may have functioned in display or as insulation.) **3.** Xenoturbellidans either retain traits that were present in the common ancestor or all deuterostomes, or they have lost many complex morphological characteristics. **4.** Independently. Birds are a highly derived lineage of reptiles, and most reptiles are not endothermic, so it is logical to infer that the common ancestor of birds and mammals was also not endothermic.

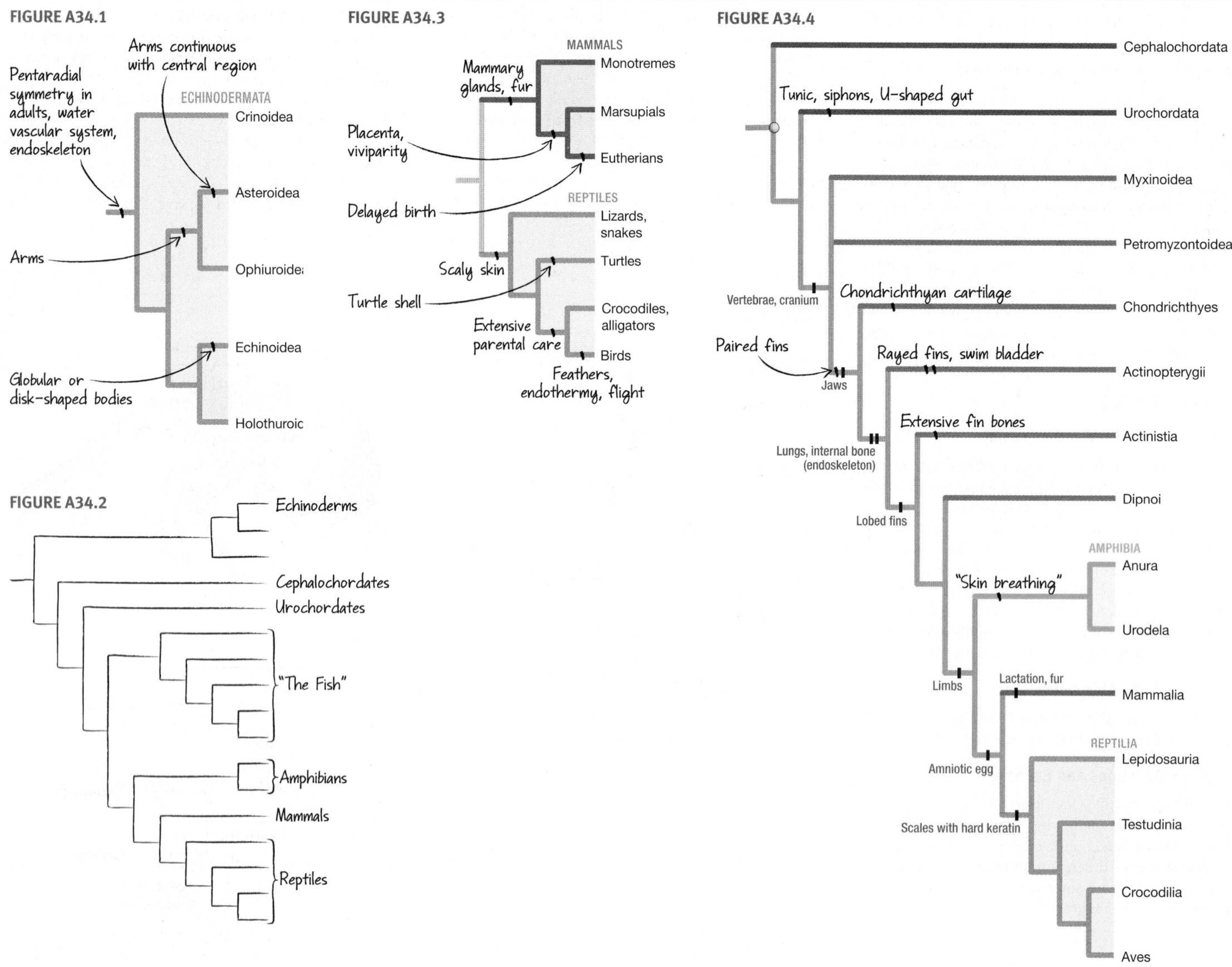

CHAPTER 35

Check Your Understanding (CYU)

CYU p. 686 HIV binds to CD4 and a co-receptor on the surface of T cells. This binding allows the viral envelope and the T-cell plasma membrane to fuse. The viral capsid then enters the cell. **CYU p. 689** (1) After the genome is copied to form positive-sense RNA, the transcript has to be copied to form a negative-sense, single-stranded genome. (2) Much more dangerous—with airborne transmission, a noninfected individual can become infected simply by inhaling virus particles released from an infected individual. This mode of transmission is much more efficient than transmission via body fluids.

You Should Be Able To (YSBAT)

YSBAT p. 680 In the lytic cycle, viral genes are transmitted to a new generation of virions. In lysogenic growth, viral genes are transmitted to daughter cells of the host cell. **YSBAT p. 686** [See Figure A35.1]

Caption Questions and Exercises

Figure 35.1 Many answers are possible: Herpes simplex 2: genitalia; sexually transmitted; HIV: immune system; sexually transmitted; Influenza virus: respiratory tract; coughing/sneezing (airborne); Chicken pox (varicella zoster virus): skin; direct contact. **Figure 35.8** Lysogenic bacteriophages are similar to transposable elements because they insert their DNA into the host cell chromosome and are passed on to daughter cells. They are unlike transposable elements because they can switch to a lytic cycle. **Figure 35.9** No—it only shows the CD4 is required. (Subsequent work showed that other proteins are involved as well.) **Figure 35.14** Budding viruses may disrupt the integrity of the host-cell plasma membrane enough to kill the cell. **Figure 35.15** You should have the following bars and labels: on branch to HIV-2 (sooty mangabey to human); on branch to HIV-1 strain O (chimp to human); on branch to HIV-1 strain N (chimp to human); on branch to HIV-1 strain M (chimp to human).

Summary of Key Concepts

KC 35.1 There is heritable variation among virions, due to random changes that occur as their genomes are copied. There is also differential reproductive success among virions in their ability to successfully infect host cells. This differential success is due to the presence of certain heritable traits. **KC 35.2** As the virus's envelope proteins are produced, they are inserted into the host cell plasma membrane. During budding, the capsid becomes surrounded by host cell membrane that includes the envelope proteins. **KC 35.3** (1) Fusion inhibitors block viral envelope proteins or host cell receptors; (2) protease inhibitors prevent processing/assembly of viral proteins; (3) reverse transcriptase inhibitors block reverse transcriptase, preventing replication of genome; (4) use of condoms or practice of monogamy prevents transmission. **KC 35.4** To replicate an ssDNA genome, a viral DNA polymerase copies the genome into a complementary strand, which then is used as a template to generate an ssDNA that is identical to the original viral genome.

Test Your Knowledge

1. b; **2.** c; **3.** d; **4.** b; **5.** b; **6.** d

Test Your Understanding

1. HIV has an envelope on its outer surface; adenovirus has only a capsid on its outer surface. A virus with an envelope exits host cells by budding. A virus that lacks an envelope exits host cells by lysis (or other mechanisms that don't involve budding). **2.** T4 bacteriophage should have a larger genome than HIV does, because T4 is much more complex morphologically—it should have genes that code for the proteins required for its head and tail regions. **3.** Growth rate is much higher during lytic growth, but viral DNA can increase during latent/lysogenic growth if the host cell is actively dividing and transmitting viral genomes to daughter cells. **4.** Viruses rely on host-cell enzymes to replicate, whereas bacteria do not. Therefore, many drugs designed to disrupt the virus life cycle cannot be used because they would kill host cells as well. Only viral-specific proteins are good targets for drug design. **5.** Each major hypothesis to explain the origin of viruses—from transposable-element-like sequences, from symbiotic bacteria, and from RNA genomes present early in Earth history—is associated with a different type of genome: single-stranded DNA or possibly single-stranded RNA, double-stranded DNA, and single- or double-stranded RNA, respectively. **6.** The phylogenies of SIVs and HIVs show that the two shared common ancestors, but that SIVs are ancestral to the HIVs. Also, there are plausible mechanisms for SIVs to be transmitted to humans through butchering or contact with pets, but fewer or no plausible mechanisms for HIVs to be transmitted to monkeys or chimps.

Applying Concepts to New Situations

1. Culture the *Staphylococcus* strain outside the human host and then add the virus to determine whether the virus kills the bacterium efficiently. Then test the virus on cultured human cells to determine whether the virus harms human cells. Then test the virus on monkeys or other animals to determine if it is safe. Finally, the virus could be tested on human volunteers. **2.** Prevention is currently the most cost-effective program but does not help people who are already infected. Treatment with effective drugs not only prolongs lives but also reduces virus loads in infected people, so that they have less chance of infecting others. **3.** Eukarya evolved from ancestors that did not contain genes encoding restriction endonucleases. Or, methylation of DNA performs other important functions in eukaryotic gene expression (see Chapter 18), so is not appropriate as a guard against endonuclease action. **4.** Viruses cannot be considered to be alive by the definition given in (a), because viruses are not capable of replicating by themselves. By the definition given in (b), it can be argued that viruses are alive because they store, maintain, replicate, and use genetic information—although they cannot perform all these tasks on their own.

FIGURE A35.1

(+)ssRNA —translation→ Viral proteins

(−)ssRNA —RNA replicase→ mRNA —translation→ Viral proteins

dsRNA —transcription→ mRNA —translation→ Viral proteins

dsDNA or ssDNA —transcription→ mRNA —translation→ Viral proteins

(+)ssRNA —reverse transcription→ dsDNA —transcription→ mRNA —translation→ Viral proteins

CHAPTER 36

Check Your Understanding (CYU)

CYU p. 704 (1) [See Figure A36.1] (2) The generalized body of a plant has a taproot with many lateral roots and broad leaves. Examples of deviations from this generalized body structure include: Modified roots: Unlike taproots, *fibrous roots* do not have one central root, and *adventitious roots* arise from stems. Modified stems: *Stolons* grow along the soil and grow roots and leaves at each node. *Rhizomes* grow horizontally underground. Both of these modified stems function in asexual reproduction. Modified leaves: The *needle-shaped leaves* of cacti lose less water to transpiration than do typical

FIGURE A36.1

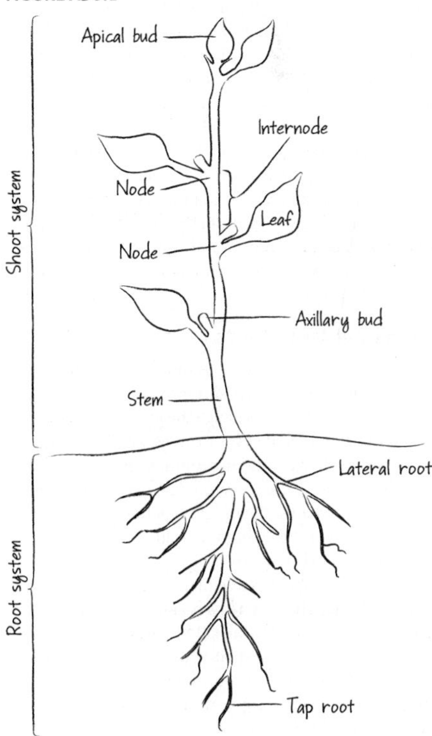

broad leaves, and they protect the plant from predation. *Tendrils* on climbing plants are modified leaves that do not photosynthesize but wrap around trees or other substrates to facilitate climbing. **CYU p. 712** (1) The apical meristem gives rise to the three primary meristems. The apical meristem is a single mass of cells localized at the tip of a root or shoot; the primary meristems are localized in distinctive locations behind the apical meristem. (2) Epidermal cells are flattened and lack chloroplasts; they secrete the cuticle (in shoots) or extend water and nutrient-absorbing root hairs (in roots) and protect the plant. Parenchyma cells are "workhorse cells" found throughout the plant body; they perform photosynthesis and synthesize and/or store materials. Tracheids are long, thin cells with pits in their secondary cell walls; they are found in xylem and are dead when mature; they conduct water and solutes up the plant. Sieve-tube elements are long, thin cells found in phloem; they are alive when mature and conduct sugars and other solutes up and down the plant. **CYU p. 715** (1) Primary growth increases the length of roots and shoots, and secondary growth increases their width. The function of primary growth is to extend the reach of the root and shoot system and thus increase a plant's ability to absorb light and acquire carbon dioxide, water, and nutrients. The function of secondary growth is to increase the amount of lateral conducting tissue and provide the structural support required for extending primary growth at the top of the plant. (2) The rings are small on the shaded side and large on the sunny side.

You Should Be Able To (YSBAT)

YSBAT p. 713a, YSBAT p. 713b, YSBAT p. 713c, YSBAT p. 714 [See Figure A36.2]

Caption Questions and Exercises

Figure 36.2 New branches and leaves should develop to the right (the plant will also lean that way); new lateral roots will develop to the left. **Figure 36.4** Lawn grasses are shallow rooted, so when the top part of soil dries out, the plants cannot grow. **Figure 36.5** For aerobic respiration, which generates the ATP needed to keep the cells alive. **Figure 36.7** The individuals are genetically identical—thus, any differences in size or shape in the different habitats are due to phenotypic plasticity and not genetic differences. **Figure 36.9** Large surface area provides more area to capture photons, but also more area from which to lose water and be exposed to potentially damaging tearing forces from wind. **Figure 36.20** Many possible answers; for example, in leaf cells, genes for forming chloroplasts would be expressed. These genes wouldn't be expressed in the root cells. **Figure 36.25** The leading hypothesis is that they lack many organelles so that the space inside the cell is available to transport nutrients. **Figure 36.27** You should have labeled a light band as early wood, and the dark band to its right as late wood; both bands together comprise one growth ring.

Summary of Key Concepts

KC 36.1 Phenotypic plasticity is more important in (1) environments where conditions vary because it gives individuals the ability to change the growth pattern of their roots, shoots, and stems to access sunlight, water, and other nutrients as the environment changes; and (2) in long-lived species because it gives individuals a mechanism to change their growth pattern as the environment changes throughout their lifetime. **KC 36.2** (1) Plants that live in warm, arid climates would have large root systems to increase their chances of accessing water, and small shoot systems because water is lost from surfaces exposed to air. (2) Plants that live in very wet climates would have small root systems because water is plentiful, and large shoot systems because the abundant water would support more aboveground growth especially in light-limiting environments such as forests. **KC 36.3** (1) In both cases, the structure would stop growing; (2) the shoot would lack epidermal cells and would die. **KC 36.4** (1) The plant would not produce secondary xylem and phloem, so its girth and transport ability would be reduced, though it would still produce bark. (2) The trunk would have much wider tree rings on one side than the other side.

Test Your Knowledge

1. b; **2.** b; **3.** a; **4.** d; **5.** c; **6.** a

Test Your Understanding

1. The general function of both systems is to acquire resources: The shoot system captures light and carbon dioxide; the root system absorbs water and nutrients. Vascular tissue is continuous throughout both the shoot and root systems. Diversity in roots and shoots enables plants of different species to live together in the same environment without directly competing for resources. **2.** Continuous growth enhances phenotypic plasticity because it allows plants to grow and respond to changes or challenges in their environment (such as changes in light and water availability). **3.** Cactus spines are modified leaves; thorns are modified stems. **4.** Cuticle reduces water loss; stomata facilitate gas exchange. Plants from wet habitats should have a relatively large number of stomata and thin cuticle. Plants living in dry habitats should have relatively few stomata and thick cuticle. **5.** Parenchyma cells in the ground tissue perform photosynthesis and/or synthesize and store materials. In vascular tissue, parenchyma cells in "rays" conduct water and solutes across the stem; other parenchyma cells differentiate into the sieve-tube members and companion cells in phloem. **6.** Cells produced to the inside of the vascular cambium differentiate into secondary xylem; cells produced to the outside of the vascular cambium differentiate into secondary phloem.

Applying Concepts to New Situations

1. Fewer sclerenchyma cells, because sclerenchyma cells have extremely tough walls and these populations have been selected to be tender. Humans select individuals with fewer sclerenchyma cells to be the parents of the next generation, so the frequency of sclerenchyma cells will decline over time. **2.** Asparagus—stem; Brussels sprouts—lateral bud; celery—petiole; spinach—leaf (petiole and blade); carrots—taproot; potato—modified stem. **3.** They grow continuously, so do not have alternating groups of large and small cells that form rings. **4.** Girdling disrupts transport of solutes in secondary phloem, and damages (but probably doesn't eliminate) the transport of water and solutes in secondary xylem. The tree starves.

CHAPTER 37

Check Your Understanding (CYU)

CYU p. 721 (1) Wet soils have almost no pressure potential, while dry soils have a highly negative pressure potential because the few water molecules present cling to soil particles. (2) Salty soils have extremely low solute potentials compared to typical soils, because the concentration of solutes is high. **CYU p. 727** (1) At the air-water interface under a stoma, hydrogen bonding is asymmetrical—the water molecules can bond only with water molecules below them. This pulls the water molecules down—creating tension at the surface. When transpiration occurs, there are few water molecules at the surface, which makes the asymmetry in bonding even more pronounced—increasing tension. (2) When a nearby stoma closes, the humidity inside the air space increases, transpiration decreases, and surface tension on menisci decreases. The same changes occur when a rain shower starts. When air outside the leaf dries, though, the opposite occurs: The humidity inside the air space decreases, transpiration increases, and surface tension on menisci increases. **CYU p. 734** (1) In early spring, the water potential of phloem sap near leaves is low because developing leaves are using sucrose much faster than they make it; root cells are releasing sucrose so the water potential of phloem cells in roots is high. (2) Midday in summer, the water potential of phloem sap near leaves is high because leaves are making much more sucrose than they consume. Root cells are storing sugar, however, so the water potential of phloem in roots is low.

You Should Be Able To (YSBAT)

YSBAT p. 719 (1) Add a solute (e.g., salt or sugar) to the left side of the U-tube, so that solute concentration is higher than on the right side. (2) Water would move from the right side of the U-tube into the left side. **YSBAT p. 720** Pull the plunger up. **YSBAT p. 723** The individual would not be able to exclude certain solutes from the xylem. If they were transported throughout the plant in high enough concentrations, they could damage tissues.

Caption Questions and Exercises

Figure 37.2 It increases (becomes less negative), which reduces the solute potential gradient between the two

FIGURE A36.2

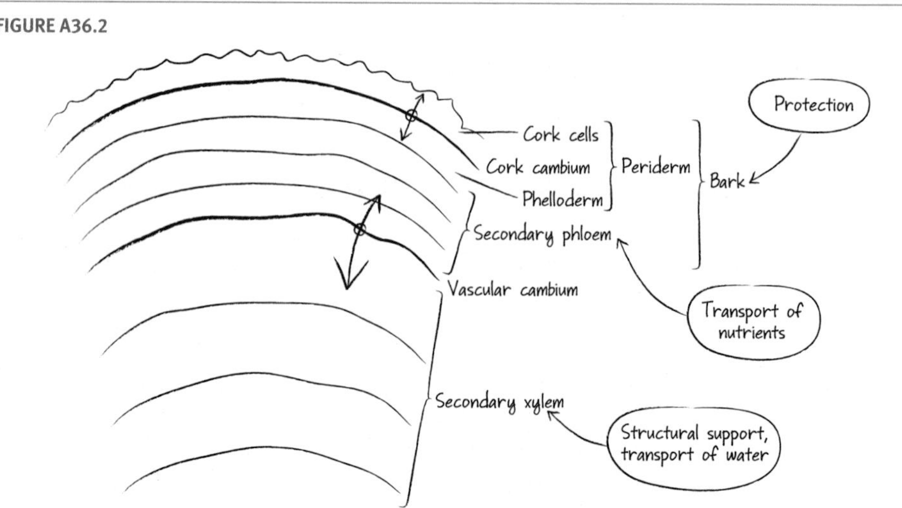

sides. **Figure 37.7** [See Figure A37.1] **Figure 37.10** Your labels should read as follows: Atmosphere: $\psi = -95.2$ MPa; Leaves: $\psi = -0.8$ MPa; Roots: $\psi = -0.6$ MPa; Soil: $\psi = -0.3$ MPa. **Figure 37.12** The line on the graph would go up, with large jumps occurring each time the light level was turned up, because increased transpiration rates would increase negative pressure (tension) at the leaf surface, and thus increase the water potential gradient. **Figure 37.17** [See Figure A37.2]

Summary of Key Concepts

KC 37.1 Because xylem cells are dead at maturity, there are no plasma membranes for water to cross and thus no solute potential. The only significant force acting on water in xylem is pressure. **KC 37.2** It drops, because transpiration rates from stomata decrease, reducing the water potential gradient between roots and leaves. **KC 37.3** To create high pressure in phloem near sources and low pressure near sinks, water has to move from xylem to phloem near sources and from phloem back to xylem near sinks. If xylem were not close to phloem, this water movement could not occur and the pressure gradient in phloem wouldn't exist.

Test Your Knowledge

1. d; **2.** c; **3.** b; **4.** a; **5.** b; **6.** c

Test Your Understanding

1. [See Figure A37.3] **2.** The force responsible for root pressure is the high solute potential of root cells at night, when they continue to accumulate ions from soil, with water following by osmosis. The mechanism is active, in the sense that ions are imported against their concentration gradient. Capillarity is passive; it is driven by adhesion of water molecules to the sides of xylem cells that creates a pull upwards, and cohesion with water molecules below. Cohesion-tension is also passive, because it is driven by the loss of water molecules from menisci in leaves, which creates a large negative pressure (tension). **3.** Water moves up xylem because of transpiration-induced tension created at the air-water interface under stomata, which is communicated down to the roots by the cohesion of water molecules. Sap flows in phloem because of a pressure gradient that exists between source and sink cells, driven by differences in sucrose concentration and flows of water into or out of nearby xylem. **4.** Aphid A is attacking a source, and Aphid B is attacking a sink. If sucrose concentrations are higher in phloem sap near sources, then aphids should be found primarily near mature leaves. (It's actually more common to find them near apical meristems, though—probably because the tissues are not as stiff and strong and difficult to pierce.) **5.** By pumping protons out, companion cells create a strong electrochemical gradient favoring entry of protons. A cotransport protein (a symporter) uses this proton gradient to import sucrose molecules *against* their concentration gradient. **6.** When it germinates—sucrose stored inside the seed is released to the growing embryo, which cannot yet make enough sucrose to feed itself.

Applying Concepts to New Situations

1. Plants have to gain CO_2 for photosynthesis to occur, which means that stomata have to be open. But transpiration occurs whenever stomata are open and the surrounding air is drier than the inside of the leaf. A similar "side effect" occurs when terrestrial animals breathe—they have to take dry air into their lungs, where water evaporates from the body and is lost. **2.** Plants would not grow as quickly or as tall, because the taller they grew, the more energy they would need to use to transport water and thus the less they would have available for growth. **3.** Because they cannot readily replace water that is lost to transpiration, they have to close stomata. During a heat wave, transpiration cannot cool the plant's tissues. They bake to death. **4.** Closing aquaporins will slow or stop movement of water from cells into the xylem and out of the plant via transpiration. Because water does not leave the cells, they are able to maintain turgor (normal solute potentials).

CHAPTER 38

Check Your Understanding (CYU)

CYU p. 748 (1) Proton pumps establish an electrical gradient across root hair membranes, with the inside of the membrane being much more negative than the outside.

FIGURE A37.1

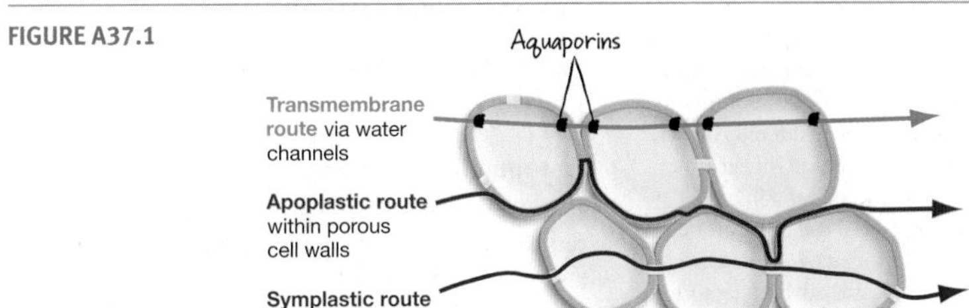

FIGURE A37.2

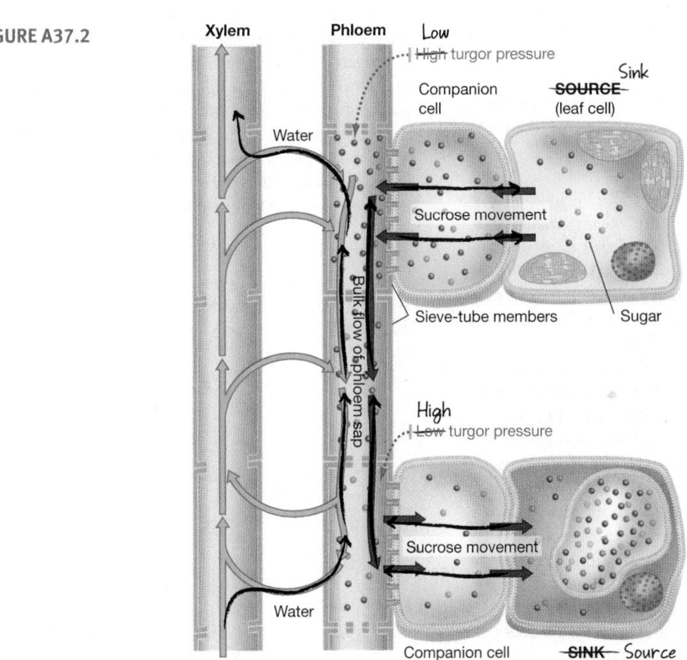

FIGURE A37.3

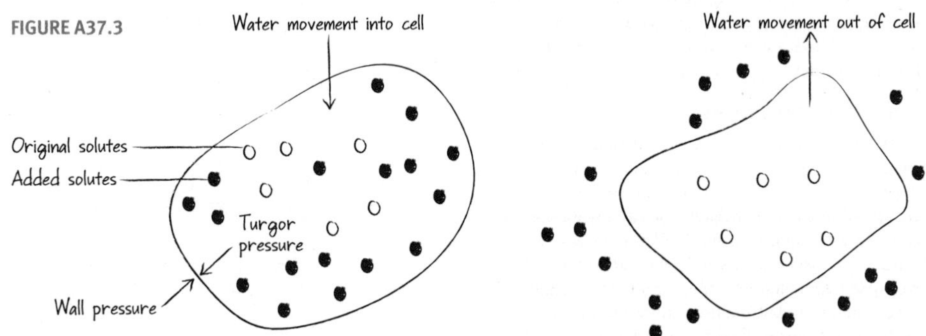

Cations follow this electrical gradient into the cell. (2) [See Figure A38.1] **CYU p. 750** (1) Many possible answers, including the following: If N-containing ions are abundant in the soil, the energetic cost of maintaining nitrogen-fixing bacteria may outweigh the benefits in terms of increased nitrogen availability. If mycorrhizal fungi provide a plant with nitrogen-containing ions, the energetic cost of maintaining nitrogen-fixing bacteria may outweigh the benefits in terms of increased nitrogen availability. Plants that grow near species with N-fixing bacteria might gain nitrogen by absorbing nitrogen-containing ions after root nodules die, or by "stealing it" (roots of different species will often grow together). There could be a genetic constraint (see Chapter 24): Plant species may simply lack the alleles required to manage the relationship. Note: To be considered correct, your hypothesis should be plausible (based on the underlying biology) and testable. (2) Many possible answers, including: Because the bacterial cell enters root cells, it is important for the plant to have reliable signals that it is not a parasitic bacterium. It is advantageous for the plant to be able to reject nitrogen-fixing bacteria if usable nitrogen is already abundant in the soil. The signals have no function.

You Should Be Able To (YSBAT)

YSBAT p. 745 (1) If there is no voltage across the root hair membrane, there is no electrical gradient favoring entry of cations, and absorption stops. (2) There is no route for cations to cross the root hair membrane, so absorption stops. **YSBAT p. 746** (1) If there is no proton gradient across the root hair membrane, there is no gradient favoring entry of protons, so symport of anions stops. (2) There is no route for anions to cross the root hair membrane, so absorption stops.

Caption Questions and Exercises

Figure 38.1 Point out that most of a plant's mass is due to carbon-containing molecules, but that water does not contain carbon. To test the hypothesis that most of a plant's mass comes from CO_2 in the atmosphere, design an experiment where plants are grown in the presence or absence of CO_2 and compare growth, and/or grow plants in the presence of labeled CO_2 and document the presence of labeled carbon in plant tissues. **Figure 38.2** Proteins and nucleic acids would be affected because they contain nitrogen. Most cell processes, including photosynthesis, would be affected because they depend on proteins. **Figure 38.3** Because soil is so complex, it would be difficult to defend the assumption that the soils with and without added copper are identical except for the difference in copper concentrations. Also, it would be better to compare treatments that had no copper versus normal amounts of copper, instead of comparing with and without added copper. **Figure 38.8** One cation (a proton) is being exchanged for another (calcium or magnesium). **Figure 38.10** The proton gradient arrow in (b) should begin above the membrane and cross the membrane pointing down. The electrical gradient arrow in (c) should begin above the membrane (marked positive) and cross the membrane pointing down (toward negative). **Figure 38.17** Mistletoe is parasitic (it extracts water and certain nutrients from host trees); bromeliads are not (they don't affect the fitness of their host plants).

Summary of Key Concepts

KC 38.1 Add nitrogen to an experimental plot and compare growth within the plot to growth of a similar plot nearby. Do many replicates of the comparison, to convince yourself and others that the results are not due to unusual circumstances in one or a few plots. **KC 38.2** A higher membrane voltage would allow cations in soil to cross root-hair membranes more readily through membrane channels and would allow anions to enter more readily via symporters. Nutrient absorption should increase. **KC 38.3** Grow pea plants in the presence of rhizobia, exposed to air with (1) N_2 containing the heavy isotope of nitrogen, and (2) radioactive carbon dioxide. Allow the plant to grow and then analyze the rhizobia and plant tissues. If the rhizobia-plant interaction is mutualistic, then ^{15}N-containing compounds and radioactive-carbon-containing compounds should be observed in both rhizobia and plant tissues. As a control, grow pea plants in the presence of the labeled compounds but without rhizobia. If the mutualism hypothesis is correct, the plant should contain labeled carbon but no heavy nitrogen.

Test Your Knowledge

1. b; **2.** d; **3.** d; **4.** c; **5.** b; **6.**d

Test Your Understanding

1. (1) Grow a large sample of corn plants with a solution containing all essential nutrients needed for normal growth and reproduction. Grow another set of genetically identical (or similar) corn plants in a solution containing all essential nutrients except iron. Compare growth, color, and seed production (and/or other attributes) in the two treatments. (2) Grow corn plants in solutions containing all essential nutrients in normal amounts, but with varying concentrations of iron. Determine the lowest iron concentration at which plants exhibit normal growth rates. **2.** If nutrients in soil are scarce, there is intense natural selection favoring alternative ways of obtaining nutrients—for example, by digesting insects or stealing nutrients. **3.** Higher in soils with both sand and clay. Clay fills spaces between sand grains and slows (1) passage of water through soil and away from roots, and (2) leaching of anions in soil water. Clay particles also provide negative charges that retain cations so they are available to plants. The presence of sand keeps soil loose enough to allow roots to penetrate. **4.** H^+-ATPases in the plasma membrane pump protons out of the cell—making the inside of the membrane less positive (more negative) and the outside of the membrane more positive. By convention, the voltage on a membrane is expressed as inside-relative-to-outside. **5.** Some metal ions are poisonous to plants, and high levels of other ions are toxic. Passive mechanisms of ion exclusion—such as a lack of ion channels that allow passage into root cells—do not require an expenditure of ATP. Active mechanisms of exclusion—such as production of metallothioneins—require an expenditure of ATP. **6.** (1) The amounts required and the mechanisms of absorption are different. Much larger amounts of C, H, and O are needed than mineral nutrients. CO_2 enters plants via stomata and water is absorbed passively along a water potential gradient; mineral nutrients enter via active uptake in root hairs or via mycorrhizae. (2) Macronutrients are required in relatively large quantities and usually function as components of macromolecules; micronutrients are required in relatively small quantities and usually function as enzyme cofactors.

Applying Concepts to New Situations

1. The water that he used may have contained many of the macro- and micronutrients that were incorporated into plant mass. To test this hypothesis, conduct an experiment of the same design but have one treatment with pure water and one treatment with water containing solutes ("hard water"). Compare the percentage of soil mass incorporated into the willow tree in both treatments. Willow may be unusual. To test this hypothesis, repeat the experiment with willow and several other species under identical conditions, and compare the percentage of soil mass incorporated into the plants.
Van Helmont's measurements were inaccurate. Repeat the experiment. **2.** Acid rain inundates the soil with protons, which bind with the negatively charged clay parti-

FIGURE A38.1

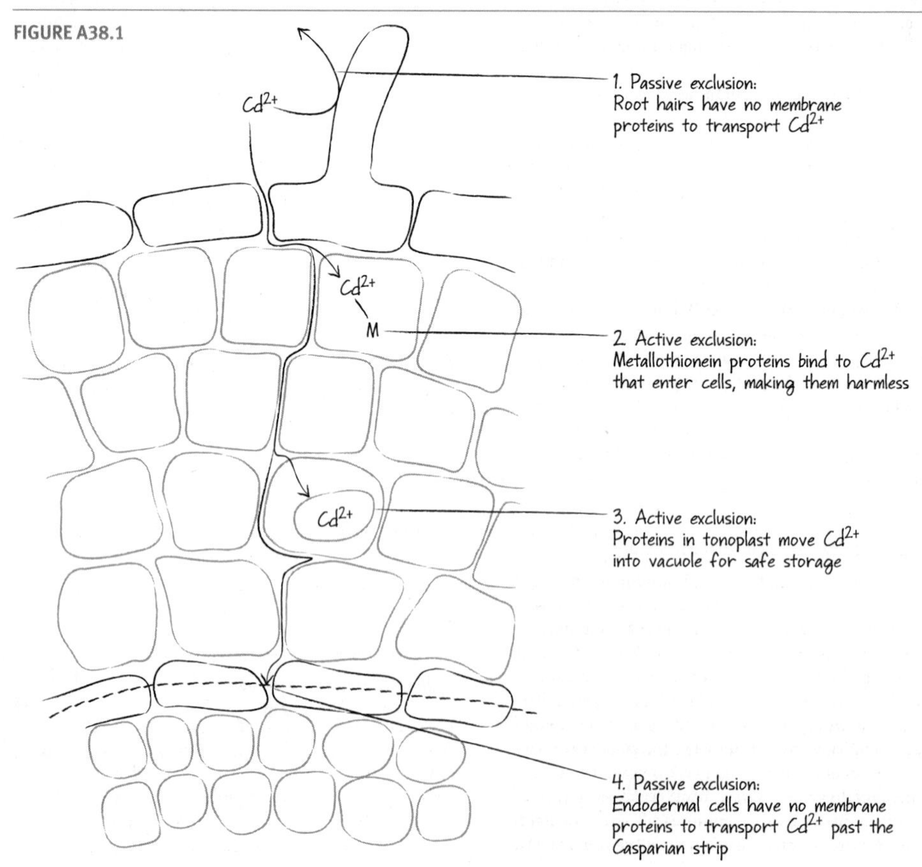

Cd2+
Cd2+
M
Cd2+
1. Passive exclusion:
Root hairs have no membrane proteins to transport Cd2+
2. Active exclusion:
Metallothionein proteins bind to Cd2+ that enter cells, making them harmless
3. Active exclusion:
Proteins in tonoplast move Cd2+ into vacuole for safe storage
4. Passive exclusion:
Endodermal cells have no membrane proteins to transport Cd2+ past the Casparian strip

cles—displacing the cations that are normally bound there. Cations may then be leached away. **3.** If the hypothesis is correct, vanadate should inhibit phosphorus uptake. To test this hypothesis, grow plants in solution with radioactively labeled phosphorus without vanadate and in solution with vanadate. Measure the amount of labeled phosphorus in root cells with and without vanadate. **4.** Alder decomposes much faster, because it provides more nitrogen to the fungi and bacteria that are responsible for decomposition—meaning that they can grow faster.

CHAPTER 39

Check Your Understanding (CYU)

CYU p. 757 (1) The only cells that can respond to a specific environmental signal or hormone are cells that have an appropriate receptor for that signal or hormone. (2) Once a receptor is activated, it triggers production of many second messengers or the activation of many proteins in a phosphorylation cascade. **CYU p. 762** (1) Using a syringe or other device, apply auxin to the west side of each stem, near the base. **CYU p. 764** (1) Lettuce grows best in full sunlight. If a seed receives red light, it indicates that the seed is in full sunlight—meaning that conditions for germination are good. But if a seed receives far-red light, it indicates that the seed is in shade—meaning that conditions for germination are poor. (2) Far-red light indicates shade, and stem-elongation gives plants a chance to grow above nearby plants into full sunlight. **CYU p. 766** (1) In both roots and shoots, auxin is redistributed asymmetrically in response to a signal. In phototropism, auxin is redistributed to the shaded side of the plant in response to blue light. In gravitropism, auxin is redistributed to the lower part of the root or shoot in response to gravity. (2) In root cells, auxin concentrations indicate the direction of gravity. The gravitropic response is triggered by changes in the distribution of auxin in root-tip cells. **CYU p. 776** (1) Short term, stomata should close and photosynthesis should stop. Long term, the plants will not grow well, compared to individuals that do not receive extra ABA, or die. (2) Their stems should elongate rapidly, compared to plants that do not receive extra GAs. **CYU p. 781** (1) Because specific *R* gene products recognize and match specific proteins produced by pathogens, having a wide array of *R* alleles allows individuals to recognize and respond to a wide array of pathogens. (2) If herbivores are not present, synthesizing large quantities of proteinase inhibitors wastes resources (ATP and substrates) that could be used for growth and reproduction.

You Should Be Able To (YSBAT)

YSBAT p. 777 Virtually all bananas have the same *R* alleles. If a pathogen evolves that is not identified by the existing *R* alleles, all bananas would be susceptible to the new disease.

Caption Questions and Exercises

Figure 39.4 These cells served as controls. If the radioactive band appeared in cells that did not have the *PHOT1* gene, it clearly could not be a *PHOT1* band. If the band appeared in cells that had not been exposed to blue light, then *PHOT1* would not be acting as a blue-light receptor. **Figure 39.5** For "Tip removed": cut the tip and replace it, to control for the hypothesis that the wound itself influences bending (not the loss of the tip). For "Tip covered": cover the tip with a transparent cover instead of opaque one, to control for the hypothesis that the cover itself affects bending (not excluding light). **Figure 39.7** Two possibilities: (1) none of the decapitated shoots would bend, or (2) the blocks from the completely divided tip would bend much less than the partially divided tip, because more cells were disrupted. **Table 39.1** The average germination rate of lettuce seeds last exposed to red light is about 99 percent. The average for seeds last exposed to far-red light is about 50 percent. Both of these values are much higher than the germination rate of buried seeds that receive no light at all; their germination rate is only 9 percent. **Figure 39.10** A signal in the form of a light wavelength is changed into a signal in the form of a protein's shape. In this way, a signal from outside the cell is changed to a signal inside the cell. **Figure 39.12** (One possibility) Direct pressure from a statolith changes the shape of the receptor protein. The shape change triggers a signal transduction cascade leading to a gravitropic cell response. **Figure 39.14** In nature, this response is most likely to occur in windy environments. A reduction in height might keep plants beneath the path of the wind, and stockier plants would be better able to resist bending when blown. **Figure 39.19** The dwarfed individuals can put more of their available energy into reproduction because they are using less energy for growth than taller individuals. **Figure 39.21** Without these data, a critic could argue that the stomata closed in response to water potentials in the leaf, not a signal from the roots. **Figure 39.24** The transported nutrients are conserved in the plant instead of being lost when leaves fall. Cessation of chlorophyll synthesis saves plants the energy and nutrients that would have gone into producing chlorophyll in leaves that soon will fall off. **Figure 39.28** The molecule travels throughout the shoot system (and root system).

Summary of Key Concepts

KC 39.1 They do not perform photosynthesis and therefore do not need to switch behavior based on being located in full sunlight versus shade. **KC 39.2** An external signal arrives at a receptor cell/protein (a person sees a barn on fire); the person calls (the receptor transduces the signal); the dispatcher rings a siren (a second messenger is produced or phosphorylation cascade occurs); the fire department members get to the pumper truck (gene expression or some other cell activity changes). **KC 39.3** Sun-adapted (high-light) plants are likely to have a significant positive phototropic response (i.e., growth toward light), whereas shade-adapted (low-light) plants are less likely to respond phototropically. **KC 39.4** They need to be transported throughout the plant (because cells throughout the body need to respond to the signal)—not in a single direction.

Test Your Knowledge

1. d; **2.** a; **3.** b; **4.** b; **5.** c; **6.** b

Test Your Understanding

1. The P_{fr}-P_r switch in phytochromes lets plants know whether they are in shade or sunlight, and triggers responses that allow sun-adapted plants to avoid shade. Phototropins are activated by blue light, which indicates full sunlight, and trigger responses that allow plants to photosynthesize at full capacity. **2.** If *PHOT1* were inserted into a plant, a critic could contend that it was being phosphorylated by a plant protein. But because the *PHOT1* was in an insect cell, this alternative hypothesis was not credible. **3.** *Transduce* means "to convert energy from one form to another." There are many examples: a touch may be converted to an electrochemical potential, light energy or pressure into a shape change in a protein, and so on. **4.** (Many possibilities) Cells or tissue involved: Auxin directs gravitropism in roots but phototropism in shoots. Developmental stage or age: GA breaks dormancy in seeds but extends stems in older individuals. Concentration: Large concentrations of auxin in stems promote cell elongation; small concentrations do not. Other hormones: The concentration of GA relative to ABA influences whether germination proceeds. **5.** Receptor cells sense changes in the availability of blue light received. They respond by changing the distribution of auxin. A change in auxin concentration causes target cells in the stem to grow, and results in the stem bending toward the source of blue light. In this way, plants can respond to changes in shading by growing toward areas where light is available. **6.** Ethylene promotes senescence. Fruits exposed to ethylene ripen faster than fruits that are not exposed to ethylene; increased ethylene sensitivity in leaves (combined with reduced auxin concentrations) causes abscission.

Applying Concepts to New Situations

1. Small-seeded plants need to perform photosynthesis early in seedling development or they will starve. Therefore, they need to germinate in direct sunlight. The food reserves in large-seeded plants can support seedling growth for a relatively long time without photosynthesis. Therefore, they do not need to germinate in direct sunlight. **2.** The experiment provides evidence that cytokinins stimulate cell division in lateral buds—suggesting that large amounts of cytokinins can overcome apical dominance. **3.** In these seeds, (a) little or no ABA is present, or (b) ABA is easily leached from the seeds, so its inhibitory effects are eliminated. Test hypothesis (a) by determining whether ABA is present in the seeds. Test hypothesis (b) by comparing the amount of ABA present before and after running water—enough to mimic the amount of rain that falls in a tropical rain forest over a few days or weeks—over the seeds. **4.** Yes—the experiments are consistent with the hypothesis that ABA must be on the surface of cells to trigger a response, meaning that the receptor must be located there.

CHAPTER 40

Check Your Understanding (CYU)

CYU p. 792 (1) Both are diploid cells that divide by meiosis to produce spores. The megasporocyte produces a megaspore (female spore); a microsporocyte produces a microspore (male spore). (2) Both are multicellular individuals that produce gametes by mitosis. The female gametophyte is larger than the male gametophyte and produces an egg; the male gametophyte produces sperm. **CYU p. 795** (1) Insects feed on nectar and/or pollen in flowers. Flowers provide food rewards to insects to encourage visitation. The individuals that attract the most pollinators produce the most offspring. (2) One product of double fertilization is the zygote, which will eventually grow by mitosis into a mature sporophyte. The other product is the endosperm nucleus, which will grow by mitosis to form a source of nutrients for the embryo.

You Should Be Able To (YSBAT)

YSBAT p. 786 (1) The microsporocyte is the male spore; the megasporocyte is the female spore. (2) In angiosperms the female gametophyte stays within the ovary at the base of the flower even after it matures and produces an egg cell. Fertilization and seed development take place in the same location. **YSBAT p. 795** The endosperm nucleus in the central cell is triploid; the zygote is diploid; the synergid and other cells remaining from the female gametophyte are haploid. **YSBAT p. 796** The protoderm gives rise to the epidermal tissue; the ground meristem gives rise to the ground tissue; the procambium gives rise to the vascular tissue.

Caption Questions and Exercises

Figure 40.5 An appropriate control treatment would be to graft a leaf from a plant that had never been exposed to a short-night photoperiod onto a new plant that also had never been exposed to the correct conditions for flowering. If the hypothesis is correct, this grafted leaf should not induce flowering. **Figure 40.9** A gametophyte is the multicellular individual that produces gametes by mitosis. The embryo sac conforms to this definition because it is a multicellular form that produces an egg by mitosis. **Figure 40.10** The pollen grain is a gametophyte because it is multicellular and produces sperm by mitosis.

Figure 40.17 Hypothesis: Fruit changes color when it ripens as a signal to fruit eaters. The color change is advantageous because seeds are not mature in unripe fruit, and fruit eaters disperse mature seeds in ripe fruit. **Figure 40.21** Beans—their cotyledons are aboveground.

Summary of Key Concepts

KC 40.1 The sporophyte is diploid, the gametophyte is haploid, spores and gametes are haploid, and zygotes are diploid. **KC 40.2** There would be four embryo sacs in each ovule, and thus four eggs. **KC 40.3** The carpel is the female portion of the flower and contains the ovary. The ovary contains ovules, which house the female gametophytes. Once fertilization has occurred, the ovary develops into the fruit and the ovules develop into seeds within the fruit.

Test Your Knowledge

1. b; **2.** c; **3.** a; **4.** d; **5.** d; **6.** c

Test Your Understanding

1. Asexual reproduction is a quick, efficient way for a plant to reproduce large numbers of offspring. The disadvantage is that offspring are genetically identical to the parent, and thus vulnerable to the same diseases and only able to thrive in habitats similar to those inhabited by the parent. **2.** Megasporocytes (female) and microsporocytes (male) undergo meiosis to generate megaspores and microspores, respectively. These divide mitotically to give rise to female and male gametophytes—the embryo sac and pollen grain, respectively. **3.** Sepals are the outermost structure in a flower and usually protect the structures within; petals are located just inside the sepals and usually function to advertise the flower to pollinators. In wind- and bee-pollinated flowers, sepals probably do not vary much in overall function or general structure. Petals are often lacking in wind-pollinated species; they tend to be broad and flat (to provide a landing site) and colored purple or blue or yellow in bee-pollinated species (often with ultraviolet markings). **4.** Outcrossing increases genetic diversity among offspring, making them more likely to thrive if environmental conditions change from the parental generation. However, outcrossing requires that cross-pollination is successful. Self-fertilization results in relatively low genetic diversity among offspring but ensures that pollination is successful. **5.** The female gametophyte develops inside the ovule; the ovary develops into the pericarp of the fruit after fertilization. A mature ovule contains both gametophyte and sporophyte tissue; the ovary and carpel are all sporophyte tissue. **6.** The endosperm of a corn seed and the cotyledons of a bean seed both contain nutrients needed for the seed to germinate. The bean cotyledons have absorbed the nutrients of the endosperm.

Applying Concepts to New Situations

1. Because the island is near the equator, photoperiod will not provide much information about when conditions are optimal for germination. Presumably, daily temperature fluctuations will also remain relatively constant. Based on the information provided, the most logical hypotheses are that flowering occurs at the onset of the rainy season and that dry, dormant seeds germinate when exposed to moisture—cues that indicate that conditions for growth are good. **2.** In response to "cheater" plants, insects should be under intense selection pressure to detect and avoid species that do not offer a food reward. Individuals would have to be able to distinguish scents, colors, or other traits that identify cheaters. In response to "cheater" insects, plants possessing unusually thick petals or sepals that prevented insects from bypassing the anthers on their way to the nectaries would be more successful at producing seed in the next generation. **3.** One possibility: Under identical growth conditions, pollinate many individuals of the same species with two pollen grains, and many individuals of this species with one pollen grain. Measure and compare the rates of pollen tube growth. **4.** Acorns have a large, edible mass and are probably animal dispersed (e.g., by squirrels that store them and forget some). Cherries have an edible fruit and are probably animal dispersed. Burrs stick to animals and are dispersed as the animals move around. Milkweed seeds float in wind. To estimate the distance that each type of seed is dispersed from the parent, (1) set up "seed traps" to capture seeds at various distances from the parent plant, (2) sample locations at various distances from the parent and analyze young individuals—using techniques introduced in Chapter 20 to determine if they are offspring from the parent being studied, (3) mark seeds, if possible, and re-find them after dispersal.

CHAPTER 41

Check Your Understanding (CYU)

CYU p. 810 *Connective tissues* consist of cells embedded in a matrix. The density of the matrix determines the rigidity of the connective tissue and its function in padding/protection, structural support, or transport. *Nervous tissue* is composed of neurons and support cells. Neurons have long projections that function as "biological wires" for transmitting electrical signals. *Muscle tissue* has cells that can contract and functions in movement. *Epithelial tissue* has tightly packed layers of cells with distinct apical and basal surfaces. They protect underlying tissues and regulate the entry and exit of materials into these tissues. **CYU p. 814** (1) As three-dimensional size increases, volume increases more quickly than does surface area, so the surface area/volume ratio decreases. (2) They are inversely proportional (negatively correlated). On a per gram basis, small animals have higher basal metabolic rates than large animals. **CYU p. 819** (1) Endotherms can remain active during the winter and at night and sustain high levels of aerobic activities such as running or flying, but require large amounts of food energy. Ectotherms need much less food and can devote a larger proportion of their food intake to reproduction, but have a hard time maintaining high activity levels at night or in cold weather. (2) [See Figure A41.1]

You Should Be Able To (YSBAT)

YSBAT p. 812 A newborn.

Caption Questions and Exercises

Figure 41.1 There would be no relationship between the variables. In all three cases, the graph would be a flat line. **Figure 41.3** Loose connective tissues have a soft matrix that is effective for cushioning and protecting organs. Dense connective tissues have a fiber-rich matrix that allows them to link adjacent structures. Structural connective tissues have a stiff or hard matrix that allows them to withstand bending or compressing forces. Fluid connective tissues are effective in transport because liquids move easily and carry solutes. **Figure 41.4** The projections allow electrical signals to be transmitted over long distances. **Figure 41.10** The dog has to eat more because it has a higher mass-specific metabolic rate than a human. **Figure 41.11** There should be no oxygen uptake on either side of the chamber.

Summary of Key Concepts

KC 41.1 Loosening tight junctions would compromise the barrier between the external and internal environments. It would allow water and other substances (including toxins) to pass more readily across the epithelial boundary. **KC 41.2** King Kong is endothermic and his huge mass would generate a great deal of heat. His relatively small surface area would not be able to dissipate the heat—especially if it were covered with fur. **KC 41.3** Individuals could acclimatize by normal homeostatic mechanisms (increased sweating, seeking shade, etc.). Individuals with alleles that allowed them to function better at higher temperatures would produce more offspring than individuals without those alleles. Over time, this would lead to adaptation via natural selection. **KC 41.4** They multiply the amount of heat or material exchanged between the two countercurrent flows, compared to a "con-current" system.

Test Your Knowledge

1. b; **2.** d; **3.** b; **4.** d; **5.** a; **6.** a

Test Your Understanding

1. It is where environmental changes are sensed first. As an effector, it has a large surface area for losing or gaining heat. **2.** A warm frog can move, breathe, and digest much faster than a cold frog, because enzymes work faster (rates of chemical reactions increase) at 35°C than at 5°C. A warm frog needs much more food, though, to support this high metabolic rate. **3.** *Absorptive regions* have numerous folds and projections, which increase their surface area for absorption via diffusion. *Capillaries* have a high surface area because they are thin and highly branched, making exchange of fluids and gases efficient. *Beaks of Galápagos finches* have sizes and shapes that correlate with the type of food. Large beaks are used to crack large seeds; long, thin beaks are used to pick insects off surfaces; etc. *Fish gills* are thin, flattened structures with a large surface area, which facilitates the efficient exchange of gases and wastes. **4.** A larger sphere has a relatively smaller surface area for its interior volume than does a smaller sphere, because surface area increases with size at a lower rate than volume does. Diffusion will be more efficient in the case of the smaller sphere. **5.** In the morning, an ant should move to a sunlit spot to gain heat by conduction from the ground and radiation striking its body directly. In midday, the ant should avoid overheating by losing heat from wet body surfaces, standing in breezy areas to lose heat by convection, or retreating to the cool burrow to lose heat by conduction and avoid gaining heat by radiation. **6.** The feedback causes a response that counters the current state of the system. It is negative in the sense of reducing the difference between the current state and the set point.

Applying Concepts to New Situations

1. Large individuals require more food, so are more susceptible to starvation if food sources can't be defended (e.g., when seeds are scattered around). They may also be slower and/or more visible to predators. **2.** You would have to show changes in the frequencies of alleles whose products are affected by temperature—for example, increases in the frequencies of alleles for enzymes that operate best at higher temperature.

FIGURE A41.1

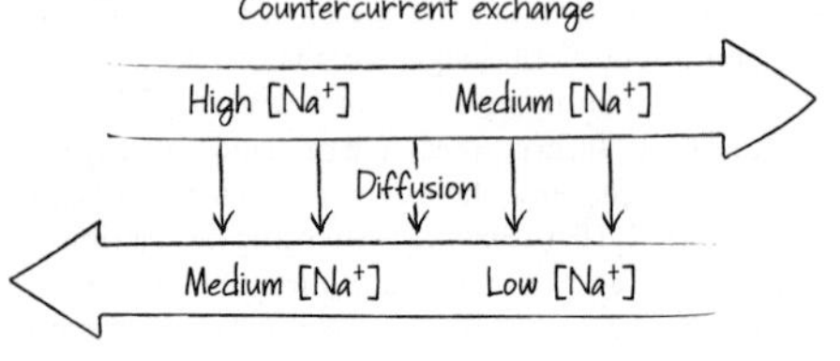

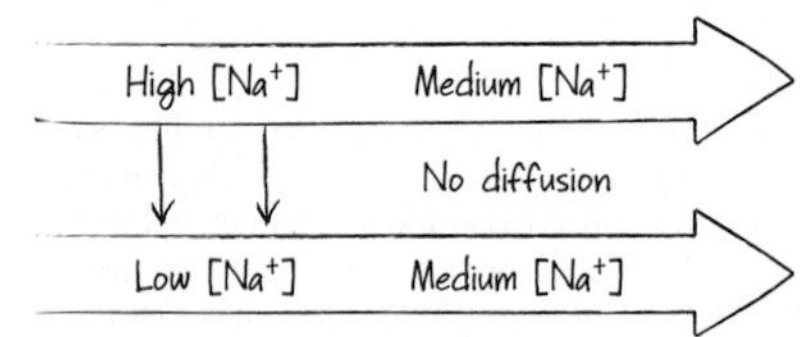

3. The heat-dissipating system should have a very high surface area/volume ratio. Based on biological structures, it should be flattened (thin) and highly folded, branched, and/or contain tubelike projections. Another approach inspired by biological systems would be to set up a countercurrent heat exchanger with fluid heated by the engine. **4.** Buy the turtle. Turtles are ectothermic, meaning they do not expend energy to regulate their body temperature. Mice are endothermic, meaning they do expend energy to regulate body temperature. Mice will therefore have a higher basal metabolic rate and consume more food.

CHAPTER 42

Check Your Understanding (CYU)

CYU p. 828 Without the "master gradient" established by the pump, sodium ions and chloride ions cannot move out of the epithelial cells into the surrounding seawater. Salt will build up in the shark's tissues. **CYU p. 831** (1) Uric acid is insoluble in water and can be excreted without much water loss. (2) When electrolytes are reabsorbed in the hindgut epithelia, water follows along an osmotic gradient. **CYU p. 838** (1) Aldosterone stimulates Na^+ reabsorption in pre-urine in the distal tubule. This increases the Na^+ level in the blood and leads to a more dilute (hypotonic) urine. (2) Water intake increases blood pressure and filtration rate in the renal corpuscle, and leads to lowered electrolyte concentration in the blood and filtrate and production of dilute urine. Eating large amounts of salt results in concentrated, hypertonic urine. Water deprivation triggers ADH release and the production of concentrated, hypertonic urine.

You Should Be Able To (YSBAT)

YSBAT p. 833 Blood contains cells and large molecules as well as electrolytes and wastes; the filtrate contains only electrolytes and wastes. **YSBAT p. 835** If sodium reabsorption is inhibited, then less water will be reabsorbed along an osmotic gradient and more urine will be produced. **YSBAT p. 837** (1) Less, because the osmotic gradient is not as steep. (2) Lower, because more water has been retained. (3) Less, because the concentration gradient is not as steep. **YSBAT p. 838** Ethanol consumption inhibits water reabsorption, leading to a larger volume of less concentrated urine. Nicotine consumption increases water reabsorption, leading to a smaller volume of more concentrated urine.

Caption Questions and Exercises

Figure 42.1 The purple and white molecules are moving down their concentration gradients, and the concentration of red molecules does not affect the concentration of purple or white molecules. **Figure 42.6** The sodium-potassium ATPase does active transport; the sodium-chloride-potassium cotransporter does secondary active transport; the chloride and potassium channels do passive transport. **Figure 42.8** The underside of the abdomen is shaded by the grasshopper's body. This location is relatively cool and should reduce the loss of water during respiration.

Summary of Key Concepts

KC 42.1 In the ocean, river otters must excrete the additional salt that is taken up from saltier marine foods and seawater. In freshwater, their bodies must conserve electrolytes and excrete the water taken in during drinking. **KC 42.2** A sodium gradient establishes an osmotic gradient that moves water. It also sets up an electrochemical gradient that can move ions against their electrochemical gradient by secondary transport through a cotransporter, or with their electrochemical gradient by passive diffusion through a channel. **KC 42.3** The desert locust—to save water, it has to actively reabsorb ions in the hindgut so that almost no water is lost during excretion. **KC 42.4** The longer the loop of Henle, the steeper the osmotic gradient in the medulla. Steep osmotic gradients allow a great deal of water to be reabsorbed as urine passes through the collecting duct.

Test Your Knowledge

1. c; **2.** a; **3.** d; **4.** d; **5.** a; **6.** c

Test Your Understanding

1. In salt water, a salmon gains NaCl by diffusion and loses water by osmosis. It replaces water by drinking and secretes ions through chloride cells in its gill epithelium. In freshwater, it gains water by osmosis and loses NaCl by diffusion. Epithelial cells in the gills take up ions, it stops drinking, and produces copious amounts of urine to rid itself of excess water. **2.** Mitochondria are needed to produce ATP. A key component of salt regulation in fish is the Na^+/K^+-ATPase, which requires ATP to function. Because ATP fuels establishment of the ion gradients necessary for the cotransport mechanisms utilized by chloride cells as well as other epithelial cells, an abundance of mitochondria would be expected. **3.** Microvilli increase the surface area available for pumps, cotransporters, and channels. **4.** The wax layer should be thinner in tropical insects, because the animals are under less osmotic stress (less evaporation due to high humidity). You could test this prediction by measuring the thickness of the wax layer in two closely related species from each habitat, or in a wide array of species from each habitat. Insects that live in extremely humid habitats may lack the ability to close the openings to their respiratory passages; animals that live in humid habitats may excrete primarily ammonia or urea and have relatively short loops of Henle. In each case, you could test these predictions by comparing individuals of closely related species that live in dry versus humid habitats. **5.** In both cases, the initial filtrate is isotonic with blood and reabsorption of certain solutes depends on Na^+/K^+-ATPase activity and is energetically costly. One difference is that the initial filtrate forms by active pumping of ions followed by osmosis in insects versus by filtration in the renal corpuscle of mammals. Another difference is that mammals make use of a countercurrent exchanger in the loop of Henle versus direct pumping of ions in the insect hindgut. **6.** Ammonia is toxic and must be diluted with large amounts of water to be excreted safely. Urea and uric acid are safer and do not have to be excreted with large amounts of water, but are more expensive to produce in terms of energy expenditure. Fish excrete ammonia; mammals excrete urea; insects excrete uric acid. You would expect the embryos inside terrestrial eggs to excrete uric acid, as it is the least toxic and is insoluble in water.

Applying Concepts to New Situations

1. The countercurrent flow removes salt from the tissues around the ascending limb, raising the osmolarity of blood, and then water from the tissues around the descending limb, along an osmotic gradient. The countercurrent arrangement maintains a concentration and osmotic gradient all along the length of the nephron. **2.** The observation that the Na^+ concentration decreased by 30 percent, even as volume also decreases, indicates that Na^+ ions were reabsorbed to a greater degree than water. The observation that the urea concentration increased by 50 percent indicates that urea was not reabsorbed. **3.** Increased salt concentrations in freshwater environments put organisms under a new kind of osmotic stress. They may not have chloride cells or other adaptations that allow them to rid themselves of additional salt. **4.** Their urine volume is much higher (actually 10 times).

CHAPTER 43

Check Your Understanding (CYU)

CYU p. 856 (1) *Mouth:* Food is taken in; teeth physically break down food into smaller particles; salivary amylase begins to break down carbohydrates; lingual lipase initiates the digestion of fats. *Esophagus:* Food is moved to the stomach via peristaltic contractions. *Stomach:* HCl denatures proteins; pepsin begins to digest them. *Small intestine:* Pancreatic enzymes complete the digestion of carbohydrates, proteins, lipids, and nucleic acids. Most of the water and all of the nutrients are absorbed here. *Large intestine:* Water is reabsorbed and feces are formed. *Anus:* Feces accumulate in the rectum and are expelled out the anus. (2) If the release of bile salts is inhibited, fats would not be digested and absorbed efficiently, and they would pass into the large intestine. The individual would likely produce fatty feces and lose weight over time. Trypsin inactivation would inhibit protein digestion and reduce absorption of amino acids, likely leading to weight loss and muscle atrophy due to protein deprivation. If the Na^+-glucose cotransporter were blocked, then glucose would not be absorbed into the bloodstream. The individual would likely become sluggish from lack of energy. **CYU p. 858** (1) Both lead to high urine volume due to reduced water reabsorption in the kidneys, but for different reasons. In diabetes mellitus, less water is reabsorbed because glucose concentrations in the filtrate from blood are high. In diabetes insipidus, less water is reabsorbed because of defects in the collecting ducts. (2) When the blood sugar of individuals with type I diabetes mellitus gets too high, they should inject themselves with insulin, which will trigger the absorption and storage of glucose by their cells. When their blood glucose gets too low, they should eat something with a high concentration of sugar (such as orange juice or a candy bar) to increase their blood glucose levels quickly.

You Should Be Able To (YSBAT)

YSBAT p. 851 Nutrient absorption occurs in the small intestine, but not in the esophagus or stomach, and the rate of absorption increases with available surface area. **YSBAT p. 854** (1) Decreased energy yield from foods due to reduced glucose absorption, (2) increased feces production due to the passing of unabsorbed glucose into the colon, (3) watery feces due to decreased water reabsorption in the large intestine (lower osmotic gradient). As an aside, also increased flatulence due to the metabolism of unabsorbed glucose by bacteria that produce methane as a waste product. **YSBAT p. 857** Individuals with type I diabetes have normal insulin receptors on their liver cells, but do not produce and release insulin from the pancreas. Individuals with type II diabetes produce and release insulin, but their insulin receptors are defective so liver cells cannot respond to insulin. In both types of diseased individuals, high glucose concentrations in the blood remain high. In individuals without disease, both insulin production and insulin receptors are normal, so liver cells respond to insulin by taking up glucose from the blood and storing it.

Caption Questions and Exercises

Figure 43.3 Your label should point to the rightmost two bones in the illustration. **Figure 43.14** Frog eggs don't normally make the sodium-glucose cotransporter protein, so the researchers could be confident that if the protein appeared, it was from the injected RNA. They would not be confident of this if the RNAs were injected into rabbit epithelial cells, where the protein was probably already present. **Figure 43.16** In type 2 diabetes mellitus, the top, black arrow from glucose (in the blood) to glycogen (inside liver and muscle cells) is disrupted. In type 1 diabetes mellitus, the green arrow on the top left, indicating insulin produced by the pancreas, is disrupted. **Figure 43.17** The leading hypothesis is an increased incidence of obesity, which is linked to development of type 2 diabetes mellitus.

Summary of Key Concepts

KC 43.1 Form two groups of mice selected from a common population, so that individuals have similar genetic characteristics and history. One group (the control group) would receive a regular diet of rat chow, and the other group (the experimental group) would receive a diet of rat chow that was deficient in magnesium. By comparing the two groups on a number of behavioral and physical measures, you could determine the effects of magnesium deficiency. **KC 43.2** Their mouth or tongue would be modified to make probing of flowers and sucking nectar efficient. They would not require teeth, and their digestive tract would be relatively simple. For example, the stomach would not have to pulverize food or digest large amounts of protein, and the large intestine would not have to process and store large amounts of bulky waste. **KC 43.3** By making the stomach smaller, gastric bypass surgery decreases the amount of food that can be comfortably contained in the stomach and digested. By bypassing a portion of the small intestine, the surgery decreases the absorption of food. (As an aside, people who have this surgery are at risk for serious nutrient deficiencies, chronic intestinal upset, ulcers, and flatulence.) **KC 43.4** The excess glucose is eventually eliminated in urine. (This takes much longer than sequestering the excess glucose in liver, muscle, or adipose cells.)

Test Your Knowledge

1. a; **2.** a; **3.** d; **4.** d; **5.** b; **6.** d

Test Your Understanding

1. The bird crop is an enlarged sac that can hold quickly ingested food; in leaf-eating species it is filled with symbiotic organisms and functions as a fermentation vessel. The cow rumen is an enlarged portion of the stomach; the elephant large intestine is enlarged relative to other species. Both structures are filled with symbiotic organisms and function as fermentation vessels. **2.** Digestive enzymes break down macromolecules (proteins, carbohydrates, nucleic acids, lipids). If they weren't produced in an inactive form, they would destroy the cells that produce and secrete them. **3.** Most biologists accept the hypothesis, for two reasons: (1) the structure is unique among fish in being effective at biting, and (2) its surface correlates with the type of food each species eats. **4.** In most terrestrial vertebrates, the primary function of the large intestine is water reabsorption; fish do not need to reabsorb water. **5.** When an individual ingests a solution of glucose and electrolytes, the solutes are absorbed. Water follows by osmosis, preventing dehydration. **6.** Insulin triggers negative feedback to high blood glucose concentrations by stimulating glucose uptake by several types of cells. Glucagon triggers negative feedback to low blood glucose concentrations by stimulating glucose release by several types of cells.

Applying Concepts to New Situations

1. Because they are feeding growing embryos and newborns, it is likely that female mammals during pregnancy and breastfeeding would require higher levels of almost every nutrient—particularly calcium, used by offspring for bone growth. **2.** Individuals with defects in pancreatic amylase would not be able to complete carbohydrate digestion and probably would be lethargic (low energy) and experience weight loss. Pepsin defects would reduce or eliminate protein digestion in the stomach, leading to severe amino acid deficiency. Defects in the fatty-acid binding protein would reduce or eliminate fatty-acid absorption in the small intestine, likely producing fatty feces, weight loss, and diarrhea. Aquaporin defects would prevent water reabsorption in the large intestine and lead to diarrhea and dehydration. **3.** The result is still valid because the injection was correlated with secretion—if lack of injection was correlated with lack of secretion. The criticism is also somewhat valid, however, because the researchers couldn't rule out the hypothesis that signaling from nerves plays some sort of role, too. **4.** Terrestrial animals are exposed to increased risk of water loss, and the large intestine is where water reabsorption occurs. In most cases fish do not need to reabsorb large amounts of water from their feces.

CHAPTER 44

Check Your Understanding (CYU)

CYU p. 864 (1) Oxygen partial pressure is high in mountain streams because the water is cold, mixes constantly, and has a high surface area (due to white water). Oxygen partial pressure is low at the ocean bottom because the area is far from the surface where gas exchange takes place and there is relatively little mixing. (2) *Warm-water species:* Large amount of air because the oxygen-carrying capacity of warm water is low. *Vigorous algal growth:* Small amount of air because algae contribute oxygen to the water through photosynthesis. *Sedentary animals:* Small amount of air because sedentary animals require relatively little oxygen. **CYU p. 870** (1) Common features include large surface area, short diffusion distance (a thin gas exchange membrane), and a mechanism that keeps fresh air or water moving over the gas-exchange surface. Only fish gills use a countercurrent exchange mechanism; only tracheae deliver oxygen directly to cells without using a circulatory system; only lungs contain "dead space"—areas that are not involved in gas exchange. (2) The anterior and posterior air sacs provide additional compartments for inhaled and exhaled air; they are arranged so that air moves through them and the lungs in one direction. **CYU p. 874** The saturation curve for Tibetans should be shifted to the left relative to people from sea level—meaning that their hemoglobin has a higher affinity for oxygen at all partial pressures. **CYU p. 883** (1) If the blood plasma that leaks out of the capillaries is not reabsorbed in capillaries, it enters lymphatic vessels that eventually merge with blood vessels. (2) [See Figure A44.1]

You Should Be Able To (YSBAT)

YSBAT p. 872 There would be an even larger change in the oxygen saturation of hemoglobin in response to an even smaller change in the partial pressure of oxygen.

Caption Questions and Exercises

Figure 44.4 External gills are ventilated passively and are efficient because they are in direct contact with water. They are exposed to predators and mechanical damage, however. Internal gills are protected but have to be ventilated by some type of active mechanism for water flow. **Figure 44.7** Using several to many different animals increases confidence that the results are true for most or all individuals in the population, and not due to one or a few unusual individuals or circumstances. **Figure 44.10** Sighing increases pressure in the lung cavity, forcing air out of the lungs more rapidly and completely than occurs during normal exhalation. When taking a deep breath, volume in the lung cavity is increased, resulting in lower pressure and causing more air to enter the lungs than during normal inhalation. **Figure 44.15** According to data in the figure, the oxygen saturation of hemoglobin is about 15 percent for blood at pH 7.2 and about 25 percent at pH 7.4. Therefore, about 85 percent of the oxygen is released from hemoglobin at pH 7.2, but only about 75 percent of the oxygen is released at pH 7.4. **Figure 44.17** In the lungs, a strong partial pressure gradient favors diffusion of dissolved CO_2 from blood into the alveoli. As the partial pressure of CO_2 in the blood declines, hydrogen ions leave hemoglobin and react with bicarbonate to form more CO_2, which then diffuses into the alveoli and is exhaled from the lungs. **Figure 44.23** Air from the alveoli mixes with air in the "dead space" in the bronchi and trachea on its way out of the body. This dead-space air is from the previous inhalation (P_{O_2} = 160 mm Hg; P_{CO_2} = 0.3), so when the alveolar air mixes with the dead-space air, the partial pressures in the exhaled air achieve levels intermediate between that of inhaled and alveolar air.

Summary of Key Concepts

KC 44.1 It is harder to extract oxygen from water than it is to extract it from air. This limits the metabolic rate of water breathers. **KC 44.2** Large animals have a relatively small body surface area relative to their volume. If they had to rely solely on gas exchange across their skin, they would not have enough skin surface area to exchange the volume of oxygen needed to meet their metabolic needs. **KC 44.3** Cold water carries more oxygen than warm water, so icefish blood can carry enough oxygen to supply the tissues with oxygen even in the absence of hemoglobin. The oxygen and carbon dioxide are simply dissolved in the blood. **KC 44.4** Their circulatory systems should have relatively high pressures—achieved by independent systemic and pulmonary circulations powered by a four-chambered heart—to maximize the delivery of oxygenated blood to metabolically active tissues.

Test Your Knowledge

1. b; **2.** c; **3.** d; **4.** d; **5.** c; **6.** a

Test Your Understanding

1. The insect tracheal system delivers O_2 directly to respiring cells, but in humans, O_2 is first taken up by the blood of the circulatory system, which then delivers it to the cells. The human respiratory system also has specialized muscles devoted to ventilating the lungs. The open circulatory system of the insect is a low-pressure system, which makes use of pumps (hearts) and body movements to circulate hemolymph. The closed circulatory system of humans is a high-pressure system, which can respond to rapid changes in O_2 demand by tissues. **2.** During exercise, P_{O_2} decreases in the tissues and P_{CO_2} increases. The increase in P_{CO_2} lowers tissue pH. The drop in P_{O_2} and pH causes more oxygen to be released from hemoglo-

FIGURE A44.1

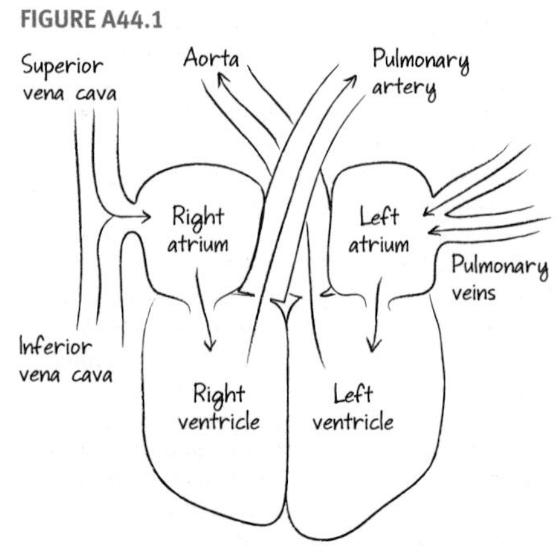

bin. **3.** In blood, CO_2 is converted to carbonic acid, which dissociates into a proton and a bicarbonate ion. Because the protons bind to deoxygenated hemoglobin, they do not cause a dramatic drop in blood pH (hemoglobin acts as a buffer). Also, small decreases in blood pH trigger a homeostatic response that increases breathing rate and expulsion of CO_2. **4.** Airflow through bird lungs is unidirectional, so it allows for continuous ventilation of gas-exchange surfaces with fresh, oxygenated air. The trachea and bronchi in a mammalian lung do not have a gas-exchange surface, and bidirectional airflow in these species means that "stale" air has to be expelled before "fresh" air can be inhaled. As a result, the alveoli are not ventilated continuously. **5.** Lungs increase the temperature of the air and are moist to allow greater solubility of gases (increasing k); alveoli present a large surface area (large A), the epithelium of alveoli is thin (small D), and constant delivery of deoxygenated blood to alveoli maintains a steep partial pressure gradient favoring diffusion of oxygen into the body ($P_2 - P_1$ is high). **6.** If the pulmonary circulation was under pressure as high as that found in the systemic circulation, large amounts of fluid would be forced out of capillaries in the lungs. There is a conflict between the thin surface required for efficient gas exchange and thick blood vessels required to withstand high pressure.

Applying Concepts to New Situations

1. In species that depend on rapid movement to chase down prey or perform other key actions, individuals with larger gill surfaces would absorb more O_2 from water and likely produce more offspring than individuals with smaller gill-surface areas. In slow-moving species, there would be no advantage to increasing the surface area in the gills, so such an adaptation would not be preferentially selected for. **2.** A shift to the right of the oxygen-hemoglobin dissociation curve represents a decrease in the affinity of hemoglobin for O_2. Thus DPG promotes the release of O_2 from hemoglobin into the tissues. **3.** Yes—the trait compensates for the small P_{O_2} gradient between stagnant water and the blood of the carp. **4.** Since O_2 cannot compete as well as CO for binding sites in hemoglobin, O_2 transport decreases. As oxygen levels in the blood drop, tissues (particularly the brain) become deprived of oxygen, and suffocation occurs.

CHAPTER 45

Check Your Understanding (CYU)

CYU p. 891 (1) The resting potential would fall (be much less negative) because K^+ could no longer leak out of the cell. (2) The size of an action potential from a particular neuron does not vary, so it cannot contain information. Only the frequency of action potentials from the same neuron varies. **CYU p. 895** (1) Once the threshold level of depolarization is attained, the probability that the voltage-gated sodium channels will open approaches 100 percent. But below threshold, the massive opening of Na^+ channels does not occur. This is why the action potential is "all or none." (2) Na^+ would continue to move into the cell as voltage-gated potassium channels opened and potassium began to diffuse out of the cell. As a result, repolarization would take much longer. **CYU p. 899** If neurons made direct electrical connections, action potentials would simply travel from one neuron to the next. The presence of synapses allows information from many different neurons to affect the activity of a postsynaptic neuron, through the process of summation. **CYU p. 904** (1) [See Figure A45.1] (2) 1. Study individuals with brain damage and correlate the location of the defect with a deficit in mental or physical function. 2. Directly stimulate brain areas in conscious patients during brain surgery and record the response.

You Should Be Able To (YSBAT)

YSBAT p. 890 The Na^+/K^+-ATPase makes the inside of the membrane less positive (more negative) than the outside, and generates a concentration gradient favoring movement of K^+ out of the cell, through the K^+ leak channel. As K^+ leaves, the resting potential becomes even more negative. **YSBAT p. 891** [See Figure A45.2] **YSBAT p. 892** (1) Na^+ channels open when the cell membrane depolarizes, and as more channels open, more Na^+ flows into the cell, further depolarizing the cell, causing more voltage-gated Na^+ channels to open. (2) After opening, the voltage-gated Na^+ channels close and temporarily cannot reopen. Also, sodium reaches its equilibrium potential, so no net force is available to drive Na^+ movement. (3) K^+ channels open in response to membrane depolarization. **YSBAT p. 897** The inside of the cell becomes more positive (depolarized). This shifts the membrane potential closer to the threshold, making it easier for the postsynaptic cell to fire an action potential.

Caption Questions and Exercises

Figure 45.3 No—as K^+ leaves the cell along its concentration gradient, the interior of the cell becomes more negative. As a result, an electrical gradient favoring movement of K^+ into the cell begins to counteract the concentration gradient favoring movement of K^+ out of the cell. Eventually, the two opposing forces balance out, and there is no net movement of K^+. **Figure 45.10** (1) Get a solution taken from the synapse between the heart muscle and the vagus nerve *without* the nerve being stimulated. Expose a second heart to this solution. There should be no change in heart rate. **Figure 45.16** In the "rest and digest" mode, pupils take in less light stimulation, the heartbeat slows to conserve energy, and liver conserves glucose and promotes digestion by stimulating release of gallbladder products. In the "fight or flight" mode, the pupils open to take in more light, the heartbeat increases to support muscle activity, and the liver releases glucose and inhibits digestion. **Figure 45.18** Point to your forehead and top of your head for the frontal lobe, the top and top rear of your head for the parietal lobe, the back of your head for the occipital lobe, and the sides of your head (just above your ear openings) for the temporal lobes. **Figure 45.19** No—for example, part (b) indicates that the size of the brain area devoted to the trunk is no bigger than the size of the brain area devoted to the thumb.

FIGURE A45.1

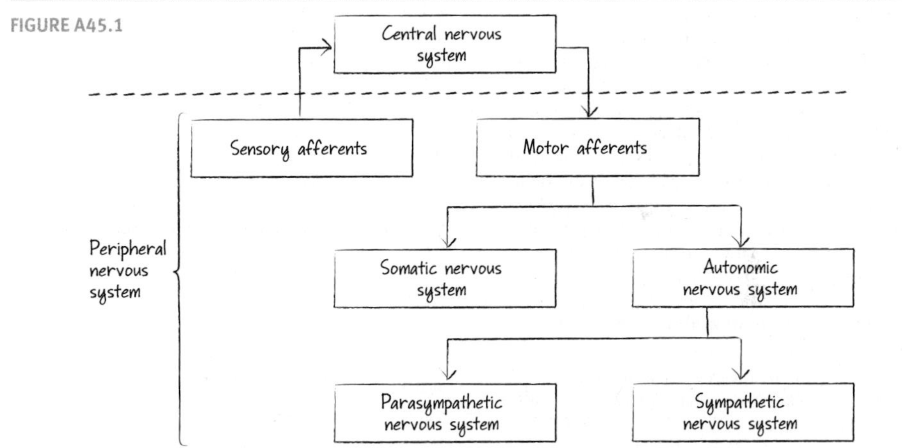

FIGURE A45.2

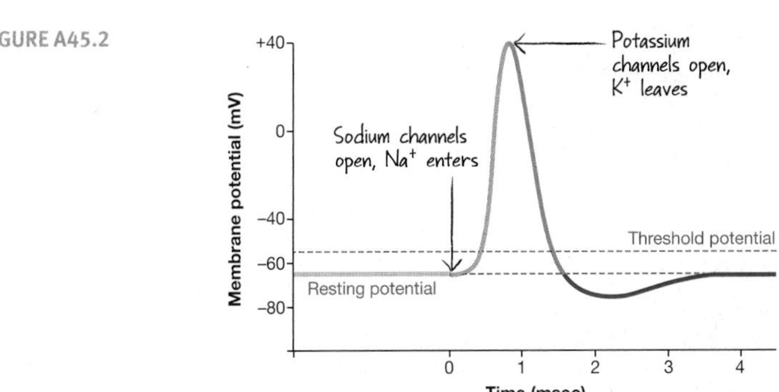

Summary of Key Concepts

KC 45.1 [See Figure A45.3] **KC 45.2** Because the behavior of voltage-gated Na^+ and K^+ channels does not change. Every time a membrane depolarizes past threshold and all of the voltage-gated Na^+ channels are open, the same amount of Na^+ will enter the cell and the cell will depolarize to the same extent. In addition, K^+ channels open on the same time course and allow the same amount of K^+ to leave the cell, meaning that it repolarizes to the same extent. **KC 45.3** Summation means that input from many neurons is involved in triggering a response from a postsynaptic cell. Thus, an action potential from a postsynaptic cell requires integration of input from many synapses. **KC 45.4** The animal's ability to learn—and perhaps remember—would be impaired or nonexistent.

Test Your Knowledge

1. b; **2.** d; **3.** c; **4.** a; **5.** c; **6.** a

Test Your Understanding

1. The Na^+/K^+-ATPase pumps 3 Na^+ out of the cell for every 2 K^+ it brings in. Since more positive charges leave the cell than enter it, there is a difference in charge on the two sides of the membrane and thus a voltage. **2.** [See Figure A45.4] **3.** The ligand-gated channel opens or closes in response to binding by a small molecule (for example, a neurotransmitter); the voltage-gated channel opens or closes in response to changes in membrane potential. **4.** EPSPs and IPSPs are based on flows of ions, which change the membrane potential in the postsynaptic cell. If a flow of ions at one point depolarizes the cell but a nearby flow of ions hyperpolarizes the cell, then in combination the flows of ions cancel each other out. But if two adjacent flows of ions depolarize the cell, then the total amount of ion movement—and thus the total change in membrane potential—sums. **5.** The somatic system responds to external stimuli and controls voluntary skeletal muscle activity, such as movement of arms and legs. The autonomic system responds to internal stimuli and controls internal involuntary activities, such as digestion, heart rate, and gland activities. **6.** Sympathetic nerves trigger responses in an array of organs and tissues that promote increased physical activity and heightened awareness. Parasympathetic nerves cause the opposite response in the same organs and tissues, promoting reduced physical and mental activity.

Applying Concepts to New Situations

1. The neurotransmitters will stay in the synaptic cleft longer so the amount of binding to ligand-gated channels will increase. Ion flows into the postsynaptic cell will increase dramatically, affecting its membrane potential and likelihood of firing action potentials. **2.** Diseased or damaged brains may not respond like healthy and undamaged brains. Also, the extent of lesions and the exact location of electrical stimulation may be difficult to determine, making correlations between the regions affected and the response imprecise. **3.** Because the current is reduced from normal levels, but not eliminated completely, there must be more than one type of potassium channel present. (The poison probably knocks out just one specific type of channel.) **4.** Record from the neuron while the individual is performing various tasks—feeding, moving, courting, etc.

CHAPTER 46

Check Your Understanding (CYU)

CYU p. 913 (1) The outer ear collects sound waves from the environment and directs them into the ear canal. The middle ear amplifies sound. The inner ear contains hair cells that transduce the information in sound waves to action potentials that are sent to the brain. (2) A punctured eardrum wouldn't vibrate correctly and would result in hearing loss at all frequencies in the affected ear. If the stereocilia are too short to come into contact with the tectorial membrane, vibration of the basilar membrane will not cause them to bend, and sound will not be detected. A loss in basilar membrane flexibility would result in the inability to hear lower-pitched sounds, such as human speech. **CYU p. 918** (1) Retinal acts like an on-off switch that indicates whether light has fallen on a rod cell. When retinal absorbs light, it changes shape. The shape change triggers events that result in a change in action potentials that signals that light has been absorbed. (2) A tear in the fovea would likely result in blurred vision in the affected eye. The mutation would produce blue-purple color blindness because this opsin responds to wavelengths in that region of the spectrum. A clouded lens would reduce the amount of light that reaches the retina, reducing visual sensitivity. **CYU p. 920** (1) Extremely hot food damages taste receptors, so proteins would no longer be able to respond to their chemical triggers. (2) Dogs have more than twice the number of odor receptors as humans. **CYU p. 926** (1) In a sarcomere, thick myosin filaments are sandwiched between thin actin filaments. When the heads on myosin contact actin and change conformation, they pull the actin filaments toward one another, shortening the whole sarcomere. (2) Increased ACh release would result in an increased rate of muscle cell contraction. Preventing conformational changes in troponin would prevent muscle contraction. Blocking the uptake of calcium ions into the sarcoplasmic reticulum would lead to sustained muscle contraction.

You Should Be Able To (YSBAT)

YSBAT p. 916 cGMP is a ligand that opens sodium channels and a second messenger in the sense that lack of it carries a signal from activated transducin. Transducin is like a G protein because it switches from "off" to "on" in response to a receptor protein, and activates a key pro-

FIGURE A45.3

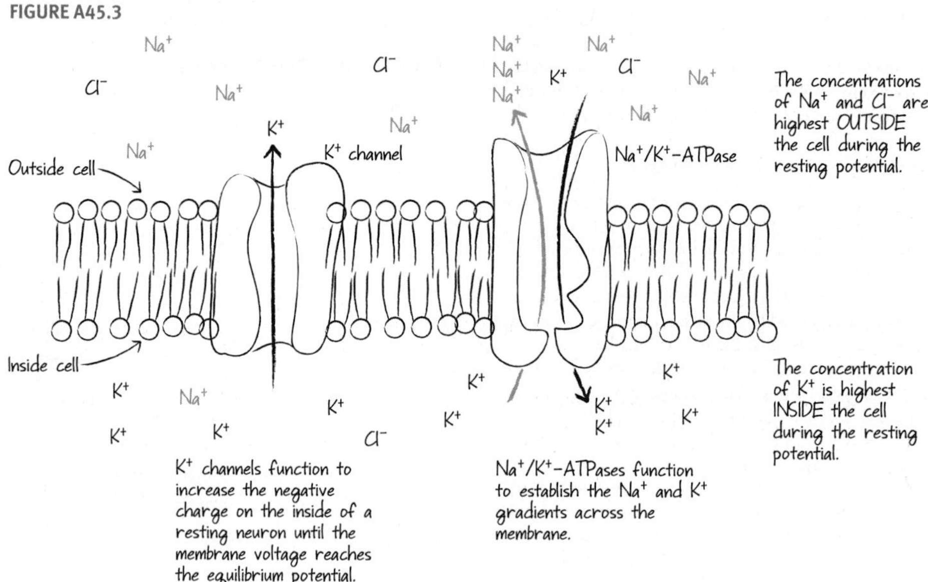
Outside cell
Inside cell
Na+
Cl−
K+
K+ channel
Na+/K+-ATPase
The concentrations of Na+ and Cl− are highest OUTSIDE the cell during the resting potential.
The concentration of K+ is highest INSIDE the cell during the resting potential.
K+ channels function to increase the negative charge on the inside of a resting neuron until the membrane voltage reaches the equilibrium potential.
Na+/K+-ATPases function to establish the Na+ and K+ gradients across the membrane.

FIGURE A45.4

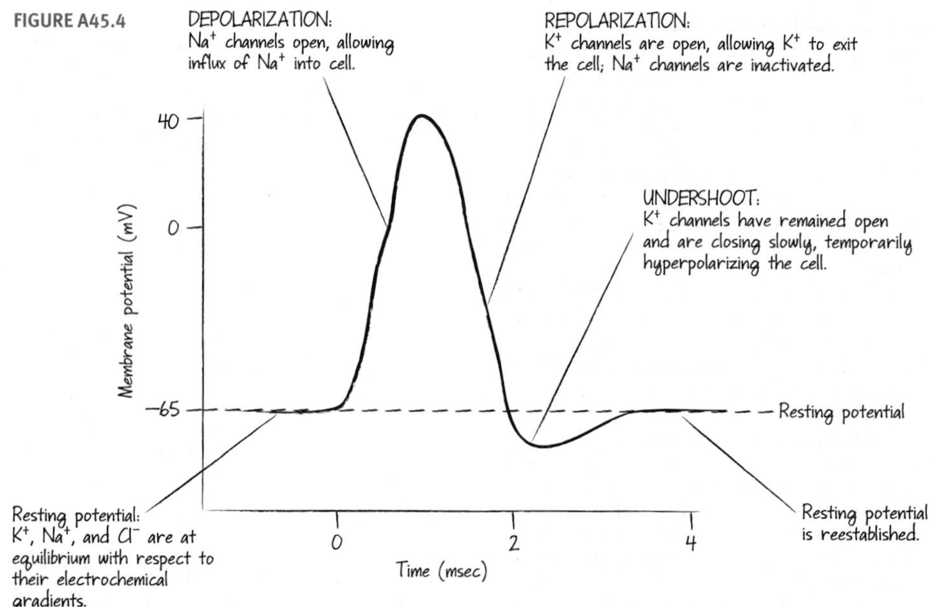
DEPOLARIZATION:
Na+ channels open, allowing influx of Na+ into cell.
REPOLARIZATION:
K+ channels are open, allowing K+ to exit the cell; Na+ channels are inactivated.
UNDERSHOOT:
K+ channels have remained open and are closing slowly, temporarily hyperpolarizing the cell.
Resting potential
Resting potential:
K+, Na+, and Cl− are at equilibrium with respect to their electrochemical gradients.
Resting potential is reestablished.
Membrane potential (mV)
40
0
−65
0
2
4
Time (msec)

tein in response. Closing Na^+ channels stops the entry of positive charges into the cell, making the inside of the cell more negative relative to the outside. **YSBAT p. 923** The trucks are the Z disks, the ropes are the thin filaments, and the burly weightlifters are the thick filaments.

Caption Questions and Exercises

Figure 46.9 The change in retinal's shape would induce a change in opsin's shape. A protein's function correlates with its shape, so opsin's function is likely to change as well. **Figure 46.10** This ion flow occurs when the photoreceptor cell is not receiving light—when the cell is in the dark. **Figure 46.12** The S, M, and L opsin proteins are each different in structure. These structural differences affect the ability of retinal to absorb specific frequencies of light and change shape. **Figure 46.18** The dark band includes thin filaments and a dense concentration of bulbous structures extending from the thick filament; the light band consists of just thin filaments.

Summary of Key Concepts

KC 46.1 The sensory neurons that used to serve that limb are cut, but may still fire action potentials that stimulate brain areas associated with awareness of pain in the missing limb. **KC 46.2** If dense, starch-filled amyloplasts exerted pressure on hair-cell-like receptors, the stereocilia would bend, causing a change in the membrane potential of the receptors and thus transducing the signal from gravity. **KC 46.3** Because different animal species have opsin molecules in their cone cells that respond to different wavelengths of light absorbed by retinal, they see different colors. **KC 46.4** Smells are part of the sensation of flavor, and people with nasal congestion cannot smell. In addition to smell being important, the small number of taste receptors can send many different frequencies of action potentials to the brain, and the taste receptors can be stimulated in many different combinations. **KC 46.5** Paralysis, because ACh has to bind to its receptors on the membrane of postsynaptic muscle fibers for action potentials to propagate in the muscle and cause contraction.

Test Your Knowledge

1. a; **2.** d; **3.** c; **4.** b; **5.** d; **6.** a

Test Your Understanding

1. The brain can distinguish sensory stimuli because the axons from different sensory neurons go to different and specific areas of the brain. For example, the axons of all the olfactory neurons that have the same receptor proteins go to the same location in the olfactory bulb, resulting in the sensation of smelling a particular odor.
2. Many possibilities, including several blue opsins allowing coelacanths to distinguish the blue wavelengths present in their deep-sea habitat, infrasound hearing allowing elephants to hear over long distances, ultrasonic hearing allowing bats to hunt by echolocation and moths to avoid bat predators, red/yellow opsins allowing fruit-eating mammals to distinguish ripe from unripe fruit. **3.** It answered the question of why so many different smells can be detected and identified separately. Each type of odorant has its own receptor. **4.** Ion channels in hair cells are thought to open in response to physical distortion caused by the bending of stereocilia. Ion channels in taste receptors, in contrast, change shape and open when certain bitter-tasting molecules bind to them. **5.** Dalton's hypothesis was reasonable because blue fluid absorbs red light and would prevent it from reaching the retina, and could be tested by dissecting his eyes. Although the hypothesis was rejected, it inspired a rigorous test and required researchers to think of alternative explanations. **6.** The key observation was that the banding pattern of sarcomeres changed during contraction. Even though the entire unit became shorter, only some portions moved relative to each other. This observation suggested that some portions of the structure slid past other portions. Muscle fibers have to have many mitochondria because large amounts of ATP are needed to power myosin heads to move along actin filaments; large amounts of calcium stored in smooth ER are needed to initiate contraction by binding to troponin.

Applying Concepts to New Situations

1. If ACh receptors are covered, they cannot bind to ACh and become activated. Muscles will have trouble contracting. **2.** Block all but a few hundred ommatidia in the center of each compound eye of many dragonflies with an opaque material, and observe any differences that might occur in their ability to detect and pursue prey compared to individuals whose ommatidia had been covered by a transparent material. **3.** Ground-dwelling birds have large amounts of "white meat" in their breasts, used for powering short, rapid flights, but "dark meat" in their legs, used for endurance running. Birds that rely on long, sustained flights need "dark meat" in the breast muscles that power flight. **4.** When a person focuses directly on an object, the image lands on the fovea, which contains relatively few rods. But if the image lands beside the fovea, there are many more rod cells. Rod cells are specialized for detecting dim light.

CHAPTER 47

Check Your Understanding (CYU)

CYU p. 940 (1) By reducing immune system function, promoting release of fatty acids from storage cells and use of amino acids from muscles for energy production, and preventing release of glucose in response to signals from insulin, cortisol conserves glucose supplies for use by the brain. (2) Increased heart rate, glucose and fatty-acid release, and blood pressure all support rapid action in response to danger. If an animal were not able to respond quickly and at the maximum level, it might not escape predators or successfully challenge rivals. **CYU p. 943** (1) ACTH triggers the release of cortisol, but cortisol inhibits ACTH release by blocking the release of CRH from the hypothalamus and suppressing ACTH production. High levels of cortisol tend to lower cortisol levels in the future. (2) Processing centers in the brain are responsible for synthesizing a wide array of sensory input. To start a response to this sensory input, they stimulate neurosecretory cells in the hypothalamus. Secretions from these cells travel to the anterior pituitary, where they trigger the production and release of hormones that control hormones that stimulate the appropriate activity. **CYU p. 947** (1) The steroid-hormone receptor complex would probably fail to bind to the hormone-response element. If so, then gene expression would not change—the arrival of the hormone would have little or no effect on the target cell. (2) The mode of action would be similar—hormone binding would have to trigger a signal transduction event that resulted in production of a second messenger or a phosphorylation cascade.

You Should Be Able To (YSBAT)

YSBAT p. 932 Room temperature is analogous to the sensory input; the thermostat, to the CNS; the thermostat signal, to the cell-to-cell signal; and the furnace, to the effector. If feedback inhibition fails, hormone production will not stop and the response will continue indefinitely. **YSBAT p. 936** The cells could have (1) different receptors, (2) different signal transduction systems, or (3) different response systems (genes or proteins that can be activated).

Caption Questions and Exercises

Figure 47.4 Injecting the dog with a liquid extract from a different part of the body, or with a solution that is similar in chemical composition to the extract from the pancreas (e.g., similar pH or ions present) but lacking molecules produced by the pancreas. **Figure 47.7** The saline injection controlled for any stress induced by the injection procedure and for introducing additional fluids. **Figure 47.9** Same—a defect in both signal and receptor has the same effect—no response—as a defective signal alone or a defective receptor alone. **Figure 47.10** Inject ACTH and monitor cortisol levels. If cortisol does not increase, adrenal failure is likely. **Figure 47.14** The arrow that joins the signal and the receptor. The subsequent events also fail to occur.

Summary of Key Concepts

KC 47.1 The production and/or release of hormones are controlled, directly or indirectly, by electrical or chemical signals from the nervous system. A good example is the hypothalamic-pituitary axis, where neuroendocrine signals regulate endocrine signals. **KC 47.2** Electrical signals are short lived and localized. All of the responses listed in the question are long-term changes in the organism that require responses by tissues and organs throughout the body. **KC 47.3** The hypothalamus would continue to release CRH, stimulating the pituitary to release ACTH, which in turn would stimulate continued release of cortisol from the adrenal cortex. Sustained, high circulating levels of cortisol have negative effects. **47.4** A cell can respond to more than one hormone at a time if the receptors for the hormones are different and the cell produces receptors for each hormone.

Test Your Knowledge

1. c; **2.** a; **3.** d; **4.** a; **5.** c; **6.** a

Test Your Understanding

1. The posterior pituitary is an extension of the hypothalamus; the anterior pituitary is independent—it communicates with the hypothalamus via chemical signals in blood vessels. The posterior pituitary is a storage area for hypothalamic hormones; the anterior pituitary synthesizes and releases an array of hormones in response to releasing hormones from the hypothalamus.
2. Steroid hormones act directly, and alter gene expression. They bind to receptors inside the cell, forming a complex that binds to DNA and activates transcription. Nonsteroid hormones act indirectly, and activate proteins. They bind to receptors on the cell surface and trigger production of a second messenger or a phosphorylation cascade, ending in activation of proteins already present in the cell. **3.** The pituitary produces hormones that regulate hormone production in other glands. **4.** This is one of the reasons that the same hormone can trigger different effects in different tissues. For example, epinephrine binds to four different types of receptors in different tissues—eliciting a different response from each. **5.** The hormonal signal is amplified through production of many copies of a second messenger, the activation of many proteins through a phosphorylation cascade, or the transcription and translation of many mRNAs. **6.** When fat stores are high, leptin release acts on the brain to reduce feeding. In this way, it helps maintain weight and energy balance—preventing obesity in mice.

Applying Concepts to New Situations

1. There are two basic strategies: (1) remove the structure from some individuals, and compare their behavior and condition to untreated individuals in the same environment, or (2) make a liquid extract from the structure, inject it into some individuals, and compare their behavior and condition to individuals in the same environment who were injected with water or a saline solution.
2. If sustained high cortisone levels suppress wound healing, individuals receiving large doses might become more susceptible to disease or heal slowly. **3.** The patient will be unable to produce oxytocin and ADH. Although the patient could manage without oxytocin, artificial

administration of ADH would be required to maintain water balance. You would need to monitor the individual's total water intake and the water content of the urine, along with symptoms of dehydration or water retention (such as swelling or high blood pressure) and increase or decrease ADH dosage accordingly. **4.** Label the hormone and introduce it into an experimental animal. Analyze fat cells by isolating cell components, testing for the presence of the label, and eventually isolating just the labeled hormone-receptor complex. To test the second messenger hypothesis, treat fat cells growing in culture with the hormone and monitor levels of cAMP, Ca^{2+}, and other known second messengers.

CHAPTER 48

Check Your Understanding (CYU)

CYU p. 957 (1) Oviparity usually requires less energy input from the mother after egg laying, and mothers do not have to carry eggs around as long—meaning that they can lay more eggs and be more mobile. Mothers have to produce all the nutrition required by the embryo prior to egg laying, however, and eggs may not be well protected after laying. Viviparity usually increases the likelihood that the developing offspring will survive until birth, but limits the number of young that can be produced to the space available in the mother's reproductive tract. If viviparous young can be nourished longer than oviparous young, then they may be larger and more capable of fending for themselves. (2) Divide a population of sperm into two groups. Subject one group of sperm to spermkillerene at concentrations observed in the female reproductive tract (the experimental group) but not the other group (the control). Document the number of sperm that are still alive over time in both groups. **CYU p. 966** (1) FSH triggers maturation of an ovarian follicle. It rises at the end of a cycle because it is no longer inhibited by progesterone, which stops being produced at high levels when the corpus luteum degenerates. (2) The drug would keep FSH levels low, meaning that follicles would not mature and would not begin producing estradiol and progesterone. The uterine lining would not thicken.

You Should Be Able To (YSBAT)

YSBAT p. 952 Asexual reproduction would be expected in environments where conditions change little over time. **YSBAT p. 965** (1) There should be an LH spike early in the cycle, followed by early ovulation. (2) LH levels should remain low—no mid-cycle spike and no ovulation. **YSBAT p. 968** (1) The uterine lining may not be maintained adequately—if it degenerates, a miscarriage is likely. (2) After the first trimester, the placenta produces enough progesterone to maintain the pregnancy, so supplementation is no longer needed.

Caption Questions and Exercises

Figure 48.3 Isolate and identify the molecules found in crowded water. Test each molecule by adding it, at the same concentration found in "crowded water," to clean water occupied by a single *Daphnia* and recording whether the female produces a male-containing brood. Repeat with many test females, and for each molecule identified. As a control, record the number of male-containing broods produced in clean water. **Figure 48.5** No—the data are consistent with the displacement hypothesis, but there is no direct evidence for it. There are other plausible explanations for the data. **Figure 48.11** (Note: There is more than one possible answer—this is an example.) Having two gonads is an "insurance policy" against loss or damage. To test this idea, surgically remove one gonad from a large number of male and female rats. Do a similar operation on a large number of similar male and female rats but do not remove either gonad. Once the animals have recovered, place them in a barn or other "natural" setting and let them breed. Compare reproductive success of individuals with paired vs. unpaired gonads. **Figure 48.12** It is similar. In both cases, the hypothalamus produces a releasing factor (GnRH or CRH) that acts on the pituitary. The releasing factor stimulates release of regulatory hormones from the anterior pituitary (LH and FSH or ACTH). These hormones travel via the bloodstream and act on the gonads or adrenals to induce the release of hormones from these glands. In both cases, the hormones are involved in negative feedback control of the regulatory hormones from the pituitary. **Figure 48.20** The birth will be more difficult, as limbs or other body parts can get caught. **Figure 48.21** A mother's chance of dying in childbirth in 1760 was a little over 1000 in 100,000 live births, or close to 1.0 percent. If she gave birth 10 times, she would have a 10 percent chance of dying in childbirth.

Summary of Key Concepts

KC 48.1 Parental care demands resources (time, nutrients) that cannot be used to produce more eggs. Sperm competition is not likely when external fertilization occurs, and most fishes use external fertilization. **KC 48.2** A surgeon can ligate (tie off) the fallopian tubes in a woman to stop the delivery of the egg to the uterus and can cut the vas deferens in a man to stop the delivery of sperm into the semen. **KC 48.3** Progesterone suppresses the release of GnRH and FSH through negative feedback. When progesterone levels are high, an LH spike does not occur, ovulation is prevented, and no egg is available for fertilization.

Test Your Knowledge

1. a; **2.** b; **3.** a; **4.** c; **5.** d; **6.** d

Test Your Understanding

1. Every offspring that is produced sexually is genetically unique; every offspring that is produced asexually is genetically identical to its parent. **2.** *Daphnia* females only produced males when they were exposed to short day lengths, *and* water from crowded populations, *and* low food levels. These conditions are likely to occur in the fall. Sexual reproduction could be adaptive in fall if genetically variable offspring are better able to thrive in conditions that occur the following spring. **3.** Spermatogenesis generates four haploid sperm cells from each primary spermatocyte; oogenesis produces only one haploid egg cell from each primary oocyte. Egg cells are much larger than sperm cells because they contain more cytoplasm. In males, the second meiotic division occurs right after the first meiotic division, but in females it is delayed until fertilization. **4.** LH triggers release of estradiol, but at low levels estradiol inhibits further release of LH. LH and FSH trigger release of progesterone, but progesterone inhibits further release of LH and FSH. High levels of estrogen trigger release of more LH. The follicle can produce high levels of estrogen only if it has grown and matured—meaning that it is ready for ovulation to occur. **5.** Ethanol can cross the placenta, enter the fetus, and adversely affect development. Ethanol abuse by pregnant women is correlated with fetal alcohol syndrome (FAS), a condition caused by loss of neurons in developing embryos. **6.** Females should be choosier about their second mate because the sperm of the second male has an advantage in sperm competition.

Applying Concepts to New Situations

1. If all sheep were genetically identical, it is less likely that any individuals could survive a major adverse event (e.g., a disease outbreak or other environmental challenge) than a population composed of genetically diverse individuals, which are likely to vary in their ability to cope with the new conditions. **2.** Compare the number of gametes produced by species that have similar body sizes, but undergo external versus internal fertilization. By comparing similar-sized animals, you would be able to rule out the possibility that differences in gamete production are due to differences in body size. **3.** Oviparous populations should produce larger eggs than viviparous populations. Because oviparous species deposit eggs in the environment, the eggs must contain all the nutrients and water required for the entire period of embryonic development. But because embryos in viviparous species receive nourishment directly from the mother, it is likely that their eggs are smaller. **4.** The shell membrane is a vestigial trait (see Chapter 24)—specifically, an "evolutionary holdover" from an ancestor of today's marsupials that laid eggs.

CHAPTER 49

Check Your Understanding (CYU)

CYU p. 977 Applying direct pressure constricts blood vessels, mimicking the effect of histamine released from mast cells. Applying bandages impregnated with platelet-recruiting compounds mimics the effect of chemokines released from injured tissues and macrophages, which attract circulating leukocytes and platelets to facilitate blood clotting and wound repair. **CYU p. 984** (1) They are identical, except that a B-cell receptor has a transmembrane domain that allows it to be located in the B-cell plasma membrane. (2) Lymphocytes that respond to self molecules would circulate in the blood, and trigger immune system responses to self cells—leading to their destruction. The mutation would probably be fatal. **CYU p. 987** Individuals who are heterozygous for MHC genes have greater variability in their MHC proteins than do individuals who are homozygous for these genes. The increased variability in MHC proteins allows a greater variety of antigens to be presented to T cells, which thus would be able to recognize, attack, and eliminate a wider range of pathogens compared with T cells in homozygotes. As a result, heterozygous individuals would likely suffer fewer infections than homozygous individuals. **CYU p. 990** A vaccination is a "false alarm" that triggers a primary response from the immune system. The memory cells that result produce the secondary response when and if an authentic infection occurs.

You Should Be Able To (YSBAT)

YSBAT p. 982 Have the lottery numbers contain a large number of digits—say 5, or even 10. To generate a number, pick a number from 0–9 for each of the 5 or 10 places. If there are 5 places, there would be $10 \times 10 \times 10 \times 10 \times 10 = 10{,}000$ different numbers. If there were 10 places, there would be 10^{10} (10 billion) different numbers. **YSBAT p. 985** Rapidly dividing B and T cells may be destroyed by chemotherapy, along with cancer cells. If clonal expansion cannot occur, antibody production and other aspects of the immune response will be dampened and possibly ineffective. **YSBAT p. 986** (1) All nucleated cells in the body express Class I MHC proteins, whereas only B cells and some other leukocytes express Class II MHC proteins. (2) Cytotoxic T cells interact only with cells that display antigens presented on Class I MHC proteins. Helper T cells interact only with antigens presented on Class II MHC proteins. (3) An MHC-peptide displayed on an infected cell means "kill me"; displayed on a dendritic cell it means "activate killer T cells"; displayed on a B cell it means "helper T cells should activate me."

Caption Questions and Exercises

Figure 49.1 Pathogens stick to mucus; contaminated mucus is then swept to a disposal point by cilia. **Figure 49.3** Blood vessels near the wound would be constricted, which restricts blood loss. Blood vessels slightly farther from the wound would be dilated, which increases blood flow and the delivery rate of platelets, neutrophils, and macrophages to the surrounding area. **Figure 49.18** [See Figure A49.1]

Summary of Key Concepts

KC 49.1 Leukocytes could not move to the damaged tissue efficiently to clean up debris, remove pathogens, and begin the repair process. **KC 49.2** The drug would inhibit the activation of T and B cells by that antigen, because it would prevent an interaction between the epitope and the appropriate TCRs and BCRs. The cell-mediated response would also fail if the drug blocked the interaction between cytotoxic T cells and infected cells displaying the antigen. **KC 49.3** Cowpox is similar enough to smallpox to be antigenic, but cannot grow effectively enough in humans to cause disease.

Test Your Knowledge

1. c; **2.** a; **3.** b; **4.** d; **5.** b; **6.** a

Test Your Understanding

1. Rubor (reddening) is due to increased blood flow to the infected or injured area. Mast cells trigger dilation of vessels by releasing histamine and other signaling molecules. Calor (heat) is the result of fever, which is activated by the release of cytokines from macrophages that are in the area of infection. Dolor (pain) occurs when tissue damage stimulates pain receptors. Tumor (swelling) occurs due to dilation of blood vessels triggered by histamines and other compounds released by macrophages. **2.** Both the BCR and TCR interact with antigens via binding sites located in the variable regions of their polypeptide chains. A TCR is like one "arm" of a BCR. [See Figure A49.2] **3.** Vaccines have to contain an antigen that can stimulate an appropriate primary immune response. Vaccines have not worked for HIV because the antigens on this virus are constantly changed through mutation, rendering it unrecognizable to memory cells generated following vaccination. **4.** "Clonal" refers to the cloning—producing many exact copies—of cells that are "selected" by the binding of their receptor to an antigen. **5.** Pattern-recognition receptors on leukocytes bind to surface molecules that are present on many pathogens. BCRs and TCRs bind to particular epitopes of antigens. Pattern-recognition-receptor binding is nonspecific; BCR and TCR binding is specific. **6.** By mixing and matching different combinations of gene segments from the variable and joining regions of the light-chain immunoglobulin gene, along with diversity regions of the heavy-chain gene, lymphocytes end up with a unique sequence for both chains in the BCR and TCR.

Applying Concepts to New Situations

1. Irrigating the wound removes dirt and debris that may contain pathogens. Scrubbing with soapy water removes more pathogens and may also kill them (soap may disrupt plasma membranes enough to be fatal). Antibiotics are toxic to pathogenic bacteria. **2.** If the antibody had a fluorescent molecule or other type of label attached, you could treat various types of cells with it, examine them under the microscope, and determine the location of the antibody-pump complexes. **3.** Natural selection favors individuals that can create a large array of antibodies, because the high mutation rates and rapid evolution observed in pathogens means that they will constantly present the immune system with new antigens. If these antigens were not recognized by the immune system, the pathogens would multiply freely and kill the individual. **4.** Tissue from the patient's own body is marked with the major histocompatibility (MHC) proteins that the patient's immune system recognizes as self. This results in the preservation, rather than the destruction, of the transplanted tissue. Tissue from a different person is marked with MHC proteins that are unique to that individual. The body of the transplant recipient will recognize the MHC proteins (and other molecules) of the donor as foreign, resulting in a full immune response and rejection of the grafted tissue.

CHAPTER 50

Check Your Understanding (CYU)

CYU p. 1001 (1) They are shallow, so a relatively large proportion of the water receives enough sunlight to support photosynthesis. In addition, nutrients are available from rivers that contribute nutrients from inland and upwellings that bring nutrients up from the ocean bottom. (2) It sinks, and the lower-density water below it rises, bringing nutrients to the surface. **CYU p. 1008** Tropical dry forests are probably less productive than tropical wet forests because they have less water available to support photosynthesis during some periods of the year. **CYU p. 1012** Your answer will depend on where you live. For example, if global warming continues at the current projected rates, several effects can be expected. If you live in a coastal area, you can expect water levels to rise. In general, plant communities will change because average temperature and moisture and variation in temperature and moisture will change.

You Should Be Able To (YSBAT)

YSBAT p. 997 The littoral zone, because light is abundant and nutrients are available from the substrate. **YSBAT p. 998** Bogs are nitrogen-poor, so plants that are able to capture and digest insects have a large advantage. This advantage does not exist in marshes and swamps, where nutrients are more readily available. **YSBAT p. 999** Cold water contains more oxygen than warm water, so it can support much more cellular respiration in fish. **YSBAT p. 1000** Species that live in estuaries must be able to tolerate variable salinity; marsh species do not. The abiotic environment (salt concentration) is so different that few species grow well in both habitats. **YSBAT p. 1001** No—the aphotic zone is lightless, so natural selection favors individuals that do not invest energy in developing and maintaining eyes. **YSBAT p. 1003** Vines and epiphytes increase productivity because they are photosynthetic organisms that fill space between small trees and large trees—they capture light and use nutrients that might not be used in a forest that lacked vines and epiphytes. **YSBAT p. 1004** Most leaves have a large surface area to capture light. Light is abundant in deserts, however, and a leaf with a large surface area would be susceptible to high water loss and/or overheating. **YSBAT p. 1005** Vegetation is continuous in a grassland, so a fire carries better. In a desert, the fire is likely to run out of fuel. **YSBAT p. 1006** There is a continuous grass cover and scattered trees. (A biome like this is called a savannah.) **YSBAT p. 1007** Boreal forests should move north. **YSBAT p. 1008** High elevations present an abiotic environment (precipitation and temperature) that is similar to the conditions in arctic tundras. As a result, the plant species present will have similar adaptations to cope with these physical conditions. **YSBAT p. 1010** [See Figure A50.1]

Caption Questions and Exercises

Figure 50.4 Freshwater, because light penetrates much deeper. **Figure 50.23** Dry, because dry air is descending here. This is consistent with the very low rainfall amounts in Barrow, Alaska, shown in Figure 50.21. **Figure 50.27** Visible wavelengths enter the chamber

FIGURE A49.1

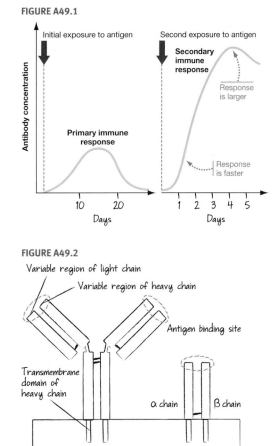

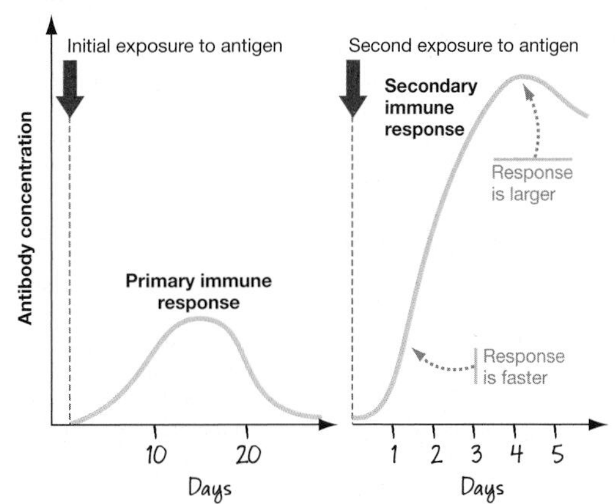

FIGURE A49.2

Variable region of light chain
Variable region of heavy chain
Antigen binding site
Transmembrane domain of heavy chain
α chain
β chain
B-cell receptor
T-cell receptor

FIGURE A50.1

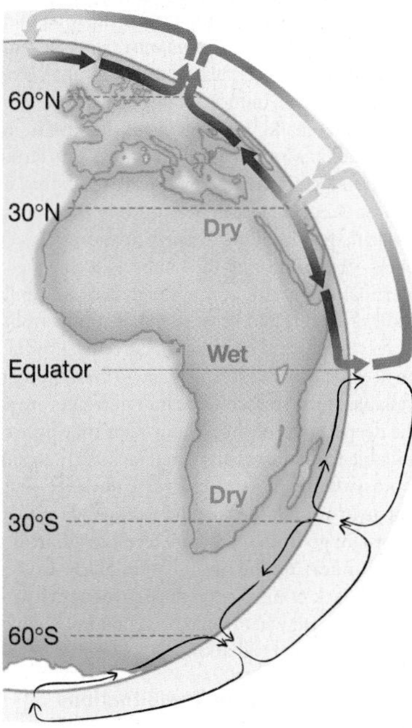

from the top and through the glass, warming the plants and ground inside the chamber. Some of this heat energy is retained within the mini-greenhouse. **Figure 50.28** They address the question posed because they capture important aspects of a plant community. NPP indicates how much biomass is available for other organisms to use; species diversity indicates how many different species are present. **Figure 50.34** (Several possible answers; one possibility is provided.) If water is short, there is not enough available to support a continuous cover of photosynthetic tissue aboveground. Roots of bunchgrasses can spread out and harvest water below spaces that are "bare" aboveground.

Summary of Key Concepts

KC 50.1 Populations are made up of individual organisms; communities are made up of populations of different species; ecosystems are made up of groups of communities (along with the abiotic environment). **KC 50.2** Increased evaporation from the oceans should increase precipitation on land—at least in some places. If precipitation does not increase, then increased transpiration rates will put plants under water stress and reduce productivity. **KC 50.3** All latitudes would get equal amounts of sunlight all year round. There would be no seasons and no changes in climate with latitude. **KC 50.4** It would increase, because cattle would no longer succumb to the disease carried by tsetse flies. (In Africa, cattle are limited by a biotic condition, not abiotic conditions.)

Test Your Knowledge

1. b; **2.** a; **3.** d; **4.** b; **5.** a; **6.** c

Test Your Understanding

1. Organismal, population, community, and ecosystem. Examples of possible questions: *Organismal:* How do humans cope with extremely hot (or cold) weather conditions? *Population:* How large will the total human population be in 50 years? *Community:* How are humans affecting prey species such as cod and tuna? *Ecosystem:* How will human-induced changes in global temperature affect sea level and thus organisms that live in the intertidal zone? **2.** More—in June the Sun's rays strike the Northern Hemisphere more directly than they do the equator. **3.** A Hadley cell just south of the equator results in warm, dry air falling to Earth at about 30 degrees south—the location of the Outback. **4.** Productivity in intertidal and neritic zones is high because sunlight is readily available and because nutrients are available from estuaries and deep-ocean currents. Productivity in the oceanic zone is extremely low—even though light is available at the surface—because nutrients are scarce. The deepest part of the oceanic zone may have nutrients available from the substrate but lacks light to support photosynthesis. **5.** As you increase in elevation, biomes change in a similar way as increasing in latitude (e.g., you might go from temperate forest to a boreal-type forest to tundra). **6.** The combination of low oxygen, low pH, and lack of nitrogen reduces productivity in bogs compared with what would be found in marshes or lakes with similar solar radiation.

Applying Concepts to New Situations

1. Yes—as Mars orbits the Sun, the part of the planet that is tilted toward the Sun is warmest because it receives the most direct sunlight. But as the planet orbits, that portion of the planet begins to tilt away from the Sun. Its temperature drops as a result, producing something like Earth's cycle of winter and summer. **2.** Mountaintops are cold because as air rises, it expands; this expansion cools the air, making the mountaintop cold. Also, there is less land surface nearby to absorb solar radiation and heat up. **3.** The hypothesis is that winds are usually from the northeast, and that higher elevations in the middle of the islands cause a rain shadow in the southwest corner. **4.** Edinburgh is surrounded by water. Because water has a high specific heat, it keeps temperatures moderate. Moscow, in contrast, is surrounded by a large landmass that does not have the same moderating effects on temperature as the ocean does.

CHAPTER 51

Check Your Understanding (CYU)

CYU p. 1022 If a sitter-like allele in bee-eaters was favored when population density is low, then it should be at high frequency in small colonies. If a rover-like allele was favored when population density is high, then it should be at high frequency in large colonies. **CYU p. 1025** (1) In the experiment, females came into breeding condition slowly if they were exposed only to springlike light conditions, but much more quickly if they were exposed to springlike light conditions *and* displaying males. Both are required for maximum effect. (2) Many possible answers, but based on the experiments with *Anolis*, it would be legitimate to hypothesize that they have to be exposed to springlike light and temperature conditions as well as courtship displays from males. **CYU p. 1031** (1) Auditory communication allows individuals to communicate over long distances but can be heard by predators. Olfactory communication is effective in the dark and scents can continue to carry information long after the signaler has left, but scents do not carry long distances. Visual communication is effective during the day but can be seen by predators. (2) The ability to detect deceit protects the individual from fitness costs (e.g., being eaten, having a mate's eggs fertilized by someone else); avoiding or punishing "liars" should also lower the frequency of deceit (because it becomes less successful). Alleles associated with detecting and avoiding or punishing "liars" should be favored by natural selection and increase in frequency. **CYU p. 1034** (1) Altruism is likely when B is high, as when a predator is about to pounce; r is high, as when a close relative is threatened; C is low, as when the caller is far from the predator. Altruism is unlikely under the opposite conditions. (2) For reciprocal altruism to work, individuals have to interact repeatedly and remember who has helped whom in the past.

You Should Be Able To (YSBAT)

YSBAT p. 1033 Humans often live near kin, are good at recognizing kin, and in many cases can give kin resources or protection that increase fitness. It is not known whether kin selection occurs in plants. In many cases, limited seed dispersal means that close relatives live together—making kin selection more likely. But it is not known whether plants can recognize kin and confer benefits preferentially to kin.

Caption Questions and Exercises

Figure 51.1 Several legitimate hypotheses are possible. Examples at the proximate level: The increased size of the antennae provide space for more sensory neurons; or large antennae allow individuals to sense a larger area. Examples at the ultimate level: large antennae help individuals navigate better and thus escape predation; or large antennae allow individuals to forage more successfully in the dark. **Figure 51.5** Bring females into the lab and give them the same food and housing conditions as the treatment groups, but expose them to artificial lighting that simulates the short daylight conditions of winter. **Figure 51.7** Because individuals were chosen for the different treatments at random, the investigator could claim that on average, the only thing that differed among individuals in the different groups was average tail length. **Figure 51.10** [See Figure A51.1] **Figure 51.14** Between first cousins, $r = \frac{1}{2} \times \frac{1}{2} \times \frac{1}{2} = \frac{1}{8}$. **Figure 51.15** The control would be to drag an object of similar size through the colony, at similar times of day. This would test the hypothesis that prairie dogs are reacting to the presence of the experimenter and the disturbance—not a predator.

Summary of Key Concepts

KC 51.1 At the proximate level, different alleles of the *for* gene affect the probability that an individual will move or stay after feeding. At the ultimate level, roving is favored when population density is high; sitting is favored when population density is low. **KC 51.2** Individuals have the capacity to act altruistically or non-altruistically toward kin. Hamilton's rule specifies the conditions—in terms of the costs and benefits of the act and the degree of relationship between the participants—under which altruism is favored. **KC 51.3** Roving is expensive energetically and could make individuals more susceptible to predation, but could allow individuals to find unexploited food sources. Sitting is inexpensive energetically, but may restrict individuals to areas where food supplies have been depleted. **KC 51.4** If only males that are well fed and free of parasites and disease are able to produce a brightly colored dewlap and do the bobbing display vigorously, then these traits would be a reliable indication of their condition. **KC 51.5** If migration does not occur when extra food is supplied, it suggests that the "decision" to migrate depends on conditions. **KC 51.6** Deceitful communication should decrease. If large males are gone, there are few nests available and more female-mimic males would compete at them. Most female-mimic males would have no (or fewer) eggs to fertilize and no male to take care of the eggs they did fertilize. There would be less natural selection pressure favoring deceitful communication. **KC 51.7** (1) Long-lived, so that extensive interactions with kin and nonrelatives are possible; (2) good memory, to record reciprocal interactions; (3) kin nearby, to make inclusive fitness gains possible; (4) ability to share fitness benefits (e.g., food) between individuals.

Test Your Knowledge

1. c; **2.** d; **3.** a; **4.** c; **5.** d; **6.** c

Test Your Understanding

1. At the ultimate level, homing to a safe den increases survival. The proximate cause for this behavior is still not fully known, but appears to involve the ability to sense changes in Earth's magnetic field. **2.** When optimal foraging occurs, an individual maximizes the benefit of foraging in terms of energy gain and minimizes the cost of foraging in terms of time, energy expended, and risk of predation. **3.** Only males that are in good shape (well-nourished and free of disease) have the resources required to produce a trait like a long tail. **4.** Longer day lengths indicate the arrival of spring, when renewed plant growth and insect activity make more food available. Courtship displays from males indicate the presence of males that can fertilize eggs. **5.** A map provides information about the spatial relationships of places on

FIGURE A51.1

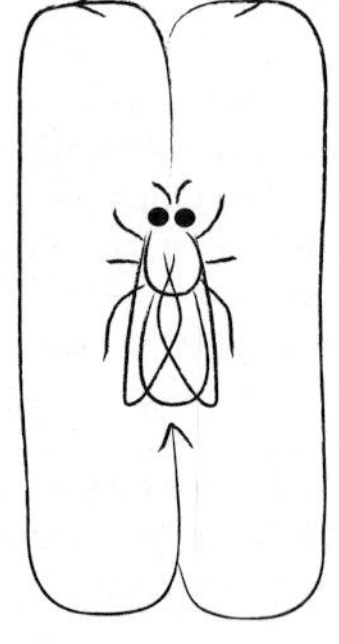

The straight run of the upward waggle dance is twice as long as in Fig. 51.9

a landscape; a compass allows you to orient the map—so that it aligns with the actual landscape. Animals have been shown to use the Sun's position, the position of the North star, and information from Earth's magnetic field as a compass (a way to find North). **6.** Kin selection can only act on kin; reciprocal altruism can occur between nonrelatives. Kin selection should be more common, because it potentially can occur any time that kin interact. Reciprocal altruism should be relatively rare because it is only possible when unrelated individuals interact repeatedly, have resources to share, and can remember the outcomes of past interactions.

Applying Concepts to New Situations

1. One possibility would be to set up a testing arena with food sources at various distances from a food source where adult flies are originally released. Test adults with a rover versus sitter genotype individually, and record how far each flies over the course of a set time interval. If the rover and sitter alleles affect foraging movements in adults, then rovers should be more likely to move and move farther than sitters. **2.** Simulate the onset of rains after a dry period, by using a hose to sprinkle water on the habitat where lizards are being maintained. **3.** Individuals have an *r* of 1/2 with full siblings and an *r* of 1/8 with first cousins. If the biologist will lose his life, he needs to save two siblings to keep the "lost copy" of his altruism alleles in the population but eight first cousins to maintain a copy. **4.** People would be expected to donate blood under two conditions: (1) They either received blood before or expect to have a transfusion in the future, and/or (2) they receive some other benefit in return, such as a good reputation among people who might be able to help them in other ways.

CHAPTER 52

Check Your Understanding (CYU)

CYU p. 1046 One possibility is competition for food. To test this idea, compare carrying capacity in identical vials that differ only in the amount of food added. (You could also test a hypothesis of oxygen limitation by adding more oxygen to one set of vials, or test the hypothesis of space limitation by doubling the volume but keeping the amount of food and oxygen the same.) **CYU p. 1053** (1) In developed nations, there are roughly equal numbers of individuals in each age class because the fertility rate has been constant and survivorship high for many years. In developing countries there are many more children and young people than older people, because the fertility rate and survivorship have been increasing. (2) Because survivorship is high in most human populations, changes in overall growth rate depend almost entirely on fertility rates.

You Should Be Able To (YSBAT)

YSBAT p. 1041 (1) Compared to northern populations, fecundity should be high and survivorship low. (2) Fecundity can be much higher if females lay eggs instead of retaining them in their bodies and giving birth to live young. **YSBAT p. 1055** The population appears to be increasing over time: from the original 1000 females, there are 2308 females in the third year.

Caption Questions and Exercises

Figure 52.2 [See Figure A52.1] **Figure 52.4** [See Table A52.1] **Figure 52.8** At high population density, competition for food limits the amount of energy available to female song sparrows for egg production. **Figure 52.12** Large-scale field experiments like this are extremely expensive and difficult to implement—it was not practical to replicate all treatments. **Table 52.2** From 1999 to 2009, $6800 = 6000\,\lambda^{10}$ (see Eq. 52.5 in Box 52.2). Solving, $\lambda = 1.013$. If $e^r = 1.013$, then $r = 0.0125$. Now use the expression $N_t = N_0e^{rt}$ (Eq. 52.7 in Box 52.2) to solve the problems.

Number of years required to add 1 billion people to 2009 level:

$$7800 = 6800e^{0.0125t}$$
$$ln(1.15) = 0.0125t$$
$$t = \sim 11 \text{ years}$$

Number of years required to double population from 2009 level:

$$13{,}600 = 6800e^{0.0125t}$$
$$ln(2) = 0.0125t$$
$$t = \sim 5.5 \text{ years}$$

Figure 52.17 [See Figure A52.2 on p. A:48]

Summary of Key Concepts

KC 52.1 A life table represents a snapshot of how a particular population is growing. As conditions change, survivorship and fecundity may change. In ancient Rome, for example, survivorship was probably low and fecundity high—women started reproducing at a young age and did not live long, on average. In Rome today, the population has high survivorship and low fecundity. **KC 52.2** [See Figure A52.3 on p. A:48] **KC 52.3** Fewer children are being born per female, but there are many more females of reproductive age due to high fecundity rates in the previous generation. **KC 52.4** Small, isolated populations are likely to be wiped out by bad weather, a disease outbreak, or changes in the habitat. For example, primrose populations may die out when a gap is shaded. But in a metapopulation, migration between the individual small populations helps reestablish subpopulations and maintain the overall size of the metapopulation.

Test Your Knowledge

1. b; **2.** d; **3.** a; **4.** a; **5.** d; **6.** c

Test Your Understanding

1. Equation 52.4: After one breeding interval, the population size is equal to the original population size times the discrete growth rate. Equation 52.5: After *t* breeding intervals, the population size is equal to the original population size times the discrete growth rate multiplied by itself *t* times (i.e., raised to the *t*th power). **2.** Species with type I survivorship curves have high survivorship until old age, so population growth is based on the number of offspring produced—not on how many of the

FIGURE A52.1

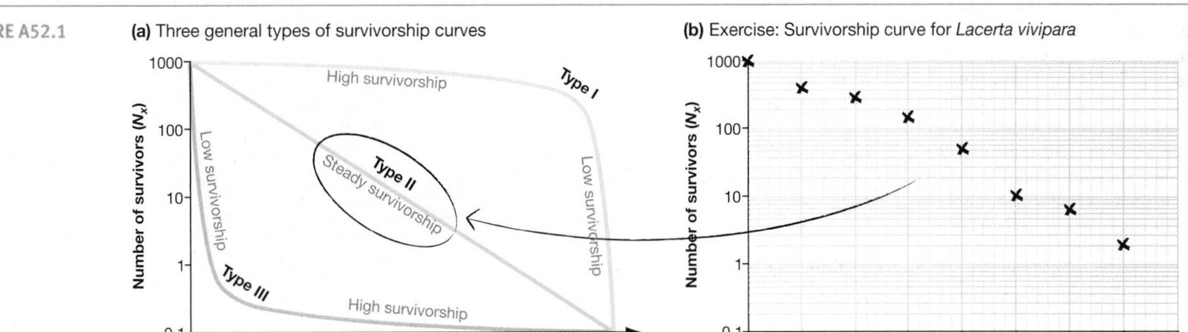

TABLE A52.1

Trait	Life-History Continuum		
	Left	Middle	Right
Growth habit	Herbaceous	Shrub	Tree
Disease- and predator-fighting ability	Low	Medium	High
Seed size	Small	Medium	Large
Seed number	Many	Moderate	Few

offspring survive. Species with type III survivorship curves exhibit very high juvenile mortality, but high survivorship of those few individuals able to reach adulthood. Because only the very few adults can reproduce, any changes to survivorship at the adult level will have a large effect on the population as a whole. **3.** The population has undergone near-exponential growth recently because advances in nutrition, sanitation, and medicine have allowed humans to live at high density without suffering from decreased survivorship and fecundity. Eventually, however, growth rates must slow as density-dependent effects increase death rates and lower birth rates. Experiments on snowshoe hares have shown that food availability and predation are density-dependent factors that influence population growth. **4.** (a) Corridors allow individuals to move between subpopulations, increasing gene flow and making it possible to recolonize habitats where populations have been lost.
(b) Maintaining unoccupied habitat makes it possible for the habitat to be recolonized. **5.** As a sexually transmitted disease, AIDS will reduce the number of sexually active adults. If the epidemic continues unabated, the numbers of both reproductive-age adults and children will decline, causing a top-heavy age distribution dominated by older adults and the elderly. [**See Figure A52.4**] **6.** Lambda is simple to calculate but only describes growth over a discrete time interval. Determining R_0 requires knowledge of all age-specific birth and death rates. r indicates the growth rate at any instant, so is independent of generation time and can be applied to any population. r_{max} gives the population growth rate in the absence of density-dependent limitation; r is the actual growth rate, which is usually affected by density-dependent factors.

Applying Concepts to New Situations

1. Fewer older individuals will be left in the population; there will be relatively more young individuals. If too many older individuals are taken, growth rate may decline sharply as reproduction stops or slows. (But if relatively few older individuals are taken, more resources are available to younger individuals and their survivorship and fecundity, and the population's overall growth rate, may increase.) **2.** The sunflowers and beetles are both metapopulations. To preserve them, you must preserve as many of the sunflower patches as possible (or plant more) and maintain corridors (which may be smaller sunflower patches) along which the beetles can migrate between the patches. **3.** Large-scale immigration into this population is projected. **4.** The growth rate of any population is female dependent, because females can produce a limited number of offspring at a time, regardless of how many males are in the population. This means that there are "extra" men in the Chinese population—they will not participate in reproduction. Growth rate should decrease, as a result.

CHAPTER 53

Check Your Understanding (CYU)

CYU p. 1070 (1) The individuals do not choose or try to have traits that reduce competition—they simply have those traits (or not). Resource partitioning just happens, because individuals with traits that allow them to exploit different resources produce more offspring, which also have those traits. (Recall from Chapter 24 that natural selection occurs on individuals, but adaptive responses such as resource partitioning are properties of populations.) (2) When species interact via consumption, a trait that gives one species an advantage will exert natural selection on individuals of the other species who have traits that reduce that advantage. This reciprocal adaptation will continue indefinitely. An example is the interaction of *Plasmodium* with the human immune system: The human immune system has evolved the

FIGURE A52.2

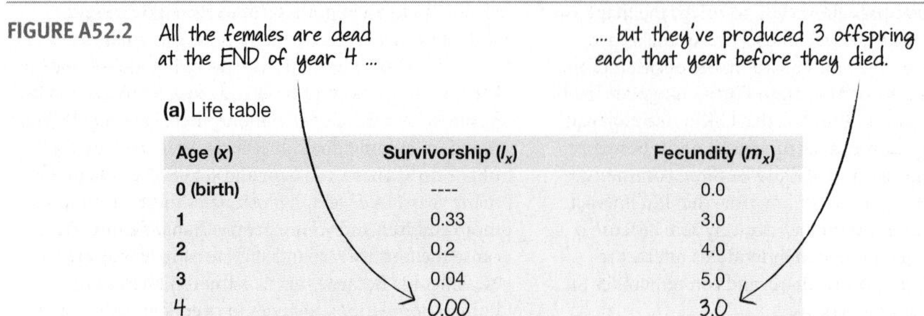

All the females are dead at the END of year 4 ...
... but they've produced 3 offspring each that year before they died.

(a) Life table

Age (x)	Survivorship (l_x)	Fecundity (m_x)
0 (birth)	----	0.0
1	0.33	3.0
2	0.2	4.0
3	0.04	5.0
4	0.00	3.0

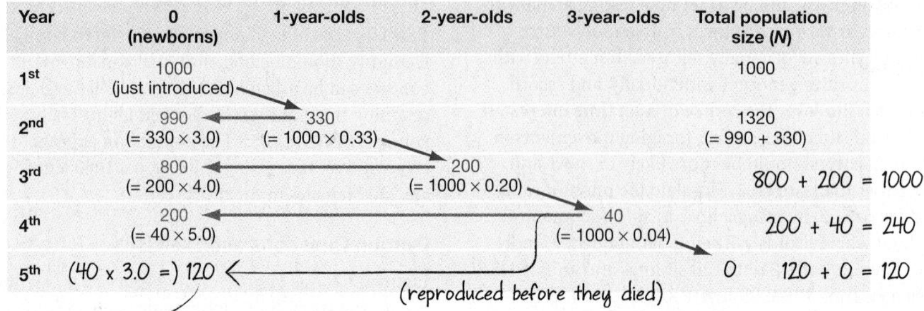

(reproduced before they died)

(b) Fate of first-generation females

Year	0 (newborns)	1-year-olds	2-year-olds	3-year-olds	Total population size (N)
1st	1000 (just introduced)				1000
2nd	990 (= 330 × 3.0)	330 (= 1000 × 0.33)			1320 (= 990 + 330)
3rd	800 (= 200 × 4.0)		200 (= 1000 × 0.20)		800 + 200 = 1000
4th	200 (= 40 × 5.0)			40 (= 1000 × 0.04)	200 + 40 = 240
5th	(40 x 3.0 =) 120			0	120 + 0 = 120

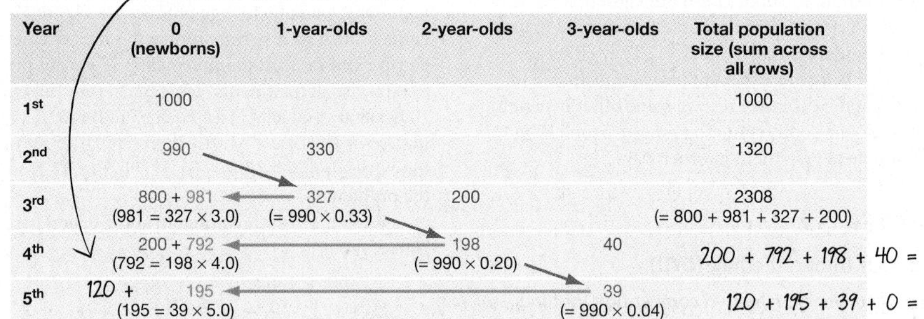

(c) Fate of first- and second-generation females

Year	0 (newborns)	1-year-olds	2-year-olds	3-year-olds	Total population size (sum across all rows)
1st	1000				1000
2nd	990	330			1320
3rd	800 + 981 (981 = 327 × 3.0)	327 (= 990 × 0.33)	200		2308 (= 800 + 981 + 327 + 200)
4th	200 + 792 (792 = 198 × 4.0)		198 (= 990 × 0.20)	40	200 + 792 + 198 + 40 = 1230
5th	120 + 195 (195 = 39 × 5.0)			39 (= 990 × 0.04)	120 + 195 + 39 + 0 = 354

FIGURE A52.3

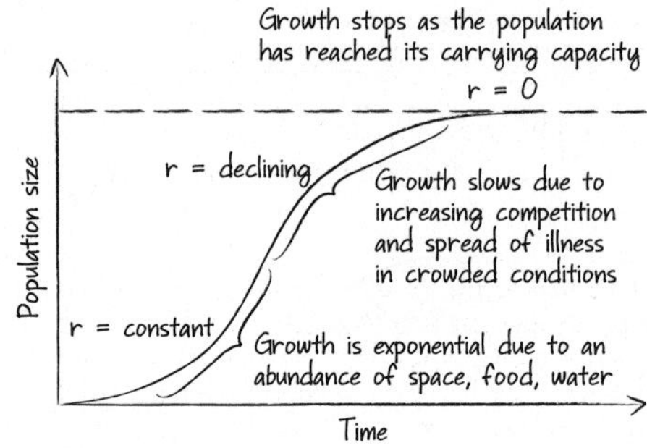

Growth stops as the population has reached its carrying capacity
r = 0
r = declining
Growth slows due to increasing competition and spread of illness in crowded conditions
r = constant
Growth is exponential due to an abundance of space, food, water
Population size
Time

FIGURE A52.4

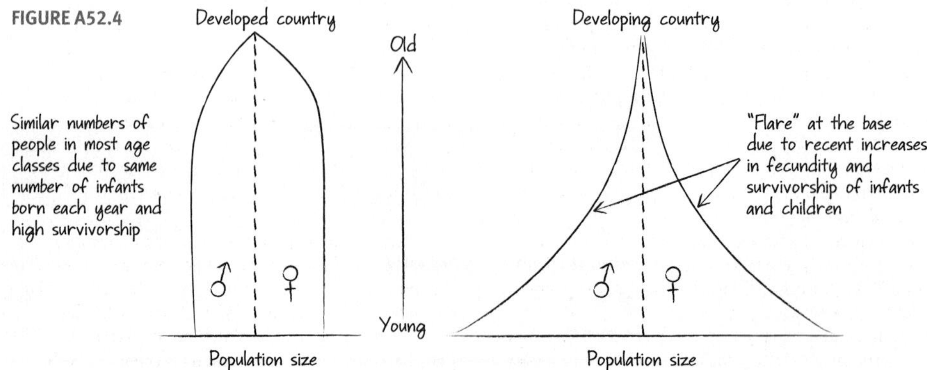

Developed country
Similar numbers of people in most age classes due to same number of infants born each year and high survivorship
♂
♀
Population size
Old
Young
Developing country
"Flare" at the base due to recent increases in fecundity and survivorship of infants and children
♂
♀
Population size

ability to detect proteins from the *Plasmodium* and kill infected cells; in response, *Plasmodium* has evolved different proteins that the immune system does not detect. **CYU p. 1077** (1) The shade provided by early successional species increases humidity, and decomposition of their tissues adds nutrients and organic material to the soil. These conditions favor growth by later successional species, which can outcompete the early successional species. (2) The presence or absence of a plant species where nitrogen-fixation occurs would dramatically alter nutrient conditions, and thus the speed of succession and the types of species that could become established. For example, species that require high nitrogen would be favored on sites where alder grew, while species that can tolerate low nitrogen would thrive on sites where alder is absent.

You Should Be Able To (YSBAT)

YSBAT p. 1079a The calculations for Community 1 are as follows:

Species A	Species B
$p_A = 10/18 = 0.55$	$p_B = 1/18 = 0.055$
$\ln 0.55 = -0.598$	$\ln 0.055 = -2.9$
$-0.598 \times 0.55 = -0.329$	$-2.9 \times 0.055 = -0.16$

Species C	Species D
$p_C = 1/18 = 0.055$	$p_D = 3/18 = 0.167$
$\ln 0.055 = -2.9$	$\ln 0.167 = -1.78$
$-2.9 \times 0.055 = -0.16$	$-1.78 \times 0.167 = -0.299$

Species E	Species F
$p_E = 2/18 = 0.11$	$p_F = 1/18 = 0.055$
$\ln 0.11 = -2.21$	$\ln 0.055 = -2.9$
$-2.2 \times 0.11 = -0.243$	$-2.9 \times 0.055 = -0.16$

Summing the values of $p \times \ln p$ for each species and multiplying by -1 gives the Shannon index of species diversity for Community 1:

$$(-1)[(-0.329) + (-0.16) + (-0.16) + (-0.299) + (-0.243) + (-0.16)] = 1.351$$

Similar calculations would give species diversity values of 1.79 for Community 2 and 1.61 for Community 3.
YSBAT p. 1079b
Community 1: Species richness = 12; Diversity = 2.04
Community 2: Species richness = 12; Diversity = 2.49
Community 3: Species richness = 10; Diversity = 2.3
YSBAT p. 1079c Species richness doubled in each community because the number of species doubled. Although the species diversity values differ from those in Figure 53.25, the communities with more equal numbers of individuals in each species still have the higher diversity values.

Caption Questions and Exercises

Figure 53.6 If he had done the treatments on different rocks, a critic could argue that differences in survival were due to differences in the nature of the rocks—not differences in competition. **Figure 53.7** The species partition, or divide up, resources. In this way, evolution leads to a reduction or elimination of competition. **Figure 53.12** This experimental design tested the hypothesis that mussels can sense the presence of crabs. If the crabs had been fed mussels, a critic could argue that the mussels detect the presence of damaged mussels—not crabs. (As it turns out, the mussels can do both. The experimenters did the same experiment with broken mussel shells in the chamber instead of a crab-fed fish.) **Figure 53.13** Beavers must represent the greater overall threat to cottonwoods, because the trees have evolved a compound that deters further beaver attack but does not deter beetle larvae. Getting the trunk cut down probably reduces fitness more than getting leaves eaten. **Figure 53.15** Put equal numbers of infected and uninfected ants in a pen that includes a bird predator. Count how many of each type are eaten over time. **Figure 53.16** Many possible answers are reasonable hypotheses. For example, acacia trees may expend energy in producing the large bulbs in which the ants live and the food they eat. The *Crematogaster* ants may expend energy and risk injury or death in repelling herbivores. The cleaner shrimp may occasionally get injured or eaten by the fish they are cleaning, or their diets may be somewhat restricted by the association. The host fish may miss meals or be at greater risk of predation when they are undergoing cleaning by the shrimp. In both examples of mutualism, however, the overall associations are positive for both parties. **Figure 53.17** These steps controlled for the alternative hypothesis that differences in treehopper survival were due to differences in the plants they occupied—not the presence or absence of ants. **Figure 53.18** Predictable, at least to a degree. **Figure 53.24** The effect would be to make the island more remote, which would lower the rate of immigration and move the whole immigration curve downward. The rate of extinction would increase, shifting the curve upward. The overall effect would be to decrease the number of species because the island would have effectively become more remote.

Summary of Key Concepts

KC 53.1 A mutualistic relationship becomes a parasitic one if one of the species stops receiving a benefit. The treehopper-ant mutualism becomes parasitic in years when spiders are rare, because the treehoppers no longer derive a benefit but pay a fitness cost (producing honeydew that the ants eat). **KC 53.2** All species at Glacier Bay must be able to survive in a cold climate with the local amount of precipitation. The species earliest in the succession must also be able to grow on rock exposed as the glacier melts. But chance, historical differences in seed sources, and presence or absence of alder (and nitrogen-fixation) created differences in the species present. **KC 53.3** Lakes are abundant in northern latitudes but rare in the tropics. The presence of much greater amounts of habitat supports higher species richness for breeding shorebirds at high latitudes.

Test Your Knowledge

1. d; **2.** a; **3.** b; **4.** d; **5.** c; **6.** d

Test Your Understanding

1. Yes—the treehopper-ant mutualism is parasitic or mutualistic depending on conditions; competition can evolve into no competition over time if niche differentiation occurs; arms races mean that the outcome of host-parasite interactions can change over time, and so forth (many other examples). **2.** *Top-down control:* Herbivore populations are kept at relatively low levels by predation and disease. *Bottom-up control:* Plant matter is low in nitrogen or defended by chemicals that are toxic to herbivores. These hypotheses are certainly not mutually exclusive. For example, resprouted cottonwoods produce compounds that make them less edible for beavers, which is a plant defense. In addition, these cottonwoods might not provide enough nitrogen for beavers to grow quickly, and species that prey on beavers or make them sick could be keeping their population under control. **3.** (a) If community composition is predictable, then the species present should not change over time. But if composition is not predictable, then there should be significant changes in the species present over time. (b) If community composition is predictable, then two sites with identical abiotic factors should develop identical communities. If community composition is not predictable, then sites with identical abiotic factors should develop variable communities. In most tests, the data best match the predictions of the "not predictable" hypothesis, though communities show elements of both. **4.** Disturbance is an event that removes biomass. Compared to low-frequency fires, high-frequency fires would tend to be less severe (less fuel builds up) and would tend to exert more intense natural selection for adaptations to resist the effects of fire. Compared to low-severity fires, high-severity fires would open up more space for pioneering species and would tend to exert more intense natural selection for adaptations to resist the effects of fire. **5.** Early successional species are adapted to disperse to new environments (small seeds) and grow and reproduce quickly (reproduce at an early age, grow quickly). They can tolerate severe abiotic conditions (high temperature, low humidity, low nutrient availability) but have little competitive ability. These species are able to enter a new environment (with no competitors) and thrive. **6.** The idea is that high productivity will lead to high population density of consumers, leading to competition and intense natural selection favoring niche differentiation that leads to speciation.

Applying Concepts to New Situations

1. Natural selection will favor orchid individuals that have traits that resist bee attack: thicker flower walls, nectar storage in a different position, a toxin in the flower walls, or other. Individuals could also be favored if their anthers were in a position that accomplished pollination, even if bees eat through the walls of the nectar-storage structure. **2.** Set fires or let natural fires burn, to match the time interval between fires recorded prior to the arrival of European settlers. Note that because fires have been suppressed for so long, the initial fires will have to be carefully controlled or fuel will have to be removed manually—for example by selective logging. The logic is that species in this habitat are adapted to a disturbance regime of frequent, low-intensity fires. For them to thrive, these conditions have to be re-created. **3.** The exact answer will depend on the location of the campus. The first species to appear must possess good dispersal ability, rapid growth, quick reproductive periods, and tolerance for very harsh and severe conditions. The two-acre plot is likely to be colonized first by pioneer species that have very "weedy" characteristics. But once colonization is under way, the course of succession will depend more on how the various species interact with each other. The presence of one species can inhibit or facilitate the arrival and establishment of another. For example, an early-arriving species might provide the shade and nutrients required by a late-arriving species. The site's history and nearby ecosystems may influence which species appear at each stage; for instance, an undisturbed ecosystem nearby could be a source for native species. The pattern and rate of this succession is also influenced by the overall environmental conditions affecting it. Only species with traits appropriate to the local climate are likely to colonize the site. **4.** One reasonable experiment would involve constructing artificial ponds and introducing different numbers of plankton species to different ponds, but the same total number of individuals. (Any natural immigration to the ponds would have to be prevented.) After a period of time, remove all of the plankton and measure the biomass present. Make a graph with number of species on the *x*-axis and total biomass on the *y*-axis. If the hypothesis is correct, the line of best fit through the data should have a positive slope.

CHAPTER 54

Check Your Understanding (CYU)

CYU p. 1092 (1) At each trophic level, most of the energy that is consumed is lost to heat, metabolism, or other maintenance activities, which leaves only a small percentage of energy for biomass production (growth and reproduction). (2) More food will be available to primary consumers, resulting in increases in biomass at each trophic level. An increase in biomass entering decomposer food chains will also increase decomposer populations. **CYU p. 1098** Perhaps the most direct impacts concern rates of groundwater replenishment. Converting biomes into farms or suburbs decreases groundwater recharge and increases runoff. Irrigation pumps groundwater to the surface, where much of it runs off into streams.

You Should Be Able To (YSBAT)

YSBAT p. 1086 To grow a kilogram of beef, you have to first grow 10 kilograms of grain or grass and feed it to the cow. Only 10 percent of this 10 kg will be used for growth and reproduction—the other 9 kg is used for maintenance or lost as heat. **YSBAT p. 1087a** Crustaceans and fish are ectothermic, so they are much more efficient at converting primary production into the biomass in their bodies than endothermic birds and mammals. **YSBAT p. 1087b** Bigger: elk to wolf, mule deer to wolf, aspen/cottonwood/willows to beaver, mice to rough-legged hawk. Smaller: mice to coyote, aspen/cottonwood/willows to elk and mule deer. **YSBAT p. 1094** The uptake arrow was removed—there were no plants to take up nutrients from the soil nutrient pool.

Caption Questions and Exercises

Figure 54.3 The percentage of energy used for maintenance would be much lower because they are ectothermic and sedentary. The percentage excreted would also be much lower because their diet contains little indigestible material. Thus the percentage of energy available for growth and reproduction would be much higher. **Figure 54.10** The diatoms' bodies sink, removing nutrients from the surface ocean and feeding organisms in the deeper ocean. **Figure 54.13** One logical hypothesis is that the total amount exported increases as tree roots and other belowground organic material decay and begin to wash into the stream; the amount exported should begin to decline as nitrate reserves become exhausted—eventually, there is no more nitrate to wash away. **Figure 54.14** Much more water should evaporate. One possibility is that more water vapor will be blown over land and increase precipitation over land. **Figure 54.21** Photosynthesis increases in summer, resulting in removal of CO_2 from the atmosphere.

Summary of Key Concepts

KC 54.1 [See Figure A54.1] **KC 54.2** A food chain consisting of only primary producers and primary decomposers would minimize energy loss. If organisms at higher trophic levels were present, the efficiency of energy transfer would be highest if they are ectothermic, move little, and have extremely efficient digestion. **KC 54.3** Decomposition is slowest in cold, wet temperatures on land and in anoxic conditions in the ocean (or in freshwater). Stagnant water often becomes anoxic as decomposers use up available oxygen and it is not replenished by diffusion from the atmosphere. **KC 54.4** Many possible answers, including: (1) reducing carbon dioxide emissions by finding alternatives to fossil fuels should decrease the amount of carbon dioxide released into the atmosphere, (2) recycling programs decrease carbon dioxide emissions from manufacturing plants and power plants, (3) reforestation can tie up carbon dioxide in wood; (4) iron fertilization could increase NPP in the open ocean—meaning that more CO_2 would be used.

Test Your Knowledge

1. a; **2.** d; **3.** a; **4.** c; **5.** c; **6.** b

Test Your Understanding

1. A food chain shows just a single connection between organisms at different trophic levels; a food web shows many or most of the connections that exist. The arrows represent the movement of chemical energy, obtained through feeding. **2.** If predation by cougars succeeds in reducing the deer population, then trees and low-growing plant species should increase. More (uneaten) biomass will remain in primary producers, meaning that more will be available for other primary consumers besides deer. **3.** Warmer, wetter climates speed decomposition; cool temperatures slow it. Lower amounts of nitrogen and oxygen also slow decomposition by limiting decomposer growth. Wood is slow to decompose because it contains lignin. Decomposition regulates nutrient availability because it releases nutrients from detritus and allows them to reenter the food web. **4.** Bleaching occurs when corals release symbiotic algae that perform photosynthesis; acidification is occurring as increased CO_2 in the atmosphere leads to increased carbonic acid in the ocean and a lower pH. Corals can starve if bleaching is extensive; acidification slows the growth of corals because it makes formation of calcium carbonate skeletons more difficult. **5.** Positive feedbacks include increased forest fires (due to drier, hotter summers) and increased rates of decomposition in tundras (due to warming); negative feedbacks include increased NPP due to warmer temperatures, increased precipitation, and/or increased access to CO_2 for photosynthesis. **6.** The open ocean has almost no nutrient input from the land and has little upwelling to supply nutrients from the deep ocean. In contrast, intertidal and coastal areas receive large inputs of nutrients from rivers as well as from upwellings from ocean depths.

Applying Concepts to New Situations

1. One possibility is to add radioactive phosphorus to the water, then follow this isotope through the primary producers, primary consumers, and secondary consumers in the system by measuring the amount of radioisotope in the tissues of organisms at each trophic level. **2.** If large amounts of organic material were produced in the open ocean and then moved to nearshore areas by currents, then decomposition of the material might use up all available oxygen and lead to creation of anoxic dead zones. **3.** Without herbivores, there is no link in the nitrogen cycle between primary producers and secondary consumers. All of the plant nitrogen would go to the primary decomposers and back into the soil. If decomposition is rapid enough, nitrogen would cycle quickly between primary producers, decomposers, the soil, and back to primary producers. **4.** Atmospheric oxygen would increase due to extensive photosynthesis, but carbon dioxide levels would decrease because little decomposition was occurring. The temperature would drop because fewer greenhouse gases would be trapping heat reflected from the Earth's surface.

CHAPTER 55

Check Your Understanding (CYU)

CYU p. 1109 Do an all-taxon survey: organize experts and volunteers to collect, examine, and identify all species present. This could include direct sequencing studies (see Chapter 29) to document the bacteria and archaea present in different habitats on campus. **CYU p. 1117** (1) Fragmentation reduces habitat quality by creating edges that are susceptible to invasion and loss of species, due to changed abiotic conditions. Genetic problems occur inside fragments as species become inbred and/or lose genetic diversity via genetic drift. (2) Using the equation $S = cA^z$, then $S = 19(100{,}000^{0.20}) = 190$ prior to habitat destruction and $S = 19(10{,}000^{0.20}) = 120$ afterwards, for a total loss of 70 species. Using the same equation for a z of 0.25 suggests that 338 species existed prior to extinction and 190 exist afterwards, a difference of 148. Twice as much diversity was lost due to the increased z. **CYU p. 1120** Establish a large number of study plots on the same hillside. Randomly assign the plots to contain 0, 1, 2, 4, 16, 32, or 64 species (or some other combination, to test a range of species richness), with several replicates for each level of species richness. Prior to planting, document soil depth and soil nutrient levels. Over time, maintain each plot by weeding to keep the original species richness intact. After several years, document soil depth and soil nutrient levels. Compare values as a function of species richness.

Caption Questions and Exercises

Figure 55.6 Nearly all endangered species are threatened by more than one factor and therefore appear on the graph more than once, leading to species totals greater than 100 percent. **Figure 55.10** Because the treatments and study areas were assigned at random, there is no bias in terms of picking certain areas that are unusual in terms of their biomass or species diversity. They should represent a random sample of biomass and species diversity in fragments of various sizes versus intact forest. **Figure 55.11** Draw a horizontal line at lambda of 1.0. **Figure 55.14** Null hypothesis: NPP is not affected by species richness or functional diversity of species. Prediction: NPP will be greater in plots with more species and more functional groups. Prediction of null: There will be no difference in NPP based on species richness or number of functional groups. **Figure 55.15** This line represents no change in biomass, which would mean the

FIGURE A54.1

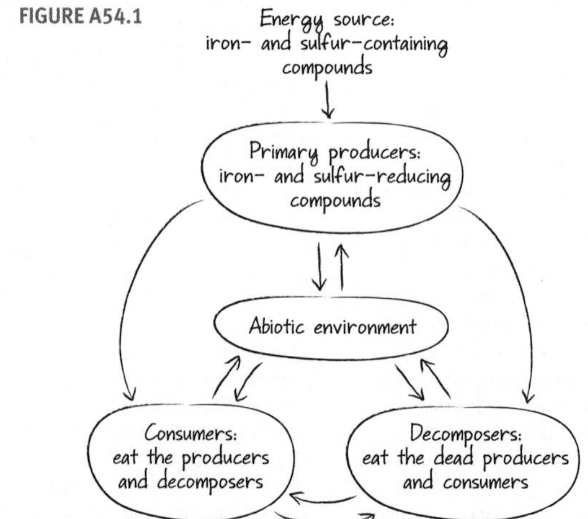

community was completely resistant to drought—that is, unaffected. **Figure 55.16** [See Figure A55.1]

Summary of Key Concepts

KC 55.1 By sequencing the genes present in a sample from a habitat, you get a broad view of the allelic diversity present—even if you don't know which species are there. **KC 55.2** Small, geographically isolated populations have low genetic diversity due to lack of gene flow and genetic drift and can suffer from inbreeding depression. Low genetic diversity makes them vulnerable to changes in their habitat, and small population size makes them susceptible to catastrophic events such as storms, disease outbreaks, or fires. **KC 55.3** If many species are present, and each uses the available resources in a unique way, then a larger proportion of all available resources should be used—leading to higher biomass production. **KC 55.4** 11percent is already protected, but a mass extinction event is still going on. This is because 11 percent is not sufficient to save most species and because many or most protected areas are not located in biodiversity or conservation hotspots.

Test Your Knowledge

1. d; **2.** a; **3.** b; **4.** d; **5.** c; **6.** a

Test Your Understanding

1. The threats have increased, and instead of most endangered species occurring on islands and being threatened by introduced species and overexploitation, now most threatened species are on continents (where habitat destruction is reducing habitats to small islands) and are in trouble due to habitat destruction and climate change. **2.** Level of resource use—because resources are only used at the rate at which they are replenished. **3.** By comparing the number of species estimated to reside in the park before and after the survey, biologists will have an idea of how much actual species diversity is being underestimated in other parts of the world—where research was at the level of Great Smoky Mountains National Park before the survey. The limitation is that it may be difficult to extrapolate from the data—no one knows if the situation at this park is typical of other habitats. **4.** Species–area curves specify how many species are expected to occur in habitats of various size. After projecting amounts of habitat loss, biologists can read a species–area curve to estimate how many species will be left in the amount of area that remains. **5.** In experiments with species native to North American grasslands, study plots that have more species produce more biomass, change less during a disturbance (drought), and recover from a disturbance faster than study plots with fewer species. **6.** Wildlife corridors facilitate the movement of individuals. Corridors allow areas to be recolonized if a species is lost and the introduction of new alleles that can counteract genetic drift and inbreeding in small isolated populations.

Applying Concepts to New Situations

1. Ecosystem services are beneficial effects that ecosystems have on the abiotic environment (soil, air, water quality, climate) for humans. No one owns or pays for these services, so no one has a vested interest in maintaining them. This is one of the primary reasons that ecosystems are destroyed, even though the services they offer are valuable. **2.** Catalog existing biodiversity at the genetic, species, and ecosystem levels. Using these data, find areas that have the highest concentration of biodiversity at each level. Protect as many of these areas as possible, and connect them with corridors of habitat. (In essence, protect and connect the biodiversity hotspots.) **3.** No "correct" answer. One argument is that conservationists could lobby officials in Brazil and Indonesia to learn from the mistakes made in developed nations, and preserve enough forests to maintain biodiversity and ecosystem services intact—avoiding the expenditures that developed countries had to make to clean up pollution and restore ecosystems and endangered species. **4.** Species with specialized food or habitat requirements, large size (and thus large requirements for land area), small population size, and low reproductive rate are vulnerable to extinction. A good example is the koala bear. Species that are particularly resistant to pressure from humans possess the opposite of many of these traits. A good example of a resistant species is the Norway rat.

THE BIG PICTURE: ECOLOGY

Check Your Understanding (CYU), p. 1126

1. Habitat loss "may eliminate" species; elimination of species (1) may disrupt community structure, (2) may reduces species richness, and (3) may reduce primary productivity. **2.** CO_2 "increases/contributes to" global warming; global warming (1) may disrupt community structure, (2) may reduce species richness, and (3) may reduce primary productivity. **3.** Populations "make up" species; ecosystems "are based on" primary productivity; water temperature, etc. "dictate species that can be found in certain" aquatic ecosystems; CO_2 "concentrations affect" climate.

APPENDIX A: BIOSKILLS

BIOSKILLS 1 (p. B:2) 1. 3.1 miles **2.** 36.96 yards **3.** 37°C **4.** Multiply your weight in pounds by 1/2.2 (0.45). **5.** 4 **BIOSKILLS 2 (p. B:4) 1.** about 18% **2.** a dramatic drop (almost 10%) **3.** No—the order of presentation in a bar chart does not matter (though it's convenient to arrange the bars in a way that reinforces the overall message). **4.** about 11 **5.** 68 inches **BIOSKILLS 3 Figure B3.1** [See Figure AB.1] **Figure B3.2** [See Figure AB.2]

FIGURE A55.1

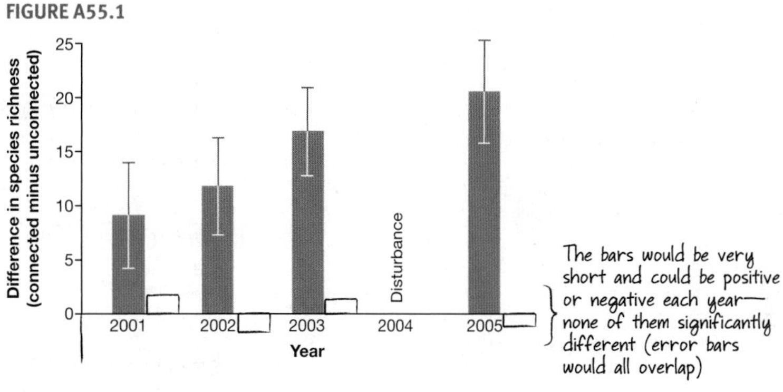

FIGURE AB.1

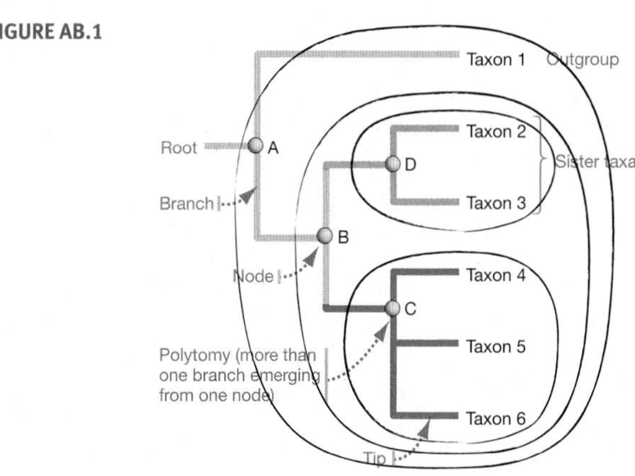

FIGURE AB.2

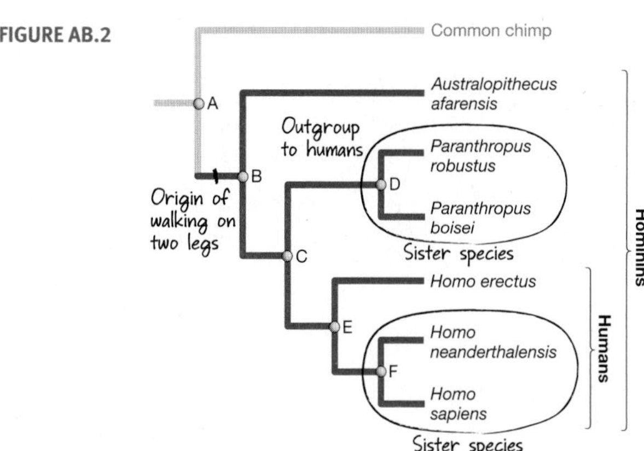

Figure B3.3 d **BIOSKILLS 4 (p. B:6) 1.** "different-yoked-together" **2.** "sugary-loosened" **3.** "study-of-form" **4.** "three-bodies" **BIOSKILLS 5 (p. B:7)** The estimate based on the larger sample—the more replicates or observations you have, the more precise your estimate of the average should be. **BIOSKILLS 6 Figure B6.1** [See Figure AB.3] **BIOSKILLS 7 (p. B:9) 1.** exponential **2.** $\ln N_t = \ln N_0 + rt$ **BIOSKILLS 8 (p. B:10) 1.** [See Figure AB.4] **2.** [See Figure AB.4] **BIOSKILLS 9 Figure B9.1** DNA and RNA are acids that tend to drop a proton in solution, giving them a negative charge. **(p. B:13)** The lane with no band comes from a sample where RNA X is not present. The same size RNA X is present in the next two lanes, but the light band has very few copies while the dark band has many. In the fourth lane, the band is formed by a smaller version of RNA X, with relatively few copies present. **BIOSKILLS 10 (p. B:16) 1.** No—it's just that no mitochondria happened to be present in this section sliced through the cell. **2.** Understanding a molecule's structure is often critical to understanding how it functions in cells. **BIOSKILLS 11 (p. B:17) 1.** size and/or density **2.** The use of high speeds generates the much higher forces needed to separate molecules that are similar in size and mass. **BIOSKILLS 12 (p. B:18) 1.** It may not be clear that the results are relevant to noncancerous cells that are not growing in cell culture—that is, that the artificial conditions mimic natural conditions. **2.** It may not be clear that the results are relevant to individuals that developed normally, from an embryo—that is, that the artificial conditions mimic natural conditions. **BIOSKILLS 13 (p. B:19) 1.** $\frac{1}{2} \times \frac{1}{2} \times \frac{1}{2} \times \frac{1}{2} = \frac{1}{16}$ **2.** $\frac{1}{6} + \frac{1}{6} + \frac{1}{6} = \frac{1}{2}$ **BIOSKILLS 14 Figure B14.1** This is human body temperature—the natural habitat of *E. coli.* **(p. B:22) 1.** *Caenorhabditis elegans* would be a good possibility, because the cells that normally die have already been identified. You could find mutant individuals that lacked normal cell death; you could compare the resulting embryos to normal embryos and be able to identify exactly which cells change as a result. **2.** Any of the multicellular organisms in the list would be a candidate, but *Dictyostelium discoideum* might be particularly interesting because cells only stick to each other during certain points in the life cycle. **3.** *Mus musculus*—as the only mammal in the list, it is the organism most likely to have a gene similar to the one you want to study.

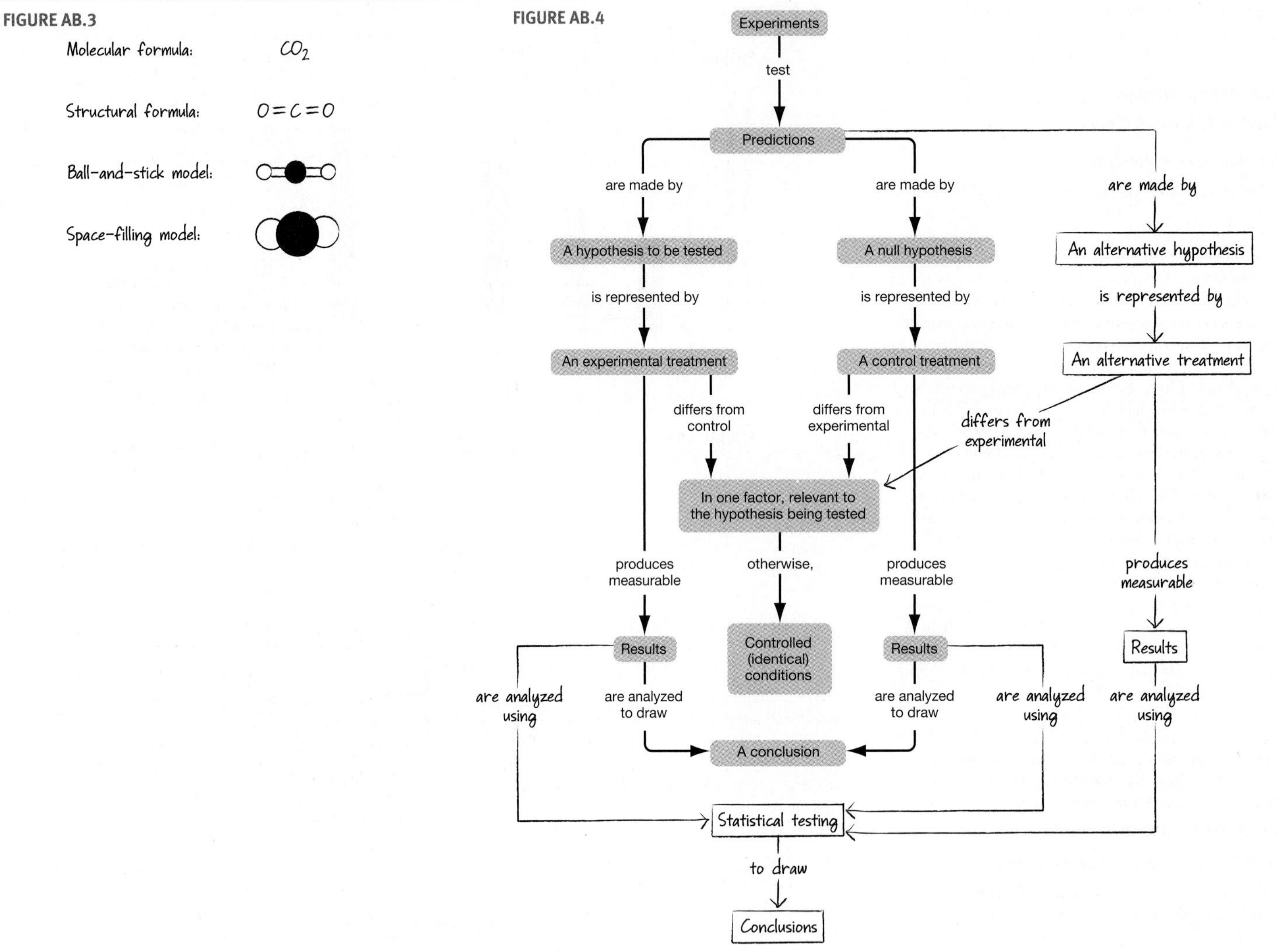

Glossary

5′ cap A chemical grouping, consisting of 7-methylguanylate and three phosphate groups, that is added to the 5′ end of newly transcribed messenger RNA molecules.

20-hydroxyecdysone See **ecdysone**.

abdomen A region of the body; in insects, one of the three prominent body regions called tagmata.

abiotic Not alive (e.g., air, water, and soil). Compare with **biotic**.

aboveground biomass The total mass of living plants in an area, excluding roots.

abscisic acid (ABA) A plant hormone that inhibits cell elongation and stimulates leaf shedding and dormancy.

abscission In plants, the normal (often seasonal) shedding of leaves, fruits, or flowers.

abscission zone The region at the base of a petiole that thins and breaks during dropping of leaves.

absorption In animals, the uptake of ions and small molecules derived from food across the lining of the intestine and into the bloodstream.

absorption spectrum The amount of light of different wavelengths absorbed by a pigment. Usually depicted as a graph of light absorbed versus wavelength. Compare with **action spectrum**.

acclimation, acclimatization Gradual physiological adjustment of an organism to new environmental conditions that occur naturally or as part of a laboratory experiment.

acetyl CoA A molecule produced by oxidation of pyruvate (the final product of glycolysis) in a reaction catalyzed by pyruvate dehydrogenase. Can enter the citric acid cycle and also is used as a carbon source in the synthesis of fatty acids, steroids, and other compounds.

acetylation Addition of an acetyl group (CH_3COO^-) to a molecule.

acetylcholine (ACh) A neurotransmitter, released by nerve cells at neuromuscular junctions, that triggers contraction of muscle cells. Also used as a neurotransmitter between neurons.

acid Any compound that gives up protons or accepts electrons during a chemical reaction or that releases hydrogen ions when dissolved in water.

acid-growth hypothesis The hypothesis that auxin triggers elongation of plant cells by inducing the synthesis of proton pumps whose activity makes the cell wall more acidic, leading to expansion of the cell wall and an influx of water.

acoelomate An animal that lacks an internal body cavity (coelom). Compare with **coelomate** and **pseudocoelomate**.

acquired immune deficiency syndrome (AIDS) A human disease characterized by death of immune system cells (in particular helper T cells and macrophages) and subsequent vulnerability to other infections. Caused by the human immunodeficiency virus (HIV).

acquired immunity Immunity to a particular pathogen or other antigen conferred by antibodies and activated B and T cells following exposure to the antigen. Is characterized by specificity, diversity, memory, and self-nonself recognition. Also called *acquired immune response.* Compare with **innate immunity**.

acrosome A caplike structure, located on the head of a sperm cell, that contains enzymes capable of dissolving the outer coverings of an egg.

ACTH See **adrenocorticotropic hormone**.

actin A globular protein that can be polymerized to form filaments. Actin filaments are part of the cytoskeleton and constitute the thin filaments in skeletal muscle cells.

actin filament A long fiber, about 7 nm in diameter, composed of two intertwined strands of polymerized actin protein; one of the three types of cytoskeletal fibers. Involved in cell movement. Also called a *microfilament.* Compare with **intermediate filament** and **microtubule**.

action potential A rapid, temporary change in electrical potential across a membrane, from negative to positive and back to negative. Occurs in cells, such as neurons and muscle cells, that have an excitable membrane.

action spectrum The relative effectiveness of different wavelengths of light in driving a light-dependent process such as photosynthesis. Usually depicted as a graph of some measure of the process versus wavelength. Compare with **absorption spectrum**.

activation energy The amount of energy required to initiate a chemical reaction; specifically, the energy required to reach the transition state.

active site The portion of an enzyme molecule where substrates (reactant molecules) bind and react.

active transport The movement of ions or molecules across a plasma membrane or organelle membrane against an electrochemical gradient. Requires energy (e.g., from hydrolysis of ATP) and assistance of a transport protein (e.g., pump).

adaptation Any heritable trait that increases the fitness of an individual with that trait, compared with individuals without that trait, in a particular environment.

adaptive immunity Immunity to a particular pathogen or other antigen conferred by antibodies and activated B and T cells following exposure to the antigen. Is characterized by specificity, diversity, memory, and self-nonself recognition. Also called *adaptive immune response.* Compare with **innate immunity.**

adaptive radiation Rapid evolutionary diversification within one lineage, producing numerous descendant species with a wide range of adaptive forms.

adenosine diphosphate (ADP) A molecule consisting of adenine, a sugar, and two phosphate groups. Addition of a third phosphate group produces adenosine triphosphate (ATP).

adenosine triphosphate (ATP) A molecule consisting of adenine, a sugar, and three phosphate groups that can be hydrolyzed to release energy. Universally used by cells to store and transfer energy.

adenylyl cyclase An enzyme that can catalyze the formation of cyclic AMP (cAMP) from ATP. Involved in controlling transcription of various operons in prokaryotes and in some eukaryotic signal-transduction pathways.

adhesion The tendency of certain dissimilar molecules to cling together due to attractive forces. Compare with **cohesion**.

adipocyte A fat cell.

adipose tissue A type of connective tissue whose cells store fats.

ADP See **adenosine diphosphate**.

adrenal glands Two small endocrine glands that sit above each kidney. The outer portion (cortex) secretes several steroid hormones; the inner portion (medulla) secretes epinephrine and norepinephrine.

adrenaline See **epinephrine**.

adrenocorticotropic hormone (ACTH) A peptide hormone, produced and secreted by the anterior pituitary, that stimulates release of steroid hormones (e.g., cortisol, aldosterone) from the adrenal cortex.

adult A sexually mature individual.

adventitious root A root that develops from a plant's shoot system instead of from the plant's root system.

aerobic Referring to any metabolic process, cell, or organism that uses oxygen as an electron acceptor. Compare with **anaerobic**.

afferent division The part of the nervous system, consisting mainly of sensory neurons, that transmits information about the internal and external environment to the central nervous system. Compare with **efferent division**.

age class All the individuals of a specific age in a population.

age structure The proportion of individuals in a population that are of each possible age.

age-specific fecundity The average number of female offspring produced by a female in a certain age class.

agglutination Clumping together of cells, typically caused by antibodies.

aggregate fruit A fruit (e.g., raspberry) that develops from a single flower that has many separate carpels. Compare with **multiple** and **simple fruit**.

AIDS See **acquired immune deficiency syndrome**.

albumen A solution of water and protein (particularly albumins), found in amniotic eggs, that nourishes the growing embryo. Also called *egg white.*

albumin A class of large proteins found in plants and animals, particularly in the albumen of eggs and in blood plasma.

alcohol fermentation Catabolic pathway in which pyruvate produced by glycolysis is converted to ethanol in the absence of oxygen.

aldosterone A hormone produced in the adrenal cortex that stimulates the kidney to conserve salt and water and promotes retention of sodium.

alimentary canal See **digestive tract**.

allele A particular version of a gene.

allergen A molecule (antigen) that triggers an allergic response (an allergy).

allergy An abnormal response to an antigen, usually characterized by dilation of blood vessels, contraction of smooth muscle cells, and increased activity by mucous-secreting cells.

allopatric speciation The divergence of populations into different species by physical isolation of populations in different geographic areas. Compare with **sympatric speciation**.

allopatry Condition in which two or more populations live in different geographic areas. Compare with **sympatry**.

allopolyploidy (adjective: allopolyploid) The state of having more than two full sets of chromosomes (polyploidy) due to hybridization between different species. Compare with **autopolyploidy**.

allosteric regulation Regulation of a protein's function by binding of a regulatory molecule, usually to a specific site distinct from the active site, causing a change in the protein's shape.

α-amylase See **amylase**.

α-helix (alpha-helix) A protein secondary structure in which the polypeptide backbone coils into a spiral shape stabilized by hydrogen bonds between atoms.

alternation of generations A life cycle involving alternation of a multicellular haploid stage (gametophyte) with a multicellular diploid stage (sporophyte). Occurs in most plants and some protists.

alternative splicing In eukaryotes, the splicing of primary RNA transcripts from a single gene in different ways to produce different mature mRNAs and thus different polypeptides.

altruism Any behavior that has a cost to the individual (such as lowered survival or reproduction) and a benefit to the recipient. See **reciprocal altruism**.

alveolus (plural: alveoli) One of the tiny air-filled sacs of a mammalian lung.

ambisense virus A virus whose genome contains both positive-sense and negative-sense sequences.

amino acid A small organic molecule with a central carbon atom bonded to an amino group ($-NH_3$), a carboxyl group ($-COOH$), a hydrogen atom, and a side group. Proteins are polymers of 20 common amino acids.

aminoacyl tRNA A transfer RNA molecule that is covalently bound to an amino acid.

aminoacyl tRNA synthetase An enzyme that catalyzes the addition of a particular amino acid to its corresponding tRNA molecule.

ammonia (NH_3) A small molecule, produced by the breakdown of proteins and nucleic acids, that is very toxic to cells. Is a strong base that gains a proton to form the ammonium ion (NH_4^+).

amnion The membrane within an amniotic egg that surrounds the embryo and encloses it in a protective pool of fluid (amniotic fluid).

amniotes A major lineage of vertebrates (Amniota) that reproduce with amniotic eggs. Includes all reptiles (including birds) and mammals.

amniotic egg An egg that has a watertight shell or case enclosing a membrane-bound water supply (the amnion), food supply (yolk sac), and waste sac (allantois).

amoeboid motion See **cell crawling**.

amphibians A lineage of vertebrates many of whom breathe through their skin and feed on land but lay their eggs in water. Represent the earliest tetrapods; include frogs, salamanders, and caecilians.

amphipathic Containing hydrophilic and hydrophobic elements.

amylase Any enzyme that can break down starch by catalyzing hydrolysis of the glycosidic linkages between the glucose residues.

amyloplasts Dense, starch-storing organelles that settle to the bottom of plant cells and that may be used as gravity detectors.

anabolic pathway Any set of chemical reactions that synthesizes larger molecules from smaller ones. Generally requires an input of energy. Compare with **catabolic pathway**.

anadromous Having a life cycle in which adults live in the ocean (or large lakes) but migrate up freshwater streams to breed and lay eggs.

anaerobic Referring to any metabolic process, cell, or organism that uses an electron acceptor other than oxygen, such as nitrate or sulfate. Compare with **aerobic**.

anaphase A stage in mitosis or meiosis during which chromosomes are moved to opposite ends of the cell.

anatomy The study of the physical structure of organisms.

ancestral trait A trait found in ancestors.

aneuploidy (adjective: aneuploid) The state of having an abnormal number of copies of a certain chromosome.

angiosperm A flowering vascular plant that produces seeds within mature ovaries (fruits). The angiosperms form a single lineage. Compare with **gymnosperm**.

animal A member of a major lineage of eukaryotes (Animalia) whose members typically have a complex, large, multicellular body, eat other organisms, and are mobile.

animal model Any disease that occurs in a nonhuman animal and has many parallels to a similar disease of humans. Studied by medical researchers in hope that findings may apply to human disease.

anion A negatively charged ion.

annelids Members of the phylum Annelida (segmented worms). Distinguished by a segmented body and a coelom that functions as a hydrostatic skeleton. Annelids belong to the lophotrochozoan branch of the protostomes.

annual Referring to a plant whose life cycle normally lasts only one growing season—less than one year. Compare with **perennial**.

anoxygenic Referring to any process or reaction that does not produce oxygen. Photosynthesis in purple sulfur and purple nonsulfur bacteria, which does not involve photosystem II, is anoxygenic. Compare with **oxygenic**.

antenna (plural: antennae) A long appendage that is used to touch or smell.

antenna complex Part of a photosystem, containing an array of chlorophyll molecules and accessory pigments, that receives energy from light and directs the energy to a central reaction center during photosynthesis.

anterior Toward an animal's head and away from its tail. The opposite of posterior.

anterior pituitary The part of the pituitary gland containing endocrine cells that produce and release a variety of peptide hormones in response to other hormones from the hypothalamus. Compare with **posterior pituitary**.

anther The pollen-producing structure at the end of a stamen in flowering plants (angiosperms).

antheridium (plural: antheridia) The sperm-producing structure in most land plants except angiosperms.

anthropoids One of the two major lineages of primates, including apes, humans, and all monkeys. Compare with **prosimians**.

antibiotic Any substance, such as penicillin, that can kill or inhibit the growth of bacteria.

antibody An immunoglobulin protein, produced by B cells, that can bind to a specific part of an antigen, tagging it for attack by the immune system. All antibody molecules have a similar Y-shaped structure and, in their monomer form, consist of two identical light chains and two identical heavy chains.

anticodon The sequence of three bases (triplet) in a transfer RNA molecule that can bind to a mRNA codon with a complementary sequence.

antidiuretic hormone (ADH) A peptide hormone, secreted from the posterior pituitary gland, that stimulates water retention by the kidney. Also called *vasopressin*.

antigen Any foreign molecule, often a protein, that can stimulate a specific response by the immune system.

antigen presentation Process by which small peptides, derived from ingested particulate antigens (e.g., bacteria) or intracellular antigens (e.g., viruses in infected cell) are complexed with MHC proteins and transported to the cell surface where they are displayed and can be recognized by T cells.

antiparallel Describing the opposite orientation of the strands in a DNA double helix with one strand running in the $5' \rightarrow 3'$ direction and the other in the $3' \rightarrow 5'$ direction.

antiporter A carrier protein that allows an ion to diffuse down an electrochemical gradient, using the energy of that process to transport a different substance in the opposite direction *against* its concentration gradient. Compare with **symporter**.

antiviral Any drug or other agent that can kill or inhibit the transmission or replication of viruses.

anus In a multicellular animal, the end of the digestive tract where wastes are expelled.

aorta In terrestrial vertebrates, the major artery carrying oxygenated blood away from the heart.

aphotic zone Deep water receiving no sunlight. Compare with **photic zone**.

apical Toward the top. In plants, at the tip of a branch. In animals, on the side of an epithelial layer that faces the environment and not other body tissues. Compare with **basal**.

apical bud A bud at the tip of a stem, where growth occurs to lengthen the stem.

apical dominance Inhibition of lateral bud growth by the apical meristem at the tip of a plant branch.

apical meristem A group of undifferentiated plant cells, at the tip of a stem or root, that is responsible for primary growth. Compare with **lateral meristem**.

apical–basal axis The long or vertical, shoot-to-root axis of a plant.

apomixis The formation of mature seeds without fertilization occurring; a type of asexual reproduction.

apoplast In plant roots, a continuous pathway through which water can flow, consisting of the porous cell walls of adjacent cells and the intervening extracellular space. Compare with **symplast**.

apoptosis Series of genetically controlled changes that lead to death of a cell. Occurs frequently during embryological development and later may occur in response to infections or cell damage. Also called *programmed cell death*.

appendix A blind sac (having only one opening) that extends from the colon in some mammals.

aquaporin A type of channel protein through which water can move by osmosis across a plasma membrane.

arbuscular mycorrhizal fungi (AMF) Fungi whose hyphae enter the root cells of their host plants. Also called *endomycorrhizal fungi*.

Archaea One of the three taxonomic domains of life consisting of unicellular prokaryotes distinguished by cell walls made of certain polysaccharides not found in bacterial or eukaryotic cell walls, plasma membranes composed of unique isoprene-containing phospholipids, and ribosomes and RNA polymerase similar to those of eukaryotes. Compare with **Bacteria** and **Eukarya**.

archegonium (plural: archegonia) The egg- producing structure in most land plants except angiosperms.

arteriole One of the many tiny vessels that carry blood from arteries to capillaries.

artery Any thick-walled blood vessel that carries blood (oxygenated or not) under relatively high pressure away from the heart to organs throughout the body. Compare with **vein**.

arthropods Members of the phylum Arthropoda. Distinguished by a segmented body; a hard, jointed exoskeleton; paired appendages; and an extensive body cavity called a hemocoel. Arthropods belong to the ecdysozoan branch of the protostomes.

articulation A movable point of contact between two bones of a skeleton. See **joint**.

artificial selection Deliberate manipulation by humans, as in animal and plant breeding, of the genetic composition of a population by allowing only individuals with desirable traits to reproduce.

ascocarp A large, cup-shaped reproductive structure produced by some ascomycete fungi. Contains many microscopic asci, which produce spores.

ascus (plural: asci) Specialized spore-producing cell found at the ends of hyphae in sac fungi (ascomycetes).

asexual reproduction Any form of reproduction resulting in offspring that are genetically identical to the parent. Includes binary fission, budding, and parthenogenesis. Compare with **sexual reproduction**.

asymmetric competition Ecological competition between two species in which one species suffers a much greater fitness decline than the other. Compare with **symmetric competition**.

atomic number The number of protons in the nucleus of an atom, giving the atom its identity as a particular chemical element.

ATP See **adenosine triphosphate**.

ATP synthase A large membrane-bound protein complex in chloroplasts, mitochondria, and some bacteria that uses the energy of protons flowing through it to synthesize ATP. Also called F_oF_1 *complex*.

atrial natriuretic hormone An animal hormone that stimulates excretion of salt from the kidneys.

atrioventricular (AV) node A region of the heart between the right atrium and right ventricle where electrical signals from the atrium are slowed briefly before spreading to the ventricle. This delay allows the ventricle to fill with blood before contracting. Compare with **sinoatrial (SA) node**.

atrium (plural: atria) A thin-walled chamber of the heart that receives blood from veins and pumps it to a neighboring chamber (the ventricle).

autocrine Relating to a chemical signal that affects the same cell that produced and released it.

autoimmunity A pathological condition in which the immune system attacks self cells or tissues of an individual's own body.

autonomic nervous system The part of the peripheral nervous system that controls internal organs and involuntary processes, such as stomach contraction, hormone release, and heart rate. Includes parasympathetic and sympathetic nerves. Compare with **somatic nervous system**.

autophagy The process by which damaged organelles are surrounded by a membrane and delivered to a lysosome to be destroyed.

autopolyploidy (adjective: autopolyploid) The state of having more than two full sets of chromosomes (polyploidy) due to a mutation that doubled the chromosome number.

autoradiography A technique for detecting radioactively labeled molecules separated by gel electrophoresis by placing an unexposed film over the gel. A black dot appears on the film wherever a radioactive atom is present in the gel. Also can be used to locate labeled molecules in fixed tissue samples.

autosomal inheritance The inheritance patterns that occur when genes are located on autosomes rather than on sex chromosomes.

autosome Any chromosome that does not carry genes involved in determining the sex of an individual.

autotroph Any organism that can synthesize reduced organic compounds from simple inorganic sources such as CO_2 or CH_4. Most plants and some bacteria and archaea are autotrophs. Also called *primary producer*. Compare with **heterotroph**.

auxin Indoleacetic acid, a plant hormone that stimulates phototropism and some other responses.

avirulence (*avr*) genes Genes in pathogens encoding proteins that trigger a defense response in plants. Compare with **resistance (*R*) genes**.

axillary bud A bud that forms in the angle between a leaf and a stem and may develop into a lateral (side) branch. Also called *lateral bud*.

axon A long projection of a neuron that can propagate an action potential and transmit it to another neuron.

axon hillock The site in a neuron where an axon joins the cell body and where action potentials are first triggered.

axoneme A structure found in eukaryotic cilia and flagella and responsible for their motion; composed of two central microtubules surrounded by nine doublet microtubules (9 + 2 arrangement).

B cell A type of leukocyte that matures in the bone marrow and, with T cells, is responsible for adaptive immunity. Produces antibodies and also functions in antigen presentation. Also called *B lymphocyte*.

BAC library A collection of all the sequences found in the genome of a species, inserted into bacterial artificial chromosomes (BACs).

background extinction The average rate of low-level extinction that has occurred continuously throughout much of evolutionary history. Compare with **mass extinction**.

Bacteria One of the three taxonomic domains of life consisting of unicellular prokaryotes distinguished by cell walls composed largely of peptidoglycan, plasma membranes similar to those of eukaryotic cells, and ribosomes and RNA polymerase that differ from those in archaeans or eukaryotes. Compare with **Archaea** and **Eukarya**.

bacterial artificial chromosome (BAC) An artificial version of a bacterial chromosome that can be used as a cloning vector to produce many copies of large DNA fragments.

bacteriophage Any virus that infects bacteria.

baculum A bone inside the penis usually present in mammals with a penis that lacks erectile tissue.

balancing selection A pattern of natural selection in which no single allele is favored in all populations of a species at all times. Instead, there is a balance among alleles in terms of fitness and frequency.

ball-and-stick model A representation of a molecule where atoms are shown as balls—colored and scaled to indicate the atom's identity—and covalent bonds are shown as rods or sticks connecting the balls in the correct geometry.

bar coding The use of well-characterized gene sequences to identify species.

bark The protective outer layer of woody plants, composed of cork cells, cork cambium, and secondary phloem.

baroreceptors Specialized nerve cells in the walls of the heart and certain major arteries that detect changes in blood pressure and trigger appropriate responses by the brain.

basal Toward the base. In plants, at the base of a branch where it joins the stem. In animals, on the side of an epithelial layer that abuts underlying body tissues. Compare with **apical**.

basal body A structure of nine pairs of microtubules arranged in a circle at the base of eukaryotic cilia and flagella where they attach to the cell. Structurally similar to a centriole.

basal lamina A thick, collagen-rich extracellular matrix that underlies most epithelial tissues (e.g., skin) in animals.

basal metabolic rate (BMR) The total energy consumption by an organism at rest in a comfortable environment. For aerobes, often measured as the amount of oxygen consumed per hour.

basal transcription complex A large multi-protein structure that assembles near the promoter of eukaryotic genes and initiates transcription. Composed of basal transcription factors, TATA-binding protein, coactivators, and RNA polymerase.

basal transcription factor General term for proteins, present in all cell types, that bind to eukaryotic promoters and help initiate transcription. Compare with **regulatory transcription factor**.

base Any compound that acquires protons or gives up electrons during a chemical reaction or accepts hydrogen ions when dissolved in water.

basidium (plural: basidia) Specialized spore-producing cell at the ends of hyphae in club fungi.

basilar membrane The membrane in the vertebrate cochlea on which the bottom portion of hair cells sit.

basolateral Toward the bottom and sides. In animals, the side of an epithelial layer that faces other body tissues and not the environment.

Batesian mimicry A type of mimicry in which a harmless or palatable species resembles a dangerous or poisonous species. Compare with **Müllerian mimicry**.

B-cell receptor (BCR) An immunoglobulin protein (antibody) embedded in the plasma membrane of mature B cells and to which antigens bind.

beak A structure that exerts biting forces and is associated with the mouth; found in birds, cephalopods, and some insects.

behavior Any action by an organism.

beneficial In genetics, referring to any mutation, allele, or trait that increases an individual's fitness.

benign tumor A mass of abnormal tissue that grows slowly or not at all, does not disrupt surrounding tissues, and does not spread to other organs. Benign tumors are not cancers. Compare with **malignant tumor**.

benthic Living at the bottom of an aquatic environment.

benthic zone The area along the bottom of an aquatic environment.

β-pleated sheet (beta-pleated sheet) A protein secondary structure in which the polypeptide backbone folds into a sheetlike shape stabilized by hydrogen bonding.

bilateral symmetry An animal body pattern in which there is one plane of symmetry dividing the body into a left side and a right side. Typically, the body is long and narrow, with a distinct head end and tail end. Compare with **radial symmetry**.

bilaterian A member of a major lineage of animals (Bilateria) that are bilaterally symmetrical at some point in their life cycle, have three embryonic germ layers, and have a coelom. All protostomes and deuterostomes are bilaterians.

bile A complex solution produced by the liver, stored in the gall bladder, and secreted into the intestine. Contains steroid derivatives called bile salts that are responsible for emulsification of fats during digestion.

biodiversity The diversity of life considered at three levels: genetic diversity (variety of alleles in a population, species, or group of species); species diversity (variety and relative abundance of species present in a certain area); and ecosystem diversity (variety of communities and abiotic components in a region).

biodiversity hot spot A region that is extraordinarily rich in species.

biofilm A hard, polysaccharide-rich layer secreted by bacterial cells, which allows them to attach to a surface.

biogeochemical cycle The pattern of circulation of an element or molecule among living organisms and the environment.

biogeography The study of how species and populations are distributed geographically.

bioinformatics The field of study concerned with managing, analyzing, and interpreting biological information, particularly DNA sequences.

biological fitness See **fitness**.

biological species concept The definition of a species as a population or group of populations that are reproductively isolated from other groups. Members of a species have the potential to interbreed in nature to produce viable, fertile offspring but cannot produce viable, fertile hybrid offspring with members of other species. Compare with **morphospecies** and **phylogenetic species concept**.

bioluminescence The emission of light by a living organism.

biomagnification In animal tissues, an increase in the concentration of particular molecules that may occur as those molecules are passed up a food chain.

biomass The total mass of all organisms in a given population or geographical area; usually expressed as total dry weight.

biome A large terrestrial ecosystem characterized by a distinct type of vegetation and climate.

bioprospecting The effort to find commercially useful compounds by studying organisms—especially species that are poorly studied to date.

bioremediation The use of living organisms, usually bacteria or archaea, to degrade environmental pollutants.

biotechnology The application of biological techniques and discoveries to medicine, industry, and agriculture.

biotic Living, or produced by a living organism. Compare with **abiotic**.

bipedal Walking primarily on two legs.

bipolar cell A cell in the vertebrate retina that receives information from one or more photoreceptors and passes it to other bipolar cells or ganglion cells.

bivalves A lineage of mollusks that have two shells, such as clams and mussels.

bladder A mammalian organ that holds urine until it can be excreted.

blade The wide, flat part of a plant leaf.

blastocoel Fluid-filled cavity in the blastula of many animal species.

blastocyst Specialized type of blastula in mammals. A spherical structure composed of trophoblast cells on the exterior and a cluster of cells (the inner cell mass), which fills part of the interior space.

blastomere A small cell created by cleavage divisions in early animal embryos.

blastopore A small opening (pore) in the surface of an early vertebrate embryo, through which cells move during gastrulation.

blastula In vertebrate development, a hollow ball of cells (blastomere cells) that is formed by cleavage of a zygote and immediately undergoes gastrulation. See **blastocyst**.

blood A type of connective tissue consisting of red blood cells and leukocytes suspended in a fluid portion called plasma.

blood pressure See **diastolic blood pressure** and **systolic blood pressure**.

body mass index (BMI) A mathematical relationship of weight and height used to assess obesity in humans. Calculated as weight (kg) divided by the square of height (m^2).

body plan The basic architecture of an animal's body, including the number and arrangement of limbs, body segments, and major tissue layers.

bog A wetland that has no or almost no water flow, resulting in very low oxygen levels and acidic conditions.

Bohr shift The rightward shift of the oxygen-hemoglobin dissociation curve that occurs with decreasing pH. Results in hemoglobin being more likely to release oxygen in the acidic environment of exercising muscle.

bone A type of vertebrate connective tissue consisting of living cells and blood vessels within a hard extracellular matrix composed of calcium phosphate ($CaPO_4$) and small amounts of calcium carbonate ($CaCO_3$) and protein fibers.

bone marrow The soft tissue filling the inside of long bones containing stem cells that develop into red blood cells and leukocytes throughout life.

Bowman's capsule The hollow, double-walled cup-shaped portion of a nephron that surrounds a glomerulus in the vertebrate kidney.

brain A large mass of neurons located in the head region of an animal, that is involved in information processing; may also be called the cerebral ganglion.

brain stem The most posterior portion of the vertebrate brain, connecting to the spinal cord and responsible for autonomic body functions such as heart rate, respiration, and digestion.

braincase See **cranium**.

branch (1) A part of a phylogenetic tree that represents populations through time. (2) Any extension of a plant's shoot system.

brassinosteroids A family of steroid hormones found in plants.

bronchiole One of the small tubes in mammalian lungs that carry air from the bronchi to the alveoli.

bronchus (plural: bronchi) In mammals, one of a pair of large tubes that lead from the trachea to each lung.

bryophytes Members of several phyla of green plants that lack vascular tissue including liverworts, hornworts, and mosses. Also called *non-vascular plants*.

budding Asexual reproduction via outgrowth from the parent that eventually breaks free as an independent individual; occurs in yeasts and some invertebrates.

buffer A substance that, in solution, acts to minimize changes in the pH of that solution when acid or base is added.

bulbourethral glands In male mammals, small paired glands at the base of the urethra that secrete an alkaline mucus (part of semen), which lubricates the tip of the penis and neutralizes acids in the urethra during copulation. In humans, also called *Cowper's glands*.

bulk flow The directional movement of a substantial volume of fluid due to pressure differences, such as movement of water through plant phloem and movement of blood in animals.

bundle-sheath cell A type of cell found around the vascular tissue (veins) of plant leaves.

C_3 photosynthesis The most common form of photosynthesis in which atmospheric CO_2 is used to form 3-phosphoglycerate, a three-carbon sugar.

C_4 photosynthesis A variant type of photosynthesis in which atmospheric CO_2 is first fixed into four-carbon sugars, rather than the three-carbon sugars of classic C_3 photosynthesis. Enhances photosynthetic efficiency in hot, dry environments, by reducing loss of oxygen due to photorespiration.

cadherin Any of a class of cell-surface proteins involved in cell adhesion and important for coordinating movements of cells during embryological development.

callus In plants, a mass of undifferentiated cells that can generate roots and other tissues necessary to create a mature plant.

Calvin cycle In photosynthesis, the set of light-independent reactions that use NADPH and ATP formed in the light-dependent reactions to drive the fixation of atmospheric CO_2 and reduction of the fixed carbon, ultimately producing sugars. Also called *carbon fixation* and *light-independent reactions.*

calyx All of the sepals of a flower.

CAM See **crassulacean acid metabolism**.

cambium (plural: cambia) See **lateral meristem**.

Cambrian explosion The rapid diversification of animal body types and lineages that occured between the species present in the Ediacaran faunas (565–542 mya) and the Cambrian faunas (525–515 mya).

camera eye The type of eye in vertebrates and cephalopods, consisting of a hollow chamber with a hole at one end (through which light enters) and a sheet of light-sensitive cells against the opposite wall.

cAMP See **cyclic AMP**.

cancer General term for any tumor whose cells grow in an uncontrolled fashion, invade nearby tissues, and spread to other sites in the body.

canopy The uppermost layers of branches in a forest (i.e., those fully exposed to the Sun).

CAP binding site A DNA sequence upstream of certain prokaryotic operons to which catabolite activator protein can bind, increasing gene transcription.

capillarity The tendency of water to move up a narrow tube due to surface tension, adhesion, and cohesion.

capillary One of the numerous small, thin-walled blood vessels that permeate all tissues and organs, and allow exchange of gases and other molecules between blood and body cells.

capillary bed A thick network of capillaries.

capsid A shell of protein enclosing the genome of a virus particle.

carapace In crustaceans, a large platelike section of the exoskeleton that covers and protects the cephalothorax (e.g., a crab's "shell").

carbohydrate Any of a class of molecules that contain a carbonyl group, several hydroxyl groups, and several to many carbon-hydrogen bonds. See **monosaccharide** and **polysaccharide**.

carbon cycle, global The worldwide movement of carbon among terrestrial ecosystems, the oceans, and the atmosphere.

carbon fixation See **Calvin cycle**.

carbonic anhydrase An enzyme that catalyzes the formation of carbonic acid (H_2CO_3) from carbon dioxide and water.

carboxylic acids Organic acids with the form R-COOH (a carboxyl group).

cardiac cycle One complete heartbeat cycle, including systole and diastole.

cardiac muscle The muscle tissue of the vertebrate heart. Consists of long branched fibers that are electrically connected and that initiate their own contractions; not under voluntary control. Compare with **skeletal** and **smooth muscle**.

carnivore (adjective: carnivorous) An animal whose diet consists predominantly of meat. Most members of the mammalian taxon Carnivora are carnivores. Some plants are carnivorous, trapping and killing small animals, then absorbing nutrients from the prey's body. Compare with **herbivore** and **omnivore**.

carotenoid Any of a class of accessory pigments, found in chloroplasts, that absorb wavelengths of light not absorbed by chlorophyll; typically appear yellow, orange, or red. Includes carotenes and xanthophylls.

carpel The female reproductive organ in a flower. Consists of the stigma, to which pollen grains adhere; the style, through which pollen grains move; and the ovary, which houses the ovule. Compare with **stamen**.

carrier A heterozygous individual carrying a normal allele and a recessive allele for an inherited trait; does not display the phenotype of the trait but can pass the recessive gene to offspring.

carrier protein A membrane protein that facilitates diffusion of a small molecule (e.g., glucose) across the plasma membrane by a process involving a reversible change in the shape of the protein. Also called *carrier* or *transporter.*

carrying capacity (*K*) The maximum population size of a certain species that a given habitat can support.

cartilage A type of vertebrate connective tissue that consists of relatively few cells scattered in a stiff matrix of polysaccharides and protein fibers.

Casparian strip In plant roots, a waxy layer containing suberin, a water-repellent substance, that prevents movement of water through the walls of endodermal cells, thus blocking the apoplastic pathway.

cast A type of fossil, formed when the decay of a body part leaves a void that is then filled with minerals that later harden.

catabolic pathway Any set of chemical reactions that breaks down larger, complex molecules into smaller ones, releasing energy in the process. Compare with **anabolic pathway**.

catabolite activator protein (CAP) A protein that can bind to the CAP binding site upstream of certain prokaryotic operons, facilitating binding of RNA polymerase and stimulating gene expression.

catabolite repression A type of positive transcriptional control in which the end product of a catabolic pathway inhibits further transcription of the gene encoding an enzyme early in the pathway.

catalysis (verb: catalyze) Acceleration of the rate of a chemical reaction due to a decrease in the free energy of the transition state, called the activation energy.

catalyst Any substance that increases the rate of a chemical reaction without itself undergoing any permanent chemical change.

catecholamines A class of small compounds, derived from the amino acid tyrosine, that are used as hormones or neurotransmitters. Include epinephrine, norepinephrine, and dopamine.

cation A positively charged ion.

cation exchange In botany, the release (displacement) of cations, such as magnesium and calcium from soil particles, by protons in acidic soil water. The released cations are available for uptake by plants.

CD4 A membrane protein on the surface of some T cells in humans. $CD4^+$ T cells can give rise to helper T cells.

CD8 A membrane protein on the surface of some T cells in humans. $CD8^+$ T cells can give rise to cytotoxic T cells.

Cdk See **cyclin-dependent kinase**.

cDNA See **complementary DNA**.

cDNA library A set of cDNAs from a particular cell type or stage of development. Each cDNA is carried by a plasmid or other cloning vector and can be separated from other cDNAs. Compare with **genomic library**.

cecum A blind sac between the small intestine and the colon. Is enlarged in some species (e.g., rabbits) that use it as a fermentation vat for digestion of cellulose.

cell A highly organized compartment bounded by a thin, flexible structure (plasma membrane) and containing concentrated chemicals in an aqueous (watery) solution. The basic structural and functional unit of all organisms.

cell body The part of a neuron that contains the nucleus and where incoming signals are integrated. Also called the *soma.*

cell crawling A form of cellular movement involving actin filaments in which the cell produces bulges (pseudopodia) that stick to the substrate and pull the cell forward. Also called *amoeboid motion.*

cell culture See **culture**.

cell cycle Ordered sequence of events in which a eukaryotic cell replicates its chromosomes, evenly partitions the chromosomes to two daughter cells, and then undergoes division of the cytoplasm.

cell-cycle checkpoint Any of several points in the cell cycle at which progression of a cell through the cycle can be regulated.

cell division Creation of new cells by division of pre-existing cells.

cell-mediated (immune) response The type of immune response that involves generation of cytotoxic T cells from $CD8^+$ T cells. Defends against pathogen-infected cells, cancer cells, and transplanted cells. Compare with **humoral (immune) response**.

cell membrane See **plasma membrane**.

cell plate A double layer of new plasma membrane that appears in the middle of a dividing plant cell; ultimately divides the cytoplasm into two separate cells.

cell sap An aqueous solution found in the vacuoles of plant cells.

cell theory The theory that all organisms are made of cells and that all cells come from preexisting cells.

cell wall A fibrous layer found outside the plasma membrane of most bacteria and archaea and many eukaryotes.

cellular respiration A common pathway for production of ATP, involving transfer of electrons from compounds with high potential energy (often NADH and $FADH_2$) to an electron transport chain and ultimately to an electron acceptor (often oxygen).

cellulose A structural polysaccharide composed of β-glucose monomers joined by β-1,4-glycosidic linkages. Found in the cell wall of algae, plants, bacteria, fungi, and some other groups.

Cenozoic era The most recent interval of geologic time, beginning 65.5 million years ago, during which mammals became the dominant vertebrates and angiosperms became the dominant plants.

central dogma The long-accepted hypothesis that information in cells flows in one direction: DNA codes for RNA, which codes for proteins. Exceptions are now known (e.g., retroviruses).

central nervous system (CNS) The brain and spinal cord of vertebrate animals. Compare with **peripheral nervous system (PNS)**.

centriole One of two small cylindrical structures, structurally similar to a basal body, found together within the centrosome near the nucleus of a eukaryotic cell.

centromere Constricted region of a replicated chromosome where the two sister chromatids are joined and the kinetochore is located.

centrosome Structure in animal and fungal cells, containing two centrioles, that serves as a microtubule-organizing center for the cell's cytoskeleton and for the spindle apparatus during cell division.

cephalization The formation of a distinct anterior region (the head) where sense organs and a mouth are clustered.

cephalochordates One of the three major chordate lineages (Cephalochordata), comprising small, mobile organisms that live in marine sands; also called *lancelets* or *amphioxi.* Compare with **urochordates** and **vertebrates**.

cephalopods A lineage of mollusks including the squid, octopuses, and nautiluses. Distinguished by large brains, excellent vision, tentacles, and a reduced or absent shell.

cerebellum Posterior section of the vertebrate brain that is involved in coordination of complex muscle movements, such as those required for locomotion and maintaining balance.

cerebrum The most anterior section of the vertebrate brain. Divided into left and right hemispheres and four lobes: parietal lobe, involved in complex decision making (in humans); occipital lobe, receives and interprets visual information; parietal lobe, involved in integrating sensory and motor functions; and temporal lobe, functions in memory, speech (in humans), and interpreting auditory information.

cervix The narrow passageway between the vagina and the uterus of female mammals.

chaetae (singular: chaeta) Bristle-like extensions found in some annelids.

channel A protein that forms a pore in a cell membrane. The structure of most channels allows them to admit just one or a few types of ions or molecules.

character displacement The tendency for the traits of similar species that occupy overlapping ranges to change in a way that reduces interspecific competition.

chelicerae A pair of clawlike appendages found around the mouth of certain arthropods called chelicerates (spiders, mites, and allies).

chemical bond An attractive force binding two atoms together. Covalent bonds, ionic bonds, and hydrogen bonds are types of chemical bonds.

chemical carcinogen Any chemical that can cause cancer.

chemical energy The potential energy stored in covalent bonds between atoms.

chemical equilibrium A dynamic but stable state of a reversible chemical reaction in which the forward reaction and reverse reactions proceed at the same rate, so that the concentrations of reactants and products remain constant.

chemical evolution The theory that simple chemical compounds in the ancient atmosphere and ocean combined via spontaneous chemical reactions to form larger, more complex substances, eventually leading to the origin of life and the start of biological evolution.

chemical reaction Any process in which one compound or element is combined with others or is broken down; involves the making and/or breaking of chemical bonds.

chemiosmosis An energetic coupling mechanism whereby energy stored in an electrochemical proton gradient (proton-motive force) is used to drive an energy-requiring process such as production of ATP.

chemokine Any of several chemical signals that attract leukocytes to a site of tissue injury or infection.

chemolithotroph An organism that produces ATP by oxidizing inorganic molecules with high potential energy such as ammonia (NH_3) or methane (CH_4). Also called *lithotroph*. Compare with **chemoorganotroph**.

chemoorganotroph An organism that produces ATP by oxidizing organic molecules with high potential energy such as sugars. Also called *organotroph*. Compare with **chemolithotroph**.

chemoreceptor A sensory cell or organ specialized for detection of specific molecules or classes of molecules.

chiasma (plural: chiasmata) The X-shaped structure formed during meiosis by crossing over between non-sister chromatids in a pair of homologous chromosomes.

chitin A structural polysaccharide composed of *N*-acetylglucosamine monomers joined end to end by β-1,4-glycosidic linkages. Found in cell walls of fungi and many algae, and in external skeletons of insects and crustaceans.

chitons A lineage of marine mollusks that have a protective shell formed of eight calcium carbonate plates.

chlorophyll Any of several closely related green pigments, found in chloroplasts and photosynthetic protists, that absorb light during photosynthesis.

chloroplast A chlorophyll-containing organelle, bounded by a double membrane, in which photosynthesis occurs; found in plants and photosynthetic protists. Also the location of amino acid, fatty acid, purine, and pyrimidine synthesis.

choanocyte A specialized flagellated feeding cell found in choanoflagellates (protists that are the closest living relatives of animals) and sponges (the oldest animal phylum).

cholecystokinin A peptide hormone secreted by cells in the lining of the small intestine. Stimulates the secretion of digestive enzymes from the pancreas and of bile from the liver and gallbladder.

chordates Members of the phylum Chordata, deuterostomes distinguished by a dorsal hollow nerve cord, pharyngeal gill slits, a notochord, a dorsal hollow nerve cord, and a post-anal tail. Include vertebrates, cephalochordata, and urochordata.

chromatid One of the two identical strands composing a replicated chromosome that is connected at the centromere to the other strand.

chromatin The complex of DNA and proteins, mainly histones, that compose eukaryotic chromosomes. Can be highly compact (heterochromatin) or loosely coiled (euchromatin).

chromatin remodeling The process by which the DNA in chromatin is unwound from its associated proteins to allow transcription or replication. May involve chemical modification of histone proteins or reshaping of the chromatin by large multi-protein complexes in an ATP-requiring process.

chromosome Gene-carrying structure consisting of a single long molecule of DNA and associated proteins (e.g., histones). Most prokaryotic cells contain a single, circular chromosome; eukaryotic cells contain multiple noncircular (linear) chromosomes located in the nucleus.

chromosome theory of inheritance The principle that genes are located on chromosomes and that patterns of inheritance are determined by the behavior of chromosomes during meiosis.

chylomicron A ball of protein-coated lipids used to transport the lipids through the bloodstream.

cilium (plural: cilia) One of many short, filamentous projections of some eukaryotic cells containing a core of microtubules. Used to move the cell and/or to move fluid or particles along a stationary cell. See **axoneme**.

circadian clock An internal mechanism found in most organisms that regulates many body processes (sleep-wake cycles, hormonal patterns, etc.) in a roughly 24-hour cycle.

circulatory system The system in animals responsible for moving oxygen, carbon dioxide, and other materials (hormones, nutrients, wastes) around the body.

cisternae (singular: cisterna) Flattened, membrane-bound compartments that make up the Golgi apparatus.

citric acid cycle A series of eight chemical reactions that starts with citrate (citric acid, when protonated) and ends with oxaloacetate, which reacts with acetyl CoA to form citrate—forming a cycle that is part of the pathway that oxidizes glucose to CO_2. Also known as the *Krebs cycle*, *tricarboxylic acid cycle*, and *TCA cycle*.

clade See **monophyletic group**.

cladistic approach A method for constructing a phylogenetic tree that is based on identifying the unique traits of each monophyletic group. Compare with **phenetic approach**.

Class I MHC protein An MHC protein that is present on the plasma membrane of virtually all nucleated cells and functions in presenting antigen to $CD8^+$ T cells.

Class II MHC protein An MHC protein that is present only on the plasma membrane of dendritic cells, macrophages, and B cells and functions in presenting antigen to $CD4^+$ T cells.

cleavage In animal development, the series of rapid mitotic cell divisions, with little cell growth, that produces successively smaller cells and transforms a zygote into a multicellular blastula, or blastocyst in mammals.

cleavage furrow A pinching-in of the plasma membrane that occurs as cytokinesis begins in animal cells and deepens until the cytoplasm is divided.

climate The prevailing long-term weather conditions in a particular region.

climax community The stable, final community that develops from ecological succession.

clitoris A small rod of erectile tissue in the external genitalia of female mammals. Is formed from the same embryonic tissue as the male penis and has a similar function in sexual arousal.

cloaca An opening to the outside used by the excretory and reproductive systems in a few mammals and many nonmammalian vertebrates.

clonal-selection theory The dominant explanation of the development of adaptive immunity in vertebrates. According to the theory, the immune system retains a vast pool of inactive lymphocytes, each with a unique receptor for a unique antigen. Lymphocytes that encounter their antigens are stimulated to divide (selected and cloned), producing daughter cells that combat infection and confer immunity.

clone (1) An individual that is genetically identical to another individual. (2) A lineage of genetically identical individuals or cells. (3) As a verb, to make one or more genetic replicas of a cell or individual.

cloning vector A plasmid or other agent used to transfer recombinant genes into cultured host cells. Also called simply *vector*.

closed circulatory system A circulatory system in which the circulating fluid (blood) is confined to blood vessels and flows in a continuous circuit. Compare with **open circulatory system**.

club mosses Common name for species in the land plant lineage Lycophyta.

cnidocyte A specialized stinging cell found in cnidarians (e.g., jellyfish, corals, and anemones) that is used in capturing prey.

cochlea The organ of hearing in the inner ear of mammals, birds, and crocodilians. A coiled, fluid-filled tube containing specialized pressure-sensing neurons (hair cells) that detect sounds of different pitches.

coding strand See **non-template strand**.

codominance An inheritance pattern in which heterozygotes exhibit both of the traits seen in either kind of homozygous individual.

codon A sequence of three nucleotides in DNA or RNA that codes for a certain amino acid or that initiates or terminates protein synthesis.

coefficient of relatedness (*r*) A measure of how closely two individuals are related. Calculated as the probability that an allele in two individuals is inherited from the same ancestor.

coelom An internal, usually fluid-filled, body cavity that is lined with mesoderm.

coelomate An animal that has a true coelom. Compare with **acoelomate** and **pseudocoelomate**.

coenocytic Containing many nuclei and a continuous cytoplasm through a filamentous body, without the body being divided into distinct cells. Some fungi are coenocytic.

coenzyme A small organic molecule that is a required cofactor for an enzyme-catalyzed reaction. Often donates or receives electrons or functional groups during the reaction.

coenzyme A (CoA) A nonprotein molecule that is required for many cellular reactions involving transfer of acetyl groups ($-COCH_3$).

coenzyme Q A nonprotein molecule that shuttles electrons between membrane-bound complexes in the mitochondrial electron transport chain. Also called *ubiquinone* or *Q*.

coevolution A pattern of evolution in which two interacting species reciprocally influence each other's adaptations over time.

coevolutionary arms race A series of adaptations and counter-adaptations observed in species that interact closely over time and affect each other's fitness.

cofactor A metal ion or small organic compound that is required for an enzyme to function normally. May be bound tightly to an enzyme or associate with it transiently during catalysis.

cohesion The tendency of certain like molecules (e.g., water molecules) to cling together due to attractive forces. Compare with **adhesion**.

cohesion-tension theory The theory that water movement upward through plant vascular tissues is due to loss of water from leaves (transpiration), which pulls a cohesive column of water upward.

cohort A group of individuals that are the same age and can be followed through time.

coleoptile A modified leaf that covers and protects the stems and leaves of young grasses.

collagen A fibrous, pliable, cable-like glycoprotein that is a major component of the extracellular matrix of animal cells. Various subtypes differ in their tissue distribution.

collecting duct In the vertebrate kidney, a large straight tube that receives filtrate from the distal tubules of several nephrons. Involved in the regulated reabsorption of water.

collenchyma cell In plants, an elongated cell with cell walls thickened at the corners that provides support to growing plant parts; usually found in strands along leaf veins and stalks. Compare with **parenchyma cell** and **sclerenchyma cell**.

colon The portion of the large intestine where feces are formed by compaction of wastes and reabsorption of water.

colony An assemblage of individuals. May refer to an assemblage of semi-independent cells or to a breeding population of multicellular organisms.

commensalism (adjective: commensal) A symbiotic relationship in which one organism (the commensal) benefits and the other (the host) is not harmed. Compare with **mutualism** and **parasitism**.

communication In ecology, any process in which a signal from one individual modifies the behavior of another individual.

community All of the species that interact with each other in a certain area.

companion cell In plants, a cell in the phloem that is connected via numerous plasmodesmata to adjacent sieve-tube members. Companion cells provide materials to maintain sieve-tube members and function in the loading and unloading of sugars into sieve-tube members.

compass orientation A type of navigation in which movement occurs in a specific direction.

competition In ecology, the interaction of two species or two individuals trying to use the same limited resource (e.g., water, food, living space). May occur between individuals of the same species (intraspecific competition) or different species (interspecific competition).

competitive exclusion principle The principle that two species cannot coexist in the same ecological niche in the same area because one species will out-compete the other.

competitive inhibition Inhibition of an enzyme's ability to catalyze a chemical reaction via a nonreactant molecule that competes with the substrate(s) for access to the active site.

complement system A set of proteins that circulate in the bloodstream and can form holes in the plasma membrane of bacteria, leading to their destruction.

complementary base pairing The association between specific nitrogenous bases of nucleic acids stabilized by hydrogen bonding. Adenine pairs only with thymine (in DNA) or uracil (in RNA), and guanine pairs only with cytosine.

complementary DNA (cDNA) DNA produced in the laboratory using an RNA transcript as a template and reverse transcriptase; corresponds to a gene but lacks introns. Also produced naturally by retroviruses.

complementary strand A newly synthesized strand of RNA or DNA that has a base sequence complementary to that of the template strand.

complete digestive tract A digestive tract with two openings, usually called a mouth and an anus.

complete metamorphosis See **holometabolous metamorphosis**.

compound eye An eye formed of many independent light-sensing columns (ommatidia); occurs in arthropods. Compare with **simple eye**.

concentration gradient Difference across space (e.g., across a membrane) in the concentration of a dissolved substance.

condensation reaction A chemical reaction in which two molecules are joined covalently with the removal of an —OH from one and an —H from another to form water. Also called a *dehydration reaction*. Compare with **hydrolysis**.

conduction (1) Direct transfer of heat between two objects that are in physical contact. Compare with **convection**. (2) Transmission of an electrical impulse along the axon of a nerve cell.

cone cell A photoreceptor cell with a cone-shaped outer portion that is particularly sensitive to bright light of a certain color. Also called simply *cone*. Compare with **rod cell**.

connective tissue An animal tissue consisting of scattered cells in a liquid, jellylike, or solid extracellular matrix. Includes bone, cartilage, tendons, ligaments, and blood.

conservation biology The effort to study, preserve, and restore threatened populations, communities, and ecosystems.

constant (C) region The portion of an antibody's light chains or heavy chains that has the same amino acid sequence in the antibodies produced by every B cell of an individual. Compare with **variable (V) region**.

constitutive Always occurring; always present. Commonly used to describe enzymes and other proteins that are synthesized continuously or mutants in which one or more genetic loci are constantly expressed due to defects in gene control.

constitutive defense A defensive trait that is always manifested even in the absence of a predator or pathogen. Also called *standing defense*. Compare with **inducible defense**.

constitutive mutant An abnormal (mutated) gene that produces a product at all times, instead of under certain conditions only.

consumer See **heterotroph**.

consumption Predation or herbivory.

continental shelf The portion of a geologic plate that extends from a continent to under seawater.

continuous strand See **leading strand**.

control In a scientific experiment, a group of organisms or samples that do not receive the experimental treatment but are otherwise identical to the group that does.

convection Transfer of heat by movement of large volumes of a gas or liquid. Compare with **conduction**.

convergent evolution The independent evolution of analogous traits in distantly related organisms due to adaptation to similar environments and a similar way of life.

cooperative binding The tendency of the protein subunits of hemoglobin to affect each other's oxygen binding such that each bound oxygen molecule increases the likelihood of further oxygen binding.

coprophagy The eating of feces.

copulation The act of transferring sperm from a male directly into a female's reproductive tract.

coral reef A large assemblage of colonial marine corals that usually serves as shallow-water, sunlit habitat for many other species as well.

core enzyme The enzyme responsible for catalysis in a multi-part holoenzyme.

co-receptor Any membrane protein that acts with some other membrane protein in a cell interaction or cell response.

cork cambium One of two types of lateral meristem, consisting of a ring of undifferentiated plant cells found just under the cork layer of woody plants; produces new cork cells on its outer side. Compare with **vascular cambium**.

cork cell A waxy cell in the protective outermost layer of a woody plant.

corm A rounded, thick underground stem that can produce new plants via asexual reproduction.

cornea The transparent sheet of connective tissue at the very front of the eye in vertebrates and some other animals. Protects the eye and helps focus light.

corolla All of the petals of a flower.

corona The cluster of cilia at the anterior end of a rotifer.

corpus callosum A thick band of neurons that connects the two hemispheres of the cerebrum in the mammalian brain.

corpus luteum A yellowish structure in an ovary that secretes progesterone. Is formed from a follicle that has recently ovulated.

cortex (1) The outermost region of an organ, such as the kidney or adrenal gland. (2) In plants, a layer of ground tissue found outside the vascular bundles and pith of a plant stem.

cortical granules Small enzyme-filled vesicles in the cortex of an egg cell. Involved in formation of the fertilization envelope after fertilization.

corticotropin-releasing hormone (CRH) A peptide hormone, produced and secreted by the hypothalamus, that stimulates the anterior pituitary to release ACTH.

cortisol A steroid hormone, produced and secreted by the adrenal cortex, that increases blood glucose and prepares the body for stress. The major glucocorticoid hormone in some mammals. Also called *hydrocortisone*.

cost-benefit analysis Decisions or analyses that weigh the fitness costs and benefits of a particular action.

cotransport Transport of an ion or molecule against its electrochemical gradient, in company with an ion or molecule being transported with its electrochemical gradient. Also called *secondary active transport*.

cotransporter A transmembrane protein that facilitates diffusion of an ion down its previously established electrochemical gradient and uses the energy of that process to transport some other substance, in the same or opposite direction, *against* its concentration gradient. Also called *secondary active transporter*. See **antiporter** and **symporter**.

cotyledon The first leaf, or seed leaf, of a plant embryo. Used for storing and digesting nutrients and/or for early photosynthesis.

countercurrent exchanger In animals, any anatomical arrangement that allows the maximum transfer of heat or a soluble substance from one fluid to another. The two fluids must be flowing in opposite directions and have a heat or concentration gradient between them.

covalent bond A type of chemical bond in which two atoms share one or more pairs of electrons. Compare with **hydrogen bond** and **ionic bond**.

cranium A bony, cartilaginous, or fibrous case that encloses and protects the brain of vertebrates. Forms part of the skull. Also called *braincase*.

crassulacean acid metabolism (CAM) A variant type of photosynthesis in which CO_2 is stored in organic acids at night when stomata are open and then released to feed the Calvin cycle during the day when stomata are closed. Helps reduce water loss and oxygen loss by photorespiration in hot, dry environments.

cristae (singular: crista) Sac-like invaginations of the inner membrane of a mitochondrion. Location of the electron transport chain and ATP synthase.

Cro-Magnon A prehistoric European population of modern humans (*Homo sapiens*) known from fossils, paintings, sculptures, and other artifacts.

crop A storage organ in the digestive systems of certain vertebrates.

cross-fertilization A mating that combines gametes from different individuals, as opposed to combining gametes from the same individual (self-fertilization).

crossing over The exchange of segments of non-sister chromatids between a pair of homologous chromosomes that occurs during meiosis I.

cross-pollination Pollination of a flower by pollen from another individual, rather than by self-fertilization. Also called *crossing*.

cross-talk Interactions among signaling pathways that modify a cellular response.

crustaceans A lineage of arthropods that includes shrimp, lobster, and crabs. Many have a carapace (a platelike portion of the exoskeleton) and mandibles for biting or chewing.

cryptic species A species that cannot be distinguished from other species by easily identifiable morphological traits.

culture In cell biology, a collection of cells or a tissue growing under controlled conditions, usually in suspension or on the surface of a dish on solid growth medium.

cup fungus See **sac fungus**.

Cushing's disease A human endocrine disorder caused by loss of feedback inhibition of cortisol on ACTH secretion. Characterized by high ACTH and cortisol levels and wasting of body protein reserves.

cuticle A protective coating secreted by the outermost layer of cells of an animal or a plant.

cyanobacteria A lineage of photosynthetic bacteria formerly known as blue-green algae. Likely the first life-forms to carry out oxygenic photosynthesis.

cyclic AMP (cAMP) Cyclic adenosine monophosphate; a small molecule, derived from ATP, that is widely used by cells in signal transduction and transcriptional control.

cyclic photophosphorylation Path of electron flow during the light-dependent reactions of photosynthesis in which photosystem I transfers excited electrons back to the electron transport chain of photosystem II, rather than to $NADP^+$. Also called *cyclic electron flow*. Compare with **Z scheme**.

cyclin One of several regulatory proteins whose concentrations fluctuate cyclically throughout the cell cycle.

cyclin-dependent kinase (Cdk) Any of several related protein kinases that are active only when bound to a cyclin. Involved in control of the cell cycle.

cytochrome *c* (cyt *c*) A soluble iron-containing protein that shuttles electrons between membrane-bound complexes in the mitochondrial electron transport chain.

cytokines A diverse group of autocrine signaling proteins, secreted largely by cells of the immune system, whose effects include stimulating leukocyte production, tissue repair, and fever. Generally function to regulate the intensity and duration of an immune response.

cytokinesis Division of the cytoplasm to form two daughter cells. Typically occurs immediately after division of the nucleus by mitosis or meiosis.

cytokinins A class of plant hormones that stimulate cell division and retard aging.

cytoplasm All of the contents of a cell, excluding the nucleus, bounded by the plasma membrane.

cytoplasmic determinant A regulatory transcription factor or signaling molecule that is distributed unevenly in the cytoplasm of the egg cells of many animals and that directs early pattern formation in an embryo.

cytoplasmic streaming The directed flow of cytosol and organelles that facilitates distribution of materials within some large plant and fungal cells. Occurs along actin filaments and is powered by myosin.

cytoskeleton In eukaryotic cells, a network of protein fibers in the cytoplasm that are involved in cell shape, support, locomotion, and transport of materials within the cell. Prokaryotic cells have a similar but much less extensive network of fibers.

cytosol The fluid portion of the cytoplasm.

cytotoxic T cell An effector T cell that destroys infected cells and cancer cells. Is descended from an activated $CD8^+$ T cell that has interacted with antigen on an infected cell or cancer cell. Also called *cytotoxic T lymphocyte (CTL)* and *killer T cell*. Compare with **helper T cell**.

dalton (Da) A unit of mass equal to 1/12 the mass of one carbon-12 atom; about the mass of 1 proton or 1 neutron.

day-neutral plant A plant whose flowering time is not affected by the relative length of day and night (the photoperiod). Compare with **long-day** and **short-day plant**.

dead space Portions of the air passages that are not involved in gas exchange with the blood, such as the trachea and bronchi.

deciduous Describing a plant that sheds leaves or other structures at regular intervals (e.g., each fall)

decomposer See **detritivore**.

decomposer food chain An ecological network of detritus, decomposers that eat detritus, and predators and parasites of the decomposers.

definitive host The host species in which a parasite reproduces sexually. Compare with **intermediate host**.

dehydration reaction See **condensation reaction**.

deleterious In genetics, referring to any mutation, allele, or trait that reduces an individual's fitness.

demography The study of factors that determine the size and structure of populations through time.

denaturation (verb: denature) For a macromolecule, loss of its three-dimensional structure and biological activity due to breakage of hydrogen bonds and disulfide bonds, usually caused by treatment with excess heat or extreme pH conditions.

dendrite A short extension from a neuron's cell body that receives signals from other neurons.

dendritic cell A type of leukocyte that ingests and digests foreign antigens, moves to a lymph node, and presents the antigens displayed on its membrane to $CD4^+$ T cells.

dense connective tissue A type of connective tissue, distinguished by having an extracellular matrix dominated by collagen fibers.

density dependent In population ecology, referring to any characteristic that varies depending on population density.

density independent In population ecology, referring to any characteristic that does not vary with population density.

deoxyribonucleic acid (DNA) A nucleic acid composed of deoxyribonucleotides that carries the genetic information of a cell. Generally occurs as two intertwined strands, but these can be separated. See **double helix**.

deoxyribonucleoside triphosphate (dNTP) A monomer that can be polymerized to form DNA. Consists of deoxyribose, a base (A, T, G, or C), and three phosphate groups; similar to a nucleotide, but with two more phosphate groups.

deoxyribonucleotide See **nucleotide**.

depolarization Change in membrane potential from its resting negative state to a less negative or a positive state; a normal phase in an action potential. Compare with **hyperpolarization**.

deposit feeder An animal that eats its way through a food-containing substrate.

derived trait A trait that is clearly homologous with a trait found in an ancestor, but which has a new form.

dermal tissue system The tissue forming the outer layer of an organism. In plants, also called *epidermis;* in animals, forms two distinct layers: *dermis* and *epidermis*.

descent with modification The phrase used by Darwin to describe his hypothesis of evolution by natural selection.

desmosome A type of cell-cell attachment structure, consisting of cadherin proteins, that binds the cytoskeletons of adjacent animal cells together. Found where cells are strongly attached to each other. Compare with **gap junction** and **tight junction**.

detergent A type of small amphipathic molecule used to solubilize hydrophobic molecules in aqueous solution.

determination In embryogenesis, progressive changes in a cell that commit it to a particular cell fate. Once a cell is fully determined, it can differentiate only into a particular cell type (e.g., liver cell, brain cell).

detritivore An organism whose diet consists mainly of dead organic matter (detritus). Various bacteria, fungi, and protists are detritivores. Also called *decomposer*.

detritus A layer of dead organic matter that accumulates at ground level or on seafloors and lake bottoms.

deuterostomes A major lineage of animals that share a pattern of embryological development, including formation of the anus earlier than the mouth, and formation of the coelom by pinching off of layers of mesoderm from the gut. Includes echinoderms and chordates. Compare with **protostomes**.

developmental homology A similarity in embryonic form, or in the fate of embryonic tissues, that is due to inheritance from a common ancestor.

diabetes insipidus A human disease caused by defects in the kidney's system for conserving water. Characterized by production of large amounts of dilute urine.

diabetes mellitus A human disease caused by defects in insulin production (type I) or the response of cells to insulin (type II). Characterized by abnormally high blood glucose levels and large amounts of glucose-containing urine.

diaphragm An elastic, sheetlike structure. In mammals, the muscular sheet of tissue that separates the chest and abdominal cavities. Contracts and moves downward during inhalation, expanding the chest cavity.

diastole The portion of the heartbeat cycle during which the atria or ventricles of the heart are relaxed. Compare with **systole**.

diastolic blood pressure The force exerted by blood against artery walls during relaxation of the heart's left ventricle. Compare with **systolic blood pressure**.

dicot Any plant that has two cotyledons (embryonic leaves) upon germination. The dicots do not form a monophyletic group. Also called *dicotyledonous plant*. Compare with **eudicot** and **monocot**.

dideoxy sequencing A laboratory technique for determining the exact nucleotide sequence of DNA. Relies on the use of dideoxynucleotide triphosphates (ddNTPs), which terminate DNA replication.

diencephalon The part of the mammalian brain that relays sensory information to the cerebellum and functions in maintaining homeostasis.

differential centrifugation Procedure for separating cellular components according to their size and density by spinning a cell homogenate in a series of centrifuge runs. After each run, the supernatant is removed from the deposited material (pellet) and spun again at progressively higher speeds.

differential gene expression Expression of different sets of genes in cells with the same genome. Responsible for creating different cell types.

differentiation The process by which a relatively unspecialized cell becomes a distinct specialized cell type (e.g., liver cell, brain cell) usually by changes in gene expression. Also called *cell differentiation.*

diffusion Spontaneous movement of a substance from a region of high concentration to one of low concentration (i.e., down a concentration gradient).

digestion The physical and chemical breakdown of food into molecules that can be absorbed into the body of an animal.

digestive tract The long tube that begins at the mouth and ends at the anus. Also called *alimentary canal, gastrointestinal (GI) tract,* or the *gut.*

dihybrid cross A mating between two parents that are heterozygous for both of the two genes being studied.

dikaryotic Having two nuclei.

dimer An association of two molecules, which may be identical (homodimer) or different (heterodimer).

dioecious Describing an angiosperm species that has male and female reproductive structures on separate plants. Compare with **monoecious**.

diploblast (adjective: diploblastic) An animal whose body develops from two basic embryonic cell layers—ectoderm and endoderm. Compare with **triploblast**.

diploid (1) Having two sets of chromosomes ($2n$). (2) A cell or an individual organism with two sets of chromosomes, one set inherited from the maternal parent and one set from the paternal parent. Compare with **haploid**.

direct sequencing A technique for identifying and studying microorganisms that cannot be grown in culture. Involves detecting and amplifying copies of certain specific genes in their DNA, sequencing these genes, and then comparing the sequences with the known sequences from other organisms.

directional selection A pattern of natural selection that favors one extreme phenotype with the result that the average phenotype of a population changes in one direction. Generally reduces overall genetic variation in a population.

disaccharide A carbohydrate consisting of two monosaccharides (sugar residues) linked together.

discontinuous strand See **lagging strand**.

discrete trait An inherited trait that exhibits distinct phenotypic forms rather than the continuous variation characteristic of a quantitative trait such as body height.

dispersal The movement of individuals from their place of origin (birth, hatching) to a new location.

disruptive selection A pattern of natural selection that favors extreme phenotypes at both ends of the range of phenotypic variation. Maintains overall genetic variation in a population. Compare with **stabilizing selection**.

distal tubule In the vertebrate kidney, the convoluted portion of a nephron into which filtrate moves from the loop of Henle. Involved in the regulated reabsorption of sodium and water. Compare with **proximal tubule**.

disturbance In ecology, any event that disrupts a community, usually causing loss of some individuals or biomass from it.

disturbance regime The characteristic disturbances that affect a given ecological community.

disulfide bond A covalent bond between two sulfur atoms, typically in the side groups of some amino acids (e.g., cysteine). Often contributes to tertiary structure of proteins.

DNA See **deoxyribonucleic acid**.

DNA cloning Any of several techniques for producing many identical copies of a particular gene or other DNA sequence.

DNA fingerprinting Any of several methods for identifying individuals by unique features of their genomes. Commonly involves using PCR to produce many copies of certain simple sequence repeats (microsatellites) and then analyzing their lengths.

DNA library See **cDNA library** and **genomic library**.

DNA ligase An enzyme that joins pieces of DNA by catalyzing formation of a phosphodiester bond between the pieces.

DNA microarray A set of single-stranded DNA fragments, representing thousands of different genes, that are permanently fixed to a small glass slide. Can be used to determine which genes are expressed in different cell types, under different conditions, or at different developmental stages.

DNA polymerase Any enzyme that catalyzes synthesis of DNA from deoxyribonucleotides.

domain (1) A section of a protein that has a distinctive tertiary structure and function. (2) A taxonomic category, based on similarities in basic cellular biochemistry, above the kingdom level. The three recognized domains are Bacteria, Archaea, and Eukarya.

dominant Referring to an allele that determines the phenotype of a heterozygous individual. Compare with **recessive**.

dopamine A catecholamine neurotransmitter that functions mainly in a part of the mammalian brain involved with muscle control. Also functions as a hypothalamic inhibitory hormone that inhibits release of prolactin from the interior pituitary; also called *prolactin-inhibiting hormone (PIH).*

dormancy A temporary state of greatly reduced, or no, metabolic activity and growth in plants or plant parts (e.g., seeds, spores, bulbs, and buds).

dorsal Toward an animal's back and away from its belly. The opposite of ventral.

double fertilization An unusual form of reproduction seen in flowering plants, in which one sperm nucleus fuses with an egg to form a zygote and the other sperm nucleus fuses with two polar nuclei to form the triploid endosperm.

double helix The secondary structure of DNA, consisting of two antiparallel DNA strands wound around each other.

Down syndrome A human developmental disorder caused by trisomy of chromosome 21.

downstream In genetics, the direction in which RNA polymerase moves along a DNA strand. Compare with **upstream**.

dynein Any one of a class of motor proteins that use the chemical energy of ATP to "walk" along an adjacent microtubule. Dyneins are responsible for bending of cilia and flagella, play a role in chromosome movement during mitosis, and can transport certain organelles.

early endosome A small membrane-bound vesicle, formed by endocytosis, that is an early stage in the formation of a lysosome.

ecdysone An insect hormone that triggers either molting (to a larger larval form) or metamorphosis (to the adult form), depending on the level of juvenile hormone.

ecdysozoans A major lineage of protostomes (Ecdysozoa) that grow by shedding their external skeletons (molting) and expanding their bodies. Includes arthropods, insects, crustaceans, nematodes, and centipedes. Compare with **lophotrochozoans**.

echinoderms A major lineage of deuterostomes (Echinodermata) distinguished by adult bodies with five-sided radial symmetry, a water vascular system, and tube feet. Includes sea urchins, sand dollars, and sea stars.

echolocation The use of echoes from vocalizations to obtain information about locations of objects in the environment.

ecology The study of how organisms interact with each other and with their surrounding environment.

ecosystem All the organisms that live in a geographic area, together with the nonliving (abiotic) components that affect or exchange materials with the organisms; a community and its physical environment.

ecosystem diversity The variety of biotic components in a region along with abiotic components, such as soil, water, and nutrients.

ecosystem services Alterations of the physical components of an ecosystem by living organisms, especially beneficial changes in the quality of the atmosphere, soil, water, etc.

ecotourism Tourism that is based on observing wildlife or experiencing other aspects of natural areas.

ectoderm The outermost of the three basic cell layers in most animal embryos; gives rise to the outer covering and nervous system. Compare with **endoderm** and **mesoderm**.

ectomycorrhizal fungi (EMF) Fungi whose hyphae form a dense network that covers their host plant's roots but do not enter the root cells.

ectoparasite A parasite that lives on the outer surface of the host's body.

ectotherm An animal that does not use internally generated heat to regulate its body temperature. Compare with **endotherm**.

effector Any cell, organ, or structure with which an animal can respond to external or internal stimuli. Usually functions, along with a sensor and integrator, as part of a homeostatic system.

efferent division The part of the nervous system, consisting primarily of motor neurons, that carries commands from the central nervous system to the body.

egg A mature female gamete and any associated external layers (such as a shell). Larger and less mobile than the male gamete. In animals, also called *ovum.*

ejaculation The release of semen from the copulatory organ of a male animal.

ejaculatory duct A short duct connecting the vas deferens to the urethra, through which sperm move during ejaculation.

elastic Referring to a structure (e.g., lungs) with the ability to stretch and then spring back to its original shape.

electric current A flow of electrical charge past a point. Also called *current.*

electrical potential Potential energy created by a separation of electric charges between two points. Also called *voltage.*

electrocardiogram (EKG) A recording of the electrical activity of the heart, as measured through electrodes on the skin.

electrochemical gradient The combined effect of an ion's concentration gradient and electrical (charge) gradient across a membrane that affects the diffusion of ions across the membrane.

electrolyte Any compound that dissociates into ions when dissolved in water. In nutrition, refers to the major ions necessary for normal cell function.

electromagnetic spectrum The entire range of wavelengths of radiation extending from short wavelengths (high energy) to long wavelengths (low energy). Includes gamma rays, X-rays, ultraviolet, visible light, infrared, microwaves, and radio waves (from short to long wavelengths).

electron acceptor A reactant that gains an electron and is reduced in a reduction-oxidation reaction.

electron carrier Any molecule that readily accepts electrons from and donates electrons to other molecules.

electron donor A reactant that loses an electron and is oxidized in a reduction-oxidation reaction.

electron microscope See **scanning electron microscope** and **transmission electron microscope**.

electron shell A group of orbitals of electrons with similar energies. Electron shells are arranged in roughly concentric layers around the nucleus of an atom, with electrons in outer shells having more energy than those in inner shells. Electrons in the outermost shell, the valence shell, often are involved in chemical bonding.

electron transport chain (ETC) Any set of membrane-bound protein complexes and smaller soluble electron carriers involved in a coordinated series of redox reactions in which the potential energy of electrons transferred from reduced donors is successively decreased and used to pump protons from one side of a membrane to the other.

electronegativity A measure of the ability of an atom to attract electrons toward itself from an atom to which it is bonded.

electroreceptor A sensory cell or organ specialized to detect electric fields.

element A substance, consisting of atoms with a specific number of protons, that cannot be separated into or broken down to any other substance. Elements preserve their identity in chemical reactions.

elongation (1) The process by which messenger RNA lengthens during transcription. (2) The process by which a polypeptide chain lengthens during translation.

elongation factors Proteins involved in the elongation phase of translation, assisting ribosomes in the synthesis of the growing peptide chain.

embryo A young developing organism; the stage after fertilization and zygote formation.

embryo sac The female gametophyte in flowering plants that exhibit alternation of generations.

embryogenesis The process by which a single-celled zygote becomes a multicellular embryo.

Embryophyta An increasing popular name for the lineage called land plants.

embryophyte A plant that nourishes its embryos inside its own body. All land plants are embryophytes.

emergent vegetation Any plants in an aquatic habitat that extend above the surface of the water.

emerging disease Any infectious disease, often a viral disease, that suddenly afflicts significant numbers of humans for the first time; often due to changes in the host species for a pathogen or host population movements.

emigration The migration of individuals away from one population to other populations. Compare with **immigration**.

emulsification (verb: emulsify) The dispersion of fat into an aqueous solution. Usually requires the aid of an amphipathic substance such as a detergent or bile salts, which can break large fat globules into microscopic fat droplets.

endangered species A species whose numbers have decreased so much that it is in danger of extinction throughout all or part of its range.

endemic species A species that lives in one geographic area and nowhere else.

endergonic Referring to a chemical reaction that requires an input of energy to occur and for which the Gibbs free-energy change (ΔG) > 0. Compare with **exergonic**.

endocrine Relating to a chemical signal (hormone) that is released into the bloodstream by a producing cell and acts on a distant target cell.

endocrine system All of the glands and tissues that produce and secrete hormones into the bloodstream.

endocrine gland A gland that secretes hormones directly into the bloodstream or interstitial fluid instead of into ducts. Compare with **exocrine gland**.

endocytosis General term for any pinching off of the plasma membrane that results in the uptake of material from outside the cell. Includes phagocytosis, pinocytosis, and receptor-mediated endocytosis. Compare with **exocytosis**.

endoderm The innermost of the three basic cell layers in most animal embryos; gives rise to the digestive tract and organs that connect to it (liver, lungs, etc.). Compare with **ectoderm** and **mesoderm**.

endodermis In plant roots, a cylindrical layer of cells that separates the cortex from the vascular tissue.

endomembrane system A system of organelles in eukaryotic cells that performs most protein and lipid synthesis. Includes the endoplasmic reticulum (ER), Golgi apparatus, and lysosomes.

endomycorrhizal fungi See **arbuscular mycorrhizal fungi (AMF)**.

endoparasite A parasite that lives inside the host's body.

endophyte (adjective: endophytic) A fungus that lives inside the aboveground parts of a plant in a symbiotic relationship. Compare with **epiphyte**.

endoplasmic reticulum (ER) A network of interconnected membranous sacs and tubules found inside eukaryotic cells. See **rough** and **smooth endoplasmic reticulum**.

endoskeleton Bony and/or cartilaginous structures within the body that provide support. Examples are the spicules of sponges, the plates in echinoderms, and the bony skeleton of vertebrates. Compare with **exoskeleton**.

endosome See **early** and **late endosome**.

endosperm A triploid ($3n$) tissue in the seed of a flowering plant (angiosperm) that serves as food for the plant embryo. Functionally analogous to the yolk in some animal eggs.

endosymbiont An organism that lives in a symbiotic relationship inside the body of its host.

endosymbiosis An association between species in which one lives inside the cell or cells of the other.

endosymbiosis theory The theory that mitochondria and chloroplasts evolved from prokaryotes that were engulfed by host cells and took up a symbiotic existence within those cells, a process termed primary endosymbiosis. In some eukaryotes, chloroplasts originated by secondary endosymbiosis, that is, by engulfing a chloroplast-containing protist and retaining its chloroplasts.

endotherm An animal whose primary source of body heat is internally generated heat. Compare with **ectotherm**.

endothermic Referring to a chemical reaction that absorbs heat. Compare with **exothermic**.

energetic coupling In cellular metabolism, the mechanism by which energy released from an exergonic reaction (commonly, hydrolysis of ATP) is used to drive an endergonic reaction.

energy The capacity to do work or to supply heat. May be stored (potential energy) or available in the form of motion (kinetic energy).

enhancer A regulatory sequence in eukaryotic DNA that may be located far from the gene it controls or within introns of the gene. Binding of specific proteins to an enhancer enhances the transcription of certain genes.

enrichment culture A method of detecting and obtaining cells with specific characteristics by placing a sample, containing many types of cells, under a specific set of conditions (e.g., temperature, salt concentration, available nutrients) and isolating those cells that grow rapidly in response.

entropy (*S*) A quantitative measure of the amount of disorder of any system, such as a group of molecules.

envelope, viral A membrane-like covering that encloses some viruses and their capsid coats, shielding them from attack by the host's immune system.

environmental sequencing The inventory of all the genes in a community or ecosystem by sequencing, analyzing, and comparing the genomes of the component organisms.

enzyme A protein catalyst used by living organisms to speed up and control biological reactions.

epicotyl In some embryonic plants, a portion of the embryonic stem that extends above the cotyledons.

epidemic The spread of an infectious disease throughout a population in a short time period. Compare with **pandemic**.

epidermis The outermost layer of cells of any multicellular organism.

epididymis A coiled tube wrapped around the testis in reptiles, birds, and mammals. The site of the final stages of sperm maturation and storage.

epigenetic inheritance Pattern of inheritance involving differences in phenotype that are not due to changes in the nucleotide sequence of genes.

epinephrine A catecholamine hormone, produced and secreted by the adrenal medulla, that triggers rapid responses relating to the fight-or-flight response. Also called *adrenaline.*

epiphyte (adjective: epiphytic) A nonparasitic plant that grows on trees or other solid objects and is not rooted in soil.

epithelium (plural: epithelia) An animal tissue consisting of sheet-like layers of tightly packed cells that lines an organ, a duct, or a body surface. Also called *epithelial tissue.*

epitope A small region of a particular antigen to which an antibody, B-cell receptor, or T-cell receptor binds.

equilibrium potential The membrane potential at which there is no net movement of a particular ion into or out of a cell.

ER signal sequence A short amino acid sequence that marks a polypeptide for transport to the endoplasmic reticulum where synthesis of the polypeptide chain is completed and the signal sequence removed. See **signal recognition particle**.

erythropoietin (EPO) A peptide hormone, released by the kidney in response to low blood oxygen levels, that stimulates the bone marrow to produce more red blood cells.

esophagus The muscular tube that connects the mouth to the stomach.

essential amino acid An amino acid that an animal cannot synthesize and must obtain from the diet. May refer specifically to one of the eight essential amino acids of adult humans: isoleucine, leucine, lysine, methionine, phenylalanine, threonine, tryptophan, and valine.

essential nutrient Any chemical element, ion, or compound that is required for normal growth, reproduction, and maintenance of a living organism and that cannot be synthesized by the organism.

ester linkage The covalent bond formed by a condensation reaction between a carboxyl group (—COOH) and a hydroxyl group (—OH). Ester linkages join fatty acids to glycerol to form a fat or phospholipid.

estradiol The major estrogen produced by the ovaries of female mammals. Stimulates development of the female reproductive tract, growth of ovarian follicles, and growth of breast tissue.

estrogens A class of steroid hormones, including estradiol, estrone, and estriol, that generally promote female-like traits. Secreted by the gonads, fat tissue, and some other organs.

estrous cycle A female reproductive cycle, seen in all mammals except Old World monkeys and apes (including humans), in which the uterine lining is reabsorbed rather than shed in the absence of pregnancy, and the female is sexually receptive only briefly during mid-cycle (estrus). Compare with **menstrual cycle**.

estuary An environment of brackish (partly salty) water where a river meets the ocean.

ethylene A gaseous plant hormone that induces fruits to ripen, flowers to fade, and leaves to drop.

eudicot A member of a monophyletic group (lineage) of angiosperms that includes complex flowering plants and trees (e.g., roses, daisies, maples). All eudicots have two cotyledons, but not all dicots are members of this lineage. Compare with **dicot** and **monocot**.

Eukarya One of the three taxonomic domains of life consisting of unicellular organisms (most protists, yeast) and multicellular organisms (fungi, plants, animals) distinguished by a membrane-bound cell nucleus, numerous organelles, and an extensive cytoskeleton. Compare with **Archaea** and **Bacteria**.

eukaryote A member of the domain Eukarya; an organism whose cells contain a nucleus, numerous membrane-bound organelles, and an extensive cytoskeleton. May be unicellular or multicellular. Compare with **prokaryote**.

eutherians A lineage of mammals (Eutheria) whose young develop in the uterus and are not housed in an abdominal pouch. Also called *placental mammals.*

evaporation The energy-absorbing phase change from a liquid state to a gaseous state. Many organisms evaporate water as a means of heat loss.

evo-devo Research field focused on how changes in developmentally important genes have led to the evolution of new phenotypes.

evolution (1) The theory that all organisms on Earth are related by common ancestry and that they have changed over time, predominantly via natural selection. (2) Any change in the genetic characteristics of a population over time, especially, a change in allele frequencies.

ex situ conservation Preserving species outside of natural areas; e.g., in zoos, aquaria, or botanic gardens.

excitable membrane A plasma membrane that is capable of generating an action potential. Neurons, muscle cells, and some other cells have excitable membranes.

excitatory postsynaptic potential (EPSP) A change in membrane potential, usually depolarization, at a neuron dendrite that makes an action potential more likely.

exergonic Referring to a chemical reaction that can occur spontaneously, releasing heat and/or increasing entropy, and for which the Gibbs free-energy change (ΔG) < 0. Compare with **endergonic**.

exocrine gland A gland that secretes some substance through a duct into a space other than the circulatory system, such as the digestive tract or the skin sufrace. Compare with **endocrine gland**.

exocytosis Secretion of intracellular molecules (e.g., hormones, collagen), contained within membrane-bounded vesicles, to the outside of the cell by fusion of vesicles to the plasma membrane. Compare with **endocytosis**.

exon A region of a eukaryotic gene that is translated into a peptide or protein. Compare with **intron**.

exoskeleton A hard covering secreted on the outside of the body, used for body support, protection, and muscle attachment. Examples are the shell of mollusks and the outer covering (cuticle) of arthropods. Compare with **endoskeleton**.

exothermic Referring to a chemical reaction that releases heat. Compare with **endothermic**.

exotic species A nonnative species that is introduced into a new area. Exotic species often are competitors, pathogens, or predators of native species.

expansins A class of plant proteins that actively increase the length of the cell wall when the pH of the wall falls below 4.5.

exponential population growth The accelerating increase in the size of a population that occurs when the growth rate is constant and density independent. Compare with **logistic population growth**.

expressed sequence tag (EST) A portion of a transcribed gene (synthesized from an mRNA in a cell), used to find the physical location of that gene in the genome.

extant species A species that is living today.

extensor A muscle that pulls two bones farther apart from each other, as in the extension of a limb or the spine. Compare with **flexor**.

extinct Said of a species that has died out.

extracellular digestion Digestion that takes place outside of an organism, as occurs in many fungi that make and secrete digestive enzymes.

extracellular matrix (ECM) A complex meshwork of proteins (e.g., collagen, fibronectin) and polysaccharides secreted by animal cells and in which they are embedded.

extremophile A bacterium or archaean that thrives in an "extreme" environment (e.g., high-salt, high-temperature, low-temperature, or low-pressure).

F_1 generation First filial generation. The first generation of offspring produced from a mating (i.e., the offspring of the parental generation).

facilitated diffusion Movement of a substance across a plasma membrane down its concentration gradient with the assistance of transmembrane carrier proteins or channel proteins.

facilitation In ecological succession, the phenomenon in which early-arriving species make conditions more favorable for later-arriving species. Compare with **inhibition** and **tolerance**.

facultative aerobe Any organism that can perform aerobic respiration when oxygen is available to serve as an electron acceptor but can switch to fermentation when it is not.

FAD/FADH$_2$ Oxidized and reduced forms, respectively, of flavin adenine dinucleotide. A nonprotein electron carrier that functions in the citric acid cycle and oxidative phosphorylation.

fallopian tube A narrow tube connecting the uterus to the ovary in humans, through which the egg travels after ovulation. Site of fertilization and cleavage. In nonhuman animals, called *oviduct.*

fat A lipid consisting of three fatty acid molecules joined by ester linkages to a glycerol molecule. Also called *triacylglycerol* or *triglyceride.*

fatty acid A lipid consisting of a hydrocarbon chain bonded to a carboxyl group (—COOH) at one end. Used by many organisms to store chemical energy; a major component of animal and plant fats.

fatty-acid binding protein Proteins that bind to fatty acids and enable them to be transported into cells.

fauna All the animals characteristic of a particular region, period, or environment.

feather A specialized skin outgrowth, composed of β-keratin, present in all birds and only in birds. Used for flight, insulation, display, and other purposes.

feces The waste products of digestion.

fecundity The average number of female offspring produced by a single female in the course of her lifetime.

feedback inhibition A type of metabolic control in which high concentrations of the product of a metabolic pathway inhibit one of the enzymes early in the pathway. A form of negative feedback.

fermentation Any of several metabolic pathways that make ATP by transferring electrons from a reduced compound such as glucose to a final electron acceptor other than oxygen. Allows glycolysis to proceed in the absence of oxygen.

ferredoxin In photosynthetic organisms, an iron- and sulfur-containing protein in the electron transport chain of photosystem I. Can transfer electrons to the enzyme $NADP^+$ reductase, which catalyzes formation of NADPH.

fertilization envelope A physical barrier that forms around a fertilized egg in amphibians and some other animals. Formed by an influx of water under the vitelline membrane.

fertilization Fusion of the nuclei of two haploid gametes to form a zygote with a diploid nucleus.

fetal alcohol syndrome A condition, marked by hyperactivity, severe learning disabilities, and depression, thought to be caused by exposure of an individual to high blood alcohol concentrations during embryonic development.

fetus In live-bearing animals, the unborn offspring after the embryonic stage, which usually are developed sufficiently to be recognizable as belonging to a certain species. In humans, from 9 weeks after fertilization until birth.

fiber In plants, a type of elongated sclerenchyma cell that provides support to vascular tissue. Compare with **sclereid**.

fibronectin An abundant protein in the extracellular matrix that binds to other ECM components and to integrins in plasma membranes; helps anchor cells in place. Numerous subtypes are found in different tissues.

Fick's law of diffusion A mathematical relationship that describes the rates of gas exchange in animal respiratory systems.

fight-or-flight response Rapid physiological changes that prepare the body for emergencies. Includes increased heart rate, increased blood pressure, and decreased digestion.

filament Any thin, threadlike structure, particularly (1) the threadlike extensions of a fish's gills or (2) the slender stalk that bears the anthers in a flower.

filter feeder See **suspension feeder**.

filtrate Any fluid produced by filtration, in particular the fluid ("pre-urine") in the nephrons of vertebrate kidneys.

filtration A process of removing large components from a fluid by forcing it through a filter. Occurs in a renal corpuscle of the vertebrate kidney, allowing water and small solutes to pass from the blood into the nephron.

finite rate of increase (λ) The rate of increase of a population over a given period of time. Calculated as the ending population size divided by the starting population size. Compare with **intrinsic rate of increase**.

first law of thermodynamics The principle of physics that energy is conserved in any process. Energy can be transferred and converted into different forms, but it cannot be created or destroyed.

fission (1) A form of asexual reproduction in which a prokaryotic cell divides to produce two genetically similar daughter cells by a process similar to mitosis of eukaryotic cells. Also called *binary fission*. (2) A form of asexual reproduction in which an animal splits into two or more individuals of approximately equal size; common among invertebrates.

fitness The ability of an individual to produce viable offspring relative to others of the same species.

fitness trade-off See **trade-off**.

fixed action pattern (FAP) Highly stereotyped behavior pattern that occurs in a certain invariant way in a certain species. A form of innate behavior.

flaccid Limp as a result of low internal pressure (e.g., a wilted plant leaf). Compare with **turgid**.

flagellum (plural: flagella) A long, cellular projection that undulates (in eukaryotes) or rotates (in prokaryotes) to move the cell through an aqueous environment. See **axoneme**.

flatworms Members of the phylum Platyhelminthes. Distinguished by a broad, flat, unsegmented body that lacks a coelom. Flatworms belong to the lophotrochozoan branch of the protostomes.

flavin adenine dinucleotide See **FAD/FADH$_2$**.

flexor A muscle that pulls two bones closer together, as in the flexing of a limb or the spine. Compare with **extensor**.

floral meristem A group of undifferentiated plant cells that can give rise to the four organs making up a flower.

florigen In plants, a protein hormone that is synthesized in leaves and transported to the shoot apical meristem where it stimulates flowering.

flower In angiosperms, the part of a plant that contains reproductive structures. Typically includes a calyx, a corolla, and one or more stamens and/or carpels. See **perfect** and **imperfect flower**.

fluid connective tissue A type of connective tissue, distinguished by having a liquid extracellular matrix.

fluid feeder An animal that feeds by sucking or mopping up liquids such as nectar, plant sap, or blood.

fluid-mosaic model The widely accepted hypothesis that the plasma membrane and organelle membranes consist of proteins embedded in a fluid phospholipid bilayer.

fluorescence The spontaneous emission of light from an excited electron falling back to its normal (ground) state.

follicle An egg cell and its surrounding ring of supportive cells in a mammalian ovary.

follicle-stimulating hormone (FSH) A peptide hormone, produced and secreted by the anterior pituitary, that stimulates (in females) growth of eggs and follicles in the ovaries or (in males) sperm production in the testes.

follicular phase The first major phase of a menstrual cycle during which follicles grow and estrogen levels increase; ends with ovulation.

food Any nutrient-containing material that can be consumed and digested by animals.

food chain A relatively simple pathway of energy flow through a few species, each at a different trophic level, in an ecosystem. Might include, for example, a primary producer, a primary consumer, a secondary consumer, and a decomposer. Compare with **food web**.

food web Any complex pathway along which energy moves among many species at different trophic levels of an ecosystem.

foot One of the three main parts of the mollusk body; a muscular appendage, used for movement and/or burrowing into sediment.

foraging Searching for food.

forebrain One of the three main regions of the vertebrate brain; includes the cerebrum, thalamus, and hypothalamus. Compare with **hindbrain** and **midbrain**.

fossil Any trace of an organism that existed in the past. Includes tracks, burrows, fossilized bones, casts, etc.

fossil record All of the fossils that have been found anywhere on Earth and that have been formally described in the scientific literature.

founder effect A change in allele frequencies that often occurs when a new population is established from a small group of individuals (founder event) due to sampling error (i.e., the small group is not a representative sample of the source population).

fovea In the vertebrate eye, a portion of the retina where incoming light is focused; contains a high proportion of cone cells.

free energy See **Gibbs free-energy change**.

free radical Any substance containing one or more atoms with an unpaired electron. Unstable and highly reactive.

frequency The number of wave crests per second traveling past a stationary point. Determines the pitch of sound and the color of light.

frequency-dependent selection A pattern of selection in which certain alleles are favored only when they are rare; a form of balancing selection.

fronds The large leaves of ferns.

frontal lobe In the vertebrate brain, one of the four major areas in the cerebrum.

fruit In flowering plants (angiosperms), a mature, ripened plant ovary (or group of ovaries), along with the seeds it contains and any adjacent fused parts. See **aggregate, multiple,** and **simple fruit**.

fruiting body A structure formed in some prokaryotes, fungi, and protists for spore dispersal; usually consists of a base, a stalk, and a mass of spores at the top.

functional genomics The study of how a genome works, that is, when and where specific genes are expressed and how their products interact to produce a functional organism.

functional group A small group of atoms bonded together in a precise configuration and exhibiting particular chemical properties that it imparts to any organic molecule in which it occurs.

fundamental niche The ecological space that a species occupies in its habitat in the absence of competitors. Compare with **realized niche**.

fungi A lineage of eukaryotes that typically have a filamentous body (mycelium) and obtain nutrients by absorption.

fungicide Any substance that can kill fungi or slow their growth.

G protein Any of various peripheral membrane proteins that bind GTP and function in signal transduction. Binding of a signal to its receptor triggers activation of the G protein, leading to production of a second messenger or initiation of a phosphorylation cascade.

G_1 phase The phase of the cell cycle that constitutes the first part of interphase before DNA synthesis (S phase).

G_2 phase The phase of the cell cycle between synthesis of DNA (S phase) and mitosis (M phase); the last part of interphase.

gallbladder A small pouch that stores bile from the liver and releases it as needed into the small intestine during digestion of fats.

gametangium (plural: gametangia) (1) The gamete-forming structure found in all land plants except angiosperms. Contains a sperm-producing antheridium and an egg-producing archegonium. (2) The gamete-forming structure of some chytrid fungi.

gamete A haploid reproductive cell that can fuse with another haploid cell to form a zygote. Most multicellular eukaryotes have two distinct forms of gametes: egg cells (ova) and sperm cells.

gametogenesis The production of gametes (eggs or sperm).

gametophyte In organisms undergoing alternation of generations, the multicellular haploid form that arises from a single haploid spore and produces gametes. A female gametophyte is commonly called an *embryo sac;* a male gametophyte, a *pollen grain*. Compare with **sporophyte**.

ganglia A mass of neurons in a centralized nervous system.

ganglion cell A neuron in the vertebrate retina that collects visual information from one or several bipolar cells and sends it to the brain via the optic nerve.

gap junction A type of cell-cell attachment structure that directly connects the cytoplasms of adjacent animal cells, allowing passage of water, ions, and small molecules between the cells. Compare with **desmosome** and **tight junction**.

gastrin A hormone produced by cells in the stomach lining in response to the arrival of food or to a neural signal from the brain. Stimulates other stomach cells to release hydrochloric acid.

gastrointestinal (GI) tract See **digestive tract**.

gastropods A lineage of mollusk distinguished by a large muscular foot and a unique feeding structure, the radula. Include slugs and snails.

gastrulation The process by which some cells on the outside of a young embryo move to the interior of the embryo, resulting in the three distinct germ layers (endoderm, mesoderm, and ectoderm).

gated channel A channel protein that opens and closes in response to a specific stimulus, such as the binding of a particular molecule or a change in the electrical charge on the outside of the membrane.

gel electrophoresis A technique for separating molecules on the basis of size and electric charge, which affect their differing rates of movement through a gelatinous substance in an electric field.

gemma (plural: gemmae) A small reproductive structure that is produced in some liverworts during the gametophyte phase and can grow into mature gametophyte.

gene A section of DNA (or RNA, for some viruses) that encodes information for building one or more related polypeptides or functional RNA molecules along with the regulatory sequences required for its transcription.

gene duplication The formation of an additional copy of a gene, typically by misalignment of chromosomes during crossing over. Thought to be an important evolutionary process in creating new genes.

gene expression Overall process by which the information encoded in genes is converted into an active product, most commonly a protein. Includes transcription and translation of a gene and in some cases protein activation.

gene family A set of genetic loci whose DNA sequences are extremely similar. Thought to have arisen by duplication of a single ancestral gene and subsequent mutations in the duplicated sequences.

gene flow The movement of alleles between populations; occurs when individuals leave one population, join another, and breed.

gene pool All of the alleles of all of the genes in a certain population.

gene therapy The treatment of an inherited disease by introducing normal alleles.

gene-for-gene hypothesis The hypothesis that there is a one-to-one correspondence between the resistance (*R*) loci of plants and the avirulence (*avr*) loci of pathogenic fungi; particularly, that *R* genes produce receptors and *avr* genes produce molecules that bind to those receptors.

generation The average time between a mother's first offspring and her daughter's first offspring.

genetic bottleneck A reduction in allelic diversity resulting from a sudden reduction in the size of a large population (population bottleneck) due to a random event.

genetic code The set of all 64 codons and the particular amino acids that each specifies.

genetic correlation A type of evolutionary constraint in which selection on one trait causes a change in another trait as well; may occur when the same gene(s) affect both traits.

genetic diversity The diversity of alleles in a population, species, or group of species.

genetic drift Any change in allele frequencies due to random events. Causes allele frequencies to drift up and down randomly over time, and eventually can lead to the fixation or loss of alleles.

genetic homology Similarities in DNA sequences or amino acid sequences that are due to inheritance from a common ancestor.

genetic map An ordered list of genes on a chromosome that indicates their relative distances from each other. Also called a *linkage map* or *meiotic map*. Compare with **physical map**.

genetic marker A genetic locus that can be identified and traced in populations by laboratory techniques or by a distinctive visible phenotype.

genetic model A set of hypotheses that explain how a certain trait is inherited.

genetic recombination A change in the combination of genes or alleles on a given chromosome or in a given individual. Also called *recombination*.

genetic screen Any of several techniques for identifying individuals with a particular type of mutation. Also called a *screen*.

genetic variation (1) The number and relative frequency of alleles present in a particular population. (2) The proportion of phenotypic variation in a trait that is due to genetic rather than environmental influences in a certain population in a certain environment.

genetics The field of study concerned with the inheritance of traits.

genitalia External copulatory organs.

genome All of the hereditary information in an organism, including not only genes but also other non-gene stretches of DNA.

genomic library A set of DNA segments representing the entire genome of a particular organism. Each segment is carried by a plasmid or other cloning vector and can be separated from other segments. Compare with **cDNA library**.

genomics The field of study concerned with sequencing, interpreting, and comparing whole genomes from different organisms.

genotype All of the alleles of every gene present in a given individual. May refer specifically to the alleles of a particular set of genes under study. Compare with **phenotype**.

genus (plural: genera) In Linnaeus' system, a taxonomic category of closely related species. Always italicized and capitalized to indicate that it is a recognized scientific genus.

geologic time scale The sequence of eons, eras, and periods used to describe the geologic history of Earth.

germ cell In animals, any cell that can potentially give rise to gametes. Also called *germ-line cells*.

germ layer In animals, one of the three basic types of tissue formed during gastrulation; gives rise to all other tissues. See **endoderm, mesoderm,** and **ectoderm**.

germ theory of disease The theory that infectious diseases are caused by bacteria, viruses, and other microorganisms.

germination The process by which a seed becomes a young plant.

gestation The duration of embryonic development from fertilization to birth in those species that have live birth.

gibberellins A class of hormones found in plants and fungi that stimulate growth. Gibberellic acid is one of the major gibberellins.

Gibbs free-energy change (ΔG) A measure of the change in potential energy and entropy that occurs in a given chemical reaction. $\Delta G < 0$ for spontaneous reactions and > 0 for nonspontaneous reactions.

gill Any organ in aquatic animals that exchanges gases and other dissolved substances between the blood and the surrounding water. Typically, a filamentous outgrowth of a body surface.

gill arch In aquatic vertebrates, curved region of tissue between the gills. Gills are suspended from the gill arches.

gill filament In fish, one of the many long, thin structures that extend from gill arches into the water and across which gas exchange occurs.

gill lamella (plural: gill lamellae) One of hundreds to thousands of sheetlike structures, each containing a capillary bed, that makes up a gill filament.

gland An organ whose primary function is to secrete some substance, either into the blood (endocrine gland) or into some other space such as the gut or skin (exocrine gland).

glia Collective term for several types of cells in nervous tissue that are not neurons and do not conduct electrical signals but provide support, nourishment, and electrical insulation and perform other functions. Also called *glial cells*.

global carbon cycle See **carbon cycle, global**.

global nitrogen cycle See **nitrogen cycle, global**.

global warming A sustained increase in Earth's average surface temperature.

global water cycle See **water cycle, global**.

glomalin A glycoprotein that is abundant in the hyphae of arbuscular mycorrhizal fungi; when hyphae decay it an important component of soil.

glomerulus (plural: glomeruli) (1) In the vertebrate kidney, a ball-like cluster of capillaries, surrounded by Bowman's capsule, at the beginning of a nephron. (2) In the brain, a ball-shaped cluster of neurons in the olfactory bulb.

glucagon A peptide hormone produced by the pancreas in response to low blood glucose. Raises blood glucose by triggering breakdown of glycogen and stimulating gluconeogenesis. Compare with **insulin**.

glucocorticoids A class of steroid hormones, produced and secreted by the adrenal cortex, that increase blood glucose and prepare the body for stress. Include cortisol and corticosterone. Compare with **mineralocorticoids**.

gluconeogenesis Synthesis of glucose from non-carbohydrate sources (e.g., proteins and fatty acids). Occurs in the liver in response to low insulin levels and high glucagon levels.

glucose Six-carbon monosaccharide whose oxidation in cellular respiration is the major source of ATP in animal cells.

glyceraldehyde-3-phosphate (G3P) The phosphorylated three-carbon compound formed as the result of carbon fixation in the first step of the Calvin cycle.

glycerol A three-carbon molecule that forms the "backbone" of phospholipids and most fats.

glycogen A highly branched storage polysaccharide composed of α-glucose monomers joined by 1,4- and 1,6-glycosidic linkages. The major form of stored carbohydrate in animals.

glycolipid Any lipid molecule that is covalently bonded to a carbohydrate group.

glycolysis A series of 10 chemical reactions that oxidize glucose to produce pyruvate and ATP. Used by all organisms as part of fermentation or cellular respiration.

glycoprotein Any protein with one or more covalently bonded carbohydrate groups.

glycosidic linkage The covalent bond formed by a condensation reaction between two sugar monomers; joins the residues of a polysaccharide.

glycosylation Addition of a carbohydrate group to a molecule.

glyoxisome Specialized type of peroxisome found in plant cells and packed with enzymes for processing the products of photosynthesis.

gnathostomes Animals with jaws.

Golgi apparatus A eukaryotic organelle, consisting of stacks of flattened membranous sacs (cisternae), that functions in processing and sorting proteins and lipids destined to be secreted or directed to other organelles. Also called *Golgi complex*.

gonad An organ that produces reproductive cells, such as a testis or an ovary.

gonadotropin-releasing hormone (GnRH) A peptide hormone, produced and secreted by the hypothalamus, that stimulates release of FSH and LH from the anterior pituitary.

grade In taxonomy, a group of species that share a position in an inferred evolutionary sequence of lineages but that are not a monophyletic group. Also called a *paraphyletic group*.

Gram stain A dye that distinguishes the two general types of cell walls found in bacteria. Used to routinely classify bacteria as Gram-negative or Gram-positive.

Gram-negative Describing bacteria that look pink when treated with a Gram stain. These bacteria have a cell wall composed of a thin layer of peptidoglycan and an outer phospholipid layer.

Gram-positive Describing bacteria that look purple when treated with a Gram stain. These bacteria have cell walls composed of a thick layer of peptidoglycan.

granum (plural: grana) In chloroplasts, a stack of flattened, membrane-bound vesicles (thylakoids) where the light reactions of photosynthesis occur.

gravitropism The growth or movement of a plant in a particular direction in response to gravity.

grazing food chain The ecological network of herbivores and the predators and parasites that consume them.

great apes See **hominids**.

green algae A paraphyletic group of photosynthetic organisms that contain chloroplasts similar to those in green plants. Often classified as protists, green algae are the closest living relatives of land plants and form a monophyletic group with them.

greenhouse gas An atmospheric gas that absorbs and reflects infrared radiation, so that heat radiated from Earth is retained in the atmosphere instead of being lost to space.

gross photosynthetic productivity The efficiency with which all the plants in a given area use the light energy available to them to produce sugars.

gross primary productivity In an ecosystem, the total amount of carbon fixed by photosynthesis, including that used for cellular respiration, over a given time period. Compare with **net primary productivity**.

ground meristem The middle layer of a young plant embryo. Gives rise to the ground tissue system.

ground tissue An embryonic tissue layer that gives rise to parenchyma, collenchyma, and sclerenchyma—tissues other than the epidermis and vascular tissue.

groundwater Any water below the land surface.

growth factor Any of a large number of signaling molecules that are secreted by certain cells and that stimulate other cells to divide or to differentiate.

growth hormone (GH) A peptide hormone, produced and secreted by the mammalian anterior pituitary, that promotes lengthening of the long bones in children and muscle growth, tissue repair, and lactation in adults. Also called *somatotropin*.

GTP See **guanosine triphosphate**.

guanosine triphosphate (GTP) A molecule consisting of guanine, a sugar, and three phosphate groups. Can be hydrolyzed to release free energy. Commonly used in many cellular reactions; also functions in signal transduction in association with G proteins.

guard cell One of two specialized, crescent-shaped cells forming the border of a plant stoma. Guard cells can change shape to open or close the stoma. See also **stoma**.

gustation The perception of taste.

guttation Excretion of water droplets from plant leaves in the early morning, caused by root pressure.

gymnosperm A vascular plant that makes seeds but does not produce flowers. The gymnosperms include four lineages of green plants (cycads, ginkgoes, conifers, and gnetophytes). Compare with **angiosperm**.

H^+-ATPase See **proton pump**.

habitat destruction Human-caused destruction of a natural habitat with replacement by an urban, suburban, or agricultural landscape.

habitat fragmentation The breakup of a large region of a habitat into many smaller regions, separated from others by a different type of habitat.

Hadley cell An atmospheric cycle of large-scale air movement in which warm equatorial air rises, moves north or south, and then descends at approximately 30° N or 30° S latitude.

hair cell A pressure-detecting sensory cell, found in the cochlea, that has tiny "hairs" (stereocilia) jutting from its surface.

hairpin A secondary structure in RNA consisting of a stable loop formed by hydrogen bonding between purine and pyrimidine bases on the same strand.

halophile A bacterium or archaean that thrives in high-salt environments.

Hamilton's rule The proposition that an allele for altruistic behavior will be favored by natural selection only if $Br > C$, where B = the fitness benefit to the recipient, C = the fitness cost to the actor, and r = the coefficient of relatedness between recipient and actor.

haploid (1) Having one set of chromosomes ($1n$). (2) A cell or an individual organism with one set of chromosomes. Compare with **diploid**.

haploid number The number of distinct chromosome sets in a cell. Symbolized as n.

Hardy-Weinberg principle A principle of population genetics stating that genotype frequencies in a large population do not change from generation to generation in the absence of evolutionary processes (e.g., mutation, migration, genetic drift, random mating, and selection).

heart A muscular pump that circulates blood throughout the body.

heart murmur A distinctive sound caused by backflow of blood through a defective heart valve.

heartwood The older xylem in the center of an older stem or root, containing protective compounds and no longer functioning in water transport.

heat Thermal energy that is transferred from an object at higher temperature to one at lower temperature.

heat of vaporization The energy required to vaporize 1 gram of a liquid into a gas.

heat-shock proteins Proteins that facilitate refolding of proteins that have been denatured by heat or other agents.

heavy chain The larger of the two types of polypeptide chains in an antibody molecule; composed of a variable (V) region, which contributes to the antigen-binding site, and a constant (C) region. Differences in heavy-chain constant regions determine the different classes of immunoglobulins (IgA, IgE, etc.). Compare with **light chain**.

helicase An enzyme that catalyzes the breaking of hydrogen bonds between nucleotides of DNA, "unzipping" a double-stranded DNA molecule.

helper T cell An effector T cell that secretes cytokines and in other ways promotes the activation of other lymphocytes. Is descended from an activated $CD4^+$ T cell that has interacted with antigen presented by dendritic cells, macrophages, or B cells.

heme A small molecule that binds to four polypeptides to form hemoglobin; contains an iron atom that can bind oxygen.

hemimetabolous metamorphosis A type of metamorphosis in which the animal increases in size from one stage to the next, but does not dramatically change its body form. Also called *incomplete metamorphosis*.

hemocoel A body cavity, present in arthropods and some mollusks, containing a pool of circulatory fluid (hemolymph) bathing the internal organs.

hemoglobin An oxygen-binding protein consisting of four polypeptide subunits, each containing an oxygen-binding heme group. The major oxygen carrier in mammalian blood.

hemolymph The circulatory fluid of animals with open circulatory systems (e.g., insects) in which the fluid is not confined to blood vessels.

hemophilia A human disease, caused by an X-linked recessive allele, that is characterized by defects in the blood-clotting system.

herbaceous Referring to a plant that is not woody.

herbivore (adjective: herbivorous) An animal that eats primarily plants and rarely or never eats meat. Compare with **carnivore** and **omnivore**.

herbivory The practice of eating plant tissues.

heredity The transmission of traits from parents to offspring via genetic information.

heritable Referring to traits that can be transmitted from one generation to the next.

hermaphrodite An organism that produces both male and female gametes.

heterokaryotic Describing a cell or fungal mycelium containing two or more nuclei that are genetically distinct.

heterospory (adjective: heterosporous) In seed plants, the production of two distinct types of spore-producing structures and thus two distinct types of spores: microspores, which become the male gametophyte, and megaspores, which become the female gametophyte. Compare with **homospory**.

heterotherm An animal whose body temperature varies markedly with environmental conditions. Compare with **homeotherm**.

heterotroph Any organism that cannot synthesize reduced organic compounds from inorganic sources and that must obtain them by eating other organisms. Some bacteria, some archaea, and virtually all fungi and animals are heterotrophs. Also called *consumer*. Compare with **autotroph**.

heterozygote advantage A pattern of natural selection that favors heterozygous individuals compared with homozygotes. Tends to maintain genetic variation in a population. Also called *heterozygote superiority*.

heterozygous Having two different alleles of a certain gene.

hexose A monosaccharide (simple sugar) containing six carbon atoms.

hibernation An energy-conserving physiological state, marked by a decrease in metabolic rate, body temperature, and activity, that lasts for a prolonged period (weeks to months). Occurs in some animals in response to winter cold and scarcity of food. Compare with **torpor**.

hindbrain One of the three main regions of the vertebrate brain; includes the cerebellum and medulla oblongata. Compare with **forebrain** and **midbrain**.

histamine A molecule released from mast cells during an inflammatory response that causes blood vessels to dilate and become more permeable.

histone One of several positively charged (basic) proteins associated with DNA in the chromatin of eukaryotic cells.

histone acetyl transferase (HAT) In eukaryotes, one of a class of enzymes that loosen chromatin structure by adding acetyl groups to histone proteins.

histone code The hypothesis that specific combinations of chemical modifications of histone proteins contain information that influences gene expression.

histone deacetylase (HDAC) In eukaryotes, one of a class of enzymes that recondense chromatin by removing acetyl groups from histone proteins.

HIV See human immunodeficiency virus (HIV).

holoenzyme A multipart enzyme consisting of a core enzyme (containing the active site for catalysis) along with other required proteins.

holometabolous metamorphosis A type of metamorphosis in which the animal completely changes its form. Also called *complete metamorphosis*.

homeosis Replacement of one body part by another normally found elsewhere in the body as the result of mutation in certain developmentally important genes (homeotic genes).

homeostasis (adjective: homeostatic) The array of relatively stable chemical and physical conditions in an animal's cells, tissues, and organs. May be achieved by the body's passively matching the conditions of a stable external environment (conformational homeostasis) or by active physiological processes (regulatory homeostasis) triggered by variations in the external or internal environment.

homeotherm An animal that has a constant or relatively constant body temperature. Compare with **heterotherm**.

homeotic gene Any gene that specifies a particular location within an embryo, leading to the development of structures appropriate for that location. Mutations in homeotic genes cause the development of extra body parts or body parts in the wrong places.

hominids Members of the family Hominidae, which includes humans and extinct related forms; chimpanzees, gorillas, and orangutans. Distinguished by large body size, no tail, and an exceptionally large brain. Also called *great apes*.

hominins Humans and extinct related forms; species in the lineage that branched off from chimpanzees and eventually led to humans.

homologous chromosomes In a diploid organism, chromosomes that are similar in size, shape, and gene content. Also called *homologs*.

homology (adjective: homologous) Similarity among organisms of different species due to their inheritance from a common ancestor. Features that exhibit such similarity (e.g., DNA sequences, proteins, body parts) are said to be homologous. Compare with **homoplasy**.

homoplasy Similarity among organisms of different species due to convergent evolution. Compare with **homology**.

homospory (adjective: homosporous) In seedless vascular plants, the production of just one type of spore. Compare with **heterospory**.

homozygous Having two identical alleles of a certain gene.

hormone Any of numerous different signaling molecules that circulate throughout the body in blood or other body fluids and can trigger characteristic responses in distant target cells at very low concentrations.

hormone-response element A specific sequence in DNA to which a steroid hormone-receptor complex can bind and affect gene transcription.

host cell A cell that has been invaded by an organism such as a parasite or a virus.

***Hox* genes** A class of homeotic genes found in several animal phyla, including vertebrates, that are expressed in a distinctive pattern along the anterior-posterior axis in early embryos and control formation of segment-specific structures.

human Any member of the genus *Homo*, which includes modern humans (*Homo sapiens*) and several extinct species.

human chorionic gonadotropin (hCG) A glycoprotein hormone produced by the human placenta from about week 3 to week 14 of pregnancy. Maintains the corpus luteum, which produces hormones that preserve the uterine lining.

Human Genome Project The multinational research project that sequenced the human genome.

human immunodeficiency virus (HIV) A retrovirus that causes AIDS (acquired immune deficiency syndrome) in humans.

humoral (immune) response The type of immune response that involves generation of antibody-secreting plasma cells from activated B cells. Defends against extracellular pathogens. Compare with **cell-mediated (immune) response**.

humus The completely decayed organic matter in soils.

Huntington's disease A degenerative brain disease of humans caused by an autosomal dominant allele.

hybrid The offspring of parents from two different strains, populations, or species.

hybrid zone A geographic area where interbreeding occurs between two species, sometimes producing fertile hybrid offspring.

hydrocarbon An organic molecule that contains only hydrogen and carbon atoms.

hydrogen bond A weak interaction between two molecules or different parts of the same molecule resulting from the attraction between a hydrogen atom with a partial positive charge and another atom (usually O or N) with a partial negative charge. Compare with **covalent bond** and **ionic bond**.

hydrogen ion (H^+) A single proton with a charge of 1+; typically, one that is dissolved in solution or that is being transferred from one atom to another in a chemical reaction.

hydrolysis A chemical reaction in which a molecule is split into smaller molecules by reacting with water. In biology, most hydrolysis reactions involve the splitting of polymers into monomers. Compare with **condensation reaction**.

hydrophilic Interacting readily with water. Hydrophilic compounds are typically polar compounds containing charged or electronegative atoms. Compare with **hydrophobic**.

hydrophobic Not interacting readily with water. Hydrophobic compounds are typically nonpolar compounds that lack charged or electronegative atoms and often contain many C—C and C—H bonds. Compare with **hydrophilic**.

hydroponic growth Growth of plants in liquid cultures instead of soil.

hydrostatic skeleton A system of body support involving fluid-filled compartments that can change in shape but cannot easily be compressed.

hydroxide ion (OH^-) An oxygen atom and a hydrogen atom joined by a single covalent bond and carrying a negative charge; formed by dissociation of water.

hyperpolarization Change in membrane potential from its resting negative state to an even more negative state; a normal phase in an action potential. Compare with **depolarization**.

hypersensitive reaction An intense allergic response by cells that have been sensitized by previous exposure to an allergen.

hypersensitive response In plants, the rapid death of a cell that has been infected by a pathogen, thereby reducing the potential for infection to spread throughout a plant. Compare with **systemic acquired resistance**.

hypertension Abnormally high blood pressure.

hypertonic Comparative term designating a solution that has a greater solute concentration, and therefore a lower water concentration, than another solution. Compare with **hypotonic** and **isotonic**.

hypha (plural: hyphae) One of the strands of a fungal mycelium (the meshlike body of a fungus). Also found in some protists.

hypocotyl The stem of a very young plant; the region between the cotyledon (embryonic leaf) and the radicle (embryonic root).

hypothalamic-pituitary axis The functional interaction of the hypothalamus and the pituitary gland, which are anatomically distinct but work together to regulate most of the other endocrine glands in the body.

hypothalamus A part of the brain that functions in maintaining the body's internal physiological state by regulating the autonomic nervous system, endocrine system, body temperature, water balance, and appetite.

hypothesis A proposed explanation for a phenomenon or for a set of observations.

hypotonic Comparative term designating a solution that has a lower solute concentration, and therefore a higher water concentration, than another solution. Compare with **hypertonic** and **isotonic**.

immigration The migration of individuals into a particular population from other populations. Compare with **emigration**.

immune system In vertebrates, the system whose primary function is to defend the body against pathogens. Includes several types of cells (e.g., lymphocytes and macrophages) and several organs where they develop or reside (e.g., lymph nodes and thymus).

immunity (adjective: immune) State of being protected against infection by disease-causing pathogens either by relatively nonspecific mechanisms (innate immunity) or by specific mechanisms triggered by exposure to a particular antigen (adaptive immunity).

immunization The conferring of immunity to a particular disease by artificial means.

immunoglobulin (Ig) Any of the class of proteins that function as antibodies.

immunological memory The ability of the immune system to "remember" an antigen and mount a rapid, effective response to a pathogen encountered years or decades earlier.

impact hypothesis The hypothesis that a collision between the Earth and an asteroid caused the mass extinction at the K-P boundary, 65 million years ago.

imperfect flower A flower that contains male parts (stamens) *or* female parts (carpels) but not both. Compare with **perfect flower**.

implantation The process by which an embryo buries itself in the uterine wall and forms a placenta. Occurs in mammals and a few other vertebrates.

in situ hybridization A technique for detecting specific DNAs and mRNAs in cells and tissues by use of labeled probes. Can be used to determine where and when particular genes are expressed in embryos.

inbreeding Mating between closely related individuals. Increases homozygosity of a population and often leads to a decline in the average fitness (inbreeding depression).

inbreeding depression In inbred offspring, fitness declines due to deleterious recessive alleles that are homozygous.

inclusive fitness The combination of (1) direct production of offspring (direct fitness) and (2) extra production of offspring by relatives in response to help provided by the individual in question (indirect fitness).

incomplete digestive tract A digestive tract that has just one opening.

incomplete dominance An inheritance pattern in which the heterozygote phenotype is a blend or combination of both homozygote phenotypes.

incomplete metamorphosis See **hemimetabolous metamorphosis**.

independent assortment, principle of The concept that each pair of hereditary elements (alleles of the same gene) behaves independently of other genes during meiosis. One of Mendel's two principles of genetics.

indeterminate growth A pattern of growth in which an individual continues to increase its overall body size throughout its life.

indicator plate A laboratory technique for detecting mutant cells by growing them on agar plates containing a compound that when metabolized by wild-type cells yields a colored product.

induced fit Change in the shape of the active site of an enzyme, as the result of the initial weak binding of a substrate, so that it binds substrate more tightly.

inducer A small molecule that triggers transcription of a specific gene, often by binding to and inactivating a repressor protein.

inducible defense A defensive trait that is manifested only in response to the presence of a consumer (predator or herbivore) or pathogen. Compare with **constitutive defense**.

infection thread An invagination of the membrane of a root hair through which beneficial nitrogen-fixing bacteria enter the roots of their host plants (legumes).

inflammatory response An aspect of the innate immune response, seen in most cases of infection or tissue injury, in which the affected tissue becomes swollen, red, warm, and painful.

inhibition In ecological succession, the phenomenon in which early-arriving species make conditions less favorable for the establishment of certain later-arriving species. Compare with **facilitation** and **tolerance**.

inhibitory postsynaptic potential (IPSP) A change in membrane potential, usually hyperpolarization, at a neuron dendrite that makes an action potential less likely.

initiation (1) In an enzyme-catalyzed reaction, the stage during which enzymes orient reactants precisely as they bind at specific locations within the enzyme's active site. (2) In DNA transcription, the stage during which RNA polymerase and other proteins assemble at the promoter sequence. (3) In RNA translation, the stage during which a complex consisting of a ribosome, a mRNA molecule, and an aminoacyl tRNA corresponding to the start codon is formed.

initiation factors A class of proteins that assist ribosomes in binding to a messenger RNA molecule to begin translation.

innate behavior Behavior that is inherited genetically, does not have to be learned, and is typical of a species.

innate immune response See **innate immunity**.

innate immunity A set of nonspecific defenses against pathogens that exist before exposure to an antigen and involves mast cells, neutrophils, and macrophages; typically results in an inflammatory response. Compare with **acquired immunity**.

inner cell mass (ICM) A cluster of cells in the interior of a mammalian blastocyst that undergo gastrulation and eventually develop into the embryo.

inner ear The innermost portion of the mammalian ear, consisting of a fluid-filled system of tubes that includes the cochlea (which receives sound vibrations from the middle ear) and the semicircular canals (which function in balance).

insulin A peptide hormone produced by the pancreas in response to high levels of glucose (or amino acids) in blood. Enables cells to absorb glucose and coordinates synthesis of fats, proteins, and glycogen. Compare with **glucagon**.

integral membrane protein Any membrane protein that spans the entire lipid bilayer. Also called *transmembrane protein*. Compare with **peripheral membrane protein**.

integrated pest management In agriculture or forestry, systems for managing insects or other pests that include carefully controlled applications of toxins, introduction of species that prey on pests, planting schemes that reduce the chance of a severe pest outbreak, and other techniques.

integrator A component of an animal's nervous system that functions as part of a homeostatic system by evaluating sensory information and triggering appropriate responses. See **effector** and **sensor**.

integrin Any of a class of cell-surface proteins that bind to fibronectins and other proteins in the extracellular matrix, thus holding cells in place.

intercalated disc A specialized junction between adjacent heart muscle cells that contains gap junctions, allowing electrical signals to pass between the cells.

intermediate disturbance hypothesis The hypothesis that moderate ecological disturbance is associated with higher species diversity than either low or high disturbance.

intermediate filament A long fiber, about 10 nm in diameter, composed of one of various proteins (e.g., keratins, lamins); one of the three types of cytoskeletal fibers. Form networks that help maintain cell shape and hold the nucleus in place. Compare with **actin filament** and **microtubule**.

intermediate host The host species in which a parasite reproduces asexually. Compare with **definitive host**.

interneuron A neuron that passes signals from one neuron to another. Compare with **motor neuron** and **sensory neuron**.

internode The section of a plant stem between two nodes (sites where leaves attach).

interphase The portion of the cell cycle between one mitotic (M) phase and the next. Includes the G_1 phase, S phase, and G_2 phase.

interspecific competition Competition between members of different species for the same limited resource. Compare with **intraspecific competition**.

interstitial fluid The plasma-like fluid found in the region (interstitial space) between cells.

intertidal zone The region between the low-tide and high-tide marks on a seashore.

intraspecific competition Competition between members of the same species for the same limited resource. Compare with **interspecific competition**.

intrinsic rate of increase (r_{max}) The rate at which a population will grow under optimal conditions (i.e., when birthrates are as high as possible and death rates are as low as possible). Compare with **finite rate of increase**.

intron A region of a eukaryotic gene that is transcribed into RNA but is later removed, so it is not translated into a peptide or protein. Compare with **exon**.

invasive species An exotic (nonnative) species that, upon introduction to a new area, spreads rapidly and competes successfully with native species.

inversion A mutation in which a segment of a chromosome breaks from the rest of the chromosome, flips, and rejoins with the opposite orientation as before.

invertebrates A paraphyletic group composed of animals without a backbone; includes about 95 percent of all animal species. Compare with **vertebrates**.

involuntary muscle Muscle that cannot respond to conscious thought.

ion An atom or a molecule that has lost or gained electrons and thus carries an electric charge, either positive (cation) or negative (anion), respectively.

ion channel A type of channel protein that allows certain ions to diffuse across a plasma membrane down an electochemical gradient.

ionic bond A chemical bond that is formed when an electron is completely transferred from one atom to another so that the atoms remain associated due to their opposite electric charges. Compare with **covalent bond** and **hydrogen bond**.

iris A ring of pigmented muscle just below the cornea in the vertebrate eye that contracts or expands to control the amount of light entering the eye through the pupil.

isotonic Comparative term designating a solution that has the same solute concentration and water concentration than another solution. Compare with **hypertonic** and **hypotonic**.

isotope Any of several forms of an element that have the same number of protons but differ in the number of neutrons.

joint A place where two components (bones, cartilages, etc.) of a skeleton meet. May be movable (an articulated joint) or immovable (e.g., skull sutures).

juvenile An individual that has adult-like morphology but is not sexually mature.

juvenile hormone An insect hormone that prevents larvae from metamorphosing into adults.

karyogamy Fusion of two haploid nuclei to form a diploid nucleus. Occurs in many fungi, and in animals and plants during fertilization of gametes.

karyotype The distinctive appearance of all of the chromosomes in an individual, including the number of chromosomes and their length and banding patterns (after staining with dyes).

keystone species A species that has an exceptionally great impact on the other species in its ecosystem relative to its abundance.

kidney In terrestrial vertebrates, one of a paired organ situated at the back of the abdominal cavity that filters the blood, produces urine, and secretes several hormones.

kilocalorie (kcal) A unit of energy often used to measure the energy content of food. A kcal of energy raises 1 g of water 1°C.

kin selection A form of natural selection that favors traits that increase survival or reproduction of an individual's kin at the expense of the individual.

kinesin Any one of a class of motor proteins that use the chemical energy of ATP to transport vesicles, particles, or chromosomes along microtubules.

kinetic energy The energy of motion. Compare with **potential energy**.

kinetochore A protein structure at the centromere where kinetochore microtubules attach to the sister chromatids of a replicated chromosome. Contains motor proteins that move a chromosome along a microtubule.

kinetochore microtubules Microtubules that form during mitosis and meiosis, and which extend from a spindle apparatus to an attachment point—the kinetochore—on a chromosome.

kinocilium (plural: kinocilia) A single cilium that juts from the surface of many hair cells and functions in detection of sound or pressure.

Klinefelter syndrome A syndrome seen in humans who have an XXY karyotype. People with this syndrome have male sex organs, may have some female traits, and are sterile.

knock-out allele A mutant allele that does not function at all, or an organism homozygous for such a mutation. Also called *null allele* or *loss-of-function allele*.

Koch's postulates Four criteria used to determine whether a suspected infectious agent causes a particular disease.

labia major (plural: labium majus) One of two outer folds of skin that protect the labia minora, clitoris, and vaginal opening of female mammals.

labia minora (plural: labium minus) One of two inner folds of skin that protect the opening of the urethra and vagina.

labor The strong muscular contractions of the uterus that expel the fetus during birth.

lactation (verb: lactate) Production of milk from mammary glands of mammals.

lacteal A small lymphatic vessel extending into the center of a villus in the small intestine. Receives chylomicrons containing fat absorbed from food.

lactic acid fermentation Catabolic pathway in which pyruvate produced by glycolysis is converted to lactic acid in the absence of oxygen.

lagging strand In DNA replication, the strand of new DNA that is synthesized discontinuously in a series of short pieces that are later joined together. Also called *discontinuous strand.* Compare with **leading strand.**

large intestine The distal portion of the digestive tract consisting of the cecum, colon, and rectum. Its primary function is to compact the wastes delivered from the small intestine and absorb enough water to form feces.

larva (plural: larvae) An immature stage of a species in which the immature and adult stages have different body forms.

late endosome A membrane-bound vesicle that arises from an early endosome and develops into a lysosome.

latency In viruses that infect animals, the ability to exist in a quiescence state without producing new virions.

lateral bud A bud that forms in the angle between a leaf and a stem and may develop into a lateral (side) branch. Also called *axillary bud.*

lateral gene transfer Transfer of DNA between two different species, especially distantly related species. Commonly occurs among bacteria and archaea via plasmid exchange; also can occur in eukaryotes via viruses and some other mechanisms.

lateral line system A pressure-sensitive sensory organ found in many aquatic vertebrates.

lateral meristem A layer of undifferentiated plant cells found in older stems and roots that is responsible for secondary growth. Also called *cambium* or *secondary meristem.* Compare with **apical meristem.**

lateral root A plant root extending from another, older root.

leaching Loss of nutrients from soil via percolating water.

leading strand In DNA replication, the strand of new DNA that is synthesized in one continuous piece, with nucleotides added to the 3′ end of the growing molecule. Also called *continuous strand.* Compare with **lagging strand.**

leaf The main photosynthetic organ of vascular plants.

leak channel Potassium channel that allows potassium ions to leak out of a neuron in its resting state.

learning An enduring change in an individual's behavior that results from specific experience(s).

leghemoglobin An iron-containing protein similar to hemoglobin. Found in root nodules of legume plants where it binds oxygen, preventing it from poisoning a bacterial enzyme needed for nitrogen fixation.

legumes Members of the pea plant family that form symbiotic associations with nitrogen-fixing bacteria in their roots.

lens A transparent, crystalline structure that focuses incoming light onto a retina or other light-sensing apparatus of an eye.

lenticels Spongy segments in bark that allow gas exchange between cells in a woody stem and the atmosphere.

leptin A hormone produced and secreted by fat cells (adipocytes) that acts to stabilize fat tissue mass in part by inhibiting appetite and increasing energy expenditure.

leukocytes Several types of blood cells, including neutrophils, macrophages, and lymphocytes, that circulate in blood and lymph and function in defense against pathogens. Also called *white blood cells.*

lichen A symbiotic association of a fungus and a photosynthetic alga.

life cycle The sequence of developmental events and phases that occurs during the life span of an organism, from fertilization to offspring production.

life history The sequence of events in an individual's life from birth to reproduction to death, including how an individual allocates resources to growth, reproduction, and activities or structures that are related to survival.

life table A data set that summarizes the probability that an individual in a certain population will survive and reproduce in any given year over the course of its lifetime.

ligand Any molecule that binds to a specific site on a receptor molecule.

ligand-gated channel An ion channel that opens or closes in response to binding by a certain molecule. Compare with **voltage-gated channel.**

light chain The smaller of the two types of polypeptide chains in an antibody molecule; composed of a variable (V) region, which contributes to the antigen-binding site, and a constant (C) region. Compare with **heavy chain.**

lignin A substance found in the secondary cell walls of some plants that is exceptionally stiff and strong. Most abundant in woody plant parts.

limiting nutrient Any essential nutrient whose scarcity in the environment significantly reduces growth and reproduction of organisms.

limnetic zone Open water (not near shore) that receives enough light to support photosynthesis.

lineage See **monophyletic group.**

LINEs (long interspersed nuclear elements) The most abundant class of transposable elements in human genomes; can create copies of itself and insert them elsewhere in the genome. Compare with **SINEs.**

linkage In genetics, a physical association between two genes because they are on the same chromosome; the inheritance patterns resulting from this association.

linkage map See **genetic map.**

lipase Any enzyme that can break down fat molecules into fatty acids and monoglycerides.

lipid Any organic subtance that does not dissolve in water, but dissolves well in nonpolar organic solvents. Lipids include fats, oils, phospholipids, and waxes.

lipid bilayer The basic structural element of all cellular membranes consisting of a two-layer sheet of phospholipid molecules whose hydrophobic tails are oriented toward the inside and hydrophilic heads, toward the outside. Also called *phospholipid bilayer.*

littoral zone Shallow water near shore that receives enough sunlight to support photosynthesis. May be marine or freshwater; often flowering plants are present.

liver A large, complex organ of vertebrates that performs many functions including storage of glycogen, processing and conversion of food and wastes, and production of bile.

lobe-finned fish Fish with fins supported by bony elements that extend down the length of the structure.

locomotion Movement of an organism under its own power.

locus (plural: loci) A gene's physical location on a chromosome.

logistic population growth The density-dependent decrease in growth rate as population size approaches the carrying capacity. Compare with **exponential population growth.**

long interspersed nuclear elements See **LINEs.**

long-day plant A plant that blooms in response to short nights (usually in late spring or early summer in the northern hemisphere). Compare with **day-neutral** and **short-day plant.**

loop of Henle In the vertebrate kidney, a long U-shaped loop in a nephron that extends into the medulla. Functions as a countercurrent exchanger to set up an osmotic gradient that allows reabsorption of water from a subsequent portion of the nephron.

loose connective tissue A type of connective tissue consisting of fibrous proteins in a soft matrix. Often functions as padding for organs.

lophophore A specialized feeding structure found in some lophotrochozoans and used in filter feeding.

lophotrochozoans A major lineage of protostomes (Lophotrochozoa) that grow by extending the size of their skeletons rather than by molting. Many phyla have a specialized feeding structure (lophophore) and/or ciliated larvae (trochophore). Includes rotifers, flatworms, segmented worms, and mollusks. Compare with **ecdysozoans.**

loss-of-function allele See **knock-out allele.**

lumen The interior space of any hollow structure (e.g., the rough ER) or organ (e.g., the stomach).

lung Any respiratory organ used for gas exchange between blood and air.

luteal phase The second major phase of a menstrual cycle, after ovulation, when the progesterone levels are high and the body is preparing for a possible pregnancy.

luteinizing hormone (LH) A peptide hormone, produced and secreted by the anterior pituitary, that stimulates estrogen production, ovulation, and formation of the corpus luteum in females and testosterone production in males.

lymph The mixture of fluid and white blood cells that circulates through the ducts and lymph nodes of the lymphatic system in vertebrates.

lymph node One of numerous small oval structures through which lymph moves in the lymphatic system. Filter the lymph and screen it for pathogens and other antigens. Major sites of lymphocyte activation.

lymphatic system In vertebrates, a body-wide network of thin-walled ducts (or vessels) and lymph nodes, separate from the circulatory system. Collects excess fluid from body tissues and returns it to the blood; also functions as part of the immune system.

lymphocytes Two types of leukocyte—B cells and T cells—that circulate through the bloodstream and lymphatic system and that are responsible for the development of acquired immunity.

lysogenic cycle A type of viral replication in which a viral genome enters a host cell, is inserted into the host's chromosome, and is replicated whenever the host cell divides. When activated, the viral DNA enters the lytic cycle, leading to production of new virus particles. Also called *lysogeny* or *latent growth.* Compare with **lytic cycle.**

lysosome A small organelle in an animal cell containing acids and enzymes that catalyze hydrolysis reactions and can digest large molecules. Compare with **vacuole.**

lysozyme An enzyme that functions in innate immunity by digesting bacterial cell walls. Occurs in saliva, tears, mucus, and egg white.

lytic cycle A type of viral replication in which a viral genome enters a host cell, new virus particles (virions) are made using host enzymes and eventually burst out of the cell, killing it. Also called *replicative growth.* Compare with **lysogenic cycle.**

macromolecule Any very large organic molecule, usually made up of smaller molecules (monomers) joined together into a polymer. The main biological macromolecules are proteins, nucleic acids, and polysaccharides.

macronutrient Any element (e.g., carbon, oxygen, nitrogen) that is required in large quantities for normal growth, reproduction, and maintenance of a living organism. Compare with **micronutrient**.

MADS box A DNA sequence that codes for a DNA-binding motif in proteins; present in floral organ identity genes in plants. Functionally similar sequences are found in some fungal and animal genes.

major histocompatibility protein See **MHC protein**.

maladaptive A trait that lowers fitness.

malaria A human disease caused by four species of the protist *Plasmodium* and passed to humans by mosquitoes.

malignant tumor A tumor that is actively growing and disrupting local tissues and/or is spreading to other organs. Cancer consists of one or more malignant tumors. Compare with **benign tumor**.

Malpighian tubules A major excretory organ of insects, consisting of blind-ended tubes that extend from the gut into the hemocoel. Filter hemolymph to form "pre-urine" and then send it to the hindgut for further processing.

mammals One of the two lineages of amniotes (vertebrates that produce amniotic eggs) distinguished by hair (or fur) and mammary glands. Includes the monotremes (platypuses), marsupials, and eutherians (placental mammals).

mammary glands Specialized exocrine glands that produce and secrete milk for nursing offspring. A diagnostic feature of mammals.

mandibles Any mouthpart used in chewing. In vertebrates, the lower jaw. In insects, crustaceans, and myriapods, the first pair of mouthparts.

mantle One of the three main parts of the mollusk body; the thick outer tissue that protects the visceral mass and may secrete a calcium carbonate shell.

Marfan syndrome A human syndrome involving increased height, long limbs and fingers, an abnormally shaped chest, and heart disorders. Probably caused by mutation in one pleiotropic gene.

marsh A wetland that lacks trees and usually has a slow but steady rate of water flow.

marsupials A lineage of mammals (Marsupiala) that nourish their young in an abdominal pouch after a very short period of development in the uterus.

mass extinction The extinction of a large number of diverse evolutionary groups during a relatively short period of geologic time (about 1 million years). May occur due to sudden and extraordinary environmental changes. Compare with **background extinction**.

mass feeder An animal that takes chunks of food into its mouth.

mass number The total number of protons and neutrons in an atom.

mast cell A type of leukocyte that is stationary (embedded in tissue) and helps trigger the inflammatory response to infection or injury, including secretion of histamine. Particularly important in allergic responses and defense against parasites.

maternal chromosome A chromosome inherited from the mother.

mechanoreceptor A sensory cell or organ specialized for detecting distortions caused by touch or pressure. One example is hair cells in the cochlea.

mediator complex regulatory proteins that form a physical link between regulatory transcription factors that are bound to DNA and the basal transcription complex

medium A liquid or solid in which cells can grow in vitro. Also called *growth medium.*

medulla The innermost part of an organ (e.g., kidney or adrenal gland).

medulla oblongata In vertebrates, a region of the brain stem that along with the cerebellum forms the hindbrain.

medusa (plural: medusae) The free-floating stage in the life cycle of some cnidarians (e.g., jellyfish). Compare with **polyp**.

megapascal (MPa) A unit of pressure (force per unit area), equivalent to 1 million pascals (Pa).

megasporangium (plural: megasporangia) In heterosporous species of plants, a spore-producing structure that produces megaspores, which go on to develop into female gametophytes.

megaspore In seed plants, a haploid (n) spore that is produced in a megasporangium by meiosis of a diploid ($2n$) megasporocyte; develops into a female gametophyte. Compare with **microspore**.

meiosis In sexually reproducing organisms, a special two-stage type of cell division in which one diploid ($2n$) parent cell produces four haploid (n) reproductive cells (gametes); results in halving of the chromosome number. Also called *reduction division.*

meiosis I The first cell division of meiosis, in which synapsis and crossing over occur, and homologous chromosomes are separated from each other, producing daughter cells with half as many chromosomes (each composed of two sister chromatids) as the parent cell.

meiosis II The second cell division of meiosis, in which sister chromatids are separated from each other. Similar to mitosis.

meiotic map See **genetic map**.

membrane potential A difference in electric charge across a cell membrane; a form of potential energy. Also called *membrane voltage.*

memory Retention of learned information.

memory cells A type of lymphocyte responsible for maintenance of immunity for years or decades after an infection. Descended from a B cell or T cell activated during a previous infection.

meniscus (plural: menisci) The concave boundary layer formed at most air-water interfaces due to surface tension.

menstrual cycle A female reproductive cycle seen in Old World monkeys and apes (including humans) in which the uterine lining is shed (menstruation) if no pregnancy occurs. Compare with **estrous cycle**.

menstruation The periodic shedding of the uterine lining through the vagina that occurs in females of Old World monkeys and apes, including humans.

meristem (adjective: meristematic) In plants, a group of undifferentiated cells that can develop into various adult tissues throughout the life of a plant.

mesoderm The middle of the three basic cell layers in most animal embryos; gives rise to muscles, bones, blood, and some internal organs (kidney, spleen, etc.). Compare with **ectoderm** and **endoderm**.

mesoglea A gelatinous material, containing scattered ectodermal cells, that is located between the ectoderm and endoderm of cnidarians (e.g., jellyfish, corals, and anemones).

mesophyll cell A type of cell, found near the surfaces of plant leaves, that is specialized for the light-dependent reactions of photosynthesis.

Mesozoic era The interval of geologic time, from 251 million to 65.5 million years ago, during which gymnosperms were the dominant plants and dinosaurs the dominant vertebrates. Ended with extinction of the dinosaurs.

messenger RNA (mRNA) An RNA molecule that carries encoded information, transcribed from DNA, that specifies the amino acid sequence of a polypeptide.

meta-analysis A comparative analysis of the results of many smaller, previously published studies.

metabolic pathway An ordered series of chemical reactions that build up or break down a particular molecule. Often, each reaction is catalyzed by a different enzyme.

metabolic rate The total energy use by all the cells of an individual. For aerobic organisms, often measured as the amount of oxygen consumed per hour.

metabolic water The water that is produced as a by-product of cellular respiration.

metabolism All the chemical reactions occurring in a living cell or organism.

metagenomics Sequencing of all or most of the genes present in an environment directly (also called environmental sequencing).

metallothioneins Small plant proteins that bind to and prevent excess metal ions from acting as toxins.

metamorphosis Transition from one developmental stage to another, such as from the larval to the adult form of an animal.

metaphase A stage in mitosis or meiosis during which chromosomes line up in the middle of the cell.

metaphase plate The plane along which chromosomes line up during metaphase of mitosis or meiosis; not an actual structure.

metapopulation A population made up of many small, physically isolated populations.

metastasis The spread of cancerous cells from their site of origin to distant sites in the body where they may establish additional tumors.

methanogen A prokaryote that produces methane (CH_4) as a by-product of cellular respiration.

methanotroph An organism that uses methane (CH_4) as its primary electron donor and source of carbon.

methyl salicylate (MeSA) A molecule that is hypothesized to function as a signal, transported among tissues, that triggers systematic acquired resistance in plants—a response to pathogen attack.

methylation The addition of a methyl ($-CH_3$) group to a molecule.

MHC protein One of a large set of mammalian cell-surface glycoproteins involved in marking cells as self and in antigen presentation to T cells. Also called *MHC molecule.* See **Class I** and **Class II MHC protein**.

microbe Any microscopic organism, including bacteria, archaea, and various tiny eukaryotes.

microbiology The field of study concerned with microscopic organisms.

microfilament See **actin filament**.

micrograph A photograph of an image produced by a microscope.

micronutrient Any element (e.g., iron, molybdenum, magnesium) that is required in very small quantities for normal growth, reproduction, and maintenance of a living organism. Compare with **macronutrient**.

micropyle The tiny pore in a plant ovule through which the pollen tube reaches the embryo sac.

microRNA (miRNA) A small, single-stranded RNA associated with proteins in an RNA-induced silencing complex. Can bind to complementary sequences in mRNA molecules, allowing the associated proteins to degrade the bound mRNA or inhibit its translation. See **RNA interference**.

microsatellite A noncoding stretch of eukaryotic DNA consisting of a repeating sequence 1- to 5-base pair long. Also called *simple sequence repeat.*

microsporangim (plural: microsporangia) In heterosporous species of plants, a spore-producing structure that produces microspores, which go on to develop into male gametophytes.

microspore In seed plants, a haploid (n) spore that is produced in a microsporangium by meiosis of a diploid ($2n$) microsporocyte; develops into a male gametophyte. Compare with **megaspore**.

microtubule A long, tubular fiber, about 25 nm in diameter, formed by polymerization of tubulin protein dimers; one of the three types of cytoskeletal fibers. Involved in cell movement and transport of materials within the cell. Compare with **actin filament** and **intermediate filament**.

microtubule-organizing center (MTOC) General term for any structure (e.g., centrosome and basal body) that organizes microtubules in cells.

microvilli (singular: microvillus) Tiny protrusions from the surface of an epithelial cell that increase the surface area for absorption of substances.

midbrain One of the three main regions of the vertebrate brain; includes sensory integrating and relay centers. Compare with **forebrain** and **hindbrain**.

middle ear The air-filled middle portion of the mammalian ear, which contains three small bones (ossicles) that transmit and amplify sound from the tympanic membrane to the inner ear. Is connected to the throat via the eustachian tube.

migration (1) In ecology, a cyclical movement of large numbers of organisms from one geographic location or habitat to another. (2) In population genetics, movement of individuals from one population to another.

millivolt (mV) A unit of voltage equal to 1/1000 of a volt.

mimicry A phenomenon in which one species has evolved (or learns) to look or sound like another species. See **Batesian mimicry** and **Müllerian mimicry**.

mineralocorticoids A class of steroid hormones, produced and secreted by the adrenal cortex, that regulate electrolyte levels and the overall volume of body fluids. Aldosterone is the principal one in humans. Compare with **glucocorticoids**.

minisatellite A noncoding stretch of eukaryotic DNA consisting of a repeating sequence that is 6 to 500 base pairs long. Also called *variable number tandem repeat (VNTR).*

mismatch repair The process by which mismatched base pairs in DNA are fixed.

missense mutation A point mutation (change in a single base pair) that causes a change in the amino acid sequence of a protein. Also called *replacement mutation.*

mitochondrial matrix Central compartment of a mitochondrion, which is lined by the inner membrane; contains the enzymes and substrates of the citric acid cycle and mitochondrial DNA.

mitochondrion (plural: mitochondria) A eukaryotic organelle that is bounded by a double membrane and is the site of aerobic respiration.

mitosis In eukaryotic cells, the process of nuclear division that results in two daughter nuclei genetically identical to the parent nucleus. Subsequent cytokinesis (division of the cytoplasm) yields two daughter cells.

mitosis-promoting factor (MPF) A complex of a cyclin and cyclin-dependent kinase that phosphorylates a number of specific proteins needed to initiate mitosis in eukaryotic cells.

mitotic (M) phase The phase of the cell cycle during which cell division occurs. Includes mitosis and cytokinesis.

model organism An organism selected for intensive scientific study based on features that make it easy to work with (e.g., body size, life span), in the hope that findings will apply to other species.

molarity A common unit of solute concentration equal to the number of moles of a dissolved solute in 1 liter of solution.

mole The amount of a substance that contains 6.022×10^{23} of its elemental entities (e.g., atoms, ions, or molecules). This number of molecules of a compound will have a mass equal to the molecular weight of that compound expressed in grams.

molecular chaperone A protein that facilitates the three-dimensional folding of newly synthesized proteins, usually by an ATP-dependent mechanism.

molecular formula A notation that indicates only the numbers and types of atoms in a molecule, such as H_2O for the water molecule. Compare with **structural formula**.

molecular weight The sum of the mass numbers of all of the atoms in a molecule; roughly, the total number of protons and neutrons in the molecule.

molecule A combination of two or more atoms held together by covalent bonds.

mollusks Members of the phylum Mollusca. Distinguished by a body plan with three main parts: a muscular foot, a visceral mass, and a mantle. Include bivalves (clams, oysters), gastropods (snails, slugs), chitons, and cephalopods (squid, octopuses). Mollusks belong to the lophotrochozoan branch of the protostomes.

molting A method of body growth, used by ecdysozoans, that involves the shedding of an external protective cuticle or skeleton, expansion of the soft body, and growth of a new external layer.

monocot Any plant that has a single cotyledon (embryonic leaf) upon germination. Monocots form a monophyletic group. Also called a monocotyledonous plant. Compare with **dicot**.

monoecious Describing an angiosperm species that has both male and female reproductive structures on each plant. Compare with **dioecious**.

monohybrid cross A mating between two parents that are both heterozygous for a given gene.

monomer A small molecule that can covalently bind to other similar molecules to form a larger macromolecule. Compare with **polymer**.

monophyletic group An evolutionary unit that includes an ancestral population and all of its descendants but no others. Also called a *clade* or *lineage.* Compare with **paraphyletic group**.

monosaccharide A small carbohydrate, such as glucose, that has the molecular formula $(CH_2O)_n$ and cannot be hydrolyzed to form any smaller carbohydrates. Also called *simple sugar.* Compare with **disaccharide** and **polysaccharide**.

monosomy Having only one copy of a particular type of chromosome.

monotremes A lineage of mammals (Monotremata) that lay eggs and then nourish the young with milk. Includes just three living species: the platypus and two species of echidna.

morphogen A molecule that exists in a concentration gradient and provides spatial information to embryonic cells.

morphospecies concept The definition of a species as a population or group of populations that have measurably different anatomical features from other groups. Also called *morphological species concept.* Compare with **biological** and **phylogenetic species concept**.

morphology The shape and appearance of an organism's body and its component parts.

motor neuron A nerve cell that carries signals from the central nervous system (brain and spinal cord) to an effector, such as a muscle or gland. Compare with **interneuron** and **sensory neuron**.

motor protein A class of proteins whose major function is to convert the chemical energy of ATP into motion. Includes dynein, kinesin, and myosin.

MPF See **mitosis-promoting factor**.

mRNA See **messenger RNA**.

mucigel A slimy substance secreted by plant root caps that eases passage of the growing root through the soil.

mucosal-associated lymphoid tissue (MALT) Collective term for lymphocytes and other leukocytes associated with skin cells and with mucus-secreting epithelial tissues in the gut and respiratory tract. Plays important role in preventing entry of pathogens into the body.

mucous cell A type of cell found in the epithelial layer of the stomach; responsible for secreting mucus into the stomach.

mucus (adjective: mucous) A slimy mixture of glycoproteins (called mucins) and water that is secreted in many animal organs for lubrication.

Müllerian inhibitory substance A peptide hormone secreted by the embryonic testis that causes regression (withering away) of the female reproductive ducts.

Müllerian mimicry A type of mimicry in which two (or more) harmful species resemble each other. Compare with **Batesian mimicry**.

multicellularity The state of being composed of many cells that adhere to each other and do not all express the same genes with the result that some cells have specialized functions.

multienzyme complex A group of enzymes that are physically attached to each other, even though each of the enzymes catalyzes a separate—but usually related—chemical reaction.

multiple allelism In a population, the existence of more than two alleles of the same gene.

multiple fruit A fruit (e.g., pineapple) that develops from many separate flowers and thus many carpels. Compare with **aggregate** and **simple fruit**.

multiple sclerosis (MS) A human autoimmune disease caused by the immune system attacking the myelin sheaths that insulate nerve axons.

muscle fiber A single muscle cell.

muscle tissue An animal tissue consisting of bundles of long, thin contractile cells (muscle fibers).

mutagen Any physical or chemical agent that increases the rate of mutation.

mutant An individual that carries a mutation, particularly a new or rare mutation.

mutation Any change in the hereditary material of an organism (DNA in most organisms, RNA in some viruses).

mutualism (adjective: mutualistic) A symbiotic relationship between two organisms (mutualists) that benefits both. Compare with **commensalism** and **parasitism**.

mycelium (plural: mycelia) A mass of underground filaments (hyphae) that form the body of a fungus. Also found in some protists and bacteria.

mycorrhiza (plural: mycorrhizae) A mutualistic association between certain fungi and most vascular plants, sometimes visible as nodules or nets in or around plant roots.

myelin sheath Multiple layers of myelin, derived from the cell membranes of certain glial cells, that is wrapped around the axon of a neuron, providing electrical insulation.

myoD A transcription factor that is critical for differentiation of muscle cells (short for "*myo*blast *d*etermination").

myofibril Long, slender structure composed of contractile proteins organized into repeating units (sarcomeres) in vertebrate heart muscle and skeletal muscle.

myosin Any one of a class of motor proteins that use the chemical energy of ATP to move along actin filaments in muscle contraction, cytokinesis, and vesicle transport.

myriapods A lineage of arthropods with long segmented trunks, each segment bearing one or two pairs of legs. Includes millipedes and centipedes.

NAD^+/NADH Oxidized and reduced forms, respectively, of nicotinamide adenine dinucleotide. A nonprotein electron carrier that functions in many of the redox reactions of metabolism.

$NADP^+$/NADPH Oxidized and reduced forms, respectively, of nicotinamide adenine dinucleotide phosphate. A nonprotein electron carrier that is reduced during the light-dependent reactions in photosynthesis and extensively used in biosynthetic reactions.

natural experiment A situation in which groups to be compared are created by an unplanned, natural change in conditions rather than by manipulation of conditions by researchers.

natural selection The process by which individuals with certain heritable traits tend to produce more surviving offspring than do individuals without those traits, often leading to a change in the genetic makeup of the population. A major mechanism of evolution.

nauplius A distinct planktonic larval stage seen in many crustaceans.

Neanderthal A recently extinct European species of hominid, *Homo neanderthalensis*, closely related to but distinct from modern humans.

nectar The sugary fluid produced by flowers to attract and reward pollinating animals.

nectary A nectar-producing structure in a flower.

negative control Of genes, when a regulatory protein shuts down expression by binding to DNA on or near the gene

negative feedback A self-limiting, corrective response in which a deviation in some variable (e.g., body temperature, blood pH, concentration of some compound) triggers responses aimed at returning the variable to normal. Compare with **positive feedback**.

negative pressure ventilation Ventilation of the lungs that is accomplished by "pulling" air into the lungs by expansion of the rib cage. Compare with **positive pressure ventilation**.

negative-sense virus A virus whose genome contains sequences complementary to those in the mRNA required to produce viral proteins. Compare with **ambisense virus** and **positive-sense virus**.

nematodes See **roundworms**.

nephron One of the tiny tubes within the vertebrate kidney that filter blood and concentrate salts to produce urine. Also called *renal tubule*.

neritic zone Shallow marine waters beyond the intertidal zone, extending down to about 200 meters, where the continental shelf ends.

nerve A long, tough strand of nervous tissue typically containing thousands of axons wrapped in connective tissue; carries impulses between the central nervous system and some other part of the body.

nerve cord A bundle of nerves extending from the brain along the dorsal (back) side of a chordate animal, with cerebrospinal fluid inside a hollow central channel. One of the defining features of chordates.

nerve net A nervous system in which neurons are diffuse instead of being clustered into large masses or tracts.

nervous tissue An animal tissue consisting of nerve cells (neurons) and various supporting cells.

net primary productivity (NPP) In an ecosystem, the total amount of carbon fixed by photosynthesis over a given time period minus the amount oxidized during cellular respiration. Compare with **gross primary productivity**.

net reproductive rate (R_0) The growth rate of a population per generation; equivalent to the average number of female offspring that each female produces over her lifetime.

neural Relating to nerve cells (neurons) and the nervous system.

neural tube A folded tube of ectoderm that forms along the dorsal side of a young vertebrate embryo and that will give rise to the brain and spinal cord.

neuroendocrine Referring to nerve cells (neurons) that release hormones into the blood or to such hormones themselves.

neuron A cell that is specialized for the transmission of nerve impulses. Typically has dendrites, a cell body, and a long axon that forms synapses with other neurons. Also called *nerve cell*.

neurosecretory cell A nerve cell (neuron) that produces and secretes hormones into the bloodstream. Principally found in the hypothalamus. Also called *neuroendocrine cell*.

neurotoxin Any substance that specifically destroys or blocks the normal functioning of neurons.

neurotransmitter A molecule that transmits electrical signals from one neuron to another or from a neuron to a muscle or gland. Examples are acetylcholine, dopamine, serotonin, and norepinephrine.

neutral In genetics, referring to any mutation or mutant allele that has no effect on an individual's fitness.

neutrophil A type of leukocyte, capable of moving through body tissues, that engulfs and digests pathogens and other foreign particles; also secretes various compounds that attack bacteria and fungi.

niche The particular set of habitat requirements of a certain species and the role that species plays in its ecosystem.

niche differentiation The change in resource use by competing species that occurs as the result of character displacement.

nicotinamide adenine dinucleotide See NAD^+/NADH.

nicotinamide adenine dinucleotide phosphate See $NADP^+$/NADPH.

nitrogen cycle, global The movement of nitrogen among terrestrial ecosystems, the oceans, and the atmosphere.

nitrogen fixation The incorporation of atmospheric nitrogen (N_2) into forms such as ammonia (NH_3) or nitrate (NO_3^-), which can be used to make many organic compounds. Occurs in only a few lineages of bacteria and archaea.

nociceptor A sensory cell or organ specialized to detect tissue damage, usually producing the sensation of pain.

Nod factors Molecules produced by nitrogen-fixing bacteria that help them recognize and bind to roots of legumes.

node (1) In animals, any small thickening (e.g., a lymph node). (2) In plants, the part of a stem where leaves or leaf buds are attached. (3) In a phylogenetic tree, the point where two branches diverge, representing the point in time when an ancestral group split into two or more descendant groups. Also called *fork*.

node of Ranvier One of the periodic unmyelinated sections of a neuron's axon at which an action potential can be regenerated.

nodule Lumplike structure on roots of legume plants that contain symbiotic nitrogen-fixing bacteria.

noncyclic electron flow See **Z scheme**.

nondisjunction An error that can occur during meiosis or mitosis in which one daughter cell receives two copies of a particular chromosome, and the other daughter cell receives none.

nonpolar covalent bond A covalent bond in which electrons are equally shared between two atoms of the same or similar electronegativity. Compare with **polar covalent bond**.

non-sister chromatids The chromatids of a particular type of chromosome (after replication) with respect to the chromatids of its homologous chromosome. Crossing over occurs between non-sister chromatids. Compare with **sister chromatids**.

non-template strand The strand of DNA that is not transcribed during synthesis of RNA. Its sequence corresponds to that of the mRNA produced from the other strand. Also called *coding strand*.

non-vascular plants See **bryophytes**.

norepinephrine A catecholamine used as a neurotransmitter in the sympathetic nervous system. Also is produced by the adrenal medulla and functions as a hormone that triggers rapid responses relating to the fight-or-flight response.

notochord A long, gelatinous, supportive rod down the back of a chordate embryo, below the developing spinal cord. Replaced by vertebrae in most adult vertebrates. A defining feature of chordates.

nuclear envelope The double-layered membrane enclosing the nucleus of a eukaryotic cell.

nuclear lamina A lattice-like sheet of fibrous nuclear lamins, which are one type of intermediate filaments. Lines the inner membrane of the nuclear envelope, stiffening the envelope and helping organize the chromosomes.

nuclear lamins Intermediate filaments that make up the nuclear lamina layer—a lattice-like layer inside the nuclear envelope that stiffens the structure.

nuclear localization signal (NLS) A short amino acid sequence that marks a protein for delivery to the nucleus.

nuclear pore An opening in the nuclear envelope that connects the inside of the nucleus with the cytoplasm and through which molecules such as mRNA and some proteins can pass.

nuclear pore complex A large complex of dozens of proteins lining a nuclear pore, defining its shape and transporting substances through the pore.

nuclease Any enzyme that can break down RNA or DNA molecules.

nucleic acid A macromolecule composed of nucleotide monomers. Generally used by cells to store or transmit hereditary information. Includes ribonucleic acid and deoxyribonucleic acid.

nucleoid In prokaryotic cells, a dense, centrally located region that contains DNA but is not surrounded by a membrane.

nucleolus In eukaryotic cells, specialized structure in the nucleus where ribosomal RNA processing occurs and ribosomal subunits are assembled.

nucleosome A repeating, bead-like unit of eukaryotic chromatin, consisting of about 200 nucleotides of DNA wrapped twice around eight histone proteins.

nucleotide A molecule consisting of a five-carbon sugar (ribose or deoxyribose), a phosphate group, and one of several nitrogen-containing bases. DNA and RNA are polymers of nucleotides containing deoxyribose (deoxyribonucleotides) and ribose (ribonucleotides), respectively. Equivalent to a nucleoside plus one phosphate group.

nucleotide excision repair The process of removing a damaged region in one strand of DNA and correctly replacing it using the undamaged strand as a template.

nucleus (1) The center of an atom, containing protons and neutrons. (2) In eukaryotic cells, the large organelle containing the chromosomes and surrounded by a double membrane. (3) A discrete clump of neuron cell bodies in the brain, usually sharing a distinct function.

null allele See **knock-out allele**.

null hypothesis A hypothesis that specifies what the results of an experiment will be if the main hypothesis being tested is wrong. Often states that there will be no difference between experimental groups.

nutrient A substance that an organism requires for normal growth, maintenance, or reproduction.

occipital lobe In the vertebrate brain, one of the four major areas in the cerebrum.

oceanic zone The waters of the open ocean beyond the continental shelf.

oil A fat that is liquid at room temperature.

Okazaki fragment Short segment of DNA produced during replication of the 5′ to 3′ template strand. Many Okazaki fragments make up the lagging strand in newly synthesized DNA.

olfaction The perception of odors.

olfactory bulb A bulb-shaped projection of the brain just above the nose. Receives and interprets odor information from the nose.

oligodendrocyte A type of glial cell that wraps around axons of some neurons in the central nervous system, forming a myelin sheath that provides electrical insulation. Compare with **Schwann cell**.

oligopeptide A chain composed of fewer than 50 amino acids linked together by peptide bonds. Often referred to simply as *peptide*.

ommatidium (plural: ommatidia) A light-sensing column in an arthropod's compound eye.

omnivore (adjective: omnivorous) An animal whose diet regularly includes both meat and plants. Compare with **carnivore** and **herbivore**.

oncogene Any gene whose protein product stimulates cell division at all times and thus promotes cancer development. Often is a mutated form of a gene involved in regulating the cell cycle. See **proto-oncogene**.

one-gene, one-enzyme hypothesis The hypothesis that each gene is responsible for making one (and only one) protein, in most cases an enzyme that catalyzes a specific reaction. Many exceptions to this hypothesis are now known.

oocyte A cell in the ovary that can undergoes meiosis to produce an ovum.

oogenesis The production of egg cells (ova).

oogonia (singular: oogonia) The diploid cells in an ovary that can divide by mitosis to create more oogonia and primary oocytes, which can undergo meiosis.

open circulatory system A circulatory system in which the circulating fluid (hemolymph) is not confined to blood vessels. Compare with **closed circulatory system**.

open reading frame (ORF) Any DNA sequence, ranging in length from several hundred to thousands of base pairs long, that is flanked by a start codon and a stop codon. ORFs identified by computer analysis of DNA may be functional genes, especially if they have other features characteristic of genes (e.g., promoter sequence).

operator In prokaryotic DNA, a binding site for a repressor protein; located near the start of an operon.

operculum The stiff flap of tissue that covers the gills of teleost fishes.

operon A region of prokaryotic DNA that codes for a series of functionally related genes and is transcribed from a single promoter into a polycistronic mRNA.

opsin A transmembrane protein that is covalently linked to retinal, the light-detecting pigment in rod and cone cells.

optic nerve A bundle of neurons that runs from the eye to the brain.

optimal foraging The concept that animals forage in a way that maximizes the amount of usable energy they take in, given the costs of finding and ingesting their food and the risk of being eaten while they're at it.

orbital The spherical region around an atomic nucleus in which an electron is present most of the time.

ORF See **open reading frame**.

organ A group of tissues organized into a functional and structural unit.

organ system Groups of tissues and organs that work together to perform a function.

organelle Any discrete, membrane-bound structure within a cell (e.g., mitochondrion) that has a characteristic structure and functions.

organic For a compound, containing carbon and hydrogen and usually containing carbon-carbon bonds. Organic compounds are widely used by living organisms.

organism Any living entity that contains one or more cells.

organogenesis A stage of embryonic development, just after gastrulation in vertebrate embryos, during which major organs develop from the three embryonic germ layers.

origin of replication The site on a chromosome at which DNA replication begins.

osmoconformer An animal that does not actively regulate the osmolarity of its tissues but conforms to the osmolarity of the surrounding environment.

osmolarity The concentration of dissolved substances in a solution, measured in moles per liter.

osmoregulation The process by which a living organism controls the concentration of water and salts in its body.

osmoregulator An animal that actively regulates the osmolarity of its tissues.

osmosis Diffusion of water across a selectively permeable membrane from a region of high water concentration (low solute concentration) to a region of low water concentration (high solute concentration).

osmotic potential See **solute potential**.

ossicles, ear In mammals, three bones found in the middle ear that function in transferring and amplifying sound from the outer ear to the inner ear.

ouabain A plant toxin that poisons the sodium-potassium pumps of animals.

outcrossing Reproduction by fusion of the gametes of different individuals, rather than self-fertilization. Typically refers to plants.

outer ear The outermost portion of the mammalian ear, consisting of the pinna (ear flap) and the ear canal. Funnels sound to the tympanic membrane.

outgroup A taxon that is closely related to a particular monophyletic group but is not part of it.

out-of-Africa hypothesis The hypothesis that modern humans (*Homo sapiens*) evolved in Africa and spread to other continents, replacing other *Homo* species without interbreeding with them.

oval window A membrane separating the fluid-filled cochlea from the air-filled middle ear through which sound vibrations pass from the middle ear to the inner ear in mammals.

ovary The egg-producing organ of a female animal, or the seed-producing structure in the female part of a flower.

oviduct See **fallopian tube**.

oviparous Producing eggs that are laid outside the body where they develop and hatch. Compare with **ovoviviparous** and **viviparous**.

ovoviviparous Producing eggs that are retained inside the body until they are ready to hatch. Compare with **oviparous** and **viviparous**.

ovulation The release of an ovum from an ovary of a female vertebrate. In humans, an ovarian follicle releases an egg at the end of the follicular phase of the menstrual cycle.

ovule In flowering plants, the structure inside an ovary that contains the female gametophyte and eventually (if fertilized) becomes a seed.

ovum (plural: ova) See **egg**.

oxidation The loss of electrons from an atom during a redox reaction, either by donation of an electron to another atom or by the shared electrons in covalent bonds moving farther from the atomic nucleus.

oxidative phosphorylation Production of ATP molecules from the redox reactions of an electron transport chain.

oxygen-hemoglobin equilibrium curve The graphical depiction of the percentage of hemoglobin in the blood that will bind to oxygen at various partial pressures of oxygen.

oxygenic Referring to any process or reaction that produces oxygen. Photosynthesis in plants, algae, and cyanobacteria, which involves photosystem II, is oxygenic. Compare with **anoxygenic**.

oxytocin A peptide hormone, secreted by the posterior pituitary, that triggers labor and milk production in females and that stimulates pair bonding, parental care, and affiliative behavior in both sexes.

p53 A tumor-suppressor protein (molecular weight of 53 kilodaltons) that responds to DNA damage by stopping the cell cycle and/or triggering apoptosis. Encoded by the *p53* gene.

pacemaker cell A specialized cardiac muscle cell in the sinoatrial (SA) node of the vertebrate heart that has an inherent rhythm and can generate an electrical impulse that spreads to other heart cells.

paleontologists Scientists who study the fossil record and the history of life.

Paleozoic era The interval of geologic time, from 542 million to 251 million years ago, during which fungi, land plants, and animals first appeared and diversified. Began with the Cambrian explosion and ended with the extinction of many invertebrates and vertebrates.

pancreas A large gland in vertebrates that has both exocrine and endocrine functions. Secretes digestive enzymes into a duct connected to the intestine and several hormones (notably, insulin and glucagon) into the bloodstream.

pancreatic lipase An enzyme that is produced in the pancreas and acts in the small intestine to break bonds in complex fats, releasing small lipids.

pandemic The spread of an infectious disease in a short time period over a wide geographic area and affecting a very high proportion of the population. Compare with **epidemic**.

parabiosis An experimental technique for determining whether a certain physiological phenomenon is regulated by a hormone, by surgically uniting two individuals so that hormones can pass between them.

paracrine Relating to a chemical signal that is released by one cell and affects neighboring cells.

paraphyletic group An evolutionary unit that includes an ancestral population and *some* but not all of its descendants. Paraphyletic groups are not meaningful units in evolution. Compare with **monophyletic group**.

parapodia (singular: parapodium) Appendages found in some annelids from which bristle-like structures (chaetae) extend.

parasite An organism that lives on or in a host species and that damages its host.

parasitism (adjective: parasitic) A symbiotic relationship between two organisms that is beneficial to one organism (the parasite) but detrimental to the other (the host). Compare with **commensalism** and **mutualism**.

parasitoid An organism that has a parasitic larval stage and a free-living adult stage. Most parasitoids are insects that lay eggs in the bodies of other insects.

parasympathetic nervous system The part of the autonomic nervous system that stimulates functions for conserving or restoring energy, such as reduced heart rate and increased digestion. Compare with **sympathetic nervous system**.

parathyroid glands Four small glands, located near or embedded in the thyroid gland of vertebrates, that secrete parathyroid hormone (PTH), a peptide hormone that increases blood calcium.

parenchyma cell In plants, a general type of cell with a relatively thin primary cell wall. These cells, found in leaves, the centers of stems and roots, and fruits, are involved in photosynthesis, starch storage, and new growth. Compare with **collenchyma cell** and **sclerenchyma cell**.

parental care Any action by which an animal expends energy or assumes risks to benefit its offspring (e.g., nest-building, feeding of young, defense).

parental generation The adult organisms used in the first experimental cross in a formal breeding experiment.

parietal cell A cell in the stomach lining that secretes hydrochloric acid.

parietal lobe In the vertebrate brain, one of the four major areas in the cerebrum.

parsimony The logical principle that the most likely explanation of a phenomenon is the most economical or simplest. When applied to comparison of alternative phylogenetic trees, it suggests that the one requiring the fewest evolutionary changes is most likely to be correct.

parthenogenesis Development of offspring from unfertilized eggs; a type of asexual reproduction.

partial pressure The pressure of one particular gas in a mixture; the contribution of that gas to the overall pressure.

particulate inheritance The observation that genes from two parents do not blend together to form a new physical entity in offspring, but instead remain separate or particle-like.

pascal (Pa) A unit of pressure (force per unit area).

passive transport Diffusion of a substance across a plasma membrane or organelle membrane. When this occurs with the assistance of membrane proteins, it is called facilitated diffusion.

patch clamping A technique for studying the electrical currents that flow through individual ion channels by sucking a tiny patch of membrane to the hollow tip of a microelectrode.

paternal chromosome A chromosome inherited from the father.

pathogen (adjective: pathogenic) Any entity capable of causing disease, such as a microbe, virus, or prion.

pattern formation The series of events that determines the spatial organization of an embryo, including alignment of the major body axes and orientation of the limbs.

pattern-recognition receptor One of a class of membrane proteins on leukocytes that bind to molecules on the surface of many bacteria. Part of the innate immune response.

PCR See **polymerase chain reaction**.

peat Semidecayed organic matter that accumulates in moist, low-oxygen environments such as bogs.

pectin A gelatinous polysaccharide found in the primary cell wall of plant cells. Attracts and holds water, forming a gel that helps keep the cell wall moist.

pedigree A family tree of parents and offspring, showing inheritance of particular traits of interest.

penis The copulatory organ of male mammals, used to insert sperm into a female.

pentose A monosaccharide (simple sugar) containing five carbon atoms.

PEP carboxylase An enzyme that catalyzes addition of CO_2 to phosphoenol pyruvate, a three-carbon compound, forming a four-carbon organic acid. Found in mesophyll cells of plants that perform C_4 photosynthesis.

pepsin A protein-digesting enzyme present in the stomach.

pepsinogen The precursor of the digestive enzyme pepsin. Is secreted from cells in the stomach lining and converted to pepsin by the acidic environment of the stomach lumen.

peptide See **oligopeptide**.

peptide bond The covalent bond (C—N) formed by a condensation reaction between two amino acids; links the residues in peptides and proteins.

peptidoglycan A complex structural polysaccharide found in bacterial cell walls.

perennial Describing a plant whose life cycle normally lasts for more than one year. Compare with **annual**.

perfect flower A flower that contains both male parts (stamens) and female parts (carpels). Compare with **imperfect flower**.

perforation In plants, a small hole in the primary and secondary cell walls of vessel elements that allow passage of water.

pericarp The part of a fruit, formed from the ovary wall, that surrounds the seeds and protects them. Corresponds to the flesh of most edible fruits and the hard shells of most nuts.

pericycle In plant roots, a layer of cells that give rise to lateral roots.

peripheral membrane protein Any membrane protein that does not span the entire lipid bilayer and associates with only one side of the bilayer. Compare with **integral membrane protein**.

peripheral nervous system (PNS) All the components of the nervous system that are outside the central nervous system (the brain and spinal cord). Includes the somatic nervous system and the autonomic nervous system.

peristalsis Rhythmic waves of muscular contraction that push food along the digestive tract.

permafrost A permanently frozen layer of icy soil found in most tundra and some taiga.

permeability The tendency of a structure, such as a membrane, to allow a given substance to diffuse across it.

peroxisome An organelle found in most eukaryotic cells that contains enzymes for oxidizing fatty acids and other compounds including many toxins, rendering them harmless. See **glyoxisome**.

petal One of the leaflike organs arranged around the reproductive organs of a flower. Often colored and scented to attract pollinators.

petiole The stalk of a leaf.

pH A measure of the concentration of protons in a solution and thus of acidity or alkalinity. Defined as the negative of the base-10 logarithm of the proton concentration: $pH = -\log[H^+]$.

phagocytosis Uptake by a cell of small particles or cells by pinching off the plasma membrane to form small membrane-bound vesicles; one type of endocytosis.

pharyngeal gill slits A set of parallel openings from the throat through the neck to the outside. A diagnostic trait of chordates.

pharyngeal jaw A secondary jaw in the back of the mouth, found in some fishes. Derived from modified gill arches.

phelloderm A component of bark; specifically, a tissue layer produced to the inside of the cork cambium in plants with secondary growth.

phelloderm In the stems of woody plants, a thin layer of cells located between the outer cork cells and inner cork cambium.

phenetic approach A method for constructing a phylogenetic tree by computing a statistic that summarizes the overall similarity among populations, based on the available data. Compare with **cladistic approach**.

phenology The timing of events during the year, in environments where seasonal changes occur.

phenotype The detectable physical and physiological traits of an individual, which are determined its genetic makeup. Also the specific trait associated with a particular allele. Compare with **genotype**.

phenotypic plasticity Within-species variation in phenotype that is due to differences in environmental conditions. Occurs more commonly in plants than animals.

phenylketonuria (PKU) A disease caused by the inability to process the amino acid phenylalanine.

pheophytin In photosystem II, a molecule that accepts excited electrons from a reaction center chlorophyll and passes them to an electron transport chain.

pheromone A chemical signal, released by one individual into the external environment, that can trigger responses in a different individual.

phloem A plant vascular tissue that conducts sugars; contains sieve-tube members and companion cells. Primary phloem develops from the procambium of apical meristems; secondary phloem, from the vascular cambium of lateral meristems. Compare with **xylem**.

phosphodiester linkage Chemical linkage between adjacent **nucleotide residues** in DNA and RNA. Forms when the phosphate group of one nucleotide condenses with the hydroxyl group on the sugar of another nucleotide. Also known as *phosphodiester bond*.

phosphofructokinase The enzyme that catalyzes synthesis of fructose-1,6-bisphosphate from fructose-6-phosphate, a key reaction (step 3) in glycolysis.

phospholipid A class of lipid having a hydrophilic head (a phosphate group) and a hydrophobic tail (one or more fatty acids). Major components of the plasma membrane and organelle membranes.

phosphorylase An enzyme that breaks down glycogen by catalyzing hydrolysis of the α-glycosidic linkages between the glucose residues.

phosphorylation (verb: phosphorylate) The addition of a phosphate group to a molecule.

phosphorylation cascade A series of enzyme-catalyzed phosphorylation reactions commonly used in signal transduction pathways to amplify and convey a signal inward from the plasma membrane.

photic zone In an aquatic habitat, water that is shallow enough to receive some sunlight (whether or not it is enough to support photosynthesis). Compare with **aphotic zone**.

photon A discrete packet of light energy; a particle of light.

photoperiodism Any response by an organism to the relative lengths of day and night (i.e., photoperiod).

photophosporylation Production of ATP molecules using the energy released as light-excited electrons flow through an electron transport chain during photosynthesis. Involves generation of a proton-motive force during electron transport and its use to drive ATP synthesis.

photoreceptor A molecule, a cell, or an organ that is specialized to detect light.

photorespiration A series of light-driven chemical reactions that consumes oxygen and releases carbon dioxide, basically reversing photosynthesis. Usually occurs when there are high O_2 and low CO_2 concentrations inside plant cells, often in bright, hot, dry environments when stomata must be kept closed.

photoreversibility A change in conformation that occurs in certain plant pigments when they are exposed to their preferred wavelengths of light and that triggers responses by the plant.

photosynthesis The complex biological process that converts the energy of light into chemical energy stored in glucose and other organic molecules. Occurs in plants, algae, and some bacteria.

photosystem One of two types of units, consisting of a central reaction center surrounded by antenna complexes, that is responsible for the light-dependent reactions of photosynthesis.

photosystem I A photosystem that contains a pair of P700 chlorophyll molecules and uses absorbed light energy to produce NADPH.

photosystem II A photosystem that contains a pair of P680 chlorophyll molecules and uses absorbed light energy to split water into protons and oxygen and to produce ATP.

phototroph An organism that produces ATP through photosynthesis.

phototropins A class of plant photoreceptors that detect blue light and initiate phototropic responses.

phototropism Growth or movement of an organism in a particular direction in response to light.

phylogenetic species concept The definition of a species as the smallest monophyletic group in a phylogenetic tree. Compare with **biological** and **morphospecies concept**.

phylogenetic tree A diagram that depicts the evolutionary history of a group of species and the relationships among them.

phylogeny The evolutionary history of a group of organisms.

phylum (plural: phyla) In Linnaeus' system, a taxonomic category above the class level and below the kingdom level. In plants, sometimes called a *division.*

physical map A map of a chromosome that shows the number of base pairs between various genetic markers. Compare with **genetic map**.

physiology The study of how an organism's body functions.

phytoalexin Any small compound produced by a plant to combat an infection (usually a fungal infection).

phytochrome A specialized plant photoreceptor that exists in two shapes depending on the ratio of red to far-red light and is involved in the timing certain physiological processes, such as flowering, stem elongation, and germination.

pigment Any molecule that absorbs certain wavelengths of visible light and reflects or transmits other wavelengths.

piloting A type of navigation in which animals use familiar landmarks to find their way.

pinocytosis Uptake by a cell of extracellular fluid by pinching off the plasma membrane to form small membrane-bound vesicles; one type of endocytosis.

pioneering species Those species that appear first in recently disturbed areas.

pit In plants, a small hole in the secondary cell walls of tracheids that allow passage of water.

pitch The sensation produced by a particular frequency of sound. Low frequencies are perceived as low pitches; high frequencies, as high pitches.

pith In the shoot systems of plants, ground tissue located to the inside of the vascular bundles.

pituitary gland A small gland directly under the brain that is physically and functionally connected to the hypothalamus. Produces and secretes an array of hormones that affect many other glands and organs.

placenta A structure that forms in the pregnant uterus from maternal and fetal tissues. Exchanges nutrients and wastes between mother and fetus, anchors the fetus to the uterine wall, and produces some hormones. Occurs in most mammals and in a few other vertebrates.

placental mammals See **eutherians**.

plankton Any small organism that drifts near the surface of oceans or lakes and swims little if at all.

Plantae The monophyletic group that includes red, green, and glaucophyte algae and land plants.

plantlet A small plant, particularly one that forms on a parent plant via asexual reproduction and drops, becoming an independent individual.

plasma The non-cellular portion of blood.

plasma cell An effector B cell, which produces large quantities of antibodies. Is descended from an activated B cell that has interacted with antigen.

plasma membrane A membrane that surrounds a cell, separating it from the external environment and selectively regulating passage of molecules and ions into and out of the cell. Also called *cell membrane.*

plasmid A small, usually circular, supercoiled DNA molecule independent of the cell's main chromosome(s) in prokaryotes and some eukaryotes.

plasmodesmata (singular: plasmodesma) Physical connection between two plant cells, consisting of gaps in the cell walls through which the two cells' plasma membranes, cytoplasm, and smooth ER can connect directly. Functionally similar to gap junctions in animal cells.

plasmogamy Fusion of the cytoplasm of two individuals. Occurs in many fungi.

plastocyanin A small protein that shuttles electrons from photosystem II to photosystem I during photosynthesis.

plastoquinone (PQ) A nonprotein electron carrier in the chloroplast electron transport chain. Receives excited electrons from pheophytin and passes them to more electronegative molecules in the chain. Also carries protons to the lumen side of the thylakoid membrane, generating a proton-motive force.

platelet A small membrane-bound cell fragment in vertebrate blood that functions in blood clotting. Derived from large cells in the bone marrow.

pleiotropy (adjective: pleiotropic) The ability of a single gene to affect more than one phenotypic trait.

ploidy The number of complete chromosome sets present. *Haploid* refers to a ploidy of 1; *diploid,* a ploidy of 2; *triploid,* a ploidy of 3; and *tetraploid,* a ploidy of 4.

podium (plural: podia) See **tube foot**.

point mutation A mutation that results in a change in a single nucleotide pair in a DNA molecule.

polar (1) Asymmetrical or unidirectional. (2) Carrying a partial positive charge on one side of a molecule and a partial negative charge on the other. Polar molecules are generally hydrophilic.

polar bodies The tiny, nonfunctional cells produced during meiosis of a primary oocyte, due to most of the cytoplasm going to the ovum.

polar covalent bond A covalent bond in which electrons are shared unequally between atoms differing in electronegativity, resulting in the more electronegative atom having a partial negative charge and the other atom, a partial positive charge. Compare with **nonpolar covalent bond**.

polar microtubules Microtubules that form during mitosis and meiosis, and which extend from a spindle apparatus and overlap with each other in the middle of the cell.

polar nuclei In flowering plants, the nuclei in the female gametophyte that fuse with one sperm nucleus to produce the endosperm. Most species have two.

pollen grain In seed plants, a male gametophyte enclosed within a protective coat.

pollen tube In flowering plants, a structure that grows out of a pollen grain after it reaches the stigma, extends down the style, and through which two sperm cells are delivered to the ovule.

pollination The process by which pollen reaches the carpel of a flower (in flowering plants) or reaches the ovule directly (in conifers and their relatives).

poly(A) tail In eukaryotes, a sequence of 100–250 adenine nucleotides added to the 3′ end of newly transcribed messenger RNA molecules.

polygenic inheritance The inheritance patterns that result when many genes influence one trait.

polymer Any long molecule composed of small repeating units (monomers) bonded together. The main biological polymers are proteins, nucleic acids, and polysaccharides.

polymerase chain reaction (PCR) A laboratory technique for rapidly generating millions of identical copies of a specific stretch of DNA by incubating the original DNA sequence of interest with primers, nucleotides, and DNA polymerase.

polymerization (verb: polymerize) The process by which many identical or similar small molecules (monomers) are covalently bonded to form a large molecule (polymer).

polymorphism (adjective: polymorphic) (1) The occurrence of more than one allele at a certain genetic locus in a population. (2) The occurrence of more than two distinct phenotypes of a trait in a population.

polyp The immotile (sessile) stage in the life cycle of some cnidarians (e.g., jellyfish). Compare with **medusa**.

polypeptide A chain of 50 or more amino acids linked together by peptide bonds. Compare with **oligopeptide** and **protein**.

polyploidy (adjective: polyploid) The state of having more than two full sets of chromosomes.

polyribosome A structure consisting of one messeng RNA molecule along with many attached ribosomes their growing peptide strands.

polysaccharide A linear or branched polymer consisting of many monosaccharides joined by glycosidic linkages. Carbohydrate polymers with relatively few residues often are called *oligosaccharides*.

polyspermy Fertilization of an egg by multiple sperm.

population A group of individuals of the same species living in the same geographic area at the same time.

population density The number of individuals of a population per unit area.

population dynamics Changes in the size and other characteristics of populations through time.

population ecology The study of how and why the number of individuals in a population changes over time.

population thinking The ability to analyze trait frequencies, event probabilities, and other attributes of populations of molecules, cells, or organisms.

pore In land plants, an opening in the epithelium that allows gas exchange.

positive control Of genes, when a regulatory protein triggers expression by binding to DNA on or near the gene.

positive feedback A physiological mechanism in which a change in some variable stimulates a response that increases the change. Relatively rare in organisms but is important in generation of the action potential. Compare with **negative feedback**.

positive pressure ventilation Ventilation of the lungs that is accomplished by "pushing" air into the lungs by positive pressure in the mouth. Compare with **negative pressure ventilation**.

positive-sense virus A virus whose genome contains the same sequences as the mRNA required to produce viral proteins. Compare with **ambisense virus** and **negative-sense virus**.

posterior Toward an animal's tail and away from its head. The opposite of anterior.

posterior pituitary The part of the pituitary gland that contains the ends of hypothalamic neurosecretory cells and from which oxytocin and antidiuretic hormone are secreted. Compare with **anterior pituitary**.

postsynaptic neuron A neuron that receives signals, usually via neurotransmitters, from another neuron at a synapse. Muscle and gland cells also may receive signals from presynaptic neurons.

post-translational control Regulation of gene expression by modification of proteins (e.g., addition of a phosphate group or sugar residues) after translation.

postzygotic isolation Reproductive isolation resulting from mechanisms that operate after mating of individuals of two different species occurs. The most common mechanisms are the death of hybrid embryos or reduced fitness of hybrids.

potential energy Energy stored in matter as a result of its position or molecular arrangement. Compare with **kinetic energy**.

prebiotic soup A hypothetical solution of sugars, amino acids, nitrogenous bases, and other building blocks of larger molecules that may have formed in [illegible]allow waters or deep-ocean vents of ancient Earth and [illegible] rise to larger biological molecules.

[illegible]**brian** The interval between the formation of the [illegible]ut 4.6 billion years ago, and the appearance of [illegible] groups about 542 million years ago. [illegible]anisms were dominant for most of this [illegible]as virtually absent for the first 2 billion [illegible]

[illegible] and eating of one organism (the [illegible]dator).

er
nd

predator Any organism that kills other organisms for food.

prediction A measurable or observable result of an experiment based on a particular hypothesis. A correct prediction provides support for the hypothesis being tested.

pressure potential (ψ_P) A component of the potential energy of water caused by physical pressures on a solution. In plant cells, it equals the wall pressure plus turgor pressure. Compare with **solute potential (ψ_S)**.

pressure-flow hypothesis The hypothesis that sugar movement through phloem tissue is due to differences in the turgor pressure of phloem sap.

presynaptic neuron A neuron that transmits signals, usually by releasing neurotransmitters, to another neuron or to an effector cell at a synapse.

prezygotic isolation Reproductive isolation resulting from any one of several mechanisms that prevent individuals of two different species from mating.

primary cell wall The outermost layer of a plant cell wall, made of cellulose fibers and gelatinous polysaccharides, that defines the shape of the cell and withstands the turgor pressure of the plasma membrane.

primary consumer An herbivore; an organism that eats plants, algae, or other primary producers. Compare with **secondary consumer**.

primary decomposer A decomposer (detritivore) that consumes detritus from plants.

primary growth In plants, an increase in the length of stems and roots due to the activity of apical meristems. Compare with **secondary growth**.

primary immune response An acquired immune response to a pathogen that the immune system has not encountered before. Compare with **secondary immune response**.

primary oocyte The large diploid cell in an ovarian follicle that can initiate meiosis to produce a haploid ovum.

primary producer Any organism that creates its own food by photosynthesis or from reduced inorganic compounds and that is a food source for other species in its ecosystem. Also called *autotroph*.

primary RNA transcript In eukaryotes, a newly transcribed messenger RNA molecule that has not yet been processed (i.e., it has not received a 5′ cap or poly(A) tail, and still contains introns). Also called *pre-mRNA*.

primary spermatocyte A diploid cell in the testis that can initiate meiosis I to produce two secondary spermatocytes.

primary structure The sequence of amino acids in a peptide or protein; also the sequence of nucleotides in a nucleic acid. Compare with **secondary, tertiary,** and **quaternary structure**.

primary succession The gradual colonization of a habitat of bare rock or gravel, usually after an environmental disturbance that removes all soil and previous organisms. Compare with **secondary succession**.

primase An enzyme that synthesizes a short stretch of RNA to use as a primer during DNA replication.

primates The lineage of mammals that includes prosimians (lemurs, lorises, etc.), monkeys, and great apes (including humans).

primer A short, single-stranded RNA molecule that base pairs with the 5′ end of a DNA template strand and is elongated by DNA polymerase during DNA replication.

prion An infectious form of a protein that is thought to cause disease by inducing the normal form to assume an abnormal three-dimensional structure. Cause of spongiform encephalopathies, such as mad cow disease.

probe A radioactively or chemically labeled single-stranded fragment of a known DNA or RNA sequence that can bind to and thus detect its complementary sequence in a sample containing many different sequences.

proboscis A long, narrow feeding appendage through which food can be obtained.

procambium A group of cells in the center of a young plant embryo that gives rise to the vascular tissue.

product Any of the final materials formed in a chemical reaction.

productivity The total amount of carbon fixed by photosynthesis per unit area per year.

progesterone A steroid hormone produced in the ovaries and secreted by the corpus luteum after ovulation; causes the uterine lining to thicken.

programmed cell death See **apoptosis**.

prokaryote A member of the domain Bacteria or Archaea; a unicellular organism lacking a nucleus and containing relatively few organelles or cytoskeletal components. Compare with **eukaryote**.

prolactin A peptide hormone, produced and secreted by the anterior pituitary, that promotes milk production in female mammals and has a variety of effects on parental behavior and seasonal reproduction in other vertebrates.

prometaphase A stage in mitosis or meiosis during which the nuclear envelope breaks down and kinetochore microtubles attach to chromatids.

promoter A short nucleotide sequence in DNA that binds RNA polymerase, enabling transcription to begin. In prokaryotic DNA, a single promoter often is associated with several contiguous genes. In eukaryotic DNA, each gene generally has its own promoter.

promoter-proximal elements In eukaryotes, regulatory sequences in DNA that are close to a promoter and that can bind regulatory transcription factors.

proofreading The process by which a DNA polymerase recognizes and removes a wrong base added during DNA replication and then continues synthesis.

prophase The first stage in mitosis or meiosis during which chromosomes become visible and the spindle apparatus forms. Synapsis and crossing over occur during prophase of meiosis I.

prosimians One of the two major lineages of primates, including lemurs, tarsiers, pottos, and lorises. Compare with **anthropoids**.

prostate gland A gland in male mammals that surrounds the base of the urethra and secretes a fluid that is a component of semen.

protease An enzyme that can degrade proteins by cleaving the peptide bonds between amino acid residues.

proteasome A multi-molecular machine that destroys proteins that have been bound to ubiquitin.

protein A macromolecule consisting of one or more polypeptide chains composed of 50 or more amino acids linked together. Each protein has a unique sequence of amino acids and, in its native state, a characteristic three-dimensional shape.

protein kinase An enzyme that catalyzes the addition of a phosphate group to another protein, typically activating or inactivating the substrate protein.

proteinase inhibitors Defense compounds, produced by plants, that induce illness in herbivores by inhibiting digestive enzymes.

proteome The complete set of proteins produced by a particular cell type.

proteomics The systematic study of the interactions, localization, functions, regulation, and other features of the full protein set (proteome) in a particular cell type.

protist Any eukaryote that is not a green plant, animal, or fungus. Protists are a diverse paraphyletic group. Most are unicellular, but some are multicellular or form aggregations called colonies.

protoderm The exterior layer of a young plant embryo that gives rise to the epidermis.

proton pump A membrane protein that can hydrolyze ATP to power active transport of protons (H^+ ions) across a plasma membrane against an electrochemical gradient. Also called *H^+-ATPase.*

proton-motive force The combined effect of a proton gradient and an electric potential gradient across a membrane, which can drive protons across the membrane. Used by mitochondria and chloroplasts to power ATP synthesis via the mechanism of chemiosmosis.

proto-oncogene Any gene that normally encourages cell division in a regulated manner, typically by triggering specific phases in the cell cycle. Mutation may convert it into an oncogene.

protostomes A major lineage of animals that share a pattern of embryological development, including formation of the mouth earlier than the anus, and formation of the coelom by splitting of a block of mesoderm. Includes arthropods, mollusks, and annelids. Compare with **deuterostomes**.

proximal tubule In the vertebrate kidney, the convoluted section of a nephron into which filtrate moves from Bowman's capsule. Involved in the largely unregulated reabsorption of electrolytes, nutrients, and water. Compare with **distal tubule**.

proximate causation In biology, the immediate, mechanistic cause of a phenomenon (how it happens), as opposed to why it evolved. Also called *proximate explanation.* Compare with **ultimate causation**.

pseudogene A DNA sequence that closely resembles a functional gene but is not transcribed. Thought to have arisen by duplication of the functional gene followed by inactivation due to a mutation.

pseudopodium (plural: pseudopodia) A temporary bulge-like extension of certain cells used in cell crawling and ingestion of food.

puberty The various physical and emotional changes that an immature animal undergoes leading to reproductive maturity. Also the period when such changes occur.

pulmonary artery A short, thick-walled artery that carries oxygen-poor blood from the heart to the lungs.

pulmonary circulation The part of the circulatory system that sends oxygen-poor blood to the lungs. Is separate from the rest of the circulatory system (the systemic circulation) in mammals and birds.

pulmonary vein A short, thin-walled vein that carries oxygen-rich blood from the lungs to the heart.

pulse-chase experiment A type of experiment in which a population of cells or molecules at a particular moment in time is marked by means of a labeled molecule and then their fate is followed over time.

pump Any membrane protein that can hydrolyze ATP to power active transport of a specific ion or small molecule across a plasma membrane against its electrochemical gradient. See **proton pump**.

Punnett square A diagram that depicts the genotypes and phenotypes that should appear in offspring of a certain cross.

pupa (plural: pupae) A metamorphosing insect that is enclosed in a protective case.

pupil The hole in the center of the iris through which light enters a vertebrate or cephalopod eye.

pure line In animal or plant breeding, a strain of individuals that produce offspring identical to themselves when self-pollinated or crossed to another member of the same population. Pure lines are homozygous for most, if not all, genetic loci.

purifying selection Selection that lowers the frequency or even eliminates deleterious alleles.

purines A class of small, nitrogen-containing, double-ringed bases (guanine, adenine) found in nucleotides. Compare with **pyrimidines**.

pyrimidines A class of small, nitrogen-containing, single-ringed bases (cytosine, uracil, thymine) found in nucleotides. Compare with **purines**.

pyruvate dehydrogenase A large enzyme complex, located in the inner mitochondrial membrane, that is responsible for conversion of pyruvate to acetyl CoA during cellular respiration.

quantitative trait A heritable feature that exhibits phenotypic variation along a smooth, continuous scale of measurement (e.g., human height), rather than the distinct forms characteristic of discrete traits.

quaternary structure The overall three-dimensional shape of a protein containing two or more polypeptide chains (subunits); determined by the number, relative positions, and interactions of the subunits. Compare with **primary, secondary,** and **tertiary structure**.

quorum sensing Cell-cell signaling in bacteria, in which cells of the same species communicate via chemical signals. It is often observed that cell activity changes dramatically when the population reaches a threshold size, or quorum.

radial symmetry An animal body pattern in which there are least two planes of symmetry. Typically, the body is in the form of a cylinder or disk, with body parts radiating from a central hub. Compare with **bilateral symmetry**.

radiation Transfer of heat between two bodies that are not in direct physical contact. More generally, the emission of electromagnetic energy of any wavelength.

radicle The root of a plant embryo.

radula A rasping feeding appendage in gastropods (snails, slugs).

rain shadow The dry region on the side of a mountain range away from the prevailing wind.

range The geographic distribution of a species.

Ras protein A type of G protein that is activated by binding of signaling molecules to receptor tyrosine kinases and then initiates a phosphorylation cascade, culminating in a cell response.

rays In plant shoot systems with secondary growth, a lateral array of parenchyma cells produced by vascular cambium.

Rb protein A tumor-suppressor protein that helps regulate progression of a cell from the G_1 phase to the S phase of the cell cycle. Defects in Rb protein are found in many types of cancer.

reactant Any of the starting materials in a chemical reaction.

reaction center Centrally located component of a photosystem containing proteins and a pair of specialized chlorophyll molecules. Is surrounded by antenna complexes and receives excited electrons from them.

reactive oxygen intermediates (ROIs) Highly reactive oxygen-containing compounds that are used in plant and animal cells to kill infected cells and for other purposes.

reading frame The division of a sequence of DNA or RNA into a particular series of three-nucleotide codons. There are three possible reading frames for any sequence.

realized niche The ecological niche that a species occupies in the presence of competitors. Compare with **fundamental niche**.

receptor tyrosine kinase (RTK) Any of a class of cell-surface signal receptors that undergo phosphorylation after binding a signaling molecule. The activated, phosphorylated receptor then triggers a signal-transduction pathway inside the cell.

receptor-mediated endocytosis Uptake by a cell of certain extracellular macromolecules, bound to specific receptors in the plasma membrane, by pinching off the membrane to form small membrane-bound vesicles.

recessive Referring to an allele whose phenotypic effect is observed only in homozygous individuals. Compare with **dominant**.

reciprocal altruism Altruistic behavior that is exchanged between a pair of individuals at different points in time (i.e., sometimes individual A helps individual B, and sometimes B helps A).

reciprocal cross A breeding experiment in which the mother's and father's phenotypes are the reverse of that examined in a previous breeding experiment.

recombinant DNA technology A variety of techniques for isolating specific DNA fragments and introducing them into different regions of DNA and/or a different host organism.

recombinant Possessing a new combination of alleles. May refer to a single chromosome or DNA molecule, or to an entire organism.

recombination See **genetic recombination**.

rectal gland A salt-excreting gland in the digestive system of sharks, skates, and rays.

rectum The last portion of the digestive tract where feces are held until they are expelled.

red blood cell A hemoglobin-containing cell that circulates in the blood and delivers oxygen from the lungs to the tissues.

redox reaction Any chemical reaction that involves the transfer of one or more electrons from one reactant to another. Also called *reduction-oxidation reaction.*

reduction An atom's gain of electrons during a redox reaction, either by acceptance of an electron from another atom or by the electrons in covalent bonds moving closer to the atomic nucleus.

reduction-oxidation reaction See **redox reaction**.

reflex An involuntary response to environmental stimulation. May involve the brain (e.g., conditioned reflex) or not (e.g., spinal reflex).

refractory No longer responding to stimuli that previously elicited a response. For example, the tendency of voltage-gated sodium channels to remain closed immediately after an action potential.

regulatory genes DNA sequences that code for regulatory proteins—products that alter gene expression.

regulatory sequence, DNA Any segment of DNA that is involved in controlling transcription of a specific gene by binding certain proteins.

regulatory site A site on an enzyme to which a regulatory molecule can bind and affect the enzyme's activity; separate from the active site where catalysis occurs.

regulatory transcription factor General term for protei[n] that bind to DNA regulatory sequences (eukaryotic enhancers, silencers, and promoter-proximal elements) not to the promoter itself, leading to an increase or de[c] in transcription of specific genes. Compare with **basal transcription factor**.

reinforcement In evolutionary biology, the natural selection for traits that prevent interbreeding between recently diverged species.

release factors Proteins that can trigger termination of RNA translation when a ribosome reaches a stop codon.

renal corpuscle In the vertebrate kidney, the ball-like structure at the beginning of a nephron, consisting of a glomerulus and the surrounding Bowman's capsule. Acts as a filtration device.

replacement mutation See **missense mutation**.

replacement rate The number of offspring each female must produce over her entire life to "replace" herself and her mate, resulting in zero population growth. The actual number is slightly more than 2 because some offspring die before reproducing.

replica plating A method of identifying bacterial colonies that have certain mutations by transferring cells from each colony on a master plate to a second (replica) plate and observing their growth when exposed to different conditions.

replication fork The Y-shaped site at which a double-stranded molecule of DNA is separated into two single strands for replication.

replisome The multi-molecular machine that copies DNA; includes DNA polymerase, helicase, primase, and other enzymes.

repolarization Return to a normal membrane potential after it has changed; a normal phase in an action potential.

repressor Any regulatory protein that inhibits transcription.

reptiles One of the two lineages of amniotes (vertebrates that produce amniotic eggs) distinguished by adaptations for reproduction on land. Includes turtles, snakes and lizards, crocodiles and alligators, and birds. Except for birds, all are ectotherms.

resilience, community A measure of how quickly a community recovers following a disturbance.

resistance, community A measure of how much a community is affected by a disturbance.

resistance (*R*) genes Genes in plants encoding proteins involved in sensing the presence of pathogens and mounting a defensive response. Compare with **avirulence (*avr*) genes**.

respiratory system The collection of cells, tissues, and organs responsible for gas exchange between an animal and its environment.

resting potential The membrane potential of a cell in its resting, or normal, state.

restriction endonucleases Bacterial enzymes that cut DNA at a specific base-pair sequence (restriction site). Also called *restriction enzymes*.

retina A thin layer of light-sensitive cells (rods and cones) and neurons at the back of a camera-type eye, such as that of cephalopods and vertebrates.

retinal A light-absorbing pigment, derived from vitamin A, that is linked to the protein opsin in rods and ones of the vertebrate eye.

ovirus A virus with an RNA genome that uces by transcribing its RNA into a DNA and then inserting that DNA into the host's replication.

riptase A enzyme of retroviruses (RNA synthesize double-stranded DNA from NA template.

izobium) Members of the nitrogen-fixing bacteria that bers of the pea family

rhizoid The hairlike structure that anchors a bryophyte (non-vascular plant) to the substrate.

rhizome A modified stem that runs horizontally underground and produces new plants at the nodes (a form of asexual reproduction). Compare with **stolon**.

rhodopsin A transmembrane complex that is instrumental in detection of light by rods and cones of the vertebrate eye. Is composed of the transmembrane protein opsin covalently linked to retinal, a light-absorbing pigment.

ribonucleic acid (RNA) A nucleic acid composed of ribonucleotides that usually is single stranded and functions as structural components of ribosomes (rRNA), transporters of amino acids (tRNA), and translators of the message of the DNA code (mRNA).

ribonucleotide See **nucleotide**.

ribosomal RNA (rRNA) A RNA molecule that forms part of the structure of a ribosome.

ribosome A large complex structure that synthesizes proteins by using the genetic information encoded in messenger RNA strands. Consists of two subunits, each composed of ribosomal RNA and proteins.

ribosome binding site In a bacterial mRNA molecule, the sequence just upstream of the start codon to which a ribosome binds to initiate translation. Also called the *Shine-Dalgarno sequence*.

ribozyme Any RNA molecule that can act as a catalyst, that is, speed up a chemical reaction.

ribulose bisphosphate (RuBP) A five-carbon compound that combines with CO_2 in the first step of the Calvin cycle during photosynthesis.

RNA interference (RNAi) Degradation of an mRNA molecule or inhibition of its translation following its binding by a short RNA (microRNA) whose sequence is complementary to a portion of the mRNA.

RNA polymerase One of a class of enzymes that catalyze synthesis of RNA from ribonucleotides using a DNA template. Also called *RNA pol*.

RNA processing In eukaryotes, the changes that a primary RNA transcript undergoes in the nucleus to become a mature mRNA molecule, which is exported to the cytoplasm. Includes the addition of a 5′ cap and poly(A) tail and splicing to remove introns.

RNA replicase A viral enzyme that can synthesize RNA from an RNA template.

RNA See **ribonucleic acid**.

rod cell A photoreceptor cell with a rod-shaped outer portion that is particularly sensitive to dim light, but not used to distinguish colors. Also called simply *rod*. Compare with **cone cell**.

root (1) An underground part of a plant that anchors the plant and absorbs water and nutrients. (2) In a phylogenetic tree, the bottom, most ancient node.

root apical meristem (RAM) A group of undifferentiated plant cells at the tip of a plant root that can differentiate into mature root tissue.

root cap A small group of cells that covers and protects the tip of a plant root. Senses gravity and determines the direction of root growth.

root hair A long, thin outgrowth of the epidermal cells of plant roots, providing increased surface area for absorption of water and nutrients.

root pressure Positive (upward) pressure of xylem sap in the vascular tissue of roots. Is generated during the night as a result of the accumulation of ions from the soil and subsequent osmotic movement of water into the xylem.

root system The belowground part of a plant.

rough endoplasmic reticulum (rough ER) The portion of the endoplasmic reticulum that is dotted with ribosomes. Involved in synthesis of plasma membrane proteins, secreted proteins, and proteins localized to the ER, Golgi apparatus, and lysosomes. Compare with **smooth endoplasmic reticulum**.

roundworms Members of the phylum Nematoda. Distinguished by an unsegmented body with a pseudocoelom and no appendages. Roundworms belong to the ecdysozoan branch of the protostomes. Also called *nematodes*.

rRNA See **ribosomal RNA**.

rubisco The enzyme that catalyzes the first step of the Calvin cycle during photosynthesis: the addition of a molecule of CO_2 to ribulose bisphosphate.

ruminants A group of hoofed mammals (e.g., cattle, sheep, deer) that have a four-chambered stomach specialized for digestion of plant cellulose. Ruminants regurgitate the cud, a mixture of partially digested food and cellulose-digesting bacteria, from the largest chamber (the rumen) for further chewing.

salivary glands Vertebrate glands that secrete saliva (a mixture of water, mucus-forming glycoproteins, and digestive enzymes) into the mouth.

sampling error The accidental selection of a nonrepresentative sample from some larger population, due to chance.

saprophyte An organism that feeds primarily on dead plant material.

sapwood The younger xylem in the outer layer of wood of a stem or root, functioning primarily in water transport.

sarcomere The repeating contractile unit of a skeletal muscle cell; the portion of a myofibril located between adjacent Z disks.

sarcoplasmic reticulum Sheets of smooth endoplasmic reticulum in a muscle cell. Contains high concentrations of calcium, which can be released into the cytoplasm to trigger contraction.

saturated Referring to fats and fatty acids in which all the carbon-carbon bonds are single bonds. Such fats have relatively high melting points. Compare with **unsaturated**.

scanning electron microscope (SEM) A microscope that produces images of the surfaces of objects by reflecting electrons from a specimen coated with a layer of metal atoms. Compare with **transmission electron microscope**.

scarify To scrape, rasp, cut, or otherwise damage the coat of a seed. Necessary in some species to trigger germination.

Schwann cell A type of glial cell that wraps around axons of some neurons outside the brain and spinal cord, forming a myelin sheath that provides electrical insulation. Compare with **oligodendrocyte**.

scientific name The unique, two-part name given to each species, with a genus name followed by a species name—as in *Homo sapiens*. Scientific names are always italicized, and are also known as Latin names.

sclereid In plants, a type of sclerenchyma cell that usually functions in protection, such as in seed coats and nutshells. Compare with **fiber**.

sclerenchyma cell In plants, a cell that has a thick secondary cell wall and provides support; typically contains the tough structural polymer lignin and usually is dead at maturity. Includes fibers and sclereids. Compare with **collenchyma cell** and **parenchyma cell**.

screen See **genetic screen**.

scrotum A sac of skin, containing the testes, suspended just outside the abdominal body cavity of many male mammals.

second law of thermodynamics The principle of physics that the entropy of the universe or any closed system increases during any spontaneous process.

second messenger A nonprotein signaling molecule produced or activated inside a cell in response to stimulation at the cell surface. Commonly used to relay the message of a hormone or other extracellular signaling molecule.

secondary active transport Transport of an ion or molecule against its electrochemical gradient, in company with an ion or molecule being transported with its electrochemical gradient. Also called *cotransport.*

secondary active transporter A transmembrane protein that facilitates diffusion of an ion down its previously established electrochemical gradient and uses the energy of that process to transport some other substance, in the same or opposite direction, *against* its concentration gradient. Also called *cotransporter.* See also **antiporter** and **symporter**.

secondary cell wall The inner layer of a plant cell wall formed by certain cells as they mature. Provides support or protection.

secondary consumer A carnivore; an organism that eats herbivores. Compare with **primary consumer**.

secondary growth In plants, an increase in the width of stems and roots due to the activity of lateral meristems. Compare with **primary growth**.

secondary immune response The acquired immune response to a pathogen that the immune system has encountered before. Compare with **primary immune response**.

secondary metabolites Molecules that are closely related to compounds in key synthetic pathways, and that often function in defense.

secondary spermatocyte A cell produced by meiosis I of a primary spermatocyte in the testis. Can undergo meiosis II to produce spermatids.

secondary structure In proteins, localized folding of a polypeptide chain into regular structures (e.g., α-helix and β-pleated sheet) stabilized by hydrogen bonding between atoms of the backbone. In nucleic acids, elements of structure (e.g., helices and hairpins) stabilized by hydrogen bonding and other interactions between complementary bases. Compare with **primary, tertiary,** and **quaternary structure**.

secondary succession Gradual colonization of a habitat after an environmental disturbance (e.g., fire, windstorm, logging) that removes some or all previous organisms but leaves the soil intact. Compare with **primary succession**.

second-male advantage The reproductive advantage of a male who mates with a female last, after other males have mated with her.

secretin A peptide hormone produced by cells in the small intestine in response to the arrival of food from the stomach. Stimulates secretion of bicarbonate (HCO_3^-) from the pancreas.

sedimentary rock A type of rock formed by gradual accumulation of sediment, as in riverbeds and on the ocean floor. Most fossils are found in sedimentary rocks.

seed A plant reproductive structure consisting of an embryo, associated nutritive tissue (endosperm), and an outer protective layer (seed coat). In angiosperms, develops from the fertilized ovule of a flower.

seed bank A repository where seeds, representing many different varieties of domestic crops or other species, are preserved.

seed coat A protective layer around a seed that encases both the embryo and the endosperm.

segment A well-defined region of the body along the anterior-posterior body axis, containing similar structures as other, nearby segments.

segmentation Division of the body or a part of it into a series of similar structures; exemplified by the body segments of insects and worms and by the somites of vertebrates.

segmentation genes A group of genes that affect body segmentation in embryonic development. Includes gap genes, pair-rule genes, and segment polarity genes.

segregation, principle of The concept that each pair of hereditary elements (alleles of the same gene) separate from each other during the formation of offspring (i.e., during meiosis). One of Mendel's two principles of genetics.

selective adhesion The tendency of cells of one tissue type to adhere to other cells of the same type.

selective permeability The property of a membrane that allows some substances to diffuse across it much more readily than other substances.

selectively permeable membrane Any membrane across which some solutes can move more readily than others.

self Property of a molecule or cell such that immune system cells do not attack it, due to certain molecular similarities to other body cells.

self molecule A molecule that is synthesized by an organism and is a normal part of its cells and/or body; as opposed to non-self or foreign molecules.

self-fertilization In plants, the fusion of two gametes from the same individual to form a diploid offspring. Also called *selfing.*

semen The combination of sperm and accessory fluids that is released by male mammals and reptiles during ejaculation.

semiconservative replication The mechanism of replication used by cells to copy DNA. Results in each daughter DNA molecule containing one old strand and one new strand.

seminal vesicles In male mammals, paired reproductive glands that secrete a sugar-containing fluid into semen, which provides energy for sperm movement. In other vertebrates and invertebrates, often stores sperm.

senescence The process of aging.

sensor Any cell, organ, or structure with which an animal can sense some aspect of the external or internal environment. Usually functions, along with an integrator and effector, as part of a homeostatic system.

sensory neuron A nerve cell that carries signals from sensory receptors to the central nervous system. Compare with **interneuron** and **motor neuron**.

sepal One of the protective leaflike organs enclosing a flower bud and later supporting the blooming flower.

septum (plural: septa) Any wall-like structure. In fungi, septa divide the filaments (hyphae) of mycelia into cell-like compartments.

serotonin A neurotransmitter involved in many brain functions, including sleep, pleasure, and mood.

serum The liquid that remains when cells and clot material are removed from clotted blood. Contains water, dissolved gases, growth factors, nutrients, and other soluble substances. Compare with **plasma**.

sessile Permanently attached to a substrate; not capable of moving to another location.

set point A normal or target value for a regulated internal variable, such as body heat or blood pH.

severe combined immunodeficiency disease (SCID) A human disease characterized by an extremely high vulnerability to infectious disease. Caused by a genetic defect in the immune system.

sex chromosome Any chromosome carrying genes involved in determining the sex of an individual. Compare with **autosome**.

sex-linked inheritance Inheritance patterns observed in genes carried on sex chromosomes, so females and males have different numbers of alleles of a gene and may pass its trait only to one sex of offspring. Also called *sex-linkage.*

sexual dimorphism Any trait that differs between males and females.

sexual reproduction Any form of reproduction in which genes from two parents are combined via fusion of gametes, producing offspring that are genetically distinct from both parents. Compare with **asexual reproduction**.

sexual selection A pattern of natural selection that favors individuals with traits that increase their ability to obtain mates. Acts more strongly on males than females.

shell A hard protective outer structure.

Shine-Dalgarno sequence See **ribosome binding sequence**.

shoot apical meristem (SAM) A group of undifferentiated plant cells at the tip of a plant stem that can differentiate into mature shoot tissues.

shoot The combination of hypocotyl and cotyledons in a plant embryo, which will become the aboveground portions of the body.

shoot system The aboveground part of a plant comprising stems, leaves, and flowers (in angiosperms).

short interspersed nuclear elements See **SINEs**.

short tandem repeats (STRs) Relatively short DNA sequences that are repeated, one after another, down the length of a chromosome. The two major types are microsatellites and minisatellites.

short-day plant A plant that blooms in response to long nights (usually in late summer or fall in the northern hemisphere). Compare with **day-neutral** and **long-day plant**.

shotgun sequencing A method of sequencing genomes that is based on breaking the genome into small pieces, sequencing each piece separately, and then figuring out how the pieces are connected.

sieve plate In plants, a pore-containing structure at one end of a sieve-tube member in phloem.

sieve-tube member In plants, an elongated sugar-conducting cell in phloem that has sieve plates at both ends, allowing sap to flow to adjacent cells.

sign stimulus A simple stimulus that elicits an invariant, stereotyped behavioral response (fixed action pattern) from an animal. Also called a *releaser.*

signal In behavioral ecology, any information-containing behavior.

signal receptor Any cellular protein that binds to a particular signaling molecule (e.g., a hormone or neurotransmitter) and triggers a response by the cell. Receptors for water-soluble signals are transmembrane proteins in the plasma membrane; those for many lipid-soluble signals (e.g., steroid hormones) are located inside the cell.

signal recognition particle (SRP) A RNA-protein complex that binds to the ER signal sequence in a polypeptide as it emerges from a ribosome and transports the ribosome-polypeptide complex to the ER membrane where synthesis of the polypeptide is completed.

signal transduction cascade See **phosphorylation cascade**.

signal transduction The process by which a stimulus (e.g., a hormone, a neurotransmitter, or sensory information) outside a cell is amplified and converted into a response by the cell. Usually involves a specific sequence of molecular events, or signal transduction pathway.

silencer A regulatory sequence in eukaryotic DNA to which repressor proteins can bind, inhibiting transcription of certain genes.

silent mutation A mutation that does not detectably affect the phenotype of the organism.

simple eye An eye with only one light-collecting apparatus (e.g., one lens), as in vertebrates. Compare with **compound eye**.

simple fruit A fruit (e.g., apricot) that develops from a single flower that has a single carpel or several fused carpels. Compare with **aggregate** and **multiple fruit**.

simple sequence repeat See **microsatellite**.

SINEs (short interspersed nuclear elements) The second most abundant class of transposable elements in human genomes; can create copies of itself and insert them elsewhere in the genome. Compare with **LINEs**.

single nucleotide polymorphism (SNP) A site on a chromosome where individuals in a population have different nucleotides. Can be used as a genetic marker to help track the inheritance of nearby genes.

single-strand DNA-binding proteins (SSBPs) A class of proteins that attach to separated strands of DNA during replication or transcription, preventing them from re-forming a double helix.

sink Any tissue, site, or location where an element or a molecule is consumed or taken out of circulation (e.g., in plants, a tissue where sugar exits the phloem). Compare with **source**.

sinoatrial (SA) node A cluster of cardiac muscle cells, in the right atrium of the vertebrate heart, that initiates the heartbeat and determines the heart rate. Compare with **atrioventricular (AV) node**.

siphon A tubelike appendage of many mollusks, that is often used for feeding or propulsion.

sister chromatids The paired strands of a recently replicated chromosome, which are connected at the centromere and eventually separate during anaphase of mitosis and meiosis II. Compare with **non-sister chromatids**.

sister species Closely related species, which occupy adjacent branches in a phylogenetic tree.

skeletal muscle The muscle tissue attached to the bones of the vertebrate skeleton. Consists of long, unbranched muscle fibers with a characteristic striped (striated) appearance; controlled voluntarily. Also called *striated muscle*. Compare with **cardiac** and **smooth muscle**.

sliding-filament model The hypothesis that thin (actin) filaments and thick (myosin) filaments slide past each other, thereby shortening the sarcomere. Shortening of all the sarcomeres in a myofibril results in contraction of the entire myofibril.

small intestine The portion of the digestive tract between the stomach and the large intestine. The site of the final stages of digestion and of most nutrient absorption.

mall nuclear ribonucleoproteins See **snRNPs**.

th endoplasmic reticulum (smooth ER) The of the endoplasmic reticulum that does not have attached to it. Involved in synthesis and lipids. Compare with **rough endoplasmic**

he unstriated muscle tissue that lines essels, and some other organs. branched cells that can sustain luntarily controlled. Compare **uscle.**

snRNPs (small nuclear ribonucleoproteins) Complexes of proteins and small RNA molecules that function in splicing (removal of introns from primary RNA transcripts) as components of spliceosomes.

sodium-potassium pump A transmembrane protein that uses the energy of ATP to move sodium ions out of the cell and potassium ions in. Also called *Na^+/K^+-ATPase.*

soil organic matter Organic (carbon-containing) compounds found in soil.

solute Any substance that is dissolved in a liquid.

solute potential (ψ_S) A component of the potential energy of water caused by a difference in solute concentrations at two locations. Also called *osmotic potential*. Compare with **pressure potential (ψ_P)**.

solution A liquid containing one or more dissolved solids or gases in a homogeneous mixture.

solvent Any liquid in which one or more solids or gases can dissolve.

soma See **cell body**.

somatic cell Any type of cell in a multicellular organism except eggs, sperm, and their precursor cells. Also called *body cells.*

somatic hypermutation Mutations that occur in the immunoglobulin genes of the immune system's memory cells, resulting in novel variation in the receptors that bind to antigens.

somatic nervous system The part of the peripheral nervous system (outside the brain and spinal cord) that controls skeletal muscles and is under voluntary control. Compare with **autonomic nervous system**.

somatostatin A hormone secreted by the pancreas and hypothalamus that inhibits the release of several other hormones.

somites Paired blocks of mesoderm on both sides of the developing spinal cord in a vertebrate embryo. Give rise to muscle tissue, vertebrae, ribs, limbs, etc.

sori In ferns, a cluster of spore-producing structures (sporangia).

source Any tissue, site, or location where a substance is produced or enters circulation (e.g., in plants, the tissue where sugar enters the phloem). Compare with **sink**.

space-filling model A representation of a molecule where atoms are shown as balls—color-coded and scaled to indicate the atom's identify—attached to each other in the correct geometry.

speciation The evolution of two or more distinct species from a single ancestral species.

species A distinct, identifiable group of populations that is thought to be evolutionarily independent of other populations and whose members can interbreed. Generally distinct from other species in appearance, behavior, habitat, ecology, genetic characteristics, etc.

species diversity The variety and relative abundance of the species present in a given ecological community.

species richness The number of species present in a given ecological community.

species–area relationship The mathematical relationship between the area of a certain habitat and the number of species that it can support.

specific heat The amount of energy required to raise the temperature of 1 gram of a substance by 1°C; a measure of the capacity of a substance to absorb energy.

sperm A mature male gamete; smaller and more mobile than the female gamete.

sperm competition Competition to fertilize eggs between the sperm of different males, inside the same female.

spermatid An immature sperm cell.

spermatogenesis The production of sperm. Occurs continuously in a testis.

spermatogonia (singular: spermatogonium) The diploid cells in a testis that can give rise to primary spermatocytes.

spermatophore A gelatinous package of sperm cells that is produced by males of species that have internal fertilization without copulation.

sphincter A muscular valve that can close off a tube, as in a blood vessel or a part of the digestive tract.

spicule Stiff spike of silica or calcium carbonate found in the body of many sponges.

spindle apparatus The array of microtubules responsible for contacting and moving chromosomes during mitosis and meiosis; includes kinetochore microtubules and polar microtubules.

spines In plants, modified leaves that are stiff and sharp and that function in defense.

spiracle In insects, a small opening that connects air-filled tracheae to the external environment, allowing for gas exchange.

spleen A dark red organ, found near the stomach of most vertebrates, that filters blood, stores extra red blood cells in case of emergency, and plays a role in immunity.

spliceosome In eukaryotes, a large, complex assembly of snRNPs (small nuclear ribonucleoproteins) that catalyzes removal of introns from primary RNA transcripts.

splicing The process by which introns are removed from primary RNA transcripts and the remaining exons are connected together.

sporangium (plural: sporangia) A spore-producing structure found in seed plants, some protists, and some fungi (e.g., chytrids).

spore (1) In bacteria, a dormant form that generally is resistant to extreme conditions. (2) In eukaryotes, a single cell produced by mitosis or meiosis (not by fusion of gametes) that is capable of developing into an adult organism.

sporophyte In organisms undergoing alternation of generations, the multicellular diploid form that arises from two fused gametes and produces haploid spores. Compare with **gametophyte**.

sporopollenin A watertight material that encases spores and pollen of modern land plants.

stabilizing selection A pattern of natural selection that favors phenotypes near the middle of the range of phenotypic variation. Reduces overall genetic variation in a population. Compare with **disruptive selection**.

stamen The male reproductive structure of a flower. Consists of an anther, in which pollen grains are produced, and a filament, which supports the anther. Compare with **carpel**.

standing defense See **constitutive defense**.

stapes The last of three small bones (ossicles) in the middle ear of vertebrates. Receives vibrations from the tympanic membrane and by vibrating against the oval window passes them to the cochlea.

starch A mixture of two storage polysaccharides, amylose and amylopectin, both formed from α-glucose monomers. Amylopectin is branched, and amylose is unbranched. The major form of stored carbohydrate in plants.

start codon The AUG triplet in mRNA at which protein synthesis begins; codes for the amino acid methionine.

statocyst A sensory organ of many arthropods that detects the animal's orientation in space (i.e., whether the animal is flipped upside down).

statolith A tiny stone or dense particle found in specialized gravity-sensing organs in some animals such as lobsters.

statolith hypothesis The hypothesis that amyloplasts (dense, starch-storing plant organelles) serve as statoliths in gravity detection by plants.

STATs See **signal transducers and activators of transcription**.

stem cell Any relatively undifferentiated cell that can divide to produce daughter cells identical to itself or more specialized daughter cells, which differentiate further into specific cell types.

stems Vertical, aboveground structures that make up the shoot system of plants.

stereocilium (plural: stereocilia) One of many stiff outgrowths from the surface of a hair cell that are involved in detection of sound by terrestrial vertebrates or of waterborne vibrations by fishes.

steroid A class of lipid with a characteristic four-ring structure.

steroid-hormone receptor One of a family of intracellular receptors that bind to various steroid hormones, forming a hormone-receptor complex that acts as a regulatory transcription factor and activates transcription of specific target genes.

sticky end The short, single-stranded ends of a DNA molecule cut by a restriction endonuclease. Tend to form hydrogen bonds with other sticky ends that have complementary sequences.

stigma The moist tip at the end of a flower carpel to which pollen grains adhere.

stolon A modified stem that runs horizontally over the soil surface and produces new plants at the nodes (a form of asexual reproduction). Compare with **rhizome**.

stoma (plural: stomata) Generally, a pore or opening. In plants, a microscopic pore on the surface of a leaf or stem through which gas exchange occurs.

stomach A tough, muscular pouch in the vertebrate digestive tract between the esophagus and small intestine. Physically breaks up food and begins digestion of proteins.

stop codon One of three mRNA triplets (UAG, UGA, or UAA) that cause termination of protein synthesis. Also called a *termination codon*.

strain A population of genetically similar or identical individuals.

stream A body of water that moves constantly in one direction.

striated muscle See **skeletal muscle**.

stroma The fluid matrix of a chloroplast in which the thylakoids are embedded. Site where the Calvin cycle reactions occur.

structural formula A two-dimensional notation in which the chemical symbols for the constituent atoms are joined by straight lines representing single (—), double (═), or triple (≡) covalent bonds. Compare with **molecular formula**.

structural gene A stretch of DNA that codes for a functional protein or functional RNA molecule, not including any regulatory sequences (e.g., a promoter, enhancer).

structural homology Similarities in organismal structures (e.g., limbs, shells, flowers) that are due to inheritance from a common ancestor.

style The slender stalk of a flower carpel connecting the stigma and the ovary.

subspecies A population that has distinctive traits and some genetic differences relative to other populations of the same species but that is not distinct enough to be classified as a separate species.

substrate (1) A reactant that interacts with an enzyme in a chemical reaction. (2) A surface on which a cell or organism sits.

substrate-level phosphorylation Production of ATP by transfer of a phosphate group from an intermediate substrate directly to ADP. Occurs in glycolysis and in the citric acid cycle.

succession In ecology, the gradual colonization of a habitat after an environmental disturbance (e.g., fire, flood), usually by a series of species. See **primary** and **secondary succession**.

sucrose A disaccharide formed from glucose and fructose. One of the two main products of photosynthesis.

sugar Synonymous with carbohydrate, though usually used in an informal sense to refer to small carbohydrates (monosaccharides and disaccharides).

sulfate reducer A prokaryote that produces hydrogen sulfide (H_2S) as a by-product of cellular respiration.

summation The additive effect of different postsynaptic potentials at a nerve or muscle cell, such that several subthreshold stimulations can cause an action potential.

supporting connective tissue A type of connective tissue, distinguished by having a firm extracellular matrix.

surface tension The cohesive force that causes molecules at the surface of a liquid to stick together, thereby resisting deformation of the liquid's surface and minimizing its surface area.

surfactant A mixture of phospholipids and proteins produced by lung cells that reduces surface tension, allowing the lungs to expand more.

survivorship On average, the proportion of offspring that survive to a particular age.

survivorship curve A graph depicting the percentage of a population that survives to different ages.

suspension feeder Any organism that obtains food by filtering small particles or small organisms out of water or air. Also called *filter feeder*.

sustainability The planned use of environmental resources at a rate no faster than the rate at which they are naturally replaced.

sustainable agriculture Agricultural techniques that are designed to maintain long-term soil quality and productivity.

swamp A wetland that has a steady rate of water flow and is dominated by trees and shrubs.

swim bladder A gas-filled organ of many ray-finned fishes that regulates buoyancy.

symbiosis (adjective: symbiotic) Any close and prolonged physical relationship between individuals of two different species. See **commensalism, mutualism,** and **parasitism**.

symmetric competition Ecological competition between two species in which both suffer similar declines in fitness. Compare with **asymmetric competition**.

sympathetic nervous system The part of the autonomic nervous system that stimulates fight-or-flight responses, such as increased heart rate, increased blood pressure, and decreased digestion. Compare with **parasympathetic nervous system**.

sympatric speciation The divergence of populations living within the same geographic area into different species as the result of their genetic (not physical) isolation. Compare with **allopatric speciation**.

sympatry Condition in which two or more populations live in the same geographic area, or close enough to permit interbreeding. Compare with **allopatry**.

symplast In plant roots, a continuous pathway through which water can flow through the cytoplasm of adjacent cells that are connected by plasmodesmata. Compare with **apoplast**.

symporter A carrier protein that allows an ion to diffuse down an electrochemical gradient, using the energy of that process to transport a different substance in the same direction *against* its concentration gradient. Compare with **antiporter**.

synapomorphy A shared, derived trait found in two or more taxa that is present in their most recent common ancestor but is missing in more distant ancestors. Useful for inferring evolutionary relationships.

synapse The interface between two neurons or between a neuron and an effector cell.

synapsis The physical pairing of two homologous chromosomes during prophase I of meiosis. Crossing over occurs during synapsis.

synaptic cleft The space between two communicating nerve cells (or between a neuron and effector cell) at a synapse, across which neurotransmitters diffuse.

synaptic plasticity Long-term changes in the responsiveness or physical structure of a synapse that can occur after particular stimulation patterns. Thought to be the basis of learning and memory.

synaptic vesicle A small neurotransmitter- containing vesicle at the end of an axon that releases neurotransmitter into the synaptic cleft by exocytosis.

synaptonemal complex A network of proteins that holds non-sister chromatids together during synapsis in meiosis I.

synthesis (S) phase The phase of the cell cycle during which DNA is synthesized and chromosomes are replicated.

systemic acquired resistance (SAR) A slow, widespread response of plants to a localized infection that protects healthy tissue from invasion by pathogens. Compare with **hypersensitive response**.

systemic circulation The part of the circulatory system that sends oxygen-rich blood from the lungs out to the rest of the body. Is separate from the pulmonary circulation in mammals and birds.

systemin A peptide hormone, produced by plant cells damaged by herbivores, that initiates a protective response in undamaged cells.

systole The portion of the heartbeat cycle during which the heart muscles are contracting. Compare with **diastole**.

systolic blood pressure The force exerted by blood against artery walls during contraction of the heart's left ventricle. Compare with **diastolic blood pressure**.

T cell A type of leukocyte that matures in the thymus and, with B cells, is responsible for acquired immunity. Involved in activation of B cells ($CD4^+$ helper T cells) and destruction of infected cells ($CD8^+$ cytotoxic T cells). Also called *T lymphocytes*.

T tubules Membranous tubes that extend into the interior of muscle cells. Propagate action potentials throughout a muscle cell and trigger release of calcium from the sarcoplasmic reticulum.

taiga A vast forest biome throughout subarctic regions, consisting primarily of short conifer trees. Characterized by intensely cold winters, short summers, and high annual variation in temperature.

taproot A large vertical main root of a plant.

taste bud Sensory structure, found chiefly in the mammalian tongue, containing spindle-shaped cells that respond to chemical stimuli.

TATA box A short DNA sequence in many eukaryotic promoters about 30 base pairs upstream from the transcription start site.

TATA-binding protein (TBP) A protein that binds to the TATA box in eukaryotic promoters and is a component of the basal transcription complex.

taxon (plural: taxa) Any named group of organisms at any level of a classification system.

taxonomy The branch of biology concerned with the classification and naming of organisms.

TBP See **TATA-binding protein**.

T-cell receptor (TCR) A transmembrane protein found on T cells that can bind to antigens displayed on the surfaces of other cells. Composed of two polypeptides called the alpha chain and beta chain. See **antigen presentation**.

tectorial membrane A membrane in the vertebrate cochlea that takes part in the transduction of sound by bending the stereocilia of hair cells in response to sonic vibrations.

telomerase An enzyme that replicates the ends of chromosome (telomeres) by catalyzing DNA synthesis from an RNA template that is part of the enzyme.

telomere The region at the end of a linear chromosome.

telophase The final stage in mitosis or meiosis during which sister chromatids (replicated chromosomes in meiosis I) separate and new nuclear envelopes begin to form around each set of daughter chromosomes.

temperate Having a climate with pronounced annual fluctuations in temperature (i.e., warm summers and cold winters) but typically neither as hot as the tropics nor as cold as the poles.

temperature A measurement of thermal energy present in an object or substance, reflecting how much the constituent molecules are moving.

template strand (1) The strand of DNA that is transcribed by RNA polymerase to create RNA. (2) An original strand of RNA used to make a complementary strand of RNA.

temporal lobe In the vertebrate brain, one of the four major areas in the cerebrum.

tendon A band of tough, fibrous connective tissue that connects a muscle to a bone.

tentacle A long, thin, muscular appendage of gastropod mollusks.

termination (1) In enzyme-catalyzed reactions, the final stage in which the enzyme returns to its original conformation and products are released. (2) In DNA transcription, the dissociation of RNA polymerase from DNA when it reaches a termination signal sequence. (3) In RNA translation, the dissociation of a ribosome from mRNA when it reaches a stop codon.

territory An area that is actively defended by an animal from others of its species.

tertiary consumers In a food chain or food web, organisms that feed on secondary consumers. Compare with **primary consumer** and **secondary consumer**.

tertiary structure The overall three-dimensional shape of a single polypeptide chain, resulting from multiple interactions among the amino acid side chains and the peptide backbone. Compare with **primary, secondary,** and **quaternary structure**.

testcross The breeding of an individual of unknown genotype with an individual having only recessive alleles for the traits of interest in order to infer the unknown genotype from the phenotypic ratios seen in offspring.

testis (plural: testes) The sperm-producing organ of a male animal.

testosterone A steroid hormone, produced and secreted by the testes, that stimulates sperm production and various male traits and reproductive behaviors.

tetrad The structure formed by synapsed homologous chromosomes during prophase of meiosis I.

tetrapod Any member of the taxon Tetrapoda, which includes all vertebrates with two pairs of limbs (amphibians, mammals, birds, and other reptiles).

texture A quality of soil, resulting from the relative abundance of different-sized particles.

theory A proposed explanation for a broad class of phenomena or observations.

thermal energy The kinetic energy of molecular motion.

thermocline A gradient (cline) in environmental temperature across a large geographic area.

thermophile A bacterium or archaean that thrives in very hot environments.

thermoreceptor A sensory cell or an organ specialized for detection of changes in temperature.

thermoregulation Regulation of body temperature.

thick filament A filament composed of bundles of the motor protein myosin; anchored to the center of the sarcomere. Compare with **thin filament**.

thigmotropism Growth or movement of an organism in response to contact with a solid object.

thin filament A filament composed of two coiled chains of actin and associated regulatory proteins; anchored at the Z disk of the sarcomere. Compare with **thick filament**.

thorax A region of the body; in insects, one of the three prominent body regions called tagmata.

thorn A modified plant stem shaped as a sharp protective structure. Helps protect a plant against feeding by herbivores.

threshold potential The membrane potential that will trigger an action potential in a neuron or other excitable cell. Also called simply *threshold*.

thylakoid A flattened, membrane-bound vesicle inside a plant chloroplast that functions in converting light energy to chemical energy. A stack of thylakoids is a granum.

thymus An organ, located in the anterior chest or neck of vertebrates, in which immature T cells generated in the bone marrow undergo maturation.

thyroid gland A gland in the neck that releases thyroid hormone (which increases metabolic rate) and calcitonin (which lowers blood calcium).

thyroid-stimulating hormone (TSH) A peptide hormone, produced and secreted by the anterior pituitary, that stimulates release of thyroid hormones from the thyroid gland.

thyroxine (T_4) A peptide hormone containing four iodine atoms that is produced and secreted by the thyroid gland. Acts primarily to increase cellular metabolism. In mammals, T_4 is converted to the more active hormone triiodothyronine (T_3) in the liver.

Ti plasmid A plasmid carried by *Agrobacterium* (a bacterium that infects plants) that can integrate into a plant cell's chromosomes and induce formation of a gall.

tight junction A type of cell-cell attachment structure that links the plasma membranes of adjacent animal cells, forming a barrier that restricts movement of substances in the space between the cells. Most abundant in epithelia (e.g., the intestinal lining). Compare with **desmosome** and **gap junction**.

tip The end of a branch on a phylogenetic tree. Represents a specific species or larger taxon that has not (yet) produced descendants—either a group living today or a group that ended in extinction. Also called *terminal node*.

tissue A group of similar cells that function as a unit, such as muscle tissue or epithelial tissue.

tolerance In ecological succession, the phenomenon in which early-arriving species do not affect the probability that subsequent species will become established. Compare with **facilitation** and **inhibition**.

tonoplast The membrane surrounding a plant vacuole.

top-down control The hypothesis that population size is limited by predators or herbivores (consumers).

topoisomerase An enzyme that cuts and rejoins DNA downstream of the replication fork, to ease the twisting that would otherwise occur as the DNA "unzips."

torpor An energy-conserving physiological state, marked by a decrease in metabolic rate, body temperature, and activity, that lasts for a short period (overnight to a few days or weeks). Occurs in some small mammals when the ambient temperature drops significantly. Compare with **hibernation**.

totipotent Capable of dividing and developing to form a complete, mature organism.

trachea (plural: tracheae) (1) In insects, one of the small air-filled tubes that extend throughout the body and function in gas exchange. (2) In terrestrial vertebrates, the airway connecting the larynx to the bronchi. Also called *windpipe*.

tracheid In vascular plants, a long, thin water-conducting cell that has gaps in its secondary cell wall, allowing water movement between adjacent cells. Compare with **vessel element**.

trade-off In evolutionary biology, an inescapable compromise between two traits that cannot be optimized simultaneously. Also called *fitness trade-off*.

trait Any heritable characteristic of an individual.

transcription The process by which RNA is made from a DNA template.

transcriptional control Regulation of gene expression by various mechanisms that change the rate at which genes are transcribed to form messenger RNA. In negative control, binding of a regulatory protein to DNA represses transcription; in positive control binding of a regulatory protein to DNA promotes transcription.

transcriptome The complete set of genes transcribed in a particular cell.

transduction Conversion of information from one mode to another. For example, the process by which a stimulus outside a cell is converted into a response by the cell.

transfer cell In land plants, a cell that transfers nutrients from a parent plant to a developing plant seed.

transfer RNA (tRNA) One of a class of RNA molecules that have an anticodon at one end and an amino acid binding site at the other. Each tRNA picks up a specific amino acid and binds to the corresponding codon in messenger RNA during translation.

transformation (1) Incorporation of external DNA into the genome. Occurs naturally in some bacteria; can be induced in the laboratory by certain processes. (2) Conversion of a normal cell to a cancerous one.

transgenic Referring to an individual plant or animal whose genome contains DNA introduced from another individual, either from the same or a different species.

transition state A high-energy intermediate state of the reactants during a chemical reaction that must be achieved for the reaction to proceed. Compare with **activation energy**.

transitional feature A trait that is intermediate between a condition observed in ancestral species and the condition observed in more derived species.

translation The process by which proteins and peptides are synthesized from messenger RNA.

translational control Regulation of gene expression by various mechanisms that alter the life span of messenger RNA or the efficiency of translation.

translocation (1) In plants, the movement of sugars and other organic nutrients through the phloem by bulk flow. (2) A type of mutation in which a piece of a chromosome moves to a nonhomologous chromosome. (3) The process by which a ribosome moves down a messenger RNA molecule during translation.

transmembrane protein Any membrane protein that spans the entire lipid bilayer. Also called *integral membrane protein.*

transmission The passage or transfer (1) of a disease from one individual to another or (2) of electrical impulses from one neuron to another.

transmission electron microscope (TEM) A microscope that forms an image from electrons that pass through a specimen. Compare with **scanning electron microscope**.

transpiration Water loss from aboveground plant parts. Occurs primarily through stomata.

transport protein Collective term for any membrane protein that enables a specific ion or small molecule to cross a plasma membrane. Includes carrier proteins and channel proteins, which carry out passive transport (facilitated diffusion), and pumps, which carry out active transport.

transporter See **carrier protein**.

transposable elements Any of several kinds of DNA sequences that are capable of moving themselves, or copies of themselves, to other locations in the genome. Include LINEs and SINEs.

tree of life A diagram depicting the genealogical relationships of all living organisms on Earth, with a single ancestral species at the base.

trichome A hairlike appendage that grows from epidermal cells of some plants. Trichomes exhibit a variety of shapes, sizes, and functions depending on species.

triglyceride See **fat**.

triiodothyronine (T_3) A peptide hormone containing three iodine atoms that is produced and secreted by the thyroid gland. Acts primarily to increase cellular metabolism. In mammals, T_3 has a stronger effect than does the related hormone thyroxine (T_4).

triose A monosaccharide (simple sugar) containing three carbon atoms.

triplet code A code in which a "word" of three letters encodes one piece of information. The genetic code is a triplet code because a codon is three nucleotides long and encodes one amino acid.

triploblast (adjective: triploblastic) An animal whose body develops from three basic embryonic cell layers: ectoderm, mesoderm, and endoderm. Compare with **diploblast**.

trisomy The state of having three copies of one particular type of chromosome.

tRNA See **transfer RNA**.

trochophore A larva with a ring of cilia around its middle that is found in some lophotrochozoans.

trophic cascade A series of changes in the abundance of species in a food web, usually caused by the addition or removal of a key predator.

trophic level A feeding level in an ecosystem.

trophoblast The exterior of a blastocyst (the structure that results from cleavage in embryonic development of mammals).

tropomyosin A regulatory protein present in thin (actin) filaments that blocks the myosin-binding sites on these filaments, thereby preventing muscle contraction.

troponin A regulatory protein, present in thin (actin) filaments, that can move tropomyosin off the myosin-binding sites on these filaments, thereby triggering muscle contraction. Activated by high intracellular calcium.

true navigation The type of navigation by which an animal can reach a specific point on Earth's surface.

trypsin A protein-digesting enzyme present in the small intestine that activates several other protein-digesting enzymes.

trypsinogen The precursor of protein-digesting enzyme trypsin. Secreted by the pancreas and activated by the intestinal enzyme enterokinase.

tube foot One of the many small, mobile, fluid-filled extensions of the water vascular system of echinoderms; the part extending outside the body is called a podium. Used in locomotion and feeding.

tuber A modified plant rhizome that functions in storage of carbohydrates.

tuberculosis A disease of the lungs caused by infection with the bacterium *Mycobacterium tuberculosis.*

tumor A mass of cells formed by uncontrolled cell division. Can be benign or malignant.

tumor suppressor A gene (e.g., *p53* and *Rb*) or the protein it encodes that prevents cell division, particularly when the cell has DNA damage. Mutated forms are associated with cancer.

tundra The treeless biome in polar and alpine regions, characterized by short, slow-growing vegetation, permafrost, and a climate of long, intensely cold winters and very short summers.

turgid Swollen and firm as a result of high internal pressure (e.g., a plant cell containing enough water for the cytoplasm to press against the cell wall). Compare with **flaccid**.

turgor pressure The outward pressure exerted by the fluid contents of a plant cell against its cell wall.

Turner syndrome A human genetic disorder caused by the presence of only one X chromosome and no Y chromosome ("XO"). Individuals with this condition are female but sterile.

turnover In lake ecology, the complete mixing of upper and lower layers of water that occurs each spring and fall in temperate-zone lakes.

tympanic membrane The membrane separating the middle ear from the outer ear in terrestrial vertebrates, or similar structures in insects. Also called the *eardrum.*

ubiquinone See **coenzyme Q**.

ulcer A hole in an epithelial layer that damages the underlying basement membrane and tissues.

ultimate causation In biology, the reason that a trait or phenomenon is thought to have evolved; the adaptive advantage of that trait. Also called *ultimate explanation.* Compare with **proximate causation**.

umami The taste of glutamate, responsible for the "meaty" taste of most proteins and of monosodium glutamate.

umbilical cord The cord that connects a developing mammalian embryo or fetus to the placenta and through which the embryo or fetus receives oxygen and nutrients.

unequal crossover An error in crossing over during meiosis I in which the two non-sister chromatids match up at different sites. Results in gene duplication in one chromatid and gene loss in the other.

universal tree The phylogenetic tree that includes all organisms.

unsaturated Referring to fats and fatty acids in which at least one carbon-carbon bond is a double bond. Double bonds produce kinks in the fatty acid chains and decrease the compound's melting point. Compare with **saturated**.

upstream In genetics, opposite to the direction in which RNA polymerase moves along a DNA strand. Compare with **downstream**.

ureter In vertebrates, a tube that transports urine from one kidney to the bladder.

urethra The tube that drains urine from the bladder to the outside environment. In male vertebrates, also used for passage of sperm during ejaculation.

uric acid A whitish excretory product of birds, reptiles, and terrestrial arthropods. Used to remove from the body excess nitrogen derived from the breakdown of amino acids. Compare with **urea**.

urochordates One of the three major chordate lineages (Urochordata), comprising sessile, filter-feeding animals that have a polysaccharide exoskeleton (tunic) and two siphons through which water enters and leaves; also called tunicates or sea squirts. Compare with **cephalochordates** and **vertebrates**.

uterus The organ in which developing embryos are housed in those vertebrates that give live birth. Common in most mammals and in some lizards, sharks, and other vertebrates.

vaccination The introduction into an individual of weakened, killed, or altered pathogens to stimulate development of acquired immunity against those pathogens.

vaccine A preparation designed to stimulate an immune response against a particular pathogen without causing illness. Vaccines consist of inactivated (killed) pathogens, live but weakened (attenuated) pathogens, or portions of a viral capsid (subunit vaccine).

vacuole A large organelle in plant and fungal cells that usually is used for bulk storage of water, pigments, oils, or other substances. Some vacuoles contain enzymes and have a digestive function similar to lysosomes in animal cells.

vagina The birth canal of female mammals; a muscular tube that extends from the uterus through the pelvis to the exterior.

valence electron An electron in the outermost electron shell, the valence shell, of an atom. Valence electrons tend to be involved in chemical bonding.

valence The number of unpaired electrons in the outermost electron shell of an atom; determines how many covalent bonds the atom can form.

valves In circulatory systems, flaps of tissue that prevent backward flow of blood, particularly in veins and between the chambers of the heart.

van der Waals interactions A weak electrical attraction between two hydrophobic side chains. Often contributes to tertiary structure in proteins.

variable (V) region The portion of an antibody's light chains or heavy chains that has a highly variable amino acid sequence and forms part of the antigen-binding site. Compare with **constant (C) region**.

variable number tandem repeat See **minisatellite**.

vas deferens (plural: vasa deferentia) A muscular tube that stores and transports semen from the epididymis to the ejaculatory duct. In nonhuman animals, called the *ductus deferens.*

vasa recta In the vertebrate kidney, a network of blood vessels that runs alongside the loop of Henle of a nephron. Functions in reabsorption of water and solutes from the filtrate.

vascular bundle A cluster of xylem and phloem strands in a plant stem.

vascular cambium One of two types of lateral meristem, consisting of a ring of undifferentiated plant cells inside the cork cambium of woody plants; produces secondary xylem (wood) and secondary phloem. Compare with **cork cambium**.

vascular tissue In plants, tissue that transports water, nutrients, and sugars. Made up of the complex tissues xylem and phloem, each of which contains several cell types.

vector A biting insect or other organism that transfers pathogens from one species to another. See also **cloning vector**.

vegetative organs The nonreproductive parts of a plant including roots, leaves, and stems.

vein Any blood vessel that carries blood (oxygenated or not) under relatively low pressure from the tissues toward the heart. Compare with **artery**.

veliger A distinctive type of larva, found in mollusks.

vena cava (plural: vena cavae) A large vein that returns oxygen-poor blood to the heart.

ventral Toward an animal's belly and away from its back. The opposite of dorsal.

ventricle (1) A thick-walled chamber of the heart that receives blood from an atrium and pumps it to the body or to the lungs. (2) One of several small fluid-filled chambers in the vertebrate brain.

venules Small veins (blood vessels that return blood to the heart).

vertebra (plural: vertebrae) One of the cartilaginous or bony elements that form the spine of vertebrate animals.

vertebrates One of the three major chordate lineages (Vertebrata), comprising animals with a dorsal column of cartilaginous or bony structures (vertebrae) and a skull enclosing the brain. Includes fishes, amphibians, mammals, reptiles, and birds. Compare with **cephalochordates** and **urochordates**.

vessel element In vascular plants, a short, wide water-conducting cell that has gaps through both the primary and secondary cell walls, allowing unimpeded passage of water between adjacent cells. Compare with **tracheid**.

vestigial trait Any rudimentary structure of unknown or minimal function that is homologous to functioning structures in other species. Vestigial traits are thought to reflect evolutionary history.

vicariance The physical splitting of a population into smaller, isolated populations by a geographic barrier.

villi (singular: villus) Small, fingerlike projections (1) of the lining of the small intestine or (2) of the fetal portion of the placenta adjacent to maternal arteries. Function to increase the surface area available for absorption of nutrients and gas exchange (in the placenta).

virion A single mature virus particle.

virulence The ability of a pathogen or parasite to cause disease and death.

virulent Referring to pathogens that can cause severe disease in susceptible hosts.

virus A tiny intracellular parasite that uses host cell enzymes to replicate; consists of a DNA or RNA genome enclosed within a protein shell (capsid). In enveloped viruses, the capsid is surrounded by a phospholipid bilayer derived from the host cell plasma membrane, whereas nonenveloped viruses lack this protective covering.

visceral mass One of the three main parts of the mollusk body; contains most of the internal organs and external gill.

visible light The range of wavelengths of electromagnetic radiation that humans can see, from about 400 to 700 nanometers.

vitamin An organic micronutrient that usually functions as a coenzyme.

vitelline envelope A fibrous sheet of glycoproteins that surrounds mature egg cells in many vertebrates. Surrounded by a thick gelatinous matrix (the jelly layer) in some species. In mammals, called the *zona pellucida.*

viviparous Producing live young (instead of eggs) that develop within the body of the mother before birth. Compare with **oviparous** and **ovoviviparous**.

volt (V) A unit of electrical potential (voltage).

voltage Potential energy created by a separation of electric charges between two points. Also called *electrical potential.*

voltage clamping A technique for imposing a constant membrane potential on a cell. Widely used to investigate ion channels.

voltage-gated channel An ion channel that opens or closes in response to changes in membrane voltage. Compare with **ligand-gated channel**.

voluntary muscle Muscle tissue that can respond to conscious thought.

wall pressure The inward pressure exerted by a cell wall against the fluid contents of a plant cell.

Wallace line A line that demarcates two areas in the Indonesian region, each of which is characterized by a distinct set of animal species.

water cycle, global The movement of water among terrestrial ecosystems, the oceans, and the atmosphere.

water potential (ψ) The potential energy of water in a certain environment compared with the potential energy of pure water at room temperature and atmospheric pressure. In living organisms, ψ equals the solute potential (ψ_S) plus the pressure potential (ψ_P).

water potential gradient A difference in water potential in one region compared with that in another region. Determines the direction that water moves, always from regions of higher water potential to regions of lower water potential.

water table The upper limit of the underground layer of soil that is saturated with water.

water vascular system In echinoderms, a system of fluid-filled tubes and chambers that functions as a hydrostatic skeleton.

watershed The area drained by a single stream or river.

Watson-Crick pairing See **complementary base-pairing**.

wavelength The distance between two successive crests in any regular wave, such as light waves, sound waves, or waves in water.

wax A class of lipid with extremely long hydrocarbon tails, usually combinations of long-chain alcohols with fatty acids. Harder and less greasy than fats.

weather The specific short-term atmospheric conditions of temperature, moisture, sunlight, and wind in a certain area.

weathering The gradual wearing down of large rocks by rain, running water, and wind; one of the processes that transform rocks into soil.

weed Any plant that is adapted for growth in disturbed soils.

wetland A shallow-water habitat where the soil is saturated with water for at least part of the year.

white blood cells See **leukocytes**.

wild type The most common phenotype seen in a population; especially the most common phenotype in wild populations compared with inbred strains of the same species.

wildlife corridor Strips of wildlife habitat connecting populations that otherwise would be isolated by man-made development.

wilt To lose turgor pressure in a plant tissue.

wobble hypothesis The hypothesis that some tRNA molecules can pair with more than one mRNA codon, tolerating some variation in the third base, as long as the first and second bases are correctly matched.

wood Xylem resulting from secondary growth. Also called *secondary xylem.*

xeroderma pigmentosum A human disease characterized by extreme sensitivity to ultraviolet light. Caused by an autosomal recessive allele that results in a defective DNA repair system.

X-linked inheritance Inheritance patterns for genes located on the mammalian X chromosome. Also called *X-linkage.*

X-ray crystallography A technique for determining the three-dimensional structure of large molecules, including proteins and nucleic acids, by analysis of the diffraction patterns produced by X-rays beamed at crystals of the molecule.

xylem A plant vascular tissue that conducts water and ions; contains tracheids and/or vessel elements. Primary xylem develops from the procambium of apical meristems; secondary xylem, or wood, from the vascular cambium of lateral meristems. Compare with **phloem**.

xylem sap The watery fluid found in the xylem of plants.

yeast Any fungus growing as a single-celled form. Also, a specific lineage of ascomycetes.

Y-linked inheritance Inheritance patterns for genes located on the mammalian Y chromosome. Also called *Y-linkage.*

yolk The nutrient-rich cytoplasm inside an egg cell; used as food for the growing embryo.

Z disk The structure that forms each end of a sarcomere. Contains a protein that binds tightly to actin, thereby anchoring thin filaments.

Z scheme Path of electron flow in which electrons pass from photosystem II to photosystem I and ultimately to $NADP^+$ during the light-dependent reactions of photosynthesis. Also called *noncyclic electron flow.*

zero population growth (ZPG) A state of stable population size due to fertility staying at the replacement rate for at least one generation.

zona pellucida The gelatinous layer around a mammalian egg cell. In other vertebrates, called the *vitelline envelope.*

zone of (cellular) division In plant roots, a group of apical meristematic cells just behind the root cap where cells are actively dividing.

zone of (cellular) elongation In plant roots, a group of young cells, located behind the apical meristem, that are increasing in length.

zone of (cellular) maturation In plant roots, a group of plant cells, located several centimeters behind the root cap, that are differentiating into mature tissues.

zygosporangium (plural: zygosporangia) The spore-producing structure in fungi that are members of the Zygomycota.

zygote The diploid cell formed by the union of two haploid gametes; a fertilized egg. Capable of undergoing embryological development to form an adult.

Credits

PHOTO CREDITS

Frontmatter p. ix Oscar Miller/SPL/Photo Researchers **p. xi** Photo by Jason Rick, supplied by Loren Rieseberg **p. xiv** Jeff Rotman/NPL/Minden Pictures **p. xvi** Frans Lanting/Minden Pictures **p. xxT** Natalie B. Fobes Photography **p. xxB** David Quillin

Chapter 1 Opener Marjorie Kibby **1.1** "Animalcules" observed by Anton van Leeuwenhoek, c1795, HIP/Art Resource, NY **1.6a** Samuel F. Conti and Thomas D. Brock **1.6b** Kwangshin Kim/Photo Researchers **1.7b** Michael Hughes/Laif/Aurora Photos **1.8R** From M. Wittlinger, R. Wehner, and H. Wolf, The ant odometer: Stepping on stilts and stumps, *Science* 312: 1965–1967, supporting online material (Jun 30 2006).

Chapter 2 Opener Frans Lanting/Corbis **2.1b** Dragan Trifunovic/Shutterstock **2.6c** Photos. com **2.11** Robert and Beth Plowes, ProteaPIX **2.14b** Dietmar Nill/Picture Press/Photolibrary **2.15c** Jan Vermeer/Minden Pictures **2.20** Shutterstock **2.20 inset** PhotoSpin/Alamy

Chapter 3 Opener Fritz Wilhelm (heisingart.com) **3.10a** Microworks/Phototake **3.10b** Walter Reinhart/ Phototake

Chapter 4 Opener SSPL/The Image Works **4.5** National Cancer Institute

Chapter 5 Opener Peter Arnold/Alamy

Chapter 6 6.5L James J. Cheetham, Carleton University **6.9a** iStockphoto **6.9b** iStockphoto **6.9c** Shutterstock **6.18** Harold Edwards/Visuals Unlimited **6.24T** David M. Phillips/Photo Researchers **6.24M** David M. Phillips/Photo Researchers **6.24B** Joseph F. Hoffman, Yale University School of Medicine

Chapter 7 Opener Torsten Wittmann/Photo Researchers **7.1** T. J. Beveridge/Visuals Unlimited **7.2** Gopal Murti/ Visuals Unlimited **7.3** Wanner/Eye of Science/Photo Researchers **7.4** Stanley C. Holt/Biological Photo Service **7.5** From David S. Goodsell, *TheMachinery of Life,* 2nd ed. (2009). © Springer-Verlag, New York. **7.7** Don W. Fawcett/ Photo Researchers **7.8** Don W. Fawcett/Photo Researchers **7.9** Don W. Fawcett/Photo Researchers **7.10** Biophoto Associates/Photo Researchers **7.11** Omikron/Photo Researchers **7.12** Don Fawcett, Daniel Friend, & Richard Wood/Photo Researchers **7.13** Gopal Murti/Visuals Unlimited **7.16** E. H. Newcomb & W. P. Wergin/Biological Photo Service **7.17** T.Kanaseki & Donald Fawcett/Visuals Unlimited **7.18** E. H. Newcomb & W. P. Wergin/Biological Photo Service **7.19** E. H. Newcomb & S. E. Frederick/ Biological Photo Service **7.20** By David S. Goodsell. Published in L. A. Moran et al., *Biochemistry* (1994). © Neil Patterson Publishers/Prentice-Hall. **7.21a** Don W. Fawcett/ Photo Researchers **7.21b** Don W. Fawcett/Photo Resarchers **7.21c** Biophoto Associates/Photo Researchers **7.21d** Dennis Kunkel/Visuals Unlimited **7.22** Don W. Fawcett/Photo Researchers **7.25a** James D. Jamieson, Yale University School of Medicine **7.25b/c** From J. D. Jamieson and G. E. Palade, *J. Cell Biol.* 34: 597–615 (1967). © The Rockefeller University Press. **7.30** Conly L. Rieder, Wadsworth Center, New York State Department of Health **7.31a/b** From *Mol. Biol. Cell* 9: cover (Dec 1998). © American Society for Cell Biology. Photo B. J. Schnapp. **7.32a** John E. Heuser, Washington University School of Medicine **7.33L** SPL/Photo Researchers **7.33R** Visuals Unlimited/Corbis **7.34a** Don W. Fawcett/Photo Researchers

Chapter 8 Opener Caroline Weight, G.I.T. Molecular Immunology, Institute of Food Research, Norwich, UK **8.1** William James Warren/Science Faction **8.2** Biophoto Associates/Photo Researchers **8.3b** Barry King, University of California, Davis/Biological Photo Service **8.5** SPL/Photo Researchers **8.6** Photo Researchers **8.7aL** Don W. Fawcett/Photo Researchers **8.7aR** Don W. Fawcett/Photo Researchers **8.8a** Don W. Fawcett/Photo Researchers **8.11a** E. H. Newcomb & W. P. Wergin/Biological Photo Service **8.11b** Don W. Fawcett/Photo Researchers **8.18** Janice Carr/CDC/Rodney M. Donlan

Chapter 9 Opener Richard Megna/Fundamental Photographs **9.7T** Shutterstock **9.7B** Shutterstock **9.13** Terry Frey, San Diego State University **9.22a** Science VU/B. Bhatnagar/Visuals Unlimited

Chapter 10 Opener Visuals Unlimited/Corbis **10.2T** John Durham/SPL/Photo Researchers **10.2M/B** Electron micrographs by W. P. Wergin, courtesy of E. H. Newcomb, University of Wisconsin, Madison **10.4b** Sinclair Stammers/ SPL/Photo Researchers **10.17L/R** James A. Bassham, Lawrence Berkeley Laboratory, UCB (retired) **10.20a** Jeremy Burgess/SPL/Photo Researchers

Chapter 11 Opener CNRI/Phototake **11.1a** From Walter Flemming, *Zellsubstanz, Kern, und Zelltheilung*. Leipzig: Verlag von F. C. W. Vogel, 1882. **11.1b** Conly Rieder, Wadsworth Center, New York State Department of Health. **11.1b** Photo Researchers **11.2T** Gopal Murti/Photo Researchers **11.2B** Biophoto Associates/Photo Researchers **11.5** Micrographs by Conly L. Rieder, Wadsworth Center, NYS Department of Health **11.6a** Ed Reschke/Peter Arnold **11.6b** Visuals Unlimited/Getty Images

Chapter 12 Opener David Phillips/The Population Council/Photo Researchers **12.6** Photos.com **12.7** Micrographs byEdward Novitski, courtesy of Charles Novitski

Chapter 13 Opener Brian Johnston **13.9a** Benjamin Prud'homme/Nicolas Gompel **13.9bL/bR** From J. Childress, R. Behringer, and G. Halder, "Learning to Fly: Phenotypic Markers in Drosophila," *Genesis* 43(1): cover illustration (2005). © Wiley-Liss. Photo Georg Halder, University of Texas. **13.15a** Robert Calentine/Visuals Unlimited

Chapter 14 Opener Gopal Murti/SPL/Photo Researchers **14.1b** Eye of Science/Photo Researchers **14.5** From M. Meselson and F. W. Stahl, *Proc. Natl. Acad. Sci. USA* 44: 671–682, fig 4 (Jul 15 1958). Photo Matthew S. Meselson, Harvard University. **14.7a** Gopal Murti/SPL/Photo Researchers

Chapter 15 Opener Alfred Pasieka/Photo Researchers **15.4a** Rod Williams/Nature Picture Library **15.4b** The Peromyscus Genetic Stock Center at the University of South Carolina. Photograph by Clint Cook. **15.9a/b** Peter Duesberg, University of California, Berkeley

Chapter 16 Opener Oscar Miller/SPL/Photo Researchers **16.5a** Bert W. O'Malley, Baylor College of Medicine **16.8a** From B. A. Hamkalo and O. L. Miller Jr., *Annu. Rev. Biochem.* 42: 379–96 (1973). © Annual Reviews. **16.9b** E. V. Kiseleva and Donald Fawcett/Visuals Unlimited

Chapter 17 Opener Stephanie Schuller/Photo Researchers **UN17.1** EDVOTEK, The Biotechnology Education Company (www.edvotek.com)

Chapter 18 18.2a Ada Olins & Don Fawcett/Photo Researchers **18.4T** Barbara Hamkalo, University of California, Irvine **18.4B** Victoria E. Foe, University of Washington

Chapter 19 Opener AJ/IRRI/Corbis **19.1b** Brady-Handy Photograph Collection (Library of Congress). Reproduction number: LC-DIG-cwpbh-02977. **19.12** Baylor College of Medicine/Peter Arnold **19.16bL/R** Golden Rice Humanitarian Board (www.goldenrice.org)

Chapter 20 Opener Sanger Institute/Wellcome Photo Library **20.7bL/R** Test results provided by GENDIA (www.gendia.net) **20.10** Agilent Technologies **20.11** Camilla M. Kao and Patrick O. Brown, Stanford University

Chapter 21 Opener Anthony Bannister/Gallo Images/Corbis **21.1a** From R. Merino et al., *Development* 126(23): 5515–5522, fig. 6 (1999). © The Company of Biologists. **21.1b** From K. Kuida et al., *Cell* 94: 325–337, figs. 2e and 2f (1998). © Elsevier Science Ltd. Photo Keisuke Kuida, Vertex Pharmaceuticals. **21.2** Roslin Institute/Phototake **21.3a** Richard Hutchings/Photo Researchers **21.3b** From S. Kulandavelu et al., Embryonic and neonatal phenotyping of genetically engineered mice. *ILAR Journal*, 47(2); 103–117, fig. 12a (2006). © National Academy of Sciences. **21.4a** F. Rudolf Turner, Indiana University **21.4b** Christiane Nusslein-Volhard, Max Planck Institute **21.5** Christiane Nusslein-Volhard, Max Planck Institute **21.6** Wolfgang Driever, University of Freiburg **21.7a/c** Jim Langeland, Stephen Paddock, and Sean Carroll, University of Wisconsin–Madison **21.7b** Stephen J. Small, New York University **21.8L** Visuals Unlimited/Getty Images **21.8M** David Scharf/Science Faction/Corbis **21.8R** Eye of Science/Photo Researchers **21.10a** F. Rudolf Turner, Indiana University **21.11** Anthony Bannister/NHPA **21.12a/b** From "The Origin of Form" by Sean B. Carroll, *Natural History* (Nov 2005); A. C. Burke, "*Hox* Genes and the Global Patterning of the Somitic Mesoderm," in C. Ordahl (ed.), *Somitogenesis, Part I* (2000). © Academic Press.

Chapter 22 Opener Yorgos Nikas/Photo Researchers **22.2** Holger Jastrow **22.5a** Michael Whitaker/SPL/Photo Researchers **22.5b** Victor D. Vacquier, Scripps Institution of Oceanography, University of California at San Diego **22.11a** Kathryn W. Tosney, University of Michigan **22.11b** Kathryn W. Tosney, The University of Miami

Chapter 23 Opener John Runions/Oxford Brookes University **23.3b** Cabisco/Visuals Unlimited **23.4b** Michael Clayton, Department of Botany, University of Wisconsin, Madison **23.8b** Ken Wagner/Phototake **23.10a/bT/bB** From M. Kim et al., The expression domain of *PHANTASTICA* determines leaflet placement in compound leaves, *Nature* 424: 438–443, fig. 1 (Jul 24 2003). © Macmillan Magazines Ltd. Photos Neelima Sinah. **23.12** Photos by John L. Bowman, University of California at Davis

Chapter 24 Opener Art Wolfe/Stone/Getty Images **24.2a** Sinclair Stammers/Photo Researchers **24.2b** From S. De Valais and R. N. Melchor, Ichnotaxonomy of bird-like footprints: an example from the Late Triassic–Early Jurassic of northwest Argentina. *J. Vertebr. Paleontol.* 28(1):145–159, fig 5c (2008). © Society of Vertebrate Paleontology. **24.2c** The Natural History Museum, London **24.3L** Gerd Weitbrecht for Landesbildungsserver Baden-Württemberg (www.schule-bw.de) **24.3R** iStockphoto **24.5aL** CMCD/ PhotoDisc/Getty Images **24.5aR** Vincent Zuber/Custom Medical Stock Photo **24.5bL** Mary Beth Angelo/Photo Researchers **24.5bR** Custom Medical Stock Photo **24.6a1** age fotostock/SuperStock **24.6a2** George D. Lepp/Photo Researchers **24.6a3** Tui De Roy/Minden Pictures **24.6a4** Stefan Huwiler/age fotostock **24.8L/M** From M. K. Richardson et al., *Anat. Embryol.* 196: 91–106, figs. 7 and 8 (1997). © Springer-Verlag. Photos Michael K. Richardson, Leiden University, and Ronan O'Rahilly, National Museum of Health and Medicine/Armed Forces Institute of Pathology. **24.8R** From M. K. Richardson et al., *Science* 280: 983 (in Letters) (May 15 1998). Photo Ronan O'Rahilly, National Museum of Health and Medicine/Armed Forces Institute of Pathology. **24.10** Walter J. Gehring, Biozentrum, University of Basel (retired) **24.15aL** G. Gerra & S. Sommazzi/www.justbirds.it **24.15aR** Tui De Roy/Minden Pictures **24.15b** Huw Rees Lewis **24.18T** From A. Abzhonov et al., *Bmp4* and Morphological Variation of Beaks in Darwin's Finches, *Science* 305: 1462–1465, fig. 1c (Sep 3

2004). **24.18B** From K. Petren et al., *Proc. R. Soc. B*, 266: 321–329, fig. 3 (1999). © The Royal Society. Photo Kenneth Petren, University of Cincinnati. **24.20** Erlend Haarberg/NHPA/Photoshot

Chapter 25 Opener Phil Savoie/Minden Pictures **25.7a** André Karwath/Wikimedia Commons. Licensed under the Creative Commons Attribution Share Alike 2.5. **25.14a** Cyril Laubscher/Dorling Kindersley **25.15** Otorohanga Kiwi House, New Zealand **25.16a** Robert Rattner **25.17TL/TR** Robert & Linda Mitchell Photography **25.17BL/BR** Fotolia

Chapter 26 Opener Photo by Jason Rick, supplied by Loren Rieseberg **26.9bL** Michael Lustbader/Photo Researchers **26.9bR** From D. E. Soltis et al., Recent and recurrent polyploidy in *Tragopogon* (Asteraceae): Genetic, genomic, and cytogenetic comparisons. *Biol. J. Linn. Soc.* 82: 485–501 (2004). Photo Douglas and Pamela Soltis, University of Florida, Gainesville; Andrew Doust, Oklahoma State University. **26.10aT/M/B** H. Douglas Pratt/National Geographic Stock

Chapter 27 Opener Merv Feick (www.indiana9fossils .com) **27.7a** Stefan Piasecki, GEUS **27.7b** iStockphoto **27.7c/d** From The Virtual Petrified Wood Museum (petrifiedwoodmuseum.org). by Mike Viney **27.11LT** Gerald D. Carr **27.11LB/LR** Forest & Kim Starr **27.12a** Eladio Fernandez/Caribbean Nature Photography **27.12b** Jonathan B. Losos and Kevin de Quieroz, Washington University in St. Louis **27.13a** Colin Keates/Dorling Kindersley, courtesy of the Natural History Museum, London **27.13b** John Davis **27.13c** Don P. Northup, www.africancichlidphotos .com **27.13d** James L. Amos/Corbis **27.14bTL** Chip Clark, National Museum of Natural History, Smithsonian Institution **27.14bTR** Ed Reschke/Peter Arnold **27.14bML** Simon Conway Morris, University of Cambridge **27.14bMR** J. Gehling, South Australian Museum **27.14bBL/BM/BR** Shuhai Xiao, Virginia Polytechnic Institute and State University **27.17bL/bR** Glen A. Izett, U.S. Geological Survey, Denver **27.17c** Peter H. Schultz, Brown University; Steven D'Hondt, University of Rhode Island Graduate School of Oceanography

Chapter 28 Opener John Hammond **28.3** Science VU/Visuals Unlimited **28.4** From S. V. Liu et al., *Science* 277:1106–1109, fig. 2 (1997). © AAAS. PhotoYul Roh, Oak Ridge National Laboratory. **28.7aT** Daniel Quilten/Visuals Unlimited **28.7aB** From H. N. Schulz, et al., *Science* 284:493–495, fig. 1b (Aug 4 1999). © AAAS. Photo Heide Schulz, Max-Planck-Institute for Marine Microbiology. **28.7bT/B** SciMAT/Photo Researchers **28.7cT** K. S. Kim/Peter Arnold **28.7cB** Ralf Wagner **28.8a** R. Hubert, Iowa State University **28.10** Bruce J. Russell/BioMEDIA Associates **28.12** Wally Eberhart/Visuals Unlimited/Getty Images **28.15** Andrew Syred/Photo Researchers **28.16** Bill Schwartz/CDC **28.17** S. Amano, S. Miyadoh, and T. Shomura/The Society for Actinomycetes Japan **28.18** From J. L. Cocchiaro et al., Cytoplasmic lipid droplets are translocated into the lumen of the Chlamydia trachomatis parasitophorous vacuole, *Proc. Natl. Acad. Sci. USA* 105(27): 9379–9384 (Jul 8 2008). © National Academy of Sciences. **28.19** Peter Siver/Visuals Unlimited **28.20a** Yves V. Brun **28.20b** George Barron, University of Guelph, Canada **28.21** Kenneth M. Stedman/Portland State University **28.21 inset** Eye of Science/Photo Researchers **28.22** Doc Searls **28.22 inset** Priya DasSarma, Center of Marine Biotechnology, University of Maryland Biotechnology Institute

Chapter 29 Opener Wim van Egmond/Visuals Unlimited/Getty Images **29.2L** D. P. Wilson/FLPA/Minden Pictures **29.2M** Mark Conlin/V&W/imagequestmarine.com **29.2R** E. R. Degginger/Animals Animals–Earth Scenes **29.4** Pete Atkinson/Photographer's Choice/Getty Images **29.4 inset** D. Anderson, WHOI **29.6** Biophoto Associates/Photo Researchers **29.11a** Steve Gschmeissner/Photo Researchers **29.11b** Biophoto Associates/Photo Researchers **29.11c** Andrew Syred/Photo Researchers **29.12a** Biophoto Associates/Photo Researchers **29.12b** Bruce Coleman/Photoshot **29.15** Wim van Egmond/Visuals Unlimited/Getty Images **29.19** Scott Camazine/Photo Researchers **29.20** Dennis Kunkel/Visuals Unlimited **29.21** Tai-Soon Yong (www.atlas.or.kr) **29.22** Image by D. Patterson, provided by the microscope web project. **29.23** Francis Abbott/NPL/Minden Pictures **29.24** Peter Parks/imagequestmarine.com **29.25** Image by D. Patterson, provided by the micro*scope web project. **29.26** Maurice LOIR (www.diatomloir.eu) **29.27** Dennis Kunkel/Visuals Unlimited/Corbis **29.28** Andrew Syred/SPL/Photo Researchers **29.29** Wim van Egmond/Visuals Unlimited **29.30** Joyce Photographics/Photo Researchers **T29.3(1)** Peter Siver/Visuals Unlimited/Getty Images **T29.3(2)** Visuals Unlimited/Corbis **T29.3(3)** Jeff Rotman/Stone/Getty Images **T29.3(4)** iStockphoto **T29.3(5)** Yuuji Tsukii, Protist Information Server (protist.i.hosei.ac.jp) **T29.3(6)** Dr. Kawachi of National Institute for Environmental Studies, Japan

Chapter 30 Opener Jean-Paul Ferrero/Auscape/Minden Pictures **30.1** Lynn Betts/USDA Natural Resources Conservation Service **30.2b** Hugh Iltis/The Doebley Lab, University of Wisconsin-Madison (teosinte.wisc.edu) **30.4L** Linda Graham, University of Wisconsin-Madison **30.4R** Lee W. Wilcox **30.5L** Lee W. Wilcox **30.5M** iStockphoto **30.5R** Rob Whitworth/GAP Photos/Getty Images **30.11a/b** Lee W. Wilcox **30.12** Biodisc/Visuals Unlimited/Alamy **30.16a/b** Lee W. Wilcox **30.18** Carolina Biological Supply Company/Phototake **30.21a** Larry Deack/Wikimedia Commons **30.21b** Shutterstock **30.21c** Andreas Lander/dpa/Corbis **30.23a** Nigel Cattlin/Holt Studios/Photo Researchers **30.23b** Shutterstock **30.25T1** Jerome Wexler/Visuals Unlimited/Getty Images **30.25T2** Keith Wheeler/SPL/Photo Researchers **30.25T3** iStockphoto**30.25T4** Noodle snacks (www.noodlesnacks.com/) via Wikipedia. This file is licensed under the Creative Commons Attribution Share Alike 3.0 and GNU Free Documentation License. **30.25B1** Shutterstock **30.25B2** ISM/Phototake **30.25B3** iStockphoto **30.25B4** Shutterstock **30.27a** Doug Allan/NPL/Minden Pictures **30.27c** Lee W. Wilcox **30.27b** Wim van Egmond/Visuals Unlimited/Getty Images **30.28** Peter Verhoog/Foto Natura/Minden Pictures **30.29** Wim van Egmond **30.30** CAIP IMAGES © 2006 University of Florida (plants.ifas.ufl.edu/slidecol.html) **30.31** Adrian Davies/NPL/Minden Pictures **30.32a** Wildlife GmbH/Alamy **30.32b** Colin McPherson/Corbis **30.33** Steven P. Lynch, Lynch Images **30.34** Lee W. Wilcox **30.35** Frans Lanting/Corbis **30.36** Peter Arnold/Alamy **30.37L** David T. Webb, Botany, University of Hawaii **30.37R** Jami Tarris/Workbook Stock/Getty Images **30.38** Lee W. Wilcox **30.39** iStockphoto **30.40a** Martin Land/SPL/Photo Researchers **30.40b** iStockphoto **30.41** Peter Chadwick/Dorling Kindersley **30.42a/b** Lee W. Wilcox **30.43** Michael & Patricia Fogden/Minden Pictures **30.44a** Brian Johnston **30.44bL/R** Lee W. Wilcox

Chapter 31 Opener Richard Shiell/Animals Animals–Earth Scenes **31.1a** David Reed, Department of Animal and Plant Sciences, University of Sheffield **31.1b** Mycorrhizal Applications Inc. (www.mycorrhizae .com) **31.2** John Cang Photography **31.3a** Visuals Unlimited/Corbis **31.3b** iStockphoto **31.4a** Eye of Science/Photo Researchers **31.4b** Tony Brain/Photo Researchers **31.5bL** George Barron, University of Guelph, Canada **31.5bR** Biophoto Associates/Photo Researchers **31.6a** Melvin Fuller/The Mycological Society of America **31.6b** James Richardson/Visuals Unlimited **31.6c** Biophoto Associates/Photo Researchers **31.6d** Nino Santamaria **31.10a/b** Mark Brundrett (http://mycorrhizas.info) **31.13** J. I. Ronny Larsson, Professor of Zoology **31.14** Thomas J. Volk (TomVolkFungi.net) **31.15a** Gregory G. Dimijian/Photo Researchers **31.15b** David M. Phillips/Photo Researchers **31.16** Jim Deacon **31.17** Michel Poinsignon/NPL/Minden Pictures **31.18b** Daniel Mosquin **31.19** N. Allen and G. Barron, University of Guelph **31.20** Michael Fogden/OSF/Photolibrary

Chapter 32 Opener Bill Curtsinger/National Geographic Stock **32.1a** Satoshi Kuribayashi/Nature Production/Minden Pictures **32.1b** Mark Moffett/Minden Pictures **32.2** Walker England/Photo Researchers **32.3a** Roberto Rinaldi/NPL/Minden Pictures **32.3b** Ingo Arndt/Minden Pictures **32.8bT** Robert Steene/imagequestmarine.com **32.8bB** Russ Hopcroft/University of Alaska Fairbanks **32.12a** J. P. Ferrero/Jacana/Photo Researchers **32.12b** Todd Husman (www.soundaquatic.com) **32.13a** Pete Oxford/Minden Pictures **32.13b** Heidi & Hans-Jurgen Koch/Minden Pictures **32.14a overlay** Tim Ridley/Dorling Kindersley, Courtesy of the Royal Veterinary College, University of London **32.14a background** Shutterstock **32.14b** Adrian Costea/Fotolia **32.14b inset** Josef Ramsauer and Prof. Dr. Robert Patzner, Salzburg University **32.15a** Satoshi Kuribayashi/OSF/Photolibrary **32.15b** Radius Images/Corbis **32.16a** Andrew Syred/Photo Researchers **32.16b** Eye of Science/Photo Researchers **32.17a** Roland Seitre/Peter Arnold **32.17b** Jeff Foott/NPL/Minden Pictures **32.18a** D. P. Wilson/FLPA/Minden Pictures **32.18b** Sue Scott **32.19L/M/R** From G. Panganiban et al., *Proc. Natl. Acad. Sci. USA* 94(10): 5162–5166, figs. 1c and 1f (May 13 1997). © National Academy of Sciences. **32.20a** Shutterstock **32.20b** D. Parer & E. Parer-Cook/Auscape/Minden Pictures **32.21a** Nigel Cattlin/FLPA/Minden Pictures **32.21b inset** Frank Greenaway/Dorling Kindersley **32.21b** Kim Taylor/NPL/Minden Pictures **32.23** Roberto Rinaldi/NPL/Minden Pictures **32.24** Gary Bell/Taxi/Getty Images **32.25** Gregory G. Dimijian/Photo Researchers **32.26** Matthew D. Hooge, University of Maine

Chapter 33 Opener Barbara Strnadova/Photo Researchers **33.3a** Peter Parks/imagequestmarine.com **33.3b** Carmel McDougall, University of Oxford **33.4a** Dave King/Dorling Kindersley **33.4b** Breck P. Kent/Animals Animals–Earth Scenes **33.8a** Amazon-Images/Alamy **33.8b** Wim van Egmond/Visuals Unlimited **33.8c** Wim van Egmond/Visuals Unlimited/Corbis **33.9a** Pankaj Bajpai **33.9b** Bates Littlehales/Animals Animals–Earth Scenes **33.9c** Jane Burton/NPL/Minden Pictures **33.11** Antonio Guillén Oterino **33.12a** Doug Anderson (doug.deep via Flickr) **33.12b** Manfred Kage/Peter Arnold **33.12c** Oliver Meckes/Photo Researchers **33.13a** J. W.Alker/Imagebroker/Alamy **33.13b** Jasper Nance (www.flickr.com/photos/nebarnix) **33.13c** Robert and Linda Mitchell **33.14a** Marevision/age fotostock **33.15a** Roger Steen/imagequestmarine.com **33.15b** Daniela Wolf (www.taucher.li) **33.16** Kjell B. Sandved/Photo Researchers **33.17** Mark Conlin/VWPICS/age fotostock **33.18a** Shutterstock **33.18b** Andrew Syred/Photo Researchers **33.19a** Eye of Science/Photo Researchers **33.19b** C. Hough/Custom Medical Stock Photo **33.20** Mark Smith/Photo Researchers **33.21a** Holger Mette/Fotolia **33.21a inset** Holger Mette/Fotolia **33.21b** Manfred Kage/Peter Arnold **33.22a** iStockphoto **33.22b** National Geographic/Getty Images **33.23** Colin Milkins/OSF/Photolibrary **T33.1(1)** Meul/ARCO/NPL/Minden Pictures **T33.1(2)** iStockphoto **T33.1(3)** iStockphoto **T33.1(4)** Trouncs/Wikimedia Commons **T33.1(5)** J. Gall/Photo Researchers **T33.1(6)** iStockphoto **T33.1(7)** J. K. Lindsey **T33.1(8)** Brand X Pictures/Jupiter Images

Chapter 34 34.2a D. P. Wilson/FLPA/Minden Pictures **34.2b** Juri Vinokurov/Fotolia **34.3b** Jeff Rotman/NPL/Minden Pictures **34.4a** Sue Scott/OSF/Photolibrary **34.4b** Gary Bell/OceanwideImages.com **34.6** Alan James/NPL/Minden Pictures **34.7a** Audra Reese Kirkland **34.7b** Jeffrey L. Rotman/CORBIS **34.9** Heather Angel/Natural Visions **34.10a** Gary Bell/Getty Images **34.10b** Sinclair Stammers/NPL/Minden Pictures **34.15** Rudie Kuiter/OceanwideImages.com **34.17** Pavel Riha **34.21a** Hiroya Minakuchi/Minden Pictures **34.21b** Shutterstock **34.23a** Tom & Theresa Stack/NHPA/Photoshot **34.23b** James L. Amos/National Geographic Stock **34.24a** Fred McConnaughey/Photo Researchers **34.24b** Paul Souders/Corbis **34.25** Oceans Image/Photoshot **34.26** Arnaz Mehta **34.27a/b** Michael & Patricia Fogden/Minden Pictures **34.28a** Tom McHugh/Photo Researchers **34.28b** Reg Morrison/Auscape/Minden Pictures **34.29** Frank Lukasseck/Photographer's Choice/Getty Images **34.30** Anup Shah/NPL/Minden Pictures **34.31** George Grall/National Geographic Stock **34.32** iStockphoto **34.33** iStockphoto **34.34a** Tui De Roy/Minden Pictures **34.34b** Ingo Arndt/Minden Pictures **34.35a** Cyril Ruoso/J. H. Editorial/Minden Pictures **34.35b** Ingo Arndt/Minden Pictures

Chapter 35 Opener Jed Fuhrman, University of Southern California **35.2** SPL/Photo Researchers **35.5** David M. Phillips/Photo Researchers **35.6a** Harold Fisher/Visuals

Unlimited/Getty **35.6b** Biophoto Associates/Photo Researchers **35.6c** Gary Gaugler/Visuals Unlimited/Getty Images **35.6d** Oliver Meckes/E.O.S./Max-Planck-Institut-Tübingen/Photo Researchers **35.12L/R** Abbott Laboratories **35.14a** Hans R. Gelderblom/Eye of Science **35.14b** From R. H. Meints et al., *Virology* 113: 698–703, fig. C (1981). Photo James L. Van Etten, University of Nebraska. **35.16** Jean Roy/CDC **35.17** Philip Leder, Harvard Medical School **35.18** Nigel Cattlin/Holt Studios/Photo Researchers **35.19** Lowell Georgia/Science Source/Photo Researchers **35.20** David Parker/SPL/Photo Researchers

Chapter 36 Opener Thomas Marent/Minden Pictures **36.5a** Matt Meadows/Peter Arnold **36.5bL** Alain Mafart-Renodier/Biosphoto/Peter Arnold **36.5bR** Rukhsana Photography/Fotolia **36.8a** Geoff Dann/Dorling Kindersley **36.8b** Dorling Kindersley **36.8c** Lee W. Wilcox **36.8d** Nigel Cattlin/FLPA/Minden Pictures **36.8e** Will Cook/Charles W. Cook **36.9a/b** Lee W. Wilcox **36.9c/d** David T. Webb **3636.10a/b/c/d** Lee W. Wilcox Victoria Firmston/GAP Photos/Getty Images **36.11** RDF/Visuals Unlimited **36.12a/d** Lee W. Wilcox **36.12b** Ivan Kmit/Alamy **36.12c** Doug Wechsler/Animals Animals–Earth Scenes **36.12e** Torsten Brehm/Nature Picture Library **36.13a** Ed Reschke/Peter Arnold **36.13b** M. I. Walker/Photo Researchers **36.16a** Keith Wheeler/SPL/Photo Researchers **36.16b** ISM/Phototake **36.18L** Ed Reschke/Peter Arnold **36.18R** Ed Reschke/Peter Arnold **36.19** Andrew Syred/SPL/Photo Researchers **36.20a** John Durham/Photo Researchers **36.20b** Lee W. Wilcox **36.21L/R** Lee W. Wilcox **36.22a/c** Lee W. Wilcox **36.22b** G. Shih and R. Kessel/Visuals Unlimited/Getty Images **36.23a** Paul Schulte, University of Nevada **36.23b** Jack Bostrack/Visuals Unlimited **36.24a** John D. Cunningham/Visuals Unlimited/Getty Images **36.24b/c** Richard Kessel & Gene Shih/Visuals Unlimited/Getty Images **36.25L/M/R** Lee W. Wilcox **36.26L** Biodisc/Visuals Unlimited/Alamy **36.26R** Michael Clayton **36.27a** Lee W. Wilcox **36.27b** Biodisc/Visuals Unlimited/Alamy **36.27c** Adam Hart-Davis/SPL/Photo Researchers

Chapter 37 Opener Creatas/Photolibrary **37.4L/R** David Cook/blueshiftstudios/Alamy **37.6L/R** Lee W. Wilcox **37.8** Bruce Peters **37.10T** Ken Wagner/Phototake NYC **37.10M** G. Shih and R. Kessel/Visuals Unlimited **37.10B** Lee W. Wilcox **37.13** Lee W. Wilcox **37.18** Martin H. Zimmerman/Harvard Forest **37.22** Michael R. Sussman/American Society of Plant Biologists

Chapter 38 Opener Angelo Cavalli/Getty Images **38.2a** Photo Researchers **38.2b** Nigel Cattlin/Photo Researchers **38.2c** Photo Researchers **38.3** Emanuel Epstein, University of California at Davis **38.6a** The Institute of Texan Cultures **38.6b** Steve Ringman/ The Seattle Times **38.9** Ed Reschke/Peter Arnold **38.13b** Eduardo Blumwald **38.14T** Hugh Spencer/Photo Researchers **38.14BL** Andrew Syred/SPL/Photo Researchers **38.14BR** E.H. Newcomb & S. R. Tandon/Biological Photo Service **38.16a/b** Ed Reschke/Peter Arnold **38.17L** Frank Greenaway/Dorling Kindersley Media Library **38.17R** Courtesy Carol A. Wilson and Clyde L. Calvin **38.18** Noah Elhardt

Chapter 39 Opener Lee W. Wilcox **39.3** Malcolm B. Wilkins, University of Glasgow **39.4bL/R** John M. Christie/AAAS **39.9T/B** Malcolm B. Wilkins, University of Glasgow **39.11a** American Society of Plant Biologists **39.14** Thomas Bjorkman **39.15** Donald Specker/Animals Animals–Earth Scenes **39.16L** Richard Shiell/Animals Animals–Earth Scenes **39.16R** blickwinkel/Alamy **39.17a/b** Lee W. Wilcox **39.19** Malcolm B. Wilkins, University of Glasgow, Glasgow, Scotland, U.K. **39.23** Adel A. Kader, University of California at Davis **39.29** Nigel Cattlin/Holt Studios International/Photo Researchers

Chapter 40 Opener Walter Siegmund/Wikimedia Commons. GNU Free Documentation License **40.3a** Dan Suzio/Photo Researchers **40.3b/c** Jerome Wexler/Photo Researchers **40.6bT** berniekasper.com **40.6bM** Rod Planck/Photo Researchers **40.6bB** Tom & Therisa Stack/Tom Stack & Associates **40.7a/b** Leonard Lessin/Photo Researchers **40.8a** iStockphoto **40.8bT/B** Lee W. Wilcox **40.11L** Frans Lanting/Corbis **40.11R** Nick Garbutt/NPL/Minden Pictures **40.17aL** Brian Johnston **40.17aR** Pablo Galán Cela/age fotostock **40.17bL** Kerstin Layer/Mauritius/Photolibrary **40.17bR** Laurie Campbell/NHPA/Photoshot **40.17cL** Nicholas and Sherry Lu Aldridge/FLPA/Minden Pictures **40.17cR** James Hardy/PhotoAlto Agency RF Collections/Getty Images **40.18** Lee W. Wilcox **40.19L** Scott Camazine/Alamy **40.19M** Jonathan Watts/OSF/Photolibrary **40.19R** Daniel Heuclin/NHPA/Photoshot

Chapter 41 Opener Joe McDonald/Corbis **41.2** The Royal Society of London **41.3aL** Educational Images/Custom Medical Stock Photo **41.3aR** Nina Zanetti **41.3b** Peter Arnold/Alamy **41.3cL/R** Nina Zanetti **41.3d** Carolina Biological Supply Company/Phototake **41.4a** Deco Images II/Alamy **41.4b** Dennis Kunkel Microscopy/Phototake **41.5a** ISM/Phototake **41.5b** Manfred Kage/Peter Arnold **41.5c** Biophoto Associates/Photo Researchers **41.6** Ed Reschke/Peter Arnold **41.8** Tom Stewart/CORBIS **41.12a** G. Kruitwagen, H.P.M. Geurts and E.S. Pierson/Radboud University Nijmegen (vcbio.science.ru.nl/en) **41.12b** Innerspace Imaging/SPL/Photo Researchers **41.12c** P. M. Motta, A. Caggiati, G. Macchiarelli/SPL/Photo Researchers **41.14** Derrick Hamrick/imagebroker.net/Photolibrary **41.15** Nutscode/T Service/Photo Researchers **41.17** From J. E. Heyning and J. G. Mead, *Science* 278: 1138–1140, fig. 1b (1997). Photo Natural History Museum of Los Angeles.

Chapter 42 Opener Frans Lanting **42.13a** Image by H. Wartenberg, © H. Jastrow, Electron Microscopic Atlas (www.drjastrow.de)

Chapter 43 Opener Jonathan Blair/Corbis **43.2a** Wallace63/Wikimedia. Creative Commons Attribution Share Alike 3.0 License. **43.2b** Kim Taylor/NPL/Minden Pictures **43.4L/M/R** Karel F. Liem **43.12** Biophoto Associates/Photo Researchers

Chapter 44 Opener Tony Freeman/PhotoEdit **44.4a** Walter E. Harvey/Photo Researchers **44.4b** John D. Cunningham/Visuals Unlimited **44.19a** Lennart Nilsson, Albert Bonniers Forlag AB **44.19b** Kessel & Kardon/Tissues & Organs/Visuals Unlimited/Getty Images

Chapter 45 Opener Marcus E. Raichle **45.9a** C. Raines/Visuals Unlimited **45.11** Dennis Kunkel/Visuals Unlimited **45.20a** Genny Anderson **45.22a** Elsevier Science Ltd.

Chapter 46 Opener Neil Hardwick/Alamy **46.1** Dmitry Smirnov **46.3a** Carole M. Hackney **46.7a** David Scharf/Peter Arnold **46.9a** Don Fawcett & T. Kwwabara/Photo Researchers **46.11L/R** David Quillin **46.17T/B** James E. Dennis/Phototake NYC **T46.1L** ISM/Phototake **T46.1M** Manfred Kage/Peter Arnold **T46.1R** Biophoto Associates/Photo Researchers

Chapter 47 Opener Ralph A. Clevenger/Corbis **47.5L** Stephen Dalton/Photo Researchers **47.5M** Duncan McEwan/Nature Picture Library **47.5R** Bernard Castelein/Nature Picture Library **47.6L** BIOS/Heras Joël/Peter Arnold **47.6M** Hans Pfletschinger/Peter Arnold **47.6R** Perennou Nuridsany/Photo Researchers **47.8** The Jackson Laboratory

Chapter 48 Opener Jurgen & Christine Sohns/FLPA/Minden Pictures **48.1a** NHPA/Photoshot **48.1b** Andrew J. Martinez/Photo Researchers **48.1c** Charles J. Cole **48.2** Bruce J. Russell/BioMEDIA Associates **48.5T/B** From C. S. C. Price et al., *Nature* 400:449–452, figs. 2 and 3 (1999). © Macmillan Magazines Ltd. Photo Jerry A. Coyne, University of Chicago. **48.6** From M. Winterbottom, T. Burke, and T. R. Birkhead, *Nature* 399: 28, fig. 1b (May 6 1999). Photo Tim R. Birkhead, The University of Sheffield. **48.7a** Suzanne L. Collins/Photo Researchers **48.8a** Johanna Liljestrand Rönn **48.8a inset** Fleur Champion de Crespigny **48.16a** Geoff Shaw, Zoology, University of Melbourne, Australia/Wikipedia. GNU Free Documentation License. **48.16b** Mitsuaki Iwago/Minden Pictures **48.16c** Mitsuaki Iwago/Minden Pictures **48.17a** Claude Edelmann/Petit Format/Photo Researchers **48.17b** Lennart Nilsson, Albert Bonniers Forlag AB **48.17c** Petit Format/Nestle/Science Source/Photo Researchers **48.19aL/R** AAAS

Chapter 49 Opener Eye of Science/Photo Researchers **49.1** Susumu Nishinaga/Photo Researchers **49.2a** Don W. Fawcett/Photo Researchers **49.2b/c** Steve Gschmeissner/SPL/Photo Researchers **49.5a/c** David M. Phillips/Visuals Unlimited **49.5b** Steve Gschmeissner/SPL/Photo Researchers

Chapter 50 Opener Jeff Rotman/NPL/Minden Pictures **50.1a** Paul Nicklen/National Geographic/Getty Images **50.1b** Sergey Gorshkov/Minden Pictures **50.1c** Werner & Kerstin Layer/Photolibrary **50.1d** Lynn M. Stone/NPL/Minden Pictures **50.6a/b** Gerry Ellis/Minden Pictures **50.6c** Tim Fitzharris/Minden Pictures **50.7T** George Gerster/Photo Researchers **50.7B** Rich Wheater/Aurora/Getty Images **50.8** David Zimmerman/UpperCut Images/Getty Images **50.12** Gerry Ellis/Minden Pictures **50.14** Michael & Patricia Fogden/Minden Pictures **50.16** D. Hartnett, Konza Prairie Biological Station **50.18L/R** Alex Hyde/Minden Pictures **50.20L** Tim Fitzharris/Minden Pictures **50.20R** Stephen J. Krasemann/NHPA/Photoshot **50.22** Pixtal Images/Photolibrary **50.27** Fairfax Photos/Sandy Scheltema/The Age **50.29T** Purestock/Getty Images **50.29B** Comstock/Getty Images **50.30** Bruce Coleman/Alamy **50.31** Michael & Patricia Fogden/Minden Pictures **50.32 inset** Steven Mihok (www.nzitrap.com) **50.33L/R** John M. Randall **50.34L** Shutterstock **50.34R** USGS Canyonlands Research Station (www.soilcrust.org)

Chapter 51 Opener Konrad Wothe/Minden Pictures **51.1** Peter Arnold/Alamy **51.3aL** Peet van Schalkwyk (www.flickr.com/photos/peetvs) **51.3aR** David Dean Peterson **51.4a** iStockphoto **51.6** Gary Woodburn (pbase.com/woody) **51.8** Operation Migration (www.operationmigration.org) **51.11a** Shinji Kusano/Nature Production/Minden Pictures **51.11b** James E. Lloyd/Animals Animals–Earth Scenes **51.12** Bryan D. Neff **51.13** Tom Vezo/Peter Arnold

Chapter 52 Opener Annie Griffiths Belt/NGS Image Collection/Getty Images **52.1** David Kjaer/NPL/Minden Pictures **52.6** Mark Trabue/USDA/NRCS **52.9L** JC Schou/Biopix.dk (www.Biopix.dk) **52.9R** Marko Nieminen, University of Helsinki, Helsinki, Finland **52.13a** John Downer/NPL/Minden Pictures

Chapter 53 Opener Scott Tuason/imagequestmarine.com **53.1a** PREMAPHOTOS/Nature Picture Library **53.1b** David Tipling/Alamy **53.8L/R** From P. R. Grant and B. R. Grant, Evolution of character displacement in Darwin's finches, *Science* 313: 224–226, fig. 1 (2006). **53.9L** blickwinkel/Alamy **53.9M** Chris Newbert/Minden Pictures **53.9R** J. Sneesby/B. Wilkins/Stone/Getty Images **53.11L** Rainer Zenz/Wikimedia Commons.GNU Free Documentation License and Creative Commons Attribution Share Alike. **53.11R** CSIRO Marine and Atmospheric Research (www.scienceimage.csiro.au) **53.13a** Thomas G. Whitham/Northern Arizona University **53.15a/b** Stephen P. Yanoviak **53.15b** Stephen P. Yanoviak **53.16a** Mark Moffett/Minden Pictures **53.16b** David B. Fleetham/OSF/Photolibrary **53.19a** Nancy Sefton/Photo Researchers **53.19b** Thomas Kitchin/Tom Stack & Associates **53.20** Raymond Gehman/Corbis **53.21a** Tony C. Caprio **53.22T** G. Carleton Ray/Photo Researchers **53.22MT** Michael P. Gadomski/Animals Animals–Earth Scenes **53.22MB** James P. Jackson/Photo Researchers **53.22B** Michael P. Gadomski/Photo Researchers **53.23L/R** Glacier Bay National Park **53.23M** Christopher L. Fastie, Middlebury College

Chapter 54 Opener Jan Tove Johansson/The Image Bank/Getty Images **54.8** MODIS/NASA **54.12a** Richard Hartnup/Wikipedia Commons **54.12b** Randall J. Schaetzl, Michigan State University **54.13** John Campbell, U.S. Forest Service, Northern Research Station **54.15** Pornchai Kittiwongsakul/AFP/Getty Images

Chapter 55 Opener Ben Simmons/Photolibrary **55.1a** DLILLC/Corbis **55.1b** Xiaoqiang Wang **55.2** Edward S. Ross, California Academy of Sciences **55.5** Theo Allofs/The Image Bank/Getty Images **55.7** Karl Ammann (KarlAmmann.com) **55.8** Lester Lefkowitz/Corbis **55.9T/B** NASA/USGS (earthobservatory.nasa.gov) **55.12a** F. Mercay, BIOS/Peter Arnold **55.16** Ellen Damschen **55.17** Photo by Ariel Poster, provided by Marlboro Productions, photographed while filming TAKING ROOT: The Vision of Wangari Maathai, A Film by Lisa Merton & Alan Dater (www.takingrootfilm.com), © 2008. **55.18a/b** Daniel H. Janzen, University of Pennsylvania **55.19a/b/c** Charles Curtin **Appendix A BS9.3** From J.P. Ferris et al., *Nature* 381:59–61, fig. 2 (1996). Photo James P. Ferris, Rensselaer Polytechnic Institute. **BS10.1a** Biology Media/Photo Researchers **BS10.1b** Janice Carr/CDC

BS10.2a/b Michael W. Davidson/Florida State University /Molecular Expressions **BS10.3** Rosalind Franklin/Photo Researchers **BS12.1aL** National Cancer Institute **BS12.1aR** E.S. Anderson/Photo Researchers **BS12.1b** Sinclair Stammers/Photo Researchers **BS14.1a** Kwangshin Kim/Photo Researchers **BS14.1b** Mark J. Grimson & Richard L. Blanton **BS14.1c** Holt Studios International/Photo Researchers **BS14.1d** Custom Medical Stock Photo **BS14.1e** Graphic Science/Alamy **BS14.1f** Sinclair Stammers/Photo Researchers **BS14.1g** iStockphoto

Illustration and Table Credits

Chapter 1 **1.3** Adapted by permission from Elsevier from S. P. Moose, J. W. Dudley, and T. R. Rocheford. 2004. Maize selection passes the century mark: A unique resource for 21st century genomics. *Trends in Plant Science* 9: 358–364, Fig 1a. © 2004. **1.7a** Adapted by permission of Wiley-Blackwell from T. P. Young and L. A. Isbell. 1991. Sex differences in giraffe feeding ecology: Energetic and social constraints. *Ethology* 87: 79–80, Figs 5a, 6a. **1.8** Adapted by permission of AAAS and the author from M. Wittlinger, R. Wehner, and H. Wolf. 2006. The ant odometer: Stepping on stilts and stumps. *Science* 312: 1965–1967, Figs 1, 2, 3.

Chapter 2 **Table 2.1** Reproduced by permission of Pearson Education, Inc., from J. McMurry and R. C. Fay, *Chemistry*, 4th ed., Table 8.1. © 2004.

Chapter 3 **Opener** PDB ID: 2DN2. S.Y. Park et al. 2006. 1.25 Å resolution crystal structures of human haemoglobin in the oxy, deoxy and carbonmonoxy forms. *J Mol Biol* 360: 690–701. **3.9a** PDB ID: 1YTB. Y. Kim et al. 1993. Crystal structure of a yeast TBP/TATA-box complex. *Nature* 365: 512–520. **3.9b** PDB ID: 2F1C. G.V. Subbarao and B. van den Berg. 2006. Crystal structure of the monomeric porin OmpG. *J Mol Biol* 360: 750–759. **3.9c** PDB ID: 2PTN. J. Walter et al. 1982. On the disordered activation domain in trypsinogen. chemical labelling and low-temperature crystallography. *Acta Crystallogr, Sect B* 38: 1462. **3.9d** PDB ID: 1CLG. J.M. Chen 1991. An energetic evaluation of a "Smith" collagen microfibril model. *J Protein Chem* 10: 535–552. **3.12bL** PDB ID: 2MHR. S. Sheriff et al. 1987. Structure of myohemerythrin in the azidomet state at 1.7/1.3 Å resolution. *J Mol Biol* 197: 273–296. **3.12bM** PDB ID: 1FTP. N.H. Haunerland et al. 1994. Three-dimensional structure of the muscle fatty-acid-binding protein isolated from the desert locust *Schistocerca gregaria*. *Biochemistry* 33: 12378–12385. **3.12bR** PDB ID: 1IXA. M. Baron et al. 1992. The three-dimensional structure of the first EGF-like module of human factor IX: comparison with EGF and TGF-alpha. *Protein Sci* 1: 81–90. **3.13a** PDB ID: 1D1L. P.B. Rupert et al. 2000. The structural basis for enhanced stability and reduced DNA binding seen in engineered second-generation Cro monomers and dimers. *J Mol Biol* 296: 1079–1090. **3.13b** PDB ID: 2DN2. S. Y. Park et al. 2006. 1.25 resolution crystal structures of human haemoglobin in the oxy, deoxy and carbonmonoxy forms. *J Mol Biol* 360: 690–701. **3.19** PDB IDs: 1Q18, 1SZ2. V.V. Lunin et al. 2004. Crystal structures of *Escherichia coli* ATP-dependent glucokinase and its complex with glucose. *J Bacteriol* 186: 6915–6927. **3.23a/b** Data from T. Hansen, B. Schlichting, and P. Schönheit. 2002. Glucose-6-phosphate dehydrogenase from the hyperthermophilic bacterium *Thermotoga maritima*: Expression of the *g6pd* gene and characterization of an extremely thermophilic enzyme. *FEMS Microbiology Letters*, 216: 249–253, Fig 1; N. N. Nawani and B. P. Kapadnis. 2001. One-step purification of chitinase from *Serratia marcescens* NK1, a soil isolate. *Journal of Applied Microbiology* 90: 803–808, Fig 3; N. N. Nawani et al. 2002. Purification and characterization of a thermophilic and acidophilic chitinase from *Microbispora* sp.V2. *Journal of Applied Microbiology* 93: 965–975, Fig 7.

Chapter 4 **4.11** PDB ID: 1GRZ. B.L. Golden et al. 1998. A preorganized active site in the crystal structure of the *Tetrahymena* ribozyme. *Science* 282: 259–264.

Chapter 6 **6.11** Data fromJ. de Gier, J. G. Mandersloot, and L. L. van Deenen. 1968. Lipid composition and permeability of liposomes. *Biochimica et Biophysica Acta* 150: 666–675. **6.25a** PDB ID: 1J4N. H. Sui et al. Structural basis of water-specific transport through the AQP1 water channel. 2001. *Nature* 414:872–878 **6.25b** PDB IDs: 1ORS, 1ORQ. Y. Jiang et al. 2003. X-ray structure of a voltage-dependent K^+ channel. *Nature* 423: 33–41.

Chapter 8 **8.7b** © 2002. From *Molecular Biology of the Cell*, 4th edition, by B. Alberts et al., Fig 19.5, p. 1069. Reproduced by permission of Garland Science/Taylor & Francis, LLC.

Chapter 9 **9.2L** PDB ID: 1IRK. S. R. Hubbard et al. 1994. Crystal structure of the tyrosine kinase domain of the human insulin receptor. *Nature* 372: 746–754. **9.2R** PDB ID: 1IR3. S. R. Hubbard 1997. Crystal structure of the activated insulin receptor tyrosine kinase in complex with peptide substrate and ATP analog. *EMBO J* 16: 5572–5581. **9.12** PDB ID: 4PFK. P. R. Evans and P. J. Hudson. 1981. Phosphofructokinase: structure and control. *Philos Trans R Soc Lond B Biol Sci* 293: 53–62.

Chapter 14 **14.5** Adapted by permission of Dr. Matthew Meselson after M. Meselson and F. W. Stahl. 1958. The replication of DNA in *Escherichia coli*. *PNAS* 44: 671–682, Fig 6. **14.17a** Data from J. E. Cleaver. 1968. Defective repair replication of DNA in xeroderma pigmentosum. *Nature* 218: 652–656. **14.17b** Adapted by permission of Macmillan Publishers Ltd from J. E. Cleaver. 1968. Defective repair replication of DNA in xeroderma pigmentosum. *Nature* 218: 652–656, Fig 5. © 1968. **14–UN02** Graph adapted by permission of the Radiation Research Society from P. Howard-Flanders and R. P. Boyce. 1966. DNA repair and genetic recombination: Studies on mutants of *Escherichiacoli* defective in these processes. *Radiation Research Supplement* 6: 156–184, Fig 8.

Chapter 16 **16.2** PDB ID: 3IYD. B. P. Hudson et al. 2009. Three-dimensional EM structure of an intact activator-dependent transcription initiation complex. *PNAS* 106:19830–19835. **16.11** PDB ID: 1ZJW. I. Gruic-Sovulj et al. 2005. tRNA-dependent aminoacyl-adenylate hydrolysis by a nonediting class I aminoacyl-tRNA synthetase. *J Biol Chem* 280: 23978–23986. **16.12** Adapted by permission of the publisher from M. B. Hoagland et al. 1958. A soluble ribonucleic acid intermediate in protein synthesis. *Journal of Biological Chemistry* 231: 241–257, Fig 6. © 1958 American Society for Biochemistry and Molecular Biology. **16.14b** PDB IDs: 3FIK, 3FIH. E. Villa et al. 2009. Ribosome-induced changes in elongation factor Tu conformation control GTP hydrolysis. *PNAS* 106: 1063–1068.

Chapter 18 **Opener** PDB ID: 1ZBB. T. Schalch et al. 2005. X-ray structure of a tetranucleosome and its implications for the chromatin fibre. *Nature* 436: 138–141.

Chapter 20 **20.4** Adapted by permission of AAAS and the author from S. J. Giovannoni. 2005. Genome streamlining in a cosmopolitan oceanic bacterium. *Science* 309: 1242–1245, Fig 1. **20.7b** Reproduced by permission of GENDIA from www.paternity.be/information_EN .html#identitytest , Examples 1 and 2. **20.9** Data from www.pantherdb.org . P. D. Thomas et al. 2003. PANTHER (Protein ANalysis Through Evolutionary Relationships): A library of protein families and subfamilies indexed by function.

Chapter 21 **21.10a/b** Adapted by permission of Macmillan Publishers Ltd after S. B. Carroll. 1995. Homeotic genes and the evolution of arthropods and chordates. *Nature* 376: 479–485, Fig 1. © 1999.

Chapter 24 **24.4** After E. B. Daeschler, Neil H. Shubin, and Farish A. Jenkins, Jr. 2006. A Devonian tetrapod-like fish and the evolution of the tetrapod body plan. *Nature* 440:757–763, Fig 6; P. E. Ahlberg and J. A. Clack. 2006. A firm step from water to land. *Nature* 440: 747–749, Fig 1; N. H. Shubin, E. B. Daeschler, and F. A. Jenkins, Jr. 2006. The pectoral fin of *Tiktaalik roseae* and the origin of the tetrapod limb. *Nature* 440: 764–771, Fig 4; M. Hildebrand and G. Goslow. 2001. *Analysis of Vertebrate Structures*, 5th ed. John Wiley and Sons, Inc. **24.11** *Indohyus* reproduced by permission of Macmillan Publishers Ltd after J. G. M. Thewissen et al. 2007. Whales originated from aquatic artiodactyls in the Eocene epoch of India. *Nature* 450: 1190–1194, Fig 5. © 2007. *Rhodocetus* reproduced by permission of AAAS after P. D. Gingerich et al. 2001. Origin of whales from early Artiodactyls: Hands and feet of Eocene Protocetidae from Pakistan. *Science* 293:2239–2242, Fig 3. *Dorudon* reproduced by permission after P. D. Gingerich et al. 2009. New protocetid whale from the middle Eocene of Pakistan: Birth on land, precocial development, and sexual dimorphism. *PLoS ONE* 4(2): e4366, Fig 1B. *Delphinapterus* reproduced by permission of Skulls Unlimited International, Inc. (www.skullsunlimited.com). **24.14** Adapted by permission from D. P. Genereux and C. T. Bergstrom. 2005. Evolution in action: Understanding antibiotic resistance, Fig 3. In J. Cracraft and R. W. Bybee (eds.), *Evolutionary Science and Society: Educating a New Generation*, pp. 145–153. Colorado Springs, CO: BSCS. **24.16** Data from P. T. Boag and P. R. Grant. 1981. Intense natural selection in a population of Darwin's finches (Geospizinae) in the Galápagos. *Science* 214: 82–85, Table 1. **24.17** Body size and beak shape graphs reproduced by permission of AAAS from P. R. Grant and B. R. Grant. 2002. *Science* 296: 707–711, Fig 1. Beak size graph reproduced by permission of AAAS from P. R. Grant and B. R. Grant. 2006. Evolution of Character Displacement in Darwin's Finches. *Science* 313: 224–226, Fig 2.

Chapter 25 **Table 25.2** Data from T. Markow et al. 1993. *HLA* polymorphism in the Havasupai: Evidence for balancing selection. *American Journal of Human Genetics* 53: 943–952, Table 3. **25.3b** Data from C. R. Brown and M. B. Brown. 1998. Intense natural selection on body size and wing and tail asymmetry in cliff swallows during severe weather. *Evolution* 52: 1461–1475. **25.4b** Data from M. N. Karn, H. Lang-Brown, H. MacKenzie, and L. S. Penrose. 1951. Birth weight, gestation time and survival in sibs. *Annals of Eugenics* 15: 306–322. **25.5b** Data from T. B. Smith. 1987. Bill size polymorphism and intraspecific niche utilization in an African finch. *Nature* 329: 717–719. **25.6** Reproduced by permission of Pearson Education, Inc., from S. Freeman and J. Herron, *Evolutionary Analysis*, 3rd ed., Figs 6.15a, 6.15c. © 2004. **25.10b** Adapted by permission of Macmillan Publishers Ltd from E. Postma and A. J. van Noordwijk. 2005. Gene flow maintains a large genetic difference in clutch size at a small spatial scale. *Nature* 433: 65–68, Fig 5b. © 2005. **25.13** Adapted by permission of Wiley-Blackwell from M. O. Johnston. 1992. Effects of cross and self-fertilization on progeny fitness in *Lobelia cardinalis* and *L. siphilitica*. *Evolution* 46: 688–702, Fig 1. **Table 25.4** Reproduced by permission of W. H. Freeman and Company from Curt Stern, *Principles of Human Genetics*, 3rd ed., Table 5.8. © 1973 by W. H. Freeman and Company. **25.14b** Data from J. D. Blount et al. 2003. Carotenoid modulation of immune function and sexual attractiveness in zebra finches. *Science* 300: 125–127, Fig 2. **25.16b, c** Data from B. J. Le Boeuf and R. S. Peterson. 1969. Social status and mating activity in elephant seals. *Science* 163: 91–93.

Chapter 26 **26.3a** Adapted by permission of AAAS and the author after J. C. Avise and W. S. Nelson. 1989. Molecular genetic relationships of the extinct dusky seaside sparrow. *Science* 243: 646–648, Fig 2. **26.3b** J. C. Avise and W. S. Nelson. 1989. Molecular genetic relationships of the extinct dusky seaside sparrow. *Science* 243: 646–648. **26.5b** Data from N. Knowlton et al. 1993. Divergence in proteins, mitochondrial DNA, and reproductive compatibility across the Isthmus of Panama. *Science* 260:1629–1632, Fig 1. **26.7** Adapted by permission of Wiley-Blackwell from H. R. Dambroski et al. 2005. The genetic basis for fruit odor discrimination in *Rhagoletis* flies and its significance for sympatric host shifts. *Evolution* 59: 1953–1964, Figs 1A, 1B. **26.10b** Data from S. E. Rohwer et al. 2001. Plumage and mitochondrial DNA haplotype variation across a moving hybrid zone. *Evolution* 55: 405–422. **26.11** Data from L. H. Rieseberg et al. 1996. Role of gene interactions in hybrid speciation: Evidence from ancient and experimental hybrids. *Science* 272: 741–745.

Chapter 27 **27.5a/b/c** Data adapted by permission of the publisher from M. A. Nikaido, P. Rooney, and N. Okada. 1999. Phylogenetic relationships among cetartiodactyls based on insertions of short and long interspersed elements: Hippopotamuses are the closest extant relatives of whales. *PNAS* 96: 10261–10266, Figs 1, 7. © 1999 National Academy of Sciences, U.S.A. **27.8, 27.9, 27.10** Data from the

International Commission on Stratigraphy, "International Stratigraphic Chart 2009" (www.stratigraphy.org/column.php?id=Chart/Time Scale). The data on this site are modified from F. M. Gradstein and J. C Ogg (eds). 2004. *A Geologic Time Scale 2004.* Cambridge, UK: Cambridge University Press; and J. G. Ogg, G. Ogg, and F. M. Gradstein. 2008. *The Concise Geologic Time Scale.* Cambridge, UK: Cambridge University Press. **27.11** B. G. Baldwin, D. W. Kyhos, and J. Dvorak. 1990. Chloroplast DNA evolution and adaptive radiation in the Hawaiian Silversword Alliance (Asteraceae–Madiinae). *Annals of the Missouri Botanical Garden* 77: 96–109, Fig 2. **27.12c** J. B. Losos, K. I. Warheitt, and T. W. Schoener. 1997. Adaptive differentiation following experimental island colonization in *Anolis* lizards. *Nature* 387: 70–73. **27.15** Data from J. W. Valentine, D. H. Erwin, and D. Jablonski. 1996. Developmental evolution of metazoan body plans: The fossil evidence. *Developmental Biology* 173: 373–381, Fig 3; R. de Rosa et al. 1999. Hox genes in brachiopods and priapulids and protostome evolution. *Nature* 399: 772–776; D. Chourrout et al. 2006. Minimal ProtoHox cluster inferred from bilaterian and cnidarian Hox complements. *Nature* 442: 684–687. **27.16** Data from M. J. Benton. 1995. Diversification and extinction in the history of life. *Science* 268: 52–58. **27.17a** Adapted by permission of AAAS and the author from W. Alvarez, F. Asaro, and A. Montanari. 1990. Iridium profile for 10 million years across the Cretaceous-Tertiary boundary at Gubbio (Italy). *Science* 250: 1700–1702, Fig 1.

Chapter 28 **28.2** Reproduced by permission of the American Medical Association from G. L. Armstrong, L. A. Conn, and R. W. Pinner. 1999. Trends in infectious disease mortality in the United States during the 20th century. *JAMA* 281: 61–66, Fig 1. © 1999 American Medical Association. All rights reserved. **28.6** After J. G. Elkins et al. 2008. A korarchaeal genome reveals insights into the evolution of the Archaea. *PNAS* 105: 8102–8107, Fig 2; F. D. Ciccarelli et al. 2006. Toward automatic reconstruction of a highly resolved tree of life. *Science* 311:1283–1287, Fig 2. **28.11** Adapted by permission of Pearson Education, Inc., from M. T. Madigan and J. M. Martinko. 2006. *Brock Biology of Microorganisms*,12th ed., Fig 5.9. © 2009.

Chapter 29 **29.1, 29.7** S. M. Adl et al. 2005. The new higher level classification of eukaryotes with emphasis on the taxonomy of protists.*Journal of Eukaryotic Microbiology* 52: 399–451; N. Arisue, M. Hasegawa, and T. Hashimoto. 2005. Root of the Eukaryota tree as inferred from combined maximum likelihood analyses of multiple molecular sequence data. *Molecular Biology and Evolution* 22: 409–420; V. Hampl et al. 2009. Phylogenomic analyses support the monophyly of Excavata and resolve relationships among eukaryotic "supergroups." *PNAS* 106: 3859–3864, Figs 1, 2, 3; J. D. Hackett et al. 2007. Phylogenomic analysis supports the monophyly of cryptophytes and haptophytes and the association of Rhizaria with chromalveolates.*Molecular Biology and Evolution* 24: 1702–1713, Fig 1; P. Schaap et al. 2006. Molecular phylogeny and evolution of morphology in the social amoebas. *Science* 314: 661–663.

Chapter 30 **30.3** Adapted by permission of Pearson Education, Inc., from O. Owen et al. 1998. *Natural Resource Conservation: Management for a Sustainable Future* 7th ed., p. 509, Fig 21.2. © 1998.**30.7, 30.10, 30. 24** Y.-L. Qiu et al. 2006. The deepest divergences in land plants inferred from phylogenomic evidence. *PNAS* 103:15511–15516, Fig 1; K. S. Renzaglia et al. 2007. Bryophyte phylogeny: Advancing the molecular and morphological frontiers. *The Bryologist* 110: 179–213; J. F. Pombert et al. 2005. The chloroplast genome sequence of the green alga *Pseudendoclonium akinetum* (Ulvophyceae) reveals unusual structural features and new insights into the branching order of chlorophyte lineages.*Molecular Biology and Evolution* 22: 1903–1918. **30.22** Data from S. D. Johnson and K. E. Steiner. 1997. Long-tongued fly pollination and evolution of floral spur length in the *Disa draconis* complex (Orchidaceae). *Evolution* 51: 45–53. **30.26** P. S. Soltis and D. E. Soltis. 2004. The origin and diversification of angiosperms. *American Journal of Botany* 91: 1614–1626, Figs 1, 2, 3.

Chapter 31 **31.7** S. M. Adl et al. 2005. The new higher level classification of eukaryotes with emphasis on the taxonomy of protists.*Journal of Eukaryotic Microbiology* 52: 399–451; N. Arisue, M. Hasegawa, and T. Hashimoto. 2005. Root of the Eukaryota tree as inferred from combined maximum likelihood analyses of multiple molecular sequence data. *Molecular Biology and Evolution* 22: 409–420; V. Hampl et al. 2009. Phylogenomic analyses support the monophyly of Excavata and resolve relationships among eukaryotic "supergroups." *PNAS* 106: 3859–3864, Figs 1, 2, 3; J. D. Hackett et al. 2007. Phylogenomic analysis supports the monophyly of cryptophytes and haptophytes and the association of Rhizaria with chromalveolates. *Molecular Biology and Evolution* 24: 1702–1713, Fig 1; P. Schaap et al. 2006. Molecular phylogeny and evolution of morphology in the social amoebas. *Science* 314: 661–663. **31.8** T. Y. James et. al. 2006. Reconstructing the early evolution of fungi using a six-gene phylogeny. *Nature* 443: 818–822, Fig 1.

Chapter 32 **32.9** A. M. A. Aguinaldo et al. 1997. Evidence for a clade of nematodes, arthropods, and other moulting animals. *Nature* 387: 489–493, Figs 1, 2, 3; C. W. Dunn et al. 2008. Broad phylogenomic sampling improves resolution of the animal tree of life. *Nature* 452: 745–750, Figs 1, 2.

Chapter 33 **33.2** A. M. A. Aguinaldo et al. 1997. Evidence for a clade of nematodes, arthropods, and other moulting animals. *Nature* 387: 489–493, Figs 1, 2, 3; C. W. Dunn et al. 2008. Broad phylogenomic sampling improves resolution of the animal tree of life. *Nature* 452: 745–750, Figs 1, 2. **33.7** T. H. Struck et al. 2007. Annelid phylogeny and the status of Sipuncula and Echiura. *Evolutionary Biology* 7: 571–11.

Chapter 34 **34.1** After C. W. Dunn et al. 2008. Broad phylogenomic sampling improves resolution of the animal tree of life. *Nature* 452: 745–750, Figs 1, 2; F. Delsuc et al. 2006. Tunicates and not cephalochordates are the closest living relatives of vertebrates. *Nature* 439: 965–968, Fig 1. **34.12, 44.21** J. E. Blair and S. B. Hedges.2005. Molecular phylogeny and divergence times of deuterostome animals. *Molecular Biology and Evolution* 22: 2275–2284, Figs 1, 3, 4. **34.16** Adapted by permission of Macmillan Publishers Ltd after E. B. Daeschler et al. 2006. A Devonian tetrapod-like fish and the evolution of the tetrapod body plan. *Nature* 440:757–763, Fig 6, © 2006; also after N. H. Shubin et al. 2006. The pectoral fin of Tiktaalik roseae and the origin of the tetrapod limb. *Nature* 440:764–771, Fig 4, © 2006. **34.22** J. E. Blair and S. B. Hedges. 2005. Molecular phylogeny and divergence times of deuterostome animals. *Molecular Biology and Evolution* 22: 2275–2284, Figs 1, 3, 4; F. R. Liu et al. 2001. Molecular and morphological supertrees for eutherian (placental) mammals. *Science* 291: 1786–1789; A. B. Prasad et al. 2008. Confirming the phylogeny of mammals by use of large comparative sequence data sets. *Molecular Biology and Evolution* 25:1795–1808. **34.36** A. B. Prasad et al. 2008. Confirming the phylogeny of mammals by use of large comparative sequence data sets. *Molecular Biology and Evolution* 25:1795–1808, Figs 1, 2, 3. **34.39** J. Z. Li et al. 2008. Worldwide human relationships inferred from genome-wide patterns of variation. *Science* 319: 1100–1104, Fig 1. **34.40** After L. L. Cavalli-Sforza and M. W. Feldman. 2003. The application of molecular genetic approaches to the study of human evolution. *Nature Genetics Supplement* 33: 266–275, Fig 3.

Chapter 35 **35.3** Adapted by permission of the publisher from J. G. Bartlett and R. D. Moore. 2003. Improving HIV Therapy. *Scientific American* July 2003: 30–37. Originally published July 1998. © 1998 Scientific American. All rights reserved. **35.4** Data compiled by the United Nations AIDS program. **35.15** F. Gao et al. 1999. Origin of HIV-1 in the chimpanzee *Pan troglodytes troglodytes. Nature* 397: 436–441, Fig 2.

Chapter 36 **36.4** Illustration adapted by permission from "Root Systems of Prairie Plants" by Heidi Natura. © 1995 Conservation Research Institute.

Chapter 37 **37.3** Data from R. G. Cline and W. S. Campbell. 1976. Seasonal and diurnal water relations of selected forest species. *Ecology* 57: 367–373, Fig 1. **37.12** Adapted by permission of the American Society of Plant Biologists from C. M. Wei, M. T. Tyree, and E. Steudle. 1999. Direct measurement of xylem pressure in leaves of intact maize plants. A test of the cohesion-tension theory taking hydraulic architecture into consideration. *Plant Physiology* 121: 1191–1205, Fig 6. © 1999 American Society of Plant Biologists. **37.22** Data from N. D. DeWitt and M. R. Sussman. 1995. Immunocytological localization of an epitope-tagged plasma membrane proton pump (H^+–ATPAse) in phloem companion cells. *Plant Cell* 7: 2053–2067, Fig 7.

Chapter 39 **39.5** After Darwin and Darwin, 1897. *The Power of Movement in Plants* (D. Appleton & Co. New York). **Table 39.1** Data from H. A. Borthwick et al. 1952. A reversible photoreaction controlling seed germination. *PNAS* 38: 662–666, Table 1.

Chapter 40 **40.12** Y.-L. Qiu et al. 2006. The deepest divergences in land plants inferred from phylogenomic evidence. *PNAS* 103:15511–15516, Fig 1; K. S. Renzaglia et al. 2007. Bryophyte phylogeny: Advancing the molecular and morphological frontiers. *The Bryologist* 110: 179–213; J. F. Pombert et al. 2005. The chloroplast genome sequence of the green alga *Pseudendoclonium akinetum* (Ulvophyceae) reveals unusual structural features and new insights into the branching order of chlorophyte lineages. *Molecular Biology and Evolution* 22: 1903–1918.

Chapter 41 **41.10** Data from K. Schmidt-Nielsen. 1984. *Scaling: Why is animal size so important?* Cambridge, UK: Cambridge University Press. **41.11** Adapted by permission of The Company of Biologists from P. R. Wells and A. W. Pinder. 1996. The respiratory development of Atlantic salmon. II. Partitioning of oxygen uptake among gills, yolk sac and body surfaces. *Journal of Experimental Biology* 199: 2737–2744, Figs 1A, 3.

Chapter 42 **42.14a/b** Adapted by permission of Elsevier from K. J. Ullrich, K. Kramer, and J. W. Boyer. 1961. Present knowledge of the counter-current system in the mammalian kidney. *Progress in Cardiovascular Diseases* 3: 395–431, Figs 5, 7. © 1961.

Chapter 43 **43.17a** Graph reproduced with permission from The American Diabetes Association from L. O. Schulz et al. 2006. Effects of traditional and western environments on prevalence of type 2 diabetes in Pima Indians in Mexico and the U.S. *Diabetes Care* 29: 1866–1871, Fig 1. © 2006 American Diabetes Association. **43.17b** Data from L. O. Schulz et al. 2006. Effects of traditional and western environments on prevalence of type 2 diabetes in Pima Indians in Mexico and the U.S. *Diabetes Care* 29: 1866–1871, Table 2.

Chapter 44 **44.7** Adapted by permission from Y. Komai. 1998. Augmented respiration in a flying insect. *Journal of Experimental Biology* 201: 2359–2366, Fig 7. **44.8** Adapted by permission from Y. Komai. 1998. Augmented respiration in a flying insect. *Journal of Experimental Biology* 201: 2359–2366, Fig 1. **44.11** W. Bretz and K. Schmidt Nielsen. 1971. Bird respiration: Flow patterns in the duck lung. *Journal of Experimental Biology* 54:103–118. **44.20** Data from E. H. Starling. 1896. On the absorption of fluids from the connective tissue spaces. *Journal of Physiology* 19: 312–326.

Chapter 45 **45.21** Adapted by permission from Elsevier from K. C. Martin et al. 1997. Synapse-specific, long-term facilitation of Aplysia sensory to motor synapses: A function for local protein synthesis in memory storage. *Cell* 91: 927–938. © 1997. **45.22b** Adapted by permission from Elsevier from K. C. Martin et al. 1997. Synapse-specific, long-term facilitation of Aplysia sensory to motor synapses: A function for local protein synthesis in memory storage. *Cell* 91: 927–938. © 1997.

Chapter 46 **46.2a/b** Data from J. E. Rose et al. 1971. Some effects of stimulus intensity on response of auditory nerve fibers in the squirrel monkey. *Journal of Neurophysiology* 34: 685–699. **46.12** Adapted by permission of AAAS and the author from David M. Hunt et al. 1995. The Chemistry of John Dalton's Color Blindness. *Science* 267: 984–988, Fig 3. **46.19** PDB ID: 1KWO. D. M. Himmel et al. 2002. Crystallographic findings on the internally uncoupled and near-rigor states of myosin: further insights into the mechanics of the motor. *PNAS* 99: 12645–12650.

Chapter 47 **47.10a/b** Data from G. V. Upton et al. 1973. Evidence for the internal feedback phenomenon in human subjects: Effects of ACTH on plasma CRF. *Acta Endocrinologica* 73: 437–443. **47.13** Data from D. Toft, and J. Gorski. 1966. A receptor molecule for estrogens: Isolation from the rat uterus and preliminary characterization. *PNAS* 55: 1574–1581. **47.15b** Data from T. W. Rall et al. 1956. The relationship of epinephrine and glucagon to liver phosphorylase. *Journal of Biological Chemistry* 224: 463–475. **47.16** Reproduced by permission of the Nobel Foundation from E. W. Sutherland. "Studies on the mechanism of hormone action." A lecture delivered in Stockholm, 11 December 1971. © 1971 Nobel Foundation.

Chapter 48 **48.3a** Reproduced by permission of AAAS from R. G. Stross and J. C. Hill. 1965. Diapause induction in Daphnia requires two stimuli. *Science* 150: 1462–1464, Fig 3. **48.3b** Data from O. T. Kleiven, P. Larsson, and A. Hobæk. 1992. Sexual reproduction in Daphnia magna requires three stimuli. *Oikos* 65: 197–206, Table 4. **48.5** Adapted by permission of Macmillan Publishers Ltd from C. S. C. Price, K. A. Dyer, and J. A. Coyne. 1999. Sperm competition between *Drosophila* males involves both displacement and incapacitation. *Nature* 400: 449–452, Fig 3c. © 1999. **48.7b** R. Shine and M. S. Y. Lee. 1999. A reanalysis of the evolution of viviparity and egg-guarding in squamate reptiles. *Herpetologica* 55: 538–549, Figs 1, 2, 3. **48.8b** Reproduced by permission of Elsevier from C. Hotzyand and G. Arnqvist. 2009. Sperm competition favors harmful males in seed beetles. *Current Biology* 19: 404–407. © 2009. **48.14, 48.15** Data from R. Stricker et al. 2006. Establishment of detailed reference values for luteinizing hormone, follicle stimulating hormone, estradiol, and progesterone during different phases of the menstrual cycle on the Abbott ARCHITECT® analyzer. *Clin Chem Lab Med* 44: 883–887, Tables 1A and 1B. **48.19b** Reproduced by permission of AAAS from C. Ikonomidou et al. 2000. Ethanol-induced apoptotic neurodegeneration and fetal alcohol syndrome. *Science* 287: 1056–1060, Fig 4. **48.21** Adapted by permission of the publisher and author from U. Högberg and I. Joelsson. 1985. The decline in maternal mortality in Sweden, 1931–1980. *Acta Obstetricia et Gynecologica Scandinavica* 64: 583–592, Fig 1. Taylor & Francis Group, www.informaworld.com.

Chapter 49 **49.6a** PDB ID: 1IGT. L. J. Harris et al. 1997. Refined structure of an intact IgG2a monoclonal antibody. *Biochemistry* 36: 1581–1597. **49.6b** PDB ID: 1TCR. K. C. Garcia et al. 1996. An αβ cell receptor structure at 2.5Å and its orientation in the TCR-MHC complex. *Science* 274: 209–219.

Chapter 50 **50.4a** Adapted by permission of the University of California Press Journals and the author from M. B. Saffo. 1987. New light on seaweeds. *BioScience* 37: 654–664, Fig 1. © 1987 American Institute of Biological Sciences. **50.28** Graphs adapted by permission of AAAS and the author from A. K. Knapp et al. 2002. Rainfall variability, carbon cycling, and plant species diversity in a mesic grassland. *Science* 298: 2202–2205, Figs 1, 2, 3.

Chapter 51 **51.2** After M. Sokolowski. 2001. *Drosophila*: Genetics meets behavior. *Nature Reviews Genetics* 2: 879–890, Fig 2. **51.3b** Data from R. E. Hegner, S. T. Emlen, and N. J. Demong. 1982. Spatial organization of the white-fronted bee-eater. *Nature* 298: 264–266. **51.4b** Adapted by permission of the author from D. Crews. 1975. Psychobiology of reptilian reproduction. *Science* 189: 1059–1065, Fig 2. **51.16** Data from M. Daly and M. Wilson. 1988. Evolutionary social psychology and family homicide. *Science* 242: 519–524, Fig 1.

Chapter 52
Table 52.1 Data reproduced with kind permission from Springer Science and Business Media from H.Strijbosch and R. C. M. Creemers, 1988. Comparative demography of sympatric populations of *Lacerta vivipara* and *Lacerta agilis*. *Oecologia* 76: 20–26. © 1988 Springer-Verlag. **52.3** Reproduced by permission of AAAS from C. K. Ghalambor and T. E. Martin. 2001. *Science* 292: 494–496, Fig 1c. **52.7b** Data from G. F. Gause. 1934. *The Struggle for Existence*. Baltimore, MD: Williams & Wilkins. **52.8a** Reproduced with kind permission of Springer Science and Business Media and the author from G. E. Forrester. 1995. Strong density-dependant survival and recruitment regulate the abundance of a coral reef fish. *Oecologia* 103: 275–282, Fig 2. © 1995 Springer-Verlag. **52.8b** Data from P. Arcese and J. N. M. Smith. 1988. Effects of population density and supplemental food on reproduction in song sparrows. *Journal of Animal Ecology* 57: 119–136, Fig 2. **52.11** Reproduced by permission of Alberta Sustainable Resource Development. 2002. From www3.gov.ab.ca/srd/fw/watch/rabb_cycles.html. © 2002 Government of Alberta. **52.12** Graphs reproduced by permission of AAAS from C. J. Krebs et al. 1995. Impact of food and predation on the snowshoe hare cycle. *Science* 269: 1112–1115, Fig 3. **52.13b** Data from T. Valverde and J. Silvertown. 1998. Variation in the demography of a woodland understorey herb (*Primula vulgaris*) along the forest regeneration cycle: Projection matrix analysis. *Journal of Ecology* 86: 545–562. **52.14a/b** Data from U.S. Census Bureau, International Data Base (IDB), www.census.gov/ipc/www/idb/.

Chapter 53 **53.2a/b, 53.3, 53.4a/b** G. F. Gause. 1934. *The Struggle for Existence*. Baltimore, MD: Williams & Wilkins. **53.8** Reproduced by permission of AAAS from P. R. Grant and B. R. Grant. 2006. Evolution of character displacement in Darwin's finches. *Science* 313, Fig 2. **53.11** Adapted by permission of Ecological Society of America from G. H. Leonard, M. D. Bertness, and P. O. Yund. 1999. Crab predation, waterborne cues, and inducible defenses in the blue mussel, *Mytilus edulis*. *Ecology* 80: 1–14. © 1999. **53.12** Adapted by permission of Ecological Society of America from G. H. Leonard, M. D. Bertness, and P. O. Yund. 1999. Crab predation, waterborne cues, and inducible defenses in the blue mussel, *Mytilus edulis*. *Ecology* 80: 1–14. © 1999. **53.13b, c** Data from G. D. Martinsen, E. M. Driebe, and T. G. Whitham. 1998. Indirect interactions mediated by changing plant chemistry: Beaver browsing benefits beetles. *Ecology* 79: 192–200. **53.17** Adapted by permission of Ecological Society of America from J. H. Cushman and T. G. Whitham. 1989. Conditional mutualism in a membracid-ant association: Temporal, age-specific, and density-dependant effects. *Ecology* 70: 1040–1047. **53.18** Adapted by permission of Ecological Society of America from D. G. Jenkins and A. L. Buikema, Jr. 1998. Do similar communities develop in similar sites? A test with zooplankton structure and function. *Ecological Monographs* 68: 421–443. **53.19c** Adapted with kind permission from Springer Science and Business Media from R. T. Paine. 1974. Intertidal community structure. *Oecologia* 15: 93–120. © 1974 Springer-Verlag. **53.21b** Data from T. W. Swetnam. 1993. Fire history and climate change in giant sequoia groves. *Science* 262: 885–889. **53.24a** Adapted by permission of Wiley-Blackwell from R. H. MacArthur and E. O. Wilson. 1963. An Equilibrium Theory of Insular Zoogeography. *Evolution*: 17: 373–387, Fig 4. **53.24b, c** Data from R. H. MacArthur and E. O. Wilson. 1963. An equilibrium theory of insular zoogeography. *Evolution*: 17: 373–387, Fig 5. **53.26** Data from W. V. Reid and K. R. Miller. 1989. *Keeping Options Alive: The Scientific Basis for Conserving Biodiversity*. Washington, DC: World Resources Institute.

Chapter 54 **54.3** Data from J. R. Gosz et al. 1978. The flow of energy in a forest ecosystem. *Scientific American* 238: 92–102. **54.7** Reproduced by permission of the publisher from C. A. Mackenzie, A. Lockridge, and M. Keith. 2005. Declining sex ratio in a first nation community. *Environmental Health Perspectives* 113: 12–96, Fig 1. **54.9a/b/c** Data from R. H. Whittaker and G. E. Likens. 1973. Primary production: The biosphere and man. *Human Ecology* 1: 357–369, Table 1; H. Lieth.1973. Primary production: Terrestrial ecosystems. *Human Ecology* 1: 303–332, Tables 2, 3; J. S. Bunt. 1973. Primary production: Marine ecosystems. *Human Ecology* 1: 333–345, Table 1; G. E. Likens. 1973. Primary Production: Freshwater ecosystems. *Human Ecology* 1: 347–356, Table 3. **54.16** Data from W. H. Schlesinger. 1997. *Biogeochemistry: An Analysis of Global Change*, 2nd ed. San Diego, CA: Academic Press. **54.17** Data from P. M. Vitousek et al. 1997. Human alteration of the global nitrogen cycle: Sources and consequences. *Ecological Applications* 7: 737–750. **54.20** Data from BP Statistical Review of World Energy 2007, pp. 6–21 **54.21** Data from NOAA.gov, www.esrl.noaa.gov/gmd/ccgg/trends/. **54.22a** After G. Beaugrand et al. 2002. Reorganization of North Atlantic marine copepod biodiversity and climate. *Science* 296: 1692–1694. **54.22b** Adapted by permission of the publisher fromN. L. Bradley et al. 1999. Phenological changes reflect climate change in Wisconsin. *PNAS* 96: 9701–9704. © 1999 National Academy of Sciences, U.S.A. **54.23** Reproduced by permission of AAAS from R. R. Nemani et al. 2003. Climate-driven increases in global terrestrial net primary production from 1982 to 1999. *Science* 300: 1560–1563, Fig 2. **54.23b** After M. J. Behrenfeld et al. 2006. Climate-driven trends in contemporary ocean productivity. *Nature* 444: 752–755.

Chapter 55 **55.1** F. R. Liu et al. 2001. Molecular and morphological supertrees for eutherian (placental) mammals. *Science* 291: 1786–1789, Fig 1; A. B. Prasad et al. 2008. Confirming the phylogeny of mammals by use of large comparative sequence data sets. *Molecular Biology and Evolution* 25:1795–1808; U. Arnason, A. Gullberg, and A. Janke. 2004. Mitogenomic analyses provide new insights into cetacean origin and evolution. *Gene* 333: 27–34, Fig 1. **55.3a/b** Adapted by permission of Macmillan Publishers Ltd after C. D. L. Orme et al. 2005. Global hotspots of species richness are not congruent with endemism or threat. *Nature* 436: 1016–1019. © 2005. **55.4** Reproduced by permission of Conservation International from *Conservation International 2008 Annual Report*, 12–13. © 2008 Conservation International (www.conservation.org). **55.6** Reproduced by permission of the University of California Press Journals and the author from O.Venter et al. 2006. Threats to endangered species in Canada. *BioScience* 56: 903–910, Fig 2. © 2006 American Institute of Biological Sciences. **55.10b** Adapted by permission of AAAS and the author from W. F. Laurance et al. 1997. Biomass collapse in Amazonian forest fragments. *Science* 278: 1117–1118, Fig 2. **55.11** Adapted by permission of The Royal Society and the author from J. T. Hogg et al. 2006. Genetic rescue of an insular population of large mammals. *Proceedings of the Royal Society B* 273: 1491–1499, Fig 1. **55.12b** Adapted by permission from D. B. Lindenmayer and R. C. Lacy. 1995. Metapopulation viability of Leadbeater's possum, *Gymnobelideus leadbeateri*, in fragmented old-growth forests. *Ecological Applications* 5: 164–182, Fig 5G. **55.13** Adapted by permission of the author fromJ. M. Diamond and E. Mayr. 1976. Species-area relation for birds of the Solomon Archipelago. *PNAS* 73: 262–266, Fig 1. **55.14** Graphs reproduced by permission of AAAS from D. Tilman et al. 1997. The influence of functional diversity and composition on ecosystem processes. *Science* 277: 1300–1302, Fig 1. **55.16** Reproduced by permission of AAAS from E. I. Damschen et al. 2006. Corridors increase plant species richness at large scales. *Science* 313: 1284–1286, Fig 2B. **55-UN01T (England maps)** Reproduced by permission of the author from D. S. Wilcove, C. H. McLellan, and A. P. Dobson. 1986. Habitat fragmentation in the temperate zone. In M. E. Soule (ed.), *Conservation Biology: The Science of Scarcity and Diversity*. Sunderland, MA: Sinauer Associates, pp. 237–256, Fig 1. **55-UN01B (U.S. maps)** After W. B. Greeley, Chief, U.S. Forest Service. 1925. Relation of geography to timber supply. *Economic Geography* 1: 1–11.

Index

Boldface page numbers indicate a glossary entry; page numbers followed by an *f* indicate a figure; page numbers followed by *t* indicate a table.

Boldface page numbers indicate a glossary entry; page numbers followed by an *f* indicate a figure; page numbers followed by *t* indicate a table.

B

Boldface page numbers indicate a glossary entry; page numbers followed by an *f* indicate a figure; page numbers followed by *t* indicate a table.

C

Boldface page numbers indicate a glossary entry; page numbers followed by an *f* indicate a figure; page numbers followed by *t* indicate a table.

Boldface page numbers indicate a glossary entry; page numbers followed by an *f* indicate a figure; page numbers followed by *t* indicate a table.

D

Boldface page numbers indicate a glossary entry; page numbers followed by an *f* indicate a figure; page numbers followed by *t* indicate a table.

E

Boldface page numbers indicate a glossary entry; page numbers followed by an *f* indicate a figure; page numbers followed by *t* indicate a table.

Boldface page numbers indicate a glossary entry; page numbers followed by an *f* indicate a figure; page numbers followed by *t* indicate a table.

G

Boldface page numbers indicate a glossary entry; page numbers followed by an *f* indicate a figure; page numbers followed by *t* indicate a table.

H

I

Boldface page numbers indicate a glossary entry; page numbers followed by an *f* indicate a figure; page numbers followed by *t* indicate a table.

J

K

L

M

Boldface page numbers indicate a glossary entry; page numbers followed by an *f* indicate a figure; page numbers followed by *t* indicate a table.

Boldface page numbers indicate a glossary entry; page numbers followed by an *f* indicate a figure; page numbers followed by *t* indicate a table.

N

O

Boldface page numbers indicate a glossary entry; page numbers followed by an *f* indicate a figure; page numbers followed by *t* indicate a table.

P

Boldface page numbers indicate a glossary entry; page numbers followed by an *f* indicate a figure; page numbers followed by *t* indicate a table.

Boldface page numbers indicate a glossary entry; page numbers followed by an *f* indicate a figure; page numbers followed by *t* indicate a table.

Q

R

Boldface page numbers indicate a glossary entry; page numbers followed by an *f* indicate a figure; page numbers followed by *t* indicate a table.

S

Boldface page numbers indicate a glossary entry; page numbers followed by an *f* indicate a figure; page numbers followed by *t* indicate a table.

T

Boldface page numbers indicate a glossary entry; page numbers followed by an *f* indicate a figure; page numbers followed by *t* indicate a table.

U

V

Boldface page numbers indicate a glossary entry; page numbers followed by an *f* indicate a figure; page numbers followed by *t* indicate a table.

W

X

Y

Z

Boldface page numbers indicate a glossary entry; page numbers followed by an *f* indicate a figure; page numbers followed by *t* indicate a table.